LIMIT STATE DESIGN OF
STEEL STRUCTURES-1

(Based on IS: 800-2007 in S.I. Units)
for Under graduate and post-graduate students (LSD: SS)

Dr. Ramchandra

B.E (Civil), M.E (Hons), Ph.D. (I.I.T, Roorkee)
Retired Professor
Department of Structural Engineering
MBM Engineering College, J.N.V. University,
Jodhpur (Raj.)

Virendra Gehlot

B.E. (Civil), M.E. (Civil) Structures
Consultant Engineer
Jodhpur, Rajasthan

Published by:

Scientific Publishers (India)
5 A, New Pali Road, P.O. Box 91
Jodhpur 342 001 (India)

E-mail: info@scientificpub.com
Website: www.scientificpub.com

Branch Office
Scientific Publishers (India)
4806/24, Ansari Road, Daryaganj
New Delhi - 110 002 (India)

Print: 2018

ISBN: 978-81-7233-614-1

Typeset by Rajesh Ojha
Printed in India

Dedicated
to
our teachers

ROLE OF STEEL STRUCTRUES IN SERVICE FOR THE SOCIETY (Chat-magni aur Patt-byah)

Role of Steel Structures is useful, significant and important inservice for the society due to following advantages of steel inconstruction Steel is strong and these structures have extremely highload-bearing capacity.

1. These may be fabricated in the work-shops quickly.
2. The construction of the structures is also quick.
3. These structures may be furbished (made new again) strengthened and remodelled, if necessary. These may be expanded or relocated.
4. These structures have long life-span.
5. These structures need few columns to stand-on, minimizing waste of space.
6. Steel is recyclable.
7. Steel is more environment friendly.
8. If necessary, steel structures are easy to dismentle.
9. These structures can withstand earthquakes to much significant degree.

Authors fully support the views of R.K.P. Singh, Director General,Institution of Steel Development and Growth, (INSDAG).

Reference: Times of India, dated 14.05.2007, page 12, dated 22.5.2007 page 1.

ACKNOWLEDGEMENT

Tables and clauses from the Indian Standard Specifications have been produced in the book with the kind permission of the Indian Standards Institution.

It is desirable that for complete detail, reference be made to the latest versions of the Standards Institution, Manak Bhavan, 9, Bahadur Shah Zafar Marg, New Delhi -I, or from its branch offices at Mumbai, Kolkata Kanpur and Chennai.

SYSTEM INTERNATIONAL d' UNITS (SI - System of Units)

In order to avoid the conversion of results obtained by engineers working with the foot pound second system (gravitational) of units in terms of centimetre-gram second absolute system of units used by the scientists, a need of common system of units was realised. The General Conference on Weights and Measures held at Paris in 1960 finalised the System International d' Units (SI). It is an absolute system of units. The mass is considered as fundamental unit and not the force. BIS has included a comment of transition in IS 3616-1966. 'Recommendation on the International System (SI) Units' that this system has begun to replace older system of units in several branches of science and technology. The SI is a universal system of units and it has been adopted in France as a legal system and it is likely to become common in many countries. SI units have the following six basic units.

Unit of length (metre, m)

The length equal to 1,650, 763.73 wave lengths, in vacuum, of the radiation corresponding to the transition between $2p^{19}$ and $5d^{5}$ levels of the krypton η atom of mass 86 is known as one metre.

Unit of mass (kilogram, kg)

The mass of platinum-iridium cylinder deposited at the international Bureau of Weights and Measures and declared as the intrnational prototype of the kilogram by the First General Conference of Weights and Measures is called as one kilogram.

Unit of time (second, s)

1131, 566, 925, 974.7 of the length of the tropical year for 1900, the year commencing at 1200 hours universal time on the first day of January, 1900 is termed as one second.

Unit of electric current (ampere, A)

The constant current which flows in two parallel straight conductors of infinite length of negligible circular cross-section and placed at a distance of one metre from each other in vacuum producing a force of 2×10^{-7} Newtons per metre length between the conductors is defined as one ampere.

Unit of thermo-dynamic temperature (degree Kelvin, °K)

The degree interval of the thermo-dynamic scale on which the temperature of triple point of water is 273.16 degrees, is known as one degree Kelvin.

Units of luminous intensity (candela, cd)

One sixtieth part of luminous intensity normally emited by one hundred millimetre square of integral radiator (black body) at the temperature of solidification of platinum is called as one candela.

Following are the derived units (derived from the basic SI units and these are relevant for the study of design of steel structures).

Force Newton (N) kg.m. s^{-2}

Work-energy Joule (J) kg.m^2 . S^{-2}=(N – m)

Power Watt (W) kg.m^2 . s^{-3}=(J . s^{-1})

Frequency Hertz (Hz) Cycles/sec.

Accereation due to gravity used is 9.81 ms^{-2}.

The force is a derived quantity and physical law connecting the quantity to the fundamental quantities or previously obtained derived quantities is force = mass x acceleration. It is defined as that force which produces unit acceleration (i.e., 1 m per sec^2 in a unit mass of 1 kg). Its unit is Newton (N). Though, the Newton is a small unit, a still larger unit kN may be used. The intensity of force (viz. stress) due to 1 Newton over a unit area of one metre quare is known as one pascal. It is denoted by symbol, Pa (1 Pa = 1 N/m^2 and 10^6. Pa = 1 N/mm^2. (viz. 1 MPa = 1 N/mm^2).

The unit of force, the Newton (N) is the force required to develop unit acceleration (m s^{-2}) to unit mass (kg). In terms of Newtons, the common force units in the foot-pound second system (with g = 9.81 ms^{-2}) are

1 lb-wt = 4.45 Newtons (N)

1 ton-wt = 9.96 x 10^3 Newtons (N)

1 ton-wt = 9.96 kN

The unit of force, the Newton (N) is used for the external loads and the internal forces, such as the shear force. Torque and bending moments are expressed as Newton-metres (N-m).

Another important unit is stress. In the foot-pound-second system the stresses are commonly expressed in lb-wt/in2 and tons/in2. In the SI system of units, these are taken as

$$1 \text{ lb. wt/in}^2 = 6.89 \times 10^3 \text{ N/mm}^2 \; 6.89 \text{ kN/m}^2$$

$$1 \text{ ton. wt/in}^2 = 15.42 \times 10^6 \text{ N/mm}^2 = 15.42 \text{ MN/m}^2$$

Yield stress of the common metallic materials are in the range 200 MN/m2 to 750 MN/m^2. Yound modulus of elasticity, E for steel

$$E_{steel} = 30 \times 10^6 \text{ lbs.wt/in}^2 = 2.07 \times 10^5 \text{N/mm}^2 = 207 \text{ GN/m}^2$$

The SI units make the use of multiples and sub-multiples 1000 times or 1/1000 times the unit quantity and in powers of 10^3 (kilo) or 10^{-3} (milli) in respect of still larger and smaller quantities respectively. The lengths are measured usually in kilometre (1 km = 1000 m), metre and millimetre (1 mm = 10^{-3} m). The symbols of units are not to be suffixed with 's' for plural.

$$(\text{kilo k } 10^{+3}), (\text{mega M } 10^{+6}), (\text{giga G } 10^{+9})$$

SI system of units have many advantages. The units are very handy. The burden of nn-decimal coefficients in foot-pound second system is avoided. It has relatively large main units in contrast to centimetre- gram- second system. At the same time, it is closely related to centimetre- gram-second sytem of units. In practice, it results in perfectly reasonable number when the value of g = 10 m/sec^2 is used instead of 9.806 m/sec^2.

References

1. (Professor V.S. Mokashi, Visvesvaraya Regional College of Engineering, Nagpur in his paper titled as Interantional System (SI) Units and their Application to Engineering published in Journal of Institution of Engineers, India Vol. 19, March 1970 has high lighted the advantages and discussed SI units. A reference has been made to this paper).
2. John Case and A.H. Chilver 'Strength of Materials and Structures' Second Edition (ELBS)/Edward Arnold, 1986, Printed of the Bath Press, Avon, Great Britain.

LIST OF THE PAPERS PUBLISHED
by Prof. Dr. Ram Chandra* and Virendra Gehlot**

I. PAPERS PUBLISHED IN INTERNATIONAL JOURNALS

*1. "Non-Linear Analysis of Steel Space Structures," Journals of American Society of Civil Engineers, Volume 116, No. 4, April, 1990, Paper no. 24525 (pp. 898-909) New York.

*2. "Elastic-Plastic Analysis of Steel Space Structures", Journal of American Society of Civil Engineers, Volume 116, No. 4, April, 1990 paper no. 24528 (PP -939-955) New York.

II. PAPERS PUBLISHED IN NATIONAL JOURNALS, CONFERENCES AND SEMINARS

*3. "Non-Linear Elastic-Plastic Analysis of Skeletal Steel Plane Frames Hinged at supports", proceedings of the seminar on Modern trend in structural analysis and design, 25-26 February, 1984, Department of Civil Engineering, Banara Hindu University, Varanasi (U.P.)

*4. "Non-Linear Elastic Plastic Analysis of Skeletal Plane Steel Frame", International conference on computer applications in Civil Engineering, October 23-25, 1979, University of Roorkee, Roorkee.

*5. "Stub Column Tests on Indian Standard H-Beams", Journal of the Institution of Engineers (India), Volume 51, No. 11, C.I. 6, July 1971.

*6. "Kani's Iteration Method I-Analysis of Continuous Beams" The Indian Engineer, Volume XIV, No. 6 June 1970.

*7. "Kani's Iteration Method II Analysis of Frames", The Indian Engineer", Volume XIV, No. 8 August, 1970.

*8. Hinge Formation in Plane Frames Considering Finite Deformations All India Conference on "Tall Buildings" The Institution of Engineers (1) Roorkee Local Centre, March 1-3, 1993.

**9. "Design, Construction and Behaviour Aspects of Concrete and Steel Structures' Recent Trends in Civil Engineering National Seminar, Jodhpur, February 22-23, 2002, Department of Civil Engineering MBM Engineering College, Faculty of Engineering, JNVyas University, Jodhpur pp 229.

**10. 'Corrosion of Reinforcing Steel Bars in Concrete' Futuristic Construction Materials and Technique (FCMT - 2004, Feb 21-22, 2005) Department of Structural Engineering Faculty of Engineering, JN Vyas University Jodhpur.

**11. 'Stresses in Reinforcing Steel Bars, HYSD of Grade Fe 415' Futuristic Construction Materials and Technique (FCMT 2004, Feb. 21-22, 2005) Department of Structural Engineering Faculty of Engineering, JN Vyas University, Jodhpur.

PREFACE

METHOD OF LIMIT STATE (Ultimate Limi State, (ULS) and Serviceability Limit State (SLS) present an improved design philosophy and makes allowance for the short-compings of working stress method (conventional and long time used in practice). This method provides basic framework, within which the performance of the steel structures may be assessed against various limiting conditions and involves some concept of probability.

Object of limit design method is to get steel structure that will remain fit for use during its life with acceptable target reliability. The probability of a limit state being reached during its life time is kept very small.

This method has been broadly adopted in many developed countries.

In our country, `Limit State Method of Design of Concrete Structures (LSD : CS) was intoduced in IS 456-1978. It was natural for Bureau of Indian Standards, to introduce this method (LSD : SS) for **Steel Structures** in its recommendations in IS : 800-2007 (Third Revised Edition).

Authors have made efforts to present the matter to understand the subject easily and conveniently. This book has been covered in eight parts in twenty six chapters and four appendices as listed in contents. Appendix A and B have been given for `shear centre for the thin walled cross-sections' and `unsymmatrical bending' being important topics for many points of view for the interest of the students.

The expressions given in IS : 800-2007 for the determination of the strengths of te structural members in bending and shear have been derived by first principle of strength of materials and / structural stability.

Inspite of careful scrutiny of the manuscript, it is possible that some typographical and computational errors are still left. Authors shall feel highly obliged to those, who bring these errors to their notice.

Authors acknowledge Shri R.K. Singhal, Shri Sher Singh Gehlot and Shri Suresh Sankhla of Department of Structural Engineering, Faculty of Engineering, Jodhpur for their encouragement and time to time consultations.

Authors express their heartly thanks to Shri Rajesh Ojha (Laser typeset) and also thanks to Shri Pawan Kumar Sharma, Proprietor, M/s Scientific Publishers (India), Jodhpur for nice get-up and publication.

Dr. Ramchandra
Virendra Gehlot

CONTENTS

PART I:
STRUCTURAL STEEL SECTIONS, STRUCTURES, CODES AND SPECIFICATIONS

PART II
LOADS (ACTIONS), METHODS OF ANALYSIS AND DESIGN, AND PROPERTIES OF STEEL, SERVICE CHARACTERISTICS

PART V
DESIGN OF COLUMNS AND COMPRESSION MEMBERS

1

Part - I

STRUCTRAL STEEL, SECTIONS, STRUCTURES, CODES AND SPECIFICATION

INTRODUCTION

1.1 INTRODUCTION

Word 'Engineer' is originated from the Latin-word 'Ingener'. In our Indian culture, (i.e., Hindu Sanskrati), a person who creates artifacts (i.e., Brahmaji/ Vishwakarma/Vidhata). An engineer uses his power of imagination and capacity, constructs basically structures ingenuously and with simple and pure thoughts. An engineer understands others constraints and their capacity to spend. The construction of such structures which is possible and practicable for use, commonly it may be said, an engineer is one who designs and manufactures engines (a machine composed of simple parts). Long back, there used to be only two categories of engineers (military engineer and civil engineer) who used to look after all expects. Now-a-days, there are different engineers for minute disciplines and their capacity and expertisation in their fields are immense.

An engineer understands the importance of finance, significance of project management and information technology. Engineer improves and updates his knowledge and quality of his design. It is essential to understand and keep collaboration with other disciplines. To-day's time is the time of inter-discipline, to work together. It is necessary to maintain communication across professional limits. Engineer is to ensure that his design is proper and suitable for use, economical, durable and safe (both for users and neighbours). Engineer ensures that his construction is aethetically pleasing.

Civil engineering is the art and science of planning, design, construction, operation, maintenance and rehabilitation of structures. Term structure includes bridges, multi-storey buildings and all types of civil engineering structures (towers, shells, ...etc.) composed of any structural material. In common language it may be said, that engineering is an art of constructing and manufacturing structures. Structural engineering is a branch of civil engineering.

The **structural engineering** is that branch of engineering which deals with structural analysis and structural design. The structural engineering covers a wide range. The structural engineering plays an important and necessary part in mechanical engineering, electrical engineering, civil engineering, naval engineering, aeronautical engineering and in all the specialized phases of engineering. The **structural analysis** deals with the development of suitable arrangement of structural elements for the structures to support the external loads or the various critical combinations of the loads which are likely to act on the structure. The analysis also deals with the determination of internal forces in the various members (viz., axial forces, bending moments and shear forces), state of stresses or critical combination of the stresses at the various points (which includes the nature, magnitude and direction of these stresses) and the external reactions due to the worst possible combination of the loads. The internal force per unit area of the original cross-section is called **stress.** The external reactions are transmitted to the foundation. The methods of structural analysis and the principles involved in them remain independent of the materials used for all types of structures, whether the structures are built of plastics, aluminium, timber, reinforced concrete or steel.

The analysis of stress, strain and the deflection characteristics of a structure is referred as **structural analysis.**

Structural design is an art and science combining the experienced engineers, feelings for the behaviours of a structure with a sound knowledge of principles of statics, dynamics, mechanics of materials and structural analysis to produce a safe economical structure which will serve its intended purpose. In the design, certain criteria must be established to evaluate whether or not an optimum has been obtained. The typical criteria may be (i) minimum cost, (ii) minimum weight, (iii) minimum construction time, (iv) minimum labour, (v) minimum cost of manufacture of owner's product and maximum efficiency of operation to owner. In most practical situations, the evaluation must be qualitative. Only simple objective criteria, such as weight or cost must be kept in mind.

The **structural design** deals with the selection of proper material, proper sizes, proportions and shape of each member and its connecting details. The selection is such that it is economical and safe. It satisfies all the stress requirements imposed by the most severe combination of the loads to which the structure is required to transmit or resist including its self- weight. The structural design further deals with the preparation of final layout of the structure. The finished design drawings with all pertinent or controlling dimensions for all the members, parts and connections are made. The finished design drawings are necessary for fabrication and construction.

The structural design in a limited sense also deals with the design of various parts or members of a structure. The structural design is governed with the standard specifications. The hand-books are used as working tools in the structural design. When the structural steel is used as the material for the structure, the structural design is known as *design of steel structures*. The procedure of design is simplified and presented for the ease of understanding and for the better appreciation of the subject.

In general building construction, all the steel structures and the structural steel components are designed as per IS : 800-1984 Code of Practice for General Construction in Steel. The provisions of the code are thoroughly followed, intelligently interpreted and carefully applied.

1.2 STEEL

Structural steel developed much later with the development of civilisation. Whereas iron was used in ancient times. After the stone age, it was the age of iron. Iron age began somewhat before 1000 BC in Western Asia and Egypt. Iron occurs in nature (as raw material) in combined state, Weapons and arms were primarily made of iron. Which were used for hunting and for fighting wars. In our country, there are many examples of use of iron which may be cited from Ramayan and Mahabharat. In the Indus Valley Civilisation, there is archaeological evidence of use of iron. In the fifth century, an iron-Isthumb was made in Mehrauli village, Delhi within a small distance, from Qutub Minar. Recently, scientists have reported (in June-July, 2002) that this Isthumb is made of molybdenum-iron. A sufficiently thick layer of molybdenum oxide exists round this Isthumb which has made this a 'Rust-less Wonder'.

Photo-Plate 1. Rust-less wonder (An Iron Isthumb) Mehrauli - Village, Delhi (Near Qutub Minar)

A Professor at the Department of Material and Metallurgical Engineering R. Bala-Subramaniam, I.I.T., Kanpur alongwith his team of students have developed a novel phosphorous containing iron which is corrosion resistant. This was in use since fourth century BC when Emperor Ashok ruled the Mauran empire. This technology was adopted in the construction of this historical iron-pillar (presently installed near Qutub Minar, Delhi). Professor Balasubramaniam and his students studied this phosphoric iron pillar and found, it is ductile and corrosion resistant to atmospheric environment.

An iron pillar in Kodachadri Village in Karnataka is seen without rusting for nearly 1500 years. In the eighteen century, famous bridge engineers started the use of iron. Initially, a number of arch bridges were built using cast-iron. First iron bridge of 30.5 m span was constructed in 1779 over the river severn in Coalbrookdale at stropshire, England by Abraham Derby and John Wilkinoon This bridge was the public demonstration of a revolution that was to change the world. Cast-iron was gradually replaced by wrought-iron dominated the field until close to the end of the nineteenth century. First major structure was built using carbon steel (St. Louis Bridge over the river Mississipi) completed in 1874.

The first major steel structure in UK was the fourth bridge completed in 1889. Which in time (mid-nineteenth century) was replaced by steel when Bessemier Process of Steel (BPS) making was introduced.

Steel was invented and patented in 1885 in England by Sir Henery Bessemier. At about the same time, the technique of making steel was developed by William Kelly of USA. In the heating process, the undesirable elements including excessive quantities of carbon were removed, subsequently, the desirable amounts of principally carbon and managanese were added. It become the beginning of the steel age. Since, then, the steel is used as a major material for buildings and bridges. The use of steel became popular in nineteenth century. Siemens Martin process of open hearth technique (oldest) was developed along with Bessemier Process, The Open-hearth process is slow and relatively uneconomical. It was largely replaced by the basic oxygen system (briefly, referred as BOS) and the electric-arc method. This method needs precise control over their chemical composition. This process was originally devised to produce high-quality steels. More recently, a new process has been devised. It is continuous-casting process (briefly referred as CONCAST). In this process, the molten metal is poured directly into a casting machine to make the initial solid shape (e.g., slabs, blooms, or billets) depending on their size and shape.

In our country, Howarah Bridge (Rabindra Sethu) in Kolkatta is a marvellous structure built in steel in 1943. It is a balanced cantilever bridge over river Hoogly having 457 m span. It is fourth largest cantilever bridge in world. Recently a cable stayed bridge Vidhyasagar Sethu (second Hoogly Bridge) has been constructed over this river Hoogly. It is completed in 1992 a huge steel construction may be seen in the Jogighopa bridge over river Brahmputra. In our country, there are many bridges built in steel for railways.

The use of iron has immersely increased with development of its common alloy steel.

1.3 STRUCTURAL STEEL

This material *steel* is essentially an alloy of iron and carbon (in small percentage) and other elements in varying percentages. It is a crystalline substance in steel, iron is predominant (more than 98 per cent) and other minerals are combined with small proportions to create an alloy that best serves the intended function. The most important constituent of steel, the element carbon which gives its strength and hardness. The strength, hardness and brittleness of steel increase and ductility of steel and malleability decreases with the increase of percentage of carbon.

For the increased strength, the carbon remains present in chemical combination with iron as a compound known as carbide of iron with entirely different properties from those of iron or carbon. Carbon, either as charcoal or graphite, is a soft, brittle substance, easily powdered, while pure iron is a soft, malleable and ductile metal. Carbide of iron, on the other hand, is intensely hard, and although brittle as compared to iron, is not so easily powdered as carbon. Carbide of iron is offen called **cementite**. Its chemical formula is Fe_3C (meaning three atoms of iron combined with one atom of carbon). Steel is really an alloy of iron and carbon, more correctly an alloy of iron and iron carbide. When the steel is heated to a certain temperature, depending on its composition, the carbide dissolves in iron of which it is largely composed to give what is called solid solution of carbide in iron.

When the steel is heated to a certain temperature, the tiny particles of carbide of iron dissolve or go into the solution in the iron, and above this temperature, although the steel is still quite solid. The particles of carbine cannot be observed under the most powerful microscope. The iron carbine and the iron form a solution, but because, this was formed, when the steel was solid, it is said to be a solid solution. This solid solution of carbide of iron in iron is called **austenite**.

Depending on the chemical composition, the different type of steels are classified as mild steel, medium carbon steel, high carbon steel, low alloy steel and high alloy steel. The mild steel, medium carbon steel and low alloy steel are generally used for steel structures. The copper bearing quality of steel contains small percentage of copper contents. The corrosive resistance of such steel is increased. The structural steel is the steel used for the manufacture of rolled structural steel sections, fastenings and other elements for use in structural steel works.

Steel may classified in different following categories depending upon its chemical composition.

(i) Mild Steel (ii) Medium carbon steel (iii) High carbon steel (iv) Low alloy steel (v) High alloy steel

Similar to carbon contents, steel having different other chemical constituents have different effects. Manganese influences similar to carbon. Increased sulphur and phosphorus (0.06 per cent) impart brittleness, affect weldability and fatigue strength. Chrome and nickel impart corrosion resistance properties to steel. Sometimes 0.5 per cent molybdenum is added to refine the lamellar spacing in pearlite (a fine mixture of ferrite and cementite). Cementite is a compound-iron (iron carbide Fe_3C) and carbon having 6.67 per cent by weight. Ferrite is practically pure iron.

In addition to the chemical composition, the mechanical properties of steel depend primarily upon the rolling process and heat treatment. Mechanical properties of the steel are also influenced by other factors (such as, rate of loading the specimen, the conditions and geometry of the specimen, cold work, and the temperature at the time of testing. These influences may produce varying results for the same grade of steel.

The function of a structure is the principal factor to decide the geometry of the structure. Using the structural geometry along with the design loads, individual components are selected to properly support and transmit loads throughout the structure. Steel members are selected from the standard rolled steel shapes.

Steel possesses **ductility** (a unique property) that no other structural material exhibits in the same way. Structural steel is able to absorb large deformation beyond the elastic limit without the risk of fracture. It is this characteristic feature of structural steel that makes possible the use of steel for structural purposes. All structural steels exhibit nearly identical mechanical behaviour in the elastic range, where the modulus elasticity of different kinds of steel does not differ by more than five per cent from the average value, $E = 2 \times 10^5$ N/mm^2.

There are many qualities of steels which are available for structural use. Some of the steels are produced by specific steel companies. These are considered as proprietary steels, which may be used with the permission of the approving agencies. Various qualities of the steels may be grouped by strength and grade, which are as under

(i) High-strength low-alloy steels, (ii) Quenched and tempered carbon-steel (iii) Construc-tional alloy steels (iv) Structural carbon steels (v) Other steels

In other steel group, some steels have been developed for specific use (e.g., for submarine hulls, missiles, space equipment... etc.) One group of these special steels are referred as maraging steels. These steels derive their high strengths from high nickel alloy content, which is then heat treated to age the iron-nickel martensite.

In America, various varieties of steels are codified by American Society for Testing and Materials (ASTM) and by American Institute of Steel Construction (AISC) American Association of State High Way Officials (AASO) incorporated many steels into their bridge specifications.

1.4 STRUCTURAL MILD STEEL

The mild steel is used for the manufacture of rolled structural steel sections, rivets and bolts. Following operations can be done easily on mild steel :

1. Cutting 2. Punching 3. Drilling 4. Machining 5. Welding and 6. Forging when heated.

The mild steel cannot be used for manufacture of cutting tools.

All structural steels used in general construction, coming within the purview of IS : 800-84 shall, before fabrication, comply with one of the following Indian Standard Specifications.

1. IS : 226-1975 structural steel (standard quality)
2. IS : 1977-1975 structural steel (ordinary quality)
3. IS : 2062-1984 weldable structural steel
4. IS : 961-1975 structural steel (high tensile)
5. IS : 8500-1977 weldable structural steel (medium and high strength qualities) Table 1.1 gives chemical composition of structural steel IS : 226-1975 as per BIS.

Table 1.1. CHEMICAL COMPOSITION

Constituents	Per cent (Max.)
Carbon (For thickness/diameter 20 mm and below)	0.23
Carbon (For thickness/diameter over 20 mm)	0.25
Sulphur	0.055
Phosphorus	0.055

1. IS : 226-1975 structural steel (standard quality) : The mild steel is designated as St 44-S for use in structural work. This steel is also available in copper bearing quality in which case it is designated as St 44-SC. The copper content is between 0.20 and 0.35 per cent. All designs and examples have been worked out for the steel, conforming to IS : 226-1975.

The physical properties of structural steel are given as under :

(a) Unit weight of steel 78.430 kN/m^3 to 79.000 kN/m^3

(b) Young's modulus of elasticity, $E = 2.04$ to 2.18×10^5 N/mm^2

(c) Modulus of rigidity, $G = 0.84$ to 0.98×10^5 N/mm^2

(d) Coefficient of thermal (alpha), $\alpha = 12 \times 10^{-6}$/C

expansion (or contraction), (alpha), $\alpha = 6.7 \times 10^{-6}$/F.

The tensile strength, yield stress and percentage elongations for IS : 226-1975 structural steel standard quality, determined in accordance with IS: 1608-1960. Method for tensile testing of steel products other than sheets, strips, wires and tubes are given in Table 1.2 as per IS : 226-1975.

Table 1.2. MECHANICAL PROPERTIES

Class of steel Product	Nominal thickness Diameter mm	Tensile Strength N/mm^2	Yield Stress Min. N/mm^2	Percentage Elongation Min.
Plates, Sections e.g. angles, tees, beams, channels etc., and flats	6 upto and including 20	410-530	260	23
	Over 20 upto and including	410-530	240	23
	40 over 40	410-530	230	23
Bars (Round, Squares and Hexagonals)	10 upto and including 20	410-530	250	23
	over 20	410-530	240	23

The steel conforming to IS : 226 is suitable for all types of steel structures subjected to static, dynamic and repeated cycles of loadings. It is also suitable for welding upto 20 mm thickness. When the thickness of elements is more than 20 mm, it needs special precautions while welding.

2. IS : 1977-1975 structural steel (ordinary quality). The steel which did not comply with IS : 226, was formerly called as steel of untested quality. The standards for such steel have been laid down in IS: 1977-1975 (Ordinary quality). There are two grades in this standard which are designated as St 44.0 and St 32.0. The steel St 44.0 is intended to be used for structures not subjected to dynamic loading other than wind loads (e.g., platform roofs, office buildings, foot overbridge). The copper bearing quality is designated as St 44.OC.

The steel conforming to IS: 1977 is not suitable for welding and for the structures subjected to high seismic forces. The steel structures using steel conforming to IS : 1977 must not be analysed and designed by plastic theory.

3. IS : 2062-1984 weldable structural steel. This structural steel is intended to be used for members in structures subjected to dynamic loading where welding is employed for fabrication and where fatigue (wide fluctuation of stress, reversal of stress), and great restraint are involved (e.g., crane gantry girder, road and rain bridges etc.). It is designated as St 42-W and copper bearing quality is designated as St 42-WC.

It is suitable for welding the elements of thickness between 28 mm and 50 mm. When the thickness of elements is less than 28 mm, it may be welded provided the limited maximum carbon content is 0.22 per cent.

1.5 HIGH TENSILE STRUCTURAL STEEL

The high tensile steel forms a specific class of steel in which enhanced mechanical properties and in most of the cases increased resistance to atmospheric corrosion are obtained by the incorporation of low proportions of one or more alloying elements, besides carbon. These steels are generally intended for application where saving in weight can be effected by reason of their greater strength and atmospheric corrosion resistance. Standard of high tensile steel have been given in IS: 961-1975. It has been classified into two grades designated as St 58-HT and St 55-HTW. St 58-HT is intended for use in structures where fabrication is done by methods other than welding. St 55-HTW is intended for use in structures where welding is employed for fabrication. The high tensile steel is also available in copper bearing quality and two grades are designated as St 58-HTC and St 55-HTWC.

There are following two varieties of high tensile steel

1. IS 961-1975, and 2. IS : 8500-1977

The steel conforming to IS : 961 is suitable for bridges and general building construction.

Following table gives chemical composition of structural steel IS : 961-1975 (High tensile Steel) as per BIS : (Bureau of India Standard)

Table 1.3(A). CHEMICAL COMPOSITION

Constituents	Steel Designation	
	St 58-HT	St 55-HTW
Carbon	0.27% max.	0.20% max.
Sulphur	0.055% max.	0.055% max.
Phosphorus	0.055% max.	0.055% max.

1. IS : 961-1975 Structural Steel (High Tensile) : The tensile strength, yield stress and percentage elongation for IS : 961-1975 high tensile steel, determined in accordance with IS : 1608-1960 method for tensile testing of steel products other than sheets, strips, wires and tubes are given in Table 1.4 (A) as per IS : 961-1975.

Table 1.4 (A). MECHANICAL PROPERTIES

Class of Steel products	Nominal thickness diameter (mm)	Tensile strength Min. (N/mm2)	Yield stress min. (N/mm2)	Percentage Elongation Min.
Structural steel (St. 58-HT)				
For Plates,	6 Upto and including 28	570	350	20
Sections,	Over 28 upto and including 45	570	340	20
Flats and	Over 45 upto and including 63	570	320	20
Bars	Over 63	540	290	20
Structural steel (St. 55-HTW)				
For Plates,	6 Upto and including 16	540	350	20
Sections,	Over 16 upto and including 32	540	340	20
Flats and	Over 32 upto and including 63	510	330	20
Bars	Over 63	490	280	20

The tensile and yield properties of steel may be increased by using increasing contents of carbon and manganese, but its ductility will be reduced. The welding is also difficult for such steel. The increased sulphur and phosphorus beyond 0.06 per cent will give brittleness and affect weldability and fatigue strength. The chrome and nickel (if added) give corrosive resistance and resistance to high temperature.

The tensile and yield strength are also increased and the ductility is maintained by other alloying elements.

The tensile strength, yield strength, ductility and weldability are the properties of steel which depend on the chemical composition of steel. The fatigue strength of steel members depends on their shape and surface characteristics. These five properties are the important properties of steel and while selecting the material, the effects, of different alloys must also be borne in mind.

Any structural steel other than those specified above may also be used provided that the permissible stresses and other design provisions are suitably modified and the steel is also suitable for the type of fabrication adopted.

IS : 8500-1977 Weldable structural steel (medium and high strength qualities)

Various medium and high strength qualities of weldable structural steel for example, Fe 440 (HT1 and HT2) Fe 540 (HT and HTA and HTB), Fe 570 HT, Fe 590 HT and Fe 640 HT have been developed. Table 1.3 (B) gives the chemical composition of various qualities of IS : 8500-1977 steel.

Table 1.3 (B). CHEMICAL COMPOSITION

Constituents Steel Designation Fe 440 HT Fe 490 HT Fe 590 HT

Constituents	Steel Designation		
	Fe440 HT	Fe490 HT	Fe590 HT
Carbon	0.25	0.25	0.25
Sulphur	0.055	0.040	0.050
Phosphorus	0.055	0.040	0.050
Manganese	1.5	1.5	1.5

Table 1.4 (B) gives the mechanical properties of various medium and high strength qualities of weldable structure steel IS : 8500-1977

Table 1.4 (B). MECHANICAL PROPERTIES (IS 8500-1977)

Class of steel products	Nominal thickness diameter (mm)	Tensile strength Min. (N/mm²)	Yield stress min. (N/mm²)	Percentage Elongation Min.
Fe 440 HT1				
Plates, sections,	upto 6	440-560	300	—
angles,	over 6	440-560	300	—
beams	upto 20			
Fe440 HT2				
channels etc	over 20, upto 40	440-560	290	—

bars, flats	over 40, upto 63	440-560	280	—
Fe540HT				
Plates, sections	upto 6	540-660	410	20-25
Fe 540 HTA				
angles,	over 6 upto 20			20-25
beams,		540-660	440	—
channels etc.		540-660	390	20-25
Fe 540 HTB				
bars, flats	over 40 upto 63	540-660	380	—
Fe 570 HT				
Plates,	upto 6	570-720	450	20-25
section	over 6			
angles	upto 20	570-720	440	20-25
beams	over 20	570-720	430	20-25
	upto 40			
channels etc.	over 40	570-720	420	20-25
bars, flats	upto 63			
Fe 590 HT	upto 6 over 6	590-740	490	20-25
	upto 20 over 6	590-740	480	20-25
Fe 640 HT	upto 6	640-790	540	
	over 6 upto 20	640-790	530	20-25

1.6 WELDABLE STEEL

Weldability is a property to adjudge the steel's suitability for welding. It is closely related to the quantity of carbon in steel. Other chemical composition also influence this property. It is important to examine that the steel used is capable of welding without which the need for special treatment such as pre-heating or post heating and therefore expensive welding procedures.

The weldability for steel is seen by determining **carbon equivalent value,** which may be expressed by the following expression.

$$C_{EQV} = \left[C + \left(\frac{M_n}{6} \right) + \left(\frac{C_r + M_o + V}{5} \right) + \left(\frac{N_i + C_u}{15} \right) \right] \tag{1.1}$$

where, C is percentage of carbon, M_n is percentage of manganese, C_r is percentage of chromium, M_o is percentage of molybdenum. V is percentage of Vanedium.

N_i is percentage of nickel and Cu is the percentage of copper (Reference : David, A. Nethercot, 'Limit state Design of Structural Steelwork, third edition, 2001, UK, pp 13).

In case, C_{EQV} is less than 0.42 per cent, steel is treated as weldable without preheating. However, C_{EQV} = 0.45 per cent, it is tolerable, in case carbon in steel is less than 0.12 per cent.

1.7 STAINLESS STEEL

Stainless steel was discovered by Harry Brearley in 1913. The first stainless steels contained 13 to 14 per cent of chromium. Later, it was found that by reducing the carbon, the chromium percentage could also be reduced and a range of stainelss steels was produced of malleable character, which would withstand the attack of many acids and ordinancs atmospheric conditions.

Stainless steel is steel which is an alloy of iron and chromium. In this quality of steel, when the chromium content is increased to about eleven percent, a thin protective passive film is formed spontaneously on the steel. This film acts as a barrier. This barrier protects the steel from corrosion. The steel becomes stainless. A more correct term for the stainless steel is write 'steels resistant to corrosion'.

In case, the chromium content is further increased, above mentioned film is further strengthened. It achieves the ability to repair itself, in case, it is damaged in the corrosive environment. Addition of nickle in stainless steel improves corrosion resistance in reducing environments' damages from sulphuric acid if present in the environment. It crystal structure changes from one form to the other. Thereby, ductility, toughness and weldability properties are improved. Addition of molybdenum increases pitting and crevice corrosion in chloride environment.

Introduction of nickle, silicon, molybdenum, titanium, tunguston, niabium (colum bind) etc. varities of stainless steels were (i) Ferritic, (ii) Martensitic, (iii) Austentie and (iv) Duplex.

1. **Fertitic stainless steels** contain 12 to 30 per cent of chromium and low carbon contents (usually less than 0.1 per cent). Their structures consist essentially of ferrite solid solution and heat treatment has little effect on their properties. These steels are used for sheets for pressing, and producing moulds by the hobbing process for use with urea and other straining plastic moulding powders. For the latter purpose, the moulds are case hardened after the hobbing process. This means that in their find condition, the surface layer consists of the hard martensite while the inner layers consist essentially of ferrite.

2. **Martensitic stainless steels** consists of martensite. They contains from 12 to 18 per cent chromium and upto 0.4 per cent of carbon. These are hardened and tempered to attain the maximum resistance to corrosion. An extremely wide range of mechanical properties may be obtained from tham. These can be used for general engineering purposes and for cutlery. For the latter purpose, these are used in the martensilic condition but tempering in the higher temperature ranges yield sorbite and the steels are then more ductile and less hard, although they still exhibit satisfactory corrosion resistance to many media.

3. **Austenitic stainless steels** when quanched from above 1000°C consist entirely of austenite and are non-magnetic. Austenitic stainless steels contain 17 to 18 per cent of chromium and 8 to 12 per cent of nickle, (usually less than 0.12 per cent carbon).

Duplex stainless steels have higher strength and wearing resistance than austenitic grade. Other elements such as silicon, tungsten, molybdenum, titanium and columbium may be present for special purpose.

Now-a-days, stainless steel is used extensively in building construction. In Kualalampur, Malaysia, about four thousand tonnes of stainless steel (made in our country, India) has been used in the World's tallest twin towers.

Despite the high cost, stainless steel is attractive to the architects. The stainless steel provides a combined effect of aesthetics, strength and durability.

1.8 PRODUCTION OF STEEL

Steel is produced by adding carbon in small quantities to iron. Iron is refined. It is smelted from basic iron ore in the blast furnace. Blast furnaces operate continuously over a period of several years producing upto eight-thousand tonnes of motten iron every 24 hours. Iron ore coke lime stone and sinter (a mixture of ore, coke and lime stone previously roasted together to remove some of the volatile matter) are fed into the top of the furnace, Air is blown through to increase the temperature, the oxygen content reacting with the hot carbon in the coke form carbon mono-oxide which in turn releases the iron. The molted metal is periodically tapped-off from near the base for subsequent use as the basic raw material used in the steel making.

There are four main processes which exist for the production of steel. Open-hearth process is the oldest. It is slow and uneconomic. It is, therefore, largely replaced by **Basic oxygen process** (BOS) and the electric arc method. Electric arc method was originally devised to produce high-quality steels. It requires higher precision control over their composition. BOS process begins with the operation known as charging. A mixture of molten iron and upto 30 percent scrap is poured into the top of BOS vessel. Oxygen of high-purity is then blown in at great speed using a water cooled lance. Oxygen combines with excess carbon and other undesired impurities. Which then float-off as slag.

Temperature and chemical composition are carefully controlled. When iron and carbon are seen in correct proportions, then, the steel is tapped onto a ladle. Molted metal is cast into moulds. The mixture solidifies into ingots. Ingots are taken to mill and rolled into structural sections, plates, bars, sheets and strips. It takes only 40 minutes as compared to ten hours in the **open- hearth process** (OHP).

Continuous casting process (CONCAST) has been developed more recently. In this system, the molted metal is poured directly into the casting machine to produce initial solid shape.

It has been introduced for the production of the structural sections. This process eliminates many defects of above mentioned production and gives a better quality final product. In this system, the production of few million of tonnes per mill per year is technically and economically sufficiently attractive.

In our country, steel is produced in the form of ingots and converted to different shapes, since 1907. First steel plant was set up at Jamshedpur by Jamsetji Nusserwanji.

After independence (i.e. 1947), realising the potential use of steel our first prime minister Pandit Jawahar Lal Nehru authorised some foreign countries to set up major steel plants, for example. by Tata Iron and steel company Indian Iron and Steel Company and Hindustan Steel. Hindustan steel has its plants at Durgapur, Rourkela and Bhilai.

Sir Mokshakundam Viseweswarayya set up the Bhadrawati Steel Plant to Karnataka.

In addition to above, there are steel plants at Bokaro, Burnpur and Vishkapatnam and 180 mini steel plants.

To-day, there are number of steel plants in our country and steel is produced in private sector. In 1990-2000 about 25 million tonnes of steel is produced in our country and its production is increasing at faster rate. In 2000-2005 the annual production of steel was 38 million tonnes.

1.9 RECENT DEVELOPMENTS IN MATERIAL

A number of developments in material such as steel have been made recently. The weldable qualities of steel (IS : 2062) designated as St 42-W) and (IS : 961 designated as St 55-HTW) are developed with the large scale use of welding.

IS : 961 has been developed with high tensile strength and there is saving in weight due to enhanced mechanical properties. Its weldable quality is advantageous for composite construction.

1.10 ADVANTAGES AND DISADVANTAGES OF STEEL MEMBERS

The steel is a versatile material for structural and other uses. It has many advantages.

The advantages of steel members are as follows :

1. The steel members have high strength. Therefore, the steel members can resist high loads with comparatively light weight and small size of members. The steel members can be conveniently handled and transported because of their small size.
2. The steel members are gas and water-tight, because of high density of steel (The unit weight of steel is 7.85 kN/m^3).
3. The steel members have long service life. This is because of high and homogenous strength and density properties of steel.
4. The steel members can be used as pre-fabricated members, because of ease of handling, fabrication and erection.
5. The steel members can be readily disassembled or replaced.
6. The existing steel structures and structural components may be strengthened by connecting additional sections or plates.
7. The steel structures may be inspected quickly and conveniently.
8. This material may be reprocessed/recycled in furnaces.
9. This material can be reused.
10. This material possesses high ductility and therefore suitable for earthquake resistant structures.

The disadvantage of steel members are as follows :

1. The steel members are susceptible to corrosion. The corrosion necessitates their painting or the use of other methods of their protection.
2. The steel members are costly

1.11 ROLLED STRUCTURAL STEEL SECTIONS

The steel sections manufactured in rolling mills and used as structural members are known as rolled structural steel section. Hot rolling process resulted in inevitably in the standardisation of sections (Shapes and sizes). The steel sections are named according to their cross-sectional shapes. Many steel sections are readily available in the market and have frequent demand. Such steel sections are known as regular steel sections. Some steel sections are rarely used. The manufacture of such steel sections is not usual. Such sections are produced on special requisition and are known as special sections.

'ISI Handbook for Structural Engineers I' gives nominal dimensions, weight and geometrical properties of various rolled structural steel sections. The handbook also gives other additional data required by the designers and architects.

The various types of rolled structural steel sections manufactured and used as structural members are given below :

1. Rolled Steel I Sections (Beam sections).
2. Rolled Steel Channel Sections.
3. Rolled Steel Tee Sections.
4. Rolled Steel Angle Sections.
5. Rolled Steel Bars.
6. Rolled Steel Tubes.
7. Rolled Steel Flats.
8. Rolled Steel Sheets and Strips.
9. Rolled Steel Plates.

1.12 ROLLED STEEL BEAM SECTIONS

The rolled steel beams are classified into the following four series as per BIS : (IS : 808-1989)

1. Indian Standard Junior Beams ...ISJB
2. Indian Standard Light Beams ...ISLB
3. Indian Standard Medium Weight Beams ...ISMB
4. Indian Standard Wide Flange Beams ...ISWB

The rolled steel columns/heavy weight beams are classified into the following two series as per BIS (IS : 808-1989)

1. Indian Standard Column Sections - ...ISSC
2. Indian Standard Heavy Weight Beams ...ISHB

The cross-section of a rolled steel beam has been shown in Fig. 1.1. The beam section consists of web and two flanges. The junction between the flange and the web is known as fillet. These hot rolled steel beam sections have sloping flanges. The outer and inner faces are inclined to each other and they intersect at an angle varying from 1.5 to 8 depending on the section and rolling mill practice. The angle of intersection of ISMB section is 8.

Abbreviated reference symbols (JB, LB, MB, WB, SC and HB) have been used in designating the Indian Standard Sections as per BIS (IS 808-1989).

Designation. The rolled steel beams are designated by the series to which beam sections belong, (abbreviated reference symbols) followed by depth (mm) of the section and weight (kN) per metre length of the beam (e.g., MB 225, @ 0.312 kN/metre). The weight per metre length of beam section should always be specified as in Indian Standard wide flange beams and Indian standard column sections - H-Beams sections of equal depths have different weights per metre length and also different properties e.g., WB 600, @ 1.340 kN/m, WB 600, @ 1.450 kN/m, HB 350, @ 0.674 kN/m, HB 350, @ 0.724 kN/m).

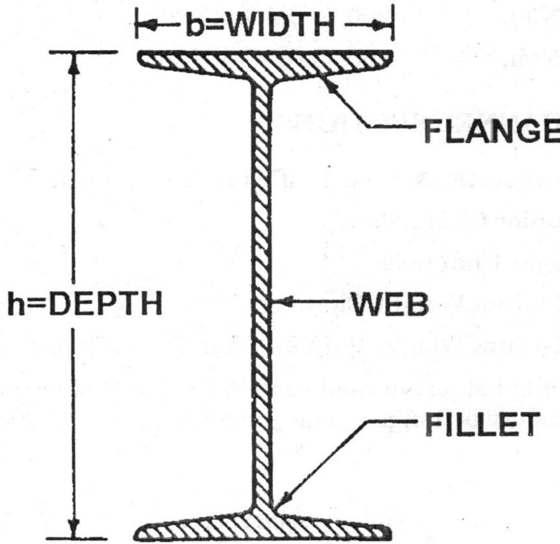

Buckling class: (i) Limits (h/b_f) > 1.2, t_f < 40 mm, Buckling about ZZ-axis buckling class (a) (ii) Limits t_f = 40 mm Buckling about yy-axis bucklingclass (b) (iii) 40 mm < mm, < t_f < 100 mm. Buckling about zz-axis, buckling class (b), yy-axis, Buckling class (c)

Fig. 1. Beam section

The rolled steel beams are mainly employed to resist bending. The rolled steel beams are used as independent sections to resist axial forces (compression or tension). These are also used in the built up sections of columns.

An efficient and a rational series of hot rolled beam and column sections for use in structures has been developed first time by our country. Out of the five above I-sections, only MB sections are being produced. Because of the rolling mill limitations, other sections have not yet been produced. Many countries are installing universal mills, which are special rolling mills and perfected to roll parallel flange I-sections as compared to conventional sloping flange sections. The beams of parallel flange sections have higher efficiency (represented by ratio of section modulus to wt.) and are easier to fabricate than the conventional sloping flange sections.

Note : In our country following sections are rolled

ISMB	600, @ 1.226 kN/m,	250, @ 0.373 kN/m
	550, @ 0.869 kN/m,	225, @ 0.312 kN/m
	450, @ 0.724 kN/m,	200, @ 0.254 kN/m
	400, @ 0.616 kN/m,	150, @ 0.149 kN/m
	350, @ 0.524 kN/m,	100, @ 0.115 kN/m
ISLB	100, @ 08 kN/m	
ISHB	250, @ 0.51 kN/m,	250, @ 0.547 kN/m
	225, @ 0.431 kN/m,	225, @ 0.468 kN/m

200, @ 0.343 kN/m, 200, @ 0.400 kN/m

150, @ 0.271 kN/m, 150, @ 0.344 kN/m

1.13 ROLLED STEEL CHANNEL SECTIONS

The rolled steel channel sections are classified in the following four series as per ISI :

1. Indian Standard Junior Channels ...ISJC
2. Indian Standard Light Channels ...ISLC
3. Indian Standard Medium Weight Channels ...ISMC
4. Indian Standard Medium Weight Parallel Flange Channels ...ISMCP

The cross-section of rolled steel channel section has been shown in Fig. 1.2. The channel section consists of web and two flanges. The junction between the flange and the web is known as fillet.

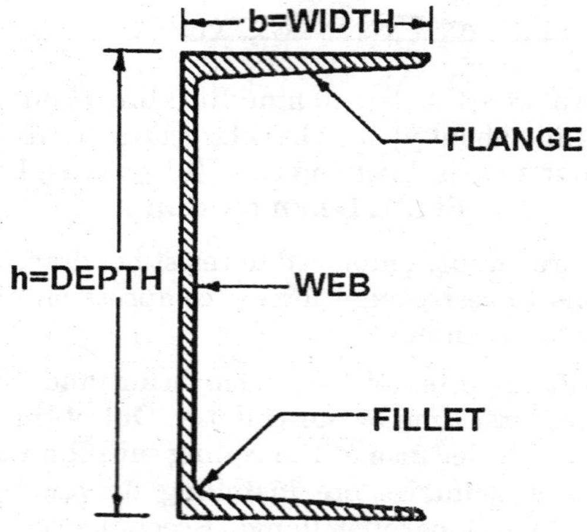

Buckling class: Buckling about any axis, Buckling class (c)

Fig. 2. Channel section

Designation. The rolled steel channels are designated by the series to which channel sections belong, (abbreviated reference symbols) followed by depth (mm) of the section and weight (kN) per metre length of the channel (e.g., MC225, 0.261 kN/m @ 0.259 kN/m).

The channels are employed as elements to resist bending (e.g., as purlins in the rolling of industrial buildings). It is to note that these are subjected to twisting or torsion because of absence of symmetry of the section with regards to the axis parallel to the web (i.e. yy-axis). Therefore, it is subjected to additional stresses. The channel sections are commonly used as members subjected to axial compression in the shape of built-up sections of two channels connected by lattices, or batten plates or perforated cover plates. The built-up channel sections are also used to resist axial tension in the form of chords of truss girders.

Note : As per IS: 808-1989, following channel sections have also been additionally adopted as Indian Standard Channel Sections

1. Indian Standard Light Channels with parallel flanges ...ISLC(P)

2. Medium weight channels ...MC
3. Medium weight channels with parallel flanges ...MCP
4. Indian Standard Gate Channels ...ISPG

In MC and MCP channel sections, some heavier sections have been developed for their intended use in wagon building industry. The methods of designating MC and MCP channel are also same as that for IS-channels described above.

In our country following sections give rolled

ISMC 400, @ 0.494 kN/m
350, @ 0.421 kN/m
300, @ 0.358 kN/m
250, @ 0.221 kN/m
225, @ 0.259 kN/m

ISMC 200, @ 0.221 kN/m
175, @ 0.191 kN/m
150, @ 0.164 kN/m
125, @ 0.127 kN/m
100, @ 0.092 kN/m
75, @ 0.057 kN/m

1.14 ROLLED STEEL TEE SECTIONS

The rolled steel tee sections are classified into the following five series as per ISI:

1. Indian Standard Normal Tee Bars ...ISNT
2. Indian Standard Wide flange Tee Bars ...ISHT
3. Indian Standard Long Legged Tee Bars ...ISST
4. Indian Standard Light Tee Bars ...ISLT
5. Indian Standard Junior Tee Bars ...ISJT

The cross-section of a rolled steel tee section has been shown in Fig. 1.3. The tee section consists of web and flange. The junction between the flange and the web is known as fillet.

Designation. The rolled steel tee sections are designated by the series to which the sections belong, (abbreviated reference symbols) followed by depth (mm) of the section and weight (kN) per metre length of the Tee (e.g. HT 125, @ 0.274 kN/m).

The tee section are used to transmit bracket loads to the columns. These are also used with flat strips to connect plates in the steel rectangular tanks. **Note :**

As per IS : 808-1984, following T-sections have also been additionally adopted as Indian Stand T-sections.

1. Indian Standard deep legged Tee bars ...ISDT

2. Indian Standard Slit medium weight Tee bars ...ISMT
3. Indian Standard Slit Tee bars from I-sections ...ISHT

It is to note that as per IS 808 (part II) 1978, H beam sections have been deleted.

Note : In our country, following sections are rolled

ISNT 80, @ 0.960 kN/m, 100, @ 0.150 kN/m

ISST 100, @ 0.081 kN/m, 150, @ 0.1996 kN/m

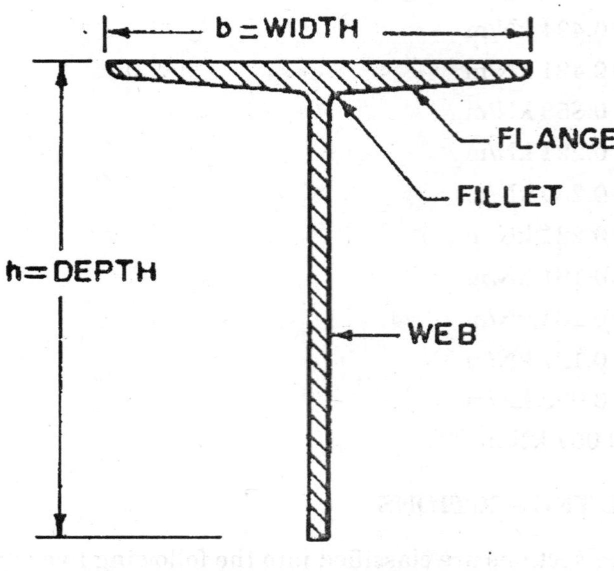

Buckling class : Buckling about any axis. Bucking class (c)

Fig. 1.3. The Tee section

1.15 ROLLED STEEL ANGLE SECTIONS

The rolled steel angle sections are classified into the following three series.

1. Indian Standard Equal Angles ...ISA
2. Indian Standard Unequal Angles ...ISA
3. Indian Standard Bulb Angles ...ISBA

The cross-section of a rolled equal angle section has been shown in Fig. 1.4 (A), unequal angle section, in Fig. 1.4 (B) and that of bulb angle section in Fig. 1.4 (C). The lengths of the legs in case of equal sections are equal and in case of unequal angle section, length of one leg is longer than the other. The thickness of legs of equal and unequal angle sections are equal. The bulb angle shown in Fig. 1.4 (C) consists of web flange and a bulb projecting from end of web. The thickness of web of bulb angle may or may not be equal to the thickness of flange.

Designation. The rolled steel equal and unequal angle sections are designated by abbreviated reference symbols (<) followed by lengths (mm) and thickness of legs, e.g. < 130

130 x 8 mm (< 130 130, @ 0.159 kN/m) < 200 100 x 10 mm (< 200 100, @ 0.228 kN/m) The rolled steel bulb angles are designated by BA, followed by depth (mm) of the section and weight (kN) per metre length of bulb angle.

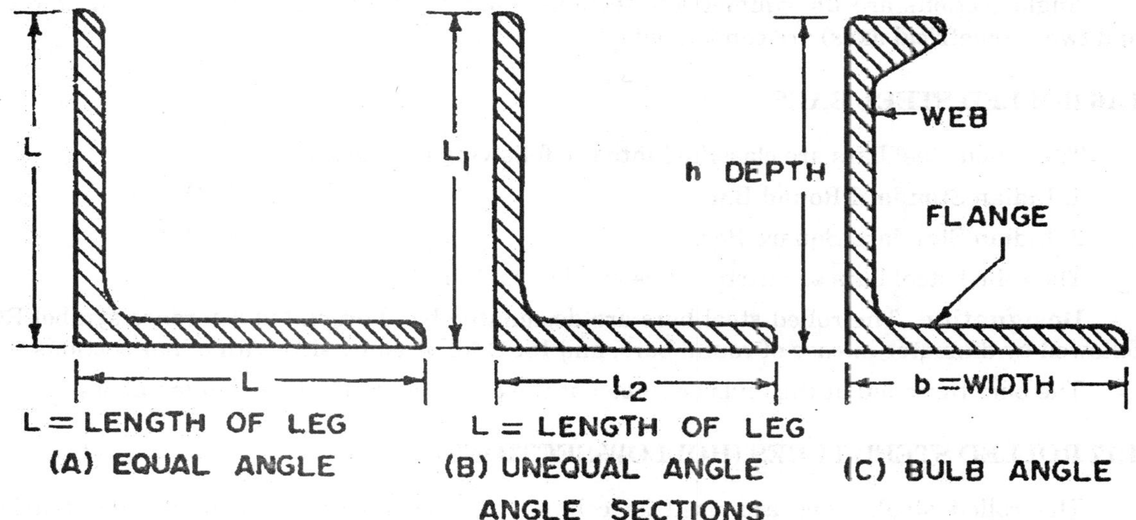

Buckling class: Bending about any axis. Buckling class : (c)
Fig. 1.4.

The Angle sections are used as independent sections consisting of one or two or four angles designed for resisting axial forces (tension and compression) and transverse forces as purlins. The angle sections are used as connecting elements to join together sheets or plates or to form a built up section. The angle sections are also used as construction elements for connecting beams to the columns, and purlins to the chords of trusses in the capacity of beam seats, stiffening ribs and cleat angles. The bulb angles are used in the ship buildings. The bulb helps to stiffen the outstanding leg when the angle is under compression. Note :

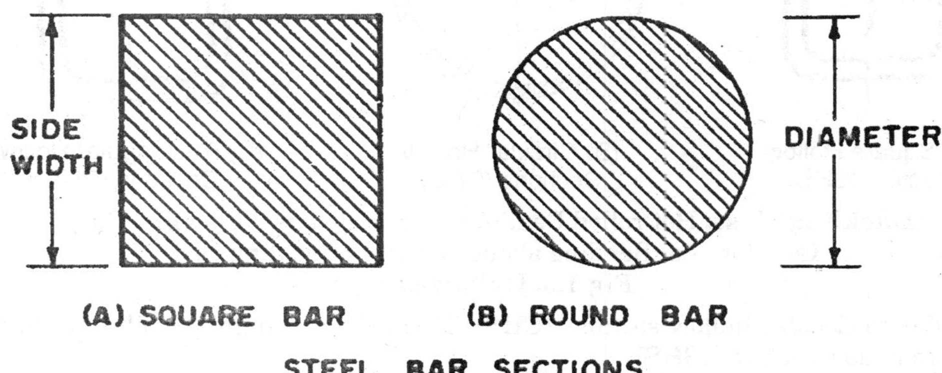

Buckling class: Buckling about any axis : Buckling class (c)
Fig. 1.5

As per IS : 808-1984, some supplementary angle sections have also additionally adopted as Indian Standard angle sections. However prefix ISA has been droped. These sections are designated by the size of legs followed by thickness (e.g. < 200 x 150 x 15).

Angle sections are un-symmetrical sections. Four axes (two principal axes, uu and vv and two geometrical axes) are considered.

1.16 ROLLED STEEL BARS

The rolled steel bars are classified into the following two series :

1. Indian Standard Round Bars ...ISRO
2. Indian Standard Square Bars ...ISSQ

The rolled steel bars are used as ties and lateral bracing.

Designation. The rolled stcel bars are designated by abbreviated reference symbol RO followed by diameter in case of round bars, and ISSQ followed by side with of bar sections.

The bars threaded at the ends or looped at the ends are used as tension members.

1.17 ROLLED STEEL TUBES (HOLLOW-SECTIONS)

The rolled steel tubes are used as columns and compression members and tension members in tubular trusses. The rolled steel tubes are efficient structural sections to be used as compression members. The steel tube sections have equal radius of gyration in all directions.

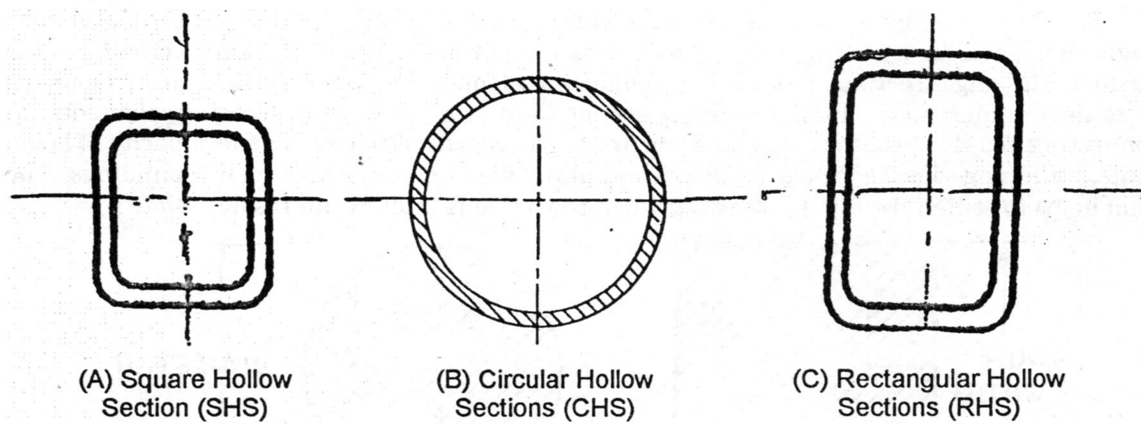

(A) Square Hollow Section (SHS)

(B) Circular Hollow Sections (CHS)

(C) Rectangular Hollow Sections (RHS)

Buckling class: Hot rolled Bending about any axis Buckling class (a)
Cold formed Bending about any axis Buckling class (b)

Fig.1.6. Hollow-sections

Similar to circular hollow sections (CHS), There are rectangular hollow sections (RHS), and square hollow sections (SHS).

1.18 ROLLED STEEL FLATS

The rolled steel flats are used for lacing of elements in built- up members, such as columns and are also used as ties.

Designation. The rolled steel flats are designated by width (mm) of the section followed by letters (abbreviated reference symbol) F, and thickness in (mm), (e.g., 50 F 8). This means a flat of width 50 mm and thickness 8 mm.

The rolled steel flats are used as lattice bars for lacing the elements of built-up columns. The rolled steel flats are also used as tension members and stays.

1.19 ROLLED STEEL SHEETS AND STRIPS

Development of sheets and plates was a further development in the production of rolled steel sections. This sheet metal was a major cladding material during the twentieth century. Steel plate allows the fabrication of non-standard elements. The plates may be used to form large plate girders and box structures plates have been used with great success in recent years in bridge design.

Designation. The rolled steel sheet is desingated abbreviated reference symbol SH followed by length (mm) x width (mm) x thickness (mm) of sheet. The rolled steel strip is designated as ISST followed by width (mm) x thickness (mm) (e.g., SH 2,000 mm x600 mm x8 mmand ST 250 mm x 2 mm).

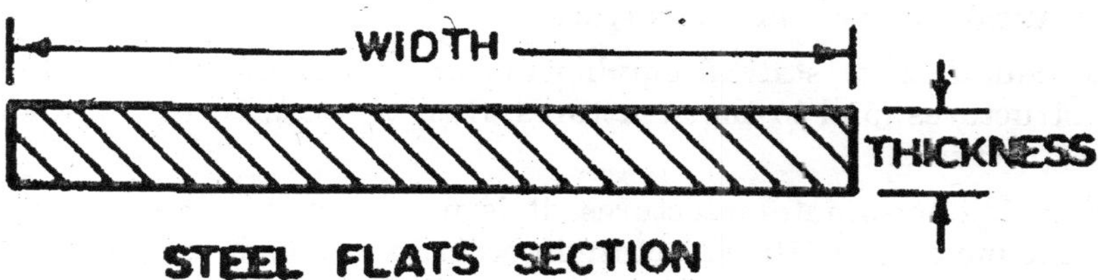

Fig. 1.7. Steel flats

1.20 ROLLED STEEL PLATES

Designation. The rolled steel plates are designated by abbreviated reference symbol PL followed by length (mm) x width (mm) x thickness (mm) of the plates (e.g., PL 2,000 mm x 1,000 mm x 6 mm).

The rolled steel sheets and plates are widely used in construction. Any sections of the required dimensions, thickness and configuration may be produced by riveting or welding the separate plates. The rolled plates are used in the web and the flanges of plate girders, plated beams and chord members and web members of the truss bridge girders. The rolled steel plates are used in special plate structures (e.g., shells, rectangular and circular steel tanks and steel chimneys).

1.21 RECENT DEVELOPMENTS IN SECTIONS

The rolled steel beam sections with parallel faces of the flanges are recently developed. These beam sections are called as parallel flange sections. These sections have increased moment of inertia, section modulus and radius of gyration about the weak axis. Such sections used as beams and columns have more stability. These sections possess ease of connections to other sections as no packing is needed as in beams of sloping flanges.

The parallel flange beam sections are not yet rolled in our country.

New welded sections using plates and other steel sections are developed because of welding. The development of beams with tapered flanges and tapered depths is also due to welding. The open web sections and the castellated beams (Fig. 1.2) were also developed with the rapid use of welding.

1.22 STRUCTURES

When any body is subjected to a system of loads and deformation takes place, and the resistance is set up against the deformation, then, the body is known as structure. The structures are means of transferring forces and moments. The structures may be classified as **statically determinate structures** and **statically indeterminate structures**.

In statically determinate structures the distribution of moments and forces may be established by applying the equations of equilibrium alone. For example, a simply supported beam is a statically determinate structure.

When the equations of statics (i.e. the equations of statical equilibrium) are enough to determine all the forces acting on the structures, and in the structures, then, the structures are known as statically determinate structures.

When the equations of statical equilibrium are not sufficient to determine all forces acting on the structures, and in the structures, then the structures are known as statically indeterminate structures.

In statically indeterminate structures, it is necessary to take into consideration, the stiffenesses of the members of the structure in order to establish the distribution of moments and forces. For example, a continuous beam is a statically indeterminate structure.

The equations of consistent deformations are added to the equations of equilibrium in order to analyse the statically indeterminate structures.

1.23 CLASSIFICATION OF STRUCTURES

The structures are characterized by their supporting systems. There are one dimensional, two dimensional and three dimensional supporting systems (i.e. line, surface and three dimensional structures). When a supporting system is subjected to only one type of stresses, then, it is known as basic system. When a system is subjected to simultaneously several types of stresses, then it is known as mixed system.

1. One dimensional supporting system can also be termed as **line supporting** *structures*. The line supporting structure is large in one dimension and small in the other two dimensions. The material remains concentrated along a straight, curved or angular line. A flexible cable is an example for such a line supporting structure.

2. Two dimensional supporting system can also be called as **surface structure**. The surface structure is large in two dimensions and small in third dimension. The material remains distributed along a surface. The surface may be plane or curved. The curved surface may possess single or double curvature. A surface structure may also consist of line elements that form a continuous structure. The frame structures are examples of surface structures made of line elements. The frame structures are built by assemblies of elongated members. The **shell structures** are rigid curved surface structures. The shell structures are built of

sheets or plates. The shell roof coverings for large buildings, aero planes, rail road cars, ship bulls, tanks etc. are the examples of shell structures.

3. Three dimensional structures are large in three dimensions and these structures may have any shape. The three dimensions structures are also called as space structures. The three dimensions frame structures are also space structures.

The structures may be further classified depending on the materials used as plastic structures, aluminium structures, timber structures, reinforced concrete structures and steel structures. The structural steel is used for the steel structures. The high strength and ductility are the properties of particular importance for the structural use of the steel.

The ductility is the unique property of steel by virtue of which it is able to deform substantially beyond the limit of yield deformation. The ratio of ultimate deformation to yield deformation is usually termed as **ductility** either in tension or compression before failure.

The structure can also be divided into the following three categories :

1. Skeletal structures. 2. Stressed skin structures. 3. Space structures.

1. The skeletal structures are those structures in which the member may be represented by a line and the intersections of lines correspond to joints of the structures.

2. Stressed-skin structures is defined as that which utilises the stiffness of the skin of the structure as part of the basis of design. Such an assumption is made in the design of certain reinforced concrete structures and this may be utilised in the design in steel frame construction where in latticed frames in surface planes simulate the solid skin. Underlying principle is that all forces which act at angles to the skin or its equivalent are resolved into components acting in the planes of the skin, the shape of which is arranged so that this may be done.

Professor E.R. Bryan of Structural Engineering, University of Salford worked for many years on the stiffening effect of steel cladding (sheeting) on steel framed structures steel sheeting makes a significant contribution to the stiffness and possibly, the strength of steel framed structures. In buildings, which are designed from the beginning as stressed-skin structures, the effect is even more.

The design method refers to the elastic design of steel framed, steel sheeted buildings which act primarily as umbrella structures. The design method leads to considerable economy of steel compared with existing design methods. It describes the true behaviour of the completed building.

There is a manual, which sets down the principles of stressed-skin construction as applied mainly to single bay single storey portal frame buildings. In the manual, the method of calculation has been given in detail. In addition, it presents tables and data relating to steel sheeting. The recommendations for the design details have also been mentioned in the manual.

Professor Bryan presented a monograph, which gives full design treatment of the composite action between steel sheeting and single storey portal frame structure and provides methods whereby the effect of such cladding on both stiffness and strength may be determined and used in design. The stresses induced in the cladding are sufficiently small for these to have no cladding to take the normal loading for which it is primarily designed.

3. Space structures: Literally, the space structures are defined as three dimensional structures. Space structures may also be referred as spatial structures which have their action in three dimensions. Working group on spatial steel structures of the International Association [Reference : International Association for shell and spatial structures (IASS), working Group on spatial structures, 1984, Analysis design and realization of space frames, Bulletin IASS No. 84/85, XXV (1/2) : 1-114] defined 'Space Structures' as under :

'**A space frame** is a structure system assembled of linear elements so arranged that the forces are transferred in a three dimensional manner. In some cases, the constituent elements may be two dimensional. Macroscopically (...a space frame often takes the form of a flat or curved surface.

ASCE Task Committee on Latticed structure prepared a report [Reference : ASCE Task Committee on Latticed Structures of the Committee on special structures of the Committee on Metals of the structural Division 1976. Latticed - structures : State of the art report. Journal of structural Division, Proceedings, ASCE 102 (ST II) : 2197-2230] above mentioned same structure was defined as a space frame, referred to as **latticed structure** as below :

'**A latticed structure** is a structure system in the form of a net- work of elements (as opposed to a continuous surface). Rolled, extruded or fabricated sections comprise the member elements. Another characteristic of latticed structural system is that their load-carrying mechanism is, three dimensional in nature.' Flat surfaces with loading perpendicular to the plane as well as curved surfaces are also included. Structural systems (e.g., common trusses or building frames, which may appropriately be divided into series of planar frame works with loading in the frame work) have been excluded in the report. Terms space frames and latticed structures are considered synonymous (word having same meaning as another in same language).

An array of single, double or multiple layers of intersecting members arrangements are considered as space frames.

A single layer space frame having a curved surface is referred as braced vault, braced dame, or latticed shell.

Space frames are referred by some authors as **double layered grid**.

The joints of a space frame are assumed as having rigid-joints. Internal torsions develop at these joints and internal moments develop in the members.

The joints of a space trusses are assumed as having hinged joints. Therefore, the internal moments do not develop in the members.

In the space frames and the space trusses, there is no difference in the member geometry. The difference between them is in the joint connection and detailing. Absolute rigid-joints or hinged joints do not exist in engineering practice.

A single layer curved surface space frame may be analysed either as having hinged connections or rigid connections. Whereas, a double layer flat surface may be analysed as hinged connections.

Both space frames and space trusses are referred by the term space structures.

Design life of a structure: is the time period for which a structure or a structural element performs its function without damage.

1.24 CLASSIFICATION OF STRUCTURAL MEMBERS

A structural member is also classified as one-dimensional, two dimensional and three-dimensional element. Actually a structural member has always three dimensions. But, in case, the width and thickness are small in comparison with length, then, such members are classified as **one dimensional elements**.

In case, the thickness is small in comparison with length and width, then such members are classified as **two dimensional elements**. Hence, the plates are considered as two dimensional elements.

As far as the ratios between length, width and thickness are concerned there is no clear cut limit to them in accordance with which members are classified as one, two, or **three dimensional elements.**

1.25 SPECIFICATIONS AND BUILDING CODES

Building codes reflect historically successful building practice and communicate to the practising engineers, those level of safety considered acceptable to society. The generic description of acceptable levels of structural performance is communicated to fresh engineers who do not have specific previous experience, so that these engineers are allowed to design and to create acceptable designs. Fresh engineers may rely heavily on historically established practice so that they may take the vast numbers of decisions needed in the process of design.

The construction of steel structures involves four parties, (*viz.,* the owners, the designers, (engineers and architects), the fabricators and the constructors). The specifications are prepared by the designers. The acceptable standard of quality of materials and of workman-ship in fabrication are defined by the specifications. The designers prepare three types of specifications, (*viz.* project specifications, material standards and design codes). The object of drafting specifications and preparing codes is to achieve economy in the use of material and the cost of construction without sacrificing the safety. The project specifications describe complete information about owner's and the engineer's precise requirements for the completed structures. The material standards describe the quality of materials and their economic use.

Important benefits of standardization are improvements of the suitability of materials, products, processes and services for their intended purposes, prevention of barrier trade and facility of technological co-operation. Standardization is an activity of establishing with regard to actual or potential problems, provisions for common and repeated use, aimed at the achivement of the optimum degree of order in a given context. Standard cover almost all aspects of economic activity of a person-engineering industry, construction and many others. Correct perspective of aims of standardization is found by defining them at the following levels :

1. **International standardization**: Relevant bodies of all countries have their involvement at this level. International standards are formulated and adopted and made available for use.

2. **Regional standardization:** Relevant bodies of some countries (e.g. from geographical, political or economic area of the world) have their involvement. These are formulated and adopted made available for use.

3. **National standardization:** These are done by one country. These are formulated by national standard body and adopted and made available for use.

In addition to above, relevant bodies of different associations and companies may have their own involvement for formulation of standards.

These standards may differ in form and type depending on the particular aspect of a subject that may be covered. These aspects may include (i) a set of nomenclature or definition of terms; (ii) a specification for the quality, composition, or performance of a material, an instrument, a machine or a structure, (iii) a method of sampling or inspection to find conformity to a specified requirement of a large batch of material by inspection of a smaller sample, (iv) a method of test or analysis to evaluate specified characteristics of a material or a chemical. (v) a scheme of simplification or rationalization, (i.e., limitation of variety of sizes, shapes or grades) designed to meet most economically the needs. This also includes dimensional freezing of component design to ensure interchangeability as also methods of grading and grade definitions for natural products, such as timber, minerals, etc. (vi) a code of practice dealing with design construction, operation, safety, maintenance of a building, an installation or a machine, and (vii) a model form of contract or agreement.

The design codes incorporate

1. General guidance as regards various loads.
2. Allowable stresses.
3. Design principles and procedures generally accepted throughout the country.

For complex and time consuming formulae and provisions of designs, the tables are included and simplified assumptions are used in the code, which reduce the design calculations and complex behaviour states to predictable performance. The various professional associations, institutes and government agencies prepare codes and specifications.

In America, the American Institute of Steel Construction specifications, briefly known as AISC specifications (Specifications for the Design, Fabrication and Erection of Structural Steel Buildings) 8th edition, Chicago, 1978, are following by steel buildings. For the special aspect of design, the following specifications are followed:

AISI Specifications	:	American Iron and Steel Institute Specifications. For the design of cold formed steel structrual members, Washington, D.C. 1980
AWS Specifications	:	American Welding Society Specifications, Structural Welding Code, Steel AWS, DI 1-84, 9th Edition, Miami, 1985.
SJI Specifications	:	Steel Joist Institute Specifications
AASHTO Specifications	:	American Association of State Highway and Transportation Officials Specifications.
ANSI	:	Specifications American National Standard Institute (Minimum Design Loads and other structures ANSI A 58.1, New York)
USBPR Specifications	:	U.S. Bureau of Public Roads Specifications
AREA Specifications	:	American Railway Engineering Association Specifications
USASI Specifications	:	U.S. American Standard Institute.

AASHO, USBPR and AREA specifications are for the highways and railway bridges.

In Britain, the British Standard Institute prepares the specifications. These specifications are known as British Standard Specifications (BSS). For structural use of steel work in building BS 5950-1990 (recently revised as BS 5950-2000) is used now-a-days. There are following nine parts : Steel structures are designed and proportioned taking into consideration the limit states upto which the structures remain fit for their intended use.

BS 5400 is an additional code available for research engineers. This code is applicable to steel concrete and composite construction. There are ten parts, which cover various aspects of bridge design. Part 3 is `code of practice for design of steel bridges.'

European Economic Community of Eurocodes intends to fulfil a similar role with `Euro-codes, EU' as done at presents by national codes within the individual member countries. EC3 is for steel structures EC4 is for the composite construction. These are available for trial use. Other codes are in course of preparation.

In Germany, the German Standard Institution (DIN) prepares the specifications. In USSR, USSR State Standards (GOST) prepares the specifications.

In our country India the following organisations have prepared and published the specifications:

1. Bureau of Indian Standard (BIS)
2. Indian Road Congress (IRC)
3. Ministry of Railways and Railway Board.

The standardization work in the field of Civil Engineering began as the part of the composite Engineering Division of the then ISI, (Indian Standards Institution). This Institution had made valuable contribution to economic and industrial development of the country and formulated many standards and specificiations and codes of practice needed by the users. The Bureau of Indian Standards Act, 1986 was passed by the Parliament, Government of India and it became functional from 1 April, 1987 and started performing all functions which were looked after by ISI. Bureau of Indian Standards (BIS) is the National Standards Body of India. Since, 1987, it is engaged in the formulation of standards in various fields.

In 1950's, the steel economy programmes was initiated in Indian Standard Institute. The object of this programme was to achieve economy in the use of structural steel. The rational, efficient and optimum standards for structural steel products and their use were established. The code of practice for 'Use of Structural Steel in General Building Construction' IS 800-1956 was published. It had been first in the series of Indian Standards. After this standard was in use for some time, the first revision, incorporating certain very important changes was prepared and published in IS : 800-1962. Considering the comments received on the standard during its usages and developments taking place in the country and abroad, the necessary modifications and additions have been incorporated in the second revision. The standard has been made more useful and IS 800-1984 with the modified title as 'Code of Practice for General Construction in steel' has been brought out. This code is applicable to all types of steel structures.

Following IS codes are useful for the design of steel structures : IS :

1	226-1975	Structural Steel (standard quality)
2	227	Galvanised (Plain and Corrugated) sheets
3	459	Unreinforced Corrugated Asbestos Cement Sheets
4	723	Mild Steel Wire nails
5	800-1984	Code of practice for General Construction in Steel
6	806	Use of Steel Tubes in General Building Construction
7	808	Rolled Steel Beams, Channels and Angle Sections
8	812-1957	Glossary of Terms Relating to Welding and Cutting Metals
9	813-1961	Scheme for Symbols of Welding
10	816-1969	Use of Metal Arc Welding for General Construction in Mild Steel
11	819-1957	Resistance Spot Welding for Light Assemblies in Mild Steel
12	875-1984	Loading Standard for Structural Safety of Buildings
13	961-1975	Structural Steel (High Tensile)
14	1148-1973	Rivets Bars for Structural Purposes
15	1149-1982	High Tensile Rivet Bars for Structural Purposes 16 1161 Steel Tubes for Structural Purposes
17	1173	Rolled Steel Sections, Tee-Bars
18	1252	Rolled Steel Sections, Bulb-Angles
19	1261-1959	Seam Welding in Mild Steel
20	1323-1962	Code of Practice for Oxy-Acetylene Welding for Structural Works in Mild Steel
21	1730	Dimensions for Steel Plates, Sheets and Strips for Structural and General Engineering Purposes
22	1731	Dimensions for Steel Flats for Structural General Engineering Purposes
23	1732	Dimensions for Round and Square Steel Bars for Structural and General Engineering Purposes
24	1911	Schedule of Unit Weight of Materials 25 1977-1975 Structural Steel (ordinary quality)
26	2062-1984	Weldable Structural Steel
27	2585	Black Square Bolts and Nuts and Black Square Screws
28	3139	Dimensions for Screw Threads for Bolts and Nuts
29	3757-1972	High Tensile Friction Grip Fasteners for Structural Engineering Purposes
30	4000-1967	Assembly for Structural Joints using High Tensile Friction Grip Fasteners
31	456-1978	Code of Practice for Plain and Reinforced Concrete (Third Revision)
32	6623-1972	High Tensile Friction Grip Nuts

33 6639-1972 Hexagon Bolts for Steel Structures

34 6649-1972 High Tensile Friction Grip Washers

In addition to above the following ISI : Handbooks are also useful for Design of Steel Structures :

1. Handbook for Structural Engineering 1. (Structural Steel Section)
2. Handbook for Structural Engineering 2. (Steel Beams and Plate Girders)
3. Handbook for Structural Engineering 3. (Steel Columns and Struts)

Following Specifications are useful for the design of timber structures :

1. 883 Use of Structural Timber in Building (Material, Grading and Design)
2. 2366 Nail Jointed Timber Construction.

Building Code (Document that recomments practices or procedures for the design manufacture, installation, maintenance or utilization of equipment, structure or products) is a broadly based document. It is a legal document and specifies legal requirement. Building codes are drafted by either state or by local bodies. It must not be dis-regarded even it may not be a perfect. Building codes generally treat all facets relating to safety, such as structural design, architectural details, fire protection, heating and air conditioning, plumbing and sanitation, and lighting.

Specifications represent agreements by individuals or groups on standard practice specifications frequently refer to rules set forth by the architect or the engineers specifications may be thought of as recommended practice.

The specifications are revised from time to time depending upon the experience, developments and research works of research institutions and laboratories.

Aims of standardization in general are (i) to achieve maximum overall economy in terms of cost (ii) to help in better understanding at contact points, (iii) to provide variety of control, (iv) to provide fitness of purpose, (v) to adjudge suitability as regards the functional aspect (vi) to provide compability to fulfill the relevant requirements, (vii) to provide safety of the environment, surrounding or the community, (viii) to protct health of consumers using items of food, (ix) to proect the environment from different kinds of pollution (such as air, water, etc.) and (x) to remove trade barriers to promote fair competition.

Many sectors of economy (*viz.* producers, consumers, trade and technologists) have accrued advantages from standardization. Technologists alone have many benefits (e.g., accepted method and materials for expediting designs, facility for introducing improvement in products and processes, enhanced knowledge of properties, possibilities and application of materials, guidance for formulation of research and development programmes etc.).

REFERENCES

1. IS:226-1975: Structural Steel (Standard Quality)
2. IS: 961-1975: Structural Steel (High Tensile)
3. IS: 1977-1975: Structural Steel (Ordinary Quality)
4. IS: 2062-1984: Weldable Structural Steel.

5. IS: 8500-1977: Weldable Structural Steel and High Strength Qualities.

6. Edwin Gregory and Eric, N. Simons. 'The Structural of Steel' Odhams Press Limited Long Acre, London, Third Edition 1957 (Stainless Steel)

7. Rangachari Narain, V. Kalyanaraman, *et al.* 'Teaching Resources Structural Steel Design Institution of Steel Development and Growth (Production of Steel)

8. IS: 808-1989: Rolled Steel Beams, Channels and Angle Sections.

9. IS: 806-.....: Use of Steel Tubes in General Building Construction.

10. IS: 1161-....: Steel Tubes for structural purposes.

11. IS: 11973- : Rolled Steel Sections -Tee Bars.

12. IS: 1252-: Rolled Steel Bulb Angles.

13. IS:1730- : Dimensions for Steel plates, Sheets and strips for structural general engineering purposes.

14. IS:1731- : Dimension for steel flats for Structural General Engineering Purposes.

15. IS:1732-: Dimensions for Round and Square Steel Bars for structural and general Engineering Purposes.

2

Part - II
LOADS (ACTIONS), METHODS OF ANALYSIS AND DESIGN AND PROPERTIES OF STEEL (CHARACTERISTIC) SERVICE

LOADS (ACTIONS), METHODS OF ANALYSIS AND DESIGN, AND PROPERTIES OF STEEL SERVICE (CHARACTERISTIC)

2.1 INTRODUCTION

The structures and structural members are designated to meet the functional and structural aspects. Both the aspects are interrelated. A building has to perform many functions satisfactorily.

The **functional aspect** takes into consideration the purpose for which the building or the structure is designed. The functional aspect includes the determination of the amount and arrangement of space necessary for the structure. The structure is able to perform satisfactorily the services for which it is designed. The functional aspect includes the determination of location and arrangement of operating utilities of the building for the intended use and occupancy, structural safety, fire safety and compliance with hygienic, sanitation, ventilation, special equipment, machinery or other features, incident, to the proper functioning of the structures. The preliminary layout of the structure is made. The preliminary layout of the structure is useful in determining the type of structure to be used, the general size, shape and arrangement of the various parts of the structure. The approximate dimensions of the structures are established. These dimensions are used for the structural design.

In the **structural aspect**, it is ensured that the building or the structure is structurally safe, strong, durable and economical. Many structures are built to shelter human beings and to protect them from the forces of nature. Occupants of the buildings or the structure should feel safe with regard to their lives and properties. The safety of many individuals at one time depends on the structural safety of the structure. The minimum requirements pertaining to the structural safety of buildings are being covered in codes dealing with loads by way of laying down minimum design loads which have to be assumed for dead loads, imposed loads, wind loads and other external loads, the structures would be required to bear. Unnecessarily, heavy loads without proper assessment should not be assumed. The structures are designed between two limits, (viz., the structural safety and economy). The structures and structural members are designed to satisfy three basic principles of design. The structures should be **strong**, **stable** and **stiff.**

All the sections of the structures should have strength at least equal to the structural effects of design loads and forces that occur during the construction and use. All the loads should be taken into consideration, which a building or a structure would be required to bear. A structure may be required to withstand any loads out of the various loads or combination of the various loads. These loads include overhead and other cranes, equipment, etc., in industrial buildings, occupancy etc. in buildings; vehicles on bridges; water in hydraulic structure; loose materials for bins and bunkers and so on. For the purpose of determining the maximum stresses in any structure or member of a structure, the various loads which are likely to act on the structure are given below : These loads are referred as **service-loads.**

1. Dead load, 2. Live load (Imposed load), 3. Wind load,
4. Snow load 5. Seismic load, 6. Temperature effects.

In addition to these above loads and effects, following forces and effects are also considered while designing the structures if these are liable to affect materially the safety and serviceability of the structures.

1. Foundation movements
2. Elastic axial shortening
3. Soil and fluid pressures
4. Vibrations
5. Fatigue
6. Impact (dynamic effects)
7. Erection loads
8. Stress concentration effect due to point load and the like.

When the loads act upon the structures the geometry of the structures charges due to action of the loads. Therefore, the loads are referred as **actions.** Loads are the externally applied forces. The internal actions produced in the various member (viz., axial force, bending moments, shear forces and torsions) are referred as **load-effects.**

According to the strength requirement, the internal actions produced in the structural members must be equal to the structural effects caused due to designed loads, forces and various effects.

2.2 DEAD LOADS (PERMENANT LOADS)

The dead loads (viz., the self-weight of the structural members), superimposed dead loads and loads due to filling materials are referred as **permanent loads** [All permanent construction and installations including the self-weight of all walls, partitions, floors, roofs and other permanent fixture, acting on a member].

Dead loads are loads which are constant in magnitude and fixed in position throughout the life time of the structure. Dead loads in a building comprise the self weight of he structure and all other super-imposed loads (*viz.,* all permanent constructions and installations including weight of walls, partitions, floors, and roofs). The self-weight of the structure being unknown in the beginning of the design, it is assumed by the designer in the light of experience and thumb rules and formula available. The super-imposed dead loads are determined for the various materials specified for the construction and unit-weight of materials. On completion of design, total dead load is calculated and checked with the dead load assumed for design and the design is revised, if necessary. Table 2.1 gives the unit-weights of materials used in building construction. The dead loads are also called as **permanent loads** as regards to the duration of their action.

Table 2.1. UNIT WEIGHT OF BUILDING MATERIALS, BUILDING PARTS AND COMPONENTS IS : 875 (PART I) - 1987
(Extracts from Tables I and II of IS:1911-1967, schedule of unit-weight of materials)

No.	Materials	Nominal size of thickness	Weight mass kN	Kg	per
(1)	(2)	(3)	(4)	(5)	(6)
	Building materials				
1	Asbestos natural	—	29.8	3040	m^3
	raw		5.90 to 8.85	600 to 900	m^3
2	Asbestos cement sheeting corrugated	6 mm	0.118 to 0.130	12	m^2
	(146 mm)	6 mm		13.3	m^2
	Semi-corrugated (240 mm)	6 mm	0.118 to 0.127	12	m^2
		6 mm		13	m^2
	plain	5 mm	0.090	9.16	m^3
3.	Bitumen	—	0.102	10.4	m^3
4	Boards (cork)				
	Compressed	10 mm	0.04	4	m^2
	Ordinary	10 mm	0.02	2	m^2
5	Bricks, chips and broken bricks				
	Fine	—	14.2	1450	m^3
	Coarse	—	9.9	1010	m^3
6	Brick dust (Surkhi)	—	9.9	1010	m^3
7	Bricks (Engineering)	—	21.2	2160	m^3
8	Cement				
	Ordinary and Aluminious	—	14.10	1440	m^3
	Rapid hardening	—	12.55	1280	m^3
9	Plain cement concrete with sand and gravel or crushed natural stone	—	22 to 23.5	2240-2400	m^3
10	Reinforced cement concrete sandstone aggregate with				
	1 Per cent steel	—	22.75-24.20	2310-2470	m^3
	2 Per cent steel	—	23.25-24.80	2370-2530	m^3
	5 Per cent steel	—	24.80-26.50	2530-2700	m^3
11	Glass	—	24.384	2480	m^3
	Glass bricks	—	24.304	2480	m^3
	Window glass and		24.304	2480	m^3
	Looking glass		24.304	2480	m^3
	Sheet	2.0 mm	0.049	5.0	m^3
		2.5 mm	0.062	6.3	m^3

	3.0 mm	0.074	7.5	m^3
	4.0 mm	0.098	10.0	m^3
	5.0 mm	0.123	12.5	m^3
	5.5 mm	0.134	13.7	m^3
	6.5 mm	0.167	17.0	m^3
12 Gypsum				
Mortar	—	11.75	1200	m^3
Powder	—	13.89-17.25	1410-1760	m^3
13 Iron				
Pig	—	70.60	7200	m^3
Gray cast	—	68.95-69.90	7030-7130	m^3
White cast	—	74.30-75.70	7530-7720	m^3
Wrought	—	75.50	7700	m^3
14 Mastic asphalt	10 mm	0.215	22	m^2
15 Metal sheeting Galvanised steel sheets (plain and corrugated)				
Case 1	1.60 mm	0.131	13.31	m^2
	1.25 mm	0.104	10.50	m^2
	1.00 mm	0.084	8.60	m^2
	0.80 mm	0.069	7.03	m^2
	0.63 mm	0.056	5.70	m^2
Case 2	1.60 mm	0.129	13.16	m^2
	1.25 mm	0.102	10.41	m^2
	1.00 mm	0.083	8.45	m^2
	0.80 mm	0.067	6.88	m^2
	0.63 mm	0.054	5.55	m^2
Case 3	1.60 mm	0.128	13.01	m^2
	1.25 mm	0.101	10.26	m^2
	1.00 mm	0.081	8.30	m^2
	0.80 mm	0.066	6.73	m^2
	0.63 mm	0.053	5.40	m^2
Case 4	1.60 mm	0.127	12.94	m^2
	1.25 mm	0.100	10.19	m^2
	1.00 mm	0.081	8.22	m^2
	0.80 mm	0.065	6.66	m^2
	0.63 mm	0.052	5.32	m^2
16 Mortar				
Cement	—	20.40	2080	m^3
Lime	—	15.70-18.05	1600-1840	m^3
17 Plaster				
Cement	—	20.40	2080	m^3
Lime	—	17.25	1760	m^3

18	Soils and gravels				
	Alluvual ground (undisturbed)	—	15.60	1600	m^3
	Clay (dry and pump)	—	10.20	1040	m^3
	Dry (compact)	—	14.10	1440	m^3
	Damp. (compact)	—	17.25	1760	m^3
	Clay (wet and compact)	—	20.40	2080	m^3
	Undisturbed	—	18.85	1920	m^3
	Undisturbed bed gravelly	—	20.40	2080	m^3
	Eatch				
	Dry	—	13.85-18.05	1410-1840	m^3
	Moist	—	15.70-19.60	1600-2000	m^3
	Gravel				
	Loose	—	15.70	1600	m^3
	Rammed	—	18.85-21.20	1920-2160	m^3
19	Stone				
	Agate	—	25.50	2600	m^3
	Aggregate	—	15.70-18.85	1600-1920	m^3
	Basalt	—	27.95-29.05	2850-2960	m^3
	Granite	—	25.90-27.45	2640-2800	m^3
	Lime	—	23.55-25.90	2400-2640	m^3
	Marble	—	26.70	2720	m^3
	Sandstone	—	27.45	2800	m^3
20	Tiles clay				
	Mangalore pattern	—			
21	Timber				
	Hard wood (e.g. Babul)	—	6.27-9.41	640-960	
	Light wood (e.g. Fir)	—	3.92-4.71	400-480	
	Medium wood (e.g. Deodar)	—	4.71-6.7	480-640	
	Building parts and components				
22	Ceiling (on wire netting)	10 m	0.265	27	m^2
23	Flooring				
	Asphalt	10 mm	0.22	22	m^2
	Terrazo paving	10 mm	0.24	24	m^2
24	Roofing				
	Allahabad tiles (including battens)				
	Single	—	0.83	85	m^2
	Double	—	1.67	170	m^2
	Country tiles (with battens)				

	Single	—	0.69	70	m^2
	Double	—	1.18	120	m^2
	Mangalore tiles with batten (tiles with mortar)	—	0.64	65	m^2
	Slates (on battens)	—	1.08	110	m^2
	Glazed with aluminium alloy bars for span upto 3 m	—	0.34-0.49	35-50	m^2
	Glazed with lead covered steel bars at	6.4 mm	0.19	19.5	m^2
	0.6 m centres	6.4 mm	0.25-0.28	26-29	m^2
27	Wall				
	Engg. bricks (in masonary)	—	23.55	2400	m^3
	Stone (masonry)				
	Granite ashlar	—	25.90	2640	m^3
	rubble	—	23.55	2400	m^3
	Lime ashlar	—	25.70	2560	m^3
	Marble dressed	—	26.5	2700	m^3
	Sand stone	—	22	2240	m^3

The dead loads are also termed as **permanent loads** as regards to the duration of their action. These loads arise due to gravitational influence. Therefore, these are also referred as **gravity loads.**

2.3 LIVE LOADS (IMPOSED/TRANSIENT/VARIABLE LOADS)

Live loads are the loads assumed to be produced by the intended use or occupancy including distributed and concentrated loads, impact and vibration. These loads which vary in magnitude and/or in positions. Live loads are also known as **imposed or transient loads**. Live loads include any external loads imposed upon the structure when it is serving its normal purpose. Live loads are assumed to be produced by the intended use of occupancy in buildings including distributed, concentrated, impact and vibration and snow loads. Live loads are expressed as uniformly distributed static loads (UDL). Live loads include the weight of materials stored, furniture and movable equipments. For buildings in most cities, the loads imposed on floors, stairs and roofs are specified in codes. For design purpose, certain live loads on floors have been assumed by many countries and which are almost in agreement. Some efforts have been made at the international level do decide live loads on floors and these have been specified in the International standards 2103 (Imposed floor loads in residential and public building and 2633 (Determination of imposed floor loads in production buildings and warehouses). These codes have been published in the International Organisation of Standardization.

Imposed load causing larger action and/or larger deflection is referred as **leading imposed load.**

Live (imposed) load acting alongwith leading imposed load and causes, smaller actions and low deflection are referred as **accompanying imposed load.**

Code IS : 875 (Part 2) - 1987 defines the **principal occupancy** for which a building or part of a building is used or intended to be used. The buildings are classified according to occupancy as under as per IS : 875 (Part 2) - 1987. *An occupancy shall be deemed to include subsidiary occupancies which are contingent upon.*

1. Assembly buildings. The assembly buildings including any building or part of a building where groups of people gather together for amusement, recreation, social, religious, patriotic, civil, travel and similar purposes (e.g., theaters, motion picture, houses, assembly halls, city halls, marriage halls, townhalls, auditorium, exhibition halls, museums, skating rinks, gymnasium, restaurants, places of worship, dance halls, club rooms, passenger stations and terminals of air, surface and other public transportation services, recreation piers and stadia, etc.).

2. Business buildings. The business buildings include any building or part of a building, which is used to conduct business (other than that covered in 6) for maintaining of accounts and records for similar purpose, offices, banks, professional establishments, court houses, and libraries shall be classified in this group so far as principal function of these is to deal with public business and the keeping and records.

The **office buildings** are primarily used for office purposes. (e.g., purpose of administration, clerical work, handling money, telephone and telegraph operating and operating computers, calculating machines). The clerical work includes writing, book-keeping, sorting papers, typing, filing, duplicating, punching cards or tapes, drawing of matter for publication and the editorial preparation of matter for publication).

3. Educational buildings. The educational buildings include any building used for school, college or day-care purposes involving assembly for instructions, education or recreation and which is not covered in 1.

4. Industrial buildings. The industrial buildings include any building or a part of a building or structure in which products or materials of various kinds and properties are fabricated, assembled or processed like assemble plants, power plants, refineries, gas plants, mills, diaries, factories, workshops, etc.

5. Institutional buildings. Institutional buildings include any building or a part thereof, which is used for purpose, such as medical or other treatment in case of persons suffering from physical and mental illness, disease or infirmity; care of infants, convalescents of aged persons and for penal or correctional detention in which the liberty of the immates is restricted. Institutional buildings ordinarily provide sleeping accommodation for the occupants. These include hospitals, sanitoria, custodial institutions or panel institutions (e.g., jails, prisons and reformations).

6. Mercantile buildings. These buildings include building or a part of a building which is used as shops, stores, market for display and sale of merchandise either wholesale or retail. Office, storage and service facilities incidental to the sale of merchandise and located in the same building shall be included under this group.

7. Residential buildings. The residential buildings include any building in which sleeping accommodation is provided for normal residential purposes with or without cooking a dining or both facilities (except building covered in 5). These buildings include one or multifamily dwellings, apartment houses (flats), lodging or rooming houses, restaurants, hostels, dormitories and residential hotels.

The *dwellings* include any building or part occupie by members of single/ multifamily units with independent cooking facilities. These also include apartment houses (flats).

8. Storage buildings. The storage buildings include any building or part of building used primarily for the storage or sheltering of goods, wares or merchandize (e.g., warehouses, cold storages, freight depots, transity sheds, store houses, garages, hangers, truck terminals, grain elevators, barns and stables).

The imposed loads are the loads assumed to be produced by the intended use or occupancy of a building, including the weight of movable partitions, distributed loads, concentrated loads, loads due to impact and vibration, and dust load but excluding wind, seismic, snow and other loads due to temperature changes, creep, shrinkage, differential settlement, etc. The imposed loads are the largest loads those probably will be produced by the intended use or occupancy, but shall not be less than the equivalent minimum loads specified in Table 2.2 subjected to any reductions allowed by code IS : 875 (Part 2) - 1987.

The floors are investigated for both the uniformly distributed loads (UDL) and the corresponding concentrated loads specified in Table 2.2 and designed for the most adverse effects but these shall not be considered to act simultaneously. The concentrated loads specified in Table 2.2 may be assumed to act over an area of 0.3 m x 0.3 m. However, the concentrated loads may not be considered where the floors are capable of effective lateral distribution of this load.

Table 2.2. IMPOSED FLOOR LOADS FOR DIFFERENT OCCUPANCIES (As per IS : 875 (Part 2) - 1987)

S.N.	Occupancy Classification	Uniformly Distributed Load (UDL)	Concentrated Load
(i)	RESIDENTIAL BUILDINGS		
	(a) Dwelling houses:		
	(1) All rooms and kitchen	2.0	1.8
	(2) Toilet and bath rooms	2.0	—
	(3) Corridors, passages, staircases including fire escapes and store rooms	3.0	4.5
	(4) Balconies	3.0	1.5 per metre run concentrated at the outer edge
	(b) Dwelling units planned and executed in accordance with IS : 8888-1989* only :		
	(1) Habitable rooms, kitchen, toilet and bathrooms	1.5	1.4
	(2) Corridors, passages and staircases including fire escapes	1.5	1.3
	(3) Balconies	3.0	1.5 per metre run concentrated at outer edges

	(c) Hotels, hostels, boarding houses, lodging houses, domitories, residential clubs :		
	(1) Living rooms, bed rooms and domitories	2.0	1.8
	(2) Kitchens and laundries	3.0	4.5
	(3) Billiards room and public lounges	3.0	2.7
	(4) Store rooms	5.0	4.5
	(5) Dining rooms, cafeterias and restaurants	4.0	2.7
	(6) Office rooms	2.5	2.7
	(7) Rooms for indoor games	3.0	1.8
	(8) Baths and toilets	2.0	—
	(9) Corridors, passages, staircases including fire escapes, lobbies - as per the floor serviced (excluding stores and the like) but not less than	3.0	4.5
	(10) Balconies	Same as rooms to which they give access but with a minimum of 4.0	1.5 per metre run concentrated at outer edges
	(d) Boiler rooms and plant room-to be calculated but not less than	5.0	6.7
	(e) Garages :	2.5	9.0
	(1) Garage floors (including parking area and repair workshop) for passenger cars and vehicles not exceeding 25 kN gross weight, including access ways and ramps to be calculated but not less than		
	(2) Garage floors for vehicles not exceeding 40 kNgross weight (including access ways and ramps) - to be calculated but not less than	5.0	9.0
(ii)	EDUCATIONAL BUILDINGS		
	(a) Class rooms and lecture rooms (not used for assembly purposes)	3.0	2.7
	(b) Dining rooms, cafeterias and restaurants	3.0*	2.7
	(c) Offices, lounges and staff rooms	2.5	2.7
	(d) Domitories	2.0	2.7
	(e) Projection rooms	5.0	—
	(f) Kitchens	3.0	4.5
	(g) Toilets and bathrooms	2.0	—
	(h) Store rooms	5.0	4.5
	(i) Libraries and archives :		
	(1) Stack rooms/stack area	6.0 kN/m2 for a minimum height of 2.2 m + 2.0 kN/m2 per metre height beyond 2.2 m	4.5

	(2) Reading rooms (without separate storage)	4.0	4.5
	(3) Reading rooms (with separatestorage)	3.0	4.5
	(j) Boiler rooms and plant rooms - to be calculated but not less than	4.0	4.5
	(k) Corridors, passages, lobbies, staircases including fire escapes as per the floor services (without accounting for storage and projection rooms) but not less than	4.0	4.5
	(l) Balconies	Same as rooms to which they give access but with a minimum of 4.0	1.5 per metre run concentrated at outer edges
(iii)	INSTITUTIONAL BUILDING		
	(a) Bed rooms, dressing rooms, domitories and lounges	2.0	1.8
	(b) Kitchens, laundries and laboratories	3.0	1.5
	(c) Dining rooms, cafeterias and restaurants	3.0	2.7
	(d) Toilets and bathrooms	2.0	—
	(e) X-ray rooms, operating rooms, general storage areas - to be calculated but not less than	3.0	4.5
	(f) Office rooms and OPD rooms	2.5	2.7
	(g) Corridors, passages, lobbies and staircases including fire escapes as per the floor services but not less than	4.0	4.5
	(h) Boiler rooms and plant rooms - to be calculated but not less than	5.0	4.5
	(i) Balconies	Same as rooms to which they give access but with a minimum of 4.0	1.5 per metre run concentrated at outer edges
(iv)	ASSEMBLY BUILDINGS		
	(a) Assembly areas :		
	(1) with fixed seats	4.0	—
	(2) without fixed seats	5.0	3.6
	(b) Restaurants (subject to 4.0 4.5 assembly), museums and art galleries and gymnasia	4.0	4.5
	(c) Projection rooms	5.0	—
	(d) Stages	5.0	4.5
	(e) Office rooms, kitchen and laudries	3.0	4.5
	(f) Dressing rooms	2.0	1.8
	(g) Lounges and billiards rooms	2.0	2.7

	(h) Toilets and bathrooms	2.0	—
	(i) Corridors, passages, staircases including fire escapes	4.0	4.5
	(j) Balconies	Same as rooms to which they give access but with a minimum of 4.0	1.5 per metre run concentrated at outer edges
	(k) Boiler rooms and plant rooms including weight of machinery	7.5	4.5
	(l) Corridors, passages, subject to loads greater than from crowds, such as wheeled vehicles, trolleys and the like. Corridors, staircases and passages in grandstands	5.0	4.5
(v)	BUSINESS AND OFFICE BUILDING (see also 3.12)		
	(a) Rooms for general use with separate storage	2.5	2.7
	(b) Rooms without separate storage	4.0	4.5
	(c) Banking halls	3.0	2.7
	(d) Business computing machine rooms (with fixed computers or similar equipment)	3.0	4.5
	(e) Records/files store rooms and storage space	5.0	4.5
	(f) Values and strong room to be calculated but not less than	5.0	4.5
	(g) Cafeterias and dining rooms	5.0	2.7
	(h) Kitchens	3.0	2.7
	(i) Corridors, passages, lobbies and staircases including fire escapes as per the floor serviced (excluding stores) but not less than	4.0	4.5
	(j) Bath and toilet rooms	2.0	—
	(k) Balconies	Same as rooms to which they give access but with a minimum of 4.0	1.5 per metre run concentrated at outer edges
	(l) Stationary stores	4.0 for each metre of storage height	9.0
	(m) Boiler rooms and plant rooms - to be calculated but not less than	5.0	5.6
	(n) Libraries	see sl. No. (ii)	
(vi)	MERCANTILE BUILDINGS		
	(a) Retail shops	4.0	3.6
	(b) Wholesale shops to be calculated but not less than	6.0	4.5

	(c) Office rooms	2.5	2.7
	(d) Dining rooms, restaurants and cafeterias	3.0	2.7
	(e) Toilets	2.0	—
	(f) Kitchens and laundries	3.0	4.5
	(g) Boiler rooms and plant rooms - to be calculated but not less than	5.0	6.7
	(h) Corridors, passages, staircases including fire escapes and lobbies	4.0	4.5
	(i) Corridors, passages, staircases subject to loads greater than from crowds, such as wheeled vehicles, trolleys and the like	5.0	4.5
	(j) Balconies	Same as rooms to which they give access but with a minimum of 4.0	1.5 per metre run concentrated at outer edges
(vii)	INDUSTRIAL BUILDINGS		
	(a) Work areas without machines/equipment	2.5	4.5
	(b) Work areas with machinery/equipment		
	(1) Light duty To be calculated but not less than	5.0	4.5
	(2) Medium duty	7.0	4.5
	(3) Heavy duty	10.0	4.5
	(c) Boiler rooms and plant rooms - to be calculated but not less than	5.0	6.7
	(d) Cafeterias and dining rooms	3.0	2.7
	(e) Corridors, passages and staircases including fire escapes	4.0	4.5
	(f) Corridors, passages, staircases subject to machine loads, wheeled vehicles - to be calculated but not less than	5.0	4.5
	(g) Kitchens	3.0	4.5
	(h) Toilets and bathrooms	2.0	—
(viii)	STORAGE BUILDINGS		
	(a) Storage rooms (other than cold storage) warehouses - to be calculated based on the bulk density of materials stored but not less than	2.4 kN/m^2 per each metre of storage height with a minimum of 7.5 kN/m^2	7.0
	(b) Cold storage - to be calculated but not less than	5.0 kN/m2 per each metre of storage height with a minimum of 15 kN/m2	9.0

(c) Corridors, passages and staircases including fire escapes as per the floor serviced but not less than	4.0	4.5
(d) Corridors, passages subject to loads greater than from crowds, such as wheeled vehicles, trolleys and the like	5.0	4.5
(e) Boiler rooms and plant rooms	7.5	4.5

* Guide for requirements of low income housing.

Where unrestricted assembly of persons is anticipated, the value of UDL should be increased to 4.0 kN/m^2.

'With fixed seats' implies that the removal of the seating and the use of the space for other purposes is improbable. The maximum likely load in this case is, therefore, closely controlled.

The loading in industrial buildings (workshops and factories) varies considerably and so three loadings under the terms 'light', 'medium' and 'heavy' are introduced in order to allow for more economical designs but the terms have no special meaning in themselves other than the imposed load for which the relevant floor is designed. It is, however, important particularly in the case of heavy weight loads, to assess the actual loads to ensure that they are not in excess of 10 kN/m^2, in case where they are in excess, the design shall be based on the actual loadings.

For various mechanical handling equipment which are used to transport goods, as in warehouses, workshops, store rooms, etc., the actual load coming from the use of such equipment shall be ascertained and design should cater to such loads.

Note 1. Where no values are given (in this table) for concentrated load, it may be assumed that the tabulation distributed load is adequate for design purposes.

Note 2. The loads specified in above table are equivalent uniformly distributed loads on the plan area and provide for normal effect of impact and acceleration. These loads do not take into consideration special concentrated loads and other loads.

Note 3. Where the use of an area or floor is not provided in the above table, the imposed load due to the use and occupancy of such an area shall be determined from the analysis of the loads resulting from :

(i) weight of the probable assembly of persons; (ii) weight of the probable accumulation of equivalent and furnishing; (iii) weight of the probable storage materials; and (iv) impact factor, if any.

While designing columns, abutments, piers, walls, their supports and foundations, the live loads on floors are reduced as in Table 2.3 as per IS : 800-84.

Table 2.3. PERCENT REDUCTION OF TOTAL LIVE LOAD

Number of floors carried by member under consideration	Percent reduction in total live load on all floors above the member under consideration
1	0
2	10
3	20
4	30
5 or more	40

Reduction in live load shall not be made in the case of ware houses, garages and other buildings used for storage purposes and for factories and workshops designed for 5 kN/m^2. However, above reductions are made for buildings such as factories and workshops designed for a live load more than 5 kN/m^2 provided that the loading assumed for any column, etc., is not less it would have been if the floors had been designed for 5 kN/m^2 with no reduction.

As per IS : 875-1984, where a single span of beam or girder supports not less than 50 m^2 of floor at one general level, the live load may be reduced in the design of beam or girder by 5 percent, for each 50 m^2 supported, subjected to a maximum reduction of 25 per cent. This reduction or that given in Table 2.3 above, whichever is greater, may be taken into account in the design of columns, supporting such a beam. These reductions are not applicable for the floors used for storage purposes, in the weight of any plant or machinery which is specifically allowed for.

Accompanying live load - Live (imposed) load of lower magnitude which may act together with **leading imposed load** in called as **accompanying live load.**

2.4 WIND LOAD (TRANSIENT LOAD)

Load experienced by member or a structure due to wind pressure acting on the surface is called as **wnd load.**

The wind loads are the **transient loads**. Relative to the surface of earth, the natural air remains in motion, (*i.e.* a current of air flows near the surface of earth). The difference in solar and terrestrial radiations setting up irregularities in temperatures (which gives rise to convection either upwards or downwards. The wind usually blows horizontal to the ground at high wind speeds. The vertical components of atmospheric motion are relative small. The term wind denotes almost exclusive the horizontal wind. The vertical winds are always identified as such. The wind speeds are assessed with the aid of anemometers or anemographs which are installed at meteorological observatories at heights generally varying from 10 m to 30 m above ground. Very weak winds (i.e., winds at low speeds say in the range of 5 kmph to 10 kmph) are called breeze. The strong winds (i.e., winds at speed say, 50 kmph) are generally associated with cyclonic storms. Very strong winds (i.e., winds at speed say 80 kmph) are also associated with cyclonic storms, thunderstorms, dust storms or vigorous monsoons. The cyclonic storms weaken rapidly after crossing the coast and move as depressions in land. A severe storm may have wind speed of 60 kmph to 120 kmph. The winds of very high speeds and very short duration (hurricanes) are called **Kal Baisaki or Norwesters** occur fairly frequently during summer months over North East India.

In addition to wind at any time, there are effects of gusts (sudden blast of wind, which may last for few seconds) which should be considered. These gusts cause increase in air pressure but their effect on the stability of the building may not be so important. Often, the gusts influence only part of the building and the increased local pressures may be more than balanced by a momentary reduction in pressure elsewhere. Short period gusts may not cause any appreciable increase in stress in main components of buildings (because of inertia of the building), although the walls, roof sheetings and individual cladding units (glass panels) and their supporting members such as purlins, sheeting rails and glazing bars may be more seriously affected. Gusts may also be extremely important for design of structures with slenderness ratios.

The liability of a building or a structure to high wind pressures depends not only upon the geographical location and proximity of other obstructions to air flow but also upon the characteristics of the structure itself. In general, wind speed in the atmospheric boundary layer increases with height from zero at ground level to maximum at a height called the **gradient height**. The variation of wind with height depends primarily on the terrain conditions. However, the wind speed at any height never remains constant and it has been found convenient to resolve its instantaneous magnitude into an average or mean value and a fluctuating component around this average value. The average value depends on the average time employed in analysing the meteorological data, which varies from a few seconds to several minutes. The magnitude of fluctuating component of the wind speed is called **gust,** it depends upon averaging time. In general, smaller the averaging interval, greater is the magnitude of the gust speed.

The wind load depends upon terrain, height of the structure, and the shape and size of structure. For pitched roof, wind load has been considered and described further in detail in chapter (Design of roof trusses). However, it is essential to know the following terms to study the new concept of wind as described in IS : 875 (Part 3) - 1987.

1. Angle of attack - It is defined as the angle between the direction of wind and a reference axis of the structure.

2. Breadth : It means horizontal dimension of the building measured normal to the direction of wind.

(Note : The breadth and depth are dimensions measured in relation to the direction of wind. Where as length and width are dimensions related to the plan of the building.)

3. Depth : It means the horizontal dimensions of the building measured in the direction of the wind.

4. Developed height : It is height of upward penetration of the velocity profile in a new terrain.

At larger fetch lengths, such penetration reaches the gradient height, above which the wind speed may be taken to be constant. **At lesser fetch lengths,** a velocity profile of a smaller height but similar to that of the fully developed profile of that terrain category has to be taken. The velocity at the top of this shorter profile equals that of the unpenetrated earlier velocity profile at that height.

5. Effective frontal area : The projected area of the structure normal to the direction of wind.

6. Element of surface area : The area of the surface over which the pressure coefficient is taken to be constant.

7. Force coefficient : It s non-dimensional coefficient such that the total wind force on a body is the product of the force coefficient, the dynamic pressure of the incident design wind speed and the reference area over which the force is required.

In the direction of wind, this coefficient is called as **'drag coefficient'**. In the perpendicular direction of wind, it is called as **lift coefficient** :

8. Ground roughness : The nature of the earth's surface as affected by small scale obstructions such as trees and buildings (as distinct from topogrpahy) is called ground roughness.

9. Gust : A positive or negative departure of wind speed from its mean value, lasting not more than say, 2 minutes over a specified interval of time.

10. Peak gust: Perk gust or peak gust speed is the wind speed associated with the maximum amplitude.

11. Fetch length : It is the distance measured along the wind from a boundary at which a change in the type of terrain occurs.

When the changes in terrain types are encountered (such as, the boundary of a town or city, forest, etc.), the wind profile changes in character but such changes are gradual and start at ground level, spreading or penetrating upwards with increasing fetch length.

12. Gradient height : It is the height above the mean ground level at which the gradient wind blows as a result of balance among pressure gradient force, coriolis force and centrifugal force.

It is taken as height above the mean ground level, above which the variation of wind speed with height need not be considered.

13. Mean ground level : It is the level of average horizontal plane of the area enclosed by the boundaries of the structure.

14. Pressure coefficient : It is the ratio of the difference between the pressure acting at a point on a surface and the static pressure of the incident wind to the design wind pressure.

where the static and design wind pressures are calculated at the height of the point considered after taking into consideration the geographical location, terrain conditions and shielding effect. Also

$$\text{Pressure coefficient} = [1 - (V_p/V_z)^2]$$

where, V_p = actual wind speed at any point on the structure at a height corresponding to that of V_z.

(*Note :* When the sign of pressure coefficient is positive, it shows that the pressure acts towards the surface and the negative sign shows that the pressure acts away from the surface).

15. Return period : It is the number of years, the reciprocal exceeding a given wind speed in any one year.

16. Shielding effect : It refers to the condition where wind has to pass along some structure(s) or the structural element(s) located on the upstream wind side, before meeting the structure or structural element under consideration.

A factor called **shielding factor** is used to consider such effects in estimating the force on the shielded structures.

17. Suction : It means the pressure less than the atmospheric (static) pressure and it is taken to act away from the surface

18. Solidity ratio : It is equal to the effective area (projected area of all the individual elements) of a frame normal to the direction of wind divided by the area enclosed by the boundary of the frame normal to the direction of wind.

(*Note :* It is to be calculated for the individual frame)

19. Terrain category : It means the characteristics of the surface irregularities of an area which arise from natural or constructed features. The categories are numbered in the increasing order of roughness.

20. Velocity profile : The variation of horizontal component of the atmospheric wind speed at different heights above the mean ground level is termed as velocity ratio.

21. Topography: The nature of the earth's surface as influenced by the hill and valley configuration.

2.5 SNOW LOAD

Load on a roof or a structure due to the accumulation of snow and ice.

The snow load depends upon latitude of place and atmospheric humidity. The snow load acts vertically and it is expressed in kilo-Newtons per square metre of plan area. The actual load due to snow depends upon the shape of the roof and its capacity to retain the snow. When actual data for snow load is not available, snow load may be assumed to be 25 N/m^2 per mm depth of snow.

It is usual practice to assume that snow load and maximum wind load will not be acting simultaneously on the structure.

2.6 SEISMIC LOAD (EARTHQUAKE FORCE)

Earthquake loads are the inertia forces produced in a structure due to ground movement during an earthquake.

Seismological forces: The word **seismology** is derived from the Greek word, **seismos'** meaning earthquake. This word was first used by Robert Mallet, an Irish scientist and engineer in about 1880. **Seismology** is the scientific study of earthquakes and the internal structure within the earth. The part of seismology is concerned with the study of seismic phenomena as applied to earthquake resistant construction, which is also known as **structural seismology**. It has to do with the location of seismic active areas and the prediction of the ground motion character. Seismic forces result from the eratic motion of the ground (i.e., earthquake. These are of kinematic origin. These forces owe their existence to vibrations caused by the movement of the earth's surface in structures during an earthquake. These forcesd are random in character, though these are usually regarded as deterministic in practical calculations to simplify the design model.

It is to note that the engineers and the scientists are in habit of using difficult and new words instead of simple words. These forces are called seismic forces instead of earthquake forces so that common people attach great significance.

Earth is a multi-layered body. It consists of a series of shells differing from one another in minearalogy, density, elastic properties, temperature and pressure. Terrestrial globe representing the Earth is 6357 km to 6378 km in radius vertically and horizontally, respectively.

The outer shell, which is caved the crust consists of several layers. These layers are inclined. The inclinations of these layers vary at different places. These layers fit together like two jigsaw pieces.

The upper layer is made up of stratified **sedimentary rock** several kilometers thick on the average. Below which there is a **granite layer**, which sometimes comes upto the surface. The granite layer is about 10 km thick in valleys and 40 km thick beneath mountain ridges. Further below, there is a layer of **basalt**, the thickness of which is 30 km under valleys and 20 km under ridges. All in all the crus is 35 km (on the average) under continental shields, being 60 km to 70 km thick under high mountains. In contrast to its continental counterpart, the oceanic crust is much thinner (6 km to 8 km thick), the basalt lying right under the sedimentary rock.

The outer crust is seperated from the underlying stratified material referred to as **mantle** by a surface called **Mohorovicic discontinuity** (or briefly Moho). At this, the propagation velocity of seismic waves abruptly changes. It is indicative of difference in density and elastic properties between crust and upper mantle materials.

The Earth's rocks, specially those forming the crust, contain radio-active elements which decay and emit heat. At a depth of 50 km to 100 km below the surface, there is a well pronounced layer of **astenosphere** (or **tectosphere**) about 200 km thick. At this, the transverse velocity suddenly drops. It shows reasons that, there is the material, which one is less harder than elsewhere. Some of this is probably in molten state. According to some specialists, the molten materials is concentrated within isolated portions in what may be called **lenses**, rather than forms a continuous shell. It is believed that these portions are directly responsible for volcanism and that these are located near the crust areas, where the intensive **folding** takes place. Below the lower mantle lies the core whose outer portion, is believed to be in molten state and whose central portion, called the inner core, is assumed to be solid.

Observation of the Earth's surface has shown that is in a state of continuous motion with some portions of the crust sinking and some going up. It is also seen that the crust executes horizontal motion. These processes are referred to as **diasterophism** or **tectonic movement**. According to the **pulsation hypothesis**, the Earth continually experiences alternate contractions and expansions. According to this theory, the contractions lead to crustal movement mainly in the areas of **foldings** where as the expansions give rise to crustal ruptures. In 1922 (a few years before the pulsation theory) it was suggested the **convection currents** take place inside the Earth's surface. By this hypothesis, the tectonic movement was accounted for by the transportation differentiation of materials of different density, caused by the difference in temperature. In plain words, it was suggested that the convention in the mantle converts heat into the mechanical energy of continental migration and orogeny.

According to the geological theory of plate tectonics the surface of the earth consists of several thick rock plates that float on the earth's molten mantle. New tectonic plates are continuously formed along the deep rift valleys of the ocean floor, where the molten material from the earth's interior is pushed upward, thus building up the edges of the oceanic plates to cause the so- called **continental drift** or **mobilism** that is, the ocean plates are pushed against the continental plates. Where the plates collide, they may be locked in place, this being temporarily prevented by the frictional resistance along the plate boundaries. The continental migration takes place in subcrustal layers and oceanic crust and owes its existence to the axial rotation of the Earth, which gives rise to forces moving the colossal contents over the oceans floor.

Theory of continental drift has given a new lease of life by the findings of geophysicists in the recent decades. Among other things it has been found that the extensive mountain ridges (upto 4 km wide and 6 km high) pass through all the central regions of the oceans. In contrast to their continental counterparts, they are made up of high basalt benches.

Between them, there are rift valleys, which are, in effect, **large faults** along the ridges. A fault is define as a zone, where the crust (and sometimes some of the uppermantle) is fractured to a great extent and may, therefore, undergo considerable deformations. The ridges are cut by quite a number of what are called **transformed faults** and deep trenches sometimes several hundred kilometers long. The surface temperature here is higher than elsewhere on the Earth.

In fact two plates are in a state of collision which creates major mountain belts as one plate over-rides the other.

Stresses are built-up along the plate edges until sudden slippage due to elastic rebound or fracture of the rock occurs, resulting in sudden release of energy that may cause the upper crust of the earth to fracture along a certain direction and form a fault. Some of the energy is propagated in the form of shock waves in all directions. It is this wave motion that is known as earthquake.

It becomes essential to consider 'seismic load' in the design of structure, if the structure is situated in the seismic areas. The seismic areas are the regions which are geologically young and unstable parts and which have experienced earthquakes in the past and are likely to experience earthquakes in future. The Himalayan region, Indo Gangetic Plain, Western India, Kutch and Kathiawar are the places in our country which experience earthquakes frequently. Sometimes these earthquakes are violent also. Seismic load is caused by the shocks due to an earthquake shocks cause movement of ground, as a result of which the structure vibrates. The vibrations caused because of earthquakes may be resolved in three perpendicular directions. The horizontal direction of vibration dominates over other directions. In some cases structures are designed for horizontal seismic forces only and in some cases horizontal seismic forces and vertical seismic forces both are taken into account. The seismic accelerations for the design may be arrived at from **seismic coefficient,** which is defined as the ratio of acceleration due to earthquakes and acceleration due to gravity. This country has been divided into seven zones for determining seismic coefficients. The seismic coefficients have also been recommen-ded for different types of soils for the guidance of designers. IS: 1893-1962 Indian Standard Recommendations for Earthquakes Resistant Design of Structure, may be referred to for actual design.

2.7 SOIL AND HYDROSTATIC PRESSURE

The pressure exerted by soil or water or both should be taken into consideration for the design of structures or parts of structures which are below ground level. The soil pressure and hydrostatic pressure may be calculated from established theories.

2.8 ERECTION EFFECTS (LOADS)

The actions (loads and deformations) experienced by the structure exclusively during erection.

The erection effects include all effects to which a structure or a part of structure is subjected during transportation of structural members, and erection of structural member by equipments. Erection effects also take into account the placing or storage of construction materials. The proper provisions shall be made, (e.g., temporary bracings, to take care of all stresses caused during erection). The stresses developed because of erection effects should not exceed allowable stresses.

Accidental loads: Loads such as those due to blast, impact of vehicles explosion or other rare loads for which the structure is considered vulnerable as per the user.

2.9 DYNAMIC LOADS AND ITS EFFECTS (IMPACT)

Dynamic loads are the loads which suddenly change in time with variations in magnitude, direction and point of application taking place either jointly or separately. Dynamic loads imparts accelerations to the bodies (structural components as complete structures) on which these are imposed. These loads give rise to inertia forces and cause system to vibrate.

Cyclic loads, impact loads, and moving loads are three types of dynamic loads.

1. Cyclic loads: Cyclic loads mean **vibration loads,** the variations in these in time loads follow a harmonic law. Such loads may arise in (for example, unbalanced mass rotating machinery).

2. Impact loads : Term impact loads means, loads of great magnitude which are applied suddenly and act for a short time.

3. Moving loads: Moving loads mean live loads caused by traffic (i.e., moving vehicles on roads and moving coaches on rails), moving cranes, etc.

Dynamic loads are classified into **deterministic** and **stochastic** or **random**. A deterministic dynamic load is always prescribed as function of time. In contrast, the variations of a stochastic or random force in time may be affected by a quite a number of random factors, so its determination always implies a certain probabilistic element.

The moving loads on a structure cause vibrations and have also impact effect. The dynamic effects resulting from moving loads are accounted for, by impact factor. The live load is increased by adding to it the impact load. The impact load is determined by the product of impact factor and live load. The effect of impact has been described in Chapter (Design of Plate Girder).

Action effect (load effect): The internal forces (axial, bending, shear and twisting moment) developed due to the external actions.

Action: An externally applied force is referred as actions. There are primary cause for stress or deformation in a structure (such as dead, live, wind seismic or temperature loads).

The expansion and contraction resulting from temperature changes are called as the **secondary effects.**

2.10 TEMPERATURE EFFECTS

The variations in temperature result in expansion and contraction of structural material. The range of variation in temperature varies from localities to localities, season to season and day to day. The temperature effects should be accounted for properly and adequately. The allowable stress should not be exceeded by stress developed because of design loads and temperature effects.

2.11 LOAD COMBINATIONS

All the parts of the steel structures shall be capable of sustaining the most adverse combination of the dead loads, prescribed live (imposed) loads, wind loads, earthquake loads where applicable and any other forces or loads to which the steel structure may reasonably be subjected without exceeding the stresses specified in this standard.

The load combinations for design purposes shall be the one that produces maximum forces and effects and consequently maximum stresses from the following combinations.

(i) Dead load + imposed (live) loads,

(ii) Dead load + imposed (live) loads + wind or earthquake loads, and

(iii) Dead load + wind load or earthquake loads.

The internal forces (axial shear bending or twisting moment) developed due to external loads (actions) and the temperature changes are referred as **action (load) effects.**

2.12 STRESSES

When a structural member is loaded, deformation of the member takes place, and resistance is set up against deformation. This resistance to deformation is known as stress. The **direct stress** is defined as force per unit cross-sectional area. The nature of stress developed in the structural member depends upon nature of loading on the member. The following are the various types of stresses :

1. Axial stress (Direct stress) :

(i) Tensile stress (ii) Compressive stress.

2. Bearing stress. 3. Bending stress. 4. Shear stress.

The bending stress and shear stress have been discussed in Chapter 'design of beams'.

A member may be subjected to combined direct and bending stresses. Such stress is known as combined stress.

Sign convention for stresses. The tensile stresses are taken as positive and compressive stresses as negative. The sign convention for stresses is convenient as a structural member elongates on application of tensile load and shortens on application of compressive load.

2.13 STRESS-STRAIN RELATIONSHIP FOR MILD STEEL

When a steel bar is pulled in a machine, holding it between the grips at each end, and pulling in the direction of the bar, the bar continues to stretch or elongates (i.e., it length is increased). The pull exerted by the machine is called **tensile load** and the complete process is known as tension test. The pull is gradually applied, till, the steel-bar is finally fractured (i.e., broken). The pull divided by the cross-section area of the bar is referred as tensile stress. The deformation per unit length or unit angle is known as **strain.**

Photo Plate 2.1. Universal Testing Machine (UTM) A steel-bar (specimen/coupen) in axial tension

When a mild steel bar is subjected to a tensile load, it elongates. The elongation per unit length is known as strain. The stress is proportional to strain within limit of proportionality. The stress-strain relationship for mild steel can be studied by plotting stress-strain curve. The stress and load on some suitable scales may be plotted on y-axis, and strain on some suitable scale may be plotted on x-axis as shown in Fig. 2.1(A).

When the tensile load increases with increases in strain, stress-strain curve follows a straight line relationship upto **'Limit of proportionality'.** The limit of proportionality is defined as stress beyond which straight line relationship ceases between stress and strain. Beyond the limit of proportionality stress approaches the **elastic limit.** The elastic limit is defined as the maximum stress upto which a specimen regains its original length and tracing the original path illustrates the growth of strain in relation to stress, known as stress-strain relationship for the mild steel bar. On the removal of the applied load, there is hardly any distinct difference in the positions of limit of proportionality and elastic limit. Practically, position of limit of proportionality coincides with the elastic limit.

When the specimen is loaded beyond the elastic limit, the specimen does not resume its original length on the removal of applied load and a little strain is left in the specimen. This little strain is known as **residual strain** or **permanent** set. The steel bar does resume original length and remains stretched. It has become irrevocable. That is, the elastic limit has passed. In steel design, it is taken as the yield stress (IS 800 : 2007)

In case, the load is not removed and when the position at which this sudden large increase of strain shows the **yield point**.

Many high tensile steels and most non-ferrous alloys dow not exhibit a definite yield point and the load-extension curve is an shown in Fig. 2.1(B). In such cases, the **proof-stress** has been evolved. Proof-stress is defined as the load, which applied for not less 15 seconds produces a plastic extension or permanent set of 0.2 percent of the original gallery length. There are different methods.

Proof-stress is defined as the stress needed to cause a given a extension in metals not showing a sudden yield point used to find the proof-stress. The obvious method is to load and unload the test piece at successive increments until eventually the pre-found permanent set is obtained. The given proof-length (0.2 per cent of the gauge length) is measured along the extension axis. Through this point a line of proportionately. This point P, Fig. 2.1(B), at which, this line intersects the load-extension curve gives the required value of the proof stress, since the difference between the total extension indicated by this point and the corresponding point on the theoretical elastic line obviously represents the pre-determined permanent set.

Proof stress is also defined as the stress to which high strength friction grip (HSFG) both are pre-tensioned.

An extension remaining after the removal of a load from a test-piece when the elastic limit has already exceeded is knwon as **permanent set.**

When the tensile load further increases the stress-reaches 'yield stress' and material starts yielding. The stress-strain curve suddenly falls showing a decrease in stress. The distinct position from where sudden fall of curve occurs marks the **upper yield point** and the position upto which fall of curve occurs is known as **lower yield point**. The material of bar stretches suddenly at constant stress. The adjustment of stress takes place in the elements of material in between upper yield point and lower yield point. On further increase of load, stress increases with the increase of strain. The phenomenon of increase in stress with increase in strain beyond yielding is called as **strain-hardening**. However, strain increase more rapidly.

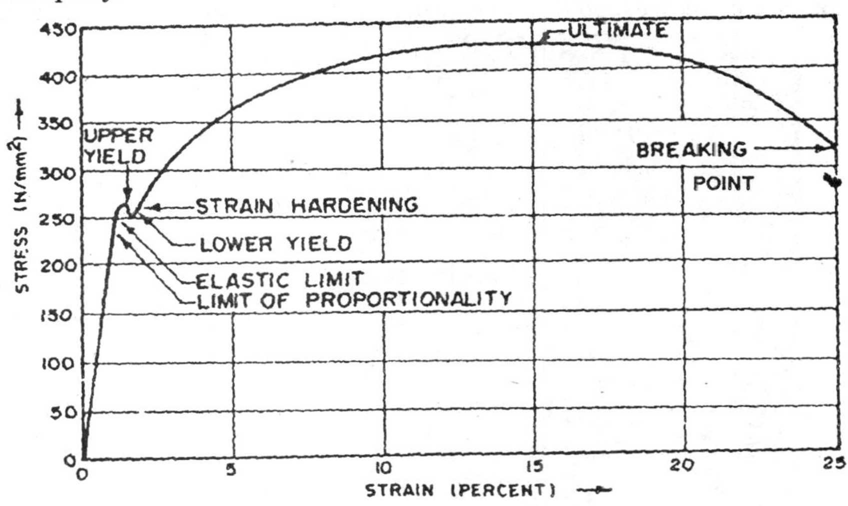

Fig. 2.1(A). Stress-strain curve

Finally the load reaches the value of **'ultimate load'**. The ultimate load is defined as maximum load, which can be placed prior to the breaking of specimen. The stress corresponding to the ultimate load is known as 'ultimate stress'. The stress- strain curve suddenly falls with rapid increase in strain and specimen breaks. The load corresponding to breaking position is nown as **'breaking load'.** The cross-section of specimen decreases. If actual breaking stress is computed on the basis of decreased (reduced) cross-sectional area, the breaking stress will be found to be more than the ultimate stress. **Strength** is defined as the resistance to failure by yielding or buckling.

Significance of upper yield point. The boundaries of grains of mild steel are composed of brittle material. The forms a rigid skeleton. The rigid skeleton prevents plastic deformation of the grains at low stress, and shows upper yield point in stress-strain curve. At upper yield point, this rigid skeleton breaks down. As a result of this, the stress in material drops down without elongation from upper yield point to lower yield point. This is followed by sudden stretching of the material at constant stress from lower yield point upto strain hardening. As shown in Fig. 2.2 (on exaggerated scale).

Note: This tensile test is carried out in an universal testing machine. A machine is called as **universal testing machine** in case a specimen may be tested in axial tension. And other specimen may be tested in axial compression. A third specimen (that is a beam) may be tested in bending (i.e. in flexure).

2.14 TENSILE STRESS

When a structural member is subject to direct axial tensile load, then the stress is known as tensile stress (σ_{at}). The tensile stress is calculated on net cross-sectional area of the member :

$$\sigma = (P_t / A_n)$$

where P_t is the direct axial tensile load and A_n is the net cross-sectional area of the member. The characteristic stress corresponding to rupture in tension is called **tensile stress.**

Photo Plate 2.2. Compression Testing Machine (with attachment for flexural test) A stub column specimen in compression Test.

2.15 COMPRESSIVE STRESS

When a structural member is subjected to direct axial compressive load, the stress is known as compressive stress, (σ_{ac}). The compressive stress is calculated on gross cross-sectional area of the member

$$\sigma_{ac} = (P_c/A_g)$$

where, P_c = direct axial compressive load, A_g = gross-sectional area of the member.

It is to note that one peculiar feature needs attention.

It is assumed that the resistance to compression would be equal to resistant to tension. It depends upon the condition of the steel.

For normalized or annealed steels, the yield-points in tension and compression are practically equal and there is no much difference.

However, for cole worked, bright-drawn steels, the compression yield-point in tension. Although, the values approach each other when the cold-drawn steel is tempered at temperature about 450°C.

Ability of a material to resist a compressive force developed by a specific load before failure is called as **compressive strength.**

2.16 BEARING STRESS

When a load is exerted or transferred by the application of load through one surface for the another surface in contact, the stress is known as 'bearing stress' (σ_π). The bearing stress in calculated on net projected area of contact)

$$\sigma_p = (P/A)$$

where, P = load placed on the bearing surface, A = net projected area of contact.

2.17 WORKING STRESS

When a structure is being designed by the working stress method the maximum stress that is permitted to be experienced in the elements, members, or structures under the nominal/service load (action).

The working stress is also termed as **allowable stress** or **permissible stress.** The working stress is evaluated by dividing yield stress by factor of safety. For the purpose of computing safe load carrying capacity of a structural member, its strength is expressed in terms of working stress. The working stress is the stress which may be developed or set up in the member without causing structural damage to it. The actual stresses resulting in a structural member from design loads should not exceed working stresses. This ensures the safety of the structural member. The maximum working stresses are adopted from IS : 800-1984.

IS : 800-1984 specifies a number of grades of steel with different yield strengths, the design parameters and the geometrical properties. There are also developments in the design of steel structures. In view of these, the permissible stresses have been revised in general. These permissible stresses have been expressed to the extent possible in terms of the yield strength of the material. However, the specific values have also been given in IS : 800-1984 for steels commonly used.

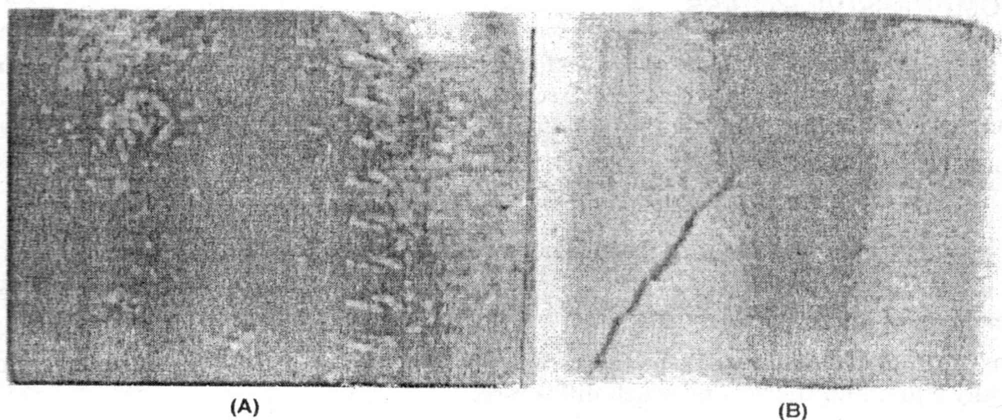

(A) (B)

Photo plate 2.3. Axial compression failures
(A) **Mild Steel specimen** Compressed (Barrel-like failure) (B) **Cast iron** Circular specimen in compression (Fracture along a diagonal plane)

2.18 INCREASE IN PERMISSIBLE STRESSES

A structure may be subjected to the different combinations of loads. The possible combinations of the loads with special reference to the roof trusses have been described in chapter 9. These loads in such combinations do not act for long period. Most of the national codes allow some increase in permissible stresses. Increases in permissible stresses as per IS : 800 is taken as follows :

(i) When the effect of wind or seismic load is taken into account, the permissible stresses in steel are increased by 33 1/3 per cent.

(ii) For rivets, bolts and tension rods, the permissible stresses are increased by 25 per cent, when the effect of wind or seismic load is taken into account.

The increased values of permissible stresses must not exceed yield stress of the material.

2.19 METHODS OF DESIGN

All parts of the steel framework of a structure shall be capable of sustaining the most adverse combination of the dead loads, the prescribed superimposed roof and floor loads, wind loads, seismic forces where applicable, superimposed roof and floor loads, wind loads, seismic forces where applicable, and any other forces or loads to which the building may be reasonably subjected without exceeding the permissible stresses.

Design of structures is done by calculating the internal forces (moments, shear forces....) acting on each component. It is followed by the selection of the appropriate cross-section for the grade of steel used and for required strength. The stresses in a structure are generally determined by the following methods, of and lysis and design of the steel framework.

1. Elastic method of design, 2. Semi-rigid design

3. Fully-rigid design, 4. Plastic method of design

These methods have been described in subsequent articles.

2.20 HISTORY OF ELASTIC (WORKING-STRESS) METHOD OF DESIGN

Until the beginning of the nineteenth century, very little design was done using any kind of calculations. Persons did not have theoretical knowledge of behaviour of the structures developed. The load carrying capacity of the structures could not be predicted. Materials and dimensions of the structural members used to be decided on the basis of experience and the safety used to be assured. There were no means to measure strength of materials. Whether, the strength of materials used was excessive or not, it could not be ascertained. These conditions prevailed upto the Industrial Revolution. Radical changes were seen in the first half of the nineteenth century. Over the next century, materials developed rapidly with wrought iron and eventually steel was developed.

In 1709 Abraham Darby discovered smelting of iron with coke. In 1780's workable wrought iron was developed. Iron-master Henry Cort patended how to roll iron into standard shapes in 1783-84.

In the later part of the eighteenth century, the large scale use of iron for the structures began in the Europe. A 30.48 m span Coal Brookadale (a large size cast-iron arch bridge) was built in 1779 in England. Abraham Darby constructed cast-iron bridge in 1790. There were radical changes in the materials as well as in the structures.

Huge number of bridges with larger spans were needed with the development of railways. Railways required to carry-out heavy- loads which were beyond then while current experience. New type of buildings were needed for the developing factories and mills. There were challenges before the designers and builders to bring new methods. Attempts to find safe designs tended to involve testing of the new structures. Brunel built his railway bridge over the Wye at Chepstow and tested it prior to the erection. Robert Stephenson used model testing extensively in the design of Britanna bridge. Efforts were made to ensure the safety. Symbolically, it was expressed

$$(CRS_s > \gamma \,.\, W_{DLA})$$

where, CRS_s is the calculated resistance of the structure, (i.e. strength of the structure), W_{DLA} is the design loading after assessment and γ is the factor of safety.

Factor of safety takes into consideration all the uncertainties in the assessment of loads and the strength of materials.

Galileo was the first person to attempt a scientific study of structures. Galileo (in his book published in 1638) had developed an analysis of the behaviour of beams. Galileo carried out his analysis by first considering a cantilever beam and then, a simply supported beam Galileo expressed that the external moment must be balanced by an equal and opposite moment. External moment is the moment due to applied loads and the internal moment is the **moment of resistance**. Mostly, his work was correct but there remained some error.

Newton (borned in the year Galileo died) stated in his third law (which states that, for static equilibrium, action and reaction must be equal and opposite). Only, when Newton had proposed his law and Hooke had devised the theory of elasticity. The development for the analysis of beams continued further.

The formulae now in practice for the calculation of stresses in beams made of materials were not clearly presented until, 1773 by Coulomb. Basically correct approach had been developed by Parent.

2.21 ELASTIC (WORKING STRESS) METHOD OF DESIGN (WSM)

Elastic design assumes elastic behaviour of the material (steel) throughout the service load range.

The stress induced in the structural component of the structure by the expected service load, or service load combination is kept equal to or less than the allowable stress. This procedure is called **allowable-stress design** (ASD) or **working-stress design** (WSD). In the **elastic methods**, the stresses due to working loads must not exceed certain permissible working stresses. The working loads are supposed to represent the maximum loads which may be expected to arise in normal usages during the life time of the structures. The values of working stresses are intended to ensure an adequate margin of safety to account for unforeseen overloads, defective workmanship etc. The values of the working loads and permissible working stresses are specified in IS : 800-1984 and IS : 875 (Part 2) - 1987. The allowable working stress σ_w is obtained by dividing the yield stress, σ_y by factor of safety, (FS). The analysis fòrces is based on Hooke's law. Today, every one is familiar with Hooke's law and its statement. (i.e., the stress in steel is directly proportional to strain produced, so long as stress does not exceed limit of proportionality). Therefore, this method is called as **elastic method of design**. This conventional elastic theory is in use for long for the analysis of the structures.

In this, the stress-strain relationship for the steel is assumed to be linear. The deformations are considered to be small. The equilibrium equations are formulated on the basis of the original geometry of the structure. In this elastic analysis, the load-deflection relation for the structure is considered linear. Therefore, this analysis is also referred as **linear analysis**. It is simple in its application and calculations. Therefore, it is also known as **simple design method**.

The linear stress-strain and load-deflection relationships are normally valid so long as the stress or strains are sufficiently below the yield point of the steel to be used. **Internal forces** (viz., axial force) bending moments, shear and torsion (if any) are found for the design.

Elastic design is based on the premise that the stress is proportional to the strain, so that the plane transverse sections before stressing remain plane after application such that for axial load, the stress is directly found as

$$\sigma_a = (P/A) \qquad \text{(i)}$$

and the corresponding deformation is

$$\Delta = (P.L/AE) = (\sigma_a \,.\, L/E) = \varepsilon L \qquad \text{(ii)}$$

The familiar mechanics of material equation for bending stresses

$$\sigma_b = (M.y/I) = (M/Z) \qquad \text{(iii)}$$

is used. As long as the stresses are in the elastic region, the principle of superposition of effects is valid and the order of the application of load is not important.

Closed-form solution may be found using slope-deflection equations. However, these are valid for the given beam, given geometry and loads.

In this method the end connections between the members are assumed to be pin-jointed. The end connections do not develop restraint moments adversely affecting the members and the structure as a whole. Following assumptions are made in this method :

(i) The beams are simply supported.

(ii) All the connections of beams, girders, or trusses are virtually flexible and are proportioned for the reaction shears applied at the appropriate eccentricity.

(iii) The members in compression are subjected to forces applied at the appropriate eccentricities.

(iv) The members in tension are subjected to longitudinal forces applied over the net area of the section.

Essential conditions for elastic design

Following essential conditions should be satisfied while using elastic method of design.

1. **Equilibrium conditions:** The conditions of static equilibrium should be satisfied (viz., (i) Algebraic sum of vertical forces should be zero, i.e., $\Sigma V = 0$,

(ii) Algebraic sum of horizontal forces should be zero, i.e., $\Sigma H = 0$, and

(iii) Algebraic sum of moments of all the forces about a point should be zero i.e $\Sigma M = 0$.

2. **Geometric conditions**: The geometric deformations should be consistent (i.e., acceptable).

3. **Boundary conditions:** Boundary conditions (viz. the conditions at supports should be satisfied).

Following **assumptions** are made in the elastic-analysis

1. Individual members are assumed to remain elastic under the action of the factored design loads for all limit states.

2. The effect of hauching or any variation of the cross-section along the axis of a member shall be considered and where significant, shall be taken into consideration while calculating the member stiffness.

2.22 SEMI-RIGID DESIGN

This method as compared with the simple design method permits a reduction in the maximum bending moment in beams suitably connected to their supports, so as to provide a degree of direction fixity. In the case of triangulated frames, it permits rotation account being taken of the rigidity of the connections and the moment of interaction of members. In cases where this method of design is employed, it is ensured that the assumed partial fixity is available and calculations based on general or particular experimental evidence shall be made to show that the stresses in any part of the structure are not in excess of those laid down in IS : 800-1984.

2.23 FULLY RIGID DESIGN

This method as compared to the methods of simple and semi-rigid design assumes that the end connections are fully rigid and are capable of transmitting moments and shears. It is also assumed that the angle between the members at the joint does not change, when it is subjected to loading. This method given economy in the weight of steel used one applied in appropriate cases. The ed connections of members of the frame shall have sufficient rigidity to hold virtually unchanged original angles between such members and the members they connect. The design should be based on accurate methods of elastic analysis and calculated stresses shall not exceed permissible stresses.

2.24 HISTORY OF PLASTIC (ULTIMATE LOAD) METHOD OF DESIGN

The basic principles of the plastic design are not new. Builders of early steel structures intuitively took advantage of the reserve strength inherent in plastic behavior.

The behaviour of beams (simply supported, continuous and fixed were loaded and tested by H. Maier-Leibnitz) Reference: Test results, their interpretation and application, Preliminary Publication, International Association for Bridge and Structural Engineering, Second Congress, Burlin, 1936) loaded beyond the elastic range had been studied much earlier apparently the first published papers on the possibility of directly utilizing the ductility of the steel to improve the design of engineering structures were those of G.V. Kazinzy (Reference : Tests with Fixed Ended Beams. Betonszemle, Volume 2, (1914). According to Professor H. Maier-Leibnitz, the possibility of the development of plastic hinges was first suggested by G.V. Kazinczy in Hungary in 1914. and N.C. Kist in Holland in 1917. Kazinczy carried out some tests on fixed ended beams, and came to the conclusion that the failure only took place when yielding had occurred at three cross-sections, at cross-sections, at which a hinge action could be said to occur. As a result of his recognition of the fundamentally important concept of the plastic hinge. Kazinczy may be fairly said to be the originator of the plastic methods.

The method was formally proposed in Germany in 1920. Considerable theoretical work was carried out by 1932. [Maier-Leibnitz : (contribution to the problem of ultimate carrying capacity of simple and continuous beams of structural steel and timber 1927]. In 1926, a small book was published in Germany. In this book, certain general results regarding the failure conditions of pin- jointed trusses were established. The analysis presented was very complicated. It was not accompanied by experimental confirmation.

In 1928 and 1929, several investigators carried out tests on continuous beams, specially by H. Maier-Leibnitz. His tests and theoretical interpretations may be said to have first put the plastic methods for continuous beams on a firm quantitative footing.

In 1931, Girkmann published a paper. An approximate method for the design on the multi-storey and multi-bay rectangular frame was presented.

In 1932, F Bleich published a treatise and a chapter on steel framed structures in Berlin. It was devoted to a review of the plastic methods developed for beams and simple portal frames.

In 1936, H. Maier-Leibnitz published a paper. This paper included the results of the principal tests on continuous beams and simple frames. This paper was considered an excellent bibliography for consultation. In 1936 an intensive investigation of inelastic behavior of structures was started in England under Professor J.F. Baker.

In 1940, J.A. van den Broek published a paper (Theory of Limit Design, Transactions, American Society of Civil Engineers, 105, 638, 1940). In this paper, the basic ideas of the plastic theory were restated. In this paper and a book published lateron, the technique of designing by the plastic methods, was termed 'Limit Design'

[J.A. Van den Broek, theory of limit design, John Wiley (N.Y.), Chapman and Hall (London), 1948].

In 1949, J.F. Baker published a paper (A review of recent Investigations into the behaviour of steel frames in the plastic range Journal of Institution of Civil Engineers, 31, 188, 1949). Baker first realized that the simple plastic theory might prove to be key to a simple and rational method for the design of complex frames By 1945, the method of Ultimate Design had received wide acceptance in many countries in Europe and South Africa.

In 1938, J.F. Baker and his associates conducted series of tests on portal frames. These portal frames failed by buckling after partially yielded. In 1956, positions reached were described in 'The Steel Skeleton Volume 2 (second), [Plastic Behaviour and Design by J.F. Baker, M.R. Horne, J. Heyman, The English Language Book Society and Cambridge University Press].

In the United States, the question of ultimate design was discussed by Van den Broek in 1940. In 1945, the Structural Steel Committee of the Welding Research Council directed its interest towards a study of elastic and plastic behavior of continuous frames and their components.

Problems of instability were studied extensively at Lehigh University.

M.R. Horne [The Stability of Elastic Plastic Structures, Progress in Solid Mechanics, Volume - Second, Amsterdam, 1961] reviewed the significance of work carried by W. Merchant. A number of plastic hinges developed in the frames.

In 1949, J.F. Baker [The design of steel frames, 27-397] described a method to calculate the failure loads.

In 1952 H.J. Greenberg and W. Prager [Limit Design of Beams and Frames, Transactions of American Society of Civil Engineering 117 447, First published as Technical Report, A 18-1, Brown University, 1949]. and M.R. Horne [Fundamental propositions in the plastic Theory of structures, Journal of Institution of civil Engineers, 34, 174 (1950)], both published independently, the formal statements and proofs of the principles.

In 1938, A.A. Gvozdev established these principles.

The plastic methods were placed on firm footing by above mentioned contributions. The basic mathematical theory of the plastic methods became complete and rigorous as the conventional methods (elastic methods) for the analysis of the structures.

Acceptance of plastic design methods in the United States came slowly. Until the adoption in 1958 of the supplementary Rules for Plastic Design incorporated in the American Institute of Steel construction specifications for Design, Fabrication and Erection of structural steel for buildings.

In 1960, Professor L.S. Beedle [on the application of plastic design, proceedings of the second symposium on Naval structural Mechanics, Brown University, pp 538, 1960] gave a review position.

In Great Britain and also in USA the Plastic methods of design are widely used.

However, the designer must be familiar with the limitations of the simple plastic design (Commentry on Plastic Design in Steel ASCE Manual of Engineering Practice 41, 1961)

Following further references are also there for the plastic analysis of the structures.

(i) P.G. Hodge : Plastic Analysis of Structures, McGraw Hill Book Company, New York, 1959.

(ii) W. Prager : Introduction to Plasticity Addison Wiley (1959).

(iii) C.E. Massonet and M.A. Save : Plastic Analysis and Design, Volume (one. Beams and Frames, Blaisdell Publishing Company, New York, Toronto, London

(iv) B.G. Neal : The plastic Methods of structural Analysis, Chapman and Hall Ltd. John Wiley and Sons, New York, 1956.

(v) Lynn S. Beedle : Plastic Design of Steel Frames, John Wiley & Sons, Inc, 1958.

(vi) Cross, H., 'Relation of Analysis to structural Design, Transactions, ASCE 101, 1936.

Design of structures on the basis of ultimate load has advantages and limitations. Maximum advantage of the ultimate load design are its rational approach, based on the actual strength of the structures.

2.25 PLASTIC (ULTIMATE LOAD) METHOD OF ANALYSIS AND DESIGN

Plastic desgin is the design against the limit state of plastic collapse.

Plastic design is an advantageous replacement for conventional elastic design as applied to statically loaded structural steel plane frames. All the members lie in one plane frame and are subjected to loads, which also lie in the same plane. Rigid jointed plane frames, continuous or restrained beams and girders, are statically indeterminate structures. An indeterminate structure (e.g., a continuous beam) has two or more points of maximum bending moments.

The load carrying capacity of a plane frame is due mainly to the ability of the joints to transmit bending moments and the resistance to bending of the members (although these members carry axial and shear forces). The basic principle of the plastic theory are concerned with the relation between bending moment and the curvature for the frame. It is assumed that the sign conventions adopted for bending moment and curvature are consistent. A positive bending moment is defined as causing tension in the outermost fibres on one side of the beam, a positive curvature will cause extension in the same fibres.

As load is applied to the structure, the cross-section at the maximum bending moments will reach the yield point. As more load is added, a zone of yielding develops there. Whereas, at elsewhere structure is still elastic. This fact serves to control deflection.

Due to ductility of the steel, the moment remains constant at the first zone of yielding. Later on, the zones of yielding develop at other points of maximum bending moments (peak points in the bending moment diagrams). On development of these yielded zones at sufficient numbers, the structure fails, These completely yielded cross-sections of the members are known as **plastic hinges** and the values of bending moments do not exceed their respective **plastic moments** (i.e. moments of resistance, M_p).

Steel possesses **ductility,** which is a unique property. No other structural material shows this in quite the same manner. Structural steel absorbs large deformations beyond the elastic limit. It is this characteristic of the structural steel that makes possible the application plastic analysis.

The stress strain diagram for the steel (i.e., $\sigma - \varepsilon$) is practically linear upto the upper yield value, σ_{yu} The stress then drops suddenly to the value of lower yield, σ_{yL} and it is followed by increase in strain under constant stress upto a certain value thereafter there occurs **strain harding.**

The flat yielded portion is the **range of large plastic strain**. It is referred as **plastic range** as shown in Fig. 2.2. The strain over this range is pure plastic, ε_p which one is about twelve to thirteen **times** elastic strain, ε_E.

A complete chapter describing **'Plastic Analysis and Design'** has been given in author's Design of steel structures, Volume second.

A set of simplifying assumptions similar to those in the simple elastic theory of bending is made. On the basis of these assumptions, the slope and deflection of a member bent so as to cause some of the material to be strained beyond the yield point. These assumptions are as below :

1. The material obeys Hooke's law until the stress reaches the upper yield value; on further straining the stress drops to the lower yield value (vertically downward) and their-after, the strain increases upto a certain point under constant stress.

2. The upper and lower yield stresses and the modulus of elasticity have the same values in compression as in tension.

3. The material is homogeneous and isotropic in both the elastic and plastic states.

4. Plane transverse sections normal to the longitudinal axis remain plane and normal to the longitudinal axis after bending. (The effect of shear being neglected).

5. There is no resultant axial force on the beam.

6. The cross-section of the beam is symmetrical about an axis through its centroid parallel to the plane of bending.

7. Every layer (fibre) of the material is free to expand and contract longitudinally and laterally under stress as if separated from the other layers.

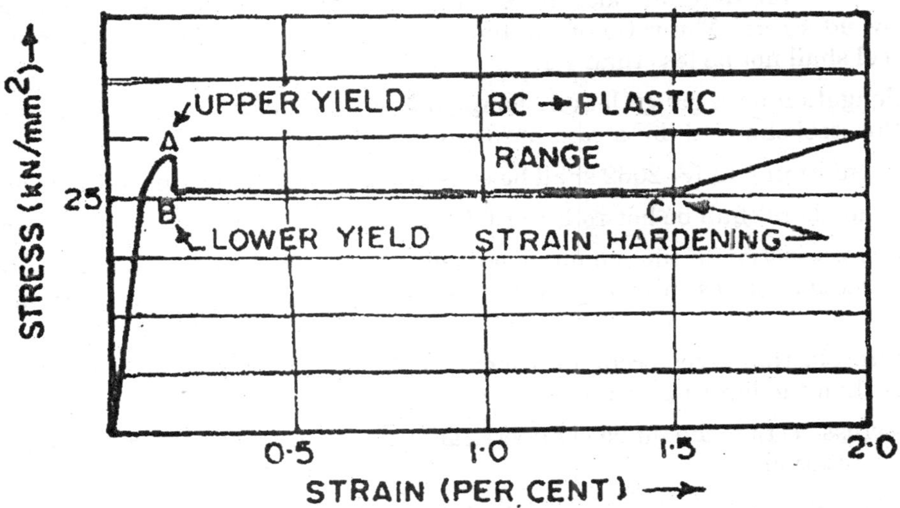

Fig. 2.2.

8. The loading on the structure is proportional (i.e. all the loads remain in constant proportion to each other during loading.

9. The deformations of the structure are small and the geometry of the undeformed structure is used in formulating the equations of equilibrium at the joints. Members are initially straight and prismatic, and instability does not develop before plastic action begins.

It is aimed to design structures, which are structurally safe, strong, durable and serviceable throughout its life.

Essential conditions for plastic analysis and design

Following essential conditions should be satisfied while using plastic analysis and design.

1. **Equilibrium conditions:** As specified for the elastic method of analysis and design.

2. **Mechansm condition:** Ultimate or collapse load is reached when a mechanism is formed. The number of plastic hinges developed should be just sufficient to form a mechanism.

3. **Plastic moment condition:** The bending moment at any section of the structure, M, should not be more than the fully plastic moment of the section, M_p (i.e. M should not more than M_p).

Requirements for plastic analysis and design

In addition to essential conditions, following requirements should also be satisfied.

1. The yield stress of the grade of the steel used shall not be more than 450 N/mm^2.

2. The stress-strain relationship of the steel shall not be significantly different from those found for steels complying with IS:2062 or equivalent and shall be such as to ensure complete plastic moment redistribution.

The stress-strain diagram shall have a plateau at the yield stress, extending for atleast **times** the yield strain. The ratio of the tensile strength to the yield-stress specified for the grade of steel shall not be less than 1.2.

The elongation on a gauge length complying with IS:2062 shall not be less than 15 per cent, and the steel shall exhibit strain-hardening capability.

Steels conforming to IS: 2062 shall be considered to satisfy the above requirement.

3. The members shall be hot-rolled or fabricated using the hot-rolled plates and sections.

4. Teh cross-section of the members not containing plastic hinges should be atleast that of compact section, unless the members meet the strength requirement from the elastic analysis.

5. Where plastic hinges occur in a member the proportions of the cross-section should not be more than the limiting values for plastic section.

6. The cross-section should be symmetrical about its axis perpendicular to the axis of the plastic hinge rotation.

7. The members shall not be subjected to impact loading, needing fracture assessment or fluating loading, needing a fatigue assessment.

Assumptions in plastic analysis and design

The design action effects shall be found using a rigid-plastic analysis. It shall be allowable to assume full strength or partial strength connections, provided the capacities of these are used in the analysis, and provided that

1. In a full strength connection, the moment capacity of the connection, shall be not less that of the member being connected.

2. In a partial strength connection the moment capacity of the connected may be less than that of the member being connected, and

3. In both cases, the behaviour of th connection shall be such as to permit all plastic hinges necessary for the collapse mechanism to develop, and shall be such that the needed plastic hinge rotation does not exceed the rotation capacity at any of the plastic hinges in the collapse mechanism.

2.26. STRUCTURAL SAFETY AND FACTOR OF SAFETY

The meaning of safety and to ensure safety are not so straight forward. It is difficult to define the concept of safety. For the purpose of structural design one needs a precise mathematically accepted definition. A balance in a rational and consistent manner is essential to be clear about the meaning of the safety. One should express and give the definition of safety as a quantity, in the absence of which, one continues to balance a high quantified concept (economy measured in money) against a vague qualitative concept of safety. The main structural safety requirement is to ensure that these is only acceptably low risk of failure during its design life. Another important requirement is that the structure is sufficiently stiff. The excessive deflection and vibrations influence the inservice performance of the structure.

Factor of safety for the structures depends upon many factors such as (i) type of structure, (ii) fire rating of the fire proofing material (iii) constructional details (iv) probability of the failure in the structural members and connections, inspection method of quality control, (vi) durability and (vii) the amount of supervision. The safety of steel structures is examined from the point of view of the uncertainties in the design considerations and structural failures.

There are many uncertainties (e.g. uncertainty about the loading, uncertainty about the strength of materials which will be used, uncertainty about the quality of the construction, uncertainty about one's ability to predict the behaviour). Inspite of all these uncertainties, one has to design and construct safe and economical structures.

Until the beginning of the nineteenth century, the safety was assumed by experience. A high value of factor of safety (around three) used to be arbitrarily adopted, which allows a margin and considers all the uncertainties in the estimation of loads and the strength of material. As a result of large value of factor of safety, the allowable stresses calculated remained small and the size of members (components of the structures) adopted in design remained large.

There were no means for the measurement of the strength of material used for the construction. There was no theoretical knowledge about the behaviour of the structure developed. The load carrying capacity of the structure could not be predicted. These conditions continued, until the industrial revolution. Lower the probabity of the failure, higher is safety.

The **factor of safety** is defined as the factor by which the yield stress of the material is divided to give the working stress (permissible stress) in the material. A greater value of factor of safety used to be adopted formerly and as such the working stresses computed were small. As a result of this a large cross- section of the member had to be adopted in design. From experience gained and experiments made in laboratories, it has been seen that such a large section of the structural member is not essential. Hence, factor of safety adopted now-a-days is comparatively small. This has resulted in appreciable saving in the material. The value of factor safety is decided keeping in view of the following considerations.

1. The average strength of materials is determined after making tests on number of specimens. The strengths of different specimens of given structural material are not identical. Some of the specimens may show relatively large deviations in strength from the average.
2. The values of design loads remains uncertain. The values of dead loads can ofcourse be determined correctly. But live load, impact load, wind load, snow load etc. cannot be determined with certainty since these depend upon statistics available. The probable values of these loads are only determined.
3. The values of internal forces in many structures depend upon the methods of analysis. The degree of precision of different method varies. The methods involving detailed analysis are more precise. In case, analysis of the structure is done precisely, a small value of factor of safety may be adopted.
4. During fabrication, structural steel is subjected to different operations. The punching of a hole in a structural element distorts the surrounding material and causes high residual stresses. The warping and buckling of elements may take place during welding. The welding leaves high residual stresses. Structural elements are subjected to uncertain erection stress.
5. The variations in temperatures and settlement of supports are uncertain. Many times, a well designed structure is damaged because of these effects. The strength of materials decreases because of corrosion. The extent of corrosion is more, when a structure is located in industrial areas and exposed to chemical wastes.
6. The failure of some small structures or some elements of a structure is less serious and less disastrous than the failure of large structure or a main element of a structure.

The minimum value of factor of safety, m_{min} may be found by defining it in a rational method. The factor of safety may be defined as the ratio of computed strength, P of the structure or the structural member to the respective computed internal load, F. Therefore,

$$m = (P / F) \tag{2.1}$$

The magnitude of lowest probable strength, P is influenced by the uncertainties in the mechanism failure, properties of material and workmanship. The magnitude of highest probable internal load. F is influenced by the uncertainties in loading conditions and the structural behaviour. The probable values deviate from the computed value because of these uncertainties. The minimum probable value of strength is given by

$$P_{min} = (P - \Delta P) \quad \text{(i)}$$

The maximum probable value of the internal load is given by

$$F_{max} = (F + \Delta F) \quad \text{(ii)}$$

where, ΔP = Maximum probable deviation of actual value from the computed value of strength

ΔF = Maximum probable deviation of actual value from the computed value of internal load.

When the minimum probable value of strength is atleast equal to or slightly more than the maximum probable value of the internal load, then the failure of structure is just prevented under most severe conditions. Therefore,

$$(P - \Delta P) > (F + \Delta F) \quad \text{(iii)}$$

or
$$P\left(1 - \frac{\Delta P}{P}\right) > F\left(1 + \frac{\Delta F}{F}\right) \quad \text{(iv)}$$

The minimum value of factor of safety is given by

$$m_{min} = \left(\frac{P}{F}\right) = \left(\frac{(1 + \Delta F / F)}{(1 - \Delta P / P)}\right) \quad (2.2)$$

Assuming the values of maximum deviations DP and DF to 25 per cent of the computed values of P and F respectively, the minimum value of factor of safety is given by

$$m_{min} = \left(\frac{P}{F}\right) = \left(\frac{1 + 0.25}{1 - 0.25}\right) = 1.67 \quad \text{(v)}$$

2.27 MECHANICAL PROPERTIES OF STEEL

Mechanical properties of steel depend upon the chemical composition, rolling process, and heat treatment of the steels. Other factors which may influence the mechanical properties are the techniques of testing, such as the rate of loading, and the temperature at the time of testing. These influences may produce varying results for the same grade of steel.

Following definitions concerning the mechanical properties of the steel are useful for the study of this subject :

1. **Modulus of elasticity** : The modulus of elasticity is defined as the ratio of longitudinal stress to the longitudinal stain within the elastic region. It is denoted by letter E, the slope of elastic portion of the stress-strain diagram gives the value of modulus of elasticity, E. ($E = 2 \times 10^5$ N/mm^2). It may also be defined as the coefficient of elasticity for stretching (e.g. the ratio of the stretching force per unit cross-sectional area to the elongation per unit length).

2. **Shear modulus of elasticity.** The shear modulus of elasticity is defined as the ratio of shearing stress to the shearing strain within the elastic range. It is denoted by letter G. The shear modulus of elasticity is also known as modulus of rigidity. It is given by

$$G = \left(\frac{E}{2(1+\mu)}\right). \; G = 0.758 \times 10^5 \text{ N/mm}^2 \tag{2.3}$$

where μ = Poisson's ratio.

3. Bulk modulus of elasticity

The bulk modulus of elasticity is defined as the ratio of hydrostatic stress to the volumetric strain within the elastic range. It is denoted by letter K. It is given by

$$K = \left(\frac{E}{3(1-2\mu)}\right) \tag{2.4}$$

It is also given by

$$K = \left(\frac{2G(1+\mu)}{3(1-2\mu)}\right) \tag{2.5}$$

4. **Tangent modulus of elasticity.** The tangent modulus of elasticity is defined as the slope of the tangent at a point on the stress-strain curve above the limit of proportionality. It is denoted by E_t.
5. **Poisson's ratio.** When the stress is applied in one direction, strains are induced not only in the direction of applied stress, but also in the other two mutually perpendicular directions. The **poisson's ratio** is defined as the ratio of transverse strain to the longitudinal strain under an axial load. It is denoted by letter μ. The value of **Poisson's ratio** for steel within the elastic region ranges from

   0.25 to 0.33. Whereas as, for the cork, poisson's ratio is zero whereas for the materials like rubber and paraffin under tension may have, maximum value of 0.5.
6. **Yield point.** The yield point on a stress-strain diagram is defined as the stress at which the strain increases to a large value without an increase of stress. For the mild steel, it is indicated by a flat portion of stress-strain diagram if the stress-strain diagram is plotted on the exaggerated scale. Some steels show upper yield point and lower yield point.
7. **Yield strength.** Some steels do not show yield plateau, (i.e., the flat portion) on the stress strain diagram at yield. The stress-strain diagram for such steels show continuous curve upto the tensile strength. Therefore, the yield strength of such steel is defined as a stress shown by a specific point on the curve obtained by drawing 0.2 per cent offset of the strain parallel to the initial elastic portion of the curve.
8. **Tensile strength.** The tensile strength of a test piece is defined as maximum axial load divided by the original area of cross-section.
9. **Fatigue strength.** The fatigue strength is defined as the stress at which steel fails under repeated applications of load. It is called as fatigue strength or **endurance limit.**
10. **Impact strength.** The impact strength of steel is the measure of ability of the steel to absorb energy at high rates of loading.

11. **Strain hardening modulus.** The strain hardening modulus is defined as the slope of the stress-strain curve in the strain-hardening range. It is denoted by E_{st}. It has the maximum value at the beginning of strain-hardening.

12. **Ductility.** The ductility is the unique property of steel. It is the property of steel by virtue of which, the deformation of structural member/structure occurs before failure. It is ability of steel to undergo large in elastic (plastic) deformation without rupture (opposite of brittleness) due to tensile force. It is to note that the adequate ductility is a pre-requisite for the plastic analysis of the steel. It is measured quantitatively by percentage of elongation of test specimen at failure in a uniaxial test. A **gauge length,** ($L_{GL} = 5.65 (A_0)^{1/2}$, where A_o is the original cross-sectional area of the bar) is marked in the middle of the bar. The bar is tested upto failure. Its elongated length, L_{EL} is measured as shown in Fig. 2.3. The strench in the length of the bar when it is subjected to tensile load is known as **elongation.**

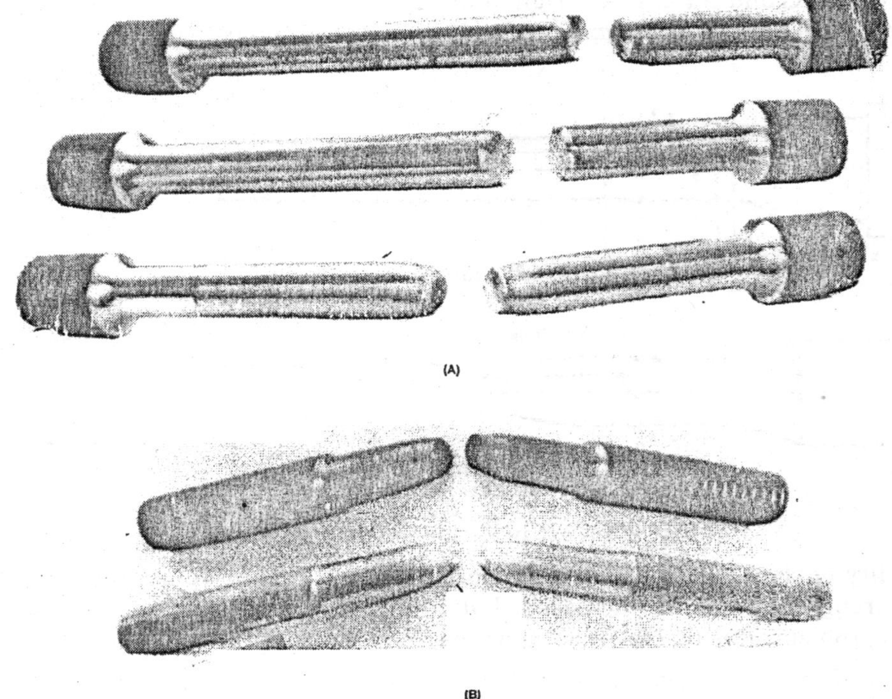

Photo Plate 2.4. Axial Tension Failures
(A) (i) Steel specimen shows necking (ii) High tensile steel shows brittle fracture
(B) (i) Mild steel: shows cup and cone (ii) Aluminium - alloy specimen shows double cup type failure.

Then, the percentage elongation $= \left(\dfrac{L_{EL} - L_{GL}}{L_{GL}}\right) \times 100$. For a good ductile steel the percentage elongation should be 20 to 33 per cent. This percentage of elongation is an acceptable measure of the ductility of the steel. However, the gauge length must always be specified. Since, the total extension is the sum of two factors: the general extension, which occurs over the total length of the test-piece, and the local extension which confined to the region in the immediate vicinity of the fractured cross-section. The local extension is a constant for the material of a given diameter, as also the general extension per unit gauge length. But the general extension is considerably less than the local extension.

The elongation of the test-piece is accompanied by a reduction in the cross-sectional area. By measuring the smallest diameter after fracture, the percentage reduction in area is determined.

However, the decrease in the length, when it is subjected to compressive load is called **shortening**.

Elongation and shortening of the length of the steel bar **deformation**. And the deformation per unit length is known as **strain**.

The ductility is also measured by percentage of reduction in area. It is given by $(A_o - A_R)/A_o \times 100$ where A_o and A_R are the original and reduced cross-sectional areas of the bars. For good ductile steel bars, the percentage of reduction may be 65 per cent.

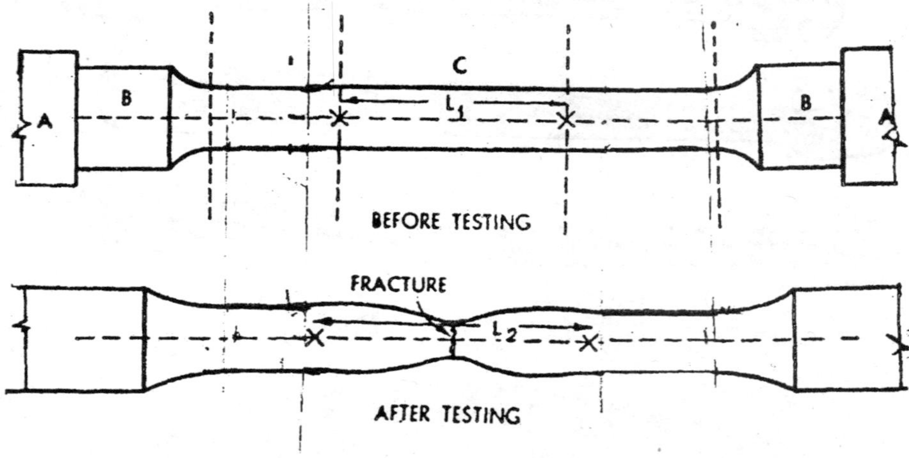

Fig. 2.3. Test-piece for tensile-test.

The ductility of steel is reduced in case more carbon is added. By increasing the carbon content, the strength of steel is increased, but it reduces ductility. The increased carbon content causes problems during the welding process.

13. **Malleability** : It is that property of steel by virtue of which it may be beaten rolled etc. and it may be given a new shape and new figure. It is ability of steel to undergo inelastic (plastic) deformation without rupture due to compressive force. This property is useful in many operations (e.g., forging, hot rolling, drop stamping etc).

Malleable iron: Malleable iron is made by prolonged heating of cast-iron in the presence of iron-oxide during which the surface is decarburised to a depth of 1.5 mm to 3 mm. Therefore, the surface becomes ductile and soft, and the general ductility and tensile strength of metal are considerably increased. On being heating to temperatures in excess of 700°C malleable iron reverts to white iron, which is extremely hard and brittle to prevent this reversion, welding must be done on the surface only, and the lowest possible current must be used in order to avoid thermal disturbance of the metal below the decarburised skin.

Other properties of the structural steel, (viz. Durability, Hardness, Brittleness, Resilience and Toughness) have been described in the subsequent sections).

2.28 DURABILITY AND CORROSION OF STEEL

Durability is the ability of a material to resist deterioration over a long period of time.

One of the earliest things one learns about iron and steel is that these materials rust. Rust is actually a result of **corrosion**. Rust itself is a hydrated oxide of iron, (that is, iron oxide containing a certain quantity of combined water).

The corrosion is the main problem which influences the durability of carbon steel.

Thin elements are more susceptible to corrosion than the conventional relative thick shapes. Increased thickness increases the durability of steel. **Durability** of the steel is the property of steel by virtue of which it lasts long and resists wear. **Corrosion** may be defined as the deterioration of a metal by chemical or electro chemical reaction with its environment.

On the basis of modern electro-chemical theory, corrosion or rusting is both an electrical and a chemical process, connected with the phenomenon known as electrolysis. When certain substances (e.g., acids and salts) are dissolved in water, the molecules of which these are originally composed are split-up, partly at any rate into negatively and positively charged particles called ions. Because these ions are electrically charged the solution will conduct electricity so that the solution is called an electrolyte.

In case, a direct electric current is passed through an electrolyte, the ions split into two groups, a it were. The negatively charged ions are driven towards the positive pole (or anode) and the positively - charged ions towards the negative pole (or cathode). The anode is the plate or point at which the electric current enters the solution, and the cathode is the plate or point at which it leaves the solution. When the ions reach the corresponding poles, these are neutralized (i.e., these are no longer charged with electricity), and those at the cathode are liberated or set free as elements. In case the element is gas, it comes out as such, if a metal, it is deposited on the cathode.

Rust on iron and steel is a state of affair similar to the formation of an electric chemical battery, where the electrolyte consists of water containing dissolved salt or of condensed moisture, which always contain carbon di-oxide (carbonic acid gas), and in industrial zones, the condensed water-vapour may contain sulphuric acid gases dissolved from the atmosphere. Moist air at the seaside also contains particles of salt. In contact with impure water or condensed water-vapour some cystals of iron and steel become anode to others and are therefore, corroded away and pitting so frequently accompanies corrosion.

Corrosion is defined as the deterioration of steel by an electron-chemical process. In corrosion, on the surface of the steel anode and cathode zones are formed in the presence of the moisture and oxygen as shown in Fig. 2.5. An electric current flows through the metal and the anode is dissolved. The situation becomes comparable to the action of battery. Exposed areas to atmospheric oxygen becomes cathode. Some areas on the surface of steel remains covered in dirt or rust scales. Such areas suffer severe attack. Such areas act as anode. As the corrosion reactions proceed, these areas shift and change. **Rust** is formed as a product of reaction as hydrated ferrous oxide. On the surface of exposed steel structures and results in gradual loss of metal and reduces the cross-section of the member. Thereby, it increases stress in the remaining material.

When the bar is corroding, there is a flow of electrons through the bar and flow of ions through the concrete. When the bar is not corroding there is no flow of electrons and ions. An isolated steel bar will rust spontaneously as shown in Fig. 2.4. In the high-quality concrete, however, corrosion of steel bars should not occur, even in the presence of moisture and oxygen. Highly alkaline concrete (pH value of about 12 to 12.5) causes passive oxide film to form on the surface of steel which prevents corrosion.

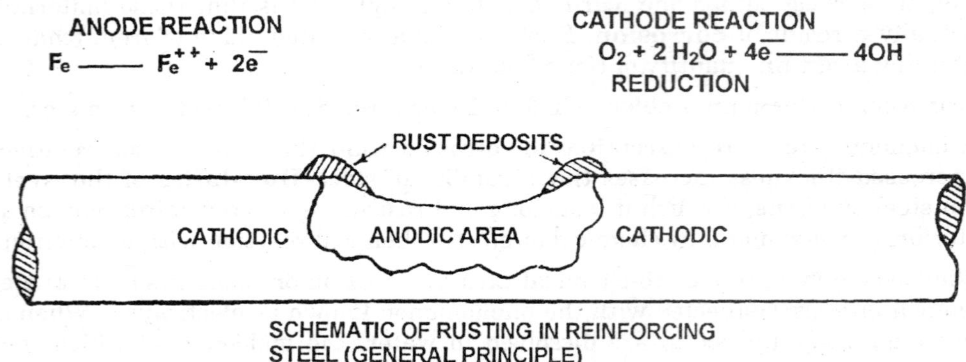

Fig. 2.4.

Only when the pH falls below 11, then, the passive iron-oxide layer is destroyed and it permits rusting to occur.

Above description shows the general principles of corrosion activity. Expression (1) indicates the process of **half-cell oxidation reaction** or the anodic reaction.

$$Fe \uparrow \left(Fe^{++} + 2_e^-\right) \tag{1}$$

Expression (2) indicates the process of **half-cell reduction reaction** or the cathodic reaction

$$\left(2\,H_2O + O_2 + 4e\right) \uparrow 4OH^+ \tag{2}$$

Half-cell potential method is used to evaluate the corrosion (i.e., an electric method) activity.

Recently, an issue of concern has been the corrosion of steel reinforcement in fly-ash concrete structures exposed to chloride ions from deicing salts or sea-water. In case, the concrete cover is sufficiently thick and impermeabile, it will normally provide adequate protection against corrosion. The protective effect of concrete cover is both a physical and a chemical nature and functions in three way as under.

(a) It provides an alkaline medium in the immediate vicinity of the steel surface.

(b) It offers a physical and a chemical barrier to the ingress of moisture, oxygen, carbon dioxide, chlorides, and other aggressive agents.

(c) It provides an electrically resistive medium around the steel members.

The reunion internale des laboratories d'essis de recherches sur les materiaux et less construction (RILEM) Technical Committee on Corrosion of Steel and Concrete (1974) made following two statements.

(a) The efficiency of the concrete cover in preventing corrosion is dependent on many factors (e.g., impermeability and a high reserve of alkalinity). Which satisfies both the physical and chemical requirements of concrete cover. In case the concrete is permeable to atmospheric gases or lean in cement corrosion of the reinforcement may be anticipated and a good protection should be attempted by the use of dense aggregate and a well compacted mix with reasonable low water/cement ratio.

(b) In case, chloride corrosion is expected, it is now usually agreed that the carbonation of concrete cover is the essential condition for corrosion of reinforcement. In polluted atmospheres (such as bad industrial areas having smoke and soot), the rate of corrosion becomes high. In the marine environment, the chloride ions remain present. These aggressive environments quicken the process.

Types of Corrosion

There are following various types of corrosion which are seen.

(i) **Pitting corrosion**: Corrosion pits are formed in anodic areas of the surface of steel bars immersed in water or soil. Due to the presence of moisture specially aided by improper detailing or due to constant exposure to alternate wetting and drying, this type corrosion occurs. It is easily handled by providing quick drainage in proper detailing and allowing free flow of air. The surface becomes dry.

(ii) **Crevice corrosion:** In this type of corrosion, long narrow openings (clinks) and fissures (thin splits) occur again due to improper detailing. The tops of splits become local anodes with respect to surrounding metal and corrosion takes place at this point. The oxygen contents of water trapped in splits is less than that of water, which is exposed to air. The corrosion begins inside the split.

(iii) **Bimetallic corrosion:** When two different metals (e.g., iron and aluminium) are in electrical contact put together in an electrolyte (normal water) an electric current passes in between them and corrosion occurs.

The basen metals become anodic and the nobler metals become cathode. The nobler metal tends to be protected at the expense of the base material.

Depending on the electro-potential, the metals may be arranged into a table known as the galvanic series. Therefore, this bimetallic corrosion is also called as galvanic corrosion. The electrical potential difference shall be large in this case. The distance between two metals becomes large.

The metallic path between these two metals may be broken by insullating the dissimilar materials from each other and this bimetallic corrosion may be prevented. Otherwise the electrolyte solution may be broken by preventing the formation of a continuous bridge of electrolyte solution between two metals. Insulating bushes and washers may be fitted in former procedure while the metals may be painted in the alternative procedure.

(iv) **Stress corrosion:** Under the simultaneous influence of a static tensile stress and specific corrosive atmosphere, at the stress concentration areas, the stress makes some spots in steel more anodic compared with the remaining. The top of a split (crack) becomes anodic part and it corrodes and the split becomes more wider. It is not common in ferrous metals. Some stainless steels are susceptible to this type of corrosion.

(v) **Fretting corrosion:** In case, two oxide coated films or rusted surfaces are rubbed together, the rust may be mechanically removed from projecting high spots (i.e., the surface in contact). These high spots become active anode and rest of the surfaces act as cathode, and corrosion begin. In mechanical components, this type of corrosion occurs.

(vi) **Bacterial corrosion:** The pipe-lines, burried structures, and off-shore structures remain in soil and water and bacterial corrosion occurs in them due to micro-biological activity.

Above types of corrosions may be grouped in atmospheric corrosion, which occurs in normal environment. Others may be grouped into chlorides corrosions.

DETERIORATION OF CONCRETE AND STEEL BY CORROSION

In concrete structures rusted steel bars swell in volume. Due to this, the cracks develop in concrete. As the rusting is progressive phenomenon, the width of cracks increases and pushes the concrete at the sides. The concrete cover breaks and the concrete cover falls down. The rate of corrosion accelerates by moisture. Alongwith fall of the concrete cover, the cross-sectional areas of the steel bars (as these are eaten-up) reduce and become unable to take-up forces for which these are designed. With progress of time, the structures may fail due to progressive corrosion alone.

Corrosion protection

Several strategies to combat steel corrosion have been developed. Concrete protects embedded steel from corrosion through its highly alkaline nature. High-pH environment (usually more than 12-5) causes a passive and non-corroding protective oxide film to form on steel. However, carbonation or the presence of chloride ions from deicers or seawater can destroy or penetrate the film. Once this happens, an electric cell is formed along the steel or between steel bars and the electro-chemical process of corrosion begins. Chloride ions alone can also penetrate the passive film on the reinforcement and combine with iron ions to form a soluble iron chloride complex that carries the iron into the concrete where it is later oxidized and forms rust.

The steel reinforcement bars should be protected from moisture and oxygen to minimise the risk of corrosion. The reinforcement bars need proper protection. Following protective measures used to protect the steel bars from corrosion have been described here.

(i) **Paint coating:** Various effective paints are available which are used to provide paint coating on the steel bars. However, the paints are also permeable. Commonly zinc, aluminium or lead based priming coat is applied first so as to provide a good base for the paints coat.

(ii) **Metallic coating:** Metallic coating including galvanizing and sherardizing (in both the process, zinc is used), is done for sheets. Electro plating is mainly confined to small items like fasteners.

(iii) **Corrosion resistant steel:** Special corrosion resistant steel are used as protective treatment. Such steel rapidly forms its own protective layer of oxide film. The rate of corrosion reduces to negligible level after few years. In Britain, such steel is knwon as weathering steel. Commerically, such steels is available as Cor-Ten. It is originally developed by the United States Steel Corporation. Elements such as copper, nickle and chromium are added in this variety of steel. There is stainless steel, which does not corrode.

(iv) **Coated reinforcement**: In IS:9077-1979, a special method of coating the individual bars with a phosphate jelly to prevent corrosion has been described. This method has been developed at the Central Electro Chemical Research Institute (CECRI) at Karaikudi, Tamilnadu. The individual bars are treated at site with derusting solution. The bars are washed with an alkaline cleaning powder. Then, the phosphate jelly is applied alongwith prohibitory solution, cement slurry, sealing solution and cement slurry again. Soon after coating, the bars are used. This process is tedious and time consuming.

(v) Electro-chemical corrosion may be suppressed by using Cathodic protection. In this technique, an electric current is applied to the rebars (which must be electrically connected in the opposite direction to the current flow in spontaneous corrosion. This makes the steel cathodic, and prevents corrosion. This technique has been used successfully but it is difficult to apply.

(vi) Various corrosion inhibiting (prohibiting) admixtures are now available commercially which serve to maintain the passive oxide film.

Epoxy-coated rebars are becoming more and more common.

For shallow concrete cover, corrosion prohibiting admixtures are often prescribed. These admixtures interfere with the galvanic process by scavenging electrons or protons near the steel. Following are three general categories of corrosion prohibitors.

(a) **Anodic or active:** These prohibitory admixtures actively interfere with the corrosion process by occupying overactive electrons, oxidizing rapidly dissolving ferrous oxide ions into an insoluble protective coating of ferric oxide on their way to the surface of the anodic steel reinforcement. Calcium nitrite is the primary material in this category. Although sodium nitrite, sodium benzoate, and sodium chromate are also used outside the United States.

(b) **Cathode or Passive:** Cathodic protection (CP) is an electro-chemical technique in which external anodes are added to the electro-chemical cell present in the seawall. The addition of external anodes to the seawall structure results in the shift of electrical potential at the reinforcing steel from anodic (actively corroding) to the cathodic values (non-corroding). Fig. 2.5 shows a schematic representation of two types of cathodic protection systems (viz, sacrificial anode and impressed current cathodic systems).

The act as protone acceptors, prohibitors indirectly inhterfere with the corrosion process by slowing down the cathodic reaction and impeding the corrosion current. These are normally highly alkaline materials that render the ferrous ions at the sjurface of the reinforcing steel insoluble in water and unable to participate in the corrosion current. The most commonly used materials in this category are sodium hydroxide, sodium bio-carbonate and aluminium hydroxide. Although organic materials like substituted forms of aniline and mercapto-benzothiazole are used.

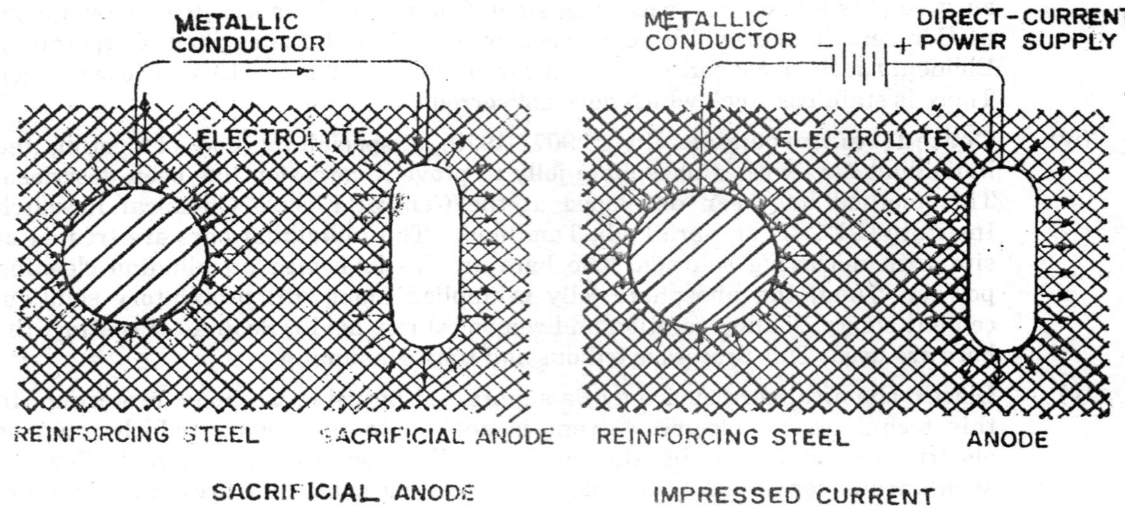

Fig. 2.5.

(c) Mixed or passive-active: This may be preferable over either of the previous types, since it attacks the problem at both ends of the reaction, interfering with the corrosion process before, it can being. These materials can be the form of more complicated organic molecules that are attracted to and tie up both cathodic and anodic sites, forming salts as these accept either the available electrons or protons. This category may also be a physical mixture of two simpler compounds one of which influence the cathodic reaction site while the other affects the anodic side.

Corrosion prohibitors are typically used at dosage rate of 1% to 4% based on the weight of cement. Concrete properties are influenced by the use of corrosion probibiliting admixtures. Inorganic salts like calcium nitrite are known accelerators, but organic corrosion prohibitory may retard set times. Workability may be improved at levels upto 2% but begins to degenerate above that. In hardened concrete strengths are slightly reduced and, the alkali-slica reaction may be aggravated by corrosion prohibitors in case reactive aggregates present in the material.

Other methods of preventing the corrosion have been described along with mentioning different types of corrosion.

Impurities in water influences the rate of corrosion. Rate of corrosion remains low in rural areas. In industrial areas, the atmosphere remains polluted, the rate of corrosion increases.

To minimise the risk of corrosion, the bare steel should be protected from moisture and oxygen. The paints which are applied to protect the steel surface from corrosion are themselves permeable. Therefore, such a system is chosen that a thickness may be maintained.

The durability is influenced by the detailing of the steel work. The moisture is trapped in details. In open and out-door locations, the intermittent welds should be avoided. At such places, the weep holes are provided so that any collected moisture is let-out.

For resistance to corrosion the steel may be chemically modified. Copper, nickle and chromium may be added so that the steel becomes rust-less and the durability of steel is increased. New corrosion resistant steel has been developed. The corrosion resistant steel has possibilities of providing an economical answer to the maintenance problem.

Corrosion in the presence of stress may produce failure called **stress corrosion cracking**. In case, the crack size, stress and temperature corresponding to the brittle fracture then the failure may occur. The failure in 1967 of the Silver Bridge across the river Ohio, with the loss of forty lives has been attributed to this effect.

Stainless steel is subjected to a type of corrosion which progresses along the grain boundaries and results in a crack. Failure of a suspended ceiling over a swimming pool in Switzerland in 1985 resulted from this type of attack. The chlorine in air above the pool produced corrosion of the stainless steel hangers supporting the suspended ceiling. Stainless steel is especially sensitive to this type of attack when halides are present in the corrosive environment.

2.29 FIRE RESISTANCE

Fire resistance is defined as ability of an element component or structure, to fulfil for a stated period of time, the required stability integrity, thermal insullation and/or other expected performance in a stand fire test.

Adequate fire resistance is necessary in buildings to allow the escape of the occupants and the reasonable safety for the fire fighters engaged in fire under control.

Fire resistance and protection are important aspects for the design in the event of fire inside the structures. An understanding of the behaviour of materials in elevated temperature is needed to ensure safety. Adequate fire resistance to structural elements is necessary to safe guard the stability of the structure Essentially, it is intended to ensure public safety than to safe guard the structure itself. The main criteria are to prevent premature collapse and the main sketelon of the building to stand up long enough so that to allow escape of the inside occupants and reasonable safety for fighters engaged in extinguishing the fire and bringing the fire under control and to limit the spread of fire, thus reducing the risk to surrounding property and their occupants. Smaller the area in which the fire may be contained, the smaller the escape and fire fighting times will need to be, although the different materials stored in the building will affect this time.

Fire resistance is obtained by ensuring that the cross-sections are of minimum thickness and that the cover reinforcement is sufficient.

In Britain, the necessary requirements form part of the Building Regulations. The minimum periods for the inside occupants to escape range from 30 minutes for a small residential buildings to 4 hours for a large store. And it is specified in building regulations. The period for which fire resistant is needed is calculated by the volume of the enclosed section. For a larger volume a greater period of fire resistance is required than for a lesser volume. The fire resistance of a concrete structural element is expressed in terms of time period in hours in accordance with IS : 1641.

All materials lose strength with increase in temperature. Some idea of the of the temperatures likely to meet may be gained by reference to an internationally accepted standard time-temperature curve shown in Fig. 2.6.

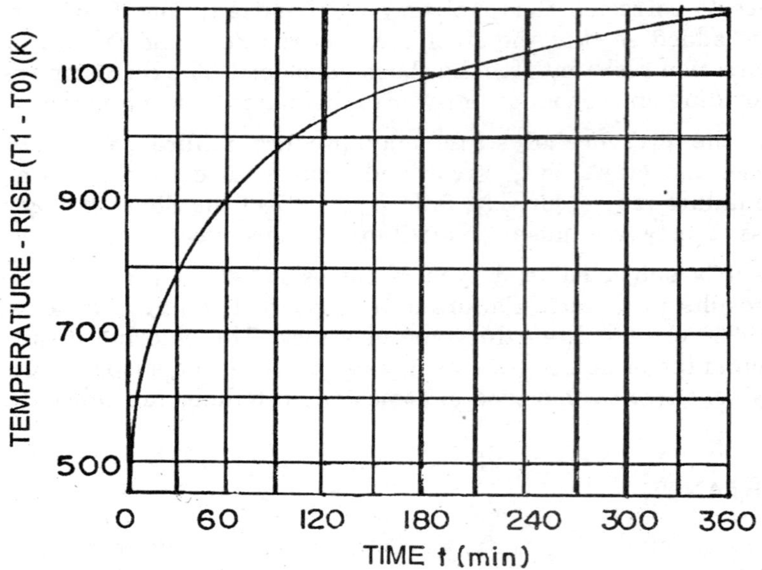

STANDARD CURVE (TIME v/s TEMPERATURE RISE.)

Fig. 2.6.

The fire resistance of a building is usually specified in hours (minutes). The fire-resistance grading period for a structural elements or system in minutes, whcih required to be attended in a standard fire test is referred as Fire Resistance Level.

A reinforced concrete structure or a structural element required to have fire resistance should be designed to possess an appropriate degree of resistance to flame penetration, heat transmission and failure. The fire resistance of reinforced concrete depends to a large extent on the control of temperature of the reinforcement and hence on the degree of insulation provided, which is obtained by concrete cover or by mixture of concrete cover, plaster and other finishes. The thickness of section also enters into the calculations. The fire resistance of concrete elements depends upon details of member size, cover to steel reinforcement and type of aggregate (normal weight or light weight). The general requirements for fire protection are given in IS: 1642. The beams should have minimum width, the floors should have minimum thickness and the columns should have minimum lateral dimensions.

The cover provided for the protection against corrosion and for the durability may not be sufficient for protection against fire. The nominal cover to meet specified period of fire resistance have been specified in IS : 456-2000 for beams, slabs, ribs and columns and these have been given in respective portions.

2.30 HARDNESS OF STEEL

Materials are described conventionally as hard or soft, as the case may be, but it is difficult to define clearly either of these terms. The difficulty is because, the hardness has no physical dimensions and cannot be measured in fundamental units. In consequence, it is not an additive property, neither does it follow the law of equivalence, which means that different materials will not be placed in the same order, when the hardness is determined by different methods.

The earliest hardness scale devised by Moh, was based on the assumption that a mineral of superior hardness scratches another that is less hard. This principle was utilised to determine the relative hardness of different metals. The hardness is regarded as resistance to wear or abrasion and machines have been devised to measure this abrasive hardness.

A more modern conception of this peculiar property that of resistance to deformation or penetration by an applied load. This is known as **indentation hardness.**

Hardness is defined as the resistance of material to indentation and scratching. For products such as armour plates, steel rail and types, hardness is an essential quality. It is tested using either indentation tests or scratching and abrasion tests Both, elastic and plastic deformation in steel take place. These test are used to find the effect of heat treatment, hardening and tampering. Hardness of a material is determined by conducting either (i) Brinell Hardness Test, or (ii) Vicker's Hardness Test or (iii) Rockwell Hardness Test. Different types of indentors are used in these tests.

A steel ball, a square based diamond pyramid of 135 included angle or a diamond core with 120 included angle are respectively used in these types of tests. Hardness numbers are expressed as the results of these tests. These numbers are related to the applied load, P to the surface area of the indentation formed.

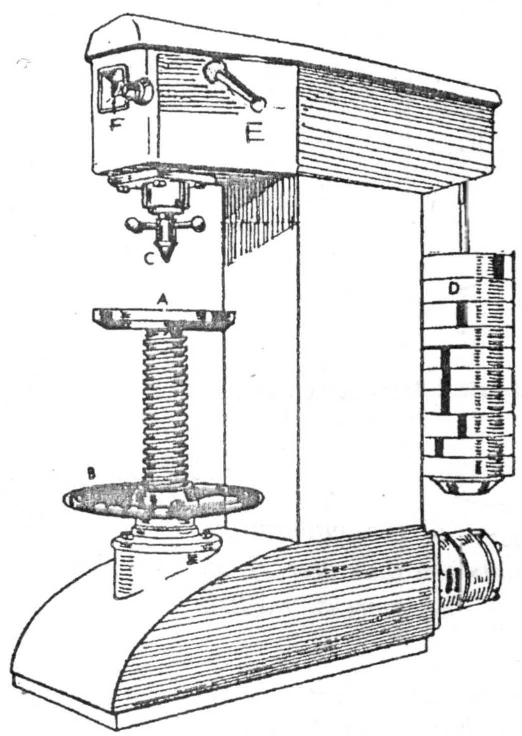

"Brinell" hardness-testing machine.
(Courtesy W. & T. Avery, Ltd.)

Fig. 2.7

(i) **Brinell hardness number,** (BHN) is given by following expression :

$$BHN = \left(\frac{\text{Applied load, P in kilograms}}{\text{Spherical area of indentation sq. mm}}\right)$$

The section of the impression as shown in Fig. 2.3 is obviously part of a great circle of the spherical ball and, therefore, by a well known property of sphere

$$\left(\frac{\text{Area of indentation}}{\text{Area of ball}}\right) = \frac{h}{d}$$

where, h is the height/depth of the indentation and area of the ball is πD^2. Therefore,

$$\text{Area of indentation} = \left(\frac{\pi\ D^2\ h}{D}\right) \quad \text{(ii)}$$

It is more convenient to measure the diameter, d of the impression instead of depth h and from the properties of the circle,

$$\left(\frac{d}{2}\right)^2 = (D-h)\ .\ h\ .\text{i.e.}$$

$$\left(h^2 - D.h + \frac{d^2}{4}\right) = 0 \quad \text{(iii)}$$

Therefore,

$$h = \left|D \pm (D^2 - d^2)^{1/2}\right| = 0 \quad \text{(iv)}$$

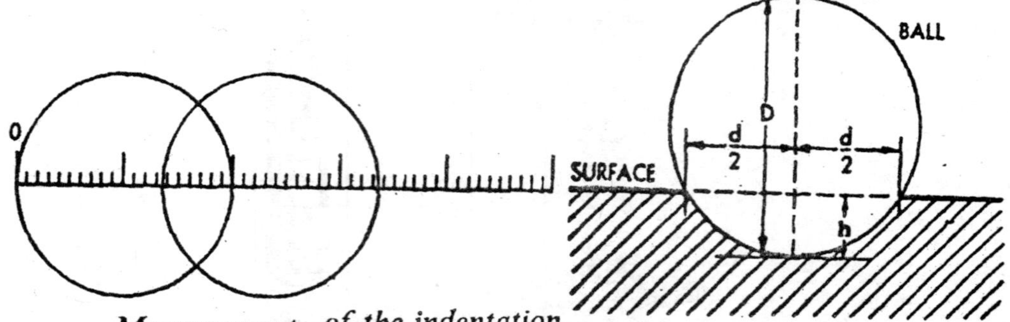

Measurement of the indentation. relationships between diameter of the ball and diameter and depth of the impression. Brinell test

Fig. 2.8

The value obtained by taking the positive sign (±) is that of the sphere (i.e., ball) outside the indentation and it is therefore, rejected. Therefore,

$$BHN = P/\left[\frac{\pi}{2}(D-(D^2-d^2)^{1/2}\right]$$

where, D is the diameter of the steel ball, and d is the diameter of indentation measured by using microscope.

Geometrically, similar impressions may obtained, if the ratio (P/D^2) is a constant. In this way, identical hardness numericals will be obtained on the same material, assuming its constituents are uniformly distributed by using

Load	Diameter of ball
30 kg	1mm
750 kg	5mm
3000 kg	10 mm

The British Standards Institution gives the following ratio

1	For iron and steel and other (P/D2) ferrous alloys of similar hardness	= 30
2	For copper alloys brass, bronzesete	= 10
3	For commerically pure copper	= 05
4	For lead and tin alloys	= 01

In the determination of hardness of metals by the identation method following precautions are to be observed.

(i) The thickness of the specimen should be more than 7 (seven) **times** depth of impression for hard materials and more than 15 **times** the depth of soft materials such as copper. The effect of load should not appear on the reverse side of the test specimen.

(ii) In case, the impression is not truly circular the greatest and the least diameters should be measured and the mean value be taken for the determination of the hardness. Although not absolutely accurate the results are then near enough for all practical purposes.

(iii) The width of the test-piece should be not less 2.5 **times** the impression.

(iv) The surface of the specimen must be flat and smooth: with balls of less than 5 mm the surface should be finished with grade 00 emery paper.

The indentation hardness of a metal bears some relation to its tensile strength, and the latter may be evaluated approximately by multiplying the Brinell number by a constant whole value depends on the nature of the alloy.

The values for this multiplying factor (k) for certain important alloys are given in Table 2.

TABLE 2.3

Alloys	$K = \left(\frac{IS}{BHN}\right)$	P/D^2
Brasses containing less than 30 per cent zinc	0.20/0.30	10
$\alpha - \beta$ Brasses	0.25/0.40	10
Monel metal	0.20/0.23	10

Copper-aluminium alloys containing less than 8 per cent of Aluminium	0.20/0.35	10
Nickle-rich alloys	0.30/0.40	10
Aluminium-copper alloys containing between 4 and 14 per cent copper	0.15/0.22	5
Magnesium alloys	0.27/0.33	5
Lead-tin and white-metal bearing alloys	0.17/0.21	5

There are certain limitations in the ordinary Brinell test. For soft materials, the plastic flow raises the surface material under test (above its original) level as shown in Fig. 2.9. The measured diameter of the impression, d' is larger than it should be and the indicated hardness less than the actual.

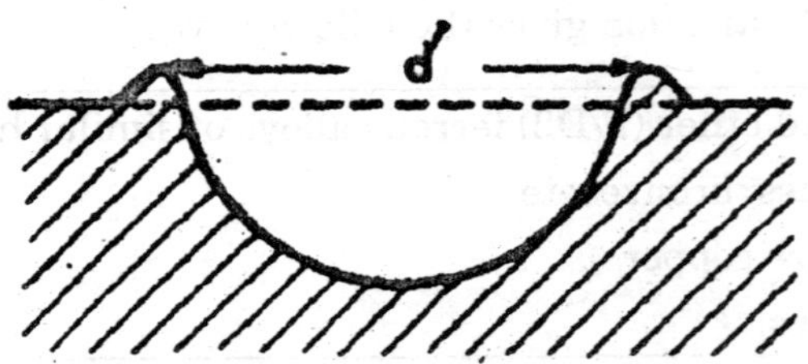

Incorret impression owing to surface flow (Brinell test)

Fig. 2.9.

In other cases, the material under test may be the harder than the material of the ball used as the indenting tool. The ball than becomes flattened and it results in erroneous hardness figures.

(ii) Vicker's hardness number (VHN) is given by the expression as under :

$$\text{VHN} = (1.854\ P/L^2)$$

where, L is the diagonal length of the indent.

Modern hardness-testing machines employ diamond indentors of conical or pyramidal in shape as shown in Fig. 2.10. The base of indentor is square of size a and having the slant height, l.

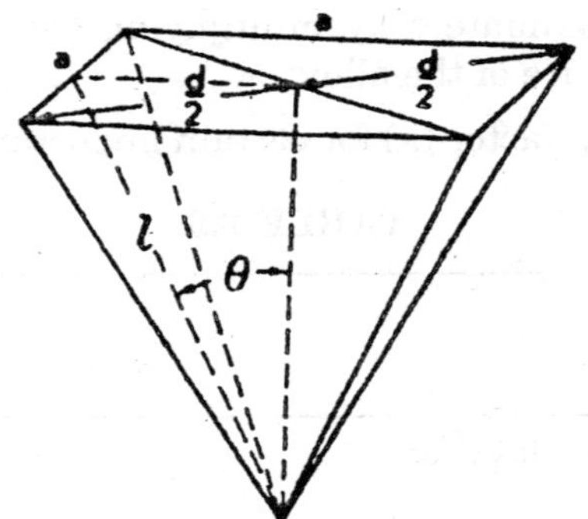

Pyramidal diamond indentor.

Fig. 2.10.

In practice, the diagonal d of the base is more conventiently and accurately measured in comparison to its side. Then,

$$d^2 = (a^2 + a^2)\ , \qquad \therefore\ a = (d/\sqrt{2})$$

$$\left(\frac{a}{2}\right) = l\,.\ \sin\theta\ . \qquad l = (a\ /\ 2\ \sin\theta)$$

where θ is the semi-vertical angle between opposite faces of the pyramid. Therefore,

$$\text{Surface area} = \frac{1}{2}\left(\frac{d^2}{\sin\theta}\right)$$

The hardness number (P.H.N.) is then given by the expression

$$\text{P.H.N.} = (2\,l\ .\ \sin\theta)\ /\ d^2$$

In the Vickers machine, the angle between the opposite faces of the pyramid to 136°, a value calculated from average diameter of ball impressions. The Vickers hardness number (V.H.N.) is then given by

$$\text{V.H.N.} = \left(\frac{2\,l.\ \sin 68}{d^2}\right)$$

$$\text{V.H.N.} = (1.854\ P\ .\ l/d^2)$$

The diagonal *d* of the impression is measured by a specially designed micrometer ocular. Looking through the eye-piece, the operator brings the left-hand extremity of the diagonal coincident with a fixed knife-edge by means of two focusing screws. The diagonal of the impression is then, measured by rotating a knurted head, which operates both a digit-counter mounted on the side of the ocular and a movable knife-edge within the eye-piece. When correctly focused and adjusted, the image of the impression appears as shown in Fig. 2.11.

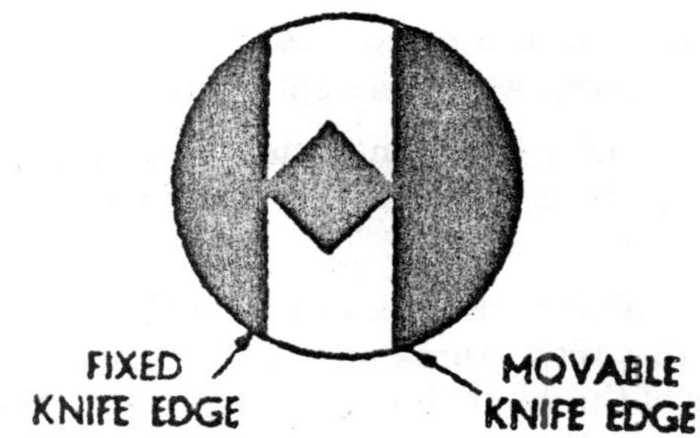

Measurement of Vickers test impression.

Fig. 2.11

It is to note that Rockwell hardness test is not generally performed for structural steels.

There is some relation between the hardness number and the tensile properties of steel.

2.31 BRITTLENESS OF STEEL

The strength of steel increases with the increase of carbon content. However, the ductility of steel decreases. High carbon steel is relatively brittle.

Brittle materials have low toughness, since, these materials have only plastic deformation before fracture. The use of such materials in structures is dangerous, since, fracture may occur suddenly without any noticeable deformation. In a brittle fracture, catastrophic low ductility failure occurs without prior plastic deformation and quickly (i.e., at extremely high speed). It may cause complete fracture. A small crack propogates very fast specially in the regions of comparatively low stress.

The structures should not be subjected to combinations of situations, which are conducive to brittle fracture (i.e., low temperature welds stressed in through thickness direction and fast rate of loading.

The brittle fracture may be avoided in steel structures, in case, there is toughness of material. This type of fracture is affected by many conditions (e.g., low temperature, loading rate, stress level plate thickness flow size, joint geometry and workmanship.

2.32 RESILIENCE OF STEEL

Resilience is defined as act of rebound. It is ability of the material to absorb elastic energy. The elastic energy to be absorbed by a unit volume of a material loaded in tension is termed as modulus of resilience. It is equal to the area under the stress-strain diagram upto the yield.

2.33 TOUGHNESS OF STEEL

Toughness is defined as the ability of the material to absorb energy during deformation (both, elastic and plastic). It relates to the total energy (mechanical energy). It is energy to be absorbed before its fracture. For uni-axial tension (i.e., in a static tensile test of steel bar), it is work done on a unit volume of a material from zero axial force upto the force corresponding to the breaking point. It is equal to the area under the complete tension stress-strain curve upto the fracture point, where the diagram terminates. This area is sometime called as **modulus of toughness**. Since, all the parts of the tensile specimen do not deform by the maximum amount, the area only represent an approximate value for the metal's toughness.

Toughness denotes the capacity of the material to resist fracture under impact loading. The area under stress-strain is a measure of toughness. Both strength and ductility contribute to toughness.

The modulus of elasticity of aluminium is only-one third that of the steel, an aluminium structural member can absorb **three times** as much energy for the same stress, as a steel member as the steel member of the same dimensions, provided the stress does not exceed the proportional limit.

In real structures, the uni-axial tension rarely exists, particularly in the region of connections. A more practical index of toughness based on the more complex stress condition (probably tri-axial) below the root of a notch is used. It is called **notch-toughness**. It is a term used to describe the resistance of a metal to the beginning and propagation of a crack at the base of a standard notch.

Notch-toughness is commonly measured by Charp V-notch (impact) test. A small rectangular simply supported beam with a V-notch at the middle length is used as a test specimen. The beam is fractured by one blow from a swinging hammer (pendulum). The energy absorbed is determined from the height, the hammer raises after breaking the specimen from which the impact value is calculated. The appearance of the fractured surface is examined (seen). A fine fibrous sharp appearance shows that the material is ductile (either, steel or copper). While, the granular appearance shows that the material is brittle (either, cast-iron or brass).

In the impact (i.e., shock)-test, the energy required to break a test-specimen (when subjected to sudden blow) is measured. It is considered as an indication of the toughness of the material. The property of toughness may be regarded as the opposite of brittleness. It follows that greater the hardness and the tensile strength, the toughness will be less as measured by the impact test.

In the impact-test, the test-piece is fractured completely and not merely bent. V-notch is made in the test-piece to ensure the fracture.

In considering the results of the impact-test, the tensile strength of the material must always be considered. High impact values are not expected from the high-tensile alloys. For a given material, the impact values are often more closely connected with the reduction of area per cent than with any other property and, in general, small-grained structures yield higher values for both these items. The fracture is generally trans-crystalline (i.e., across the crystals) although in abnormal cases, where an extremely brittle constituent exists along the crystal boundaries, the fracture may be inter-crystalline, resulting in low impact value.

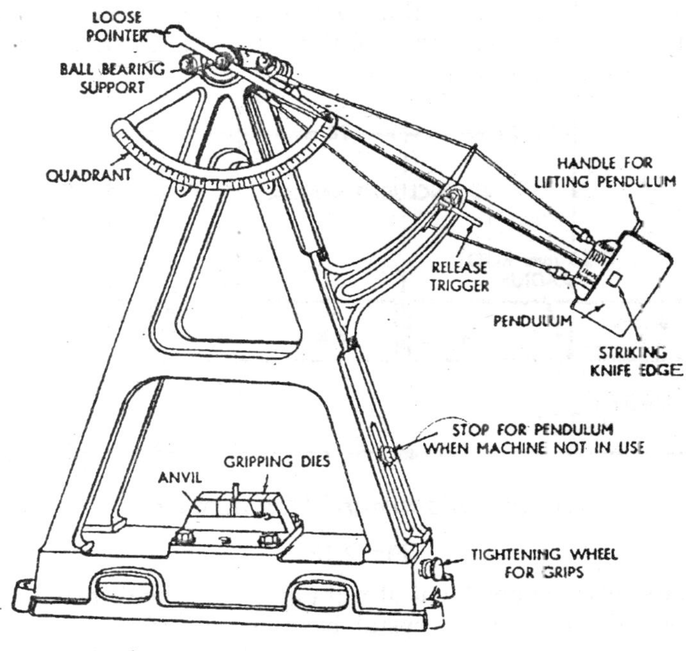

Izod impact-testing machine
(Courtesy W. & T. Avery, Ltd.)

Fig. 2.12

In Great Britain, the impact tests are usually carried-out **Izod-testing** made by Messrs Avery, Ltd. as shown in Fig. 2.12. It will be seen that the machine is essentially a compound swinging predulum released from a fixed height. The height of centre of gravity of the pendulum above the level of vice, multiplied by its mass, determines its potential energy and therefore, the capacity of the machine.

In the standard-machine, the pendulum weighs 60 lbs (27.273 kg), and the vertical height of fall of its centre of gravity is 2 ft (0.6096 m). The lower extremity of the pendulum terminates in a hammer fitted with a hardned knife-edge, which strikes the test-piece of a fixed distance above the root of the notch. The notch is made to lie in the same horizontal plane as the upper-surface of the vice by adjusting the wedge-shape dies by means of the hand-wheel, into the holes of which a tommy-ber is inserted for tightening. The upper part of the pendulum terminates in a pointer which actuates an idle or loose pointer.

At the beginning of a test, the pendulus is raised to the necessary height and the two pointers are brought into contact. When the pendulum is raised, the idle pointer is moved over into a stationary position in the right-hand quadrant which is graduated to indicate the energy absorbed in breaking the test-piece. In fact, the machine measures the residual energy of the pendulum, but the graduation of the quadrant backwards indicate the absorbed energy, because the sum of these quantities is equal to the capacity of the machine.

The energy required to bring about fracture is sum of several factors. The most important being (i) that required to start a crack in the material, and (ii) that required to propagate the crack across the whole section. It accounts for the fact, previously mentioned that the small-grained structures yield relatively high impact values, since the crack changes direction and transverse a longer path than in a similar material consisting of larger but fewer crystals. Crystal boundaries are stronger than the crystals themselves, and thus offer a greater resistance to crack propagation.

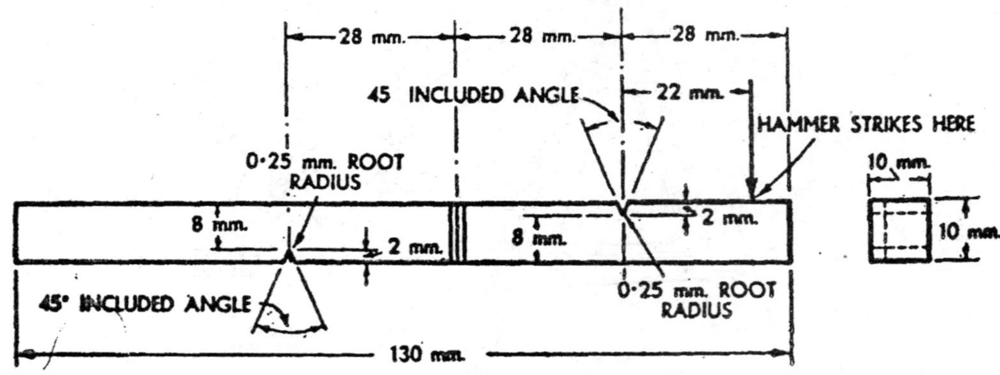

Details of standard Izod notched test-piece.

Fig. 2.13.

In rolled materials with the notch cut at right angles to the direction of rolling, the crack tends to proceed along the elongated crystal boundaries and drawn-out particles of non-metallic matter. This involves successive crack regeneration, and the crack traverses a longer path. The absorbed energy is then greater than when the notch is out parallel to the direction of rolling, because, in this instance, the crystal boundaries or inclusion of foreign matter may form a more or less straight path for the crack. Whereever possible, therefore, test-pieces should be cut at right-angles to each other. At least, three tests should be carried out, and the average result is taken, on each piece cut in a certain direction. For this the test-pieces are prepared in lengths having three notches as shown in Fig. 2.13, so arranged that the second notch is at right angles to the first and last, which are on opposite sides.

Izod V-notch test is the second test to obtain the notch-toughness. A notch-test piece rectangular in shape (i.e., a cantilever) gripped vertically with the bottom of the notch in the same way as the upper surface at a fixed position on the face having the notch.

It is to note that the V-notch in both the tests is kept on the tension side. And the energy absorbed is obtained as the difference of energy of the hammer before and after the striking and it is taken as energy needed to smash the specimen.

It may be concluded that in order to have high toughness, the material must have high strength and at the same time, large ductility.

Brittle materials have low toughness, as they have small plastic deformation before fracture.

2.34 RELATION BETWEEN YOUNG'S MODULUS OF ELASTICITY, E AND MODULUS OF RIGIDITY, G

The deformation of an element, ABCD as shown in Fig. 2.14 may be considered. There are no normal stresses action on the sides of this element, AB, AD, BC and CD. As such, their lengths will not change due to deformation. However, the diagonal BD will be stretched and the diagonal AC will be shortened due to shear stresses acting along these sides.

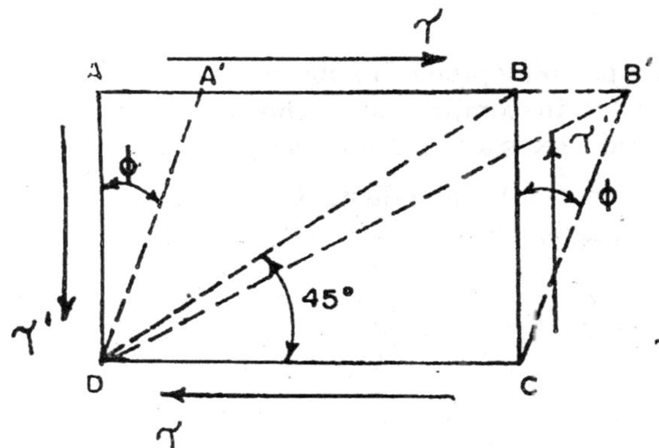

Fig. 2.14.

The square ABCD will change to a rbonbus as indicated in Fig. 2.14 by dotted lines. The angle at B before deformation ($\pi/2$), will now be become less than ($\pi/2$) [Say, ($\pi/2$) -φ]. And at the same time, the angle at A increases and becomes equal to more than ($\pi/2$), [i.e. ($\pi/2$) + φ]. Small angle φ is the angle of distorsion due to shear. This distorsion of element ABCD is called **shearing strain.**

This strain due to shear φ is the small angle. It may be considered as ratio of (AA_1/AD).

Which is equal to sliding AA_1 of the side AB with respect to the side DC, divided by the distance between these sides. In case, the material follows Hooke's law, this sliding is proportional to the stress, T. Therefore,

$$\text{Shear strain}, \phi = \left(\frac{\text{Shear stress} . \tau}{\text{Modulus of rigidity, G}}\right) \text{ where } G = (\tau / \phi)$$

where, G is known as modulus of elasticity in shear, or modulus of rigidity.

Modulus of rigidity, G may be expressed in terms of the modulus of elasticity, E in axial tension and poisson's ratio, (μ = 1/m) as below.

In diagonal BD, there is a tensile stress, p and in diagonal AC there is a compressive stress, p. The linear strain, ε of the diagonal BD due to these two stresses mutually perpendicular direct stresses. Therefore

$$\varepsilon = \left(\frac{P}{E}\right) - \left(\frac{-P}{mE}\right) = \frac{P}{E}\left(1 + \frac{1}{m}\right)$$

But, p = τ, and linear strain, $= \frac{1}{2}\frac{\tau}{G}$

$$\varepsilon = \frac{\tau}{E}\left(1 + \frac{1}{m}\right). \quad \therefore \frac{1}{2}\frac{\tau}{G} = \frac{\tau}{E}(1 + \mu)$$

$$E = 2G(1 + \mu), G = [E/(2)(1 + \mu)]$$

2.35 RELATION BETWEEN YOUNG'S MODULUS OF ELASTICITY, E AND BULK MODULUS, K

In case a rectangular parallelepiped is subjected to either uniformly distributed forces Px, Py and Pz in tension (or in compression), the normal stresses over the cross-sections perpendicular to the x, y and z axes are as under respectively.

It is assumed that ($\sigma_x > \sigma_y > \sigma_z$). The effects of these forces may be combined. The strains ε_x, ε_y and ε_z may be written as below

$$\varepsilon_x = \left[\left(\frac{\sigma_x}{E}\right) - \left(\mu\frac{\sigma_y}{E}\right) - \left(\mu\frac{\sigma_z}{E}\right)\right] \quad \text{(i)}$$

$$\varepsilon_y = \left[\left(\frac{\sigma_y}{E}\right) - \left(\mu\frac{\sigma_x}{E}\right) - \left(\mu\frac{\sigma_z}{E}\right)\right] \quad \text{(ii)}$$

$$\varepsilon_z = \left[\left(\frac{\sigma_z}{E}\right) - \left(\mu\frac{\sigma_x}{E}\right) - \left(\mu\frac{\sigma_y}{E}\right)\right] \quad \text{(iii)}$$

Volume of above rectangular parallel piped increases in the ratio

$$\left|(1 + \varepsilon_x).\ (1 + \varepsilon_y)\ (1 + \varepsilon_z) : 1\right| \quad \text{(iv)}$$

Neglecting the quantities of higher order, above expression reduces as under :

$$\left[\left(1+\varepsilon_x+\varepsilon_y+\varepsilon_z\right):1\right] \tag{v}$$

Therefore, it is seen that volume increases by δ_v where,

$$\delta_v = \left(\varepsilon_x+\varepsilon_y+\varepsilon_z\right) \tag{vi}$$

By adding the expressions (i), (ii) and (iii) above

$$\delta_v = \left(\varepsilon_x+\varepsilon_y+\varepsilon_z\right) = \left(\frac{1-2\mu}{E}\right)\left(\sigma_x+\sigma_y+\sigma_z\right) \tag{vii}$$

In case of uniform hydro-static pressure,

$$\sigma_x = \sigma_y = \sigma_z = -p \tag{viii}$$

$$\varepsilon_x = \varepsilon_y = \varepsilon_z = \varepsilon = \left(\frac{-P}{E}\right)(1-2\mu) \tag{ix}$$

From the expression (vii)

$$\frac{E}{3(1-2\mu)} = K \quad \text{and} \quad K = \left(\frac{E}{3(1-2\mu)}\right) \tag{x}$$

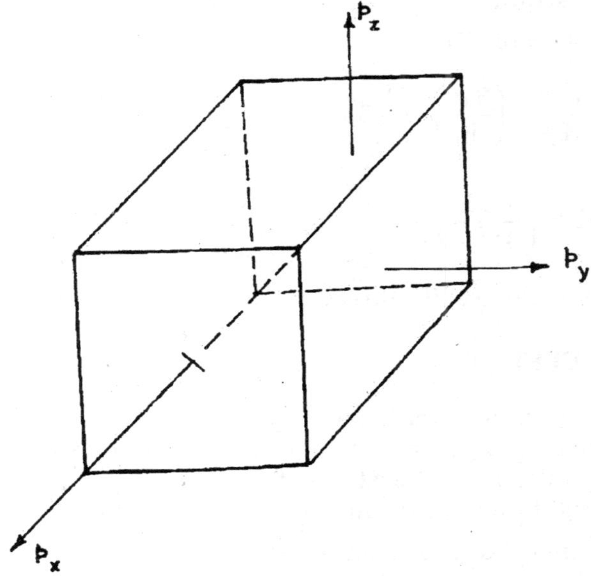

Fig. 2.15.

The unit volume contraction is proportional to the compressive stress, p and inversely proportional to K, which one is known as bulk modulus of elasticity.

From the expression (vii) and the expression (ix) epsilon

$$\left(\varepsilon_x + \varepsilon_y + \varepsilon_z\right) = \frac{(1-2\mu)}{E}\left(\sigma_x + \sigma_y + \sigma_z\right)\ 3\,\varepsilon = \left(\frac{1-2\mu}{E}\right)(-3P) \qquad \text{(xi)}$$

In case, p is tensile then, all the values shall be tensile. Then (1-2 μ) must also be positive or at the most zero poisson's ratio shall be 0.5.

Materials like rubber and paraffin may reach this value.

Alternatively :

In case, the sides of rectangular parallel pipe are equal and each side is L [i.e., length = width = Height = L]. Then, the volume of parallelpipe

$$V = (L.\ L.\ L) = L^3 \qquad \text{(i)}$$

Its partial derivative with respect to L

$$\left(\frac{\delta_v}{\delta_L}\right) = 3L^2\ .\ \delta V = (3L^2 . \delta L) \qquad \text{(ii)}$$

$$\text{or} \quad \left(\frac{\delta V}{V}\right) = \left(\frac{3L^2 . \delta L}{L^3}\right) = 3\left(\frac{\delta L}{L}\right) = 3\,\varepsilon \qquad \text{(iii)}$$

where, ε is the linear strain.

Since, (δv = V) K, and ε = (σ/E)

$$\left(\frac{\sigma}{K}\right) = \left(\frac{3\,\sigma}{E}\right)(1-2\mu) \qquad \text{(iv)}$$

$$K = \left(\frac{E}{3(1-2\mu)}\right)$$

where, the parallelpipe carries like direct stress, σ on its all the three respective faces.

2.36 BAUSCHINGER'S EFFECT

The stress-strain diagram for a mild steel bar (test-specimen) tested in axial tension is shown in Fig. 2.16. When the specimen attains yield at that instant, in case, the bar is unloaded, then, during unloading, the stress-strain diagram follows a straight line, CD parallel to the initial straight line, OB during loading as shown in Fig. 2.13.

In case, after unloading, the specimen is loaded second time, it is seen that the yield point stress is raised as shown in Fig. 2.16.

When the load is applied to the bar, a second time, the material again follows approximately Hooke's law as shown by line DF. At point F, which corresponds to the previous loading point, C, the curve abruptly change the behaviour. Instead of tracing the portion, FG (which one is prolongation of the curve, BC, the curve raises to point F' (i.e. the yield point, F is raised to F') due to previous stretching of the material.

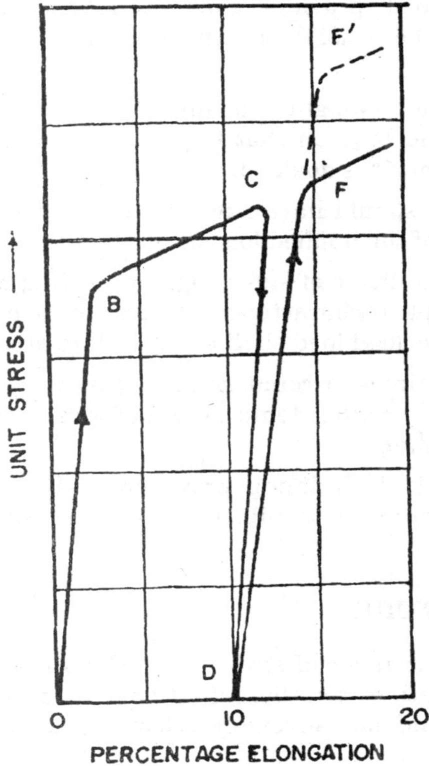

Fig. 2.16.

In case, several days are allowed to lapse after the first, unloading then, upon reloading, a still higher yield point may be obtained.

More complete investigations have shown that the time which lapses between unloading and reloading is of great influence on the stress-strain curve during reload.

This raising of yield-point stress due to unloading and reloading is known as Bauschinger's effect (Reference : S. Timoshenko, Strength of Materials' Part Second, Affiliated East-West Press Pvt. Ltd, New Delhi, P416-422, 1965).

2.37 STABILITY OF STRUCTURE

According to the stability requirement, the stability of a structure as a whole against overturning is ensured so that the restoring moment is greater than the maximum overturning moment.

1. **Overturning.** The stability of a structure as a whole against overturning is ensured such that the restoring moment, M_R shall be not less than the sum of 1.2 times the maximum overturning moment due to the characteristic dead load, $M_{o.CDL}$ and 1.4 times the maximum overturning moment due to characteristic imposed loads $M_{o.CIL}$. That is,

$$M_R < (1.2\, M_{o.CDL} + 1.4\, M_{o.CIL})$$

In cases, where dead load provides the restoring moment, only 0.9 times the characteristic dead load shall be considered. The restoring moment due to imposed loads shall be neglected.

The anchorages or counterweights against for overhanging members (during construction and service) should be such that the static equilibrium should be maintained, even when the overturning moment is doubled.

2. **Sliding.** The structure should have adequate factor of safety against sliding due to the most adverse combination of the applied loads.

The structure shall have a factor of safety against sliding not less 1.4 under the most adverse combination of the applied characteristic forces. In case only dead loads are acting, only 0.9 times the characteristic dead load shall be taken into account.

To ensure stability at all times, account shall be taken of probable variations in dead load during construction, repair or other temporary measures. The wind and seismic loading shall be treated as imposed loading.

In designing the framework of a building, provisions shall be made by adequate moment connections or by a system of bracings to effectively transmit all the horizontal forces to the foundations.

2.38 STIFFNESS OF STRUCTURE

In order to meet the requirement of stiffness for the structure as a whole and for the structural members, the limitations for the deflections are specified. The deflection of the structure or part thereof should not adversely affect the appearance or efficiency of the structure or finishes or partitions.

REFERENCES

1. IS: 875 (Part 1), 1987: Code of Practice for Design Loads (Other than Earthquake) for Buildings and Structures (second revised edition). Part 1: Dead Loads (Unit weight of Building Materials and Stored Materials), Part - 2: Imposed Loads; Part - 3:Wind Loads; Part -4: Snow loads.
2. IS: 1893-Recommendations for Earthquake Resistant Design of Structures.
3. David A. Nethercot – Limit States design of Structural Street Works Third Edition, published Span Press, 11 New Fetter Lane, London, EC4P4EE, simulataneously published by Spon Press, 29 West 35th Street, New York, NY 10, 001 USA. (pp 17).
4. European Coal and Steel Community (ECSC-Guide), 1982 BS : 5950 : Part 1, 2000, Amendment. The Structural Use of Steelwork in Building : Code of Practice for Design in Simple and Continuous Construction : Hot-Rolled Sections, 1990.
5. BS: 5950: Part 1. 2000. Amendment. The Structural Use of Steelwork in Building : Code of Practice for Design in Simple and Continuous Construction: Hot-Rolled Sections, 1990.
6. R.S. Narayanan and A.W. Beeby. Introduction to Design for Civil Engineers as in Reference (1) (pp 116-117)
7. R.S. Narayan and *et al.* Teaching Resource on Structural Steel Design. Volume : 1-3 Prepared jointly by Indian Institute of Technology, Madras, Anna University, Chennai,

Structural Engineering Research Centre, Madras Organised by Institute for Steel Development and Growth (INS. DAG) Kolkata (India)

8. Edward G. Nawy (Editor-in-chief). Concrete Construction Engineering Handbook' CRC Press Boca Raton, New York.

9. Ramchandra and Virendra Gehlot. `Corrosion of Reinforcing Steel bars in Concrete'. A key note presented in UGC sponsored National Seminar, `Futuristic construction materials and Techniques' (FCMT 2005) February, 21-22, 2005, organized by Department of Structural Engineering, Faculty of Engineering, M.B.M. Engineering College, Jai Narain Vyas University, Jodhpur - 342 011 (Rajasthan, Inida).

3

Part - III

CONCEPT OF LIMIT STATE METHOD OF ANALYSIS AND DESIGN

SOURCES OF UNCERTAINTIES INFLUENCING SAFETY OF STEEL STRUCTURES

3.1 INTRODUCTION

Civil engineers ensures the safety of the structures. The structures designed are: (i) safe (ii) fit for their purpose, and (iii) durable and economical. Safety is one of the paramount requirement for the structures. It should be ascertained at the design stage.

In limit state design, there are various levels of reliability analysis to obtain the risk involved and to find a probability based estimate of structural safety. There are various uncertainties in the design and reliability methods. These uncertainties have been described briefly here.

However, there are many uncertainties, which influence both safety and economy. There are many uncertainties about loading (e.g., the prediction of the loading, which a structure may be subjected to on a future occasion), natural variability in the strength of material and about the structural dimensions and behaviour. Following uncertainties influence the safety of a structure.

A realistic, rational and quantity representation of safety must be based on statistical and probabilistic analysis which provides for both over-load and under-strength.

1. Uncertainties about loadings

In the service life of a structure, there may be the variation in the magnitudes of the maximum loads and the load effects (e.g., bending moments, shear force, axial force and torsion), which the different structural elements might meet. Uncertainties both due to variability of the load applied to the structure and also due to variability of load distribution through the structure. In case, a particularly weak structural component is subjected to a heavy load, which exceeds the strength of the component clearly, failure is likely to occur.

2. Uncertainties about the strength of materials

It is necessary to look at the possibility of weakness possible in the material. The strength of every piece of material is tested and the material is used for the construction. One may even guarantee the strength provided. During testing a specimen is destroyed the material used (actual material incorporated in the structure), remains untested. Some material is tested and it is hoped that the remaining material which is included in the structure possesses the similar strength.

3. Uncertainties about the structural dimensions and behaviour

The exact sizes of the component members are mentioned on the drawings. There are recognized tolerances. The measurements show that these are not infrequently exceeded. In practice it would be seen that neither the spans for the beams nor the sizes of the cross section will be exactly what were specified.

3.2 NORMAL (OR GAUSSIAN PROBABILITY) DISTRIBUTION CURVE

There are following three standard forms of the frequency distribution used in statistics. These are referred as distribution functions/slash density functions.

1. Poisson density. It is expressed by equation

$$y = e^{-\mu . x\left(\frac{\mu x}{r_1}\right)} \qquad \text{(i)}$$

2. Gauss density : It is expressed by equation

$$y = \frac{1}{\sigma\sqrt{2\pi}} \cdot e^{-\left(\frac{x - \mu_u}{2\sigma^2}\right)^2} \qquad \text{(ii)A}$$

$$y = \frac{1}{\sigma\sqrt{2\pi}} \cdot e^{-0.5\left(\frac{x - \mu)}{\sigma}\right)^2} \qquad \text{(ii) B}$$

Where, y = frequency

x = Value of ordinate at which, y is to be calculated (a random variable)

μ = mean value

σ = standared deviation of the distribution

3. Maxwell Bolizmann density

It is expressed by equation

$$y = 4a\sqrt{\frac{a}{\pi}}\ x^2 . e^{-ax^2} \qquad \text{(iii)}$$

These density functions may be used for the calculations to be carried out.

Gaussian distribution is used. The normal density function is the special case $\sigma = 1,\ \mu = 0$ of the Gauss density, in which

$$y = \frac{1}{\sqrt{2\pi}}\ e^{-0.5\,x^2} \qquad \text{(iv)}$$

When, it is plotted, it is a symmetrical bell-shaped distribution as shown in Fig. 3.1. In case, the data exist for each of the variables and in case frequency distribution is approximated to Normal Gaussian-distribution, then there is standard procedure to calculate the probability.

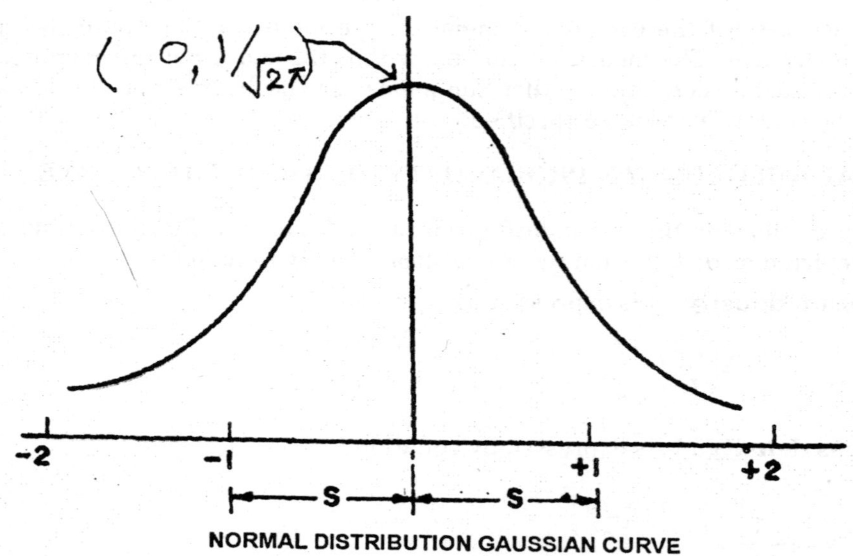

NORMAL DISTRIBUTION GAUSSIAN CURVE

Fig. 3.1

In many mathematical handbooks, this distribution has been tabulated in the form of standard variable `Z', so that

$$Z = \left[\frac{x - \mu}{\sigma}\right] \quad \text{or} \qquad (3.1)$$

$$f(z) = \frac{1}{\sigma\sqrt{2\pi}} \cdot e^{-0.5\,Z^2} \qquad (3.2)$$

Area under the curve by obtained by numerically integrating Eq. 3.1 with appropriate limit. From the standard tables for the curve, it may be seen that 95 per cent of the area lies between the limits (–1.64) and (infinity). This means that, there is only a 5 per cent probability that a result will fall below $\left(\bar{x} - 1.64\,\sigma\right)$. Following are properties of **Normal Distribution Curve**

(i) It is symmetric about the vertical axis which passes through $Z = 0$

(ii) It's highest point is at $\left[0,\ 1/\sqrt{2\pi}\right]$

(iii) The curve is concave downward between the $Z = \pm 1$ and concave upward for the values of Z outside the range.

(iv) Total area under the curve and Z-axis is unity. Shape of the distribution changes with the change in standard deviation (discribed in Art. 3.3).

As shown in Fig. 3.2.

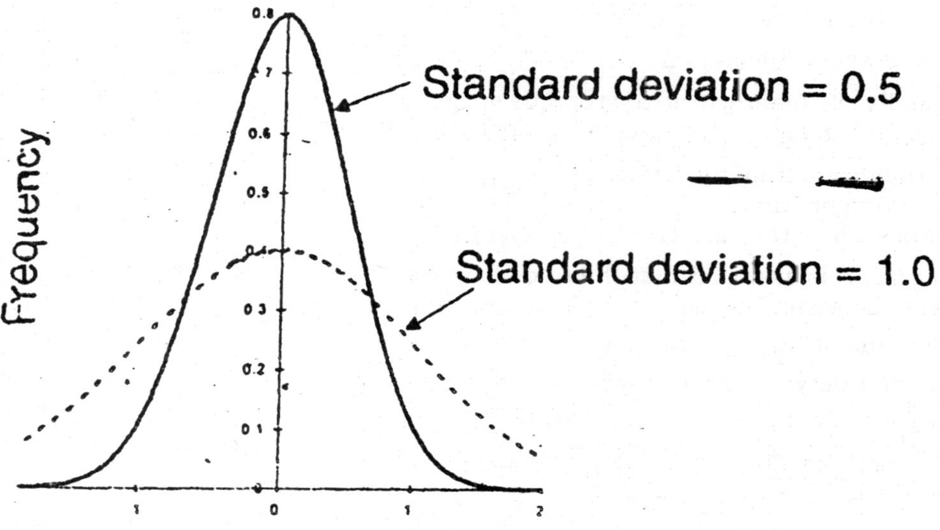

SHAPE OF A FREQUENCY DISTRIBUTION CURVE AS A FUNCTION OF THE STANDARD DEVIATION

Fig. 3.2.

3.3 STANDARD DEVIATION

The standard deviation is a measure of the spread of the distribution. For any frequency distribution built-up from n measurements. The **Standard-deviation** is given by

$$\sigma = \sqrt{\frac{\Sigma x^2}{n} - \left(\frac{\Sigma x}{n}\right)^2} \tag{3.3}$$

For calculating, f_{ck}, it is known from Equ. 3.1

$$Z = \left[\frac{x - \bar{x}}{\sigma}\right] \tag{i}$$

$$\sigma . Z = (x - \bar{x}) \tag{ii}$$

For 5 percentage probability, that for the value of 0.05 from Table 3.1 for safety index (β), β = 1.64 (i.e, z = 1.64). Substituting z = 1.64 and x = f_{ck}.

$$1.64 = \left[\frac{x - f_{ck}}{\sigma}\right]$$

$$(\pm 1.64\,\sigma - x) = -f_{ck}\ ,\ f_{ck} = (x - 1.64\,\sigma) \tag{3.4}$$

where,

f_{ck} = Characteristic strength of steel, and

σ = Standard-deviation

Usually, f_{ck} (characteristic strength of material) is adopted such that below which not more than 5 percent of test results are expected to fall.

Variability of many common natural factors (e.g, the variation in height of the population) approximates to a normal distribution and it is often found to give a reasonable fit to data such as the variation in the strength of materials.

In case, it is assumed that the frequency distribution of R and S is to be calculated and these may be assumed to be normal then, the distribution of (R – S) will also be normal with a mean value of (R_{mean} – S_{mean}) and a standard deviation of $(\sigma_R^2 + \sigma_S^2)^{1/2}$. It is convenient to express the safety in terms of the safety index, β.

where,

$$\beta = \left[\frac{R_{mean} - S_{mean}}{\sqrt{\sigma_R^2 + \sigma_S^2}}\right] \quad (3.5)$$

$$\beta = \left[\frac{R_{mean} - S_{mean}}{\left(\sigma_R^2 + \sigma_S^2\right)^{1/2}}\right] \quad (3.6)$$

In case, the distribution is normal, there is an exact relationship between β and the probability of failure. [(R – S) < 0].

Table 3.1 shows the probability of failure as a function of the safety index β for normal distribution.

TABLE 3.1. PROBABILITY OF FAILURE AS A FUNCTION OF SAFETY INDEX, β FOR A NORMAL DISTRIBUTION

β	Probability failure
1.64	0.05
2.0	0.023
3.0	0.003
4.0	0.0009
5.0	0.0000

A very simple example shows how a design approach works.

The science of reliability based design has been developed with the aim to provide a rational solution for sufficient safety. This method has been described in the next chapter 4.

Example 3.1. It is required to support a load of 100 kN (an average value) with a standard deviation of 40 kN by suspending it from a bar. The tensile strength of bar is 400

N/mm² with a standard deviation of 40 N/mm². Find the cross-sectional area of the bar. Take the value of a safety index 5 to be achieved.

Solution : Strength of hanger

= strength of the material x cross-sectional area of the hanger, A

The cross-sectional area is assumed to be subjected to no uncertainty, then, it may be substituted R_n Eq. (3.5), and Eq. 3.6 for the safety index, β

$$\beta = (R_{mean} - S_{mean}) / \sqrt{\sigma_R^2 + \sigma_S^2}$$

$$\beta = (R_{mean} - S_{mean}) / \left(\sigma_R^2 + \sigma_S^2\right)^{1/2}$$

Substituting the respective values.

$$5 = \frac{400\ A - 100 \times 1000}{\left[(40\ A)^2 + (40 \times 1000)^2\right]^{1/2}}$$

where, A is the cross-sectional area. It is only unknown. Solving above expression, A = 934 mm²

Example 3.2. 10 steel coupens (samples cut from an I-section were tested in an axial tension, in Universal Testing Machine. Following results were noted. (σ_y noted in N/mm²)

249.00 254.50 257.50 256.50 254.00

257.60 240.00 254.40 249.60 246.00

Take characteristic strength of steel σ_y = 250 N/mm²

Solution :

Step 1. Mean strength for the above tests = (2519.50/10) = 251.95 N/mm²

Step 2. Standard deviation

$$\sigma = \left(\frac{\Sigma \Delta^2}{n-1}\right)^{1/2}$$

Step 3. The values of Δ and Δ^2 have been calculated and Listed as under :

S.No.	$\Delta = (x - x_{mean})$	Δ	Δ^2
1.	[(249.00–251.95)]	2.95	8.763
2.	[(254.50–251.95)]	2.55	6.503
3.	[(257.50–251.95)]	5.55	30.803
4.	[(256.50–251.95)]	4.55	20.703
5.	[(254.50–251.95)]	2.55	6.503
6.	[(257.60–251.95)]	5.55	30.803
7.	[(240.50–251.95)]	11.95	142.800

8.	[(254.40–251.95)]	2.45	6.001
9.	[(249.60–251.95)]	2.35	5.523
10.	[(246.00 – 251.95)]	5.95	35.403
n = 10	(n–1) = 9	$\sum \Delta^2$ =293.745	

Step 4. σ Standard deviation

$$\sigma = \left(\frac{293.745}{9}\right)^{1/2} = 5.712$$

Step 5. Characters strength

$$f_{ck} = (\bar{x} - 1.64\,\sigma)\ \text{N/mm}^2$$

$$f_{ck} = (251.95 - 1.64 \times 5.712)$$

$$= 242.58\ \text{N/mm}^2$$

REFERENCES

1. **Nethercot David A.** `Limit State Design of Structural Steel-work' Third edition, 2001, published by Spon-Press 11 New Fetter Lane, London, EC 4 P 4 EE Simultaneously published in The USA and Canada, by Spon Press, 29 West 35th Street, New York, NY - 10001.

2. **Sokulnikoff I.S. and Redheffer R.M.** `Mathematics of Physics and Modern Engineering' Second-edition (International Student Edition McGraw-Hill Book-Company, Ltd. New York, St. Louis, San Francisco, Toronto, London, Sydney (PP 615).

4

Part - IV

LIMIT STATE DESIGN

METHOD OF ULTIMATE LIMIT STATE (ULS) AND SERVICEABILITY LIMIT STATE (SLS) AND PARTIAL SAFETY FACTORS FOR LOADS AND STRENGTH OF MATERIALS

4.1. INTRODUCTION

An improved design philosophy to make allowance for the short comings of the **`working stress method of design'** **`Ultimate load-method of design'** (method of plastic analysis of design) of steel structures, was developed in the late 1970's and it has been extensively incorporated in design standards and codes formulated in all the developed countries. Although, there are many variations between practices adopted in different countries, the basic concept is broadly similar. The probability of operating conditions not reaching failure conditions form the basis of **`Limit State Design (LSD)'** of steel structures adopted in many countries.

A limiting condition beyond which the structures ceases to fulfil its intended function. The structure no longer satisfies the performance requirements specified.

'Limit States' are the various conditions in which a structure would be considered to have failed to fulfil the purpose for which it was built.

'Limit State Method' provides the basic framework, within which the performance of the structure may be assessed against various limiting conditions. When formulating procedures, now-a-days, it is customary to do so in a way which recognizes the inherent variability of loads, the strength of materials, construction practices and approximations made in design. This usually involves the use of some concept of probability.

In the limit state design method, the structures are designed to with stand safely all the loads likely to act on it throughout its life. It will not suffer total collapse under accidental loads such as from explosions or impact or due to consequences of human error to an extent beyond the local damages.

The object of design is to get a structure that will remain fit for use during its life with acceptable target reliability. In other words, the probability of a limit state being reached during its life time should be very small.

The acceptable limit for the safety and serviceability requirements before failure occurs is called a limit state.

In general, the structures are designed on the basis of the most critical limit state and these are checked for other limit states.

4.2 BRIEF HISTORY OF LIMIT STATE METHOD

As early as 1930, after understanding various uncertainties and draw-backs of working-stress method, a new formulation, **Limit-State Philosophy** in Soviet-union, it is believed to have developed the **limit state approach** and it was embodied in the Russian Code in 1954, (Zalesov 1973).

In 1966, Bakers Statement that for absolutely, unforseenable events, a zero probability of failure has to be demanded from the designer. But to quote, Professor Freude that "The difference between the safe and unsafe design is in the degree of risk considered acceptable, not in the delusion, (false belief) that such a risk may be completely eliminated". **For the first time,** a probabilistic approach to the structural safety, taking into consideration, an expected variation in calculating the loads and finding the strength of steel was introduced in the structural design in the air-craft industry in the late 1940s.

Following the attempts by Freudenthal (1947, 1948), to derive the safety factor from statistical variation of the design parameters, the concept of safety factor was extensively considered by Pugsley (1951) and Johnson (1953).

A composite international practicing body, research workers and practicing designers, known as Euro-pien Committee of Concerete (CEB), comite'-Europeen du Beton was set-up in 1953, with the object of providing a sound philosophy of design, together with the recommendations on the detailed aspects of design based upon experimental and theoretical research. With the inception of CEB, more serious attention began to be devoted to the complete subject of stuctural design. The limit state philosophy has been adopted and elaborated by into its present form in the Recommendations of CEB (1964).

Cornell (1969) proposed a code format for necessary safety factor in the structural design. It was suggested incorporating the measures of uncertainty in predicting the strength and applied loads and a measure of degree of safety in determining the safety factors.

The permissible stress formulation was generally used in that there is not necessarily a direct relationship between both stress calculated elastically and failure load. The safety factors are thus notional and do not necessarily relate closely to the true safety of the structure. During the middle years of the 20th centuries, much research was done on the prediction of the strength of the members and in considering safety, it seemed more logical to many people, if the safety problem was formulated in terms of ultimate strength rather than service conditions. In a number of design British Codes (e.g., CP114 (1969), ultimte strength equations were introduced.

Kemp in 1973 gave a number of valid reasons why classical reliability theory cannot be applied, in its purest form.

In 1967, it was decided to recast the documents, detailed procedure of design of steel structures into most progressive limit state shape. In 1972, in U.K., CP110 (The structural use of concrete Part 1; Design materials and workmanship British Standard Institution, now revised as BS8110. A code of practice for concrete was first published. Now this is revised BS

: 8110. Moreover, BS 5400, the Bridge Code including Part 3, relating to the Design of Steel Bridges was prepared at the same time as BS : 5950. CP110 introduced 'Limit State Design' in combination with characteristic values and partial safety factors. CP110 treated each aspect of design seperately and logically and recognized the inherent variability of both loads and material properties in a more formal way. Limit State Design was considered a revolutionary design approach and improved design philosophy.

In 1974, in Canada, the Limit State Steel work code was first published and in the other parts of world, these codes appeared gradually. Code BS : 5950 was published in nine parts in 1990, 1992, 1994, 1995 and 1998.

TABLE 4.1. BS : 5950 : THE STRUCTURAL USE OF STEEL WORK IN BUILDING

Part-1	Code of practice for design in simple and continuous construction Hot Rolled Section. (1990)
Part-2	Specifications for materials and erection : Hot Rolled Sections (1992)
Part-3	Code of practice for design in composite construction.
Part-4	Code of practice for design of floors with protiled steel sheeting (1994).
Part-5	Code of practice for design of cold formed sections (1998)
Part-6	Code of practice for design of Light gauge sheeting, decking and cladding (1995).
Part-7	Specifications for materials and workmanship : cold formed section and sheeting (1991).
Part-8	Code of practice for fire protection of structural steel work
Part-9	Code of practice for stressed skin design

These codes of practice for the design of steel structures provide detailed necessary information. A person who designs structural components and steel structures, it is clearly necessary to be familiar with the provisions of this code. It is equally important that the persons use these codes in an intelligent manner.

The trend towards the general introduction of this more rational approach (i.e, Limit State Design) for the structural Steel design, that is, taking place for the major construction materials on a World-wide basis is reflected by BS : 5950 (The Structural use of steel-work in Building).

The European Economy Community produced Euro-codes (EC1, EC2, EC3, ..., EC9) in nine parts. The European Commission finances the developments as parts of its approach towards a Europe-wide system for the control of construction products. The European stand and Organisation (CEN), Euro-Code 3 is the Code of practice for the design of steel structures.

These Euro-Codes are under development for the last twenty years. It is likely that these Euro-codes will replace national codes.

The Institution for Steel Development and Growth (INSDAG) conducted workshops at Mumbai, Calcutta, and Delhi to introduce and to familiarize engineers with the proposed Limit State Design. Code of Practice for our country BIS [BUREAU OF INDIAN STANDARDS] Delhi issued a draft for approval and suggestions and finally code IS : 800-2007 has been issued.

4.3 REQUIREMENTS FOR LIMIT STATE DESIGN METHOD

Following design requirements with regard to strength, stability, serviceability, brittle fracture, durability, fatigue and fire should be satisfied to design steel structures. Following requirements should be met.

1. The structures remain fit with adequate reliability and they should be able to sustain all loads (actions) and other effects experienced during construction and use.
2. The structures should have sufficient durability under normal maintenance.
3. The structures do not suffer overall damage or collapse dis-proportionately under accidental events like explosions, vehicle impact or due to consequences of human error to an extent beyond local damage.

The potential for catas-trophic damage shall be limited or avoided by appropriate choice of one or more of the following :

(a) Avoiding, eliminating or reducing exposure to hazards, which the structure is likely to sustain.

(b) Choosing structural forms, layouts and details and designing such that

(i) the structure has small sensitivity to hazardous conditions; and (ii) the structure survives with only local damage even after serious damage to any one individual element by the hazard.

(c) Selecting suitable material, design and detailing procedure, construction specifications, and control procedures for shop fabrication and field construction as relevant to the particular structure.

4.4 CONDITIONS FOR LIMIT STATE DESIGN

Following conditions should be satisfied to avoid a dis-proportionate collapse

(1) The structure should be effectively tied together at each principal floor-level and each column should be effectively held in position by means of continuous ties (beams) nearly orthogonal, except where the steel-work supports only cladding weighing not more than 0.7 kN/m^2 along with imposed and wind loads.

These ties must be steel members such as beams, which may be designed for other purposes, steel bar reinforcement anchoring the steel frame to concrete floor or steel mesh reinforcement in composite slab with steel profiled sheeting directly connected to beam with shear connectors.

These steel ties and their end end connections should be able of resisting factored tensile force not less than the factored dead and imposed loads acting acting on the floor area tributary to the tie nor less than 75 kN.

Such connections of ties to edge column should also be able to resist 1 per cent of the maximum axial compression in the column at the level due to factored dead and imposed loads.

All column splices should be able to resist a tensile force equal to the largest of a factored dead and live load reaction from a single floor level located between that column and next column splice below that splice. Lateral load system to resist notional horizontal loads should be distributed throughout the structure in nearly orthogonal directions so that no substantial portions is connected at only point to such a system.

(1) Pecast concrete or other heave floor or roof units should be effectively anchored in the direction of their span either to each other over the support or directly to the support.

(2) When the above conditions to tie the columns to the floor adequately are not satisfied, each storey of the structure should be checked to ensure that dis-proportionate collapse would not precipitate by the notional removal, one at a time[of each column.

(3) When each floor is not laterally supported by more then one system, it should be checked at each storey by removing one such lateral system at a time to ensure that disproportionate collapse would not occur.

The collapse is considered dis-propotionate, in case more than 15 per cent of the floor or roof area of 70m^2 collapse at that level and at one adjoining level either above or below it, under a load equal to 1.05 or 0.9 times the dead load, 0.33 times temporary or full imposed load of permanent nature (as in storage structures) and 0.33 times wind load acting together.

Structures designed for unusual or special functions shall comply with any other relevant additional limit state considered appropriate to that structure.

Generally, structures and elements shall be designed by limit state method. In case the limit state method cannot be conveniently adopted, the working stress design may be used.

4.5 ELEMENTS AND SUB-ELEMENTS OF LIMIT STATE DESIGN

A limit state is the state or a condition beyond which the structure would become less than completely fit for its intended use. If this happens the structure is said to have entered a limit state.

There are following four elements of limit state design.

1. Ultimate Limit State (ULS). Also referred as Limit State of strength.
2. Serviceability Limit State (SLS)
3. Limit State of Durability.
4. Fatigue Limit State

1. Ultimate Limit State (ULS)

The Limit State of strength are those associated with failures (or imminent failure), under the action of probable and most unfavourable combination of the loads on the structure using the appropriate partial safety factors, which may endanger the safety of life and property.

It is the State, which if exceeded may cause collapse of a part or the whole of the structure (or safety limit state (inability to sustain any increased load. The attainment of one or more ultimate limit states (ULS) may be regarded as an inability to sustain any increase in load.

The calculation of ultimate strength may need the consideration of the following (i) Loss of equilibrium of the structure or any part of it, considered as a rigid body, and (ii) Failure by excessive deformation, rupture or loss of stability of the structure or any part of it including support and foundation.

Ultimate Limit States are those catastrophic (great disaster) states, which require a large reliability in order to reduce the probability of its occurrence to a very low level. It includes:

(i) **Strength limits,** (Generally, including yielding, rupture, buckling transformation into a mechanism, buckling). A limit State of collapse or loss of structural integrity. It includes loss of equilibrium of the structure as a whole or any of its parts or components.

(ii) Stability (Overall loss of equilibrium (overturning and sway where appropriate and overturning) or any of its parts including supports and the foundation.

(iii) Elasting and plastic instability.

(iv) Fatigue (leading to fracture).

(v) Brittle fracture.

Ultimate Limit State is related to the maximum design load capacity under the extreme conditions. The partial load-factors are chosen to reflect the probability of extreme conditions, when load acts alone or in combination. In Limit State design methods, the structure shall be designed to withstand safely all loads likely to act on it throughout its life.

The limit state of collapse of the structure or part of the structure could be assessed from rupture of one or more critical sections and from buckling due to elastic or plastic instability (including the effects of sway where appropriate or overturning. The resistance to bending, shear, torsion and axial loads at every section shall not be less than the appropriate value at that section produced by the probable most unfavourable combination of loads or the structure using the appropriate partial safety factors. **Ultimate Limit State (ULS) are the conditions be avoided.**

Following are the **sub-elements** of Ultimate Limit State of collapse (i.e, Limit State of collapse (ULS) :

1. Limit State of Collapse in axial compression
2. Limit State of Collapse in Flexure (Bending)
3. Limit State of Collapse in Shear
4. Limit State of Collapse in Torsion.

Details of above stated sub-elements have been described in respective chapters which deal these elements individually.

2. Serviceability Limit State (SLS)

Serviceability Limit State is related to the criteria governing normal use. Loss of utility and/or requirement for remedial action or some other loss of utility.

It is a Limit State of those acceptable service condition exceedence of which causes serviceability failure. Serviceability Limit State refers to limits on acceptable performance of the structure. Serviceability checks against the need for remedial action or some other loss of utility. Serviceability Limit State is the State beyond which service criteria, specified below are no longer met : Serviceability Limit State (SLS) is considered merely undesirable

Following are the sub-elements of Limit State of Serviceability (SLS) :

(i) Excessive deflection

(ii) Excessive vibration (Oscillations induced due to wind floor vibrations, etc)

(iii) Corrosion Durability consideration

(iv) Fire-resistance

Note : Partial safety factors γ_{fL} of value equal to 1.00 shall be used for all loads to check the adequacy of the structure under the serviceability limit state, unless specified otherwise.

(a) Excessive deflection (deformation)

The deflection under serviceability loads of a building or a building component should not damage the strength of the structure or components or cause damage to finishing. Deflections are to be checked for the most adverse but realistic combination of service loads and their arrangement, by elastic analysis, using a load factor of 1.00 (unity).

Circumstances may arise where the greater or lesser values would be more appropriate depending upon the nature of material in element to be supported and intended use of structure. Table 4.2 gives the recommended limit of deflections for certain structural member systems.

Table 4.2 (A). DEFLECTION LIMITS OTHER THAN FOR PITCHED ROOF PORTAL FRAME [FOR INDUSTRIAL BUILDINGS] [FOR VERTICAL DEFLECTIONS]

Design load	Member	Supporting	Maximum deflection
Live-load	Purlin and Girts	Elastic cladding	Span/150
		Brittle cladding	Span/180
Live-load	Simple span	Elastic cladding	Span/240
		Brittle cladding	Span/300
Live-load	Cantilever	Elastic cladding	Span/120
		Brittle cladding	Span/150
Live-load	Simple span	Elastic cladding	Span/240
Live-load	Cantilever	Elastic cladding	Span/120
Live-load	Simple span	Floor (Brittle cladding)	Span/300
Live-load	Cantilever	Floor (Brittle cladding)	Span/150
Live-load/Wind load	Rafter supporting	Profield	Span/180
		Metal sheeting	
		Plastered sheeting	Span/240
Crane load (Manual operation)	Gantry girder	Crane	Span/500
Crane load (Electric operation) upto 500 kN	Gantry girder	Crane	Span/750
Crane load (Electric operation) over 500 kN	Gantry girder	Crane	Span/1000
Crane load (Vertical) + Roof load	Gantry girder	Crane	Inward 12mm
Moving load (Charge Cars) etc.	Gantry girder	Crane	Span/600

TABLE 4.2 (B). DEFLECTION LIMITS OTHER THAN FOR PITCHED ROOF PORTAL FRAME [FOR INDUSTRIAL BUILDING] (FOR LATERAL CRANE+WIND)

Design Load	Member	Supporting	Maximum Deflection
No Crane	Column	Elastic Cladding	(Height/150)
No Crane	Column	Masonry/Brittle Cladding	(Height/240)
Crane + Wind	Gantry (Lateral)	Crane absolute	(Span/400)
		Relative between rails	(10 mm)
	Column Frame	Gantry (Pendent operated)	(Height/200)
Crane	Column Frame	Gantry (Cab-operated)	(Height/400)

TABLE 4.2 (C). DEFLECTION LIMITS OTHER THAN FOR PITCHED ROOF PORTAL FRAME [FOR OTHER BUILDINGS]

(i) For Vertical Deflection

Design Load	Member	Supporting	Maximum Deflection
Live load	Floor & Roofs	Elements not susceptible to cracking	(Span/300)
Live load	Floor & Roofs	Elements susceptible to cracking	(Span/360)

(ii) For Lateral Deflection

Wind	Building	Elastic Cladding	(Height/300)
		Brittle cladding	(Height/500)
Wind	Inter storey Drift		Storey Height/300

Note : (a) The deflection of the members shall be calculated without considering the impact factor or dynamic effects of the a deflection.

(b) Roofs, which are very flexible, shall be designed to withstand any additional load, that is, likely to occur as a result of ponding of water as the accumulation of snow.

In a simply supported beam, carrying total uniformly distributed load, W = w.L, the maximum bending moment occurring at the centre is M = [WL/8], and the maximum deflection, Δ.

$$\Delta = \left[\frac{5}{384} \times \frac{WL^3}{EI}\right] = \left[\frac{5}{48}\right] \times \left[\frac{1}{8}\frac{WL}{EI}\right] L^2$$

$$\Delta = \left(\frac{5}{48}\right)\left[\frac{ML^2}{EI}\right] \qquad \text{(i)}$$

From theory of simple bending, substituting

$$\left[\frac{M}{I}\right] = \left[\frac{\sigma_b}{d/2}\right]$$

where, σ_b is the bending stress at the extreme fibre

$$\Delta = \left[\frac{5}{24}\right]\left[\frac{\sigma_b . L^2}{E . d}\right] \quad \text{(ii)}$$

This expression may be written in terms of (L/d) as

$$\left[\frac{L}{d}\right] = \left[\frac{24}{5}\right]\left[\frac{E . \Delta}{\sigma_b . L}\right] \quad (4.1)$$

Eq. (4.1) represents the (length/depth) ratio for the beam corresponding a specific ratio of deflection to span.

In the code of practice, IS : 800-1984 (Δ/L) is restricted as (1/325) using $\sigma_b = 0.6\,\sigma_y$, it gives necessary (L/d) ratio as below : Take ($E = 2 \times 10^5$ N/mm^2)

$$\left(\frac{L}{d}\right) = \left[\frac{24}{5} \times \frac{2 \times 10^5}{0.6\,\sigma_y} \times \frac{1}{325}\right] \quad \text{(iii)}$$

$$\left[\frac{L}{d}\right] = \left[\frac{4923}{\sigma_y}\right]$$

For $\sigma_y = 250$ N/mm^2

$$\left[\frac{L}{d}\right] = \left[\frac{4923}{250}\right] = 19.692 = 20 \quad \text{(iv)}$$

Hence, in case, a beam is selected for the design, for which (L/d) ratio is less than 19, then, the deflection criterion is automatically satisfied.

Similarly, for a simply supported beam carrying a concentrated load at the centre, the ratio (L/d) works out to be 24.

Ratio (L/d) provides a guidance parameter to satisfy the limit state of serviceability. It is not mendated (compulsory) in design, so long as, the check for the serviceability is applied seperately.

Where the deflection due to the combination of dead load and live load is likely to be excessive, consideration should be given to pre-camber beams, trusses and girders. The value of desired camber shall be specified in design drawing.

Generally, for spans more than 25 m, a camber approximately equal to the deflection due to dead loads plus half the live load may be used. The deflection of a member shall be determined without considering the impact factor or dynamic effect of the load on deflection. Roofs which are very flexible, shall be designed to with stand any additional load, that is likely to occur as a result of ponding of water or accumulation of the snow or ice.

(b) Vibration

Suitable provisions in the design shall be made for the dynamic effects of the live loads, impact loads and vibrations due to machinery operating loads. In the severe cases, possibility of resonance, fatigue or unacceptable vibrations shall be investigated. Un-usually, the flexible structures shall be investigated for lateral vibrations under the dynamic wind loads; Structures subjected to large number of cycles of loading shall be designed against fatigue failure, as specified in the code.

(c) Durability and corrosion of steel

Durability is the ability of the material to resist deterioration over long period of time. Durability of the steel is the property of steel by virtue of which it lasts long and resists wear. Factors which affect the durability of the structures under conditions relevant to their intended life are (i) environment, (ii) degree of exposure, (iii) shape of the member and the structural detail, (iv) protective measures and (v) ease of maintenance.

As per IS : 800, a durable steel structure is one that performs satisfactorily the desired function in the working environment under the anticipated exposure conditions during its service life, without deterioration of the cross-sectional area and loss of strength due to corrosion. A durable steel structure is one that performs satisfactorily the desired function in the working environment under the anticipated exposure condition during its service life without deterioration of the cross- sectional area and loss of strength due to corrosion.

General environment to which a steel structure is exposed during its working life, is classified into following five categories :

1. **Mild exposure :** The surfaces normally protected against exposure to weather or aggressive condition is in interior of buildings, except when located in coastal areas

2. **Moderate exposure** : The surfaces of the structural steel exposed to

   (i) Condensation and rain

   (ii) Continuously under water

   (iii) Non-aggressive soil/ground water

   (iv) Sheltered from saturated salt air in coastal areas.

3. **Severe exposure** : The surfaces of the structural steel exposed to

   (i) Severe frequent rain

   (ii) Alternate wetting and drying

   (iii) Severse condensation

   (iv) Completely immersed in sea-water

   (v) Exposed to saturated salt air in coastal area.

4. **Very severe exposure** : The surface of the structural steel exposed to

   (i) Sea-water spray

   (ii) Corrosive fumes

   (iii) Aggressive sub-soil or ground water

5. **Extreme exposure** The surfaces of the structural steel exposed to

(i) Tidal zones and splash zones in the sea

(ii) Aggressve liquid or solid chemicals

Corrosion may be defined as the deterioration of steel by the chemical or electro chemical reaction with its environment. It is an electro-chemical process over surface of steel leading to oxidation of the metal.

One of the earliest things one learns about iron and steel is that, these materials rust. Rust is actually a result of corrosion. Rust itself is a hydrated oxide containing a certain quantity of combined water. Rust is formed as a product of reaction as hydrated ferrous oxide (Fe_2O_3 – yellowish loose powder).

This rust becomes 2.50 times in volume than original volume as steel. The rate at which surface is lost, may reach 0.075 mm/year. It results in gradual loss of metal and reduces the cross-sectional area of the member. Thereby it increases the stress in the remaining member and decreases the strength. The material used, the detailing, fabrication erection and surface protection measures should all address.

1. Corrosion reduction and

2. Durability requirements

Requirements for durability take into consideration of the following

1. Shape size, orientation of the members, connections and details

2. Exposure conditions : (General environment has been classified into five levels of severity (viz., mild, moderate, severe, very severe and extreme).

3. Exposure to sulphate attack.

Corrosion protection methods are governed by actual environment conditions as per IS : 9077 and IS : 9172. Main corrosion protection methods are given below :

(i) Controlling the electrode - potential

(ii) Inhibitors

(iii) Inorganic/metallic coatngs or organic/pain systems.

(iv) Surface protection

Surface Protection as suggested by BIS (Bureau of Indian Standards)

(i) In the case of mild exposure, a coat primer after the removal of any looe mill scale may be adequate. As the exposure condition becomes more critical, more elaborate surface preparations and coatings become necessary, as per literature.

(ii) Steel surfaces shall be provided with at least one coat of primer immediately after surface preparations, such as sand blasting to remove all mill scales and rust to expose steel

(iii) Steel without protective coating shall not be stored for long duration in out door environment.

(iv) Surfaces to transfer forces by friction as in HSFG connection shall not be painted. However, it shall be ensured that moisture shouldnot be trapped on such surfaces after pre-tensioning of bolts by proper protective measures.

(v) Members to be assembled by welding shall not be pre-painted at the adjacent to the location of such welds. However, after welding, appropriate protective coating is shall be applied in the region as required by the exposure conditions. If the contact surfaces cannot be properly protected against ingress of moisture by surface coating, these may be completely sealed by appropriate welds.

(vi) Pre-painted members shall be protected against abrasion of the coating during transportation handling and erection.

TABLE 4.3 (A). SPECIFICATION FOR DIFFERENT COATING SYSTEMS

Coating systems	1	2	3	4	5	6
Surface preparation	Blast Clean	Blast Clean	Blast Clean	Blast Clean	Girt Blast	Blast Clean
Pre Fabrication Primer	Zinc Phosphate Epoxy 20 μm	2-Pack Zinc-Rich Epoxy 20μm	-	2- Pack Zinc Rich Epoxy 20 μm	-	Ethyl Zinc Silkate 20 μm
Intermediate coat	-	High building Zinc Phosphate 25 μm	-	2-Pack Epoxy Micaceous Iron oxide	Sealer	Chlorinated Rubber Alkyol 35 um
Top Coat	-	-	-	2-Pack Epoxy Iron Oxide 85 μm	Sealer	-

TABLE 4.3 (B). SPECIFICATION FOR DIFFERENT COATING SYSTEMS

(B) SIDE APPLIED TREATMENTS

Coating systems	1	2	3	4	5	6
Surface preparation	As necessary	Touch in	No site Treatment	As necessary	No site treatment	As Necessary
Primer	Touch in	Touch in	-	-	-	Touch in
Intermediate	-	Modified Alkyd Micaceous Iron-oxide 50 μm	-	Touch in	-	High build Micaceous Iron Oxide Chlorinated Rubber 75 μm
Top Coat	High Build Alkyd Finish 60 um	Modified Alkyd Micaceous Iron oxide 50 μm	-	High Build Chlorinated Rubber	-	High Build Iron oxide Chlorinated Rubber 75 μm

Note: Steels with special alloying elements and production process to obtain better corrosion resistance may be used.

(d) Fire resistance

Fire resistance of a steel member is a function of its mass, its geometry, the actions to which it is subjected, its structural support condition. Fire protection measures are adopted and the fire to which it is exposed.

It is the ability of an element, component or structure to fulfil for a stated perior of time, the required stability integrity, Thermal insulation, and/or other expected performance specfied in standard fire test.

Serviceability Limit State (SLS) is related to the criterion controlling normal use. Unfactored loads are used to check the adequacy of the structure. Serviceability (SLS) checks against need for remedial action or some other loss of utility. SLS could be considered as undesirable.

Limit State of serviceability although includes both excess deflection and excessive vibration, the codes usually provide only limits for the deflection. A beam designed to have sufficient strength will be suitable only if there is no excessive deflection due to service loads.

Fatigue limit state is important where distress (subjected to severe strain) to the structure by repeated loading is possible.

Stability limit state

A limit state corresponding to the loss of static equilibrium of a structure by excessive deflection transverse to the driection of predominant loads.

4.6 ACTIONS (LOADS)

Euro-codes introduced a new word **actions** which refers to loads.

When the loads act on the structure, the deformations occur due to the actions of loads and the geometry of the structure changes. The action (loads) to be considered in the design include direct actions (loads) experienced by the structure due to self-weight, external actions... etc. and impsed deformations such as that due to temperature and settlements. These actions (loads) are classified according to their variation with time as given below.

1. Permanent actions [Dead loads/Gravity loads], W_{PL}

The structures support materials which have mass. The gravitational forces of these masses (the weight) act on the structure. These actions are permanent (which do not change their magnitude and their position. These are due to self-weight of structural and non-structural components, fittings ancillaries (additional equipments, [e.g., computer, fixed equipments etc.].

2. Variable actions [Live loads/Imposed loads], W_{LL}

Loads dut to occupany of a building are other major sources of loading, commonly referred as **super-imposed (Live) Loads or Live loads** (crane loads, snow loads), wind loads and earthquake loads etc. or **Variable action.**

In general, the magnitudes of these loads may change. Although the portions in buildings may look permanent when built, these may be modified. The weight of the portion should strictly be classified as live loads. Variable actions due to construction and service stage loads as **imposed (live loads)** [Crane loads].

The climatic loads are other major source of loads. These loads include **wind loads** and snow loads. The wind may act in any direction, and this load may act normally on the structure or the tangentially to the surfaces, producing horizontal and vertical loading.

The roof-trusses (provided over sloping-roofs) are also designed for the wind load, and the weight of the snow. Snow is likely to be deposited in the winter and its possible accumulation caused by drifting.

Another form of climate induced load arises when movement caused by temperature changes are restrained. Actions caused by movement are called indirect.

3. Accidental Actions, W_{AL}

Actions due to explosions, impact (dynamic effect) of moving vehicles.

4.7 CHARACTERISTIC (SERVICE) LOAD ACTIONS), W_{CK}

In order, to obtain the design object, the design shall be based on characteristic values for applied loads and for material strength, which take into consideration, the probability of variations in the loads to be supported and the material strength.

Generally a single value for each load and for each material property are used with a margin to take care of all uncertainties - such a value is referred as characteristic load.

The characteristics values should based on statistical data, in case, these are available.

Where such data is not available, these should be based on experience. The design values are derived from the characteristic values through the use of partial safety factors, both for the loads and the materials strength. In the absence of special considerations, these factors shall have the values given in IS : 800-2007 according to the material the type of load and the limit state being considered.

These loads have an acceptability small probability of not being exceeded during the life of the structure.

The characteristic/service loads are defined as the value of specified percentage (usually 5 percent) load (action), above which not more than specified percentage (usually 5 per cent) of corresponding stresses of samples tested are expected to occur. (as shown in Fig. 4.1 and Fig. 4.3) Characteristic actions (loaded) denoted by W_{CK} are the values of different actions that are not expected to be exceeded, with more than 5 per cent probability, during the service life of structure. The prescribed pertage is normally adopted as 95 per cent. These are the values of loads while the structures are in service. These are referred as service (working) loads. The values of these loads are adopted as under :

(a) The self-weight in most cases may be calculated on the basis of nominal dimensions and unit weights as specified in IS : 875 : Part I.

(b) For the variable loads, the values specified in relevant code IS:875-Part II, 1987, IS: 1893.

(c) The upper limit with a specified (usually 5 per cent) probability of non-exceedance during some reference period (design life of a structure).

(d) Specified by client, or designer in consultantion with the client provided. Which satisfy the minimum provisions of the relevant loading standard.

(e) The accidental load generally corresponds to the value specified by relevant code, standard or client.

Characteristic yield/ultimate stress The minimum value of stress below which not more than a specified percentage (usually 5 percent) of the corresponding stresses of samples tested are expected to occur.

4.8 DESIGN (FACTORED) LOADS, W_{fd}

Design (factored) load, W_{fd} is defined as the product of the characteristic load and the relevant partial safety factor for loads, γ_{fL}. Since the characteristic load is multiplied by partial safety factor for loads, it is referred as **factored load** Factored load is obtained as below :

$$W_{fd} = (\Sigma\, W_{ck} \cdot \gamma_{fL}) \qquad (4.1)A$$

where,

W_{fd} = Design (factored load)

W_{ck} = Characteristic load

γ_{fL} is partial safety factor for load. Its value is generally greater than one. It has been described in the subsequent article.

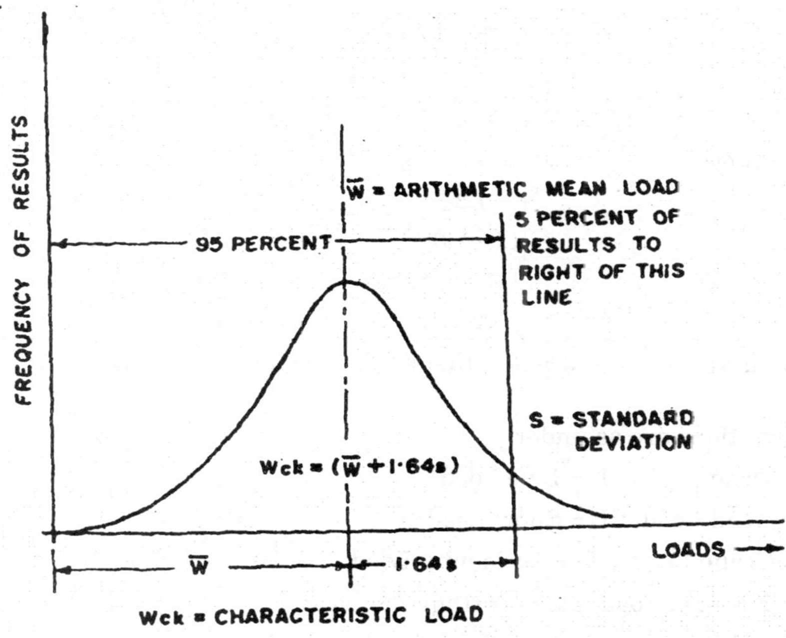

Fig. 4.1.

Factored load (design load) is used for calculating the internal forces (viz, axial forces, bending moments, and torsion). These internal forces are referred as **Load effects.**

4.9 PARTIAL SAFETY FACTORS FOR LOADS, γ_{fL}

The factor normally greater than unity by which the loads (actions) are multiplied to obtain the design values.

Design (factored) load, is found by multiplying the characteristic (service) loads, W_{CK} and partial safety factor for loads, r_{fL}. It is expressed as under :

$$W_{fd} = (\Sigma\, W_{ck} \cdot \gamma_{fL}) \qquad 4.1\,(B)$$

Partial safety factor for load accounts for the following uncertainties :

(i) The possibility of inaccurate assessment of the load (i.e, the magnitude of load).

(ii) The possibility of unfavourable derivation of the load from the characteristic value.

(iii) The uncertainty of assessment of effects of the load.

(iv) The uncertainty in the assessment of the limit state being considered.

(v) The uncertainity as regards the distribution of loads.

The values of γ_{fL} have been listed in Tables 4.4 (A) and 4.4 (B) [For Ultimate Limit Ŝtate, ULS and for the limit state of serviceability, SLS, respectively.

TABLE 4.4 (A). PARTIAL SAFETY FACTORS FOR LOADS r_{fL}, for ULS

Combination of Loads	Limit States for Loads			
	DL	LL	WL/EL	AL
	Leading		Accompanying CL, SL, etc.	
DL + LL + CL	1.5	1.5	1.05 -	-
DL + LL + CL + WL/EL	1.2	1.2	1.05 -0.6	-
	1.2	1.2	0.53 -1.2	
DL + WL/EL	1.5 (0.9)*	-	- 1.5	-
DL + ER	1.2 (0.9)*	1.2	- -	-
DL + LL + AL	1.0	0.35	0.35 -	-

This value is to considered when the stability against overturning or stress reversal is critical.

The abbreviations are as under:

DL = Dead load, L.L = Live Load

WL = Wnd Load, SL = Snow Load

CL = Crane Load, AL = Accidental Load

ER = Erection Load, EL = Earthquake Load.

Note : Usually the values of partial safety factor, γ_{fL} is more than one.

However, looking to Table 4.4 (A) it is seen that some values which have been mentioned are less than unity.

TABLE 4.4 (B). PARTIAL SAFETY FACTORS FOR LOADS γfL1 for SLS

Combination of Loads	Limit State of Serviceability			
	DL	LL		WL/EL
		Leading	Accompanying	
DL + LL + CL	1.0	1.0	1.0	–
DL + LL + CL + WL/EL	1.0	0.8	0.8	0.8

DL + WL/EL	1.0	-	-	1.0
DL + ER	-	-	-	-
DL + LL + AL	-	-	-	-

Note : Abbreviations used have been mentioned under Table 4.4 (A).

Note : (1) In general, the values of load-factors, γ_{fL} is more than unity, since, over-estimation gives improved safety.

(2) Exception is sometimes, $\gamma_{D.fL}$ for dead load is adopted as 0.9. Whenever dead load increases stability against overturning or sliding, or while considering reversal of stresses when dead loads are combined with wind/earthquake loads. Then, $\gamma_{D.fL}$ = 0.9 for counter acting influence of D.L results in increased safety.

(3) For dead loads, γ_{fL} is taken less than unity when the different nature of loads (DL, LL and EL) are considered to act simultaneously at their peat values. The probability of all loads acting simultaneously at peak values is less.

4.10 CHARACTERISTIC STRENGTH OF MATERIALS (STEEL), f_{ck}

The characteristic strength of material (steel) is defined as the value of the yield stress or proof strength below which five percent of all possible test results would be expected to fall. As shown in Fig. 4.2 and Fig. 4.3. The prescribed pertage is normally adopted as 95 per cent.

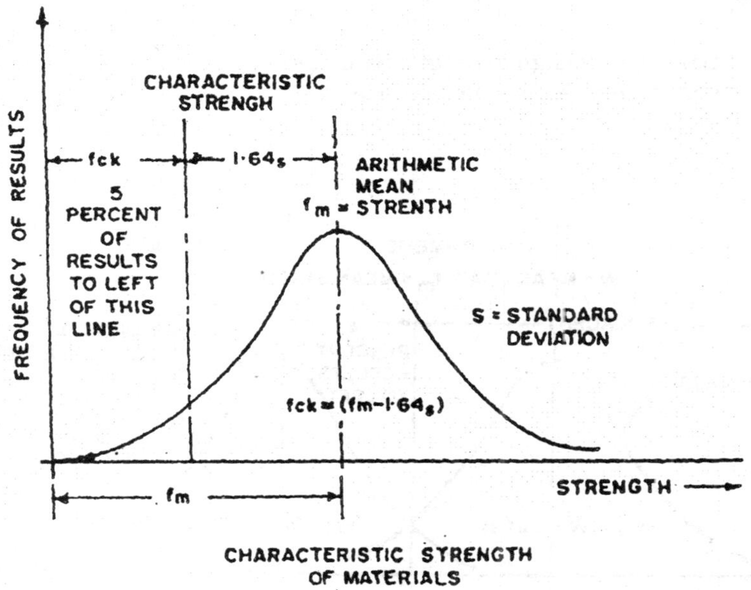

Fig. 4.2.

Yield stress (the characteristic stress of the material in tension before the elastic limit of the material is exceeded.

Some qualities of steel (e.g, high-tensile steel) do not show distinctly the yield stress. At stresses above the limit of proportionality, σ_p, the behaviour of the materials is non-linear.

On unloading and reloading, the behaviour of the material is linear and elastic. These steels do not have a plastic plateau and exhibit strain hardening throughout inelastic range. For such materials in the absence of clearly defined yield-point, the proof stress is used for the purpose of calculations.

Proof stress is a stress corresponding to a pre-assigned residual strain, normally which is taken as a strain 0.002 for most steel structure is used.

Strength calculation may require considerations of the following :

Note : In general and specially in plastic analysis and design, the effect of strain-hardening is neglected. In fact, the strain hardening provides a margin of safety.

4.11 PARTIAL SAFETY FACTORS FOR STRENGTH OF MATERIALS, γ_{ms}

The partial safety factor for the strength of materials, γ_{ms} is defined as the factor which is used to divide the characteristic value of resistance for calculating the strength of the material (design values). Therefore, the design strength of material.

$$f_d = \left[\frac{\text{Characteristic value of Resistance}}{\text{Partial safety factor for the strength of materials}}\right]$$

$$f_d = \left[\frac{f_{ck}}{\gamma_{ms}}\right] \qquad (4.2)$$

Design strength : The design strength S_d is determined as given below from ultimate strength, S_u and partial safety factors for material, γ_{ms}.

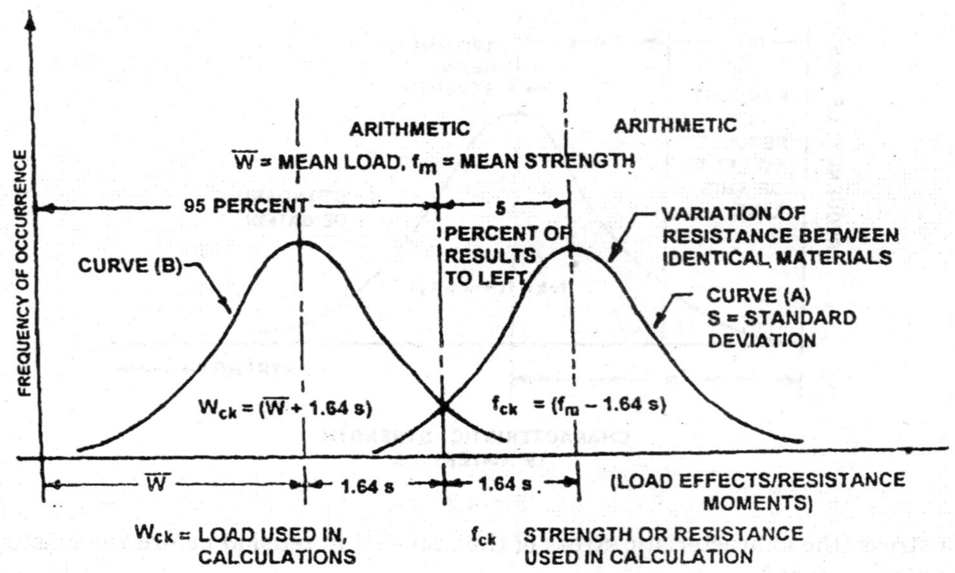

CHARACTERISTIC LOADS AND CHARACTERISTIC STRENGTH / CHARACTERISTIC RESISTANCE

Fig. 4.3.

The partial safety factor for the strength of material γ_{ms} covers the following (i) Uncertainty about the strength of the material tested. (i.e, possibility of unfavourable reduction in member strength due to fabrication and tolerance.

$$S_d = (S_u/\gamma_{ms})$$

(ii) Uncertainty about the difference between, the material as tested and the material as incorporated in the structure (i.e, possibility of unfavourable deviation of material strength from the characteristic value.)

(iii) Uncertainity about the structural dimensions, so far as this affects the calculation of the resistance (i.e., the possibility of unfavourable variation of member sizes).

(iv) Uncertainity about the accuracy of the methods of prediction of the strength. (i.e, uncertainty in the computation of strength of members).

The values of partial safety factor for the strength of materials are given in Table 4.5. Its values is generally more than one.

TABLE 4.5. PARTIAL SAFETY FOR STRENGTH OF MATERIALS, γ_{ms}

No.	Definition	Partial Safety Factor, γ_{ms}
1.	Resistance, governed by yielding (γ_{mso})	1.10*
2.	Resistance governed by buckling (γ_{mso})	1.10*
3.	Resistance governed by ultimate stress (γ_{ms1})	1.25
	Shop Fabrication	Field Fabrication
4.	Resistance of connections, γ_{ms1}	
	(i) Bolts-friction type 1.25	1.25
	(ii) Bolts-bearing type 1.25	1.25
	(iii) Rivets 1.25	1.25
	(iv) Welds 1.25	1.50

Note : Authors started to write this book much earlier than the issue of IS : 800-2007. As such, this value has been used as 1.15 or recommended in IS : 456-1978.

4.12 EFFECT OF LOADS/LOAD EFFECTS (INTERNAL FORCES)

It is also a new word introduced to refer the internal forces developed due to action of loads (i.e, axial forces, bending, shear and torsion).

When any load or combination of loads acts/act on a structure, a set of internal forces (i.e, axial forces, bending moment, shear forces and torsion) are produced in the various members of the structure.

The members of the structure are designed for the design axial forces, design bending moment, design shear forces and designed torsion. These provide resistance from the structure and these forces are kept more than the load effects.

4.13 REQUIREMENTS OF LIMIT STATE AS REGARDS SAFETY OF STRUCTURES

Knowing the characteristic loads and the characteristc property of the strength of the material, the requirements of limit state as regards safety of the structure may be described.

In case, S_d is considered as calculated factored load effect on the components of the structure (e.g, bending moments, shear forces axial forces, which are the function of the product of characteristic loads W_{ck} and γ_{fL} and design resistance, R_d is calculated from the factored resistance of the components being checked. And it is function of nominal value of the material yield strength. S_d is the function of the combined effects of factored dead, live and wind loads.

Other loads, if applicable are also considered.

In accordance with above concepts, the safety format used. In the limit State code is based on probable maximum load and probable minimum strength, so that, a consistant level of safety is achieved. Thus, the design requirements are expressed as below.

Calculated factored resistance of the component,

$$R_d = f_n \left\{ \frac{f_{ck1}}{\gamma_{ms1}}, \frac{f_{ck2}}{\gamma_{ms2}}, \frac{f_{ck3}}{\gamma_{ms3}} \text{ etc.} \right\} \quad (4.3)$$

where, R_d is the factored resistance of the component (design strength) and f_n indicates a function of the quantities within the parentheses. Calculated factored load effect on the component.

$$S_d = f_n \left\{ W_{ck1}, \gamma_{fL1}, W_{ck2}, \gamma_{fL2}, W_{ck3}, \gamma_{fL3}, \ldots \text{ etc.} \right\} \quad (4.4)$$

where, S_d is the calculated factored load effects and f_n is a function of quantities in the parentheses.

$\gamma_{fL}, \gamma_{fL}, \gamma_{fL}, \ldots\ldots$ are the partial safety factors for loads.

In this approach, the ultimate strength is considered. The individual loads are multiplied by the partial safety factors for loads. Ultimate design loads have been defined

$$W_{fd} = \Sigma \left(W_{ck} \cdot \gamma_{fL} \right) \quad (4.5)$$

The design requirement for a safe structure may now be written as under

(Design Action) < (Design Strength) (4.6) (A)

$$\left(R_d > S_d \right) \quad (4.6)\ (B)$$

$$\left(S_d < R_d \right) \quad (4.6\ C)$$

Therefore, the ultimate limit state is related to the maximum design load capacity under the extreme conditions. The partial load factors are chosen to reflect the probability of the extreme conditions, when the loads act alone or in combination.

4.14 STABILITY OF STRUCTURES

The structure as a whole or any part of it shall be designed to prevent instability due to overturning, uplift or sliding under the factored load as under

1. Stability against Overturning uplift or sliding

The stability of the structure is checked for the factored load as below.

(a) The actions (loads) shall be divided into components aiding instability and components resisting instability.

(b) The permanent and variable actions (loads) and their effects causing instability shall be combined using appropriate load factors as per limit state requirements to obtain maximum destabilizing effects.

(c) The permanent loads and effects causing resistance shall be multiplied with a partial safety factor 0.9 and added together with design resistance after multiplying with appropriate partial safety factor. Variable actions and their effects contributing to resistance shall be disregarded.

(d) The resistance effect shall be greater than or equal to be destabilizing effect. Combination of imposed and dead loads should be such as to cause most severe effect on overall stability.

2. Sway stability

The whole structure including portion between expansion joints, shall be adequately stiff against sway. To ensure this, in addition to designing for applied horizontal loads a seperate check should be carried out for notional horizontal loads as to evaluate the sway under gravity loads.

3. Fatigue

Generally fatigue need not be considered unless a structure or its element is subjected to numerous significant fluctuations of stress changes due to fluctuations in wind loading also need not be considered. When designing for fatigue, the load factors for action, r_{fL} equal to unity shall be used for the load causing stress fluctuation and stress range. Fatigue is the damage caused by repeated fluctuations of stress, leaking to progressive and consequent cracking of structural element.

4. Plastic collapse

Plastic analysis and design may be used if the conditions under the plastic method of analysis are satisfied. The plastic collapse is the failure stage at which sufficient number of plastic hines have formed due to actions (loads) in a structure leading to a failure mechanism.

4.15 CLASSIFICATION OF CROSS-SECTIONS

Due to compressive-stresses, the plate elements of a cross-section may buckle locally. Before the limit-state is achieved, the local buckling may be avoided by limiting the width to thickness ratio of each element of a cross-section subjected to compression due to axial force, moment or shear.

The members shall be capable of forming plastic hinges with sufficient rotation capacity (ductility) without local buckling to enable the redistribution of bending moment needed before formation of the failure mechanism in plastic analysis.

The members shall be capable of developing the yield stress under compression without local buckling in case, elastic analysis is used.

The cross-sections of the members have been defined on the basis of above description as under :

1. **Plastic**: The member cross-sections are called **`plastic-type'** in which the plastic-hinges may develop and have the rotation capacity needed for failure of the structrure due to formation of plastic mechanism. The width to thickness ratio of plate elements shall be less than specified in Table 4. These sections are also referred as Class-1, type
2. **Compact** : The member cross-sections shall be referred as the **`compact type'**, in which the sections may develop plastic moment of resistance, but have inadequate plastic hinge rotation due to local buckling. The width to thickness ratio of plate elements shall be less than that specified in Table 4, but greater than that specified in Table 4. These type of cross-sections are also referred as class-2 type.
3. **Semi-compact** : The member cross-sections may be referred as **`semi-compact'** in which the extreme fibre in compression may reach yield stress, but cannot develop the plastic moment of resistance, due to local buckling. The width to thickness ratio of plate elements shall be less than that specified in Table 4, but greater than that specified in Table 4. These types of cross-sections are also called as Class-3 type.
4. **Slender**. The member cross-sections may be referred as **`slender'** in which the elements buckle locally even before reaching the yield stress. The width to thickness ratio of plate element shall be greater than that specified in Table 4. In such cases, the effective sections for the design shall be calculated either by following the provisions of IS: 801 to consider for the post-local buckling strength or by deducting width of the compression plate element in excess of the semi-compact section limit. These types of cross-sections are also called as Class-4 type.

4.16 TYPES OF ELEMENTS

In case, the cross-section of a member is considered, as if, it consists of flange elements, web elements, stem of a tee-section and seperate legs of an angle section, then, these elements are referred as below.

1. Internal elements : Those elements which are connected along both longitudinal edges to other elements or to longitudinal stiffeners attached at suitable intervals to transverse stiffeners, for example web of I-section and flanges and web of a box section are called as **internal elements.**

2. Outside elements or outstands: Those elements which are connected along only one of the longitudinal edges to an adjacent element, the other edge kept free to displace out of plane (for example flanges of an I-section overhanging, stem of a tee-section and different legs of an angle section) are called as **outside elements or out stands.**

3. Tapered elements: Tapered elements may be considered as flat elements having average thickness.

4.17. COMBINATIONS OF LOADS

Combinations of loads for purpose of design shall be those that gives maximum forces and effects and consequently results in maximum stresses and deformations.

In Art 2.2, the self-weight of the structural members, super-imposed dead loads and loads due to filling materials have been referred **as permanent loads.**

In Art 2.3, loads assumed to be produced by the intended use or occupancy (i.e., live load, including distributed and concentrated loads, impact and vibration (which vary in magnitude and/or in position) are referred as **Variable loads.**

In Art 2.4, Art 2.5 and Art 2.6 Wind load, snow load and the seismic (earthquake forces) loads have been referred as **other variable loads.**

Let W_{PL}, W_{VL} and W_{OVL} represent the characteristic values of the permanent loads, the variable loads and other variable loads, respectively similarly, γ_{FL}, $\gamma_{fL.VL}$ and γ_{fLOVL} represent partial safety factors for permanent loads, variable loads and other variable loads. The values of these partial safety factors have been found statistically using the theory of premeability.

Euro-code (EC3 for steel structures, 1993 specifies a general expression for the combinations of appropriate values of above categories of loads

$$\text{Combinations of load} = \Sigma\, W_{PL} \cdot \gamma_{fL.PL} + \Sigma\, W_{VL} \cdot \gamma_{fL.VL} + \Sigma\, W_{OVL} \cdot \gamma_{fL.OVL} \quad (4.7)$$

In case of accidental loads, above combination is expessed as below

$$= \Sigma\, W_{PLA}\, \gamma_{fLPLA} + \Sigma\, W_{OVLA} \cdot \gamma_{fLOVLA} + A_d + W_{OVL.11} \cdot \gamma_{fL.OVL} \quad (4.8)$$

where, A_d are the design values for accidental design events and suffix A represents the respective load factors for the accidental events.

As it is nicely known that the factored (i.e, ultimate) load (Action) is used for the design purpose (which one is the product of characteristic load and the appropriate corresponding partial safety factor).

As in the above expressions, more than one variable load (action) are used. In such cases, the second variable load (action) is multiplied by a partial safety factor, g_{fL}, the value of which shall be smaller than its main value used for its being single load (Action) is to be used. For the combinations of more loads, the total value is effectively reduced. The fact is that the probability of all the variable loads (Actions) attainng their maximum value at the same time is very low.

IS 800-2007, in its Table 4 recommends the values of partial safety factors for the different combinations as below in Table 4.

TABLE 4. VALUES OF FACTORED LOAD AS THE PRODUCT OF PARTIAL SAFETY FACTORS AND THE RESPECTIVE LOADS

Combinations	Factored values	
First	1.5 DL + 1.5 IL + 1.05 (CL or SL)	(4.9)
Second	1.2 DL + 1.2 IL + 1.05 (CL or SL) ± 0.6 (WL or EL)	(4.10A)
Alternatively	1.2 DL + 1.2 IL ± WL or EC + 0.53 (CL or SR)	(4.10B)
Third	1.5 DL ± 1.5 WL or + 1.5 EL	(4.11A)

Alternatively	0.9 DL ± 1.5 WL or + 1.5 EL	(4.11B)
Fourth	1.2 DL + 1.2 ER	(4.12A)
Alternatively	0.9 DL ± 1.2 ER	(4.12B)
Fifth	1.0 DL + 0.35 IL + 3.5 CL or SL + AL	(4.13)

1. **Abbreviations** of above table are as under

DL = Dead load, LL = Live load (Imposed); WL = Wind load, ER = Erection load

AL = Accidential load, SL = Snow load CR = Crane load (Vertical/Horizontal)

EL = Earthquake (Seismic) loads

2. **Different loads** (Simultaneously considered): The leading live load shall be considered to be the one causing to higher load effects in the member/section.

3. (i) **Dead load** contributes to stability against overturning is critical or

(ii) **Dead load** causes reduction in stress due to other loads

Wind loads are the horizontal loads and these loads may change their direction. Therefore, these are referred as reversible. These loads have two components, WL_x and WL_y.

Earthquake loads/forces have two horizontal components EL_x, EL_y and in addition to these, a third components, EL_z. IS 1893 (Part 4) - 2005, Clause 7.3 (Sub-clause 7.3.2.1) recommends that all possible combinations of these three components (EL_x, EL_y and EL_z) **including the variation in sign** (plus or minus) has to be considered.

[**Note** : In this way, x and y are two horizontal orthogonal directions and Z-is the vertical direction]. In buildings, Wind load and Earthquake loads resisted either by walls frames builts in orthogonal directions (x and y directions).

Again, wind load and earthquake do not attend their peak values simultaneously. Therefore, only one is considered.

In case, the crane loads, CL, snow load, SL, and accidental loads, AL are not taken into consideration, then above mentioned combination (Table 4.) reduce as below

I (1)	(1.5 DL + 1.5 IL)	(4.14)A
(2)	(1,5 DL + 1.5 EL_x)	(4.14)B
(3)	(1.5 DL − 1.5 EL_x)	(4.14)C
(4)	(1.5 DL + 1.5 EL_y)	(4.14)D
(5)	(1.5 DL - 1.5 EL_y)	(4.14)E
II (1)	(1.2 DL + 1.2 IL + 1.2 EL_x)	(4.15)A
(2)	(1.2 DL + 1.2 IL − 1.2 EL_x)	(4.15)B
(3)	(1.2 DL + 1.2 IL + 1.2 EL_y)	(4.15)C
(4)	(1.2 DL + 1.2 IL − 1.2 EL_y)	(4.14)D
III (1)	(0.9 DL + 1.5 EL_x)	(4.16)A
(2)	(0.9 DL − 1.5 EL_x)	(4.16)B

(3) $(0.9\ DL + 1.5\ EL_y)$ (4.16)C

(4) $(0.9\ DL - 1.5\ EL_y)$ (4.16)D

In some cases. the designed is controlled by wind or earthquake loads minus the vertical loads. In such cases, the value of partial safety factor for loads, g_{fL} more than 1.00 for vertical loads make the calculations unconservative. Therefore, γ_{fL} is adopted as 0.9.

In micro-wave and other tower strcutures, the imposed/live load is very small. In such structures, the combination of load (1.5 D.L + 1.5 W.L) controls the design.

In industrial building structures the crane loads combined with wind load become most critical. Then, the product of 1.5 CL becomes leading load and the roof live load is considered as the accompanying live load. In case of earthquake loads, EL only the dead load of the crane is taken into account. Load lifted is not likely to experience the inertia force due to earthquake.

As compared to the live loads, various countries adopt smaller value of γ_L is adopted for the dead loads as these loads may be estimated more accurately.

4.18 CRITERIA FOR FAILURE/COLLAPSE OF STEEL

The mechanical properties of structural materials are normally obtained by conducting tests. The test specimen are subjected to comparatively simple stress conditions. Tensile tests are conducted to know most of information concerning the strength of metals Compression tests are used for studying brittle materials like cast iron.

Load displacement behaviour until fracture occurs suddenly with little or no plastic deformation as shown by the cast iron is linear. Such a fracture is **brittle fracture.**

Where as steel shows a linear load-displacement relationship upto yield with large plastic deformation (e.g, necking and cup and cone formation) such a fracture is **ductile**.

For shear loading, a small information is available relating the strength of materials.

In actual structures, the materials are subjected to more complicated stress conditions and the yielding is not defined as defined in simple uniaxial tensile test. For which various strength theories have been developed to predict when the failure will occur (that is, which decide the criteria for failure). By failure of the failure is meant yielding on actual rupture, which occurs first. It is assumed that yielding is reached when one component of stress becomes equal to the uniaxial value, σ_y.

Generally, following four theoretical criteria for yield are followed.

(1) The maximum stress theoryb

This is the oldest theory of failure and it is sometimes referred as Rankine's theory.

In this theory, it is stated that the yield occurs when the maximum principal stress reaches the uniaxial tensile stress. Thus, the conditions for yielding are

$$(\sigma_1)_{yp} = \sigma_{yp} \quad (4.16)A$$

$$|(\sigma_3)_{yp}| = \sigma_{yp} \quad (4.16)B$$

in which σ_{yp} and σ_{yp} are the yield point stresses in simple tension and compression, respectively.

(2) The maximum strain theory

This theory is attributed to st. – Venant

In this theory, it is stated that a ductile material begins to yield either

(i) When the maximum strain (elongation) equals the yield point strain in simple tension

$$\left(\frac{\sigma_1}{E}\right) - \frac{\mu}{E}(\sigma_2 + \sigma_3) = \epsilon_1 \quad (4.17)$$

or

(ii) When the minimum strain (shortening) equals the yield point strain in simple compression

$$\left(\frac{\sigma_3}{E}\right) = \frac{\mu}{E}(\sigma_1 + \sigma_2) = \epsilon_3 \quad (4.18)$$

(3) The maximum shear stress theory

This theory gives better agreement with experiments, at least for ductile materials (which have yield in tension = yield in compression, $\sigma_{yp} = \sigma_{yp}$)

This theory states that the maximum shear stress in the material becomes equal to maximum shear stress at the yield point in a simple tension.

Since, the maximum shear in the material is equal to the half the difference between the maximum and minimum principal stresses and since the maximum shear stress in tension test is equal to half the normal stress, the condition for yielding.

$$(\sigma_1 - \sigma_3) = \sigma_{yp} \quad (4.19)$$

(4) The maximum energy theory

(Distorsion or shear strain energy theory). This theory was first proposed by Beltrami, Rendiconti.

In this theory, it is assumed that the quantity of strain energy per unit volume of the material becomes equal to the shear strain energy in simple axial tension.

$$\frac{1}{2E}\left(\sigma_1^2 + \sigma_2^2 + \sigma_3\right) - \frac{\mu}{E}(\sigma_1\sigma_2 + \sigma_2\sigma_3 + \sigma_1\sigma_3) \quad (4.20)$$

$$= \frac{1}{2E}.\sigma_{yp}^2$$

Huber, Von-Mises and Hencky gave the most commonly accepted theory. Uniaxial yield stress has been expressed in terms of the principal stresses.

$$(\sigma_1 - \sigma_2)^2 + (\sigma_2 - \sigma_3)^2 + (\sigma_3 - 1)^2 = 2\sigma_{ys}^2 \quad (4.21)$$

where, stresses, s_1, s_2 and s_3 are stresses which are in three mutually perpendicular planes of zero shear and σ_{ys} is the yield stress comparable with uniaxial value, σ_y.

Above expression may be expressed in terms of direct stresses σ_b, σ_{bc} σ_{bt} and the shear stresses τ on two mutually perpendicular planes. The principal stresses may be written from Mohr's circle.

$$\sigma_1 = \frac{(\sigma_b + \sigma_{bc})}{2} + \left[\left(\frac{\sigma_x - \sigma_y}{2}\right)^2 + \tau^2\right]^{1/2} \tag{4.22)(i}$$

$$\sigma_2 = \frac{(\sigma_b - \sigma_{bc})}{2} - \left[\left(\frac{\sigma_x - \sigma_y}{2}\right)^2 + \tau^2\right]^{1/2} \tag{4.23}$$

Substituting from Eqs. 4.22 and 4.23 in Eq. (4.21) with $\sigma_3 = 0$ and σ_{ys} is equal to the design stress, σ_y

$$\sigma_{bc}^2 + \sigma_b^2 - \tau_{bc}\ \sigma_b + 3\tau_q^2 = \sigma_y^2 \tag{4.24}$$

In case σ_{bc} is replaced by σ_{bt} with change in sign

$$\sigma_y{}^2 + \sigma_{bc}{}^2 + 3\ \tau_q{}^2 \tag{4.25)A}$$

$$\sigma_y = \left(\sigma_{bc}{}^2 + 3\ \tau_q{}^2\right)^{1/2} \tag{4.25)B}$$

REFERENCES

1. **Nethercot David** - `Limit State Design, of Structural Steel-Work'. Third-edition, 2001. Published by Spon Press. 11 New Fetter Lane, London EC, 4P, 4EE Simultaneously published in the USA and Canada, Spon Press 29 West 35th Stress, New York NY, 10001.

2. `Teaching Resource on Structural Steel Design' Compiled by V. Kalyanaraman Rangachari Narayanan, A.R. Santhakumar, S. Seetharaman, S.R. Satish Kumar, R. Senthil, S. Arul Jayachandran: INSDAG : (Institute for Steel Development And Growth)

3. Draft for Approval and Suggestion - Issued by (BIS) Bureau of Indian Standards Manak Bhawan, 9 Bahadur Shah Zafar Marg, New Delhi 110, 002.

4. IS : 800-2007: General Construction Steel Code of Practice, BIS-2007. Bureau of Indian Standards manak Bhawan, 9 Bahadur Shah Zafar Marg, New Delhi 110, 002.

5. BS : 5950 (British Standards) Structural use of Steel-Work in Buildings.

6. **N. Subramanian** `Design of Steel Structures' published by Oxford University Press, First Published 2008.

7. **S. Timoshenko** Strength of Materials (Part I) `Elementary Theory and Problems, Third Edition. An East-West Edition, D. van Nostrand Company, Inc. Affiliated East-West Press Pvt. Ltd., New Delhi.

8. **S. Timoshenko** Strength of Materials (Part II) Advanced Theory and Problems. Third Edition. An East West Edition. D. Van Nostrand Company Inc. Affiliated East-West Press Pvt. Ltd., New Delhi.

5

Part - V
DESIGN OF COLUMNS AND COMPRESSION MEMBERS

LIMIT STATE OF COLLAPSE IN COMPRESSION DESIGN OF AXIALLY LOADED COLUMNS AND COMPRESSION MEMBERS

5.1 INTRODUCTION

A column is defined as a structural member in upright (vertical position) subjected to compressive force in a direction parallel to its longitudinal axis. In general, a column in a structural member in upright (vertical) position which supports a roof or a floor system and predominately subjected to compression. The terms stanchions and posts are also used for columns. The columns, stanchions and posts are general terms used in building construction. In the simply supported at both the ends truss bridge girders, carrying vertical loads acting downward, the top chord members are compression members and the end compression members are termed as end posts. In building frames, columns transfer the loads from roof slabs and beam girders to the foundation by compression.

Stocky/short columns

Very short columns (*viz.* members with low slenderness) carrying gradually increasing equal and opposite forces at the ends, fail by crushing (*i.e.*, by yielding or even into the strain hardening range).

A **strut** is defined as a structural member subjected to compression in a direction parallel to its longitudinal axis. The term strut is commonly used for compression members in roof trusses. A strut may be used in a vertical position or in an inclined position in roof trusses. The compression members may be subjected to both axial compression and bending.

In struts, it results in compressive strain and shortening. The shortening of struts continues until the struts squeeze (i.e., crush).

Slender/Long columns terms long and very long column are used for slender columns

In long columns, axial shortening is seen only at the initial stages. As the axial load increases, the deformation in the direction normal to the axis of loading develops. After that the columns become unstable.

Very long columns (viz. members with high slenderness ratio) fail by elastic buckling in the Euler mode on attaining the crippling load. Generally, such columns fail by inelastic buckling (i.e., failure occurs. by buckling after a portion of the cross-section has already yielded). The behaviour of the columns does not conform to the assumptions made in the Euler theory. The behaviour does not remain linearly elastic upto failure.

Most of the compression members used in construction, have **intermediate slenderness ratios Practical Columns** (i.e, the columns which do not conform to the assumptions made in Euler's theory are referred as practical columns) fail by inelastic buckling. The columns do not normally remain linearly elastic upto failure, unless these are very slender.

In general, a column is a structural member in upright (vertical) position). Which supports a roof or a floor system and predominately subjected to compression.

Term **stocky column** is used for the columns of short height and strongly built-columns. The stocky-columns have high buckling resistance (i.e., high bending stiffness). The appropriate basis for the design of stocky-columns of the compact cross-section is the **squash load,** $\sigma_y.A$ (i.e., theoretical yield stress multiplied by the cross-sectional area of the column).

Very stocky columns (for example, stub columns) resist loads more than squash load ($\sigma_y.A$) Product of theoretical yield stress and the area of the column).

When compression members are overloaded then their failure may take place because of one of the following :

1. Direct compression, 2. Excessive bending, 3. Twisting and 4. Bending combined with twisting.

The failure of column depends upon its slenderness ratio. The load required to cause above mentioned failures decreases as the length of compression member increases, the cross- sectional area of the member being constant. Therefore, columns are commonly classified as short and long columns. This classification is arbitrary and there is no absolute way to determine the exact limits for each classification.

Note : It is not possible to apply the load exactly axially (i.e, exactly at the centroid of the column). There remains always a small eccentricity, e, while the load is applied. Applied load induces a bending moment (P.e) at every cross-section of the column. As a result of which the column deflects laterally. There develops small additional bending stresses, which are added to the direct compressive stress due to axial load.

5.2 BASIC CONCEPT OF STABILITY

In case, the equilibrium of a ball on a concave spherical surface is considered as shown in Fig. 5.1(A) and if it is displaced slightly, the ball returns to its original position on the removal of the disturbance. It is said to be in the **state of stable equilibrium**. In case, the ball is on the horizontal plane as shown in Fig. 5.1 (B) and if it is displaced, it remains at the position to which the small disturbance has taken. This is called a **state of neutral equilibrium**. Whereas, in case, the ball is in equilibrium on the convex spherical surface, and it is displaced and the disturbing force is removed, the ball continues to move farther away from the original position. This is known as an **unstable equilibrium**.

Instability in the phenomenon which disables an element, member or structure to carry further load due do excessive (deflection lateral) to the direction of loading and vanishing stiffness.

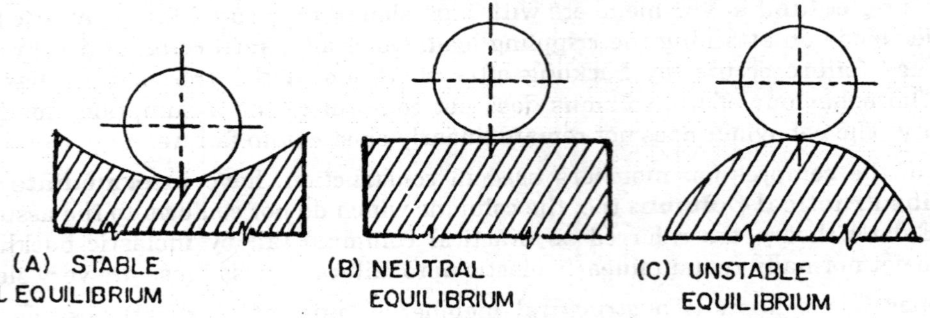

Fig. 5.1. Three states of Equilibrium

These are the basic concepts of three states of equilibrium.

Above mentioned basic concepts may be further illustrated by considering a rigid bar and a linear spring of stiffness, k. A vertical load, P is applied at the top of the bar as shown in Fig. 5.2. Let a small rotation, 0 be given to the rigid-bar at C. Then,

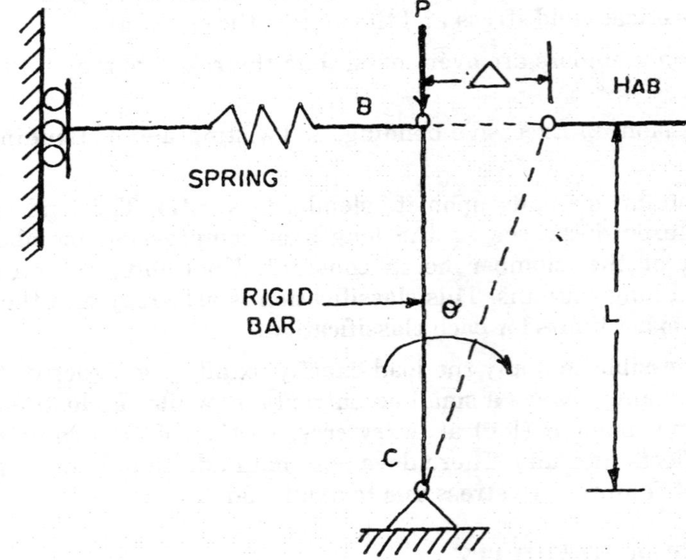

Fig. 5.2.

Moment about point, C

$$M_C = (P.\Delta - H_{AB}.BC) = [P.L.\sin\theta - (K\Delta)(L.\cos\theta)]$$

$$= [P.L.\sin\theta - (K.L.\sin\theta)(L.\cos\theta) \quad \text{(i)}$$

Angle θ is infinitesimal. Then (sin θ = 0) and (cos 0 =1). Substituting in the expression (i)

$$M_C = [(P.L.\theta) - (K.L^2\,\theta) = \theta L(P - kL) \quad \text{(ii)}$$

$$\left(\frac{M_C}{\theta}\right) = [L(P - KL)] \quad \text{(iii)}$$

It may be stated that in case (i) **moment, M_c acts in the direction opposite to that of rotation, θ [i.e., (P – kL) < 0] then the system is in stable equilibrium.**

(ii) moment M_c is zero [i.e., (P – kL) = 0], **then, the system is in the neutral equilibrium.**

(iii) moment M_c **acts in the some direction, as the rotation, θ, [i.e., (P – kL)>0], the system is in the state of unstable equilibrium.**

Mathematically, it may be stated that

(i) For stable equilibrium

(P – kL) is less than zero (i)

(ii) For neutral equilibrium

(P – kL) is equal to zero (ii)

(iii) For unstable equilibrium

(P – kL) is more than zero (iii)

These expressions imply that **as P increases, the state of the system changes from stable equilibrium to unstable equilibrium. At the critical load at which multiple equilibrium** positions are possible (i.e, θ is zero, **and θ is not equal to zero). The critical load serves also as** a bifurcation point of equilibrium path.

The load at such a bifurcation is called the **buckling load.**

For the present system considered, the **buckling load for length kL is stability limit** as well as neutral equilibrium.

In general, the buckling load **corresponds to a state of neutral equilibrium but not** necessarily to stability limit. **Nevertheless, the buckling load is often associated with the** characteristic change of structure before **and therefore, may be considered as the Limit State** of **Serviceability.**

5.3 STRUCTURAL INSTABILITY

Structural instability is defined as **condition in which is no tendency for the structure to return** to its initial position.

There are three classes of structural instability phenomenon **which are observed in the** structures namely (i) **bifurcation,** (ii) **snap-through,** and (iii) **softening.**

A schematic load-displacement diagram **is shown in Fig. 5.3 (A). At point A, the tangent to the actual behaviour of the load displacement diagram bifurcates. Then, the point where** the **bifurcation** of tangent **takes place, represents that the structure has become unstable. It** is an important measure **to note the structural instability. Due to imperfection (such as, the** initial crookedness of a member), **and the eccentricity of loading, the bifurcation is rarely** observed.

The bifurcation point may be evaluated **by the linear buckling analysis.**

In some structures, it is seen that **the displacement increases abruptly at a certain load level.** It occurs at point A as shown in **Fig. 5.3 (B). The displacement Δ_A increases to Δ_B at** load, P_A as shown by a faint **arrow. This phenomenon is referred as snap-through.** In shell-like structures, **and shallow arches, the path of equilibrium is typical. It may be traced** only by the finite displacement analysis.

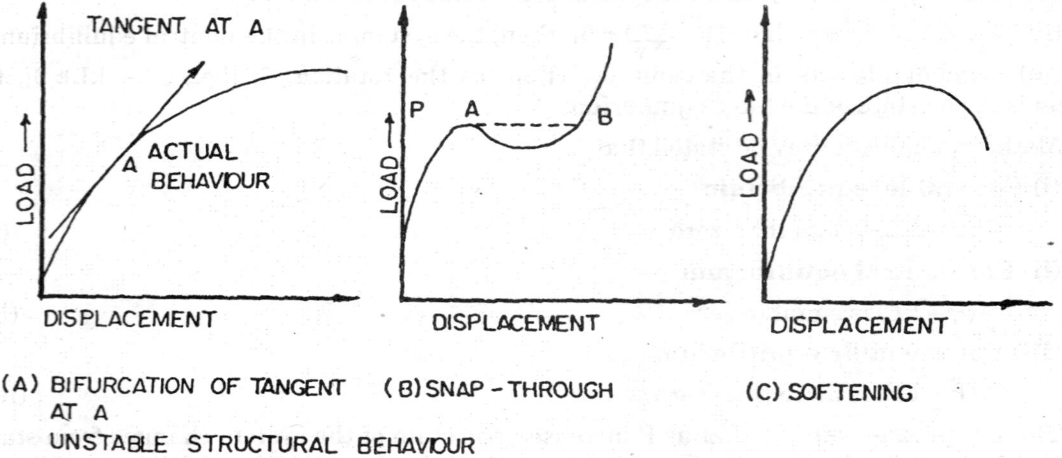

Fig. 5.3. Unstable structural behavious

In many structures, the actual load-displacement relationship shows a peak point after which the structural strength reduces as shown in Fig. 5.3(C). This phenomenon may be observed in case, the yielding takes place. This phenomenon is referred as the **softening**. Non-linear analysis is necessary to trace the path of associated equilibrium.

Non-linear analysis is really difficult, complicate and costly.

5.4 ASSUMPTIONS FOR DETERMINING BASIC COLUMN STRENGTH

Professor Euler made certain assumptions for the determination of basic column strength of an ideal column [Reference : Bruce G. Johnston, 'A Survey of Progress, 1944-51', Bulletin No. 1, Column Research Council, January, 1952].

Following assumptions were made as regards to material

1. The material is perfectly homogeneous and isotropic. The material is linearly elastic and limit of proportionality does not exceed anywhere. The compressive stress-strain properties are same throughout the complete cross-section. The elastic moduli in tension and in compression are equal.

2. There are no initial internal stresses, which exist (as those due to cooling after rolling and those due to welding (i.e. the member is free from residual stresses).

3. The column is perfectly straight and prismatic.

4. The column load acts through the centroidal axis of the member until the member begins to bend.

5. The end conditions are determinate so that a definite equivalent pinned length may be established. The axial shortening is not restrain.

Further assumptions were made as regards the buckling of the column.

6. Small deflection theory of ordinary bending remains applicable and the shear is neglected and its elements do not undergo local buckling.

7. Twisting or distortion of the cross-section does not occur during bending.

It is well known truth that the long columns and the compression members fail by **elastic buckling** (i.e., no portion of the cross-section reaches yield). The short stubby compression members may be loaded until the material yields or perhaps even into the strain-hardening range. However, in the vast majority of usual situations, the failure occurs by **inelastic-buckling** (i.e., buckling after a portion of the cross-section has already yielded).

The practical use of the term buckling is that it is boundary between stable and unstable deflections of a compression member, rather than the instantaneous condition which occurs in the isolated slender elastic column.

As a general approach, Euler elastic buckling controls the strength for large slenderness ratios. Yield stress σ_y is used for short columns. A transition curve is used for inelastic buckling.

These assumptions are not usually realized in practice. Because of various imperfections which occur in practice, the line of action of the resultant thrust does not coincide with the geometrical axis of the member. It results in inevitable (unavoidable) flexure (bending) of the column.

The compressive strengths of the practical columns are significantly influenced by the following

1. Initial imperfection, 2. Eccentricity of loading,
3. Lack of well defined yield-point, and
4. Strain hardening, 5. Residual stresses

5.5 EULER'S ELASTIC BUCKLING LOAD FOR AN AXIALLY LOADED (PRISMATIC MEMBER) COLUMNS

In an axially loaded column, it is assumed that the load is applied at the centroid of the section and in a direction parallel to the longitudinal axis of the column. It is also assumed that the column is perfectly straight. The terms centrally loaded and concentrically loaded are also used for axially loaded columns.

An axially loaded column (as defined by the structural engineers) transmits a compressive force without an explicit design requirement to carry lateral loads or end moments. There exist accidental end eccentricity, initial crookedness, initial curvature and the residual stresses with the tolerance limits in the columns. By having appropriate modifications in the analytical expressions and using suitable factor of safety, the design formula makes the account for these factors. Due to these factors, the strength of an actual column used to be less than that of a perfectly axially loaded column.

Consider a column vertically fixed at the base and free at the top (i.e., as a cantilever column) as shown in Fig. 5.4 Euler's buckling load for this column may be evaluated as under:

In general, the length L of a compression member in any plane is taken as the distance between the points at which, it has effective positional or directional restraint against buckling in that plane.

One end of the column, B is fixed and other end A is free to take up any position from A to A'. Let AB be the axis of the column in the undeflected shape when an axial load (i.e., a thrust), P is applied. The end A of the column moves by δ and the column deforms laterally as shown in Fig. 5.4. Then a fixing moment develops at the end A'.

$$M = (P.\ AA') = (P.\delta) \qquad \text{(i)}$$

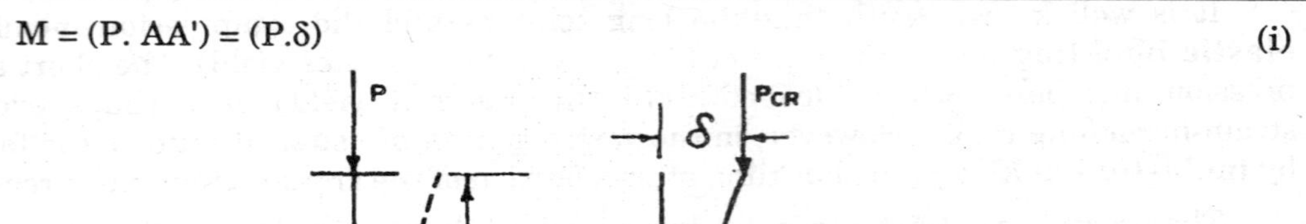

Fig. 5.4.

The bending moment at any point x, Fig. 5.4 (B) at a distance, y from the original vertical axis of the column.

$$M_x = P.\ (\delta - y)\ ...\text{(hogging)} \qquad \text{(ii)}$$

where, x is measured along the axis of the column and y is measured in the direction perpendicular to the axis.

When the value of load P is less than critical load, and stress is within the limit of proportionality, the column remains straight. The column is in **stable equilibrium** that is, if a small lateral load is applied at the free end, the column deflects. On withdrawal of the lateral load, the column resumes its vertical position and deflection vanishes. When the axial load P is gradually increased, a stage will be reached when the vertical position of the column is in the **unstable equilibrium** (i.e., if a small lateral load is applied, a deflection will be produced, which will not vanish on withdrawal of lateral load). The problem of column from stable equilibrium is converted to that of unstable equilibrium, (i.e., the problem of elastic instability). The axial load which is sufficient to keep the column in such a slight deflected shape is called **critical load**. The critical load is the load at which the slender columns and compression members fail by buckling (i.e. instability). Critical load is also called as **buckling load** or **crippling load**. The buckling load is defined as the load at which a member of a structure as a whole collapses in service (or buckles in a load test). **The buckling** is defined as the sudden bending, warping, curling or crumpling of the elements or members under compressive stresses. The direction of buckling of a column depends upon flexural rigidity. E*I*, of the column. It buckles in a direction perpendicular to the axis, about which the moment of inertia of the section is minimum.

The load at which an element or a member of a structure as a whole, either collapses in service or buckles in a load test and develops excessive lateral (out of plane) deflection or instability.

In about 1759 Prof. Leonhard Euler (a Swiss mathematician) derived the most popular column formula. The critical load for the column as shown in Fig. 5.4 was determined as under. Euler dealt with a member, built-in at one end and simply supported at the other end.

The differential equation of the deflected shape of the column is

$$\left(EI\frac{d^2y}{dx^2}\right) = +M_x \text{ (Hogging moment +ve)} \quad \text{(i)}$$

The bending moment at any point on the deflected shape, $M = +P(\delta - y)$

Therefore, $$\left(EI\frac{d^2y}{dx^2}\right) = P(\delta - y) \quad \text{(ii)}$$

$$\left[\left(\frac{d^2y}{dx^2}\right) - \frac{P}{EI}(\delta - y)\right] = 0 \quad \text{(iii)}$$

Let $n = \left(\frac{P}{EI}\right)^{1/2}$

Then the differential equation becomes

$$\left[\frac{d^2y}{dx^2} + n^2y - n^2\delta = 0\right] \quad \text{(iv)}$$

The general solution of this differential equation is

$$y = A.\sin(nx) + B.\cos(nx) + \delta \quad \text{(v)}$$

At $\left(x = 0, y = 0, \frac{dy}{dx} = 0\right)$. $B = -\delta A = 0$

$\therefore\ y = \delta(1 - \cos nx)$ at $x = l = \delta$

This condition is satisfied when, $\delta \cos(n\,l) = 0$

From this, either $\delta = 0$ or $\cos nl = 0$ (vi)

If δ is zero, buckling of column does not occur and then, there is no deflection. If $\cos nl = 0$, then

$$nl = \frac{\pi}{2}(2\pi - 1) \text{ where } n = 1,2,3 \text{ ....}$$

For $n = 1$, the values of nl is smallest and the column is unstable. It is equal to $\pi^2/4$

$$P_{cr} = \left(\frac{\pi^2 EI}{4l^2}\right) \quad (5.1)$$

This load is sustained by the column in a deflected shape. It is not necessary to consider higher values of P_{cr}. In this column, it is seen that there are no deflection and no slope at the fixed end, so also, there are no external moment and no external shear force at the free end. Therefore, the boundary conditions are

$$y = 0, y' = 0 \text{ at } x = 0 \qquad \text{(i)}$$

$$\text{and} \quad y'' = 0, \; y'' + K^2 y' = 0 \text{ at } x = L$$

It is to note that since, the column is considered in the slightly deformed shape in the linear buckling analysis the axial force has a transverse, component. Therefore, P comes in the boundary condition at x = L.

This condition at x = L eliminates A_3. With this and second condition at x = 0, A_1 is obtained as zero. Remaining two conditions results to

$$\begin{bmatrix} 1 & 1 \\ K^2 . \cos kL & 0 \end{bmatrix} \begin{Bmatrix} A_2 \\ A_4 \end{Bmatrix} = \begin{Bmatrix} 0 \\ 0 \end{Bmatrix}$$

The smallest eigenvalue and the corresponding eigen vector coincide with the buckling load obtained earlier.

A column simply supported at top and bottom, Fig. 5.8 (A)

The moment equilibrium in a slightly deformed geometry

$$\frac{d^2y}{dx^2} + k^2 y = 0 \qquad \text{(i)}$$

$$\text{i.e. } y'' + k^2.y = 0 \qquad \text{(ii)}$$

where, (k^2 = P/EI), and

EI = flexural (bending) rigidity.

General solution of this equation.

$$y = A_1 \sin(kx) + A_2 \cos(kx) \qquad \text{(iii)}$$

Arbitry constants A_1 and A_2 are obtained from the boundary conditions

$$y = 0 \text{ at } x = 0 \qquad \text{(iv)}$$

$$y = 0 \text{ at } x = L \qquad \text{(v)}$$

From the expression (iv), $A_2 = 0$ and from the expression (v)

$$A_1. \sin(kL) = 0 \qquad \text{(vi)}$$

Note : $A_1 = 0$ is a solution of the characteristic equation above. It is the trivial solution corresponding to a perfectly straight column. It is of no interest. Therefore,

$$P_{cr} = \left(\frac{n^2 \pi^2 EI}{L^2}\right) \qquad \text{(vii)}$$

From the practical point of view, the lowest critical buckling load is found when

$$n = 1, \; P_{cr} = \left(\frac{\pi^2 EI}{L^2}\right) \qquad \text{(viii)}$$

The deflected shape is

$$y = A_1 . \sin\left(\frac{\pi x}{L}\right) \qquad \text{(ix)}$$

Upto the Euler's load, the column remains straight. At the Euler load, the state of column becomes the neutral equilibrium.

The boundary conditions are as follows. There are no deflection and no external moment at each end of the column.

$$y = 0,\ y'' = 0 \text{ at } x = L$$

$$y = 0,\ y'' = 0 \text{ at } x = L$$

From the conditions at x = 0, it is seen that $A_2 = A_4 = 0$. Using this result and the conditions at x = L,

$$\begin{bmatrix} \sin kL & L \\ -K^2 \sin kL & 0 \end{bmatrix} \begin{Bmatrix} A_1 \\ A_3 \end{Bmatrix} = \begin{Bmatrix} 0 \\ 0 \end{Bmatrix}$$

For the non-trivial solution to exist, the determinant of the coefficent matrix, leads

$$k^2L.\ \sin (kL) = 0$$

It gives the critical load for the buckled shape, $P_{cr} = \left(\dfrac{\pi^2 EI}{L^2}\right)$

In general, the buckling load for the columns with any other sets of boundary conditions, the buckling load may be expressed in the form.

$$P_{cr} = \left(\frac{\pi^2 EI}{(KL)^2}\right)$$

where, kL is called **effective length.** Since, $I = A.r^2$, A is the area of column cross-section and r is the radius of gyration.

$$P_{cr} = \left(\frac{\pi^2 . E. A. r^2}{(KL)^2}\right) \qquad \sigma_c = \left(\frac{P_{cr}}{A}\right) = \left(\frac{\pi^2 . E}{(KL/r)^2}\right)$$

For a column of perfectly plastic material, the stress does not exceed the value of yield stress, σ_y. The ratio

$$\left(\frac{\sigma_c}{\sigma_y}\right) = \left(\frac{\pi^2 E}{(KL/r)^2}\right) \quad \text{or} \quad \left(\frac{\sigma_c}{\sigma_y}\right) = \left(\frac{\sigma_y}{\pi^2 E / (KL/r)^2}\right)$$

$$\left(\frac{\sigma_y}{\sigma_c}\right)^2 = \left(\frac{1}{\pi^2}\right) . \left(\frac{kL}{r}\right)^2 . \left(\frac{\sigma_y}{E}\right)$$

$$\left(\frac{\sigma_y}{\sigma_c}\right) = \left(\frac{1}{\pi}\right) . \left(\frac{kL}{r}\right) . \left(\frac{\sigma_y}{E}\right)^{1/2}$$

When (σ_y / σ_c) is more than 1-0, elastic buckling occurs. When (σ_y / σ_c) is less than 1.0, it fails plastically.

This load is referred as **'Euler's buckling load'** or the critical load. For the column in a deflected shape. It is not necessary to consider the higher values of Eq. 5.1.

The values of critical loads for other end condition can be determined from this case. For a column, hinged at both ends slenderness.

$$P_{cr} = \left(\frac{\pi^2 EI}{l^2}\right) \tag{5.2}$$

where P_{cr} = Critical load (or Buckling load).

Since each half of the column is in the same position as the whole of the column. This is called **fundamental case** of buckling of a bar. Eq. 5.2 is usually called as Euler's expression for critical load, P_{cr}.

It is interesting to observe from Eq. 5.2 that the Euler's load is independent of the strength of the material and it depends on the stiffness, E.I.

When the length of a column in very short, the Euler's formula gives load-carrying capacity more than strength of material (i.e., yield stress multiplied by the cross-sectional area). It is seen that it is clearly unreasonable. Euler's formula is applied only where, it gives load-carrying capacity smaller than its strength.

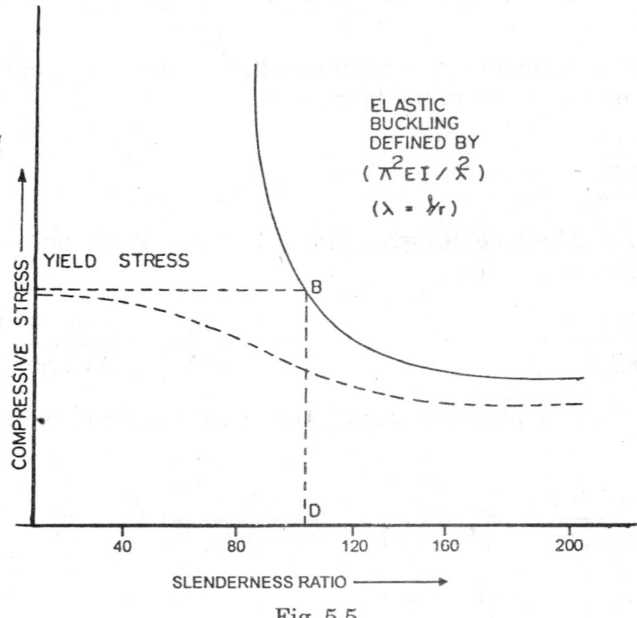

Fig. 5.5

When the columns are gradually loaded (i.e. the load is applied in small increments). It remains straight only at the initial stages. When the loads are further increased, the column begins to deflect. The deformation develops in the direction normal to the axis of loading. The column deflects because the applied load is not exactly through the centroid of column-section and the column itself is not perfectly straight. The imperfections present introduce moment, which causes some deflections. After occurrence of deflection, a large moment develops. It is seen that the middle point along the height of column carries moment equal to the product of applied load and the deflection at mid-height. The moment reduces the load-carrying capacity of the column. The deflection in column further increases with the increases in the applied load.

From the experiments it is seen that the slender steel columns with (l = l/r) more than 100-120, test results agree closely with the values obtained from Eq. 5.2. For the smaller slenderness values results deviate from the ideal elastic buckling because load stresses exceeds the limit of proportionality.

It is seen that a column which remains perfectly stiff under the load, is **stable.** After reaching the critical load, the column becomes weak and stays deflected. If pushed aside. The column buckles in the shape of a half-sine-wave; the amplitude of which is indeterminate. The stresses in the column also increase until the material yields. The column does not remain as a structural member.

Limitations of Euler's formula

The stress corresponding to Euler's load at failure (for the column pinned at both the ends) is given by

$$P_E = \left(\frac{P_{cr}}{A}\right) = \left(\frac{\pi^2 E.A.r^2}{A.l^2}\right) = \left(\frac{\pi^2 E}{(l/r)^2}\right) \qquad \text{(i)}$$

It is seen from the expression (i) that the critical stress, P_E is inversely proportional to the square of the slenderness ratio.

For mild steel, σ_y = 250 N/mm² and E = 2 x 10⁵ N/mm². Substituting in the expression (i)

$$(l/r)^2 = \left(\frac{\pi^2 \times 2 \times 10^5}{250}\right) \cdot \left(\frac{l}{r}\right) = 78.956$$

It shows that for Euler's formula. For pin-ended column (l/r) >> 78.956. The Euler's formula is more nearly expressed by a curve shown by dotted curve.

5.6. STUB COLUMNS (COMPRESSION TEST SPECIMEN)

A **stub column** is a short compression test specimen, sufficiently long for use in measuring the stress-strain relationship for the complete cross-section, but short enough to avoid buckling in the elastic range. The length of a stub column is long enough to contain the same residual stress pattern that exists in the column cross-section itself. Professor Bruce G. Johnston Guide to Design Criteria for Metal Compression Members, John Wiley and Sons, New York, 1966 have specified the length of **stub-column compression test specimen**. It should be (twice the depth of section, d + 250) mm or three times the depth minimum. The maximum length of the test specimen may be twenty times the radius of gyration about weak axis of the section or five times the depth of section.

Professor Lynn S. Beedle and Lambert Tall, ['Basic column strength' Proceedings of the American Society of Civil Engineers, Journal of structural Division, Vol. 86 No. ST 7, July 1960 (pp 139-173)] gave the following expressions or the critical stresses for strong axis, σ_{xx} and weak axis, σ_{yy}, respectively as under :

For (l/r) < π (E/σ_p)$^{1/2}$

$$\sigma_{xx}\left[\sigma_y - \left(\frac{\sigma_p}{\pi^2 E}\right)(\sigma_y - \sigma_p)\left(\frac{l}{r}\right)^2\right] \qquad (5.3)$$

For $(l/r) > \pi\,(E/\sigma_p)^{1/2}$

$$\sigma_{yy} = \left[\sigma_y - \left(\frac{\sigma_y - \sigma_p}{\pi}\right)\left(\frac{\sigma_p}{E}\right)^{1/2} \cdot \left(\frac{l}{r}\right)\right] \quad (5.4)$$

$$\sigma_{xx} = \sigma_{yy} = \left|\pi^2\,E/(l/r)^2\right| \quad (5.5)$$

where, σ_y is the yield stress of steel σ_p is the stress at limit of proportionality, E is young's modulus of elasticity, 2 x 10^5 N/mm^2 and r is the approximate radius of gyration. Above equations include yield stress and stress at the limit of proportionality, which are the actual properties of steel. Yield stress, σ_y of the material is a major factor, which influences the ultimate strength of axially loaded columns.

Author, of this book has conducted `stub-column tests on Indian standard H-Beam sections' for his M-E Thesis, August, 1970 and submitted to University of Jodhpur and has published a paper with this title along with Professor S. Divakaran as superviser [`Stub column Tests on Indian Standard H-Beams, Journal of the Institution of Engineers (India), Vol. 51, Number II, Part CI 6, July, 1971 (pp 313-316)]. Complete details of tests have been reported in this paper. The stub column test specimen were tested upto yield. Six coupons (four from the flanges and two from the web) were tested. Young's modulus of elasticity, yield stress of the steel and the limit of proportionality were found as E = 2 x 10^5 N/mm^2, σ_y = 247.2 N/mm^2 and σ_p = 187 N/mm^2, respectively. The critical stresses, σ_{xx} and σ_{yy} were calculated with factor of safety, m = 1.68 (as per IS : 800- 1962) and the stress-strain curves were plotted for both strong axis and weak axis of the section. The recommendations and the conclusions were reported in above cited reference.

For axial loading, compressive force is to be axial through the centroid and parallel to the longitudinal axis of the test specimen. Four strain gauges were affixed at the four corners as shown in Fig. 5.6 to check the alignment. The strain-gauge bridge readings were noted (which should be within five per cent). Then, it is assumed that the application of load is through the centroid and parallel to the vertical axis

The residual stresses present across the cross-section and their effects have been described and discussed in subsequent-section. From above equations, the value of critical slenderness ratio $\left[\left(\frac{l}{r}\right) = \pi\left(\frac{E}{\sigma_p}\right)^{1/2} = \pi\left(\frac{2.1 \times 10^5}{187}\right) = 105\right]$ representing the point of transition from elastic buckling to inelastic buckling was found as 105. The value of E was found as 2.1 x 10^5 N/mm^2 and σ_p as 187 N/mm^2.

5.7 SLENDER (LONG) COLUMNS AND COMPRESSION MEMBERS

Theory of elastic stability of perfectly straight columns hinged at both ends is sometimes referred to as the Euler's Theory. This theory explains to some extent the behaviour slender (i.e. long) columns and compression members. The compressive strength of practical columns are significantly influenced by (i) The initial imperfections, (ii) eccentricity of loading, (iii) the presence of residual stresses, (iv) lack of well defined yielding point and (v) strain hardening. These factors are often clubbed under the general title of imperfections. These columns generally fail by inelastic buckling and do not conform to the assumptions made in Euler's theory. Very slender columns fail at loads which are close to the elastic critical loads. Whereas, the columns of intermediate slenderness collapse at loads someway between the elastic critical loads and squash loads (i.e. at loads at which the columns yield).

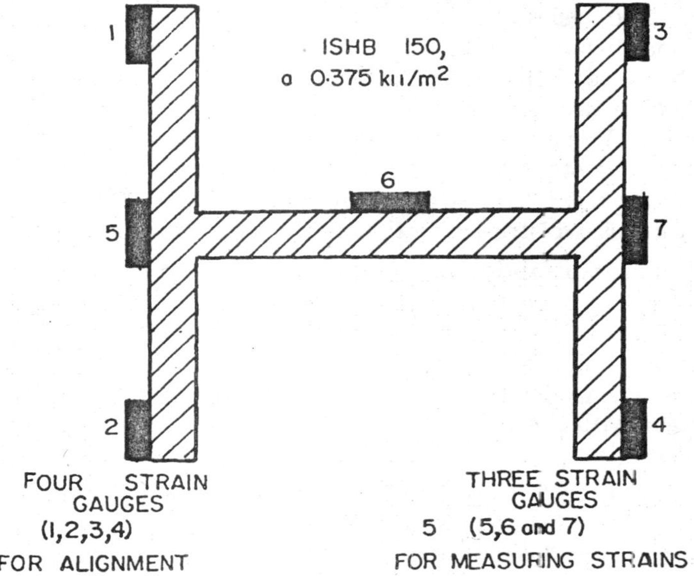

Fig. 5.6

The design of columns is based on the concept of one or more column curves.

Columns are usually an integral part of a structure and as such, their behaviour is not independent.

5.8 RESIDUAL STRESSES IN HOT-ROLLED STEEL SECTIONS

The Column Research Council (now, it is known as Structural Stability Research Council) has made significant contribution for the development of column strength curves including the effect of residual stresses. [Reference : Bruce G. Johnston, ed `Structural Stability Research Council, Guide to Stability Design Criteria for Metal Structures, 3rd edition, New York, John Wiley & Sons, Inc, 1976]. The logic behind this SSRC basic column strength is essentially as presented by A.W. Huber and L.S. Beedle [Reference : `Residual Stresses and the compressive strength of steel', Welding Journal (December, 1954), pp 589-614]. There are two general methods for determining the column strength. Analytically the distribution of residual stresses are found along with the stress-strain diagram for the material. An alternate method is to obtain experimentally an average stress- strain relationship from a short length of rolled steel shape containing the residual stress. Lynn S. Beedle and Lambert Tall [`Basic Column Strength' Journal of Structural Division, ASCE, 1986, ST7 (July, 1960, pp 139-173].

In general, the effect of residual stresses in a steel member is two fold. Firstly, the residual stresses cause an initiation of yield at loads lower than predicted by usual stress analysis. Secondly, the residual stresses may lower the ultimate capacity by inducting either local or general buckling of a compression member.

Eqs. 5.3, 5.4 and 5.5 are the expressions for the critical stresses about strong axis and weak axis of the steel column sections.

The **residual stresses** develop in the hot rolled structural steel sections due to differential cooling. The residual stresses result from cooling after hot rolling, welding or fabrication operations such as plane cutting, cold bending or combining. The residual stresses develop in a member after it has been formed into its finished shape. The flange tips and interior web parts being thinner and more exposed, always cool more quickly than the other parts. The junctions of flange and web are the thickest and most protected parts and always cool last. Residual tensile stresses develop in those parts of the section that cool last as the metal tends to contract but it is restrained by colder metal. These residual tensile stresses produce residual compressive stresses in the adjacent metal which has cooled earlier.

Earlier cooled material tends to shrink first. However, it continues to remain integral part of the rest of the cross- section. The thicker elements also cool off and shrink. Thicker elements remain composite part with the elements at the edges. The differential shrinkages induce compression at the outer edges. The cross-section remains in equilibrium. Therefore, the compressive stresses are balanced by the tensile stresses at the inner locations (i.e., near the junction of flanges with the web). The residual stresses developed across the cross-section are shown in Fig. 5.19.

Compressive stress, (σ_c), then, the nett stress-distribution across the cross-section becomes non-uniform due to the presence of the residual stresses. At the outer edges of the cross-section the total stress becomes (σ_{cc} + σ_{rt}) and at the inner elements, the total stresses becomes (σ_{cc} + σ_{rt}). In case, the total stress in compression, no where, has reached yield, the section continues to deform elastically. For the gradually increasing load, the flange tips attain yield first. When the gradual load is further increased, the yield spreads inward and the junction of the flange and the web yields.

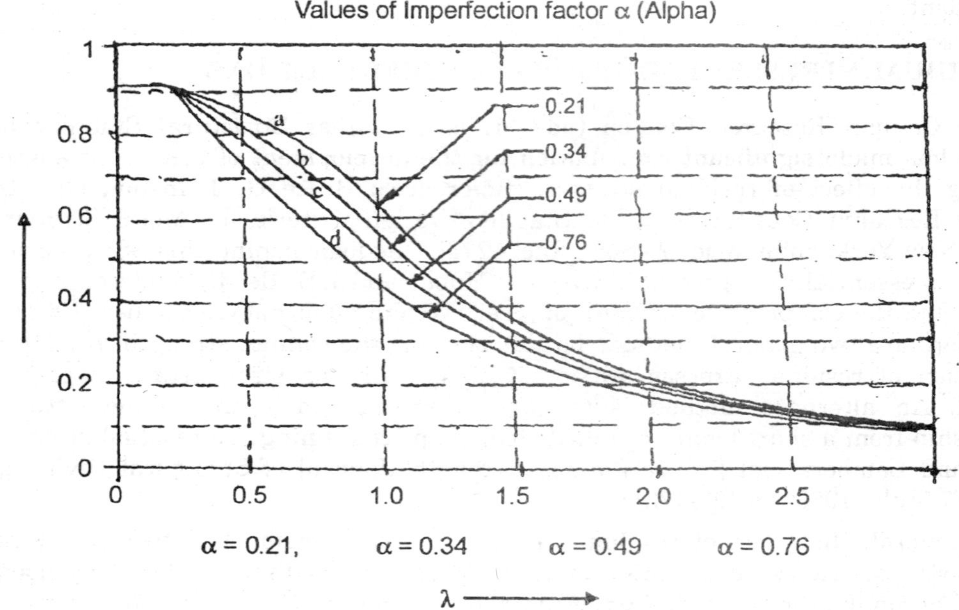

Fig. 5.7 Column Buckling Curves

Studies on columns of various types were carried out by the European Community. This community recommended that a family of design curves may be adopted Four column curves as shown in Fig. 5.6 (B) have been advised in British and European codes for the different types of sections, which are commonly used as compression members.

In these curves,

$\eta = \alpha\ (\lambda - \lambda_o)$, where η a parameter, which permits the effect of lack of straightness and eccentricity of loading.

$$\lambda_o = \pi\ (E / \sigma_y)^{1/2}$$

The values of α have been depending upon various sections.

Note : These four buckling curves a, b, c and d are for the different values of α [0.21, 0.34, 0.49 and 0.76, respectively.

The classification of different sections under different class a, b, c or d has been done as mentioned in Table 5.1.

Table 5.1. BUCKLING CURVES FOR COLUMN SECTIONS

Sections	Buckling about axis	Buckling curve
1. Rolled I-sec		
A. $(h/b_f) > 1.2$		
$(t_f < 40$ mm)	xx-axis	(a)
	yy-axis	(b)
	$(40$ mm $< t_f < 100$ mm)	
	xx-axis	(b)
	yy-axis	(c)
B. $(h/b_f) < 1.2$		
$(t_f < 100$ mm)	xx-axis	(b)
	yy-axis	(c)
$(t_f > 100$ mm)	xx-axis	(d)
	yy-axis	(d)
2. Channel, angle tee and solid-section		
	Any	(c)
3. Built-up section	Any	(c)
4. Hollow-section Circular, square, rectangular		
(A) Hot-rolled	Any	(a)
(B) Cold-formed	Any	(b)
5. Welded I-sec		
(A) No additional plate	xx-axis	(b)
$(t_f < 4.0$ mm)	yy-axis	(c)
$(t_f >$ mm)	xx-axis	(c)
	yy-axis	(d)
(B) Additionally/welded plate	xx-axis	(b)

(t_f < 40 mm)	yy-axis	(c)
(t_f > 40 mm)	xx-axis	(c)
	yy-axis	(d)
6. Welded Box sec		
(A) Generally	Any	(b)
(B) Thickwelds and (b_f/t_f) < 30	xx-axis	(c)
(d_w/t_w) < 30	yy-axis	(c)

Author of this book, conducted stub-column tests on Indian Standard H-beams (M.E Thesis, University of Jodhpur, Jodhpur in August, 1970). His paper along with Dr. S. Divakaran on this title was published in the Journal of the Institution of Engineers (India) [Vol. 51, No. 11, Part CI 6, July, 1971]. It has been reported that (σ_y) = 247.2 N/mm^2 and (σ_p) = 187.0 N.mm^2. Hence, the magnitude of the residual stress [$\sigma_r = (\sigma_y - \sigma_p)$]. i.e. (247.2 − 187.0) = 60.2 N/mm^2 present in H-section were found.

The residual stresses present in the rolled steel sections are the primary cause for the non-linear portion of the average stress-strain curve for the axially loaded compression member [Reference: A.W. Huber and L.S. Beedle, `Residual stress and the compressive' strength of steel' pp. 589-614]. The initial curvature and the accidental eccentricity have a relatively minor effect. The residual stresses are essentially independent of yield stress. The residual stresses depend on the cross-sectional dimensions and configuration (shape). These factors control the cooling rate [Reference: C.H. Yang, L.S. Beedle and B.G. Johnston, `Residual stresses and the Yielding Strength of Steel Beams Welding Journal (April, 1952), pp. 205-229].

The welding also produces residual stresses as the hot metal and flame cutting in the vicinity is restrained from contraction by the cooler-to-cold surrounding metal.

Under the ordinary conditions, those residual stresses which result from uneven cooling and welding are the most important. Actually, the important residual residual stresses due to welding are really the result of uneven cooling. The effect of residual stresses on compression structural members is summarized by Huber and Beedle.

[A.W. Huber and L.S. Beedle, `Residual Stresses and the Compressive Strength of Steel', Welding Journal (December, 1954) pp 589-614] as well as Beedle and Tall [Reference : Basic Column Strength' Journal of Structural Division, ASCE, 86, ST 7, (July , 1960) pp 139-173]. The mechanism of residual stresses due to cooling has been discussed in Welding Handbook [Reference : Vol. 1, American Welding Society, 7th edition, 1976, Chapter 6].

Welding of the built-up shapes is an even greater contributor to the residual stresses than cooling of hot-rolled H-shapes [References : N.R. Nagaraja Rao, F.R. Estuar and L-Tall `Residual Stresses in Welded Shapes', Welding Journal, (July, 1964), pp. 295-306]. Plate themselves generally have small residual stresses because of relatively uniform cooling.

5.9 EFFECTIVE LENGTH OF COMPRESSION MEMBER

Effective defined as restraint that produces sufficient resistance to prevent deformation in the lateral direction. The effective length of a compression member depends upon end restraint conditions (i.e. the rotational and relative translational conditions). A restraint should have sufficient strength and stiffness to forbid movement of the restraint point in position or direction as appropriate. The end restraint conditions are of two types as given below :

1. Position restraint. 2. Direction restraint.

1. Position restraint. In position restraint end of the column is not free to change its position but rotation about the end of the column can take place (e.g., hinged end of column as shown in Fig. 5.8(A)).

2. Direction re-straint. In direction restraint, end of the column is free to change its position but rotation about the end of the column cannot take place.

End conditions: The end conditions of the columns influence their effective lengths which influence slenderness ratios and hence the values of compressive stresses.

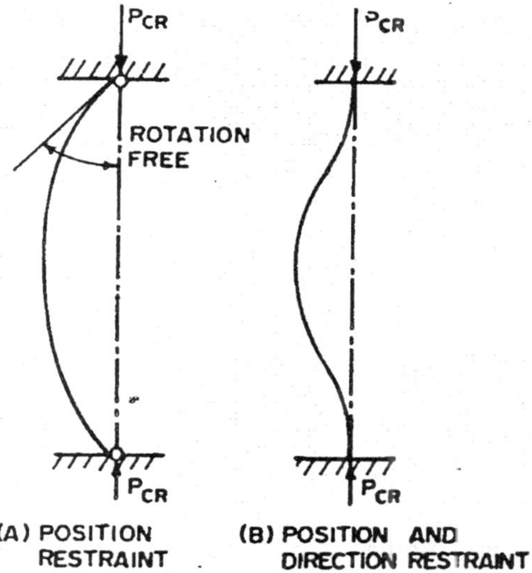

Fig. 5.8(A).

When an end of a column is having restraint in position and direction both then end is not free to change its position, and the rotation about the end of the column also cannot take place as shown in Fig. 5.8(B).

There are various possible combinations of restraints about either or both axes. The restraint conditions at the two ends of a column may be different or may be same. Following are the ideal cases of the end conditions :

1. Both ends of column hinged, 2. Both ends of column fixed, 3. One end of column fixed and the other end hinged, 4. One end of column fixed and the other end free.

The deflected shapes of columns under critical loads have been shown in Figs. 5.8 (A), (B), (C) and (D), respectively. The actual lengths (i.e. length from centre to centre of intersections with supporting members in the plane of the buckling deformation or in the case of a member with free end the free standing length from the centre of interesting member at the supported end) have been indicated by 'L'. Cases one (i.e., both ends of the column hinged has been considered as standard case. The effective length (l) of a column is expressed in terms of **equivalent length** of compression members, hinged at both the ends.

It is the length of column between two adjacents points of zero moments, and it is represented by `l'. Where, (l = k. L) is the effective length and the factor, k is formed as factor for effective length. It is also called as **unsupported length**.

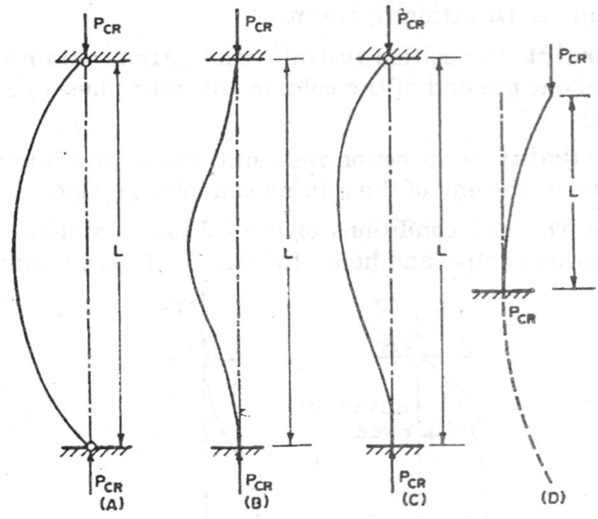

Fig. 5.8(B).

Effective length: Actual length of a member between the points of effective restraint and free and multiplied by a factor to take account of the end conditions in buckling strength calculations.

The ideal end conditions cannot be achieved in actual practice. The effective length of a compression member is adopted as per Table 5.1 as recommended by BIS in IS : 800-1984 for different types of compression members. The effective length as given in Table 6.1 will be adequate in most of the cases. The effective length as given in this table may also be adopted where the columns directly form part of the frame structures.

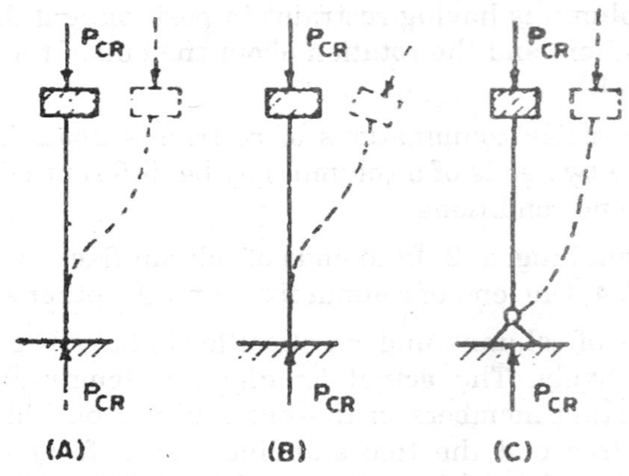

Fig. 5.9.

TABLE 5.2. EFFECTIVE LENGTH OF COMPRESSION MEMBERS BS : 5950 Part I recommends the following effective length for design purpose

S.N.	Type	Effective length of member (l)
1	Effectively held in position and restrained against rotation at both ends [Fig. 5.8 (i) (B)]	0.70 L
2	Effectively held in position at both ends and restraint against rotation at one ends [Fig. 5.8 (i) (C)]	0.80 L
3	Effectively held in position at both ends but not restrained against rotation [Fig. 5.8 (i) (A)]	1.00 L
4	Effectively held in position and restrained against rotation at one end and at the other end restrained against rotation but not held in position [Fig. 5.9 (ii) (A)]	1.00 L
5	Effectively held in position and restrained against rotation at one end and at the other end partially restrained against rotation but not held in position	1.2 L
6	Effectively held in position at one end but not restrained against rotation, at the other end restrained against rotation but not held in position.	2.00 L
7	Effectively held in position and restrained against rotation at one end but not held in position [Fig. 5.8 (i) (D)] or restrained against rotation at the other end.	2.0 L

Note : For battened struts, the effective length 'l' is increased by 10 per cent.

Where the exact frame analysis is not done, the effective length of columns in the frame structures may be found from the ratio of effective length of there unsupported length (l/L) from Fig. 5.10 when the relative displacements of the column is prevented (i.e. when there is no away) and from Fig. 5.11 when the relative lateral displacement of the ends is not prevented (i.e., without restraint against sway viz., the sway occurs), when sway occurs, IS : 800-1984 recommends that the effective length ratio, (l/L) may not be taken to be less than 1.2.

In Figs. 5.10 and 5.11 (iii) and (iv)

$$\beta_1 = \left(\frac{\Sigma k_c}{\Sigma k_c + \Sigma k_b}\right) \quad \text{and} \quad \beta_2 = \left(\frac{\Sigma k_c}{\Sigma k_c + \Sigma k_b}\right)$$

where, the summation is to be done for the members framing into a joint at top and bottom, respectively, and

k_c = flexural stiffness of the column, and k_b = flexural stiffness of the beam

Figs. 5.10 and 5.11 are from the paper titled as Effective Length of columns in multistorey buildings by Professor R.H. Wood, published in the Structural Engineer Vol. 52, No. 7 July 1974.

It is worthwhile to note that IS : 800-1984 'code of practice for general construction in steel' and IS : 456-1978 'code of practice for plain and reinforced concrete' have recommended the same effective length for the column with similar support conditions.

The cross-sectional area calculated from the specified size of the member or part there of or read from appropriate table slenderness ratio is a major factor which influences ultimate strength of an axial loaded column.

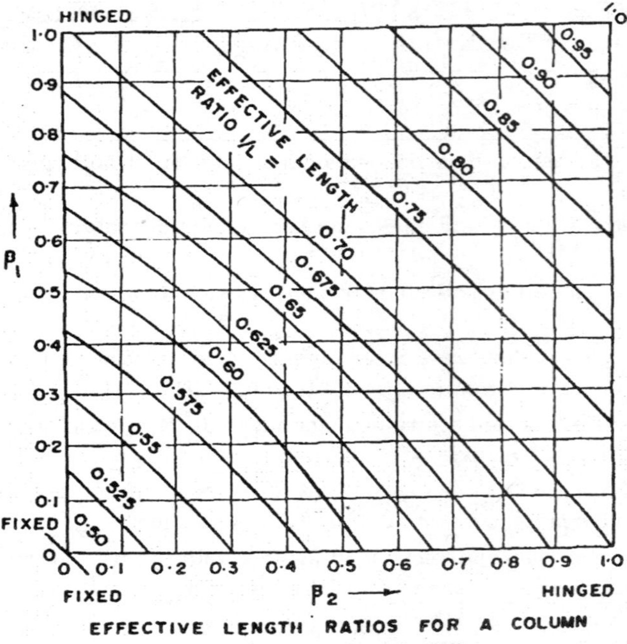

Fig. 5.10.

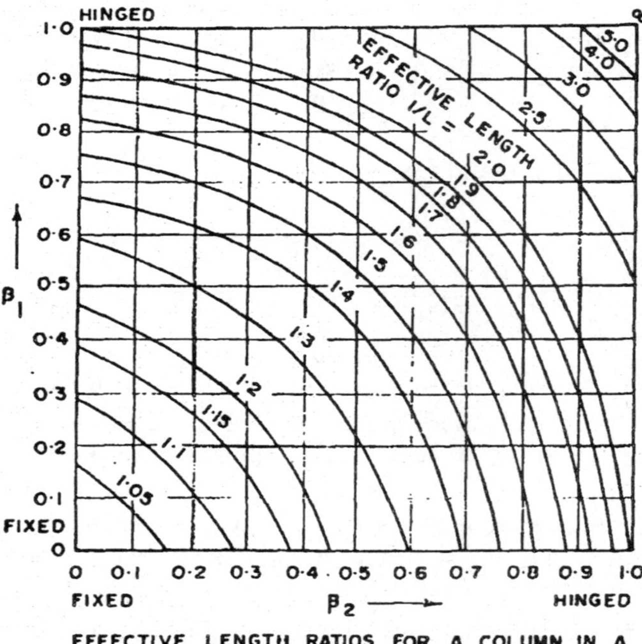

Fig. 5.11.

5.10 EFFECTIVE SECTIONAL AREA (OR CROSS-SECTIONAL AREA)

The gross cross-sectional area is the area as calculated from the specified size of the member or part thereof.

The effective sectional area of a compression member is the gross cross-sectional area of the member. The deduction is not made for members connected by rivets, bolts and pins. If the holes are not filled by the fastening material, modified when the ratio of the outstand to thickness exceeds the limits specified by BIS. This point has been discussed further in Article 6.20. The deduction is also made from the gross cross-sectional area for excessive effective plate width (if any) to determine the effective sectional area.

The usual approach for design consists of reducing the actual case under consideration to an equivalent pinned-end case by means of a factor for the effective length. It is found from comparison of elastic critical loads. This process assumes that the influence of imperfections will be broadly similar for all forms of restraints being the function of effective slenderness only.

5.11 RADIUS OF GYRATION OF CROSS-SECTIONAL SHAPE

The radius of gyration of a section is a geometrical property of the section and it is denoted by r

$$r = (I/A)^{1/2}$$

where I = moment of inertia of the section about the axis.

r = radius of gyration of the section about the axis.

A = effective sectional area of the section.

Along with the cross-sectional shapes of various columns used in practice as shown in Fig. 5.22, approximate radii of gyration for sections are given. In the cases of the rectangular and circular cross-sections, the values indicated are closely approximate to the correct values, but for the built up sections, there may be considerable fluctuation because of the variation in relative cross-sectional dimensions.

5.12 SLENDERNESS RATIO OF COMPRESSION MEMBERS

The slenderness ratio of a compression member is defined as ratio of effective length of compression member (l) to appropriate radius of gyration (r) of the cross-section about the axis under consideration.

Slenderness ratio is a major factor which influences ultimate strength of an axially loaded column.

Slenderness ratio, $\lambda = l/r$.

The radii of gyration about various axes of rolled steel sections can be obtained from structural steel section tables (ISI Handbook No. 1). The minimum radius of gyration is used for computing the maximum slenderness ratio. For built-up compression members, value of radius of gyration is calculated. The slenderness ratio for a compression member should be as small as possible so that the material may be stressed to its greatest possible limit. The maximum slenderness ratio of compression members should not exceed the values given in Table 5.3. These limits have been laid down by BIS in IS : 800-1984.

The end restraints of columns are often different in the two principal planes. The different moments of inertia of the column cross-section in these planes are sometimes desirable to achieve approximately equal slenderness ratios. The intermediate supports are provided to the columns for this purpose.

TABLE 5.3. MAXIMUM SLENDERNESS RATIO OF COMPRESSION MEMBER $\lambda = l/r$

S.N.	Type of member	$\lambda = l/r$
1	A member carrying compressive loads resulting from dead loads and imposed loads	180
2	A member subjected to compression forces resulting from wind/earthquake forces provided that the deformation of such members does not adversely affect the stress in any part of the structure	250
3	A member normally acting as a tie in a roof truss or a bracing system but subjected to possible reversal of stress resulting from the action of wind or earthquake forces.	350

The intermediate supports reduce the unsupported length of the columns. When the unsupported lengths of columns are reduced, then, the smaller sections may be used at a higher average stress.

Sometimes, the intermediate supports are furnished only in one direction. For example, a rolled steel I-section column is having its continuous length upto two storeys. At the level of one storey, intermediate support is provided by connecting beams with the web. The radius of gyration, r_{yy} of the section, about yy-axis (axis parallel to the web) is much smaller than the radius of gyration r_{xx} of the section, about xx-axis. By providing the intermediate support, the effective lengths of the column become different in two different direction. The effective length of columns, l_{yy} for bending about yy-axis is found by considering the length of column between one-storey only. The effective length of column l_{xx} for bending about xx-axis is found by considering the length of column between two storeys. It is seen that the effective length of column l_{yy} is much smaller than that of l_{xx}. The values of slenderness ratio (l_{yy}/r_{yy}) and (l_{xx}/r_{xx}) about two directions may be made approximately equal. As such the use of sections with different values of radii of gyration in two directions may be made economical. When a column is subjected to different bending moments in two directions then, the greater value of r may be kept in the direction of greater moment. The intermediate supports in the weak direction make the use of I-section and channel-section economical.

5.13 POSSIBLE MODES OF FAILURE FOR AXIALLY LOADED COLUMNS

The possible modes of failure for axially loaded columns depend on its physical weakness to bending. These are described as under :

Primary (over all) lateral buckling

Primary lateral buckling (i.e., overall flexural buckling failure) occurs by excessive deflection in the plane of the weaker principal axis, (in case, otherwise the column is restraint). **Buckling** is characterized by deformations developed in a direction or plane normal to that of loading. It is bending without twisting. That is, the sections displace from their original position by translation without rotation. It occurs mainly in the members subjected to compressive force. The load at which buckling occurs becomes progressively less useful as the column's slenderness increases. Overall buckling of the members may occur in one of the three different modes. These are as follows: 1. Flexural (bending) buckling, 2. Torsional buckling, 3. Flexural torsional buckling.

5.14. FLEXURAL BUCKLING

Flexural buckling (bending about principal axis without twisting, that is, the section deforms from the original position by translation rotation) occurs in the members with doubly symmetric or doubly antisymmetric cross-sections (e.g, I-sections or Z-sections) and in the members with singly symmetric sections (e.g, channels, tees, equal legged angles, double angle sections. These sections buckles about an axis, that is perpendicular to the axis of symmetry.

In case, the bending stiffness of a member is high, its resistance to buckling reduces as the length of the member increases (provided the cross-sectional area remains constant). It is of two types as mentioned above : (i) Elastic buckling (ii) Inelastic buckling

(i) Elastic buckling : The failure column in bending due to excessive deflection prior to its any portion yields. The practical use of the term buckling is that it is the boundary between stable equilibrium and unstable equilibrium of the compression member rather than the instantaneous condition that occurs in the isolated slender elastic bar.

Elastic flexural buckling stress is given by Euler's expression

$$\sigma_E = \left(\frac{\pi^2 . E}{(KL/r)}\right)$$

(ii) Inelastic buckling : In general, the majority of column fail when the failure occurs by buckling after any portion of the cross-section has yielded. That is known as inelastic buckling.

As a general approach Euler's elastic buckling controls the strength for large slenderness ratios and yield stress, σ_y is used for short columns. A transition curve is used for inelastic buckling.

5.15. TORSIONAL BUCKLING

Torsional buckling occurs in members with doubly symmetric sections such as cruciform or built-up shapes with very thin walls (such as, angles, tees and channels with relatively low torsional stiffness, which may under axial compression buckle torsionally while the longitudinal axis remains straight. It occurs when the torsional rigidity of the member is appreciably smaller than its bending rigidity.

Consider a doubly symmetric section in the shape and cross-section as shown in Fig. 5.12. In this section the shear centre and the centroid coincide. The member is of cruciform section with four identical flanges of width, b and thicknes, t Axes x and y are the symmetric axes of the section. Torsion buckling occurs due to compression. Longitudinal axis of the member straight, while each flange buckles by rotating about z-axis. Consider buckling the deflection of the flanges during column is pinned at its both the ends. This column cross-section is shown in Fig. 5.13 in plan.

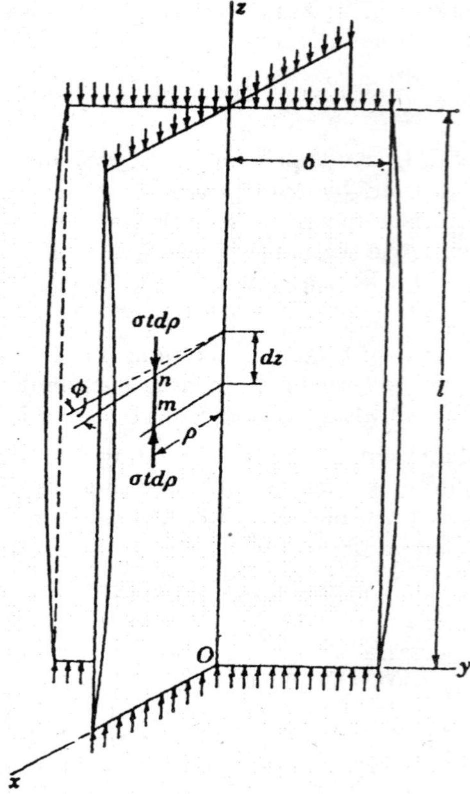

Fig. 5.12. Torsional buckling

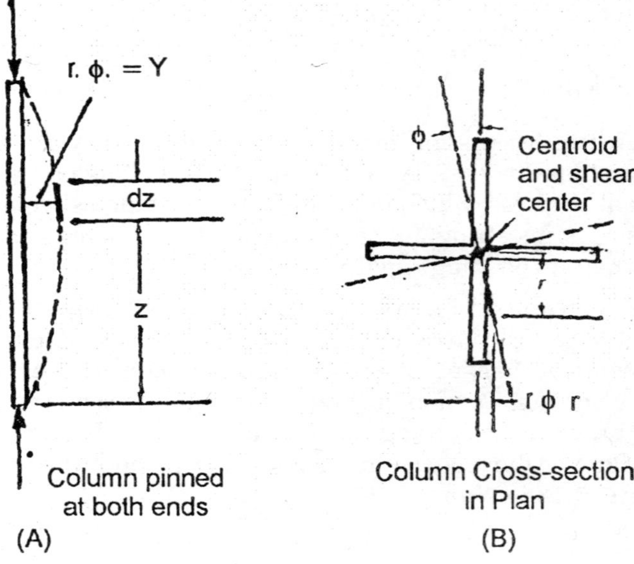

Fig. 5.13.

Consider a pin-ended compression member. The member is initially straight and subjected to the centrally applied force, P force, P reaches its critical value and the member may have a slightly deflected shape and it is in equilibrium. Due to this deflection, the bending stresses super-impose on the initial uniformly distributed compressive stresses. The initial compressive stresses are acting on slightly rotated cross-section. Euler's expression is

$$EI \frac{d^2y}{dz^2} + P.y = 0 \qquad \text{(i)}$$

Differentiating the expression (i) twice,

$$EI \left(\frac{d^4y}{dz^4}\right) = -P \left(\frac{d^2y}{dz^2}\right) \qquad \text{(ii)}$$

EI. (d^4y/dz^4) is the rate of loading. The Euler's column may be considered as a beam lateraly loaded with the arbitrary loading – P. (d^2y/dz^2). This expression may be used for calculating the critical value of the compressive force.

It may be stated that at the critical state, the buckled form of equilibrium is sustained by the compressive stresses acting on rotated cross-section of the member.

Consider an element in the form of a thin strip of length, dz, located at a distance, r from z-axis and having cross-sectional area t.dr. Owing to torsional buckling, the deflection of this element in the y-direction is

$$= r\,\phi \qquad \text{(iii)}$$

where, ϕ is the small angle of twist of the cross-section. (It is assumed that the shape of the cross-section does not change during twisting). The compressive forces are statically equivalent to lateral load of intensity

$$-(\sigma_z . t . dr)\left(\frac{d^2\phi}{dz^2}\right) \qquad \text{(iii)}$$

The increment of torsional moment about the z-axis tributory to the length length, dz equals the load *times* the moment arm, r, thus

$$dm_z = -\,(\sigma_z . dr . t . r)\left(r\frac{d^2\phi}{dz^2}\right) dz \qquad \text{(iv)}$$

Total torsional moment for the slice, dz of the column is

$$m_z = -\left(\sigma_z \frac{d^2\phi}{dz^2} . dz\right)\left(r^2 t . dr\right) \qquad \text{(v)}$$

It shows the contribution to the torsional moment, M_z tributory to the element, dz.

$$dM_z = m_z.dz \qquad \text{(vi)}$$

Differential equation for the torsion on I-shaped section is

$$M_z = G.J \frac{d\phi}{dz} - E.C_w \frac{d^3\phi}{dz^3} \qquad \text{(vii)}$$

Differentiating with respect to Z

$$\left(\frac{dM_z}{dz}\right) = G.J.\left(\frac{d^2\phi}{dz^2}\right) - E.C_w\left(\frac{d^4\phi}{dz^4}\right) \quad \text{(viii)}$$

Positive moment M_z at the section, z gives a clockwise rotation. Where as at the section, Z, there as at the section, z, there is an anticlockwise rotation. Therefore, the relationship (vi) requires a minus sign. Differential equation for torsional buckling is then

$$\sigma_z\left(\frac{d^2\phi}{dz^2}\right)\int_A r^2.t.dr = G.J\left(\frac{d^2\phi}{dz^2}\right) - E.C_w\left(\frac{d^4\phi}{dz^4}\right) \quad \text{(ix)}$$

$$E.C_w\left(\frac{d^4\phi}{dz^4}\right) - (G.J - \sigma_z)\int_A r^2.t.dr.\left(\frac{d^2\phi}{dz^2}\right) = 0 \quad \text{(x)}$$

In which $\int_A r^2t.\ dr = I_p$, the polar moment of inertia about the shear centre when the centroid of the section coincides with the shear centre, the expression (x) alone determines the buckling condition.

It is to note that the warping rigidity, $E.C_w$ is zero for shapes consisting of thin rectangular elements intersecting at a common point. Therefore, the expression (x) may be written as below $\left(\therefore \int_A r^2.t.dr = I_p\right)$.

$$E.C_w\left(\frac{d^4\phi}{dz^4}\right) + (\sigma_z - G.J).\ I_p\left(\frac{d^2\phi}{dz^2}\right) = 0$$

$$\left(\frac{d^4\phi}{dz^4}\right) + \left(\frac{\sigma_z - G.J.}{E\,C_w}\right).I_p.\left(\frac{d^2\phi}{dz^2}\right) = 0 \qquad \left(\frac{d^4\phi}{dz^4}\right) + \left(\frac{\sigma_z.I_p - G.J.I_p}{E\,C_w}\right).\left(\frac{d^2\phi}{dz^2}\right) = 0$$

$$\left(\frac{d^4\phi}{dz^4}\right) + p^2.\left(\frac{d^2\phi}{dz^2}\right) = 0 \quad \text{(xi)}$$

General solution of this differential equation is

$$\phi = A_1.\sin(pz) + A_2\cos(\phi z) + A_3 Z + A_4 \quad \text{(xii)}$$

For pin-ended column, with rotation about z-axis restrained at each and, but with warping not restricted at the ends, then

$A_2 = A_3 = A_4 = 0$

Since, A_1 cannot be zero,

$\sin(pL) = 0 \qquad (pL = n\pi)$

$$p^2 = \left(\frac{n\pi}{L}\right)^2, \quad (\text{for } n = 1)$$

Therefore,

$$\left(\frac{\pi^2}{L^2}\right) = (p^2) \cdot \left(\frac{\sigma_z I_p - G.J}{E.C_w}\right)$$

$$\sigma_z = \left(\frac{E.C_w}{I_p} \cdot \frac{\pi^2}{L^2}\right) + \left(\frac{G.J}{I_p}\right) \qquad \text{(xiii)}$$

For doubly symmetric section

The shear centre and centroid of the section (e.g, I-section and Z-shaped section coincide).

The expression (xiii) is accurate. It gives **elastic-buckling stress** for which torsional buckling stress.

For the common Sections (e.g, single angle strut, since, the distance from the centroid is small, the expression (xiii) gives a reasonable approximation for the torsional buckling.

Note: For the design purpose, the torsional buckling is usually not considered. A check may be applied to find whether an equivalent radius of gyration is less than r_x and r_y.

In the expression (xiii), from Euler's expression

$\sigma_z = \pi^2 E/(L/r_E)^2$, then, the expression, for equivalent radius of gyration, r_e may be defined.

$$\frac{\pi^2 E}{(L/r_E)^2} = \left(\frac{E.C_w.\pi^2}{I_p.L^2} + \frac{G.J}{I_p}\right)$$

$$r_E^2 = \left(\frac{E.C_w.\pi^2}{I_p.L^2.\pi^2 E}.L^2 + \frac{G.J.L^2}{E.I_p.\pi^2}\right)$$

$$r_E^2 = \left(\frac{C_w}{I_p} + \frac{G.J.L^2}{E.I_p.\pi^2}\right)$$

$$r_E = \left(\frac{C_w}{I_p} + \frac{G.J.L^2}{E.Ip.\pi^2}\right)^{1/2} \qquad \text{(xiv)}$$

For steel (E/G) = 2.6, substituting in the expression (xiv)

$$r_E = \left(\frac{C_w}{I_p} + \frac{0.04.J.L^2}{I_p}\right)^{1/2} \qquad \text{(xv)}$$

For doubly symmetric sections, it is seen that only for short lengths, r_E shall be less than r_x and r_y.

For the sections, which are symmetrical to only about y-axis

$$\left(\frac{1}{r_e^2}\right) = \left(\frac{1}{2\,r_E^2}\right) + \left(\frac{1}{2\,r_y^2}\right) + \left[\left(\frac{1}{2\,r_E^2} - \frac{1}{2\,r_y^2}\right) + \left(\frac{y_o}{r_E \cdot r_y \cdot r_p}\right)\right]^{1/2} \qquad \text{(xvi)}$$

where, r_e = Equivalent radius of gyration for torsional-flexural buckling

r_E = Equivalent radius of gyration for torsional buckling as in the expression (xv)

r_y = radius of gyration for axis of symmetry

r_p = Polar radius of gyration

I_p = Polar moment of inertia.

y_o = distance from the centre of gravity of cross-section to its shear centre.

This type of failure occurs by twisting about shear centre, about the longitudinal axis. For fabricated sections of normal proportions and for hot-rolled steel-sections, this type of failure is unlikely. However, in light gauge cold formed members (particularly unsymmetrical shapes) may fail in torsional buckling.

Thin walled sections, with open cross-sectional shape unlike box or thick walled sections are sometimes weak in torsion rather than bending. Torsional buckling occurs when the torsional rigidity of the member is appreciably smaller than its bending rigidity. [Timoshenkos' `Theory of Bending, Torsion and Buckling of Thin-walled members of Open Cross-Sections, J. Franklin Inst. 239, No S3, 4 and 5, 1945].

In addition to bending and torsional modes of buckling, some sections fail in a combined torsion and bending mode of buckling.

5.16 COMBINED TORSIONAL FLEXURAL BUCKLING

In the primary (i.e., **flexural buckling**), columns and compression members bend in the plane of one of the principal axes without twisting (i.e., in a plane of symmetry of the cross-section). The column-sections displace from their original geometry by translation without rotation.

Thin-walled members with open cross-sectional shape are weak in torsion. Angle sections (which are unsymmetrical to both the axes), and tee-sections and channel-sections which are symmetrical/unsymmetrical about one axis. In such sections, the shear centre and the centroid do not coincide. These sections have low torsional stiffness, and buckle by twisting rather than by bending. The torsional rigidity of the column is appreciably smaller than bending rigidity. [Reference : Timoshenko's, Theory of Bending, Torsion and Buckling of Thin-walled members of Open cross-sections' Franklin. Inst. 239, Nos. : 3, 4 and 5, (1945)].

Singly symmetrical/unsymmetric sections fail in **combined torsion and flexural buckling**. Because of the bent shape of the column axis the vertical load does not remain, perpendicular to the cross-sectional plane. There are a normal and a shear component.

This centroidal shear component does not pass through shear-centre. It results in translation of the shear centre. The complete phenomenon (which one is referred as **combined torsional flexural buckling**) occurs.

The critical load for the columns which buckle by twisting or by combined twisting and bending may be found. The critical load, P_{cr} for such a combined mode of buckling is smaller than the Euler's critical load, $P_{E.xx}$ for the lateral buckling about xx-axis. The difference

between P_{cr} and $P_{E.xx}$ may be small for long columns with adequate torsional rigidity. But for torsionally weak columns of the intermediate length, this difference is appreciable.

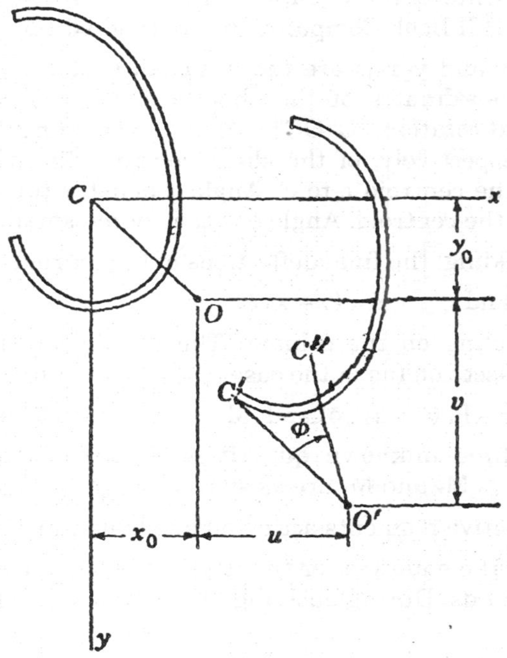

Fig. 5.14A. Combined Flexural-Torsional Buckling

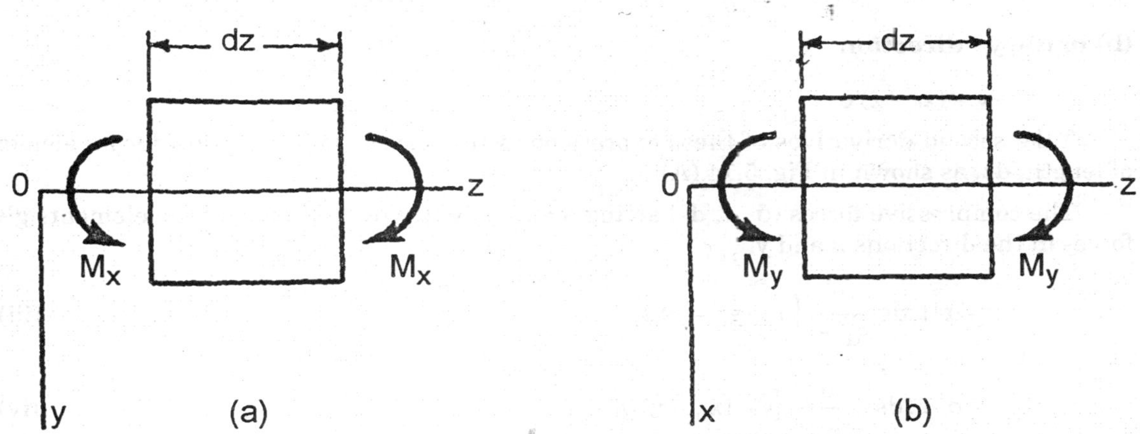

Fig. 5.14B. Sign convention for bending moments

An unsymmetrical cross-section as shown in Fig. 5.14 A may be considered Professor Wagner. H obtained the solution for torsional and flexural buckling combined. Three critical values of Euler's axial loads (viz, the bending about principal axes x and y and twisting about the longitudinal axis are needed. Therefore, three differential equations [viz, (i) buckling by lateral bending about x-axis, (ii) buckling by lateral bending about y-axis and (iii) twisting about the shear centre) are considered. These are interdependent. Three simultaneous differential equations may be solved to obtain the critical buckling loads for the columns by making the use of an energy principle (which uses the Rayleigh-Ritz energy method) or by integrating the governing differential equations [Ref.: Timoshenko, S.P. & Gere, `Theory of Elastic Stability, McGraw-Hill Book Company, Inc., Second-edition, 1961 (pp. 230-236)].

In Fig. 5.14(A), x-axis and y-axis are the principal centroidal axes of the cross-section and x_o and y_o are the co-ordinates of the shear centre, o. The cross-section undergoes translation (OO' = CC') and rotation, (C' OC" = ϕ) due to buckling. The deflections u and v in the x and y directions, respectively, of the shear centre define the translation. The shear centre o moves to o' and the centroid c to c'. Angle ϕ denotes the rotation (i.e, twisting). C" shows the final position of the centroid. Angle ϕ is considered small.

Therefore, during buckling, the final deflections of th centroid

$$x = (u + y_o \, . \, \phi) \text{ and }, \qquad Y = (v - x_o \, \phi)$$

Only the load P is acting on the column. The bending moments with respect to the principle axes at any cross-section (as in the case of the column pinned at both the ends),

$$M_x = (-P.\, Y) = -P.\,(v - x_o \, . \, \phi) \text{ and } M_y = (-P.X) = -P\,(u + y_o\, \phi) \qquad \text{(iii)}$$

where, u, v and ϕ are three unknown quantities in these, expressions (i) and (ii) sign conventions for the moments M_x and M_y are as shown in Fig. 5.14(B).

Fig. 5.14(B) shows positive moments acting on an elementary length, d_z of the column.

In order to determine the equation for the angle of twist, ϕ, consider a longitudinal strip of the column cross-section t.ds. During buckling, the components of its deflection

(a) in the x-direction

$$= u + (y_o - y)\, \phi \qquad \text{and}$$

(b) in the y - direction

$$v - (x_o - x)\, \phi$$

Take second derivatives of these expressions with respect to Z and consider an element of length, dz, as shown in Fig. 5.14 (A).

The compressive forces (σ . t. ds) acting on the slightly rotated ends of the elements give forces in the directions x and y.

$$-\sigma.\text{t.ds}.\frac{d^2}{dz^2}\left[u + (y_o - y)\,\phi\right] \qquad \text{(iii)}$$

$$-\sigma.\text{t.ds}\,.\,\frac{d^2}{dz^2}\left[v - (x_o - x)\,\phi\right] \qquad \text{(iv)}$$

A third equation is obtained by considering the twisting of the column.

Consider the moment of above forces about the axis of the shear centre, torque (torsion) per unit length of the bar for the one longitudinal strip is determined.

$$dm_z = -(\sigma.\text{t.ds})\,(y_o - y)\left[\frac{d^2u}{dz^2} + (y_o - y)\frac{d^2\phi}{dz^2}\right] + (\sigma.\text{t.ds})\,(x_o - x)\left[\frac{d^2v}{dz^2} - (x_o - x)\frac{d^2\phi}{dz^2}\right] \qquad \text{(v)}$$

Integrating the expression (v) over the complete cross-sectional area, it is seen that

$$\sigma \int_A t.ds = P, \int_A x.t.ds = \int_A y.t.ds = 0$$

Since, $I_x = \int_A y^2 \, t\,ds$ and $I_y = \int_A x^2 \, tds$

and $I_o = I_x + I_y + A\left(x_o^2 + y_o^z\right)$

Now, it may be stated that at the critical condition, the buckled form of equilibrium is sustained by the compressive stresses acting on the rotated cross-section of the longitudinal element. The element in the form of a thin strip of length dz, located at a distance, r, from the z-axis and having cross-sectional area, t-dr. Due to torsional buckling, the deflection of this element in the y-direction is v = (r – ϕ), where ϕ is the small angle of twist. It is to note that the geometry of the cross-section does not change during twisting.

The compressive forces acting on the rotated ends of the element dz and σ.t.dr, where, σ = (p/A'). These compressive forces are statically equivalent to a lateral load of intensity (–σ.t.dr) (d^2v/dz^2) which may be written in the above form.

$$(-\sigma.t.r.dr)\left(d^2\phi / dz^2\right)$$

Moment about z-axis of this arbitrary lateral load acting on the element, dz is then

$$-\sigma\left(\frac{d^2\phi}{dz^2}\right) dz.t.r^2\,dr.$$

Adding the moments for the entire cross-section, the torque acting on an element of the buckled column included between two consecutive cross-sections. Hence, the torque is

$$-\sigma\left(\frac{d^2\phi}{dz^2}\right) dz \int_A t.dr.r^2 = -\sigma\left(\frac{d^2\phi}{d}\right).$$

where, I_o is the polar moment of inertia of the cross-section about the shear centre, O. Let m_z be the torque per unit length

$$m_z = -\sigma\left(\frac{d^2\phi}{dz^2}\right).I_o \quad (5.6A)$$

This expression holds for any, shape (in case, the shear centre and the centroid coincide). For a column of thin-walled open section

$$\left(\frac{dM_t}{dZ}\right) = C\left(\frac{d^2\phi}{dz^2}\right) - C_1\left(\frac{d^2\phi}{dz^4}\right) \quad (5.6B)$$

Considering the equilibrium of this element, dz

$$m_z = -\left(\frac{dM_t}{dz}\right)$$

Three simultaneous differential equations for buckling by bending about two axes and torsion are used to obtain the critical loads.

Angle of rotation appears in all the three equations to represent torsional buckling combined with bending about two axes simultaneously.

Expression for the torque is as under

$$m_z = \int_A dm_z$$

$$= P\left[x_o \frac{d^2 v}{dZ^2} - y_o \frac{d^2 u}{dZ^2}\right] - \frac{I_o}{A}.P.\frac{d^2\phi}{dZ^2} \quad (5.6C)$$

where, I_x and I_y are the principal centroidal moments of inrtias of the column cross-section and I_o is the polar moment of inertia about the shear, centre r_o is polar radius of gyration of the cross section with respect to shear centre.

$$r_o^2 = y_o^2 + x_o^2 + (I_y + I_x)/A$$

Since, the expression for the torque is as under

$$m_z = \int_o dm_z = C_1\left(\frac{d^4\phi}{dZ^4}\right) - C\left(\frac{d^2\phi}{dZ^2}\right) \quad (5.6D)$$

Positive directions of m_t and m_z are given by the right hand rule.

where, C, is the **warping rigidity** = E.C_w and C is the torsional constant = GJ

C_1 = E.C_w (C_w is the **warping constant**)

C = G.J. (Torsional constant)

where, G is the modulus of shear and J is the torsional constant

The units of C_w is in sixth power

Following expression is obtained by substituting the expression (vi) in the expression (vii) for non-uniform torsion

$$C_1\left(\frac{d^4\phi}{dZ^4}\right) - \left(C - \frac{I_o}{A}.P\right)\frac{d^2\phi}{dZ^2} - Px_o\left(\frac{d^2 v}{dZ^2}\right) + P.y_o\frac{d^2 u}{dZ^2} = 0 \quad (5.6E)$$

Expressions Eq. 5.6 (B), Eq. 5.6 (C) and Eq. 5.6 (D) are the three simultaneous differential equations for buckling due to combined bending and torsion. These expressions may be used to obtain the critical loads. It may be observed that the angle of rotation appears in all the three expressions. As such in the general case, torsional buckling and bending of the axis occur simultaneously.

Consider a general case : It is assumed that the bar has simple supports, so that, the ends of bar are free to warp and free to rotate about the x and y axes. But it cannot rotate about z-axis or deflect in the x and y directions (i.e, the displacements in the x and y directions are zero and the moments about these axes disappear at the ends of the members. In such a case, the end conditions (i.e, **boundary conditions are u = o,** v = o and ϕ = o. z = o and z = l. Therefore, **at z = o and z = l.**

$$\left(\frac{d^2u}{dZ^2}\right) = \left(\frac{d^2v}{dZ^2}\right) = \left(\frac{d^2\phi}{dZ^2}\right) = 0$$

Above stated end conditions (i.e, boundary conditions) are satisfied in case, the deflected shape is assumed, in the following form.

$$u = A_1 . \sin\left(\frac{\pi z}{l}\right), \quad v = A_2 . \sin\left(\frac{\pi z}{l}\right)$$

$$\phi = A_3, \sin\left(\frac{\pi z}{l}\right)$$

These values of u, v and ϕ are substituted in these three Eqs. 5.6 (B), 5.6 (C) and 5.6 (D) to obtain the values of the constants A_1, A_2 and A_3

$$\left(P - E I_y . \frac{\pi^2}{l^2}\right) A_1 + P. y_o A_3 = 0 \qquad (5.7)\ (A)$$

$$\left(P - E I_x . \frac{\pi^2}{l^2}\right) A_2 - P. x_o A_3 = 0 \qquad (5.7)\ (B)$$

$$P. y_o . A_1 - P. x_o A_2 - \left(C_1 \frac{\pi^2}{l^2} + C - \frac{I_o . P}{A}\right) A_3 = 0 \qquad (5.7)\ (C)$$

In case, the column remains straight in the equilibrium, then, $A_1 = A_2 = A_3 = 0$

However, when the column remains in the buckled form in the equilibrium, these constants A_1, A_2 and A_3 must not disappear (i.e, should not become zone simultaneously. It is possible only in case, the determinant of Eqs. 5.7 (A), 5.7 (B) and 5.7 (C) disappear. In order to simplify these expressions, let

$$P_{E.xx} = \left(\frac{\pi^2 E. I_{xx}}{l^2}\right) \qquad (5.8)A$$

$$P._{E.yy} = \left(\frac{\pi^2 E. I_{yy}}{l^2}\right) \qquad (5.8)B$$

$$\text{and } P_\phi = \left(\frac{A}{I_o}\right)\left(C + C_1 \frac{\pi^2}{l^2}\right) \qquad (5.8)C$$

Where, $P_{E.xx}$ and $P_{E.yy}$ are the Euler critical loads for buckling about the x and y axes, respectively and ϕ is the critical load for purely torsional buckling. Determinant representing Eqs. 5.7 (A) 5.7 (B), and 5.7(C) may be written as under and it may be equated to zero.

$$\begin{vmatrix} (P - P_{E.yy}) & 0 & P . y_o \\ 0 & (P - P_{E.xx}) & -P . x_o \\ P . y_o & -P. X_o & (P - P_\phi) \end{vmatrix} \qquad (5.9)$$

In its expanded shape, determinant reduces to the following cubic equation

$$\frac{I_o}{A} (P - P_{E.yy}) (P - P_{E.xx})(P - P_\phi) - P^2 . y_o^2 (P - P_{E.xx}) - P^2 . x_o^2 (P - P_{E.yy}) = 0$$

$$\left(\frac{I_C}{I_o}\right).P^3+\left[\frac{A}{I_o}\left(P_{E.xx}\right).y_o^2+P_{E.yy}.x_0^2\right]-\left(P_{E.xx}+P_{E.yy}+P_\phi\right)P$$

$$-\left(P_{E.xx}.P_{E.yy}+P_{E.xx}.P_\phi+P_{E.yy}.P_\phi\right)-P_{E.xx}.P_{E.yy}.P_\phi=0 \quad (5.10)$$

where, $I_c = (I_x + I_y)$ is the polar moment of inertia about the centroideof the cross-section.

This is a cubic equation in P. These are three roots of this equation which corresponds to three buckled mode shapes. Sh of three will be critical load.

In order to obtain the critical load in any particular case, the numerical values of the coefficients in Eq. 5 are calculated. Then, three values of the critical loads are found. The smallest value of the load is used in practical application as a critical load.

Substitution of three values of the critial loads in Eqs 5.8 (A), 5.8 (B) and 5.8 (C) gives the ratios of (A_1/A_3) and (A_2/A_3) for each of the corresponding three forms of buckling. The relations between rotation and translation of the cross-sections are established by these ratios. These ratios also define the deflected shape of the axis of shear centre.

Conclusion: Regarding the relative magnitudes of the critical loads from Eq. 5, an important conclusion may be obtained.

Consider the left hand side of this equation as a function f(P). In order to obtain the information concerning the values of P for which f(P) disappears, the sign of this function is found for the following different conditions.

1. For the very large values of P : The function f(P) takes the sign of the term with the highest power (i.e, P^3. I_o/A) and it is positive.

2. In case P is equal to zero : The value of f(P) is negative

$$\left(-P_{E.xx}.P_{E.yy}.P_\phi\right)\left(I_o / A\right)$$

2.(i) **In case $P_{E.xx} < P_{E.yy}$** That is, smaller Euler load corresponds to bending in the y-z plane, if $P = P_{E.xx}$ then, $f(P) = -P_{E.xx}\, x_o\, (P_{E.xx} - P_{E.yy})$ which is positive.

2(ii). **In case P = P,** Then,

$$f_{(P)} = -P_{E.yy}^{2'}.y_o^2\left(P_{E.yy}-P_{E.xx}\right), \text{ which is negative.}$$

As such, it is seen that; Eq. 5.9 and Eq. 5.10, have three positive roots. One root is less than $P_{E.xx}$. Second root is greater than $P_{E.yy}$ and third root is between $P_{E.xx}$ and $P_{E.yy}$. A similar result is found, in case, it is assumed.

3. **$P_{E.xx}$ is greater than $P_{E.yy}$:** It may be proved that the smallest value of P is less than P_ϕ in case, P_ϕ is less than $P_{E.xx}$ and less than $P_{E.yy}$, both. It is seen that f(P) is positive for $P = P_\phi$.

Likewise, the largest root must always be greater than P_ϕ. In this way, finally, it is concluded that

(i) One critical load is less than $P_{E.xx}$, $P_{E.yy}$ or P_ϕ and one is greater. Third critical load is always intermediate between $P_{E.xx}$ and $P_{E.yy}$.

It means that when the possibility of torsion during buckling is considered, the critical load obtained is smaller than either the Euler load or the purely torsional buckling load. This solution may be taken in a more general form and it may be assumed that

$$u = A_1 . \sin\left(\frac{n\pi z}{l}\right), v = A_2 \sin\left(\frac{n\pi z}{l}\right) \text{ and } \quad \phi = A_3 . \sin\left(\frac{n\pi z}{l}\right)$$

Then, it corresponds to the assumption that during buckling, the member is divided intop n-half sine waves. Previous conclusions will hold in this case, in case, the values $\left(n^2 \pi^2/l^2\right)$ are substituted for $\left(\pi^2/l^2\right)$ in Eq. 5.10. The corresponding critical loads are larger than those obtained for n-equal to unity and are of practical interest only if, the member has intermediate equidistant lateral supports.

Following two cases of interest are important for the studics.

1. When the cross-section of column member is symmetrical to both the axes.

In such cases, the shear center coincides with the centroid. Then, $x_o = y_o = 0$ (Zero).

Then, Eqs 5.8 (A), 5.8 (B), and 5.8 (C), reduce to

$$E.I_y.\left(\frac{d^2u}{dZ^2}\right) = P.u \quad \text{(i)}$$

$$E.I_x.\left(\frac{d^2v}{dZ^2}\right) = -P.v \quad \text{(ii)}$$

$$\text{And } C_1.\left(\frac{d^4\phi}{dZ^4}\right) - \left(C - \frac{I_o}{A}.P\right)\left(\frac{d^2\phi}{dZ^2}\right) \quad \text{(iii)}$$

Each of these above expressions, have only one unknown quantity, and these may be solved individually, so that torsional buckling is independent of buckling by bending. First two expressions give the values of Euler's critical loads for buckling in the two principal planes. Third expression is same as Eq. 5, and gives the critical load for pure torsional buckling. Only, the lowest of these three values of the critical loads is of interest in practical applications.

2. When the column cross-section is symmetrical about one axis

It is assumed that x-axis is an axis of symmetry. A channel section is shown in Fig. 5.15. Deformations due to combined torsional and flexural Buckling.

In this case, when $y_o = 0$ (Zero), Eqs. 5. and 5. reduce to

$$E.I_{yy}.\left(\frac{d^2u}{dZ^2}\right) = -P.u \quad \text{(5.11A)}$$

$$E.I_{xx}.\left(\frac{d^2v}{dZ^2}\right) = -P.\left(v - x_o . \phi\right) \quad \text{(5.11B)}$$

$$C_1.\left(\frac{d^4\phi}{dZ^4}\right) - \left(C - \frac{I_o}{A}.P\right)\left(\frac{d^2\phi}{dZ^2}\right) - P.x_o\left(\frac{d^2v}{dZ^2}\right) = 0 \quad \text{(5.11C)}$$

In the expression (i) there is no term ϕ and it shows that the buckling in the plane of symmetry is independent of torsion. The corresponding critical load is given by Euler's formula.

CHANNEL SECTION

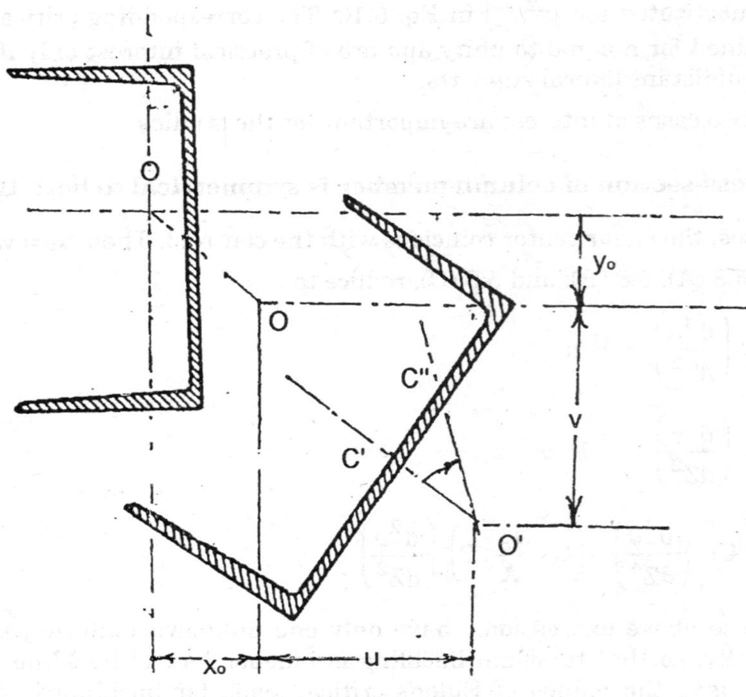

Fig. 5.15

Buckling perpendicular to the plane of symmetry is combined with torsion. It is given by Eqs. 5., and 5. The ends of the column member are simply supported (i.e, pinned at both ends) that is, free to warp and free to rotate about x-axis but restrained about x-axis but restrained about. x-axis but restrained against rotation about z-axis, then,

$$v = \phi = 0 \text{ (zero)} \quad \text{at } z = 0 \text{ and } z = l$$

$$\left(\frac{d^2v}{dZ^2}\right) = \left(\frac{d^2\phi}{dZ^2}\right) = 0 \text{ (zero)} \quad \text{at } z = o \text{ and } z = l$$

Let the form of deformations be

$$v = A_2 . \sin\left(\frac{\pi z}{l}\right) \text{ and } \phi = A_3 . \sin\left(\frac{\pi z}{l}\right)$$

The expressions for the critical loads may be written in the determinant form

$$\begin{vmatrix} (P - P_{E.xx}) & -P.x_o \\ -P.x_o & \frac{I_o}{A}(P - P_\Phi) \end{vmatrix} = 0 \quad (5.12)A$$

In the expanded form, which may be written as under

$$\frac{I_o}{A}(P - P_{E.xx})(P - P_\phi) - P^2.x_o^2 = 0 \tag{5.12B}$$

or $$\frac{I_c}{I_o}.P^2 - (P_{E.xx} + P_\phi)P + P_{E.xx}.P.\phi = 0 \tag{5.12C}$$

Above expression gives two values of the critical loads. One of which is smaller than $P_{E.xx}$ or P_ϕ, while the other is greater than either (both).

- Smaller of these two roots or the Euler's load for the buckling in the plane of the symmetry represents the critical load for the column.

From the determinant,

$$(P - P_{E.xx})(P - P_\phi) - \frac{A}{I_o}P^2.x_o^2 = 0 \qquad (\because I_o = A.r^2)$$

$$(P - P_{E.xx})(P - P_\phi) - \left(\frac{P^2.x_o^2}{r_o^2}\right) = 0$$

$$P^2\left[1 - \left(\frac{x_o}{r_o}\right)^2\right] - P(Px_{E.xx} + P_\phi) + P_{E.xx}.P_\phi = 0$$

$$P^2\left[\left(\frac{r_0^2 - x_0^2}{r_0^2}\right)\right] - P(P_{Exx} + P\varphi) + P_{Exx}.P_\varphi = 0$$

$$k.P^2 - (P_{E.xx} + P_\phi)P + P_{E.xx}.P_\phi = 0$$

$$P_{cr.tf} = (P_{E.xx} + P_\phi) - \left[(P_{E.xx} + P_\phi)^2 - 4k\,P_{E.xx}.P_\phi\right]^{1/2} = 0$$

$$P_{cr.tf} = \frac{1}{2k}\left[(P_{E.xx} + P_\phi) - \left\{(P_{E.xx} + P_\phi)^2 - 4k\,P_{E.xx}.P_\phi\right\}^{1/2}\right]$$

All axially loaded columns have three distinct buckling loads [(i) Euler's load for buckling about x-axis, (ii) Euler's load for buckling about y-axis, and (iii) Torsional buckling load] in which at least one corresponds to torsional or torsional-flexural mode in doubly symmetric section.

Flexural buckling load about the weak-axis is almost always the smallest. Therefore, in doubly symmetric sections, torsional buckling load is disregarded.

When the sections are symmetric, the buckling shall always be in torsional flexural mode irrespective of its shape and dimensions. However, such sections are rarely used.

Note – Influence of eccentricity of load

Mode of buckling (i.e, pure bending, pure torsion or combined torsion-bending) depends on the eccentricity of the load, position of shear center and the symmetry of the cross-section. Effects of yielding or buckling may be neglected.

In case, the column simply suppored at both the ends carries an **eccentric load,** the column shall buckle at the critical load, P_{cr} due to bending or torsion or combined torsion-bending. The critical load shall be given by the smallest of three roots of the expression given below :

$$(P_{cr} - P_{x.cr})\,(P_{cr} - P_{y.cr})\,(\alpha . P_{cr} - P_{z.cr}) - P_{cr}^2\left[\beta_y\,(P_{cr} - P_{x.cr}) + \beta_x\,(P_{cr} - P_{y.cr})\right] = 0 \quad (5.12D)$$

where, α, β_x and β_y are the coefficients which depends on the geometrical proerties of the cross-section, the values of which are as follows :-

$$\alpha = \left[1 - \left(\frac{e_x . x_o - e_y . y_o}{r^2 + x_e^2 + y_e^2}\right)\right] \quad (5.13A)$$

$$\beta_x = \left[\frac{(x_e - e_x)^2}{r_o^2 + x_e^2 + y_e^2}\right] \quad (5.13B)$$

$$\beta_y = \left[\frac{(y_e - e_y)^2}{r_o^2 + e_x^2 + y_e^2}\right] \quad (5.13C)$$

where, e_x and e_y are the eccentricities of the load with respect to the x and y principal axes, respectiveiy and x_e and y_e are the co-ordinates of shear center with respect to x and y principal axes; and x_o and y_o are the co-ordinate axes of a point characteristic of the cross-sectional shape to be defined.

$$x_o = 2x_e - \int_A \frac{x(x^2 + y^2)}{I_y}\,dA$$

$$y_o = 2y_e - \int_A \frac{y(x^2 + y^2)}{I_x}\,dA$$

Following special cases show the importance of eccentricity of the load, position of shear center and symmetry of the cross-sectional shape.

Case I. Shear center coincides with the centroid (i.e, the section is symmetrical with two axes of symmetry. For the simply supported column,

$x_c = y_c = 0$, $x_o = y_o = 0$, $\alpha = 1$

$\beta_x = (e_x / r_o)^2$ and $\beta_y = (e_y / r_o)^2$

In case, the load is axial, then $e_x = e_y = 0$, then

$$(P_{cr} - P_{x.cr})\,(P_{cr} - P_{y.cr})\,(P_{cr} - P_{\phi.cr}) = 0 \quad (5.14)$$

which one is same as Eq. (5.12D)

Case 2. Section is symmetrical about one axis of symmetry (i.e, x – axis)

In such a case, the end end conditions are as follows.

$$y_e = y_o = 0, \quad \alpha = 1 - \left(\frac{e_x . x_o}{r_o^2 + x_c^2}\right)$$

$$\beta_x = \left(\frac{x_c - e_x}{(r_o^2 + x_c^2)}\right)^2, \quad \beta_y = \left(\frac{e_y^2}{r_o^2 + x_c^2}\right)$$

In case, it is an **axial load,** $e_x = e_y = 0$, then

$$\alpha = 1, \quad \beta_y = 0, \quad \beta_x = \left(\frac{x_c^2}{r_o^2 + x_c^2}\right)$$

In this case, it may be shown that the smallest value of the root is either $P_{cr} = P_{y.cr}$ or smaller than either $P_{x.cr}$ or $P_{z.cr}$.

In case, the load passes through center, $e_x = x_c$, $e_y = y_c$, then,

$$\beta_x = \beta_y = 0, \quad \alpha = \left[1 - \left(\frac{e_x \cdot x_o}{r_o^2 + x_c^2}\right)\right]$$

and the critical load P_{cr} is the smallest value of three roots $P_{x.cr}$, $P_{y.cr}$ and $(P_{z.cr}/\alpha)$. It is a critical buckling load only when it is less than $(P_{z.cr}/\alpha)$ and when either the shear center coincides with the centroid or the load is applied at the shear center

5.17. LOCAL BUCKLING IN COLUMNS

Most structural members may be considered as assemblage of flat plates, that is, the flat plate elements have been placed together to form the structural members.

I-sections, channel sections and angle sections are such elements in which, as if, the flat plates have been placed together. These sections are referred as open sections. Rectangular square and circular tube's are referred as closed-sections. The strength of such built-up compression members depends upon slenderness.

Buckling of complete length of the column has been discussed so far. When the member is made up of thin element, it is quite possible that the buckling of one or more element may occur at loads which are less than the buckling load for the complete column. Such buckling is known as **local-buckling**. Local buckling depends on the width-to-thickness ratio of the elements of compression members that make up the column and the edge supports.

When the plate elements are subjected to direct compression, bending or shear stresses or to combination of these stresses, then, these flat elements buckle locally before the entire member fails.

AISC-Specifications specify limits on width-to-thickness ratios to the extent to which these sections may be built thin walled under two categories (i) unstiffened elements and (ii) stiffened elements. The element is called as unstiffened in case only its one edge is supp-orted. The element is known as stiffened in case both the edges supported. (while the edges in both the categories are parallel to the direction of the compress stress.

The load carrying capacity of columns reduce due to local buckling. It happens due to decrease in stiffness and the strength of locally buckled flat elements. Local buckling of the compression members before the column as a whole buckles is prevented. Either the allowable stress for the column is reduced and/or the effective area of the element is reduced.

Local buckling before yielding of the member should be avoided. Hot-rolled steel sections have sufficient thickness in their elements so that local buckling before yielding does not occur.

5.18 COLUMN FORMULAE FOR AXIAL STRESS IN COMPRESSION (A HISTORICAL REVIEW)

The strength of a column depends upon large number of variables. The efforts are made to obtain a design formula by fitting a curve to experimentally found buckling loads for the intermediate range of the slenderness ratio. It is tried to draw a curve which may merge with the Euler hyperbola in the very slender column range on one side and with the material yield strength for the zero length on the other side.

A perfectly straight column of perfectly homogenous material (i.e., an ideal column) is subjected to an axial load. The primary object is to find the average axial stress in compression, which corresponds to the allowable load. The average axial stress is uniform across the section. It is given by

$$\sigma_{ca} = (P_a / A) \quad \text{(i)}$$

where P_a = allowable load and A = cross-sectional area of the column

The required cross-sectional area for a given design load may be found conveniently in case σ_c is known

$$A_{regd.} = (P / \sigma_c) \quad \text{(ii)}$$

where, P is the design load.

It is observed that the average axial stress in the column at the time of failure of column is less than the yield strength of the material. The difference in the stress depends on the type of failures of the column. The various types of failures of the column other than the direct compression or direct crushing are as follows :

1. Primary laterally buckling

(a) Inelastic buckling, (b) Elastic buckling

2. Gradual yielding due to excessive local stress

3. Local buckling, and 4. Combined Torsional and Flexural buckling.

The failure by **local buckling** and by **torsional buckling** are prevented by selecting the rolled steel sections of proper shape and proportions. Therefore, the failure of a column may occur by lateral buckling or by gradual yielding due to an excessive local stress. The average allowable axial stress for use in the design is influenced by amount of lateral buckling. It depends upon the **slenderness ratio** and the **compressive strength** of the material. Therefore, the columns may be classified in two categories viz., (i) those columns which fail in elastic buckling, and (ii) those columns which fail in inelastic buckling. The columns having very large slenderness ratios fail in elastic buckling. Leonhard Euler predicted that the lateral buckling of the column occurs when the value of load, P, exceeds the critical or ultimate load, P_{cr} either in the elastic range or in the inelastic range, the critical stress in the column for **elastic buckling** is given by **Euler's formula** as follows :

$$\sigma_c = \left(\frac{P_{cr}}{A}\right) = \left(\frac{\pi^2 E}{(KL/r)^2}\right) \text{ whereas} \quad (5.15)$$

K = Factor, which depends upon the end conditions of the column

E = Modulus of elasticity

L = Actual length of the column

r = Radius of gyration of the section.

Eq. 5.6 may be written as

$$\sigma_c = \left(\frac{P_{cr}}{A}\right) = \left(\frac{\pi^2 E}{(l/r)^2}\right) \qquad (5.16)$$

where, l = K.L. = Effective length of the column.

It is concluded that the columns remain straight for all values of P. The value of P reaches [$P_{cr} = P_E = (\pi^2 EI/l^2)$] when the load is gradually increased. the column buckles in the shape of a half-sine wave. the magnitude of buckling deflection is indeterminate. Other sinusoidal buckled shapes (n half-ways) are possible for the higher values of loads (for n = 4, 9, ...).

$$P_{cr.n} = (n^2 . \pi^2 . EI/l^2) \qquad (i)$$

$$P_{cr4} = (4\,\pi^2 . EI/l^2) \qquad (ii)$$

$$P_{cr.9} = (9\,\pi^2 . EI/l^2) \qquad (iii)$$

The column deflects away from its original position due to very small disturbance.

1. Euler's formula is suitable for large slenderness ratio (i.e., l/r >200). It was considered appropriate to apply a variable factor of safety to account for this stress level and additionally to consider the effect of eccentricity, residual stresses and the several other factors which complicate the column theory. The maximum average allowable stress in the axial compression may be found by dividing σ_{cr} by factor of safety (F.S.). Therefore,

$$\sigma_a = \left(\frac{\sigma_{cr}}{F.S.}\right) = \frac{1}{F.S.} \cdot \left(\frac{\pi^2 E}{(l/r)^2}\right) \qquad (5.17)$$

The relationship between Euler's formula and A.I.S.C. (American Institute of Steel Construction) 1953 Specification formula may be established.

The critical stress by Euler's formula may be expressed as

$$\sigma_{cr} = \left(\frac{\pi^2 E}{(l/r)^2}\right) \qquad (i)$$

Differentiating (i) with respect to (l/r)

$$\left(\frac{d(\sigma_{cr})}{d(l/r)}\right) = \left(\frac{-2\pi^2 . E}{(l/r)^3}\right) \qquad (ii)$$

The maximum value of Euler's stress or any critical buckling stress is limited to yield stress σ_y. The Euler's formula is not valid for small values of slenderness ratio. For such a region, the critical stress may be obtained by assuming the following expression (which one is also based on experimental column test data).

$$\sigma_{cr} = \left[\sigma_y - m . (l/r)^2\right] \qquad (iii)$$

Differentiating (iii) with respect to (l/r)

$$\left(\frac{d(\sigma_{cr})}{d(l/r)}\right) = (-2m.(l/r)) \quad \text{(iv)}$$

At $l/r = 0$, the slope may be assumed as zero. The two curves defined by the expressions (i) and (iii) will have a common tangent. At that point, the slopes given by (ii) and (iv) will be equal. Equating the two slopes at a common point.

$$(-2m.(l/r)) = \left[\frac{-2\pi^2.E}{(l/r)^3}\right] \quad \text{(v)}$$

The slenderness ratio at this common point be arbitrary defined by a parameter

$(l/r) = C_c$ (vi)

Substituting this in (v),

$$(-2m.(C_c)) = \left[\frac{-2\pi^2.E}{(C_c)^3}\right].m = \pi^2 E/C_c^4 \quad \text{(vii)}$$

Thus, the expression (ii) may be written as

$$\sigma_c = \left(\sigma_y - \frac{\pi^2.E}{C_c^4}(l/r)^2\right) \quad \text{(viii)}$$

By substituting, the values of σ_{cr} by Euler's formula (on left hand side)

$$\left(\frac{\pi^2.E}{(l/r)^2}\right) = \left[\sigma_y - \frac{\pi^2 E}{C_c^4}.(l/r)^2\right] \text{At} \quad (l/r) = C_c \quad \text{(ix)}$$

$$\left(\frac{\pi^2 E}{(C_c)^2}\right) = \left[\sigma_y - \frac{\pi^2 E}{C_c^4}.C_c^2\right] \quad . \quad C_c = \left(\frac{2\pi^2 E}{f_y}\right)^{1/2} \quad \text{(x)}$$

Substituting the value of C_c in (x)

$$\sigma_c = \left[\sigma_y - \frac{\pi^2 E}{2\pi^2 E}\sigma_y\right] = 0.5\,\sigma_y \quad \text{(xi)}$$

Thus, the critical buckling stress becomes $0.5\,\sigma_y$ at $l/r = C_c$.

In general, the expression (viii) may be written as

$$\sigma_{cr} = \sigma_y\left[1.0 - \frac{0.5\,(l/r)^2}{C_c^2}\right] \quad (5.18)$$

The lowest of slenderness ratio upto which the elastic buckling occurs is denoted by C_c.

The minimum slenderness ratio ($l/r_{min} = C_c$) distinguishes the elastic buckling from inelastic buckling. When the slenderness ratio is greater than C_c, then failure of column occurs by elastic buckling. A.I.S.C. Specifications recommended the following for the determination of allowable stress in axial compression for (l/r) C_c.

$$\sigma_c = \left[\frac{\pi^2 E}{(l/r)^2 \times (F.S.)}\right] = \left[\frac{9.87 \times 2.047 \times 10^6}{(l/r)^2 \times 1.92} \times \frac{10}{100}\right] N/mm^2$$

$$\sigma_c = \left(\frac{10.52 \times 10^5}{(l/r)^2}\right) = \left(\frac{10.52.000}{(l/r)^2}\right) N/mm^2 \tag{5.19}$$

When the slenderness ratio for the column is less than or equal to C_c, then, the failure of column occurs by inelastic buckling. The allowable stress in axial compression for $(l/r) \le C_c$ is given by

$$\sigma_c = \frac{1}{F.S.}\left[1 - 0.5\left(\frac{1}{r}\right)^2 \cdot \left(\frac{1}{C_c^2}\right)^2\right] \cdot \sigma_y \tag{5.20}$$

where, $$F.S. = \left[\frac{5}{3} + \frac{3}{8}\left(\frac{1}{r}\right) \cdot \left(\frac{1}{C_c}\right) - \left(\frac{1}{r}\right)^3 \cdot \left(\frac{1}{C_c}\right)^3\right] \tag{5.21}$$

when $\left(\frac{l}{r}\right) = 0 .\ F.S. = 1.67$ and $\left(\frac{1}{r}\right) = C_c .\ F.S. = 1.92$

The factor of safety of 1.92 remains constant for (l/r) greater than C_c. Eqs. 5.10 and 5.11 and are used for main compression members. For bracing and secondary compression members for which (l/r) exceed 120, the factor of safety is reduced by multiplying it by a factor Ψ equal to $[1.6 - (l/r)(1/200)]$. A more conservative value of factor of safety is used (F.S. = 2.12) for the members of the bridge structures since these members are in a hostile atmosphere than those in buildings.

All the columns have some initial curvature and the accidental eccentricity, the factor of safety takes these into consideration. Short members are negligibly influenced by small eccentricity of load or by residual stresses. Therefore, the factor of safety may not be more than it is desired for tension members (i.e., 1.67 under the AISC specifications). The influence of accidental eccentricity is more for the columns having large value of slenderness ratio. Therefore, AISC prescribes an increased factor of safety with slenderness ratio to a maximum of 15 per cent over the basic value (i.e., a maximum factor of safety = (23/12) = 1.92). Arbitrarily to get a smooth transition from FS = 1.67 for $(kL/r) = 0$ to FS = 1.92 for $(kL/r) = C_c$, a cubic equation approximation as Eq. 5.12 to a quarter sine wave is used.

For columns with slenderness ratio more than $(C_c = kL/r)$, elastic buckling governs the strength according to the Euler's formula, which gives stress on gross area for long columns.

2. Engesser's basic tangent modulus theory In elastic buckling of axially loaded (prismatic members) columns

F. Engesser ['Ueber die knickfestigkeit gerader stable' Zeitschrift fur Architektur und Ingenieurwesen der Bauverwaltung, Berling, (December 5, 1891, pp. 483)] and A. Considere ['Resistance des pieces compromise' conges International des Proceeds de Construction, Paris, 1891, Vol. v pp. 371)] were first to utilize the possibility of a variable modulus of elasticity. In Engesser's tangent modulus theory, the column remains straight upto the instant of failure. The modulus of elasticity at failure is the tangent to stress-strain curve which is smaller than the initial modulus E. The critical load, P_{cr} based on the assumption of linear elastic behaviour is no longer valid when the critical stress exceeds the limit of proportionality.

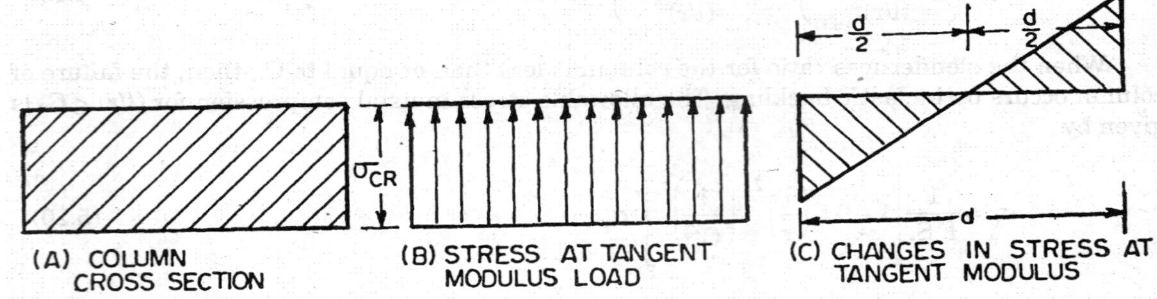

Fig. 5.16A.

The relationship for the column pinned at both the ends are shown in Fig. 5.16B. The deformation at σ_{cr} is controlled by $E_t = (\sigma_d / d\varepsilon)$ that is, no unloading of fibres

Engesser formula suggested the tangent modulus formula for the buckling load of an axially load column.

$$P_{cr} = \left(\frac{\pi^2 E_t I}{l^2}\right) \tag{5.22}$$

$$\text{or} \quad \sigma_{cr} = \left(\frac{P_{cr}}{A} = \frac{\pi^2 E_t}{(l/r)^2}\right) \tag{5.23}$$

where E_t = Tangent modulus of the elasticity

Engesser in 1989 modified Euler's formula

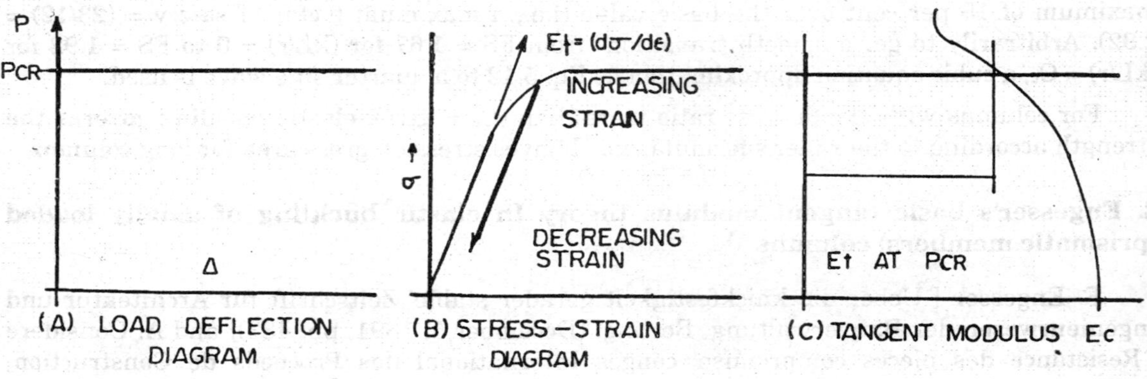

Fig. 5.16B.

This relationship is based on the assumption that the deformation of the fibres of the cross-section is controlled by the law ($d\sigma/d\varepsilon$), that is the modulus, E is replaced by E_t no unloading of the fibres take place.

This theory, however, still did not satisfy the test results. The calculated loads found were lower than the measured ultimate load capacity. The tangent modulus load theory remained erroneous due to the member changes from straight to bent shape. No strain reversal takes place.

However, in case, the column is slightly curved, an increase in curvature corresponds to increase in compression on the concave side and decrease in compression on the convex side. The unloading on the convex side will follow a linear stress-strain law whereas the increase in compression will follow the non-linear law ($d\sigma/d\varepsilon$) = E_t. This concept was proposed by Considere and developed by Karman, leads leads to the so-called reduced modulus theory $E_t < E_r < E$).

Therefore, $\sigma_{cr} = \left[\pi^2 E_t / \left(kL/r)^2\right)\right]$

In 1985, Engesser argued that during bending, some fibres are under-going increased strain (lowered tangent modulus) and some fibres are being unloaded (higher modulus at reduced strain). Therefore, a combined modulus should be used.

Engesser's double modulus (or reduced modulus theory) theory

Engesser considered the cross-section of a column, in an arbitrary deflected position under unstable equilibrium and the process of column bending at stresses beyond the elastic limit as shown in Fig. 5.17. From which, the double modulus (or reduced modulus) theory was developed. There was a logic to this concept. It was also accepted. The values of loads calculated were found more than measured ultimate loads.

At the neutral axis-xx at unstable equilibrium, the stress remains equal to that it was prior to Δ. With the increase of load on one side, the strain increases and therefore, the stress also increases proportional to E_t = ($d\sigma$ / $d\varepsilon$). On the other side, the strain reduces from the value of elastic strain in proportion to the elastic modulus, E. On both the sides, (i.e., on the column cross-section), the strains vary linearly.

At the extreme relieving on the side of strain, as per Hooke's law, the stress as shown in Fig. 5.14.

$$\sigma_{2(max)} = (\text{Unit strain} \times E) = \left(\frac{\Delta\ (d\varepsilon)}{d\varepsilon}\right) E \qquad \text{(i)}$$

On the strain increasing side

$$\sigma_{1(max)} = \left(\frac{\Delta\,(d\varepsilon)}{d\varepsilon}\right) \cdot \left(\frac{d_1}{d_2}\right) . E_t \qquad \text{(ii)}$$

$$\text{Let} \quad \left[\Delta\ (d\varepsilon) = d_2\right] = d\,\phi \qquad \text{(iii)}$$

Substituting this in the expressions (i) and (ii)

$$\sigma_{2(max)} = \left(\frac{d\,\phi}{d\varepsilon}\right) d_2 . E \quad \text{and,} \quad \sigma_{1(max)} = \left(\frac{d\,\phi}{d\,\varepsilon}\right) d . E_t \qquad \text{(iv)}$$

$$\left(\frac{M}{E_t\ I}\right) = \frac{1}{(\text{Radius of curvature})} = \left(\frac{d\,\phi}{d\varepsilon}\right) = \left(\frac{d^2y}{d\varepsilon^2}\right) \quad (v)$$

where, E_r is Engesser's double modulus

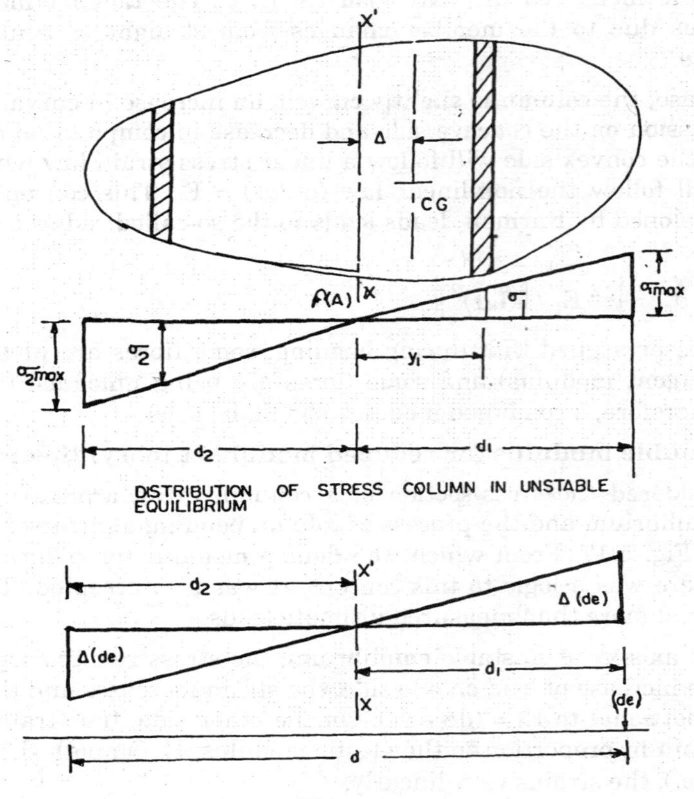

Fig. 5.17.

3. **Shanley and Von-Karman** defined the tangent modulus load as the smallest value of the axial load at which the bifurcation of the equilibrium position can occur regardless of whether the transition to the bent position requires an increase of the axial load.

Moment of resistance for the condition of the stress as stated above

$$M = (-P.y)\int_0^{d_1} \sigma_1\, dA\ (y_1 - \Delta) + \int_0^{d_2} \sigma_2\, dA_2\ (y_2 + \Delta) \quad (vi)$$

From stress distribution diagram, Fig. 5.17.

$$\left(\frac{\sigma_{1max}}{d_1}\right) = \left(\frac{\sigma_1}{y_1}\right). \quad \therefore \sigma_1 = \sigma_{1\,max}\cdot\left(\frac{y_1}{d_1}\right) \quad (vii)$$

$$\left(\frac{\sigma_{2\,max}}{d_2}\right) = \left(\frac{\sigma_2}{y_2}\right). \quad \therefore \sigma_2 = \sigma_{2\,max}\cdot\left(\frac{y_2}{d_2}\right) \quad (viii)$$

Substituting for $\sigma_{1\,max}$ and $\sigma_{2\,max}$ from the expressions (v) and (iv)

$$\sigma_1 = E_t \cdot d_1 \left(\frac{d\,\phi_1}{d\varepsilon_1}\right) = E_t\, d_1 \cdot \left(\frac{y_1}{d_1}\right)\left(\frac{d^2y}{d\,\varepsilon_1^2}\right) \quad \text{(ix)}$$

$$\sigma_2 = E_t \cdot d_2 \left(\frac{d\,\phi_2}{d\varepsilon_2}\right) = E\, d_2 \cdot \left(\frac{y_2}{d_2}\right)\left(\frac{d^2y}{d\,\varepsilon_2^2}\right) \quad \text{(x)}$$

$$\therefore\ (-P.y) = E_t \left(\frac{d^2y}{d\,\varepsilon^2}\right) \int_0^{d_1} y_1 (y_1 - \Delta)\, dA_1 + E \left(\frac{d^2y}{d\,\varepsilon^2}\right) \int_0^{d_2} y_2^2\, dA_2 \quad \text{(xi)}$$

Since, the column cross-section is in equilibrium

$$\int_0^{d_1} \sigma_1\, dA_1 = \int_0^{d_2} \sigma_2 \cdot dA_2 \quad \text{(xii)}$$

$$E_t \cdot \left(\frac{d^2y}{d\varepsilon^2}\right) \int_0^{d_1} y_1\, dA_1 = E \left(\frac{d^2y}{d\varepsilon^2}\right) \int_0^{d_2} y_2\, dA_2 \quad \text{(xiii)}$$

$$(-P.y) = E_t \left(\frac{d^2y}{d\,\varepsilon^2}\right) \int_0^{d_1} y_1^2\, dA_1 + E \left(\frac{d^2y}{d\,\varepsilon^2}\right) \int_0^{d_2} y_2^2\, dA_2 \quad \text{(xiv)}$$

$$\therefore \left(\frac{d^2y}{d\varepsilon^2}\right) \cdot \left[E_t \cdot \int_0^{d_1} \cdot y_1^2\, dA_1 + E \int_0^{d_2} y_2^2\, dA_2\right] + P.y = 0 \quad \text{(xv)}$$

It is obviously of the same form as the elastic buckling equation (i.e., differential equation).

Therefore, for the double modulus theory

$$P_{cr} = \frac{\pi^2}{L^2}\left[E_t \cdot \int_0^{d_1} y_1^2\, dA_1 + E \int_0^{d_2} y_2^2\, dA_2\right] \quad (5.24)$$

3. Shanley's Concept : The actual behaviour of the column was explained by Shanley, F.R. [('The Column Paradox' Journal of the Aeronautical Sciences, 13, 5 (December 1946), pp 678-679].

A column pinned at both the ends with a rectangular cross-section as shown in Fig. 5.18(A) was considered. At the onset of bending (infinitesimal curvature), there will be a small (infinitesimal) increase in axial strain and stress, $\Delta\sigma_1$. By the time, the curvature becomes finite (i.e., the point B moves to x_1), some reversal of strain occurs and the cross-section develops a resisting moment to maintain equilibrium with the moment due to the external load, P.Δ.

For small finite values of curvature, the increment of load represented by the stress on the area of increasing strain exceeds the increment of load represented by the stress on the area of decreasing strain. Therefore, the axial load is increased by dP (point D, Fig. 5.18). At every increment of curvature takes place, P further increases as long as dC > dT. The increased compressive force, dC is calculated by using tangent modulus, E_t. dT is found for the region of reversal of strain using the elastic modulus E. Double modulus theory, which similarly treated loading, which similarly treated loading getting additional pressure and unloading (relieving fibres), did not agree (dC > dT), but rather only considered equilibrium positions near the perfectly straight one.

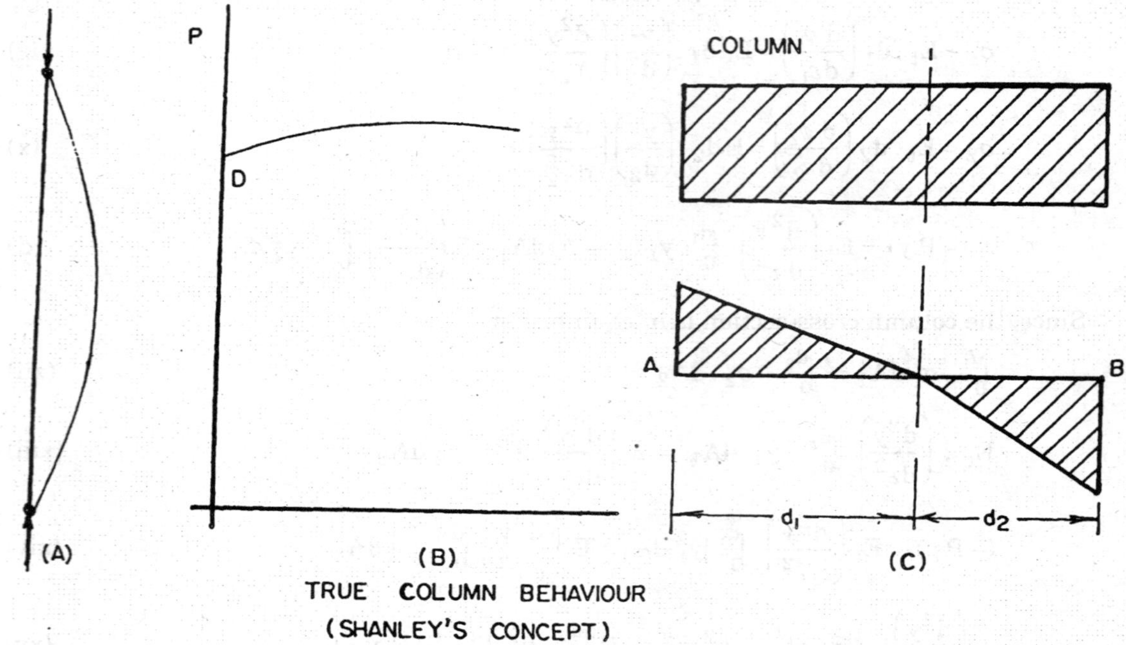

Fig. 5.18.

For the practical purpose the increase of capacity, P_T to P_s, Fig. 5.15(B) may be neglected. Therefore, the tangent modulus load may be considered as the critical load (i.e., the load at which bending begins).

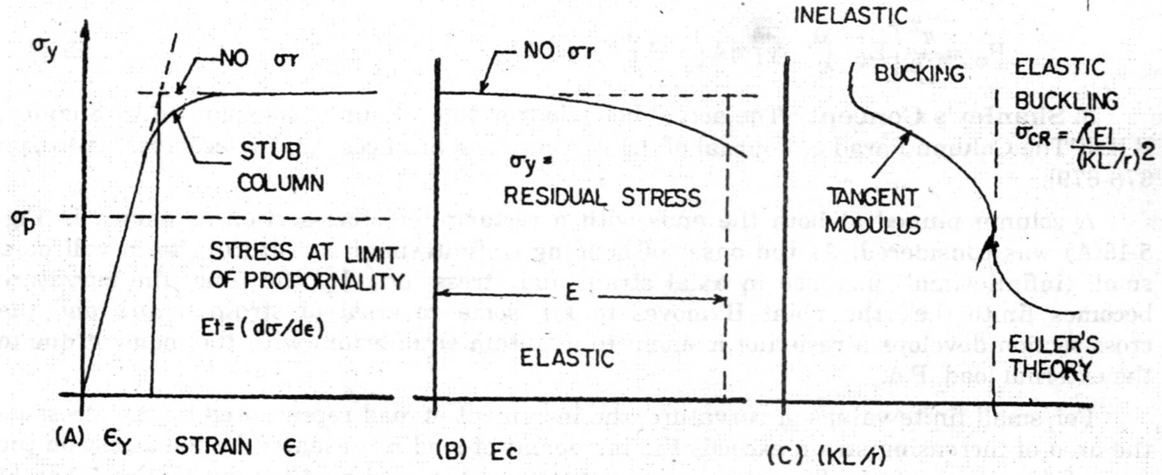

Fig. 5.19.

Beedle and Tall pointed out that Eq. cannot be applied to a steel column if the stress-strain relationship is determined from a small coupon from the section. The buckling strength of the column is reduced due to the existence of residual stress in the cross-section, since there is early localized yielding in the certain portions of the cross-section where the compressive residual stresses prevails. The basic equation for the critical strength of a column containing residual stresses is given by

$$P_{cr} = \left(\frac{\pi^2 E . I_e}{t^2}\right) \quad \text{and} \quad \sigma_{cr} = \left(\frac{P_{cr}}{A}\right) = \left(\frac{\pi^2 E . I_e/I}{(l/r)^2}\right) \qquad (5.25)$$

4. Steel remains perfectly elastic until strain becomes equal to that at yield, E_y and then, it is plastic (i.e., constant stress with increasing strain). In case, a coupon is cut from the web of a rolled shape, shows a behaviour as shown by dotted lines, Fig. 5.19. The behaviour of an H-shaped rolled steel section including residual stress is indicated by solid.

Due to presence of residual stress, early yielding occurs. To account for this effect, consider one fibre at a distance x from the axis of zero strain due to bending. It is taken as an infinitesimal amount consistent with the equilibrium at the tangent modulus load. Let Δ M be the bending moment contribution from stress on the fibre. Then, Δ M = (stress). (area). (Moment arm) = $(\phi . E_t . x)$ (dA). (x).

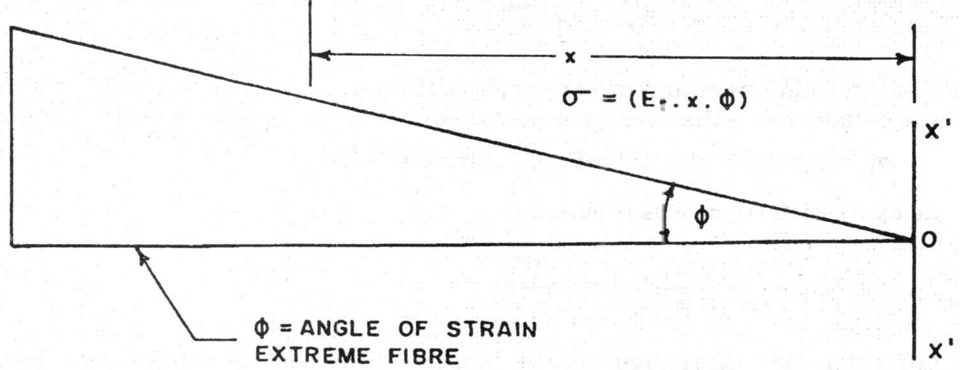

(X'X' — AXIS OF ZERO STRAIN DUE TO BENDING ASSUMED TO COINCIDE WITH C.G FOR INFINITESIMAL BENDING

Fig. 5.20.

For the complete cross-section

$$M(\Sigma \Delta M) = \int_A \phi E_t x^2 d_A = \phi \int_A E_t x^2 dA$$

From simple theory of bending

$$\left(\frac{M}{EI}\right) = \left(\frac{1}{R}\right) = \phi$$

where, R is the radius of curvature. Then

$$\phi = \left(\frac{1}{R}\right) = \left(\frac{M}{\text{Equivalent} . EI}\right) = \left(\frac{M}{EI}\right)$$

Assumed to coincide with C.G. for infinitesimal bending

$$E' \, I = \left(\frac{M}{\phi}\right) = \int_A E_t \, . x^2 \, dA$$

$$E' = \left(\frac{1}{I}\right) \int_A E_t \, . x^2 \, . dA \tag{5.26}$$

where, E' is called **effective modulus** and it is used as an equivalent value for E_t.

4. SSRC Basic Column Strength Curve

Structural Stability Research Council (SSRC) adopted the basic column strength curve based on the parabolic equation proposed by Fredrich Bleich ['Buckling Strength of Metal Structures', Mc-Graw-Hill Book Company, Inc, 1952, New York] Test results have generally shown essentially parabolic variation for buckling in the inelastic range, it seems reasonable. Equation proposed by Bleich is as under :

$$\sigma_{cr} = \sigma_y - \left(\frac{\sigma_p}{\pi^2 E}\right) . (\sigma_y - \sigma_p) \left[\frac{kL}{r}\right]^2 \tag{i}$$

where, σ_y is the yield stress and σ_p is the stress at the limit of proportionality. The deviation from elastic behaviour on the average stress-strain curve is considered due to residual stress.

$$\sigma_r = (\sigma_y - \sigma_p), \text{ and therefore, } \quad \sigma_p = (\sigma_y - \sigma_r)$$

In the expression (i), σ_p was replaced.

$$\sigma_{cr} = \sigma_y - \left(\frac{\sigma_y - \sigma_r}{\pi^2 E}\right) . \sigma_y \left(\frac{kL}{r}\right)^2 \tag{ii}$$

In order for the expression (ii) to properly represent a compromise between the strengths for the weak-axis and strong-axis of bending, the residual stress must be assumed than 0.3 σ_y. A value of 0.5 σ_y was taken by the SSRC, so as to give a smooth transition from the Euler's curve representing inelastic buckling. Two curves are tangent at ($\sigma_{cr} / \sigma_y = 0.5$). The SSRC curve then becomes.

$$\sigma_{cr} = \sigma_y \left[1 - \left(\frac{\sigma_y}{4 \pi^2 E}\right)\left(\frac{kL}{r}\right)^2\right] \tag{iii}$$

In case, the idealized elastic-plastic $\sigma - \varepsilon'$ curve, Fig. 5.16 (as shown by dotted line) is used (for $\sigma < \sigma_y$, E_t = E and for $\sigma = \sigma_y$, E_t = 0), the bending stiffness of yielded portions become zero; however, the buckling strength will be same as a column whose moment of inertia, Ie, is the moment of inertia of the portion remaining elastic. Eq. 5.17 then, becomes

$$E' = \left(\frac{E}{I}\right) \int_{AEP} x^2 \, dA = (E \, I_e / I) \tag{5.27}$$

The column begins to bend at the stress, Eq. 5.28.

$$\sigma_{cr} = \left(\frac{P}{A}\right) = \left(\frac{\pi^2 (I_e/1)}{(kL/r)^2}\right) \tag{5.28}$$

Eq. 5.28 needs the relationship between σ_{cr} and I_e. For this, following two cases of buckling of the column about weak-axis and about strong-axis are considered.

(i) Buckling of the column about weak-axis

Consider a rolled steel section as shown in Fig. 5.21. It is assumed that the flange tips becomes fully plastic before the yielding of web.

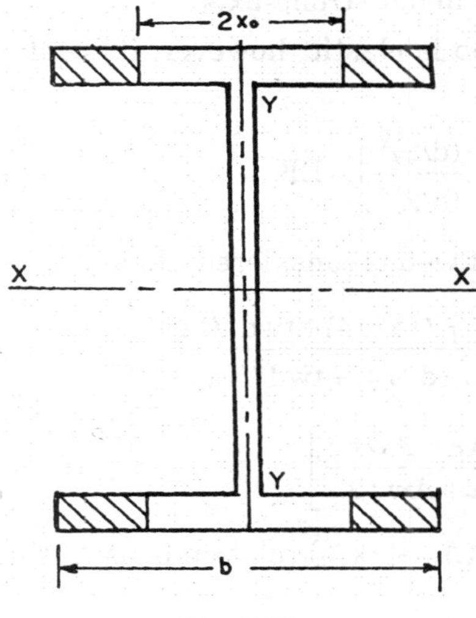

Fig. 5.21.

Then, Eq. 5.29A may be written as under.

$$\left(\frac{E.\,Ie}{I}\right) = \left(\frac{E.\,t_f\,.(2x_0)}{12} \times \frac{12}{t_f\, b^3}\right) = E\,K^3 \tag{5.29A}$$

where, $K = (2\,x_o / b) = (2\,x_o\, t_f / b\, t_f) = (Ae/A)$

In case, the web is ignored in calculating, I and applying the definition of tangent modulus.

$$E_t = \left(\frac{d\,\sigma}{d\,\varepsilon}\right) = \left(\frac{\text{Normal incremental stress}}{\text{Incremental elastic strain}}\right)$$

$$E_t = \left(\frac{d\,(P/A)}{\dfrac{d\,(P/A_e)}{E}}\right) = (E\,A_e / A)$$

$$(E_t . A) = (E\,A_e) = E\,(2\,k\,A_f + A_w) \tag{5.29B}$$

where, A_f is the area of the flange, (6.) A_w is the area of web and A is the total gross-area of the section. From Eq. 5.29B.

$$k = \left(\frac{E_c}{2E}\frac{A}{Af} - \frac{Aw}{2A_f}\right) \quad (5.29C)$$

$$\sigma_{cr} = \left(\frac{\pi^2 E K^3}{(kL/r)^2}\right) = \left(\frac{\pi^2 E}{(kL/r)^2}\right)\left[\frac{A E}{2A_f E} - \frac{A_w}{2A_f}\right] \quad (5.29D)$$

Eq. 5.29(D) includes the effect of elastic web, for buckling with respect to weak-axis, yy.

(ii) Buckling of the column about strong-axis

It is assumed that the web is elastic, however, its contribution towards the moment of inertia is neglected. Therefore,

$$\left(\frac{E.I_e}{I}\right) = \left(\frac{E.2A_c}{2A_f}\frac{(d/2)^2}{(d/2)^2}\right) = EK \quad (5.30A)$$

In case, the contribution of web is considered, then

$$\left(\frac{E.I_e}{I}\right) = E\left[\frac{2k.A_f/(d^2/4) + twd^3/12}{2A_f(d^2/4) + twd^3/12}\right] \quad (5.30B)$$

$$\left(\frac{E.I_e}{I}\right) = E\left[\frac{2k.A_f + Aw/3}{2A_f + Aw/3}\right] \quad (5.30C)$$

In case, $(E_t A) = (A_e.E) = E(A_w + 2kA_f)$ relation is used, then

$$2kA_f = \left(\frac{E_t - A}{E} - A_w\right)$$

Substituting for $(2k.A_f)$ in Eq. 5.23

$$\left(\frac{E.I_e}{I}\right) = \left[\frac{E_t.A/E - (2A_w/3)}{2A_f + (A_w/3)}\right] \quad (5.30D)$$

And $$\sigma_{cr} = \left(\frac{\pi^2.E.k}{(kL/r)^2}\right) \quad (5.31A)$$

Eq. 5.31A is the approximate relation. More exact relation for buckling about the strong axis shall be

$$\sigma_{cr} = \left(\frac{\pi^2 E}{(kL/r)^2}\right)\left[\frac{E_t.A/E - (2A_w/3)}{2A_f + (A_w/3)}\right] \quad (5.31B)$$

Above analysis shows that two equations are necessary to find column strength (one for the strong-axis bucking and other for the bucking about the weak-axis)

The value of (I_e/I) is not itself a function the residual stress distribution provided that it satisfies geometric requirement. Nevertheless, the critical stress, σ_{cr} calculated as the

buckling load divided by the gross area, has relationship with (kL/r) that does depend on the residual stress.

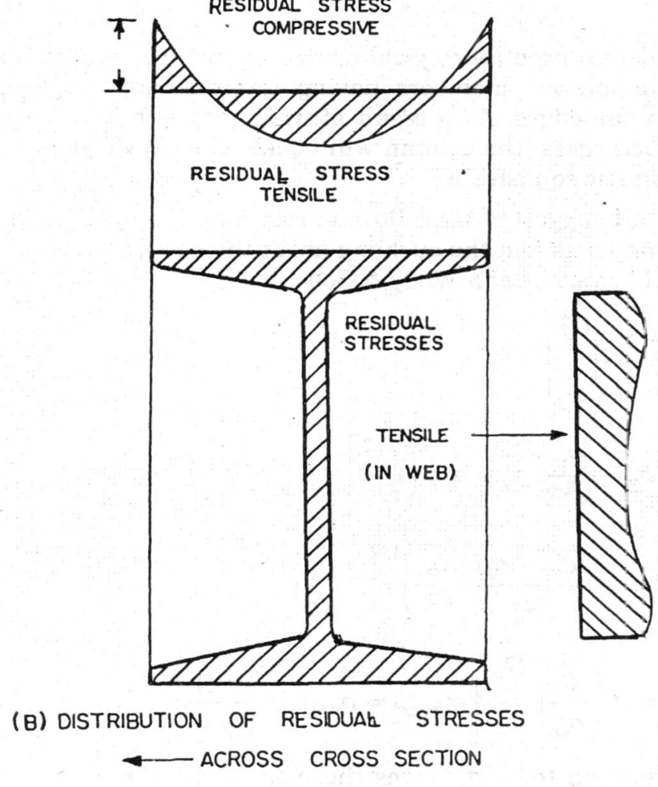

Fig. 5.22.

Residual stresses in welded box steel-sections

As mentioned in Art. 5.8, that earlier cod/ed portion of a section tends to shrink first and thicker elements cool last.

Middle portions of the plate elements of the box steel sections cool earlier. As such, the compressive residual stresses develop in the middle portions.

Where as, at the junctions (i.e, where, the web elements and flange elements meets (i.e, corners of the closed portions) cool last) the tensile stresses are produced.

However, the complete section remains in the equilibrium. Therefore, the compressive residual stresses and the tensile residual stresses remain balanced.

Effect of residual residual stresses

In case, an I-section is subjected to an uniform compressive stress, σ_{cc} the net stress distribution across the cross-section becomes non-uniform because of the presence of residual stresses. The maximum compressive stress shall be at the edges of flanges (i.e, ($\sigma_{cc} + \sigma_r$)).

In case, the total stress does not reach yield stress, the section continues to deform elasti-cally. With the incremental loading, the edges of the flanges yield first. [i.e, ($\sigma_{cc} + \sigma_r$)]. When the load further increases, the yield spreads inwards, eventually, yield enters the web-portion.

In very stocky-column, premature yielding may occur. A very slender struct may fail by buckling. For the compressive members having intermediate slenderess, the premature yielding may occur at the edges. As a result of which the effective bending stiffness of the column reduces. In such cases, the column will buckle elastically at a load below the elastic critical load and the plastic squash load.

5. **Beedle and Tall** suggested the following relationships for the critical stresses for the bucking about the strong axis and the buckling about the weak axis respectively. [Reference : Plastic Design of Steel Frames, John Wiley & Sons, Inc.].

For $\left(\frac{l}{r} < \left(\frac{E}{\sigma_p}\right)^{1/2}\right)$

$$\sigma_{cr(xx)} = \left[\sigma_y - \frac{\sigma_p}{\pi^2 E}(\sigma_y - \sigma_p).\left(\frac{l}{r}\right)^2\right] \quad (5.32A)$$

$$\sigma_{cr(yy)} = \left[\sigma_y - \frac{(\sigma_y - \sigma_p)}{\pi}\left(\frac{\sigma_p}{E}\right)^{1/2} . \left(\frac{l}{r}\right)\right] \quad (5.32B)$$

$$\text{For } \left(\frac{l}{r}\right) > \pi\left(\frac{E}{\sigma_p}\right)^{1/2} . \left[\sigma_{cr(xx)} = \sigma_{cr(yy)} = \frac{\pi^2 E}{(l/r^2)}\right] \quad (5.32C)$$

The difference between ($\sigma_y - \sigma_p$) gives the magnitude of **residual stress**, where σ_p is stress at the **limit of proportionality** and σ_y is yield stress for the steel.

The effect of stresses was recognised as an important factor in calculating the strength of a centrally loaded column in a report published by Column Research Council.

6. **AREA (American Railway Engineering Association)** recommends the following for the determination of allowable stress in axial compression.

$$\text{For } \frac{1}{r} > 140 \text{ (Rivet ends)}, \quad \sigma_c = 15000 - \frac{1}{4}\left(\frac{l}{r}\right)^2 \text{ .... (psi units)} \quad (5.33A)$$

$$\text{For } \frac{1}{r} > 140 \text{ (Pin ends)}, \quad \sigma_c = 15000 - \frac{1}{3}\left(\frac{l}{r}\right)^2 \text{ .... (psi units)} \quad (5.33B)$$

7. **AASHTO (American Association of State Highway and Transport Officials)** also recommends the above mentioned expressions. In bridges, the compression members may have one end riveted and the other end pin connected. For example, the end post of a riveted bridge truss is sometimes pin-connected at its lower end. Bridge specifications do not have any provisions for this condition. In the absence of the specific instructions, the allowable stress in axial compression for the member pinned at one end and riveted at the other end may be found from the following:

$$\sigma_c = \left[15000 - \frac{7}{24}\cdot\left(\frac{l}{r}\right)^2\right] \text{ (psi unit)} \tag{5.34A}$$

Eqs. and are the forms of **Johnson's parabolic** formula

$$\sigma_c = [\, f - k\,(l\,/\,r)^2 \,] \tag{5.34B}$$

where, σ_c = Permissible stress in axial compression

f = Crushing stress at failure and k = Constant.

AISC specifications recommend Johnson's parabolic formula for $(l/r) > 120$

$$\sigma_c = [17000 - 0.485\,(l/r)^2] \text{ (psi units)} \tag{5.35A}$$

The allowable stress in axial compression may be found by a straight line formula

$$\sigma_c = [\, f - k\,(l\,/\,r)] \tag{5.35B}$$

AREA specifications recommend this straight formula for $(l/r) > 120$

The equations of AASHTO for working stress design of axially loaded columns follow the logic of the structural stability Research Council. It used higher factor of safety than AISC to account for the dynamic nature of the loads. These equations are expressed in terms of C_c, as defined previously.

$$\text{For } (kL/r) < C_c,\ \sigma_a = \left(\frac{\sigma_y}{2.12}\right)\left[1 - \frac{(kL/r)^2}{2C_c^2}\right] \tag{i}$$

$$\text{For } (kL/r) > C_c,\ \sigma_a = \left(\frac{\pi^2 E}{2.12\,(kL/r)^2}\right) \tag{ii}$$

The factor of safety for columns (fS = 2.12) is used as 16 per cent more than the basic value of 1.82 used for the tension members.

$$\sigma_c = 18000 - 70.\,(l/r) \text{ (psi units)} \tag{5.36}$$

8. Rankine-Gorden formula or simply known as Rankine formula is adopted for (l/r) greater than 120 and less than 200

$$\sigma_c = \left(\frac{f}{1 + a\,.\,(l/r)^2}\right) \tag{5.37}$$

where a = constant. Its value depends on the material.

It is a semi-empirical formula. AISC recommended this formula in 1949 for $\frac{l}{r}$ >120 and <200.

$$\sigma_c = \left[\frac{18000}{1 + \frac{1}{18000}\cdot\left(\frac{l}{r}\right)^2}\right] \text{ (psi units)} \tag{5.38}$$

Rankine-Gorden formula, straight line formula and Johnson parabolic formula were extensively used in the past. The curves given by Rankine-Gordon formula and Johnson parabolic formula and Johnson parabolic formula when plotted are seen to be the same curve. Johnson parabolic formula is simple and it was more generally used.

9. Perry-Roberston formulas - British Standard Specifications for steel and girder bridges recommended Perry-Roberston formula for the determination of allowable stress in axial compression. This formula deals with **imperfect** such columns have minor crookedness of shape and there is small eccentricity of loading.

Bucling loads have been derived so far, by idealizing the column. It has been assumed that the columns are perfectly straight and the loading is concentric at every cross-section. These assumption simplify the derivation.

In practice, columns are not mathematically straight. Perfect columns do not exit therefore, it became desirable to derive formulae assuming slight initial deflection due to crookedness of the column.

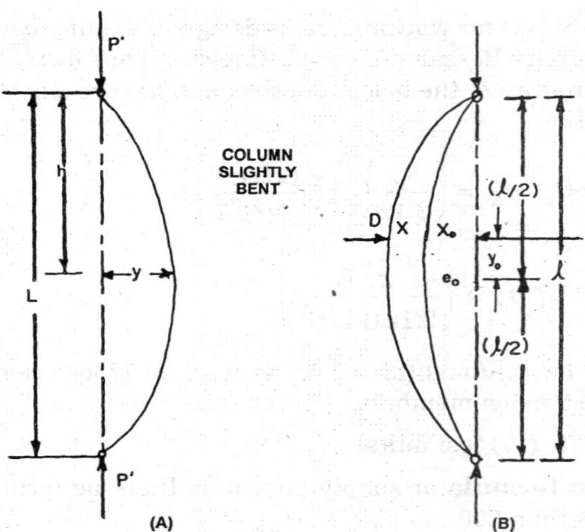

Fig. 5.23A.

Consider a column pinned at both the ends, (i.e., standard case in which the ends are hinged), as shown in Fig. 5.23(A). For such a column, the analysis is simple. It is easy to express equivalent length. It is assumed that the column is initially slightly bent. Further deflection due to bending moment is calculated. The bending stress at the centre of column is found and then added to the stress due to direct load. The maximum stress in the column is obtained for the given load and equated to the yield stress, σ_y of the material. As such an expression is derived to get the ultimate stress in the column.

After the application of load, the deflection increases. It becomes (x+xo) to the form as shown in Fig. 5.23B. **Let the initial deflection at the centre (or the eccentricity)** is e_o and that the initial curve is of the sine form. This shape is assumed to assist the mathematician. (i.e., for the simplicity of calculation). At the same time, half sine wave corresponds to the lowest node shape and it represents maximum influence on the actual behaviour. Then, at the point D, initial eccentricity is

$$x_o = e_o \cdot \cos(\pi y/l) \qquad \text{(i)}$$

where, e_o represents small eccentricity at mid-point consider a point D. Let the deflection from this initial position caused by the load applied by x.

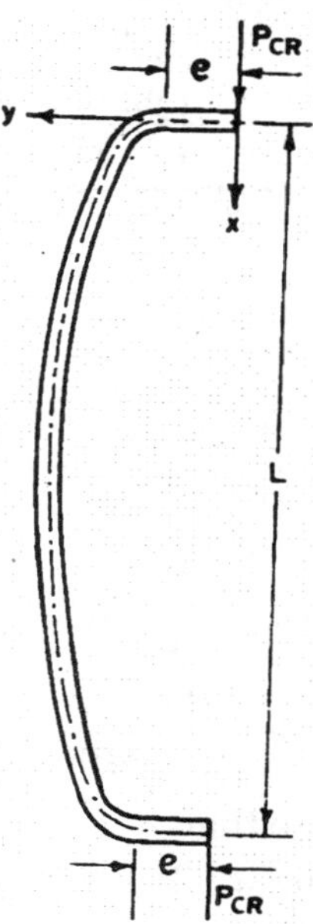

Fig. 5.23B.

The bending moment at D, after the application of load

$$M_D = P(x + x_o) = P[x + e_o \cos(\pi y/l)]$$

Since, the deflection is small, therefore

$$\left(-EI\frac{d^2x}{dy^2}\right) = M_D \text{ and Let } \left(\frac{P}{EI}\right) = n^2 \qquad \text{(ii)}$$

$$\left(\frac{d^2x}{dy^2}\right) = \left(\frac{-P}{EI}\right)[x + e_o \cos(\pi y/l)] \qquad \text{(iii)}$$

$$\text{Hence, } \left(\frac{d^2x}{dy^2}\right) = -n^2[x + e_o \cos(\pi y/l)] \qquad \text{(iv)}$$

The general solution of this differential equation is

$$x = A \sin(ny) + B.\cos(ny) + \left(\frac{e_o n^2 l^2}{\pi^2 - n^2 l^2}\right)\cos(\pi y/l) \qquad \text{(v)}$$

Note : Expression (v) may be expressed in the alternative shape as under by substituting $n^2 = (P/EI)$ and $P_E = (\pi^2 EI/l^2)$ as below

$$X = A \sin(ny) + B. \cos(ny) + \frac{e_o \left(\frac{P}{EI}\right).l^2}{\pi^2 - \frac{P}{EI}.l^2} \cos\left(\frac{ny}{l}\right)$$

$$x = A \sin(ny) + B \cos(ny) + \frac{\left(\frac{e_o . P.l^2}{EI}\right)}{\left(\frac{\pi^2 EI - P.l^2}{EI}\right)} \cos\left(\frac{ny}{l}\right)$$

$$x = A \sin(ny) + B \cos(ny) + \frac{l^2 (P.e_o)}{l^2\left(\frac{\pi^2 EI}{l^2} - P\right)} \cos\left(\frac{ny}{l}\right)$$

$$x = A \sin(ny) + B \cos(ny) + \frac{P.e_o}{(P_E - P)} . \cos\left(\frac{ny}{l}\right)$$

$$x = A \sin(ny) + B \cos(ny) + \frac{(P/PE)}{(1 - P/PE)} . e_o \cos\left(\frac{n.y}{l}\right)$$

For particular solution, it is to note that

$x = 0$, for $y = +(l/2)$ and $(-l/2)$ $\therefore A = 0$

Then, by taking $y = (l/2)$

$$0 = B \cos\left(\frac{nl}{2}\right) + \left(\frac{e_o . n^2 l^2}{\pi^2 - n^2 l^2}\right). \cos\left(\frac{\pi}{2}\right) \tag{vi}$$

In this $\left(\frac{nl}{2}\right)$ may be $= 0$, if π, (i.e. $n^2 l^2 = \pi^2$)

$$\left(\frac{P}{EI}\right).l^2 = \pi^2 \text{ . or } P = \left(\frac{\pi^2 EI}{l^2}\right) \tag{vii}$$

From Euler's formula, the value of P is same as obtained from the expression (vii). For all other cases, the value of P shall be less than that, so that

$\cos(nl/2)$ is not $= 0$, $\therefore B = 0$

Eq. 5.38 has two roots. The smaller root gives the value of compressive stress σ_c.

The expression (v) then becomes

$$x = \left(\frac{e_o . n^2 l^2}{\pi^2 - n^2 l^2}\right). \cos(\pi y / l) \tag{viii}$$

The value of x shall be maximum, when y = 0. Therefore

$$x_{max} = \left(\frac{e_o n^2 l^2}{\pi^2 - n^2 l^2}\right) \tag{ix}$$

Total deflection at the centre of column is $e = (e_o + x_{max})$. Therefore,

$$e = \left[e_o + \left(\frac{e_o\, n^2\, l^2}{\pi^2 - n^2\, l^2} \right) \right] = \left(\frac{1}{\frac{1}{\pi^2}(\pi^2 - n^2 . l^2)} \right) \quad e_o \left(1 - \frac{n^2 . l^2}{\pi^2 - n^2\, l^2} \right)$$

$$e = e_o \left(\frac{\pi^2}{\pi^2 - n^2 . l^2} \right) = \left(\frac{e_o}{\frac{1}{\pi^2}\left(\pi^2 - n^2\, l^2\right)} \right) = \left(\frac{e_o}{\left(1 - \frac{n^2\, l^2}{\pi^2}\right)} \right) \tag{x}$$

Let P_E be the Euler's buckling load and σ_c be (P/A), actual intensity of load. Since (n^2 = P/EI) then,

$$\left(\frac{n^2\, l^2}{\pi^2} \right) = \left(\frac{P}{EI} \right) . \left(\frac{l^2}{\pi^2} \right) = \left(\frac{P}{EA} \right) \left(\frac{l^2}{\pi^2} \right)$$

$$\left(\frac{P}{A} \right) \left(\frac{l^2 / r^2}{\pi^2\, E} \right) = \left(\frac{P}{A} \right) \left(\frac{1}{\pi^2\, E / (l/r)^2} \right) = \left(\frac{\sigma_c}{\sigma_E} \right)$$

The resultant compressive stress at the centre of column, in case, d_c is the distance from the compression edge to the neutral axis

$$\left(\frac{P}{A} \right) + \left(\frac{M}{I} d_c \right) = \frac{P}{A} \left(1 + \frac{e.d_c}{r^2} \right) \tag{xii}$$

The column fails, when the resultant stress in a ductile material (steel) becomes equal to σ_y. Therefore

$$\sigma_y = \sigma_c \left[1 + \frac{e_o . d_c}{r^2 \left(1 - \sigma_c / \sigma_E\right)} \right] \tag{xiii}$$

In case, ($e_o\, d_c / r^2$) = α, then, this expression (xiii) may be written as under

Rearranging the terms, it gives Ayrton and Perry formula, 1886 as in Eq. (5.39)A.

$$\sigma_y = \sigma_c \left[1 + \frac{\alpha_1 . \sigma_E}{\left(\sigma_E - \sigma_c\right)} \right] \tag{xiv}$$

$$\sigma_y = \left(\sigma_c + \frac{\alpha_1\, \sigma_c . \sigma_E}{\left(\sigma_E - \sigma_c\right)} \right)$$

or $$\left(\sigma_y - \sigma_c\right) = \left[\frac{\alpha_1 . \sigma_c . \sigma_E}{\left(\sigma_E - \sigma_c\right)} \right] \tag{xv}$$

or $$\left(\sigma_y - \sigma_c\right) \left(\sigma_E - \sigma_c\right) = \alpha_1 . \sigma_E . \sigma_c \tag{5.39A}$$

where, α_1 is the imperfection factor (Perry's - factor) and ($\alpha_1 = e_o . d_c / r_y^2$)

British Codes had used this column strength curve developed on the basis of the studies of **Ayrton and Perry** [The Engineer, December, 10, 1886]. From Eq. 5.39A.

$$\sigma_c^2 - \left[\sigma_y + (1+\alpha)\,\sigma_E\right]\sigma_c + \sigma_E \cdot \sigma_y = 0 \qquad (5.39B)$$

The primary root of this expression (xvi) may be written on the basis of quadratic equation using Shridhar Achary's Sootra (i.e., Hindu method). [See Reference given at the end of this chapter]. It is worth to note that this method of determining roots of quadratic equation was known world-wide.

$$\left[Ax^2 + Bx + C = 0\right] \quad \text{for which} \quad x = \left[\frac{-B \pm (B^2 - 4A.\,C)^{1/2}}{2A}\right]$$

As such the value of critical stress in a column, σ_c may be expressed as under:

$$\sigma_c = \left[\frac{\sigma_c + (1+\alpha_1)\,\sigma_E}{2}\right] - \left[\left\{\frac{\sigma_y + (1+\alpha_1)\,\sigma_E}{2}\right\} - \frac{4\,\sigma_y \cdot \sigma_E}{4}\right]^{1/2} \qquad (5.39C)$$

From eq. 5.39(C) $\phi = \left[\sigma_y + (1+\alpha_1)\,\sigma_E\right]$

$$\sigma_c = \phi - \left[\left\{\phi^2 - \sigma_y\,\sigma_E\right\}^{1/2}\right], \quad \left[\left\{\phi^2 - \sigma_y\,\sigma_E\right\}^{1/2}\right] = (\phi - \sigma_c)$$

$$\left(\phi^2 - \sigma_y\,\sigma_E\right) = (\phi - \sigma_c)^2 \qquad \phi^2 - (\phi - \sigma_c)^2 = \sigma_y\,\sigma_E$$

$$\left[\phi - (\phi - \sigma_c)^2\right] = \sigma_y\,\sigma_E \qquad \sigma_c = (2\,\phi - \sigma_c) = \sigma_y\,\sigma_E$$

$$\sigma_c = \left[\frac{\sigma_y\,\sigma_E}{(2\,\phi - \sigma_c)}\right] = \left[\frac{\sigma_y\,\sigma_E}{\phi + \phi - \sigma_c}\right]$$

Substituting the value, this gives compressive strength

$$\sigma_c = \left(\frac{\sigma_y\,\sigma_E}{\phi + \phi - \phi + \left[\left\{\phi^2 - \sigma_E\,\sigma_y\right\}^{1/2}\right]}\right)$$

Perry strut formula

$$\sigma_c = \left[\frac{\sigma_y\,\sigma_E}{\phi + \left(\phi^2 - \sigma_E\,\sigma_y\right)^{1/2}}\right] \qquad (5.40)$$

Note : Eq. 5.40 is the smaller root of Eq. (5.39)A/Eq. (5.39) B. BS : 5950 (Part 1) specifies the value of compressive strength in this form. Appendix C, page 98

Alternatvely

Consider a column hinged at both the ends. It carries a compressive force, P acting at the centroid of each end. Axis of y passes through the centroid of every-cross section (i.e, it is longitudinal vertical axis of the member). The cross-section translates and rotates. The equilibrium conditions may be derived for the column deformed slightly by the **torsional flexural buckling.**

$$E\,I_y\,.\,v^{IV} + P.\,v'' \quad + \; P_{zs}\;\phi'' = 0 \qquad \text{(i)}$$

$$E\,I_x\,.\,w^{IV} + P.\,w'' \quad - \; P_{ys}\;\phi'' = 0 \qquad \text{(ii)}$$

$$E\,I_w\,.\,\phi^{IV} + (Pr_s^2\;.\,\phi'' - gJ)\;\phi'' + Pxs\,.\,v'' - P.\,y_s\;w'' = 0 \qquad \text{(iii)}$$

where,

f = Rotation

E Cw = Warping rigidity

GJ = Torsional rigidity

v and w = Displacements in the y, x directions, respectively

ys, xs = Co-ordinates of the shear-centre and

$$E\,I_y = \int y^2\;dA \qquad \text{(iv)}$$

$$E\,I_x = \int x^2\;dA \qquad \text{(v)}$$

$$r_s^2 = (I_s / A) \qquad \text{(vi)}$$

where,

Is = Polar moment of inertia about shear centre

A = Cross-sectional area

The buckling load may be obtained by solving the eigen-value problem controlled by the expressions (i), (ii) and (iii) and the boundry conditions of y

At x = O, L

$$v = v'' = w = w'' \quad = \phi = \phi'' = \;0 \qquad \text{(vii)}$$

Consider two cases of the beam, cross-section.

Case 1. When the cross-section of beam is doubly symmetric

The shear centre and the centroid coincide. As such, ys, xs and rs disappear. Three expressions (i), (ii) and (iii) become independent of each other. The critical lods may be found as under

For flexural buckling

$$P_{y.cr} = \left(\pi^2\,E\,I_y / L^2\right) \qquad \text{(viii)}$$

$$P_{x.cr} = \left(\pi^2\,E\,I_x / L^2\right) \qquad \text{(ix)}$$

Torsional buckling

$$P_{\phi.cr} = \frac{1}{r_s}\left(GJ + \pi^2\,E\,C_w/L^2\right) \qquad \text{(x)}$$

For all the cases, the buckling mode is in the shape of sin (πy/L). The critical load shall be smallest of (viii), (ix) and (x) of practical importance for a relatively short column with low GJ and EC_w, the torsional buckling may take place.

Case 2 : When the cross-section of beam is symmetrical only to the y-axis (one-axis)

$$E.I_y.\ v^{IV} + P.\ v'' = 0 \quad \text{(xi)}$$

$$E.I_x\ w^{IV} + P.\ w'' - P_{ys} \cdot \phi'' = 0 \quad \text{(xii)}$$

$$EC_w\ \phi^{IV} + \left(P.\, r_s^2 - GJ\right)\phi'' - P_y\ w'' = 0 \quad \text{(xiii)}$$

First expression (xi) shows that the flexural buckling in the x-y plane occurs independently and the corresponding critical load is given by $P_{y.cr}$.

While, the flexural buckling in the y-z plane and the torsional buckling are coupled. It is assumed that the buckling modes are described by $w = A_1 \sin(\pi y/L)$ and $\phi = A_2 \sin(\pi y/L)$. From the expressions (xii) and (xiii)

$$\begin{bmatrix} (P - P_{y.cr}) & -P_{ys} \\ -P_{ys} & r_s^2\left(P - P_{\phi.cr}\right) \end{bmatrix} \begin{Bmatrix} A_1 \\ A_2 \end{Bmatrix} = \begin{Bmatrix} 0 \\ 0 \end{Bmatrix} \quad \text{(xiv)}$$

This eigen problem results to (xiv)

$$f_{(P)} = r_s^2\left(P - P_{\phi.cr}\right)\left(P - P_{y.cr}\right) - \left(P_{ys}\right)^2 = 0 \quad \text{(xv)}$$

Since, $f_{(0)} = r_s^2\ P_{\phi.cr} \cdot P_{x.cr}$ is more than zero, and

$f_{(P_{\phi.cr})} = -\left(P_{ys}\right)^2$ is less than, zero, and

$f\left(P_{y.cr}\right) = -\left(P_{ys}\right)^2$ is less than zero.

It is obvious that the critical load is lower than $P_{y.cr}$ and $P_{\phi.cr}$. In case, this load is smaller than $P_{y.cr}$ than, the torsional flexural buckling will take place.

In case, there is no axis of symmetry in the cross section, all the three expressions are coupled. Torsional-flexural buckling occurs in this case, since, the critical load for this teckling mode is ower than any of the three loads stated above.

Allowable stress in axial compression

$$\sigma_a = \left(\frac{1}{FS}\right)\left[\frac{1}{2}\left\{\sigma_y + \left(1 + \alpha_1\right)E\right\} - \frac{1}{2}\left[\left\{\sigma_y + \left(1 + \alpha_1\right)\sigma_E\right\}^2 - 4\,\sigma_y \cdot \sigma_E\right]\right]^{1/2} \quad \text{(5.40A)}$$

where, FS is the factor of safety, 1.70 and σ_y is the yield stress for steel.

It is interesting to consider that in Eq. (5.39)B that, in case, original eccentricity $e_o = 0$ $\therefore\ \alpha_1 = 0$. Then, σ_c gives the value equal to that, of Euler's value.

$$\sigma_c = \left[\frac{\left(\sigma_y + \sigma_E\right) - \left[\left(\sigma_y + \sigma_E\right)^2 - 4\,\sigma_y \cdot \sigma_E\right]^{1/2}}{2}\right] \quad \text{(5.40B)}$$

$$\sigma_y = \left(\frac{(\sigma_y + \sigma_E) - [(\sigma_y - \sigma_E)]}{2}\right) = \sigma_E = \left(\pi^2 \text{ E}/(l/r)^2\right) \qquad (5.40C)$$

Professor Perry used the value of α_1 as 0.0015. Later on, Roberston (1925) examined the experimental data and suggested that the initial eccentricity, e_o may be adopted as (l/1000), where, l is the length of the column. Then, the **imperfection factor** (a parameter which permits the influence of lack of straightness and eccentricity). Value of α depends upon shape of th cross-section.

$$\alpha_1 = \left(\frac{e_o \cdot d_c}{r^2}\right) = \left(\frac{1}{1000} \text{x} \frac{d_c}{r^2}\right) = \left(\frac{1}{1000} \cdot \frac{d_c}{r}\right)\left(\frac{l}{r}\right) = \left(\frac{4000\, l}{r^2}\right)$$

Let, α_1 be equal to $\alpha \cdot \lambda$, ($\therefore \lambda = l / r$).

For beams and columns, bending about the strong-axis and weak-axis, this value (0.001 dc/r) may be taken as 0.0012 and 0.0020, respectively. Whereas, for rectangular hollow section, it is adopted as 0.0013 and for tee-section, bending in the plane of stem, it is taken as 0.0028. The value of α = 0.003 (l/r) was considered appropriate. For the design of columns.

Values of imperfection factors (α - Alpha)

The values of hot-rolled and welded sections for the buckling about xx-axis and yy-axis are as follows.

TABLE 5.4(A). SHAPE OF COLUMN SECTION VALUES OF ROBERSTON CONSTANT (a/1000) BUCKLING AXES & CURVES

A. Hot-rolled sections		
α - Alpha for Buckling about		
xx-axis		yy-axis
1. I - section	0.0020	0.0035
2. Hollow sections	0.0020	0.0020
3. I-sections with welded cover plates	0.0035	0.0035
4. Angle, channel, tee-sections compound sections Two sections bac to back (battened/laced)		
Backling about any-axis		

TABLE 5.4(B). SHAPE OF COLUMN SECTION VALUES OF ROBERSTON CONSTANT (a/1000) BUCKLING AXES & CURVES

B. Welded sections	xx-axis	yy-axis
1. Plates I-sections		
upto 40 mm thickness	0.0035	0.0055
Over 40 mm thickness	0.0035	0.0080
2. Box-secton		
Upto 40 mm thickness	0.0035	0.0035.
Over 40 mm thickness	0.0055	0.0055

Above values of α are from British Codes.

Residual stresses in remain present in the hot rolled and welded sections. Their distribution and magnitude are different.

In welded sections, these stresses are of high magnitude and reduce the ultimate capacity.

20 N/mm² reduction has been recommended in British Codes.

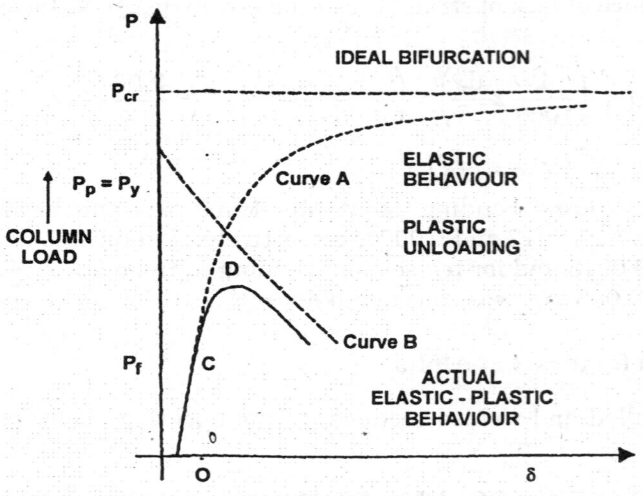

Fig. 5.24. Actual load deflection and theoretical load deflection behaviour

Modification to Perry-Roberston Formula

It has been mentioned earlier that the very stocky columns resist loads larger than their squash loads, (σ_y . A), because of the strain-hardening for low values of slenderness ration. Whereas, the column strength found by using Eq. 5.40B is lower than (σ_y . A) for low values of slenderness ratio. There remains a discrepancy in two values found separately. Recently, while finalising British and Eurocodes, some modification in Euler's formula was suggested.

$$\sigma_E = \left(\frac{\pi^2 E}{(l/r)^2}\right) = \left(\frac{\pi^2 E}{\lambda^2}\right)$$

where, λ is (l/r), [slenderness ratio] for the column. In this expression, the modification was made as under *limiting slenderness ratio*, λ_o is introduced in above expression.

$$\sigma_E = \left(\frac{\pi^2 E}{(l/r)^2}\right) = \left(\frac{\pi^2 E}{(\lambda - \lambda_o)^2}\right) \qquad (5.41)$$

where, $\lambda_o = 0.20 . (\pi) (E/\sigma_y)^{1/2}$

Perry's factor a_1 for flexural buckling under load is obtained from

$$\alpha_1 = \frac{1}{1000}.\alpha.(\lambda - \lambda_o) \text{ but not less than zero}$$

where, α = Roberston constant. The values of $0.001\ \alpha = 0.0020, 0.0035, 0.0055$ and 0.0080

Reference BS : 5950 (Part I) Page 98, Appendix, C2.

By modifying the slenderness ratio λ to $(\lambda - \lambda_o)$ a plateau to the design curve at low slenderness values is introduced.

In the British Design Code, (BS : 5950-Part I), the column curve used is based on Eq. 5.41. It fits appropriately for the observed test data for all values of $(\lambda > \lambda_o)$. It is to note that for $(\lambda < \lambda_o)$, the calculations for (σ_E) is not required. The column fails by squashing at (σ_y). there is no drop in observed strengths in very short columns.

From the expression (xi), since $[n^2 = P/EI]$ and axially loaded column hinged at the ends, $P_E = (\pi^2\ EI/l^2)$

$$\left[\frac{n^2 l^2}{\pi^2}\right] = \left(\frac{P}{EI}.\frac{l^2}{\pi^2}\right) = \left(\frac{P}{\pi^2\ EI/l^2}\right) = \left(\frac{P}{P_E}\right) \text{ or } \left(\frac{nl}{2}\right) = \frac{\pi}{2}.\left(\frac{P}{P_E}\right)^{1/2} \quad (5.42A)$$

The expression (ix) shows that due to the influence of the end load, the deflection at the centre becomes $[P_E/(P_E - P)]$ times its initial value. In case, there is originally, an eccentricity, e_o, the central deflection becomes

$[(P_E + 0.26\ P) / (P_E - P)]$ times the eccentricity. Then, it may be written as under:

$$\left(\frac{\text{Max B.M. per unit deflection}}{\text{Max B.M. per unit eccentricity}}\right) = \left(\frac{P_E}{P_E + 0.26\ P}\right) = \left(\frac{1}{1 + 0.26\ P/P_E}\right) \quad (5.42B)$$

The maximum value of this may be $[(1/1.26) = 0.7936 = (0.8) = (4/5)$, in the worst case when P approaches P_E.

In case, the material remains elastic, it may be proved that the applied load increases the initial deflection at every point along the length of the column by a factor, $[1/(1 - P/P_E)]$. This factor is known as multiplier-factor or magnification factor. The value of $[1/(1 - P/P_E)]$ is greater than unity.

As the deflection increases, the bending moment on the cross-section also increases.

The struts are considered as secondary members. These members do not carry primary loads. These members are used to brace other structural members or the structures as a whole. AISC specification 1963 recommended that for the struts having slenderness ratio (l/r) greater than or equal to 120 and less than 200, the allowable stress is increased by factor, K.

$$K = \left[\frac{1}{1.6 - \frac{1}{200}\left(\frac{l}{r}\right)}\right] \quad (5.43)$$

The factor *K* for end-conditions is adopted as unity.

In the above column formulae, the columns were considered perfectly straight and uniform. The load was considered perfectly axial. In actual practice, the ideal column and loading do not exist. The load supported by the column is eccentric with respect to the longitudinal axis of the column. The initial crookedness and other imperfections of column and imperfectness of the service load distribution give small eccentricity. The secant formula accounts for the initial crookedness of the column and imperfectness of the axial loading. Indian Standard Institution (ISI) adopted the secant formula for the determination of allowable stress in axial compression in 1962. The secant formula was considered as design formula in IS : 800-1962 as per Indian Standard Specification.

12. Secant formula

Consider an initially straight column. Suppose a load P is applied to it at a distance e from the centre line of the column. The deflected shape of the column would be as shown in Fig. (5.20).

The bending moment at any point on the deflected shape of the column

$$M = (P.y) \quad \text{(i)}$$

The negative sign in adopted since the bending moment is sagging. The differential equation to the elastic curve is as under :

$$EI\frac{d^2y}{dx^2} = M = -P.y \quad \text{(ii)}$$

$$\text{or} \quad \left(\frac{d^2y}{dx^2}\right) + \left(\frac{P}{EI}\right)y = 0, \quad \left(\frac{d^2y}{dx^2}\right) + n^2 y = 0 \quad \text{(iii)}$$

The general solution of this differential equation is

$$y = A.\sin(nx) + B.\cos(nx), \text{ where } n = \left(\frac{P}{EI}\right)^{1/2} \quad \text{(iv)}$$

The constants of this equation can be found by applying the boundary values.

At $x = 0$, $y = e$ and $x = l$, $y = e$, $A = \left(\frac{1-\cos(nl)}{\sin(nl)}\right).e$ and $B = e$

The expression (v) reduces to

$$y = e.\left[\frac{1-\cos(nl)}{\sin(nl)}.\sin(nx) + \cos(nx)\right] \quad \text{(iv)}$$

Note : The deflection at every point along the length of column is enhanced due to increase in the applied load P, by a factor.

$\{1/(1-P/P_E)\}$. This factor is known as the **magnification factor** (or **amplification factor**)

This deflectin tends to infinity as P increases to P_{cr} as shown in Fig. 5.20 (C).

As the deflection increases, the bending moment on the cross-section also increases.

Resulting bending stress (M.x/I), on the concave face of the column is compressive and it gets added to the axial compressive force of (P/A).

With the increase in the axial force, P, the stress an the concave face reaches yield stress, σ_y.

Let the load P be designated as P_y at first yield at point C, Fig. 5.20 (C). When the load P_y is further increased, then, the yield spreads across the cross-section. The bending stiffness is deteriorated when the maximum load reaches P_E, then, the column collapses the slenderness ratio influences the extent of the post-first load increase and section plastification.

Theoretical rigid plastic curve B, is shown in Fig. 5.20 (C). It is assumed that [$P_E > (P_p = A \cdot \sigma_y)$] P_E and P_P are the upper bounds to the loads P_y.

In case, the column is **stocky,** P_E shall be very large, but P_P may be expacted close to P_y.

For **very stocky members,** (i.e, a column having very low slenderness value) the initial out of straightness

(Which one is more a function of length than cross-sectional dimensions) has very negligible effect and the failure occurs by **Plastic squash load.**

In every slender columns, the collapse begins by elastic buckling. The lower bound curve is close to the elastic critical stress (σ_{cr}) curve.

For the intermediate values of slenderness, the **effect of initial out of straightness,** the lower bound curve is significantly below (σ_y) line and (σ_{cr}) line.

The maximum deflection in the column occurs at centre, i.e, when **x** is equal to $l/2$

$$y_{max} = e.\left[\frac{1-\cos(nl)}{\sin(nl)} . \sin\frac{(nl)}{2} + \cos\frac{(nl)}{2}\right] = e\left[\frac{2\sin\frac{nl}{2}}{2\sin\frac{nl}{2}\cos\frac{nl}{2}}\sin\frac{nl}{2} + \cos\frac{nl}{2}\right] \quad \text{(v)}$$

$$y_{max} = e.\left[\frac{\sin^2\frac{nl}{2} + \cos^2\left(\frac{nl}{2}\right)}{\cos\frac{nl}{2}}\right] = e.\sec\left(\frac{nl}{2}\right)\left[\because \sin^2\left(\frac{nl}{2}\right) + \cos^2\left(\frac{nl}{2}\right) = 1\right] \quad \text{(vi)}$$

The maximum bending moment in the column

$$M_{max} = P.e\sec\left(\frac{nl}{2}\right) \quad \text{(vii)}$$

The maximum stress in the column (combined direct stress plus bending stress)

$$\sigma_{max} = \left(P_A + \frac{M_{max}}{I}.c\right) \quad \text{(viii)}$$

where c = Distance to the extreme fibre from centre line of column

$$\sigma_{max} = \left[\frac{P}{A} + \frac{P}{A}.\left(\frac{ec}{r^2}\right)\sec\frac{nl}{2}\right] \quad \text{(ix)}$$

$$\text{where, } I = Ar^2 \quad n = \left(\frac{P}{EI}\right)^{1/2}, \quad \text{and}\left(\frac{P}{A} = \sigma_c\right)$$

In the expression (x), it is inconvenient to obtain the value of P, since, n itself is a function of P. In arithmetical cases, a solution may only be found by trial. In such cases, the following procedure will prove very useful. For axially loaded column pinned at both the ends, Euler's load is $P_E = \left(\pi^2 EI/l^2\right)$. Then,

$$\theta = \left(\frac{nl}{2}\right) = \left(\frac{1}{2}\right)\left(\frac{P}{EI}\right)^{1/2} = \frac{\pi}{2}\left(\frac{Pl^2}{\pi^2 EI}\right) \quad \text{(x)}$$

$$\text{or} \quad \theta = \frac{\pi}{2}\left(\frac{P}{\pi^2 EI/l^2}\right)^{1/2} = \frac{\pi}{2}\left(\frac{P}{P_E}\right)^{1/2} \quad \text{(xi)}$$

Now, an extremely close approximation to the value of sec. θ, for all values of θ between zero and (π/2) is as under :

$$\sec(\theta) = \frac{1+\left(\frac{4\theta^2}{\pi^2} \text{x}\, 0.26\right)}{1-\left(\frac{4\theta^2}{\pi^2}\right)} = \left(\frac{P_E + 0.26\,P}{P_E - P}\right) \quad \text{(xii)}$$

Substituting this value of θ, (i.e., sec (n*l*/2) in the expression (xii) and let (ec/r^2) be equal to K (xiii)

$$\sigma_{max}\,.\,A = \left[P + P.K.\left(\frac{P_E + 0.26\,P}{P_E - P}\right)\right]$$

$$\left(\sigma_{max}\,A\,.\,P_E - \sigma_{max}\,.\,A\,.\,P\right) = \left|\left(P\,.\,P_E - P^2\right) + \left(P\,.\,P_E\,K + 0.26\,P^2\,K\right)\right|$$

$$P^2\left(1 - 0.26\,K\right) - P\left[P_E\left(1+K\right) + \sigma_{max}\,.\,A\right] + \sigma_{max}\,.\,A + P_E = 0 \quad (5.44)$$

Eq. 5.44 may be easily solved in any case. In case, initial eccentricity e = 0, then, k is also equal to zero. Then, Eq. 5.44 may be written as below

$$P^2 - P\left(P_E + \sigma_{max}\,A\right) + \sigma_{max}\,.\,A\,.\,P_E = 0$$

$$P = \frac{\left(P_E + \sigma_{max}\,A\right) \pm \left(P_E - \sigma_{max}\,A\right)}{2} \quad (5.45)$$

Eq. 5.45 gives $P = P_E$, that is, Euler's value and $P = \sigma_{max}\,.\,A$, that to a short column, for which bending is negligible and the direct stress will reach the limiting value of σ_{max}. In case, nl = π, (i.e., $P = P_E$) the deflection for the extreme fibre on the axis becomes infinite and the column fails by buckling.

$$\text{Then,} \quad \sigma_{max} = \frac{P}{A}\left[1 + \frac{ec}{r^2}\sec\frac{l}{2r}\left(\frac{P}{EA}\right)^{1/2}\right]$$

$$\sigma_c = \left(\frac{\sigma_{max}}{1 + \frac{ec}{r^2}\sec\left[\frac{l}{2r}\left(\frac{P}{EA}\right)^{1/2}\right]}\right) \quad \text{(x)}$$

This expression is known as secant formula.

Let m be the factor of safety, and therefore, $\sigma_c = m \cdot \sigma_a$.

where σ_a = Allowable average stress in axial compression, as it was recommended in IS: 800 -1962.

The maximum stress σ_{max} reached can be yield stress, f_y,

Then

$$\sigma_a = \left(\frac{(\sigma_y / m)}{1 + \frac{ec}{r^2} \sec \frac{l}{r}\left(\frac{m\sigma_a}{4E}\right)^{1/2}} \right) \quad \text{(xi)}$$

Indian Standard Code of Practice for general construction steel IS : 800-1984 has dropped this secant formula for calculating the allowable average stress in axial compression. In secant formula, the value of (ec/r^2) was taken as 0.20. The formula takes into account any initial crookedness of the column and imperfectness of axial loading.

Therefore, for (l/r) = 0 to 160

$$\sigma_a = \left(\frac{(\sigma_y/m)}{1 + 0.20 \sec\left(\frac{l}{r}\left(\frac{m\sigma_a}{4E}\right)^{1/2}\right)} \right) \quad (5.46)$$

where σ_y = Guaranteed minimum yield stress

m = Factor of safety taken as 1.68 on the yield point

(l/r) = Slenderness ratio

l = Effective length of the member

r = Appropriate radius of gyration

E = Young's modulus of elasticity, 20,47,00 N/mm^2

For values of (l/r) = 160 and above

$$\sigma_a' = \sigma_a \cdot \left(1.2 - \frac{l}{800\,r}\right) \quad (5.47)$$

where σ_a' = Allowable average axial stress in compression as it was recommended in IS : 800-1962 for $(l/r) > 160$.

The secant formula for axial compression as it was recommended to determine permissible stress in IS : 800-1962 has been dropped in IS : 800-1984.

Professor Ashok K. Jain of University of Roorkee made the comparison of permissible compressive stresses as given by different national standards in his paper titled as `compressive stresses in columns' published in International Journal of structures, Vol. 4, No. 2, April 1984. The disadvantages of secant formula were pointed out. The secant formula provides a conservative estimate of the average column stress at maximum load for an eccentrically loaded steel column. The secant formula is not convenient for hand calculations. It needs number of iterations to calculate the permissible axial stress as this term appears on both the sides of the equation.

The secant formula does not show an yield plateau for the columns of slenderness ratio upto 30. The secant formula gives very conservative values of columns of slenderness ratio up to 90 as compare to those given by AISC, BS 449, IS 802 and AS : 1250. Further, this secant formula gives higher values of permissible stresses for the columns having slenderness ratio beyond 90.

Secant formula remained a long-time rational formula. It could be applied to the complete range of slenderness ratio. Such a formula gives the correct relationship in case, one assumes the deviation from elastic buckling is totally due to initial curvature and accidental eccentricity. However, it is now agreed that the main cause of inelastic buckling is due to residual stress. The secant formula has lost much of its significance. The value of limit load P obtained from secant formula and the value of critical load found from eccentrically loaded column represent two basically different design criteria. The difference between these values ranges widely, it may be small or large depending on shape of the cross-section slenderness, relative eccentricity and stress-strain characteristic. It is realized that the loads obtained from this criterian are not directly related to the actual buckling load.

Instead of the secant formula, Merchant-Rankine formula has been specified in IS : 800-1984.

Andrews formula : Andrew derived this formula on the basis that, in practice, there is no column, which remains mathematically straight. This formula was derived by assuming slight initial deflection due to **crookedness** of the column.

A column is shown in Fig. 5.20 A (pinned at both ends). It is assumed that the pillar is initially slightly bent. After the application of load, the deflection increases. It is assumed that the initial central deflection or the eccentricity is e_o and the initial curve is of sine form. This shape is adopted as it helps in mathematics.

Consider a point D (Fig. 5.20A) at which the initial eccentricity is given by

$$x_o = e_o \cos\left(\frac{\pi y}{l}\right) \text{ and} \qquad \text{(i)}$$

let the deflection from the initial position due to load W, then, the bending moment at the point D is given by

$$M_D = P\,(x + x_o) \qquad \text{(ii)}$$

$$M_D = P \cdot \left[x + e_o \cos\left(\frac{\pi y}{l}\right)\right] \qquad \text{(iii)}$$

$$\text{Since, } EI\left(\frac{d^2x}{dy^2}\right) = -M_D \qquad \text{(iv)}$$

$$EI\left(\frac{d^2x}{dy^2}\right) = -P\left[x + e_o \cdot \cos\left(\frac{\pi y}{l}\right)\right] \qquad \text{(v)}$$

Let $(P/EI) = n^2$, then

$$\left(\frac{d^2x}{dy^2}\right) = -n^2\left[x + e_o \cdot \cos\left(\frac{\pi y}{l}\right)\right] \qquad \text{(vi)}$$

General solution of this differential equation is

$$x = A,\ \sin(ny) + B,\ \cos(ny) + \left(\frac{e_o\, n^2 . l^2}{\pi^2 - n^2 . l^2}\right) \cos\left(\frac{\pi y}{l}\right) \qquad \text{(vii)}$$

For the particular solution, for

$$x = 0,\ \ y = +\left(\frac{l}{2}\right) \text{ and } -\left(\frac{l}{2}\right)$$

then, A = O. By considering $y = +(l/2)$

$$0 = B\cos\left(\frac{nl}{2}\right) + \frac{e_o\, n^2\, l^2}{(\pi^2 - n^2\, l^2)} \cos\frac{\pi}{2} = B\cos\left(\frac{nl}{2}\right) + 0 \qquad \text{(viii)}$$

$\cos\left(\frac{nl}{2}\right)$ may only be zero, in case $nl = \pi$, i.e. $n^2 l^2 = \pi^2$

$$\text{i.e., } \left(\frac{P}{EI}\right) . l^2 = \pi^2 \text{ or } P = \left(\frac{\pi^2 . EI}{l^2}\right) \qquad \text{(ix)}$$

It is Euler's value for the critical load. Here, the cases which shall be considered the value of P will be less than that, so that, $\cos\left(\frac{nl}{2}\right)$ is not = 0. Therefore, B = 0

Therefore, Expression reduces to

$$x = \left(\frac{e_o . n^2\, l^2}{(\pi^2 - n^2\, l^2)}\right) \cos\left(\frac{\pi y}{l}\right) \qquad \text{(x)}$$

The value of x shall be maximum when y = 0,

$$x_{max} = \left(\frac{e_o . n^2 . l^2}{\pi^2 - n^2\, l^2}\right) \qquad \text{(xi)}$$

Total deflection or effective eccentricity at the centre of the pillar is given by

$$e = e_o + x_{max},\ e = e_o\left(1 + \frac{n^2\, l^2}{\pi^2 - n^2\, l^2}\right) \qquad \text{(xii)}$$

$$e = e_o\left(\frac{\pi^2 - n^2\, l^2 + n^2\, l^2}{\pi^2 - n^2\, l^2}\right) \qquad \text{(xiii)}$$

$$e = e_o\left(\frac{\pi^2}{\pi^2 - n^2\, l^2}\right), \qquad e_o = \left(\frac{e_o}{1 - \left(\frac{n^2 . l^2}{\pi^2}\right)}\right),$$

Now, let σ_E be the Euler's critical stress and σ_c be the actual intensity of load = (P/A), then

$$\left(n^2 . \frac{l^2}{\pi^2}\right) = \left(\frac{P}{E.I} . \frac{l^2}{\pi^2}\right) \quad (\therefore I = A . r^2) \tag{xiv}$$

$$\left(n^2 . \frac{l^2}{\pi^2}\right) = \left(\frac{P}{A} . \frac{l^2}{r^2} . \frac{1}{\pi^2 E}\right) \tag{xv}$$

$$\left(n^2 . \frac{l^2}{\pi^2}\right) = \left(\frac{P}{A} . \frac{1}{\pi^2 E / (l/r)^2}\right) = \left(\frac{\sigma_c}{\sigma_E}\right)$$

Substituting in above Eqs. 5.

This equation may be written as

$$e = \left[\frac{e_o}{\left(1 - \frac{\sigma_c}{\sigma_E}\right)}\right] \tag{xvi}$$

In case, x_d is the distance from the compression edge to the neutral axis, then, the resulting compressive stress at the centre of the pillar,

$$= \left(\frac{P}{A} + \frac{M . x_d}{I}\right) \qquad (\therefore M = Pe \text{ and } I = A.r^2) \tag{xvii}$$

$$= \left(\frac{P}{A} + \frac{P. e . x_d}{A . r^2}\right) = \frac{P}{A} . \left(1 + \frac{e . x_d}{r^2}\right) \tag{xviii}$$

In case, this resulting stress reaches the value of yield-stress, σ_y then, the ductile material shall fail. Then

$$\sigma_y = \sigma_c . \left(1 + \frac{e . x_d}{r^2}\right)$$

Substituting the value of e from the expression (xvi)

$$\sigma_y = \sigma_c \ . \left[1 + \frac{e_o}{\left(1 + \frac{\sigma_c}{\sigma_E}\right)} \frac{x_d}{r^2}\right] \tag{xix}$$

In case, $\alpha = \left(\frac{e_o . x_d}{r^2}\right)$, then

$$\sigma_y = \sigma_c \left[1 + \frac{\alpha . \sigma_E}{(\sigma_E - \sigma_c)}\right] \tag{xx}$$

$$\left(\sigma_y - \sigma_c\right) = \left(\frac{\alpha \cdot \sigma_E \cdot \sigma_c}{\sigma_E - \sigma_c}\right) \quad \text{(xxi)}$$

$$\left(\sigma_y - \sigma_c\right)\left(\sigma_E - \sigma_c\right) = \alpha \cdot \sigma_E \cdot \sigma_c \quad \text{(xxii)}$$

$$\sigma_E \cdot \sigma_y - \sigma_y \cdot \sigma_c - \sigma_c \, \sigma_E + \sigma_E^2 - \alpha \, \sigma_E \cdot \sigma_c = 0 \quad \text{(xxiii)}$$

$$\sigma_c^2 - \sigma_c \left\{\sigma_y + \left(1 + \alpha_1\right) \sigma_E\right\} + \sigma_E \cdot \sigma_c = 0 \quad \text{(5.48A)}$$

This is a quadratic expression (like, Ax² + Bx + c = 0), which may be solved by Hindu method, as suggested by Shridhar Acharya. There are two roots (two values)

$$\sigma_c = \left\{\frac{\sigma_y + (1 + \alpha_1)\,\sigma_E}{2} - \left[\left\{\frac{\sigma_y + (1 + \alpha)\,\sigma_E}{2}\right\}^2 - \sigma_y \, \sigma_E\right]^{1/2}\right\} \quad \text{(5.48B)}$$

Let $\phi = \{\sigma_y + (1 + \alpha)\,\sigma_E\}^2$

then $\sigma_c = \phi - \left[\left\{\phi^2 - \sigma_y \cdot \sigma_E\right\}^{1/2}\right]$ (5.49)

For any given material σ_y (yield stress) and σ_E (Euler's critical stress) are known. The values of the resulting stress, for the column at the critical state may then be found for any column.

Note : The values found by **E.S. Andrews** and those by **Ayroton and Perry (Eq. 5.39) (A)** are same (equal). At the same time, it is interesting to see that when the original eccentricity $e_o = 0$, then, $\alpha = 0$ then, substituting this in above Eq. (5.48)

$$\sigma_c = \left\{\frac{\sigma_y + \sigma_E - \left(\sigma_y - \sigma_E\right)}{2}\right\} = \sigma_E$$

Other root shall be $\sigma_c = \sigma_y$. This agrees with the correct interpretation of Euler's expression.

5.19 MERCHANT-RANKINE FORMULA

Indian Standard Code of Practice for general construction in steel IS : 800-1984 has recommended Merchant-Rankine formula as below

$$\left(\frac{1}{\sigma_{ac}} \cdot \frac{1}{\text{F.S.}}\right)^n = \left(\frac{1}{\sigma_{cc}}\right)^n + \left(\frac{1}{\sigma_y}\right)^n \quad \text{(i)}$$

or

$$\sigma_{ac} = \left(\frac{\sigma_E \, \sigma_y}{\text{F.S.}\left[\left(f_{cc}\right)^n + \left(f_y\right)^n\right]^{1/n}}\right) \quad \text{(ii)}$$

The direct stress in compression on the gross-sectional area of axially loaded compression member was found with a factor of safety of 10/6. It should not exceed 0.60 f_y not the permissible stress σ_{ac} calculated using the following formula.

$$\sigma_{ac} = 0.6 \cdot \left(\frac{\sigma_{cc}\,\sigma_y}{\left[(\sigma_{cc})^n + (\sigma_y)^n\right]^{1/n}} \right) \tag{5.50}$$

where, σ_{ac} = Permissible stress in axial compression in N/mm² (MPa)

σ_y = Yield stress in steel in N/mm² (MPa)

σ_{cc} = Elastic critical stress in compression, and $\left(f_{cc} = \frac{\pi^2 E}{(l/r)^2} \right)$

E = Young's modulus of elasticity of steel 2 x 10^5 N/mm² (MPa)

n = Imperfection factor. It is assumed as 1.4.

Professor D. Allen suggested the use of this formula for the design of columns in his paper title as `Merchant-Rankine Approach to Member Stability' a technical note published in Journal of Structural Division, ASCE, Vol. 104. No. ST 12, December 1978. Depending on the degree of residual stresses, the value of parameter takes the value of 1, 1-4 or 2. The value of *n* is taken as 1.4 for riveted and normal welded members. Its value is adopted as 1 where the high residual stresses are expected.

A constant stress for the members of very small slenderness ratio is not indicated by Merchant-Rankine formula. A cut-off point below which the permissible stress is constant, is specified by other standards to include this feature. A correction factor in the Merchant-Rankine expression was suggested by Maquoi and J. Rondal, in their discussions on `Merchant-Rankine Approach to Member Stability' by D. Allen published in Vol. 105, No. ST 11, November, 1979 as below

$$\sigma_{ac} = \left(\frac{1}{\text{F.S.}} \right) \left(\frac{\sigma_{cc} \cdot \sigma_y}{\left[(\sigma_y)^n + (\sigma_{cc})^n \left(1 - 0.15^{2n}\right)\right]^{1/n}} \right) \tag{iv}$$

This expression (iv) includes the constant yield plateau for the columns of slenderness ratio less than about 30. It gives about one to four per cent difference in values given by the expressions (iii) and (iv). IS : 800-1984 does not specify the expression (iv).

Ashok K. Jain in his paper referred above has mentioned in the conclusion that Merchant-Rankine formula is very simple. It has lot of potential for use in different situations. It is more convenient for use in case the multiple column curves are to be developed. The value of imperfection factor is simply changed. Only three parameters, namely, yield stress, critical stress and imperfection factor appear in Merchant-Rankine formula. A computer-program may easily be written for use of Merchant-Rankine formula. The values of allowable stresses for columns of slenderness ratio less than 90 obtained by Merchant-Rankine formula are higher than those found by the secant formula. However, these higher values closely agree with those determined by other national standards.

5.20 ROLLED STEEL SECTIONS USED AS COMPRESSION MEMBERS

Rolled steel sections used as compression members are shown in Fig. 5.24 with their approximate radii of gyration.

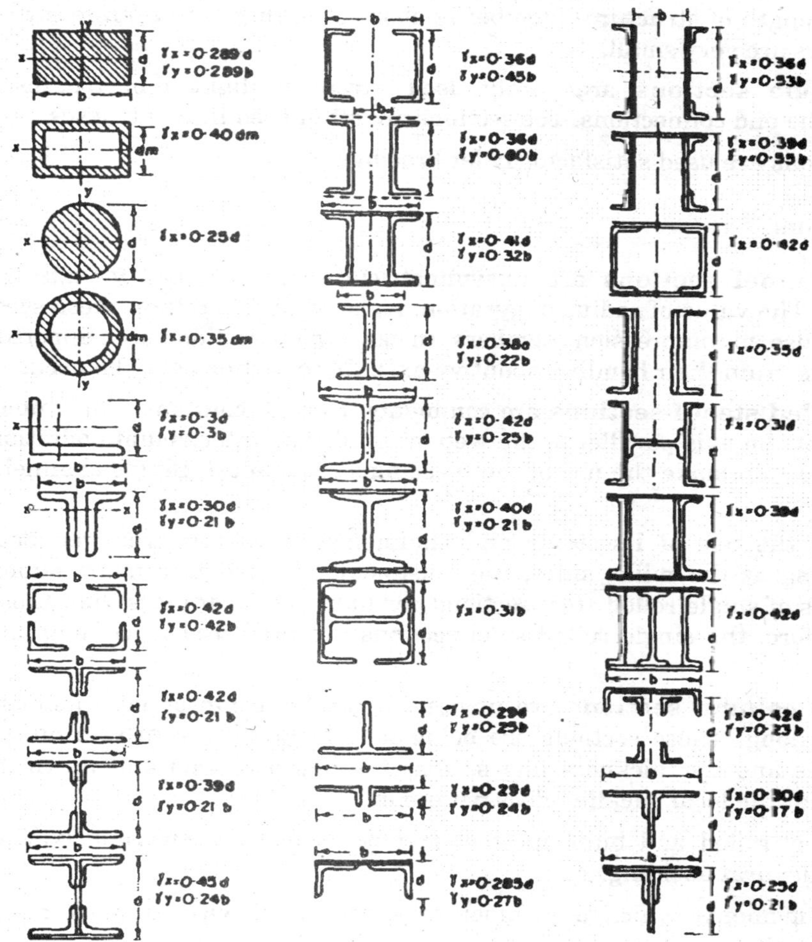

APPROXIMATE RADIUS OF GYRATION FOR COMMON SECTIONS FOR COMPRESSION MEMBERS

Fig. 5.24.

(Reference : J.A.L. Waddell `Bridge-engineering' John Wiley and Sons, Inc. New York, 1925).

A column or a compression member may be made of many different sections to support a given load. Few sections satisfy practical requirement in a given case. A **tubular section** (also referred as circular section) is most efficient and economical for the column free to buckle in any direction. The radius of gyration r for the tubular section in all the directions remains same. The tubular section has high local buckling strength. The tubular sections are suitable for medium loads. However, it is difficult to have their end connections. Rectangular and square hollow sections have been widely used in multi-storey buildings satisfactorily. These are increasingly becoming popular. A solid round bar having a cross-sectional area equal to that of a tubular section has radius of gyration, r much smaller than that of tube. The solid round bar is less economical than the tubular section. The solid round bar is better than the thin rectangular section or a flat strip. The radius of gyration of flat strip about its narrow direction is very small. Theoretically, the rods and bars do resist some compression.

When the length of structural member is about 3 m, then, the compressive strengths of the rods and bars are very small.

Single angle sections are rarely used except in light roof trusses, because of eccentricity at the end connections. Tee sections are often used in roof trusses.

These sections are used satisfactorily for bracings.

Channel sections

Single channel sections are unsymmetrical sections about yy-axis (i.e, one axis parallel to web. The value of radius of gyration, r_{yy} is small. Therefore, these sections are not satisfactory for use as compression members. In case, these sections are supported externally as which may restrain their bending about yy-axis, these sections may be used.

Single rolled steel I-sections are commonly used, as columns. The value of radius of gyration r, about the axis parallel to the web is small. The intermediate additional supports in the weak direction make the use of these sections economical. Single channel sections are generally not used.

Sometimes the use of I-sections and channel sections are preferred because of the method of rolling at the mills, since, the out-to-out dimensions remain same for a given depth. The costs of single rolled steel sections per unit weight are less than those of built-up sections. Therefore, the single rolled steel sections are preferred since long as their use is feasible.

Universal column sections are in use as suitable compression members in Britain and USA since long. These sections are being manufactured recently in our country. The width of flanges and its thickness and so also the depth of web and its thickness are so proportioned that the local buckling does not occur.

Common hot-rolled and built-up steel members used for carrying axial compression, usually fail by flexural buckling.

Forces or moments which a member may withstand without buckling is known as **buckling strength/resistance**.

Universal beams : These sections are used when the columns which carry transverse loads (i.e, these are subjected to bending) in addition to the axial loads. These are designed as beam-columns.

Hollow closed sections

Square hollow sections, Rectangular sections are used in multi-storey buildings very widely.

In the **circular hollow sections,** all the axes are equally strong. As these sections have equal strengths about these axes.

At present, depending upon the type of sections used in the columns, British Codes have specified the values of **imperfection factor**, α as 0.0020, 0.0035, 0.0055 and 0.0080.

The weight of hollow sections needed is only 50 per cent of that required for open sections [I-section channel section].

Heavily welded sections

There are lock-in residual stresses in the heavily welded sections, the distribution and magnitude of which differ significantly as compared to those in hot rolled sections. The magnitude of these stresses are very high. These reduce the ultimate capacity. It is recommended in the British codes that the design strength value be reduced by 20 N/mm². The resulting curves would be sufficient to account for the reduced buckling capacity.

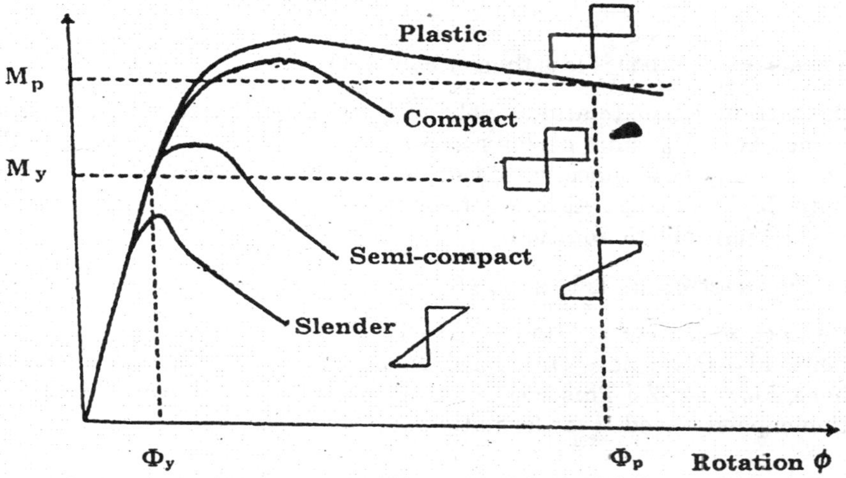

SECTION CLASSIFICATIION BASED ON MOMENT-ROTATION CHARACTERISTICS

Fig. 5.25A

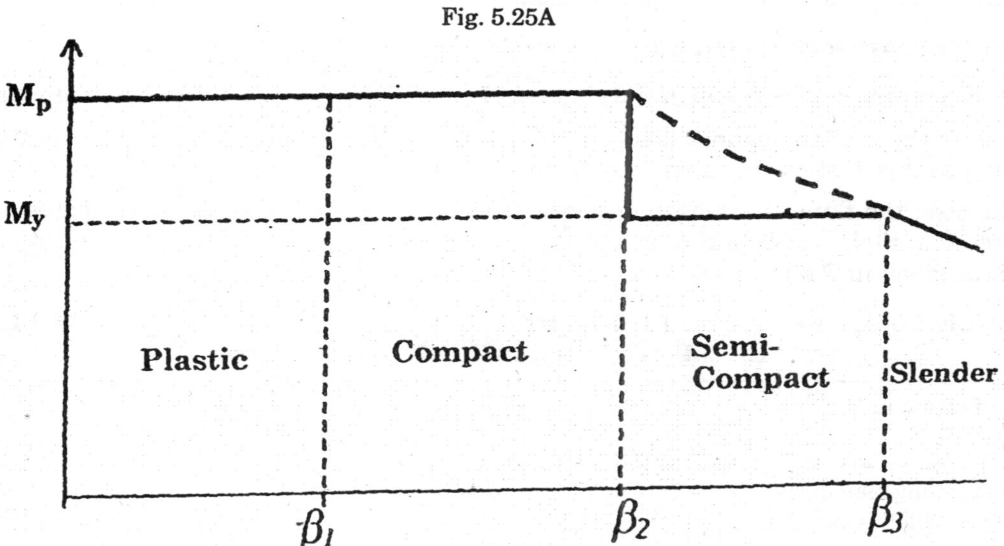

MOMENT CAPACITIES OF BEAM SECTION

Fig. 5.25B

5.21 CLASSIFICATION OF COLUMN CROSS-SECTIONS

Steel-sections (rolled or welded) are classified as (i) plastic (ii) compact, (iii) semi-compact and (iv) slender depending on their moment rotation characteristics as shown in Fig. 5.25A. The limiting width thickness ratios $\beta = (b/t)$ for the component plate elements have been specified and accordingly the sections are classified as under. The values of β are expressed as a constant *times* ε th· t is, $\varepsilon_2\ (250/\sigma_y)^{1/2}$. For the, quality of steel, $\sigma_y = 250$ N/mm², $\varepsilon = 1.00$.

1. Plastic cross-section [(b/t) is less than or equal to β_1].

When the section is able to attend yield with sufficient plastic plateau (i.e, able to attend fully plastic moment, M_p) and permits sufficient rotation, so that the re-distribution of bending moments may take place within the structure until complete failure mechanism is formed. [Where, $M_p = \sigma_y . Z_p$]. At this section develops a plastic hinge and sustains plastic moment over sufficient plastic rotation.

2. Compact cross-section [β_1 is less than (b/t) and (b/t) is less than β_2]

When the section is able to attain yield with sufficient plateau (i.e, able to develop full-plastic moment, M_p) but where the local buckling prevent the required rotation at this moment [where, $M_p = \sigma_y . Z_p$] This section can develop plastic moment, but it has inadequate plastic rotation capacity needed for formation of a plastic collapse mechanism of member of a structure.

3. Semi-compact cross-section [β_2 is less than (b/t) and (b/t) is less than β_3]

When the section is able to attain yield but local buckling limits available plastic plateau so that, at these the sections full plastic moment, M_p cannot be attained. [Where, $M_p = \sigma_y . Z$]

4. Slender cross-section [(b/t) ratio is more than β_3]

It is a cross-section in which the elements buckle locally before reaching yield moments.

When the sections cannot attain yield at the extreme fibres because of premature local buckling in the elastic range. [where $M_p < \sigma_y Z$]

In case, the flanges and web of a cross-section may be considered to be plates and web compression, their width thickness ratios are calculated and compared with the ratios B_1, B_2 and B_3 as given in Tables 5. (A) and 5. (B) as under.

TABLE 5.5(A). ($\varepsilon = (250/\sigma_y)$ LIMITS ON WIDTH-THICKNESS RATIOS OF PLATE ELEMENTS (Rolled Steel Sections, RSS)

Types of elements	Classification of cross-section		
	Plastic (β_1)	Compact (β_2)	Semi-compact (β_3)
1. Out standing element of compression flange	$\left(\frac{b}{t_f}\right) < 9.4\,\varepsilon$	$\left(\frac{b}{t_f}\right) < 10.5\,\varepsilon$	$\left(\frac{b}{t_f}\right) < 15.7\,\varepsilon$
2. Internal element of compression flange	$\left(\frac{b}{t_f}\right) < 29.3\,\varepsilon$	$\left(\frac{b}{t_f}\right) < 33.5\,\varepsilon$	$\left(\frac{b}{t_f}\right) < 42\,\varepsilon$

3. Web of an I-sec H or box section or section Neutral-axisat mid-depth	* r_1 negative $*\left(\frac{d}{t_w}\right) < \left(\frac{84\,\varepsilon}{1+r_1}\right)$ ** < 42 ε **r_1 positive	$\left(\frac{105\,\varepsilon}{1+r_1}\right)$ $\left(\frac{105\,\varepsilon}{1+1.5\,r_1}\right)$		$\frac{126\,\varepsilon}{1+2\,r_1}$ but < 42 ε
4(A). Web under Axial compression	$\left(\frac{d}{t_w}\right)$	Not applicable		42 ε
(B) Web of channel	$\left(\frac{d}{t_w}\right)$	42 ε		42 ε
5. Angle (Compr.) due to bending (Both criteria should be satisfied.	$\left(\frac{b}{t}\right)$ 9.4 ε $\left(\frac{d}{t}\right)$ 9.4 ε	10.5 ε 10.5 ε		15.7 ε 15.7 ε
6. Single angle or Double angle with components seperated (All) the three	$\left(\frac{b}{t}\right)$ $\left(\frac{d}{t}\right)$ $\left(\frac{b+d}{t}\right)$	Not applicable		15.7 ε 15.7 ε 25 ε
7. Circular hollow Tube, subjected to (a) Moment (b) Axial compr.	D/t D/t	$42\,\varepsilon^2$ NA	$52\,\varepsilon^2$ NA	$146\,\varepsilon^2$ $88\,\varepsilon^2$

Notes : NA : Not applicable

1. Slender - cross-section (Elements which exceed semi-compact limit)

2. Stress-ratio $r_1 = \frac{\text{Actual average axial stress (*)}}{\text{Design compressive stress of web}}$

$$r_2 = \frac{\text{Actual average axial stress (*)}}{\text{Design compressive stress of over-all section}}$$

(*if tensile- negative)

TABLE 5.5 (B) ($\varepsilon = (250/\sigma_y)^{1/2}$) RATIO LIMITS ON WIDTH-THICKNESS OF PLATE ELEMENTS (Welded Steel sections, WSS)

Types of elements		Classification of cross-section		
		Plastic (β_1)	Compact (β_2)	Semi-compact (β_3)
1. Outstanding Elements of compression flange		$\left(\frac{b}{t_f}\right) < 8.4\,\varepsilon$	9.4 ε	13.6 ε
2. Internal Element of compression flange (A) compression due to bending		< 29.3 ε	< 33.5	42 ε
3. Web of an I-section, H or box section Neutral axis at mid-depth		If r_1, is negative $\left(\frac{d}{t_w}\right)\left(\frac{84\,\varepsilon}{1+r_1}\right)$	$\left(\frac{105.0\varepsilon}{1+r_1}\right)$	$\left(\frac{126.0}{1+2r_1}\right)\varepsilon$
		If r_1 is +ve $\left(\frac{d}{tw}\right) < 42\,\varepsilon$	$\left(\frac{105.0\,\varepsilon}{1+1.5r_1}\right)$	< 42 ε
4. Web under axial compression		NA	NA	42 ε
5. Single/Double Angles, T-stems		NA	NA	15.7 ε 15.7 ε 25.0 ε
6. Circular Tube outer Diameter	DE	42 ε^2	52 ε^2	146 ε^2
	D/t	NA	NA	88 ε^2

Notes : For the flanges of I-section NA = Not applicable

$(b/t) = 0.5\,(b_f/t_f)$ and

For the Web of I-section

$(d/t) = (d_w/t_w)$, $d_w = h_o = (h - 2\,t_f)$

Flat elements in a cross-section are called either

(i) **Internal elements** attached on both logitudinal edges to other elements or to longitudinal stiffeners which are connected a suitabe intervals to transverse stiffeners, or

(ii) **Outstand elements** attached only on one horizontal edge to an adjacent element, the other edge being free.

In order to consider the influence of residual stresses, initial bow and accidental eccentricities of the load, buckling curves, a, b, c and (d) as shown in Fig. 5.7 in the non-dimensional form (σ_{cd}/σ_y), are used.

The values of the imperfection factors α (alpha) for the different buckling curves are as under

Buckling Curves	α	Buckling Curves	α
a	0.21	c	0.49
b	0.34	d	0.79

Table 5.6 A. BUCKLING CURVES FOR A CROSS SECTION (ROLLED STEEL SECTIONS) (RSS)

Cross section	Limits	Buckling about axis	Buckling curve
Rolled I-Sections	h/b ≥ 1.2 : $t_f \le$ 40 mm	y-y	a
			b
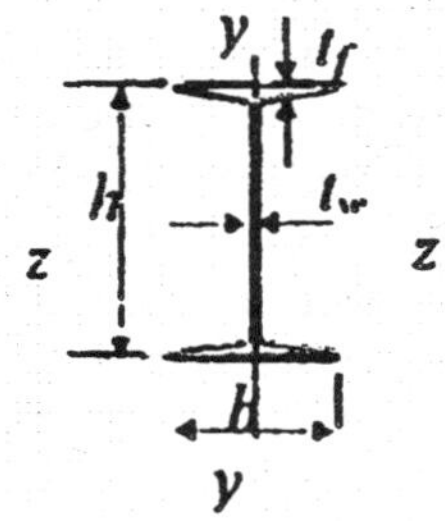		z-z	b
	40 mm ≤ t_f < 100 mm	y-y	c
	h/b ≤ 1.2 : t_f < 100 mm	z-z	b
		y-y	c
		z-z	d
	t_f ≥ 100 mm		
Channel, Angle, T and Solid Sections 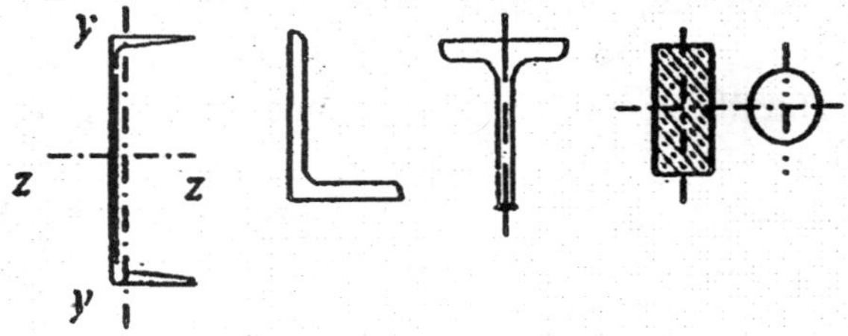		Any	c
Built-up Member 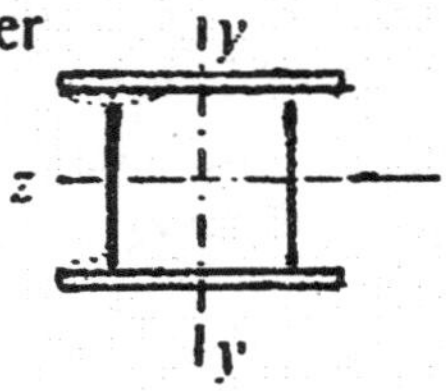		Any	c
Hollow Section	Hot rolled	Any	a
	Cold formed	Any	b

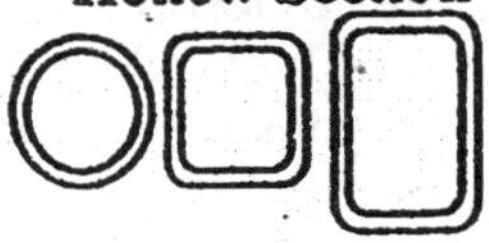

$t_f \leq 40$ mm	y-y	b
$t_f \geq 40$ mm	y-y	c
Generally (Except as below)	Any	b
Thick welds and	z-z	c
$b/t_f < 30$	y-y	c
$d/t_w < 30$		

5.22 STRENGTH/RESISTANCE OF COMPRESSION MEMBERS

Common-hot rolled and built-up steel members used to carry axial compression usually fail by flexural buckling. Force or moments which a member can withstand without buckling is known as **buckling strength** or **resistance.**

The strength of a compression member is defined as its safe load carrying capacity. The strength of a centrally loaded straight steel column depends on the effective cross- sectional area, radius of gyration (viz., shape of the cross-section), the effective length, the magnitude and distribution of residual stresses, annealing, out of straightness and cold straightening. The effective cross-sectional area and the slenderness ratio of the compression members and the design strength, σ_y or reduced design strength for the slenderness and the relevant strut curves are the main features, which influence its strength. In case, the allowable stress is assumed to vary parabolically with the slenderness ratio, it may be proved that the efficiency of a shape of a compression member is related to (A/r^2). The efficiency of a shape is defined as the ratio of the allowable load for a given slenderness ratio to that for slenderness ratio equal to zero. The safe load carrying capacity of compression member of known sectional area may be determined as follows :

From ISI structural Engineer's Handbook No. 1, following geometrical properties of the given section are noted : ($h_o = h - 2t_f$)

(i) Cross-sectional area, A, (ii) Self-weight and (iii) overall depth of I-section, h-width and thickness of flange (bf, tf), thickness of web, tw, radius of gyrations about xx-axis and yy-axis, (r_{xx} and r_{yy}), clear depth of section,

Ratios of flange and thickness (0.5 bf/tf) and depth of web to thickness of web (ho/tw) are found.

Note : The dimensions b_f, t_f, h, h_o and t_w for I-sections to be used as columns are decided prior to their manufacture and it is ensured that the phenomenon of local buckling does not occur for the elements of the sections.

Hot-rolled and built-up steel members, with cross-sections commonly used for carrying axial compression, usually failby flexural buckling. The compressive strength Pc depends or (i) the slenderness ratio, λ. (ii) effective sectional area, Ae and (iii) the design stress in compression. In addition to these, the strength of member is influenced by the presence of (i) residual stresses, (ii) initial bow and the accidental eccentricities of the load. The buckling curves a, b, c and d as shown in Fig. 5.7 are used to consider the influences of these factors. The values of **imperfections factor** α (alpha) are noted. Knowing the axis of buckling shape of the member cross-section and thickness of the steel, the design compressive stress, σ_{cd} is obtained as follows. IS 800-2007, clause 7.1.2.1 (design compressive stress of `axially loaded compression member)

$$\sigma_{cd} = \frac{(\sigma_y/\gamma_{ms})}{\phi + \left[\phi^2 - \lambda^2\right]^{1/2}}, \text{ where, } \chi = \left(1 / \left[\Phi + \left\{\Phi^2 - \lambda^2\right\}^{1/2}\right]\right) \quad (5.51)$$

$$\sigma_{cd} = \chi \cdot (\sigma_y / \gamma_{ms}) \quad (5.52)A$$

$$\phi = \frac{1}{2}\left[1 + \alpha(\lambda - \lambda_o) + \lambda^2\right]$$

$$\Phi = 0.5\left[1 + \alpha(\lambda - 0.2) + \lambda^2\right]$$

$$\lambda_e^2 = \sigma_y \left(\frac{1}{\left(\pi^2 E / \frac{kL}{r}\right)^2}\right) = \sigma_y\left(\frac{1}{\sigma_E}\right) \quad (i)$$

$$\lambda_e^2 = \left[\sigma_y \left(\frac{kL}{r}\right)^2 / \pi^2 E\right], \text{ where} \quad (ii)$$

λ = Non-dimensional effective slenderness ratio

$$\sigma_E = \left(\frac{\pi^2 E}{(kL/r)^2}\right) = \text{Euler's buckling stress}$$

α = Alpha, imperfection factor [given in Table 7, IS : 800-2007]

χ = (chi), stress-reduction factor [given in Table 8, IS : 800-2007 for different buckling class, slenderness ratio and yield stress]

$$\chi = \left(1/\left[\phi + \left\{\phi^2 - \lambda^2\right\}^{1/2}\right]\right) \quad (5.52)B$$

$\gamma_{ms.y}$ = Partial safety factor for material strength.

Note : The values of stress-reduction factor, χ (Chi) for the different buckling curves, a, b, c and d are given in Table 5.6 (a), (b), (c) and (d). The values of χ (chi) are noted from these tables. The design compressive stress, σ_{cd} is calculated from Eq. 5.51.

Alternatively, the compressive strength (Resistance) of a column may be found from **Perry-Roberston formula**.

The multiple column curves as shown in Fig. 5.7 are similar to those used in the British code BS : 5950 (Part I) - 2000 and these curves are based on the Perry-Roberston Formula Bureau of Indian Standards (BIS), first circulated the draft for approval (DFA) and then recommended and specified the some expressions in IS : 800-2007 issued later-on.

5.23 COMPARISON OF TWO PROCEDURES (DETERMINATION OF STRENGTH OF COMPRESSION MEMBERS)

(As suggested in draft for approval and derived in Perry-Roberston formula)

Let ϕ be defined in draft for approval be denoted by ϕ_{DFA} and ϕ be defined in the expression derived in Perry-Roberston formula be denoted by ϕ_{PRF}

ϕ_{DFA} has been defined as under

$$\phi_{DFA} = \frac{1}{2}\left[1+\alpha\left(\lambda-\lambda_o\right) + \lambda^2_e\right] \quad \text{(i)}$$

Where, $\lambda_e^2 = \sigma_y \cdot \left(\frac{1}{\pi^2 E/(KL/r)^2}\right) = \sigma_y \cdot \left(\frac{1}{\sigma_E}\right)$

$$\lambda_e^2 = \left[\sigma_y \cdot \left(\frac{KL}{r}\right)^2 / \pi^2 E\right]$$

σ_E = Euler's buckling stress $= \left(\frac{\pi^2 E}{(kL/r)^2}\right)$

Substituting the value of λ^2

$$\phi_{DFA} = \frac{1}{2}\left[1 + \alpha\left(\lambda - \lambda_o\right) + \left(\frac{\sigma_E}{\sigma_y}\right)\right] \quad \text{(ii)}$$

$$\sigma_E \cdot \phi_{DFA} = \frac{1}{2}\left[\sigma_y + \left[\left\{1+\alpha\left(\lambda-\lambda_o\right)\sigma_E\right\}\right]\right] \quad \text{(iii)}$$

$$\sigma_E \cdot \phi_{DFA} = \frac{1}{2}\left[\sigma_y + \left\{1+\alpha\left(\lambda-\lambda_o\right)\sigma_E\right\}\right] \text{[Right side As per IS : 800-2007 clause 7.1.2.1]} \quad \text{(iv)}$$

where, $\alpha_1 = \alpha\left(\lambda-\lambda_o\right)$

$\sigma_E\, \phi_{DFA} = \phi_{PRF}$

Note : The expression appearing on right-side is same as it appears for Perry-Roberston formula. So also, it is true for Chi `χ' function. [Stress-reducton factor (Table 18, 800-2007, clause 7.1.2.1 for different buckling class, slenderness ratio and yield stress]

Chi, $\chi = \frac{1}{\phi + \left[\phi^2 - \lambda^2\right]^{1/2}}$ that is ϕ Substituting the value of λ^2 in the expression appearing in the denominator, (Deno). [λ is non-dimensional effective slenderness ratio]

$$\text{(Deno)} = \phi_{DFA} + [\phi^2_{DFA} - \lambda^2]^{1/2} \quad \left[\because \lambda^2 = \left(\sigma_y / \sigma_E\right)\right] \quad \text{(v)}$$

$$= \phi_{DFA} + \left[\phi^2_{DFA} - \left(\frac{\sigma_y}{\sigma_E}\right)\right]^{1/2} \quad \text{(vi)}$$

$$= \phi_{DFA} + \left[\frac{\sigma_E \cdot \phi_{DFA}^2 - \sigma_y}{\sigma_E}\right]^{1/2} \quad \text{(vii)}$$

Multiplying in the numerator and denominator by $\sigma_E^{1/2}$ with second term

$$= \phi_{DFA} + \frac{\sigma_E^{1/2}}{\sigma_E^{1/2}} \cdot \left[\frac{\sigma_E \cdot \phi_{DFA}^2 - \sigma_y}{\sigma_E^{1/2}}\right]^{1/2} \quad \text{(viii)}$$

$$= \phi_{DFA} + \frac{1}{\sigma_E}\left[\sigma_E^2 \cdot \phi_{DFA}^2 - \sigma_E\,\sigma_y\right]^{1/2} \quad \text{(ix)}$$

$$\sigma_E\,(\text{Deno}) = \sigma_E \cdot \phi_{DFA} + \left[(\sigma_E \cdot \phi_{DFA})^2 - \sigma_E\,\sigma_y\right]^{1/2} \quad \text{(x)}$$

$$= \phi_{PRF} + \left[\phi_{PRF}^2 - \sigma_E \cdot \sigma_y\right]^{1/2}$$

Note : 1. The design compressive-stress by Perry-Roberston formula is determined by using the following expression Eq. 5.51. 2. The design compressive stress, σ_{cd} as per IS : 800-2007, clause 7.1.2.1 and that by Perry-Roberston Formula are **same** though expressed slightly differently (as proved in this Art 5.23). 3. IS : 800-2007, clause 7.1.2.1 recommends and specifies the same expressions as in its `Draft for approval (DFA)

$$\sigma_{cd} = \phi_{PRF} \pm \left[\phi_{PRF}^2 - \sigma_y\,\sigma_E\right]^{1/2} \quad (5.53)$$

Design strength of a compression member P_d is calculated as follows :

$P < P_d$ where, $P_d = \sigma_{cd} \cdot A_{ed}$

Step 1 : Types of cross-section

Ratios of width of flange and thickness. of flange [for an I-section $0.5\, b_f/t_f$)] and depth of web to the thickness of web, [$d = h_o$, (d/t_w)] are found. The value of epsilon $\varepsilon = (250/\sigma_y)^{1/2}$ is determined. The value of yield stress, σ_y is adopted as per the grade of steel.

Step 1. Type of the section is decided by comparing above two ratios with the ratios given for the classification of the section, [viz, plastic, compact or semi-compact. Slender-sections are not used.]

Step 2. Effective length

From the actual length of the compression member and given support/end conditions, the effective length is calculated.

Step 3. Values of λ_{xx}, λ_{yy} and λ_e

Values of slenderness ratios $\lambda_{xx} \left(= l_{e.xx}/r_{ax}\right)$ and $\lambda_{yy} = \left(l_{e.yy} / r_{yy}\right)$ are determined. The value of limiting slenderness ratio.

$\lambda_o = 0.2\,(E/\sigma_y)^{1/2}$ is found.

Step 4. Values of Perry's factor/ Imperfection factor (λ_1) for buckling about xx-axis and yy-axis are calculated.

$$\alpha_{1.xx} = \alpha_{xx}\left(\lambda_{xx} - \lambda_o\right) \text{ and } \alpha_{1.yy} = \alpha_{yy} \cdot \left(\lambda_{yy} - \lambda_o\right)$$

Step 5. Values of Euler's buckling stress about xx-axis and yy-axis are obtained

$$\sigma_{E.x} = \pi^2 \, E / (\lambda_{xx} - \lambda_o)^2 \text{ and } \sigma_{E.y} = \pi^2 . E / \left(\lambda_{yy} - \lambda_o\right)^2$$

Step 6. Values of ϕ_x and ϕ_y for Ayrton and Perry's formula are found

$$\phi_x = \left[\sigma_y + \left(1 + \sigma_{1.xx}\right)\sigma_{E.x}\right] / 2 \text{ and } \phi_y = \left[\sigma_y + \left(1 + \sigma_{1.yy}\right)\sigma_{E.y}\right] / 2$$

Where σ_y is the yield stress for steel.

Step 7. Values of the design compressive stresses, $\sigma_{c.x}$ and $\sigma_{c.y}$

$$\sigma = \left(\frac{\sigma_y + \left(1 + \alpha_1\right)\sigma_E}{2}\right) - \left[\left\{\left(\sigma_y + (1 + \alpha_1\right)^2 . \sigma_E - \sigma_y . \sigma_E\right\}^{1/2}\right]$$

$$\sigma_{c.x} = \phi_x - \left[\phi_x^2 - \sigma_y . \sigma_{Ex}\right]^{1/2} \text{ and } \sigma_{c.y} = \phi_y - \left[\phi_y^2 - \sigma_y . \sigma_{Ey}\right]^{1/2}$$

In case, the value of ϕ is substituted, then **Aryton and Perry's values** for compressive stress are obtained.

The ultimate compressive stress is that whichever is less. The limiting value of compressive/ design stress ($\sigma_{cd} / \gamma_{ms}$) is obtained. γ_{msy} is the partial safety for the strength of the materal for yield.

Step 8. Load-carrying capacity (i.e, compressive strength) of the column is calculated

Factored load, $W_{Fd} = \left(\sigma_c . A\right)$ = Equ. and

Service load, $W_{ck} = \left(\frac{\sigma_c . A}{\gamma_{ms}}\right)$

Note : (i) for **plastic, compact and semi compact section** the compressive strength of a given section.

$P_c = \left(\frac{\sigma_c . A}{\gamma_{ms}}\right)$, **where for steel. r_{ms} = 1.15**

(ii) **For slender sections**

$P = \left(\frac{\sigma_{cs} . A}{\gamma_{ms}}\right)$

where, σ_{cs} Reduced compressive stress for the slender-section

The compressive strength depends on the following

(i) Slenderness ratio, λ

(ii) Effective sectional area, A

(iii) Yield strength of steel, σ_y

or the reduced design strength of the slender member.

Example 5.1 A rolled steel beam section HB 350,@ 0.674 kN/m is used as a column. In case, the column is a short find the compressive strength of the column.

Solution : Properties of given rolled steel beam section HB 350, @ 0.674 kN/m, width of the flange, b_f = 250 mm. Thickness of the flange, t_f = 11.6 mm

Overall depth of the section, h = 350 mm. Thickness of the web, t_w = 8.3 mm, Cross-sectional area, A = 8591 mm²

Step 1. Type of the section

Ratio, $(0.5\ b_f/t_f) = (0.5 \times 250/11.6) = 10.775$

$d = (h_o - 2 \times t_f) = (350 - 2 \times 11.6) = 326.8$ mm

Ratio, $(d/t_w) = (326.8 / 8.3) = 39.37$

Epsilon, $\varepsilon = (250/\sigma_y)^{1/2} = (250/260)^{1/2} = 0.98$

(i) For outstanding element of the compression flange (for rolled steel sec.)

	Plastic	Compact	Semi-compact
$(b/t) \leq$	$9.4\ \varepsilon = 9.212$	$10.5\ \varepsilon = 10.29$	$15.7\ \varepsilon = 15.386$

(ii) Web with neutral axis at mid-depth

$(d/t) \leq$	$84\ \varepsilon = 82.32$	$105\ \varepsilon = 102.9$	$126\ \varepsilon = 123.48$

Hence, the section is a **semi-compact** due to **proportions of the flange.**

Step 2. Design compressive stress, σ_{cd} (Ayrton and Perry's formula)

Short column shall fail by yielding

$$\sigma_{cd} = (\sigma_y / \gamma_{ms}) = (260/1.15) = 226.087 \text{ N/mm}^2$$

Step 3. Design compressive strength (i.e, load carrying capacity of the member)

$$\text{Factored load, } W_{Fd} = (\sigma_{cd} \times A) = \left(\frac{226.087 \times 8591}{1000}\right) = \mathbf{1942.31\ kN}$$

$$\text{Service load, } W_{cK} = \left(\frac{1942.31}{1.5}\right) = \mathbf{1294.87\ kN}$$

Example 5.2. A roll-steel beam section HB 350, @ 0.674 kN/m is used as a stanchion. If the unsupported length of the stanchion is 4 m, determine safe load carrying capacity (i.e, compressive strength) of the section.

Solution:

Step1(A) Properties of I-section, from ISI Handbook No.1, Section HB350, @ 0.674 kN/m

Width of flange, b_f = 250 mm

Thickness of flange, t_f = 11.6mm

Overall depth of the section, ho = 350 mm.

Thickness of web, t_w = 8.3 mm

Cross-sectional area, A = 8591 mm²

Radius of gyration, r_{xx} = 149.3 mm

Radius of gyration, r_{yy} = 53.4 mm

Step 1(B) Type of the section.

Ratio, $(0.5\ b_f/t_f) = (0.5 \times 250/11.6) = 10.775$

$d = (h_o - 2t_f) = (350 - 2 \times 11.6) = 326.8$ mm

Ratio $(d/t_w) = (326.8/8.3) = 39.37$

Epsilon, $\varepsilon = (250/260)^{1/2} = 0.98$

For **rolled steel I-section :**

For outstanding element of the compression flange ($\varepsilon = 0.98$)

	Plastic	Compact	Semi-compact
(b/t) ≤	$9.4\ \varepsilon = 9.212$	$10.5\ \varepsilon = 10.29$	$15.7\ \varepsilon = 15.386$
(ii) Internal element of the compression flange			
(d/t) ≤	$29.3\ \varepsilon = 28.714$	$33.5\ \varepsilon = 32.83$	$42\ \varepsilon = 40.33$
(iii) Web with neutral axis at mid depth			
(d/t) ≤	$84\ \varepsilon = 82.32$	$105\ \varepsilon = 102.9$	$126\ \varepsilon = 123.48$

By comparing the actual values with the requirement, it is seen that the section is a Semi compact section

Step 2 : Slenderness ratios Unsupported length = 4000 mm

$$\lambda_{xx} = \left(\frac{4000}{149.3}\right) = 26.79$$

$$\lambda_{yy} = \left(\frac{4000}{53.4}\right) = 74.9$$

$$\lambda_o = 0.2\,\pi\left(E/\sigma_y\right)^{1/2}$$

$$= 0.2 \times 3.14 \times (2 \times 10^5/260)^{1/2}$$

$$= 0.628 \times 10 \times (2000/260)^{1/2} = 17.41$$

Step 3 : Imperfection factors, α_1 For rolled steel I-section

$\alpha_{xx} = 0.0020$, $\alpha_{yy} = 0.0035$

About xx-axis, $\alpha_{1\text{-}xx} = \alpha_{xx}(\lambda_{xx} - \lambda_o) = 0.0020\,(26.79 - 17.41) = 0.01876$

About yy-axis, $\alpha_{1\text{-}yy} = \alpha_{yy}(\lambda_{yy} - \lambda_o) = 0.0035\,(75 - 17.41) = 0.20156$

Step 4: Euler's stress, σ_E

About xx-axis

$$\sigma_{EX} = \pi^2 E/(\lambda_{xx} - \lambda_o)^2$$

$$= 9.8596 \times 2 \times 10^5/(26.79 - 17.41)^2$$

$$= 19.7192 \times 10^5/(9.38)^2 = 22413\ \text{N/mm}^2$$

About yy-axis

$\sigma_{EY} = \pi^2 E / (\lambda_{yy} - \lambda_o)^2$

$= 9.8596 \times 2 \times 10^5/(75 - 17.41)^2$

$= 19.7192 \times 10^5/(57.59)^2$

$= 594.523$ N/mm²

Step 5 : Values of ϕ_x and ϕ_y

$\phi_x = [\sigma_y + (1 + \alpha_{1.xx})\, \sigma_{EX}]/2$

$= [260 + (1 + 0.01876)22413]/2$

$= 11546.222$

$= [\sigma_y + (1 + \sigma_{1.yy})\, \sigma_{EY}]/2$

$= [260 + (1 + 0.20156)\ 594.467]/2$

$= 487.1472$

Step 6. Design Compressive stresses, σ_{cx}, σ_{cy} (As per Ayrton & Perry-Formula)

$$\sigma_{cx} = \phi_x - [\phi_x^2 - \sigma_y \cdot \sigma_{EX}]^{1/2} = 11546.222 - [11546.222^2 - 260 \times 22413]^{1/2}$$

$= [\ 11546.222 - 11290.599] = 255.623$ N/mm²

It is not greater than 260 N/mm²

$$\sigma_{cy} = \phi_y - [\phi_y^2 - \sigma_y \cdot \sigma_{Ey}]^{1/2}$$

$$\sigma_{cy} = 487.1474 - [487.1474^2 - 260 \times 594.523]^{1/2} \text{ N/mm}^2$$

$$\sigma_{cy} = 200.5074 \text{ N/mm}^2$$

It is also not more than $\sigma_y = 260$ N/mm².

Ultimate compressive stress is that whichever is less.

Limiting compressive stress

$$(\sigma_{cy} / \gamma_{msy}) = \left(\frac{200.5074}{.1.15}\right) = 174.354 \text{ N/mm}^2$$

Step 7. Load carrying capacity (compressive strength strength of column))

Factored load = $W_{Fd} = (\sigma_{cd} \cdot A)$

$$W_{Fd} = \left(\frac{174.354 \times 8591}{1000}\right) = 1498.878$$

Service load, $W_{Fd} = (1498.878 / 1.5) = 998$ kN.

Example 5.2 (A). In Example 5.2 determine the compressive strength of the member using (i) stress-reduction factor χ (chi) and (ii) noting.

Solution : The buckling curve for compressive stress directly corresponding the slenderness ratio class `a' shall be used.

Unsupported length of column = 4000 mm

Radius of gyration

r_{xx} = 149.3 mm

Radius of gyration

r_{yy} =53.4 mm

Slenderness ratio

$$\lambda_{xx} = \left(\frac{4000}{149.3}\right) = 26.79$$

Slenderness ratio

$$\lambda_{yy} = \left(\frac{4000}{53.4}\right) = 74.9$$

The compressive stress shall be minimum for maximum value of the slenderness ratio. From Table 5. for stress-reduction factor for a class of steel with yield stress 260N/mm².

Stress reduction factor Chi, χ, for slenderness ratio

70 χ – 0.794, 80, χ = 0.772, 74.9 χ = 0.759

Design compressive stress [IS : 800-2007, clause 7.1.2.1]

$$\sigma_{cd} = \chi(\sigma_y / \gamma_{ms})\ \text{N/mm}^2$$

$= 0.759 \times (260/1.15)\ \text{N/mm}^2$

$= (197.34/1.15) = 171.6\ \text{N/mm}^2$

Load carry capacity (i.e, the compressive strength of the column)

$$\text{Factored load} = (\sigma_{cd} \times A_e) = \left(\frac{171.6 \times 85.91}{1000}\right) = 1474.21\ \text{kN}$$

$$\text{Service load} = \left(\frac{1474.21}{1.5}\right) = 982.8\ \text{kN}$$

Note : Using Perry-Roberston formula compressive stress was 174 N/mm². Using stress-reduction factor, χ, it is 171.6 N/mm². As such it seems satisfactory.

Alternatively : Noting the value of compressive stress for class a steel from Table 5.

σ_{cc} for slenderness ratio

70 σ_{cc} = 188 N/mm², 80 σ_{cc} = 171 N/mm²

$$74.9\ \ \sigma_{cc} = \left[188 - \frac{(188-171)}{10} \times 4.9\right] = [(188 - 8.33)] = 179.67\ \text{N/mm}^2$$

$$\text{Factored load, } W_{Fd} = \left(\frac{\sigma_{cd} \times A_e}{\gamma_{msy}}\right) = \left(\frac{179.67 \times 8591}{1000}\right) = \left(\frac{1543.5\ \text{kN}}{1.15}\right) = 1342.21\ kN$$

$$\text{Service load, } W_{ck} = \left(\frac{1342.21}{1.5}\right) = 894.81\ \text{kN}$$

Note: This gives higher value (179.67 N/mm²) as compared to Perry-Roberston value 174.354 N/mm². The difference is small.

Example 5.2(B). In Example 5., in case, a standard column, section SC 250, @ 0.856 kN/m is used as a column, determine the safe load carrying capacity (i.e, compressive strength) of the section.

Solution:

Step 1 (A). Properties of an I-section from IS: 808-89 (Third revision).

Width of flange, b_f = 250 mm

Thickness of flange, t_f = 17.0 mm

Depth of the section, h_o = 250 mm

Thickness of web, t_w = 10 mm

Radius of gyration, r_{xx} = 107 mm

Radius of gyration, r_{yy} = 54.6 mm

Cross-sectional area, A = 10900 mm²

Step 1 (B): Type of the section

Ratio, $(0.5\, b_f/t_f) = (0.5 \times 250/17) = 7.353$

$d = (h_o - 2t_f) = (250 - 2 \times 17) = 216$ mm

Ratio, $(d/t_w) = (216/10) = 21.6$

Epsilon, E = (250/260)1/2 = 0.98

For rolled steel I-section

(i) For outstanding element of the compression flange

	Plastic	Compact	Semi-Compact (b/t)
$(b/t) \leq$	$9.4\ \varepsilon = 9.212$	$10.5\ \varepsilon = 10.084$	$15.7\ \varepsilon = 15.386$

(ii) Internal element of the compression flange

$(b/t) \leq$	$29.3\ \varepsilon = 28.714$,	$33.5\ \varepsilon = 32.83$	$42\ \varepsilon = 41.16$

(iii) Web with neutral axis at the mid-depth

$(d/t) \leq$	$84\ \varepsilon = 82.32$	$105\ \varepsilon = 102.9$	$126\ \varepsilon = 123.48$

By comparing the actual values with the requirements, it is seen that the section is a **plastic section.**

Unsupported length = 4000 min

Step 2 : Slenderness ratio,

$\lambda_{xx} = (4000/107) = 37.38$

$\lambda_{yy} = (4000/54.6) = 73.26$

$\lambda_o = 0.2\,\pi\,(E/\sigma_y)^{1/2}$

$= 0.2 \times 3.14\ [2 \times 10^5/260]^{1/2}$

= 0.628 x 10 [2000 / 260]$^{1/2}$ = 17.41

Step 3 : Imperfection factors, α_1

For rolled steel sections

$$\alpha_{xx} = 0.0020 \text{ and } \lambda_{yy} = 0.0035$$

About xx-axis

$$\alpha_{1-xx} = \alpha_{xx}\left(\lambda_{xx} - \lambda_o\right) = 0.0020\ (37.28 - 17.41) = 0.03994$$

About yy-axis

$$\alpha_{1-yy} = \alpha_{yy}\left(\lambda_{yy} - \lambda_o\right) = 0.0035\ (73.26 - 17.41) = 0.1955.$$

Step 4 : Euler's stress, σ_{EX}, σ_{EY}

About xx-axis

$$\sigma_{EX} = \pi^2 E / \left(\lambda_{xx} - \lambda_o\right)^2$$

= 9.8596 x 2 x 10^5 / (37.38 – 17.41)^2

= (19.7192 x 10^5/398.80) = 4944.63 N/mm^2

About yy-axis

$$\sigma_{EY} = \pi^2 E / \left(\lambda_{yy} - \lambda o\right)^2$$

= 9.8596 x 2 x 10^5/(73.26 –17.41)^2

= (19.7192 x 10^5/ 3119.2225) = 632.176 N/mm^2

Step 5 : Values of ϕ_x and ϕ_y

$$\phi_x = \left[\sigma_y + \left(1 + \alpha_{1-xx}\right) 4944.6\ \right]/2$$

= [260 + (1 = 0.03994) 4944.63]/2

= [260 + 5142.1185]/2 = 2701.06 N/mm^2

$$\phi_y = \left|\sigma_y + \left(1 + \alpha_{1-yy}\right) 632.176\right| / 2$$

= [260 + (1 + 0.1955) 632.176]/2

= [260 + 755.766]/2 = 507.88 N/mm^2

Step 6 : Compressive stresses σ_{cx}, σ_{cy} (Ayrton and Perry's formulae)

$$\sigma_{cx} = \phi_x - \left[\phi_x^2 - \sigma_y \cdot \sigma_{Ex}\right]^{1/2}$$

= 2701.06 – [2701.06^2 – 260 x 4944.63]$^{1/2}$

= 2701.06 – 1000 [7.296 – 1.288]$^{1/2}$

= 2701.06 – 1000 [6.01]$^{1/2}$

= (2701.06 – 2451.61) = 249.45 N/mm^2

$$\sigma_{cy} = \phi_y - \left[\phi_y^2 - \sigma_y \cdot \sigma_{Ey}\right]^{1/2}$$

= 507.88 – [507.88^2 – 260 x 507.88]$^{1/2}$

$= 507.88 - 100\ [25.794 - 13.205]^{1/2}$

$= 507.88 - 100\ [12.589]^{1/2}$

$= [507.88 - 354.80) = 153.08\ \text{N/mm}^2$

Ultimate compressive stress is that whichever is less.

Limiting compressive stress

$$\sigma_{cd} / \gamma_{ms} = (153.28 / 1.15) = 133.183\ \text{N/mm}^2$$

Step 7 : Load carrying capacity (i.e, compressive strength) of the column section

$$\text{Factored load, } W_{Fd} = \left(\frac{133.113 \times 10900}{1000}\right) = \mathbf{1450.93\ kN}$$

Service load, W_{cK} = (1452.83/1.5)

$$W_{cK} = \left(\frac{1450.93}{1.5}\right) = \mathbf{967.288\ kN}$$

5.24 ANGLE SECTIONS USED AS COMPRESSION MEMBERS

The compression members consisting of angle sections are of two types :

1. Continuous members 2. Discontinuous members.

1. Continuous members. The compression members (consisting of single or double angles) which are continuous over a number of joints, are known as continuous members. The top chord members of truss girders and principal rafters of roof trusses are continuous members.

The effective length of such compression members are adopted between 0.7 and 1.0 times the distance between the centres of intersections, depending upon degree of end restraint provided. When the members of trusses buckle in the plane perpendicular to the plane of the truss, the effective length shall be taken as 1.0 times the distance between the points of restraint.

The ultimate compressive stresses for such compression members are adopted from IS : 800-2007 corresponding to the slenderness ratio of the member, and yield stress for steel.

2. Discontinuous members. The compression members which are not continuous over a number of joints, (i.e., which extent between two adjacent joints only), are known as discontinuous members. The discontinuous members may consist of

(i) Single angle section, or (ii) Double angle section.

When an angle strut is connected to a gusset plate or to any structural member by one leg, the load transmitted through the strut, is eccentric on the section of the strut. As a result of this, bending stress is developed along with direct stress. While designing or determining strength of an angle strut, the bending stress developed because of eccentricity of loading is accounted for as follows :

(i) Single angle sections used as strut

(A) When single angle discontinuous strut is connected to a gusset plate with one rivet as shown in Fig. 5.26(a) its effective length is adopted as centre to centre of intersection at each end. However, the slenderness ratio of such single angle strut should not exceed 180. These are designed for 1.25 times the factored load (i.e. 1.25 W_{fd}).

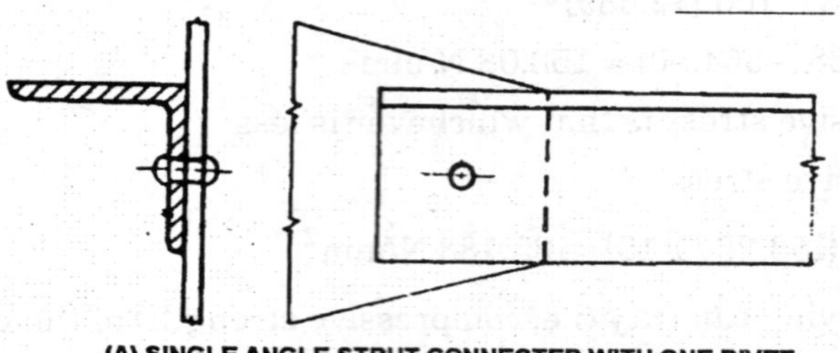

(A) SINGLE ANGLE STRUT CONNECTED WITH ONE RIVET

Fig. 5.26(A) Single angle strut

(B) When a single angle discontinuous strut is connected with two or more number of rivets or welding as shown in Fig. 5.26(b) its effective length is adopted as 0.85 times the length of strut centre to centre of intersection of each end. The member is designed for the factored load 1.05 times W_{Fd}.

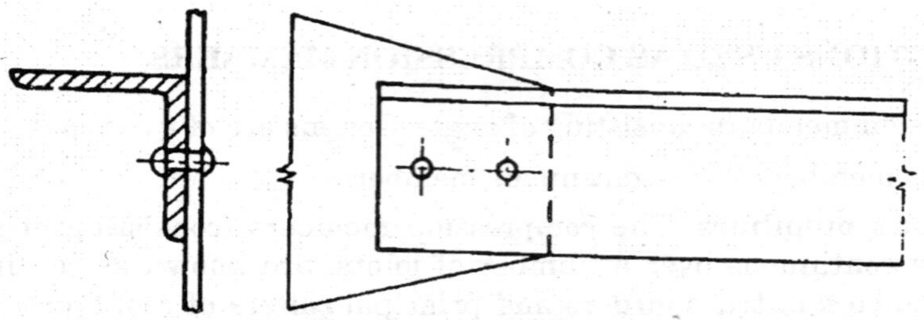

(B) SINGLE ANGLE STRUT CONNECTED TWO OR MORE RIVETS

Fig. 5.26(B) Single angle strut

(ii) Double angle sections used as strut

(A) A double angle discontinuous strut with angles placed back to back and connected to opposite sides of a gusset or any rolled steel section by not less than two rivets or bolts or in line along the angles at each end or by equivalent in welding as shown in Fig. 5.27, can be regarded as an axially loaded strut. Its effective length is adopted between 0.7 to 0.85 times the distance between intersections, depending on the degree of restraint provided and in the plane perpendicular to that of the gusset, the effective length 'l' shall be taken as equal to the distance between centres of intersections.

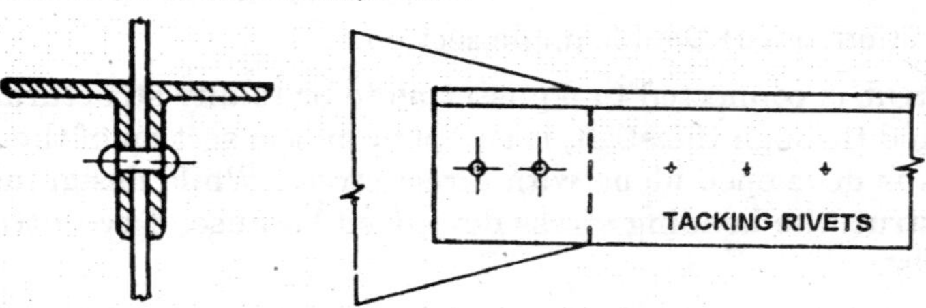

DOUBLE ANGLE STRUT (ANGLES PLACED BACK TO BACK) CONNECTED TO BOTH SIDES OF GUSSET

Fig. 5.27 (A)

The tacking rivets should be provided at appropriate pitch at sufficient intermediate points that the load for buckling of one angle between fasteners exceeds the load for overall buckling of the compound (built-up) section. It is of course, necessary to ensure that the two sections function together as one compound member. Welds may be used at suitable intervals with a spacer bar between the connecting legs.

(B) The double angles, back to back connected to one side of gusset plate or a section by one or more rivets or bolts or welds as shown in Fig. 5.27(B), these are designed as single angle discontinuous strut connected by single rivet or bolt.

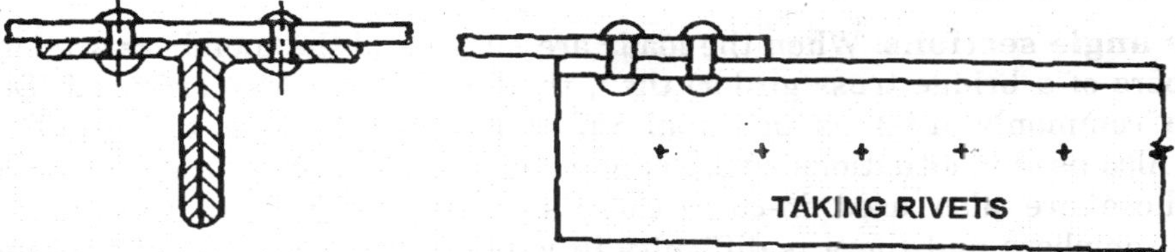

DOUBLE ANGLE STRUT (ANGLES PLACED BACK TO BACK) CONNECTED TO ONE SIDE OF GUSSET

Fig. 5.27b.

A member fabricated by inter connecting more than one element to form acompound section acting as a single member is referred as **built-up member.**

If the struts carry, in addition to axial loads, loads which cause transverse bending, the combined bending and axial stresses shall be checked as described for the columns subjected to eccentric loading.

The tacking rivets/bolts should be provided at appropriate pitch.

The **tacking rivets** are also termed as stitching rivets bolts. In case of compression members, when the maximum distance between centres of two adjacent rivets bolts exceeds 12 t to 200 mm whichever is less, then tacking rivets bolts are used. The tacking rivets are not subjected to calculated stress. The tacking rivets bolts are provided throughout the length of a compression member composed of two components back to back. The two components of a member act together as one piece by providing tacking rivets bolts at a pitch in line not exceeding 600 mm, and such that minimum slenderness ratio of each member between the connections is not less than 40 more than 60.

In case where plates are used, the tacking rivets are provided at a pitch in line not exceeding 32 times the thickness of outside plate or 300 mm, whichever is less. Where the plates are exposed to weather the pitch in line shall not exceed 16 times the thickness of the outside plate or 200 mm, whichever is less. In both cases, the lines of rivets shall not be apart at a distance greater than these pitches. M16 diameter bolts are used for the members upto 10 mm thicker otherwise, M20 diameter bolts are used. Spacing of tacking bolts or welds should not be less than 600 mm.

Single angle sections are used to transmit small forces for the compression members for light trusses and bracing. The equal angle sections are more desirable usually. The unequal angle sections are also used. The minimum radius of gyration about one of the principal axis is adopted for calculating the slenderness ratios. The minimum radius of gyration of the single angle section is much less than the other sections of same cross-sectional area. Therefore, the single angle sections are not suitable for the compression member of long lengths. The single angle sections are commonly used in the single plane trusses (i.e., the trusses having gusset plates in one plane). The angle sections simplify the end connections. In case of a single angle it is important to note that the weakest plane will be in the direction of the minor principal axis, which one is different from rectangular axis. These sections are commonly used as discontinuous struts in roof trusses. These sections are connected to the gusset plates either by single/more rivets/bolts in line along the length of the member.

These sections are designed for 1.25 times factored load, connections at the ends are designed for factored load.

Double angle sections. When the loads are heavy and the members are long (e.g., top chord members of a bridge truss girder) then, the built-up sections are used. Double angle sections are commonly used as principal rafter members in roof-trusses. Double angle sections are also used in situations where single angles cannot provide sufficient compressive strength. These are also used in-case, there is disparity in size between tension and compression members and there is difficulty in jointing. These consist of two angles placed back to back, generally with space between them to allow the connections at their ends with gusset plate in such a manner that there is minimum of the eccentricity at the joints. When two angle sections are used, then these are tacked together using stitching rivets at appropriate spacing so that these section act as one unit.

These sections are connected to gusset plates at their ends. Two or more rivets/bolts are used in line along the length of member. These members are designed for factored load. These sections may also be used as discontinuous struts.

The spacing of tacking rivets should be the maximum slenderness ratio is not more than sixty (60) and not less than forty (40).

Following values may be used as guide - lines for selecting initial sizes.

(a) Single angle size

(1/30)th of the length with $(l/r_{min}) = 150$

(b) Double angle size

(1/35)th of the length with $(l/r_{min}) = (100 - 120)$

(c) Circular hollow sections, Diameter = (1/40)th of length with l/r = 100

M16 bolts are used for thickness of the member upto 10 mm and

M20 bolts are used for thickness of members more than 10 mm

Spacing of stitching rivets should not be less than 600 mm

The **tee-sections** are suitable for the compression members for small trusses. The tee-sections are more suitable for welding.

5.25 COMPRESSIVE STRENGTH OF ANGLE SECTIONS

The compressive load in single angles may be transferred as follows :

1. Concentrically to its centroid through end gusset: The member may be designed by determining its strength using general procedure and class C curve may be used.

2. **Eccentrically** by connecting one of its legs to a gusset or an adjacent member.

Generally in the angle sections, compressive forces are transferred through one leg to a gusset or an adjacent member.

In single angle sections since these are eccentrically loaded (loaded through one leg) flexural torsional buckling occurs flexural torsional buckling strength is obtained by using the **equivalent slenderness** ratio, λ_e, calculated as under. [As per IS : 800-2007, clause 7.5.1.2]

$$\lambda_e = \left(k_1 + k_2 \cdot \lambda_{vv}^2 + k_3\, \lambda_\phi^2\right)^{1/2} \quad (5.54)$$

where, k_1, k_2, and k_3 are the constants which depend upon end conditions as mentoned below

$$\lambda_{vv} = \left[\frac{l}{r_{vv}}\right] / \left[\varepsilon\left(\pi^2 E/250\right)^{1/2}\right] \quad \text{and} \quad \lambda_{\phi\phi} = [b_1 + b_2] / \left[\varepsilon\left(\pi^2 E/250\right)^{1/2} .2t\right]$$

Where,

L = Laterally unsupported length of the member

r_{vv} = Radius of gyration about the minor axis

b_1, b_2 = Width/length of two legs of the angle

t = thickness of the leg

ε = Yield stress ratio $(250 / \sigma_y)^{1/2}$

Substituting the value of E, above two expressions may be written as under

$$\lambda_{vv} = \left(\frac{l}{r_{vv}}\right) \Bigg/ \left[\left(\frac{\pi^2 E}{\sigma_y}\right)^{1/2}\right] \quad (5.55)$$

$$\lambda_{\phi\phi} = (b_1 + b_2) \Bigg/ \left[\left(\frac{\pi^2 E}{\sigma_y}\right)^{1/2} .2t\right] \quad (5.56)$$

Table 5.7 VALUES OF CONSTANTS, k_1,k_2 and k_3

Constants	Number of bolts at each end ≥ 2			
	Gussets connecting member fixity			
	Fixed	Hinged	Fixed	Hinged
k_1	0.20	0.70	0.75	1.25
k_2	0.35	0.60	0.35	0.50
k_3	20.00	5.00	20.00	60.00

Substituting the values of k_1, k_2 and k_3

λ_{vv} and $\lambda_{\phi\phi}$ in Eq. 5.55 and 5.56 the equivalent slenderness λ_e is determined.

The values of design compressive stress, σ_{cd} is obtained either by using the formulae for stress reduction factor or it is noted directly using, λ_e.

5.25 COMPRESS STRENGTH OF BUILT-UP MEMBERS

When the loads are heavy and the members are long (e.g., top chord members of a bridge truss girder) then, the built-up sections are used. The built-up compression members are needed when the single rolled steel sections are not sufficient to furnish the required cross-sectional area.

A built-up compression member may consist of two or more rolled structural steel sections connected together effectively and acts as one compression member. The built-up compression members are given effective column cross-sections.

The built-up sections are used for one or more of the reasons mentioned below:

(i) The built-up sections provide large cross-sectional area which cannot be furnished by single rolled steel sections.

(ii) The built up sections provide special shape and depth. The special shape and depth facilitate connections between the different members.

(iii) The built-up sections provide sufficient large radius of gyration of more desirable ratio of the radii of gyration in two different directions. In the single rolled steel section, the ratio of radii of gyration in two directions cannot be altered.

(iv) The built-up column sections become necessary particularly when the loads are heavy and the members are long as in top-chord members of bridge truss-girders.

For built-up section (1), Fig. 5.28.

Buckling class : Bending about any axis, buckling class (C).

The element of built-up members are placed at farthest possible distance from centroid of the section. It gives the largest radius of gyration and the smallest slenderness ratio for constant cross-sectional area. The material of the column is used at the greatest possible stress. As the size of compression member increases with constant area, the slenderness ratio becomes less and less due to increase of radius of gyration. The compressive stress increases accordingly. At the higher compressive stress the local buckling of the member may take place. The local buckling of the member limits the size of the compression member. The material of the column is used most economically when radii of gyration of the compression members about both the axes are kept approximately equal. The various shapes of built-up members are shown in Fig. 5.28.

These built-up sections may be classified into two categories (namely, built-up columns consisting of solid webs and those consisting of open webs), which are described as under :

(i) Built-up columns consisting of solidwebs

Built-up columns as shown in Fig. 5.28(B), (D), (F), (G), (K), (L) and (M) consist of solid webs.

Double angle sections with plates at their top (or laced/battened) are used for light axial forces in the bridges.

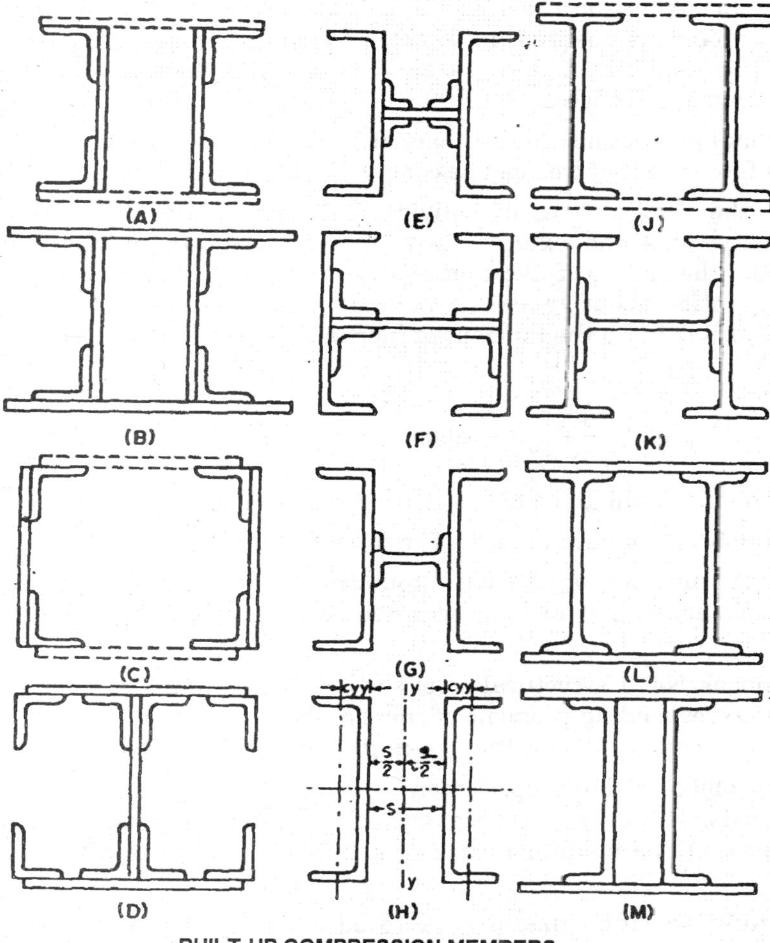

BUILT-UP COMPRESSION MEMBERS

Fig. 5.28.

Double channel sections with plates at their top (or laced/battened) are used for light axial forces in the bridges.

The webs in these sections are solid and continuous. The flange plates or channel sections may be used in combination with the rolled sections.

(ii) Built-up columns consisting of open webs

Built-up columns as shown in Fig. 5.28(A), (C) and (J) consist of open webs. These sections are connected together by lacing or batten plates as shown by dotted lines. Lacing elements tie the main elements and keep them together to act as one unit. Laced built-up columns are flexible and the main elements may adjust by lateral movements when these are loaded. These laced built-up columns carry ten percent more load as compared to battened columns. Therefore, the effective length of battened built-up columns (i.e., their slenderness ratio) is increased by ten percent.

All the built-up columns are tied at the ends by tie plates or end batten plates. Thus, a satisfactory performance is ensured. In top chord members, two channels are connected by cover plates on the top side and these are laced on the bottom side gusset plates are connected on the inner side of channels.

When two channel sections are used back to back as shown in Fig. 5.28(H) the maximum spacing between face to face of web can be computed as given below :

yy-axes are the vertical axes of individual channel sections, and C_{yy} represents the distance between yy-axis and external face of web as shown in Fig. 5.28(H). For most economical section, the radius of gyration about xx-axis (r_{xx}) should be approximately equal to the radius of gyration about yy-axis (r_{yy}) of built-up section (i.e., $I_{xx} = I_{yy}$). The moment of inertia about XX-axis is twice the moment of inertia of individual sections and therefore it is known.

$$I_{YY} = 2I_{yy} + 2A\left(\frac{S}{2} + C_{yy}\right)^2 \text{ approx.} = I_{xx}$$

where, A = effective sectional areas of one component member

S = spacing between back to back of channel section.

The spacing (S) may be computed and rounded-off in millimeters which can be adopted practically. the moment of inertia I_{yy} is recomputed for the built-up section for the adopted spacing.

In ISI Handbook No. 1 structural steel sections, tables are given for various values of spacing and the corresponding moment of inertia about yy-axis and radius of gyration r_{yy}. Comparing the values of I_{yy} with I_{xx}, that spacing (S) is adopted which gives $I_{yy} = I_{xx}$.

The primary object of built up section is to furnish large radius of gyration and to support heavy loads. However, the arrangement of different sections must allow ease of fabrication, connection and maintenance (e.g., painting). The lattice material used should be small.

The two-angle sections placed back to back are most frequently used in the roof trusses. The built-up sections consisting of such sections are particularly economical when single gusset plates are used. The angle sections are tied by rivets at suitable spacings with filler plates between the angles. The unequal angle sections are preferred for such built-up section, in order to obtain radii of gyration about two centroidal axes approximately equal.

Four angle sections are used in a built-up section to maintain the over-all depth of the section. In one case, two pairs of four angles are formed, each pair is made of two angles. Two pairs are connected by lacing, when the load is small. Two pairs are connected with a solid plate forming a built-up I-section when the load is large. The long legs of angle sections are kept outstanding. It gives higher value of radius of gyration. The four angle sections are also used to form a box section. It provides large value of radius of gyration. Such a built-up section is economical to carry small load and when the length of the member is long. Such a built-up section needs large amount of lacing. A section built-up of four angles is used for compression members having large length (e.g., masts, crane boom etc.) It provides equal stiffness in both the directions.

A built-up section consisting of **two channels back to back** is occasionally used. Such a built-up section has small radius of gyration about y-axis. A built-up section consisting of two channel sections placed at a distance apart is more frequently used. The flanges of channels are kept outward. It gives large value of radius of gyration. When the large value of radius of gyration is not necessary, then, the flanges are kept inward. It reduces the amount of lacing. A built-up section is also made of two channel sections with a solid plate on one side or on both the sides. The built-up section with solid plate on one side permits the access for inspection and painting. It is not possible, when solid plates are used on both the sides. Two channel sections with solid plates constituting built-up section, provide large cross-sectional area. The bracing between the parts is also stiff.

The built-up sections are also made of solid plates and angle sections to have very large cross-sectional area. When the channels of maximum size do not provide sufficient area, then, such built-up sections are used. Such built-up sections are used in the bridges.

It is to note that the cost of built-up section per unit weight is high because of fabrication and additional material needed for lacing the compression members. The single rolled steel sections are more economical for the equal area and equal slenderness ratio, if the normal connections may be designed for the member.

The lateral dimensions of the column is generally kept about (1/10th) to (1/15th) of the height of column.

TABLE 5.7A. STRESS REDUCTION FACTOR? FOR COLUMN BUCKLING CURVE "A" (SECTION 7.1.2.1)

KL/r	Yield stress, f_y																	
	200	210	220	230	240	250	260	280	300	320	340	360	380	400	420	450	480	510	540
10	1.000	1.000	1.000	1.000	1.000	1.000	1.000	1.000	1.000	1.000	1.000	1.000	1.000	1.000	1.000	1.000	1.000	1.000	1.000
20	1.000	0.999	0.998	0.997	0.995	0.994	0.993	0.992	0.990	0.988	0.986	0.984	0.983	0.981	0.979	0.977	0.975	0.972	0.970
30	0.977	0.975	0.974	0.972	0.970	0.969	0.967	0.964	0.961	0.957	0.954	0.951	0.948	0.946	0.943	0.938	0.934	0.930	0.925
40	0.952	0.949	0.947	0.944	0.942	0.939	0.937	0.931	0.926	0.921	0.916	0.911	0.906	0.901	0.896	0.888	0.881	0.873	0.865
50	0.923	0.919	0.915	0.911	0.908	0.904	0.900	0.892	0.884	0.876	0.867	0.859	0.851	0.842	0.834	0.820	0.807	0.794	0.780
60	0.888	0.883	0.877	0.871	0.865	0.859	0.853	0.841	0.824	0.816	0.803	0.790	0.77	0.763	0.750	0.730	0.710	0.690	0.671
70	0.846	0.837	0.829	0.820	0.811	0.803	0.794	0.776	0.758	0.740	0.722	0.703	0.686	0.668	0.651	0.626	0.602	0.579	0.557
80	0.793	0.781	0.769	0.757	0.746	0.734	0.772	0.698	0.675	0.653	0.631	0.610	0.589	0.570	0.551	0.525	0.501	0.478	0.458
90	0.730	0.715	0.700	0.685	0.671	0.657	0.643	0.615	0.590	0.565	0.542	0.520	0.500	0.481	0.463	0.439	0.416	0.396	0.377
100	0.661	0.644	0.627	0.610	0.594	0.579	0.564	0.536	0.510	0.486	0.463	0.443	0.424	0.407	0.390	0.368	0.348	0.331	0.314
110	0.591	0.573	0.555	0.538	0.522	0.507	0.492	0.465	0.440	0.418	0.397	0.379	0.362	0.346	0.332	0.312	0.295	0.279	0.265
120	0.525	0.507	0.489	0.473	0.458	0.443	0.429	0.404	0.381	0.361	0.343	0.326	0.311	0.297	0.284	0.267	0.252	0.238	0.226
130	0.466	0.448	0.432	0.416	0.402	0.388	0.376	0.353	0.332	0.314	0.298	0.283	0.269	0.257	0.246	0.231	0.217	0.206	0.195
140	0.413	0.397	0.382	0.368	0.355	0.342	0.331	0.310	0.291	0.275	0.260	0.247	0.235	0.224	0.214	0.201	0.189	0.179	0.170
150	0.368	0.353	0.339	0.326	0.314	0.303	0.293	0.274	0.257	0.243	0.229	0.218	0.207	0.197	0.189	0.177	0.166	0.157	0.149
160	0.329	0.316	0.303	0.291	0.280	0.270	0.261	0.244	0.229	0.215	0.204	0.193	0.184	0.175	0.167	0.157	0.147	0.139	0.132
170	0.296	0.283	0.272	0.261	0.251	0.242	0.233	0.218	0.204	0.192	0.182	0.172	0.164	0.156	0.149	0.140	0.131	0.124	0.117
180	0.267	0.255	0.245	0.235	0.226	0.218	0.210	0.196	0.184	0.173	0.163	0.155	0.147	0.140	0.134	0.125	0.118	0.111	0.105

190	0.242	0.231	0.222	0.213	0.205	0.197	0.190	0.177	0.166	0.156	0.147	0.140	0.133	0.126	0.121	0.113	0.106	0.100	0.095
200	0.220	0.210	0.202	0.193	0.186	0.179	0.172	0.161	0.151	0.142	0.134	0.127	0.120	0.115	0.109	0.102	0.096	0.091	0.086
210	0.201	0.192	0.184	0.177	0.170	0.163	0.157	0.147	0.137	0.129	0.122	0.115	0.110	0.104	0.099	0.093	0.087	0.083	0.078
220	0.184	0.176	0.169	0.162	0.155	0.149	0.144	0.134	0.126	0.118	0.111	0.106	0.100	0.095	0.091	0.085	0.080	0.075	0.071
230	0.170	0.162	0.155	0.149	0.143	0.137	0.132	0.123	0.115	0.108	0.102	0.097	0.092	0.088	0.083	0.078	0.073	0.069	0.065
240	0.157	0.149	0.143	0.137	0.132	0.127	0.122	0.114	0.106	0.100	0.094	0.089	0.085	0.081	0.077	0.072	0.068	0.064	0.060
250	0.145	0.138	0.132	0.127	0.122	0.117	0.113	0.105	0.098	0.092	0.087	0.082	0.078	0.074	0.071	0.068	0.062	0.059	0.056

TABLE 5.8A. STRESS REDUCTION FACTOR? FOR COLUMN BUCKLING CURVE "B" (SECTION 7.1.2.1)

KL/r	Yield stress, f_y																		
	200	210	220	230	240	250	260	280	300	320	340	360	380	400	420	450	480	510	540
10	1.000	1.000	1.000	1.000	1.000	1.000	1.000	1.000	1.000	1.000	1.000	1.000	1.000	1.000	1.000	1.000	1.000	1.000	1.000
20	1.000	0.998	0.996	0.994	0.993	0.991	0.990	0.986	0.983	0.981	0.978	0.975	0.972	0.970	0.967	0.963	0.960	0.956	0.953
30	0.963	0.961	0.958	0.955	0.953	0.50	0.948	0.943	0.938	0.933	0.929	0.924	0.920	0.915	0.911	0.904	0.898	0.892	0.886
40	0.925	0.921	0.917	0.913	0.909	0.906	0.902	0.895	0.887	0.880	0.873	0.866	0.859	0.852	0.845	0.835	0.825	0.815	0.805
50	0.883	0.877	0.872	0.866	0.861	0.855	0.850	0.839	0.829	0.818	0.808	0.798	0.787	0.777	0.767	0.752	0.737	0.722	0.708
60	0.835	0.827	0.820	0.812	0.805	0.798	0.790	0.775	0.761	0.746	0.732	0.718	0.704	0.691	0.677	0.657	0.638	0.620	0.602
70	0.781	0.771	0.761	0.751	0.742	0.732	0.722	0.703	0.685	0.667	0.649	0.632	0.615	0.599	0.584	0.561	0.540	0.520	0.502
80	0.721	0.709	0.697	0.685	0.673	0.661	0.650	0.627	0.606	0.585	0.566	0.547	0.529	0.512	0.496	0.474	0.453	0.434	0.416
90	0.657	0.643	0.629	0.615	0.602	0.589	0.576	0.552	0.530	0.508	0.488	0.470	0.452	0.436	0.421	0.400	0.380	0.363	0.346
100	0.593	0.577	0.562	0.548	0.534	0.520	0.507	0.483	0.461	0.440	0.421	0.403	0.387	0.372	0.358	0.339	0.321	0.306	0.291
110	0.531	0.515	0.500	0.485	0.471	0.458	0.445	0.422	0.401	0.381	0.364	0.348	0.333	0.319	0.306	0.289	0.274	0.260	0.247
120	0.474	0.458	0.443	0.429	0.416	0.403	0.391	0.370	0.350	0.332	0.316	0.301	0.288	0.276	0.265	0.249	0.236	0.223	0.212
130	0.423	0.408	0.394	0.380	0.368	0.356	0.345	0.325	0.307	0.291	0.276	0.263	0.251	0.240	0.230	0.217	0.204	0.194	0.184
140	0.378	0.364	0.350	0.338	0.327	0.316	0.306	0.287	0.271	0.256	0.243	0.231	0.221	0.211	0.202	0.190	0.179	0.169	0.161
150	0.339	0.325	0.313	0.302	0.291	0.281	0.272	0.255	0.241	0.227	0.215	0.205	0.195	0.186	0.178	0.167	0.158	0.149	0.142
160	0.305	0.292	0.281	0.271	0.261	0.252	0.243	0.228	0.215	0.203	0.192	0.182	0.174	0.166	0.158	0.149	0.140	0.133	0.126
170	0.275	0.264	0.253	0.244	0.235	0.227	0.219	0.205	0.193	0.182	0.172	0.163	0.155	0.148	0.142	0.133	0.125	0.118	0.112
180	0.249	0.239	0.229	0.220	0.212	0.205	0.198	0.185	0.174	0.164	0.155	0.147	0.140	0.133	0.128	0.120	0.113	0.106	0.101
190	0.227	0.217	0.208	0.200	0.193	0.186	0.179	0.168	0.157	0.148	0.140	0.133	0.127	0.121	0.115	0.108	0.102	0.096	0.091
200	0.207	0.198	0.190	0.183	0.176	0.169	0.163	0.153	0.143	0.135	0.128	0.121	0.115	0.110	0.105	0.098	0.092	0.087	0.083
210	0.190	0.182	0.174	0.167	0.161	0.155	0.149	0.140	0.131	0.123	0.117	0.110	0.105	0.100	0.096	0.090	0.084	0.080	0.075
220	0.174	0.167	0.160	0.154	0.148	0.142	0.137	0.128	0.120	0.113	0.107	0.101	0.096	0.092	0.088	0.082	0.077	0.073	0.069
230	0.161	0.154	0.147	0.141	0.136	0.131	0.126	0.118	0.111	0.104	0.098	0.093	0.088	0.084	0.080	0.075	0.071	0.067	0.063
240	0.149	0.142	0.136	0.131	0.126	0.121	0.117	0.109	0.102	0.096	0.091	0.086	0.082	0.078	0.074	0.070	0.065	0.062	0.058
250	0.138	0.132	0.126	0.121	0.117	0.112	0.108	0.101	0.095	0.089	0.084	0.080	0.076	0.072	0.069	0.064	0.060	0.057	0.054

TABLE 5.9A. STRESS REDUCTION FACTOR? FOR COLUMN BUCKLING CURVE "C" (SECTION 7.1.2.1)

KL/r	Yield stress, f_y																		
	200	210	220	230	240	250	260	280	300	320	340	360	380	400	420	450	480	510	540
10	1.000	1.000	1.000	1.000	1.000	1.000	1.000	1.000	1.000	1.000	1.000	1.000	1.000	1.000	1.000	1.000	1.000	1.000	1.000
20	0.999	0.997	0.994	0.992	0.990	0.987	0.985	0.981	0.976	0.972	0.968	0.964	0.961	0.957	0.953	0.948	0.943	0.938	0.933
30	0.948	0.944	0.941	0.937	0.933	0.930	0.926	0.920	0.913	0.907	0.901	0.895	0.889	0.883	0.877	0.869	0.861	0.853	0.845
40	0.896	0.891	0.885	0.880	0.875	0.870	0.866	0.856	0.847	0.838	0.829	0.820	0.812	0.803	0.795	0.783	0.771	0.760	0.748
50	0.841	0.834	0.827	0.821	0.814	0.807	0.801	0.788	0.776	0.763	0.752	0.740	0.729	0.717	0.706	0.690	0.675	0.660	0.645
60	0.783	0.774	0.765	0.757	0.748	0.740	0.732	0.716	0.700	0.685	0.670	0.656	0.642	0.628	0.615	0.596	0.578	0.561	0.544
70	0.722	0.711	0.700	0.690	0.680	0.670	0.660	0.641	0.623	0.605	0.588	0.572	0.557	0.542	0.528	0.508	0.489	0.471	0.454
80	0.659	0.646	0.634	0.622	0.611	0.600	0.589	0.568	0.548	0.529	0.512	0.495	0.479	0.464	0.450	0.430	0.412	0.395	0.379
90	0.596	0.583	0.569	0.557	0.544	0.533	0.521	0.499	0.749	0.460	0.443	0.426	0.411	0.397	0.383	0.365	0.348	0.332	0.318
100	0.536	0.522	0.508	0.495	0.483	0.471	0.459	0.438	0.418	0.400	0.384	0.368	0.354	0.341	0.328	0.311	0.296	0.282	0.269
110	0.480	0.466	0.453	0.440	0.428	0.416	0.405	0.385	0.366	0.349	0.333	0.319	0.306	0.294	0.283	0.268	0.254	0.242	0.230
120	0.430	0.416	0.403	0.391	0.379	0.368	0.358	0.339	0.321	0.306	0.291	0.278	0.267	0.256	0.246	0.232	0.220	0.209	0.199
130	0.385	0.372	0.360	0.348	0.337	0.327	0.317	0.299	0.283	0.269	0.256	0.244	0.234	0.224	0.215	0.203	0.192	0.182	0.173
140	0.346	0.333	0.322	0.311	0.301	0.291	0.282	0.266	0.251	0.238	0.227	0.216	0.206	0.197	0.189	0.178	0.168	0.160	0.152
150	0.311	0.300	0.289	0.279	0.269	0.261	0.252	0.237	0.224	0.212	0.202	0.192	0.183	0.175	0.168	0.158	0.149	0.141	0.134
160	0.281	0.270	0.260	0.251	0.242	0.234	0.227	0.213	0.201	0.190	0.180	0.172	0.164	0.156	0.150	0.141	0.133	0.126	0.120
170	0.255	0.245	0.236	0.227	0.219	0.212	0.205	0.192	0.181	0.171	0.162	0.154	0.147	0.140	0.134	0.126	0.119	0.113	0.107
180	0.232	0.223	0.214	0.206	0.199	0.192	0.186	0.174	0.164	0.155	0.147	0.139	0.133	0.127	0.121	0.114	0.107	0.102	0.096
190	0.212	0.203	0.195	0.188	0.181	0.175	0.169	0.158	0.149	0.140	0.133	0.126	0.120	0.115	0.110	0.103	0.097	0.092	0.087
200	0.194	0.186	0.179	0.172	0.166	0.160	0.154	0.144	0.136	0.128	0.121	0.115	0.110	0.105	0.100	0.094	0.089	0.084	0.079
210	0.178	0.171	0.164	0.158	0.152	0.146	0.141	0.132	0.124	0.117	0.111	0.105	0.100	0.096	0.092	0.086	0.081	0.076	0.072
220	0.164	0.157	0.151	0.145	0.140	0.135	0.130	0.122	0.114	0.108	0.102	0.097	0.092	0.088	0.084	0.079	0.074	0.070	0.066
230	0.152	0.145	0.140	0.134	0.129	0.124	0.120	0.112	0.105	0.099	0.094	0.089	0.085	0.081	0.077	0.073	0.068	0.065	0.061
240	0.141	0.135	0.129	0.124	0.120	0.115	0.111	0.104	0.098	0.092	0.087	0.082	0.078	0.075	0.071	0.067	0.063	0.060	0.056
250	0.131	0.125	0.120	0.115	0.111	0.107	0.103	0.096	0.090	0.085	0.081	0.076	0.073	0.069	0.066	0.062	0.058	0.055	0.052

TABLE 5.10A. STRESS REDUCTION FACTOR? FOR COLUMN BUCKLING CURVE "D" (SECTION 7.1.2.1)

KL/r	Yield stress, f_y																		
	200	210	220	230	240	250	260	280	300	320	340	360	380	400	420	450	480	510	540
10	1.000	1.000	1.000	1.000	1.000	1.000	1.000	1.000	1.000	1.000	1.000	1.000	1.000	1.000	1.000	1.000	1.000	1.000	1.000
20	0.999	0.995	0.991	0.988	0.984	0.980	0.977	0.970	0.964	0.958	0.952	0.946	0.940	0.935	0.930	0.922	0.915	0.908	0.901

30	0.922	0.916	0.911	0.906	0.901	0.896	0.891	0.881	0.872	0.863	0.885	0.847	0.839	0.831	0.823	0.813	0.802	0.792	0.782
40	0.848	0.841	0.834	0.828	0.821	0.815	0.808	0.796	0.784	0.773	0.762	0.751	0.741	0.731	0.721	0.707	0.694	0.681	0.668
50	0.777	0.768	0.760	0.752	0.744	0.736	0.728	0.713	0.699	0.685	0.672	0.659	0.647	0.635	0.624	0.608	0.592	0.577	0.563
60	0.707	0.697	0.687	0.678	0.668	0.659	0.651	0.634	0.617	0.602	0.587	0.573	0.560	0.547	0.535	0.517	0.501	0.486	0.471
70	0.640	0.629	0.618	0.607	0.597	0.587	0.578	0.559	0.542	0.526	0.510	0.496	0.482	0.469	0.456	0.439	0.423	0.408	0.394
80	0.576	0.564	0.553	0.542	0.531	0.521	0.511	0.492	0.474	0.458	0.442	0.428	0.414	0.402	0.390	0.373	0.358	0.344	0.330
90	0.517	0.505	0.493	0.482	0.471	0.461	0.451	0.432	0.415	0.399	0.384	0.370	0.357	0.345	0.334	0.319	0.304	0.292	0.280
100	0.434	0.451	0.440	0.428	0.418	0.408	0.398	0.380	0.363	0.348	0.334	0.321	0.309	0.298	0.288	0.274	0.261	0.249	0.239
110	0.416	0.404	0.392	0.381	0.371	0.361	0.352	0.335	0.319	0.305	0.292	0.281	0.270	0.259	0.250	0.237	0.226	0.215	0.206
120	0.373	0.361	0.350	0.340	0.330	0.321	0.313	0.297	0.282	0.269	0.257	0.246	0.236	0.227	0.219	0.207	0.197	0.187	0.179
130	0.336	0.325	0.314	0.305	0.295	0.287	0.279	0.264	0.251	0.239	0.228	0.218	0.209	0.200	0.193	0.182	0.173	0.164	0.157
140	0.303	0.292	0.283	0.274	0.265	0.257	0.250	0.236	0.224	0.213	0.203	0.194	0.185	0.178	0.171	0.161	0.153	0.145	0.138
150	0.274	0.264	0.255	0.247	0.239	0.231	0.224	0.212	0.201	0.190	0.181	0.173	0.165	0.159	0.152	0.144	0.136	0.129	0.123
160	0.249	0.240	0.231	0.223	0.216	0.209	0.203	0.191	0.181	0.171	0.163	0.155	0.149	0.142	0.137	0.129	0.122	0.116	0.110
170	0.227	0.218	0.210	0.203	0.196	0.190	0.184	0.173	0.164	0.155	0.147	0.140	0.134	0.128	0.123	0.116	0.110	0.104	0.099
180	0.207	0.199	0.192	0.185	0.179	0.173	0.167	0.157	0.149	0.141	0.134	0.127	0.122	0.116	0.111	0.105	0.099	0.094	0.089
190	0.190	0.183	0.176	0.169	0.164	0.158	0.153	0.144	0.136	0.128	0.122	0.116	0.111	0.106	0.101	0.095	0.090	0.085	0.081
200	0.175	0.168	0.162	0.156	0.150	0.145	0.140	0.132	0.124	0.118	0.112	0.106	0.101	0.097	0.093	0.087	0.082	0.078	0.074
210	0.161	0.155	0.149	0.143	0.138	0.134	0.129	0.121	0.114	0.108	0.102	0.097	0.093	0.089	0.085	0.080	0.075	0.071	0.068
220	0.149	0.143	0.138	0.133	0.128	0.123	0.119	0.112	0.105	0.100	0.094	0.090	0.086	0.082	0.078	0.074	0.069	0.066	0.062
230	0.138	0.133	0.128	0.123	0.118	0.114	0.110	0.104	0.097	0.092	0.087	0.083	0.079	0.075	0.072	0.068	0.064	0.061	0.058
240	0.129	0.123	0.119	0.114	0.110	0.106	0.103	0.096	0.090	0.085	0.081	0.077	0.073	0.070	0.067	0.063	0.059	0.056	0.053
250	0120	0.115	0.110	0.106	0.102	0.099	0.095	0.089	0.084	0.079	0.075	0.071	0.068	0.065	0.062	0.058	0.055	0.052	0.049

TABLE 5.11A. DESIGN COMPRESSIVE STRESS, F_{CD} FOR COLUMN BUCKLING CURVE "A" (SECTION 7.1.2.1) IS 800-2007

KL/r	Yield stress, f_y																		
	200	210	220	230	240	250	260	280	300	320	340	360	380	400	420	450	480	510	540
10	182	191	200	182	218	227	236	255	273	291	309	327	345	364	382	409	436	464	491
20	182	191	200	182	217	226	235	252	270	287	305	322	339	357	374	400	425	451	476
30	178	186	195	178	212	220	229	245	262	279	295	311	328	344	360	384	408	431	454
40	173	181	189	173	205	213	221	237	253	268	283	298	313	328	342	363	384	405	425
50	168	176	183	168	198	205	213	227	241	255	268	281	294	306	318	336	352	368	383
60	162	169	175	162	189	195	202	214	226	237	248	258	268	278	286	299	310	320	329
70	154	160	166	154	177	182	188	197	207	215	223	230	237	243	249	256	263	268	274
80	144	149	154	144	163	167	171	178	184	190	195	199	204	207	210	215	219	222	225
90	133	137	140	133	146	149	152	157	161	164	168	170	173	175	177	179	182	184	185

100	120	123	125	120	130	132	133	136	139	141	143	145	146	148	149	151	152	153	154
110	107	109	111	107	114	115	116	118	120	121	123	124	125	126	127	128	129	129	130
120	95.5	96.7	97.9	95.5	100	101	101	103	104	105	106	107	107	108	109	109	110	110	111
130	84.6	85.5	86.3	84.6	87.7	88.3	88.8	89.8	90.6	91.3	92.0	92.5	93.0	93.5	93.9	94.4	94.9	95.3	95.7
140	75.2	75.8	76.4	75.2	77.4	77.8	78.2	78.9	79.5	80.0	80.5	80.9	81.3	81.6	81.9	82.3	82.6	83.0	83.2
150	67.0	67.4	67.9	67.0	68.6	68.9	69.2	69.7	70.2	70.6	70.9	71.2	71.5	71.8	72.0	72.3	72.6	72.9	73.1
160	59.9	60.3	60.6	59.9	61.1	61.4	61.6	62.0	62.4	62.7	62.9	63.2	63.4	63.6	63.8	64.0	64.3	64.5	64.6
170	53.8	54.1	54.3	53.8	54.8	55.0	55.1	55.5	55.7	56.0	56.2	56.4	56.6	56.7	56.9	57.1	57.3	57.4	57.6
180	48.6	48.8	49.0	48.6	49.3	49.5	49.6	49.9	50.1	50.3	50.5	50.6	50.8	50.9	51.0	51.2	51.3	51.5	51.6
190	44.0	44.2	44.3	44.0	44.6	44.7	44.9	45.1	45.3	45.4	45.6	45.7	45.8	45.9	46.0	46.2	46.3	46.4	46.5
200	40.0	40.2	40.3	40.0	40.5	40.7	40.7	40.9	41.1	41.2	41.3	41.4	41.5	41.6	41.7	41.8	41.9	42.0	42.1
210	36.6	36.7	36.8	36.6	37.0	37.1	37.2	37.3	37.4	37.6	37.7	37.8	37.8	37.9	38.0	38.1	38.2	38.3	38.3
220	33.5	33.6	33.7	33.5	33.9	34.0	34.0	34.2	34.3	34.4	34.5	34.5	34.6	34.7	34.7	34.8	34.9	35.0	35.0
230	30.8	30.9	31.0	30.8	31.2	31.2	31.3	31.4	31.5	31.6	31.6	31.7	31.8	31.8	31.9	31.9	32.0	32.1	32.1
240	28.5	28.5	28.6	28.5	28.7	28.8	28.8	28.9	29.0	29.1	29.1	29.2	29.3	29.3	29.4	29.4	29.5	29.5	29.6
250	26.3	26.4	26.5	26.3	26.6	26.6	26.7	26.7	26.8	26.9	26.9	27.0	27.0	27.1	27.1	27.2	27.2	27.3	27.3

TABLE 5.12A. DESIGN COMPRESSIVE STRESS, F_{CD} FOR COLUMN BUCKLING CURVE "B" (SECTION 7.1.2.1) IS 800-2007

KL/r	Yield stress, f_y																		
	200	210	220	230	240	250	260	280	300	320	340	360	380	400	420	450	480	510	540
10	182	191	200	209	218	227	236	255	273	291	309	327	345	364	382	409	436	464	491
20	182	190	199	208	217	225	234	251	268	285	302	319	336	353	369	394	419	443	468
30	175	183	192	200	208	216	224	240	256	271	287	302	318	333	348	370	392	414	435
40	168	176	183	191	198	206	213	228	242	256	270	283	297	310	323	342	360	378	395
50	161	167	174	181	188	194	201	214	226	238	250	261	272	283	293	308	322	335	347
60	152	158	164	170	176	181	187	197	207	217	226	235	243	251	259	269	279	287	295
70	142	147	152	157	162	166	171	179	187	194	201	207	213	218	223	230	236	241	246
80	131	135	139	143	147	150	154	160	165	170	175	179	183	186	190	194	198	201	204
90	120	123	126	129	131	134	136	141	144	148	151	154	156	159	161	163	166	168	170
100	108	110	112	114	116	118	120	123	126	128	130	132	134	135	137	139	140	142	143
110	96.5	98.3	100	101	103	104	105	107	109	111	112	114	115	116	117	118	119	121	121
120	86.2	87.5	88.6	89.7	90.7	91.7	92.5	94.1	95.4	96.6	97.7	98.6	100	100	101	102	103	104	104
130	76.9	77.8	78.7	79.5	80.3	81.0	81.6	82.7	83.7	84.6	85.4	86.1	86.8	87.3	87.9	88.6	89.2	89.8	90.3
140	68.7	69.4	70.1	70.7	71.3	71.8	72.3	73.1	73.9	74.6	75.2	75.7	76.2	76.6	77.1	77.6	78.1	78.5	78.9
150	61.6	62.1	62.6	63.1	63.6	64.0	64.3	65.0	65.6	66.1	66.6	67.0	67.4	67.7	68.1	68.5	68.9	69.2	69.5
160	55.4	55.8	56.2	56.6	56.9	57.3	57.5	58.1	58.5	59.0	59.3	59.7	60.0	60.3	60.5	60.9	61.2	61.5	61.7

170	50.4	50.3	50.7	51.0	51.2	51.7	51.7	52.2	52.5	52.9	53.2	53.5	53.7	53.9	54.1	54.4	54.7	54.9	55.1
180	45.3	45.6	45.9	46.1	46.3	46.7	46.7	47.1	47.4	47.7	47.9	48.1	48.3	48.5	48.7	48.9	49.2	49.3	49.5
190	41.2	41.5	41.7	41.9	42.1	42.4	42.4	42.7	42.9	43.2	43.4	43.6	43.7	43.9	44.0	44..2	44.4	44.6	44.7
200	37.6	37.8	38.0	38.2	38.3	38.5	38.6	38.9	39.1	39.3	39.5	39.6	39.8	39.9	40.2	40.2	40.3	40.5	40.6
210	34.5	34.7	34.8	35.0	35.1	35.2	35.3	35.5	35.7	35.9	36.0	36.2	36.3	36.4	36.5	36.6	36.8	36.9	37.0
220	31.7	31.9	32.0	32.1	32.2	32.3	32.4	32.6	32.8	32.9	33.0	33.1	33.2	33.3	33.4	33.6	33.7	33.8	33.9
230	29.2	29.4	29.5	29.6	29.7	29.8	29.9	30.0	30.1	30.3	30.4	30.5	30.6	30.7	30.7	30.8	30.9	31.0	31.1
240	27.1	27.2	27.3	27.3	27.4	27.5	27.6	27.7	27.8	27.9	28.0	28.1	28.2	28.3	28.3	28.4	28.5	28.6	28.7
250	25.1	25.2	25.3	25.3	25.4	25.5	25.6	25.7	25.8	25.9	26.0	26.0	26.1	26.2	26.2	26.3	26.4	26.5	26.5

TABLE 5.13A. DESIGN COMPRESSIVE STRESS, F_{CD} FOR COLUMN BUCKLING CURVE "C" (SECTION 7.1.2.1) IS 800-2007

KL/r	Yield stress, f_y																	
	200	210	220	230	240	250	260	280	300	320	340	360	380	400	420	450	480	510	540
10	182	191	200	209	218	227	236	255	273	291	309	327	345	364	382	409	436	464	491
20	182	190	199	207	216	224	233	250	266	283	299	316	332	348	364	388	412	435	458
30	172	180	188	196	204	211	219	234	249	264	278	293	307	321	335	355	376	395	415
40	163	170	177	184	191	198	205	218	231	244	256	268	280	292	304	320	337	352	367
50	153	159	165	172	178	183	189	201	212	222	232	242	252	261	270	282	295	306	317
60	142	148	153	158	163	168	173	182	191	199	207	215	222	228	235	244	252	260	267
70	131	136	140	144	148	152	156	163	170	176	182	187	192	197	202	208	213	218	223
80	120	123	127	130	133	136	139	145	149	154	158	162	165	169	172	176	180	183	186
90	108	111	114	116	119	121	123	127	131	134	137	140	142	144	146	149	152	154	156
100	97.5	100	102	104	105	107	109	112	114	116	119	120	122	124	125	127	129	131	132
110	87.3	89.0	90.5	92.0	93.3	94.6	95.7	97.9	100	102	103	104	106	107	108	110	111	112	113
120	78.2	79.4	80.6	81.7	82.7	83.7	84.6	86.2	87.6	88.9	90.1	91.1	92.1	93.0	93.8	94.9	95.9	96.8	97.6
130	70.0	71.0	71.9	72.8	73.5	74.3	75.0	76.2	77.3	78.3	79.2	80.0	80.7	81.4	82.0	82.9	83.6	84.3	84.9
140	62.9	63.6	64.4	65.0	65.6	66.2	66.7	67.7	68.6	69.3	70.0	70.7	71.2	71.8	72.3	72.9	73.5	74.1	74.6
150	56.6	57.2	57.8	58.3	58.8	59.2	59.7	60.4	61.1	61.7	62.3	62.8	63.3	63.7	64.1	64.6	65.1	65.5	65.9
160	51.1	51.6	52.1	52.5	52.9	53.3	53.6	54.2	54.8	55.3	55.7	56.1	56.5	56.9	57.2	57.6	58.0	58.4	58.7
170	46.4	46.8	47.1	47.5	47.8	48.1	48.4	48.9	49.3	49.8	50.1	50.5	50.8	51.1	51.3	51.6	52.0	52.3	52.6
180	42.2	42.5	42.8	43.1	43.4	43.6	43.9	44.3	44.7	45.0	45.3	45.6	45.8	46.1	46.3	46.6	46.9	47.1	47.3
190	38.5	38.8	39.0	39.3	39.5	39.7	39.9	40.3	40.6	40.9	41.1	41.4	41.6	41.8	42.0	42.2	42.5	42.7	42.9
200	35.3	35.5	35.7	35.9	36.1	36.3	36.5	36.8	37.0	37.3	37.5	37.7	37.9	38.1	38.2	38.4	38.6	38.8	39.0
210	32.4	32.6	32.8	33.0	33.1	33.3	33.4	33.7	33.9	34.1	34.3	34.5	34.7	34.8	34.9	35.1	35.3	35.4	35.6
220	29.9	30.1	30.2	30.4	30.5	30.6	30.8	31.0	31.2	31.4	31.5	31.7	31.8	31.9	32.1	32.2	32.4	32.5	32.6
230	27.6	27.8	27.9	28.0	28.2	28.3	28.4	28.6	28.8	28.9	29.1	29.2	29.3	29.4	29.5	29.7	29.8	29.9	30.0

240	25.6	25.7	25.9	26.0	26.1	26.2	26.3	26.4	26.6	26.7	26.9	27.0	27.1	27.2	27.3	27.4	27.5	27.6	27.7
250	23.8	23.9	24.0	24.1	24.2	24.3	24.4	24.5	24.7	24.8	24.9	25.0	25.1	25.2	25.3	25.4	25.5	25.6	25.7

TABLE 5.14A. DESIGN COMPRESSIVE STRESS, F_{CD} FOR COLUMN BUCKLING CURVE "D" (SECTION 7.1.2.1) IS : 800-2007

KL/r	Yield stress, f_y																		
	200	210	220	230	240	250	260	280	300	320	340	360	380	400	420	450	480	510	540
10	182	191	200	209	218	227	236	255	273	291	309	327	345	364	382	409	436	464	491
20	182	190	198	206	215	223	231	247	263	279	294	310	325	340	355	377	399	421	442
30	168	175	182	189	197	204	211	224	238	251	264	277	290	302	314	332	350	367	384
40	154	161	167	173	179	185	191	203	214	225	235	246	256	266	275	289	303	316	328
50	141	147	152	157	162	167	172	182	191	199	208	216	224	231	238	249	258	268	277
60	129	133	137	142	146	150	154	161	168	175	182	188	193	199	204	212	219	225	231
70	116	120	124	127	130	133	137	142	148	153	158	162	167	171	174	180	184	189	193
80	105	108	111	113	116	118	121	125	129	133	137	140	143	146	149	153	156	159	162
90	94.1	96.4	98.6	101	103	105	107	110	113	116	119	121	123	126	128	130	133	135	137
100	84.3	86.2	87.9	89.6	91.1	92.6	94.0	96.7	99.1	101	103	105	107	108	110	112	114	116	117
110	75.6	77.0	78.4	79.7	81.0	82.1	83.2	85.3	87.1	88.8	90.4	91.8	93.1	94.4	95.5	97.1	98.5	100	101
120	67.8	69.0	70.1	71.1	72.1	73.0	73.9	75.5	77.0	78.3	79.5	80.6	81.7	82.6	83.5	84.7	85.8	86.9	87.8
130	61.0	62.0	62.8	63.7	64.5	65.2	65.9	67.2	68.3	69.4	70.4	71.2	72.1	72.8	73.5	74.5	75.4	76.2	76.9
140	55.0	55.8	56.5	57.2	57.8	58.4	59.0	60.0	61.0	61.8	62.4	63.3	64.0	64.6	65.2	66.0	66.7	67.3	67.9
150	49.8	50.4	51.0	51.6	52.1	52.6	53.1	53.9	54.7	55.4	56.0	56.6	57.2	57.7	58.1	58.8	59.3	59.9	60.4
160	45.2	45.7	46.2	46.7	47.1	47.5	47.9	48.6	49.3	49.9	50.4	50.9	51.3	51.7	52.1	52.7	53.1	53.6	54.0
170	41.2	41.6	42.1	42.4	42.8	43.1	43.5	44.1	44.6	45.1	45.5	45.9	46.3	46.7	47.0	47.4	47.8	48.2	48.6
180	37.7	38.0	38.4	38.7	39.0	39.3	39.6	40.1	40.5	41.0	41.3	41.7	42.0	42.3	42.6	43.0	43.3	43.6	43.9
190	34.5	34.9	35.2	35.4	35.7	35.9	36.2	36.6	37.0	37.4	37.7	38.0	38.2	38.5	38.7	39.1	39.4	39.6	39.9
200	31.8	32.0	32.3	32.5	32.8	33.0	33.2	33.6	33.9	34.2	34.5	34.7	35.0	35.2	35.4	35.7	35.9	36.2	36.4
210	29.3	29.6	29.8	30.0	30.2	30.4	30.5	30.9	31.2	31.4	31.7	31.9	32.1	32.3	32.5	32.7	32.9	33.1	33.3
220	27.1	27.3	27.5	27.7	27.9	28.0	28.2	28.5	28.7	29.0	29.2	29.4	29.6	29.7	29.9	30.1	30.3	30.5	30.6
230	25.2	25.3	25.5	25.7	25.8	26.0	26.1	26.4	26.6	26.8	27.0	27.1	27.3	27.5	27.6	27.8	27.9	28.1	28.2
240	23.4	23.6	23.7	23.9	24.0	24.1	24.2	24.5	24.7	24.8	25.0	25.2	25.3	25.4	25.5	25.7	25.9	26.0	26.1
250	21.8	22.0	22.1	22.2	22.3	22.5	22.6	22.8	22.9	23.1	23.2	23.4	23.5	23.6	23.7	23.9	24.0	24.1	24.2

5.27 WIDTH THICKNESS RATIO OF ELEMENTS IN COMPRESSION MEMBERS

In addition to the slenderness ratio and the cross-sectional shape, the behaviour of the compression members is influenced by the relative thickness of the component elements that constitute the cross section the relative thickness of the component element is quantified by the width-thickness ratio (b/t) of the element.

When any plates element is subjected to direct compression bending, or shear stresses or to combination of these stresses, the plate element may buckle locally before the entire member fails. In local buckling, waves (single wave or series of waves) or wrinkles are formed in the elements of compression members. In case, the flanges and the web of a rolled steel member cross-section are considered to plate elements carrying compression, their thicknesses may be found by equating the critical buckling stress to the yield stress of material. In this approach the various factors (i.e., the actual support restraint provided by the adjoining plate elements, the residual stresses and the initial imperfections) are not considered.

The dimensions of rolled steel member cross-sections (viz, width of flange, b_f thickness of flange, t_f the depth of web, d_w and the thickness of web, t_w) are so proportioned that the local buckling does not occur. The ratios of $(0.5\ b_f/t_f)$ and (d_w/t_w) are determined and certain limits are imposed over these width- thickness ratios. These limits are decided by testing these sections in the laboratory and after examining their validations. The yield stress, σsub y> of the steel is also considered.

ISI in IS : 800-62 recommended the thickness of element in compression members in terms of outstanding lengths of compression members. These thicknesses have been recommended to avoid the failure of elements owing to local buckling.

The outstanding length of compression member in case of plated I-section is measured from the free edge to the first row of rivet, as shown in Fig. 5.29(a).

In case of channels, angles and stem of tee-section, it is measured as nominal width of sections as shown in Fig. 5.29(b) and in case of flanges of beams and tee-section, it is measured as half the nominal width, as shown in Fig. 5.29(c)

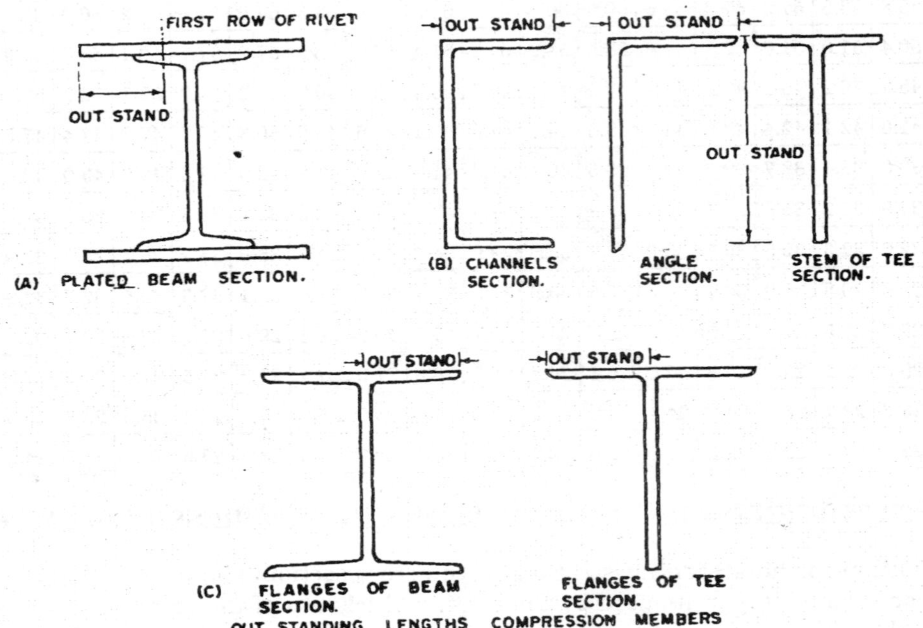

Fig. 5.29.

The thickness is taken as 1/16 of outstand of any member in compression for unstiffened outstand and 1/20 of outstand of any member in compression for stiffened outstand. The thickness is taken as 1/90 of outstand for unsupported width of unstiffened plate forming any part of a compression member, measured between adjacent line of rivets. But for the purpose of calculating effective sectional area and radius of gyration only 50 times the thickness of plate will be taken instead of 90 times the thickness in case single plate is used. However, it is total thickness of two or more plates effectively tacked together.

All hot rolled steel sections as per IS : 808 (beams, channels and angle sections) conform to outstand requirements as per IS : 800-62 described above.

As per the recommendations in IS : 800-1984, the projection of a plate or flange beyond its connection to a web, or other line of support or the like, exceeds the relevant values given in (i), (ii) and (iii) below, the area of excess flange shall be neglected when the effective geometrical properties of the section are determined :

(i) Flanges and plates in compression with unstiffened edges.

The excess projection if exceeds $256t/(f_y)^{1/2}$ subjected to a maximum of 16t shall be neglected.

(ii) Flanges and plates in compression with stiffened edges.

The excess projection if exceeds 20t to the innermost face of the stiffening.

(iii) Flanges and plates in tension the excess projection if exceeds 20t.

Note. 1. the stiffened flanges shall include flanges composed of channels or I-sections or of plates with continuously stiffened edges.

2. *t* denotes the thickness of the flange of a section or of a plate in compression, or the aggregate thickness of the plates, if connected together appropriately.

Example 5.3 A single angle dis-continuous strut ISA 150 mm x 150 mm x 12 mm (ISA 150 150, @ 0.272 kN/m with single riveted/bolted connection is 1.60 m long calculate safe load carrying capacity (i.e, compressive strength) of the section. This angle strut is loaded through one leg. Take $\sigma_y = 260$ N/mm^2 and $E = 2 \times 10^5$ N/mm^2.

Solution :

Step 1 : Properties of angle section

From ISI HB No. 1 for ISA 150 mm x 150 mm x 12 mm

(ISA 150, 150, @ 0.272 kN/m)

Sectional area, A = 3459 mm^2

$r_{xx} = r_{yy} = 46.1$ mm

$r_{uu} = 58.3$ mm, $r_{uu} = 29.3$ mm

Length of each leg, $b_1 = b_2 = 150$ mm

Thickness of each leg t = 12 mm

Step 2 : Type of section

Yield stress ratio $\sigma_y = 260$ N/mm^2

Epsilon, $\varepsilon = (250/\sigma_y)^{1/2} = (250/260)^{1/2} = 0.98$

Ratio, $(b/t) = (150/12) = 12.5$

For semi-compact section

(b/t) ≤ [(15.7 x 0.98)] = 15.386

Hence, the given section is semi-compact section.

Step 3 : Slenderness ratio

Effective length, l = (1.60 x 1000) mm

$$\left(\frac{l}{r_{uu}}\right) = \left(\frac{1600}{58.3}\right) = 27.44$$

$$\left(\frac{l}{r_{vv}}\right) = \left(\frac{1600}{29.3}\right) = 54.61$$

Yield stress ratio

Epsilon, $\varepsilon = (250/\sigma_y)^{1/2}$ = (250/260)$^{1/2}$ = 0.98

From IS : 800-2007, Clause 7.5.1.2, Ratio, (b/t) = (150/12) = 12.5

For semi-compact section

(b/t) < (15.7 x 0.98) = 15.386.

Hence, the given section is semi-compact type.

$$\lambda_{vv} = \left(\frac{l/r_{vv}}{\varepsilon.\left(\frac{\pi^2 E}{250}\right)^{1/2}}\right)$$

$$= \left(\frac{l/r_{vv}}{\left(\frac{\pi^2\,E}{\sigma_y}\right)^{1/2}}\right) = \left(\frac{54.61}{\left(\frac{9.8596 \times 2 \times 10^5}{260}\right)^{1/2}}\right) = 0.627$$

$$\lambda_{\phi\phi} = \left(\frac{b_1 + b_2}{\varepsilon\left(\frac{\pi^2 . E}{250}\right)^{1/2}}\right) = \left(\frac{b_1 + b_2}{\left(\frac{250}{\sigma_y}\right)^{1/2}\left(\frac{\pi^2 E}{250}\right)^{1/2} . 2t}\right)$$

$$\lambda_{\phi\phi} = \left(\frac{b_1 + b_2}{\left(\frac{\pi^2 E}{\sigma_y}\right)^{1/2} . (2t)}\right) = \left(\frac{150 + 150}{\left(\frac{9.8596 \times 2 \times 10^5}{260}\right)^{1/2} \times (2t)}\right)$$

$$\lambda_{\phi\phi} = \left(\frac{300}{87.08 \times 24}\right) = \left(\frac{12.5}{87.08}\right) = 0.1435$$

Step 4 : Equivalent slenderness ratio

Member is connected to a gusset plate by one rivet/one bolt at each end. As such, it shall act as a hinged member. From IS : 800–2007 (clause 7.5.1.2)

$k_1 = 1.25$, $k_2 = 0.50$, $k_3 = 60$

Substituting these values

$$\lambda_e = \left(k_1 + k_2 \cdot \lambda_{vv}^2 + k_3 \cdot \lambda_{\phi\phi}^2\right)^{1/2}$$

$$= (1.25 + 0.50 \times 0.627^2 + 60 \times 0.1435^2)^{1/2} = 1.659$$

Step 5 : Design compressive stress, σ_{cd}

For buckling curve `C', and $\sigma_y = 260$ N/mm²

For $\lambda_e = 0$, $\sigma_{cd} = 260$ N/mm²

$\lambda_e = 10$, $\sigma_{cd} = 236$ N/mm² $\lambda_e = 1.651$, $\sigma_{cd} = 256.04$ N/mm²

Cross-sectional area, A = 3459 mm

$$\text{Factored load, } W_{Fd} = \left(\frac{3459 \times 256.04}{1000 \times 1.15}\right) = 770.12 \text{ kN}$$

$$\text{Service load, } W_{cK} = \left(\frac{770.12}{1.5}\right) = 513.417 \text{ kN}$$

Example 5.3A. A discontinuous strut ISA 150 mm x 150 mm x 12 mm (ISA 150 150, @, 0.272 kN/m). With two more than rivets/bolts in a line along the angle at each end. The length of strut is 3.5 m calculate safe load carrying capacity (compressive strength of the section).

This angle is loaded through one leg. Take $\sigma_y = 260$ N/mm², and $E = 2 \times 10^5$ N/mm²

Solution

Step 1 : Properties of angle section

From ISI HB No. 1

Sectional area, A = 3459 mm²

$r_{yy} = r_{xx} = 46.1$ mm

$r_{uu} = 58.3$ mm, $r_{vv} = 29.3$ mm

Length of each leg = 150 mm

Thickness of each leg = 12 mm

Step 2 : Type of section, $\sigma_y = 260$ N/mm²

Yield stress ratio

Epsilon, $\varepsilon = (250/\sigma_y)^{1/2} = (250/260)^{1/2} = 0.98$

Ratio, (b/t) = (150/12) = 12.5

For semi-compact section

$$(b/t) < [(15.7 \times \varepsilon) = (15.7 \times 0.98) = 15.386]$$

Hence, the given section is a semi-compact section.

Member is connected to a gusset plate by two/more than two rivets/ bolts at each end. Its such, it shall act as a fixed member From IS : 800–2007 (Clause 7.5.1.2)

$k_1 = 0.20$, $k_2 = 0.35$ and $k_3 = 20$

$$\text{Factor, } \varepsilon = \sqrt{\frac{\pi^2 E}{250}} = 0.98\sqrt{\frac{9.8596 \times 2 \times 10^5}{250}} = 87.04$$

Step 3 : Slenderness ratio

Effective length l = (0.85 x 3.50 x 1000) mm

$$\left(\frac{l}{r_{uu}}\right) = \left(\frac{0.85 \times 3.5 \times 1000}{58.3}\right) = (0.85 \times 60.034) = 51.029$$

$$\left(\frac{l}{r_{vv}}\right) = \left(\frac{0.85 \times 3.5 \times 1000}{29.3}\right) = (0.85 \times 119.45) = 101.536$$

From IS 800–2007, Clause 7.5.1.2,

$$\lambda_{vv} = \left(\frac{(l/r_{vv})}{\varepsilon\sqrt{\frac{\pi^2 E}{250}}}\right) = \left(\frac{101.536}{8.704}\right) = 1.1665$$

$$\lambda_{\phi\phi} = \left(\frac{(b_1 + b_2)}{\varepsilon\sqrt{\frac{\pi^2 E}{250}} \times 2t}\right) = \left[\frac{(150+150)}{8.704 \times 2 \times 12}\right] = \left(\frac{300}{8.704 \times 24}\right) = 1.436$$

Step 4 : Equivalent slenderness ratio (From IS 800–2007, Clause 7.5.1.2)

$$\lambda_e = \left[k_1 + k_2 \cdot \lambda_{vv}^2 + k_3 \cdot \lambda_{\phi\phi}^2\right]^{1/2}$$

$= [0.20 + 0.35 \times 1.1665^2 + 20 \times 1.436^2]^{1/2}$

$= [0.20 + 0.35 \times 1.3608 + 20 \times 2.062]^{1/2}$

$= [0.20 + 4.76 + 41.24]^{1/2}$

$= (41.916)^{1/2} = 6.474$

Step 5 : Design compressive stress

For buckling curve `C' and

$\sigma_y = 260\ N/mm^2$

For $\lambda_e = 0, \quad \sigma_{cd} = 260 \quad N/mm^2$

$\lambda_e = 10 \quad \sigma_{cd} = 236 \quad N/mm^2$

$\lambda_e = 6.474, \quad \sigma_{cd} = \left(260 - \frac{6.474}{10} \times 24\right) = (260 - 15.54) = 244.46 \text{ N/mm}^2$

Cross-sectional area of the angle-section, A = 3459 mm².

Factored compressive strength.

$$W_{Fd} = \left(\frac{244.46 \text{ x } 3459}{1000 \times 1.15}\right) = 735.293 \text{ kN}$$

Service (Characteristic) compressive strength

$$W_{cK} = \left(\frac{735.293}{1.5}\right) = 490.195 \text{ kN}$$

Example 5.4 A double angle discontinuous strut, 2ISA 125 mm x 95 mm x 10 mm (ISA 125, 95 @ 0.165 kN/m) long legs connected back to back on opposite sides of a gusset plate 10 mm thick by two rivets/bolts in a line along the length of strut at each end.

The length of strut between centre to centre of intersections is 1.8 m. Determine the safe load carrying capacity (compressive strength) of the section. Take (σ_y = 260 N/mm² and E = 2 x 10⁵ N/mm².

Solution :

Step 1 : Properties of angle sections (From ISI HB No. 1)

Double angle discontinuous strut 2 ISA 125 mm x 95mm x 10 mm (2 ISA 125, 95, @ 0.165 kN/m) is shown in Fig. 5.30. It is assumed that the tacking rivets/bolts shall be used along the length at appropriates spacing. Sectional area, A = 4204 mm².

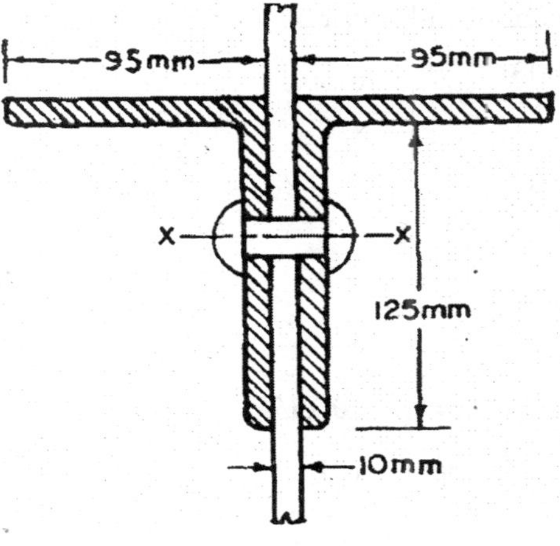

Fig. 5.30.

Radius of gyration (with gusset plate 10 mm thick)

r_{xx} = 39.4 mm, r_{yy} = 40.1 mm

Step 2 : Type of the section

b_1 = 125 mm, b_2 = 95 mm, t = 10 mm

Ratios, (b_1/t) = 12.5, (b_2/t) = 9.5

For semi-compact section

Epsilon, $\varepsilon = (250/260)^{1/2} = 0.98$

Requirements for the semi-compact section are as follows :

$(b_1/t) \leq [(15.7 \times \varepsilon) = (15.7 \times 0.98) = 15.388]$

$(b_2/t) \leq [(9.5 \times \varepsilon) = (9.5 \times 0.98) = 9.3]$

These satisfy the requirements. Hence, the section is a semi-compact section.

Step 3 : Slenderness ratio

Length of the strut between centre to centre of intersection

L = 1.80 m

Effective length in the plane of gusset

kL = (0.85 x 1.80) = 1.53 m

$$\lambda_{xx} = \left(\frac{kL}{r_{xx}}\right) = \left(\frac{1530}{39.4}\right) = 38.83$$

$$\lambda_{yy} = \left(\frac{kL}{r_{yy}}\right) = \left(\frac{1530}{40.1}\right) = 38.15$$

$$\lambda_o = 0.2\,\pi \left(E/\sigma_y\right)^{1/2}$$

$$= 0.2 \times 3.14\ (2 \times 10^5/260)^{1/2} = 17.417$$

Step 4 : Imperfection factor, α_1

For angle-sections and compound sections (two **sections back** to back)

For buckling about xx-axis

$\alpha_{xx} = 0.0055$

For buckling about yy-axis

$\alpha_{yy} = 0.0055$

About xx-axis

$$\alpha_{1\text{-}xx} = \alpha_{xx}\left(\lambda_{xx} - \lambda_o\right)$$

$$= 0.0055\ (38.83 - 17.417) = 0.1177$$

About yy-axis

$$\alpha_{1\text{-}yy} = \alpha_{yy}\left(\lambda_{yy} - \lambda_o\right)$$

$$= 0.0055\ (38.15 - 17.417) = 0.1140$$

Step 5 : Euler's stress, σ_e

About xx-axis

$$\sigma_{Ex} = \pi^2 \text{ E} / (\lambda_{xx} - \lambda_o)^2$$

$= 9.8596 \text{ x } 2 \text{ x } 10^5 / (38.83 - 17.417)^2$

$= 4300.6$

About yy-axis

$$\sigma_{Ey} = \pi^2 \text{ E} / (\lambda_{yy} - \lambda_o)^2$$

$$\sigma_{Ey} = \left(\frac{9.8596 \times 2 \times 10^5}{(38.15 - 17.417)^2}\right) = \left(\frac{19.7192 \times 10^5}{(20.733)^2}\right)$$

$= [(19.719 \text{ x } 10^5) / 429.857] = 4587.386$

Step 6 : Values of ϕ_x and ϕ_y

$$\phi_x = \lfloor \sigma_y + (1 + \alpha_{1.xx}) . \sigma_{Ex} \rfloor \ / \ 2$$

$= [260 + (1 + 0.1177) \text{ x } 4330.6]^{1/2} \quad = 2550.155$

$\phi_y = [260 + (1 + 0.1140)4587.388]^{1/2}$

Step 7 : Design Compressive stresses, σ_{cx} σ_{cy} (Ayrton and Perry's Formula)

$$\sigma_{cx} = \phi_x - \left|\phi_x^2 - \sigma_y . \sigma_{Ex}\right|^{1/2}$$

$2550.155 - [2550.155^2 - 260 \text{ x } 43306]^{1/2}$

$= (2550.155 - 2318.91) = 118.65 \text{ N/mm}^2$

$$\sigma_{cy} = \phi_y - \left[\phi_y^2 - \sigma_y . \sigma_{Ey}\right]^{1/2}$$

$= 2555.175 - [2555.175^2 - 260 \text{ x } 4587.358]^{1/2}$

$= 2555.175 - 100 \ [652.89 - 119.27]^{1/2}$

$= 2555.175 - 100 \ [533.62]^{1/2} = 2555.175 - 2301.02 = 245.15 \text{ N/mm}^2$

Design compressive stress, σ_{cd} **shall be that whichever is minimum.**

Step 8 : Load carrying capacity (i.e, the compressive strength) of the

$$W_{Fd} = \left(\frac{118.65 \times 4204}{1000 \times 1.15}\right) = \left(\frac{498.80}{1.15}\right) = 433.74 \text{ kN}$$

$$W_{cK} = \left(\frac{433.74}{1.5}\right) = 289.16 \text{ kN}$$

Example 5.5. A built-up column consists of three rolled steel beam sections WB 450, @ 0.794 kN/m, connected effectively to act as one column as shown in Fig. 5.31. Determine the safe load carrying capacity (compressive strength) of the built-up section. The unsupported length of column is 4.25 m. Take $\sigma_y = 260 \text{ N/mm}^2$, $E = 2 \text{ x } 10^5 \text{ N/mm}^2$.

Solution :

Step 1 : Properties of built-up section

From ISI Handbook No.1 (steel section tabels)

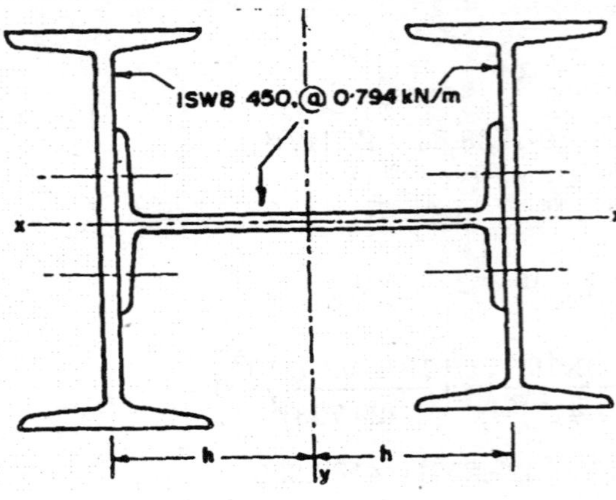

Fig. 5.31.

The built-up column consists of

3 WB 450,@ 0.794 kN/m

Area of section = (3 x 101.15 x 100) mm^2 = 30345 mm^2

Moment of inertia of built-up section about xx-axis

I_{xx} = (2 x 35057.6 + 1706.7) x 10^4 mm^4

= 71821 x 10^4 mm^4

Moment of inertia of built-up section about yy-axis

I_{yy} = [35057.6 + 2 x 1706.7 + Ab^2] x 10^4 mm^4

where, A is the sectional area of one I - section.

I_{yy} = [38471.0 + 2 x 101.15 (22.5 + (1/2) x 0.92)^2] x 10^4 mm^4

= 145115.79 x 10^4 x mm^4

Step 2 : Type of section

Width of flange, b_f = 200 mm

Thickness of flange, t_f = 15.4 mm

Overall depth of section t_o = 450 mm

Thickness of web, t_w = 9.2 mm

d = (450 – 2 x 15.4) = 419.2 mm

All the three sections are equal

Ratio, (0.5 b_f / t_f) = (0.5 x 200 / 15.4) = 6.494

Ratio, (d/t_w) = (419.2/9.2) = 45.565

For the section having web with neutral axis at mid-depth, **for the plastic section.**

$$(d/t_w) < [(83.0 \times \varepsilon)]$$

$$\text{where, } \varepsilon = \left(\sqrt{\frac{250}{\sigma_y}}\right) = \left(\frac{250}{260}\right)^{1/2} = 0.98$$

$(d/t_w) < [(83 \times 0.98) = 81.34]$

Hence, the section is a **plastic section**

Step 3 : Slenderness ratio

Unsupported (Effective) length = 4.25m

Radius of gyration

$$r_{xx} = (I_{xx}/A)^{1/2} = \left(\frac{71821.9 \times 10^4}{30345}\right)^{1/2}$$

$$= (100) \times (2.367)^{1/2} = 153.8$$

$$r_{yy} = (I_{yy}/A)^{1/2} = \left(\frac{14115.79 \times 10^4}{30345}\right) = 100 \times (4.78)^{1/2} = 218.68$$

Slenderness ratio for the column

$$\lambda_{xx} = \left(\frac{l}{r_{xx}}\right) = \left(\frac{4.25 \times 1000}{153.8}\right) = 27.633$$

$$\lambda_{yy} = \left(\frac{l}{r_{yy}}\right) = \left(\frac{4.25 \times 1000}{218.68}\right) = 19.435$$

Step 4 : Imperfection factors

For rolled steel sections

$\alpha_{xx} = 0.0020, \qquad \alpha_{yy} = 0.0035$

$\lambda_o = 0.2\,\pi\,(E/\sigma_y)^{1/2} = (0.2 \times 3.14)\,(2 \times 10^5/260)^{1/2}$

$= 0.628 \times 10\,(2000/260)^{1/2} = 17.41$

$\alpha_{1\text{-}xx} = \alpha_{xx}\,(\lambda_{xx} - \lambda_o)$

$= 0.0020\,(27.633 - 17.41) = (0.0020 \times 10.223) = 0.0204$

$\alpha_{1\text{-}yy} = \alpha_{yy}\,(\lambda_{yy} - \lambda_o)$

$= 0.0035\,(19.435 - 17.41) = (0.0035 \times 2.025) = 0.0071$

Step 5 : Euler's stress

About xx-axis, $\sigma_{Ex} = \pi^2 E/(\lambda_{xx} - \lambda_o)^2$

$\sigma_{Ex} = [3.14^2 \times 2 \times 10^5/10.223^2]$

$= [9.8596 \times 2 \times 10^5/104.51] = 18868.29\ \text{N/mm}^2$

About yy-axis

$\sigma_{Ey} = \pi^2 E/(\lambda_{yy} - \lambda_o)^2$

$= [(3.14^2 \times 2 \times 10^5)/(19.435 - 17.41)^2] = [9.8596 \times 2 \times 10^5/(2.025)^2]$

$= [19.7192 \times 10^5/4.101] = 482658\ \text{N/mm}^2$

Step 6. Values of ϕ_x and ϕ_y

$$\phi_x = \left[\sigma_y + \left(1+\alpha_{1\text{-}xx}\right)\sigma_{Ex}\right]/2$$

= [260 + (1 + 0.0204) x 18868.29]/2

= [260 + 19253.20]/2 = [19513.20/2] = 9756.60

$$\phi_y = \left[\sigma_y + \left(1+\alpha_{1\text{-}yy}\right)\sigma_{Ey}\right]/2$$

= [260 + (1 + 0.0071) x 482658]/2

= [260 + 489536.07]/2 = 244898.034

Step 7. Design Compressive stresses $(\sigma_{cx}$ and $\sigma_{cy})$ [Ayrton and Perry's Formula]

$$\sigma_{cx} = \phi_x - \left[\phi_x^2 - \sigma_y \cdot \sigma_{Ex}\right]^{1/2}$$

= 9756.60 – [9756.60² – 260 x 18868.29]$^{1/2}$

= 9756.60 – [95.19 x 10⁶ – 4.90 x 10⁶]$^{1/2}$

= (9756.60 – 9.502 x 10³) = 254.6 N/mm²

less than 260 N/mm²

$$\sigma_{cy} = \phi_y - \left[\phi_y^2 - \sigma_y \cdot \sigma_{Ey}\right]^{1/2}$$

= 244898.034 – [244898.034² – 260 x 482658]$^{1/2}$

= 244898.034 – [244.898 x 10⁶ – 2.60 x 48.2658x 10⁶]$^{1/2}$

= 244898.034 – [59975.03 x 10⁶ – 125.49 x 10⁶]$^{1/2}$

= 244898.034 – [244.6416 x 10³]

= 256.375 N/mm² less than 260 N/mm²

Ultimate compressive stress is that whichever is less. Therefore,

σ_{cu} = 254.64 N/mm²

Step 8. Load carrying capacity (Compressive strength) of the column

$$\text{Factored load, } W_{Fd} = \left(\frac{\sigma_{cu} \times A}{\gamma_{ms}}\right)$$

$$= \left(\frac{254.64 \times 303.45}{1.15 \times 1000}\right) = 6719.175\text{kN}$$

Service load, $W_{cK} = W_{Fd}/\gamma_{FL}$

= (6719.175/1.5) = **4479.44 kN**

5.28 DESIGN OF AXIALLY LOADED COLUMNS AND COMPRESSION MEMBERS

When a column or compression member is designed, for given load, actual length of the member and its support conditions, the cross-sectional shape of the member is determined.

The cross-sectional shape of axially loaded compression member depends largely on whether the compression member is long or short and whether it carries a small load or a large load. It is difficult to decide, whether a column is short or long. It is arbitrarily decided.

When the slenderness ratio of a column is less than 60, then, it may be considered as a short column. When the slenderness ratio is between 60 and 180, the column may be considered as long column. Following are the length and load categories arbitrarily made for design of compression members :

1. Short compression members with small loads.
2. Short compression members with large loads.
3. Long compression members with small loads.
4. Long compression members with intermediate load.

The strength of axially loaded compression member depends upon slenderness ratio (l/r_{min}). For the design of axially loaded compression member load to be carried, the length of compression member and end conditions are known. The effective length of the compression member for the given end conditions is computed. The radius of gyration of compression member is not known as the cross-sectional shape of the compression member is not known. The allowable working stress in compression can be found when the slenderness ratio is known. There is no direct method of designing a compression member. The compression member is designed by trial and error method. The design of compression member is also done by using safe load tables, if available.

ISI Handbook No. 1 provides tables for safe concentric loads on rolled steel column sections (HB-sections) for bending about xx-axis and yy-axis. The effective length of column is determined knowing the end conditions. The value of safe concentric loads corresponding to respective effective lengths are given for various sizes of HB-sections. A column section having safe axial load equal to or slightly greater than the required load on the column is selected.

Design strength. Following are the usual steps in design of compression members. σ_{cd} is the design compressive stress for axially loaded compression members.

Step 1. The slenderness ratio for the compression member and the value of yield stress for the steel are assumed.

For the rolled steel beam section compression members, the slenderness ratio varies from 70 to 90. For struts, the slenderness ratio varies from 110 to 130. For compression members carrying large loads, the slenderness ratio is about 40.

Step 2. The effective sectional area (A) required for compression member is determined, $A = (W_{Fd} / \sigma_{cd})$

where, W_{Fd} / σ_{cd} = Design load to be carried by the member.

Step 3. From the steel section tables, section for the compression member of the required area is selected.

The section for the compression member is selected such that it has the largest possible radius of gyration for the required sectional area. It should also be most economical section.

Step 4. Knowing the geometrical properties of the section, slenderness ratio is computed and allowable axial stress in compression is found from IS : 800-1984 for the quality of steel assumed.

Step 5. The safe/service load carrying capacity of the compression member is determined.

The section selected for the compression member is revised in case the safe load carrying capacity of the compression member is less than or much larger than the load to be carried by it.

Axially loaded columns are designed as follows :

Step 1. A suitable trial section may be assumed by following the working stressmethod of design. This trial section is classified as per (b_f/t_f) and (d_w/t_w) ratios as recommended in IS800-2007. It is to note that a slender section is not used for columns.

Whether this trial section is adequate as per limit state method as recommended in IS: 800-2007. Or not. The adequacy of the section is checked.

Step 2. Actual length of column and the support conditions are given.

The values of slenderness ratios (λ) about weak axis and strong axis, both are calculated. So also, the value of critical slenderness ratio

$$\lambda_o = 0.2\,\pi\left(E/\sigma_y\right)^{1/2} \tag{5.57}$$

is also calculated. The value of yield stress σ_y is assumed for the grade (quality) of the steel to be used.

Step 3. Slenderness ratios Values of buckling class (that is, a, b, c or d) are noted keeping in view the grade of steel, the cross-section of the column and likely, axis of bending for x-direction and y-direction (λ_{yy}).

The values of **imperfection factors** (tht is, Perry's factor, $\alpha_{xx} = \eta_{xx}$ and $\alpha_{1\text{-}yy} = \eta_{yy}$ are noted.

$$\alpha_1 = \alpha\ \left[\left(\lambda_{xx} - \lambda_o\right)\right] \tag{5.58}$$

Step 4. The value of equivalent slenderness ratio

$$\sigma_e = \left[\pi^2 E/(\lambda - \lambda_o)^2\right]$$

for, both the axes (weak and strong) are also found.

Onward, using the value of σ_e, and the respective tables for design compressive stress for grade of steel (a, b, c, or d), σ_{cd} is noted.

Otherwise, the general procedure is used to determine, the value of σ_{cd} (design compressive stress).

Step 5. In general procedure, **the values of ϕ are calculated** (about both the axes) values of Euler's stresses, σ_{Exx} and $\sigma_{E.yy}$ are found.

$$\phi = \left(\frac{\sigma_y + (1+\alpha_1)\,\sigma_E}{2}\right)$$

Step 6. The value of σ_{cd} is obtained as under :

$$\sigma_{cd} = \phi \pm \left[(\phi)^2 - \sigma_y\,.\,\sigma_E\right]^{1/2}$$

where, σ_E = Euler's critical load

$$= \left(\pi^2 E/\lambda^2\right)$$

Step 7. The cross-sectional area, A is found by equating the strength (resistance) of the column with the factored load (design load)

$$W_{fd} = \left(\frac{\sigma_{cd} \cdot A}{\gamma_{ms}}\right)$$

and, the design or factored load

$W_{fd} = (\gamma_{fL} \times W_{ck})$

γ_{fL} = partial safety factor for loads (1.5)

W_{cK} = characteristic or service load.

Example 5.6 Design a rolled steel beam section column to carry an axial load 1100 kN. The column is 4 m long and adequately restrained in position byt not in direction at both the ends.

Use limit state method of design. Take $\sigma_y = 260$ N/mm², and $E = 2 \times 10^5$ N/mm².

Solution :

Step 1 : Selection of trial section. For guidance, working stress method is used.

Allowable stress, (as per IS : 800-1984), $\sigma_{ac} = 103$ N/mm² is used Effective sectional area required,

$$A = \left(\frac{1100 \times 1000}{103}\right) = 10679.61 \text{ mm}^2$$

From ISI Handbook No.1 (Steel section tables), the beam section HB 450, @ 0.872 kN/m is selected for trial.

Geometrical properties of the section sectional area = 11,114 mm²

$r_{xx} = 187.8$ mm, $r_{yy} = 51.8$ mm

Step 2 : Type of the section

Width of flange, $b_f = 250$ mm

Thickness of flange, $t_f = 13.7$ mm

Depth of the section, $h_o = 450$ mm

Thickness of web, $t_w = 11.3$ mm

$d = (h_o - 2\,t_f) = (450 - 2 \times 11.3) = 427.4$ mm

Ratio, $(0.5\, b_f/t_f) = (125/13.7) = 9.124$

Ratio, $(d/tw) = (427.4/11.3) = 37.823$

$$\text{Factor, } \varepsilon = \left(\frac{250}{\sigma_y}\right)^{1/2} = \left(\frac{250}{260}\right)^{1/2} = 0.98$$

For semi-compact section

Ratio, $(0.5\, b_f / t_f) < [(15.7\, \varepsilon) = (15.7 \times 0.98)] = 15.386$

Ratio, $(d/t_w) < [(41.0\, \varepsilon) = (41.0 \times 0.98)] = 37.50$ mm

Hence, the section is a semi-compact section.

Step 3 : Slenderness ratio

Effective length of given column = 4 m

Slenderness ratios about x and y-axes

$$\lambda_{xx} = \left(\frac{l}{r_{xx}}\right) = \left(\frac{4000}{187.88}\right) = 21.290$$

$$\lambda_{yy} = \left(\frac{l}{r_{yy}}\right) = \left(\frac{4000}{51.8}\right) = 77.22$$

Critical slenderness ratio

$$\lambda_o = 0.2\,\pi\,(E/\sigma_y)^{1/2} = (0.2 \times 3.14)\,(2 \times 10^5 / 260)^{1/2}$$

$$= 0.628 \times 10 \times (2000/260)^{1/2} = 17.41$$

Step 4 : Imperfection factor

For rolled steel section

$$\alpha_{xx} = 0.0020, \quad \alpha_{yy} = 0.0035$$

$$\alpha_{1\text{-}xx} = \alpha_{xx}\,(\lambda_{xx} - \lambda_o) = 0.0020\,(21.20 - 17.41) = 0.00776$$

$$\alpha_{1\text{-}yy} = \alpha_{yy}\,.\,(\lambda_{yy} - \lambda_o)$$

$$= 0.0035\,(77.22 - 17.41) = 0.209$$

Step 5 : Euler's stresses σ_{Exx} and $\sigma_{E.yy}$

$$\sigma_{Ex} = \pi^2 E / (\lambda_{xx} - \lambda_o)^2$$

$$= [9.8596 \times 2x\ 10^5/(21.29 - 17.41)^2]$$

$$= (19.7192 \times 10^5/3.88^2) = 130986.29\ \text{N/mm}^2$$

$$\sigma_{Ey} = \pi^2 E / (\lambda_{yy} - \lambda_o)^2$$

$$= [19.7192 \times 10^5 / (77.22 - 17.41)^2]$$

$$= [19.7192 \times 10^5 / (59.81)^2]$$

$$= 551.24\ \text{N/mm2}$$

Step 6 : Values of ϕ_x and ϕ_y

$$\phi_x = [\sigma_y + (1 + \alpha_{1-2x})\,\sigma_{Ex}] / 2$$

$$= [260 + (1 + 0.00776)\ 130986.29]^{1/2}$$

$$= [260 + 130002.74]^{/2} = 66131.37$$

$$\phi_y = [\sigma_y + (1 + \alpha_{1\text{-}yy})\,551.24] / 2$$

$$= [260 + (1.209)\ 551.24]^{/2}$$

$$= [260 + 666.44916]^{/2} = 463.224$$

Step 6: Design Compressive stresses, $\sigma_{c.xx}$ and $\sigma_{c.yy}$ (Ayrton & Perry's formula)

$$\sigma_{cx} = \phi_x - \left[\phi_x^2 - \sigma_y\ \sigma_{Ex}\right]^{1/2}$$

$$= 66131.37 - [66131.37^2 - 260 \times 130986.29]^{1/2}$$

$$= 255.70 \text{ N/mm}^2 \text{ less than } 260 \text{ N/mm}^2$$

$$\sigma_{cy} = \phi_y - \left[\phi_y^2 - \sigma_y \cdot \sigma_{Ey}\right]^{1/2}$$

$$= 463.224 - [463.224^2 \times 10^4 - 260 - 551.24]^{1/2}$$

$$\sigma_{cd} = 196.247 \text{ N/mm}^2 \text{ less than } 260 \text{ N/mm}^2$$

Ultimate compressive stress is that which is ever is less. $\sigma_{cy} = 196.247$ N/mm²

Step 7 : Load carrying capacity (Compressive strength) of the column

$$W_{fd} = \left(\frac{\sigma_{cd} \times A}{\gamma_{msy}}\right) = \left(\frac{196.247 \times 10679.61}{1.15 \times 1000}\right) = 1822.47 \text{ kN}$$

Service load, $W_{ck} = \left(\frac{1822.47}{1.5}\right) = 1214.98$ kN >1100 kN. Hence safe

Alternatively : Following

Step 2 : Trial section may be tried.

From IS : 808-1989, SC 250, @ 0.872 kN/m (Standard column section), properties of this trial sections are noted.

Cross-sectional area, $A = 109 \times 100$ mm²

Radius of gyrations,

$r_{xx} = 107.0$ mm, $r_{yy} = 54.6$ mm

Width of flange, $b_f = 250$ mm

Thickness of flange, $t_f = 17$ mm

Overall depth, $h_o = 250$ mm

Thickness of web, $t_w = 10$ mm

Yield stress, $\sigma_y = 260$ N/mm²

$$\varepsilon = (250 / \sigma_y)^{1/2} = (250/260)^{1/2} = 0.98$$

Step 3 : Type of the section

Ratio, $(0.5\ b_f/t_f) = (0.5 \times 250/17) = 7.353$

$d = (h_o - 2t_f)\ (250 - 2 \times 17) = 216$ mm

Ratio, $(d/t_w) = (216/10) = 21.6$

For plastic-section

Ratio, $(0.5\ b_f / t_f) < [(8.9\ \varepsilon) = (8.9 \times 0.98)] = 0.908$

Ratio, $(d/t_w) < [(273\,\varepsilon) = (27.3 \times 0.98)] = 27.8$

Trial section is a plastic section

Critical slenderness ratio

$$\lambda_o = 0.2 \times \pi \left[2 \times 10^5 / 260\right]^{1/2}$$

$= (0.2 \times 3.14)\ [20 \times 10^4/260]^{1/2}$

$= 0.628 \times 10 \times [2000/260]^{1/2} = 17.41$

Step 4 : Imperfection factors

For rolled steel section

$\alpha_{xx} = 0.0020, \quad \alpha_{yy} = 0.0035$

About xx - axis, $\alpha_{1\text{-}xx} = \alpha_{xx}\left(\lambda_{xx} - \lambda_o\right)$

$\alpha_{1\text{-}xx} = 0.0020\ (107 - 17.41) = 0.17912$

About yy-axis, $\alpha_{1\text{-}yy} = \alpha_{yy}\left(\lambda_{yy} - \lambda_o\right)$

$\alpha_{1\text{-}yy} = 0.0035\ (54.6 - 17.41) = 0.1302$

Step 5 : Euler's stresses, $\sigma_{E.xx}$ and $\sigma_{E.yy}$

About xx-axis, $\sigma_{Ex} = \pi^2\ E/\left(\lambda_{xx} - \lambda_o\right)^2$

$\sigma_{Ex} = (9.8596 \times 2 \times 105)/(37.38 - 17.41)^2$

$= (19.7192 \times 105)/(19.97)^2$

$= (19.7192 \times 105/398.80) = 4944.64\ N/mm^2$

About yy-axis $\sigma_{Ey} = \pi^2\ E/\left(\lambda_{yy} - \lambda_o\right)^2$

$\sigma_{Ey} = (9.8596 \times 2 \times 10^5)/(73.26 - 17.41)^2$

$= (19.7192 \times 10^5)/(55.85^2)$

$= (19.7192 \times 10^5)/(3119.2225)$

$= 632.176\ N/mm^2$

Step 6 : Values of ϕ_x and ϕ_y

$\phi_x = \left[\sigma_y + \left(1 + \alpha_{1\text{-}xx}\right)\sigma_{Ex}\right]/2$

$= [260 + (1 + 0.17918) \times 4944.62]/2$

$= [260 + 5830.597]/2 = 3045.2985$

$\phi_y = \left[\sigma_y + \left(1 + \alpha_{1\text{-}yy}\right)\sigma_{Ey}\right]/2$

$= [260 + (1 + 0.1302)\ 632.176]/2$

$= [260 + 714.485]/2 = 487.242\ N/mm^2$

Step 7 : Design Compressive stress σ_{cxx} and σ_{cyy} (Ayrton and Perry's Formula)

$$\sigma_{cxx} = \phi_x - \left[\phi_x{}^2 - \sigma_y.\sigma_{Ey}\right]^{1/2}$$

$= 3045.2985 - [3045.2985^2 \times 4994.62]^{1/2}$

$= (3045.2985 - 2846.96) = 198.335 \text{ N/mm}^2$

$$\sigma_{cyy} = \phi_y - \left[\phi_y{}^2 - \sigma_y.\sigma_{Ey}\right]^{1/2}$$

$= 487.2427 - [487\text{-}2427^2 - 260 \times 632.176]^{1/2}$

$= [487.2427 - 270.2591]$

$= 216.9835 \text{ N/mm}^2$

$\sigma_{cx} = 198.335 \text{ N/mm}^2$ is not more than 260 N/mm^2

$\sigma_{cy} = 216.9855 \text{ N/mm}^2$ is not more than 260 N/mm^2

Ultimate compressive stress is that whichever is less. Limiting compressive stress

$\sigma_{cd} / \gamma_{msy} = (198.335/1.15) = 172.465 \text{ N/mm}^2$

Step 8 : Load carrying capacity

(Compressive strength) of the column

Factored load, $W_{fd} = (\sigma_{cd} \times A)$

$$\gamma_{cd} = \left(\frac{172.465 \times 109 \times 100}{1000}\right) = 1879.871 \text{ kN}$$

$$\text{Service load, } W_{cK} = \left(\frac{1879.871}{1.5}\right) = 1253.247 \text{ kN}$$

Hence, the section is satisfactory this trial section.

SC 250, @ 0.872 kN may be used.

Example 5.7. A single-angle section ISA 120 mm x 120 mm x 10 mm (i.e, ISA 120, 120 @) is used as a discontinuous strut. The length of the strut (centre to centre of intersections) is 3.25 m.

Use limit state method of design. The load is to be transferred through one leg. Take $\sigma_y = 260 \text{N/mm}^2$ and $E = 2 \times 10^5 \text{ N/mm}^2$. Find the compressive strength of the strut.

Solution

It is assumed that the angle strut is connected to the gusset plate at both the ends with two or more than two rivets/bolts.

Effective length of the strut

$l = 0.85 \text{ L} = (0.85 \times 3.25 \times 1000)$

$= 2762.5 \text{ mm}$

Step 1 : Geometrical properties of strut are as under

Thickness of angle, t = 10 mm

Sectional area, A = 2320 mm²

Radius of gyrations

$r_{xx} = r_{yy} = 36.7$ mm

$r_{uu} = 46.3$ mm and $r_{vv} = 23.6$ mm

Length of each leg $b_1 = b_2 = 120$ mm.

Step 2 : Slenderness ratio of strut

$$(l/r_{vv}) = \left(\frac{2762.5}{23.6}\right) = 117.055$$

Yield stress ratio, epsilon ε

$$\varepsilon = (250/260)^{1/2} = 0.98$$

$$\varepsilon = \left(\frac{\pi^2 E}{250}\right)^{1/2} = 0.98\left(\frac{9.8596 \times 2 \times 10^5}{250}\right)^{1/2} = 87.04$$

From IS 800-2007, Clause 7.5.1.2,

$$\lambda_{vv} = \left(\frac{(l/r_{vv})}{\in\left(\frac{\pi^2 E}{250}\right)}\right)$$

$$\lambda_{vv} = \left(\frac{117.055}{87.04}\right) = 1.3448$$

$$\lambda_{\varphi\varphi} = \left(\frac{(b_1 + b_2)2t}{t\left(\frac{\pi^2 \varepsilon}{250}\right)}\right) = \left(\frac{(120 + 120)/20}{87.04}\right) = 0.1378$$

Step 4 : Equivalent slederness ratio, λ_e, (IS 800-2007, Clause 7.5.1.2)

$$\lambda_e = \left[k_1 + k_2.\lambda^2_{vv} + k_3.\lambda^2_{\varphi\varphi}\right]^{1/2}$$

Given angle section is connected at its two ends with the gusset plate by more than one rivet bolt along its length. It may be assumed as fixed at eachend. The values of k_1, k_2 and k_3 are noted from IS : 800-2007. (Clause 7.5.1.2)

$k_1 = 0.20$, $k_2 = 0.35$ and $k_3 = 20$ substituting these values

$$\lambda_e = \left[0.20 + 0.35 \times 1.8086^2 + 20 \times 0.1380^2\right]^{1/2}$$

$$= [1.21388]^{1/2} = 1.102$$

Step 5 : Design compressive stress for strut σ_{cd}

Yield stress for steel, $\sigma_y = 260\ N/mm^2$

From IS : 800-2007

Angle section may buckle about any axis. The buckling class is `c'

For $\left(\frac{l}{r}\right) = 0.0 \qquad \sigma_{cd} = 260\ N/mm^2$

$\left(\frac{l}{r}\right) = 10.0 \qquad \sigma_{cd} = 236 N/mm^2$

For (l/r) = 1.102

$$\sigma_{cd} = [260 - \left(\frac{1.102 x 24}{10}\right) = [260 - 2.6448] = 257.355\ N/mm^2$$

Step 6 : Factored design compressive strength (resistance)

$$W_{fd} = \left(\frac{257.355 \times 2320}{1000 \times 1.15}\right) = 519.18 kN$$

Step 7 : Service (characteristic) strength

$$W_{cK} = \left(\frac{W_{fd}}{r_{fL}}\right) = \left(\frac{519.18}{1.5}\right) = 346.124 kN$$

5.29 DESIGN OF ANGLE STRUTS

1. Single angle struts

Struts (whether dis-continuous or continuous) are the members which carry compression. As such, the design of angle struts is design of compression members. That is, a trial section is taken using working stress method, (which one remained in practice since long and so also, so far). The geometrical properties of this section are noted. This section is classified keeping in view local buckling and section classification.

In general, the end conditions are given (otherwise, these are assumed suitably). The effective length of the strut is calculated. The angle struts are designed as under.

Step 1. The slenderness ratio (l/r_w) is calculated. As it is known that there are two geometrical reference axes (x-axis and y-axis) and in additions to these axes, there are two principal axes (uu-axis and vv-axis). Its respectivc radii of gyrations are known and r_{vv} (radius of gyration about vv-axis used to be minimum).

Step 2. The critical slenderness ratio

$\lambda_o = 0.2\pi(E/\sigma_y)$ is also found.

Step 3 : The value of yield stress ratio ε (epsilon) is determined as under.

$$\varepsilon = (250/\sigma_y)^{1/2}$$

where, σ_y = Yield stress of steel, being used

Step 4. Equivalent slenderness ratios are obtained as defined in IS : 800-2007.

$$\lambda_{vv} = \left(\frac{\left(\frac{l}{r_{vv}}\right)}{\in . \left(\frac{\pi^2 E}{250}\right)^{1/2}} \right) \quad \text{(i)}$$

Substituting the expression for ε (epsilon)

$$\lambda_{vv} = \left(\frac{\left(\frac{l}{r_{vv}}\right)}{\left(\frac{250}{\sigma_y}\right)^{1/2} \left(\frac{\pi^2 E}{250}\right)^{1/2}} \right) \quad \text{(ii)}$$

$$\lambda_{vv} = \left(\left[\frac{(l/r_{vv})}{\left(\pi^2 E/\sigma_y\right)^{2}} \right] \right) \quad \text{(5.59)}$$

And the other equivalent slenderness ratio

$$\lambda_{\varphi\varphi} = \left(\frac{[(b_1 + b_2)/2t]}{\in \left(\frac{\pi^2 E}{250}\right)^{1/2}} \right) \quad \text{(iii)}$$

$$\lambda_{\varphi\varphi} = \left(\frac{[(b_1 + b_2)/2t]}{\left(\frac{250}{\sigma_y}\right)^{1/2} \left(\frac{\pi^2 E}{250}\right)^{1/2}} \right) \quad \text{(i)}$$

$$\lambda_{\varphi\varphi} = \left(\frac{[(b_1 + b_2)/2t]}{\left(\pi^2 E/\sigma_y\right)^{1/2}} \right) \quad \text{(5.60)}$$

where,

l = Effective length of the member

r_{vv} = Radius of gyration about the minor principle axis

b_1, b_2 = Width of the two legs of the angle

t = Thickness of the angle section

ε = Yield stress ratio

Angle strut is loaded through one leg. The flexural-torsional buckling occurs in such member. Strength of the strut may be obtained by calculated equalent slenderness ratio as per IS : 800-2007, (Clause 7.5.1.2)

$$\lambda_e = \left[k_1 + k_2 .\lambda^2_{vv} + k_3 .\lambda^2_{\varphi\varphi}\right]^{1/2} \tag{5.61}$$

where, k_1, k_2, and k_3 are the constants. Their values depend upon the end connections and these may be noted from IS : 800-2007.

Step 5. Design compressive stress, σ_{cd} may be noted from IS: 800-2007 corresponding to the equivalent slenderness ratio and the buckling class (a, b, c and d) and bending axis. Angle section may bend about any axis. It belong to the buckling class, `c'.

Step 6. Factored compressive strength (resistance) may be calculated

$$W_{fd} = \left(\frac{\sigma_{cd} \text{ x } A}{1.15}\right) \qquad (\gamma_{ms} = 1.15)$$

Step 7. Service compressive strength may then be determined.

$$W_{cK} = (W_{Fd}/1.5)$$

This should be more by 25 per cent of the service load.

Double angle struts

(A) Double angle discontinuous struts connected back to back, on opposite/sides of the gusset plate or a section, by not less than two rivets or bolts along the length of the strut may be considered as compressive load applied axially.

Its effective length, kL in the plane of end gusset shall be taken as between 0.7 and 0.85 *time* 'istance between intersections, depending on the degree of the restraint provided.

The effectve length, kL, in the plane perpendicular to that of the end gusset shall be taken as equal to the distance between centres of intersections.

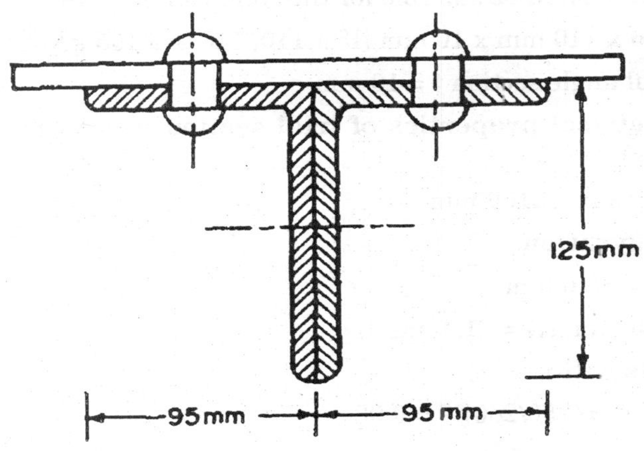

Fig. 5.32.

Average compressive stress is calculated using the following expression (general procedure).

$$\sigma_{cd} = \left(\frac{\sigma_y / \gamma_{ms}}{\varphi + \left[\varphi^2 - \lambda^2\right]^{1/2}} \right)$$

(B) Double angle discontinuous struts connected back-to-back to one-side of a gusset plate is designed as single strut as shown in Fig. 5.32.

(C) Double angle continuous struts such as those forming the flanges, chords of the truss girders or the legs of towers shall be designed as axially loaded compression members.

Example 5.8 : Design a single angle discontinuous strut to carry 88 kN load. The length of the strut (centre to centre of intersections is 3.25 m.

Use limit state method of design. The load is transferred through one leg. Take, $\sigma_y = 260$ N/mm² and E = 2 x 10⁵ N/mm²

Design : Selection of a trial section

The procedure for design by limit state method needs a trial section. A trial section may be selected taking guidance from `working stress method', which one is as follows.

It is assumed that the angle strut is connected to the gusset plates at both the ends with two or more than two rivets/bolts.

Effective length of the strut

$$l = 0.85\,L = (0.85 \times 3.25 \times 1000) = 2762.5 \text{ mm}$$

The slenderness ratio for the single angle strut may be assumed as 130. For which the allowable stress from IS : 800-1984 may be taken as 57 N/mm².

Effective sectional area required for the compressive force 1.25 *times* the given load (=1.25 x 88 = 110 kN)

$$A = \left(\frac{110 \times 1000}{57} \right) = 1929.29 \text{ mm}^2$$

Equal angle section shall be suitable for the single angle strut. It suggests a trial section.

ISA 110 mm x 110 mm x 10 mm [ISA 110, 110 @ 0.165 kN/m]

Thickness of trial angle section t = 10 mm

Step 2 : Geometrical properties of trial section may be noted from ISI Handbook No. 1. (Section Tables)

Sectional area A = 2106 mm²

Radius of gyration

$r_{xx} = r_{yy} = 33.6$ mm

$r_{uu} = 42.5$ mm, $r_{vv} = 21.4$mm

Yield stress ratio,

Epsilon, $\varepsilon = (250/260)^{1/2} = 0.98$

Step 3 : Maximum slenderness ratio

$$\frac{l}{r_{vv}} = \left(\frac{2762.5}{21.4} \right) = 129.088$$

From IS : 800 - 2007 as per Clause 7.5.1.2

$$\lambda_{vv} = \left(\frac{(l/r_{vv})}{\varepsilon\left(\frac{\pi^2 E}{250}\right)^{1/2}} \right) = \left(\frac{129.088}{\frac{0.98(9.8596 \times 2 \times 10^5)}{250}} \right) = 1.483$$

$$\lambda_{\varphi\varphi} = \left(\frac{(b_1 + b_2)/2t}{\varepsilon\left(\frac{\pi^2 E}{250}\right)^{1/2}} \right)$$

$$\lambda_{\varphi\varphi} = \left(\frac{(110 + 110)/2 \times 10}{0.98\left(\frac{9.8596 \times 2 \times 10^5}{250}\right)^{1/2}} \right) = 0.1265$$

Step 4 : Equivalent slenderness ratio, λ_e

$$\lambda_e = \left[k_1 + k_2 . \lambda^2_{vv} + k_3 . \lambda^2_{\varphi\varphi}\right]^{1/2}$$

The angle section is connected at its two ends with the gusset plate by more than one rivet/bolt along, its length. The values of k_1, k_2 and k_3 are noted from IS : 800-2007.

$k_1 = 0.20$, $k_2 = 0.35$ and $k_3 = 20$

$\lambda_e = [0.20 + 0.35 \times 1.483^2 + 20 \times 0.1265^2]^{1/2}$

$= [0.20 + 0.35 \times 2.1993 + 20 \times 0.016]^{1/2}$

$= [0.20 + 0.76975 + 0.32]^{1/2} = [1.28975]^{1/2} = 1.3567$

Step 5 : Design compressive stress

from buckling class `c', $\sigma_y = 260$ N/mm²

for $(l/r) = 0$, $\sigma_{cd} = 260$ N/mm²

For $(l/r) = 10$, $\sigma_{cd} = 236$ N/mm²

For $(l/r) = 1.1356$

$$\sigma_{cd} = \left[260 - \left(\frac{1.1356 \times 24}{10}\right)\right] = [260 - 2.725] = 237.275 \text{ N/mm}^2$$

Step 6. Factored (design) compressive strength

$$W_{fd} = \left(\frac{237.275 \times 2106}{1000 \times 1.15}\right) = \mathbf{434.522\ kN}$$

Step 7 : Service (characteristic) compressive strength

$$W_{ck} = \left(\frac{434.522}{1.5}\right) = \mathbf{289.68\ kN}$$

The capacity of trial section checked by limit state method of design is much more than given compressive load. However, it is safe.

A lighter section may be used and checked.

5.30 LACED AND BATTENED COLUMNS

Two channels are commonly used for laced columns. These channels have a tendency to buckle independently.

In built-up columns, when rolled steel sections are not connected by plates (viz. load sharing elements) suitable lateral system is needed to connect different load carrying elements of column. The lateral system holds the load carrying elements of the built-up column in their relative positions and it does not share the load. The object of providing lateral system is to carry the transverse shear force which occurs when the column deflects. The following lateral systems are used :

1. Lacing. 2. Batten plates. 3. Perforated plates.

1. Lacing

The lacing is also termed as `latticing' and it is most commonly used. The rolled steel flats, angles and channels are used for lacing. The rolled steel sections or tubes of equivalent length may also be used instead of flats. The lacing is of two types:

1. Single lacing 2. Double lacing.

Lacing provides a tying force and checks the tendency of the channels buckling.

1. Single lacing. In single lacing as shown in Fig. 5.34(A), the lacing flats are placed to form a single-triangular system. The single lacing on opposite sides of the main components shall preferably be in the same direction so that one may be shadow of the other, instead of being mutually opposed in direction.

The effective slenderness ratio $(kL/r)_e$ of a laced columns shall be taken as 1.05 times the (kL/r). [Actual slenderness ratio considering shear deformation influence].

2. Double lacing. In double lacing as shown in Fig. 5.34(B), the lacing flats are used to intersect one another and connected at the points of intersections by rivets.

The compression members comprised of two main components laced and tied should, where practicable, have a radius of gyration about the axis perpendicular to the plane of the lacing not less than the radius of gyration at right angles to that axis (i.e., r_{yy}, should not be less than r_{xx}, see Fig. 5.34).

As far as, it is practicable, the lacing system should not be varied throughout the length of strut. The tie plates are attached at the top and bottom of a laced column.

The single laced system and double laced system on opposite sides of the main components should not be combined with cross-members (except the plates) perpendicular to

the longitudinal axis of the strut unless all forces resulting from deformation of the strut are calculated and accounted for in the lacing and its fastening.

When the components of built-up column are connected by a lateral system, the effect of shearing forces to the deflection of the column is more. The reduction in buckling strength due to the shear deflection is more than that of solid built-up columns. The buckling strength of laced or battened column is given by

$$P = \left(\frac{\pi^2 \bar{E}.I}{(kL)^2}\right) = \left(\frac{\pi^2 E_t\, I}{(KL)^2}\right)$$

$\bar{E}$ = **Effective modulus of elasticity**

E_t = **Tangent modulus**

I = **Moment of inertia of the column section,** $(I_o + 2I_1)$, $\left(I_o + 2I_1\right) I_o = \frac{1}{2} Ah^2$

I_1 = **Moment of inertia of each component of column about their own axes**

A = **Cross-sectional area of each component of column**

h = **Distance between yy-axis of components of column**

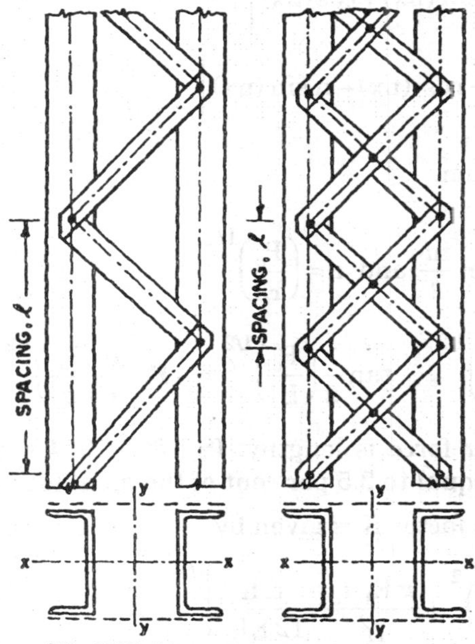

(A). SINGLE LACING (B) DOUBLE LACING
LACING OF COMPRESSION MEMBERS.

Fig. 5.34.

For single lacing the factor K is given by

$$K = \left(1 + \frac{\pi^2 E_t\, I_o}{L^2} \cdot \frac{d^2}{Ec \,.\, h^2 \,.\, A_d}\right)^{1/2} \quad \text{and} \qquad (5.48)$$

For double lacing

$$K = \left(1 + \frac{\pi^2 E_t I_o}{L^2} \cdot \frac{d^2}{2Ec \cdot h^2 \cdot A_d}\right)^{1/2} \tag{5.49}$$

where c = Panel length, as shown in Fig. 5.34

n = Number of panels

d = Length of lacing bars as shown in Fig. 5.34

A_d = Cross-sectional area of lacing bars.

The bending moment at any point on the deflected shape of the column at buckling

$$M_a = -P_c.y$$

The shear force in built-up column is given by

$$V = \left[\frac{dM_x}{dx}\right] = -P_c\left[\frac{dy}{dx}\right]$$

From Article 3.7,

$$y = e\left[\left(\frac{1-\cos nl}{\sin nl}\right)\sin(nx) + \cos(nx)\right]$$

$$\left(\frac{dy}{dx}\right) = e\left[n \tan\frac{nl}{2}\cos(nx) - n\sin(nx)\right]$$

At x = 0, $\left(\frac{dy}{dx}\right)$ is maximum.

$$\left(\frac{dy}{dx}\right)_{max} = e.n\tan\frac{nl}{2} \text{ and } n = \left(\frac{P_c}{EI}\right)^{1/2}$$

Therefore, $V = P_c \cdot e\left(\frac{P_c}{EI}\right)^{1/2} \cdot \tan\frac{1}{2}\left(\frac{P_c}{EI}\right)^{1/2} = -P_c \cdot \frac{e}{r}\left(\frac{P_c}{EI}\right)^{1/2} \cdot \tan\frac{l}{2r}\left(\frac{P_c}{EI}\right)^{1/2}$

This expression for shear force is lengthy. IS : 800-1984 recommended that lacing bar resists a transverse shear, S equal to 2.5 per cent of the axial load in the member.

For battened column, the factor K is given by

$$K = \left[1 + \frac{\pi^2 I_o}{24\, l_1} \cdot \left(\frac{C}{L}\right)^2 + \frac{\pi E_1 I_o}{L^2} \cdot \frac{c.h}{12\, EI_b}\right]^{1/2}$$

where, I_b = Moment of inertia of batten plate about the centroidal axis of the cross-section of column perpendicular to plane of buckling.

For simplification (E_t/E) is considered as unity. Therefore,

$$K = \left[1 + \frac{\pi^2 I_o}{24\, l_1} \cdot \left(\frac{C}{L}\right)^2 + \frac{\pi E_1}{12} \frac{I_o}{I_b} \cdot \frac{c.h}{L^2}\right]^{1/2}$$

The second term of the square root depends on flexibility of the component moments of built-up column, and last term depends upon flexural rigidity of batten plates. The last term is small as compared with the sum of first two terms for a properly designed column, and hence neglected. The expression for K becomes as under :

$$K = \left[1 + \frac{\pi^2 I_o}{24 I_l} \cdot \left(\frac{c}{L}\right)^2\right]^{1/2} \tag{5.50}$$

The design of lacing or batten plates of built-up column is done in such a way that the lacing or batten plates do not fail before the load carrying capacity of whole column is reached. A laced or battened column is considered to collapse if any element of the column begins to yield locally prior to the buckling load, P_{cr} is reached. Therefore, IS: 800-1984 specified rules for design of these elements of built-up column to avoid premature failure of the column.

When the perforated plates are used as lateral system then, the factor K for the column is given by

$$K = \left[1 + \frac{\pi^2}{12 (L/r)^2} \cdot \left(\frac{a}{r_2}\right)^2\right]^{1/2} \tag{5.51}$$

where, a = Spacing of perforations

r = radius of gyration of over-all section of the column

L = Length of the column between supports

rx = Radius cf gyration of main segment about its centroidal axis.

Critical stress for the column

On the basis of studies of **Ayrton and Perry** and so also as per studies of **Andrews, E.S.,**

$$\sigma_c^2 - [\sigma_y (1+\alpha_1)\sigma_E] . \sigma_c + \sigma_E \sigma_y = 0$$

From which, the value of critical stress in a column, Eq. 5.39 (A) σ_c is expressed as under

$$\sigma_c = \left(\frac{\sigma_y + (1+\alpha_1)\sigma_E}{2}\right) - \left[\left\{\frac{\sigma_y + (1+\alpha_1)\sigma_E}{2}\right\}^2 - \sigma_y \sigma_E\right]^{1/2} \tag{5.52)A}$$

$$\sigma_c = \varphi - [\varphi^2 - \sigma_y . \sigma_E]^{1/2}$$

where,

$$\varphi = \left\{\frac{\sigma_y + (1+\alpha_b)\sigma_E}{2}\right\}$$

$\alpha_b = 0.0055 (\alpha)$ = Factor for the imperfection of the column (i.e, factor which allows for it influence of lack of straightness and eccentricity, that is initial crookedness of the column).

$\alpha_b = \left(\lambda_f^2 + \lambda^2\right)^{1/2}$, where,

λ_f = Lower value of the slenderness of the individual vertical member between the columns

λ = slenderness of the overall column, using radius of gyration of the whole built-up section.

Design of lacing. Following are the usual steps required for the design of lacing as per IS : 800-2007.

Step 1. For lacing, the components of built-up column, adopt single or double lacing. The **angle of inclination of lacing bars** with the longitudinal axis of the component member shall not be less than 40 and not more than 70 to the axis of the member. (IS : 800-2007, Clause 7.6.4)

Step 2. The spacing of lacing (l) as shown in Fig. 5.34 is computed. The maximum spacing should be such that the following condition is fulfilled :

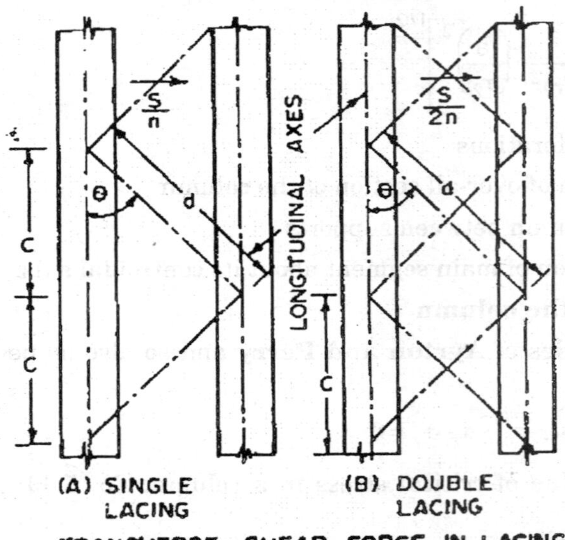

Fig. 5.35

(a) Compression members either consist of two I-sections or two channels (back to back)

When the compression members are composed of two I-sections, these shall be connected together by riveting so that. Spacing As per IS : 800-2007, Clause 7.6.5)

Minimum slenderness ratio $\lambda = (l/r)$ of compo-nent of column between consecutive connections	>	0.7 times most unfavourable slenderness ratio of the ratio of the column as a whole.
or	>	50 whichever is less

where, 'l' is the distance between the centres of connection of the lattice bars to each component.

Step 3. Transverse shear

As a compression member bends, a component of the axial load arises, which one is known as transverse shear. This shear effect reduces the column strength in proportion to the amount of deformation that may be attributed to shear. Solid web sections have less shear deformation as compared to latticed columns (using lacing bars and/or batten plates). In solid web sections, the shear has insignificant effect on reducing the column strength and it may be neglected safely. The effect of shear should be considered for latticed columns.

The curvature due to shear is added to that due to buckling in order to include the effect of the shear. The moment at any point on the deformed column (pinned at both the ends) $M_h = (P.y)$

$$\therefore \; V = \left(\frac{d\,M_n}{dh}\right) = P\left(\frac{dy}{dh}\right) \qquad \text{(i)}$$

Slope θ due to shear deformation

$$\theta = \left(\frac{\text{Shear stress}}{\text{Shear modules}}\right) = \beta\left(\frac{V}{AG}\right) \qquad \text{(ii)}$$

where, β is a factor to correct for non-uniform stress across various cross-sectional shapes. Curvature due to shear

$$\left(\frac{d\theta}{dh}\right) = \left(\frac{\beta}{A.G}\right)\left(\frac{dV}{dh}\right) = \left(\frac{P}{AG}\right).\beta\left(\frac{d^2y}{dh^2}\right) \qquad \text{(iii)}$$

Total curvature is equal to the sum of curvature due to bending and that due to shear. Therefore,

$$\left(\frac{d^2y}{dh^2}\right) = \left[-\left(\frac{P.y}{EI}\right) + \frac{P.\beta}{AG}\left(\frac{d^2y}{dh^2}\right)\right] \qquad \text{(iv)}$$

$$\therefore \left(\frac{d^2y}{dh^2}\right) + \frac{P}{EI}\left(\frac{1}{1 - P.\beta/A.G}\right) y = 0 \qquad \text{(5.52)B}$$

The form of this differential Eq. 5.52 B is same as

$$\therefore \left(\frac{d^2y}{dh^2}\right) + \left(\frac{P}{EI}\right) y = 0 \; \left(\therefore \; n^2 = P/EI\right)$$

$$\left(\text{i.e.}\left(\frac{d^2y}{dh^2}\right) + n^2 y = 0\right)$$

Therefore, the modified form of the Euler's critical load considering the effect of shear

$$P_{cr} = \left(\frac{\pi^2 E.I}{L^2}\right)\left[\frac{1}{1 + \left(\frac{\beta}{A.G}\right)\left(\frac{\pi^2 EI}{L^2}\right)}\right] \qquad \text{(5.53)}$$

The modulii, G. and E may be replaced by the tangent modulii values, G_t and E_t. Since G = E/2 (1 + μ).

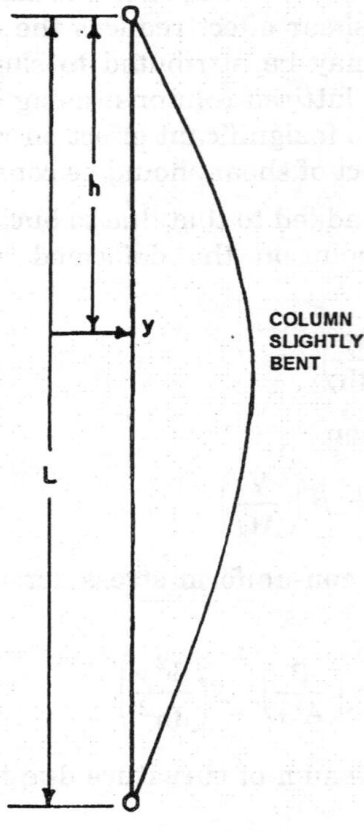

Fig. 5.36.

The length, L may be replaced by the effective length, kL. The effect of shear may be combined with kL.

$$\sigma_{cr} = \left(\frac{P_{cr}}{A}\right) = \frac{\pi^2 . E_t}{(\alpha . kl/r)^2} \qquad \text{(vi)}$$

where, $\alpha^2 = \left|1 + 2(1+\mu)\pi^2 \beta(kL/r)^2\right|$

As such, the effect of shear may be included for an adjustment for the effective length. The average value of β is about 2 (two) for W-shapes when bending about the weak-axis. For steel, pisson's ratio is 0.3. Typical values of α are as under:

(kL/r) =	50	70	100
α =	1.01	1.005	1.003

For slenderness ratio less than about 50, the failure of column is controlled by the yielding. Therefore, the effect of shear on solid H-shaped column is equivalent to an increase in effective length of less than one percent, which may be safely neglected.

The main longitudinal element of the compression members are tied to-gether by using lacing or batten plates. These elements are also subjected to deformation.

Consider one panel of a laced/battened column of height, l and c- g to c-g spacing, S. The length of diagonal is d as shown in Fig. 5.37.

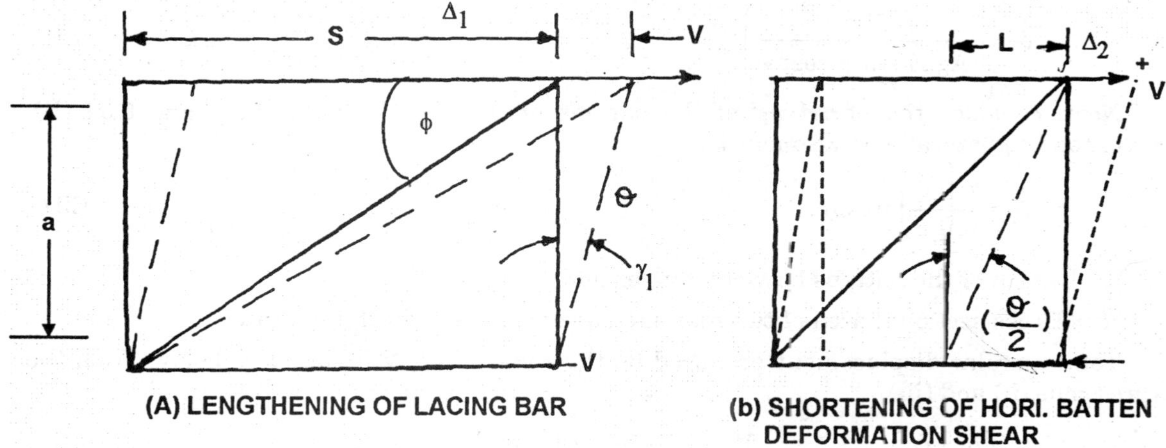

(A) LENGTHENING OF LACING BAR

(b) SHORTENING OF HORI. BATTEN DEFORMATION SHEAR

LACED BATTENED COLUMN

Fig. 5.37

The deformation of these bars are neglected, when the deflections produced by the bending moment alone is considered.

The deformation of column panel due to shear is shown in Fig. 5.37. Slopes θ_1 and θ_2 occur due to lengthening of the diagonal and shortening of the batten over a panel height, l. The cross-sectional area of the lacing bar is comparatively small. The stiffness of lacings to resist the transverse shear component is smaller than that of solid web. members.

It is assumed that the joints are hinged. The **elongation** of the diagonal produced by the shear force, V (Fig. 5.37) is

$$= \left(\frac{\text{Force}}{\text{Area}}\right)\left(\frac{1}{E}\right)(\text{Length})$$

$$= \left(\frac{V}{\cos\varphi}\right)\left(\frac{1}{A_d E}\right)\left(\frac{a}{\sin\varphi}\right) \qquad \text{(i)}$$

$$= \left(\frac{V.a}{A_d E.\ \sin\varphi.\ \cos\varphi}\right), \text{ where}$$

ϕ = Angle between diagonal and horizontal bar

$\left(\frac{V}{\sin\varphi}\right)$ = Tensile force in the diagonal

$\left(\frac{a}{\sin\varphi}\right)$ = Length of the diagonal

A_d = Cross-sectional area of two diagonals (one on each side)

Corresponding **lateral displacement**

$$\delta_1 = \left(\frac{V.a}{A_d\, E.\sin\varphi\, \cos^2\varphi} \right) \quad \text{(ii)}$$

Next, consider the **shorting** of the battens or horizontal lacing bars Fig. 5.34 (B), [corresponding lateral displacement is]

$$\delta_2 = \left(\frac{V.s}{A_b.E} \right) \text{ where} \quad \text{(iii)}$$

s = Length of the battens between hinges and

A_b = Cross-sectional area of two battens (one on each side of the column)

Total angular displacement produced by the shear force, V is found as below (from the expressions (ii) and (iii))

$$\gamma = \left(\frac{\delta_1 + \delta_2}{a} \right) = \left(\frac{V}{A_d\, E\, \sin\varphi\, \cos^2\varphi} + \frac{V.b}{a\, A_b\, E} \right) \quad \text{(iv)}$$

where, $\gamma = \left(\frac{V}{P_d} \right)$

$$\left(\frac{1}{P_d} \right) = \left(\frac{1}{A_d.E.\sin\varphi \cos^2\varphi} + \frac{b}{aA_bE} \right) \quad \text{(v)}$$

For built-up column (laced column)

$$P_{cr} = \left(\frac{P_E}{1 + (P_E/P_D)} \right) \quad \text{(vi)}$$

Critical load for a column with hinged end (substituting from the expression (v) in the expression (vi)

$$P_{cr} = \left(\frac{\pi^2 EI}{l^2} \right) \left(\frac{1}{1 + \left(\frac{\pi^2 EI}{l^2} \right) \left(\frac{1}{A_d E \sin\phi \cos^2\phi} + \frac{b}{aA_b E} \right)} \right)$$

where, I is the moment of inertia of the column.

In case, the cross-sectional areas A_b and A_d are small as compared with the areas of channels or other main elements of the member, the critical load found from the above expression shall be considerably less than the Euler value. As such, it is seen that the laced column shall be considerably weaker than the solid built-up column. With same EI. However, the quantity of material used is less the laced column shall be more economical.

When, two diagonal lacing bars are used in each panel, then, there shall be tensile stress in diagonal and compressive stress in the other, under the action of shear force. The battens, if any, do not share shear force. The critical load is determined from the above expression by neglecting the term containing, A_b, and the term containing, A_d is doubled. Therefore,

$$P_{cr} = \left(\frac{\pi^2 EI}{l^2}\right)\left(\frac{1}{1+\frac{\pi^2 .EI}{l^2}.\frac{1}{A_d E.\ \sin\varphi\ \cos^2\varphi}}\right)$$

where, A_d is the cross-sectional area of four diagonal bars (two on each side of the column in the same panel). This expression may also be used by considering A_d as the area of two diagonals and ϕ is measued with the horizontal.

Friedrich Bleich [Buckling strength of Metal structures New York, McGraw Hill Book, Company, Inc, 1952] suggested a conservative estimate of the influence of 60 or 45 lacing as generally specified in bridge design practice, may be made modifying the effective length factor, k to a new factor as under.

For (kL/r) > 40, $\alpha = [1 + 300\ (kL/r^2)]$

and For (kL/r) < 40, $\alpha = 1.0$

The lacing bars resist a total transverse shear `V' at any point in the length of the member where V equal to 2.5 per cent of the axial load in the member.

The transverse shear `V' is equally distributed in all transverse systems of parallel planes. Any other shear due to end moment or transverse loading is respectively accounted for.

Step 4. The force in the lacing bar is computed as given below :

In single lacing as shown in Fig. 5.34(A) transverse shear in each lacing is V/n

$$\text{Force in each lacing, } F = \left(\frac{V}{n}\times\frac{d}{h}\right).\ F = (V/n).\operatorname{cosec}\theta \qquad (5.54)$$

where, θ = Inclination of lacing with the longitudinal axis of member

n = Number of transverse systems in parallel planes

h = Horizontal distance between rivet lines.

In the double lacing as shown in Fig. 5.34(B) the transverse shear is to be resisted by two lacing bars.

$$\text{Force in each lacing} = \frac{1}{2}.\left(\frac{V}{n}\operatorname{cosec}.\theta\right)$$

Step 5. The section for all lacing elements is adopted as below :

Let t be the **thickness** of flat lacing : (IS : 800-2007, Clause 7.6.3)

(i) In the single lacing t shall not be $< \frac{1}{40}$th (one-fortieth) length between inner end rivets.

(ii) In the double lacing t shall not be $< \frac{1}{60}$th (one-sixtieth) length between inner end rivets.

IS : 800-1984 specifies the width of lacing flats as given in Table 5.15.

TABLE 5.15. WIDTH OF LACING BARS = Minimum width = 3 ×Nominal diameter

Nominal diameter of rivets	Width of lacing bars (Clause 7.6.2)
22 mm	65 mm
20 mm	60 mm
18 mm	55 mm
16 mm	50 mm

The diameter of the rivets connecting lacing bars with components of column is assumed.

Step 6. The minimum radius of gyration for lacing bar is computed. The slenderness ratio of the lacing bar for compression member shall be not exceed 145. (Clause 7.6.6.3)

In the riveted construction, the effective length of lacing bar in single lacing is adopted as the length between inner end rivets, and in double lacing as 0.7 of this length. In the welded construction, this effective length is taken as 0.7 times the distance between the inner ends of welds connecting the lacing bars to the members.

Step 7. The rivet value for the rivet adopted is computed and should be greater than load coming to the rivet (twice the transverse shear in each lacing).

Step 8. The lacing bar designed is checked for compressive strength and tensile strength. These should be more than the respective forces in each lacing bar.

The end tie plates are provided at the ends of lacing system in the laced compression members. These tie plates are also provided at the points where the lacing system is interrupted.

The lacing bars are attached to the main members by rivets or welding. The riveting or welding of lacing bars to the main members shall be sufficient to transmit the load in the bars. Where the welded lacing bars overlap the main members, the amount of lap measured along either edge of the lacing bar shall be not less than **four times** the thickness of the bar or the members, whichever is less. The welding should be sufficient to transmit the load in the bar and shall, in any case, be provided along each side of the bar for the full length of lap.

Where the lacing bars are fitted between the main members, these shall be connected to each member by fillet welds on each side of the bar or by full penetration butt welds. The lacing bars shall be so placed as to be generally opposite the flange or stiffening elements of the main members.

Example 5.12. A steel column 12 m long carries an axial load of 1000 kN. The column is hinged at both the ends. Design an economical built-up section with double lacing. Design the lacing also. Use limit state method of design.

Design : Step 1 : Design of built-up column (column is to be laced)

1. Selection of trial section (by working stress method)

The length of built-up column, L is (12 x 1000) = 12000 mm

Effective length of built-up column, l = L = (12 x 1000) = 12000 mm (ii)

Axial load carried by the column, P = 1000 kN (iii)

The slenderness ratio for the laced built-up column and the value of yield stress for steel may be assumed as 80 and 260 N/mm², respectively.

Allowable stress in axial compression for the column, σ_{ac} = 103 N/mm² (iv)

$$\text{Effective sectional area required, } A = \left(\frac{1000 \times 1000}{103}\right) = 9708.74 \text{ mm}^2 \quad \text{(v)}$$

2. Properties of trial section

From ISI Handbook No. 1 double channels laced used as columns, try 2 MC 350, @ 0.421 kN/m

$A = 2 \times 5366 = 10732 \text{ mm}^2$ (vi)

$I_{xx} = 20016 \times 10^4 \text{ mm}^4$ (vii)

$r_{xx} = 136.6 \text{ mm}$ (viii)

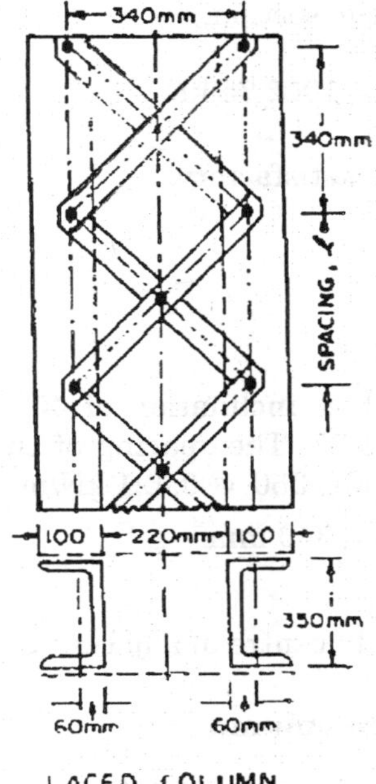

Fig. 5.38.

3. Spacing of channels

For the built-up column, the spacing between back to back of the channels should be such that $I_{yy} = I_{xx}$ approx.

From steel section tables adopt spacing of the channels

S = 200 mm (ix)

$I_{yy} = 20246.8 \times 10^4 \text{ mm}^4$

$r_{yy} = 137.4 \text{ mm}$

Moment of inertia and radius of gyration about yy-axis (I_{yy} and r_{yy}) have been kept slightly larger than I_{xx} and r_{xx}, so taht the bending occurs about xx-axis. However, the column may buckle about any axis, buckling class `c' shall be used.

4. Effective slenderness ratio (Clause 7.6.1.5)

$$1.05 \times \left(\frac{l}{r_{min}}\right) = 1.05 \times \left(\frac{12000}{136.6}\right) = (1.05 \times 87.848) = 92.24 \quad \text{(x)}$$

5. Safe load carrying capacity

From IS: 800-2007, maximum permissible stress in axial compression for the steel having yield stress as 260 N/mm²

$$\sigma_{ac} = 94.367 \text{ N/mm}^2 \quad \text{(xi)}$$

Safe load carrying capacity of the section,

$$P = \left(\frac{94.367 \times 10732}{1000}\right) = 1012.75 \text{ kN}$$

> 1000 kN. Hence the design is satisfactory.

Step 2 : Design of lacing

1. Spacing of lacing bar

The double lacing is provided at inclination of 45° with the longitudinal axis of the column member as shown in Fig. 5.37. The spacing of channels S is equal to 220 mm. The gauge distance, g for rivet lines for MC 350, @ 0.421 kN/m is 60 mm.

$$(S + 2g) = (220 + 2 \times 60) = 340 \text{ mm} \quad \text{(xiv)}$$

Inclination of lacing is 45°.

Spacing of the lacing is centre to centre of adjacent rivets, = 340 mm (xii)

2. Slenderness ratio of component column

Minimum radius of gyration of one channel MC 350, @ 0.421 kN/m,

$$r_{min} = 28.3 \text{ mm} \quad \text{(xiii)}$$

The column is composed of two channels Slenderness ratio of component of the column

$$\left(\frac{l}{r_{min}}\right) = \left(\frac{340}{28.3}\right) = 12.014$$

< 40 and $< 0.6 \left(\frac{l}{r_{min}}\right)$ of column as a whole (i.e. 0.6 x 87.8)

22 mm nominal diameter power driven rivets are used for the connection of lacing. Width of lacing flat = 65 mm (xvii)

Length of lacing = 340 $\sqrt{2}$ mm (xiv)

Thickness of lacing = $\frac{1}{60} \times 340\sqrt{2} = 8.1$ mm (xv)

Adopt lacing flat 65 F 10. Sectional area of flat 65 x 10 = 650 mm^2 (xvi)

Part A. (Example 5.) Design of built-up (to be laced) two channel-sections using '**Limit State Method**'.

Step 2 : Type of the column section

For trial section 2ISMC 350, @ 0.421 kN/m

Width of flange, b_f = 100 mm

Thickness of flange, t_f = 13.5 mm

Depth of the section, h_o = 350 mm

Thickness of Web, t_w = 8.1 mm

For two channels,

Cross-sectional area, A = 2 x 5366 mm^2

Epsilon, $\varepsilon = (250/260)^{1/2} = 0.98$

Depth of web, d = $(h_o - 2t_f)$

(350 – 2 x 13.5) = (350 – 27) = 323 mm

Ratio, (b_f/t_f) = (100/13.5) = 7.07 < 9.5

Ratio, (d_w/t_w) = (323/8.1) = 39.876

39.876 < (42 ε = (42 x 0.98) = 41.16

Hence, the section is **semi-compact**

Step 3 : Slenderness ratio of the column

Effective length of the column = 6 m

$$\lambda_{xx} = \left(\frac{l}{r_{xx}}\right) = \left(\frac{6000}{136.6}\right) = 43.924$$

$$\lambda_{yy} = \left(\frac{l}{r_{yy}}\right) = \left(\frac{6000}{137.4}\right) = 43.688$$

Effective slenderness ratio of laced columns is taken as 1.05 times the actual maximum slenderness ratio, in order to account for shear deformation effects. (Clause 7.6.1.5)

λ_{xx} = (1.05 x 43.924) = 46.12

λ_{yy} = (1.05 x 43.688) = 45.87

Critical slenderness ratio

$$\lambda_o = 0.2\pi\left(E/\sigma_y\right)^{1/2}$$

λ_o = (0.2 x 3.14) (2 x 10^5/260)$^{1/2}$

$\lambda_o = (0.628 \text{ x } 10)\ (2000/260)^{1/2} = 17.41$

Step 4 : Imperfection factors

For rolled steel sections

$\alpha_{xx} = 0.0020,\ \alpha_{yy} = 0.0035$

$\alpha_{1-xx} = \alpha_{xx}(\alpha_{xx} - \lambda_o)$

$= 0.0020\ (46.12 - 17.41)$

$= (0.0020 \text{ x } 28.72) = 0.057$

$\alpha_{1-yy} = \alpha_{yy}.(\lambda_{yy} - \lambda_o)$

$= 0.0035\ (45.87 - 17.4)$

$= (0.0035 \text{ x } 28.47) = 0.0996$

Step 5. Euler's stress

About xx-axis

$$\sigma_{E.xx} = \left[\pi^2 E / (\lambda_{xx} - \lambda_o)^2\right]$$

$= [3.14^2 \text{ x } 2 \text{ x } 10^5 / (28.72 \text{ x } 28.72)]$

$= [19.719 \text{ x } 10^5 / 824.838]$

$$= \left[\frac{1971.9 \times 1000}{824.838}\right] = [2390.65 \text{ N/mm}^2]$$

About yy-axis

$$\sigma_{E.yy} = \left[\pi^2 E / (\lambda_{yy} - \lambda_o)^2\right]$$

$= [3.14^2 \text{ x } 2 \text{ x } 10^5 / (28.47 \text{ x } 28.47)]$

$$= \left[\frac{1971.9 \text{x} 1000}{810.541}\right] = 2432.819 \text{ N/mm}^2$$

Step 6 : Values of ϕ_{xx} and ϕ_{yy} (from Ayrton and Perry's formula)

$$\phi_{xx} = [\sigma_y + (1 + \alpha_{1-xx})\,\sigma_{Exx}]/2$$

$\phi_{xx} = [260 + (1 + 0.057)\ 2390.65]/2$

$= 1393.46 \text{ N/mm}^2$

$$\phi_{yy} = \left[\sigma_y + (1 + \alpha_{1.yy})\,\sigma_{E.yy}\right]/2$$

$= 1467.564 \text{ N/mm}^2$

Step 7 : Design Compressive stresses, $\sigma_{c.xx}$ and $\sigma_{c.yy}$ (from Ayrton and Perry's Formula)

$$\sigma_{c.xx} = \phi_{xx} - \left[\phi_{xx}^2 - \sigma_y.\sigma_{E.xx}\right]^{1/2}$$

$= 1393.46 - [1393.46^2 - 260 \times 2390.65]^{1/2}$

$= 248.46$ N/mm² (Not more than 260 N/mm²)

$$\sigma_{c.yy} = \phi_{yy} - \left[\phi_{yy}^2 - \sigma_y.\sigma_{E.yy}\right]^{1/2}$$

$= 1467.564 - [1467.564^2 - 260 \times 2432.819]^{1/2}$

$= 234.217$ N/mm² (Not more than 260.00 N/mm²)

Ultimate compressive stress is that whichever is less $\sigma_{c.yy} = 234.217$ N/mm²

Step 8 : Load carrying capacity

Strength of the column

$$W_{fd} = \left(\frac{\sigma_{cd}.A}{Y_{ms}}\right) = \left(\frac{234.217 \times 2 \times 5366}{1000 \times 1.15}\right)$$

$$= \left(\frac{2513.62}{1.15}\right) = 2185.75 \text{ kN}$$

Strength of the column at service level

$$W_{cK} = \left(\frac{2185.75}{1.5}\right) = 1457.16 \text{ kN}$$

It is more than 1000 kN

Part B (Example 5...) Design of lacing

Step 1. Spacing of lacing bar

Double lacing is provided at an inclination of 45° with the longitudinal axis of the column, as shown in Fig. 5.35. The spacing of channels is kept equal to 220 mm. The gauge distance, g for rivet lins for ISMC 350, @ 0.421 kN/m is 60 mm.

$(S + 2g) = (220 + 2 \times 60) = 340$ mm

Inclination of lacing is kept 45° spacing of the lacing (centre to centre distance between rivets is kept 340 mm.

Step 2 : Slenderness ratio of component of column

Minimum radius of gyration of one channel ISMC 350, @ 0.421 kN/m is $r_{min} = 9.5$ mm. r_{vv}

Column consists of two channels slenderness ratio of the component of the column

$$\left(\frac{l}{r_{min}}\right) = \left(\frac{340}{9.5}\right) = 35.79$$

which one is not more than 50 or 0.7 times the most unfavourable slenderness ratio of the column as a whole.

22 mm nominal diameter power driven rivets are used for the connection of lacing. Use equal angle ISA 65 mm x 45 mm x 8 mm. A = 817 mm²

Length of lacing angle = $340\sqrt{2}$ mm

Thickness of angle = $\left(\frac{1}{60} x 340\sqrt{2}\right) = 8$ mm

Sectional area of angle A = 817 m²

Step 3 : Check for the strength

Step 3.1 Force in the lacing angle

Transverse shear to be resisted is 2.5 percent of the axial load

$$V_{cK} = \left[\frac{2.5x1000}{100}\right] = 25 \text{ kN}$$

Force in each lacing (since, double lacing angles are provided) at service load level.

$$V_{cK} = \left[\frac{1}{2} x \frac{25}{2} x \operatorname{cosec}\varphi\right] = \left[\frac{1}{2} x \frac{25}{2} x \sqrt{2}\right] = 8.84 \text{ kN}$$

Force in each lacing at ultimate load level

V_{fd} = (1.5 x 8.84) = 13.26 kN

Step 3.2 Compressive strength of lacing angle slenderness ratios

Effective length of the lacing angle = $340\sqrt{2}$ mm

r_{vv} = 9.5 mm

$$\text{Slenderness ratio} = \left(\frac{l}{r_{vv}}\right) = \left(\frac{340\sqrt{2}}{9.5}\right) = 50.614$$

$$\text{Epsilon, } \varepsilon = \left(\frac{250}{260}\right)^{1/2} = 0.98$$

$$\varepsilon\left(\frac{\pi^2 E}{250}\right)^{1/2} = 0.98 x \left(\frac{3.14^2 x 2x10^5}{250}\right)^{1/2} = 87.04$$

From IS : 800-2007, Clause 7.5.1.2

$$\lambda_{vv} = \left(\frac{(l/r_{vv})}{\varepsilon\left(\frac{\pi^2 E}{250}\right)^{1/2}}\right) = \left(\frac{50.614}{87.04}\right) = 0.582$$

$$\lambda_{\varphi\varphi} = \left(\frac{(b_1+b_2)/2t}{\varepsilon\left(\frac{\pi^2 E}{250}\right)^{1/2}}\right) = \left(\frac{(65+45)/2x8}{87.04}\right) = \left(\frac{110}{16x87.04}\right) = \left(\frac{1.264}{16}\right) = 0.079$$

Step 3 : Equivalent slenderness ratio (IS : 800-2007, Clause 7.5.1.2)

$$\lambda_e = \left[k_1 + k_2.\lambda_{vv}^2 + k_3.\lambda_{\phi\phi}^2\right]^{1/2}$$

The angle section used for lacing is connected with the channels at its ends by more than one rivet/bolt. Values of k_1, k_2 and k_3 are noted from IS : 800-2007 (Clause 7.5.1.2).

$k_1 = 0.20$, $k_2 = 0.35$ and $k_3 = 20$

$$\lambda_e = \left[0.20 + 0.35 x 0.582^2 + 20 x 0.079^2\right]^{1/2}$$

$$= (0.4434)^{1/2} = 0.666$$

Step 3.4 Design compressive stress

from buckling class `c',

$\sigma_y = 260$ N/mm²

For $(l/r) = 0$, $\sigma_{cd} = 260$ N/mm²

For $(l/r) = 10$, $\sigma_{cd} = 236$ N/mm²

For $(l/r) = 0.666$

$$\sigma_{cd} = \left[260 - \left(\frac{0.666 x 24}{10}\right)\right] = [260 - 1.5984] = 258.40 \text{ N/mm}^2$$

Step 3.5 Factored (design) compressive strength

$$W_{fd} = \left(\frac{258.40 x 817}{1000 x 1.15}\right) = 183.576 \text{ kN}$$

Step 3.6 : Service (characteristic) compressive strength

$$W_{ck} = \left(\frac{183.576}{1.5}\right) = 122.38 \text{ kN}$$

Which one is more than the force in the lacing angle.

5.31 BATTEN PLATES

Compression members composed of two main components battened should have preferably individual members of same cross-section and symmetrically placed about their major axis.

The batten plates are also called as the plates, and these are also used in lateral system. The angle sections, channels and I-sections are also used as battens. The components of built-up column sharing the load are connected together by batten plates. The battening of columns shall not be done where the columns are subjected in the plane of battens to eccentric loading.

In the case, the battened compression members are subjected, in the plane of battens, to eccentric loading, applied moments or lateral forces shall be designed according to the exact theory of elastic stability or empirically from the verification tests, so that, these have a load factor of not less than 1.7 in the actual structure. It is note that if the column section is subjected to eccentricity or other moments about yy-axis (axis perpendicular to the plane of battens), the battens and the column section should be specially designed for such moments.

The design includes design of built up (main) column and design of batten plates. The batten plates are designed as follows :

Design of batten plates

Following are the usual steps in design of batten plates. The specifications for the design of plates as per IS : 800-2007 have also been given. The batten plates used for connecting the component members of a built-up column have been shown in Fig. 5.39. The batten shall be placed opposite to each other at each end of the member and at points where the member is stayed in its length and shall, as far as practicable be spaced and proportioned uniformly throughout.

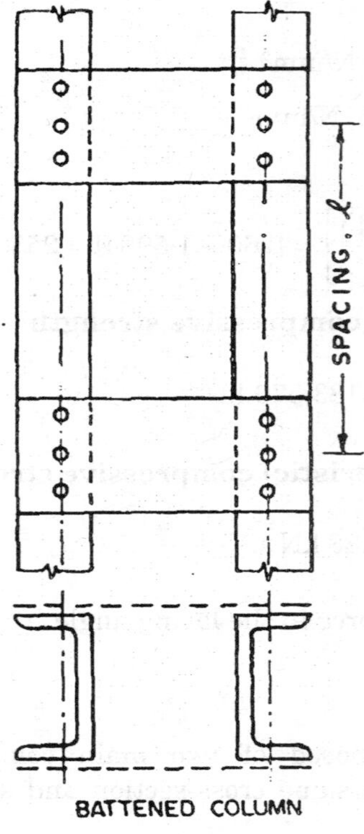

Fig. 5.39.

The number of battens shall be such that the member is divided into not less than three bays within its actual length from centre to centre of connection (i.e. the effective slenderness ratio $(kL/r)_e$ shall be taken as 1.10 times the (kL/r). The maximum axial slenderness ratio considering shear deformation influence.

Step 1. A battened column is designed with its effective length ten per cent in excess of the usual standard cases.

Step 2. The compression members composed of two main components battened should preferably have their two main components of the same cross-section and symmetrically disposed about their xx-axis. The compression members composed of two main components battened, should, where practicable, have a radius of gyration about the axis perpendicular to the plane of batten not less than radius of gyration at right angles to that axis (i.e. r_{yy} < r_{xx}).

Step 3. The spacing of battens is centre to centre distance between adjacent battens as shown in Fig. 5.39.

The spacing of battens is kept equal to length of lesser main component of built-up column. The lesser main component of built up column is the portion of column between consecutive connections centre to centre of end fastenings. It is computed as below :

(a) Compression member is composed of two I-sections

(i) In the battened compression members not specifically checked for shear stress and bending moment as specified above, the spacing of battens centre to centre of end fastenings shall be such that slenderness ratio `λ' of the lesser main component over that distance shall be not greater than 50 or greater than 0.7 times the slenderness ratio of the main members as a whole about its x-axis (axis parallel to the battens).

(b) Compression member is composed of two channel (back to back) Sections

When the compression members composed of two angles, channels, or tees back to back in contact or separated by a small distance shall be connected together by riveting, bolting or welding so that the ratio of slenderness of each member between the connections is not greater than 40 or greater than 0.6 times the most unfaourable ratio of slenderness of the strut as a whole whichever is less.

The minimum radius of gyration of the component member of column is known. The slenderness ratio of the lesser main component is equated to the minimum value as the case may be, length of the lesser main component is computed. The spacing of batten plates is taken equal to this length. The number of batten plates should be such that it divides the column longitudinally in at least three parts.

In no case shall the ends of the strut be connected together with less than two rivets or bolts or their equivalent in welding, and there shall be not less than two additional connections spaced equidistant in the length of strut. Where the members are separated back-to-back, the rivets or bolts through these connections shall pass through solid washers or packings, and where the legs of the connected angles or tables of the connected these are 125 mm wide or over, or where the webs of channels are 150 mm wide or over, not less than two rivets or bolts shall be used in each connection, one on line of each gauge mark.

The battens shall be designed to carry the bending moments and shears arising from the transverse shear force `V' of 2.5 per cent of the total axial force on the whole compression member, at any point in the length of the member divided equally between parallel planes of the battens. The main members shall also be checked for the same shear force and bending moments as for the battens.

Step 4. The batten plates are designed to resist simultaneously the moments and longitudinal force arising due to transverse shear force. The expression for which may be derived as under :

A battened column carrying an axial load may deform as shown in Fig. 5.40. The transverse shear, S at any point in the length of member is takes as 2.5 per cent of the total axial force on the whole compression member. It is divided equally between n parallel planes of the battens. It may be assumed that there are points of contraflexures, where the curvatures of the column change. Thus, a transverse shear force $\left(\frac{S}{n}.\frac{1}{2}\right)$ shall be acting as shown in Fig. 5.40(A). Let S be equal to V and n be equal to N.

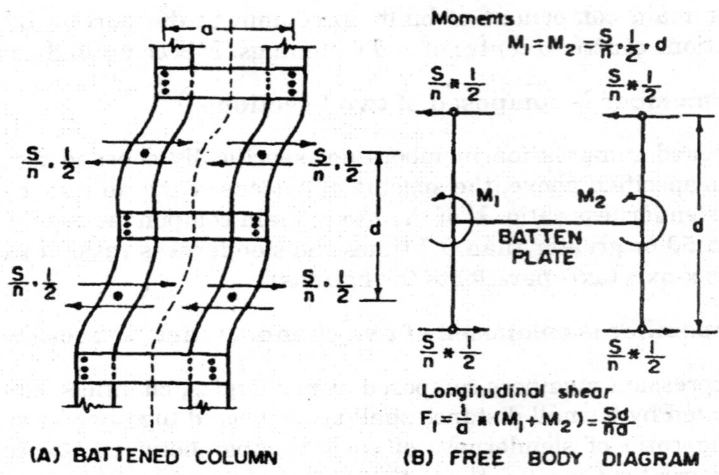

Fig. 5.40.

Consider the free body diagram of the portion of battened column as shown in Fig. 5.40(B). The bending moments arising at each end of the batten plate due to transverse shear

$$M_1 = M_2 = \left(\frac{S}{n}.\frac{1}{2}\right) x\, d = \left(\frac{S.d}{2n}\right) \tag{5.55}$$

$$\therefore M_1 = M_2 = \left(\frac{V.C}{2n}\right) \qquad (C = d) \tag{5.56}$$

The longitudinal shear arising due to these moments

$$V_1 = \frac{1}{a}(M_1 + M_2) = \left(\frac{S.d}{n.a}\right) \tag{5.57}$$

$$\therefore \; V_1 = \left(\frac{V \times C}{N \times S}\right) \qquad (a = s) \tag{6.58}$$

where C= Distance centre to centre of battens longitudinally

N = Number of parallel planes of batten, and

S = Minimum distance across between c.g. of rivets or welding.

Step 5. The tie plates shall be designed by the same method as the other sections used for battens. In no case shall a tie plate and its fastenings be incapable of carrying the forces for which the lacing are designed. When the plates are used as battens, then, the size of batten plates is computed as below :

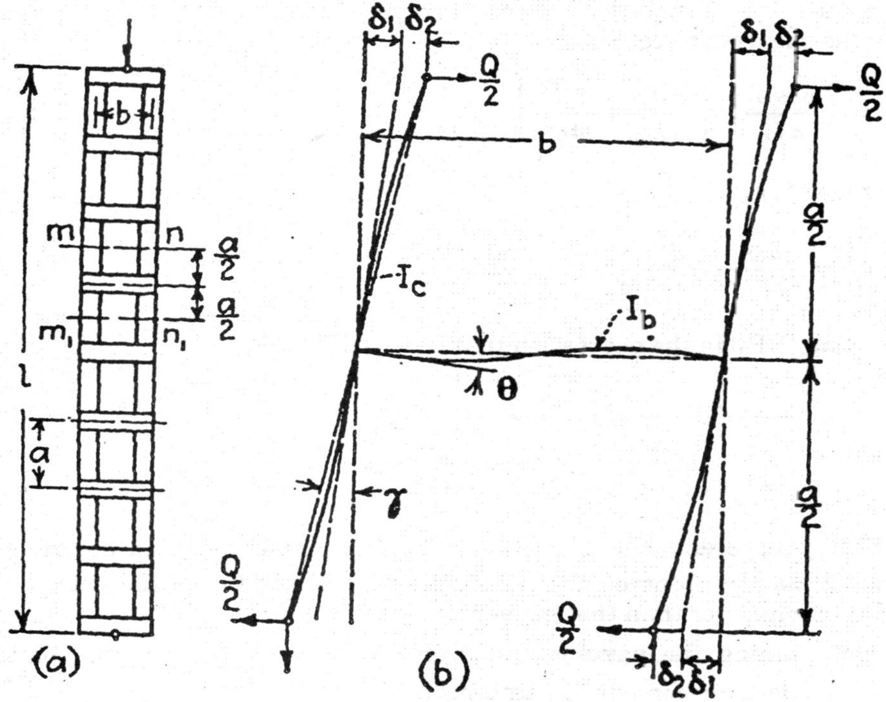

Fig. 5.40C.

Design of battens

Consider the deformation of an element of the column between the sections mm and m_1n_1, Fig. 5.40 (B). The deflection curves of the channels shows the points of inflection at these sections (assumed). The bending of the element is as shown in Fig. 5.40(C). The lateral deflection consists of two parts. It is sum of the deflection, δ_1, due to bending of the batten, and δ_2 due to bending of the channels. There are two couples (S.d/2) acting at the ends of the batten, and the angle at each end of the batten is

$$\theta=\left(\frac{S.d.a}{12EI_b}\right) \tag{i}$$

where, a is the length of the batten and EI_b is its flexural rigidity. The lateral displacement, δ_1 produced by this bending of the batten is

$$\delta_1=\left(\frac{S.d^2.a}{24EI_b}\right) \tag{ii}$$

The displacement δ_2 may be obtained from the deflection of cantilever beam. Therefore,

$$\delta_2=\frac{S}{2}\cdot\left(\frac{d}{2}\right)^2\frac{1}{3EI_c} \tag{iii}$$

where, EI_c is the flexural rigidity of one of the vertical channels. Total angular displacement produced by the shearing force, S is

$$\gamma=\left(\frac{\delta_1+\delta_2}{d/2}\right)=\left(\frac{S.d.a}{12EI_b}+\frac{S.d^2}{24EI_c}\right) \quad \text{(iv)}$$

Since, $\gamma=(S/P_d)$

$$\frac{1}{P_d}=\left(\frac{d.a}{12EI_b}+\frac{d^2}{24EI_c}\right) \quad \text{(v)}$$

Substituting ($1/P_d$) in the expression that for laced column,

$$P_{cr}=\left(\frac{\pi^2 EI}{l^2}\right)\frac{1}{\left(1+\frac{\pi^2 EI}{l^2}\left(\frac{d.a}{12EI_b}+\frac{d^2}{24EI_c}\right)\right)}$$

where, $(\pi^2 EI/l^2)$ represents the critical load for the complete column determined as for a solid column. It may be observed that when the flexural rigidity of the batten is small, the critical load is much lesser than that given by Euler's formula.

While determining the angular displacement, γ, the shear in the batten may be considered also. The shear force in the batten shall be

$$=\left(\frac{S.d.}{a}\right), \text{ and} \quad \text{(i)}$$

the corresponding shear strain is

$$=\left(\frac{n.s.d}{a.A_b.G}\right). \quad \text{(ii)}$$

where, A_b is the cross-sectional area of two battens and n = 1.2, since the battens have a rectangular cross-section. This may be added to the expression (ii), the critical load is found as

$$P_{cr}=\left(\frac{\pi^2 EI}{l^2}\right)\left(\frac{1}{1+\frac{\pi^2 EI}{l^2}\left(\frac{d.a}{12EI_b}+\frac{d^2}{24EI_c}+\frac{nd}{d.A_bG}\right)}\right)$$

In case, the vertical channels of the built-up column are very flexible, or, in case the distance between battens is large, the collapse of column may occur as a result of local buckling of the channels between two consecutive battens. In order to consider, local buckling into consideration, element between two battens may be considered. It is assumed that the rigidity of the battens is very large. The critical value of the compressive force at which the assumed buckling will occur

$$P = \left(\frac{2\pi^2 E.I_c}{d^2}\right)$$

The effect of axial load P/2 on the bending of the vertical channel may now be taken into consideration by writing

$$\delta^2 = \left(\frac{Sd^3}{48EI_c}\right)\left(\frac{1}{1-\alpha}\right)$$

$$\alpha = \left(\frac{P_{cr}}{2\pi^2 EI_c/d}\right)$$

Critical load for the column, then, be expressed as

$$P_{cr} = \left(\frac{\pi^2 EI}{l^2}\right)\left(\frac{}{1+\frac{\pi^2 EI}{l^2}\left[\frac{d.a}{12EI_b}+\frac{d^2}{24EI_c}\frac{1}{(1-\alpha)}+\frac{n.d}{a.A_b.G}\right]}\right)$$

This expression may be solved by trial and error. It is seen also that the critical load for the column between the battens is always less.

1. Effective depth Intermediate patterns (Clause 7.7.2.3, IS 800-2007)

The effective depth shall be taken as the longitudinal distance between end rivets or welds.

(a) The effective depth of end battens and those at points where the member is stayed in its length shall be as below :

It shall not be less than the perpendicular distance between the centroids of the main members. (Clause 7.7.2.3)

(< the distance between c.g. of component members of column).

(b) The effective depth of intermediate battens shall be not less than three quarters of this distance. (Clause 7.7.2.3)

(< $\frac{3}{4}$ the distance between c.g. of component members of column).

But in no case the effective depth of any batten shall be not less than twice the width of one member in the plane of the battens. (Clause 7.7.2.3)

(shall not be < twice the width of one component member of column).

The total depth is found by adding edge distance on both the sides.

2. Thickness of batten plates (Clause 7.7.2.3, IS : 800-2007)

Thickness of battens (t) should not be less than 1/50th (one-fiftieth) the distance between innermost connecting rivets.

(This clause does not apply for angles, channels, I-sections, used as batten their legs or flanges perpendicular to the main (member) and sufficient to resist longitudinal shear (F_1) and moment (M) arising from transverse shear. However, it should be ensured that the ends of the compression members are tied to achieve adequate rigidity.

3. Spacing of battens (Clause 7.7.3, IS : 800-2007)

In battened compression members where the individual members are not specifically checked for shear stress and bending moments the spacing of battens (centre to centre) of its end fastening) shall be such that the slenderness ratio (kL/r) of any component over that distance shall be neither greater than 50 not greater than 0.7 times the slenderness ratio of the member as a whole about axis parallel to the battens.

Step 6. The rivets are provided to connect batten plates with components of column to resist shear force (F_1) an moment (M) computed above.

The rivets, bolts or welds in these connections shall be sufficient to carry the shear force and moments, if any, specified for the battened struts. The diameter of rivets shall not be less than 16 mm for members upto the including 10 mm thickness; 20 mm diameter for members upto and including 16 mm thick; and 22 mm diameter for members over 16 mm thick.

The compressive members connected by such riveting, bolting or welding shall not be subjected to transverse loading in plane perpendicular to the washer riveted, bolted or welded surface.

Example 5.15. Design a built-up column consisting of two channels connected by battens to carry an axial load of 800 kN. The effective length of the column is 6 metres. Design suitable batten system, also.

Design

Part A (Example 5.)

Design of built up (two channel sections to be battened) using `Limit State Method'

Step 1 : Selection of a trial section (Clause 7.7.1.4, IS : 800-2007)

Effective length of column is 6 m. The effective length of column is increased by 10 percent to account for shear deformation. Therefore, the effective length of column shall be 6.6 m value of yield stress for the steel may be assumed 260 N/mm². Maximum slenderness ratio for the built-up battened column may be assumed 80.

Allowable stress in axial compression

$$\sigma_{ac} = 103\ \text{N/mm}^2$$

Effective sectional area required

$$A = \left(\frac{800 x 1000}{103}\right) = 7766.99\ \text{mm}^2$$

From steel section tables,

Double channels to be battened to be used as column, try 2 ISLC 300, @ 0.331 kN/m

Effective sectional area, A = 8422 mm²

I_{xx} = 12095.8 x 10^4 mm^4

r_{xx} = 119.8 mm

I_{yy} should be approximately equal to I_{xx} and r_{yy} (not less than) r_{xx}.

From steel section tables, the spacing between back to back of channels, S is adopted as 200 mm. r_{yy} = 128.7 mm

Step 2 : Type of the column section

For trial section 2 ISLC 300, @ 0.331 kN/m

Width of flange, b_f = 100 mm

Thickness of flange, t_f = 11.6 mm

Overall depth of section, h_o = 300 mm

Depth of web, d_w = (h_o – 2 x t_f)

(300 – 2 x 11.6) = 276.8 mm

Thickness of web, t_w = 6.7 mm

$$\text{Ratio, } (b_f/t_f) = \left(\frac{100}{11.6}\right) = 8.62 < 9.5$$

$$\text{Ratio, } (d_w/t_w) = \left(\frac{276.8}{6.7}\right) = 41.3 \text{ about } 41.16$$

Hence, the section is taken as **semi-compact.**

Step 3 : Slenderness ratio of the column

Effective length of the column = 6.6 m

$$\lambda_{xx} = \left(\frac{l}{r_{xx}}\right) = \left(\frac{6600}{119.8}\right) = 55.09$$

$$\lambda_{yy} = \left(\frac{l}{r_{yy}}\right) = \left(\frac{6600}{128.7}\right) = 51.28$$

Critical slenderness ratio

$$\lambda_o = 0.2\pi(E/\sigma_y)^{1/2}$$

$$\lambda_o = (0.2x3.14)\left(2x10^5/260\right)^{1/2} = 17.41$$

Step 4 : Imperfection factors

For rolled steel sections

$$\alpha_{xx} = 0.0020, \alpha_{yy} = 0.0035$$

$$\alpha_{1-xx} = \alpha_{xx}(\alpha_{xx} - \lambda_o)$$

= 0.0020 (55.09 –17.41)

= (0.0020 x 37.68) = 0.0754

$$\alpha_{1-yy} = \alpha_{yy}(\lambda_{yy} - \lambda_o)$$

= 0.0035 (51.28 – 17.41)

= (0.0035 x 33.87) = 0.1185

Step 5 : Euler's stress

About xx-axis

$$\sigma_{E.xx} = \left[\pi^2 E / (\lambda_{xx} - \lambda_o)^2\right]$$

= [3.14² x 2 x 10⁵/(37.68)²]

$$= \left[\frac{19.719 \times 100 \times 1000}{1419.78}\right] = 1388.87 \text{ N/mm}^2$$

About yy-axis

$$\sigma_{E.yy} = \left[\pi^2 E / (\lambda_{yy} - \lambda_o)^2\right] = [3.14^2 \times 2 \times 10^5/(33.87)^2]$$

$$= \left[\frac{19.719 \times 100 \times 1000}{1147.177}\right] = 1718.9 \text{ N/mm}^2$$

Step 6 : Values of ϕ_{xx} and ϕ_{yy} (Ayrton and Perry's formula)

$$\phi_{xx} = \left[\sigma_y + (1 + \alpha_{1-xx})\sigma_{Exx}\right] / 2$$

= [260 + (1 + 0.0754) 1388.87] / 2

= 876.800 N/mm²

$$\phi_{yy} = \left[\sigma_y + (1 + \alpha_{1-yy})\sigma_{E.yy}\right] / 2$$

= 1091.295 N/mm²

Step 7 : Compressive stress, $\sigma_{c.xx}$ and $\sigma_{c.yy}$

$$\sigma_{c.xx} = \phi_{xx} - \left[\phi_{xx}^2 - \sigma_y . \sigma_{E.xx}\right]^{1/2}$$

= 876.8 – [876.8² – 260 x 1388.87]^1/2

= 238.286 N/mm² (Not more than 260 N/mm²)

$$\sigma_{c.yy} = \phi_{yy} - \left[\phi_{yy}^2 - \sigma_y . \sigma_{E.yy}\right]^{1/2}$$

= 1091.295 – 100 [10.91295² – 260 x 17.189]^1/2

= 228.745 N/mm²

Ultimate compressive stress is that whichever is less. $\sigma_{c.yy} = 228.745$ **N/mm²**

Step 8 : Load carrying capacity

Strength of the column

$$W_{fd} = \left(\frac{228.745 \times (2 \times 4211)}{1.15 \times 1000}\right) = 1456.7 \text{ kN}$$

Strength of the column at service level

$$W_{cK} \doteq \left(\frac{1456.7}{1.5}\right) = 971.13 \text{ kN. It is more than 800 kN.}$$

Part B (Example 5.)

Design of batten plates

Step 1 : Spacing of batten plates (centre to centre distance of end, fastenings) (Clause 7.7.3, IS : 800-2007)

The battened column is composed of two channels back to back seperated by a small distance, S = 200 mm

slenderness ratio of column about xx-axis

$$\left(\frac{l}{r_{min}}\right) = \left(\frac{6.6 x 1000}{119.8}\right) = 55.09$$

Minimum radius of gyration of one channel (one component) is r_{xx} = 119.8 mm

r_{yy} = 28.7 mm

Slenderness ratio of component member of battened column between the connections

$\left(\frac{l}{r_{min}}\right)$ should be less 50

$\left(\frac{l}{r_{min}}\right)$ for battened column as a whole 0.7 x

(0.7 x 119.8) = 83.86 or

$\left(\frac{l}{28.7}\right) = 83.86$, l = (83.86 x 28.7) mm = 2406.782 mm

Adopt the spacing of battens as 1000 mm

Transverse shear due to axial load at service level (Clause 7.7.2.1, IS : 800-2007)

$$V_{TS} = \left(\frac{2.5}{100} x 800\right) = 20 \text{ kN}$$

Transverse shear at ultimate level

V_{TU} = (20 x 1.5) = 30 kN

Longitudinal shear arising from transverse shear at service level

$$V_{LS} = \left[\frac{V.C}{NS}\right] = \left[\frac{20x1000}{2x250}\right]$$

where, V_T = Transverse shear

C = Distance between centre to centre of battens, longitudinally

N = Number of parallel planes of battens

S = Minimum transverse distance between centroid of rivet/bolt connecting the batten to the main member

$$= \left[\frac{10x1x1000}{250}\right] = 40 \text{ kN}$$

Longitudinal shear arising from transverse shear at ultimate level

V_{LU} = (1.5 x 40) = 60 kN

Moment arising from transverse shear at service load level

$$M_s = \left(\frac{V.C}{2N}\right) = \left(\frac{20x1000}{2x2}\right) = 5000 \text{ mm-kN}$$

M_u = (1.5 x 10,000) = 75000 mm-kN

Step 2 : Size of batten plates (Clause 7.7.2.3, IS : 800-2007)

2 (A) : Effective depth of end batten (Clause 7.7.2.3)

= Distance between c.g of component members

= (200 + 2 x 25.5) = 251 mm

It should not be less than twice the width of one component of column.

Assume 22 mm nominal diameter of rivets shall be used for connection

Overall depth of end batten

= (251 + 2 x 33) = 284 mm

Effective depth of intermediate batten (Clause 7.7.2.3)

= three-fourth distance between c.g of component members

$$= \left(\frac{3}{4}x251\right) = 188.25 \text{ mm}$$

It should not be less than twice the width of one component member of the column, i.e, (2 x 100) = 200 mm

Overall depth of intermediate battens

= (200 + 2 x 33) = 266 mm

Step 4 : Thickness of batten plates (Clause 7.7.2.3, IS : 800-2007)

Thickness should not be less than one-fiftith distance between inner most connecting lines of rivets

$$t = \left(\frac{1}{50}x250\right) = 5 \text{ mm}$$

Adopt 6 mm thickness. Let the depth of intermediate batten required for resisting the moment be d. Then

Moment of resistance

$$MR = \left(\frac{1}{6}t.d^2 x\sigma_c\right)$$

$$\frac{1}{6}\text{x}\left(\frac{6\text{x}d^2\text{x}260}{1.15\text{x}1000}\right)=7500$$

$$d^2 = \left(\frac{7.5\text{x}1000\text{x}1000}{260}\right)\text{x}1.15$$

$$d = \left(\frac{1000}{5.888}\right)1.07=169.84 \text{ mm} = 182.133$$

Adopt the size of end batten plate as 300 mm x 6 mm and the size of intermediate batten plate as 230 mm x 6 mm.

5.32 PERFORATED COVER PLATES

The perforated cover plates may be used instead of lacing, as a lateral system with the columns. The perforated cover plates are particularly suitable for a built-up box section consisting of four angle sections. The interior of the column remains accessible for painting. Bureau of Indian Standards has not given specifications for the design of perforated plates. Guide to design criteria for metal compression members published by the Column Research Council, U.S.A. has given the following specifications for the design of perforated cover plates.

(1) The perforation of plate should be of ovaloid shape (i.e. two semicircles connected with straight sides) or they may be elliptical or circular. The long axis of the perforations for the first two cases should be in the direction of the axis.

(2) The spacing of the perforations (i.e., the centre to centre distance between the perforations) should not be less than 1.5 times the length of perforation, C.

(3) The clear distance between perforations should be not less than the distance between the nearest lines of longitudinal fasteners [i.e., $(a - c) > d$].

(4) The net section of the column (defined as the section at the perforations) should be used in computing the axial rigidity, EA, and the column moments of inertia about x-axis and y-axis.

The net area of perforated plate is used as the part of cross- sectional area of the column. It is the advantage of the perforated plates as compared with the lacing.

(5) If the slenderness ratio c/r_f (length of perforation divided by radius of gyration of flange, see Fig. 5.41 is 20 or less, and also not greater than one-third of the slenderness ratio l/r_x, the appropriate specification for column stress, applied to the column net area can be used to find the allowable load.

(6) For columns built-up of plates, the net area of each web at the perforation should be sufficient to resist l/n times the transverse shear force, where n is the number of perforated plates. Perforated plates designed in accordance with provision (3) need not be checked for shear introduced as a specified percentage of the column load.

(7) The transverse distance from the edge of a perforation to the nearest line of longitudinal fastener, divided by the plate thickness, i.e., the b/t ratio of plate adjacent to a perforation should conform to minimum specification requirements for plates in main compression members.

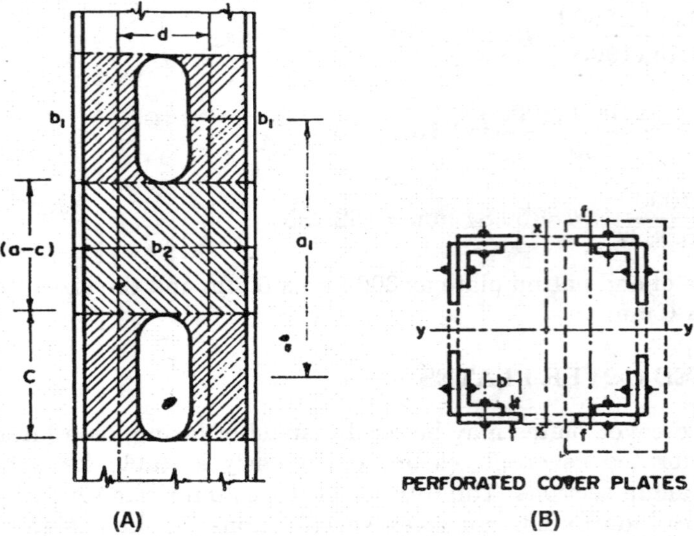

Fig. 5.41

A member with staggered perforations should be treated as if the perforations where opposite to each other. The perforated cover plates should be detailed in such a way that the axial compressive stresses do not cause the failure by local buckling before the strength of column is reached.

The lateral displacement due to shear force, V may be found, by considering an element from the column. This element is similar to the element from the column with batten plates, except that, instead of a narrow batten plate, there is the partion of the cover plate between performations. In this idealized element, the horizontal cross-member is infinitely rigid. The lengths of the vertical projections may be treated as cantilever beams, will be somewhere between (c/2) and (a_1/2), where, C is the length of a perforation. The value (3c/4) is reasonable. It gives results which agree with the experiments.

The expression for a column with batten plate may now be modified for this case. Since, the cross-member (analogus to a batten) is infinitely regid, I_b may be substited as and δ_2 is obtained as the deflection of a cantilever.

$$\delta_2 = \frac{V}{2}\left(\frac{3c}{4}\right)^2 \cdot \left(\frac{1}{3EI_f}\right) = \frac{9V.c^3}{128EI_f}$$

where, I_f represents the moment of inertia of the column flange that is, the complete area of the column on one side of x-axis, taken about the centroid of the flange (axis 1-1). The angular displacement due to V is

$$\gamma = \left(\frac{\delta_1 + \delta_2}{\frac{1}{2}a_1}\right) = \left(\frac{9Vc^2}{64a_1EI_f}\right)$$

and therefore,

$$\frac{1}{P_d} = \left(\frac{9c^3}{64a_1EI_f}\right)$$

From which, it gives the critical load for a column with perforated cover plates

$$P_{cr} = \frac{\pi^2.EI}{l^2}.\left(\frac{1}{1+\frac{\pi^2.EI}{l^2}\left(\frac{9C^3}{64a_1EI_f}\right)}\right)$$

The perforated cover plates are designed by the criterion obtained by equating the plate buckling stress $(f_{cr})_{plane}$ and the column strength $(f_{cr})_{column}$. The plate buckling stress is given by

$$(f_{cr})_{plate} = k.E.\ (t/b)^2 \qquad (5.59)$$

where, E = Effective modulus of elasticity corresponding to the critical stress

(t/b = Thickness to width ratio of plate, k = Coefficient).

The coefficient k depends upon the type of loading, the conditions of edge support and the proportion of plate.

The length of column depends upon the slenderness ratio, yield stress, f_y, and effective modulus of elasticity, E. Therefore, the proportion of perforated plate may be expressed as a function of t/r. E and f_y. The function is of complex form. The following expressions give the design criteria for all types of steel. These functions approximate to the complex function with sufficient accuracy.

$$\text{For } l/r < 60,\ b/(t.\sqrt{k}) = 19.2 \qquad (5.60)$$

$$\text{For } l/r > 60,\ b/(t.\sqrt{k}) = 0.32\,(l/r) \qquad (5.61)$$

When the values of k are known, then, the width thickness ratio (b/t) may be found. Two portions of the perforated plates are of interest. The portion b_1 is free along one edge and restrained along the other edge. The other edge of portion b_1 is riveted or welded to the segment of the column. The restraint action along the edge varies between the two cases. In one case, it is equivalent to a longitudinal hinge, which allows the plate to rotate about the axis of the edge. In the other case, it is equivalent to that of continuous clamp, which does not allow rotation. The value of k for above two cases are 0.38 and 0.15, respectively. In practice, the type of restraint gives the lower value out of two values.

The portion of plate with width b_2 is restraint along both the edges. The restraint of edges is also equivalent to two above mentioned cases. The values of k for which vary from 3.6 to 5.3. In practice, the type of restraint gives the lower value.

PROBLEMS

5.1 A rolled steel beam HB 300, @ 0.588 kN/m is used as a column. The column is fixed in position but not in direction at both ends. Determine the safe load carrying capacity of the column if the length of column is 4.50 m.

5.2 A single angle discontinuous strut ISA 130 mm x 130 mm x 10 mm (ISA 130 130, @ 0.197 kN/m) with single riveted connection is 2.5 m long. Calculate safe load carrying capacity of section.

5.3 In Problem 5.2, if single discontinuous strut is double riveted, determine safe load carrying capacity of section.

5.4. A double angle discontinuous strut ISA 150 mm x 75 mm x 10 mm (ISA 150 75, @ 0.165 kN/m) long legs back to back, is connected to both sides of a gusset plate 10 mm thick with 2 rivets. The length of strut between centre to centre of intersections is 3.50 m. Determine the safe load carrying capacity of the section.

5.5 In Problem 5.4, if double angle discontinuous strut is connected to one side of a gusset, determine safe load carrying capacity of the strut.

5.6 In Problem 5.4, if double angle strut is continuous and connected with the gusset plate with single rivet, determine safe load carrying capacity of strut.

5.7 A built-up column consists of two MC 400, @ 0.494 kN/m and two plates 500 mm x 500 mm x 10 mm. The clear distance between back to back of channels is 200 mm. One plate is connected to each flange side. Determine the safe load carrying capacity of built-up section if the effective length of column is 5 m.

5.8 Design a rolled steel beam section column to carry an axial load of 1120 kN. The column is 6.20 m long and adequately restrained in position and direction at both ends.

5.9 Design a single angle discontinuous strut to carry 80 kN load. The length of the strut between centre of intersection is 2.75 m.

5.10 Design a double angle discontinuous strut to carry 120 kN load. The length of strut between centre to centre of intersection is 3.80 m.

5.11 A built-up column consists of two channels MC 250, @ 0.304 kN/m placed back to back with 147 mm gap between them. Determine the safe load on the column when (i) channels are effectively laced together, (ii) channels are effectively battened together. Take effectively length of the column to be 6000 mm.

5.12 Design a built-up column to support 1200 kN axial load. The length of column is 18.5 m. The column is fixed at both ends. Design (i) angle lacing, (ii) double lacing for the built-up column.

5.13 Design a built-up column of effective length 5 m to carry an axial load of 1000 kN. Also, design a suitable lacing system for the above column.

5.14 An axially loaded built-up strut is made up of two channels back to back with a gap of 200 mm. The column is 3 metres long and is pinned at both ends. The individual channels of the column are MC 300, @ 0.858 kN/m. Design a suitable lacing for the column. The properties of the channels are

Area	= 4564 mm^2
Depth of section	= 300 mm
Width of flange	= 90 mm
Thickness of flange	= 13.6 mm
Thickness of web	= 7.6 mm

Centre of gravity $C_{yy} = 23.6$ mm

$I_{xx} = 6362.6 \times 10^4$ mm^4 $I_{yy} = 310.8 \times 10^4$ mm^4

$r_{xx} = 118.1$ mm, $r_{yy} = 26.1$ mm.

5.15 Design a mild steel built-up column to carry an axial load of 2000 kN with an effective height of 6.10 m. Use 4 m.s. angle sections laced together to form the column. Design also a suitable system of lacing bars.

5.16 A column 400 mm x 400 mm consists of 4 angles of ISA 80 mm x 80 mm x 10 mm (ISA 80 80, @ 0.118 kN/m) as shown in Fig. 5.42. The column is 8 m long and is hinged at both ends. Find the maximum safe load for the column and design a suitable lacing system for it. Draw a neat sketch of the column showing connection details (Properties of each angle are Area = 1505 mm^2, c.g. line $c_{xx} = c_{yy} = 23.6$ mm, $I_{xx} = I_{yy} = 87.7 \times 10^4$ mm^4, $r_{min} = 15.5$, rivet gauge distance = 45 mm.

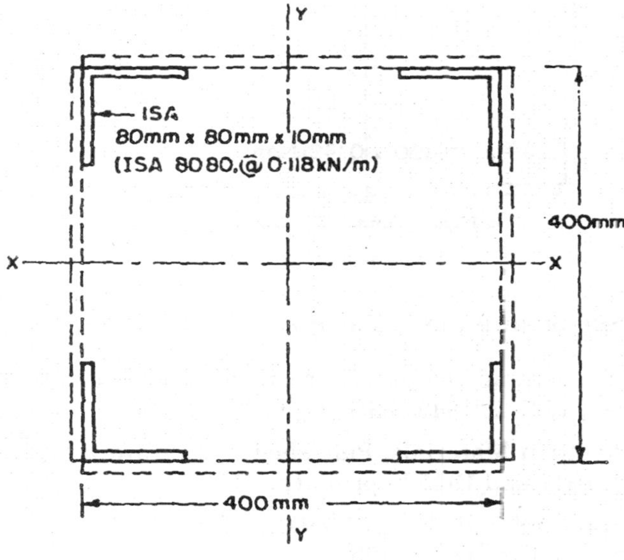

Fig. 5.42.

5.17 Design a built-up column to carry an axial load of 3600 kN. The effective length of column is 5 m. Provide two channel sections and plates.

5.18 Design a steel column 3 metre long carrying an axial load of 500 kN. Assume both the ends to be pin-jointed. Only channel sections are available.

5.19 Design a battened column to carry an axial load of 1100 kN. The effective length of column is 700 m.

5.20 The ground floor column of a multistorey building is 5 metres high and is subjected to an axial load of 1000 kN. The column height of the floor next to the ground floor is 4 metres. The beams connected to the flanges of the column have an eccentric load of 10 kN. The beams connected to the web of the column have an eccentric load of 6 kN. Design and sketch the section of the column. Adopt bending stress as per IS : 800-1984.

5.21 The cross-section of a 6 m long, pin ended column consists of 4 ISA 100 mm x 100 mm x 10 mm (4 ISA 100 100, @ 0.149 kN/m) suitably connected with lacing bars. The angles face inwards and the outside dimensions of the cross-section are 350 mm x 350 mm as shown in Fig. 5.43.

(i) Determine the safe axial compressive load for the column.

(ii) Also design the lacing bars and their connection to angles.

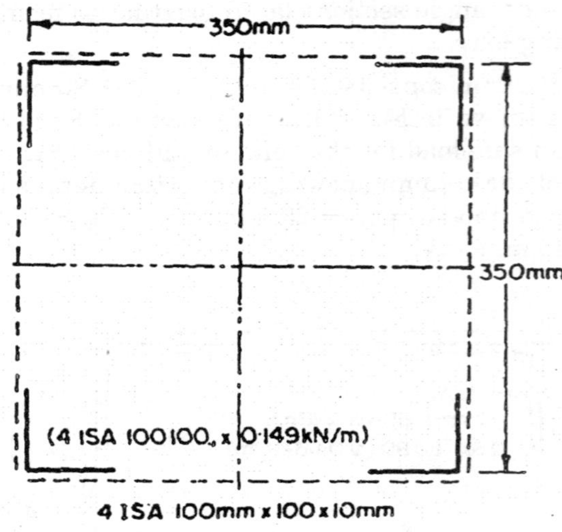

Fig. 5.43.

REFERENCES (For Ayrton and Perry Formula)

1. Ewart S. Andrews : Theory and Design of Structures. Published by Chapman & Hall, Ltd. II, Henrie TTA Street, W.C. 2, 1942 Edition (pp. 370-371).

2. John, Case: The Strength of Materials, Published by Edward Arnold & Co., London, 1938 Third Edition, 1940, 1941 and 1945 Reprint, (pp. 245-310).

3. Boris Brester, T.Y. Lin, John B. Scalzi. Design of Steel Structures, Second Ed., Wiley Eastern Private Limited, New Delhi, 1968.

4. Stephen P. Timoshenko, James, M. Gere. `Theory of Elastic Stability'. Second Ed. 1961. McGraw Hill Book Company, Inc., New York, Toronto, London.

5. Charles, G. Salmon, John E. Johnson. Steel Structures - Design and Behaviour Second Ed. 1980, Harper International/ Edition Harper and Row, Publishers, New York.

6. Bruce, G. Johnston, C. Structural Stability Research Council Guide to stability Design Criteria for Metal Structures, Third Edition, New York , John Wiley & Sons, 1976.

7. Lynn. S. Beedle. Plastic Design of Steel Frames' John Wiley and Sons, Inc. 1958.

6

LIMIT STATE OF COLLAPSE IN COMPRESSION AND BENDING COMBINED DESIGN OF ECCENTRICALLY LOADED COLUMNS

6.1 INTRODUCTION

In practice, columns do not carry concentic axial load. All members in a structure carry both bending moment and axial force (either tension or compression). When the magnitude of one or the other is comparatively small, its effect is usually neglected and the member is designed as a beam, as an axially loaded column, or as a tension member. For many situations neither influence may properly be neglected. Therefore, the behaviour under the combined loading has to be considered in the design.

Column may bend about one or both the axes along with axial compression. Loads may be eccentric. In **braced rigid** portal frames beams carry lateral loads. These beams transfer bending moments to the columns in addition to axial load. Rotations occur at the ends of beams. The connections of beams and columns may be rigid. As such, these rotations (so also moments) are also transferred to the ends of the columns along with axial loads. In unbraced frames, the frames are subjected to sway deflections and bending. At corners, the beams frame perpendicularly from two directions and columns may be bending about both the principal axes.

A beam-column may bend (along its complete length) in **single curvature**. The nature of bending stress (compressive or tensile) at a point in the cross-section and sign of bending moment diagram over its length of the beam-column remain same. The curvature has also the same sign over the complete length. Whereas in sway frames, the columns may bend in **double curvature** (that is, in reversed curvature). It results in variation of the nature (positive or negative of the bending moment and curvature of the length of the member).

A beam-column is defined as a structural member, which carries an axial compression and bending due to simultaneously acting lateral (transverse loads.) For combined bending and axial compression the possibility of instability is increased. A secondary bending moment arises, when axial compression is present. [Secondary moment = axial compression times the deflection].

Likely mode of failure for combined bending and axial compression may be stated as under.

1. For bending about one axis

Instability failure in the plane of bending, without twisting may take place.

2. For bending about strong axis

Lateral torsional buckling failure may occur.

3. For bi-axial bending (Sections torsionally stiff)

Instability failure in one of the principal direction.

3 (ii) For bi-axial bending (thin walled open sections)

Combined twisting and bending failure may take place

3 (iii) For bi-axial bending and torsion

Combined bending and twisting (when plane of bending does not contain shear centre) failure may occur.

Semi-empirical interaction procedures based on ultimate strength may be used for the design.

When both axial compression and transverse loads act simultaneously on a structural member, the conditions are completely different. The bending moments, shear forces, stresses, and the deflections produced will not be proportional to the magnitude of the axial compression.

The **principle of super-position** is widely used when the transverse loads act alone on a beam. However, this principle cannot be applied in case of combined action of transverse loads and axial compression.

Eccentrically loaded columns are the examples, where their cross-sections are subjected to both bending stresses and stresses due to axial compression. These columns have been described in subsequent article.

6.2. ECCENTRICALL LOADED COLUMNS AXIALLY LOADED MEMBERS WITH MOMENTS

When a load is applied at an eccentric distance from the centroid of the column section, as shown in Fig. 6.1, the column is called eccentrically loaded.

A column may carry an eccentric load in addition to axial load. The eccentrically loaded column is subjected to bending stress besides the axial compression. The strength of the column is very much reduced.

When the column is subjected to axial loading only, then, the cross-sectional area, A_a required is given by

$$A_a = (P / \sigma_{ac}) \quad \text{(i)}$$

where, σ_{ac} = Allowable working trees in compression on the member subjected to axial load only.

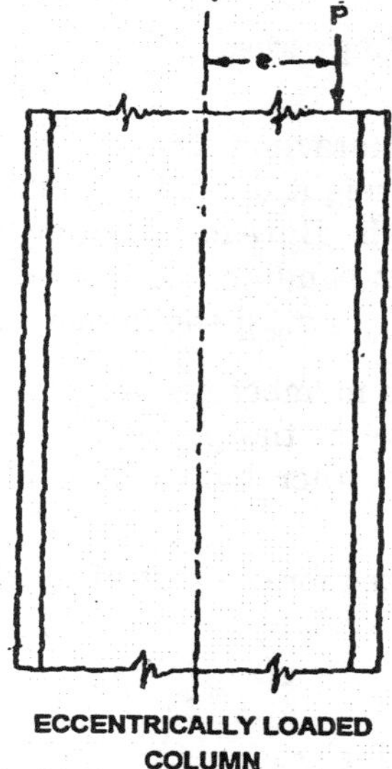

Fig. 6.1. Eccentrically loaded column

When the column is subjected to bending only, then, cross-sectional area required is given by

$$A_b = \left(\frac{P.e.y}{\sigma_{bc}.r^2}\right) \quad \text{(ii)}$$

where, σ_{bc} = Maximum allowable bending compressive stress on the extreme fibre

y = Distance to the extreme fibre of the section measured from the neutral axis

$A_b.r^2$ = Moment of inertia of the section about the appropriate axis

r = Radius of gyration of the section about the appropriate axis

$P.e$ = Moment due to the eccentricity, e of the load P, and

$$\left(\therefore \frac{P.e}{I} = \frac{\sigma_{bc.cal}}{y}\right) \therefore \left(b_{c.cal} = \left(\frac{P.e.y}{I}\right) = \frac{P.e.y}{A_b.r^2}\right)$$

When the column is subjected to both axial load and bending, then, the total area required

$$A = (A_a + A_b) = \left(\frac{P}{\sigma_{ac}} + \frac{P.e.y}{\sigma_{bc}.y^2}\right)$$

Dividing both the sides of expression (iii) by A and writing the expression as below

$$\left[\left(\frac{P/A}{\sigma_{ac}}\right) + \left(\frac{P.e.y/Ar^2}{\sigma_{bc}}\right)\right] = 1, \left(\frac{\sigma_{ac.cal}}{\sigma_{ac}} + \frac{\sigma_{bc.cal}}{\sigma_{bc}}\right) = 1 \quad \text{(6.1)}$$

where, $\sigma_{ac.cal} = \left(\frac{P}{A}\right)$ (calculated average axial compressive stress)

$$\sigma_{bc.cal} = \left(\frac{P.e.y}{Ar^2}\right) \text{ (calculated bending stress in the extreme fibre)}$$

Eq. 6.1 gives the fundamental interaction formula for the compression member subjected to the axial load and bending. It is straight line when plotted graphically, left ($\sigma_{ac.cal}/\sigma_{ac}$) along y-axis and left ($\sigma_{bc.cal}/\sigma_{bc}$) along x-axis. Bureau of Indian Standards adopted this interaction formula. The column subjected to both bending and axial compression shall be so proportioned that the quantity $[(\sigma_{ac.cal}/\sigma_{ac})+(\sigma_{bc.cal}/\sigma_{bc})]$ should not exceed unity.

When a column or a compression member carries axial load with moment, (beam-column) it should be checked for local capacity at the points of maximum bending moment and axial load. This capacity may be limited either by yielding or local buckling depending on the properties of the cross-section.

The member should also be checked for **overall buckling** by either

(i) simplified approach or

(ii) more exact approach.

For semi-compact and the slender cross-secion local capacity is checked. For plastic or compact cross-sections, as an alternative simplified approach.

$$\left(\frac{P}{A_g \, .\sigma_{yo}}\right)+\left(\frac{M_x}{M_{cx}}\right)+\left(\frac{M_y}{M_{cy}}\right) <= 1, \textit{ where}$$

Note : 1. For a slender cross-section, a reduced value of σ_y should be used

2. **Local buckling** may become the design criterion when the proportions of elements in a cross-section exceed those given for semi-compact elements in IS : 800-2007, Clause 3.7, Table 2.

P = Applied (axial) compressive load on the member.

M_x = Bending moment about strong-axis

M_y = Bending moment about weak-axis

A_g = Gross-area of the cross-section

A_e = Effective cross-section (slender section)

Z_{px}, Z_{py} = Plastic moduli of the cross-section about strong and weak axes, respectively.

σ_{yd} = Design yield strength (= σ_y / γ_{msy})

γ_{msy} = 1.15 Partial safety factor for the strength of material

Normally, actual bending moments at one of the two ends of the member in compression would control. The moments found from the linear-elastic analysis would be sufficient for buildings with small number of storeys and smaller axial cor pressive compressive force. For very tall buildings with large axial compression and large lateral sway, for calculating the end-moments, the P – Δ effects are to be included.

Overall buckling failure (Simplified approach)

Following interaction equation should be satisfied, when the capacity of the member is controlled by overall buckling.

$$\left(\frac{P}{\sigma_{cd} \cdot A_g}\right) + kx\left(\frac{M_x}{M_{cx}}\right) + ky\left(\frac{M_y}{M_{uy}}\right) < = 1.0, \text{ where}$$

P = Applied (actual) axial compressive force

M_x, M_y = Actual bending moment about strong and weak axes, respectively.

$P_{cd} = (\sigma_{cd} \cdot A_g)$ = Design compressive strength

M_{ux} = Design bending strength (the buckling resistance moment capacity) about strong axis

$M_{u.y} = (\sigma_y \cdot Z)$ Design bending strength about weak axis

Z = Elastic-section modulus about weak axis

σ_{yd} = Design yield stress. These, above interaction may be written as

$$\left(\frac{P}{P_{cd}}\right) + kx\left(\frac{M_x}{M_{xd}}\right) + ky\left(\frac{M_y}{\sigma_{yd}.Z}\right) <= 1.0$$

k_x, k_y = Moment amplification factor which include the effect of moment gradient over the length of member instead of uniform moment over the complete length. The values of k_x and k_y are obtained from

$$k = 1 - \mu.\frac{P}{P_{cd}} < = 1.5, \text{ where}$$

$$\mu = \bar{\lambda}\,(2\,\beta_m - 4) + \left(\frac{Z_p - Z}{Z}\right) < = 0.9$$

for plastic and compact sections

$\beta_m = 1.8 - 0.7\ (M_{min} / M_{max})$

= 1.3 (Plane lateral uniformly) distributed load over the member

= 1.4 (Plane lateral concentrated load over the member)

Z_p = Plastic modulus of the section

Z = Elastic modulus of the section.

More exact approach: Non-linear P–Δ analysis may be adopted for more exact approach.

When the bending occurs about both the axes of the member, then the interaction formula may be written as below :

$$\left(\frac{\sigma_{ac.cal}}{\sigma_{ac}} + \frac{\sigma_{bcx.cal}}{\sigma_{bcx}} + \frac{\sigma_{bcy.cal}}{\sigma_{bcy}}\right) < 1.0 \qquad \text{(iv)}$$

The value of σ_{cbx} and σ_{cby} to be used in the above formulae shall be lesser of the values of the maximum permissible stresses, σ_{bc} for bending about the appropriate axis.

At a support and using the values σ_{bcx} and σ_{bcy} at the support.

$$\left(\frac{\sigma_{ac.cal}}{0.60\ f_y} + \frac{\sigma_{bcx.cal}}{\sigma_{bcx}} + \frac{\sigma_{bcy.cal}}{\sigma_{bcy}}\right) < 1.0$$

IS and BS codes give the equivalent stress formula for combined bending, bearing and shear stress as under. It is based on von Mises-Hencky's theory.

$$\sigma_{e.cal} = \left(\sigma_{bt}^2 \text{ cal} + 3\,\tau_{vm\,.cal}^2\right)^{1/2} \quad \text{or} \quad \sigma_{e.cal} = \left(\sigma_{bc}^2 \text{ cal} + 3\,\tau_{vm.\,cal}^2\right)^{1/2}$$

Where a bearing stress is combined with tensile stress or compressive stress, bending and shear stresses under the most unfavourable condition of loading the equivalent stress $\sigma_{e.cal}$ obtained from the following formulae, shall not exceed. ($\sigma_e = 0.9\ \sigma_y$)

$$\sigma_{c.cal} = \left[\sigma_{bt}^{\ 2}{}_{.cal} + \sigma_{p\ .cal}^2 + \sigma_{bt.cal}\ \sigma_{p.cal} + 3\tau_{vm.cal}^2\right]^{1/2}$$

$$\text{or} \qquad \sigma_{c.cal} = \left[\sigma_{bc.cal}^2 + \sigma_{p.cal}^2 - \sigma_{bc.cal}\ \sigma_{p.cal} + 3\tau_{vm.cal}^2\right]^{1/2}$$

where, $\sigma_{ac.cal}$ = calculated average axial compressive stress

$\sigma_{bc.cal}$ = calculated bending compressive stress in the extreme fibre

$\sigma_{bt.cal}$ = calculated bending tensile stress in the extreme fibre

$\tau_{vm.cal}$ = calculated shear stress

$\sigma_{bc.cal}$; $\sigma_{bc.cal}$; $\tau_{vm.cal}$ and $\sigma_{p.cal}$ are the numerical values of the co-existing bending (compressive or tensile stress) shear and bearing stresses. When the bending occurs about both the axes of the member, $\sigma_{bt.cal}$ and $\sigma_{bc.cal}$ shall be taken as the sum of the two calculated fibre stresses, σ_e is the maximum permissible equivalent stress. ($\sigma_e = 0.9\ \sigma_y$).

The compression flange of the column subjected to bending is laterally unrestrained. As such the allowable bending compressive stress is reduced. The column section is adopted depending upon the effective length of column. Allowable bending compressive stress for compression flange of a section unrestrained, has been discussed in chapter (15), in design of laterally unrestrained beams.

For design of a column, the eccentricity of loading shall be taken as the distance from the assumed point of application of load to the centroid of the column. For the purpose of determining the stress in a stanchion or column section, the beam reactions or similar loads shall be assumed to be applied 100 mm from the face of the section or at the centre of the bearing whichever dimension gives the greater eccentricity except in the following two cases.

(i) In the case of cap connections, the load shall be assumed to be applied at the face of the column shaft or stanchion section; or edge of packing if used, towards the span of the beam; and

(ii) In the case of a roof truss beating on a cap, no eccentricity need be taken for single bearing without connections capable of developing an appreciable moment.

In continuous columns, the bending moment due to eccentricities of loading on the columns at any floor may be taken as:

(i) ineffective at the floor levels above and below that floor and

(ii) divided equally between the column's length above and below that floor level, provided that the moment of inertia of either column section, divided by its effective length does not exceed 1.5 times the corresponding value of the other column. In case where this ratio is exceeded, the bending moment shall be divided in proportion to the moment of inertia of the column sections divided by their respective effective length.

The load was assumed to be applied as given in Table 6.1 as per IS : 800-1962. The connections are designed as simple connections to transmit shear only.

TABLE 6.1. ASSUMED ECCENTRICITY OF LOADS IN COLUMN

Types of connections	Assumed point of application of load
(i) Stiffened bracket	Mid point of stiffened seating
(ii) Unstiffened bracket	Other face of vertical leg of bracket
(iii) Cleats to the web	Face pf strut

The type of connections have been shown in Figs. 6.2 (A), (B) and (C) respectively.

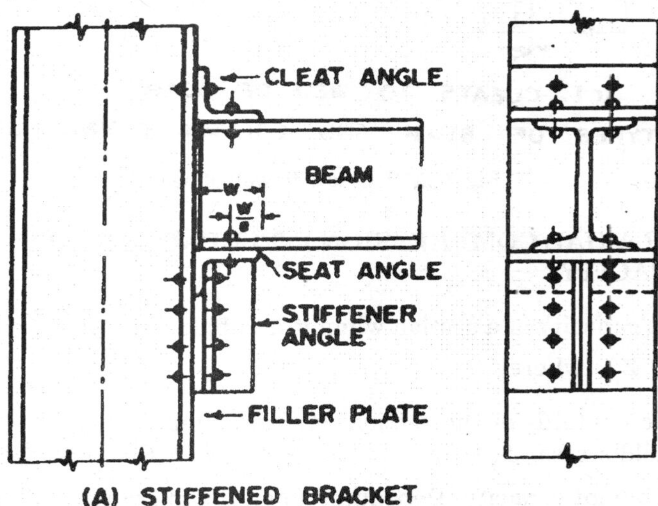

Fig. 6.2(A). Stiffened Bracket

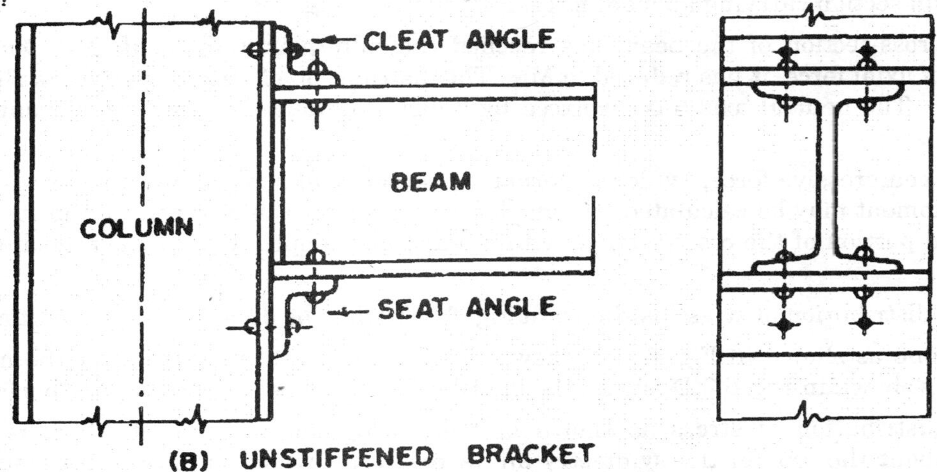

Fig. 6.2(B). Unstiffened bracket

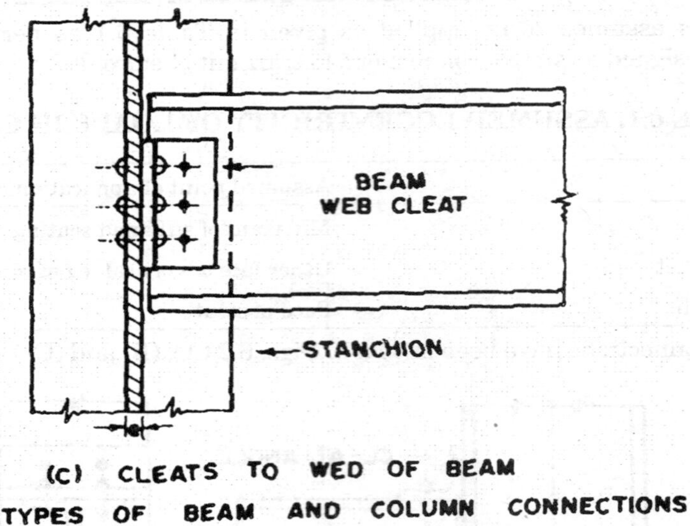

Fig. 6.2(C). Cleats to web of beam

6.3 PRESENCE OF AXIAL COMPRESSION INFLUENCES PLASTIC MOMENT CAPACITY

Plastic moment capacity of a section when it is subjected to pure bending is M_p, where

$M_p = \sigma_y . Z_p$, where,

Z_p is the plastic modulus of the section. Its value depends on the shape and size of the cross-section and yield stress, σ_y of the material. However, the presence of axial compression influences plastic moment capacity. Depending upon the magnitude of the axial compression, the plastic moment capacity is reduced.

A beam section carrying an axial force, P is shown in Fig. 6.3.

The cross-section of the beam has reached the fully plastic moment, M_p. Due to the presence of axial force, it has reduced to M_{PC}. The distribution of stress is shown in shown in Fig. 6.3(B). The neutral axis has displaced by a distance (depth) e from the elevant axis of symmetry.

Total compressive force, (which is normal or perpendicular to the member cross-section) and the moment may be calculated by considering seperately the influence of the stresses on the central portion of the cross-section, within a distance (depth) e from the relevant axis of symmetry.

Total distribution of stress as shown in Fig. 6.3 (B) may be divided into two portions.

Part one as shown in Fig. 6.3(C) shows the distribution of the stress for the moment. Part two as shown in Fig. 6.3(D) shows the distribution of stress for the compressive force, P.

The distribution of stress as shown in Fig. 6.3(B) has a resultant which is a pure moment about the axis of the symmetry and a resultant, which is pure compressive force through the centroid of the cross-section.

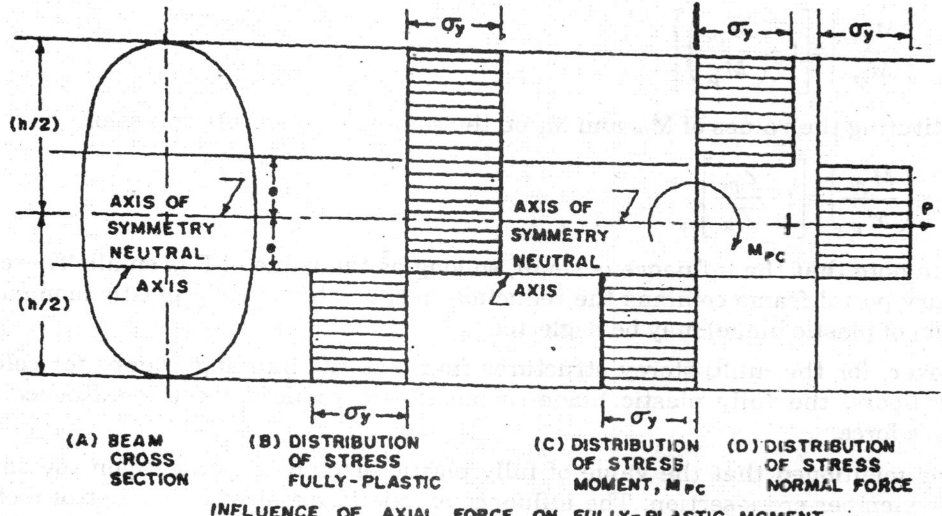

Fig. 6.3

The central portion of the cross-section of depth, 2e may be considered as carrying the compressive force, P. The remaining portion develops the reduced moment, M_{PC}. This reduction corresponds to fully plastic moment of the central portion.

Let the area of above central portion be denoted by A_e and the fully plastic moment of this portion of depth, 2e be M_{pe}. Then,

$$P = (\sigma_y . A_e) \quad \text{(i)}$$

$$M_{pc} = (M_p - M_{p.e}) \quad \text{(ii)}$$

The value of compressive force at which the cross-section would become fully plastic in the absence of any bending moment may be conveniently defined by P_y. In case, the complete cross-sectional area is A, then

$$P_y = (\sigma_y . A) \quad \text{(iii)}$$

From the expressions (i) and (iii)

$$\left(\frac{P}{P_y}\right) = \left(\frac{A_e}{A}\right) \quad \text{(iv)}$$

Fully plastic moment of the cross-section in the absence of compressive force is

$$M_p = (\sigma_y . Z_p) \quad \text{(v)}$$

where, Z_p is the plastic modulus of the cross-section of beam. Let M_{pe} be equal $(\sigma_y . Z_{pe})$. Then, using the expression (ii)

$$M_{pc} = (M_p - M_{pe}), \quad M_{pc} = M_{pe} \left(\frac{M_p}{M_{pe}} - 1\right)$$

$$\left[\frac{M_{pc}}{M_p}\right]=\left[1-\left(\frac{M_{pe}}{M_p}\right)\right] \quad \text{(vi)}$$

Substituting the values of M_{pe} and M_p on the right side of above expression

$$\left(\frac{M_{pc}}{M_p}\right)=\left[1-\frac{Z_{pe}}{Z_p}\right] \quad \text{(vii)}$$

It is to note that the influence is small, so long as the value of P is small. Therefore, for the ordinary portal frame columns the reduction in the value of fully plastic moment (i.e, at the position of plastic hinge) may be neglected.

However, for the multi-storey structures (multi-storey building frames for columns in the lower floors, the fully plastic hinge moments are reduced for the influence of axial compressive force.

It was mentioned that the value of fully plastic moment depends upon the shape and size of the member cross-section. The influence of axial force for the member of rectangular shape and I-shape have been distinctly described as under.

1. Rectangular shape

A beam of rectangular shape, (breadth b and depth d) as shown in Fig. 6.4, bends about an axis parallel to the breadth of the cross-section.

From the expression (iv), and substituting the respective values

$$\left[\frac{P}{P_y}\right]=\left[\frac{A_e}{A}\right]=\left[\frac{b.2e}{b.h}\right]=\left[\frac{2e}{h}\right] \quad \text{(viii)}$$

From the expression (vii)

$$Z_{pe}=\frac{1}{4}b.(2e)^2=\frac{1}{4}\left(b.e^2\right)$$

$$Z_p=\frac{1}{4}\left(b.h^2\right)=\frac{1}{4}\left(b.h^2\right)$$

$$\left[\frac{M_{pc}}{M_p}\right]=\left[1-\frac{Z_{pe}}{Z_p}\right]$$

$$=\left[1-\frac{b.e^2}{\frac{1}{4}b.h^2}\right]=\left[1-\frac{4e^2}{h^2}\right] \quad \text{(ix)}$$

From the expression (viii)

$$\left[\frac{P}{P_y}\right]=\left[\frac{2e}{h}\right], i.e., \left[\frac{4e^2}{h^2}\right]=\left[\frac{P}{P_y}\right]^2 \quad \text{(x)}$$

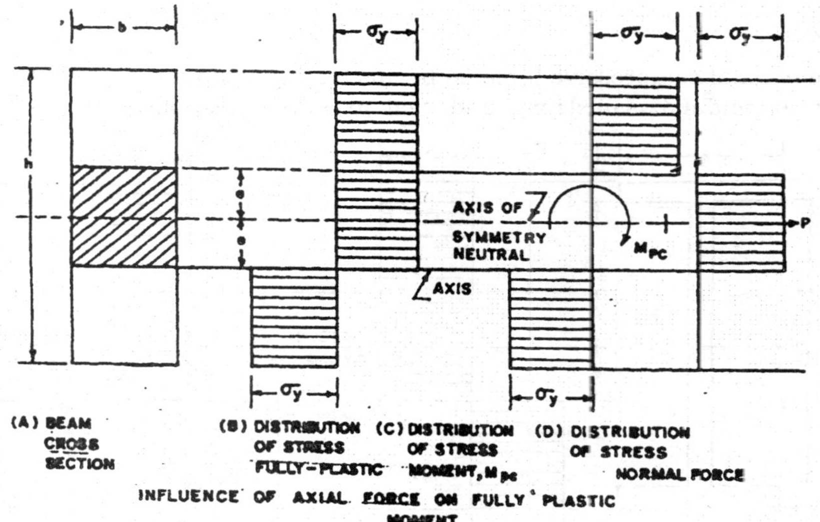

Fig. 6.4.

Substituting this in the expression (ix)

$$\left[\frac{M_{pc}}{M_p}\right]=\left[1-\left(\frac{p}{P_y}\right)^2\right] \tag{xi}$$

or

$$\left[\frac{M_{pc}}{M_p}\right]+\left[\frac{P}{P_y}\right]^2=1 \tag{xii}$$

In cartesian co-ordinates, (P/P_y) may be plotted as ordinate and (M_{pc}/M_p) may be plotted as abscissa, the expression (xii) represents a parabola with horizontal axis as shown in Fig. 6.5.

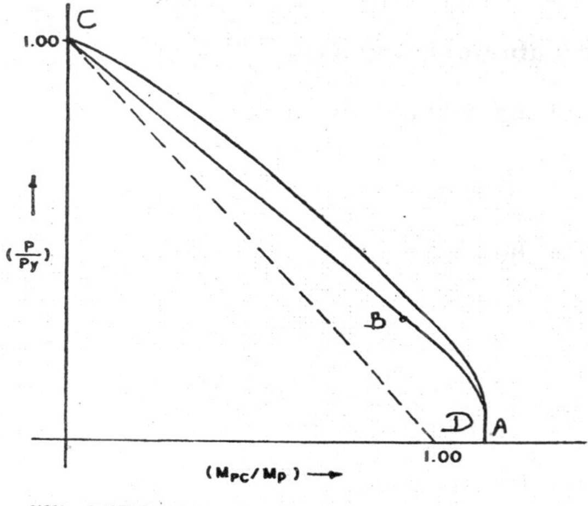

Fig. 6.5.

2. I-shaped

Influence of axial compressive force an beams having I-shapes subjected to bending with respect to the centroidal axis of strong and weak axes have been described as under.

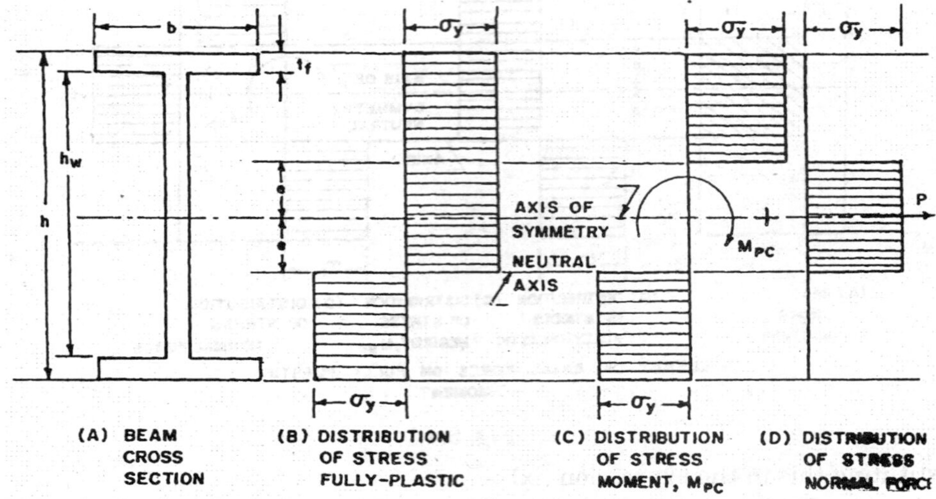

Fig. 6.6.

An I-shape beam having breadth of flange, b, and the overall depth, h is shown in Fig. 6.6.

Following two cases of I-shape beams subjected to bending with respect to its centroidal axis with

(i) Neutral axis in the web

(ii) Neutral axis in the flange have been considered

(i) I-shape beam bending about strong axis

Let the neutral axis of the beam be in the Web. Then, the axial compressive force, P is given by

$$P = (\sigma_y . 2et_w), \quad (P < \sigma_y . A_w) \tag{i}$$

$$P_y = (\sigma_y . A_w), \quad \text{then} \tag{ii}$$

$$\left(\frac{P}{P_y}\right) = \left[\frac{2t_w . e}{A}\right] \tag{iii}$$

$$\left(\frac{P}{P_y}\right) < \left(\frac{A_w}{A}\right) \tag{iv}$$

where, A is the total cross-sectional area of I-section, and A_w is the area of web ($A_w = t_w . h_w$)

The relation as in the expression holds true, so long as (t_w.2e) does not exceed the area of web. The value of Z_{pe} is that appropriate to a rectangular section of breadth, t_w and depth, 2e.

$$Z_{pe} = \frac{1}{4}\left(t_w.(2e)^2\right) = \left(t_w.e^2\right) \tag{v}$$

An I-shape beam may be considered as if it consists of three rectangles without any loss of accuracy). The elastic section modulus, Z may be expressed as below

$$Z = \frac{1}{4}\left[\frac{1}{3}b.t_f^3 + b_f.t_f\left(h-t_f\right)^2 + \frac{1}{6}t_w\left(h-2t_f\right)^2\right] \tag{vi}$$

And the plastic modulus of the section, Z_p may be expressed as under

$$Z_p = \left[\left(b.t_f\right)\left(h-t_f\right) + \frac{1}{4}t_w\left(h-2t_f\right)^2\right] \tag{vii}$$

where, t_f is the thickness of the flange and A_f is the area of both the flanges, A_f = (2.b.t_f)

h_w = depth of web = (h-2t_f), then

A_w = area of web = (t_w.h_w)

$$Z_p = \left[\left(b-t_f\right)\left(h-t_f\right) + \frac{1}{4}t_w h_w^2\right] \tag{viii}$$

$$Z_p = \left[\frac{1}{2}A_f.\left(h-t_f\right) + \frac{1}{4}h_w\ t_w.h_w\right] \tag{ix}$$

$$Z_p = \frac{1}{2}A_f\left(h-t_f\right) + \frac{1}{4}A_w\left(h-2t_f\right) \tag{x}$$

And

$$M_{pc} = \sigma_y.\left(Z_p - t_w e^2\right) \tag{xi}$$

$$M_{pc} = \sigma_y\left(Z_p\right) - \sigma_y\left(t_w.e^2\right) \tag{xii}$$

Since, P = ($\sigma_y.2e.t_w$)

$$P^2 = \sigma_y^2.4e^2.t_w^2$$

$$\left[\frac{P^2}{\sigma_y.4t_w}\right] = \sigma_y.t_w.e^2 \tag{xiii)A}$$

Substituting this in the expression (xii)

For $\quad P < (\sigma_y.t_w.h_w)$

$$M_{pc} = \left[M_p - \frac{P^2}{\sigma_y.4t_w}\right] \tag{xiii) B}$$

In the non-dimensional form

$$\left[\frac{M_{pc}}{M_p}\right]=\left[1-\frac{P^2}{\sigma_y.4t_w\,\sigma_y Z_p}\right]$$ and

$$P_y = \sigma_y.A,\ \left(\frac{A}{P_y}\right)^2=(1/\sigma_y)^2$$

$$\left[\frac{M_{pc}}{M_p}\right]=\left[1-\frac{A^2}{4t_w\,Z_p}\left(\frac{P}{P_y}\right)^2\right] \quad \text{(xiii)C}$$

2(ii) When the neutral axis lies in one of the flange instead of web

$$P = \sigma_y\,[A-b(h-2e)] \quad \text{(i)}$$

$$\left[\frac{P}{\sigma_y}\right]=[A-b(h-2e)] \quad \text{(i)A}$$

$$b.\,(h-2e) = (A-P/\sigma_y) \quad \text{(i)B}$$

$$(h-2e) = \frac{1}{b}\left(A-\frac{P}{\sigma_y}\right) \quad \text{(i)C}$$

$$2e = \left[-\frac{1}{b}\left(A-\frac{P}{\sigma_y}\right)+h\right] \quad \text{(ii)A}$$

$$M_{pc} = \sigma_y.b\left[\frac{h^2}{4}-e^2\right]$$

For $(A_w/A) < (P/P_y) < 1.00$

$$M_{pc} = \frac{\sigma_y.b}{4}(h^2-4e^2)$$

$$M_{pc} = \frac{\sigma_y}{4}[b(h-2e)\,(h+2e)]$$

$$M_{pc} = \frac{\sigma_y}{4}[b(h-2e)\,(h+2e)]$$

$$M_{pc} = \frac{\sigma_y}{4}\left[\left(A-\frac{P}{\sigma_y}\right)h+\left(A-\frac{P}{\sigma_y}\right)2e\right]$$

$$M_{pc} = \frac{\sigma_y}{4}\left[\left(A-\frac{P}{\sigma_y}\right)h+\left(A-\frac{P}{\sigma_y}\right)\left(-\frac{1}{b}\left(A-\frac{P}{\sigma_y}\right).h\right)\right]$$

$$M_{pc} = \frac{\sigma_y}{4}\left[2\left(A-\frac{P}{\sigma_y}\right)h-\frac{1}{b}\left(A-\frac{P}{\sigma_y}\right)^2\right]$$

$$M_{pc} = \frac{\sigma_y}{2}\left[\left(A-\frac{P}{\sigma_y}\right)h-\frac{1}{2b}\left(A-\frac{P}{\sigma_y}\right)^2\right] \quad \text{ii(B)}$$

It may also be expressed as under :

$$M_{pc} = \frac{\sigma_y}{2}\left[A\left(1-\frac{P}{\sigma_y . A}\right)h-\frac{A^2}{2b}\left(1-\frac{P}{\sigma_y A}\right)^2\right]$$

$$M_{pc} = \frac{\sigma_y A}{2}\left[\left(1-\frac{P}{\sigma_y}\right)\left\{h-\frac{A}{2b}\left(1-\frac{P}{P_y}\right)\right\}\right]$$

$$\left[\frac{M_{pc}}{M_p}\right]=\frac{\sigma_y A}{2\sigma_y Z_p}\left[\left(1-\frac{P}{P_y}\right)\left(h-\frac{A}{2b}\right)\left(1-\frac{P}{P_y}\right)\right]$$

$$\left[\frac{M_{pc}}{M_p}\right]=A\left(1-\frac{P}{P_y}\right)\left(h-\frac{A}{2b}\right)\left(1-\frac{P}{P_y}\right) 2\,Z_p \quad \text{(iii) B}$$

In Fig. 6.5, the expression (xiii) c is represented by AB and the expressions (xiii) A and (xiii) B are represented by line BC.

The values of (M_{pc}/M_p) and (P/P_y) are such that the maximum stress intensity in the elastic stressed sections equals the yield limit, which satisfies the following relation

$$\left[\frac{P}{P_y}\right]+\left[\frac{M_{pc}}{M_p}\right]=1.00 \quad \text{(iv)}$$

This expression (iv) is represented by the dashed line CD, Fig. 6.5 from the expressions (i) and (iii) B

$(h_w = h - 2t_f)$

For $\left|(\sigma_y . t_w . h_w) < P < P_y\right|$

$$\left[\frac{M_{pc}}{M_p}\right]=\frac{\sigma_y}{2}\left[h\left(A-\frac{P}{\sigma_y}\right)-\frac{1}{2b}\left(A-\frac{P}{\sigma_y}\right)^2\right]$$

$$\left[\frac{M_{pc}}{M_p}\right]=\frac{\sigma_y}{2}\left[hA\left(1-\frac{P}{\sigma_y . A}\right)-\frac{A^2}{2b}\left(1-\frac{P}{\sigma_y A}\right)^2\right]$$

$$M_{pc} = (\sigma_y . A/2)\left[h\left(1-\frac{p}{P_y}\right)-\frac{A}{2b}\left(1-\frac{P}{P_y}\right)^2\right]$$

In non-dimensional form

$$\left[\frac{M_{pc}}{M_p}\right]=\left[\frac{\sigma_y A/2}{\sigma_y . Z_p}\right]\left[h\left(1-\frac{P}{P_y}\right)-\frac{A}{2b}\left(1-\frac{P}{P_y}\right)^2\right]$$

$$\left[\frac{M_{pc}}{M_p}\right]=\left[\frac{A}{2Z_p}\right]\left[h\left(1-\frac{P}{P_y}\right)-\frac{A}{2b}\left(1-\frac{P}{P_y}\right)^2\right] \quad \text{(v)}$$

From the expressions (iii) C and (v) an interaction as shown in Fig. 6.7 had been plotted. When the value of axial compressive force, P is zero, $M_{pc} = M_p$. When this axial force attained the value ($P = P_y = \sigma_y A$), then the moment capacity, M_p becomes zero.

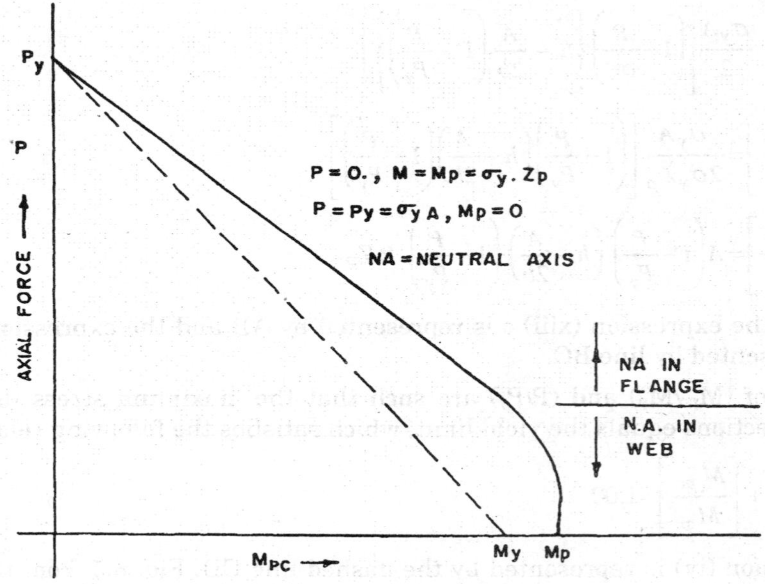

Fig. 6.7.

6.4 BEHAVIOUR AND STRENGTH OF SHORT BEAM-COLUMNS

Short beam-columns fail in case strength of material has reached. Then, the failure is referred as **material failure.**

A short column (i.e, stub column, which may be called as non-slender) cross-section may be plastic, compact or semi-compact. Such a column may fail by yielding (i.e, due to large deformation) at a squash load.

$P_{cs} = (A_{gross}\ x\ \sigma_y)$, where,

σ_y is the yield strength and A_g is the gross area of the cross-section.

Slender cross-section of such a column fail by local buckling before attending the yield stress. The effective cross-sectional area reduces and becomes less than the gross-area. Such column fails at a load below P_{cs}.

A short beam-column have a cross-section of plastic or compact and subjected to only bending member fails, when the bending moment reaches plastic-moment capacity, M_p,where

$$M_p = Z_p . \sigma_y, \quad \text{where}$$

Z_p is the plastic modulus of the cross-section.

Failure of a semi-compact section carrying bending moment alone fails by buckling of the plate element before reaching the plastising of the complete cross-section. However, after the (stress at the extreme fibre reaches the yield stress).

Therefore, only plastic, compact and semi-compact cross-sections (types) are used for beam-columns and the slender cross-sections are not used.

As it is seen from Fig. 6.3 and Fig. 6.4, at smaller values of axial compression, a smaller area of the cross-section closer to the neutral axis-corresponds to axial compression. This area equilibrates the axial compression. This area (closer to the neutral axis) contributes very little to the plastic moment capacity, M_p, the reduction in the moment capacity, M_p, is negligible when the axial compression is small.

A typical failure envelop diagram of a stub-column having I-section and subjected to axial compression P and bending moment, M is shown in Fig. 6.7, in a non-dimensional form.

6.5 BEHAVIOUR AND STRENGTH OF LONG COLUMNS

In practice, the columns are usually long and slender. Such columns may fail by instability (which may be either due to local buckling, lateral buckling or column buckling) when the columns carry axial compression, then, these may fail by column buckling rather than yielding. In case, slender I-sections when these are not laterally supported, may fail by lateral-torsional buckling. However, when I-sections bend about weak-axis (y-axis) the plastic and compact sections may reach the plastic moment capacity, M_{py} without occurring of premature and lateral buckling or local buckling.

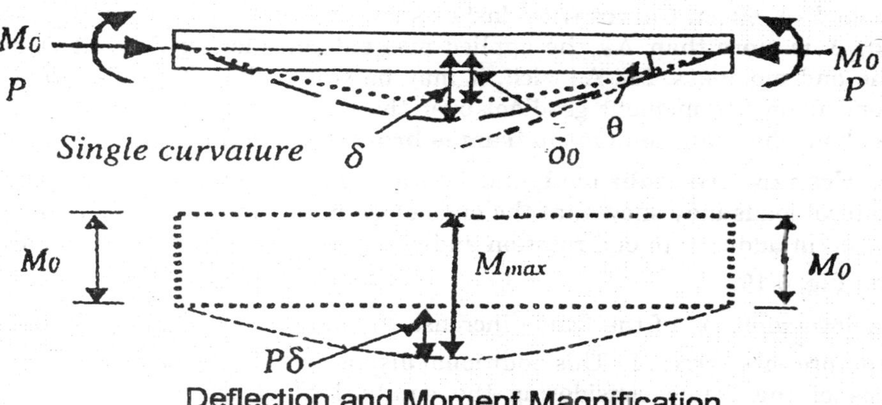

Fig. 6.8.

Such long slender members carrying combined axial compression and bending may fail in different modes of failure of instability or material failure. Following cases, have been discussed. When the beam-columns carry equal and opposite end moment, M_o, then, these may bend into a **single curvature** as shown in Fig. 6.8 with maximum deflection, Δ_o.

When the axial compression is now applied on the deformed shape. It results in additional deflection due to additional bending moment. Let the final deflection be Δ, when, the equilibrium is reached. This is called as **P-Δ effect.** The deflected shape and total bending moment are shown by dotted curves. It is observed that due to P-Δ effect, the maximum bending moment, M_{max} is more than that due to externally applied moment, M_o.

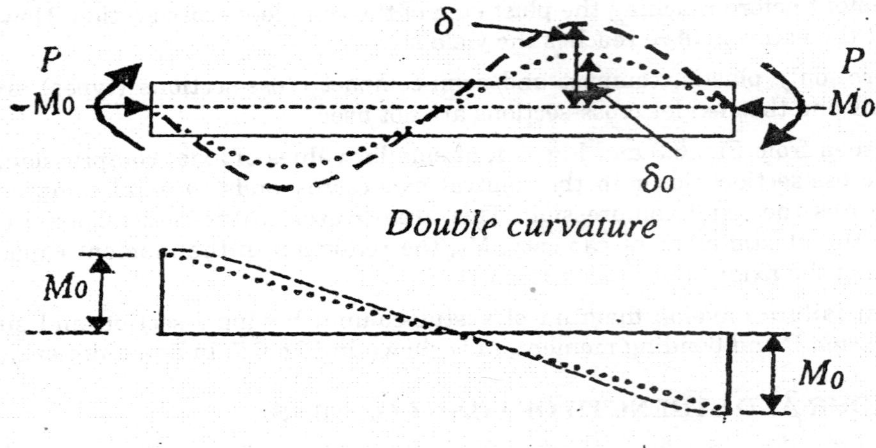

Deflection and Moment Magnification

Fig. 6.9.

However, in case, the beam-column carry equal end moments which act in the same direction, then, the member bends in **doubler curvature** bending in addition to axial compression as shown in Fig. 6.9.

Dotted curves show the deflected shape and the bending moment diagram for the member without P-Δ-effect. Curves with dashes show considering P-Δ-effect. It is observed that although, Δ is more than Δ_o, the applied moment considering P-Δ effect may not be more than the end moments, M_o. As such, it may be stated that P-Δ effect and amplified moment depend upon the moment gradient over the complete length of the member. This description is about the beam-columns in **frames braced against sway.**

In case, Frames are unbraced, the lateral sway is not prevented. Linear elastic analysis for lateral loads may show that the ends of column translate relative to one another by a distance, δ_o in addition to end rotation and reverse curvature deformation as shown by dotted curve in Fig. 6.10.

Sway displacement, δ_o of the frame increase the sway to δ as shown by dashed curve due to axial compressive forces, P. This additional displacement may be determined by a non-linear analysis of the frame considering the equilibrium of the frame in the deformed geometry. The beam moment also increase. Increase in sway deformation and bending moments is due to P-δ effect.

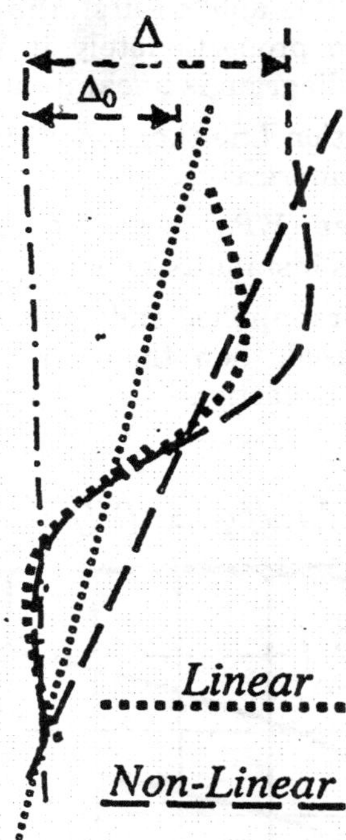

Fig. 6.10.

The load-deformation behaviour of a beam-column may be observed when the axial compression and end moments are applied one after the another.

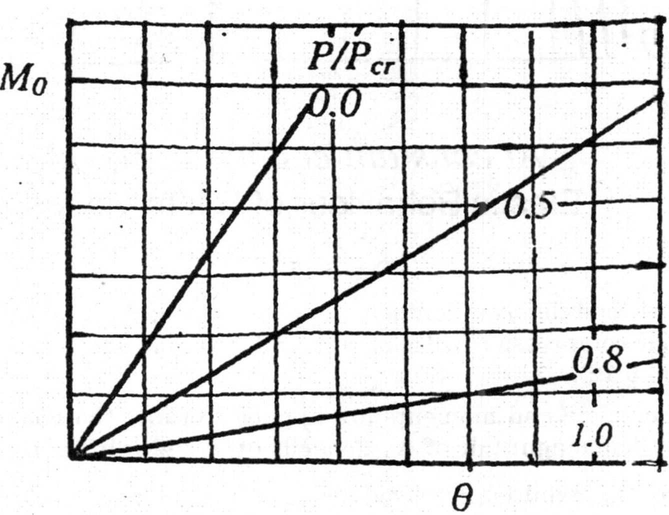

Fig. 6.11.

In case, an axial compression, P is applied first, then (P/P_{cr}) is maintained constant and the moments at the two ends are proportionately increased in the elastic range of the material, the moment end rotation diagram is linear as shown in Fig. 6.11.

The slope of the moment rotation line (i.e, the stiffness of beam-column) reduces with increase in the first applied axial compression.

Let the end rotation be θ when (P/P_{cr}) = 1.0, Fig. 6. This end rotation, θ increases to infinity even for M = 0, which shows instability under axial compression.

However, in case the ratio of first applied end moments, (M_o/M_P) is maintained constant, and the axial compression is increased, then, the compressive force and lateral deformation, behaviour of the beam column in the elastic range is plotted, then, it is seen as shown in Fig. 6.12.

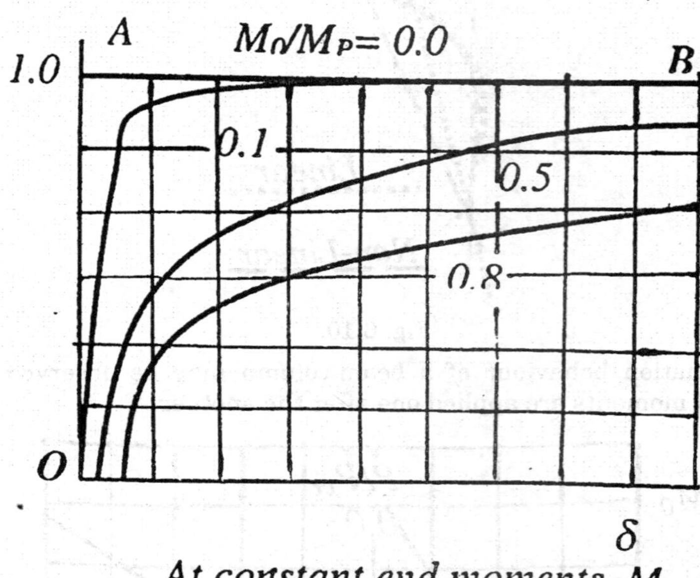

Fig. 6.12.

It is seen that the load-deformation behaviour is non-linear. It is observed that a perfect column carrying axial compression without any end rotation undergoes bifurcation type of buckling (OAB in Fig. 6.12).

However, in case, some end moments are applied initially the beam column undergoes initial deflection, Δ_o. The magnitude of Δ_o depends upon the value of end moments.

Subsequently, as the axial is increased gradually, the lateral deflection increases even from the very beginning. The rate of increase is slow, but reaching to critical load, it increases rapidly before failure. Commonly the failure commences by yielding under combined effect of axial compression and maximum moment. It is observed that the end moments change the behaviour of an axially loaded column in a manner similar to the initial bow like imper-fection.

These additional bending moments in beam-columns due to P–Δ and P –δ effects are not shown by the conventional first order linear elastic analysis since, the equilibrium equations in these analyses methods are derived for the un-changed geometry of the structure.

In the linear elastic analysis of a frame, the axial force and the bending moments in a beam column increase linearly, with the increase in the load frame as observed by the straight line OA, Fig. 6.12.

In case, the equlibrium of the beam-column in the changed geometry is seen, the bending moments are amplified (increases) by P-δ and P-Δ effects in a non-linear manner, as the load increases.

Fig. 6.12 shows non-linear variation of the maximum bending moment, M_{max} due to increase in the value of axial compression, P acting at a constant end eccentricity (M_o/P) of a long column.

At the intersection of non-linear P and M curve with the failure envelope at B, the long beam-column fails. It is observed that if the moment amplification of the long beam-column due to P-Δ and P-δ effects were notconsidere, the compressive strength ofthe long beam column would be as P_o, where as the actual compressive strength is decreased to P_{ce}, due to the moment amplification effect.

When the actual amplified moment is M_{max}, the linear analysis shows the moment, M_o. The values of enhanced deflection and bending moment due to the axial compression acting on the chaged geometry are as under.

$$\delta = \left(\frac{\delta_o}{(1 - P / P_E)} \right)$$

$$\Delta = \left(\frac{\Delta_o}{(1 - P / P_E)} \right)$$

$$M_{max} = \left(\frac{C_m M_o}{(1 - P / P_E)} \right)$$

Where,

C_m = a coefficient that accounts for the moment gradient effects

P_E = Euler's buckling load of the column in the plane bending

It is observed that as the value of applied axial compression reaches the value of Euler's buckling load, both, the deflection Δ and amplified moment, M_{max} increases rapidly and tends to approach infinity. As P reaches P_E, the failure will occur soon.

Above expressions have been derived while dealing with beam-columns subjected to lateral loads in the subsequent sections.

6.6 DIFFERENTIAL EQUATION FOR AXIAL COMPRESSION AND BENDING

Consider a **general case** of a column pin ended at both the ends carrying transverse load w(z) in combination with any end moments M_1 and M_2, which constitute the primary bending moment M_o and it is a function of Z, as shown in Fig. 6.3. It is assumed that the failure occurs by instability in the plane of bending.

There is a deflection, y. A secondary moment P.y arises due to axial compression. The moment at any section at a distance, Z may be expressed as under

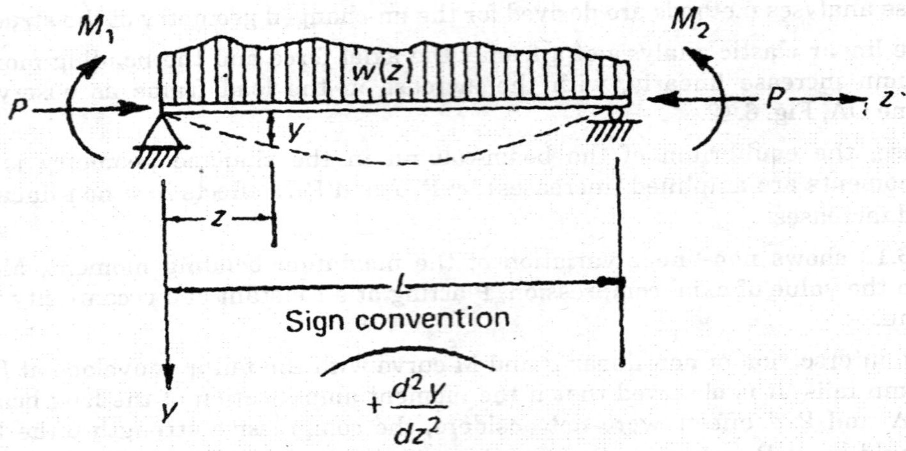

Fig. 6.13.

$$M_Z = (M_i + P.y) = -EI\left(\frac{d^2y}{dz^2}\right) \quad \text{(i)}$$

In case, the complete member has sections with EI, constant throughout the length, L, then, divide the expression (i) by EI

$$\frac{d^2y}{dz^2}+\frac{P}{EI}y=\left(\frac{-M_i}{EI}\right) \quad \text{(ii)}$$

Differenting the expression (ii) twice

$$\frac{d^4y}{dz^4}+\frac{P}{EI}.\left(\frac{d^2y}{dz^2}\right)=\frac{-1}{EI}\frac{d^2Mi}{dz^2} \quad \text{(iii)}$$

From the expression (i)

$$\left(\frac{d^2y}{dz^2}\right)=-\left(\frac{Mz}{EI}\right) \text{ and } \left(\frac{d^4y}{dz^4}\right)=\frac{-1}{EI}\frac{d^2Mz}{dz^2} \quad \text{(iv)}$$

Substituting these in the expression (iii)

$$-\frac{1}{EI}\left(\frac{d^2Mz}{dz^2}\right)+\frac{P}{EI}\left(\frac{-Mz}{EI}\right)=\frac{-1}{EI}\frac{d^2Mi}{dz^2} \quad \text{(v)}$$

Let P/EI be equal to n^2, then

$$\frac{d^2Mz}{dz^2}+n^2Mz=\left(\frac{d^2Mi}{dz^2}\right) \quad \text{(vi)}$$

The differential expression (vi) is of the some form as the differential equation for deflection (i)

Homogeneius solution for the expression (vi) is

$$M_Z = A \sin (nz) + B \cos (nz) \quad \text{(vii)}$$

A particular solution is added, so that it satisfies the right hand side of the differential equation. Since $M_i = f(z)$, where, f(z) is usually a polynomial in Z. The particular solution will be of the same form. Therefore, the complete solution will be

$$M_Z = A \sin (nz) + B \cos (nz) + f(z) \quad \text{(viii)}$$

where, f_1 (z) is a value of M_z which satisfies the expression (vi)

When M_Z is a continuous function, the maximum value of M_Z may be obtained by differentiation

$$\left(\frac{dMz}{dz}\right) = 0 = An\cos(nz) - B.n.\sin(nz) + \frac{df_1(z)}{dz} \quad \text{(ix)}$$

For most ordinary loading cases (e.g, concentrated loads, uniform loads, end moments, or the combinations, there of..), it may be shown that

$$\left(\frac{df_1(Z)}{dz}\right) = 0 \quad \text{(x)}$$

In which case, a general expression for maximum Mz may be established. From the expression (ix)

$$A.n.\cos(nz) = B.n.\sin(nz)$$

$$\tan(nz) = (A/B)$$

In a right-angled triangle, perpendicular (opposite to angle nz) shall be A and the base shall be B. The hypotenue of the triangle shall be $(A^2 + B^2)^{1/2}$. At maximum, Mz

$$\sin(nz) = A/(A^2+B^2)^{1/2}$$

$$\cos(nz) = B/(A^2+B^2)^{1/2}$$

Substituting these values in the expression (viii)

$$M_{Z.max} = \frac{A^2}{\left(A^2+B^2\right)^{1/2}} + \frac{B^2}{\left(A^2+B^2\right)^{1/2}} + f_{1(z)}$$

$$M_{Z.max} = \left(A^2+B^2\right)^{1/2} + f_{i(z)} \quad (6.2)$$

It may be noted that when $df_{1(z)}/dz$ is not equal to zero, then the expression (ix) may be solved for (nz) and result maybe substituted into the expression (viii).

6.7 BEAM-COLUMN SUBJECTED TO UNEQUAL END MOMENTS WITHOUT TRANSVERSE LOADING

An initially straight pin-ended column subjected to axial load P and end moments M_1 and M_2, with M_1 less than or equal to M_2 is shown in Fig. 6.4. Since, the member is pinned at both the ends, there is no translation at the joints. The moments applied to the member are usually of the opposite sense.

Fig. 6.14

The basic differential equation, which describes the behaviour of the beam column has been developed in Art. 6.3. Its solution form, is simplified in the applied end moment case by the fact that f(z) is zero because $(d^2Mi/dz^2) = 0$. According to the homogeneous solution to the differential equation.

$$Mz = A \sin(nz) + B.\cos(nz) \tag{i}$$

The maximum value of M_z may be obtained by differentiating Mz with respect to Z and setting this equal to zero.

$$0 = A.n.\cos(nz) - B.n.\sin(nz)$$

and $\tan(nz) = (A/B) = \text{Perpendi/Base}$

$$\text{Hypotenuse} = (A^2+B^2)^{1/2}$$

$$\sin(nz) = A/(A^2+B^2)^{1/2}$$

$$\cos(nz) = B/(A^2+B^2)^{1/2}$$

Therefore, $M_{z.max} = (A^2+B^2)^{1/2}$

Using the applied loads and the boundary conditions, coefficients

At $z = 0$, $M_z = M_1$, $\therefore B = M_1$ and $Z = 1$, $M_z = M_2$

$$\therefore M_2 = A \sin(n.L) + M_1 \cos(nL)$$

$$A = \left(\frac{M_2 - M_1\cos(nL)}{\sin(nL)}\right)$$

so that $Mz = \left(\frac{M_2 - M_1.\cos(nL)}{\sin(nL)}\right)\sin(nz) + M_1.\cos(nz)$

$$M_{z.max} = \left[\left(\frac{M_2 - M_1.\cos(nL)}{\sin(nL)}\right)^2 + M_1^2\right]^{1/2}$$

$$M_{z.max} = M_2\left[\frac{1 - 2(M_1/M_2)\cos(nL) + (M_1/M_2)^2}{\sin^2(nL)}\right]^{1/2} \tag{6.3}$$

In case, M_2 is the (absolute) larger of the applied end moments $M_{z.max}$ may be recast in a form that contains the maximum applied (primary) moments, M_2 and an amplification factor (i.e, magnification factor).

$M_{z.max}$ = (Primary moment) (Magnification factor)

When the moment along the beam is constant (i.e, uniform)

$M_2 = M_1 = M$ **(Equal moment)**

The uniform moment case reduces to

$$M_{z.max} = M\left(\frac{2(1-\cos(nL))}{\sin^2(nL)}\right)^{1/2} \quad (6.4)$$

M_z = (Uniform moment) (magnification factor)

Since, M is the uniform moment along the beam, the **amplification factor** may be developed

For equal end moments without transverse loading

$$M_{z.max} = M\left(\frac{2(1-\cos(nL))}{\sin^2 nL}\right)^{1/2}$$

$$M_{z.max} = M\left(\frac{2(1-\cos(nL))}{1-\cos^2(nL)}\right)^{1/2}$$

$$M_{z.max} = M\left(\frac{2(1-\cos(nL))}{\{1+\cos(nL)\}\{1-\cos(nL)\}}\right)^{1/2}$$

$$M_{z.max} = M\left(\frac{2}{1+\cos nL}\right)^{1/2}$$

$$M_{z.max} = M\left(\frac{2}{2\cos^2(nL/L)}\right)^{1/2}$$

$$M_{z.max} = M\left(\frac{1}{\cos(nL/2)}\right)$$

$$M_{z.max} = M.\sec\left(\frac{nL}{2}\right) \quad (6.5)A$$

$$M_{z.max} = M.Am \quad (6.5)B$$

Where, Am = Amplification (magnification) factor

$$Am = \text{Sec}\,(nL)^{1/2} = \left(\frac{1}{1-\alpha}\right)\left(\therefore nL = \pi\sqrt{\alpha}\right)$$

In case, **a beam-column is subjected to a constant moment over its entire length, it results in** a constant compression along one entire flange and constitutes the most severes loading on such a member. It would always be conservative to multiply the maximum primary/uniform moment due to any loading by sec $\left(\frac{nL}{2}\right)$ which one is excessively conservative in most cases.

It is seen that Eq. 6.3 describes the maximum experienced moment in terms of largest (absolute) of the applied end moment and **an amplification factor.** There are two terms in Eq. 6.3, an amplification term, which for the uniform moment case is sec (nL/2) and a source of quantifying an equivalent moment.

A further simplification of the amplification term is possible given the fact that sec (nL/2), especially for low values of α, is comparable with [1/(1 – 2)], the values of which have been tabulated in Table 6.2.

Since nL = $\pi(\alpha)^{1/2}$,

TABLE 6.2 : EXPRESSION (A & B)

α	sec (nL/2)	Expr (A)	Expr. (B)
0.1	1.137	1.114	1.114
0.2	1.310	1.257	1.257
0.3	1.533	1.441	1.441
0.4	1.832	1.686	1.685
0.5	2.252	2.030	2.028
0.6	2.884	2.546	2.542
0.7	3.941	3.405	3.399
0.8	6.058	5.125	5.112
0.9	12.419	10.284	10.253

Note : For the differential equation,

Am = **Magnification factor**

$$= \frac{2}{(nL/2)^2}\left(\sec\left(\frac{nL}{2}\right)-1\right) \qquad \text{(A)}$$

where, $\frac{nL}{2} = \frac{1}{2}\sqrt{\frac{P}{EI}} = \frac{\pi}{2}\sqrt{a}$

For the approximation Am = $\left(\frac{Cm}{1-\alpha}\right), \delta_o = \left(\frac{5wL^4}{384}\right), M_o = \left(\frac{W}{8}\right)$

$$\left(\frac{\delta_o}{M_o}\right) = \left(\frac{5L^2}{48EI}\right), C_m = 1 + \left(\frac{\pi^2 EI}{L^2}.\frac{5L^2}{48EI} - 1\right)$$

Cm = $1 + 0.028\alpha, \therefore A_m = (1 + 0.028\alpha)(1-\alpha)$

Inspite of the fact that $[1/(1-\alpha)]$ is notas conservative as sec (nL/2). It is specially attractive form because it is same **magnification** term developed to deal transverse loads.

Now the interior amplified design moment may be restated as a function of the maximum applied end moment; (a term that converts this maximum end moment to an equivalent **uniform moment,** C_m and a magnification term $[1/(1-\alpha)]$

From a codification perspective, when a beam is subjected to end moments, the largest absolute end moment is designated. M_2 and the smaller absolute end moment

M_1 (1 M_2/>1M_1) (6.5) c

$$M_{Z.max} = M_2 . \left(\frac{C_m}{(1-\alpha)}\right)$$

C_m is found from Eqs. (6.6) and (6.7)

$$C_m = \left[\frac{+1-2\left(\frac{M_1}{M_2}\right)\cos(nL)+\left(\frac{M_1}{M_2}\right)^2}{2(1-\cos(nL))}\right]^{1/2} \quad (6.6)A$$

$$C_m = (M_E/M_2) \quad (6.6)B$$

Eq. 6.6 is complex but it does permit for a transformation into a practical form which quantities, C_m with sufficient accuracy for design purpose

$C_m = 0.6 - 0.4\ (M_1/M_2)$

Positive values are related with **double curvature** Codified version of C_m are

$$C_m = [0.6 - 0.4\ (M_1/M_2)] \quad (6.7)$$

more than or equal to 0.4 (ASDE).

This ASDE (Allowable stress design elastic) version is adopted by the LRFD (Load and resistance factor design) and ASD (Allowable stress design. But the identification of 0.4 as a lower bound has been deleted.

$$C_m = [0.6 - 0.4\ (M_1/M_2)] \quad (6.8)$$

A conceptional understanding of codification objectives and procedures (LRFD and ASD)

It becomes easy to apply the design equations which contain (i) magnification and (ii) equivalence terms.

It is to remember that first C_m approximately convert any end moment configuration to an equivalent uniform or constant moment along the length of the member

$[1/(1-\alpha)]$ or $[1/(1 - P_u/P_e)]$

will then magnify this equivalent uniform moment so as to include the axial (secondary) moment on the now uniform moment

It is to be observed that the moment produced through implementation of

$$M_{Zmax} = M_2 \left(\frac{C_m}{(1-\alpha)}\right) \quad (6.9)$$

may be and quite often is less than maximum applied end moment M_2.

The moment used to identify the **strength limit state** for the member must. However, be the larger of the magnified equivalent uniform $[C_m.M_2/(1-\alpha)]$ or the maximum applied end moment, M_2.

The **stability limit state** (to be now developed) for the equivalent uniform moment case and then modified to consider for moment gradient effects (C_b). According to $M_{Z.max} = [M_2./C_m/(1-\alpha)]$ converts M_2 into its appropriate form $[C_m.M_2/(1-\alpha)]$ and one need not be concerned when it is less than an applied end moment as long as the design concern is exclusively with stability.

It is to be seen that C_b is in this case 1.0 for moment gradient effects have been incorporated directly into the expression for maximum moment, (M_2).

ASDE and ASDF `Design Specifications resolve the conflict in maxima (M_2-M_1) by requiring the designer to consider both strength and stability. The strength requirement is satisfied if the imposed axial load and maximum end moment is satisfied.

ASDE - strength limit State

$$\left(\frac{Fa}{0.6F_y}\right)+\left(\frac{f_b}{F_b}\right) \text{ is less than or equal to unity}$$

ASDE stability criterion is complied in case

$$\frac{fa}{Fa}+\frac{f_b}{F_b}\left(\frac{C_m}{1-fa/F_{ex}^1}\right) \text{ is less than or equal to unity.}$$

Where,

f_b is now derived from the maximum applied end moment M_2, and C_m is defined as

$$C_m = \left[0.6-0.4\left(\frac{M_1}{M_2}\right)\right] \tag{6.11}$$

for both limit states, the bending stress that would be allowed on the beam in the absence of axial load (M_{cr}/Z_x). F_{ex} is the Euler's buckling stress about the axis of bending divided by the adopted factor of safety (1.92).

6.8 BEAM COLUMN SUBJECTED TO TRANSVERSE UNIFORM LOADING

From Art. 6.4, it is seen that the primary moment, M_i may be expressed as

$$M_i = \frac{W}{2}z(L-z) \tag{i}$$

$$\text{Since, } \left(\frac{d^2M_i}{dz^2}\right) = -W,\ f_1(z) \tag{ii}$$

$f_{1(z)}$ is not equal to zero. Then, the particular solution for the differential equation is needed. Let $f_{1(z)} = C_1 + C_2.Z$; i.e, any polynomial. Substitute the particular solution in the expression

$$\left(\frac{d^2M}{dz^2}\right)+n^2.M_z=\left(\frac{d^2M_i}{dz^2}\right), \text{ where, } (n^2 = P/EI)$$

$$\left(\frac{d^2[f_1(z)]}{dz^2}\right)=0$$

$$0 + n^2.(C_1 + C_2.Z) = -w \quad \text{(iii)}$$

$C_1 = -(w/n^2)$, and $C_2 = 0$

Then,

$$M_z = A\sin(nz) + B\cos(nz) - (w/n^2)$$

Using the boundary conditions

(i) at $z = 0$, $M_z = 0$, $0 = B - (w/n^2)$

$\therefore B = (w/n^2)$

(ii) at $z = l$, $M_z = 0$

$$0 = A\sin(nL) + \left(\frac{w}{n^2}\right)\cos(nL) - \left(\frac{w}{n^2}\right) \quad \text{(iv)}$$

$$A = \frac{w}{n^2}\left(\frac{1-\cos(nL)}{\sin(nL)}\right) \quad \text{(v)}$$

$$\text{Since, } \left(\frac{df_1(z)}{dz}\right)=0 \quad \text{(vi)}$$

Maximum moment

$$M_{z.max} = \left(\frac{w}{n^2}\right)\left(\frac{1-\cos(nL)^2}{\sin nL}\right)^2 - \left(\frac{w}{n^2}\right) \quad \text{(vii)}$$

$$M_{z.max} = \left(\frac{w}{n^2}\right)\left[\left(\frac{\sin^2(nL/2)}{\sin\left(\frac{nl}{2}\right)\cos\left(\frac{nL}{2}\right)}\right)^2 + 1\right]^{\frac{1}{2}} - \left(\frac{w}{n^2}\right) \quad \text{(viii)}$$

$$M_{z.max} = \left(\frac{w}{n^2}\right)\left[\frac{\sin^2(nL/2)+\cos^2 nL/2}{\cos^2\left(\frac{nL}{2}\right)}\right]^{1/2} - \left(\frac{w}{n^2}\right) \quad \text{(ix)}$$

$$M_{z.max} = \left(\frac{w}{n^2}\right)\left[\frac{1}{\cos^2\left(\frac{nL}{2}\right)}\right]^{1/2} - \left(\frac{w}{n^2}\right) \quad \text{(x)}$$

$$M_{z.max} = \left(\frac{w}{n^2}\right)\left(Sec\left(\frac{nL}{2}\right)\right) - \left(\frac{w}{n^2}\right) \quad \text{(xi)}$$

$$M_{z.max} = \left(\frac{w}{n^2}\right)\left(Sec\left(\frac{nL}{2}\right)-1\right) \quad \text{(xii)}$$

$$M_{z.max} = \left(\frac{wL^2}{8}\right)\left(\frac{8}{(nL)^2}\right)\left(Sec\left(\frac{nL}{2}\right)-1\right) \quad \text{(6.13)A}$$

$$M_{z.max} = \left(\frac{wL^2}{8}\right)(Magnification\ factor\ due\ to\ axial\ compression) \quad \text{(6.13)B}$$

6.9 BEAM COLUMN IN SINGLE CURVATURE WITHOUT END TRANSLATION (MOMENT MAGNIFICATION)

Instead of, the differential equation approach, a simple approximate procedure is satisfactory for many common situation.

It is assumed that the beam-column carries lateral loading w(z) that gives a deflection δ_o at mid-span as shown in Fig. 6.15. The secondary bending moment (i.e, moment $P.y_{max}$ due to axial load, P) may be considered to vary as a sine wave, which is nearly correct for the members with no end restraint whose primary moment and deflection are maximum at mid-span.

The portion of the mid-span deflection y_1, due to the secondary bending moment, equals the moment of the M/EI diagram between the support and midspan (shaded portion of Fig.6) taken about the support. According to the moment-area principle.

$$y_1 = \frac{P}{EI}(y_1+\delta_o)\left(\frac{L}{2}\right)\left(\frac{2}{\pi}\right)\left(\frac{L}{\pi}\right) \quad \text{(6.14)A}$$

$$= (y_1+\delta_o)\left(\frac{P.L^2}{\pi^2 EI}\right) \quad \text{(6.14)B}$$

$$y_1 = (y_1+\delta_o)\left(\frac{P}{PE}\right) \quad \text{(6.15)A}$$

where, $P_E = (\pi^2 EI / L^2)$

solving for y_1

$$y_1 = \delta_o\left(\frac{P/P_E}{1-\frac{P}{P_E}}\right) \quad \text{(i)}$$

$$y_1 = \delta_o\left(\frac{\alpha}{1-\alpha}\right) \quad \text{(6.15) B}$$

where, $\alpha = (P/P_E)$ (ii)

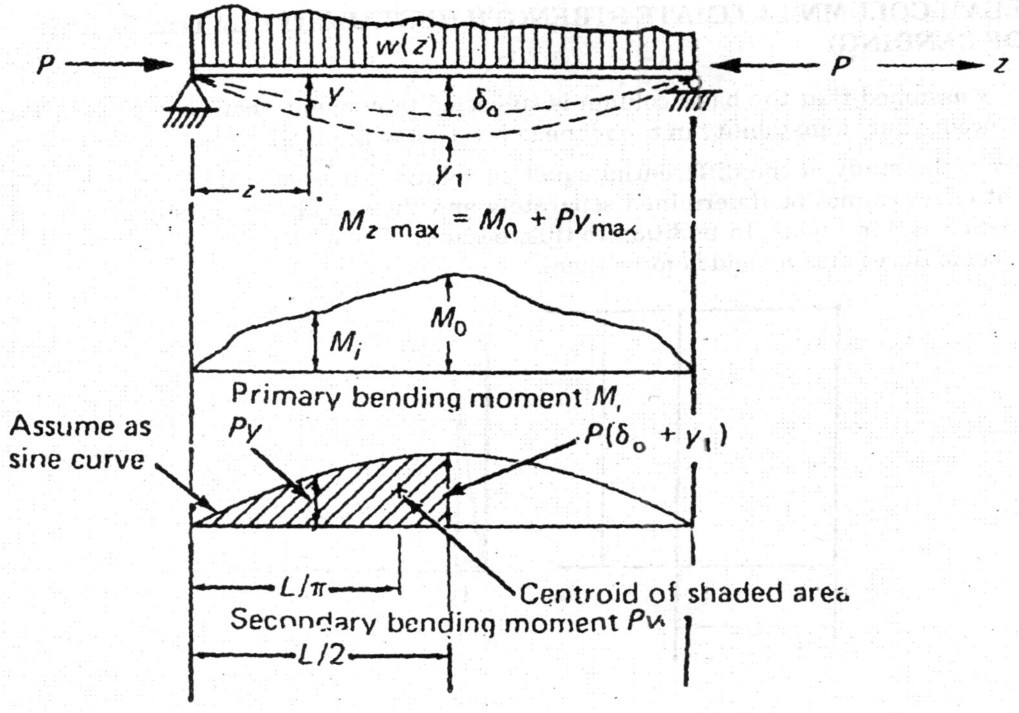

Fig. 6.15

Since, y_{max} is the sum δf.

δ_o and y_1

$$y_{max} = (\delta_o + y_1) = \left[\delta_o + \delta_o\left(\frac{\alpha}{1-\alpha}\right)\right]$$

$$= \delta_o / (1-\alpha) \qquad (6.16)$$

The maximum bending moment including the effect of axial load becomes

$$M_{z.max} = M_o + P.y_{max} \qquad (6.17)$$

Substituting the expression for y_{max} and setting P = $\alpha.P_E = \alpha.\pi^2 EI/2$ as the primary moment, M_o multiplied by a magnification factor Am. Therefore,

$M_{z.max}$ = Mo. Am (6.18) A

where, $A_m = C_m / (1-\alpha)$ (6.18)B

and

$$C_m = 1 + \left(\frac{\pi^2 EI\,\delta_o}{M_o.L^2} - 1\right)\alpha \qquad (6.19)A$$

$$\therefore C_m = (1 + \psi.\alpha) \qquad (6.19)B$$

6.10 BEAM COLUMN-ULTIMATE STRENGTH (INSTABILITY IN THE PLANE OF BENDING)

It is assumed that the basic column is subjected to moment about one principal and the failure occurs due to instability in the plane of bending without twisting.

From the study of the differential equation solution, it is seen that the axial effect and moment effect cannot be determined separately and then combined by super-position, since the relation is non-linear. In addition to this, because of bending, as well as due to residual stress, some fibres attain yield before others.

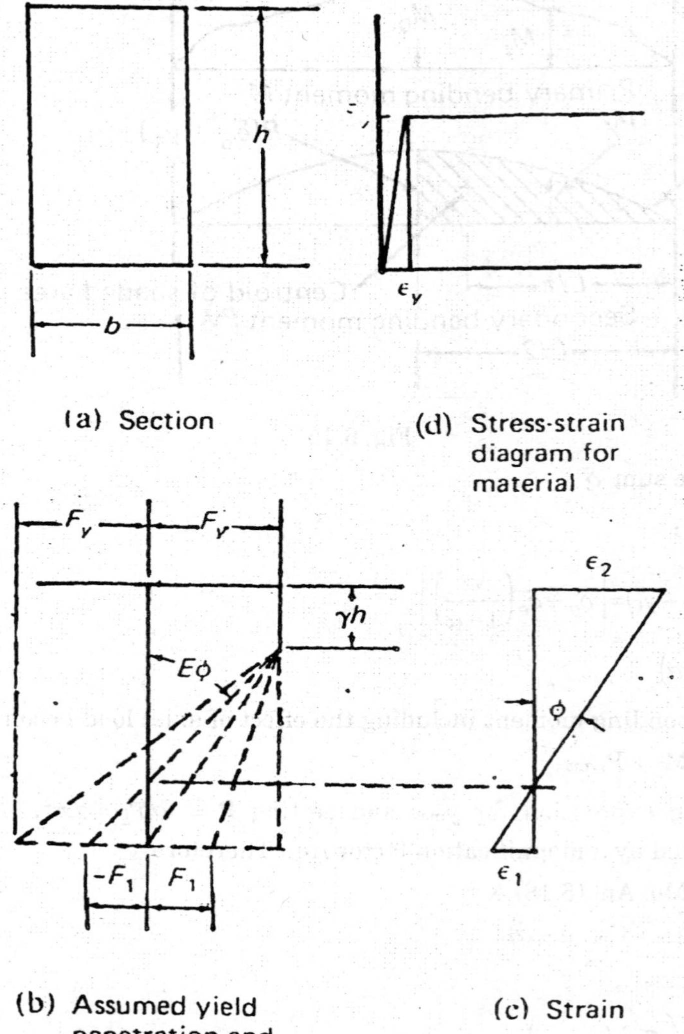

Fig. 6.16

When the stiffness of the members reduce because of yielding under combined stress, to the state where instability occurs, then the **ultimate capacity** is reached. The ultimate axial load is the critical load on the partially yielded member.

The moment-curvature-axial compression (M-ϕ-P) relationship is needed to find the actual strength. The yield penetration at various proportions of the depth of the member measured from the extreme fibre having the maximum compressive strain is assumed.

Let, yield penetration in a member having a rectangular cross-section subjected to axial compression and bending be γh. It is to note that strain ϵ_1 is less than or equal to ϵ_y.

The stress distribution representing complete range of M-ϕ– P is shown by dashed lines as shown in Fig. 6.16 Any position of the dashed line represents a value ϕ, for which the volume of the stress solid gives the correspond axial load. The summation of the moments of the stress solid about a convenient axis gives the corresponding bending moment M. It results in an M-ϕ-P function for each yield penetration depth, γh. This series of the curves for a rectangular cross-section is given in Fig. 6.17.

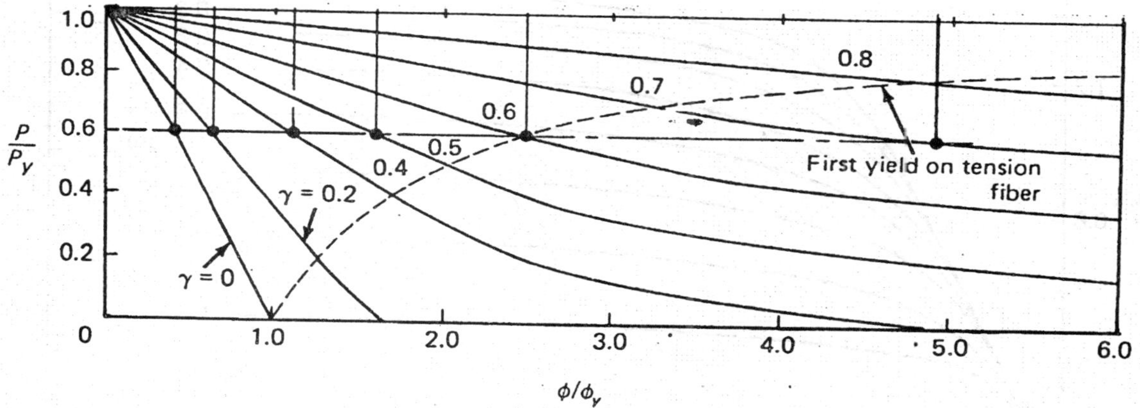

Fig. 6.17A. Auxiliary curves of M-P-ϕ for constant values of yield penetration at compression face of rectangular beam.

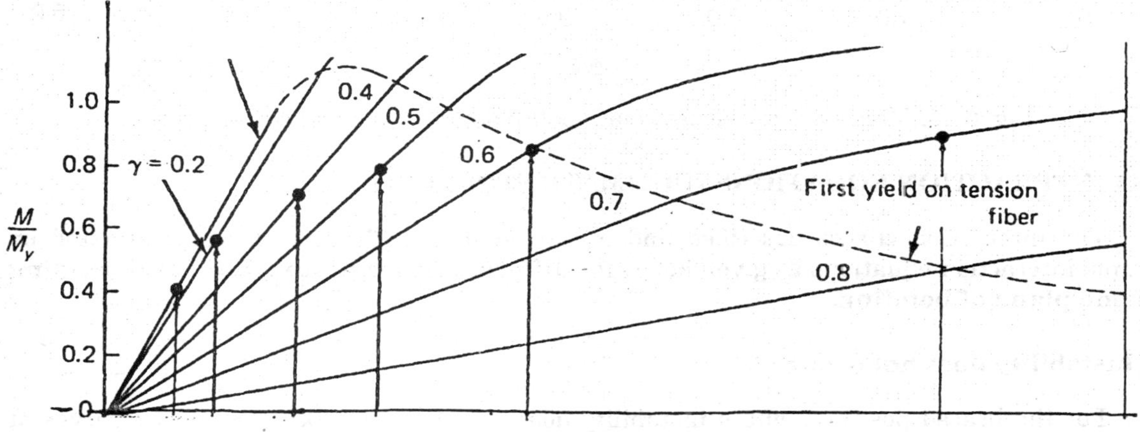

Fig. 6.17B.

From, the auxillary curves, Fig. 6.17, a series of M-ϕ curves may be obtained for constant values of (P/P_y). The graphical procedure is shown n Fig. 6. For (P/P_y) = 0.6.

Begin at ($P=P_y$) = 0.6, proceed to the right to intersect a line corresponding to a value of γ, say γ = 0.2, proceed upward to the γ = 0.2 line in the (M/M_y) portion of the curves and read off the value of (M/M_y).

This may be repeated for the for sufficient values of γ, until a complete (M/M_y) v/s (ϕ / ϕ_y) curve for the given (P/P_y) may be plotted Fig. 6.18 shows the results, which gives the beam-column strength for a short-member, where, yielding controls.

After obtaining curves, such as shown in Fig. 6.18, a specific value of (P/P_y), and slenderness ratio (kL/r) is investigated to establish the value of (M/M_y), at which the member becomes **unstable**.

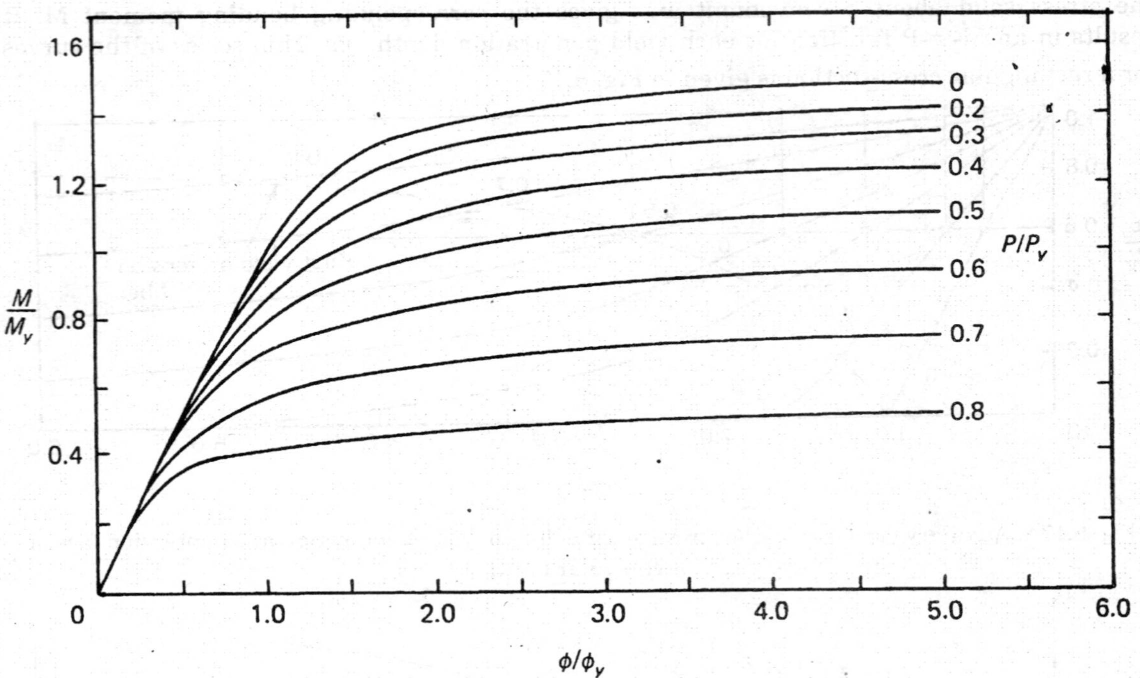

Fig. 6.18. Moment - curvature relationship for rectangular beam-column

6.11 INTERACTION EQUATIONS (ULTIMATE STRENGTH)

The interaction curves described and explained in Art. 6.7 may be approximated by simple interaction equations as given below for **different cases of instabiliy and bending in one plane of bending.**

1. Instability does not occur

For the braced position, where instability does not occur (i.e, (kL/r) = 0), the curves at the top, Fig. 6.18 apply and may be approximated by

$$\left(\frac{P}{P_y}\right)+\left(\frac{M}{1.18M_P}\right)=1.0 \qquad (6.20)$$

where, ratio (M/M_p) is less than or equal to 1.0 (unity)

$P_y = (A_{gross} \cdot \sigma_y)$, and

M_p = Maximum moment capacity of the member in the absence of axial load. (It equal to the plastic moment for all cases where premature local buckling is prevented).

2. Instability occurs in the plane of bending

The curves for various combinations of moments and values of (L/r_x) may be approximated by

$$\left(\frac{P}{P_{cr}}\right)+\left(\frac{M_E}{M_P(1-P/P_E)}\right)=1.0 \qquad (6.21)A$$

where,

P_{cr} = Buckling strength under axial load based on slenderness ratio for the axis of bending

$$M_E = M_2 \cdot \left(\frac{+1-2(M_1/M_2)\cos nL+(M_1/M_2)}{2(1-\cos nL)}\right)^{1/2} \qquad (6.21)B$$

= Equivalent moment (or alternates from)

$P_E = \left(\pi^2 EI/L^2\right)$ Euler's buckling load.

For primary bending moment from lateral (transverse) loading

In general, the interaction equation may be written for all the cases of instability in the plane of bending

$$\left(\frac{P}{P_{cr}}\right)+\left(\frac{C_m \cdot M_i}{M_P(1-P/P_E)}\right)=1. \qquad (6.22)A$$

3. Bi-axial bending

Simple plastic theory becomes inadequate when moments exist about two principal axes. In case, only one moment exists, plastic behaviur is exhibited, irrespective of the magnitude of axial compression Effective plastic moment reduces as the axial compression increases. However, the plastic behaviour does occur.

When there is bi-axial bending that an additional moment about the other principal axis, one might consider an interaction surface relating P, M_x and M_y. Even for the ideal plastic material however, present plastic analysis theorems neglect the information of deformation on the equilibrium.

While, few designers concern themselves greatly about the sequence of load application, nevertheless loading sequence influences ultimate strength. This is also true for uniaxial bending and compression, but it has less effect on that case than for the biaxial loading.

The ultimate strength of the compressions under biaxial bending is not sufficiently well known to make use of it for plastic design of rigid space frames; therefore, the plastic design should be restricted to planar structures, or ones for which planar behaviour represents a reasonable approximation.

Complete interaction equation may be written

$$\left(\frac{P}{P_{cr}}\right)+\left(\frac{M_x C_{mx}}{M_{mx}(1-P/P_{EX})}\right)+\left(\frac{M_y . C_{my}}{M_y\left(1-P/P_y\right)}\right) \tag{6.22B}$$

is less than or equal to unity (1.00).

6.12 INTERACTION EQUATIONS FOR MEMBERS SUBJECTED TO AXIAL COMPRESSION AND BENDING AS RECOMMENDED BY CODE

The strength of a member carrying axial compression and bending is controlled by material failure and the strength of member is controlled by buckling failure. This depends upto type of section (i.e, (i) Plastic and compact sections, and (ii) semi-compact sections. As it is known that the slender sections are not used for columns. Further, the strength of a member depends on whether the section is rolled or welded and so also, on the shape of the cross-section.

Strength of type of section

A. Plastic and compact sections

Strength of such sections is checked by following interaction equation

$$\left[\frac{M_{yy}}{M_{ndy}}\right]^{\alpha}+\left[\frac{M_{zz}}{M_{ndz}}\right]^{\alpha} \text{ is less than or equal to unity (1.00)} \tag{6.23}$$

conservatively

$$\left[\frac{P}{P_d}\right]+\left[\frac{M_{yy}}{M_{dyy}}\right]+\left[\frac{M_{zz}}{M_{dzz}}\right] \text{ is less than or equal to unity (1.00)} \tag{6.24}$$

where,

M_{yy}, M_{zz} = Factored applied moments about weak and strong axes of the cross-sections, respectively

M_{ndyy}, M_{ndzz} = Design reduced flexural strength under axial compression and respective uniaxial moment alone.

P = Factored applied axial compression

$P_d = (A_{gross} . \sigma_y)/\gamma_{ms}$, design strength in compression due to yielding.

M_{dyy}, M_{dzz} = Design strength under corresponding moment acting alone.

A_{gross} = Gross area of the cross-section.

α_1 and α_2 = constants as given in Table 6.

γ_{ms} = Partial factor of safety in yielding for the strength of material.

In case, there are no bolt holes following approximations may be used for evaluating M_{ndyy} and M_{ndzz}.

B. Rolled steel sections ($n = P/P_d$)

(1) Rolled steel I or H sections

(a) For n less than or equal to 0.2

$M_{ndyy} = M_{dyy}$ (i)

(b) For n more than 0.2

$M_{ndyy} = 1.56\,(1 - n)\,(n + 0.6)\,M_{dyy}$ (6.25) (ii)

$M_{ndzz} = 1.11\,(1 - n)\,M_{dzz}$ (iii)

at the same time less than or equal to M_{dzz}. (6.26)

(2) Circular hollow tubes without bolt holes

$M_{nd} = 1.04\,(1 - n^{1.7})\,M_d$ so also less than or equal to M_d (iv)

(3) Plates

$M_{nd} = M_d\,(1 - n^2)$ (6.27) (v)

C. Welded steel sections

(1) Welded I or H-sections

$$M_{ndyy} = M_{dyy}\left[1-\left(\frac{n-a}{1-a}\right)^2\right] \quad (6.28)\text{ A}$$

less than or equal to M_{dyy},

Where, n is more than or equal to a

$M_{ndzz} = M_{dzz}\left(\dfrac{1-n}{1-0.59}\right)$ is less than or equal to M_{dzz}, where,

$$n = (P/P_d) \text{ and } a = \left(\frac{a-(2b_f)}{A}\right) < 0.5 \quad (6.28)\text{B}$$

(2) Rectangular hollow sections and Welded box sections

When the section is symmetrical about both axes and there are no bolt holes

$M_{ndyy} = M_{dyy}\left(\dfrac{1-n}{1-0.5a_f}\right)$ is less than or equal to M_{dyy} (6.29)A

$M_{ndzz} = M_{dzz}\left(\dfrac{1-n}{1-0.5a_w}\right)$ is less than or equal to M_{dzz}, where, (6.29)B

a_w = (A – 2b.t_f)/A is less than or equal to 0.5

$a_f = \left(\frac{A - 2ht_w}{}\right)$ / A is less than or equal to 0.5.

B. Semi-compact section

In case, there is no high shear force, **semi-compact section design** is satisfactory under combined axial force and bending in case, the maximum longitudinal stress under combined axial force and bending, f_{abx} satisfies

f_{abx} is less than or equal to

$$(\sigma_y / \gamma_{ms}) \quad (6.30)A$$

When there are no holes, above criterial reduces to

$$\left(\frac{P}{P_d}\right) + \left(\frac{M_{yy}}{M_{dyy}}\right) + \left(\frac{M_{zz}}{M_{dzz}}\right) \text{ is less than or equal to unity (1.00)} \quad (6.30)B$$

TABLE 6.3 VALUES OF CONSTANTS α_1 and α_2

S.No.	Section	α_1	α_2
(1)	(2)	(3)	(4)
(i)	I and Channel	5n > 1	2
(ii)	Circular tubes	2	2
(iii)	Rectangular tubes	$\frac{1.66<6}{(1-1.13n^2)}$	$\frac{1.66<6}{(1-1.13n^2)}$
(iv)	Solid rectangles	$(1.73 + 1.8\ n^2)$	$(1.73 + 1.8\ n^2)$

6.13 OVERALL STRENGTH OF BEAM-COLUMN

Beam-column subjected to axial compression and bending shall comply with the following interaction relationships

$$\left(\frac{P}{P_{dyy}}\right) + K_{yy}\left(\frac{C_{my} M_{yy}}{M_{dyy}}\right) + K_{LT}\left(\frac{M_{zz}}{M_{dzz}}\right) \text{ is less than or equal to unity (1.0)} \quad (6.31)A$$

$$\left(\frac{P}{P_{dzz}}\right) + 0.6K_{yy}\left(\frac{C_{my} . M_{yy}}{M_{dyy}}\right) + K_z\left(\frac{C_{mz}.M_{zz}}{M_{dzz}}\right) \text{ is less than or equal to unity (1.0)} \quad (6.31)B$$

where,

C_{my}, C_{mz} = Equivalent uniform moment factors as per Table 6.

P = Applied axial compression under factored load

P_{dyy}, P_{dzz} = Design strength under axial compression as controlled by buckling about weak (yy-axis) and strong (zz-axis), respectively.

M_{yy}, M_{zz} = Maximum factored applied bending moments about yy and zz-axes of the members, respectively.

M_{dyy}, M_{dzz} = Design bending strength about yy (weak-axis) or zz (strong-axis) considering laterally unsupported length of the member.

$$K_{yy} = 1 + (\lambda_y - 0.2)n_y \text{ is less than or equal to } (1 + 0.8\, n_y) \quad (6.32)A$$

$$K_{zz} = 1 + (\lambda_z - 0.2)n_z \text{ is less than or equal to } (1 + 0.8\, n_z) \quad (6.32)B$$

$$K_{LT} = 1 - \left(\frac{0.1\lambda_{LT}.n_y}{C_{mLT} - 0.25}\right) \text{ is more than or equal to } \left(1 - \left(\frac{0.1\lambda_y}{C_{mLT} - 0.25}\right)\right) \quad (6.32)C$$

where,

η_y, η_z = Ratio of actual applied force to the design axial strength for buckling about the yy and zz axes, respectively.

C_{mLT} = Equivalent uniform moment factor for lateral torsional buckling as per Table 6.

Corresponding to the actual moment gradient between lateral support against torsional deformation in the critical region under consideration.

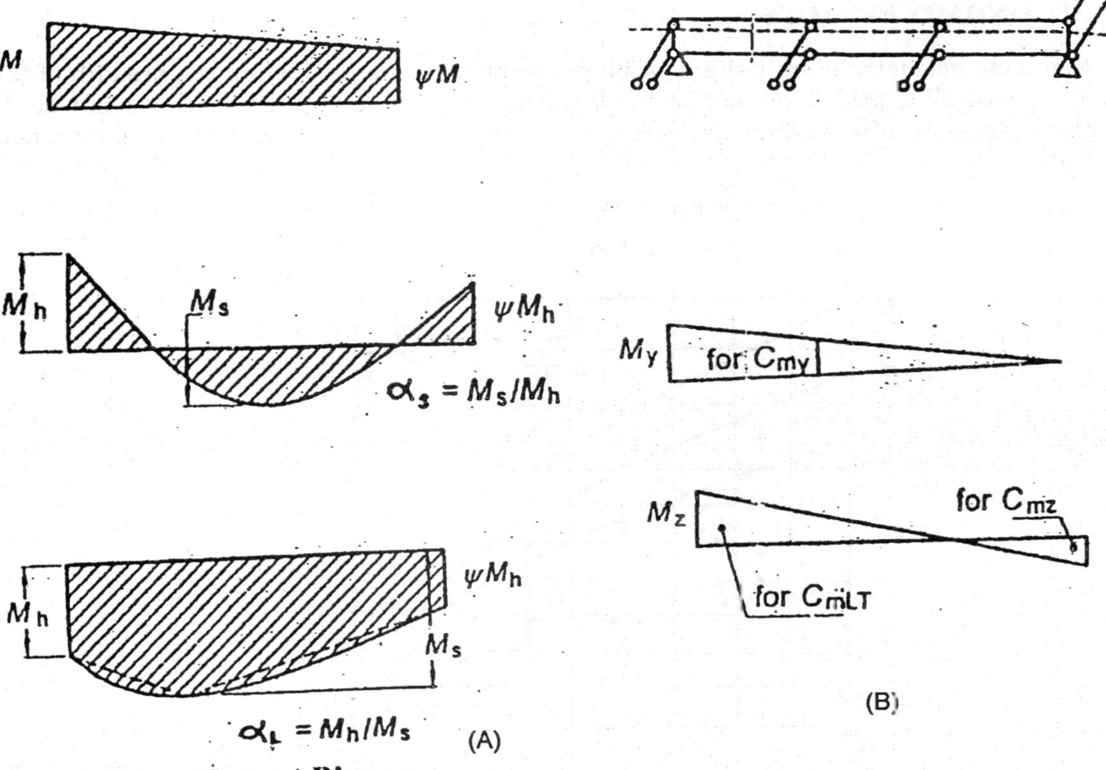

Fig. 6.19.

TABLE 6.4. EQUIVALENT UNIFORM MOMENT FACTOR

S.No.	Figure	Range		C_{my}, C_{mz}, C_{mLT}	
(1) A	1 (B)	(2)		(3)	(4)
1.	Fig. (6.19) A	$(-1 \le \psi \le 1)$		Uniform loading 0.6 + 0.4	Concentrated load $\psi \ge 0.4$
2.	Fig. (6.19) B	$0 \le a_g \le 1$	$-1 \le \psi \le 1$	$0.2 + 0.8\ a_g \ge 0.4$	$0.2 + 0.8\ a_g \ge 0.4$
		$-1 \le a_g \le 0$	$0 \le \psi \le 1$	$0.1 - 0.8\ a_g \ge 0.4$	$-0.8\ a_g \ge 0.4$
			$-1 \le \psi \le 0$	$0.1\,(1-\psi) - 0.8 a_g \ge 0.4$	$0.2\ (1-\psi)0.8 - a_g \ge 0.4$
3.	Fig. (6.19) C	$0 \le a_g \le 1$		$0.095 + 0.05\ a_h$	$0.90 \div 0.10\ a_h$
		$-1 \le a_g \le 0$		$0.095 + 0.05\ a_h$	$0.90 + 0.10\ a_h$
				$0.095 + 0.05\ a_h\ (1+2\psi)$	$0.90 + 0.05\ a_h\ (1+2\psi)$

For members with sway buckling mode, equivalent uniform moment factor $C_{my} = C_{mz} = 0.9$

6.14 BEAM COLUMN ULTIMATE STRENGTH CONSIDERING SECONDARY EFFECTS

Complete geometry of an I-shape beam cross-section is defined when the breadth of the flange, b_f, over-all depth of the section, h, thickness of flange, t_f and the thickness of web, t_w are given. Presence of axial force as in Art. 6. does not show the effects of these dimensions directly.

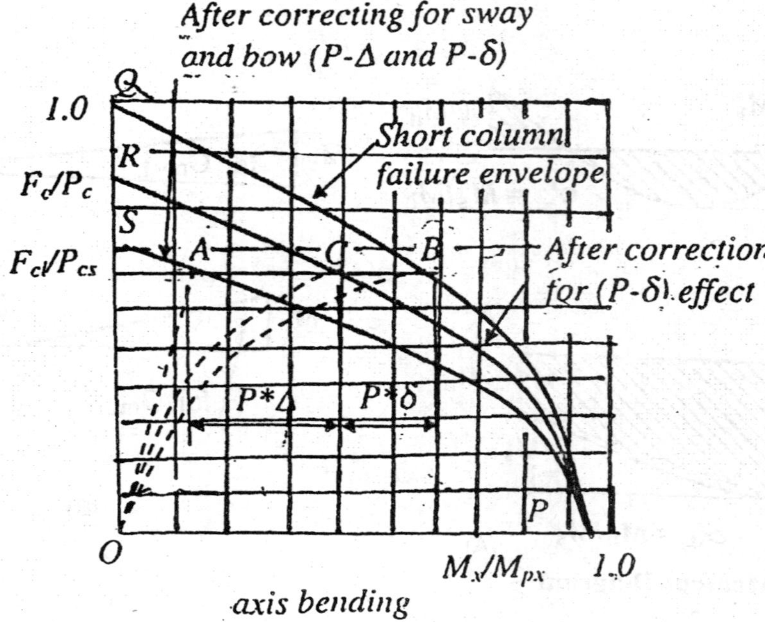

Fig. 6.20A. Uniaxial Bending of Slender Beam-Columns

When the eccentricity of applied compressive force is equal to zero, it becomes an axially loaded column. However, in case, the eccentricity of the applied compression is very large (which may tend to be infinity, the member tends to behave like a beam, as the effects of axial compression is negligible. As such two limits for beam column are defined. In between these two limits beam-column shows various combinations of axial compression and bending moment. Buckling and plastic failures are shown by the beam column depending on the relative values of P and M and the respective buckling strength and the bending strength of the member. Then, the bending of the member may be about strong axis or weak axis. In case of bending occurs about weak axis then failure may be flexural yield type. However, if the bending occurs about strong axis, failure of torsional flexural type or a combination of moments about.

Two cases of strength of an I-section beam columns subjected to axial compression and bending about.

(i) Strong axis, Fig. 6.20A.

(ii) Weak-axis, Fig. 6.20B.

shall be considered.

(i) Bending about strong axis

Curves shown by solid lines show the strength envelope of beam-column in a frame considering the different long columns and sway effect in Fig. 6.20A and Fig. 6.20B.

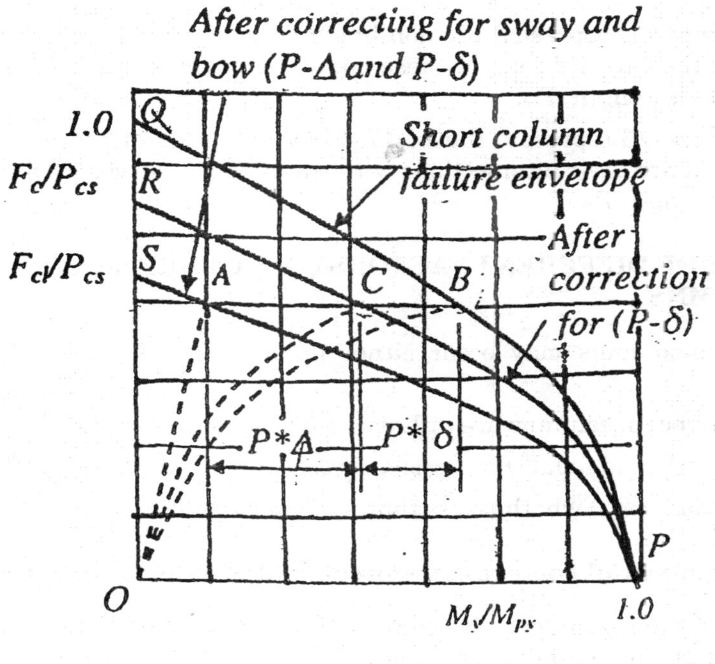

axis bending

Fig. 6.20b. Uniaxial Bending of Slender Beam-Columns

Curves PCQ in Fig. 6.20 and Fig. 6.20B represent the strength envelopes of stub columns without considering P-Δ and P-δ effects.

Therefore, in case short column strength envelopes are used, the actual bending moments and axial forces used in determining the strength should be based on non-linear analysis, considering for the P-δ and P-Δ effects.

Then the loading path from such an analysis would be represented by the curves OC.

Similarly the strength envelopes PBR accout for P-Δ effects only. Therefore, in case, these strength envelopes are used, the actual axial force and moment determination should be based on an analysis method that accounts for P-δ effects. The curve shown by dashed line OB for P v/s M represent such an analysis.

The strength envelopes PAS account for both P-Δ and P-δ effects. Therefore, a linear analysis is sufficient to find the axial forces and moments, while these strength curves are used.

It is observed from Fig. 6.20 and Fig. 6.20 that the strength under axial compression reduces when P-Δ and P-δ effects are accounted. It is also observed that under pure bending, bending strength about the strong axis is influenced due to lateral buckling and that pure bending strength may be smaller than the plastic moment capacity, M_p.

Curves shown by dashed lines represent the loading curve corresponding to different levels of analysis. The dashed curves OA represent the linear path and hence their inter-secti-ons with the strength envelopes corrected for P-Δ and P-δ effects give member strength, P_{ce}.

The dashed curves OC represent the non-linear analysis curves, considering P-Δ and P-δ effects and hence the intersections of these curves with the short column strength curve PCQ give the member strength, P_{ce}.

The dashed curves OB represent the loading curves from non-linear analysis, accounting P-δ effects only and hence their intersections with the strength curves PBR corrected for P-Δ effect the member strength P_{cl}.

6.15 INFLUENCES OF DIFFERENT FACTORS ON FAILURE MODES OF BEAM-COLUMNS

Failure of beam-columns may occur either in flexural yielding or flexural torsional buckling modes.

Failure modes of beam columns are influenced by the following factors.

1. Slenderness ratio and axial compressive forces, and
2. Axial load ratios (P/P_y) whether less than or more than 0.15.

Slenderness ratio and axial compressive forces, PLUS Axial Load ratios

The magnitude of axial compressive force and eccentricity and slenderness ratio of the beam-column influence the actual failure mode including the influence of axial load ratios (P/P_y).

For (P/P_y) ratio less than 0.15

The linear interaction as under may be regarded as sufficient

$$\left[\frac{M_{pc}}{M_p}\right]=1.18\left[1-\frac{P}{P_y}\right] \quad \text{(i)}$$

$$\left[\frac{P}{P_y}\right]+\frac{1}{1.18}\left[\frac{M_{pc}}{M_p}\right]=1.00 \quad \text{(ii)}$$

since, (1/1.18) = 0.84745 (i.e, approximately may also be written as below

$$\left[\frac{P}{P_y}\right]+0.85\left[\frac{M_{pc}}{M_p}\right]=1.00 \quad \text{(iii)}$$

The co-relation of experimental results is shown in Fig. 6.21A. Theoretical M-ϕ relationship represented by the dotted line In this Fig. 6.21A. The results of test on a member (stub-column 12 WF) loaded eccentrically is shown in Fig. 6.21B. The magnitude of the actual reduced plastic moment, M_{pc} agrees the value obtained theoretically.

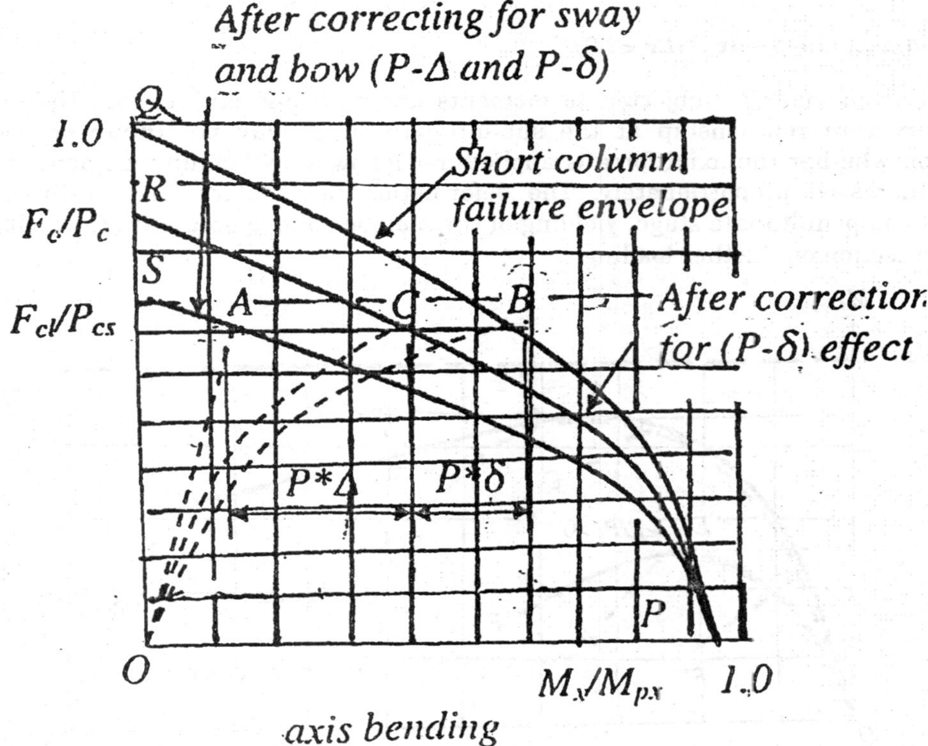

Fig. 6.21A. Inter-Action Curve for Load V/s Moment (Design approximation).

The value of plastic modulus, Z_p needed in the design may be obtained by multiplying the trial value Z_{Pt} calculated in the initial design by the ratio (M_{pc}/M_p).

For the bending about the strong axis, the influence of axial force may not be considered and M_{pc} may be taken equal to M_p in case ($P/P_y < 0.15$) with the error less than 5 per cent.

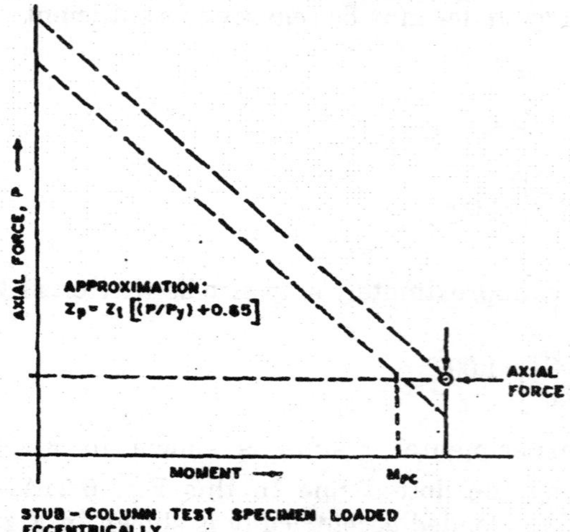

Fig. 6.21B. Stub-column test specimen loaded eccentrically

When the slenderness ratio (l/r < 50)

When the member is subjected to moments about strong axis at both the ends, the moment curvature relationship at the sub-ultimate stage may be **linear** or non-linear depend upon whether the axial force is appliedfirst followed by bending moment or both of them are increased proportionately. The deformation takes place only in the plane of bending. At the penultimate stage, yielding of the member flange occurs first, which spreads through the section on further loading.

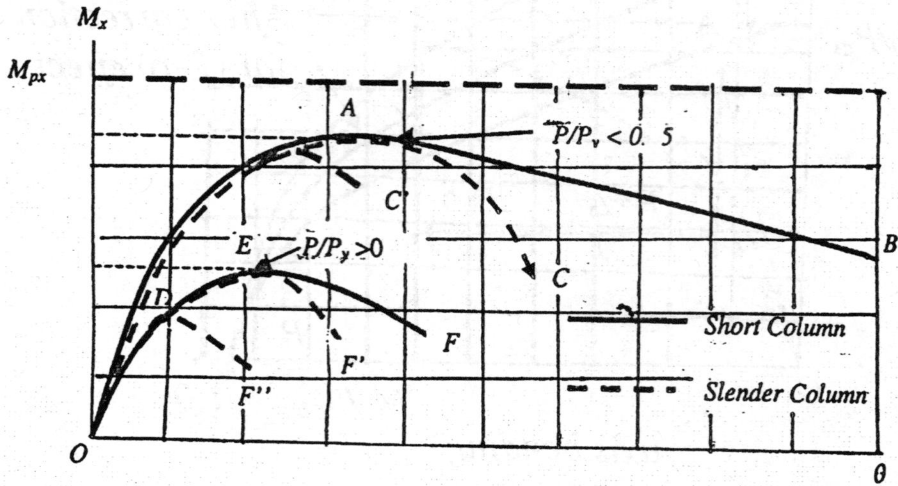

Fig. 6.22. Beam-column Moment Rotation Behaviour

The ultimate strength is reached when the plastic hinge is formed at one or both the ends. Therefore, the failure is due to the section strength being reached at one or both the ends.

Under proportionate loading, the loading curve is shown by the line OAB (Fig. 6.22). It is observed that there would be a small reduction in the moment capacity (below M_P) due to the presence of the axial compression in the member. Further, these would be reduction in the moment (unloading) beyond point A with increasing and rotation due to the spread of yielding from the end sections to other sections and the bending.

Slenderness ratio (l/r > 80)

When the slenderness ratio is more, under smaller compression combined with end moments as before, would fail by buckling out of the plane of the bending. In case, the bending is pre-dominant lateral buckling of compression flange asin unrestrained beams occur as shown in Fig. 6.22.

Due to the axial force, the deformation may occur about weak-axis and hence, the failure may be by weak-axis bending and twisting (OC') at moments less than the full in plane strength found in the case of short/stocky columns.

The moment rotation behaviour at the ultimate stage is shown by the dashed line in Fig. 6. The failure would be after plastic hinge formation unless the slenderness ratio is very large.

For (P/P_y) ratio more than 0.15

The value of plastic modulus, Z_p, is found from the following expression

$$Z_P = \left(\frac{0.85 Z_{pt}}{1 - P/P_y}\right) \quad \text{(i)}$$

This expression calculates the value of plastic modulus, that is too great. In redesign, the ratio (P/P_y) will be less. Therefore, the reduction in the value of M_p will also be less than that calculated in the first trial.

In case, it is assumed that the area of the cross-section varies proportionally to Z_p, then

$$\left[\frac{Z_{pt}}{Z_p}\right] = \left[\frac{At}{P}\right] \quad \text{(ii)}$$

Let $P_y = (\sigma_y . A_c)$, the axial force corresponding in the first approximate design

$$\left[\frac{P}{P_y}\right] = \left[\frac{P}{P_{yt}}\right] x \left[\frac{P_{yt}}{P_y}\right] = \left[\frac{P}{P_y}\right] x \left[\frac{Z_{pt}}{Z}\right] \quad \text{(iii)}$$

Substituting this in the expression (i) it gives [(for P/P_y) > 0.15]

$$Z_p = Z_t\,[0.85 + (P/P_{yt})] \quad \text{(iv)}$$

For bending about weak-axis, M_{pc} approximately equals to M_p so long as [(P/P_y) < 0.4]

$$\left[\frac{M_{pc}}{M_p}\right] = 1.19 \left[1 - \left(\frac{P}{P_y}\right)^2\right] \quad \text{(v)}$$

$$\left[\frac{P}{P_y}\right]=\frac{1}{1.19}\left[\frac{M_{pc}}{M_p}\right]=1.00$$

$$\left[\frac{P}{P_y}\right]+0.84\left[\frac{M_{pc}}{M_p}\right]=1.00$$

For slenderness ratio [l/r < 50]

For large value of axial compression combined with bending moment, the yielding may occur over large portion of the member due to combined axial stress, bending stress and residual stress. For single curvature bending deflection caused by equal end moments, the moment amplification may occur at the mid-depth of the column.

In the case of short/stockey member, the failure is due to the yielding strength reached at a section over the length of the member under combined axial force and corresponding curve is shown by line ODEF in Fig. 6.22.

The main differences in the behaviour of stockey beam-column under larger compression, compared to smaller axial compression are the moment amplification larger reduction in moment capacity due to larger axial compression and the drastic unloading in the penultimate state (EF in Fig. 6.22).

For high slenderness ratio [(l/r) > 80]

For the slender beam column with a large axial compression, the P- Δ effect is large both in the plane of and out of plane of the moment. In such members, the moment may drop drastically when the yielding begins under the combined axial compression and the amplified moment (OEF). The weak axis buckling/flexural torsional buckling causing out-of-plane deformation, could occur earlier than that, corresponding to the sec tion strength under combined axial force and amplified moment (ODF') Fig. 6.22.

As such, design of beam-columns having larger slenderness ratio needs investigation of in-plane bending failure by flexural yielding and out-of-plane buckling failure. The behaviour of beam-columns subjected to bending about the weak-axis is similar that subjected to strong axis bending.

Example 6.1. As ISHB 350, @ 0.674 kN/m column section 4.20 m high has been used as a non-sway intermediate column in a building frame. Its joints are flexible. The column section is subjected to the following.

(i) Factored axial load = 540 kN, and

(ii) Factored, moments

	About strong axis (M_{xy})	About weak axis (M_{yy})
Top end	18 kN-m	+0.80 kN-m
Bottom end	8 kN-m	-1.20 kN-m

Take σ_y = 260 N/mm², E = 2 x 10⁵ N/mm² check the adequacy of the section.

Step 1 : Geometrical properties of the section : From ISI, Handbook No. 1.

For ISHB 350, @ 0.674 kN/m

Gross cross-sectional area

A_g =8591 mm^2

Width of flange, b_f = 250 mm

Thickness of flange, t_f = 11.6 mm

Overall depth of section h_o = 350 mm

Thickness of web, t_w = 8.3 mm

Clear depth of wew, $d_w = (h_o - 2t_f)$

$= (350 - 2 \times 11.6) = 326.8$

Moment of inertia

About strong axis, I_{xx} = 19159.7 x 10^4 mm^4

About weak axis, I_{yy} = 2451.4 x 10^4 mm^4

Radius of gyration, r_{xx} = 149.3 mm

About weak axis r_{yy} = 53.4

Elastic Section modulus

About strong axis, Z_{xx} = 1094.8 x 10^3 mm^3

About weak axis, Z_{yy} = 196. x 10^3 mm^3

Plastic section modulus

About strong axis, Z_{px}

About weak axis, Z_{py}

Yield stress ratio Epsilon, $\varepsilon = (250/260)^{1/2} = 0.98$

Step 2: Classification of section

0.5 Width of flange : thickness of flange

$(0.5\, b_f/t_f) = (0.5 \times 250/11.6) = 10.776$

$10.776 < (0.98 \times 15.7 = 15.386)$

Clear depth do web : thickness of web

$(d_w/t_w) = (326.8/8.3) = 39.37$

$39.37 < (0.98 \times 42 = 41.16)$

Cross-section is `Class-3' type **semi-compact.**

Step 3 : Check for resistance of the cross-section for the combined effect

$$\text{Design compressive stress } \sigma_{cd} = \left(\frac{\sigma_y}{\gamma_{ms}}\right) = \left(\frac{260}{1.15}\right) = 226.086 \text{ N/mm}^2$$

Inter-action equation for combined effect

$$\left(\frac{P_{cy}}{\sigma_{cd} \cdot A_g}\right) + \left(\frac{M_{xx}}{\sigma_{cd} \cdot Z_{xx}}\right) + \left(\frac{M_{yy}}{\sigma_{cd} \cdot Z_{yy}}\right) \leq 1.00$$

$$\left(\frac{540x10^3}{226.086x8591}\right)+\left(\frac{18}{226.086x1094.8x10^3}\right)+\left(\frac{1.20x10^6}{226.086x196.1x10^3}\right)$$

$$\left(\frac{540}{226.086x8.591}\right)+\left(\frac{18}{226.086x1.0948}\right)+\left(\frac{1.20}{226.086x1.961}\right)$$

(0.278 + 0.072 + 0.071)

0.421 < 1.00 section is satisfactory against combined effects.

Step 4 (A) Check for resistance of cross-section for combined effects of buckling

Column is non-sway flexible intermediate member

Effective length = (0.85 x 4.20) = 3.57 m

= 3.57 x 1000 =3570 mm

Slenderness ratios about

Strong axis $\lambda_{xx} = \left(\frac{3570}{149.3}\right) = 23.91$

Weak axis $\lambda_{yy} = \left(\frac{3570}{53.4}\right) = 66.854$

Critical slenderness ratio

$$\lambda = \pi\left(\frac{E}{\sigma_y}\right)^{1/2}$$

$$= 3.14\left(\frac{200000}{260}\right)^{1/2}$$

= (3.14 x 27.735) = 87.088

Step 4(B): Slenderness ratio in non-dimensional form

$$\overline{\lambda_{xx}} = \left(\frac{\lambda_{xx}}{\lambda}\right) = \left(\frac{23.91}{87.088}\right) = 0.2745$$

$$\overline{\lambda_{yy}} = \left(\frac{\lambda_{yy}}{\lambda}\right) = \left(\frac{66.854}{87.088}\right) = 0.7676$$

Step4(C) Calculation for ϕ values (Imperfection factors for ISHB Section)

$\alpha_{xx} = 0.21$, $\alpha_{yy} = 0.3500$

$$\phi = 0.5\left[1+\alpha\left(\overline{\lambda}-0.2\right)+\overline{\lambda}^2\right]$$

$\phi_{xx} = 0.5\ [1 + 0.21\ (0.2745 - 0.2) + (0.2745)^2]$

$\phi_{xx} = 0.5\ [1 + (0.21 \times 0.0745) + (0.2745)^2]$

$= 0.5\ [1 + .015645 + 0.0754] = (0.5 \times 1.09) = 0.545$

$\phi_{yy} = 0.5\ [1 + 0.35\ (0.7676 - 0.2) + (0.7676)^2]$

$\phi_{yy} = 0.5\ [1 + (0.35 \times 0.5676) + (0.7676)^2]$

$= 0.5\ [1 + 0.19866 + 0.5892]$

$= 0.5\ [\ 1.19866 + 0.5892] = \left(\frac{1.787}{2}\right) = 0.8939$

Step 4D: Calculation of CHI-values

$$\chi = \left[\frac{1}{\phi + \left(\phi^2 - \overline{\lambda}^2\right)^{1/2}}\right] < 1.0$$

$$\chi_{xx} = \left[\frac{1}{0.545 + \left(0.545^2 - 0.2745^2\right)^{1/2}}\right] = \left[\frac{1}{0.545 + (0.297 - 0.0754)^{1/2}}\right]$$

$$= \left[\frac{1}{0.545 + 0.2216^{1/2}}\right]$$

$$= \left[\frac{1}{0.545 + 0.471}\right] = \left(\frac{1}{1.016}\right) = 0.984 < 1.00$$

$$\chi_{yy} = \left[\frac{1}{0.8939 + \left(0.8939^2 - 0.7676\right)^{1/2}}\right]$$

$$= \left[\frac{1}{0.8939 + (0.799 - 0.589)^{1/2}}\right]$$

$$= \left[\frac{1}{0.8939 + (0.2098)^{1/2}}\right] = \left(\frac{1}{1.352}\right) = 0.7396$$

Step 4(E). Check for interaction equation

$$\frac{P_c}{P_{cl}} + \frac{k_x\ M_x}{M_{ux}} + \frac{k_y\ .\ M_y}{M_{uy}} < 1.00$$

Moment at top = 18 kN-m

Moment at bottom = 8 kN-m

Moment gradient for strong axis

$$\Psi_x = \left(\frac{M_z}{M_1}\right) = \left(\frac{8}{18}\right) = 0.4444$$

$$\beta_{Mx} = 1.8 - \Psi_x$$

$$= (1.8 - 0.7 \times 0.4444) = (1.8 - 0.311) = 1.4889$$

$$\mu_x = \bar{\lambda}\left(2 . \beta_{Mx} - 4\right)$$

= 0.2745 (2 x 1.4889 – 4)

= 0.2745 (2.9778 – 4)

= – (0.2745 x 1.0222) = 0.2806

Moment gradient for weak axis Ψ_y

Moment at top = 0.8 kN-m

Moment at bottom = 1.20 kN-m

$$\Psi_y = \left(\frac{0.8}{-1.20}\right) = -0.667$$

$$\beta_{My} = (1.8 - 0.7\ \Psi_y)$$

= [1.8 + 0.7 x 0.667] = [1.8 + 0.4669] = 2.2669

Step 4(F): Substitution of various values in the interaction equation

$$\chi_{xx} = 0.984\ ,\quad \chi_{yy} = 0.7396$$

Note:

$$P_{cl} = \chi_{min}\ A_{gross} \cdot \sigma_{cd}$$

$$= \left(\frac{0.7396 \times 8591 \times 226.086}{1000}\right) = 1436.528\ \text{kN}$$

$$M_{ux} = \sigma_{cd} \times Z_{xx}$$

$$= \left(\frac{226.084 \times 1094.8 \times 10^3}{1000 \text{ x } 1000}\right) = (226.086 \text{ x } 1.0948) = 247.519 \text{ kN-m}$$

$$M_{uy} = \left(\sigma_{cd}\ x Z_{yy}\right)$$

$$= \left(\frac{226.086 x 196.1 x 10^3}{1000 x 1000}\right) = 44.335\ \text{kN-m}$$

Substituting the respective values in interaction equation

$$\frac{P_c}{P_{cl}} + \frac{k_x . M_{xx}}{M_{ux}} + \frac{k_y\, M_{yy}}{M_{uy}} < 1.0$$

$$\left(\frac{540}{1436.528}\right) + \left(\frac{1.0689 x 18}{247.579}\right) + \left(\frac{0.8661 x 1.20}{44.335}\right)$$

(0.3759 + 0.0777 + 0.0234) = 0.477 < 1.00

Hence, given section is adequate for combined effects for buckling.

$$\mu_y = \bar{\lambda}(2 . \beta_{my} - 4.0)$$

= 0.7676 (2 x 2.2669 – 4)

$= 0.7676 \text{ x } (4.5338 - 4)$

$= (0.7676 \text{ x } 0.5338) = 0.4097$

Step 4 (F) Calculation for k-coefficients

$$k_{xx} = (1 - \frac{\mu_x . P_c}{P_{cx}})$$

$$k_{xx} = \left(1 - \frac{\mu_x P_c}{\chi_{xx} A \sigma_y}\right)$$

$$= \left[1 - \frac{\left(-0.2806 x 540 x 10^3\right)}{0.984 x 8591 \; x 260}\right] = \left[1 + \frac{151.524}{0.984 x 8.591 \; x 260}\right] = [1.0689]$$

$$k_y = \left(1 - \frac{l_{iy} . P_c}{P_{cy}}\right)$$

$$= \left(1 - \frac{0.4097 x 540 x 10^3}{0.7396 x 8591 x 260}\right) = \left(1 - \frac{221.238}{0.7396 x 8.591 x 260}\right) = [1 - 0.1339] = 0.8661$$

Step 4(G) : Substitution of various values in the interaction equation

$\chi_{xx} = 0.984, \quad \chi_{yy} = 0.7396$

Note :

$$P_{cl} = x_{\min} A_{gross} \cdot \sigma_{cd}$$

$$= \left(\frac{0.7396 \times 8591 \times 226.086}{1000}\right) = 1436.528 \text{ kN}$$

$$M_{ux} = \sigma_{cd} \times Z_{xx}$$

$$= \left(\frac{226.086 \times 1094.8 \times 10^3}{1000 \times 1000}\right)$$

$$= (226.086 \times 1.094.8) = 247.519 \text{ kN-m}$$

$$M_{uy} = (\sigma_{cd} \times Z_{yy})$$

$$= \left(\frac{226.088 \times 196.1 \times 10^3}{1000 \times 1000}\right) = 44.335 \; kN - m$$

Substituting te respective values in interaction equation

$$\frac{P_c}{P_{cl}} + \frac{k_x . M_{xx}}{M_{ux}} + \frac{k_y M_{yy}}{M_{uy}} < 1.0$$

$$\left(\frac{540}{1436.528}\right) + \left(\frac{1.0689 \times 18}{247.519}\right) + \left(\frac{0.8661 \times 1.20}{44.335}\right)$$

$(0.3759 + 0.0777 + 0.0234)$

$= 0.477 < 1.00$

Hence, given section is adequate for combined effects for buckling.

PROBLEMS

6.1 An ISHB 400, @ kN/m column section 4.40 m high has been used as a non-sway intermediate column in a building frame. Its joints are flexible. This column section is subjected to the following

(i) Factored axial load 680 kN, and

(ii) Factored moments

About	strong axis (M_{xx})	Weak axis (M_{yy})
Top end	22 kN-m	+ 1.20 kN-m
Bottom end	12 kN-m	1.60 kN-m

Take, $\sigma_y = 260$ N/mm^2, $E = 2 \times 10^5$ N/mm^2 check adequacy of the section.

REFERENCES

1. **Charles G. Salmon, John E. Johnson**: `Steel Structures Design and Behaviour' Second edition, 1980. Harper International Edition. Harper & Row, Publishers, New York 10 East, 53 street, New York, 10022.

2. **Robert Englekirk** : `Steel structures (Controlling Behaviour Through Design)' John Wiley & Sons, Inc (1994 Edition) Singapore.

3. **Boris Bresler, T.Y. Lin John, B. Scalzi**: `Design of Steel Structures' Second Edition. Wiley Eastern Private Limited New Delhi.

4. **Stephen P. Timoshen ko James M Gere**: `Theory of Elastic Stability' 1961, Second Edition (International Student Mcgraw-Hill Book Company Inc. New York, Toronto, London.

5. **Rangachari Narayanan, V. Kalyanaraman *et al.*** Institution of steel Development and Growth.

6. **IS : 800-2007**: Indian Standard for General Construction in Steel Code of Practice (Third Revision). Bureau of Indian Standards Mark Bhawan, 9 Bahadur Shah Zafar Marg, New Delhi, 110 002.

7

LIMIT STATE OF COLLAPSE IN COMPRESSION DESIGN OF STEEL CONCRETE COMPOSITE (ENCASED) COLUMNS

7.1 INTRODUCTION

Structural members made from two or more materials are referred as **composite members.** A large number of composite sections used for building constructions are made from steel and concrete. Steel provides strength and concrete provides rigidity. Combination of two materials results in the efficient load-carrying members. Both steel and concrete resist the ext load by interacting together by bond and fining Lighter weight and higher strength of steel-allow the use of smaller and lighter foundation. When hot-rolled steel sections are encased in the cement concrete, then concrete is cast around the steel sections, these are called as **steel concrete (encased) composite columns** as shown in Fig. 7.1. In addition to enhance strength and providing rigidity, the concrete acts as fire-proofing materials to the steel shapes and serves as a corrosion barrier shielding the steel under adverse environmental conditions. Other types of composite members, concrete is filled in hollow steel tubular sections as shown in Fig. 7.2. then, these are known as **concrete filled composite sections**. In both, these types, the rigidity of the concrete often eliminates the problem of local buckling (likely to be experienced by some slender steel sections).

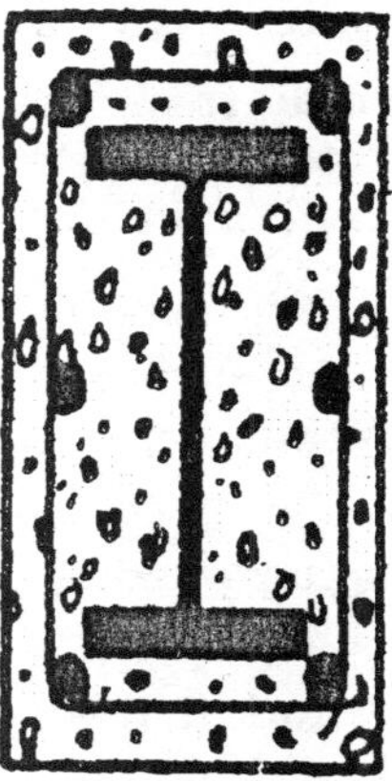

Fig. 7.1. Concrete Encased Composite Section

Concrete creep and shrinkage associate some **dis-advantage**. The design of column is complicated because of uncertainties as regards the mechanical bond to be developed between steel shapes and concrete.

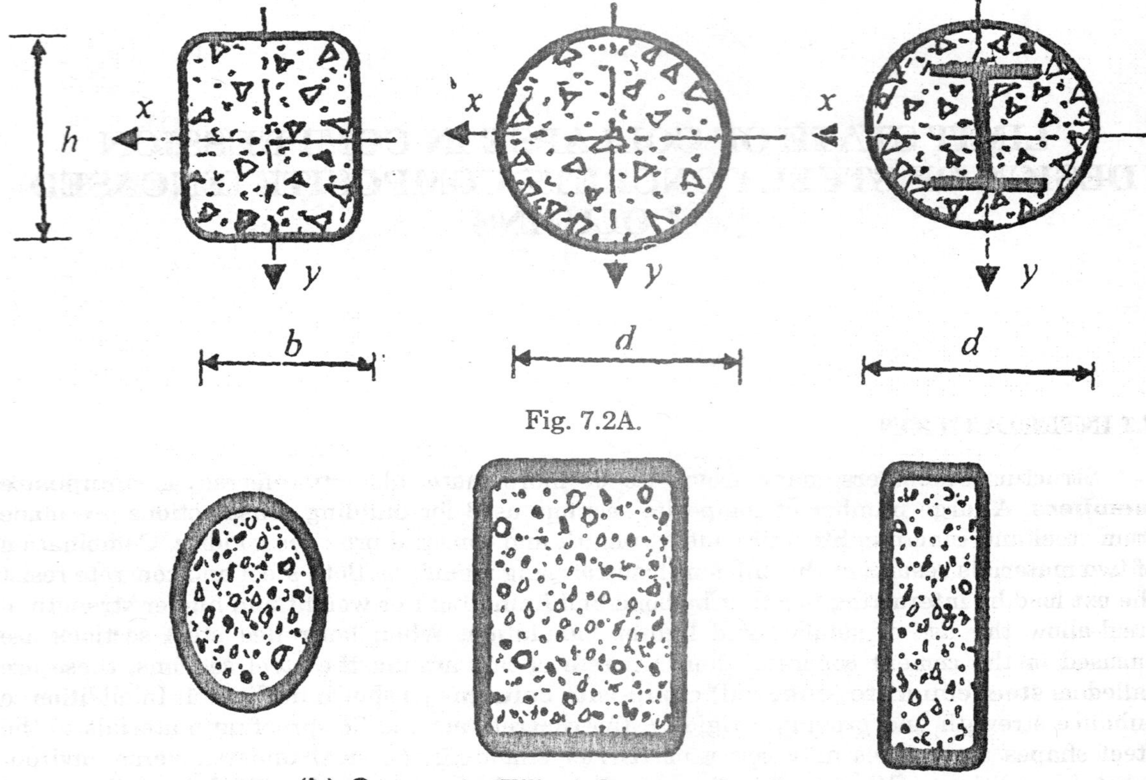

Fig. 7.2A.

(b) Concrete Filled Composite Sections

Fig. 7.2B.

The advantages of composite columns include (i) increased strength (ii) increased stiffness, (iii) good-fire resistance, (iv) protection from the corrosion.

As compared to complete structural steel shapes, these columns given significant economic advantages. In concrete filled tubular sections, the form-work is not needed. High rise building erection may be carried in efficient manner.

Supplementary reinforcement in the concrete encasement, prevents spalling of concrete.

During composite construction, bare steel sections support the intial construction loads, which include the weight of structure during construction.

Later on, the concrete is cast around steel sections or filled inside tubular sections. Subsequent concrete addition enables the building frame to limit the sway and lateral deflections easily.

Formerly, **in working stress method**, it was specified in IS : 800-1984 that encased column when the following conditions are fulfilled.

(a) A member is of symmetrical I-shape or a single I-beam or channels back to back, with or without flange plates.

(b) The overall dimensions of the steel do not exceed 750 mm x 450 mm over plating wen used, the larger dimensions being measured prallel to web.

(c) The column is unpainted and it is solidly encased in ordinary concrete with 20 mm aggregate (unless solidly can be obtained with a larger aggregate) and grade designation of concrete minimum M20.

(d) The minimum width of solid casong in equal to (bo + 100) mm where, bo, is the width of the steel flanges in mm.

(e) The surface and edges of the steel column have a concrete cover of not less than 50mm.

(f) The casing is effectively reinforced with steel-wires. The wires shall be at least 5 mm of diameter and the reinforcement shall be in the form of stirrups or bending at not more than 150 mm pitch so arranged as to pass thrugh the centre of the covering of the edges and the outer faces of the flanges and supported by longitudinal spacing bars not less than four in number.

(g) The steel cores in encased columns shall be accurately machined at splices and provisions shall be made for the alignment of column. At the column base, the provisions shall be made to transfer the load to the footing at safe unit stresses in accordance with IS : 456-2000.

(h) The axial load an the cased column should not exced two times that would be allowed on the un cased section. The slenderness ratio of the uncased section for its complete length centre to centre of connections should not exceed 250.

The steel column section shall be considered as carrying the entire load but allowance shall be made for stiffened effect. This allowance may be made by assuming the radius of gyration, r of the column section about the axis in the plane of web to be 0.2 (bo + 100) mm where, bo is the width of the steel, flanges in mm. The radius of gyration doubt its other axis shall be taken as that of the uncased section.

The slenderness ration of the uncased section for its length centre to centreof connection shouldnot exceed 250.

In the LRFD specification (18) [Load Resistance Factor Design], Certain conditions have to be satisfied for composite columns.

(a) The cross-sectional area of the steel shape is at least 4 (four) percent of the total compsite area.

(b) Longitudinal reinforcement and lateral ties are provided for concrete - encased member. The cross-sectional area of the reinforcing bars shall be 0.1778 mm^2 per mm.

To avoid spalling, lateral ties shall be placed at a spacing not greater than 2/3 the least dimension of the composite cross-section.

Where as in IS 456-2000, it is specified that the cross-sectional area of the **metal core** hot rolled steel section) shall not exceed 20 per cent of the gross-area of the column. For **hollow metal core** (hot rolled tubular sections) shall be filled with concrete.

The quantity of longitudinal and spiral reinforcement and the requirements as to spacing of bars, details of splices and thickness of protective shell outside the spiral, shall conform to requirements as per IS 456-1978.

A clearance of atleast 75 mm shall be maintained between the spiral all the metal core at all points, except that when the core consists of structural H-columns, the minimum clearance may be reduced to 50 mm.

Further, it is specified in IS : 456-2000 that the metal core of composite columns shall be designed to carry safely any construction or other loads to be placed upon them prior to their encasedment in concrete.

Splices if necessary and the connections of metal core have been specified. Transfer of loads to the metal core shall be provided by the use of bearing members, (such as (i) billets, brackets or other positive connections). These shall be provided at the top of the metal core and at the intermediate floor level where needed.

In the limit state method, a composite column is also designed for the ultimate load (i.e, factored load. A member is considered adequate, when the internal forces and moments due to most unfavourable load combination should not be more than the design resistance of the cross-section.

Local buckling of steel sections may be eliminated. However, the reduction in the compression resistance due to overall buckling should be considered together. With the influences of residual stresses and initial imperfection. At the same time the seconder order effects in the slender columns, creep and shrinkage concrete under long time loading are also considered in case, these are significant.

Reduction in flexural stiffness due to cracking of the concrete in the tension area is also considered.

7.2 TYPES OF STEEL CONCRETE COMPOSITE COLUMNS

In Art. 7.1, following two types of composite columns have been mentioned.

1. Steel concrete composite columns

Hot rolled steel sections (One or more) are encased in cement concrete as shown in Fig. 7.3. In addition to supporting a proportion of the load acting on the column the cement-concrete around the steel section, increases the stiffening and makes the steel section mare effective against local buckling as well as overall buckling. In order to avoid spalling of concrete cover specially in the case of eccentrically loaded columns, lateral reinforcement (ties or spiral) is needed. Cement concrete provides fire-proofing of the steel sections.

TYPES OF COMPOSITE COLUMNS

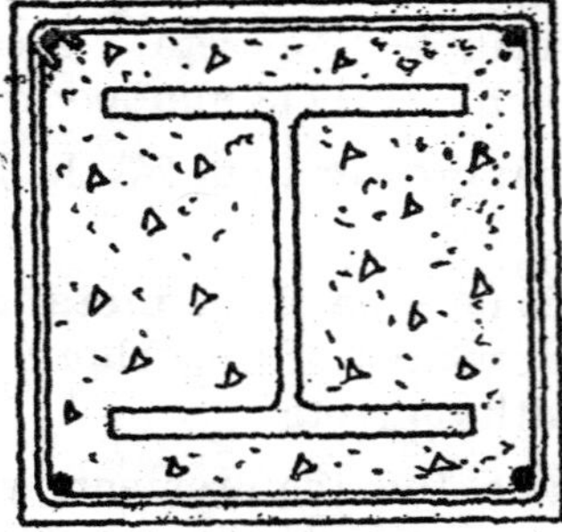

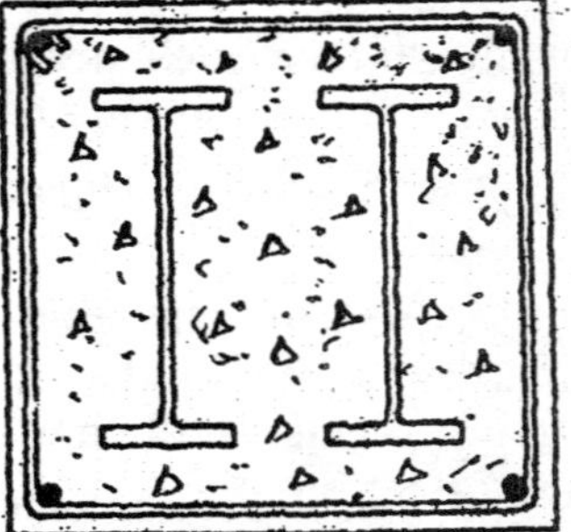

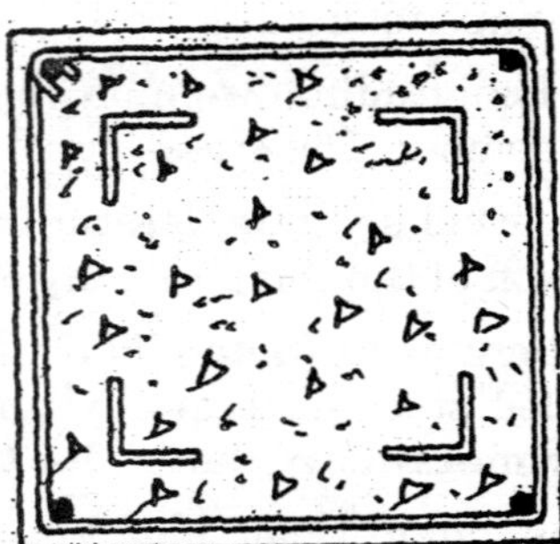

Fig. 7.3. Concrete-encased composite columns

2. Concrete filled hot rolled steel tubular sections

Concrete filled hot rolled steel sections composite columns as shown in Fig. 7.4, do not need reinforcement and form-work. Hollow square rectangular lar and circular tubes ae used and filled with cement concrete as shown in Fig. 7.4. The strength of concrete is increased since it is axially contained within the tube. However, steel tubes do not get protection against fire. There is difficulty of effective connections.

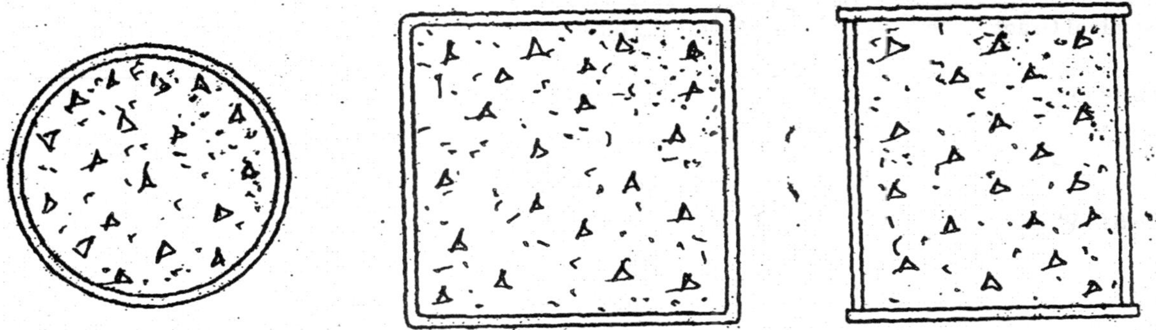

Fig. 7.4. Concrete filled composite columns

Two steel channels facing each other and connected together by batten plates at the intermediate and end positions as shown in Fig. 7.5.

The space in between the channels is filled with cement concrete. This does not require reinforcement. There are advantages when channel sections are used.

1. The load carrying capacity is increased by increasing the distance between channels.

2. An easy approach is available for inside core. This type of composite columns, need simple and cheap shuttering. Since the structural steel is placed on the outer faces (i.e, at the distance) and therefore, it is used most efficiently steel and cement concrete (two materials) are used in both these types of composite columns. These a materials have been described in the subsequence sections.

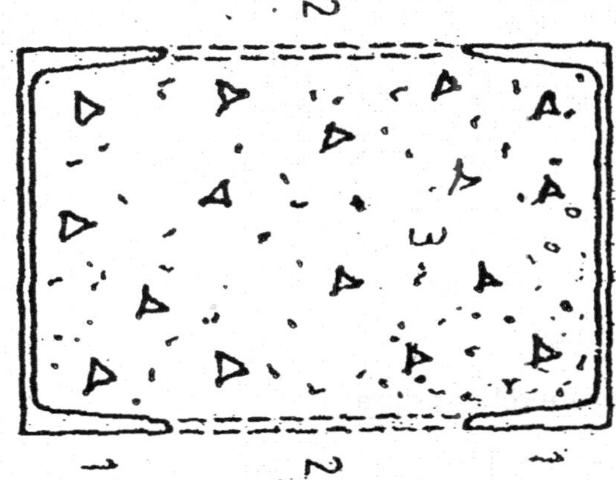

Fig. 7.5. Battened Composite Column
(1) Steel Channels; (2) End and intermediate batten plates; (3) Plain concretecore

7.3 STRUCTURAL STEEL

Structural steel used (before fabrication) confirm to IS : 1977-1975 IS : 8500-1977 and IS: 2062-1992 as appropriate. Some of the grades of commonly used in construction IS : 961-1975 and IS : 1977-1975 have been listed in Tables 7.1 (A) and 7.1 (B)

TABLE 7.1 (A). YIELD STRENGTH, σ_Y OF STEEL SECTIONS

Nominal Grade	Nominal Thickness/Diameter (mm)	Yield Stress, σ_y N/mm²
Fe 570-HT	t < 6 mm	350
	6 < t < 28 mm	350
	28 t < 45 mm	340
Fe 540 W-HT	t < 6	350
	6 < t < 16	350
	16 < t < 32	340
Fe 410-0	t < 6	250
	6 < t < 20	250
	20 < t < 40	240

TABLE 7.1 (B). YIELD STRENGTH, σ_Y OF STEEL SECTIONS AS PER IS : 2062-1992

Nominal Grade	Nominal Thickness/ Diameter (mm)	Yield Stress, σ_y N/mm²
Fe 410 WA	< 20	250
	20- 40	240
	>40	230
Fe 410 WB	<20	250
	20-40	240
	>40	230
Fe 410 WC	<20	250
	20-40	240
	>40	230

Steel used for reinforcement

Hot rolled steel composite columns are encased in cement concrete. These columns need steel reinforcement bars.

However, hot rolled tubular steel sections are filled with cement concrete. These columns do not require reinforcement.

In composite beams, dutility property of reinforcement bars has a significant influence. Whereas in the composite columns ductility has little effect.

Grades of steels used in construction have been listed in Table 7.2 (A).

TABLE 7.2 (A). CHARACTERIS STRENGTH OF REINFORCING STEEL BARS

Grade of Steel	Indian Standard	Nominal size (mm)	Yield stress s_y (N/mm²)
Mild steel Grade I (Plain bars)	IS : 432-(Part-I) - 1982	$d < 290$	250
		$20 < d < 50$	240
Mild Steel Grade II (Plain bars)	IS : 432-(Part-I) - 1982	$d < 20$	225
		$20 < d < 50$	215
Medium tensile steel (Plain bars)	IS : 432 - (Part-I) 1982	$d < 16$	540
		$16 < d < 32$	540
		$32 < d < 50$	510
Medium tensile steel (Hot-rolled Defomed bars and Cold-twisted Deformed bars	IS 1986	Bars of all sizes	415
	1985		500
			550

7.4 CEMENT CONCRETE

Value of strength of the material below which not more than 5 percent of the the test results are expected to fall is referred as **characteristic strength**

The design strength of the cement concrete shall be

$$f_d = (f_{ck}/\gamma_{ms}) \qquad \text{(i)}$$

where, f_{ck} = Characteris strength of cement concrete

γ_{ms} = Partial safety factor for the strength of cement concrete (γ_{ms} = 1.5).

Characteristic compressive strength is 150 mm cube strength at 28 days expressed in N/mm². For the different grades of concrete, the characteristic compressive strength has been listed in Table 7.2 (B).

TABLE 7.2 (B). CHARACTERISTIC COMPRESSIVE CUBE STRENGTH

Grade Designation	Characteristic compressive strength at 28 days (in/mm²)
M20	20
M25	25
M30	30
M35	35
M40	40

TABLE 7.2 (C). CHARACTERISTIC COMPRESSIVE (CYLINDER) STRENGTH, AND MEAN TENSILE STRENGTH AND VALUES OF YOUNG'S MODULUS OF ELASTICITY (E_{CM}- SHORT TIME)

Grade Designation	f_{ck} cylinder strength (N/mm²)	f_{ct}/m mean tensile (N/mm²)	$E_{cm} = 5000 \times (f_{ck})^{1/2}$ (N/mm²)
M25	20	2.2	25 x 10^3
M30	24	2.6	27.386 x 10^3
M35	35	2.8	29 .580 x 10^3
M40	32	3.3	31.622 x 10^3

For light weight concrete

The values of $E_{cm}$is found as under

= E_{cm} (as per Table 7.2 (c))

multiplied by p 2400 where, ρ is unit mass (kg/m³).

7.5 DEFINITIONS OF SOME TERMS USED

Limit State design method need definition of some terms used for design of steel and concrete composite columns whether encased with cement concrete or filled with cement concrete.

1. Partial safety factors for loads γ_{fL} for different combination of loads.

TABLE 7.3

Combination of loads	γ_{fL}		
	DL	LL	WL
Dead load (unfavourable effects)	1.35	-	-
Dead load (restraining uplift or overturning)	1.00	-	-
Dead Load + Imposed load	1.35	1.5	-
Dead load + Wind load	1.35	-	1.5
D.L. + Imposed Load + WL(mg/r)	1.35	1.05	1.5
D.L. +I.L. (major) + W.L.	1.35	1.5	1.05

2. Partial safety factorfor strength of materials, γ_{ms}

TABLE 7.4

Materials	γ_{ms}
1. Hot rolled steel sections.	1.15
2. Cement Concrete	1.50
3. Steel reinforcement bars	1.15

3. Squash load

A short compression test specimen, sufficiently long for use in measuring the stress strain relationship for the complete cross-section is referred as a **stub column**. However, it is short enough to avoid buckling in the elastic range.

The length of a stub column is long enough to contain the same residual stress pattern that exists in the column cross-section.

Professor Bruce G. Johnston (Reference : Guide to Design Criterion for Metal Compression Members. John Wiley and Sons, New York, 1966) has specified the length of stub column as under.

(i) (2 x d + 250) mm or

(ii) 3d minimum

(iii) Maximum length = 25 x radius of gyration about weak axis of the section, or

(iv) 5d (depth).

A perfectly straight short composite column is loaded axially. The strain values in the two structural materials remain equal, such that the the steel yields before the concrete reaches its full strength. The residual stress pattern in the steel section is retained.

The concrete prevents the steel column from buckling.

Further increase in the applied load will be supported by the concrete until the concrete finally attains ultimate strength.

The final value of load at this level is the ultimate strength of axially loaded short column. This load is referred as **squash load.**

7.6. ULTIMATE LOAD CARRYING CAPACITY (PLASTIC RESISTANCE) OF COMPOSITE COLUMNS

Simplified design method based on Euro-code 4 is used for design of Isolated symmetric columns of uniform cross-section in braced or non-sway frames. Euro-pean buckling curves for steel are adopted as the basis for design.

In case, a sufficiently stiff frame is subjected to in-plane horizontal forces, the additional internal forces and moments due to the consequent horizontal displacement of the joints are neglected and frame is classified as non-sway.

It is to note that this method is not applicable to sway-columns.

Two casesof prismatic composite columns having doubly symmetrical cross-sectionshave been considered here.

Case 1. Squash (Ultimate) Load

Axial (ultimate) load P_u on a composite column consisting of structural steel or cast-iron thoroughly encased in concrete reinforced with both longitudinal and spiral reinforcement is obtained as the sum of the resistances of it components is given by

$$P_u = \sigma_c \cdot \left(\alpha_c \cdot \frac{f_{ck} \cdot cy}{\gamma_{ms.c}}\right) + A_{sc} \cdot \left(\frac{f_{sc.}}{\gamma_{ms.sc}}\right) + A_{ms} \cdot \left(\frac{\sigma_y}{\gamma_{ms.s}}\right) \quad (7.1)$$

The designer should check for each of the principal axes of the composite column

P is less than or equal to χ (chi) times the plastic resistance in compression that is,

$$P \leq (\chi \cdot P_u), \text{ where} \tag{i}$$

χ is the reduction factor due to column buckling. It is a function of non-dimensional slenderness of the composite column.

$$\chi = \left(1 / \left[\varphi + \left(\varphi^2 - \bar{\lambda}^2\right)\right]^{1/2}\right) \text{ and} \tag{ii}$$

However, χ should be less than or and equal to 1.00

$$\phi = 0.5\left[1 + \alpha\left(\bar{\lambda} - 0.2\right) + \bar{\lambda}^2\right] \tag{iii}$$

$$\bar{\lambda} = \left(\frac{\sigma_y}{\pi^2 E}\right)^{1/2} \cdot \left(\frac{l}{r}\right). \text{ It is} \tag{iv}$$

defined subsequently in Eqs. 7.7(A) and 7.7(B), and α = Perr-Roberston's coefficien.It is also referred as imperfection factor.

The effects of imperfection and residual stresses onthe column are accounted for the values of α. Seperate values of α for the three different European buckling curves a, b, and C as 0.21, 0.35 and 0.49, respectively. It may be noted.

Curve a : For concrete filled tubular sections.

Curve b : For completely spartiallencased steel I-sections buckling about.

Curve c : For buckling about minor axis that the second order moment due to imperfection has been incorporated in this procedure by making the use of multiple curves. Therefore, no other considerations are needed.

Note : In IS 456-2000, the value of eccentricity has been specified as 200 mm irrespective of geometry of the column. Imperfection factor α permits for the eccentricity of the applied load. Further provision is not required for steel composite columns.

By making the use of $\bar{\lambda}$ (non-dimensional slenderness/and χ (chi), the design buckling resistance of the steel composite columns in compression P_u or $P_{buckling}$ may be obtained.

In case, any one of the following condition is satisfied, then effect of isolated non-sway check is not applied.

1. In case, the axial force in the column is less than to percent of P_{cr}.

2. Value of $\bar{\lambda}$ is less than 0.2 . Fig. 7.6 shows European buckling curves proposed for composite columns. These are adopted χ as per grade of steel and axis of bending.

α_c = strength coefficient for concrete (α_c = 0.85 for completely or partially concrete encased steel sections) and α_c = 1.00 for concreted filled tubular sections).

A_c = Net area of concrete section ($A_c = A_{gross} - A_{sc} - A_{ms}$)

A_{sc} = Cross-sectional area of longitudinal bar reinforcement

f_{sc} = Yield strength of the reinforcing steel

A_{ms} = Cross-sectional area of the steel (metal core)

σ_y = Yield strength of steel-section

$f_{ck.cy}$ = characteristic compressive strength (cylindrical), for the concrete

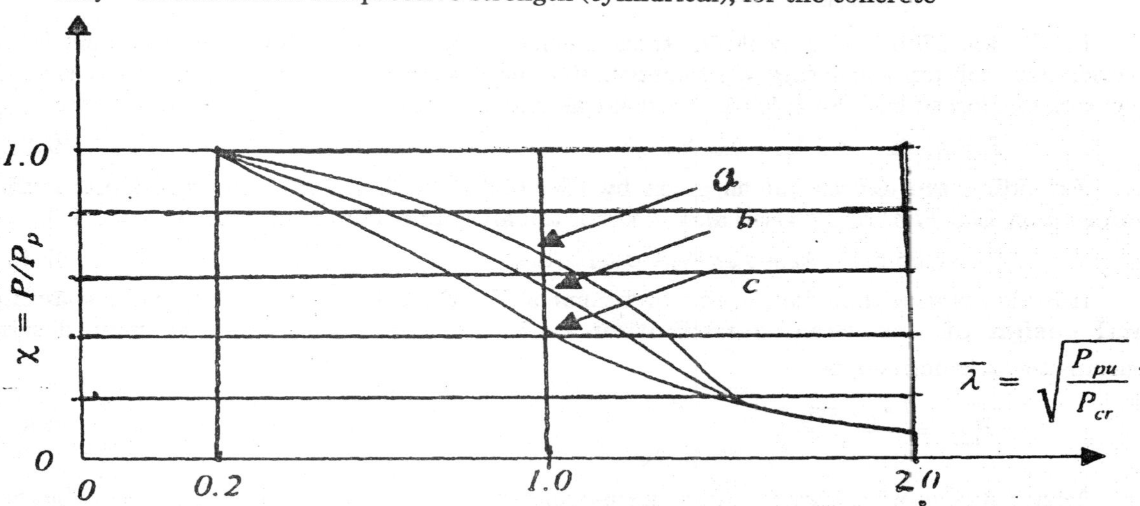

European buckling curves

Fig. 7.6

1. Ultimate axial load For tubular sections with filled concrete (σ_c = 1.00)

$$P_u = 1.00.\ A_c\ .\left(\frac{f_{ck.cy}}{\gamma_{ms.c}}\right) + A_{sc}\left(\frac{f_{sc}}{\gamma_{ms.sc}}\right) + A_{ms}\cdot\left(\frac{\sigma_y}{\gamma_{ms.s}}\right) \quad (7.2)$$

2. Ultimate axial load For steel (Metal core) sections encased in concrete (σ_c = 0.85)

$$P_u = 0.85\ A_c.\left(\frac{f_{ck.cy}}{\gamma_{ms.c}}\right) + A_{sc}\cdot\left(\frac{f_{sc.}}{\gamma_{ms.sc}}\right) + A_{ms}\cdot x\left(\frac{\sigma_y}{\gamma_{ms.s}}\right) \quad (7.3)$$

For ease of expression

$$\sigma_{my.1} = \left(\frac{\sigma_y}{\gamma_{ms.s}}\right) = \left(\frac{\sigma_y}{1.15}\right) = 0.87\,\sigma_y \quad (7.4)A$$

As it is well established that the cylinder strength is approximately equal to 0.8 times cube strength. Considering the factors influencing the strength of concrete and mainly the effect of sustained load, the maximum compressive stress is taken equal to 0.85 cylinder strength. Therefore, ($\gamma_{ms.conc}$ = 1.5).

$$\sigma_{cc.1} = \left(\frac{0.85x0.80\,x\,f_{ck.(cube)}}{1.5}\right)$$

$$= \left(\frac{0.67 f_{ck}}{1.5}\right) = 0.446 f_{ck} = 0.4 f_{ck} \quad (7.4)\ (ii)$$

$$\sigma_{sc.1} = \left(\frac{\sigma_y}{\gamma_{ms.reinf.}}\right) = \left(\frac{1}{1.15} x\, \sigma_y\right) = 0.87 \sigma_y \quad (7.4)\ (iii)$$

In IS : 456-2000 (using working stress method of design), the allowable axial load, P for a composite column consisting of structural steel or cast-iron column thoroughly encased in concrete reinforced both longitudinal and spiral reinforcement, shall not exceed the value

$$P = \sigma_{cc.} A_c + \sigma_{sc.} A_{sc} + \sigma_{mc.} A_m \quad (7.5)$$

Extending the expression as given by IS : 456-2000 (Eq. 7.5), and substituting the values from Eqs. (7.4) (i), (7.4) (ii) and (7.4) (iii) the allowable axial load. (Since, $\alpha_c = 0.85$).

$$P_u = 0.85\, \sigma_{cc.1} . A_c + \sigma_{sc.1} A_{sc} + \sigma_{ms.1} . A_{m.s} \quad (7.6)$$

It is also to note that the recommendations of IS : 456-2000 is also used for the **spacing and design of ties. Steel contribution ratio,** β_a is used as an important design parameter. It is defined as

$$\beta_{a.S.T} = \left(\frac{A_{m.s.}\, x \sigma_y}{P_u}\right) \quad (7.6)$$

Note : Above procedure for determining squash (i.e, ultimate load is also applicable for the rectangular and square hollow (tubular) sections filled with cement concrete.

However, for **circular tubular sections,** filled with cement concrete, there is some additional provision as described in step 1 (B).

Case 2: Additional provision for circular tubular (hollow sections) filled with cement concrete

Composite columns which consists of circular tubular (hollow) sections some confining influence is provided. Due to this confining effect, its resistance is increased. However, this influence is significant only in **Stocky (Strongly built) columns.**

For composite columns with non-dimensional slenderness ratio

$$\bar{\lambda} = \left(\frac{P_u}{P_{cr}}\right)^{1/2}, \text{ where,} \quad (7.7)A$$

Plastic resistance of the section in compression

P_{cr} = Elastic buckling load of the as in Eq. 7.2 or Eq. 7.6 column.

$$= \pi^2 \left(\frac{(EI)_e}{l^2}\right) = \pi^2 . \left(\frac{E.A.r^2}{l^2}\right) = \left(\frac{\pi^2 .EA}{(l/r)^2}\right)$$

$(EI)_e$ = Effective elastic flexural stiffness of the column

$$\text{Then, } \bar{\lambda} = \left(\frac{P_u}{\pi^2 EA/(l/r)^2}\right)^{1/2}$$

$$\bar{\lambda} = \left(\frac{P_u}{A} . \pi^2 . E\right)^{1/2} . \left(\frac{l}{r}\right) \qquad \left(\therefore \frac{P_u}{A} = \sigma_y\right)$$

$$\bar{\lambda} = \left(\frac{\sigma_y}{\pi^2 E}\right)^{1/2} . \left(\frac{l}{r}\right) \tag{7.7)(B}$$

After obtaining $\bar{\lambda}$, the buckling resistance to compression of the column may be found. Where, the eccentricity e of the applied load is not more than 0.1 d (where d is the outer diameter of the circular tubular (section).

Effect of eccentricity, e

The effect of eccentricity, e (when its value is more than 0.1 d) is to be considere, e is defined as

$$e = \left(\frac{M}{P}\right) < 0.10 \text{ d} \tag{7.8}$$

where, M = the maximum applied design moment (The second order effects are neglected).

P = Applied design moment

Plastic resistance of the cross-section of the column (circular tubular section) filled with cement concrete is obtained by using two coefficient η_1 and η_2.

$$P_u = \sigma_{cc.1} A_c \left(1 + \eta_1 \frac{t}{d} \cdot \frac{\sigma_y}{f_{ck}}\right) + \sigma_{sc.1} . A_{sc.1} + \sigma_{ms1} A_{ms} \tag{7.9}$$

where, t is the thickness of circular tube section and η_1 and η_2 are two coefficients, which are obtained as under.

$$\eta_1 = \eta_{10}\left(1 - \frac{10e}{d}\right), \text{ and} \tag{7.10)A}$$

$$\eta_2 = \eta_{20} \left(1 - \eta_{20}\right) \frac{10e}{d} \tag{7.10)B}$$

Basic values of η_{10} and η_{20} to permit the effectof tri-axial confinement in circular tube section filled with cement concrete is applicable for concrete mixes with $(f_{ck})_{cy}$ = 25 to 55 N/mm² (as per EC4).

Plastic resistance to compression of a composite cross-section, P_u represents the maximum load,which may applied to a short column. For long columns with low elastic, critical load, overall buckling may be critical.

For an ideal column as shown in Fig. 7.7 and Fig. 7.8 and the horizontal line represents, P_u in a typical buckling curve. These buckling curves are the function of the slenderness of the column.

These two curves limit the compressive resistance of ideal column.

Fig. 7.8 shows the buckling resistance of a column. For the convenience, column strength curve is plotted in a non-dimensionalised form. Chi, χ represents the ratio (P/P_u) plotted as an ordinate an y-axis, Chi, χ is called the reduction faction. Ratio $(P_u/P_{cr})^{1/2}$ is plotted as absicca.

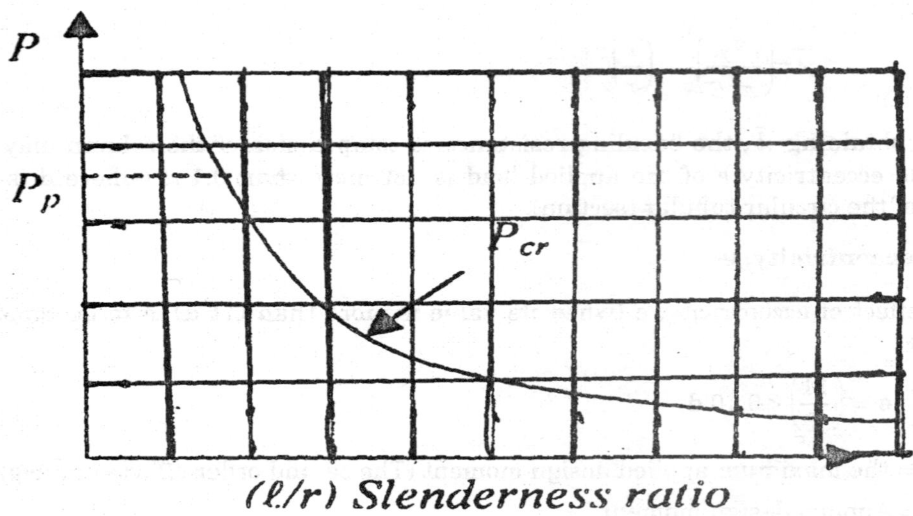

Fig. 7.7. Idealised column buckling curve.

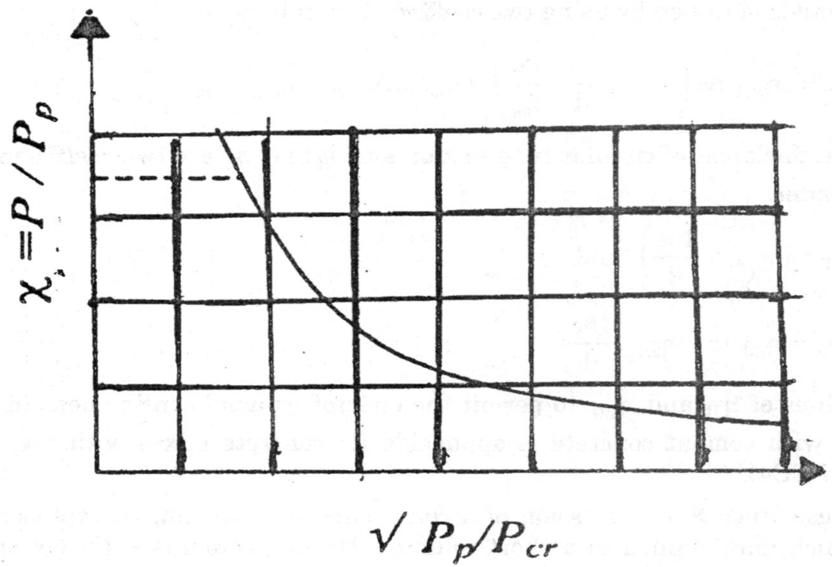

Fig. 7.8. Non-dimensionalised column buckling curve.

Due to presence of residual stresses and geometric imperfections these strength curvesfor the practical columns are different.

After incorporating the effects of both (residual stresses and geometric imperfections) the the European buckling curves are drawn. These curves form the basis of column buckling design for both steel and composite columns in codes EC3 and EC4). $\bar{\lambda}$-non dimensional slenderness is needed. It is found first as above.

TABLE 7.5. VALUES OF η_{10} AND η_{20}, WHICH DEPENDS ON $\bar{\lambda}$ FOUND AS ABOVE

Values of $\bar{\lambda}$	0.0	0.1	0.2	0.3	0.4	Equal to more than 0.5
η_{10}	4.90	3.22	1.88	0.88	0.22	0.00
η_{20}	0.75	0.80	0.85	0.90	0.99	1.00

15 per cent increase in the resistance of circular tube filled with cement concrete subjected to axial load when the effect of confinement due to tri-axial effect is considered.

It is to be noted from Table 7.5, when the value of eccentricity, e is more than 0.1 d or in case non-dimensional slenderness ratio is more than 0.5, then, η_{10} is zero and η_{20} is unity.

Step 2 : Local buckling of hot rolled steel sections

In case, the local buckling loes not occur, Eq. (7.2) and (Eq. 7.6) hold true. Hot rolled steel sections are so manufactured and produced that the prepature local buckling is prevented.

For cement concrete filled circular columns (d/t) ratio is kept equal to or less than 85 $\in^2$.

For concrete filled rectangular tubular sections, the ratio (h/t) is kept equal to or less than 50$\in$.

For the partially encased I-sections, the ratio (b_f/t_f) is 43 $\in$.

Where, yield ratio, $\in=(250/\sigma_y)^{1/2}$ and σ_y yield strength of the steel-section in N/mm².

This type of verification for local buckling is not required for the completely encased steel sections. Local buckling does not occur in such columns since the column is effectively surrounded by the concrete.

For the completely encased steel sections, concrete cover to the flange should not be less 40 mm, nor less than one-sixth of the flange width.

7.7 EFFECTIVE ELASTIC FLEXURAL STIFFNESS

Effective elastic flexural stiffness is an important factor which influences buckling load. Euler's buckling (critica) load for a column with both ends pin ended is given by

$$P_E = P_{cr} = \left(\frac{\pi^2 (EI)_e}{l^2}\right) \qquad \text{(i)}$$

$$P_E = P_{cr} = \left(\frac{\pi^2 E.A.r^2)}{l^2}\right) \qquad \text{(ii)}$$

$$P_E = P_{cr} = \left(\frac{\pi^2 E.A)}{(l/r)^2}\right) \qquad \text{(iii)}$$

$$\sigma_y = \left(\frac{P_E}{A}\right) = \left(\frac{P_{cr}}{A}\right) = \left(\frac{\pi^2 E}{(l/r)^2}\right) \qquad (7.11)$$

where,

(EI_e) = Effective flexural stiffness and

(l/r) = Slenderness ratio. Ratio of effective length to the appropriate radius of gyration.

Creep and shrinkage properties of concrete reduce the flexural stiffness. Creep and shrinkage are time dependent phenomena. The effective flexural stiffness, $(EI)_e$ is to be studied and determined seperately for

1. Short term loading, and
2. Long term loading.

1. Short term loading

Hot rolled steel sections, cement concrete and the steel reinforcement are three components of composite columns. Sum of the flexural stiffness of these individual components of the cross-sections given the effective flexural stiffness $(EI)_e$, of the composite column

$$(EI)_e = E_s I_s + 0.8\ E_c + E_{sc}.I_{sc} \tag{7.12}$$

where,

E_s and E_{sc} = Moduli of elasticity of steel and steel reinforcement bars, respectively.

I_s and I_{sc} = Second moment of areas of the steel section, and steel reinforcement bars about the axis of bending considered.

$0.8\ E_{cd}$ = Effective stiffness of the concrete. It is multiplied by an empirical factor 0.8, which one is obtained by calibration excise to give good agreement with the results.

I_c = Moment of inertia of uncracked column section about the centroid.

$E_{cd} = (E_{cm}/\gamma_{ms.c})$

E_{cm} = Secant modulus of the concrete.

Instead of 1.5 $\gamma_{ms.c}$ is reduced to 1.35 according to Eurocode EC2.

Note : It is unusual to divide E_{cm} by $\gamma_{ms.c}$ of width coefficient good results.

2. Long term loading

The effective flexural stiffness of the long composite column decreases under long-term loading due to shrinkage and creep. As a result of this, resistance to buckling also decreases. For long-columns only. In case, the buckling length to depth ratio of a composite column is more than 15, then, effect of long term loading should be considered.

In case, the eccentricity of loading (e = M/P) is more than two times the cross section dimension, the effect of bending moment distribution resulting due to increased deflection due to creep and shrinkage of concrete will be very small. Hence, it is neglected. Then, it is not required to keep provision for long term loading.

At the same time, provision is also not needed, in case, the non-dimensional slenderness, $\bar{\lambda}$ of the column is less than the limiting values listed in Table 7.6.

TABLE 7.6. LIMITING VALUES OF $\bar{\lambda}$ FOR LONG TERM LOADING

Composite column concrete	Braced Non-sway system	Unbraced and/or sway system
Encased steel columns	0.8	0.5
Concrete filled cross-section	$\left(\dfrac{0.8}{1-B_{a.ST}}\right)$	$\left(\dfrac{0.5}{1-B_{a.ST}}\right)$

Note : $B_{a.ST}$ steel contribution ratio $B_{a.ST} = \left(\sigma_y . A_s / P_u\right)$ as defined earlier Eq. 7.6.

In case, the values of $\bar{\lambda}$ becomes more than the limits listed in Table 7.6, and (eccentricity/depth) ratio is less than 2 (two) the effect of creep and shrinkage of concrete should be permitted by using the modulus of elasticity of the concrete $E_{c.w}$ instead of E_{cd}. Where,

$$E_{c.w} = E_{cd}\left(1-\frac{0.5P_d}{P}\right) \tag{7.13}$$

where, P is applied design load and P_d is the portion of applied design load permanently acting on the column.

The effect of long term loading is neglected for concrete filled tubular sections with $\bar{\lambda} < 2.0$ permitted $B_{a.ST}$ is more than 6.6 for braced or non-sway column and 0.75 for unbraced sway columns.

7.8 LOAD CARRYING CAPACITY/RESISTANCE STEEL COMPOSITE COLUMNS SUBJECTED TO AXIAL LOAD

The procedure for determination of safe axial load (ultimate), that is, plastic resistance of a steel column ISHB 400, @ 0.822 kN/m is explained by simultaneously carrying out an illustrative example by limit state design. Let the effective length of the given composite column be 4m. The column is hinged at both the ends.

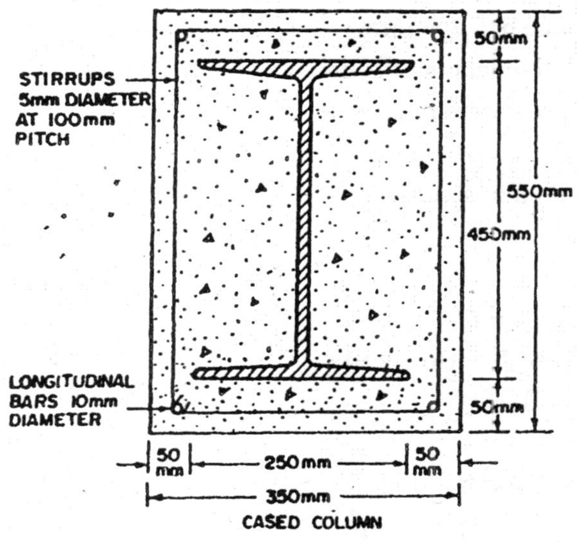

Fig. 7.9.

Solution : Given hot rolled steel section to be encased in cement concrete is ISHB, 400 @ 0.822 kN/m.

Step 1(A) : Following material and strength properties are noted

Young's modulus of steel $E_s = 2 \times 10^5$ N/mm^2

Yield stress for steel, $\sigma_y = 250$ N/mm^2

HYSD Steel bars of grade, Fe 415 shall be used, this is $f_{sc} = 415$ N/mm^2

Value of $\pi = 3.14159$

$f_{ck.cube} = 30$ N/mm^2

Characteristic value of the grade of concrete, $f_{ck.cy1} = 24$ N/mm^2

Young's modulus of elasticity for concrete

$E_{cm} = 27386$ N/mm^2

Partial safety for strength of steel

$\gamma_{ms.ST} = 1.15$

Concrete $\gamma_{ms.con} = 1.5$

Concrete (reduced) = $\gamma_{ms.con} = 1.35$

E_{cm} (Secant modulus) of elasticity

$E_{cm} = 5000\,(f_{ck.cube})^{1/2} = 5000 \times (30)^{1/2} = 27386.13$ N/mm^2

Perry-Roberston (imperfection) coefficients

$\alpha_{xx} = 0.35$

$\alpha_{yy} = 0.49$, and $\alpha_c = 0.85$

Step 1 (B) From ISI Handbook No. 1 following geometrical properties of the steel section, ISHB 400, @ 0.822 EN/m are as under :

Width of the flange $b_f = 250$ mm

Thickness of the flange, $t_f = 12.7$ mm

Overall depth of the section, $h_o = 400$ mm

Thickness of Web, $t_w = 10.6$ mm

Cross-sectional area of steel section = A = 10466 mm^2

Radius of gyration

About strong-axis, $r_{xx} = 166.1$ mm

About weak-axis, $r_{yy} = 51.6$ mm

Clear depth of web, $d_w = (h_o - 2\,t_f)$

$d_w = (400 - 2 \times 127)$

$= 374.6$ mm

Step 1(C). Effective length = 4 m

Following are the cross-sectional properties of steel section ISHB 400, @ 0.822 kN/m.

Moment of inertia

About strong axis = $28823.5 \times 10^{+4}$ mm^4

About weak axis = $2783.0 \times 10^{+4}$ mm^4

Steel reinforcement bars

HYSD Fe 415 - 4 (four) in number

$\sigma_{y.STB}$ = 415 N/mm^2

Reinforcement 4 bars 12 mm diameter

$$A_{sc} = 4 \times \frac{\pi}{4} x (12)^2 = 452.3 \text{ mm}^2$$

Cement concrete cover upto vertical face each 50mm

= Width of flange +2 side cover of concrete)

Width of composite column = (250 + 2 x 50) = 350 mm

Clear cover (at the top and bottom) over flanges (each) =75 mm

Depth of composite column

= (400 + 2 x 75) = 550 mm

Gross area of composite column

= (350 x 550) = 192500 mm^2

Cross-sectional area of steel-section = 10466 mm

Area of concrete A_c = (192500 –10466 – 452.39)

= 181581.61 mm^2

Step 2 : Classification of Section

$$\text{Ratio } (0.5\, b_f/t_f) = \left(\frac{0.5 \times 250}{12.7}\right) = 9.843$$

$$\text{Ratio } (d_w/t_w) = \left(\frac{374.6}{10.6}\right) = 35.34$$

Yield stress ratio, epsilon

$$\epsilon = \left(\frac{250}{250}\right)^{1/2} = 1.00$$

As per IS : 800-2007, for rolled steel section, for compact section

$(0.5\, b_f/t_f) = (10.5 \times 1.00) > 9.843$

$(d_w/t_w) = (42 \times 1.00) > 35.34$

Hence, the given section ISHB 400, @ 0.822 kN/m is a **compact section**

Step 3 : Ultimate/Plastic resistance, P_u of the composite column (From Eq. 7.1)

$$P_u = \alpha_c\, A_c \left\{\frac{f_{ck}}{\gamma_{ms.c}}\right\} + A_{sc} \cdot \left(\frac{f_{sc}}{\gamma_{ms.sc}}\right) + A_{ms} \cdot \left(\frac{\sigma_y}{\gamma_{ms.s}}\right) \qquad \text{(i)}$$

Since,

$$\sigma_{ccl} = \left(\frac{0.85 x 0.80 x f_{ck.cube}}{1.5}\right) = 0.4 f_{ck} \quad \text{(ii)}$$

$$\sigma_{sc.1} = \left(\frac{\sigma_y}{\gamma_{ms.sc}}\right) = \left(\frac{1}{1.15}\sigma_y\right) = 0.87\,\sigma_y \quad \text{(iii)}$$

$$\sigma_{ms.1} = \left(\frac{\sigma_y}{\gamma_{ms.s}}\right) = \left(\frac{\sigma_y}{1.15}\right) = 0.87\,\sigma_y \quad \text{(iv)}$$

Substituting in the expression

$P_u = \alpha_c \,.\, A_c\, \sigma_{cc1} + 0.87\, A_{sc}\, \sigma_{sc1} + 0.87\, A_{ms.1}\, \sigma_y$

Area of concrete

$= (A_{gross} - A_{ms} - A_{sc})$

$= [350 \text{ x } 550 - 10466. - 452.3]$

$= 192500 - 10466 - 452.3]$

$= 181581.17 \text{ mm}^2$

For completelyor partially encased steel column, $\alpha = 0.85$

$P_{ult} = 0.85, A_{c.} \text{ x } (0.4\, f_{ck}) + 0.87\, A_{sc.}\sigma_y + 0.87\, A_{ms.}\sigma_y$ (vi)

$$P_{ult} = \left[\frac{0.85 x 181581.17 x 0.4 x 30}{1000}\right] + \left(\frac{0.87\ \ 452.3 \times 415}{1000}\right) + \left(\frac{0.87 \times 10466 \times 260}{1000}\right)$$

$P_{ult} = (1852.128 + 163.303 + 2367.41) = 4382.84\text{kN}$

Step 4 : Effective elastic flexural stiffness of the section

$I_{cx} = Ah^2$

$$= 452.3 \text{ x } \left(\frac{550}{2} - 25 - 10.6\right)^2 = 16.153 \text{ x } 10^6 \text{ mm}^4$$

$$I_{sc} = \left(\frac{350 \text{ x } 550^3}{12}\right) - [2.8823.5 + 1615.3] \times 10^4 = 566.13 \times 10^6\ mm^4$$

$$E_{cm} = \left(\frac{27386}{1.35}\right) = 2.0285.926 \times 10^4$$

About strong axis

$(EI)_{ex} = E_{ms}.\, I_{ax} + 0.8\, E_{cd}.I_{cx} + E_s\, I_{sc}$

$= 2 \text{ x } 10^5 \text{ x } 2882.35 \text{ x } 10^5$

$2.0285926 \times 10^4 + 0.800 \text{ x } 10^4 \text{ x } 16.153 \text{ x } 10^6 + 2 \text{ x } 10^5 \text{ x } 566.13 \text{ x } 10^6$

$(EI)_{ex} = 57.647 \text{ x } 10^{12} + 0.262 \text{ x } 10^{12} + 113.200 \text{ x } 10^{12}$

$= 171.109 \text{ x } 10^{12} \text{ Nmm}^2$

About weak axis

$(EI)_{ey} = 123.77 \times 10^{12}$ Nmm²

Step 5 : Non-dimensional slenderness

Effective length = 4000 mm

Step 5(A): Euler's (critical) axial load

$$P_{EX} = \left(\frac{\pi^2 (EI_x)_{eff}}{l^2}\right) = \left(\frac{3.14159^2 \, x 171{:}109 \, x 10^{12}}{4000 \, x 4000}\right) = 105.548 \text{ kN}$$

$$P_{Ey} = \left(\frac{\pi^2 (EI)_{ey}}{l^2}\right) = \left(\frac{3.14159^2 \, x 123.77 \, x 10^{12}}{4000 \, x 4000 \, x 1000}\right) = 76.347 \text{ kN}$$

Step 5(B) Axial load for the values of strength of materials at service load

$\gamma_{ms.c} = \gamma_{ms.s} = \gamma_{ms.sc} = 1.00$

$P = \alpha_c . A_c . f_{ck.\ cyL} + A_{ms.s.}\ \sigma_y + A_{sc.}\ \sigma_{sc.}$

$= \alpha_c . A_c . 0.80\ f_{ck.cube} + A_{ms.s}\ \sigma_y + A_{sc.}\ \sigma_{sc}$

$$= \left[\left(\frac{0.85 \, x 181581.17 \, x 30}{1000}\right) + \left(\frac{10466 \, x 260}{\cdot 1000}\right) + \left(\frac{452.3 \, x 415}{1000}\right)\right]$$

$= (4630.32 + 2721.164 + 187.70) = 7539.19$ kN

Step 5(C): Non-dimensional slenderness

$$\bar{\lambda}_{xx} = \left(\frac{P}{P_{Ex}}\right)^{1/2} = \left(\frac{7539.19}{105548}\right)^{1/2} = 0.267$$

$$\bar{\lambda}_{yy} = \left(\frac{P}{P_{Ey}}\right)^{1/2} = \left(\frac{7539.19}{76347}\right)^{1/2} = 0.314$$

Step 6 : Buckling resistance of the composite column

Buckling load, P_b should be less than χP_u. Where

χ = Reduction factor for the column buckling

Step 6(A) : Value of Chi (χ), about strong axis, $\alpha_x = 0.35$

$$\chi = \frac{1}{\left\{\phi_{xx} + \left(\phi_{xx}^2 - \bar{\lambda}_{xx}^2\right)^{1/2}\right\}}$$

$$\phi_{xx} = \frac{1}{2}\left[1 + \alpha_x\left(\bar{\lambda}_x - 0.20\right) + \bar{\lambda}_{xx}^2\right]$$

$$= \frac{1}{2}\left[1 + 0.35(0.267 - 0.20) + (0.267)^2\right] = 0.547$$

Chi, $\chi = (1/\{0.547 + (0.547^2 - 0.0717)^{1/2}\}) = 0.9766$

Step 6 (B) Value of Chi, χ about weak axis ($\bar{\lambda}_{yy} = 0.314$)

$$\varphi_{yy} = \frac{1}{2}\left[1 + \alpha_{yy}\left(\bar{\lambda}_{yy} - 0.20\right) + \bar{\lambda}_{yy}^2\right]$$

$$= \frac{1}{2}\left[1 + 0.5\left(0.314 - 0.20\right)^{1/2} + 0.314^2\right] = 0.578$$

$$\text{Chi, } \chi = \left[1/\left\{\varphi_{yy} + \left(\varphi_{yy}^2 - \bar{\lambda}_{yy}^2\right)^{1/2}\right\}\right]$$

$$= [1/\{0.578 + (0.578^2 - 0.314^2)^{1/2}\}] = 0.940$$

Step7 (A) Buckling load due to compression

$(P_b)_{xx} = \chi$, P `(Plastic resistance of the section)

= (0.9766 x 4382) = 4279.46 kN

$(P_b)_{yy} = \chi$. P (Plastic resistance of the section)

= 0.940 x 4382 = 4119.08 kN

Buckling load shall be lower value (i.e, whichever less). Therefore, P_b = 4119.08 kN

7.9 DESIGN PROCEDURE FOR STEEL COMPOSITE AXIALLY LOADED COLUMNS

Hot rolled steel sections are used for both the types of axially loaded columns (namely, (i) steel columns encased in cement concret,and (ii) cement concrete filled tubular columns.A trial section for the steel section is needed, which can carry complete load, while, it is uncased and the same shall be encased in the cement concrete later on. Under the circumstances, working stress method is used only to **select the trial section** of the steel.

Axial load to be carried by the column is given. The actual length of column and its support conditions at both the ends are given. Hence, the effective length of the column is known.

Step 1 : Trial section for steel column

Yield stress for the steel to be used is assumed. The slenderness ratio is also assumed as 70. Corresponding to this, the allowable stress in axial compression is noted as per IS : 800-1984 as Jac. Effective sectional area needed for hot rolled steel section, A is obtained

$$A = \left(\frac{\text{Given Compressive load}}{\sigma_{ac}}\right)$$

From ISI Handbook No. 1 (geometrical properties), a trial section is selected (preferrably, some ISHB column section).

Following geometrical properties of this section are noted

(i) Width of flange, b_f

(ii) Thickness of flange, t_f

(iii) Total depth of section, h_o

(iv) Thickness of web, t_w

(v) Moment of inertia about strong axis, I_{xx}

(vi) Moment of inertia about weak-axis, I_{yy}

(vii) Cross-sectional area, A

Step 2 : In case, for a rectangular composite column, width and depth are B_c parallel to strong axis (i.e, xx-axis) and D_c parallel to weak axis, (i.e, yy-axis, axis parallel to web). The moment of inertial, I_{yy} for this composite rectangular column.

$$I_{yy} = \frac{1}{12} D_c . B_c^3 \text{ and}$$

Area = D_c . B_c. Then, the radius of gyration, r_{yy} (parallel to the plane of web) is adopted as

$$r_{yy} = (b_o + 50 + 50) \text{ mm}$$

$$= (b_f + 100) \text{ mm, where}$$

concrete cover is kept 50 mm and $b_o = b_f$.

Step 3 : Cement concrete cover over the flanges has been specified 50 mm minimum. However, it may be taken as 75 mm each. On the top and bottom side.

Step 4 : Reinforcement is kept more than 4 per cent of the gross concrete area = (4 $B_c.D_c$/100).

Step 5 : Complete composite column encasing hot rolled steel section in cement concrete is checked. Whether, this section is adequate or not by. Limit State design as described in Art. 7.8.

Ultimate axial load carrying capacity (i.e, plastic resistance of the section), P_u is determined. The buckling resistance of the section, P_u is obtained.

Lower value of plastic resistance (i.e, whichever P_u or P_b is less) is found corresponding axial load at service condition/level is found. Which should be more than axial load over the column.

PROBLEMS

7.1 Design a suitable column using a hot rolled ISHB section encased in cement concrete to support an axial load of 1800 kN. The column is 4 m long and adequately restrained in position, but not in direction at both the ends.

Use `Limit State Method of Design', Take yield stress s_y for steel 260 N/mm² and $E_s = 2 \times 10^5$ N/mm². Use f_{ck} 30 grade of concrete and HYSD Steel bars of grade Fe 415 for longitudinal reinforcement.

REFERENCES

1. WF. CHEN (Editor in Chief) - `Handbook of Structural Engineering' CRC Press Boca Raton, New York.

2. R. Narayanan, V. Kalyanaraman, *et al.* Teaching Resources on structural steel Design Volume 2 (Institution of Steel Development and Growth).

3. IS : 800-2007 - Code of Practice for General Construction published by Bureau of Indian Standards Manak Bhawan, 9 Bahadur Shah Zafer marg New-Delhi, 110002.

4. IS : 456-2000 - Indian Standard Code of Practice for Plain and Reinforcement concrete (Fourth Revision), Published by Bureau of Indian Standards Manak Bhawan, 9 Bahadur Shah Zafer marg, New Delhi, 110002.

8

ULS/LSD METHOD OF DESIGN OF COLUMN BASES SLAB AND GUSSETED BASES FOR COLUMNS

8.1 INTRODUCTION

The columns are supported on the column bases. Column bases are steel placed at the bottom of columns. The column bases transmit the column load to the concrete **Pedestals** are the concrete columns, the effective length of which does not exceed three **times** the least lateral dimension or masonry foundation blocks. The column load is spread over large area on concrete or masonry blocks. The intensity of bearing pressure on concrete or masonry is kept within the maximum allowable bearing pressure. The safety of the structure depends upon stability of foundation. The column bases should be designed with utmost care and skill. In the column bases, intensity of pressure on concrete block is assumed to be uniform as shown in Fig. 8.1. The column bases shall be of adequate strength and stiffness to transfer the axial load, bending moments and shear and shall have area to spread the load upon the concrete, masonry, other foundation or other supports without exceeding the allowable stress on such foundation under any combination of the load and bending moments. Anchor bolts and shear keys are provided wherever necessary. Resistance to shear at the proper contact surface between steel base and concrete/ground may be found using a friction coeffieicnt of 0.45. The column bases are of two types :

1. Slab bases; and 2. Gusseted bases.

Nominal bearing pressure under the base plate may be found on the basis of linearly varying distribution of pressure. Maximum bearing pressure shall not exceed the bearing strength = 0.6 f_{ck}. Where f_{ck} is the smaller of the characteristic cubes strength of the concrete or bedding material. Size of base plate is kept larger than that needed to limit the bearing pressure on the base support. The projection \`C' of the base plate beyond the face of the column and the gusset are kept equal. These proejctions are considered as effective in transferring the column load as shown in Fig. 8.2 (B), such that the bearing pressure on the effective area does not exceed bearing capacity of concrete base.

The designs of column base (slab and gusseted bases) have been described in this chapter.

8.2 DESIGN OF SLAB BASES FOR COLUMNS

The slab base as shown in Fig. 8.1 consists of cleat angles and base plate. The column end is faced for bearing over the whole area.

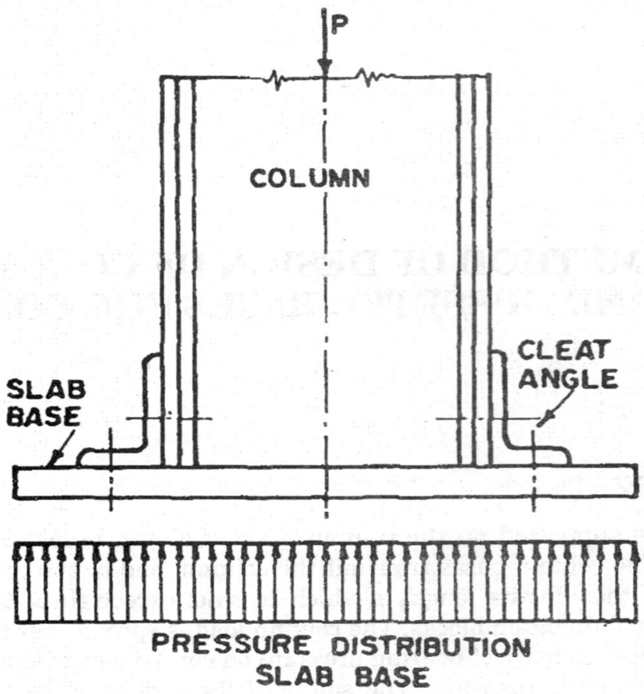

Fig. 8.1

The gussets (gusset plates and gusset angles) are not provided with the column with the slab bases. The sufficient fastenings are used to retain the parts securely with the plate and to resist all moments and forces, other than the direct compression. The forces and moments arising during transit, unloading and erection are also considered.

Design of column base steel plates requires two main steps (i) to obtain the size of the base plate and (ii) to obtain the thickness, *t* of the plate. Generally, the size of the plate is found based on the limit state of bearing on concrete and the thickness of plate is obtained based on the limit state of plastic bending of critical sections in the plate. Depending on the types of forces (viz, axial force, bending moment, shear force), the plate will be subjected to, the plate will be subjected to, the design procedures, which differ slightly. In all cases, a layor of grout should be placed between the base plate and its support. Grouting serves the purpose of levelling.

Base plates supporting concentrically loaded columns in frames in which the column bases are assumed pinned are designed with the assumption that the column factored load, P_u is distributed uniformly on the area of the concrete, pedestal under the base plate.

1. Size of base plate is found from the limit state of bearing on concrete.

In case of plain concrete pedestal IS: 456-2000 recommends that the thickness at edge shall be not less than 150 mm for footings or soils, nor less than 300 mm above the tops of piles (for footings on piles).

The angle α between the plain passing through the bottom edge of the pedestal and the corresponding junction edge of the column with pedestal and the horizontal plain as shown in Fig. 8.2 (A) It is controlled by the following expression

$$\tan \alpha \text{ not less than } 0.9\left(\frac{100q_o}{f_{ck}}+1\right)^{\frac{1}{2}} \tag{8.1}$$

where,

q_o = calculated maximum bearing pressure at the base of the pedestal in N/mm², and

f_{ck} = characteristic strength of the concrete at 28 days in N/mm².

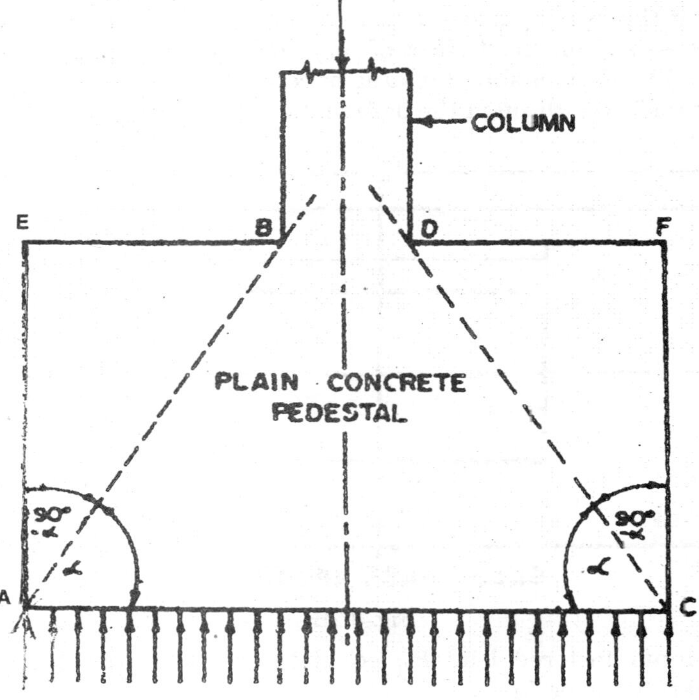

Fig. 8.2A.

After deciding the width of projection BE/DE, the depth of pedestal AE/CF may be calculated as under.

tan (90°-α) = (BE/AE), i.e, AE = BE. tan α.

The design bearing strength of concrete is given by the expression as under

$$\left(\phi_c P_p\right)=0.60\left[0.85 f_{ck}\, A_1\sqrt{\frac{A_2}{A_1}}\right] \tag{8.2}$$

where,

f_{ck} = Compressive strength of concrete

A_1 = Area of base plate (i.e, area of the column.

A_2 = Area of concrete pedestal that is, geometrically similar to and concentric with the loaded area.

A_1 is less than or equal to A_2 (which one is less than or equal to 4 A_1)

Transfer of load at the base of column

It is specified in IS : 456-1978 that the compressive stress in the concrete at the base of the column or pedestal shall be considered as being transferred by bearing to the top of the supporting pedestal footing. The bearing pressure on the loaded area shall not exceed to allowable bearing stress in direct compression multiplied by a value equal to $(A_1/A_2)^{1/2}$ but not greater than two (2).

It is to note that the bearing capacity increases when the area of concrete is more than the area of plate. It is the beneficial effect of confinement. The upper limit of the bearing is obtained when $A_2 = 4A_1$. Presumably, the area of concrete in excess of $4A_1$ is not effective in resisting the load transferred through the base plate.

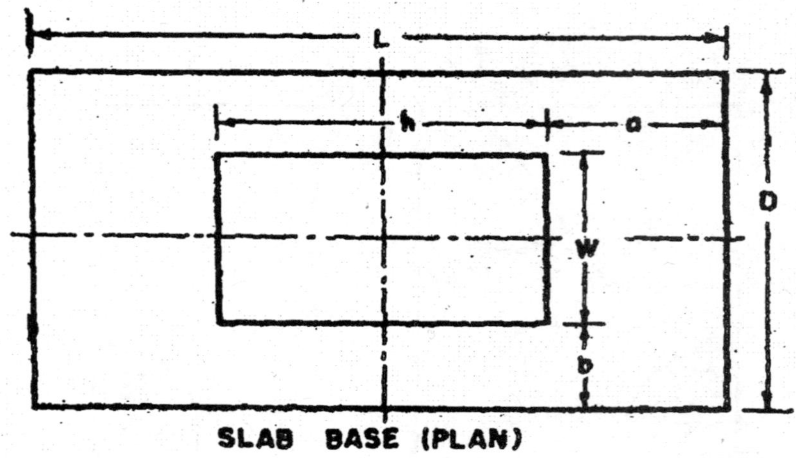

Fig. 8.2B.

Setting the column factored load P_u, equal to the bearing capacity of the concrete pedestal, $(\phi_c P_p)$ and solving for A_1,

$$A_1 = \frac{1}{A_2}\left(\frac{P_u}{0.6\left(0.85 f_c'\right)}\right)^2 \tag{8.3}$$

The size of the base plate L x D (length, L and width D) should be established such that L x D is greater than A_1. For an efficient design, the length, L may be obtained from the expression

$$L = \left(\sqrt{A_1} + 0.50\left(0.95d - 0.80b_f\right)\right) \tag{8.4}$$

where, 0.95d and $0.80b_f$ define the so-called **effective load bearing area** shown by cross-hatched. Once, the length, L is found and then D may be determined $D = (A_1/L)$ Both L and B may be rounded up to the nearest full millimetre.

Fig. 8.2C.

2. Thickness of slab base

Required thickness of slab base, t_{sb} is to be obtained from the limit state of yield line formation along the most severely stressed section. When the value of moment capacity becomes equal to its plastic moment capacity, then a yield line develops.

The size of the column relative to the plate and the value of the factoed axial load yield lines may form in various patterns on the plate.

Three patterns of yield line formations of plate failure in axially loaded plates are shown in Fig. 8.3 (A), (B) and (C). These three-cases have been described are as under.

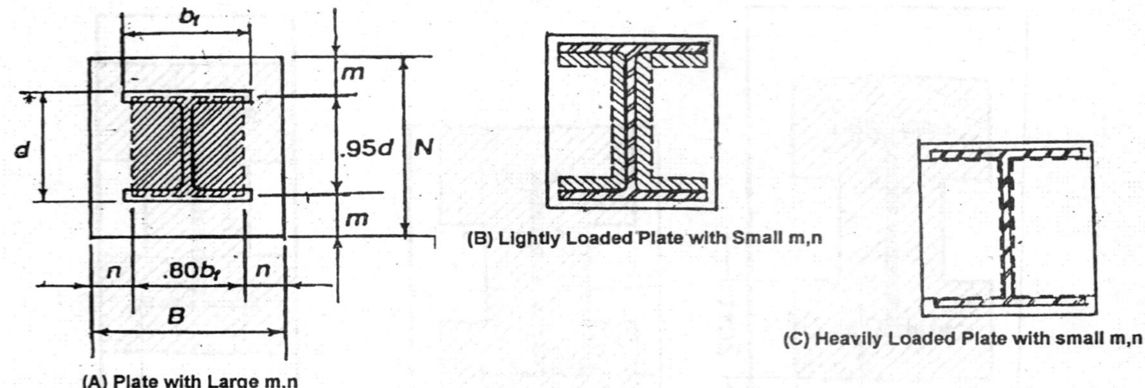

Fig. 8.3.

(i) Plate is large (Fig. 8.3 (A))

Yield lines are assumed to develop around the perimeter of the effective load bearing area (the cross-hatched area) as shown in Fig. 8.3 (A).

(ii) Plate is small

(A) In case, the plate is small and factored column load is **light**, yield linesa re assumed to develop around the inner perimeter of the I-shaped area as shown in Fig. 8.3 (B).

(B) In case, the factored column load is **heavy**, yield lines are assumed to form around the inner edge of the column flange and both sides of the column web as shown in Fig. 8.3 (C).

The required thickness of slab base, t_{sb}, may be found from the following expression

$$t_{sb} = l\left(\frac{2P_u}{0.90\,\sigma_y\,.B.N}\right)^{\frac{1}{2}} \qquad (8.5)\text{ A}$$

where, l is the larger of m, n, and λn'

$$m = \frac{1}{2}(N - 0.95\,d) \qquad (8.5)\text{ B}$$

$$n = \frac{1}{2}(B - 0.80bf) \qquad (8.5)\text{ C}$$

and $$n' = \frac{1}{4}(d.bf)^{\frac{1}{2}} \qquad (8.)\text{ D}$$

$$\lambda = 2(X)^{1/2} / \left|1 + (1 - x)^{1/2}\right| \qquad (8.5)\text{ E}$$

which should be less than or equal to unity (1.00)

$$X = \left(\frac{4\,d\,t_f}{(d + b_f)^2}\right) \cdot \left(\frac{P_u}{\phi_c P_p}\right) \qquad (8.5)\text{ F}$$

IS : 800-2007 Recommendation for obtaining thickness of slab base

When the slab alone distributes the load uniformly, then minimum thickness of a rectangular slab is derived as below :

The column is carrying an axial load P. Consider the load distributed over area h x w and under the slab; over the L X D as shown in Fig. 8.2(B).

Let t_f = Flange thickness of the compression member.

w = Pressure or loading on the underside of the base under the factored compressive axial load

a = Larger projection beyond column

b = Smaller projection beyond column

$\sigma_{sb.y}$ = Yield in N/mm² stress in the slab bases for all steels, but not more than 270 N/mm²

Consider a strip of unit width. (Fig. 8.2) B

Along the xx-axis, and $M_{xx} = \left(\frac{wa^2}{2}\right)$ **Along the yy-axis,** $M_{yy} = \left(\frac{wb^2}{2}\right)$

If Poisson ratio is adopted as 0.3, the effective moment for width D,

$$= \frac{w}{2}.\left(a^2 - 0.3\, b^2\right) \qquad (8.6)A$$

Effective moment for width L, $= \frac{w}{2}.\left(b^2 - 0.3\, a^2\right)$ (8.6)B

A is the greater projection from the column. Effective moment for width D is more. moment of resistance of the slab base of unit width

$$M.R. = (Z_p) = \left(\sigma_{SB.y}\right) = \left(\frac{1}{4} \times 1 \times t_{sb}^2 \cdot \sigma_{sby}\right)$$

Therefore, $\left(\frac{1}{4} \times 1 \times t_{sb}^2 \cdot \sigma_{sby}\right) \cdot \frac{1}{\gamma_{ms}} = \frac{w}{2}\left(a^2 - 0.3\, b^2\right)$

Thickenss of slab $t_{sb} = \left[2\, w \frac{\left(a^2 - 0.3\, b^2\right)\gamma_{ms}}{\sigma_{SB.y}}\right]^{1/2}$ as per IS:800-2007, Clause 7.4.3.1(8.7)A

where, σ_{sBy} is the yield stress for the material of slab base.

Z_p is the plastic modulus of the section for rectangular strip of unit width.

However, IS:800-2007, clause 7.4.3.1 recommends

$$t_{sB} = \left(\frac{2.5\, w \left(a^2 - 0.3\, b^2\right)\gamma_{ms}}{\sigma_{SB.y}}\right)^{1/2} \text{ but not less} \qquad (8.7)B$$

than the flange thickness, t_f when only the effective area of base plate is used, c^2 may be used in stead of ($a^2 - 0.3\, b^2$).

For solid round or hollow steel column, where the load is distributed over the whole area, the minimum thickness of square base BS: 5950 part I, 1990.

$$t = \left[\frac{w}{2.4\,\sigma_{Sb.y}} . D_p\left(D_p - 0.9D\right)\right]^{1/2} \tag{8.8}$$

where, t = Thickness of plate in mm

w = Pressure on the underside of plate.

D_p = Length of the side of base diameter of the cap or base plate not less than 1.5 (D+75) mm cap in mm

D = Diameter of the column in mm

σ_{bsy} = Design strength of the plate but not more than 270 N/mm²

The allowable intensity of pressure on concrete may be assumed as 4 N/mm². When the slab does not distribute the load uniformly or where the slab is not rectangular, separate calculation shall be made to show that stresses are within the specified limits.

When the load on the cap or under the base is not uniformly distributed or where end of the column shaft is not machined with the cap or base, or where the cap or base is not square in plan, the calculations are made on the allowable stress of $\sigma_{sB.y}$.

The cap or base plate shall not be less than 1.50 (d_o + 75) mm in length or diameter.

The area of the shoulder (the annular bearing area) shall be sufficient to limit the stress in bearing, for the whole of the load communicated to the slab to the maximum value 0.75 f_y and resistance to any bending communicated to the shaft by the slab shall be taken as assisted by bearing pressures developed against the reduced end of the shaft in conjunction with the shoulder.

The bases for bearing upon concrete or masonry need not be machined on the underside provided the reduced end of the shaft terminate short of the surface of the slab, and in all cases the area of the reduced end shall be neglected in calculating the bearing pressure from the base.

In cases where the cap or base is fillet welded direct to the end of the column without boring and shouldering, the contact surfaces shall be machined to give a perfect bearing and the welding shall be sufficient to transit the forces specified above. Where the full length T-butt welds are provided no machining of contact surfaces shall be required.

8.3 BRIEF PROCEDURE FOR DESIGN OF SLAB BASE FOR AXIAL LOAD ONLY (PINNED CONNECTIONS)

The procedure for the design of a slab base plate carrying axial load may be described briefly as under.

1. Ultimate/factored axial load from the column is determined.
2. Maximum allowable bearing strength is considered as 0.40 f_{ck}, where f_{ck} is the 28 days characteristic cube strength of concrete.
3. Actual bearing pressure is taken either lecs than or equatic 0.4 f_{ck}.
4. Size of base plate is found.
5. For rolled steel I-section, H-section, channel section or rectangular hollow section columns, the thickness of slab base t_{sb} is found as under : As per IS : 800-2007, Clause 7.4.3.1,

$$t_{sb} = \left(\frac{2.5w}{\sigma_{sB.y}}.\left(a^2 - 0.3b^2\right)\right)^{1/2} \qquad (8.9)$$

where,

w = Pressure from under-side of the base-plate in N/mm²

a = Greater plate projection from the column

b = smaller plate projection from the column

$\sigma_{sB.y}$ = Design strength of the plate, but not more than

The lengths and widths of column base plates are provided in multiples of 10 mm. The thickness provided should conform to the rolled steel plates manufactured.

The area of the concrete support the base plate is kept larger than the base plates so as to distributed the load uniformly under the base plate.

Example 8.1 A column section HB 250, @ 0.510 kN/m carries an axial load of 800 kN. Design a slab base for the column. M20 grade of concrete has been used under the slab. Use limit state method of Design. Take σ_y = 260 N/mm².

Design :

Step 1. Area of slab base required

Axial load of column = 800 kN

Factored/ultimate axial load

= (1.5 x 800) = 1200 kN

It is assumed uniformly distributed under the load.

Maximum allowable bearing strength of concrete = (0.4 f_{ck})

= (0.4 x 20) = 8 N/mm²

Area of slab base required

$$= \left(\frac{1200\,x\,1000}{8}\right) = 15\,x\,10^4 \text{ mm}^2$$

Step 2A. Size of Slab base

The length and width of slab base are proportioned so that the projections on either side beyond the column are approximately equal.

Size of column section HB 250, @ 0.510 kN/m = (250 mm)

Area of slab base

= (250 + 2a) (250 + 2b) mm²

where, a and b are the projections.

Let the projections be equal.

Area of slab base = (250 + 2a)²

(250 + 2a)² = 15 x 10⁴ (250 + 2a) = 387.298

2a = (387.29 – 250) = 137.29

a = 68.645 mm

Let a = 70 mm.

Size of base plate = (250 + 2 x 70) (250 + 2 x 70) = 390 mm x 390 mm

Area of base plate provided = 15.21 x 10^4 mm²

Intensity of pressure from concrete under the slab

$$w = \left(\frac{1200 x 1000}{15.21 x 100^4}\right) = 7.8947 \text{ N/mm}^2$$

Step 3 : Thickness of slab

$\sigma_{sB.y}$ = (260/1.15) = 226.09 N/mm²

Thickness of slab, as per IS : 800-2007, Clause 7.4.3.1,

$$t_{sB} = \left(\frac{2.5\, w\left(a^2 - 0.3b^2\right)}{\sigma_{sB.y}}\right)^{1/2}$$

$$t_{sB} = \left(\frac{2.5 x 8 x \left(70^2 - 0.3 x 70^2\right)}{226.09}\right)^{1/2}$$

$$= \left(\frac{20 x 1430}{226.09}\right)^{1/2} = (126.498)^{1/2} = 11.247 \text{ mm.}$$

Provide 16 mm thickness of slab. It should not be less than thickness of flange.

Example 8.1 (A) A column section SC 250, @ 0.856 kN/m carries an axial load of 800 kN. Design a slab base for the column. M20 grade of concrete has been used under the slab. Use limit state method of design.

Take σ_y = 260 N/mm².

Design :

Step 1. Area of slab base required

Axial load of column = 800 kN

Factored/Ultimate load = (1.5 x 800) = 1200 kN

It is assumed uniformly distributed under the slab.

Maximum allowable bearing strength of the concrete

= (0.4 f_{ck}) = (0.4 x 20) = 8 N/mm².

Area of slab base required

$$= \left(\frac{1200 x 1000}{8}\right) = 15 x 10^4 \text{ mm}^2$$

Step 2 : Size of slab base

Size of column section

SC 250, @ 0.856 kN/m = (250 mm x 250 mm)

The length and width of the slab base are proportioned so that the projectiond as that the projections on the either side beyond the column are kept approximately equal

Area of slab base = (250 + 2a) (250 + 2b)

Projections a and b are equal

$$(250 + 2a)^2 = 15 \times 10^4$$

$$(250 + 2a) = 387.298$$

$$2a = (387.298 - 250)$$

$$a = \left(\frac{137.298}{2}\right) = 68.649 \text{ mm}$$

Provide projections a = b = 70 mm

Provide slab base

$$(250 + 2 \times 70)(250 + 2 \times 70) = 390 \text{ mm} \times 390 \text{ mm}$$

Area of slab base provided.

$$= 15.21 \times 10^4 \text{ mm}^2$$

Intensity of bearing pressure In concrete below the slab

$$w = \left(\frac{1200 x 1000}{15.21 x 10^4}\right) = 7.889 \text{ N/mm}^2$$

Step 3 : Thickness of slab base, t_{sB}

$$t_{sB} = \left(\frac{2.5 \times w \times (a^2 - 0.3b^2)}{\sigma_{sB.y}}\right)^{1/2}$$

$$\sigma_{sB.y} = \left(\frac{260}{1.15}\right) = 226.09 \text{ N/mm}^2$$

$$t_{sB} = \left(\frac{2.5 x 8 x (70^2 - 0.3 x 70^2)}{226.09}\right)^{1/2}$$

$$\left(\frac{20 x (4900 - 0.3 x 4900)}{226.09}\right)^{1/2}$$

$$= \left(\frac{20 x 1430}{226.09}\right)^{1/2} = (126.498)^{1/2} = 11.247 \text{ mm}$$

Provide 16 mm thick slab base. It should not be less than thickness of flange.

8.4 DESIGN OF GUSSETED BASE FOR A COLUMN

Gusset plates and gusset angles are the structural components attached to strength the members or elements (or to enlarge) so as to accommodate connections.

In the gusseted base, gusset plates and gusset angles are used on either side of the column. In addition to the gusset plates, cleat angles are used to connect the column with the slab base. The gusseted base plate has been shown in Fig. 8.4. The columns with gusseted bases, the gusset plates, cleats, angle stiffeners, etc., in combination with the bearing area of the shaft should be sufficient to take the loads, bending moments and reactions to the base plate without exceeding the specified stresses. All the bearing surfaces are machined to ensure perfect contact.

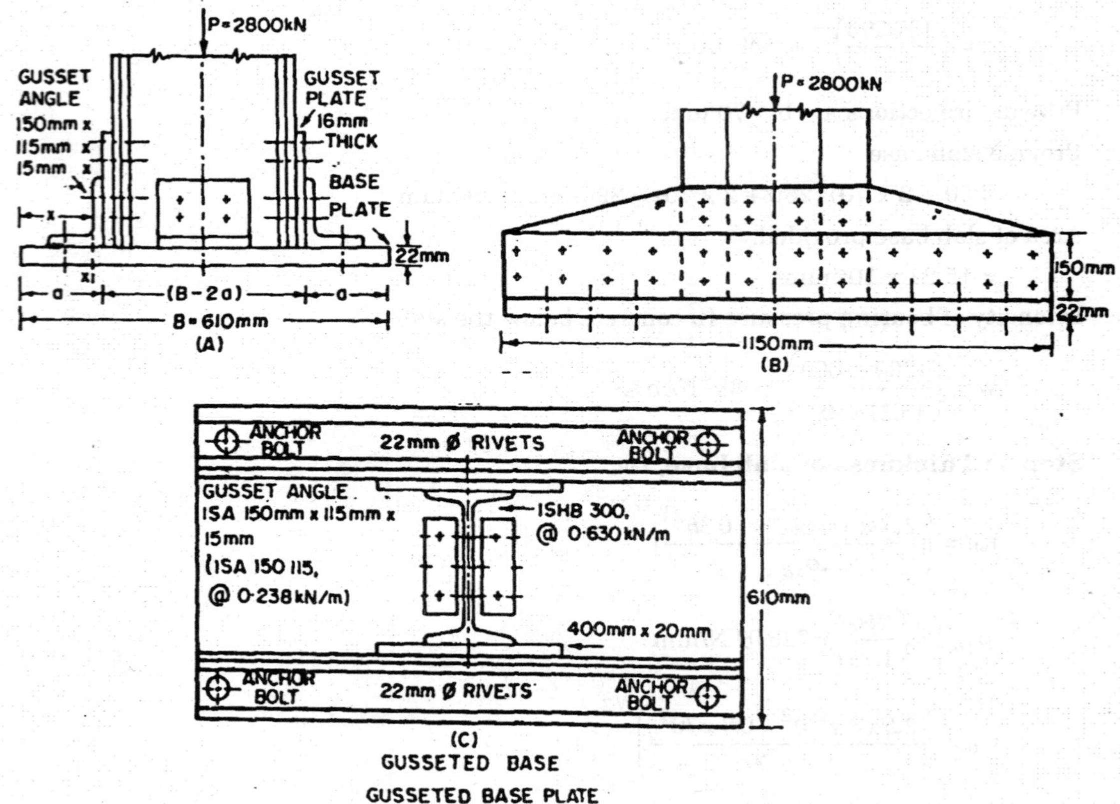

Fig. 8.4. Gusseted base plate

When the ends of the columns and gusset plates are not faced for complete bearing, the fastenings connecting them to the base plate shall be sufficient to transmit all the forces to which the base plate is subjected.

Following are the usual steps for the design of gusseted base.

Step 1. The load carried by the column, and intensity of bearing pressure on concrete or masonry block are known. The area required by the gusseted base plate is computed by dividing the load by allowable intensity of pressure in concrete.

Step 2. The gusset materials used in the gusseted base are assumed. The thickness of gusset plate is assumed as 16 mm. The size of gusset angle is assumed such that its vertical leg can accommodate two rivets in one verticals line and corresponding to that leg length, other leg is assumed in which one rivet can be provided. The thickness of cleat angle is kept approximately equal to the thickness of gusset plate (i.e., ISA 150 mm x 115 mm x 15 mm is assumed to be used as cleat angle).

Step 3. The depth of column section, thickness of gusset plate and length of leg of angle being known, the width of gusset plate for these distances as shown in Fig. 8.4 (A) is known.

Step 4. The length of gusset plate is computed by dividing area required for gusset plate by the width of the gusset plate.

Step 5. The fastenings of the gusseted base and the end of column faced for complete bearing over whole area shall be sufficient to take loads, bending moments and reactions to the base plate without exceeding specified stresses (As advised in IS : 800-56 fastenings having flush ends may be designed for 50 per cent axial load and for other forces).

Where ends of the column section and gusset plates are faced for complete bearing, the fastenings connecting them to the base plate shall be sufficient to transmit all the forces to which the base plate is subjected.

A 22 mm nominal diameter is assumed for rivets to be used and rivet value and thus number of rivets to transmit the forces as specified above are computed. The horizontal shear component is equal to the vertical shear. The number of rivets connecting the gusset plate and gusset angle is adopted equal to the number of rivets connecting the gusset plate and the column as shown in Fig. 8.4 (B).

Step 6. The thickness of base plate is computed by equating moment of resistance with the moment due to intensity of pressure at the underside of base plate at two critical sections X and Y shown in Fig. 8.4(A).

The thickness whichever is more and safe at both the sections is adopted. The thickness computed is reduced by the thickness of the gusset angle in case the thickness at XX section governs the value. It is usual to make the base plates at least as thick as the gussets or as angles.

In the above computation, base plate is assumed to have supports only at gusset plates, and bending of plate takes in one plane only. But the bending of central portion actually occurs in two planes mutually at right angles. It is also having some support from web of column. The thickness of the plate is made one-half of that calculated. The computation of thickness with the said assumption is on safer side. The thickness of base plate determined at the section XX or that determined at the section YY (whichever is maximum) is adopted as the thickness of base plate.

When the end of the column is connected directly to the base plate by means of full penetration butt welds, the connection should be deemed to transmit to the base all the forces and moments to which the column is subjected.

Example 8.2 A column section HB 300, @ 0.630 kN/m with one cover plate 400 mm x 20 mm on either side is carrying an axial load of 1800 kN inclusive of self-weight of base and column. Design a gusseted base. Use limit state method of design. M20 grade of concrete is to be used under the base slab. Take $\sigma_y = 260$ N/mm^2.

Design :

Step 1 (A) Area of base slab required

Axial load of column = 1800 kN

Factored/ultimate load of column

$(\gamma_{fL} P) = (1.5 \times 1800) = 2700$ kN

It is assumed uniformly distributed under the slab. M20 grade of concrete is to be used under the base plate. Maximum allowable bearing pressure in concrete

$= (0.40 \times f_{ck}) = (0.40 \times 20) = 8$ N/mm^2

Area of slab base required

$$= \left(\frac{2700 \times 1000}{8}\right) = 33.75 \times 10^4 \text{ mm}^2$$

Assume gusset plate 16 mm thick and gusset angles ISA 150 mm x 115 mm x 15 mm (ISA 150 115, @ 0.295 kN/m) shall be used as the gusset materials.

Step 1.B Width of gusseted base

(In the direction parallel to the web of column section)

$= [300 + 2 \times 20 + 2 \times 16 + 2 \times 115]$ mm $= 602$ mm

Adopt width of base plate = 610 mm

Step 1 (C). Length of gusset plate

$$= \left(\frac{33.74 \times 10^4}{610}\right) = 553.278 \text{ mm}$$

Adopt the length of gusseted base 580 mm

Size of base plate = 610 mm x 580 mm

Area of gusseted base plate to be provided = (610 x 580) mm^2

$= 35.38 \times 10^4$ mm^2

Intensity of bearing pressure

$$w = \left(\frac{2700 \times 1000}{35.38 \times 10^4}\right) = 7.63 \text{ N/mm}^2$$

Step 2. Thickness of base plate, t_{sB}

Thickness of base plate is designed for maximum allowable bearing pressure, **8 N/mm^2.** Consider the cantilever portion upto the critical section, xx. The length of cantilever portion

$a = (115 - 15 + 4) = 104$ mm

Consider one mm width of cantilever. Bending moment at the critical section XX

$$M_{xx} = \left(\frac{8 \times 104^2}{2 \times 1000}\right) = 43.264 \text{ kN-mm}$$

Let t_{sB} be the thickness of base plate inclusive of thickness of gusset angle

Ultimate moment of resistance (Yield stress x Plastic modulus of section)

$$= 260 \text{x} \frac{1}{4} \text{x} 1 \text{x} t_{sb}^2 = (43.264 \text{x} 1000)$$

$\sigma_{sb.y} = (260/1.15) = 226.09 \text{ N/mm}^2$

$$t_{sb}^2 = \left(\frac{4 \text{x} 43.264 \text{x} 1000}{226.09}\right) = 765.429$$

$t_{sB} = 27.666$ mm

Thickness of base plate shall be (t_{sB} - gusset angle of thickness) = (27.666 – 15) = 12.666 mm

Thickness of flange of the column section HB 300, @ 0.630 kN/m is t_f = 10.6 mm.

Consider the central portion of unit width under the Web. Bending moment at the critical section, yy

(B-2a) = (610 – 2 x 115 – 2 x 4) = 372 mm

$$M_{yy} = \left[\frac{w(B-2a)^2}{8} - \frac{w.a^2}{2}\right]$$

$$= \frac{1}{1000} \text{x} \left[\frac{8(372 \text{x} 372)}{8} - \frac{8 \text{x} (104+15)^2}{2}\right]$$

$$= \frac{1}{1000} \text{x} [138384 - 56644] = 81.74 \text{ kN-mm}$$

Let $t_{sB.2}$ be the thickness of base plate. Equaling plastic moment of resistance to that of ultimate moment

$$\frac{1}{4} \text{x} 1 \text{x} t_{sb.2}^2 \text{x} \left(\frac{260}{1.15}\right)^2 = 81.74 \text{x} 10^3$$

$$t_{sb.2}^2 = \left(\frac{4 \text{x} 81.74 \text{x} 1000}{226.009}\right) = 1446.149$$

$t_{sB.2} = 38.028$ mm

The critical portion of the slab base plate bends in two directions and it is supported at web. The thickness of slab base may be made about one-half. Therefore,

$$t_{sb.2} = \left(\frac{1}{2} \text{x} 38.028\right) = 19.014 \text{mm}$$

$t_{sB.1} = 12.666$ mm.

Provide 20 mm thick base plate, 2ISA 150 mm x 115 mm x 15 mm (2 ISA 150 115, @ 0.295' kN/m) gusset angles and 16 mm thick gusset plate.

Cleat angles shall be used to connect the base plate and the column. The design of gusseted base has been shown in Fig. 8.4.

Note : Design of connections shall be described later on.

8.5 COLUMN BASES SUBJECTED TO MOMENT ECCENTRICALLY LOADED BASE PLATES

Consider a column carrying an axial load `P' and moment `M' as shown in Fig. 8.5.

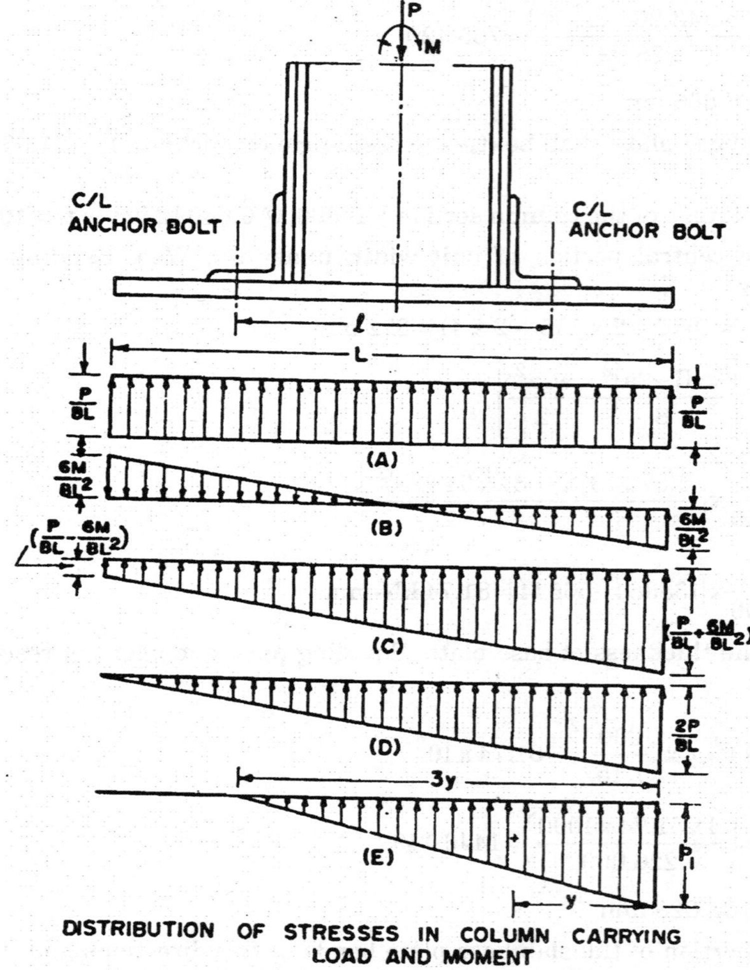

Fig. 8.5

The intensity of bearing pressure between column base and concrete (footing) because of axial load `P' has been assumed to be uniform as shown in Fig. 8.5(A). The intensity of pressure is given by

$$p = (P/BL)$$

where, B = Width of the column base

L = Length of the column base

The stress developed between column base and concrete (footing) because of moment `M' as shown in Fig. 8.5 (B) is given by $f_b = \pm (M/Z)$

where, Z = section modulus of column base, $\left(\frac{1}{6}BL^2\right)$

Depending upon direction of moment, stress due to moment acts upward on one side and downward on other side. The combined stress due to axial load `P' and moment `M' is given by.

$$(P \pm f_b) = \left(\frac{P}{BL} + \frac{6M}{BL^2}\right)$$

The combined stress varies linearly. When intensity of bearing pressure due to axial load is greater than the stress developed due to moment, (i.e., $p > f_b$)

$$\text{or} \quad \left(\frac{P}{BL} > \frac{6M}{BL^2}\right) \qquad \left(\frac{M}{P} < \frac{L}{6}\right)$$

then, the intensity of bearing pressure between column base and concrete is compressive throughout the length of column base as shown in Fig. 8.5(C).

When the intensity of bearing pressure due to axial load is equal to the stress developed due to moment, i.e.,

$$p = f_b \quad , \quad \left(\frac{P}{BL} = \frac{6M}{BL^2}\right)$$

$$\text{or} \quad M = \left(\frac{PL}{6}\right) \quad \text{or} \quad \left(\frac{M}{P} = \frac{L}{6}\right)$$

then, intensity of bearing pressure between column base and concrete is compressive, throughout the length of column base, and it is zero at one edge, as shown in Fig. 8.5(D).

When the intensity of bearing pressure due to axial load is less than the stress developed due to moment, i.e.,

$$p < f_b \quad , \quad \left(\frac{P}{BL} < \frac{6M}{BL^2}\right)$$

$$\text{or} \quad M > \left(\frac{PL}{6}\right) \quad \text{or} \quad \left(\frac{M}{P} > \frac{L}{6}\right)$$

then, intensity of bearing pressure between column base and concrete is maximum at one edge and it is zero at a point (o) within the base. The portion `mo' will not remain in contact with the base and the intensity of bearing pressure, will, therefore, be zero for all points between `m' and `o'. The intensity of bearing pressure acting upward (compressive) has been shown in Fig. 8.5 (E). The resultant passes at a distance greater than L/6 from centre of base (i.e., outside the middle third portion). It acts at the c.g. of the triangle at a distance y from the edge, and is equal to an coincides with the line of action of total vertical load acting downward. Length of base of triangle is 3y. The ordinate of triangle (p_1) represents maximum intensity of bearing pressure. This should not exceed allowable bearing pressure of concrete (footing).

Area of base giving upward pressure, = 3y x B.

Total upward pressure (resultant) (But R = P)

$$R=\left(\frac{1}{2}\times p_1\times 3y\times B\right),\quad P=\left(\frac{1}{2}\times p_1\times 3y\times B\right) \tag{8.3}$$

and taking moment about centre of the base

$$R\left(\frac{L}{2}-y\right)=M,\quad \left(\frac{L}{2}-y\right)=\frac{M}{R} \qquad (\because\ R=P)$$

$$\left(\frac{L}{2}-y=\frac{M}{P}\right),\quad y=\left(\frac{L}{2}-\frac{M}{P}\right) \tag{8.4}$$

Length of column base is determined by trial and error and the distance y can be computed from Eq. 8.4. Width of column base can be found from Eq. 8.3. Thickness of the base plate is found by equating the moment of resistance of base plate with the moment due to intensity of bearing pressure.

Example 8.3. Design a suitable base for a column to carry an axial load of 400 kN and a moment 54 m-kN in the plane of the Web. The column section is HB 300, @ 0.588 kN/m. M20 grade of concrete shall be used for the footing. Use limit state method of design. Take σ_y = 260 N/mm².

Design :

Step 1 : Force/load transmitted

Axial load over the column = 400 kN. Ultimate/factored load = (1.5 x 400) = 600 kN.

Moment to be carried = 54 m-kN

Ultimate/factored moment = (1.5 x 54) = 81 m-kN

Provide a gusseted base plate a gusset plate 16 mm thick, and two gusset angles (2ISA 150 mm x 115 mm x 15 mm (2ISA 150 115, @ 0.295 kN/m)

It is assumed that the moment is transferred to the gusset base plate by the flanges only. The column section to be used is HB 300, @ 0.588 kN/m. Overall depth of the section is 300 mm. Thickness of flange = 10.6 mm. C.G. to C.G. distance of flanges is Lever arm between the flanges

$$= (300 - 10.6) = 289.4 \text{ mm}$$

Force in each flange due to moment

$$=\left(\frac{\text{Moment}}{\text{Lever arm between the flanges}}\right)$$

$$=\left(\frac{54\times 1000}{289.4}\right)=186.59\text{ kN}$$

It is assumed that the gusset plates, gusset angles and the end of the column are having flush end (i.e, complete bearing over whole area). Therefore, 50 per cent of axial load is transferred through bearing and 50 per cent of axial load is transferred through both the faces.

$$\text{Load transferred through one face} = \left(\frac{1}{4}\times 400+186.59\right)\text{kN} \quad = 286.59 \text{ kN}$$

Step 2 : Base-plate

When the anchor bolts are not resisting over-turning moment

Length of base plate (dimensions parallel to web)

$$L = [300 + 2 \times 16 + 2 \times 115] = 562 \text{ mm}$$

Factored/ultimate load

$$= (1.5 \times 400) = 600 \text{ kN}$$

Factored/ultimate moment

$$= 54 \times 1.5 = 81 \text{ m-kN}$$

Provide length of base plate = 570 mm.

From Eq. 8.4

$$\left(\frac{L}{2} - y\right) = \frac{M}{P}, \left(\frac{570}{2} - y\right) = \left(\frac{81000}{600}\right)$$

$$285 - y = 135$$

$$y = 150 \text{ mm}$$

$$3\,y = (3 \times 150) = 450 \text{ mm}$$

Maximum allowable bearing pressure in concrete over footing

$$= 0.4\, f_{ck} = (0.4 \times 20) = 8 \text{ N/mm}^2$$

From Eq. 8.3

$$\frac{1}{2}\, p \times 3y \times B = P$$

$$\left(\frac{1}{2} \times 8 \times 450 \times B\right) = (600 \times 1000)$$

$$B = \left(\frac{600 \times 1000}{4 \times 450}\right) = 333.33 \text{ mm}$$

Provide width of base plate = 400 mm

Actual maximum intensity of pressure, p_1, then

$$\frac{1}{2}\, p_1 \times 450 \times 400 = 600 \times 1000$$

$$p = \left(\frac{600}{200} \times \frac{1000}{450}\right) = 6.67 \text{ N/mm}^2$$

Length of the cantilever portion

$$= (115 - 15) + 4 = 104 \text{ mm}$$

Intensity of pressure reduces to zero in a length 450 mm. from right side.

Intensity of pressure at aXX, at a distance 104 mm

$$p_2 = 6.67 \times \left(\frac{450 - 104}{450}\right) = 5.128 \text{ N/mm}^2$$

The bending moment at a section xx

$$\left[\frac{1}{2} \text{ x } 5-128\text{x}104^2+\frac{1}{2} \text{ x } (6.67-5.128) \text{ x } \frac{2}{3} \text{ x } 104\right] \text{ mm-N}$$

$$= [27732.224 + 3605.33 \text{ x } 1.542]$$

$$= [27732.224 + 5559.42] = 33291.644 \text{ mm-N}$$

Moment of resistance of 1 mm strip

$$\left(\frac{\sigma_y}{\gamma_{ms}} . Z_p\right) = M$$

$$\left(\frac{260}{1.15} . \frac{1}{4} \times 1. t_{sB}^2\right) = 33291.644$$

$$\left(226.09 . \frac{1}{4} . t_{sB}^2\right) = 33291.664$$

$$\left(56.522 x t_{sB}^2\right) = 33291.644$$

$$t_{sB}^2 = 589.0033 \text{ mm}^2$$

$$t_{sB} = 24.269 \text{ mm}$$

Step 3 : Thickness of base plate, t_{sB}

$$t_{sB} = (24.269 \text{ - thickness of gusset-angle})$$

$$= (24.269 - 15) = 9.269 \text{ mm}$$

Provide 570 mm x 400 mm x 16 mm thick base plate. It is fixed to the column. Provide two anchor bolts of 20 mm diameter each.

8.6 USE OF ANCHOR BOLTS TO RESIST OVER-TURNING MOMENT

In case, the column carries an axial load only, in that case, plain steel slab base plate is attached to the column. Such an arrangement is sufficient.

When the column bases carries moments in addition to the axial load, then, the uplift or overturning forces also act on the slab base. In such a case, a more positive attachment is required. Either, the base plates are fixed to the column by direct welding to the columns, or the base plates are fixed by means of **anchor bolts** or by using welding lug angles.

A column is welded directly to the base plate as shown in Fig. 8.6.

The base plates are shop-welded to the columns of small size.

Sometimes, the base plates are required to transport separately and set then to the correct postions. In such cases, the columns are attached to the footing with **the anchor bolts.** These anchor bolts pass through the lug angles as shown in Fig. 8.6. Lug angles have been shop welded to the columns.

When the value of moment is large as compared to the vertically applied load, a gusseted base plate may be provided. In this case, the lever arm distance is increased to holding-down bolts. It gives large efficiency. At the same time, the thickness of base-plate is kept acceptable minimum.

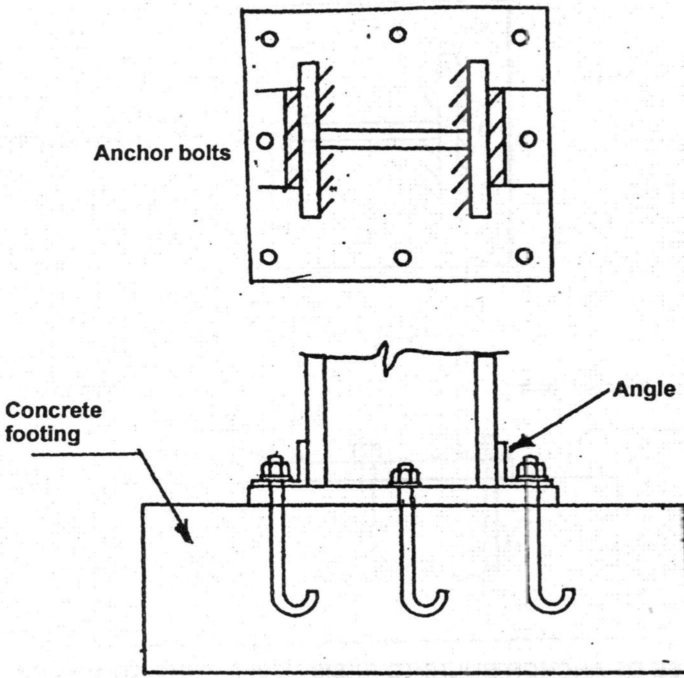

Fig. 8.6

During erection of steel building, the proper positioning of column base plates is a critical phase. Otherwise, it will result in serious stress changes in the beams and columns of the steel frame. In many cases, the levelling plates of the same size as the base plate are carefully grouted in place to the proper elevations first and then, the columns with attached base plates are set on the levelling plates.

Analysis as described in Art. 8.4 and explained in Example 8.3 is for the base plate fixed to the column, and anchor bolts are provided to keep the column base in position.

When the anchor bolts also resist overturning moment, then, the resultant upward pressure

$$R = (P + P_1) \tag{8.5}$$

where, P_1 is the tension induced in the anchor bolts on one side due to overturning moment.

Taking moment about c.g. of the triangle in Fig. 8.5 (E)

$$P_1 (a-y) + P \left(\frac{1}{2} - y\right) = P \tag{8.6}$$

where, a is the distance of anchor bolt from the edge under maximum pressure.

The position of anchor bolt is known known. Value of P_1 is substituted from Eq. 8.5 and the distance y is found.

The length and width of column base are determined by trial and error.

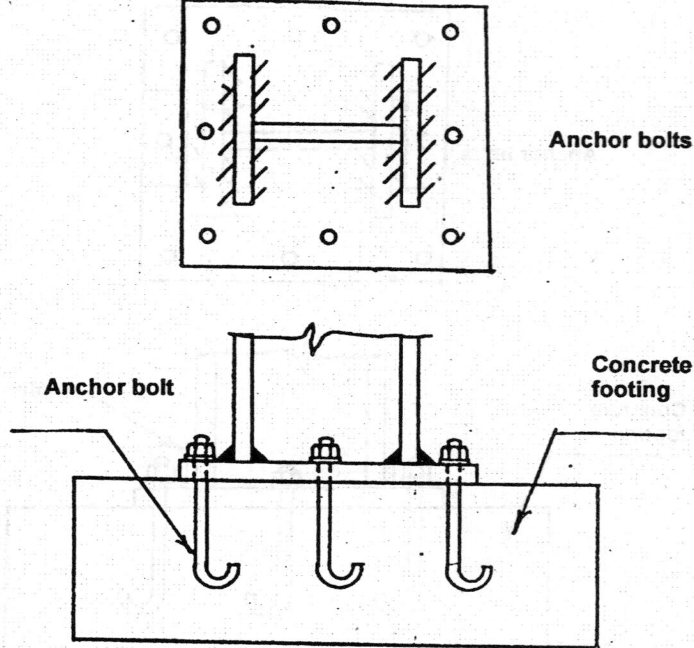

USE OF ANCHOR BOLTS TO RESIST OVER-TURNING MOMENT

Fig. 8.7

Example 8.4 In Example 8.3, design the slab base when the anchor bolts also resist the overturning moment.

Design :

From **Example 8.3**, provide a gusseted base plate 570 mm x 400 mm size. Provide two anchor bolts in a gusset angle on each side at a distance 54 mm from the edge of the base plate upto the centre line of anchor bolt.

Step 1 : Forces transmitted

From Eq. (8.5), $(P + P_1) = R$

$$[600 \times 10^3 + P_1] = \left(\frac{1}{2} P \times 3y \times B\right)$$

$$P_1 = \left(\frac{1}{2} \,.\, 8 \,.\, 3y \,.\, 400 - 600 \,.\, 10^3\right)$$

$$P_1 = (4800y - 600 \times 10^3)\ N$$

From Eq. (8.6),

$$P_1 = (a\text{-}y) + P\left(\frac{1}{2} - y\right) = M$$

Substituting the values of P_1, P a, L and M

$$(4800y - 600 \times 10^3)(570 - 54 - y) + 600 \times 10^3 \left(\frac{570}{2} - y\right) = 81 \times 10^6$$

$$y = 113.75 \text{ mm}$$

$$P_1 = \left(\frac{4800 \times 113.75 - 600 \times 10^3}{1000}\right) \text{kN} = 54 \text{ kN}$$

This shows that the vertical load is larger than that due to overturning.

The anchor bolts each of 20 mm diameter is provided to maintain the position of the column.

PROBLEMS

8.1 A steel column HB 300 @ 0.630 kN/m resting on a steel base plate is supported on concrete foundation. The column carries a central load of 3000 kN. Design the base plate and gusseted connection. The safe bearing capacity of concrete is 0.4 f_{ck} . M20 grade of concrete shall be used under the slab base. Take $\sigma_y = 260$ N/mm^2

8.2 Design the base plate and its anchor bolts for a 300 mm x 140 mm @ 0.442 kN/m I-section steel stanchion under an axial load of 480 kN and a moment of 5600 kN-mm in the plane of web.

8.3 Design a suitable riveted base plate connection for WB 200 column subjected to an axial load for 400 and a moment of 600 kN- mm at the bottom in the plane of web.

8.4 Design a suitable base for a column carrying an axial load of 320 kN and a moment of 30,000 kN-mm in the plane of the web. The section of the column is HB 250, @ 0.547 kN/m

REFERENCES

1. W.F. CHEN, Editor in Chief `Handbook of Structural Engineering' CRC Press, Boca Raton, New York, 1997.
2. BS 5950: Part 1: 1990. British Standard `Structural Usse of Steel work Building Code of Practice for dsign in simple and continuous construction : Hot rolled steel sections.
3. IS: 800-2007. Indian Standard `General Construction in Steel, Code of Practice Published by Bureau of Indian Standards, Manak Bhawan, 9 Bahadur Shah Zafar Marg, New Delhi 110 002.
4. A.B. Clarke, S.H. Coverman. `Structural Steel Work Limit State Design' Chapman and Hall Ltd., New Fetter Lane, London, EC4PEE.
5. Teaching Resources on Structural Steel Design Institution of Steel Development and Growth compiled by Rangachari Narayan, V. Kalyanraman and *et al.*

9

ULS/LSD METHOD OF DESIGN OF COLUMN FOOTINGS DESIGN OF INDEPENDENT AND COMBINED COLUMN FOOTINGS

9.1 INTRODUCTION

Footings are defined as the structural members used to support columns and walls and to transmit and distribute their loads to the soil in such a way that (i) the load bearing capacity of the soil is not exceeded, (ii) excessive settlement, differential settlement and rotat-ion are prevented, and (iii) sufficient safety against overturning or sliding is maintained. When a column load is transmitted to the soil by the footing, the soil becomes compressed. Footings or the foundations are the lower part of a structure provided below the column slab base to support and transmit and distribute the load to the soil.

The magnitude of settlement depends on many factors such as (i) type of the soil, (ii) the intensity of load, (iii) the depth below the ground level and (iv) the type of footing. In case, the different types of footings are provided under the same structure, new type of the stresses develop in the structure, and the structure have different settlements. Excessive differential settlement may lead to the damage of non-structural components in the buildings, even failure of the influenced parts.

Vertical loads are generally applied at the centroid of the footing. In case, the resultant of the applied loads does not coincide with the centroid of the bearing area, a bending moment develops. In such a case, the pressure on one side of the footing will be larger than the pressure on the other side, causing higher settlement on one side and a possible rotation of the footing.

In case, the bearing soil capacity is different under the footings, a differential settlement will occur. For example, if the footings of a building are partly on soil and partly on rock. It is customary in such cases to provide a joint between the two parts to seperate them, and allowed for independent settlement.

Footings are designed to carry column loads and transmit them to the soil safely. The code requirements must be satisfied. The area of footing is based on the allowable bearing capacity of the soil.

The independent footing is used to distribute the load over large area from individual column. When the bearing capacity of the soil is not poor, then concrete block is used as column footing. When the bearing capacity of the soil is poor, then grillage footing is used.

The column footings are designed to sustain the applied loads, moments and forces and the induced reactions. The column load is spread over large area, so that the intensity of bearing pressure between the column footing and soil does not exceed the safe bearing capacity of the soil. It is ensured that any settlement which may occur shall be as nearly uniform as possible and limited to an accepted small amount. The column load is first transmitted to the column footing through the column base. It is then spread over the soil through the column footing.

9.2 INDEPENDENT COLUMN FOOTING

(Grillage footing) The steel grillage is a frame-work of rolled steel beams forming foundation in the soil of poor bearing capacity. In the grillage footing, the steel beams are used in one tier or more depending upon spread of the base. In case, two or more tiers are used, each tier is placed perpendicular to the tier above it. The beams are embedded in cement concrete and are protected from corrosion. As far as possible, square base is provided in the independent footing. These footings are also called as **single column footing.**

Footings usually supports loads from (i) dead loads from the sub-structure and super-structure, (ii) live loads resulting from materials or occupany (iii) Weight of materials used in back-filling and wind loads.

Each footing in a building is designed to support the maximum load, which may come on any column due to the critical combination of loadings, using the permissible soil pressure.

The dead load; and may be a small portion of live load, may act continuously on the structure. The rest of the live load may act at intervals and on some parts of the structure only, causing different loads on columns.

Consequently, the pressure on the soil under different loadings will vary according to the loads on the different columns. Therefore differential settlement will occur under the various footings of one structure. Since, partial settlement is inevitable, the problem is defined by the magnitude of differential settlement that the structure may tolerate.

The magnitude of differential settlement depends upon on the variation in the compressibility of the soils, the thickness of compressible material below the foundation level, and the stiffness of the combined footing and the super-structure.

Excessive differential settlement results in cracking of concrete and damage to claddings, partitions, and finishes.

For practical purpose, it may be assumed that the soil pressure under the effect of sustained loadings is the same for all footings and therefore results in equal settlement.

The sustained load (or the usual load) may be assumed equal to dead load plus a percentage of the live load, which occur very frequently on the structure.

Footings are proportioned for these sustained loads to produce the same soil pressue under all footings. The allowable soil bearing capacity should not be exceeded in any case.

The column load `P' is distributed through a column base of length `a' to the grillage beam of length `L' as shown in Fig. 9.1.

It is assumed that each tier distributes the load uniformly to the lower tier and intensity of bearing pressure is also uniform. A load P/2 at a distance a/4, from the centre line acts downward, and resisting force P/2 at a distance L/4 from centre line acts upward.

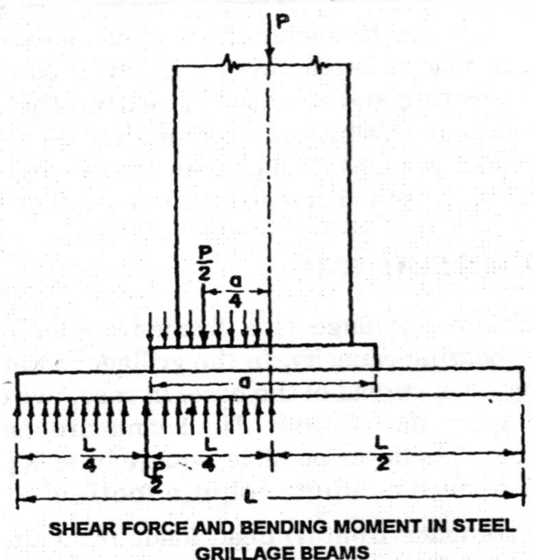

Fig. 9.1

The maximum bending moment occurs at the centre of beam. **Maximum bending** moment

$$M = \frac{P}{2}\left(\frac{L}{4} - \frac{a}{4}\right) = \frac{P}{8}(L - a) \qquad (9.1)$$

The maximum shear force occurs at the edge of the base plate.

Maximum shear force,

$$F = \frac{P}{L}\left(\frac{L - a}{2}\right) = \frac{1}{2}\frac{P}{L}(L - a) \qquad (9.2)$$

The grillage beams are kept unpainted and solidly encased in ordinary dense concrete with 100 mm aggregate.

The pipe separators or equivalent are used to keep the grillage beams properly spaced and the beams are spaced apart so that the distance between edges of adjacent flanges is not less than 75 mm.

The thickness of the concrete cover on the upper flange, at the ends, and the outer edges of the sides of the outermost beam is not less than 100 mm.

The concrete is properly compacted solid around all beams.

The distribution of load, soil pressure, shear force and bending moment diagrams are shown in Fig. 9.2.

9.3 DESIGN OF INDEPENDENT COLUMN FOOTINGS

Beams are arranged either in one tier or in two tiers (which are referred as **Grillage footings**). The beams in lower tier are placed perpendicular to those in upper tier. Axial load to be carried by the column and the footing is given.

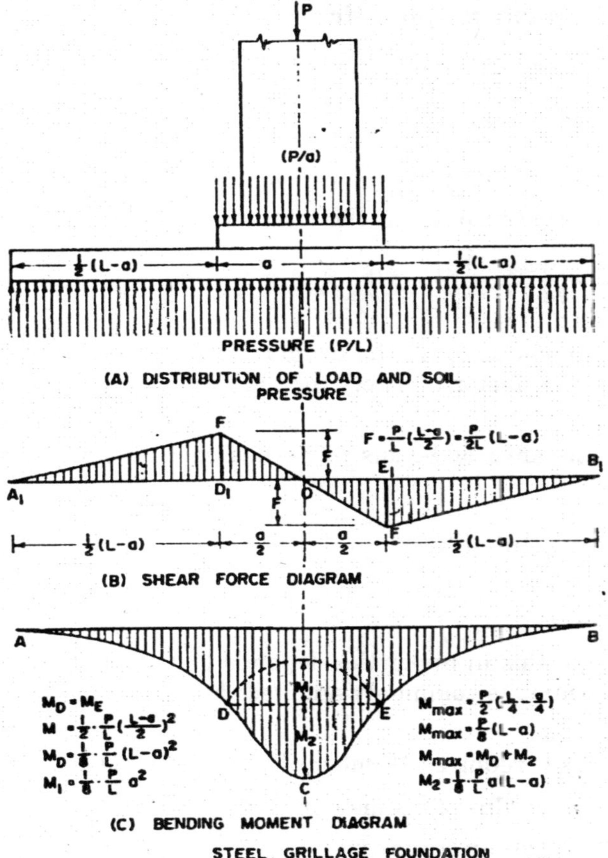

Fig. 9.2.

Ten percent of direct axial load (for the self-weight of the foundation) is added to the axial load. Following procedure is used for the design of independent column footing.

Step 1. Factored/ultimate/design axial is found

$$W_{fd} = (\gamma_{fL} \cdot W_{ck})$$

$$W_{fd} = (1.5 \times W_{ck}) \qquad \text{(i)}$$

Step 2. Area of footing required is then obtained. Safe bearing capacity, q_o of the soil is given. The ultimate bearing capacity, q_u is taken as twice the safe bearing capacity.

$$q_u = (2 \times q_o) \text{ assumed} \qquad \text{(ii)}$$

Area of footng required

$$A = \left(\frac{\text{Factore / design column load}}{2\,q_o}\right)$$

Usually, the square area is provided.

Step 3. Design of beams in upper tier

Factored/ultimate/design maximum moment is obtained from the ultimate load

$$M_{Fd} = \frac{P}{8}(L-a) \qquad \text{(iii)}$$

a = size of base plate, provided.

Step 4. Moment of resistance (i.e, the total strength of beams) required is found.

$$MR = \left(\frac{\sigma_y}{\gamma_{ms}}.Z_p\right) \qquad \text{(iv)}$$

where, Z_p is the plastic modulus of all the beams to be provided. σ_y is the yield stress of the steel used. γ_{ms} is the partial factor of safety for the strength of the material (For steel γ_{ms} = 1.15)

The moment of resistance of beams to be designed is equated to the factored/design moment.

$$\left(\frac{\sigma_y}{\gamma_{ms}}.Z_p\right) = M_{fd}.10^3.10^3$$

Step 5. Size of beams

Number of grillage beams in upper tier depends upon minimum clearance spacing of 75 mm. Three number of beams is commonly adopted. For I-section, a shape factor, S = 1.10 is assumed.

Elastic section modulus required Number of beams x $Z_E = (Z_P/S)$

Total elastic section modulus of beams required is provided.

Step 6. Design of beams in lower tier

The procedure for the design of beams in lower tier is similar to that described for the beams in upper tier.

The beams designed in upper tier and in lower tier are checked, Whether the beams are adequate for shear.

Step 1: Factored/Ultimate maximum shear force is determined for the factored/ Ultimate load, W_{fd}

$$P_u = \frac{1}{2}\frac{P_{ck}}{L}(L-a).\frac{1}{n}$$

where, n is the number of beams and a is the size of base slab provided under the column

Step 2 Shear resistance/strength of beams

$$V = \left(\frac{\sigma_y}{\sqrt{3}}.\frac{1}{\gamma_{ms}}.d_u.t_w\right)$$

where,

d_w = depth of the web

t_w = thickness of the web

γ_{ms} = 1.15, partial factor of safety for the strength of material steep used.

$$\left(\frac{1}{\sqrt{3}} = 0.577 = 0.60\right) \qquad \text{(approximately).}$$

The beams in upper tier are also checked for web buckling and web crippling.

Step 1. Effective length of Web, which acts as a strut, L_e = 0.7d Depth upto root of flange, h_2 for the flange of I-section is noted from ISI HB No.1

Depth of web

$$d = (D - 2h_2)$$

Sectional area of web

$$A_w = (d_w \,.\, t_w)$$

where, t_w thickness of web is noted from ISI HB No.1

Moment of inertia of web about yy-axis $I_y = \left(\frac{1}{12} d \,.\, t_w^3\right)$

Radius of gyration of web about yy-axis, r_y

$$r_y^2 = \left(\frac{1}{12} x \frac{d.t_w^3}{d.t_w}\right) = \left(\frac{t_w^2}{12}\right)$$

$$r_y = \left(\frac{1}{2\sqrt{3}} . t_w\right)$$

Step 2. Slenderness ratio of the web is obtained as (Le/ry). It is adopted as (2.5 d/t_w)

$$\left(\frac{Le}{r_y}\right) = \left(0.7 x 2\sqrt{3}\, \frac{d}{t_w}\right)$$

= (2.4248 d/tw) = (2.5 d/tw) Approximately

Step 3. Compressive stress for web buckling

Web is considered as fixed ended column. Stiff bearing length (width of the base plate at the end i.e near point A reaction is assumed 100 m. Angle of dispersion of load is taken as 450. Thickness of flange, t_f is noted from ISI Handbook No. 1.

(B + D/2) = (B + 1/2 x Overall depth)

Step 4. Load bearing capacity of Web

$$P_{web} = \left(B + \frac{D}{2}\right).\ t_w \,.\, \sigma_c.$$

In case, P_{web} is more than the factored/design shear, then the beam section is considered adequate for web buckling.

Further the beam is checked for Web crippling.

Step 1. Radius at the root of flange r_1, thickness of Web, t_w and the thickness of flange, t_f are noted from ISI Handbook No. 1.

Dispersion length is assumed as 1 in 2.5

Resistance to crippling

$$P_{crippling} = \left(\frac{(r_1 + t_f)\, t_w.\, \sigma_y}{\gamma_{ms}}\right)$$

In case, it is more than factored shear force, P_u it is consided adequate against web crippling.

Example 9.1 Design two tier grillage footing for flexure to carry a stanchion, designed to carry 1600 kN. The base plate of the stanchion is 600 mm square in size. The bearing pressure on earth is limited to a value of 200 kN per square metre. Use limit State method of design.

Design : Area of footing required

Direct axial load on the stanchion P = 1600 kN

Add 10 percent of direct axial load for self weight of the grillage foundation = 160 kN

Total service/characteristic load

$W_{ck} = (1600 + 160) = 1760$ kN

Factored/ultimate design/ axial load

$= (1.5 \times 1760) = 2640$ kN

Allowable bearing pressure on earth

$q_{ck} = 200$ kN/m²

Ultimate bearing pressure of soil (assumed)

$q_u = (2 \times 200) = 400$ kN/m²

Area of footing required in plan

$$= \left(\frac{2640}{2 \times 200}\right) = 6.6 \text{ m}^2 \quad = (6.6 \times 10^3 \times 10^3) \text{ mm}^2$$

Area to be provided (for square area)

$= (2.5690 \text{ m} \times 2.5690 \text{ m})$

Provide area = 2600 mm x 2600 mm

$= (6.76 \times 10^3 \times 10^3)$ mm²

Step 2 : Design of beams in Upper-Tier

The beams are designed for P = 1760 kN

Length of base plate of stanchion = 800 mm

Length of beams in upper tier, L = 2600 mm

Factored/ultimate/design axial load

$P = W_{fd} = (1.5 \times 1760) = 2640$ kN

Maximum bending moment occurs at the centre of beam

$$M_{fd} = \frac{P}{8}(L-a)$$

$$M_{fd} = \frac{2640}{8}\left(\frac{2600-600}{1000}\right) \text{ kN-m} = 660 \text{ kN-m}$$

Yield stress for the grade of steel

$\sigma_y = 260$ N/mm²

Step 3. Moment of resistance of steel beams

$$MR = \left(\frac{\sigma_y . Z_p}{\gamma_{ms}}\right)$$

where, Z_P = Plastic modulus of beams

$$MR = \left(\frac{260 \, x \, Z_p}{1.15}\right) \text{ kN-m}$$

Equating moment of resistance of beams to the design bending moment

$$\left(\frac{260 \text{x} Z_p}{1.15}\right) = 660 \text{x} 10^3 \text{x} 10^3$$

Plastic modulus of beams required

$$Z_p = \left(\frac{660 \text{x} 1.15 \text{x} 10^6}{260}\right) \text{mm}^3$$

$Z_p = 2919.236 \times 10^3$ mm³

Step 4. Size of beams

Number of grillage beams in upper tier depends upon minimum clear spacing of 75 mm between the flanges which are accommodated in the width of the slab base plate provide three (03) beams.

Elastic section modulus required

$$Z_E = \frac{1}{3}\left(\frac{Z_P}{\text{Shape factor}}\right)$$

Shape factor for I-section may be taken as 1.10

$$Z_E = \frac{1}{3} \text{x} \left(\frac{2919.230 \text{x} 10^3}{1.10}\right) \text{mm}^3 = 884.615 \times 10^3 \text{ mm}^3$$

Provide three MB 400, @ 0.616 kN/m

Width of flange = 140 mm

Width of base plate = 600 mm

Clear spacing between the flanges of beams $\frac{1}{2}(600-3 \times 140)= 90$ mm

Elastic section modulus provided

$(3 \times 1022.9 \times 10^3) = 3068.7 \times 10^3 \text{ mm}^3$

$= 2919.230 \times 10^3 \text{ mm}^3$

Note : Beams shall be checked for shear force and, crippling stress in Example 9.

Step 5. Design of beams in lower tier

These beams are designed for factored/ultimate/design axial load

$P = W_{Fd} = 2640$ kN-m

Maximum bending moment occurs at the centre of beam

$$M_{Fd} = \frac{P}{8}(L - a)$$

$$M_{Fd} = \left(\frac{2640 \times (2600-600)}{8 \times 1000}\right) = 660 \text{ kN-m}$$

Yield stress for the grade of steel

$\sigma_y = 260 \text{ N/mm}^2$

Step 6. Moment of resistance of steel beams

$$MR = \left(\frac{\sigma_y \, x Z_p}{\gamma_{ms}}\right) \text{ kN-m}$$

$$MR = \left(\frac{260 x Z_p}{1.15}\right) \text{ kN-m}$$

Equating moment of resistance of beams to the design bending moment

$$\left(\frac{260 x Z_p}{1.15}\right) = 660 x 10^3 x 10^3$$

Plastic modulus of beams required

$$Z_p = \left(\frac{660 x 1.15}{260} x 10^3 x 10^3\right), \qquad Z_P = 2919.23 \times 10^3 \text{ mm}^3$$

$$\text{Elastic section modulus required} = \left(\frac{2919.23 x 10^3}{1.10}\right) = 2653.84 \times 10^3$$

Step 7. Size of beams

Number of beams depends upon length of beams in upper tier and minimum clear spacing 75 mm

Provide 9 number of beams

LB 250, @ 0.279 kN/m in lower tier

Section modulus of are beam

$= 297.4 \times 10^3\ mm^3$

Total section modulus provided

$= 9 \times 297.4 \times 10^3\ mm^3 \qquad = (2876.6 \times 10^3) > 2653.84 \times 10^3\ mm^3$

Width of flange of this beam section is clear distance between the flanges

$$= \frac{1}{8}(2600 - 8 \times 125) = 200\text{mm}$$

Design of grillage footing/foundation for flexure is shown in Fig. 9.3.

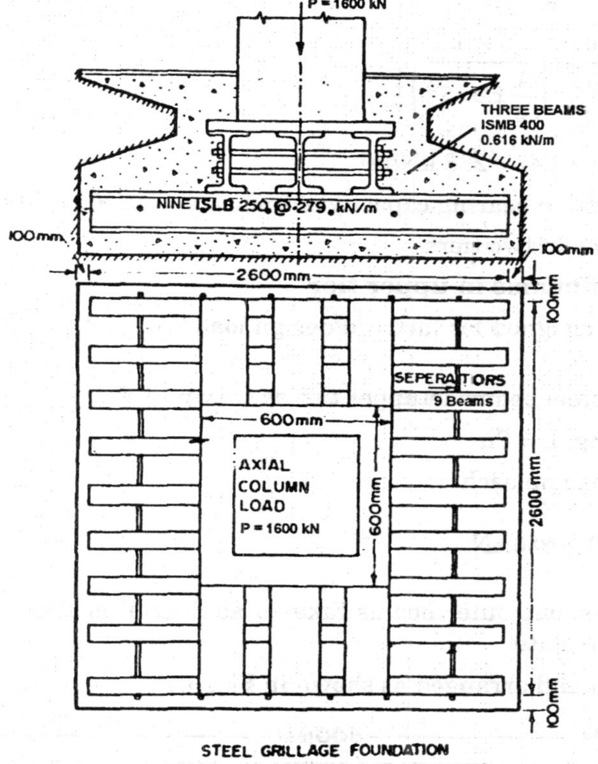

Fig. 9.3.

Example 9.2 Design a foundation footing for flexure for a column to carry on axial load 600 kN along with a moment in the same vertical plane equal to 200 m-kN. The bearing capacity of soil is 200 kN/m². The width of the foundation should not exceed 2 m. Use limit state method of design.

Design : Axial vertical column load = 1600 kN

Add 10(ten) percent for the self-weight for the foundation = 160 kN

Total vertical load at service load level W_{ck} = (1600 + 160) = 1760 kN

Factored/ultimate/design load

$W_{fd} = (1.5 \times 1760) = 2640$ kN

Moment acting at service load level M_{ck} = 200 kN-m

Factored/ultimate/design moment

$M_{fd} = (1.5 \text{ x } 200) = 300$ kN-m

Step 1 : Area of footing required

Bearing capacity of the soil

$q_o = 200$ kN/m²

Let a grillage footing 2 m wide x 5 m long shall be provided. Intensity of bearing pressure

$$q_1 = \left[\left(\frac{2640}{2x5}\right) + \left(\frac{300}{\frac{1}{4}x2x5^2}\right)\right]$$

$$= [264 + 24] = 288 \text{ kN/m}^2$$

It is less than ultimate bearing capacity of soil $q_u = (2 \text{ x } 200) = 400$ kN/m (assumed)

Provide steel beams in two tiers

Step 2. Design of beams in upper tier

These beams are designed for factored/design load, W_{fd} and the factored/design moment, M_{fd}

Provide 3 (three) steel beams in upper tier as shown in Fig. 9.4.

Beams are 2 m long, L = 2m

Factored/design load on each

$$= \left(\frac{1}{3}x2640\right) = 880 \text{ kN}$$

In addition to this, one outer beams take an additional load because of design moment Assume size of the base plate

800 mm x 600 mm and arranged as shown in Fig. 9.4.

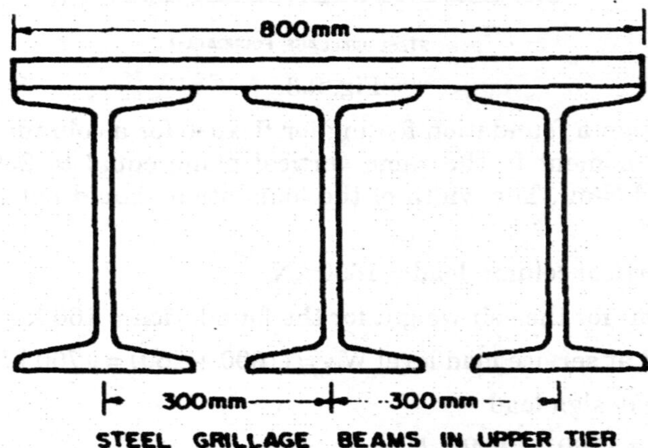

Fig. 9.4. THREE BEAMS, MB 400 @ 0.616 kN/m

$$\text{Additional load} = \left(\frac{300 \, x \, 1000}{600}\right) = 500\,\text{kN}$$

Maximum load on one outer beam

= (880 + 500) = 1380 kN

Maximum moment to be resisted

$$M_{Fd} = \left(\frac{W_{Fd}\,(L-a)}{8}\right) = \left(\frac{1380(2000-600)}{8\,x\,1000}\right) = 241.50\,\text{m-kN}$$

Step 3. Moment of resistance of the beams

$$M.R = \left(\frac{\sigma_y}{\gamma_{ms}}.Z_p\right) = \left(\frac{260}{1.15}.Z_p\right)$$

Equating bending moment to factored design moment

$$\left(\frac{260}{1.15}x\,Z_p\right) = 241.5.10^3.\,10^3$$

$$Z_P = \left(241.5\frac{10^3\,x\,10^3\,x\,1.15\,x\,1.10}{260}\right)$$

= (0.9288 x 1.15 x 10^3 x 10^3)

= 1068.173 x 10^3 x 10^3 mm^3

Elastic section modulus of one beam

$$Z_E = \left(\frac{1068.173\,x\,10^3}{1.10}\right) = 971.066\,x\,10^3\,mm^3$$

Provide MB 400, @ 0.616 kN/m

Section modulus provided = 1022.9 x 10^3 mm^3

Step 3. Design of beams in lower tier

The beams are designed for above factored/design load and factored/design moment

Length of beams = 5 m

Moment due to vertical load

$$= \left(\frac{2640(5000-800)}{8\,x\,1000}\right) = \left(\frac{2640\,.\,4200}{8\,x\,1000}\right) = 1386\,\text{m-kN}$$

Factored/ultimate/design moment

= 300 m-kN

Total moment = (1386 + 300) = 1686 kN-m

Provide 8 (eight) beams in lower tier.

Moment shared by each beam

$$= \left(\frac{1688}{8}\right) = 211\,\text{kN-m}$$

$$\left(\frac{\sigma_y}{\gamma_{ms}} \cdot Z_p\right) = 211 x 10^6$$

Plastic section modulus

$$Z_p = \left(\frac{211 x 10^6 \, x 1.15}{260}\right) = 933.26 \text{ x } 10^3 \text{ mm}^3$$

Elastic section modulus assuming shape-factor 1.10 for I-beams

$$Z_E = \left(\frac{933.26 x 10^3}{1.10}\right) = 848.426 x 10^3 \text{ mm}^3$$

Provide LB 400, @ 0.569 kN/m

Section modulus = 965.3 x 10^3 mm^3

Example 9.3 : In Example 9.1, the beams ingrillage footing have been designed for flexure using limit state method of design. Check the beam for shear.

Step 1. Check the beams for shear

Factored/Ultimate maximum shear force

$$P_u = \frac{1}{2} \cdot \frac{P}{L}(L-a) x \frac{1}{n}$$

where, n is the number of beams in upper tier and P = W_{fd} = Design/factored load. Number of beams in upper tier = 03

$$\text{Maximum Factored/ultimate shear force} = \left(\frac{1}{2} \text{x} \frac{2640(2600-600)}{2600} \text{x} \frac{1}{3}\right)$$

P_u = 338.46 kN

Depth of beams MB 400, @ 0.616 kN/m

$d_w = h_o$ = 400 mm

Thickness of web, t_w = 8.9 mm

$$\left(1/\sqrt{3}\right) = 0.577 = 0.60\,(Approx.)$$

Shear resistance of beams

$$V = \left(0.60 \frac{\sigma_y}{\gamma_{ms}} \cdot d_w \cdot t_w\right)$$

$$= \left(\frac{0.60 x 260}{1.15} \cdot \frac{400 x 8.9}{1000}\right) = 482.92 \text{ kN}$$

Shear resistance of the beam is more than maximum shear force. Here, the beams are safe in shear. In the lower tier, the number of beam is eight. It is not necessary to check the beams in shear.

Example 9.4. In Example 9.1, Step 2 **check the adequacy of MB 400, @ 0.616 kN/m for web buckling and web crippling**

(i) Effective length of web, which acts as a strut

L_e = 0.7 d, and

Depth upto root of flange, h_2 = 32.80 mm

d = (D – 2h_2)

d = (400 – 2 x 32.80) = 334.4 mm

Sectional area of web

A_w = (d_w . t_w)

= (334.4 x 8.9) = 2976.16 mm²

Moment of inertial of web (web is rectangular in section)

$$I_y = \left(\frac{1}{12}.d.t_w^3\right)$$

Radius of gyration of web about yy-axis, r_y

$$r_y^2 = \left(\frac{1}{12} x \frac{d\,x t_w^3}{d\,x t_w}\right) = \frac{t_w^2}{12}$$

$$r_y = \left(\frac{1}{2\sqrt{3}}.t_w\right)$$

(ii) Slenderness ratio of the web

$$\left(\frac{L_e}{r_y}\right) = \left(\frac{0.7\,x\,d\,x\,2\sqrt{3}}{t_w}\right) = \left(\frac{2.4248\,d}{t_w}\right)$$

It is adopted as (2.5 d/t_w)

d = (400 – 2 x 32.40) = 334.44 mm

$$\left(\frac{2.5d}{t_w}\right) = \left(\frac{2.5\,x\,334.44}{8.9}\right) = 93.944$$

Step 3. Compressive stress for axial buckling

For hot-rolled steel I-section

α_{xx} = 0.0020, α_{yy} = 0.0035

For slenderness ratio 93.944 from curves, compressive stress, σ_{cd} = 164 N/mm²

Note : Web is considered as fixed ended column. Stiff bearing length (width of the base plate at the end (the point of reaction = 100 mm assumed)

Angle of dispersion = 450

Thickness of flange, t_f = 16 mm

Thickness of web = 8.9 mm

Half the depth of I-section

$$\text{(i.e, upto N.A.)} = \left(\frac{D}{2}\right) = \left(\frac{400}{2}\right) = 200mm$$

$$\left(B + \frac{D}{2}\right) = (100 + 200) = 300 \text{ mm}$$

Step 4. Load bearing capacity of web

$$P_{web} = \left(\frac{300 x 8.9 x 164}{1000}\right) = 437.88 kN$$

It is more than factored shear force P_u = 338.46 kN. Therefore, the buckling web does not occur. The beam section MB 400, @ 0.616 kN/m is adequate against web buckling.

Step 5. Adequacy of web for crippling

Beam section MB 400, @ 0.616 kN/m has been proposed

From ISI HB No. 1, radius at the root of flange, r_1 = 14 mm

Thickness of web, t_w = 8.9 mm

Thickness of flangem, t_f = 16 mm

$(r_1 + t_f)$ = (14 + 16) = 30 mm.

Dispersion length 1 in 2.5

= (2.5 x 30) = 75 mm

Resistance to crippling

$$P_{crippling} = \left[\frac{(100+75).\,9.4\,.\,260}{1000 x 1.15}\right] = \left[\frac{175\,.\,9.4\,.\,260}{1000 x 1.15}\right]$$

= 371.91 kN

It is more than factored shear force P_u = 338.46 kN. Therefore, the crippling of the web does not occur. The beam section MB 400, @ 0.616 kN/m is adequate against web crippling.

Example 9.5. In Example 9.2 check the beams designed for flexure and shear force

Design : Factored/design load for each beam

= 880 kN

Additional load resisted/transmitted by one outer beam (i.e, due to moment = 500 kN

Total load to be transmitted by one beam

W_{fd} = (880 + 500) = 1380 kN

Maximum/ultimate shear force in upper tier

$$V_u = \frac{W(L-a)}{2L} = \left(\frac{1380 x (2000-600)}{2 x 2000}\right)$$

$$= \left(\frac{1380 x 1400}{4000}\right) = 489 kN$$

Shear resistance for the web of the beam MB 400, @ 0.616 kN/m

$$V_u = \left(\left(\frac{\sigma_y}{\sqrt{3}}\right) t_w \text{ . depth of web}\right)$$

$$\frac{1}{\sqrt{3}} = 0.577 \text{ approximately} = 0.60$$

Depth of web, D = 400 mm

Thickness of web, t_w = 8.9 mm

$$V_u = \left(\frac{0.6 \,.\, 8.9 \,.\, 400}{1000}\right) = 555.36 \text{ kN}$$

It is more than factored shear force = 489 kN

In lower tier, number of beams is eigth. These is no necessity to check the beams for shear. The beams are adequate for shear force.

9.4 DESIGN OF COMBINED COLUMN FOOTING

Spread footings which support more than one column or wall are known as **combined footings** combined footings become necessary in two situations (i) in case, the columns are so close to the property line that single-column cannot be provided without projecting beyond that line, and (ii) in case some adjacent columns are so close to each other that their bearing areas overlap each other.

The combined footing is used to distribute loads from two or more columns over large area. The steel grillage footing used as combined footing to support two column loads, has been described here. Two column loads may be equal and may be unequal.

(i) Steel grillage footing to support two equal column loads. A steel grillage footing supporting two equal column loads has been shown in plane in Fig. 9.5.

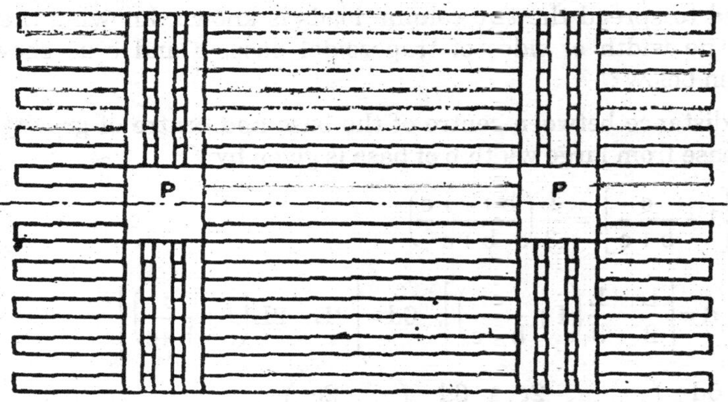

STEEL GRILLAGE FOOTING SUPPORTING TWO EQUAL LOADS

Fig. 9.5

The size of the column bases, and centre to centre distance between columns are known. The rectangular shape for the base of the footing is used. The projections of beams in lower tier are kept such that bending moments under the columns and centre of span of beams are approximately equal. It gives economical design. Length of base of footing is thus known. The width of base of footing is computed from area of base required to spread the two column loads. In case width of base of footing is restricted comparative more projections are provided in beams in lower tier, and beams are designed for maximum bending moment.

(ii) **Steel grillage footing to support two unequal column loads.** A steel grillage footing supporting two unequal column loads P_1 and P_2 has been shown in plan in Fig. 9.6.

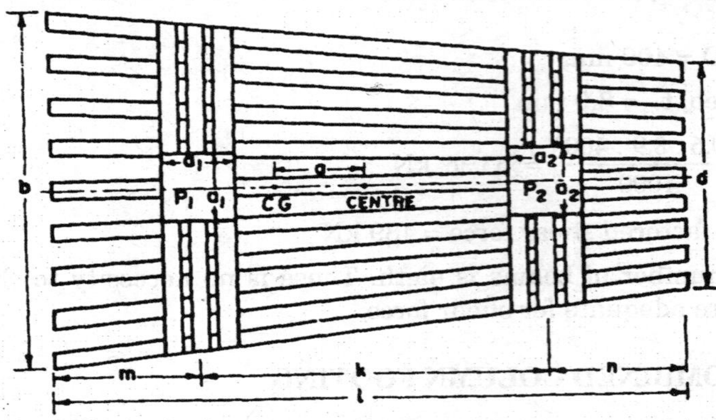

STEEL GRILLAGE FOOTING SUPPORTING TWO UNEQUAL LOADS

Fig. 9.6

The line of action of the resultant of the two unequal column loads is made to coincide with the centre of gravity of base of the footing. The trapezoidal shape is used for the base of the footing. The centre to centre distance between the two columns K is known. The projections of beams in lower tier m and n are assumed, such that, as far as possible, bending moments under the columns and between two columns are approximately equal. However, it is by trial to fix up projection most satisfactory. In case, there are restrictions for projections, the projections are fixed accordingly. The length of the base of the footing L is known, knowing the centre to centre distance of columns K and fixing the projections of beams in lower tier. The area of base of footing, A, required to spread the two column loads is known, knowing the allowable bearing pressure of soil. The width of sides of trapezoidal base, b and c as shown in Fig. 9.6 are determined as given below :

Let d be the distance between centre of the base and centre of gravity of base (c.g.). The distance of c.g. of base from large width b of base is given by

$$\bar{x} = \frac{L}{3}\left[1+\frac{c}{b+c}\right] = \frac{L}{3}\left(\frac{b+c+c}{b+c}\right)$$

then, $$d = \left[\frac{L}{2} - \frac{L}{3}\left(1+\frac{c}{b+c}\right)\right],\quad 6d = \left[3L - 2L\left(1+\frac{c}{b+c}\right)\right]$$

or $$\frac{6d}{L} = \left(3-2-\frac{2c}{b+c}\right).\ \frac{6d}{L} = \left(1-\frac{2c}{b+c}\right)$$

or $$\frac{6d}{L} = \left(\frac{b+c-2c}{b+c}\right),\quad \frac{6d}{L} = \frac{(b-c)}{(b+c)}$$

Multiply both sides by (A/L)

$$\left(\frac{6d}{L}\times\frac{A}{L}\right)=\frac{(b-c)}{(b+c)}\times\frac{A}{L},\quad \text{(but } (b+c)\,L=2A)$$

$$\therefore\ \left(\frac{6d}{L}\times\frac{A}{L}\right)=\left(\frac{(b-c)A}{2A}\right)$$

$$\text{or}\quad \frac{6A.d}{L^2}=\frac{(b-c)}{2},\quad \text{and}\quad \frac{A}{L}=\frac{(b+c)}{2} \qquad \text{(i)}$$

then from (i) and (ii) $b=\left(\frac{A}{L}+\frac{6A.d}{L^2}\right)$ (9.3)

and $c=\left(\frac{A}{L}-\frac{6A.d}{L^2}\right)$ (9.4)

or b and c are given by $\frac{A}{L}\left(1\pm\frac{6.d}{L}\right)$

where plus sign gives b, and minus sign gives c.

In case the combined grillage footing is of two tiers, then maximum bending moments for beams in upper tier under the heavy column load P_1 and the other column load P_2 are given below.

Maximum bending moment for beams in upper tier under heavy column load :

$$M_1=\frac{P_1(L_1-a_1)}{8}$$

where, L_1 = length of the beam under the heavy column

a_1 = Width of the base plate under the heavy column along the length of beams

Maximum bending moment for beams in upper tier under other column load :

$$M_2=\frac{P_2(L_2-a_2)}{8}$$

where, L_2 = Length of the beam under the other column

a_2 = Width of the base plate under the other column along the length of beams

The beams are designed for maximum bending moments and checked for shear.

For the design of beams in lower tier maximum bending moments are determined at three sections as below :

Let x be the distance from large end of trapezoidal base upto corresponding section as shown in Fig. 9.7. Then, bending moment due to intensity of pressure of soil w acting upward is

From Fig. 9.7

$$\frac{y}{(L-x)}=\left(\frac{b-c}{2}\right)\cdot\frac{1}{L}\ .\ y=\left(\frac{L-x}{L}\right)\left(\frac{b-c}{2}\right),\ 2y=(L-x)\left(\frac{b-c}{L}\right) \qquad \text{(i)}$$

Therefore, $CD = (c + 2y) = C + (L - x)\left(\frac{b-c}{L}\right) \quad \therefore \; CD = b - x\left(\frac{b-c}{L}\right)$ (ii)

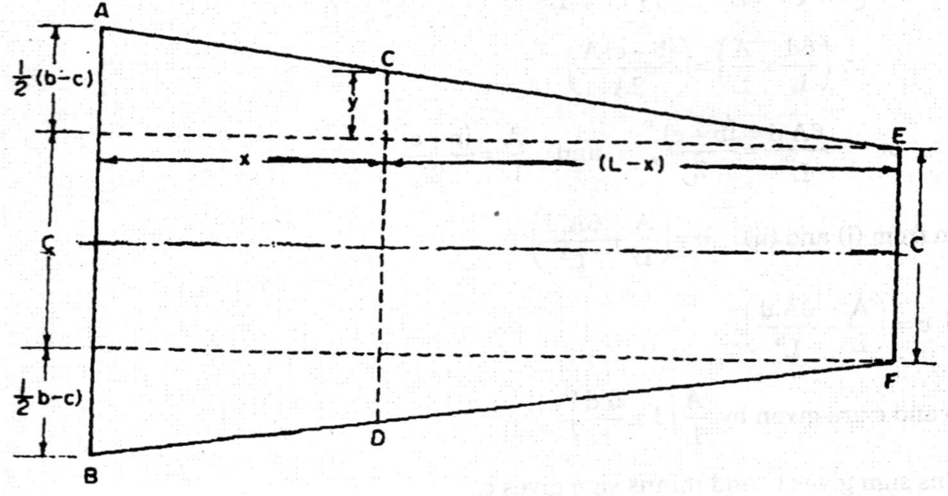

Fig. 9.7

Area of trapezium ABCD = $\left(\frac{AB+CD}{2}\right).x$ (iii)

$$\frac{1}{2}\left[b+b-x\left(\frac{b-c}{L}\right)\right].x = \frac{x}{2}\left[2b-\frac{x}{L}(b-c)\right]$$

Total load = (intensity) x area

$$= w.\frac{x}{2}\left[2b-\frac{x}{L}(b-c)\right] \quad \text{(iv)}$$

Distance of centroid from right side

$$= \frac{\frac{x}{3}\left[2b+b-\frac{x}{L}(b-c)\right]}{\left[b+b-\frac{x}{L}(b-c)\right]} \quad \text{(v)}$$

Moment = Load x distance of centroid

$$\frac{wx}{2}\left[2b-\frac{x}{L}(b-c)\right]\times\frac{\frac{x}{3}\left[3b-\frac{x}{L}(b-c)\right]}{\left[2b-\frac{x}{L}(b-c)\right]}$$

$$\frac{w.x^2}{2}.\left[b-\frac{x}{3}\left(\frac{b-c}{L}\right)\right] = \frac{wx^2}{2}\left[b-\frac{x}{3}\frac{(b-c)}{L}\right] \quad \text{(9.5A)}$$

Bending moment under **the heavy column** load

$$M_1' = \frac{wx^2}{2}\left[b - \frac{x}{3}\left(\frac{b-c}{L}\right)\right] - \frac{P_1}{a_1} \times \frac{\left[x - \left(m - \frac{a_1}{2}\right)\right]^2}{3} \quad (9.5)\,(B)$$

Bending moment between the two columns

$$M_2' = \frac{wx^2}{2}\left[b - \frac{x}{3}\left(\frac{b-c}{L}\right)\right] - P_1\,(x - m) \quad (9.6)$$

Bending moment under the other column load

$$M_3' = \frac{wx^2}{2}\left[b - \frac{x}{3}\left(\frac{b-c}{L}\right)\right] - P_1\,(x - m) - \frac{P_2}{a_2}\frac{\left[x - \left(m + K - \frac{a^2}{2}\right)\right]^2}{2} \quad (9.7)$$

M_1', M_2' and M_3' are differentiated with respect to x, and are equated to zero separately. The values of x are found for maximum bending moments in the above expressions. The maximum bending moments from the above expressions are determined by substituting values of x. The maximum of maximum bending moment is found and it is used for the design of beams in lower tier.

In the trapezoidal shape of the base of the footing, the beams in lower tier are arranged at close spacing at smaller end than that at the other end.

The rectangular shape is also adopted for combined footing to support two unequal column loads. The line of action of the resultant of two column loads is made to coincide with centre of gravity of the base. The cantilever projection near the other column. The length of the base is kept such that bending moments under the heavy column load and between two columns (though these are of opposite sign) are approximately equal.

9.5 APPROXIMATE METHOD OF DESIGN OF COMBINED FOOTING

The exact method of design of combined footing for two columns has been discussed in Art. 9.3. The approximate method of design of combined footings in discussed here. The first point will be kept in view in dealing the combined grillage footing is that the line of action of the resultant of the two column loads is made to coincide with the centre of gravity of the base of footing.

The size of column bases and centre to centre distance between columns are found by the least projection required from a consideration of the connection of a column base to the grillage. The beams in the lower tier in the grillage may be considered to be inverted beams loaded with the uniformly distributed load equal to the intensity of pressure under the grillage footing and supported at two points (i.e., at the columns). Thus, the beams have overhanging portions at both the ends. The beams are subjected to the cantilever moment at each end and the span moment in between the two columns. The cantilever moments and the span moments are of opposite sign. The cantilever moments are assumed to be maximum at the **centre line of the** columns. The maximum span moment is assumed to occur at the point, where the shear force changes sign. The maximum cantilever moments and the maxi-mum span moments are determined. The beams are designed for the absolute maximum bending moment. The shear force and the bending moment diagrams are drawn. The beams are suitably spaced such that the clear distance the flanges is not less tan 75 mm.

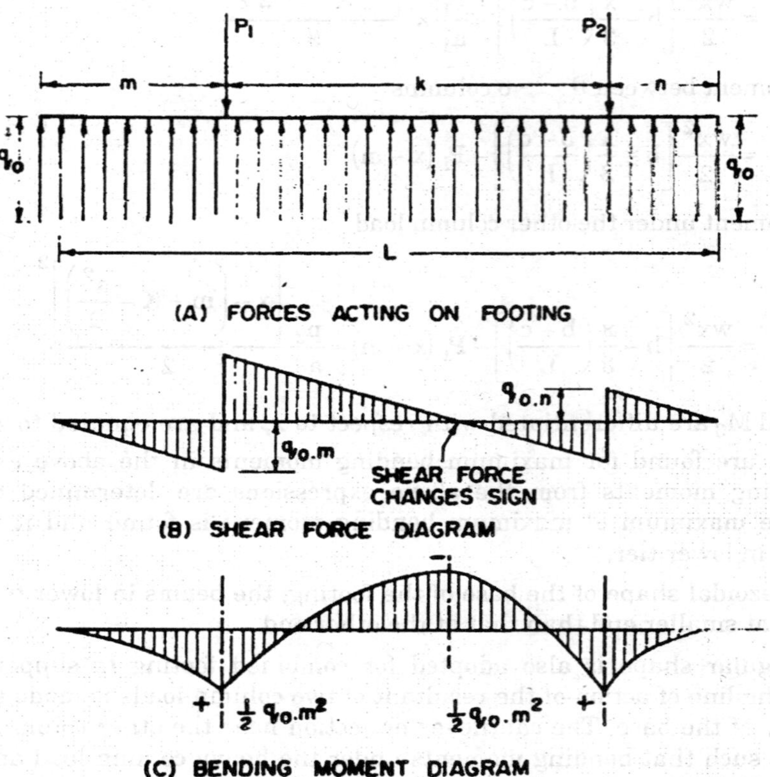

Fig. 9.8

The beams designed are checked for shear, web buckling and web crippling.

Example 9.6. Design a combined footing foundation for two columns, to carry axial 1200 kN and 1600 kN, respectively for flexure. The distance between to n centre of columns is 4 m. The safe bearing capacity of soil is 200 kN/m².

Design : A combined steel grillage foundation shall be provided.

Load from column 1, to carry heavy axial load = 1600 kN.

Load from column 2, to carry comparatively less heavy load = 1200 kN.

Factored/ultimate/design load 2400 from column 1 = (1.5 x 1600 = 2400 kN)

Factored/ultimate/design load from column, 2 = (1.5 x 1200 = 1800 kN-m

Total ultimate load = 4200 kN

Step 1. Area of base required

Add ten percent for the self-weight of the foundation

$$= \left(\frac{10}{100} x 4200\right) = 420 \text{ kN}$$

Total load inclusive of self-weight of the foundation

= (4200 + 420) = 4620 kN

Let the ultimate bearing capacity of soil, $q_u = (2 \times q_o) = (2 \times 200)$

= 400 kN/m²

Area of base required

$$= \left(\frac{4620}{400}\right) = 11.55 \text{ m}^2$$

Step 2. Length and Width of grillage footing

The distance of line of action of resultant of two column loads from 1600 kN

$$\bar{x} = \left(\frac{1200 x 4}{1200+1600}\right) = 1.714 m$$

Provide rectangular base of the and cantilever projection of 1.556 m

Centre of gravity of base is kept coinciding with the line of action of the resultant of column loads

Length of column base

= 2 (1.714 + 1.556) = 6.64 m

Width of grillage base

$$= \left(\frac{11.52}{6.64}\right) = 1.735 \text{ m}$$

Provide width of grillage base 1.88 m

Provide grillage beams in two tiers.

Step 3A. Design of beams in upper tier

Length of beams = 1.88 m = 1880 mm

Assume size of steel base plate under as 650 mm x 650 mm.

Factored design bending moment for beams under column 1

$$\frac{2400}{8}\left(\frac{1880-650}{1000}\right) = 369 \text{ kN-m}$$

Moment of resistance of beams to be provided MR = $\left(\frac{\sigma_y}{1.15} x Z_p\right)$

$$\frac{260}{1.15} . Z_p = 369 . 10^3 . 10^3$$

$$Z_p = \left(\frac{369 . 1.15}{260} . 10^3 . 10^3\right)$$

$= 1632.115 \times 10^3\ mm^3$

Elastic section modulus of beams to provided assuming shape factor 1.10

$$Z_E = \left(\frac{Z_p}{1.10}\right) = \left(\frac{1632.115x10^3}{1.10}\right) = 1483.74 \times 10^3\ mm^3$$

Provide 3 MB, 300, @ 0.442 kN/m

$(3 \times 573.6) = 1720.8 \times 10^3\ mm^3$

Width of flange of MB 300, @ 0.442 kN/m

$b_f = 140$ mm

Clear distance between flanges

$$= \left(\frac{650-3x140}{2}\right) = 115mm > 75mm$$

Step 3B. Design of beams under the column 2

Assume size of steel plate base under column 2 = 600 mm x 600 mm

Factored/design bending moment for beams under column2

$$\frac{1800}{8}\left(\frac{1880-600}{1000}\right) = 28.8\,m/kN$$

Moment of resistance for the beams required under column 2

$$\left(\frac{\sigma_y}{1.15} \cdot Z_p\right) = (288 \,.\, 10^3 \,.\, 10^3)$$

$$Z_p = \left(\frac{288}{260} x \frac{1.15x10^3\,x10^3}{}\right) = 1273.8 \times 10^3\ mm^3$$

Elastic section modulus required for beams under column 2, assuming shape factor 1.10 for

I-section

$$Z_E = \left(\frac{1273.8}{1000} x10^3\right) = 1158.04x10^3\,x10^3$$

Provide 3 LB 275, @ 0.330 kN/m

Section modulus provided

$3 \times 392.4 = 1177.2 \times 10^3\ mm^3$

It is more than $1158.04 \times 10^3\ mm^3$ required.

Step 4. Design of beams in lower tier

Safe bearing capacity of soil acting upward $q_o = 200\ kN/m^2$

Ultimate bearing capacity of soil assuming $q_u = 2q_0 = (2 \times 200) = 400\ kN/m^2$

Both the column loads as uniformly distributed load through the column bases.

The beams in lower tier act as inverted overhanging beams. Let x be the distance from left end of the base. **Factored/design bending moment under column**

$$M'_1 = \left(\frac{400\ .\ 1.88\ .\ x^2}{2} - \frac{2400}{0.650}\left(\frac{x-1.395}{2}\right)^2\right)$$

Differentiating w.r.t x and equate to zero

$M_1' = 918.84$ kN-m

Factored/design/bending moments between two columns

$$M_2' = \left[\frac{400x1.88\ .\ x^2}{2} - 2400(x-1.752)\right] \text{kN-m}$$

Differentiating w.r.t x and equating to zero,

x = 3.19 m

Substituting the value of x

$$M_2' = \left[\frac{400x1.88x3.19^2}{752} - 2400-(3.19-1.752)\right]$$

M_2' = (negative) 298.2 kN-m

Factored/design bending bending moment under column

$$M_3' = \left[\frac{400x1.88xx^2}{2} - 2400(x-1.72) - \frac{1800}{0.600}\left(\frac{x-5.42}{2}\right)^2\right]$$

Differentiating w.r.t x and equate to zero

x = 6.165 m

Substituting the value of x

$$M_3' = \left(\frac{400\ .\ 1.88\ .\ 6.165^2}{2}\right) - 2400(6.165-1.72) - \frac{1800}{0.600}\left(\frac{6.165x5.42}{2}\right)^2$$

M_3' = 2790.132 kN-m

Maximum factored/desgin bending moment occurs under the column 2 (i.e, M_3' = 2790.192 kN-m

Step 5. Moment of resistance of beams in lower tier

$$= \left(\frac{\sigma_y}{\gamma_{ms}}\ .\ Z_p\right)$$

where, Z_p is the total plastic section modulus.

Equating total moment of resistance to factored bending moment

$$\left(\frac{\sigma_y}{1.15}\ .\ Z_p\right) = \left(2790.192\ .\ 10^3\ .\ 10^3\right)$$

$$Z_P = \left(\frac{1.15x2790.192x10^3 \, x \, 10^3}{260}\right)$$

$= 12341.23 \times 10^3 \times 10^3$

Step 6. Elastic section modulus, Z_E

Let the shape factor for I-section beams be 1.10

$$Z_E = \left(\frac{Z_p}{1.10}\right) = \left(\frac{12341.23x10^3}{1.10}\right) = 11219.30 \times 10^3 \times 10^3$$

Let 8 (eight) number of beams be provided.

$$Z_{E.1} = \left(\frac{11219.30x10^3}{8}\right) = 1402.4$$

Provide LB 500, @ 0.750 kN/m with Z_{xx} 1543.2 x 10^3 mm^3 x 10^3 mm^3

Example 9.7. In Example 9.6, the beams have been designed for bending for the combined footing. Check the adequacy of these beams sections for shear.

Design : From ISI HB No. 1, the geometrical properties of the beams are as follows :

(i) **For upper tier under column1**

3MB 300, @ 0.442 kN/m have been proposed.

Thickness of web, t_w = 7.5 mm

Thickness of flange, t_f = 12.4 mm

Depth of section, h_o = 300 mm

Radius at the root of flange

r_1 = 14 mm

(ii) **For upper tier under column2**

3LB 275, @ 0.330 kN/m have been proposed

Thickness of web, t_w = 6.4 mm

Thickness of flange, t_f = 8.8 mm

Depth of section, h_o = 275 mm

Radius at the root of flange

r_1 = 14 mm

(iii) **For lower tier**

8 Beams LB 500, @ 0.750 kN/m have been proposed

Thickness of web, t_w = 9.2 mm

Thickness of flange, t_f = 14.1 mm

Depth of section, h_o = 500 mm

Radius at the root of the flange

r_1 = 17 mm.

Step 1. Factored/ultimate/design load

(i) Under column 1, for each beam

$$W_{fd} = \frac{1}{3} x 2400 = 800 kN$$

Ultimate (Maximum)/factored/design shear force for one beam

Length of beam = 1880 mm, a = 650 mm

$$V_u = \frac{W_{fd}}{2L}(L-a)$$

$$= \frac{800(1880-650)}{2x1880} = \left(\frac{800x1230}{3760}\right) = 261.70 \text{ kN}$$

Shear resistance of the beam

MB 300, @ 0.442 kN/m

$$V_u = \left(\frac{\sigma_y}{\sqrt{3}} \cdot \frac{t_w \cdot d_w}{\gamma_{msy}}\right)$$

$$\left(\frac{1}{\sqrt{3}}\right) = 0.577 = 0.60 \text{ approximately}$$

$$= \left(\frac{260x0.60x7.5x300}{1.15x1000}\right) = 305.217 \text{ kN.}$$

It is more than 261.70 kN. Proposed beam section is adequate for the shear.

Step 2 : Under column 2. for each beam

Factored/design/ultimate load for one beam

$$W_{fd} = \left(\frac{1}{3} x\ 1800\right) = 600 \text{ kN}$$

Ultimate (maximum)/factored/design shear force for one beam a = 600 mn

$$V_u = \frac{W_{fd}\ (L-a)}{2L}$$

$$= \frac{600\ (1880-600)}{2x1880} = \left(\frac{600x1280}{3760}\right) = 204.255 \text{ kN}$$

Shear resistance of the beam section

LB 275, @ 0.330 kN/m

$$V_u = \left(\frac{\sigma_y}{\sqrt{3}} x \frac{1}{\gamma_{msy}} x t_w x\ 275\right)$$

$$= \left(\frac{0.6x260}{1000} x \frac{1}{1.15} x\ 6.4\ x275\right) = 238.74 \text{ kN}$$

It is more than 204.255 kN.

Beams in upper tier under column 1 and under column 2, are adequate.

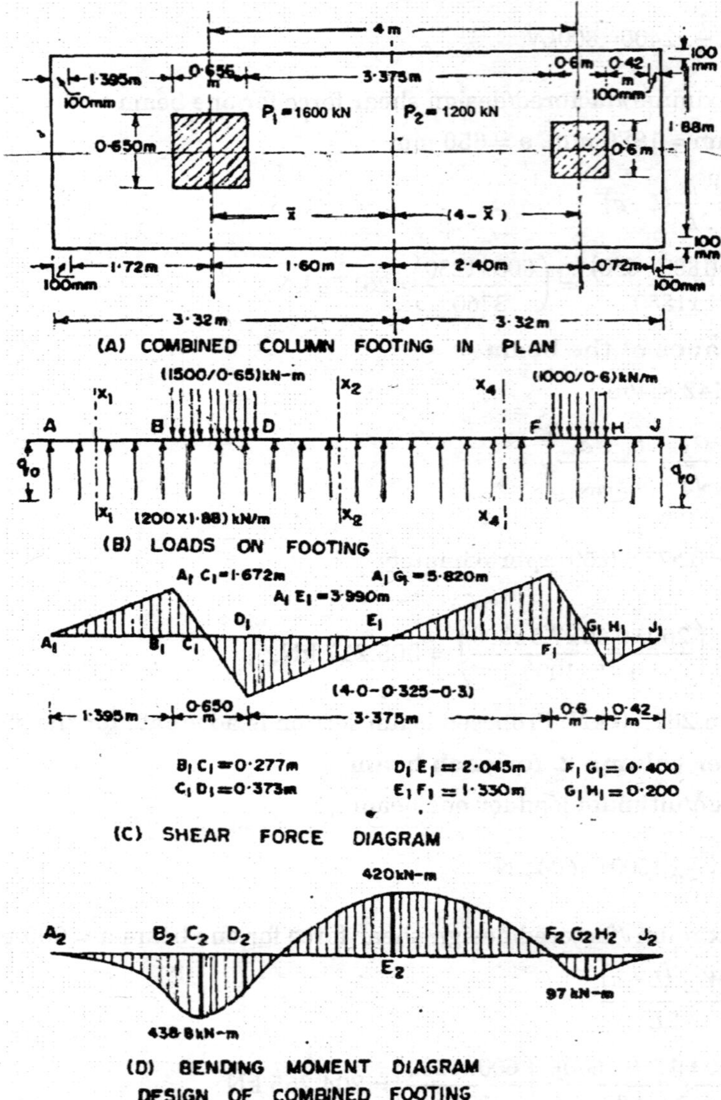

Fig. 9.9.

Step 3. Beams in lower tier

LB 500, @ 0.750 kN/m.

(A) Ultimate/factored/design shear **Shear force under the column1**

(i) Ultimate bearing capacity of soil acting upward ($q_u = 2q_o$ assumed) for the width of footing 1.88 m shear force at the left hand edge of base slab = (400 x 1.88 x 1.395)

= 1049.04 kN

(ii) Factored column load under column1, 2400 kN acts as uniformly distributed downward = $\left(\frac{2400}{0.650}\right) = 3692.31\,\text{kN}$

Ordinate of negative shear force

= (+1049.04 - 3692.31) = 2643.27 kN

(iii) Ultimate bearing capacity of the soil acting upward ($q_u = 2q_o$ assumed) shear force at the right hand edge of base slab = (400 x 1.88 x 0.42)

= 315.84 kN

(iv) Factored column load under column 1800 kN acts downward as uniformly, distributed downward = (1800/0.600) = 3000 kN

Ordinate of shear force = (315.84 – 3000) = 2684.16 kN

Ultimate (maximum) factored/design shear force is that whichever is maximum = 2684.16 kN

It is resisted by eight (08) beams. Ultimate/factored/design shear force resisted by one beam = $\left(\frac{1}{8} \text{x} 2684.16\right)$

= 335.52 kN

Shear resistance of the beam section LB 500, @ 0.750 kN/m. Its properties have been noted above. $1/\sqrt{3}$ = 0.577 treated as 0.60 approximately

$$= \left(\frac{0.60\,\sigma_y}{\gamma_{ms}} \text{x}\, t_w \,\text{x depth of section}\right)$$

$$= \left(\frac{0.60 \text{x} 260}{1.15} \text{x} \frac{9.2 \text{x} 500}{1000}\right) = 623.99\,\text{kN}$$

It is more than the factored/ultimate shear force. Therefore, the beams in lower tier are adequate for shear force.

The positions of maximum bending moments may be located by writing the expressions for shear forces and equating them to zero. The shear force under the column 1 (see Fig. 9.6).

$$F_1 = 200 \times 1.88 \times - \frac{1500}{0.65}(x - 1.395) = 0, \;\therefore\; x = 1.672\,\text{m}$$

The shear force between the two columns

$$F_2 = 200 \text{ x } 1.88 \text{ x} - 1500 = 0, \; x = 3.99\text{ m}$$

The shear force under the column 2

$$F_2 = \left(200 \times 1.88\,x - 1500 - \frac{1000}{0.60}(x - 5.42)\right) = 0, \;\therefore x = 5.82\,\text{m}$$

Step 2 : Bending moments

The values of bending moment at these locations have been found as above.

Maximum of maximum bending moments occur under column 1, M_1 = 438.8 kN-m

Step 3 : Selection of section

Total section modulus required, $\left(\dfrac{438.8 \times 1000 \times 1000}{171.6}\right) = 2557.11 \times 10^3\ mm^3$

Provide 8 LB 275, @ 0.330 kN/m. Total section modulus provided

$(8 \times 392.4 \times 10^3) = 1319.2 \times 10^3\ mm^3$

Width of flange of LB 275, @ 0.330 kN/m = 125 mm.

Clear distance between flanges, (1/7) (1880 – 8 x 140) mm = 108.57 mm 75 mm

The web crippling does not occur in lower tier. The beams in lower tier may be check for shear and these would be found safe.

The design of combined footing for two columns is shown in Fig. 9.10.

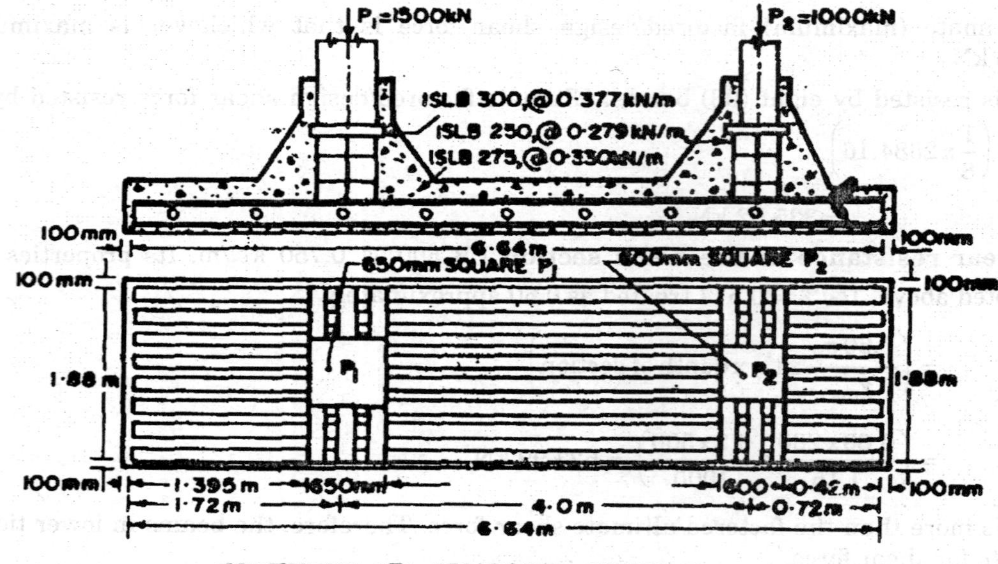

Fig. 9.10. Rectangular combined footing.

Example 9.8. Design beams for flexure in a combined trapezoidal steel grillage foundation for two columns to carry axial load 2800 kN and 2240 kN spaced 3m apart. The available space by the side of 2240 kN column load is 1400 mm. Safe allowable soil pressure is 200 kN/m². Sketch the arrangement of upper tier and lower tier with the concrete encasing the beams. Use limit state method of design.

Design : Steel grillage trapezoidal foundation is to be designed for bending (flexure) to support following two column loads

Column load, P_1 = 2800 kN,

Column load, P_2 = 2240 kN.

Factored/ultimate column load

$P_1 = (1.5 \times 2800) = 4200$ kN

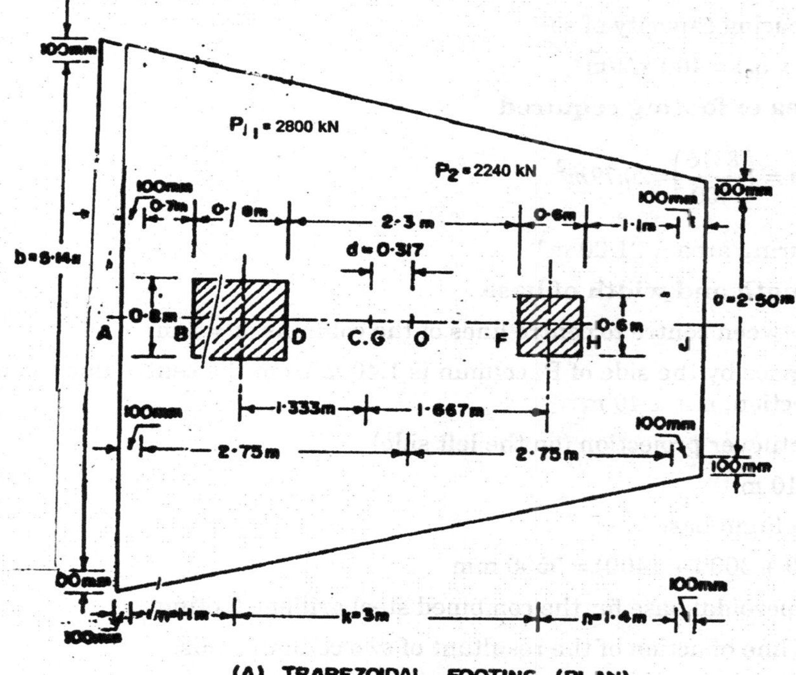

(A) TRAPEZOIDAL FOOTING (PLAN)

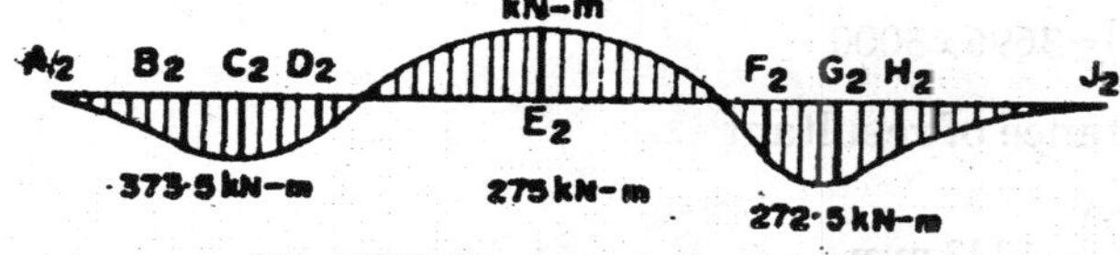

(B) BENDING MOMENT DIAGRAM

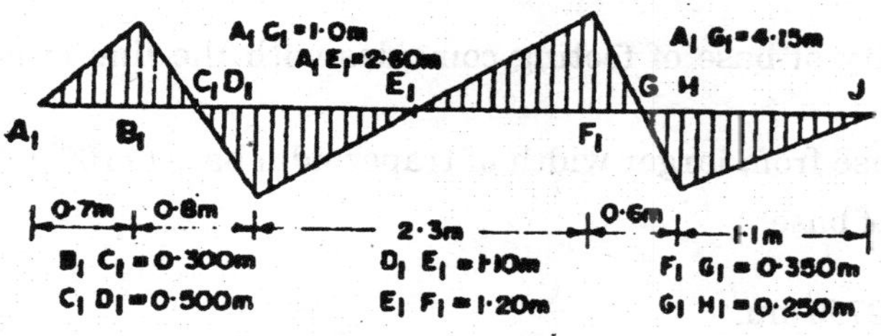

(C) SHEAR FORCE DIAGRAM
COMBINED (TRADEZODAL) FOOTING

Fig. 9.11.

Factored/ultimate column load

$P_2 = (1.5 \times 2240) = 3360$ kN

Add 10 (ten) percent for the self-weight of foundation.

$P_1 = (1.10 \times 4200) = 4620$ kN

$P_2 = (1.10 \times 3360) = 3696$ kN

Total column loads = 8316 kN

Ultimate bearing capacity of soil

$q_u = (2 \times q_o) = 400$ kN/m^2

Step 1. Area of footing required

$$\text{Area} = \left(\frac{8316}{400}\right) = 20.79\, m^2$$

Provide bearing area = 21.00 m^2

Step 2. Length and width of base

Distance between centre to centre lines of the columns, k = 3 m.

Available space by the side of P_2 column is 1.40 m from the centre line (on the right side) Cantilever projection, n is 1.40 m

Provide cantilever projection (on the left side)

m = 1.10 m

Length of column base

= (1100 + 3000 + 1400) = 5500 mm

Provide trapezoidal base for the combined steel grillage footing.

Distance of line of action of the resultant of two column loads.

Take moment about centre line of column P_1 [$P_1 + P_2 = 8316$ kN]

$$\bar{x}\,.\,(4620+3696) = 3696\, x\, 3000$$

where, $\bar{x}$ is the distance of resultant

$$\bar{x} = \left(\frac{3696\,.\,3000}{4620+3696}\right) = 1333\,\text{mm}$$

The centre of gravity of base of footing coincides with the line of action of resultant of two column loads

Distance of c.g of base from larger width of trapezoidal base (1100 + 1333) = 2433 mm

Distance of centre of base

$$\left(\frac{L}{2}\right) = \left(\frac{5500}{2}\right) = 2750\,\text{mm}$$

Distance between centre of base and c.g of base

d = (2750 – 2433) = 317 mm

Larger width of trapezoidal base is given by

$$b = \left(\frac{A}{L}\right)\left(1+\frac{6d}{L}\right)$$

$$= \left(\frac{21x10^3\, x10^3}{5500}\right)\left(1+\frac{6x\,317}{5500}\right) = 5140\text{ mm}$$

Smaller width of trapezoidal base is given by

$$C = \left(\frac{A}{L}\right)\left(1+\frac{6d}{L}\right) = \left(\frac{21x10^3\ x10^3}{5500}\right)\left(1-\frac{6x317}{5500}\right)$$

$= 2500$ mm

Let y_1 be the width of base at the centre of column P_1

$$y_1 = \left[C+(b-c)\left(\frac{k+n}{L}\right)\right]$$

$$= \left[2500+(5140-2500)\left(\frac{3000+1400}{5500}\right)\right] = 4610 \text{ mm}$$

Let y_2 be the width of base of the centre of column P_2

$$y_2 = \left[C+(b-c)\left(\frac{n+L}{L}\right)\right]$$

$$= \left[2500+(5140-2500)\left(\frac{1400}{5500}\right)\right] = 3172 \text{ mm}$$

Step 3 : Design of beams in upper tier under column, P_1 = 2800 kN

Ultimate value of this load p_1 = 4620 kN.

Length of middle beam, y = 4610 mm

Assume size of base plate under this column as 800 mm x 800 mm.

Therefore, a_1 = 800 mm

Factored/Ultimate/design bending moment

$$= \frac{P_1}{8}(y-a_1)$$

$$= \frac{4620}{8}(4610-800)$$

$$= \left(\frac{4620}{8}x\frac{3810}{1000}\right) = 2200.275 \text{ kN-m}$$

Bending resistance/strength of beams = $\left(\frac{\sigma_y}{\gamma_{ms}}.Z_p\right)$

Equating this with factored bending moment.

$$\left(\frac{260xZ_p}{1.15}\right) = (2200.275x10^3\ x10^3)$$

$$Z_p = \left(\frac{2200.275x10^3\ x10^3\ x1.15}{260}\right)$$

$Z_p = 9730.769 \times 10^3\ mm^3$

Elastic section modulus required assuming shape factor for I-section 1.10

$$Z_E = \left(\frac{9730.769 x 10^3}{1.10}\right) = 8846.15 x 10^3\ mm^3$$

Provide three beams. Elastic section modulus required for one beam = 2948.718 x 10^3 mm^3.

Provide 3 (three) MB 600, @ 1226 kN/m

$Z_{xx} = 3060.4 \times 10^3\ mm^3$

Width of flange, b_f = 210 mm.

Clear distance between the flanges

$$= \frac{1}{2}(800 - 3\ x 210)$$

$$= \frac{1}{2}(800 - 630) = 85 mm > 75 mm$$

Total section modulus provided. = 3 x 3060.4 x 10^3 = 9181.2 x 10^3 mm^3

It is more than 8846.15 x 10^3 mm^3.

Step 4. A Design of beams in upper tier under column, P_2 = 2240 kN

Ultimate value of this load

= 3696 kN

Length of middle beam under this column, p_2 is 3172 mm

Assume size of base plate under this column as 800 mm x 800 mm.

Factored/ultimate/design bending moment. Therefore, a_2 = 800 mm.

$$M_2 = \frac{P_2}{8} x (y_2 - a_2)$$

$$= \frac{3696(3170 - 800)}{8 x 1000} = \left(\frac{3696 x 2370}{8 x 1000}\right) = 1094.94 \text{ kN-m}$$

Bending resistance/strength of the beams

$$= \left(\frac{\sigma_y}{\gamma_{ms}} . Z_p\right)$$

$$= \left(\frac{260}{1.15} . Z_p\right)$$

Equating this is design Bending moment

$$\left(\frac{260}{1.15} . Z_p\right) = 1094.94 \times 10^3 \times 10^3$$

$$Z_p = \frac{1094.94x10^3\,x10^3\,x1.15}{260}$$

Z_p = 4843.00 x 10^3 mm^3

Elastic section modulus required assuming shape factor = 1.10 for I-section

$$Z_E = \left(\frac{4843.00x10^3}{1.10}\right) = 4402.73 \text{ x } 10^3 \text{ mm}^3$$

Provide three beams. Elastic section modulus for one beam = 1467.576 x 10^3 mm^3

LB 500, @ 0.750 kN/m. Z = 1543.2 x 10^3

Width of flange, b_f = 180 mm

Clear distance between the flanges = $\frac{1}{2}(800-3x180) = 130mm > 75mm$

Total section modulus provided = 3 x 1543.2 x 10^3 = 4629.6 x 10^3 mm^3

It is more than 4402.73 x 10^3 mm^3.

Step 4.B Design of beams in lower tier

Factored/Ultimate/design load inclusive of weight of foundation, P_1 = **4620 kN**

Factored/Ultimate/design load inclusive of weight of foundation, P_2 = **3696 kN**

Ultimate bearing capacity of the soil, q_u = 2 x safe bearing capacity

q_u = (2 x 200) = 400 kN/m^2 (assumed)

Determination of bending moments

Let x be the distance from larger width of the base.

Ultimate bending capacity of the soil acts upward as uniformly distributed pressure. Where, the column load loads act downward over the base plates.

(i) **Bending moment under column, P_1**

The bending moment from both the cantilever projections side are assumed to be maximum at under the centre line of columns' load as described in Art. 9.4 (Approximate method of design of combined fating).

$$M_1' = \frac{W.x^2}{2}\left[b - \frac{x}{3}\left(\frac{b-c}{L}\right)\right] - \frac{4620}{0.800}\left(\frac{x-0.7}{2}\right)^2$$

$$= \frac{400xx^2}{2}\left[5.14 - \frac{h}{3}\left(\frac{5.14-2.50}{5.50}\right)\right] - \frac{4620}{0.800}\frac{(x-0.7)^2}{2}$$

Distance of centre-line of column load P_1 from left hand-side is 1.1 m substituting x = 1.10 m,

$$M_1' = \frac{400x^2}{2}\left[5.14 - \frac{x}{3}\left(\frac{5.14-2.50}{5.50}\right)\right] - \frac{4620}{0.80}\frac{(x-0.7)^2}{2}$$

= 630 kN-m. Hogging

(ii) **Bending moment under the column, P_2**

Distance of centre line of column load P_2 from left Hand side is $(1.10^2 + 3.0) = 4.10$ m. Bending moment M_3'

$$M_3' = \frac{400x^2}{2}\left[b - \frac{x}{3}\frac{(b-c)}{L}\right] - \frac{4620}{0.800}(x-1.10) - \frac{3696}{0.800}\left(\frac{x-3.70}{2}\right)^2$$

Substituting a = 4.10 m

$$M_3' = \frac{400x4.10^2}{2}\left[5.14 - \frac{4.10}{3}x\left(\frac{5.14-2.50}{5.50}\right)\right] - \frac{4620}{0.800}(4.10-1.10) - \frac{3696}{0.800}\left(\frac{4.10-3.70}{2}\right)^2$$

= 2619.392 kN-m

Width of trapezoidal footing at the extreme edge on left hand side

= 5.14 m

Width of trapezoidal footing through the centre-line of column load, p_1

y_1 = 4.610 m

Distance between these two lines

= 1.10 m.

Area of above footing

$$= \left(\frac{5.14x1.10}{2} + \frac{4.610x1.10}{2}\right)$$

$$= \frac{1.10}{2}(5.14+4.610) \quad = 0.55\,(9.75) = 5.36\ m^2$$

Intensity of ultimate soil pressure

$q_u = 2q_o$ = 400 kN/m² (assumed)

Shear force from left side cantilever portion = 400 x area of the portion

= (400 x 5.36) = 2144 kN

Column load P_1 acting downward

= 4620 kN

Ordinate of shear force below datum

= (4620 - 2144) = 2476 kN

Similarly width of trapezium at the extreme edge an right side

= 2.50 m

Width of trapezium footing through, the centre line of column P_2

= 3.172 m.

Distance from right side upto the centre line of column, P_2 = 1.400 m

Area of this portion $= \frac{1}{2}x\,3.172x1.4 + \frac{1}{2}x2.50x1.40$

$$\frac{1}{2}(3.172+2.50)=0.7x5.672=3.97m^2$$

Intensity of ultimate soil pressure

$q_u = 2q_o = 400$ **kN/m² (assumed).**

Shear force from right side cantilever projection

= (400 x 3.97) = 1588 kN

Ordinate of the shear force above datum line

= (3696 – 1588) = 2108 kN

Distance between centre lines of two columns P_1 and P_2 = 3 m

Shear force changes sign at a distance y from column, p_1. Therefore,

$$\left(\frac{2476}{y}\right)=\left(\frac{2108}{3-y}\right)$$

y = 1.7025

From extreme left side edge, shear force changes at a distance

= (1.7025 + 1.10) = 2.8025 m

(iii) Bending moment between two columns, P_1 and P_2

Shear force changes sign at a distance 2.80 m

Factored/ultimate/design bending moment between two columns at a distance x = 2.80 m from extreme left.

$$M_2' = \frac{wx^2}{2}\left[b-\frac{x}{3}\left(\frac{b-c}{L}\right)\right]-4620(x-1.10)$$

$$=\frac{400x2.8^2}{2}\left[5.14-\frac{2.8}{3}\left(\frac{5.14-2.50}{5.50}\right)\right]-4620(2.80-1.10)$$

$$=\left(\frac{400x7.84}{2}\right)\left[5.14-\frac{2.8}{3}x\left(\frac{2.64}{5.50}\right)\right]-(4620x1.70)$$

$$=(200x7.84)\left[5.14-\left(\frac{2.8x0.88}{5.50}\right)\right]-7854$$

= 1560 [5.14 – 0.448] – 7854

= (1560 x 4.692) – 7.854

= 7319.52 – 7854

= 534.48 kN-m (sagging)

Step 5. Design of beams in lower tier

Factored/ultimate/design bending moments

Under column P_1 = 630 kN-m.

Under column P_2 = 2619.392 kN-m

In between the two columns P_1 and P_2

Beams are designed for = 534.48 kN-m

Maximum value of bending moment

= 2619.392 kN-m

Moment of resistance of the beams

$$= \left(\frac{\sigma_y}{\gamma_{ms}} \cdot Z_p \right)$$

$$\left(\frac{260}{1.15} \cdot Z_p \right) = 2619.392 \cdot 10^3 \ mm^3$$

$$Z_p = 11585.77 \times 10^3 \ mm^3$$

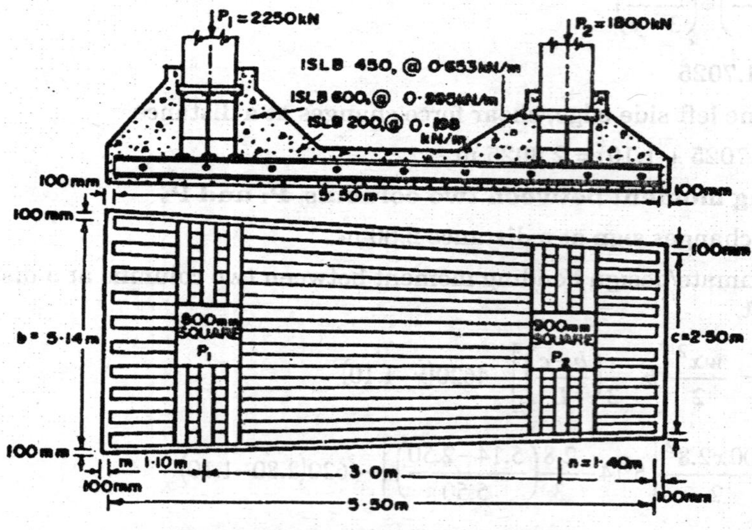

TRAPEZOIDAL COMBINED FOOTING

Fig. 9.12

Provide 10 (ten) beams. Elastic section modulus of beam, assuming shape-factor 1.10 for I-section

$$Z_E = \left(\frac{11585.77 x 10^3}{10 x 1.10} \right) mm^3$$

$$Z_E = 1053.25 \times 10^3 \ mm^3$$

Provide 10 LB, @ 0.653 kN/m, elastic section modulus, for each beam

$$Z_{xx} = 1223.8 \times 10^3 \ mm^3$$

In uppter tier, beams may be checked in shear and web crippling. In lower tier, it is not necessary to check the beams in shear and web crippling.

The complete design is shown in Fig. 9.12.

PROBLEMS

9.1 Design a grillage foundation for a compound column consisting of HB 400, @ 0.822 kN/m with flange plate 300 mm x 18 mm one on each flange and carrying a load of 2300 kN. Draw a neat drawing.

9.2 Design a grillage foundation for a column carrying an axial load of 1500 kN inclusive of self-weight. The bearing capacity of soil is 150 kN/m². The column base plate resting on the grillage is 600 mm x 600 mm.

9.3 Design a two tier grillage foundation to carry a stanchion designed to carry 1040 kN. The base plate of the stanchion is 700 mm square. The bearing of earth is limited to a value of 150 kN/m².

9.4 A combined steel grillage foundation for two columns carrying loads of 1500 kN and 1000 kN spaced at 6100 mm centres is to be designed. The available space by the side of 1000 kN column is only 460 mm.

The maximum soil pressure is limited to 110 kN/m².

(a) Draw the bending moment and shear force diagrams for the upper tier of the grillage foundation.

(b) Work out the sections for the upper and lower tiers of the grillage.

(c) Sketch the arrangement of upper and lower tiers with concrete encasing.

9.5 A column is subjected to an axial load of 2000 kN. Design a grillage foundation for the column assuming that the bearing capacity of the soil is 100 kN/m². The size of the column base plate is 800 mm x 600 mm.

9.6 Design a combined grillage foundation for the two columns HB 450, @ 0.925 kN/m and HB 400, @ 0.774 kN/m spaced at centre to centre distance of 4 m. The columns carry axial loads of 2000 kN and 1500 kN respectively. The safe bearing capacity of the soil is 150 kN/m². Give neat sketches to show design details.

9.7 A column made of an HB rolled steel section carries an axial load of 400 kN and a bending moment of 36000 kN-mm at the bottom end acting about the major principal axis of the column section. The base plate is connected to the column flange by two 150 mm x 75 mm x 12 mm (2 ISA 15075, @ 0.201 kN/m) angles with 20 mm rivets, the 75 mm legs being outstanding from the column flanges. The end of the I-section is faced for complete bearing on the base plate. Using the data given, find (a) the required horizontal dimensions of the base plate (b) the required thickness of the base plate and (c) the number of 20 mm rivets necessary to connect each angle to the column flange as governed by the 3600 kN-mm moment.

Properties of sections :

ISHB 250 column section : depth = 250 mm, flange width = 250 mm. flange thickness = 9.7 mm, area = 6971 mm².

Rivets : Nominal diameter 20 mm, diameter of rivet holes = 21.5 mm.

Gauge distance of the rivet holes :

Column flange = 140 mm, 150 mm leg of angles distance from back to first row = 55 mm, from first row to second row = 65 mm.

Adopt permissible stresses as per IS : 800-1984.

9.8 It is required to design a two tier grillage foundation for a steel column carrying a load of 1000 kN. The dimensions of the base plate are about 800 mm x 850 mm. The footings is to be supported on concrete base which in turn rests on solid rock. The allowable pressure on the concrete base is 4 N/mm². Design the grillage using I-section girder built-up of suitable rectangular plates. Plates 12 mm and 20 mm thick and of any width and length may be assumed to be available. Assume standard working stresses as per IS : 800-1984.

Sketch the arrangement of the built-up beams forming the grillage.

9.9 Design a two tier grillage foundation footing for a HB 350 @ 0.674 κN/m column supporting a load of 1000 kN. The footing is to be supported on soil which has a bearing capacity of 100 kN/m².

9.10 Design a grillage foundation for a column consisting of HB 450 section with flange plate 400 mm x 40 mm each carrying an axial load of 4500 kN. The concrete can take a safe bearing pressure of 4 N/mm². Sketch the arrangement.

9.11 A column consisting of HB 200, @ 0.373 kN/m with cover plates 300 mm x 20 mm (one on each flange) is fixed at the base and free at the top. It carries a load of P at an eccentricity of 200 mm from the centroidal axis of member giving rise to bending moment about the major axis of the member. If the height of column is 4 m, design a suitable attached base for the column. Permissible bearing pressure on concrete is 4 N/mm².

REFERENCES

1. A.B. Calrk, S.H. Coverman, `Structural Steel Work, Limit State Design. Chapman and Hall Ltd. 11, New Fetter Lane, London, EC 4P 4EE.

2. Teaching Resources on Structural Steel Design, Institution of Steel Development and Growth Compiled by Rangachari Narayan, V. Kalyanaraman and et al.

3. IS 800-2007. Indian Standard `General Construction in Steel' Code of Practice Published by Bureau of Indian Standards. Manak Bhawan, 9 Bahadur Shah Zafar Marg, New Delhi, 110 002.

10

Part - VI

DESIGN OF TENSION MEMBERS

LIMIT STATE OF FAILURE DUE TO YIELD IN TENSION/RUPTURE OF CRITICAL SECTION/BLOCK SHEAR DESIGN OF TENSION MEMBERS

10.1 INTRODUCTION

A tension member is defined as a structural member subjected to tensile force in the direction parallel to its longitudinal axis. A tension member is also called as a tie member or simply a tie. The term is commonly used for tension members in the roof trusses. When a tension member is subjected to axial tensile force, then the distribution of stress over the cross-section is uniform. The complete net area of the member is effectively used at the maximum permissible uniform stress. Therefore, a tensile member subjected to axial tensile is said to be efficient and economical member. Tension members elongate when the tensile force is applied. At the ultimate load these members may fail either due to yield by excessive elongation and the members may become non-functional or by rupture at a critical section or by block shear plates and other rolled steel sections in tension fail by block shear at ends bolted bolted region. In the building design, the interest in the use of tension members is increasing. The tension members are used as hangers for floors and cables for roof.

Suspenders (i.e., hangers) in suspension bridges and cables in cable-stay bridges are also tension elements. Some building which are used in earthquake sensitive areas are built with a central core. From the top of central core, suspenders are attached diagonally at top. From these members, the suspenders are hanged vertically supporting the floor-beams of all the storeys.

Tension members are linear members. These members elongate due to axial forces acting over the them. These members may be reach upto the ultimate load. At which these members fail by rupture at a critical section.

In case, the gross-area of the member yields over a major portion of its length before the rupture load is reached, the member may become non-functional due to excessive elongation.

Plates and other rolled sections in tension may also fail by block shear of end bolted region.

The various members used as tension members are not perfectly straight. The initial crookedness and imperfections of the members result in the small eccentricity. The distribution of stress in the eccentrically loaded tension member is not uniform. The member is subjected to combined stress, (i.e., direct tension and bending). Besides this, a tension member may be subjected to axial tension and bending.

Suspenders (i.e, hangers) in suspension bridges, and cables in cable-stay bridges are also tension elements. Some buildings which are used in earthquake sensitve areas are built with a central core. From the top of central core, suspenders are attached diagonally at top. From these members, the suspenders are hanged vertically supporting the floor-beams of all the storeys.

In the industrial building frames, the cross-bracings are provided to lateral to resist lateral loads. In these cross-bracings, some members remain in tension, when the lateral forces act in one directons. And the other members remain in compression. When the lateral loads change opposite direction, then the nature of the forces in both the above types of members reverse.

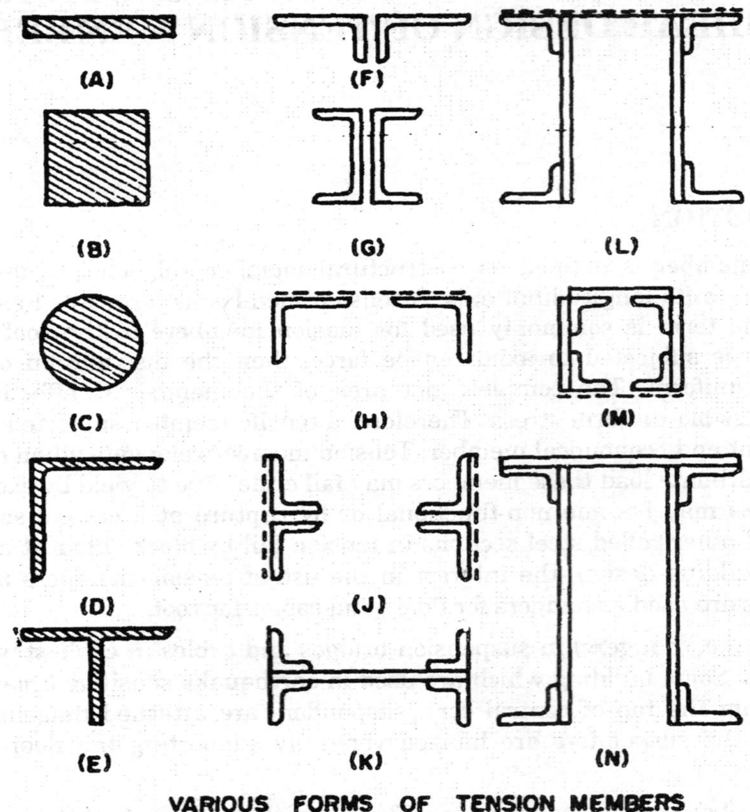

Fig. 10.1. Various forms of tension members

Former compression members act as tension members and the former tension members act as compression members. Such members are designed to resist tension as well as compression members.

Sag ties in roof trusses are also tension members, which are provided to reduce the deflection in the roof trusses.

Single angle members are attached with gusset plates in the roof trusses.

The tension in the member becomes eccentric. The distribution of stress over the section becomes non-unform.

Double angles connected to on the opposite sides of the plate also experience the effect of shear lag.

The channel sections as shown Fig. 10.1(L) are used as tension members (i.e, bottom chord members of the bridge truss girders.

10.2 TYPE OF TENSION MEMBERS

The types of structure and method of connection decide the type of tension members. The compact tension members are generally used. As far as possible, a large portion of the tension member is connected directly to the gusset plates or splice plates. This minimizes stress concentration in the members. The various types of tension members used may be grouped into following four groups:

(i) Wires and cables (ii) Rods and bars

(iii) Single structural shapes and plates (iv) Built-up members.

(i) Wires and cables. The wire ropes are used for hoists, derricks, rigging slings, guy wires and hangers for the suspension bridges. The untwisted parallel wires are used for the manufacture of main cables of the suspension bridges. The wire ropes have the advantages of flexibility and strength. The wire ropes require special fittings for proper end connections. The wire ropes are used as guy wires or guy lines with the steel towers. The wire ropes are given initial tension. The initial tension prestresses the frame. The effectiveness of frame in resisting the external load is increased when the frames are prestressed.

(ii) Rods and bars. The square and round bars as shown in Fig. 10.1 (B) and (C) are quite often used for the small tension members as the tie rods in the rectangular water tanks. These tie rods restrain the opposite faces of tanks outward deflection. The round bars are threaded at the ends are used with nuts. The area of cross-section at the ends and are used with nuts. The area of cross-section at the root of the thread is the net area which resists the tension. The round bars with threaded ends are used with pin-connections at the ends with standard clevises. Sometimes, the bars are bent to form loops at the ends instead of threads. The round bars with clevises and loop rods are made in two portions. These two portions are joined with turn buckles. The tension members are tightened with these turn buckles. Such members are commonly used for lateral bracing. The ends of rectangular bars or plates are enlarged by forging and then, these are bored to form eye bars. The eye bars are used with pin-connections. The rods and bars have disadvantages of inadequate stiffness resulting in noticeable sag under the self-weight.

(iii) Single structural shapes and plates. The single structural shapes, (viz. angle sections and tee-sections) as shown in Fig. 10.1 (D) and (E) are used as tension members. The angle sections are considerably more rigid than the wire ropes, rods and bars. When the length of tension member is too long, then, the single angle section is also too flexible. The equal angle sections are commonly used when the angles are used as single structural members. The single angle members are also used for bracing. These members can resist small compression if reversal of stress takes place. The single angle sections have the disadvantages of eccentricity in both the planes in the riveted connections.

Sometimes, the single channel sections are also effectively employed as the tension members. The channel sections have eccentricity only in one direction. The single channel sections have high rigidity in the direction of web and low rigidity in the direction of flange.

Occasionally, the I-sections are also used as tension members. I- sections have more rigidity. The single I-sections are more economical than the built-up sections.

(iv) Built-up members. Two or more than two members are used to form built-up members. When the single rolled steel sections cannot furnish the required area, then, the built-up sections are used. The built-up sections may be made more rigid and more stiff than the single structural shapes. The moment of inertia of the built-up sections may be increased. The built-up sections can resist compression if the reversal of stresses takes place.

The double angle sections of unequal legs shown in Fig. 10.1 (F) are extensively used as tension members in the roof trusses. The angle sections are placed back to back on two sides of a gusset plate. When both the angle sections are attached on the same sides of the gusset, then, built-up section has eccentricity in one plane. The built-up section is subjected to tension and bending simultaneously. The two angles sections may be arranged in the star shape (i.e., the angles are placed diagonal opposite to each other with legs on outer sides). The star shape angle sections may be connected by batten plates. The batten plates are alternately placed in two perpendicular directions. The star arrangement provides a symmetrical and concentric connection. Two angle sections as shown in Fig. 10.1(H) are used in the two plane trusses, where two parallel gussets are used at each connection. Two angle sections as shown in Fig. 10.1(H) have the advantage of adjusting the distance between them. Four angle sections as shown in Fig. 10.1(J) are also used in the two plane trusses. The angles are connected to two parallel gussets. Four angle sections connected by plates as shown in Fig. 10.1(K) are used as tension members in the bridge truss girders.

A built-up section may be made of two channels placed back to back with a gusset plate in between them. Such sections are used for medium loads in a single plane truss. In two plane trusses, two channels are arranged at a distance with their flange turned inward. It simplifies the transverse connections. It also minimizes lacing. The flanges of two channels are occasionally turned outward in order to have greater lateral rigidity.

The heavy built-up tension members in the bridge truss girders are made of angles and plates. Such members can resist compression if the reversal of stresses occurs.

Single angle members are attached with gusset plates in the roof trusses.

The tension in the member becomes eccentric. The distribution of stress over the section becomes non-uniform.

Double angles connected to on the opposite sides of the plate also experience the effect of shear lag.

The channel sections as shown Fig. 10.1(L) are used as tension members (i.e, bottom chord members of the bridge truss girders).

10.3 BEHAVIOUR OF TENSION MEMBERS

When a mild steel bar or a flat conforming to IS: 226 and IS: 2062 is tested in a laboratory using universal testing machine upto breaking, then, the bar is loaded in tension, as shown in Fig. 2.1 (A).

Upto elastic range, the stress-strain curve exhibits initially a linear relationship and obeys Hooke's law. The slope of this straight line portion represents Young's modulus of elasticity, E. As strain is increased a point is reached at which the curve tends to depart from the linearity. The stress at which this occurs is termed as **Limit of proportionality,** σ_y.

Further straning results in yielding of the steel bar at a stress equal to the upper yield, σ_{uy}. The stress drops suddenly to lower yield stress, σ_{Ly}. The difference between σ_{uy} and σ_{Ly} depends on the type of steel and also on the speed of loading during conducting the test.

After the drop of stress to lower yield, it remains sensibly constant for the considerable increase in strain called, **plastic range**. During this the plastic flow of the material occurs. The extent of which is a measure of the ductility of the material. The structural steel shows **yield plateau.** During the yield plateau, the load remains constant as the elongation increases to nearly ten (10) to twelve (12) times the yield strain, ε_y (elastic strain).

When the material is further stretched, the material shows a need of smaller increase in tension for the further increase in elongation. The initial slope of this portion of the curve is termed as the **strain hardening modulus,** E_{st}. It is much lesser steeper than the elastic range with [E_{st}/E] being typically between (1/30) and (1/100) **Reference** Alpsten GA, 1973 `Variation in mechanical and Cross-sectional properties of steel, Swedish Institute of steel construction, Publication No. 42.

The behaviour of a steel bar in tension is in many respect very similar. Yet there is difference when the member is attached to the other parts of the structure. Irrespective of the method of joining, it is influenced in the manner in which the load is transferred. When the mechanical means are used, the presence of holes has the direction effect on the strength of the member. The effect of a hole in a tension member is more than the simply the absence of some material. In the immediate surrounding of the hole, stres concentration develops.

For the ductility of steel, it is customary in design to neglect these effects. Simply the net sectional area is calculated by subtracting the area of holes.

The tension member may sustain loads upto the ultimate load without failure. At this load, nearly 10 to 15 percent of elongation occurs. The structure would become unserviceable.

After the **ultimate tensile stress,** σ_{ult}, the stress appears to decrease until the fracture occurs finally.

However, the actual stress continues to increase (Load) actual reduced cross-sectional area). The area of original cross-section is reduced once ultimate stress, σ_{ult} reaches. This area reduces rapidly because of **necting** phenomenon as shown in Fig. 10.2.

10.4 NET SECTIONAL AREA FOR STAGGERED HOLES

Usually, there are more holes (at least more than one hole) for the purpose of connections. The net cross-sectional area of the plate with holes depends upon the arrangement of the holes.

The arrangement of holes may be either in parallel lines parallel to the longitudinal axis of the member (i.e, parallel to the line of action of the tensile load) that is in (i) **chain riveting** or (ii) **diamond riveting** (i.e, staggered arrangement). The calculations of net area of the plate with holes for these two different arrangement are as follows.

In **zig-zag riveting** (staggered arrangement) the rivets used to connect the tension members are staggered as shown in Fig. 10.2.

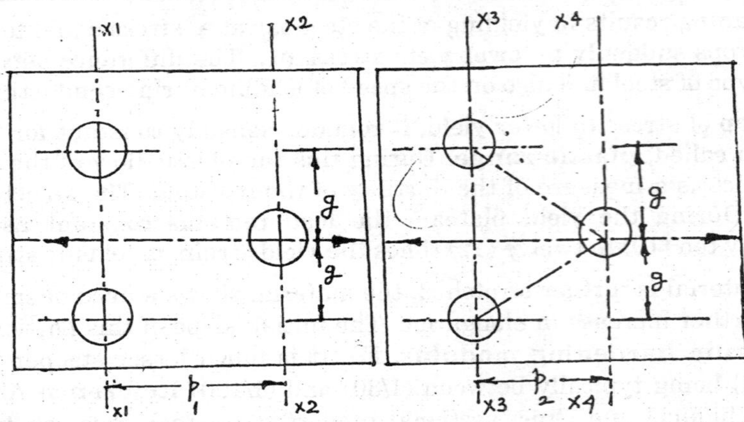

Fig. 10.2.

Consider the arrangements of holes as shown in Fig. 10.2 (A) and Fig. 10.2 (B). In Fig. 10.2 (A), the distance x_1x_2 is comparatively more than the distance x_3x_4.

In Fig. 10.2 (A), in case, the failure occurs along $x_1 - 1\text{-}2 - x_1$, then, the gross-area of the plate is reduced for two holes to find the net area.

$$A_{net.1} = (b - 2d) \cdot t \qquad \text{(i)}$$

In case, the failure occurs along $x_2 - 3 - x_2$, then, the gross area is reduced for one rivet hole.

$$A_{net.2} = (b - d) \cdot t$$

The failure of a plate may occur on a zig-zag section passing through all the three rivets instead of a straight right angle section passing through first two rivets. In case all the three rivets are in one line along a right angle section then, the deduction is made for all the three rivet holes in determining the net area. When one or more than one rivet hole is off the line, the failure of the plate depends upon the staggered pitch, p the gauge distance g, and the diameter of the hole, d. When the gauge distance g, is large as compared to the staggered pitch, p then, the failure may occur in zig-zag line. When the gauge distance is small, as compared to the staggered pitch, p then, the failure may occur along a straight right angle section passing through rivet holes in Fig. 10.2. When the staggered pitch, p and the gauge, g are fixed, then, the zig-zag failure become more likely as the diameter of hole, d increases. From the theoretical investigations and laboratory tests, it is found that the net area of the plate for Fig. 10.2 have been found as below :

$$A_{net} = \left[A_{gross} - 2dt - d\,t\left(1 - 2 \times \frac{p^2}{4\,g.d}\right)\right] \qquad \text{(i)}$$

or $$A_{net} = \left[bt - 2dt - d\,t\left(1 - 2 \times \frac{p^2}{4\,g.d}\right)\right] \qquad \text{(ii)}$$

or $$A_{net} = \left[b - \left(3d - 2 \times \frac{p^2}{4\,g.d}\right)\right] . t \qquad \text{(iii)}$$

In general, as per IS : 800-2007, Clause 6.3.1,

$$A_{net} = \left[b - \left(n\ d - n'\ \frac{p^2}{4g}\right)\right].t \tag{10.1A}$$

where, b = width of plate, and t = thickness of plate

p = staggered pitch, i.e., the distance between any two consecutive rivets in zig-zag chain measured parallel to the direction of stress in the number as shown in Fig. 10.2.

g = gauge space, that is, the distance between the same two consecutive rivets in a zig-zag chain at right angles to the direction of stress in the number as shown in Fig. 10.2. It is transverse spacing between parallel adjacent lines of fasteners.

n = number of rivet holes in zig-zag line.

n = number of gauge spaces. and d = gross diameter of rivet.

Net sectional area of plate as shown in Fig. 10.2

$$A_{net} = \left[b - \left(3d - 2\frac{p^2}{4g}\right)\right].t \tag{iv}$$

Eq. 10.1 is known as **Steinman's formula.**

In case, the rivets are connected in different staggered pitches, p_1, p_2, etc. and the gauge distances g_1, g_2, then, the deductions shall be as below :

$$\text{Deduction} = \left[(\text{Sum of sectional areas of holes}) - \left(\frac{p_1^2}{4g_1} + \frac{p_2^2}{4g_2}\right)t\right]$$

In case, the staggered rivets are used in the two legs of an angle section, then, the net area can also be determined by Eq. 10.1. Then, the riveted connections are in two different planes, (viz., in non-planer sections), such as angles with the holes in both the legs. The gross width of angle is the sum of lengths of two legs less the thickness of angle. The gauge distance for the rivets in two legs is equal to the sum of gauge distance for the rivets in each legs less the thickness. It is to note that the standard rivet gauge distances for legs of angles are given is ISI Handbook No. 1.

In the design of tension members, the tensile design strength resistance is taken as the yield load (as the limiting load)

$$T_d = \left(\frac{\sigma_y \cdot A_{gross}}{\gamma_{msy}}\right)$$

where, σ_y is the yield stress of the material in N/mm². A_{gross} area of the cross-section in mm². And γ_{msy} is the partial safety factor. For failure in tension by yielding. As per IS : 800-2007, the value of γ_{msy} is 1.15.

10.5 DESIGN TENSILE STRENGTH/RESISTANCE DUE TO YIELDING OF CROSS-SECTION

Factored/ultimate/design tension in the member shall be obtained by multiplying the axial tension by partial safety factor, γ_{fL} for loads and the axial tension in the member at the service load level.

$$T_{udt} = T_{fd} = (\gamma_{fL} * T_{ck}) \tag{10.2}$$

where, T_{ck} is the axial tension in the member at the service load level. It may also be referred as characteristic axial tension. The values of partial safety factor, γ_{fL} have been given in IS : 800-2007, Table 5.

The requirement is that, the factored/ultimate/design tension, $T_{udt} = T_{fd}$ in the members shall be less than the design tensile strength/resistance.

As per IS : 800-2007, Clause 6.1, factored design tension T_{fd} should be less than the design strength of the members

$$(T_{fd} < T_{ds}) \tag{10.3}$$

The design tensile strength/resistance of a member, T_{ds} under axial tension is the lowest of the following :

(i) Design strength due to yielding of the gross-section, T_{dg}

(ii) Rupture strength/resistance of the critical section, T_{dnrsc}, and

(iii) Design tensile strength/resistance due to block shear.

These have been defined and described in the subsequent sections.

The design tensile strength/resistance of members due to yielding of gross-section, A_{gross} is controlled by yielding of the gross area. It is found as below :

$$T_{dg} = \left(\frac{A_{gr} \cdot \sigma_y}{\gamma_{msy}}\right) \tag{10.4}$$

where,

σ_y = Yield stress of the steel.

A_{gr} = Gross-area of the cross-section.

γ_{msy} = Partial safety factor for the strength of material against failure in tension by yielding.

10.6 DESIGN TENSILE STRENGTH/RESISTANCE DUE TO RUPTURE OF CRITICAL SECTION

The dictionary meaning of rupture is breach (breaking or bursting, that is, the state of being broken).

In tension test of steel bar, the highest stress related to the original cross-sectional area of the specimens, when the material of the bar many withstand is called, **`the ultimate (tensile) strength'**.

Reaching the magnitude of the ultimate strength, a sharp local reduction of the cross-section occurs, which is called **necking.** Along with this, the length of the specimen elongates and yields.

Two stresses exist (namely, normal and tangential). The tangential stress remains tangential the plane/surface of necking. When this tangential stress becomes maximum (one of the two principal stresses), then the normal principal stress becomes zero.

The direction of this maximum tangential stress measured from the vertical axis (90-2ϕ) becomes zero. Then $(90 - 2\phi) = 0$, $2\phi = 90°$, $\phi = 45°$. The yielding and necking occur under maximum tangential stress.

The cross-sectional area of the specimen during necking reduces rapidly. The stress obtained using actual area becomes large. **The rupture of the specimen** takes place at the narrowest section of the neck (which one called as the critical section). The complete phenomenon has been referred as **'rupture of the specimen at the critical'** since, the rupture takes place at the critical stress, it is also called as **stress rupture.**

The design tensile strength due to rupture of the critical section for the following different elements have been described in the subsequent sections as given in IS : 800-2007

(i) Threaded rods

(ii) Plates

(iii) Single angle sections

(iv) built-up section consisting of two angles

10.7 THREADED RODS AND BARS

Threaded rods and bars subjected to tension may fail by rupture at the root of the area. Therefore, the net area at the root of threaded section is found.

The effective net area at the root of the threaded tension member may be obtained from the following expression (approximately). As per AISC specification.

$$A_{eff\,net} = \frac{\pi}{4}\left(D - \frac{0.9743}{n}\right)^2 \text{mm}^2 \qquad (10.5)A$$

where, D is the nominal diameter of threads in mm and n is the number of threads per mm.

The design tensile strength/resistance of the threaded rods in tension is given by

$$T_{d.rods} = \left(\frac{0.9\ A_{net.eff}\ \sigma_u}{\gamma_{msu}}\right) \qquad (10.5)B$$

where, $A_{eff.net}$ is the effective net area at the root of the threaded rods and bars

γ_{msu} = Partial safety factor for failure at the ultimate stress as given in IS : 800-2007 (=1.25)

σ_u = Ultimate stress of the material.

10.8 PLATES UNDER TENSION (NET SECTIONAL AREA)

The plates are connected by bolt holes. Bolt-holes are made slightly larger than the diameter of the bolt. Upto 24 mm diameter, the holes are made 2 mm larger than the bolt- diameter. Holes allow clearance, so that, the bolts may be placed inside the holes conveniently.

The net sectional area is used, which one is the gross-sectional area of the members less the maximum deduction for the holes. The projection for the holes. The projected area of each hole is a rectangle (one side of which is the thickness of the member and the other side is the diameter of the hole).

The deduction for the area of holes depends upon the types of arrangement in which the holes are made. The arrangement may be chain riveting or in zig-zag (staggered riveting.

(i) **Chain riveting** : The holes are made in parallel lines parallel to the longitudinal axis of the member (i.e, parallel to the line of action of the load. While, these holes are in one line in the directon perpendicular to the line of action of the load. The net sectional area of a member at a section is the gross-area *minus* the area obtained from the product of thickness of plate and the sum of the diameters of bolt holes in that line.

(ii) Zig-zag riveting (staggered riveting) :

The tensile stress in a plate at the cross-section of a hole is not uniformly distributed in the elastic range. It exhibits (shows) **stress concentration** adjacent to the hole as shown in Fig. 10.3.

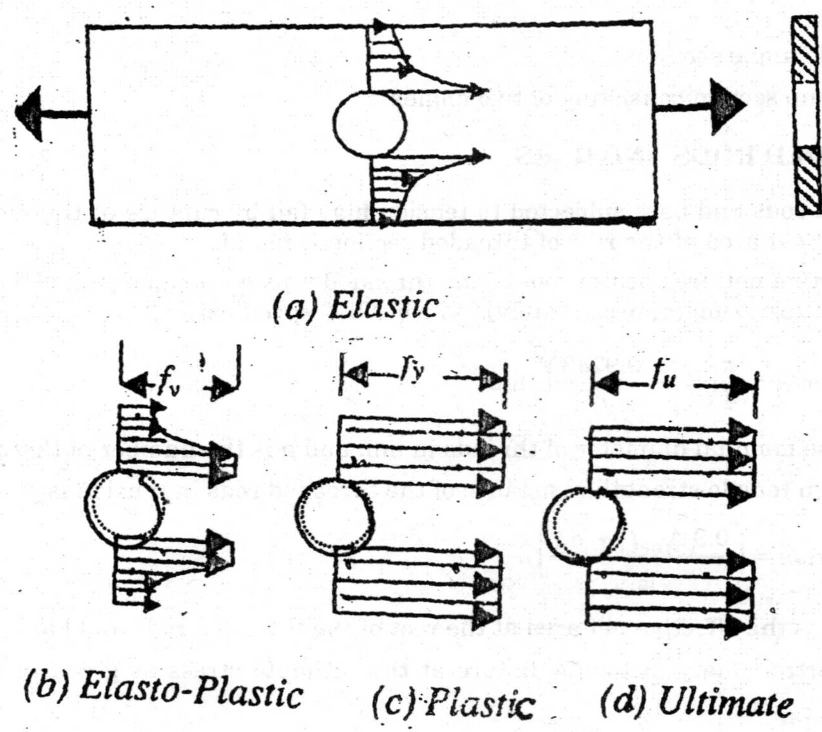

Fig. 10.3

The maximum elastic stress adjacent to the hole is nearly 2 to 3 *times* more than the average stess. The ratio of this maximum elastic stress to the average stress is called as **stress concentration**. The value of this factor depends upon the ratio of the diameter of hole to the width of plate normal to the direction of stress.

In the statically loaded plates with holes in tension, the region surounding the hole attains yield stress, σ_y as shown in Fig. 10.3 when the load is further increased the yield stress at the inner points remain constant and the yield travels and plastifies progressively until the complete net section at the hole attains the yield stress on the outer side from the hole.

Lastly, the tension member fails and the stress across the complete net section attains the ultimate stress, σ_y as shown in Fig. 10.3 (D)

Only small portion of the plate adjacent to the smallest cross-section stretches much at the ultimate stress. The elongation of the overall length of the plate shall not be large so long as the gross-cross-section does not reach yield stress.

The design tensile strength/resistance in tension of the plate is controlled by the rupture of the net cross-sectional area, A_{net} at the holes at ultimate stress is given by [IS : 800-2007, Clause 6.3.1]

$$T_{ust} = \left(\frac{0.9\, A_{net}.\ \sigma_u}{\gamma_{msu}} \right) \qquad (10.6)$$

where, γ_{msu} is the partial safety factor against ultimate tensile failure by rupture. (γ_{msu} = 1.25).

Note: Determination of design tensile strength/resistance due to rupture at the critical section of a single angle, double angles, tee-sections and other built-up section needs the study of **shear lag** (**shear flow**).

10.9 SHEAR LAG (OR SHEAR FLOW)

The problem of shear flow or shear lag as load is transfered to gusset plates from the main member has been qualitatively illustrated in Fig. 10.4.

Any case, where tension members have a profile with not all the segments in a common plane (as the outstanding lag of an angle, webs of beams with flange connections, flanges of channels when the web is connected, and similarly for built-up members) will have an effective area (A_{eff}) calculated as under :

(1) Equal or unequal leg angles connected by one leg to a gusset plate (singly or in pairs [A_{eff} = 0.90 A_{net}) in roof trusses. The tension in the member becomes eccentric. The distribution of stress over the section becomes non-uniform. There is a shear lag at the end-connections Kulak and Wu, `Shear lag in Bolted Angle Tension Members', ASCE Journal of structural Engineering, Vol. 123, No. 9 September, 1997 (pp 1144-1152) have reported that when the ratio of area of outstanding leg to the total area of the cross-secton increases, the effect of shear lag and hence, the reduction of strength is more.

Further it is reported that the double angle's connected on the opposite sides of the gusset plates experience the effect of shear lag. Kulak and Wu, 1997, have been that the shear lag due to connection by one leg only, at the ultimate stage, the stress in the outstanding lag remains closer only to yield stress, even though the stress at the net area of section of connected leg may have attained ultimate stress.

(2) For W, M or S shapes with b_f more than or equal to 0.67 d or the structural tees cut from these shapes and fastened by the flanges only [A_{eff} = 0.90 A_{net}].

(3) For all other shapes, induding built-up shapes, with at least three fasteners in a line

[A_{eff} = 0.85 A_{net}]

(4) Any tension members with only two fasteners in a line

[A_{eff} = 0.75 A_{net}]

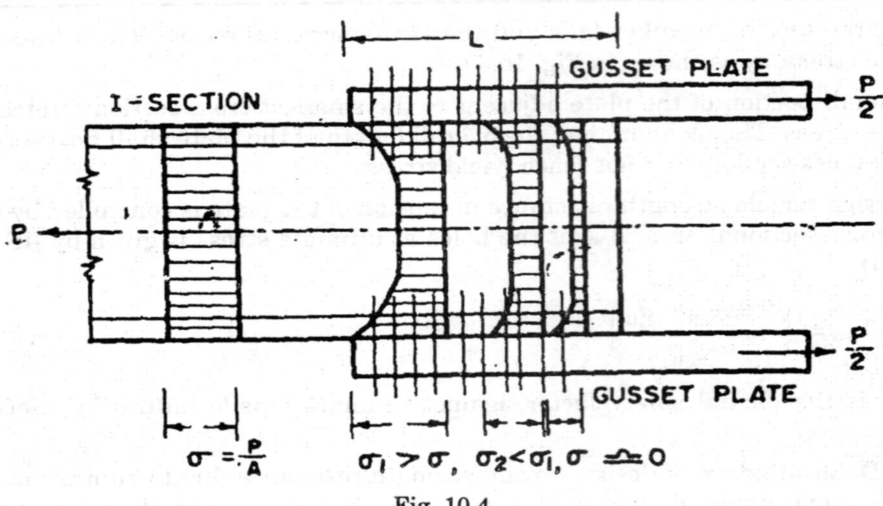

Fig. 10.4.

It is noted that the effective area is alway less than the net area, when the shear lag is to be considered.

Tests on large number of fabricated tension connections have shown that the welded connections may be close to 100 percent efficiency [A_{eff} is nearly equal to A_{net}].

No connections using mechanical fasteners were found to be much over 90 percent efficient so as to provide an adequate statistical coverage of the tests.

Note : AISC specifications limit the efficiency of any connection using mechanical fasteners to a maximum net cross-sectional area of A_{net} less than or equal to 0.85 A_{gross}.

With welded connections, without holes, the net area, A_{net} is equal to the total gross area, A_{gross}, then, the effective area.

$$A_{eff} = 0.90\ A_{net}$$

be used for tension members which are continuous through the joint, to calculate the effective tension area for double angles, channels and structural tees used as bottom chord of trusses and continuous across several panels.

In case, a tension member (angle section or channel section) is connected to the gusset plates, it is necessary to keep length of the joint sufficiently long (viz, equal to the length of transfer or length of shear lag), so that, the stresses, in the connected member at the beginning remain uniform over its complete cross-section. If the length of joint is smaller, then there will be stress concentration near the edges of the member and its effectiveness will be reduced as shown in Fig. 10.4.

The axial force in the tension member shown legs the transfer to the gusset plate, because of it is distributed across the member cros-section. If the length of joint is too large, the portion of axial tensile force transmitted by rivet/bolt shall be equal.

In case, N is the number of rivets/bolts used in the connection, the rivets/bolts in the first line carry more than (P/N) and the rivet/bolt in the last line will carry practically zero force. Due to high concentration of stress, the web plate may fail in tearing and tearing of plate progresses across the section. In case, the web plate is adequate in the section, the tearing of plate does not occur, but stretches (PL/AE) depending upon the value of tensile

strain. The rivets/bolts in the first line will either undergo compatible shear strain or will shear-off, if the strain and the resulting forward displacement of the hole is too large. The failure of rivets/bolts in second line will also fail and so on. As such, it produces a progressive joint failure. This process is called as **unbuttoning.** It is note that when the axial force in tension to be transmitted is large, this process is almost instantaneous.

If the length of joint is short, then all rivets/bolts carry load. The rivets or bolts in first line strain with the plate. The strain in these rivets increases corresponding to the shear yield stress. On further increase of strain, the rivets/bolts continue to strain with no increase of load and the rivets/bolts in the second line will take up the transferred load.

The ultimate load fa the joint will reach when all the rivets/bolts have yielded.

The analysis of joint for strain compatibility is rarely done. The factor of safety used together with the properly of steel ductility, are such that except for long joint, only the rivets/bolts in the first line in connection may yield or reach close to yield.

The proportioning of a tension member to find the required net area economically depends mainly on the type of end connections.

The components which form a tension member should be so taken and arranged that as large part of it as practicable may be directly connected to the gusset plates.

The rivets/bolts should be arranged such that the net area provides as large proportion of the gross-area as possible.

As the members are designed earlier that the connections, a suitable allowance is made for deduction due to rivet/bolt holes. A too liberal allowance will be uneconomical and a small allowance may result in excessively long connection.

Excessively long connections are undesirable. Long connections may develop more secondry stresses. The efficiency is reduced by 85 percent when the length of joint is more than 406 mm. But, in cases, where, W, S, or C shapes are used with the gusset plates on the flanges as shown in Fig. 10.4, it is necessary to produce a joint sufficiently long that the stress at the section AA may be transferred to and through the flanges and into the gusset plates.

The axial force in the tension member shown, lags (go too slow hang back, fall behind) the transfer to the gusset plate, because of its distribution across the tension member.

In case, the length of joint is too large, the tension member may fail from a large concentration of stresses in the web members producing a tear and a resulting a progressive failure across the section.

A measure of shear lag efficiency is based on the distance from the gravity axis of the member to the fastener (i.e, the gusset plate) plane Munge and chesson (Reference : "Riveted and Bolted Joints, Net Section Design" proceedings ASCE, ST I, February, 1963) have been an expression for shear lag efficiency.

$$\eta = E_{SL} = \left[\left(1 - \bar{x}/L\right) \times 100\right] \qquad (10.7)A$$

where, η is the efficiency of the joint

$\bar{x}$ = Distance from the gravity axis to the fastener plane

L = Length of the joint

For most well designed section, this expression gives a value of $\eta = E_{SL}$ = 85 percent to 90 percent

Professor Bendigo, Hansan and Rmp in their paper, titled as "Long bolted joints", proceedings ASCE, Vol. 84, ST 6, December, 1963 has mentioned that the efficiency of a joint is given by the following expression (assuming spacing) as 3 (three) *times* diameter

$$\eta = 0.85 - C_1 (L - C_2) \tag{10.7)B}$$

where, C_1 = 0.00275 in S.I units

and C_2 = 406 mm (L > C)

Eq. 10.7(B) shows that the connections with joint length upto 406 mm, have an efficiency of 85 percent. The effect of shear lag may be accounted by calculating the effective area

$$A_{eff} = A_{net} (1 - \bar{x}/L) \tag{10.7)C}$$

For a single plane connection, $\bar{x}$ is the centroidal co-ordinate of the entire cross-section, while for a symmetrical doube plane connection, it is the distance from each connector plane to the centroid of the half cross-section as shown in Fig. 10.4.

This is to note that the factor of safety for the connection is kept higher than that used for the member. This so because, a failure of a member (if it occurs) will be earlier than the failure of a joint. A joint failure used to be catastrophic, where as the failure of member permits time for safety measures to be under taken.

In the design of riveted/bolted connections for the wide plates in tension, certain precautions are necessary, in the absence of which, the strength of such plates becomes less due to the effect of shear lag.

The effect of shear lag may be seen by applying two tensile forces p_1 and p_1 at the edges of a wide rubber sheet as shown in Fig. 10.4(C). The stretches at the edges AA_1 and CC_1 are large as compared to that at the middle line, BB_1. Except at A and A_1, and C and G, the stresses in the longitudinal direction at all the points on edges ABC and $A_1B_1C_1$ are zero.

The small elements at D and D_1 on the edges ABC and $A_1B_1C_1$, respctively, deform as shown. It shows that the shear stresses are developed in the material of the sheet. These shear stresses transmit the tensile forces from the edges towards the middle line in a certain length approximately equal to the width of the sheet.

The stresses in tension in this length shall not be equal. However, the stresses in tension beyond this length may be treated as uniform over the complete cross-section of the rubber sheet, the forces transfer from the edges to the complete width through shear stresses is termed as shear lag.

10.10 DESIGN TENSILE STRENGTH/RESISTANCE DUE TO RUPTURE OF CRITICAL SECTION (NET EFFECTIVE SECTION FOR ANGLES AND TEES IN TENSION)

When a single angles section are used as tension member, these are connected with the gusset plate by one leg. In such cases, the rivets connecting tension member and the gusset plate do not lie on the line of action of the load. This results in an eccentricity moment on the connection. The distribution of the stress becomes non-uniform because of eccentric connection and effect of shear leg (as described in Art. 10.6). The net sectional area of an angle is reduced to account for this non-uniform distribution of stress. The reduced net sectional area of such member is called **net effective area**. The net effective section for angles are adopted as specified in IS : 800-2007, Clause 6.3.3.

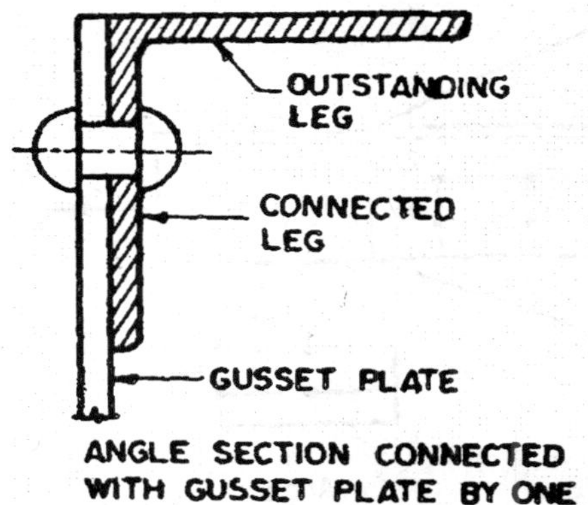

Fig. 10.5

When the tensile load in a single angle section connected lag reaches the ultimate stage, then, the stress in the outstanding lag remains close to the yield stress, σ_{sty} (as reported by Professor Kulak and Wu (1997). At the connections, the are holes, which remain at the ultimate stress, σ_{stu}.

The design tensile strength/resistance T_{dn} is controlled by rupture at the net section (critical section). It is obtained as sum of

1. Ultimate strength over net area of the connected leg T_{stu} and
2. Yield strength strength over gross area of the outstanding, T_{sty}

$$T_{dn} = (T_{stu} + T_{sty})$$

$$T_{dn} = \left(\frac{0.9A_{ncL} \cdot \sigma_u}{\gamma_{msu}} + \beta \frac{A_{groL} \cdot \sigma_y}{\gamma_{msy}}\right) \qquad (10.8)A$$

where, $$\beta = 1.4 - 0.076\,(w/t)\left(\frac{\sigma_y}{\sigma_u}\right)\left(\frac{b_s}{L_c}\right) \qquad (10.8)B$$

At the same time, this value of β should be less than or equal to

$$[(\sigma_u\,\gamma_{msy}/\sigma_y \cdot \gamma_{msu})] = \left[\left(\frac{\sigma_u}{\sigma_y}\right)\left(\frac{\gamma_{msu}}{\gamma_{msu}}\right)\right] \qquad (10.8)C$$

So also, its value should be more than or equal to 0.7

where, γ_{msu} and γ_{msy} are the partial safety factors for the strength of materials against ultimate stress and yield stress, respectively.

w = width of outstanding lag

b_s = shear lag width as shown in Fig. 10.6.

L_c = Length of end connection (i.e, the distance between outermost bolts in the end joint measured along the direction of load of weld along the direction of load.

The factor β accounts for the end restraint effect of the fasteners. The value of β depends upon number of bolt, N. These values are as follow

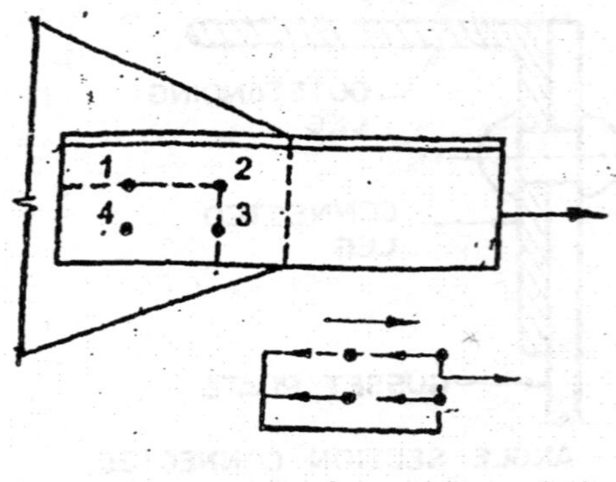

Fig. 10.6. Block shear failure

Number of bolts, N	1	2	3	4	>4
Value of β	0.50	0.50	0.75	1.00	1.00

A_{ncL} = Net area of connected leg

A_{groL} = Gross area of outstanding lag

For preliminary sizing, the rupture strength/resistance of net section may be approximately, taken as

$$T_{dn} = \alpha\left(\frac{A_{net} \cdot \sigma_u}{\gamma_{msu}}\right) \tag{10.9}$$

where,

α = 0.6 for one or two bolts

= 0.7 for three bolts and

= 0.8 for four or more bolts along the length in the end connections or equivalent weld length

A_{net} = Net area of the total cross-section

A_{ncL} = Net area of the connected leg, and

t = Thickness of the leg

Net sectional area of total cross-section

In the case of single angle in tension connected by one leg, net sectional area of total cross-section is taken as

$$A_{net} = (A_1 + k \cdot A_2) \tag{4.}$$

where,

A_1 = Net area of cross-section of the connected lag

$$= \left[\left(Length\,of\,leg - d - \frac{1}{2}t\right) . \, t\right]$$

d = Diameter of hole

A_2 = Gross cross-sectional area of outstanding leg

$$A_2 = \left[\left(Length\, of\, leg - \frac{1}{2}t\right)^t\right]$$

and $$k = \left(\frac{3A_1}{3A_1 + A_2}\right) \qquad (10.10)$$

Net area of connected leg

$$A_{ncL} = \left[\left(Length\, of\, leg - nd - \frac{1}{2}t\right) . t\right]$$

n = number of holes in one or two rivet/bolt lines in the direction perpendicular to the tension force in the member [For one rivet/bolt hole, n = 1 and for two rivet/bolt holes, n = 2]

Gross area of outstanding leg

$$A_{groL} = \left[\left(\text{Length of outstanding leg} - \frac{1}{2}\text{t}\right) . t\right]$$

It is to note that rivet gauge distances in legs of angle are given in Table 31 in ISI Handbook No. 1.

10.11 DESIGN TENSILE STRENGTH/RESISTANCE OF AN ANGLE SECTION DUE TO BLOCK SHEAR

The failure of the end connection of an angle section and a plate may occur along the **plane of block shear** as shown in Fig. 10.7. The tensile strength as controlled by **block-shear** is calculated as follows. [IS : 800-2007, Clause 6.4.1]

Bolted connection : The block shear tensile strength shall be adopted as the smaller of

$$= \left[\frac{1\, A_{vg} \cdot \sigma_y}{\sqrt{3}\, \gamma_{msy}} + \frac{0.9\, A_{net.c} \cdot \sigma_u}{\gamma_{msu}}\right]$$

Since, $\left(1/\sqrt{3}\right) = 0.5773 = 0.6\ (approximately)$

$$T_{dBS} = \left[\left(\frac{0.6\, A_{vg}\, \sigma_y}{\gamma_{msy}}\right) + \left(\frac{0.9\, A_{netc.}\, \sigma_u}{\gamma_{msu}}\right)\right] \qquad (10.11)B$$

or

$$T_{d.BS} = \left[\left(\frac{0.9\, A_{vn}\, \sigma_u}{\sqrt{3}\, \gamma_{msu}}\right) + \left(\frac{A_{tg}\, . \sigma_y}{\gamma_{msy}}\right)\right] \qquad (10.11)C$$

where,

A_{vg}, A_{vn} = Minimum gross and net area in shear along bolt line parallel to external force, respectively (1-2 and 3-4) as shown in Fig. 10.7 for A_{tg}, A_{tn} = Minimum gross and net area in tension from the bolt hole to the toe of the angle, end bolt line perpendicular to the line of force, respectively 2-3 as shown in Fig. 10.7 and

σ_u, σ_y = Ultimate and yield stress of the material, respectively

γ_{msu}, γ_{msy} = Partial safety factor for the ultimate strength (1.25) and yield strength (1.15)

In ISI Handbook No. 1, Table 31, gauge distances (i.e, distance from the root; upto centre line of bolt hole) for the single hole and for the double holes are given. These gauge distances one noted.

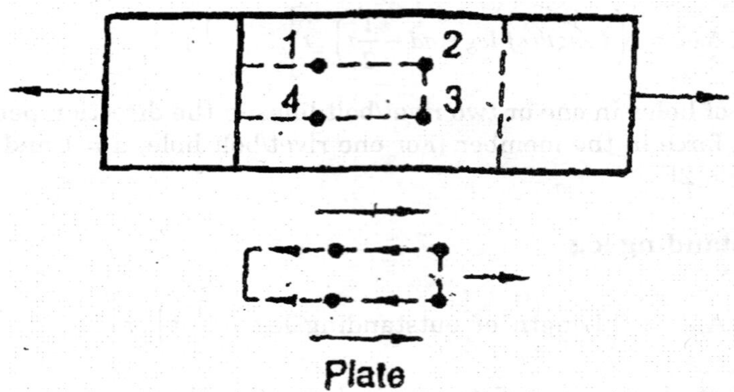

Fig. 10.7. Block shear failure

Then, Av_{gross} is product of minimum distance and the thickness is calculated.

And $A_{v.net}$ is the product of length of line AB, Fig. 10.7 minus deduction for the bolt holes diameter and the thickness of angle whereas, $A_{T.gross}$ and $A_{T.net}$ are the gross area and net area in tension along the line perpendicular to the line transmitting the force (i.e, along the transverse line BC, Fig. 10.7.

Gauge distance (measured from the centre line of bolt-hole upto the toe of the angle multiplied by the thickness of the angle gives $A_{t\text{-}gross}$.

Whereas, half the bolt hole distance in deducted from this gauge distance. Then it is multiplied by the thickness of angle, which gives $A_{t\text{-}net}$.

Efficiency of angle sections in tension

The efficiency of an angle section in tension is determined as follows :

$$\text{Efficiency (Etta)}, \eta = \left[\frac{T_{VBS1}, T_{VBS2}, \text{ or } T_{STY}}{T_{STY}}\right] \times 100 \qquad (10.12)$$

The efficiency depends upon the type of end connection and the configuration of the built-up member.

Its value remains between 0.85 and 1.00.

When, the double angles are connected on the opposite sides of the gusset plate and ends are welded, then, it gives higher efficiency.

In bolted single angle tensin member, usually, the efficiency remains on lower side.

For the threaded members, the efficiency remains about 0.85.

Double angles Fig. 10.8 channels, I-sections and other rolled steel sections have been referred as other sections when any one of these sections is connected by one or move elements to an end gusset, then, he connection is affected by shear lag. The design tensile strength/resistance in such connections is also controlled by shear lag. Which one is ruled by tearing of net sections. And, it is obtained by following expression.

$$T_{dn} = \left[\left(\frac{0.9\,A_{nc}\,\sigma_u}{\gamma_{msu}}\right) + \left(\frac{\beta\,A_{groL}\,\sigma_y}{\gamma_{msy}}\right)\right] \qquad (10.13)$$

In this expression, b is determined based on the shear lag distance distance (shear lag width, b_s) taken from the farthest edge of the outstanding leg of the cross-section.

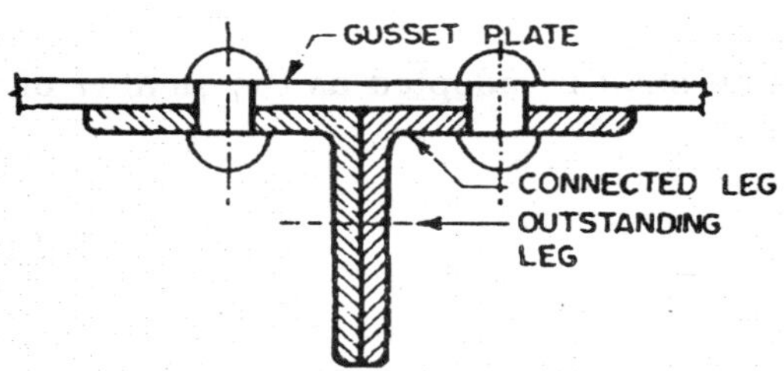

PAIR OF ANGLES (BACK TO BACK) CONNECTED BY ONE LEG OF EACH ANGLE TO SAME SIDE OF GUSSET PLATE

Fig. 10.8.

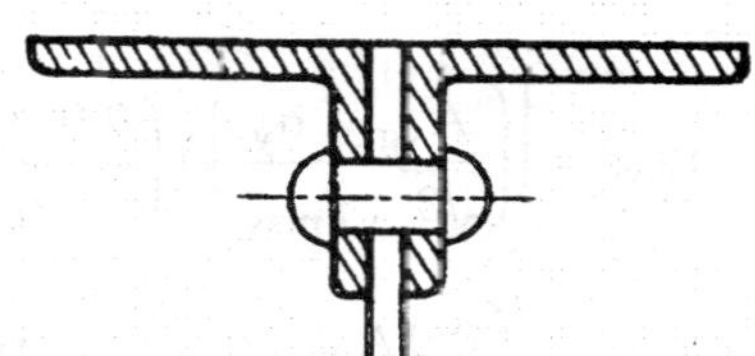

TWO ANGLES CONNECTED TO EACH SIDE OF GUSSET PLATE

Fig. 10.9

In the case of a pair of angles back to back (or a single tee) in tension connected by only one leg of each angle (or by th flange of a tee) to the same side of a gusset, the net area of total cross-section is taken as

$$A_{net} = (A_1 + k_2\,A_2)$$

where, A_1 = effective sectional area of the connected legs (or flange of the tee)

A_2= gross cross-sectional area of the unconnected (outstanding) legs (or web of the tee), and

$$k = \left(\frac{5A_1}{5A_1 + A_2}\right) \qquad (10.12)$$

The tacking rivets are used with the appropriate spacing to connect the angles along their length.

(iii) For double angles or tees carrying direct tension placed back to back and connected to each side of a gusset as shown in Fig. 10.5 or to each side of a rolled section, sectional area to determine the mean tensile stress shall be the full gross area less the deductions for holes, provided the tacking rivets are used with the appropriate spacing to connect the angles along their length.

The area of the leg shall be taken as the product of the thickness by the length from the outer corner minus half thickness, and the area of the leg of a tee as the product of the thickness by the depth minus the thickness of the table.

When the angles are back to back and the tacking rivets are not used to connect the angles along their length, then, each angle shall be designed as single angle connected through one leg only as described above.

When two tee sections are placed back to back but the tacking rivets are not used to connect them along the length, then each tee-section shall be designed as single tee-section connected to one side of a gusset as specified above.

10.12 DESIGN TENSILE STRENGTH/RESISTANCE DUE TO BLOCK SHEAR

The tensile strength/resistance due to **block-shear** at an end connection of plate and angles is calculated as follows :

Bolted connections

The block shear strength, $T_{d.BS}$ of a connection shall be adopted as the smaller of two below

$$T_{d.BS} = \left[\left(\frac{A_{vgr} \cdot \sigma_y}{\sqrt{3} \cdot \gamma_{msy}}\right) + \left(\frac{0.9\ A_{tn}\ \sigma_y}{\gamma_{msu}}\right)\right] \qquad (10.14)A$$

$$T_{d.BS} = \left[\left(\frac{0.9 A_{vn} \cdot \sigma_{cu}}{\sqrt{3} \cdot \gamma_{msu}}\right) + \left(\frac{A_{tgr} \cdot \sigma_y}{\gamma_{msy}}\right)\right] \qquad (10.14)B$$

where,

$A_{v.gr}$, $A_{v.n}$ = Minimum gross-area and net area in shear along bolt-line parallel to external force, respectively (1-2 and 3-4 as shown in Fig. 10.6 and 1-2 as shown in Fig. 10.7)

$A_{t.gr}$, A_{tn} = Minimum gross and net area in tension from the bolt hole to the toe of the angle, end-bolt line, perpendicular to the line of force, respectively (2-3 as shown in Fig. 10.7) and

σ_u, σ_y = Ultimate and yield stress of the material, respectively

γ_{msy} = 1.15 Partial safety factr for strength of material corresponding to yield stress

γ_{msu} = 1.25 partial safety factor for strength of material corresponding to ultimate stress

Welded connection

The block shear strength, $T_{d.BS}$ shall be checked for welded end connections by taking an appropriate section in the member around the end weld, which may shear off as a block.

Example 10.1. An ISA 150 mm x 75 mm x 10 mm (ISA 150, 75 @ kN/m) is connected to a gusset plate by 20 mm diameter rivets/bolts in two rows. The long leg of the angle section has been connected. The gauge distance, g is 55 and the staggered pitch, p. is 40 mm.

Determine the net area of the total angle section.

Solution : The rivet/bolt connection is shown in Fig. 10. The long leg of the angle has been connected.

Gauge distance, g = 55 mm

Staggered pitch, p = 40 mm

Step 1. Net area of total cross-section

Zig-Zag chain-line 1-2-3-4 for the rivet/bolt in shown in Fig. 10. Rivet/bolt diameter = 20 mm.

Net width Gross-diameter = (20 + 2) = 22 mm

$$b_1 = \left[b_1 - \left(nd - n'd\,\frac{p^2}{4g}\right)\right]$$

$$b_1 = \left[150 - \left(2 \times 22 - \frac{1 \times 40^2}{4 \times 55}\right)\right]$$

$$= \left[150 - \left(44 - \frac{1600}{220}\right)\right]$$

$$= [150 - (44 - 7.273)] = 113.273 \text{ mm}^2$$

Net area of connected leg

$$A_1 = \left[\left(113.273 - \frac{10}{2}\right)10\right] = 1082.73 \text{ mm}^2$$

Gross area of outstanding leg

$$A_2 = \left[\left(75 - \frac{10}{2}\right)10\right] = 700 \text{ mm}^2$$

It is single angle connected to the gusset plate.

Net area of total cross-section

$$A_{net} = \left[\frac{3A_1}{3A_1 + A_2}\right]$$

$$= \left[\frac{3 \times 1082.73}{3 \times 1082.73 + 700}\right] = \left(\frac{3248.19}{3248.19 + 700}\right) = 0.8227$$

Net area of total cross-section of the angle section

$$A_{net} = (A_1 + A_2 \,.\, k)$$

$$= (1082.73 + 0.8227 \times 700) = \mathbf{1658.622 \text{ mm}^2}$$

Example 10.2. Two plates 300 mm x 16 mm have been connected with 22 mm diameter bolts. The bolts have been arranged in diamond staggered arrangement as shown in Fig. 10.10. Determine the tensile strength/resistance of the plate. Take yield stress 250 N/mm² and ultimate stress 420 N/mm² of the steel.

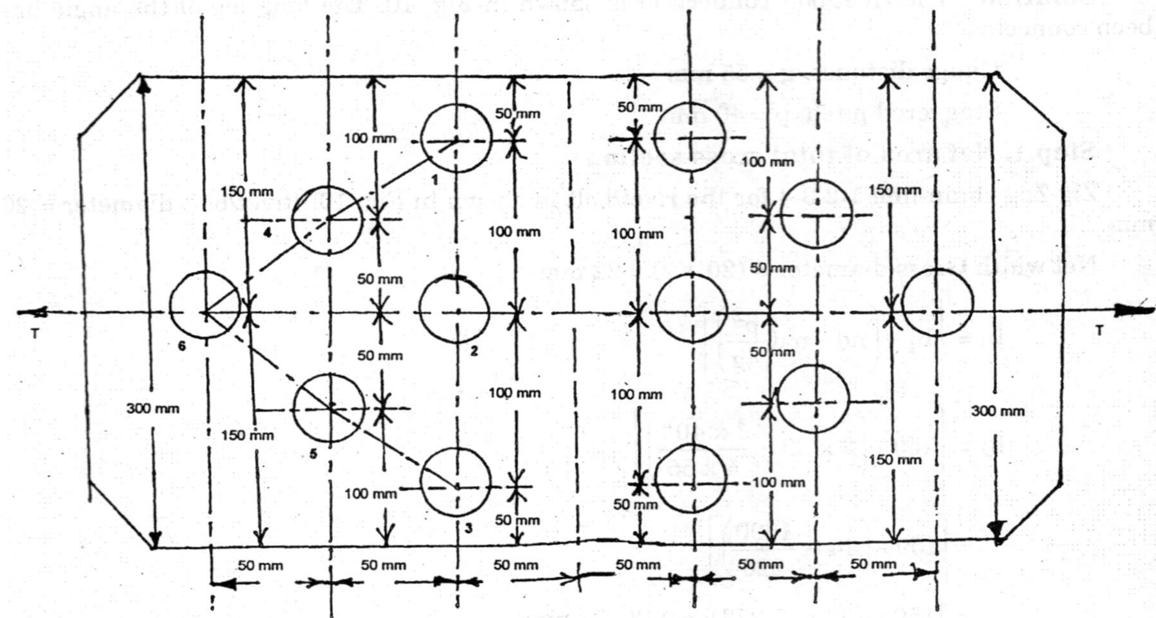

Fig. 10.10

Solution:

Yield stress of steel, $\sigma_y = 250$ N/mm²

Ultimate stress of steel, $\sigma_u = 420$ N/mm²

Step 1: Net cross sectional area of the plates

The arrangement of bolt-holes is as shown in Fig. 10.10.

Nominal diameter of botl = 22 mm

Gross-diameter of bolt holes = (22 + 2) = 24 mm

(i) Plates may fail along $x_1 = 1 - 2 - 3 - x_1$

Minimum cross-sectional area [As per IS : 800-2007, Clause 6.3.1]

$= (300 - 3 \times 24) \times 16 = 3648$ mm²

(ii) Plates may fail along $x_1 - 1 - 4 - 6 - 5 - 3 - x_1$

Minimum cross-sectional area

$$= \left(300 - 4 \times 24 + \frac{2 \times 50^2}{4 \times 50}\right) x16$$

$= (300 - 96 + 25) \times 16 = 3664$ mm²

(iii) Else, plate may fail along

$X_1 - 4 - 6 - 5 - 3 - x_1$

Minimum cross-sectional area

$$= \left(300 \ - 5 \text{ x } 24 + \frac{4 \text{ x } 50^2}{4 \text{ x } 50}\right) \text{x}16 \text{ mm}^2$$

$$= (300 - 120 \ + 50) \text{ x } 16 = 3680 \text{ mm}^2$$

Gross area, A_{gr} = (300 x 16) = 4800 mm²

Therefore, the net cross-sectional area of plates shall be

A_{net} (minimum of above three) = 3648 mm²

Step 2 : Tensile strength/resistance of the joint

$$T_{sty} = \left(\frac{\sigma_y \cdot A_{gross}}{\gamma_{msy}}\right) = \left(\frac{250 \text{ x } 4800}{1000 \text{ x } 1.15}\right) = 1043.48\text{kN}$$

$$T_{stu} = \left(\frac{0.9 \ A_{net} \cdot \sigma_u}{\gamma_{msu}}\right) = \left(\frac{0.9 \text{ x } 3648 \ \text{ x } 420}{1000 \text{ x } 1.25}\right) = 1103.155\text{kN}$$

The design tensile strength/resistance of plate in controlled by rupture of net cross-sectional area, A_{net} at ultimate stress, σ_y or rupture of gross area, A_{gross} at yield-stress, σ_y whichever less

$$= 1043.48 \text{ kN}$$

Note : Reference :

`Teaching Resources, INSDAG' by R. Narayanan and *et al.* (solved Page 5-11).

Tensile strength shall be T_{sty} or T_{stu} whichever is less **seems more reasonable.**

Efficiency of plates with holes

$$= \left(\frac{1043.48 \text{ x } 100}{1103.155}\right) = 94.59 \text{ percent}$$

Example 10.3. A single angle section ISA 150 mm x 75 mm x 10 mm (ISA 150, 75, @ 0.169 kN/m) is connected to 12 mm thick gusset plate. 5 (five) bolts of 18 mm diameter one used to transmit tension. Find the tensile strength of the angle, in case, its strength is controlled by tearing. For the mild steel, σ_y = 250 N/mm² and σ_u = 420 N/mm².

Solution :

Diameter of bolt hole = (18 + 2 - 0) = 20 mm

Angle section is connected to 12 mm thick gusset plate and long leg is kept vertical as shown in Fig. 10.11.

Step 1 : Area of connected leg (Net area)

$$A_{net.c} = \left(150 - \frac{10}{2} - 20\right) \text{x } 10 = 1250 \text{ mm}^2$$

Area of outstanding leg (Gross - area)

$$A_{gross.c} = \left[\left(150 \ - \frac{10-20}{2}\right)\right] \text{x } 10 \ mm^2 \ = 1250 \ mm^2$$

$$A_{gross.o} = \left(75 - \frac{10}{2}\right).10 = 700\ mm^2$$

$$= (1450 + 700) = 2150\ mm^2$$

Fig. 10.11

Step 2 : Strength of angle section is controlled by tearing (Given) as shown

$$T_{STU} = \left(\frac{\sigma_u \,.\, A_{net.c}}{\gamma_{msu}}\right) = \left(\frac{420 \times 1250}{1000 \times 1.25}\right)$$

$$= 420\ kN$$

Number of bolts = 5 (more than 4)

Therefore, k = 1.00

Strength of outstanding leg

$$T_{STY} = \left(\frac{k \,.\, A_{gross.o} \,.\, \sigma_y}{\gamma_{msy}}\right)$$

$$= \left(\frac{1 \times 700 \times 250}{1.15}\right) = 152.174\ kN$$

Step 3. Strength of angle section in tearing

$$T_{ST} = \left[\left(\frac{\sigma_{cu} \,.\, A_{gross}}{\gamma_{msu}}\right) + \left(\frac{\sigma_y \,.\, A_{net.c}}{\gamma_{msy}}\right)\right]$$

$$= [(421.68 + 152.174)]$$

$$= 573.854\ kN$$

Example 10.4. In Example 10.3 in case, the short leg (i.e, 75 mm) is connected to the gusset plate. Find the tensile strength of angle section. Also find the efficienciency of the joint.

In case two angles are attached to the opposite side of the gusset plate through 75 mm leg. Find the strength of angles.

Solution : In case, the tensile strength of the angle section is controlled by yielding of gross-area

Gross-area of angle section, $A_{gross} = 2150 \text{ mm}^2$

Step 1. Tensile strength of angle section

$$T_{STY} = \left[\frac{\sigma_y . A_{gross}}{\gamma_{msu}}\right] = \left(\frac{250 \ x \ 2150}{1000 \ x \ 1.15}\right) = 467.39 \text{ kN}$$

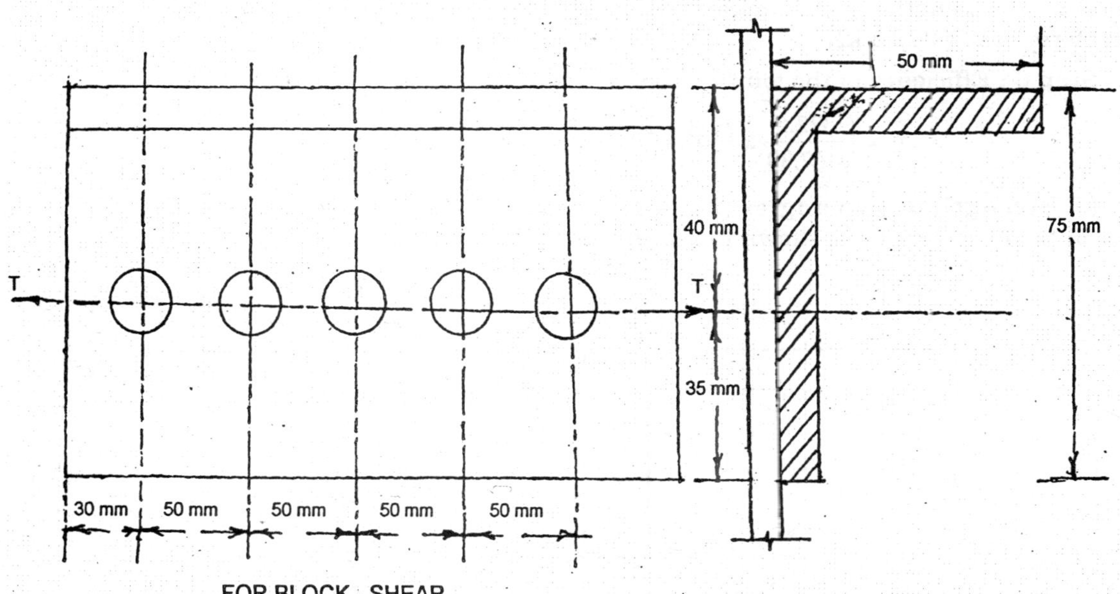

Fig. 10.12

Step 2. In case, the tensile strength of angle section is controlled by **block-shear,** then, the block shear failure shall occur along the cross-section line ABC. From Eq. 10.

$A_{v.net} = (4 \text{ x } 50 + 30 - 4.5 \text{ x } 20) \text{ x } 12 = 1600 \text{ mm}^2$

$A_{T.gross} = (35 \text{ x } 10) = 350 \text{ mm}^2$

Step 3. Strength of angle section in block shear (IS : 800-2007, Clause 6.3.3)

$$T_{V.BS.2} = \left[\left(\frac{1}{\sqrt{3}}\frac{A_{v.net} . \sigma_{cu}}{\gamma_{msu}}\right) + \left(\frac{A_{T\ gross} . \sigma_y}{\gamma_{msy}}\right)\right]$$

$$= \left[\left(\frac{0.577 \ x \ 1600 \ x \ 420}{1000 \ x \ 1.25}\right) + \left(\frac{350 \ x \ 250}{1000 x \ 1.15}\right)\right] = 386.282 \text{ kN}$$

Step 4. **Strength of angle section in block shear**

$$T_{v.gross} = (4 \times 50 + 30) \times 10 = 2300 \text{ mm}^2$$

$$A_{T.net} = \left(35 - \frac{20}{2}\right) x 10 = 250 \text{ mm}^2$$

Step 5. **Strength of angle section in block shear (IS : 800-2007, Clause 6.3.3)**

$$T_{v.BS.1} = \left[\left(\frac{1}{\sqrt{3}} \frac{A_{v.gross} \cdot \sigma_y}{\gamma_{msy}} + \frac{A_{T.net} \cdot \sigma_u}{\gamma_{msu}}\right)\right]$$

$$= \left[\left(\frac{0.577 \ x\ 2300 \ x 250}{1000 \ x\ 1.15} + \frac{2.50 \ x\ 420}{1000 \ x\ 1.25}\right)\right] = 372.5 \text{ kN}$$

Tensile strength of the angle section shall be minimum of T_{STY}, T_{UBS1}, and T_{VBS2}. Therefore, T_{VBS} = 372.5 kN

Step 6 : Efficiency of the joint

$$\eta = \left(\frac{372.5}{467.39} x\ 100\right) = 79.68 \text{ per cent}$$

(ii) **Strength of two angles** (2 ISA 150 mm x 75 mm x 10 mm) attached to the opposite sides of the gusset plate as shown in Fig. 10. through 75 mm leg

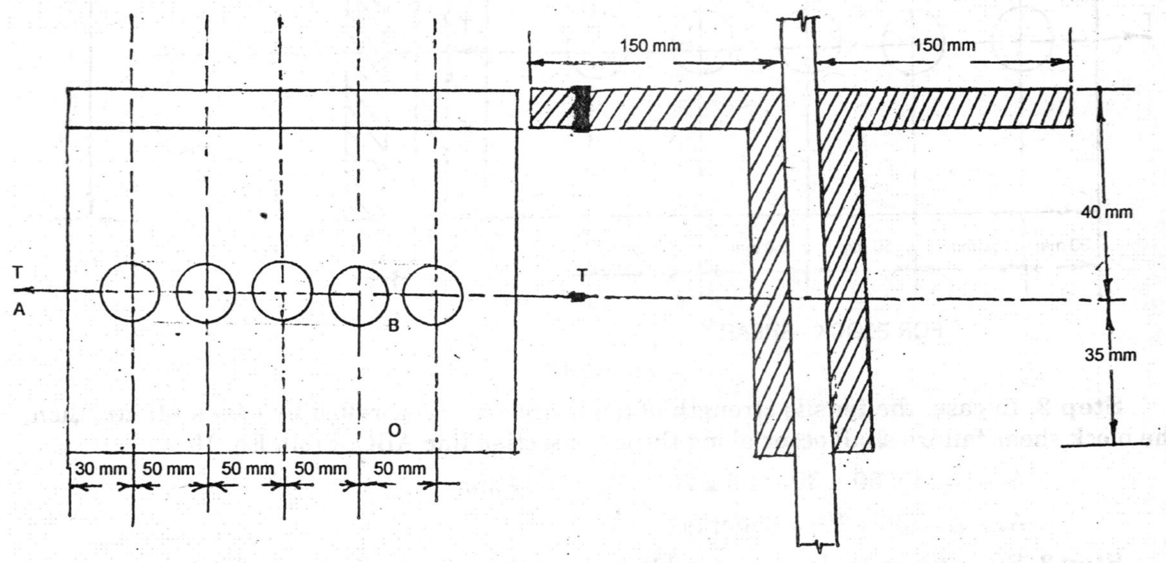

Fig. 10.13

Strength of two angles (2 ISA 150 mm x 75 mm x 10 mm) attached to the opposite sides of a gusset plate as shown in Fig. 10. shall be **twice** the strength of single angle attached to the gusset plate through 75 mm leg as shown in Fig. 10.13.

$$T_{VBS} = (2 \times 372.500) = 745 \text{ kN}$$

10.13 DESIGN OF A TENSION MEMBER

In case, a tension member is to be designed, then, the dead load to be carried and the imposed load are given. The partial safety factors for dead load, $\gamma_{fL.DL}$ and imposed load, $\gamma_{fL.IL}$ are noted from IS : 800-2007. Following procedure is used for the design of a tension member (specially, in case, of angle section.

Step 1. Factored/design load

Factored/design load is calculated

$$W_{fd} = \left[\gamma_{fL.DL}(DL) + \gamma_{fL.IL}.(L.L.)\right] \quad \text{(i)}$$

Step 2. A trial section

A trial section is needed for the design. **Net area** required for the angle section is obtained

$$A_{net} = \left[W_{fd} / \sigma_u . \gamma_{msu}\right] \quad \text{(ii)}$$

where, σ_u = Ultimate stress in the material. It is adopted as 420 N/mm²

γ_{msu} = Partial safety factor for the strength of the material in case, it is concerned with ultimate stress. Its value is used as 1.25.

Gross area required for the angle section is found as below

$$A_{gross} = \left[W_{fd} / (\sigma_y . \gamma_{msy})\right] \quad \text{(iii)}$$

where, σ_y is the yield stress for the material

γ_{msy} = Partial safety factor for it is 1.15 the strength of material corresponding to yield

From ISI Handbook No. 1, a trial section for the angle is selected. This section is checked for (i) its strength in tearing, and

(ii) its tensile stress in block shears (T_{BS1} and T_{BS2})

Diameter of bolt hole is taken as diameter of bolt hole is taken as diameter of bolt plus 2 mm. Above mentioned strengths of the angle section are called calculated as under.

Step 3. Strength of angle section in tearing

Tensile strength of angle section in tearing

$$T_{ST.T} = \left[\left(\sigma_{cu}.\frac{A_{net.c}}{\gamma_{msu}}\right) + k\left(\frac{A_{gross}.\sigma_y}{\gamma_{msy}}\right)\right] \quad \text{(iv)}$$

where, $A_{net.c}$ = Net area of connected lag.

$A_{gross.o}$ = Gross area of outstanding value of k depends upon number of bolts. In case, it is four or more than four, the value of k is adopted as unity (1.00)

Note : In the above expression, it is to be noted that when the net area of connected leg reaches ultimate stress, σ_{cu}, then, the gross-area of outstanding leg reaches, yield stress, σ_y.

Step 4. Strength of angle section in block shear : $T_{TS.BS.1}$ and $T_{TS.BS2}$

The failure of the anglection in block shear occurs along AB and BC as shown in Fig. 10. The portion AB is part of the connected leg and the portion BC is part of the outstanding leg. There are following two possibilities of failing of the angle section in block shear.

Possibility 1: The block shear failure may occur along AB (gross-area of outstanding leg) and BC (net area of connected leg as shown in Fig. 10.

Possibility 2 : In this possibility, the block shear failure may occur along AB (Net area of connected leg) and BC (gross area of outstanding leg)

Strength of angle section shall be minimum of the following

(i) Tensile strength of angle section in tearing, $T_{ST.T}$

(ii) Tensile strength of angle section in block shear, T_{TSBS1}

(iii) Tensile strength of angle section in block shear, T_{TSBS2}

The minimum value out of these three T_{STT}, T_{TSBS1} and T_{TSBS2} and this minimum value is more than the factored/design load acting over the angle section reasonably, the design is considered satisfactory. Otherwise, the design is revised suitably.

Example 10.5 Design a single angle section for a tension member to carry a pull of 60 kN due to dead load and 180 kN due to imposed load. The member is likely to have reversal of stress due to action of wind. The length of member from centre to centre of intersection is 1.20 m. Use limit state method of design.

Design :

Given tension member carries a pull of 60 kN due to dead load and 180 kN due to imposed load. The partial safety factors for dead load and imposed load are 1.35 and 1.50, respectively.

Step 1 : Factored load for tension member

$$W_{fd.T} = (\gamma_{fL.DL} \cdot D.L. + \gamma_{fL.LL} \cdot L.L.)$$

$$= (1.35 \times 60 + 1.50 \times 180) = 351 \text{ kN}$$

Step 2 : Selection of trial section

Considering $\sigma_u = 420 \text{ N/mm}^2$

$$\gamma_{msu} = 1.25$$

Net effective required for angle section

$$A_{net} = \left(\frac{W_{fd.T}}{\sigma_u / \gamma_{msu}}\right)$$

$$= \left(\frac{351 \times 1000}{420/1.25}\right) = 1044.64 \text{ mm}^2$$

Taking $\sigma_y = 250 \text{ N/mm}^2$

$$\gamma_{msy} = 1.15$$

Gross-area required for angle section

$$A_{gross} = \left(\frac{W_{fd.T}}{\sigma_y / \gamma_{msy}}\right)$$

$$A_{gross} = \left(\frac{351 \times 1000}{250/1.15}\right) = 1614.6 \text{ mm}^2$$

From ISI Handbook No. 1, select single angle section ISA 100 mm x 75 mm x 10 mm (i.e, ISA 100, 75, @ 0.130 kN/m).

This section is checked for tearing strength and so also for the tensile strength for block-shear. Long leg of angle section is attached to the gusset plate. Five bolts of 20 mm diameter are provided for transmitting total tension.

Diameter of bolt-holes = (20 + 2) = 22 mm

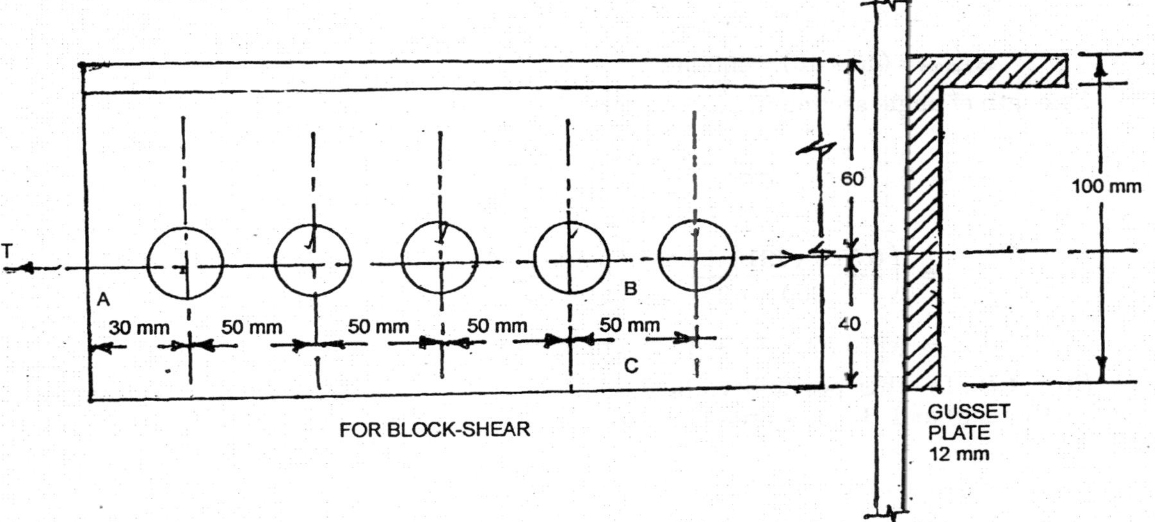

Fig. 10.14

Step 3 : Strength of angle section in tearing

$$T_{TST} = \left[\left(\frac{\sigma_u \cdot A_{net.c}}{\gamma_{msu}}\right) + \left(\frac{k \cdot A_{gross.o} \cdot \sigma_y}{\gamma_{msy}}\right)\right]$$

Number of bolt-holes = 5,

which one is more than 4, therefore,

k = 1.00 strength of angle section in tearing

$$A_{gross.o}\text{, for outstanding leg } = \left(75 - \frac{10}{2}\right) \times 10 = 700 \text{ mm}^2$$

Net area for the connected leg

$$A_{net.c} = (100 - 22 - \frac{10}{2})\ 10 = 730 \text{ mm}^2$$

Strength of angle section tearing

$$T_{STT} = \left[\left(\frac{420 \times 730}{1000 \times 1.25}\right) + 1.00\left(\frac{700 \times 250}{1000 \times 1.15}\right)\right]$$

$$= [(245.28 + 152.174)] = 397.454 \text{ kN}$$

It is more than 351 kN.

Hence, it is satisfactory.

Step 4A : Strength of angle section in block-shear

In its one possibility, the block-shear failure may occur along AB (net area of connected leg) and BC (gross-area of out standing leg).

$$A_{v.net} = (4 \times 50 + 30 - 4.5 \times 20) \times 10 \text{ mm}^2$$

$$= 1310 \text{ mm}^2$$

$$A_{v.gross} = (40 \times 10) = 400 \text{ mm}^2$$

Strength of angle section in block-shear [IS 800-2007, Clause 6.4]

$$T_{VBS2} = \left[\left(\frac{1}{\sqrt{3}} \cdot \frac{A_{v.netc}\,\sigma_u}{\gamma_{msu}} + \frac{A_{t.gross} \cdot \sigma_y}{\gamma_{msy}}\right)\right]$$

$$= \left[\left(\frac{0.577 \times 1310 \times 420}{1000 \times 1.25}\right) + \left(\frac{400 \times 250}{1000 \times 1.15}\right)\right]$$

$$= [(253.97 + 86.956)] = 340.926 \text{ kN}$$

Step 4 B.: In other possibility, the block-shear failure may occur along AB (gross-area of outstanding leg) and BC (net area of connected leg). These areas as shown in Fig. 10.14 are as below :

$$A_{v.gross.o} = (30 + 4 \times 50) \times 10 = 2300 \text{ mm}^2$$

$$A_{T.net.c} = \left(35 - \frac{22}{2}\right) \times 10 = 240 \text{ mm}^2$$

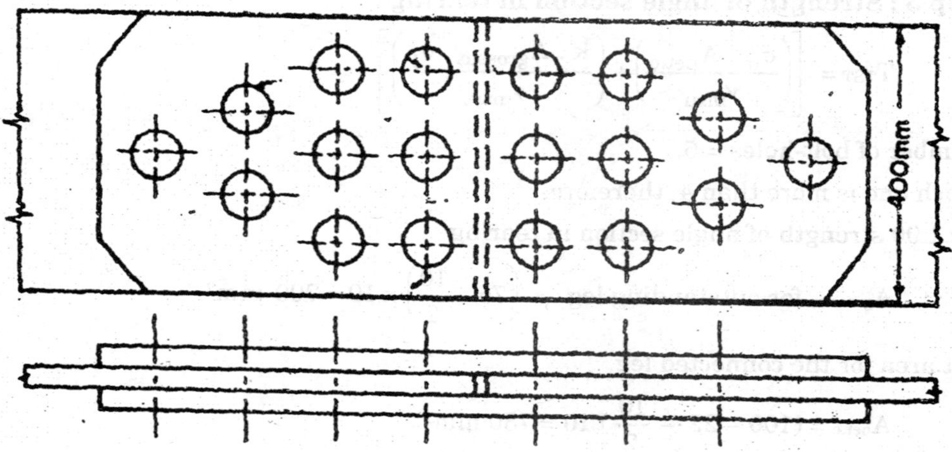

Fig. 10.15

Strength of angle section in block-shear

$$T_{VBS.1} = \left[\left(\frac{A_{Tnet} \cdot \sigma_u}{\gamma_{msu}}\right) + \left(\frac{1}{\sqrt{3}} \times \frac{A_{v.gross} \cdot \sigma_y}{\gamma_{msy}}\right)\right]$$

$$= \left[\left(\frac{240 \times 420}{1000 \times 1.25}\right) + \left(\frac{0.577 \times 2300 \times 250}{1000 \times 1.15}\right)\right]$$

$$= [(80.64) + (288.5)] = 369.14 \text{ kN}$$

Strength of angle section shall be minimum of the following

$T_{ST.T} = 397.454$ kN

$T_{V.BS.2} = 340.926$ kN

$T_{V.BS.1} = 369.14$ kN

Strength of angle section is that whichever is minimum (i.e, 340.926 kN).

This may be revised with angle section t = 12 mm, which will be found safe.

10.14 SPLICE (JOINT) IN TENSION MEMBERS

A joint in a member is known as splice. When a joint is to be provided in a tension member, then splice plates are used. Splice plates and rivets are designed for the pull required to be transmitted by the tension member. If the tension members are of unequal thickness, then, packings are used to have surfaces of tension members in one level. The rivets or bolts carrying calculated shear stress through a packing greater than 6 mm thick shall be increased about the number required by normal calculations by 2.5 per cent for each 2.0 mm thickness of packing. For double shear connections packed on both sides, the number of additional rivets or bolts required shall be determined from the thickness of the thicker packing. The additional rivets or bolts should preferably be placed in an extension of the tacking.

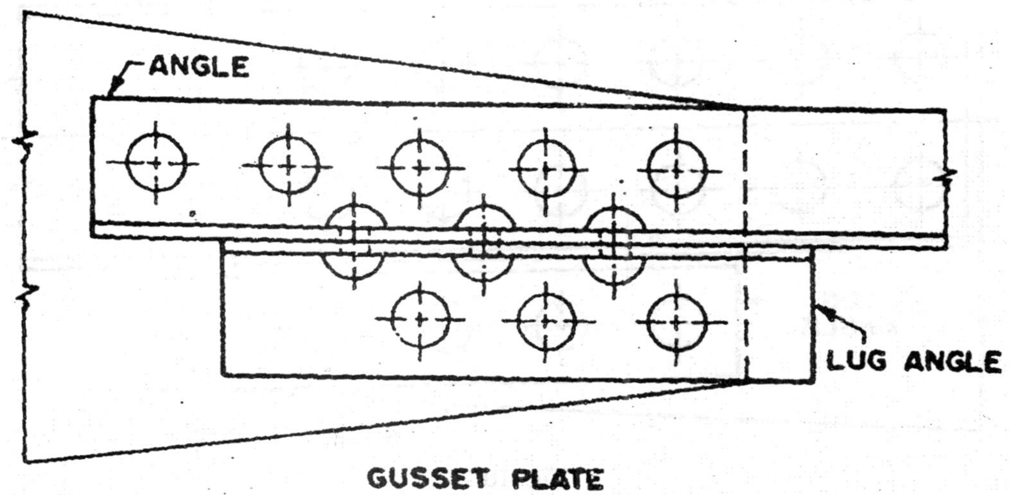

Fig. 10.16

10.15 LUG ANGLES

Length of the end connections of angle or channel section to a gusset plate is reduced by using lug angles. Lug angle used with single angle member is shown in Fig. 10.16. Rivets connecting lug angle to the gusset plates are less effective because of possible deformation of outstanding leg of the lug angle. As such, rivets are provided in excess (as per specification) of the member of rivets required by computation.

Specification : The following specifications for the design of lug angle may be followed :

1. In the case of angle members, the lug angles and their connections to the gusset or other supporting member shall be capable of developing strength not less than 20 per cent in excess of the force in the outstanding leg of the angle, and the attachment of the lug angle to the angle member shall be capable of developing 40 per cent in excess of that force.

2. The effective connection of lug angle shall be far as possible terminate at the end of the member connected and the fastening of the lug angle to the member preferably start in advance of the direct connection of the member to the gusset or other supporting member.

3. Where lug angles are used to connect an angle member, the whole area of the member shall be taken as effective rather than net effective section for angles as per Art. 10.3 (The whole area of the member is the gross area less deducton for rivet and bolt holes).

4. In no case, less than two rivets or bolts shall be used for attaching the lug angle to the gusset or other supporting member. The lug angle used with channels section has been shown in Fig. 10.17.

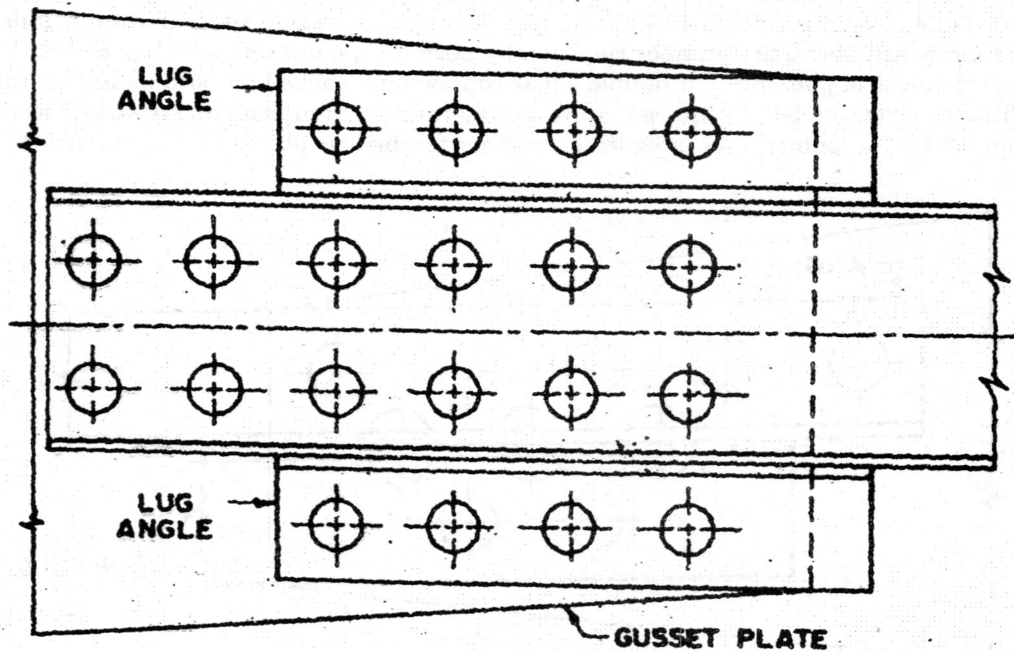

Fig. 10.17

5. In the case of channel members and the like, the lug angles and their connection to th gusset or other supporting member shall be capable of developing a strength of not less than 10 per cent in excess of the force not accounted for by the direct connection of the member and the attachment of the lug angles to the member shall be capable of developing 20 per cent in excess of that force.

PROBLEMS

10.1 An ISA mm x mm x mm (ISA, , @ kN/m is connected to a gusset plate by 20 mm diameter rivets/bolts in two rows. The long leg of the angle section has been connected. The gauge distance, g is 50 mm and the pitch is 40 mm. The rivets/bolts have been arranged in staggered manner. Find the net area of the total angle section.

10.2 Two plates 250 mm x 12 mm have been connected in diamond (staggered) arrangment. The pitch of rivets/bolts is 40 mm and the gauge distance is 60 mm. Determine the tensile strength/resistance of the plate. Take yield stress, $\sigma_y = 250$ N/mm^2 and ultimate stress, $\sigma_u = 420$ N/mm^2.

10.3 A single angle section ISA mm x mm x mm (ISA, @ kN/m) is connected to 10 mm thick gusset plate. Four bolts of 20 mm diameter are used to transmit tension. Find the tensile strength of the angle, in case, its strength is controlled by traning. For the mild steel σ_y = 250 N/mm^2 and σ_u = 420 N/mm^2.

10.4 Design a single angle section for a tension member to carry a pull of 80 kN due to dead load and 200 kN due to imposed load. The length of member from centre to centre of intersection is 1.20 m. Use limt state method of design.

REFERENCES

10.1. W.F. Chen (Editor-in-chief) - `Hand-book of structural Engineering' Published by CRC Press, Boca Raton, New York, 1997.

10.2. Rangachari Narayan, V. Kalyanaramal, *et al.* – `Teaching Resource on structural Steel Design' Published by Institute of Development of steel and Growth, Structural Engineering Research centre, Madras.

10.3. IS : 800-2007 (Third-Revision) Indian Standard General Construction in Steel, Code of Practice, Published by Bureau of Indian Standards Manak Bhawan, 9 Bahadur Shah Zafar Marg, New Delhi 110 002.

11

Part - VII

DESIGN OF BEAMS

LIMIT STATE OF COLLAPSE IN FLEXURE DESIGN OF LATERALLY SUPPORTED (RESTRAINED) STEEL BEAMS [BENDING/FLEXURAL STRENGTH] (SB-1)

11.1 INTRODUCTION

A beam is most commonly used structural member. A **beam** is defined as a structural member subjected to transverse loads and predominantly to bending. The beam shall have adequate design strength to resist bending moment, shear force, and concentrated forces imposed upon and their combination. The plane of transverse load is perpendicular to the longitudinal axis of the beam and parallel to the plane of symmetry of the cross-section of the beam and it passes through bending axis of beam and the shear centre, so that the simple bending occurs. The term transverse load includes end moments. The transverse loads produce bending moments and shear forces in the beams at all the section of the beam. The main purpose of beams is to transfer loads from the slabs supported and secondary beams, (if any) to the columns providing supports to those beams. The design of flexural members (i.e., beams) should satisfy the following criteria : (i) Flexural strength criterion, (ii) shear strength criterion (described in chapter 12) (iii) criterion for concentrated loads (also described in chapter 12) and (iv) deflection criterion (described in chapter 14).

The term **joist** is used for beams of light sections. Joists support floor construction; they do not support other beams. The term subsidiary beam or **secondary beam** is also used for the beams supporting floor construction. **Main beams** are the supporting joists for subsidiary beams : These are called floor **beams in buildings**. The term **girder** is most commonly used in buildings. Any major beam in a structure is known as a girder. In rectangular building frames, the beams span between adjacent columns.

In the roof trusses, horizontal beams spanning between the two adjacent trusses are known as **purlins**. The beams resting on the purlins are known as common rafter or simply rafters. In the buildings the beams spanning over the doors, windows and other openings in the walls are known as **lintels**. The beams at the outside wall of a building, supporting its share of the floor and also the wall upto the floor above it are known as **spandrel beams**. The beams framed to two beams at right angles to it, and usually supporting joists on one side of it; used at openings such as stair wells are known as **headers**. The beams supporting the headers are termed as **trimmers**. The beams supporting the stair steps are called as **stringers**. In the bridge floors, **the longitudinal beams** supported by the floor beams are also called as stringers. In the mill buildings, the horizontal beams spanning between the wall columns, and supporting wall covering are called as **girts**. Girts resist bending due to wind on the side and frequently support corrugated siding.

The beams are also called **simply supported, overhanging cantilever, fixed** and **continuous** depending upon nature of supports and end conditions.

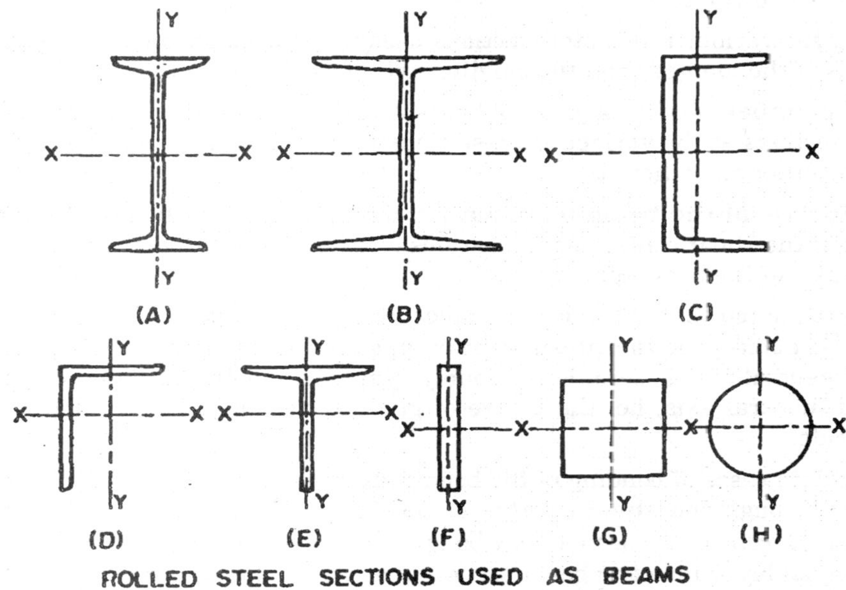

Fig. 11.1.

The rolled steel I-sections, channel sections, angle sections, tee-sections, flat sections and bars as shown in Fig. 11.1 are the regular sections, which are used as beams. The rolled steel I-sections as shown in Fig. 11.1(A) are most commonly used as the beams, and as such these sections are also termed as beam sections. The rolled steel I-sections are symmetrical sections. In these sections, more material is placed near top and bottom faces, (i.e., in the flanges) as compared to the web portion. The rolled steel I-sections provide large moment of inertia about xx- axis with less cross-sectional area. The rolled steel I-sections provide large moment of resistance as compared to the other sections and as such these are most efficient and economical beams sections. The rolled steel wide flange beams as shown in Fig. 11.1(B) provide additional desirable features. As the name indicates, the flanges of the sections are wide. These sections provide greater lateral stability and facilitate the connections of flanges to other members. I-sections and wide flange beam sections have excellent strength. The rolled steel channel sections as shown in Fig. 11.1(C) are used as purlins and other small structural members. The channel sections have reasonably good lateral strength and poor lateral stability. The channel sections are unsymmetrical sections about yy-axis. When the channel sections are loaded and supported by vertical forces passing through the centroid of the channel, then the channel sections bend and twist if these are laterally unsupported, except for the special case, wherein the loads act normal to the plane of web, causing bending in the weakest direction. The rolled steel angle sections as shown in Fig. 11.1 (D) are also used as purlins and so other small structural members. The angle sections act as unsymmetrical sections about both xx-axis and yy-axis. The rolled steel tee-sections as shown in Fig. 11.1(E) are used as beams in the rectangular water tanks. The angles and tee-sections are used for light loads. The rolled steel flats and bars as shown in 11.1(F), (G) and (H) are very rarely used. There sections are weak in resisting bending. Most commonly the beams are loaded in the direction perpendicular to xx-axis, so that the bending of beams occurs about strong (xx-axis) and xx-axis becomes neutral axis. The beams are very rarely loaded in the direction perpendicular to yy-axis. In such cases, yy-axis becomes neutral axis.

Knowing the various rolled steel sections used for the beams, one should distinctly learn about the axis of the member and the bending axis of the beam section.

Axis of member is the longitudinal axis which passes through the centre of gravity of the member sections at the various cross-sections perpendicular to the longitudinal axis from one end to the other end whereas.

Bending axis of the beam member is the axis parallel to the longitudinal axis which passes through the shear centres at the various cross-sections perpendical (transverse) to the longitudinal axis of the beam member.

It is also be noted that the axis of member and the bending axis of member are parallel. It may also be noted that the **neutral axis** and the bending axis are same. The bending stress at the neutral axis is zero. For a doubly symmetric sections, the longitudinal axis and bending axis (neutral axis, i.e, the centre of gravity, c.g. and shear centre, s.c. at a cross-section) coincide.

In general, in cases of bending of the beams about one (single) axis, the load is considered to be applied through the **shear centre** of the beam sections. In case, the loading passes through the shear centre, the section may be analysed for simple bending and shear. The shear centre for the doubly symmetric beam section is at the centre of area, and this load position produces simple bending about either axis.

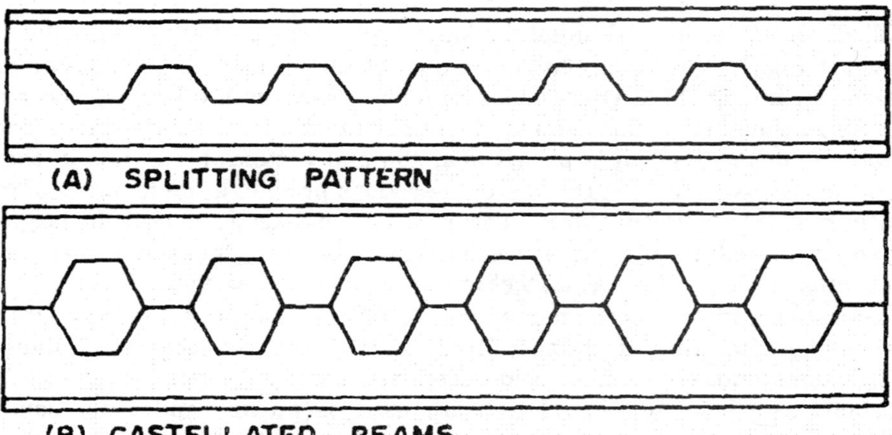

Fig. 11.2

When the load does not pass through the shear centre as in channels, angles and some built up sections (symmetric about one axis and unsymmetrical sections.), a torsional moment is produced along with the bending moment and both are considered to avoid over stressing of the member. For such sections, a special load device may be used so that the load passes through shear centre of the section and the torsional moment may be avoided.

In addition to the above, **expanded** or **castellated beams** as shown in Fig. 11.2 (B) are used. The castellated beams are light beams and these are economically used for the light construction. The castellated beams are made by splitting the web of rolled steel I-sections in a predetermined pattern as shown in Fig 11.2 (A). The splitted portions are rejoined in such a manner as to produce a regular pattern of opening in the web. The modulii of sections of castellated beams increase without the increase of material and weight.

In order to develop complete moment capacity for a beam section, **local buckling** of the elements of beam (i.e., flanges and web) should not occur. Local buckling failure of the beam section occurs by buckling (i.e., bending of a flange in compression or of the web due to shear or combined shear and bending or where, the concentrated loads are applied as a result of vertical compression. At the same time **lateral buckling** of the beam as a whole should not occur.

In case, the flanges and the web of a beam cross-section selected for a trial, do not allow local buckling. The ratios of half the width of flange to the thickness of flange, ($0.5\ b_f/t_f$) and the depth of the web to the thickness of web, (i.e., d_w/t_w) are sufficiently more than their limitations specified in codes of practice. So also, in case, appropriate restrained are provided in the horizontal plane of compression flange of the beam then the lateral buckling of beams also does not occur.

Laterally supported beams (IS : 800-2007, Clause 8.2.1)

A beam may be assumed to be sufficiently supported at the supported, provided the compression flange has full lateral restraint and nominal torsional restraint at supports supplied by web cleats, partial depth end plates, fish plates or continuity, with adjacent span full lateral restraint to compression flange may be assumed to exist in case, the frictional or other positive restraint of a floor connection to the compression flange of the member is capable of resisting a lateral force not less than 2.5 percent of the maximum force in the compression flange.

11.2 CLASSIFICATION OF STEEL BEAM SECTIONS (TYPES OF CROSS-SECTIONS)

In our country, the flanges and outstands of I-beams are rolled and so proportioned that these attained yield stress and the local buckling does not occur. The beam sections are classified depending on their moment-rotation characteristics as shown in Fig. 11.3. The width-thickness ratios $\beta = b/t$ (i.e, $0.5\ b_f/t_f$ and (d_w/t_w) are specified by the code IS : 800-2007 for the plate components (i.e, flange and web) of the beams. In this classficiation, it is assumed that the web slenderness ratio, d/t is such that the local buckling does not occur before yielding. It may be noted thatthe complete web may not be in uniform compression. In case, the neutral axis lies in the web, a portion of the web may be in tension. The beam sections are classified as (i) plastic, (ii) compact, (iii) semi-compact and (iv) slender, depending on the design strength in bending (flexure) requirements.

1. **Plastic cross-sections** : In case, the beam is requred to have significant ductility, the plastic cross-sections are invariably used. These sections satisfies the requirements of preventing local buckling. These sections can develop their full plastic moment (M_p) and allow sufficient rotation at or above this moment so that the redistribution of bending moments can take place in the structure until the failure mechanism is formed (b/t is less

than or equal to β_1) Fig. 11.3. This section may develop a plastic hinge and sustain plastic moment over sufficient plastic rotation required for the formation of the failure mechanisms of the member or the structure. The plastification of the member under bending has been shown in Fig. 11.4 and Fig. 11.5.

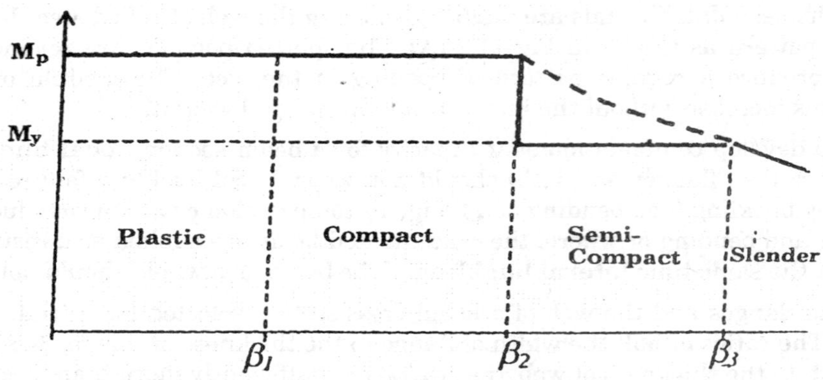

MOMENT CAPACITIES OF BEAM SECTION

Fig. 11.3

2. **Compact cross-sections** : The compact cross-sections are those which can develop their full-plastic moment (M_p) but where the required rotation at this moment, M_p is restrained. It is prevented by the local buckling. ($\beta_1 < (b/t) < \beta_2$). The cross-section of the beam is such that the effect of local buckling may be neglected. It is in adequate to have plastic-rotations capacity needed for the formation of a plastic collapse mechanism of the member or the structure as shown in Fig. 11.4.

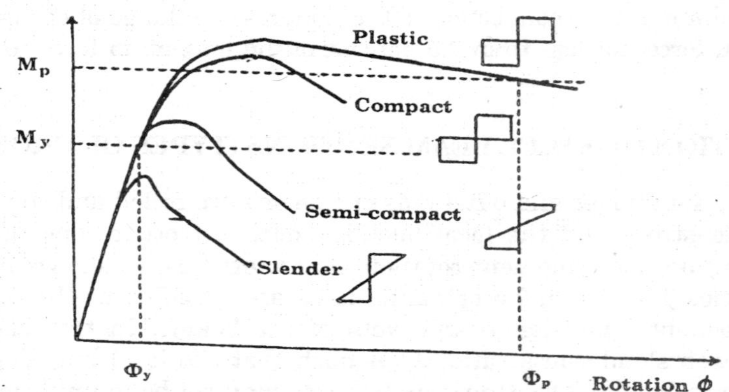

SECTION CLASSIFICATIION BASED ON MOMENT-ROTATION CHARACTERISTICS

Fig. 11.4

3. **Semi-compact cross-section**. In these cross-sectins of the beams, in which the extreme fibres remain limited to yield stress and the full plastic moment, M_p does not develop because of the local buckling. Such beam cross-section can develop only the yield moment, M_y ($\beta_2 < (b/t) < \beta_3$). These beam sections are called as semi-compact.

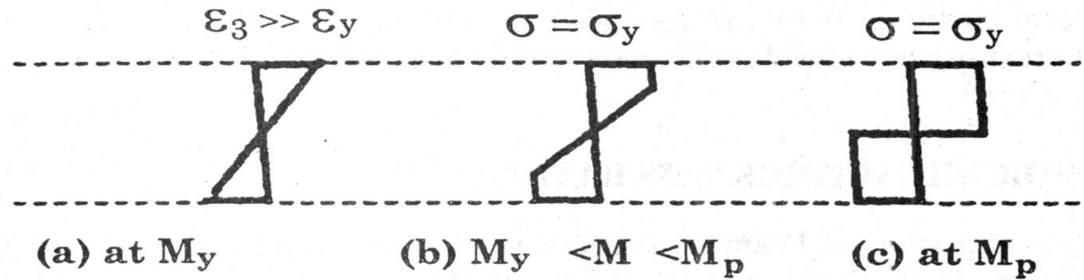

PLASTIFICATION OF CROSS-SECTION UNDER BENDING

Fig. 11.5

4. **Slender cross-sections**. In these cross-sections, in which, yield in the extreme fibres cannot be attained because of premature local buckling in the elastic range. ($\beta_3 < (b/t)$). These sections are also referred as **non-compact.** In the slender section of a beam cross-sections, the elements buckle locally before reaching yield moment.

Some strain hardening effects are seen in some qualities of steel, which have a large yield plateau and the maximum moment is likely to be more than M_p for the plastic and the compact cross-sections. In such cases, the rotation capacity considered is the ratio of the rotation when the moment capacity becomes equal to M_p, while reaching to the rotation at yield.

The moment capacity, M_p and the compression flange (b/t) ratio relationship shows the β-limits as in Fig. 11.3. Conservatively, the value of M for the semi-compact section is considered as M_y.

Above classifications of the beam section is based upon bending. It may not be applied to the compression members.

11.3 LOCAL BUCKLING EFFECTS IN STEEL BEAMS

Local buckling is defined as failure which occurs by buckling of a flange on the compression side or on the web due to shear or combined shear and bending or where the loads are applied, as a result of vertical compression. It is necessary to check for each of the following forms of instability.

Buckling of the compression flange (The level of strain which the design moment applies is noted carefully).

Buckling of web in shear and/or shear and bending combined.

Vertical buckling of a portion of the web under concentrated loads or over reaction (i.e, web crippling).

Two classes of plate elements are identified by Clause 1.3.5.1 of BS : 5950.

Part 1: As an internal element which would correspond to the flange of a box-section, and

Part 2: As outstand elements correspondingly to the flange of more commonly used I-section.

For both these types, four diferent ranges of compactness, each corresponding to a different performance requirement, are specified. In addition to the failure due to excessive bending and local buckling, there are several possibilities of local failure (e.g, shear yield of the web, local crushing of the web, excessive curing of thin flanges, local failure due around the web opening etc.)

11.4 LIMITS FOR WIDTH-THICKNESS RATIOS

The flanges and webs of beam cross-sections may be considered to be plate elements under compression. Their limiting width-thickness ratios may be obtained by equating the critical buckling stress to the yield stress.

The limits of width-thickness ratios β_1, β_2 and β_3 are useful for the designers and are normally arrived at the testing laboratory.

The limits for the width-thickness ratios for rolled steel sections unwelded and welded steel sections are given in Tables 11.1(A) and 11.1(B) as per IS : 800-2007. Tables 11.1(A) and 11.1(B) are applicable for steel with yield stress 250 N/mm² in which ease $\varepsilon = 1.0$. For other grades of steel, the yield stress ratio, $\varepsilon = (250/\sigma_y)^{1/2}$ is calculated and used.

Note: For higher strengths of steel the values of yield stress ratios, ε is less than 1.0, and hence, the limiting ratios are reduced.

In our country, various of values of yield stress for different grades of steel are 220, 230, 240, 250, 260, 280, 300, 320, 340, 360, 380, 400, 420, 450, 480, 510 and 540 N/mm² as per IS : 800-2007 (third revision).

In Tables 11.1 (A) and 11.1 (B) width to thickness ratios have been mentioned for above values. Phenomenon of local buckling controls the strength of compression bending members, only plastic and compact sections may be used in case of Limit State Design. Plastic sections may be used in mechanism forming in indeterminate structures.

Slender sections are to be avoided the elastic design. But these sections are invariably used in cold-formed construction for ecomic reasons.

TABLE 11.1 (A)
LIMITS OF WIDTH-THICKNESS RATIOS OF PLATE ELEMENTS (ROLLED STEEL SECTIONS, RSS)

Compression Elements	Class/Type of Sections			
	Ratio	Plastic Class-1	Compact Class-2	Semi-compact Class-3
1	2	3	4	5
Out-standing Elements of Compression	(b/t_f)	$9.4\,\varepsilon$	$10.5\,\varepsilon$	$15.7\,\varepsilon$
Internal Element of Compression Flange compression				
(i) due to bending	(b/t_f)	$29.3\,\varepsilon$	$33.5\,\varepsilon$	$42\,\varepsilon$
(ii) Axial Compr.	(b/t_f)	Not	applicable	$42\,\varepsilon$
Web of An I, H or box section Neutral axis at mild-depth	(d/t_w)	$84\,\varepsilon$	$105\,\varepsilon$	$126\,\varepsilon$
Generally				

(i) if, r_1 is negative (ii) if r_1 is positive	(d/t_w) (d/t_w)	$\left(\frac{84\in}{1+r_1}\right)$ but less than equal to 42 e	$\left(\frac{105\in}{1+r_1}\right)$ $\left(\frac{105\in}{1+1.5\,r_1}\right)$ but less than equal to 42 ε	$\left(\frac{126\in}{1+2\,r_1}\right)$ but less than equal to 42 ε
(iii) Axial compression	(d/t_w)	Not	applicable	42 ε
Web of a channel	(d/t_w)	42 ε	42 ε	42 ε
Angle compression due to bending (both should be satisfied)	(b/t) (d/t)	9.4 ε 9.4 ε	10.5 ε 10.5 ε	15.7 ε 15.7 ε
Single or Double agles comp. seperated	(b/t) (d/t)	Not applicable		15.7 ε 15.7 ε
All the three should be satisfied	(b + d)t			25 ε
Out standing leg of an angle in contact back to back in double angle member	(d/t)	9.4 ε	10.5 ε	15.7 ε
Out standing leg of an angle with its back with in continuous contact with another component	(d/t)	9.4 ε	10.5 ε	15.7 ε
Stem of a Tee-sec., rolled or cut from a rolled I-or H sec.	(D/t_f)	8.4 ε	9.4 ε	18.9 ε
Circular hollow tube				
(i) Moment	D/t	42 ε^2	52 ε^2	146 ε^2
(ii) Axial compression	D/t	Not applicable		88 ε^2
D = Circular tube outer Diameter				

5. The stress ratio r_1 and r_2 are defined as

$$r_1 = \left(\frac{\text{Actual average axial stress (negative if tensile)}}{\text{Design compressive stress of web alone}}\right)$$

$$r_2 = \left(\frac{\text{Actual average axial stress (negative, if tensile)}}{\text{Design compressive stress of overall section}}\right)$$

TABLE 11.1 (B).
LIMITS OF WIDTH-THICKNESS RATIOS OF PLATE ELEMENTS
(Welded Steel Sections, WSS)

Compression Elements	Class/Type of Sections			
	Ratio	Plastic	Compact	Semi-compact
1	2	3	4	5
Out-standing elements of Compression flange	(b/t_f)	8.4 ε	9.4 ε	13.6 ε
Internal element of compression Flange	(b/t_f) less than	24.0 ε	26.0 ε	29 ε

	equal to			
Web with neutral axis at mild-depth	(d/t_w) less than equal to	83.0 ε	103 ε	126 ε
Web under uniform compr.	(d/t)	-	-	29.4 ε
Single/Double angles T-stems	(d/t)	-	-	-
Circular tube with outer diameter D	-	-	-	-

Notes : 1. Elements which exceed semi-compact limits are to be taken as of slender cross-section

2. Yield stress ratio $\varepsilon = (250/\sigma_y)^{1/2}$

3. Webs shall be checked for shear buckling. when (d/t) > 67 ε, where b, is the width of the elements), may be taken as clear distance between lateral supports or between support and free edge as appropriate), t is the thickness of the element, d is the depth of the web, D is the outer diameter of element.

4. Different elements of a cross-section can be in different class/type. In such cases the section in classified based on least favourable class.

11.5 BENDING (FLEXURAL) STRESS

The bending stress is also termed as *flexural stress*. When beams are loaded, they bend and bending stresses are set up at all the sections. The established theory of bending is expressed in the following formula :

$$\left(\frac{M}{I} = \frac{\sigma}{y} = \frac{E}{R}\right) \tag{11.1}$$

where, M = Bending moment

I = Moment of inertia

σ_b = Bending stress at any point

R = Radius of curvature of the beam.

y = Distance from the neutral axis to the point under consideration

Flexural stiffness of a member against rotation is evaluated by the value of bending deformation moment require, to cause a unit rotation while all other degree of freedom of joints of the member except the rotation-one are assumed to be restrained.

Eq. 11.1 holds good when the plane of bending coincides with the plane of symmetry of the beam section. The bending of beam occurs in the principal plane of the beam section. The simple bending of beam occurs (i.e., the bending is produced by the application of pure couples at the ends of the beam). In such bending the deflection of beam does not occur due to shear. In Eq. 11.1, further, it assumed that the vertical sections of the beams plane before bending remain plane after bending. Then stress in any given fibre is proportional to its strain (i.e., Hooke's law holds good). For the material of beam, the value of E is same for the complete beam.

When the load is acting downward in a simply supported beam, then the distribution of bending stress, for any section of a beam is as shown in Fig. 11.6.

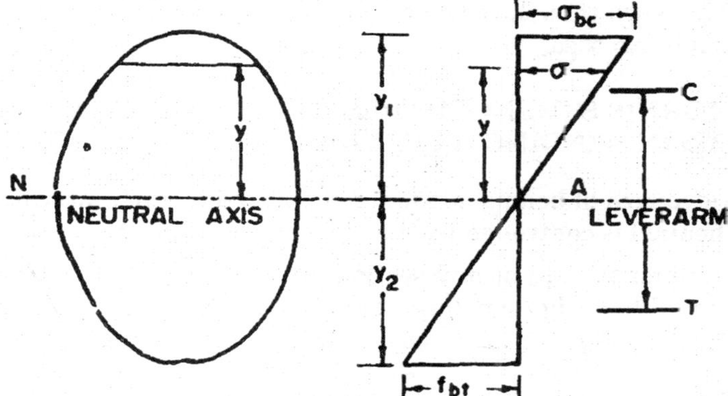

(A) SECTION OF BEAM (B) DISTRIBUTION OF BENDING STRESS

Fig. 11.6

The bending stress varies linearly. The bending stress is zero at the neutral axis. When the load is acting downward, the bending stress is compressive above the neutral axis of the section and tensile below it and these are denoted by $\sigma_{bc.cal}$ and $\sigma_{bt.cal}$, respectively. The bending stress is maximum at the extreme fibre.

$$\sigma_{b.max} = \left(\frac{M}{I} \times y_{max}\right) = \left(\frac{M}{I/y_{max}}\right) \tag{11.2}$$

or $$\sigma_{b.max} = \left(\frac{M}{Z}\right) \tag{11.3}$$

where, Z is the **section modulus** ($Z = I/y_{max}$), y_{max} is the distance from the neutral axis to the extreme fibre, and σ_{max} is the maximum bending stress.

The maximum bending stress in the beam section (if compressive) should be less than the allowable bending compressive stress and (if tensile); should be less than the allowable bending tensile stress. When the section of beam is symmetrical about the neutral axis then the value of y_{max} is equal to half the depth of section, and the maximum bending stress in compression and in the tension at the extreme fibres are equal. When the beam section is not symmetrical about the neutral axis, then there are two distances y_1 and y_2 to the two extreme fibres from the neutral axis. The bending stresses at the extreme top and bottom are not equal. Then, the values of $z_1 = (I/y_1)$ and $z_2 = (I/y_2)$ both should be calculated and compared with the section modulus, Z of the beam section provided.

The total compressive force `C' above the neutral axis is equal to the tensile force `T', for the beam in equilibrium. These two forces act in opposite directions, and form a couple. This couple resists the bending moment and this moment is known as **moment of resistance `M_r'**.

The moment of resistance of a beam section is the moment of the couple which is set up at the section by the longitudinal forces C and T created in the beam due to bending.

$$M_r = (C \times \text{Lever arm}) = (T \times \text{Lever arm})$$

For the beam in equilibrium, the moment of resistance `$M_r$' would be equal to the maximum bending moment `M' at any section ($M_r = M$).

Beams are designed such that the calculate stresses under the action of service load do not exceed specified allowable stresses.

11.6 DESIGN (FACTORED) STRENGTH / RESISTANCE OF LATERALLY SUPPORTED BEAMS IN BENDING (FLEXURE)

Design bending strength of beams adequate supported against lateral torsional buckling (laterally supported beams) is controlled by the yield stress of steel.

Design (factored) moment, M_{fd} at any section, in a beam due to external actions (loads), shall be less than or equal to design bending strength (IS : 800-2007, Clause 8.2)

M_{fd} less than or equal to $M_{db.S}$

$$M_{fd} < M_{db.S} \tag{11.4}$$

where, $M_{db.S}$ = Design bending strength of the section (in which $d/t_w \leq 67$) obtained as under.

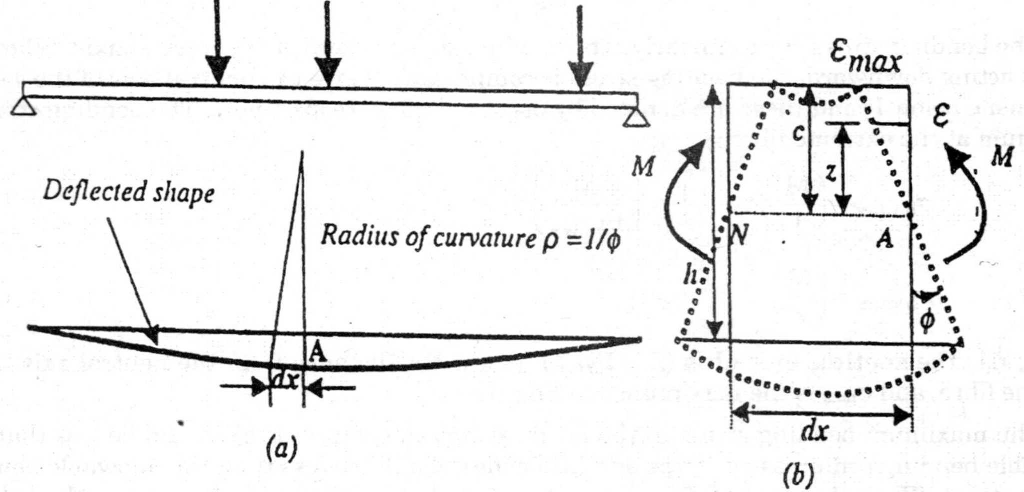

Fig. 11.7. Curvature of Bending for Steel Beams

While describing the classification of beam section, it has already been noted that **only plastic sections** can be used in **plastic/ultimate** design of frames, where, the moment redistribution is required throughout the frames. The flexural behaviour of steel beams at the instant of collapse may be studied.

In case, a flexural member is progressively loaded, it deflects and the curvature of such bending varies along the length as shown in Fig. 11.7.

Initially, the beam is elastic throughout its length.

The idealized stress-strain curve for steel is shown in Fig. 11.8.

In this Fig. 11.8, the strain, ε_1 is the less than the strain, ε_y. The strain at the extreme fibre at point B is maximum, $\varepsilon_{max.} = \varepsilon_2$. It is equal to the strain at yield ε_y and the stress is equal to yield stress, σ_y. Upto this, the strain is proportional to strain. At this extreme fibre, the beam section is stressed within the elastic range throughout its length.

The moment corresponding to this is called as **yield moment**, M_y.

$$M_y = (\sigma_y \,.\, Z) \tag{11.5}$$

Z = Elastic modulus of the section At, yield moment, yield just begins. Beyond the yield stress, the stress does not increase.

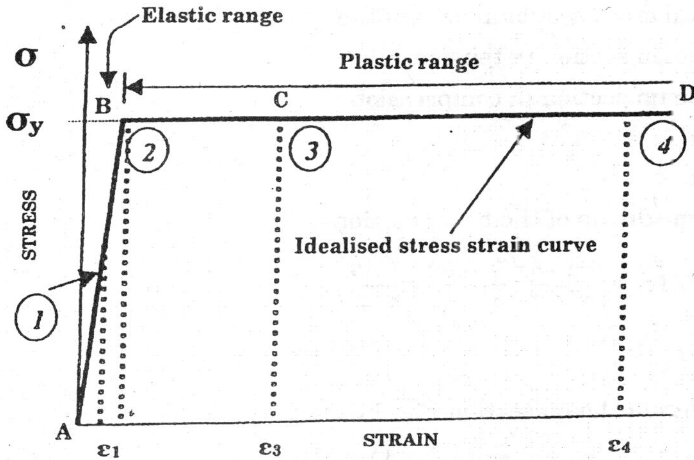

IDEALISED ELASTO-PLASTIC STRESS-STRAIN CURVE FOR STEEL

Fig. 11.8

When the load on the beam is further increased, then, the moment, M also increases. The maximum strain, ε_{max} becomes more than strain, ε_y.

This point is referred as the point 3 and $\varepsilon_3 >> \varepsilon_y$ as shown in Fig. 11.8. At this stage, the strain is in the plastic range. And when, the stress σ becomes equal to σ_y, the redistribution of the stresses begin and the yield spreads on the inner fibres.

When the load is further increased the strain increases rapidly upto **equal area axis** as shown in Fig. 11.8 and the complete cross-section of the beam is yielded. A **plastic hinge** develops and the moment M becomes equal to M_p (**Plastic moment**). The curvature of the beam increases enormously.

By considering the moment of total tensile force, T and total compressive force, C about equal area axis, the **plastic moment,** M_p is obtained as below :

$$M_p = T.\left(\frac{h}{4}\right) + C\left(\frac{h}{4}\right)$$

$$M_p = \left[\sigma_y \,.\, A_t\left(\frac{h}{4}\right) + \sigma_y . A_c\left(\frac{h}{4}\right)\right]$$

Since, ($A_t = A_c = A/2$)

$$M_p = \sigma_y\left[\left(\frac{A}{2}.\frac{h}{4}\right) + \left(\frac{A}{2}.\frac{h}{4}\right)\right]$$

$$M_p = \sigma_y\,[Z_t + Z_c]$$

$$M_p = \sigma_y\, A.\left(\overline{y_1} + \overline{y_2}\right)$$

$$M_p = \left(\sigma_y\, .\, Z_p\right) \tag{11.6}$$

where, A is the total cross-sectional area of the beam.

A_t = Area of beam section in tension.

A_c = Area of beam section in compression.

$y_1 = y_2 = (h/4)$

From Eq. 11.6

Z_p = Plastic modulus of the beam section.

$$Z_p = \frac{1}{2}A\left(\overline{y_1} + \overline{y_2}\right) = \frac{A}{2}\left(\frac{h}{4} + \frac{h}{4}\right) = \left(\frac{Ah}{4}\right) \tag{11.7}$$

$$M_p = \left(\sigma_y\, .\, Z_p\right) \tag{11.8}$$

For the rectangular beam section Fig. 11.5

$$Z_p = \left(\frac{1}{4}\,A.h\right) = \left(\frac{1}{4}\,.\,b.h.h\right) = \left(\frac{bh^3}{4}\right) \tag{11.9}$$

The shape factor, S may be introduced here. It is defined as the ratio of the plastic modulus of the section, Z_p to the elastic section modulus, Z.

$$S = \left(\frac{Z_p}{Z}\right) = \left(\frac{\frac{1}{4}bh^2}{\frac{1}{6}bh^2}\right) = 1.5 \tag{11.10}$$

It is also defined as the ratio of plastic moment to the yield moment

$$S = \left(\frac{\sigma_y\, .\, Z_p}{\sigma_y\, .\, Z}\right) = \left(\frac{M_p}{M_y}\right) \tag{11.11)A}$$

For an I-section, the value of shape factor, S varies between 1.07 to 1.20 and for the most practical cases of I-section, S is taken as 1.12. This value may be found as below

$Z_p = \sigma_y\, .\, A_f\, .\, h\, [1 + (A_w/4A_f)]$ and

$Z_p = \sigma_y\, .\, A_f\, .\, d.\, [1 + (A_w/6A_f)]$

$$S = \left(\frac{Z_p}{Z}\right) = 1.04 \text{ for } (A_f = A_w) \text{ and}$$

$$S = \left(\frac{Z_p}{Z}\right) = 1.12 \text{ for } (A_f = 0.5A_w)$$

Design bending strength, M_d shall be taken as (IS : 800-2007, Clause 8.2.1.2)

$$M_d = (\beta_b\, Z_p \,.\, \sigma_y / \gamma_{msy}) \quad (11.11)B$$

$\beta_b = 1.0$ for the plastic and compact section

$\beta_b = (Z_E / Z_p)$ for the semi-compact section

To avoid irreversible deformation under serviceability loads, M_d shall be taken less than (1.2 $Z_E . \sigma_y / \gamma_{msy}$) in case of simply supported beam and (1.5 $Z_E\, \sigma_y / \gamma_{msy}$) in cantilever beams.

When the design shear force (factored), V exceed 0.6 V_d, where V_d is the design shear strength of the cross-section.

$$M_d = M_{dv} \text{ (Design bending strength under high shear)} \quad (11.11)C$$

11.7 MOMENT CURVATURE RELATIONSHIP

Consider a simply supported beam. The beam is subjected to gradually increasing couple, M applied at the ends. From the theory of simple bending

$$\left(\frac{M}{I}\right) = \left(\frac{\sigma}{y}\right) = \left(\frac{E}{R}\right)$$

$$\text{or} \quad \phi = \left(\frac{I}{R}\right) = \left(\frac{M}{EI}\right) \quad (11.12)$$

where, R is the radius of curvature of the beam $\phi = (1/R)$ is called as **curvature** of beam.

When the value of moment M increases, the curvature of beam also increases. The curvature of beam is directly proportional to the Moment, M upto the limit of proportionality. The curvature of beam, ϕ increases with the increase of value of moment M. When the value of moment, M reaches yield moment, M_y, then the bending stress σ becomes σ_y.

$$\left(\frac{M_y}{I}\right) = \left(\frac{\sigma_y}{y}\right) = \left(\frac{E}{R}\right)$$

In case, the depth of beam section is h, then, y = (h/2)

$$\left(\frac{\sigma_y}{h/2}\right) = \left(\frac{E}{R}\right)$$

$$\left(\frac{2\sigma_y}{h}\right) = \left(\frac{E}{R}\right)$$

$$\phi = \left(\frac{1}{R}\right) = \left(\frac{2\sigma_y}{E.h}\right) \quad (11.13)$$

When the value of moment, M increases further, the section becomes elastic plastic. The section is partially still, there remains depth of elastic core. The curvature of the beam is still controlled by the elastic core

$$\phi = \left(\frac{1}{R}\right) = \left(\frac{2\sigma_y}{E.R}\right) \qquad (11.14)$$

The linear relation between curvature and the moment does not hold good further. When the value of moment, M reaches M_p = fully plastic moment of the section, the depth of elastic core becomes zero (e = 0). As such, the curvature of the beam increases infinitely. The moment curvature relationship is shown in Fig. 11.9.

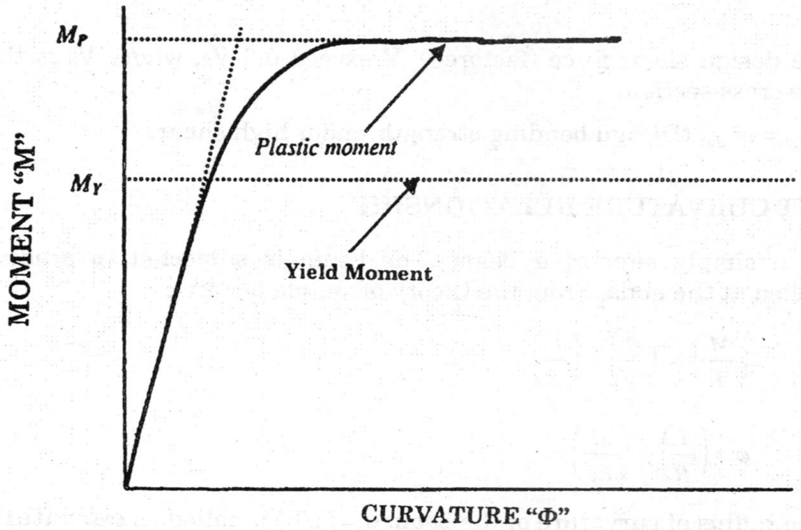

Fig. 11.9. Moment curvature characteristics of a simply supported beam

Then, the finite change of slope takes place over an infinitely small length of the beam, at this section. This section behaves a hinge, which is called as **plastic hinge**. The moment curvature relationship may be plotted in non-dimensional form as shown in Fig. 11.10.

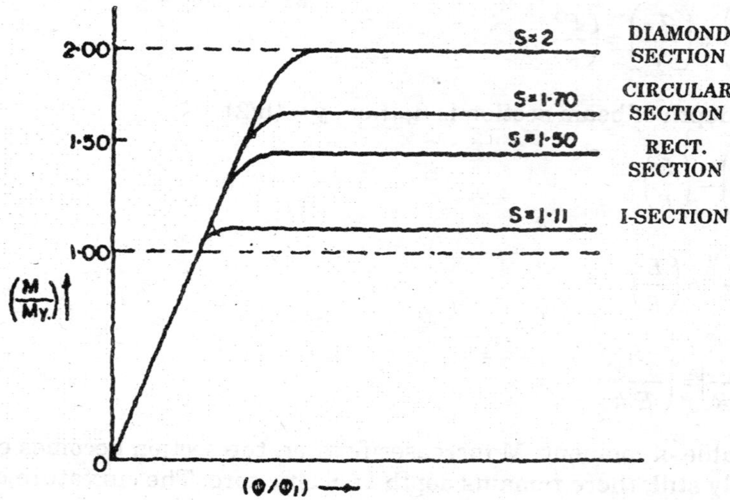

Fig. 11.10. Shape factors for different sections

Then, $\left(\frac{M_p}{M_y}\right) = \left(\frac{\phi_p}{\phi_y}\right)$

The shape of the curve depends upon the shape factor, S = (M_p/M_y)

For a rectangular beam section Simply supported : When the beam section is in elastic-plastic range.

$$M = M_p\left(1 - \frac{e^2}{3h^2}\right) \quad \text{or} \quad \left(\frac{M}{M_p}\right) = \left(1 - \frac{e^2}{3h^2}\right) \tag{11.15}$$

The value of e may be substituted from

$$\left(\frac{M}{M_p}\right) = \left(1 - \left(\frac{2\sigma_y}{E}.R\right)^2\right)$$

$$\left(\frac{1}{3h^2}\right) = \left[1 - \frac{1}{3}\left(\frac{2\sigma_y}{E}\right)\left(\frac{1}{h/R}\right)\right]$$

At the first yield,

$$\phi_y = \left(\frac{1}{R}\right) = \left(\frac{2\sigma_y}{E.h}\right) \tag{11.16}$$

$$\left(\frac{M}{M_p}\right) = \left[1 - \frac{1}{2} = \frac{h/R_y}{(h/R)}\right] \tag{11.17}$$

11.8 A PLASTIC HINGE

When the fully plastic moment is developed at any section, then, the section is completely yielded or plasticized. The curvature of the section becomes infinitely large, and the section of the beam continues to rotate at constant moment. The fibres on one side of the equal area axis of the section are in compression and the fibres on the otherside of the equal area axis of the section are in tension. The fibres carry bending compressive stress equal to yield stress tend to contract (shrink). Whereas, the fibres carrying bending tensile stress equal to yield stress tend to expand. The section acts like a hinge. This hinge is known as **a plastic hinge**. At the plastic hinge, an infinite rotation (large rotation over a small length can occur, under a constant moment equal to plastic moment, M_p of the beam section.

The plastic hinge is defined as a yielded zone due to bending in a structural member, at which infinite rotation can take place (i.e, a large rotation over a small length, at a constant plastic moment, M_p of the section). In contradiction to a mechanical hinge, at which the value of bending moment is zero. At plastic moment free rotation occurs under constant moment.

Theoretically, the plastic hinges are assumed to form at discrete (separate) points at which the plastic rotation occurs. The length of a plastic hinge is considered as zero.

11.9 LENGTH (SPREAD) OF A YIELDED ZONE

The values of moment at the adjacent section of the yielded zone are more than the yield moment, M_y upto a certain length, ΔL of the structural member. This length, ΔL is known as the **length of the hinge**, (i.e, the length of the spread of plasticity). This length of the hinge depends upon the type of loading and the geometry of the cross-section of the structural member. This region of length of hinge is known as **region of yield or plasticity**.

Two cases of a simply supported rectangular beam section have been considered.

1. **A concentrated at the centre as shown in Fig. 11.11** : The variation of bending moment is shown in Fig. (11.11) B. The maximum bending moment occurs at the centre (i.e, under the concentrated load). The concentrated load, W increases gradually. The value of moment, M reaches the value of fully plastic moment, M_p of the section and a plastic hinge is formed at the centre (under the load).

$$M_p = \left(\frac{WL}{4}\right)$$

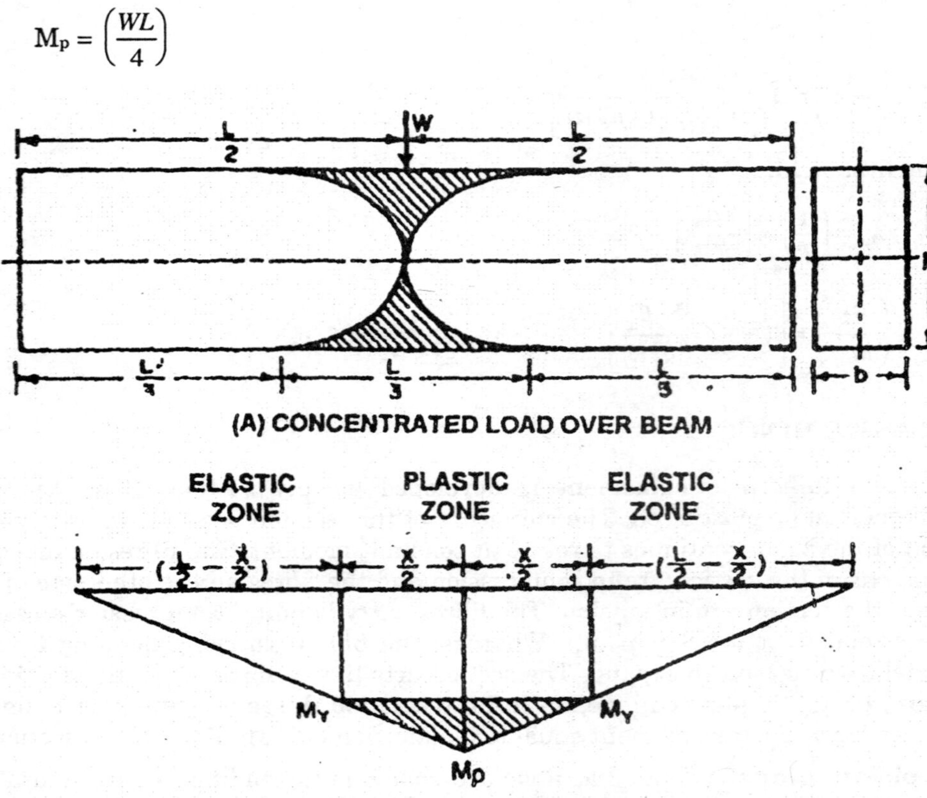

Fig. 11.11. Length (spread) of a yielded zone for a Concentrated Load.

Let the breadth of the rectangular beam be b and its depth be h. The moment $M_y = (\sigma_y \,.\, Z)$ is first reached, when the extreme fibres attain the value of yield stress

$$M_y = \sigma_y \,.\, \left(\frac{1}{6} bh^2\right)$$

$$= \sigma_y \left(\frac{1}{6} x 4 \, x \frac{bh^2}{4}\right) = \left(\frac{2}{3}\sigma_y . Z_p\right)$$

Since, $Z_p = \left(\frac{bh^2}{4}\right)$. Let the length of yield zone be x. From the bending moment diagram as shown in Fig. 11.11 (B)

$$\left(\frac{M_p}{M_y}\right) = \left(\frac{L/2}{(L/2)-(x/2)}\right)$$

or $(L - x), M_p \; M_y . L$

Substituting the value of M_y

$$(L - x) M_p = \frac{2}{3} M_p . Z$$

or $$(L - x) = \left(\frac{2}{3} L\right), \therefore x = \left(\frac{L}{3}\right) \qquad (11.18)$$

The length of yielded zone, (i.e, length of a plastic hinge) is equal to one-third of the span.

Now consider a section XX in the region of plasticity at a distance, x_1 from the centre

$$M = M_p \left[\left(\frac{L/2 - x_L}{L/2}\right)\right]$$

Let the depth of elastic core at this section be e. Since

$$M = \sigma_y . \left(\frac{bh^2}{4}\right)\left(1 - \frac{e^2}{3h^2}\right)$$

$$M = M_p \left(1 - \frac{e^2}{3h^2}\right) \qquad (11.19)$$

From the above two expressions

$$\left(1 - \frac{e^2}{3h^2}\right) = \left[\frac{(L/2)-(x_L)}{(L/2)}\right]$$

$$\left(1 - \frac{e^2}{3h^2}\right) = \left[1 - \left(\frac{2x_1}{L}\right)\right]$$

$$e^2 = \left[\frac{6h^2}{h}\right] . x_1 \qquad (11.20)$$

This shows that depth of elastic core varies parabolically with x_1.

2. **The simply supported rectangular beam subjected to uniformly distributed load, W = w.L,** as shown in Fig. 11.12. The maximum bending moment, M occurs at the centre and attains the value of fully plastic moment, M_p

$$M_p = \left(\frac{WL}{8}\right)$$

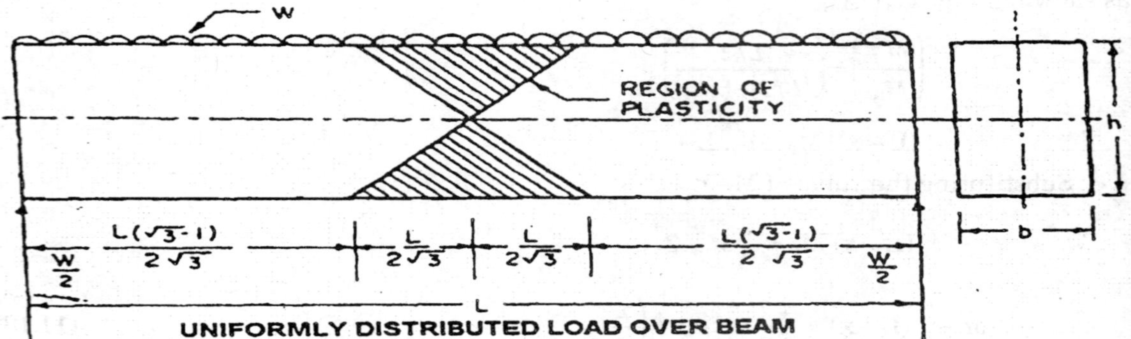

Fig. 11.12. Length (spread) of a yielded zone for a uniformly distributed load.

Consider a section at a distance x_1 from the centre, with the region of plasticity. The moment at this section

$$M = \frac{W}{8L}\left(L^2 - 4x^2\right) = \frac{8M_p}{8L}\left(L^2 - 4x_1^2\right)$$

$$M = M_p\left[1 - \frac{4x_1^2}{L}\right]$$

Let the depth of the elastic core be e. And

$$M = M_p\left[1 - \frac{e^2}{3h^2}\right] \tag{11.21}$$

Therefore,

$$\left[1 - \frac{e^2}{3h^2}\right] = \left[1 - \frac{4x_1^2}{L^2}\right] \text{ and}$$

$$e^2 = 12\,h^2 \cdot \left[\frac{x_1^2}{L^2}\right]$$

$$e = 2\sqrt{3}\left(\frac{h}{L}\right).x \tag{11.22}$$

This shows that the variation of depth of the elastic core, e is linear. The regions of plasticity are triangular in shape. When the depth of elastic core, e = h, then

$$h = 2\sqrt{3}\left(\frac{h}{3}\right) x$$

$$x = \left[\frac{L}{2\sqrt{3}}\right] \quad (11.23)$$

as shown in Fig. 11.12. Total length of region of the plasticity is equal to ($L/\sqrt{3}$).

Similarly, it may be worked out than the length of region of plasticity of an I-section (simply supported beam) subjected to a concentrated load at the centre is about **one-eight** of the span.

11.10 BEHAVIOUR OF BEAMS IN ELASTIC RANGE

A simply supported beam gradually loaded is shown in Fig. 11.13. The beam bends and deflects along its complete length. The beam remains elastic over whole length in the beginning.

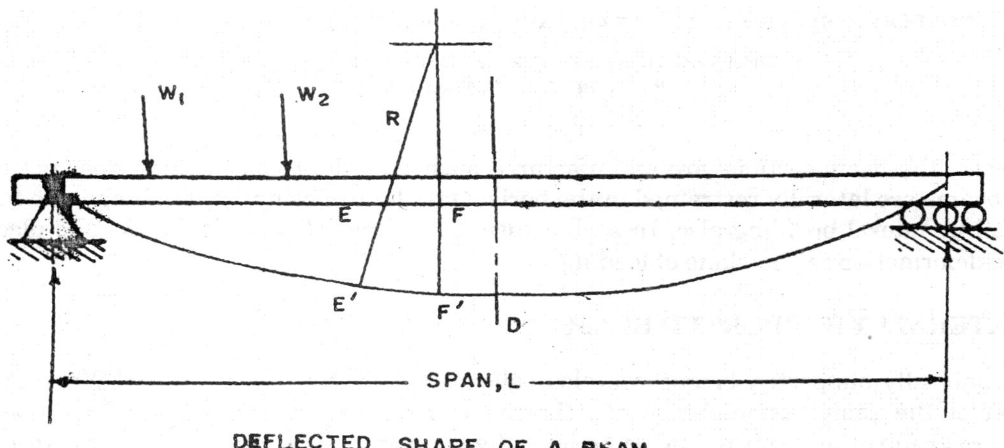

Fig. 11.13.

Consider a small length EF = dx. OE' and OF' show the radius of curvature, R. The fibres of beam at the bottom bending downward are in tension and those at the top are in compression. The variation of strains in bending over complete depth may be determined geometrically. Let the maximum strain at the top of beam be $E_{max} = E_1$ and it is less than yield strain, E_y. The strain upon this stage is directly proportional to the strain as shown in Fig 11.14.

From simple theory of bending, the **moment of resistance of the beam** (i.e., **the flexural strength**)

$$MR = \left(\sigma_{bci.}\ I/y_{c.max}\right) \quad \text{(i)}$$

where, σ_{bc1} in the bending stress in compression in the extreme fibre from the neutral axis, I is the moment of inertia of complete cross-section about neutral axis and $y_{c.max}$ is the distance from the neutral axis to the extreme fibre.

$$\text{M.R.} = (\sigma_{bc1} \cdot Z_E) \qquad \text{(ii)}$$

where, $Z_E = (I/y_{c.max})$ is the elastic section modulus. I and Z are the geometrical properties of the beam-section.

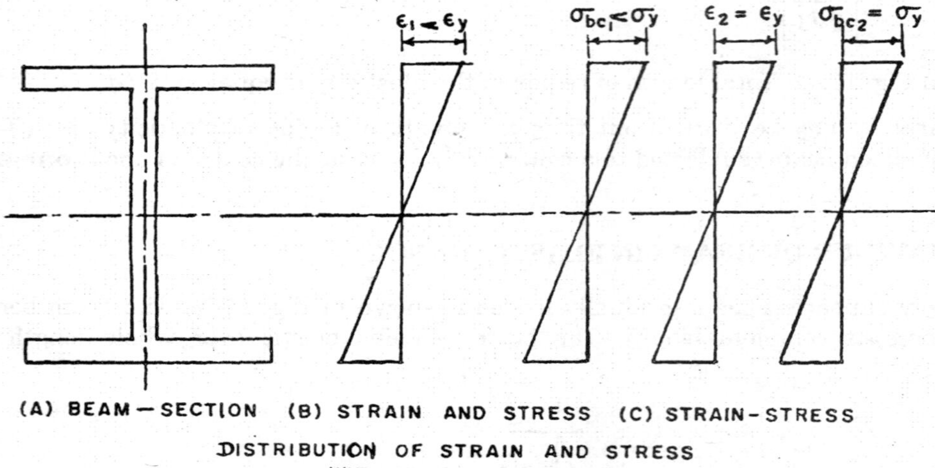

Fig. 11.14.

In case, the beam-sections are manufactured such that the local buckling does not occur and the beams are laterally restrained in the horizontal plane of compression flange, the beams do not fail by lateral buckling also. In such a case, the failure of the beam may take place by excessive deformation in the plane of loading.

11.11 LATERALLY SUPPORTED BEAM

The laterally supported beams are also called *laterally restrained beams*. When lateral deflection of the compression flange of a beam is prevented by providing effective lateral support, (restraint) the beam is said to be laterally supported. The *effective lateral restraint* is the restraint which produces sufficient resistance in a plane perpendicular to the plane of bending to restrain the compression flange of a beam from lateral buckling to either side at the point of application of the restraint. The concrete slab encasing the top flange (so that the bottom surface of the concrete slab is flush with the bottom of the top flange) is shown in Fig. 11.14(A). It provides a continuous lateral support to the top flange of the beam. When other beams frame at frequent intervals into the beam in questions as shown in Fig. 11.14(B), lateral support is provided at each point of connection but main beam should still be checked between the two support.

A beam may be assumed to be adequately **laterally supported** at the supports, provided the compression flange has full lateral restraint and nominal torsional restraint at the supports supplied by web cleats, partial depth end plates, flange plates or continuity with the adjacent span.

Complete lateral supports to compression flange may be assumed to exist if the frictional or other positive support of a floor connection to the compression flange of the member is capable of resisting a lateral force not less than 2.5 percent of the maximum force in the compression flange of the member.

This may be treated as uniformly distributed along the flange, provided gravity loads constitute the dominant loading on the member and the floor construction is capable of resisting this lateral force.

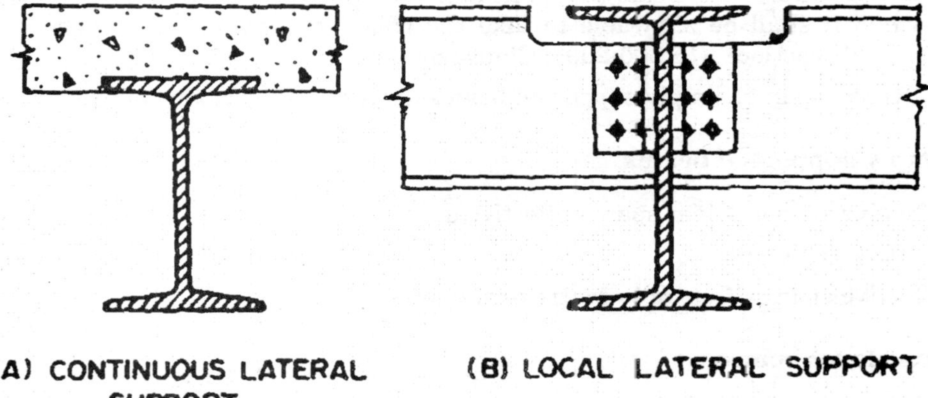

Fig. 11.15. Adequate lateral support

The design bending strength of a section [which is not susceptible to web buckling under shear yielding, where, (d/t_w) shall be less than or equal to 67 ε] is obtained as under : [Since, $(1/\sqrt{3}) = 0.577$, approximately equal to 0.6]

$$M_{bS} = (\beta_b \cdot Z_p \cdot \sigma_y / \gamma_{msy})$$

In order to avoid irreversible deformation under serviceability loads, shall be less than

$(1.2\ Z_e \cdot \sigma_y / \gamma_{msy})$ in case of simply supported and $(1.5\ Z_e \cdot \sigma_y / \gamma_{msy})$ in cantilever beams, where

β_b = 1.0 for plastic and compact section

$\beta_b = (Z_e/Z_p)$ for semi-compact section

Z_p = Plastic modulus of the cross-section

Z_e = Elastic modulus of the cross-section

σ_y = Yield stress of the material

γ_{msy} = Partial safety factor at yield

When the design (factored) shear force V exceeds 0.6 V_{bS}, where, V_{bS} is the shear strength of the cross-section, **the design bending strength, M_{bs}** shall be taken as below

$$M_{bS} = M_{dhvs}, \text{ where} \tag{11}$$

M_{dhvs} = Design bending strength under high shear

2. When the design (factored) shear force does not exceed 0.6 V_d, where, V_d is the design shear strength of the cross-section, the design bending strength shall be taken as

$$M_{bds} = \left(\beta_b \cdot Z_p \cdot \sigma_y / \gamma_{msy}\right) \quad (11.)$$

11.12 EFFECTIVE SPAN OF LATERALLY SUPPORTED BEAMS

The **effective span** of a beam shall be taken as the distance between centres of the supports except where the point of application of reaction is taken as eccentric at the supports, when it shall be allowable to take the effective span as the length between the assumed lines of reactions. (IS 800-2007, Clause 8.1.1)

The effective span of simply supported beams and cantilever beams are taken as under :

(i) For simply supported beams

(a) Effective span = (Clear span + effective depth)

$$ES = (C.S + d) \quad (i)$$

(b) Effective span = (Centre to centre to supports)

(ii) For cantilever beams

ES = (Length from free end of the cantilever to the face of the supports + half its effective depth)

$$ES = (\text{Clear projection} + \frac{1}{2}\ d)$$

(iii) For continuous beams

The effective span shall be taken as the length between the centre of supports, except in cases where the point of application of the reaction is taken as eccentric to the supports, then, it shall be permissible to take the effective span as the length between the assumed points of application of reaction.

Example 11.1 The effective length of compression flange of a simply supported beam ISMB 500, @ 0.869 kg/m is 8 m. Determine the safe uniformly distributed load per metre length, which may be placed over the beam having an effective span of 8 m. The ends of beam are restrained against rotation at the bearing.

Use limit state method of design.

Solution :

The beam section is ISMB 500, @ 0.869 kg/m. The beam is considered as laterally supported.

Step 1. Geometrical properties of ISMB 500, @ 0.869 kg/m from ISI Handbook No. 1.

Overall depth, D = 500 mm

Width of flange, b_f = 180 mm

Thickness of flange, t_f = 17.2 mm

Thickness of web, t_w = 10.2 mm

Moment of inertia

About strong axis, I_{xx} = 45218.3 x 10^4 mm^4

About weak axis, I_{yy} = 1369.8 x 10^4 mm^4

Section modulus

About strong axis, Z_{xx} = 1808.7 x 10^3 mm^3

About weak axis, Z_{yy} = 152.2 x 10^3 mm^3

Radius at the root of flange, V_1 = 17.0 mm

Step 2 : Classification/type of the section

Yield stress for proposed beam section

(Assumed), σ_y = 250 N/mm^2

Yield stress ratio

Epsilon, $\varepsilon = (250/\sigma_y)^{1/2}$

$= (250/250)^{1/2} = 1.00$

Criterion for flange (IS : 800-2007)

Actual ratio, $(0.5\ b_f/t_f) = (0.5 \times 180/17.2) = 5.23$

Requirement of flange

Ratio $(8.9\ \varepsilon) = (8.9 \times 1.0) = 8.9$

Actual ratio, (5.23 < 8.9) requirement

Criterion for web (IS : 800-2007)

$$\text{Actual ratio} \left[\left(\frac{D-2t_f}{t_w}\right)\right] = \left(\frac{500-2\,x\,17.2}{10.2}\right) = 45.64$$

Requirement for the web

Ratio $(83\ \times \varepsilon) = (83 \times 1.0) = 83$

Actual ratio (45.64 < 83) requirement

Section ISMB 500, @ 0.869 kg/m is classified as **Plastic section.**

Step 3 : Moment capacity/resistance of the section

Elastic section modulus of ISMB 500, @ 0.869 kg/m

Z_{xx} = 1808.7 x 10^3 mm^3

Shape factor, S for I-secton 1.12 (assumed)

Plastic section modulus

Z_p = (1808.7 x 10^3 x 1.12) mm^3

= 2025.744 x 10^3 x mm^3

Moment carrying capacity/resistance of the section for plastic section, $\beta_b = 1.00$

$$M_c = \left(1.0\ \frac{\sigma_y \cdot Z_p}{\gamma_{ms}}\right) = \left(1.0 \times \frac{250\,x\,2025.744\,x\,10^3}{1000\,x\,1000\,x\,1.15}\right)$$

$= 440.37$ kN-m

Step 4 : Let the design (factored) load be W_{Fd} Effective span, l = 8m

Bending moment for uniformly distributed load $= \left(\frac{W_f \cdot L}{8}\right) = \left(\frac{W_f \times 8}{8}\right) = 440.37$ kN-m

Step 5 : Uniformly distributed characteristic load, L

$$w.L = (W_{fd} / \gamma_{fL}) = \left(\frac{440.37}{1.5}\right)$$

$$w = \left(\frac{440.37}{1.5 \times 8}\right) = 36.698 \text{ kN/m}$$

This load may be placed from the requirement of bending. However, this load may also be calculated from the shear. The load, whichever is less may be taken as required load.

11.13 DESIGN OF LATERALLY SUPPORTED/RESTRAINED BEAMS

Design of laterally supported (Restrained beam) by limit State Method needs a trial section. The elastic-section modulus required for the beam section is obtained from the plastic-section modulus divided by the shape factor. Whether the beam-section proposed is adequate for the requirement of moment carrying capacity/resistance of the beam section.

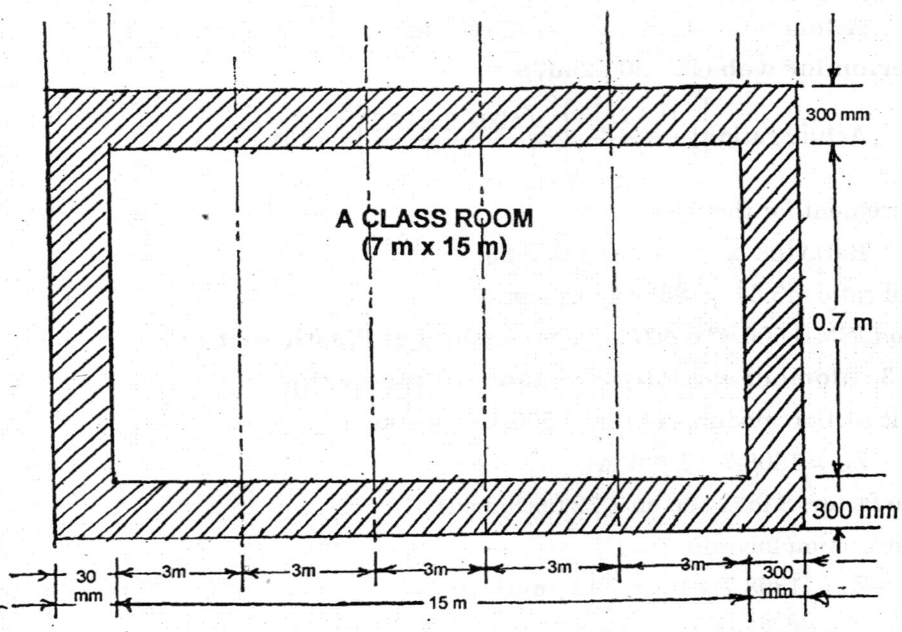

Fig. 11.16.

The factored design moment, M at any section in a beam due to external actions should be less than or equal to design bending strength/resistance of the beam section, M_d.

$(M \leq M_d)$

The design of beams deal with proportioning of size of the beam members, the determination of effective section modulus, maximum deflection and the shear. In general the rolled steel sections have flanges and webs of sufficient thicknes, such that the criterion for the design is seldom controlled by shear. Following are the usual steps in the design of laterally supported beams.

Step 1 : For the design of beams, characteristic loads to be carried by the beam, and the effective span of the beam are known. The value of yield stress, σ_y for the proposed grade/ quality of steel is decided.

Step 2 : Total design (factored) load

$W_{fd} = (\gamma_{fL.DL} . (D.L.) + \gamma_{fL. LL.} (L.L.))$ is found.

Maximum design (factored) moment is calculated.

Step 3 : Plastic-section modulus, Z_p required for the factored load is obtained

$$Z_p = \left(\frac{M_{fd} / \sigma_y}{\gamma_{msy}} \right)$$

Step 4 : Elastic-section modulus, Z_{xx} and found by assuming the shape-factor, S for I-section (S = 1.06 to 1.12)

$$Z_{xx} = (Z_p / S)$$

A trial section is proposed, and which one is checked whether, it adequate for its moment capacity. The properties of the beam sectioned is.

Step 5 : The section of proposed beam is classified and its type is determined. The actual ratios ($0.5\, b_f/t_f$) for the flanges and $[(D - 2t_f)/t_w]$ for the web are found and compared with those recommended in IS : 800-2007 whether the beam section proposed is (i) plastic (ii) compact (iii) semi-compact or (iv) slender.

Step 6 : The moment carrying capacity (i.e, the design strength/resistance) of the beam section) is calculate.

It is that M_d is more than the factored moment, M due to external action. Then, it is considered that the beam section i adequate. Further, whether the beam section proposed is adequate for the requirement of shear, web buckling and web crippling. These have been checked as described in chapters 12, 13 and 14, respectively.

Example 11.2. A class room is 7 m x 15 m. It is provided with 120 mm thick stone patties over rolled steel beams spaced 3 m centre to centre. A wearing coat of 20 mm thick cement concrete is provided over 120 mm thick lime concrete. The effective length of compression flange would be equal to the effective span. Design rolled steel beams. The ends of beams shall not be free to rotate at the bearing. Use **Limit State Method of Design.**

Design :

Unit weight of stone patties = 24 kN/m³

Unit weight of plane cement concrete = 24 kN/m³

Unit weight of lime concrete = 24 kN/m³

From IS : 875

Live load for the class room = 4 kN/m²

Step 1 : Design (Factored) load supported and the corresponding bending moment

The rolled steel beam would carry a load from a strip of 3 m width. Uniformly distributed load per metre length.

(A). Dead (permanent) load

Weight of stone patties = $\left(3 \, x \, 1 \, x \dfrac{120}{1000} x24\right)$ = 8.64 kN/m

Weight of cement concrete = $\left(3 \, x \, 1 \, x \dfrac{20}{1000} \, x \, 24\right)$ = 1.44 kN-m

Weight of lime concrete $\left(3 \, x \, 1 \, x \dfrac{120}{1000} \, x \, 24\right)$ = 8.64 kN-m

Assume self weight of beam = 0.60 kN/m

Total dead (permanent) load = 19.32 kN/m

(B) Live (imposed load) = (1 x 3 x 4) = 12 kN/m

(i) Partial safety factor for dead

(permanent) load = (γ_{fL}) = 1.35

(ii) Partial safety factor for live (imposed) load (γ_{fL}) =1.50

(c) Total factored (Design) load, W_{fd}

[(1.35 x 19.32) + (1.5 x 12)] = 44.082 kN/m

(D) Maximum factored (Design) moment, M_{fd}

Clear span of beams = 7.0 m

Assume 300 mm wide bed blocks would be provided.

The effective span = (7000 + 300) mm = 7.30 m

Maximum factored (design) moment occursat the centre

$$M_{fd} = \left[\frac{w.l^2}{8}\right] = \left[\frac{44.082 \times 7.30 \times 7.30}{8}\right] = 293.641 \text{ kN-m}$$

Step 2 : Yield stress for steel σ_y = 250 N/mm²

Plastic-section modulus required for the design load. For plastic-section, $\beta_b = 1.00$

M = (σ_y . Z_p) [IS : 800-2007, Clause 8.2.1.2]

$$Z_p = \left(\frac{293.641 \, x \, 1000 \, x \, 1000}{250}\right) = 1174.565 \text{ mm}^2$$

Partial factor of safety for the strength of material

$\gamma_{ms} = 1.15$

Design section (plastic) required for the strength = $\left(\frac{1174.565\ x10}{1.15}\right) = 1021.361\ x\ 10^3\ mm^3$

Step 3 : Elastic-section modulus required

$$Z = \left(\frac{Plastic - section}{Shape\ factor}\right)$$

$$= \left(\frac{1021.361\ x\ 10^3}{1.10\ (assumed)}\right) = 928.51\ x\ 10^3\ \ mm^3$$

Following rolled steel beams may be provided for the required section modulus

(i) ISLB 450, @ 65.3 kg/m, Z_{xx} = 1223.8 x 10^3 mm^3

(ii) ISMB 450, @ 72-4 kg/m, Z_{xx} = 1350.7 x 10^3 mm^3

(iii) ISWB 400, @ 66.7 kg/m, Z_{xx} = 1171.30 x 10^3 mm^3

Note : Trial section may be checked. Whether the proposed section is adequate for its moment capacity.

Step 4. Trial section (proposed)

ISLB 450, @ 65-3 kg/m

Section modulus, Z_{xx} = 1223.8 x 10^3 mm^3

Its weight is minimum as compared to other two sections mentioned

ISLB 450, @ 65-3 kg/m

Step 5 : Properties of trial section

Overall depth, D = 450 mm

Width of the flange, b_f = 170 mm

Thickness of the flange, t_f = 13.4 mm

Thickness of web, t_w = 8.6 mm

Moment of inertia

About xx-axis, I_{xx} = 27536.1 x 10^4 mm^4

About yy-axis, I_{yy} = 853.0 x 10^4 mm^4

Radius at the root of the flange

r_1 = 16.0 mm

Step 6 : Classification/type of beam section

Yield stress, σ_y = 250 N/mm^2

Yield stress, ratio

Epsilon, $\varepsilon = (250/\sigma_y)^{1/2} = 1.0$

(A) Criterion for flange

Actual ratio ($0.5\, b_f/t_f$) = (0.5 x 170/13.4) = 6.34

Requirement for flange

(8.9 x ε) = (8.9 x 1.00) = 8.9

Actual ratio (6.34 < 8.9) requirement.

(B) Criterion for web

Actual ratio [$(D-2t_f)/t_w$]

$$= \left(\frac{450 - 2\,x 13.40}{8.6}\right) = 49.209$$

Requirement for web

(83 x ε) = (83 x 1.0) = 83

Actual ratio (49.209 < 83) requirement

The section ISLB 450, @ 65.3 kg/m.

Elastic section modulus provided

Z_{xx} = 1223.8 x 10^3 mm^3

Shape factor for I-section

S = 1.12 (assumed)

Plastic section modulus of the section

Z_p = (1223.8 x 10^3 x 1.12) = 1370.656 x 10^3 mm^3

Moment carrying capacity the section (i.e, resistance of the beam section)

$$M_c = \left(\frac{Z_p\ x\ \sigma_y}{\gamma_{ms}}\right)$$

$$M_c = \left(\frac{1370.656\ x\ 250\ x\ 10^3}{1000\,x\ 1.15}\right) = 297.969 \text{ kN-m}$$

This moment carrying capacity/resistance should be more than design moment (external action) 293.641 kN-m. Hence, the beam section ISLB 450, @ 65.3 kg/m is adequate to resist external bending moment.

Note : In addition to this, the proposed trial section has to be checked for the following :

(i) Whether, it is adequate for shear strength (It has been done in Chapter 12)

(ii) Whether, it is adequate for **Web** buckling and web crippling (It has been done in chapter 13), and

(iii) Whether, the beam is satisfactory for deflection for limit of serviceability (It has been seen in chapter 14).

Example 11.3 Check, whether the trial section proposed ISMB 450, 72.4 kg/m is adequate.

Proposed trial section ISMB 450, @ 72.4 kg/m has been checked as follows :

Step 1. Geometrical proposed of ISMB 450, @ 72.4 kg/m ISI Handbook No. 1.

Overall depth, D = 450 mm

Width of flange, b_f = 150 mm

Thickness of flanges, t_f = 17.4 mm

Thickness of web, t_w = 9.4 mm

Moment of inertia

About strong axis, I_{xx} = 30,390.80 x 10^4 mm^4

About weak axis, I_{yy} = 834.0 x 10^4 x mm^4

Section modulus

About strong axis, Z_{xx} = 1350.70 x 10^3 mm^3

About weak axis, Z_{yy} = 111.20 x 10^3 mm^3

Radius at the root of flange

r_1 = 15.00 mm.

Step 2 : Classification/type of the section

Yield stress for steel propoed

(Assumed), σ_y = 250 N/mm^2

Yield stress ratio

Epsilon, $\varepsilon = (250/\sigma_y)^{1/2} = (250/250)^{1/2} = 1.00$

Criterion for flange

Actual ratio (0.5 b_f/t_f) $= \left(\dfrac{0.5 \, x \, 150}{17.4}\right) = 4.31$

Requirement of flange

Ratio (8.9 ε) = (8.9 x 1.00) = 8.9

Actual ratio (4.3 < 8.9) requirement

Criterion for web

Actual ratio $= \left[\left(\dfrac{D-2t_f}{t_w}\right)\right] = \left[\dfrac{450-2x\,17.4}{9.4}\right] = 44.170$

Requirement for the web

Ratio (83 ε) = (83 x 1.00) = 83

Actual ratio (40.1702 < 83) requirement

Section ISMB 450, @ 72.4 kg/m is classified as **Plastic section.**

Step 3 : Moment capacity/resistance of the section. For plastic-section, $\beta_b = 1.00$

Elastic section modulus of proposed beam section

Z_{xx} = 1350.70 x 10^3 mm^3

Z_{yy} = 111.20 x 10^3 mm^3

Shape factor, S for I-section, 1.12 (assumed)

Plastic section modulus

$$Z_p = (1350.70 \times 10^3 \times 1.12) = 1512.784 \times 10^3 \text{ mm}^3$$

Moment carrying/capacity resistance ofthe section, M_d

$$M_d = \left(\frac{\sigma_y \cdot Z_p}{\gamma_{ms}}\right) = \left(\frac{250 \ x \ 1512.784 \ x \ 10^3}{1000 \ x \ 1.15 \ x \ 1000}\right) = 328.866 \text{ kN-m.}$$

This moment capacity/resistance is more than factored (design moment due to load) 293.641 kN-m

Therefore, ISMB 450, @ 72.4 kg/m is adequate.

Note : The adequacy of this beam section for shear web buckling and web crippling and deflection should be checked.

11.14 COMPUTER PROGRAMME FOR ANALYSIS AND DESIGN OF LATERALLY SUPPORTED BEAMS

To-day, it is **computer-age**. High speed and large memory computers are available are these are used in every step of practical life.

One category of the persons prepare **flow-charts** and then, they write **computer programes.** While, other category of the persons write computer programmes and then, they prepare flow-charts

For laterally supported beams, two computer programmes have been written1

1. **ALSSB :** Computer programme deals with analysis and design and determination of elastic section modulus.

2. **BLSSB :** Computer programme deals with to find the adequacy of the beam-section in bending, shear and deflection of beams.

1. Flow-chart for AL SSB

Brief step by step description of above two programmes have been given with respect to these two programmes. Figs. 11.17, 11.18 show flow charts. Every step (block) include many computer intructions.

Size of a room (class-room or a hall) width and length and the yield stress of the grade of the steel proposed are read.

In case, the length of room is les than 8 m, only one beam is provided alongthe short side (i.e, width of room plus 0.30 m). However, in case, the length is greater than 8 m and upto 15 m, four beams are provided at a spacing equal to length divided by 5.

Unit weight of stone patties and thickness, unit weight of cement concrete and its thickness, unit weight of lime concrete and its thickness, self-weight of the beam (assumed as 0.60 kN/m) and the intensity of live (imposed) load as per the 2 likely use of room are read.

Corresponding dead loads and live loads likely to act over the beam per unit length are calculated.

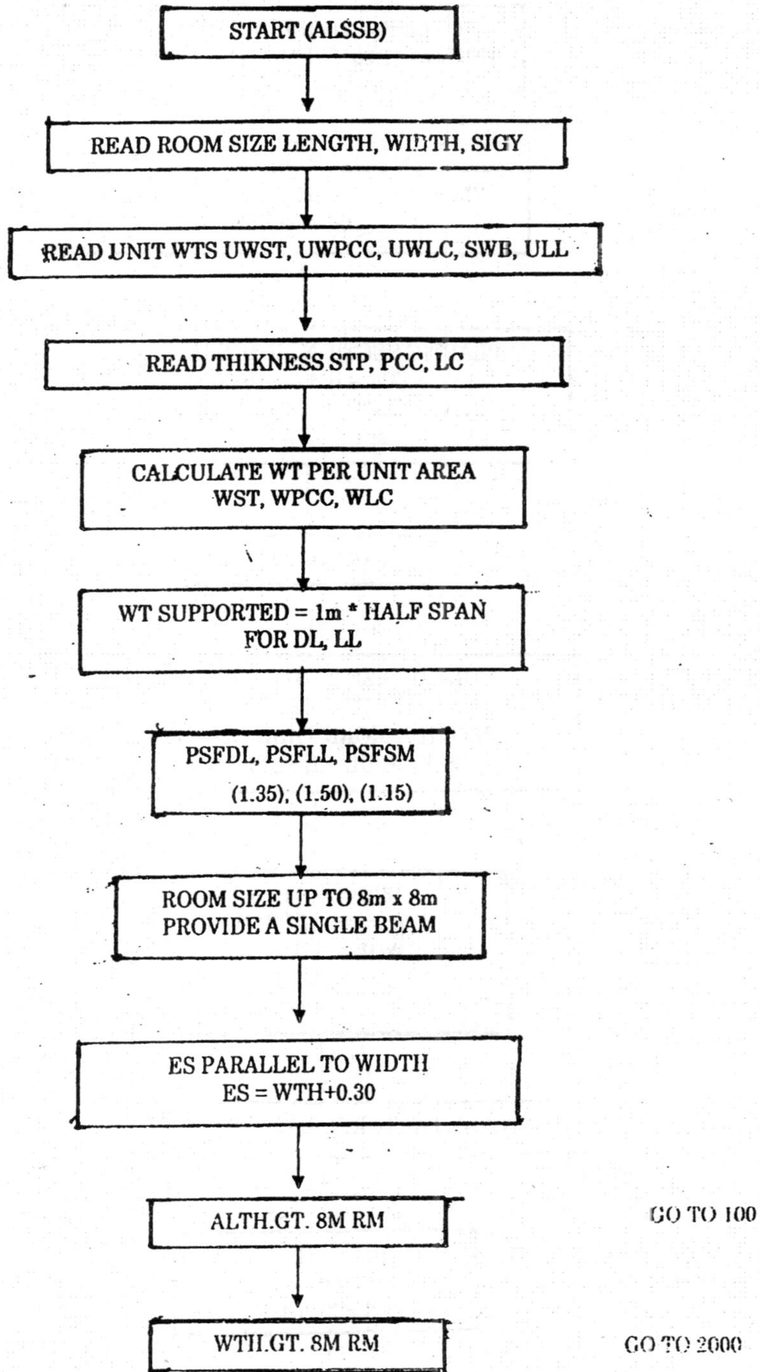

Fig. 11.17. Flow chart for ALSSB (contd..)

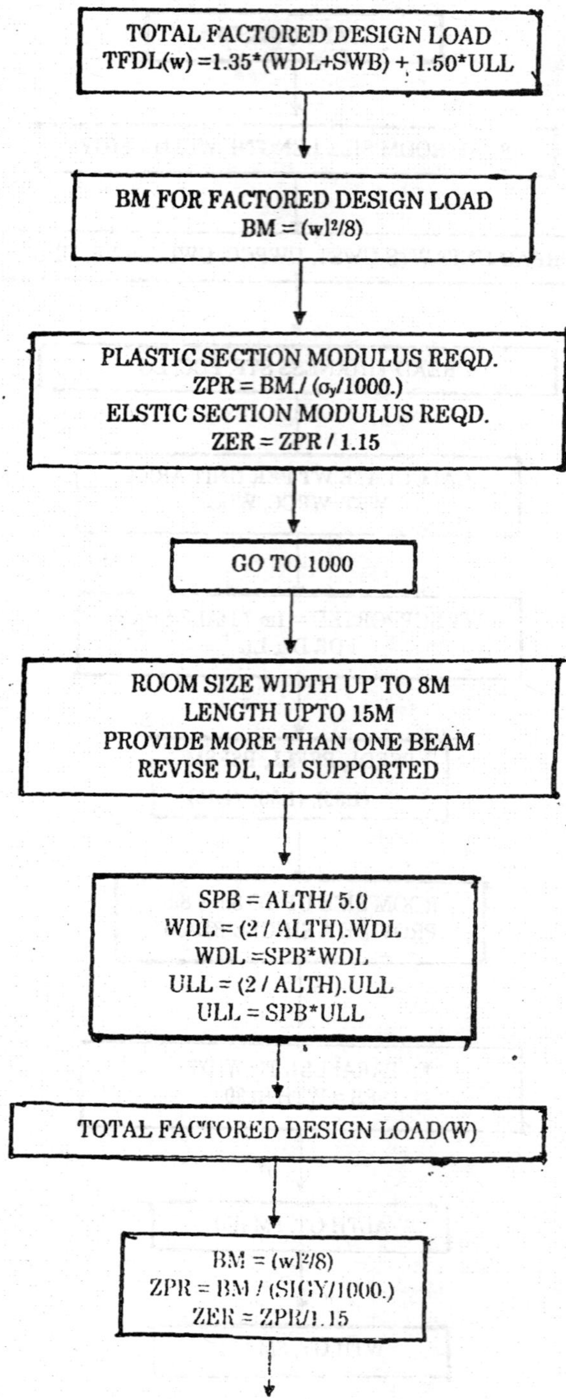

Fig. 11.17. Flow chart for ALSSB (contd..)

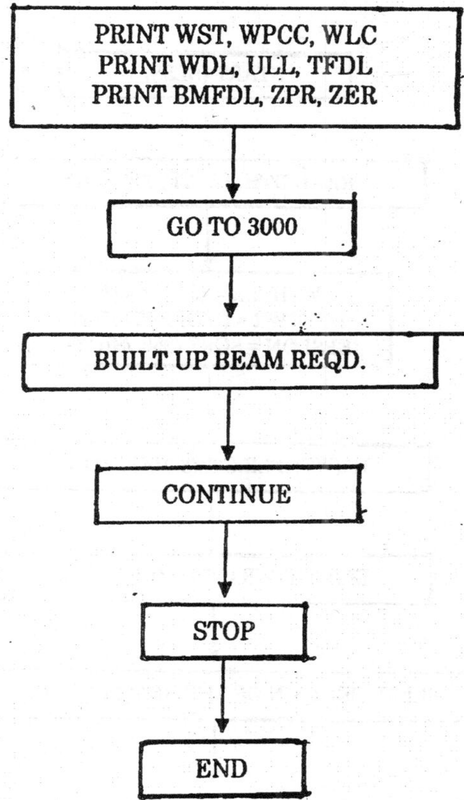

Fig. 11.17. Flow chart for ALSSB

Partial factor of safety for dead loads and the live loads and that for the strength of material are speficied. Total design (factored) external loads bending moment, and shear force are found. Plastic section modulus, Z_p = (M_d/σ_y) are obtained. Elastic section modulus required is found (Z_p / Shape factor) S = 1.06 to 1-12 (assumed).

In-case, the width of room is more than 8 m then, the built-up beam has to be provided.

2. Flow chart for BL SSB

This programme deals to check the adequacy of the beams in bending, in shear and deflection.

The width and thickness of flanges of I-sections proposed as trial section, their depth and thickness of the web, radius at the root of the flange and yield stress are read.

The proposed (trial section is classified and its type (plastic, compact, semi-compact and slender) are checked Actual ratios for the flanges (0.5 x width of flange/its thickness) and the clear depth of web/its thickness) are determined. These ratios are compared with respective specified values in IS : 800-2007. The beam section is accordingly declared (plastic or compact or does not belong to these two categories).

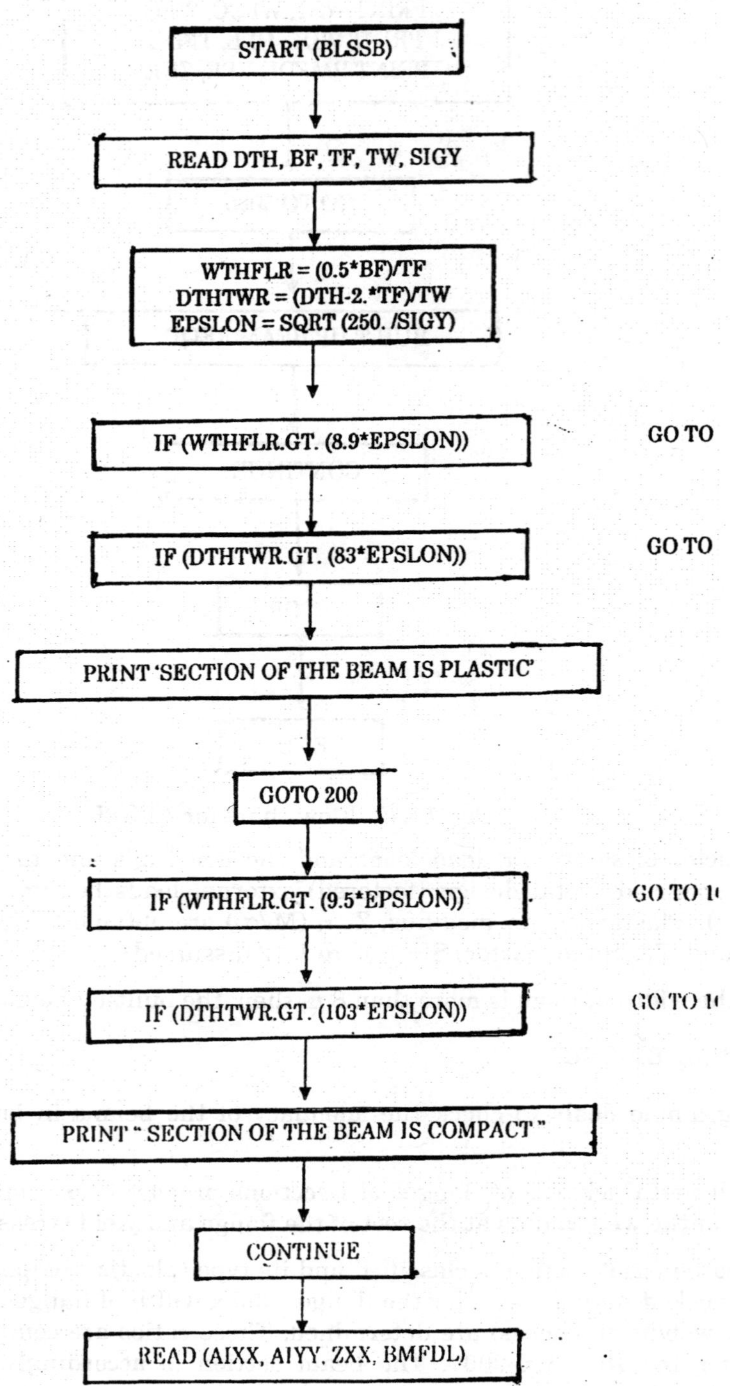

Fig. 11.18. Flow chart for BLSSB (contd..)

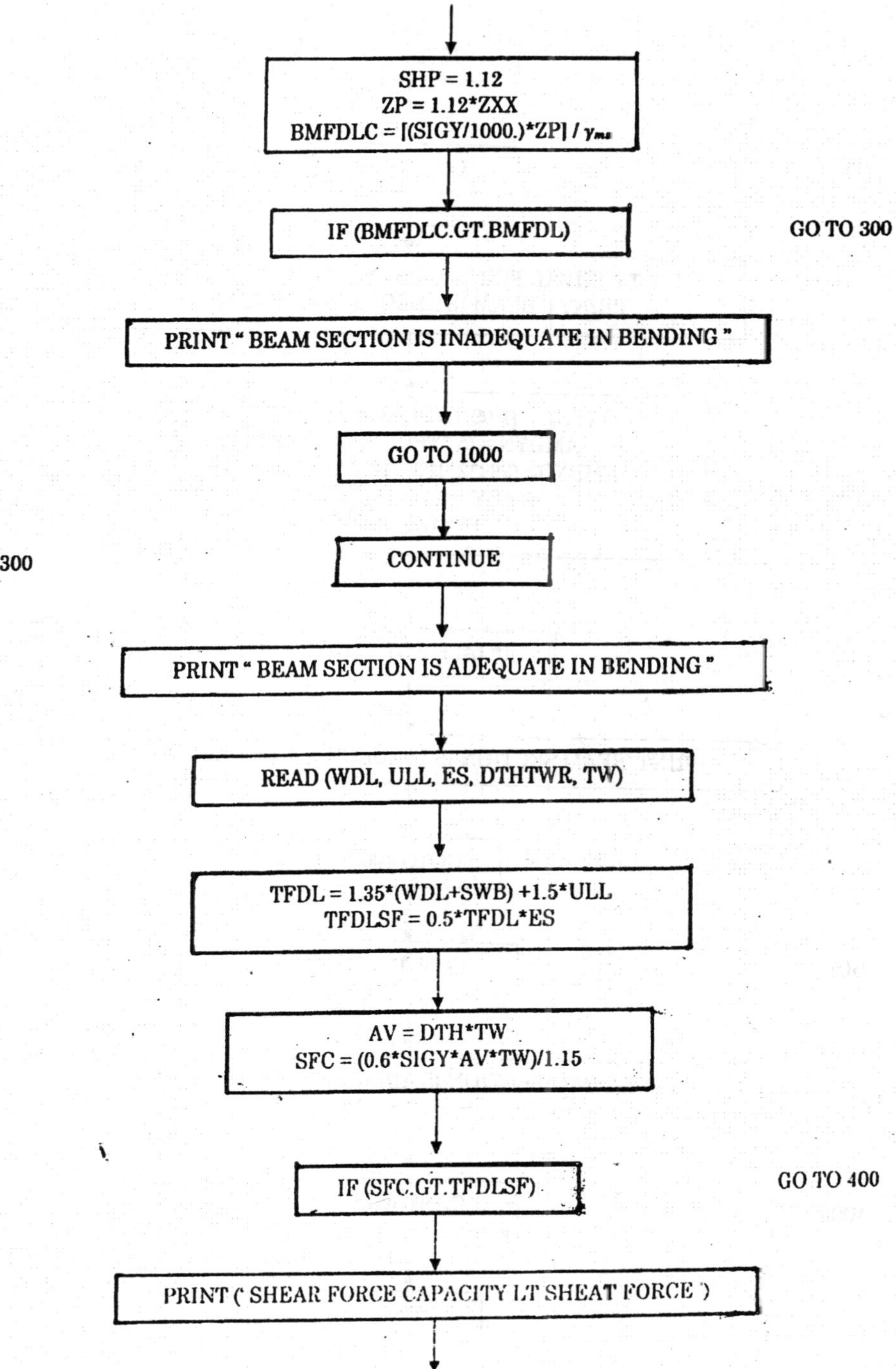

Fig. 11.18. Flow chart for BLSSB (contd..)

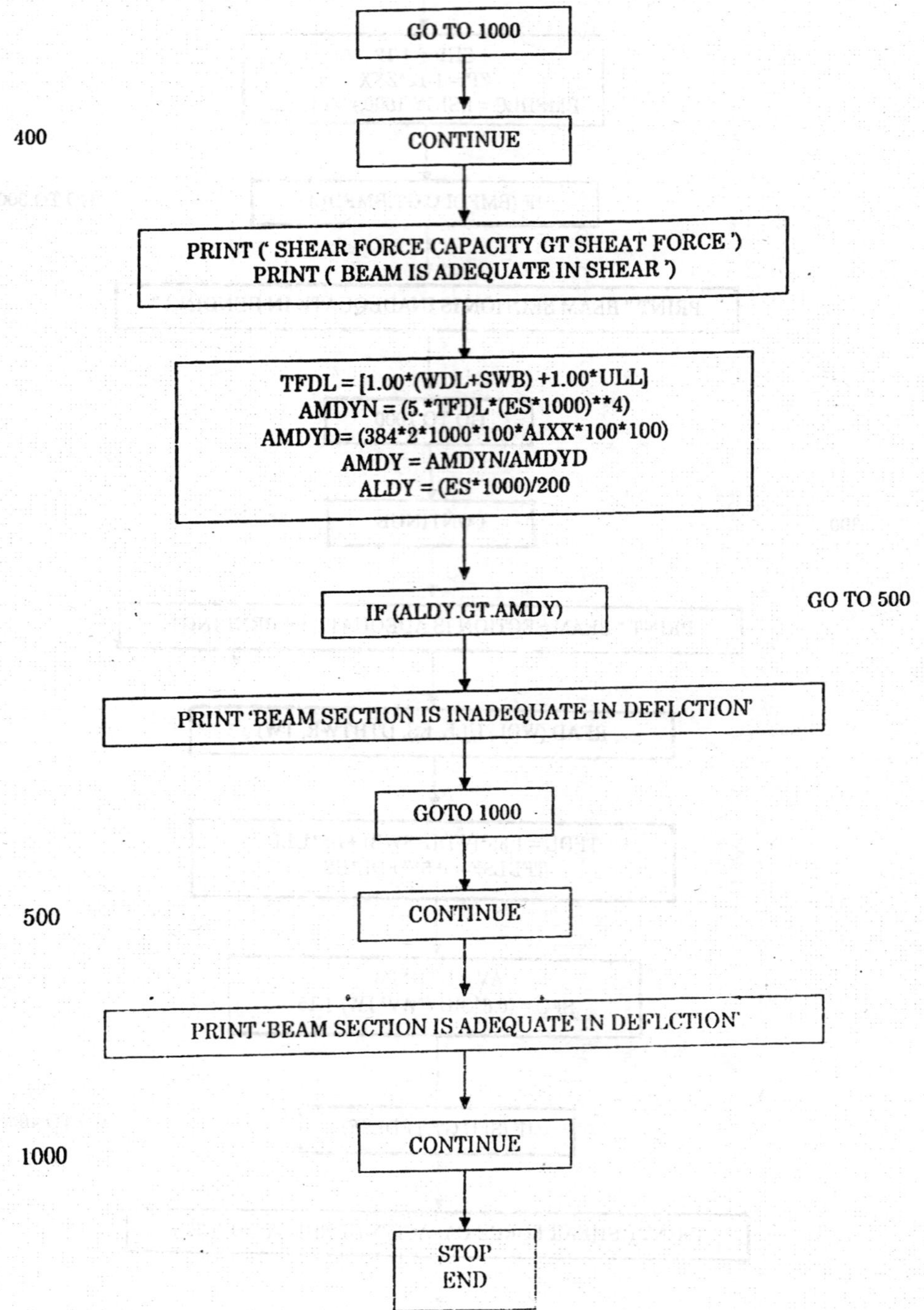

Fig. 11.18. Flow chart for BLSSB

The values of moment of inertias and section modulii about strong and weak axes are read.

The value of maximum bending moment is noted from previous programme, ALSSB. Shape factor, S = 1.12 for I - section is assumed. Plastic modulus of beam section, section (Z_p= 1.12 x elastic section modulus, Z_{xx}) is obtained.

The strength of beam section/resistance (i.e, capacityof beam section in bending is calculated (M_{fd} = s_y . Z_p) and compared with the value of maximum bending moment. M due to external moment. M due to external load should be less than M_{fd} (strength of the beam section. Accordingly it is declared that the beam section is adequate in bending.

The values of weight of dead load, live load,effective span, depth of the section and thickness of web are read from the previous programme ALSSB. Total factored design is calculated as (1.35 DL + 1.50 ULL). The clear depth of the section is determined (DTH - 2 x TF) shear strength/resistance/capacity, SFC = [(0.6 SIG Y x DTH)]. Thickness of web)/(Partial factor of safety for the strength of material] is found. It is compared with the maximum shear force due to external load. The shear force due to external load should be less than the shear strength of the beam section. Then, it is declared that the beam section is adequate in shear.

For SLS, total factored (design) load is obtained by considering partial safety factor for load as unity. Actual maximum deflection AMDY is determined and allowable deflection ALDY = [(ES x 1000)/200] are calculated. In case AMDY is less than ALDY, it is declared thatthe beam section is adequate in deflection.

In this manner, the adequacy of the beam section in bending, shear and deflection is checked. It is declared that the beam section is adequate.

11.15 INPUT AND OUTPUT STATEMENTS FOR COMPUTER PROGRAMMES

Input data are given by using READ statements and the values ofvarious variables are read. READ*, List of the variables statements reads the values of the variables. This statement remains free from the device used and format is not needed. In case, the user desires to read the variable as per his desired format, then READ format number statement is used and format with its label number is mentioned. The real variables are read in floating mode and integer values are read in integer mode. Some values remain standing and they do not change. Such variables are specified in the computer programmes. These values work as built-up data.

Similarly, outputs are noted by using PRINT statements. PRINT *, list of the variables writes the values of desred variables. The title, if given, may be written by using PRINT *, ``Title" desired. The values of variables may be printed in the desired formats by using statement PRINT format label number, list of the variables. In such cases, the values may be written in desired formats. The desired format is mentioned with its respective label number. Intermediate outputs are taken to know the correctness of programme and so also if needed. Final output shows the results.

Inputs and outputs of these three programmes ALSSB, BLSSB and CLUSB have been given in the subsequent sections.

INPUT-OUTPUT STATEMENTS FOR ALSSB (i.e, for analysis and design of laterally supported steel beams)

I. Input statements :

Following are the input data used in programme ALSSB.

1. READ *, ALTH, WTH, SIGY

   Length of room, ALTH =15.0 m

   Width of the room, WTH = 7.0 m

   Grade of the Steel, SIGY = 250 N/mm^2

2. READ *, UWST, UNPCC, UWLC, SWB, ULL

   Unit weight of the Stone Patties, UWST = 24 kN/m^3

   Unit weight of Plain Cement Concrete, UWPCC = 24 kN/m^3

   Unit weight of Lime Concrete, UWLC = 24 kN/m^3

   Self-weight of the Beam, SWB = 0.60 kN/m

   Live-load, ULL = 4 kN/m^2

3. READ *, TST, TPCC, TLC

   Thickness of Stone Patties, TST =120 mm

   Thickness of Plain Cement Concrete, TPCC = 20 mm

   Thickness of Lime Concrete, TLC = 120 mm

4. Specified input data

   Partial safety factor for loads (dead loads), PSFDL = 1.35

   Partial safety factor for loads (live-loads), PSFLL = 1.5

   Partial safety factor for strength of materials, PSFMS = 1.15

II. OUTPUT STATEMENTS

Following are output statements noted for ALSSB.

1. PRINT *, ``ALSSB" (Name of the file)
2. PRINT *, ``WST WPCC WLC"
3. PRINT *, WST, WPCC, WLC

Weight of the Stone Patties, WST (kN/m)

Weight of the Plain Cement Concrete.

WPCC (kN/m) Weight of the lime concrete, WLC (kN/m)

WST	WPCC	WLC
2.88	0.48	2.88

4. PRINT *, ``WDL ULL TFDL"
5. PRINT *, WDL, ULL, TFDL

Dead load, WDL (kN/m),

Live-load, ULL (kN/m),

Total factored design load,TFDL (kN/m)

WDL	ULL	TFDL
18.72	12.00	44.08

6. PRINT *, ``BMFDL ZPR ZER"
7. PRINT *, BMFDL, ZPR, ZER

Bending moment for factored design load, BMFDL, (kN-m)

Plastic modulus of beam section, ZPR, (cm^3)

Elastic modulus of beam section, ZER, (cm^3)

BMFDL	ZPR	ZER
293.641	1174.56	1021.361

INPUT-OUTPUT STATEMENTS FOR BLSSB, (i.e., to check adequacy of selected beam section for bending, shear and deflection for SLS)

1. Input statements :

Following are the input data used in programme BLSSB

1. READ *, BF, TF, DTH, TW, SIGY

    Breadth of the flange, BF = 170.0 mm

    Thickness of the flange, TF = 13.4 mm

    Depth of the section, DTH = 450 mm

    Thickness of the web, TW = 8.60 mm

    Yield stress of grade of the Steel, SIGY = 250 N/mm^2

2. READ *, AIXX, AIYY, ZXX, BMFDL

    Moment of inertia, AIXX = 27536.10 cm^4

    Moment of inertia, AIYY = 853.0 cm^4

    Moment of the section, ZXX = 1223.80 cm^3

    Bending moment for factored design load, BMFDL = 293.641 kN-m

    (It is read from output of ALSSB).

3. READ *, WDL, ULL, ES, DTH, TW

    Dead load, WDL = 18.72 kN/m

    Live-load, ULL = 12.00 kN/m

    Effective-span, ES = 7.30 m

    Depth of the section, DTH = 450 mm

    Thickness of the web, TW = 8.60 mm

II. Output statements of BLSSB :

Following are the output statements used and noted from BLSSB. Some of the intermediate outputs are noted to verify and to check the correctness of BLSSB.

1. PRINT *, ``WTHFLR, DTHTWR, EPSLON"
2. PRINT *, WTHFLR, DTHTWR, EPSLON

Width of flange/its thickness ratio, (WTHFLR)

Clear depth of the web/its thickness ratio. (DTHTWR).

EPSLON = SQRT (250/SIGY)$^{1/2}$

WTHFLR	DTHTWR	EPSLON
6.3432	49.2093	1.000000

3. PRINT *, ``SECTION OF THE BEAM IS PLASTIC"
4. PRINT *, ``SECTION OF THE BEAM IS COMPACT"
5. PRINT *, ``ZP, ZXX BMFDLC BMFDL"
6. PRINT *, ZP, BMFDLC, BMFDL

ZP (cm^3)	ZXX (cm^3)	BMFDLC (kN-m)	BMFDL (kN-m)
1370.565	1223.80	297.969	293.641

7. PRINT *, ``BEAM SECTION IS ADEQUATE IN BENDING"
8. PRINT *, ``SFC TFDLST"
9. PRINT *, SFC, TFDLSF

Shear force capacity, SFC (kN)

Total factored design load shear force, TEDLSF (kN)

SFC (kN)	TFDLSF (kN)
504.7827	160.8993

10. PRINT *, ``SHEAR FORCE CAPACITY GT SHEAR FORCE"
11. PRINT *, ``BEAM SECTIONS IN ADEQUATE IN SHEAR"
12. PRINT *, ``AMDY, ALDY, TFDL"
13. PRINT *, AMDY, ALDY, TFDL

Actual Maximum deflection, (AMDY (mm)

Allowable deflection, ALDY (mm)

Total factored design load, TFDL

Partial factor of safety for loads is equal to unity.

AMDY (mm)	ALDY (mm)	TFDL (kN)
19.05372	36.50000	31.32

PROBLEMS

11.1 A laterally supported beam LB 350, @ 0.495 kN/m has been placed on two supports. Determine the safe uniformly distributed load per metre length which can be placed over the beam for an effective span of 6.30 m.

11.2 Design a beam of 5 metres span carrying a uniformly distributed load of 30 kN per metre including self-weight and a concentrated load of 10 kN at the centre. The beam is simply supported at both ends.

11.3 A simply supported beam of 12 metres span is made of MB 600, @ 1.226 kN/m.

The associated properties of the beam are

Depth = 600 mm

Width of flange = 210 mm

Thickness of flange = 20.8 mm

Thickness of web = 12 mm

I_{xx} = 91813 x 10^4 mm^4, I_{yy} = 2651 x 10^4, Z_{xx} = 3060.4 x 10^3 mm^3

Determine that total uniformly distributed load that this beam can carry. The top flange is fully restrained laterally.

11.4 A beam carries a uniformly distributed load 44.00 kN/m. The effective span of beam is 6.40 m. The ends of beam are restrained against torsion and not restrained against lateral bending. Design a suitable section of a rolled steel beam.

11.5 Two LB 600, @ 0.995 kN/m joists at 300 mm with one 36 mm thick plate, 600 mm wise on each flange is used as a simply supported beam on a span on 12 m. Find the maximum uniformly distributed load per metre run which can be allowed on this built-up girder. Neglect rivet holes.

11.6 Design a simply supported plated rolled steel beam section to carry a uniformly distributed load 36.00 kN/m inclusive of self-weight of beam. The effective span of beam is 7.20 m, the depth of beam is limited to 500 mm. The beam is restrained against lateral deflection.

11.7 Design a beam of 6 m span carrying uniformly distributed load of 10 kN/m and two concentrated loads of 80 kN each at 2 m and 4 m, respectively from the left support. Only available rolled steel section is ISMB 350, @ 0.524 kN/m. The compression flange of the beam is laterally supported.

11.8 Design a beam of 5.4 m span carrying a uniformly distributed load of 30 kN/m and two concentrated loads of 50 kN each at third points. The depth of the section is restricted to 350 m. Assume the compression flange flange to have lateral restraint.

11.9 A freely supported beam spanning 12 m is made up of an I- section MB 500, @ 0.869 kN/m with two plates in each flange of 200 mm x 12 m each. Calculate the maximum uniformly distributed load the beam can carry if the compression flange is fully restrained. Assume the flange connections to be made with 18 mm ϕ rivets. Show calculations for the curtailment of the outermost plate in each flange.

11.10 A plated steel beam of span 12 metres effective is to carry a uniformly distributed load of 20 kN per metre inclusive of its own weight. Design a suitable plated beam including minimum spacing of rivets using any one of the three section listed below. Plate of 10 mm thickness is available in any width. The flange of beam should not be wider than 300 mm. Assume that 22 mm diameter rivets are used for the connection.

MB 500, @ 0.869 kN/m. LB 600, @ 0.995 kN/m, LB 500, @ 0.750 kN/m.

11.11 A simply supported beam of effective span 8 metres is required to carry. a.u.d.; 1. of 50 kN/m including self-weight over the entire span. Check whether the LB 450, @ 0.653 kN/m with the following properties is suitable for the purpose. (Area = 8314 mm²; b_f = 170 mm; t_f = 13.4 mm; t_w = 8.6 mm; I_{xx} = 27566.1 x 10^4 mm⁴). If not, design a safe section using 150 mm x 10 mm thick plates.

Design a suitable welded connection between top plate and flange of the joint.

Assume that the compression flange is adequately restrained laterally.

11.12 Design a jack arch roof for a hall of size 8 m x 4 m. The self-weight of the roof may be taken as 8.5 kN/m². The permissible compressive stress in the brick masonry is 0.3 N/mm².

11.13 A jack arch roof is to be constructed for a room measuring 5.5 m x 4.2 m. The live loaded acting is 4 kN/m². Design the roof assuming appropriate self-weight for the roof.

11.14 A beam of simple framed building is made up of LB 550, @ 0.863 kN/m (b_f = 190 mm, t_f = 15 mm, A = 10.97 mm², t_w = 9.9 mm) and carries a total load of 400 kN uniformly distributed over the entire span. It is connected to the flange of column at ends of section WB 600, @ 0.481 kN/m b_f = 200 mm, t_f = 10 mm, A = 6133 mm², t_w = 7.4 mm). Using 20 mm ϕ rivets, design a suitable connection for the joints. Draw a neat sketch of the joint.

11.15 A beam has an effective span of 10 m and carries a uniformly distributed load of 100 kN. Design the cross-section, of the beam if its depth is limited to 45 mm.

11.16 The cross-section of a built up beam consists of an LB 325 with a flange plate of 1200 mm x 10 mm provided at the top as well as at the bottom. It uses 18 mm diameter rivets. It is simply supported over an effective span of 6 m. Calculated the maximum uniformly distributed load the beam can carry, if the beam is laterally supported throughout its length along the compression flange. Consider bending moment only.

REFERENCES

1. David, A. Nethercot `Limit State Design of Structural Steel work' Third Edition (Based on revised BS : 5950 : Part 1, 2000), second Edition published in 1991 by Chap and Hall Reprinted 1994, 1995, 1996.

2. R.S. Narayanan and A.W. Beeby `Introduction to Design for Civil Engineers'. Simultaneously published in the USA and Canada by SPON PRESS (London and New York), First published in 2001 by Spon Press, 11 New Feter Lane, London, EC 4P 4 EE. 29 West 35th Street, New York, NY 10001.

3. Rangachari Narayanan, V. Kalyanaraman *et al.* (INSDAG) `Teaching Resource on structural steel Design', Vol. : 1, Organised by Institute for Steel Development And Growth (INSDAG) Delhi College of Engineering, Delhi, (June-July, 2001), Sponsored by Ministry of Steel, Government of India.

4. Draft for the approval for IS : 800 - (new revised code yet to be issued, circulated by Bureau of Indian Standards (BIS), Manak Bhawan, New Delhi.

5. A.B. Clark and S.H. Coverman - Structural Steel Work (Limit State Design), First Published in 1987, by Chapman and Hall Ltd. 11 New Fetter Lane, London, EC 4P 4EE, 29 West 35th Stress, New York 10001.

6. Adams, P.F. Kulak, G.L.and Gilmor, M. (1990), - ``Limit State Design in Structural Steel'', 4th Edition, Canadian Institute of Steel Construction.

7. British Standards Institution - ``BS : 5950, Part - 1, Structural Use of Steel-work in Buildings", British Standards Institution, London, 1985.

8. AISC-LRFD. -``Load and resistance Factor Design Specification for Structural Steel Buildings". American Institute of Steel Construction (AISC), Chicago, III, 1993.

9. British Standards Institutions - (1996, 1997, 1998) BS :6399 : Parts 1, 2 and 3. ``Design Loading for Buildings".

10. British Steel (1996). - ``The Prevention of Corrosion on structural Steel-work".

11. Ramchandra and Virendra Gehlot - ``Design of Steel Structures - 1" Twelfth Edition, 2007, published by M/s Scientific Publishers (India), 5-A, New Pali Road, Jodhpur.

12. Ramchandra and Virendra Gehlot -``Limit State Design of Concrete Structures" Third Edition, 2007, published by M/s Scientific Publishers (India), 5-A, New Pali Road, Jodhpur.

13. Eurocode 3. - ``Design of Steel Structures", British Standards Institute 1992.

14. Ramchandra- ``Stub Column Tests on Indian Standard H-Beams". A Dissertation (M.E.) submitted to University of Jodhpur, August, 1970.

15. Virendra Kumar Gehlot (M.E. Student) - A Seminar Report submitted to Head of Department of Structural Engineering, MBM. Engineering College, JNV University of Jodhpur, Jodhpur.

12

LIMIT STATE OF COLLAPSE IN SHEAR DESIGN OF STEEL BEAMS SB-II (SHEAR STRENGTH/RESISTANCE OF BEAMS)

12.1 INTRODUCTION

When the beams are subjected to loads, then, these are also required to transmit large shear forces either at supports or at concentrated loads. For simply supported beams, the shear force is maximum at the supports. The values of shear force at the concentrated loads also remain large. Due to shear force, the shear stresses are set up along with the bending stresses at all sections of the beams. The shear stress at any point of the cross-section is given by

$$\tau_v = \left(\frac{V.Q}{I.t}\right) \qquad (12.1)$$

where, τ_v is the shear stress and V is the shear force at any cross- section

Q = Static moment about the neutral axis of the portion of cross- sectional area beyond the location at which the stress is being determined.

I = Moment of inertia of the section about the neutral axis

t = Thickness of web (Width of section at which the stress is being determined).

The distribution of shear stresses for rectangular section of beam and I-beam section are shown in Fig. 12.1 (A) and Fig. 12.2 (B), respectively.

The maximum shear stress occurs at the neutral axis of the section.

The maximum shear stress in a member having regard to the distribution of stresses in conformity with the elastic behaviour of the member in flexure (bending) should not exceed the value of maximum permissible yield condition shear stress, τ_y found as follows :

$$\tau_y = \sigma_y / \sqrt{3} = 0.5773\, \sigma_y \approx 0.58\, \sigma_y \qquad (i)$$

where, σ_y is the yield stress of structural steel to be used.

It is to note that in the case of rolled beams and channels, the design shear is to be found as the average shear.

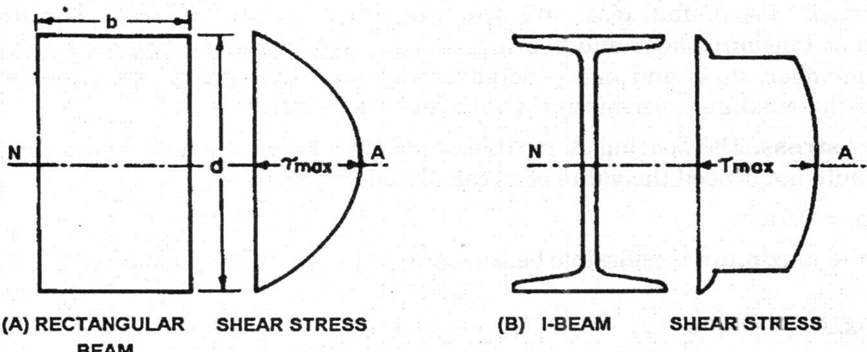

Fig. 12.1

The average shear stress for rolled beams or channels calculated by dividing the shear force at the cross-section by the gross-section of the web. *The gross-cross-section* of the web is defined as the depth of the beam or channel multiplied by its web thickness.

Average shear stress for rectangular beam is given by

$$\tau_v = \left(\frac{V}{b \times d}\right) \qquad (12.2)$$

Average shear stress for I-beam is given by

$$\tau_v = \left(\frac{V}{h \times t_w}\right) \qquad (12.3)$$

For rolled steel beams and channels, it is assumed that shear force is resisted by web only. The portion of shear resisted by the flanges is neglected.

When the beams are subjected to co-existent bending stresses (tension or compression) and shear stress, then the equivalent stress, $\sigma_{e.cal}$ is obtained from the following formula

$$\sigma_{e.cal} = \left[\sigma_{bt.cal}^2 + 3\,\tau_{vm.cal}^2\right]^{1/2} <= \sigma_u / \left(\sqrt{3}\,\gamma_{msu}\right) \qquad (iii)$$

or

$$\sigma_{e.cal} = \left[\sigma_{bc.cal}^2 + 3\,\tau_{vm.cal}^2\right]^{1/2} <= \sigma_u / \left(\sqrt{3}\,\gamma_{msu}\right) \qquad (iv)$$

The equivalent stress $\sigma_{e.cal}$ due to co-existent bending (tension or compression) and shear stresses should not exceed the maximum permissible equivalent stress, σ_e found as under

$$\sigma_e = 0.90\ \sigma_y \qquad (v)$$

When the bearing stress σ_p is combined with tensile or compressive bending and shear stresses under the most unfavourable conditions of loading, the equivalent stress $\sigma_{e.cal}$ obtained as below should not exceed

$$\sigma_e = 0.9\ \sigma_y \qquad (vi)$$

$$\sigma_{e.cal} = \left[\sigma_{bt}^2 + \sigma_{p.cal}^2 + \sigma_{bt.cal} \cdot \sigma_{p.cal} + 3\,\tau_{vm.cal}^2\right]^{1/2} \qquad (vii)$$

$$\sigma_{e.cal} = \left[\sigma_{bc}^2 + \sigma_{p.cal}^2 + \sigma_{bc.cal} \cdot \sigma_{p.cal} + 3\,\tau_{vm.cal}^2\right]^{1/2} \qquad (viii)$$

$\sigma_{bc.cal}$, $\sigma_{bt.cal}$, $\tau_{vm.cal}$ and $\sigma_{p.cal}$ are the numerical values of the co-existent bending (compression or tension), shear and bearing stresses. When the bending occurs about both the axes of the member, $\sigma_{bt.cal}$ and $\sigma_{bc.cal}$ should be taken as the sum of the two calculated fibre stresses, σ_e is the maximum permissible equivalent stress.

Bearing stress. The bearing stress in any part of a beam when calculated on the net area of contact should not exceed the value of σ_p calculated as below

$$\sigma_p = 0.75\ \sigma_y \qquad \text{(ix)}$$

where, σ_p is the maximum permissible bearing stress and σ_y is the yield stress.

Shear strength criterion

For practically all structural shapes commonly used in constructions, the shear resistance for the flexural members is normally determined on the basis of the webs only. Web- shear resistance is dependent on the depth of web-thickness of web ratio (h/tw). In case (h/tw) is small, the failure mode is **web yielding**. In case, (h/tw) is large, the failure mode is **web buckling**. To avoid the web shear failure the calculated shear stress, τ_v shall not exceed the allowable shear stress, τ_{va}.

12.2 SHEAR BEHAVIOUR OF STEEL BEAMS IN ELASTIC STAGE

In a simply supported steel I-beams subjected to maximum shear force at the supports, the shear force, V varies along the longitudinal axis of the beam with the bending moment, M, such that

$$V = (d_m/d_x) \qquad (12.5)$$

while the beam is in the elastic stage. The shear stresses, τ, which resists the external shear, V may be expressed as below :

$$\tau = (V.Q/I.t_w) \qquad (12.6)$$

where,

V = shear force at the section

I = Moment of inertia of the cross-section about the neutral axis

Q = Moment about the neutral axis of the area that is, beyond the fibre at which τ is calculated.

t_w = Thickness of web portion at which τ is being found.

The shear stresses in the elastic range are shown in Fig. 12.2. It is to note that a significant proportion of the shear force is resisted by the web and the distribution of the shear stress over the web area is nearly uniform. Therefore, for the purpose of the design, it may be assumed without much error that the average shear stress is

$$\tau_{av} = (V/d_w.t_w) \qquad (12.6)$$

where, d_w is the depth of the web and τ_w is the thickness of web.

The nominal shear yielding strength of the web is based on the Von Mises yield criterion, which states that for a steel beam, the width-thickness ratio is comparatively small, (so that, the buckling failure of the web does not occur the shear strength may be taken as

$$\tau_y = \left(\frac{1}{\sqrt{3}}.\sigma_y\right) = (0.577\sigma_y) = 0.60\sigma_y$$

where, σ_y is the yield stress of steel.

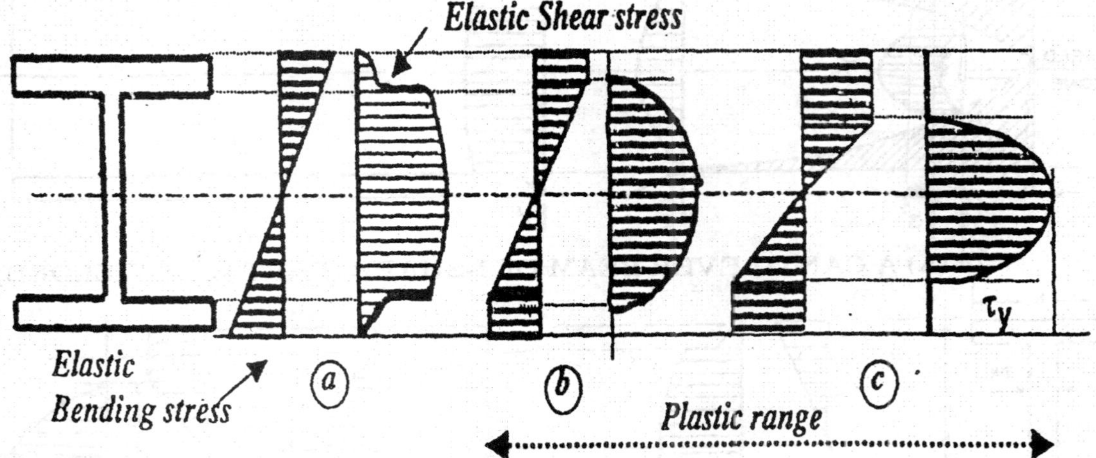

Fig. 12.2. Combined bending and shear in beams.

The shear capacity, (shear strength) of rolled steel I-beams may be obtained as under :

$$V_{bS} = (0.60\ \sigma_y\ d_w\ .\ t_w) \quad (12.7)$$

In case, the shear capacity is exceeded, the shear failure occurs by the excessive shear yielding of the gross-area of the web. The shear yielding occurs very rarely in the rolled steel beams.

12.3 SHEAR BEHAVIOUR OF STEEL BEAMS IN PLASTIC STAGE

A cantilever beam AB (with its fixed support A and free end B) of length, L carries a concentrated load, W at its fress end as shown in Fig. 12.3. The cross-section of the beam is an I-section. It bends about the major axis. The value of shear force, is V the distribution of the shear stress at the various section 1-1', 2-2', and 3-3' along the beam has also been shown in Fig. 12.3. It is assumed that the gradually increased shear force may reach the limiting value, V_u corresponds to the development of the yield stress in shear over the full area of web, ($A_w = d_w t_w$). For the Von-Mises yield criterion, the value of V_u would not exceed.

$$V = \left(\left(\frac{1}{\sqrt{3}}\right).\sigma_y.d_w.t_w\right) \quad \text{(i)}$$

$$= (0.57735\ \sigma_y\ .\ d_w\ .\ t_w) = (0.6\ \sigma_y\ d_w\ .\ t_w) \quad (12.8)$$

At the section 3-3', the shear stress reaches its yield value at the centre line. There are two following two possibilities of pre-mature failure due to the pressure of shear.

(1) In general yield may occur in the presence of higher shear to moment ratios (sections 1-1' and 2-2') Fig. 12.3. The maximum possible shear according to the Von-mises criterion as given by Eq. 12.8.

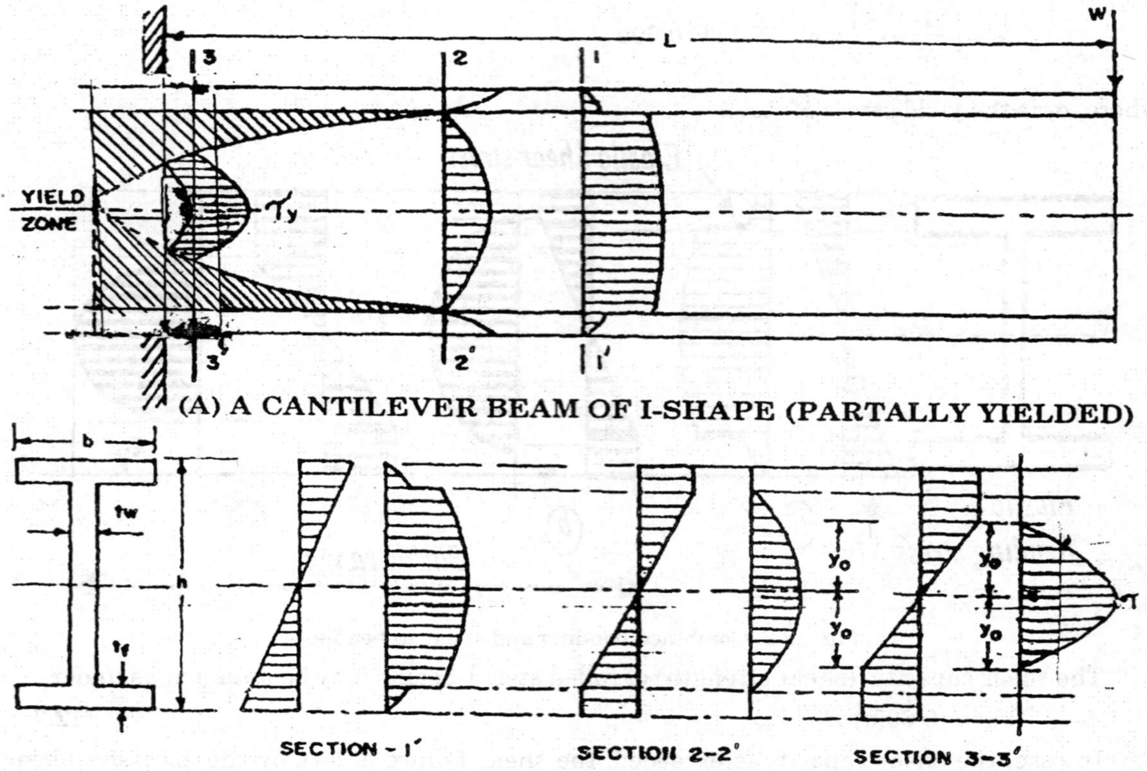

(A) A CANTILEVER BEAM OF I-SHAPE (PARTALLY YIELDED)

(B) DISTRIBUTION OF STRESS IN BENDING AND SHEAR

Fig. 12.3

(2) It is seen approximately, by considering the distribution of shear and bending stresses at section 3-3'. The maximum shear stress for a parabolic distribution is 1.5 times the average shear. That is,

$$\tau_{max} = \frac{3}{2}\left(\frac{V}{t_w.2y_o}\right) = \frac{3}{4}\left(\frac{V}{y_o t_w}\right) \tag{12.9}$$

where, y_o is the distance from the mid-height to the elastic plastic boundary. The maximum value of the shear stress is taken as the yield value.

$$\tau_y = \left[\left(\frac{1}{\sqrt{3}}\right).\sigma_y\right]$$

$$V = \left[\frac{\sigma_y}{\sqrt{3}}.d_w.t_w\right] \tag{12.10}$$

This expression may be written as

$$\left(\frac{\sigma_y}{\sqrt{3}}\right) = \frac{3}{4}.\left(\frac{M_{ps}}{L.y_o.t_w}\right) \tag{12.11}$$

where, M_{ps} denotes the value of moment at which the limiting condition of yield due to shear force at the centre line is found. Then, M_{ps} may also be obtained directly from the bending stress distribution at section 3-3', Fig. 12.3

$$M_{ps} = \left[M_p - \sigma_y . \frac{t_w . y_o^2}{3} \right] \tag{12.12}$$

In case, y_o is eliminated between these two expressions and in case, both the sides of the resulting expressions are divided by (M_y – s_y . Z_e), then

$$\frac{9}{16}\left[\frac{Z_p}{L^2 . t_w}\right]\left[\frac{M_{ps}}{M_p}\right] + \left[\frac{M_{ps}}{M_p}\right] = 1.00 \tag{12.13}$$

This shows the influence of the shear on the plastic moment.

Sections with webs susceptible to shear buckling before yielding

When the flanges are plastic, compact or semi-compact but the web is susceptible to shear buckling before yielding [(d/t_w) is less than or equal to 67 ε], the design bending strength shall be calculated using one of the following two methods.

(1) The bending moment and axial force acting on the section may be assumed to be resisted by flanges only and the web is designed only to resist shear force.

(2) The bending moment and axial force acting on the section may be assumed to be resisted by the whole section.

In such a case, the web shall be designed for combined shear and shall be designed and normal stresses using simple elastic theory in case of semi-compact webs and simple plastic theory, in case of compact, and plastic webs.

12.4 EFFECTS OF HOLES IN TENSION ZONE

The design bending strength is influenced by the presence of holes in the tension zone.

The effect of holes in the tension flange, on the design bending strength need not be considered if

[(A_{nf}/A_{gf})] is more than or equal to

$$\left[\left(\sigma_y / \sigma_u\right)\left(\gamma_{msu} / \gamma_{msy}\right)/0.9\right] \tag{12.14}$$

where,

(i) (A_{nf}/A_{gf}) is the ratio of net area to the gross area of the tension flange.

(ii) (σ_y/σ_u) is the ration of yield stress and the ultimate stress of the material, and

(iii) ($\gamma_{msu}/\gamma_{msy}$) is the ratio of partial safety factors against ultimate to yield stress (1.25/1.15).

When the (A_{nf}/A_{gf}) does not satisfy above requirement, the reduced effective flange area, A_{ef} satisfying the above equation may be taken as the effective flange area in tension, instead of A_{gf}

(2) The effect of holes in the tension region of the web on the design flexural strength need not be consider, if the limit given in about is satisfied for the complete tension zone of the cross-section comprising the tension flange and the tension region of the web.

(3) Fastener holes in the compression zone of the cross-section need not be considered in the design bending strength calculation except for oversize and slot holes or holes without fasteners.

12.5 EFFECTS OF SHEAR LAG

The effects of shear lag may be neglected provided.

(i) For outstand elements (supported along one edge)

b_o is less than or equal to $L_o/20$ and

(ii) For internal elements (supported along two edges)

b_i is less than or equal to $L/10$ where, b_o is the width of the flange with outstand and b_i is the width of the flange as an internal element.

Where these limits are exceeded the effective width of flange for the design may be calculated using specialist literature, or conservatively taken as the value satisfying above given limit.

12.6 DESIGN (FACTORED) SHEAR RESISTANCE / STRENGTH OF BEAMS

The (design) factored shear force, V in a beam due to external actions (loads) shall satisfy the following requirement.

V is less than or equal to V_{fd}

where, V_{fd} is the factored (design) strength. It is obtained as below :

$$V_{fd} = (V_n / \gamma_{msy}) \text{, where} \quad (12.15)$$

γ_{msy} = Partial safety factor for the material strength of steel (1.15)

Nominal shear strength of a cross-section is controlled by plastic shear resistance, V_p or the strength of the web in the post buckling range. Nominal plastic shear resistance under pure shear is given by ($V_n = V_p$)

$$V_p = \left(\frac{1}{\sqrt{3}} . (A_v . \sigma_{yw})\right) = \left(\frac{1}{1.732} A_v . \sigma_{yw}\right) \quad (12.16)$$

V_P = (0.5773 A_w . σ_{yw}) adopted as

$$V_p = (0.60\, \sigma_{yw} . A_v) \text{, where} \quad (12.17)$$

A_v = Area in shear

$\sigma_{y.w}$ = Yield stress in the web

The shear capacity of rolled steel beams, V_p is found as under

$$V_p = 0.60\, \sigma_{y.w} (d_w . t_w) \quad (12.18)$$

The area in shear may be calculated as below in the follow Table 12.1

TABLE 12.1. AREAS IN SHEAR FOR DIFFERENT SECTIONS

S.No.	Sections	Axis	
1.	I-section	Major axis	Minor axis
2.	Channel section		
	Hot rolled	$(h.t_w)$	$(2b.t_f)$
	Welded	$(d.t_w)$	$(2b.t_f)$
3.	Rectangular hollow sections of uniform thickness	Loaded parallel to depth (h)	Loaded parallel to width (b)
	-	A.h/(b+h)	A.b/(b+h)
4.	Circular hollow tubes of uniform thickness	$(2A/\pi)$	$(2\ A/\pi)$
5.	Plates and solid bars	A	A

where,

A = Area of cross-section

d = Clear depth of web between flanges

t_f = Thickness of flange

t_w = Thickness of Web

b = Overall breadth of the tubular section

(Flange with of I-section)

h = Overall depth of the section

Note : Fasteners holes need not be a considered for in plastic design shear strength calculations provided that A_{vn}

A_{vn} is more than or equal to

$$(\sigma_y/\sigma_u)\ .\ (\gamma_{msu}/\gamma_{msy})/0.9 \qquad (12.19)$$

In case, A_{vm} does not satisfy above condition the effective shear area may be taken as satisfying above limit $[(\sigma_y/\sigma_u).\ \gamma_{msu}/\gamma_{msy}]/0.9$. Block shear failure criteria may be verified at the end connections.

12.7 RESISTANCE TO SHEAR BUCKLING IN LATERALLY SUPPORTED BEAMS

Resistance to shear buckling shall be verified as specified when

For a web without stiffeners,

$$\text{(i)}\ (d/t_w) > 67\ \varepsilon\ \text{(epsilon)} \qquad (12.20)$$

For a web with stiffeners,

$$\text{(ii)}\ (d/t_w) > 67 \in \left(\frac{k_v}{5.35}\right)^{1/2} \qquad (12.21)$$

where, k_v is **shear buckling** coefficient and

Yield stress ratio

$$\in \;= (250/s_y)^{1/2} \tag{12.22}$$

12.8 DESIGN METHODS CONSIDERING SHEAR BUCKLING

The nominal shear strength, V_n of webs with or without intermediate stiffeners as controlled by **shear buckling** may be evaluated using one of the following methods

1. Simple post-critical method

The simple post critical method, based on **the shear buckling strength** may be used for webs of I-section beams, with or without intermediate transverse stiffeners at the supports. The nominal shear strength is given by

$$V_n = V_{cr} = A_v \,.\, \tau_b \tag{12.23}$$

where, V_{cr} is the shear force corresponding to web buckling

τ_b = shear stress corresponding to web buckling calculated as follows :

1. When λ_w is less than or equal to 0.8

$$\tau_b = (\sigma_{y.w}/\sqrt{3}\,) \tag{12.24}$$

2. When $0.8 < \lambda_w < 1.2$

$$\tau_b = [1-0.8\,(\lambda_w - 0.8)]\left(\frac{\sigma_{yw}}{\sqrt{3}}\right) \tag{12.25}$$

3. When λ_w is greater than or equal to 1 (unity)

$$\tau_b = \left(\frac{\sigma_{yw}}{\sqrt{3}\,.\,\lambda_w^2}\right) \tag{12.26}$$

where, λ_w is non-dimensional web slenderness ratio for the shear buckling stress given by

$$\lambda_w = \left(\sigma_{yw}/\sqrt{3}\,.\,\tau_{cr.e}\right) \tag{12.27}$$

$$\text{where,}\quad \tau_{cr.e} = \frac{K_y\,.\,\pi^2\,.E}{12\left(1-u^2\right)[d/t_w]^2} \tag{12.28}$$

μ = poisson's ratio and

k_y = 5.35 when the transverse

stiffeners are provided only at supports (i.e, only bearing stiffeners are provided at supports)

$k_y = 4.0 + 5.35\,/\,(c/d)^2$ for $c/d < 1.00$

(12.29)

$k_y = 5.35 + 4.0\,/\,(c/d)^2$ for $c/d \geq 1.00$ (12.30)

where, c, d are the spacing of the transverse stiffeners and depth of the web, respectively.

2. Tension field method

This method, based on the post shear buckling strength, may be used for webs with intermediate transverse stiffeners in addition to the transverse stiffeners at supports in chapter.

This method has been described for the design of plate girders.

Example 12.1 In Example 11.2, three trial sections have been proposed and these were found to be adequate for their strength in bending.

Check the adequacy of beam section ISMB 450, @ 72.4 kg/m. Find the shear strength/ resistance and check its adequacy in shear

Use limit state method of design.

Step 1. From Example 11.2, total design (factored) load is 44.082 kN/m. Effective span = 7.30 m

Design (Factored) shear force due to external load

$$S_{fd} = \left(\frac{44.082 \times 7.30}{2}\right) = 160.90 \text{ kN}$$

Shear is resisted by the web.

Overall depth of beam = 450 mm

Thickness of web, t_w = 9.4 mm

Area of web in shear

$$A_v = (D \times t_w) = (450 \times 9.4) = 4230 \text{ mm}^2$$

Step 2 : Shear strength/resistance of the beam section

$$\left(\sigma_{yw} = \sigma_y / \sqrt{3}\right), \sigma_y = 250 \text{ N/mm}^2$$

$$S_{vd} = \left(\frac{\sigma_{yw} \times A_v}{\gamma_{msy}}\right) = \left(\frac{\sigma_y}{\sqrt{3}} \times \frac{A_v}{\gamma_{msy}}\right)$$

$$= \left(\frac{0.6 \times 250 \times 4230}{1.15 \times 1000}\right) = 551.739 \text{ kN}$$

Note : The factored shear force S_{fd} 160.90 kN is less than the design shear strength/ resistance. 551.739 kN. The requirement for the shear ($S_{fd} < S_{vd}$) is satisfied. The beam section is adequate for shear requirement.

Example 12.2, In Example 11.2, find the shear strength/resistance of the beam section, ISLB 450, @ 65.3 kg/m and check the beam section for its adequacy for shear.

Step 1. From Example 11.2, total design (factored) load is 44.082 kN/m Design (factored) **shear force due to external load**

Effective span = 7.30 m

$$V_{fd} = \left(\frac{44.082 \times 7.30}{2}\right) = 160.90 \text{ kN}$$

Shear is resisted by the web. Overall depth of beam beam, D = 450 mm.

Thickness of web, t_w = 8.6 mm

Area of web in shear

$$A_v = (D \times t_w) = (450 \times 8.6) = 3870 \text{ mm}^2$$

Step 2 : Shear strength/resistance of the beam section

$$\sigma_y = 250 \text{ N/mm}^2,\ \sigma_{yw} = (\sigma_y / \sqrt{3})$$

$$S_{vd} = \left(\frac{\sigma_{yw} \times A_v}{\gamma_{msy}}\right) = \left(\frac{\sigma_y}{\sqrt{3}} \times \frac{A_v}{\gamma_{msy}}\right)$$

$$= \left(\frac{0.6 \times 250 \times 3870}{1.15 \times 1000}\right) = 504.78 \text{ kN}$$

Note : The factored shear force S_{fd} = 160.90 kN is less than the design shear strength/ resistance 504.78 kN. The requirement for the shear ($S_{fd} < S_{vd}$) is satisfied. The beam section is adequate for shear requirement.

Example 12.3. In Example 11.2, find the shear strength/resistance of the beam section, ISWB 400, @ 0.660 kN/m and check the beam for its adequacy for shear.

Use limit state method of design.

Step 1. From Example 11.2, **total design (factored) load is 44.082 kN/m.**

Effective span = 7.30

Design (factored shear force due to external load)

$$S_{fd} = \left(\frac{44.082 \times 7.30}{2}\right) = 160.90 \text{ kN}$$

shear is resisted by the web. Overall depth of beam section, D = 400 mm.

Thickness of web, t_w = 8.60 mm

Area of web $A_v = (D \times t_w)$

$$A_v = (400 \times 8.6) = 3440 \text{ mm}^2$$

Step 2 : Shear strength/resistance of the beam section

$$\sigma_y = 250 \text{ N/mm}^2,\ \sigma_{yw} = (\sigma_y / \sqrt{3})$$

$$S_{vd} = \left(\frac{\sigma_{yw} \cdot A_v}{\gamma_{msy}}\right) = \left(\frac{\sigma_y}{\sqrt{3}} \times \frac{A_v}{\gamma_{msy}}\right)$$

$$= \left(\frac{0.6 \times 250 \times 3440}{1.15 \times 1000}\right) = 440.69 \text{ kN}$$

Note : The factored shear force S_{fd} = 160.90 kN is less than the design shear strength/ resistance, S_{vd} = 440.69 kN. The requirement for the shear is satisfied. The beam section is adequate for shear requirement.

PROBLEMS

12.1. In Problem 11.1, design the beam in shear.

12.2. In Problem 11.2, design the beam in shear.

12.3. In Problem 11.3, design the beam in shear.

12.4. In Problem 11.4, design the beam in shear.

12.5. In Problem 11.5, design the beam in shear.

12.6. In Problem 11.6, design the beam in shear.

12.7. In Problem 11.7, design the beam in shear.

12.8. In Problem 11.8, design the beam in shear.

12.9. In Problem 11.9, design the beam in shear.

12.10. In Problem 11.10, design the beam in shear.

12.11. In Problem 11.11, design the beam in shear.

12.12. In Problem 11.12, design the beam in shear.

12.13. In Problem 11.13, design the beam in shear.

12.14. In Problem 11.14, design the beam in shear.

12.15. In Problem 11.15, design the beam in shear.

12.16. In Problem 11.16, design the beam in shear.

REFERENCES

1. David, A. Nethercot `Limit State Design of Structural Steel work' Third Edition (Based on revised BS : 5950 : Part 1, 2000), second Edition published in 1991 by Chap and Hall Reprinted 1994, 1995, 1996.
2. R.S. Narayanan and A.W. Beeby `Introduction to Design for Civil Engineers'. Simultaneously published in the USA and Canada by SPON PRESS (London and New York), First published in 2001 by Spon Press, 11 New Feter Lane, London, EC 4P 4 EE. 29 West 35th Street, New York, NY 10001.
3. Rangachari Narayanan, V. Kalyanaraman *et al.* (INSDAG) `Teaching Resource on structural steel Design', Vol. : 1, Organised by Institute for Steel Development And Growth (INSDAG) Delhi College of Engineering, Delhi, (June-July, 2001), Sponsored by Ministry of Steel, Government of India.
4. Draft for the approval for IS : 800 - (new revised code yet to be issued, circulated by Bureau of Indian Standards (BIS), Manak Bhawan, New Delhi.
5. A.B. Clark and S.H. Coverman - Structural Steel Work (Limit State Design), First Published in 1987, by Chapman and Hall Ltd. 11 New Fetter Lane, London, EC 4P 4EE, 29 West 35th Stress, New York 10001.
6. Adams, P.F. Kulak, G.L. and Gilmor, M. (1990), - ``Limit State Design in Structural Steel", 4th Edition, Canadian Institute of Steel Construction.
7. British Standards Institution - ``BS : 5950, Part - 1, Structural Use of Steel-work in Buildings", British Standards Institution, London, 1985.

8. AISC-LRFD. -``Load and resistance Factor Design Specification for Structural Steel Buildings". American Institute of Steel Construction (AISC), Chicago, III, 1993.

9. British Standards Institutions - (1996, 1997, 1998) BS : 6399 : Parts 1, 2 and 3. ``Design Loading for Buildings".

10. British Steel (1996). - ``The Prevention of Corrosion on structural Steel-work".

11. Ramchandra and Virendra Gehlot - ``Design of Steel Structures - 1" Twelfth Edition, 2007, published by M/s Scientific Publishers (India), 5-A, New Pali Road, Jodhpur.

12. Ramchandra and Virendra Gehlot -``Limit State Design of Concrete Structures" Third Edition, 2007, published by M/s Scientific Publishers (India), 5-A, New Pali Road, Jodhpur.

13. Euro code 3. - ``Design of Steel Structures", British Standards Institute 1992.

14. Ramchandra - ``Stub Column Tests on Indian Standard H-Beams". A Dissertation (M.E.) submitted to University of Jodhpur, August, 1970.

15. Virendra Kumar Gehlot (M.E. Student) - A Seminar Report submitted to Head of Department of Structural Engineering, MBM. Engineering College, JNV University of Jodhpur, Jodhpur.

13

LIMIT STATE OF FAILURE DUE TO WEB BUCKLING AND WEB CRIPPLING, DESIGN OF STEEL BEAMS - (SBIII)

13.1 INTRODUCTION

The steel beams may fail due to web-buckling and web crippling. While the steel beams are designed, then, there are checked and it is seen that the beam sections are adequate and do not fail in web buckling and web crippling. These phenomena have been described in the subsequent sections.

13.2 WEB BUCKLING OF STEEL BEAMS

The failure of web in which the web vertically above the bearing plate and the lower flange at reaction or below the concentration load (bearing plate and the upper flange) is subjected to column action and tends to or buckle under the load is known as **web buckling.** It is out of plane distortion of web of a beam. It is also called as **vertical web-buckling** as shown in Fig. 13.1. Compressive stresses produced in the web either at the support or under the load become high. It results due to combination of large value of ratio (h_1/t_w) and the bending stress. The unbraced length of the compression flange may also contribute to web buckle depending upon the restraint of the flanges against lateral deflection and rotation. The web may buckle depending on the type of support, the top flanges receive from the surrounding construction.

The bottom flange is restrained against lateral deflection and rotation in all the cases. The restraint conditions. for the top flange are as below

(1) Restraint against laeral deflection and rotation as shown in Fig. 13.1 (A).

(2) Restraint against lateral deflection and not against rotation a shown in Fig. 13.1 (B).

(3) Restraint against rotation but not against lateral deflection as shown in Fig. 13.1 (C), and

(4) Restraint against neither of lateral deflection nor rotation as shown n Fig. 13.1 (D).

The bearing stiffeners are provided at the points of concentrated loads and the points of supports where the concentrated, reaction exceeds the value (σ_{ac} . B. t_w), where,

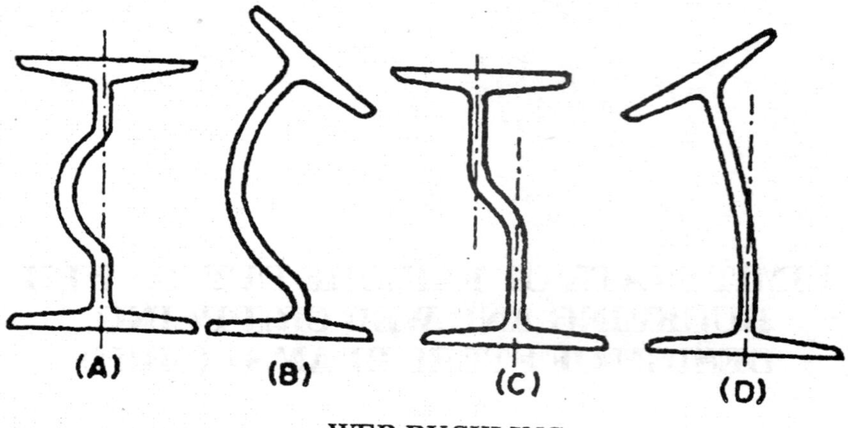

WEB BUCKLING

Fig. 13.1

σ_{ac} = Axial stress in the column for the slenderness ratio $\left(\sqrt{3}\, d_1 / t_w\right)$

$d_1 = (D - 2 h_2)$

t_w = Thickness of web

h_2 = Depth h at the roof of the flange

B = Length of the stiff portion of the bending plus the additional length by dispersion at 45° to the level of the neutral axis, *plus* the thickness of the seating angle, if any.

(The stiff portion of a bearing is that length which cannot deform appreciably in bending and shall not be taken more than half the depth for the simply supported beams and the full depth of beams continuous over bearing).

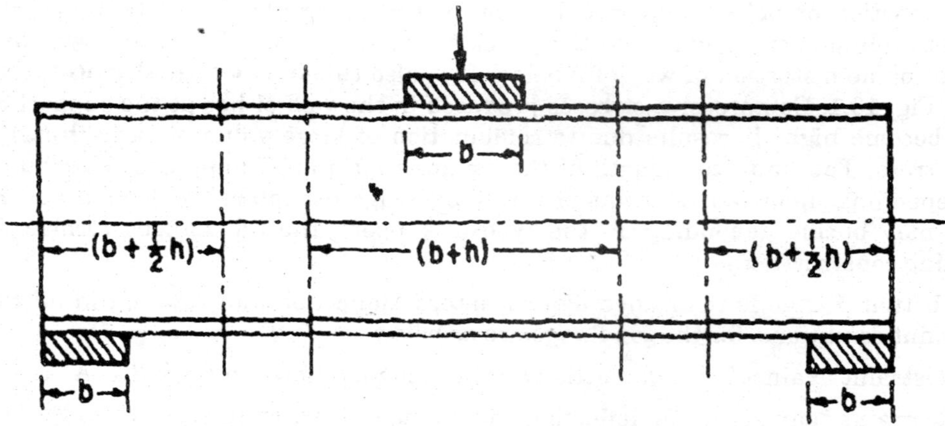

LENGTH OF BEARING FOR A CONCENTRATED LOAD

Fig. 13.2

For the concentrated load at any intermediate position, B is the length of bearing plate under the load *plus* depth of beam plus thickness of flange plates, if any. [e.g, in plated beams (i.e, B = (b + h)] as shown in Fig. 13.2.

where,

d_1 = Clear depth of web between root fillet

$d_1 = (h - 2h_2)$ and

h_2 = Depth of root of the fillet from the outer surface of the flanges.

The slenderness ratio for the portion of web acting as a column when the two flanges as restrained against lateral displacement and rotation may be found as follows :

$$\text{Slenderness ratio} = \left(\frac{\text{Effectivel length}}{\text{Minimum radius}}\right) = \left(\frac{0.5 d_1}{r_y}\right)$$

It is taken as $(0.7\ d_1/r_y)$

$$\left(\frac{d_1/2}{r_y}\right) = \left(\frac{(d_1/2)}{(I_{yy}/A)^{1/2}}\right) = \frac{d_1/2}{\left[\left(\frac{1}{12}\ B.\,t_w^3/B_1\,t_w\right)\right]^{1/2}} = \left(\frac{d_1\sqrt{3}}{t_w}\right) \qquad (13.1)$$

The slenderness ratio for the flange of a web acting as a column, when the top flange is held in position only while the bottom flange is restrained against lateral displacement and rotation both.

$$\text{Slenderness ratio} = \frac{d_1/2}{\left[\left(\frac{1}{12}\ Bt_w^3/B.t_w\right)\right]^{1/2}} = \left[\frac{d_1}{t_w}\right]\sqrt{6}$$

$$\lambda = \left[\frac{L_e}{r_y}\right] = \left[0.7\ d\ \frac{2\sqrt{3}}{t_w}\right] = 2.5\ (d/t) \qquad (13.2)$$

13.3 WEB CRIPPLING OF STEEL BEAMS

The webs of rolled steel beams are usually thin. The web is relatively weak when placed under direct compression.

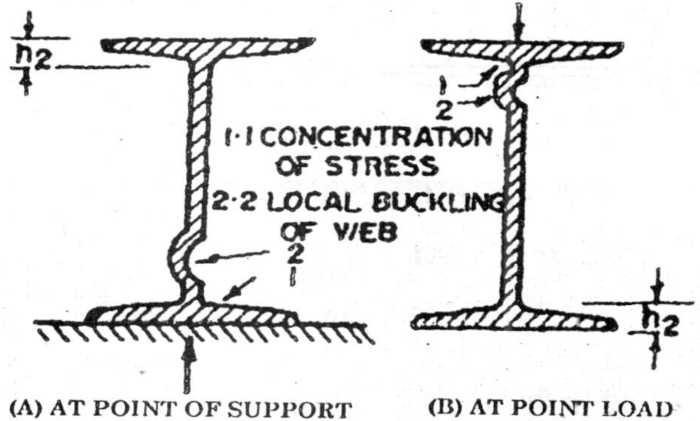

Fig. 13.3

In rolled steel beams, the failure of the web may occur at concentrated loads and at supports, where, there is reaction. **The failure of web in-direct crushing under concentrated loads in known as web crippling or web crippling.**

The web crippling may occur it (b + 2k) distance (Fig. 13.3) used to deliver load from the web to the flanges is too small.

The web crippling may also occur if the uniformly load on the flanges is too large for the thickness of web.

The web crippling measn the concentration of the stress due to bottle neck condition at the web toe of the fillet undr or over heavy concentration of the load. The web crippling is the localized bearing stress caused by the transmission of compression from the comparatively wide flange to the narrow web. The crippling may occur by failure in bearing of the metal at the web toe of the fillet and the tendency of the flange and web to fold over on each other at the plane.

In the web crippling, local buckling of web occurs immediately adjacent to a concentration of stress as shown in Fig. 13.3.

For safety against web crippling, adequate bearing length along the beam may be provided so that the bearing stress, $\sigma_b = 0.75\ \sigma_y$ (0.75 times yield stress). As per IS : 800 - 1962, the dispersion of the load or reaction through the flange to the web occurs uniformly at an angle of 30° to the horizontal as shown in Fig. 13.4.

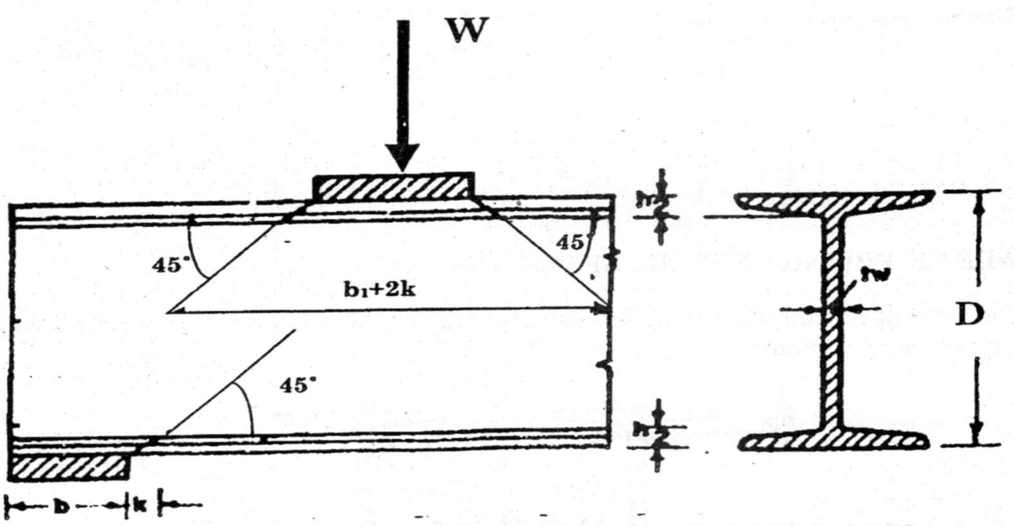

DISPERSION OF CONCENTRATED LOADS AND REACTIONS FOR EVALUTING WEB BUCKLING

Fig. 13.4

In IS : 800-2007, in case of web crippling the dispersion through the flange to the web connection at a slope of 1 : 2.5 to the plane of flange (k_z = 1.5 d) as shown in Fig. 13.4.

As earlier, the crippling strength of the web at support is calculated as below

$$P_{crippling} = [(b + k_2) \cdot t_w\ \sigma_{yw}]$$

where, σ_{yw} is the design yield strength of the web.

At an interior point, where the concentrated load is acting, the crippling strength is given by

$$P_{crippling} = [(b + 2k) . t_w . \sigma_{yw}]$$

Note : A beam, that is safe in web crippling will usually safe in web buckling as well

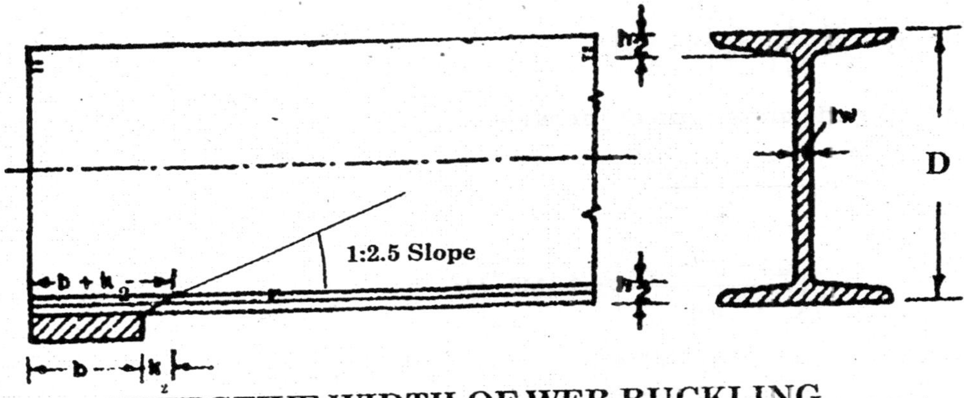

EFFECTIVE WIDTH OF WEB BUCKLING

Fig. 13.5

The web crippling and the web buckling may be avoided by adopting increased thickness of web or by adopting the increased length of bearing plates.

The web buckling is controlled by limiting either (d_1/t_w) ratio or the stress that can be used for the given (d_1/t_w) ratio. The web buckling may be prevented by stiffening the web. The bearing stiffeners may be provided under the heavy loads.

Example 13.1 : Check the adequacy of beam section ISMB 450, @ 0.724 kN/m for web-buckling and web-crippling.

Step 1. Geometrical properties of beam section.

Thickness of web, t_w = 9.4 mm

From ISI Handbook No. 1

Depth of section, D = 450 mm

Depth of flange upto web, h_2 = 35.40 mm

Effective length of the web, acting as a strut, L_e = 0.7 d, where d is the depth of the strut between the flanges

$$(D - 2 . h_2) = (450 - 2 \times 35.4) = 379.20 \text{ mm}$$

Sectional area of web = ($d \times t_w$)

Moment of inertia of web

$$I_y = \left(\frac{1}{12} x \; d \; x \; t_w^3\right)$$

Let r_y be the radius of gyration about the weak-axis-yy

$$r_y^2 = \left(\frac{Moment\ of\ inertia}{Sectional\ area\ of\ Web}\right) = \left(\frac{I_y}{A}\right)$$

$$= \left[\frac{1}{12} x \frac{d\ x\ t_w^3}{d\ x\ t_w}\right] = \left(\frac{t_w^2}{12}\right)$$

$$r_y = \left[\frac{1}{2\sqrt{3}} . t_w\right]$$

Step 2 : Slenderness ratio of the web

$$\left(\frac{L_e}{r_y}\right) = \left(\frac{0.7 x\ d\ x\ 2\sqrt{3}}{t_w}\right)$$

$$= \left(\frac{0.7 x\ d\ x\ 2\sqrt{3}}{t_w}\right) = 2.4248\ (d / t_w)$$

It is adopted as 2.5 (d/t_w). Therefore, the slenderness ratio

$$\left(\frac{2.5 d}{t_w}\right) = \left[\frac{2.5\ x\ 379.20}{9.4}\right] = 100.85$$

Step 3 : Compression stress (for axial buckling)

For hot-rolled steel section

Curve for strong axis and weak axis

a = 0.002

a = 0.0035, respectively

Corresponding design compressive stress for slenderness ratio 100.85

$\sigma_y = 151$ N/mm²

Note : Web has been considered as fixed ended column

Step 4 : Stiff bearing length

Width of the base plate at the end Near the point of reaction = 100 mm (assumed).

Angle of dispersion = 45°

Thickness of web = 9.40 mm

Half the depth of I-section

(i.e, upto neutral axis) $= \frac{D}{2}$

$$= \left(\frac{450}{2}\right) = 225\ \text{mm}$$

(B + D/2) = (100 + 225) = 325 mm

Step 5 : Load bearing capacity of web

$$P = \left(\frac{325 \times 9.4 \times 151}{1000}\right) = 461 \text{ kN}$$

$>> 160.90$ kN

Therefore, the web-buckling does not occur. ISMB 450, @ 0.724 kN/m is adequate against web buckling.

Step 6 : Adequacy of beam-section against web-crippling

From ISI Handbook No. 1, the radius at the root of flange

$r_1 = 15$ mm

Thickness of web, $t_w = 9.4$ mm

Thickness of flange, $t_f = 17.4$ mm

$(r_1 + t_f) = (15 + 17.4) = 32.4$ mm

Dispersion length 1 in 2.5

$= (25 \times 32.4) = 81$ mm

Step 7 : Resistance to crippling

Bearing length = (100 + 81) = 181 mm

$P_{crippling} = (\sigma_y \times \text{Bearing length} \times t_w)$

$$= \left(\frac{250 \times 181 \times 9.4}{1.15 \times 1000}\right) = 369.87 \text{ kN} >> 160.9 \text{ kN}$$

Therefore, web crippling does not occur. Beam section ISMB 450, @ 0.724 kN/m is adequate against web-crippling.

14

LIMIT STATE OF SERVICE ABILITY (STIFFNESS AND DEFLECTIONS) DESIGN OF STEEL BEAMS (SB-III)

14.1 INTRODUCTION

Limit State of serviceability is related to the criteria controlling the normal use. It is the limit state beyond which the service criteria in below (chapter 3) are no longer met (i) Deflection limit, (ii) Vibration limit, (iii) Durability consideration, and (iv) Fire resistance.

Deformations and deflections adversely influence the appearance or effective use of the structure. These may cause improper functioning of equipment or services or may cause damages to finishes and non-structural members.

The value of partial safety factor, γ_{fL} is adopted as **unity** for all the loads leading to serviceability limit state.

Deflections

When the beams are used on long spans, the ratio of the span length to the depth of beam section becomes large. The beams may then be referred as shallow beams, for which high strength steels are used.

There are number of methods available for calculating deflections on uniform and variable moment of inertia sections in statically determinate and indeterminate structures. In general, the maximum deflection in an elastic member may be expressed as under :

$$\Delta_{max} = \alpha.(WL^3/EI) \tag{14.1}$$

where, W is the total load over the span and L is the span length. E is Young's modulus of elasticity (2 x 10^5 N/mm^2) and I is the moment of inertia. For the simply supported beams, for uniformly distributed load over complete span Eq. 14.1 may be written as below.

$$\Delta_{max} = \left(\frac{5}{384}\right)\left(\frac{w.L^4}{EI}\right) \tag{14.2}$$

By substituting, $M = (WL^2/8)$ and $\sigma_b = (M.y_{max}/I)$ and $y_{max} = (h/2)$, Eq. 14.2 becomes

$$\Delta_{max} = \left(\frac{10\,\sigma_b\,L^2}{48\,E.h}\right) \tag{14.3A}$$

$$\left(\frac{\Delta_{max}}{L}\right) = \frac{5}{24}\left(\frac{\sigma_b}{E}\right)\left(\frac{L}{h}\right) \tag{14.3B}$$

Beams and girders supporting floors and roofs are proportioned with due regard to the deflection produced by the loads. In addition, live-load deflection (where plastered ceilings) are supported is limited to (L/360), (E = 2 x 10^5 N/mm^2)

$$\left(\frac{L}{h}\right) < \left(\frac{48 \times 2 \times 10^5}{10 \times 360\, \sigma_b}\right) = \left(\frac{2667}{\sigma_b}\right) \tag{14.4}$$

where, σ_b is the allowable stress in bending in compression.

Dead-load deflections may be commonly considered for during construction by either cambering (negative bending) or thickening the slab or floor top.

It is only the deflection, that occurs due to loads applied after construction is completed that may crack ceiling, partitions or walls Deflection restrictions may be specified to control the design.

The stiffness of a beam is a major consideration in the selection of a beam section. The allowable deflections of beams depend upon the purposes for which the beams are designed.

The large deflections of beams are undesirable for the following reasons :

(i) When the loads are primarily due to human occupants especially in the case of public meeting places, large deflections result in noticeable vibratory movement. This produces an uncomfortable sensation to the occupants.

(ii) The large deflections may result in cracking of ceiling plaster, floors or partition, walls.

(iii) The large deflections indicate the lack of rigidity. It may cause vibrations and over-stresses under dynamic loads.

(iv) The large deflections may cause the distorsions in the connections. The distorsions cause secondary stresses.

(v) The large deflections may cause poor drainage, which will lead to ponding of water, and therefore increase the loads.

14.2 EFFECTS OF DEFLECTIONS

The present trend in design is to use steel of higher strengths. As a result of which, the design of more slender members is becoming possible. Modern building structures often lack substantial walls and portions. Non-structural members are prone to damage caused by the deflections of the structural members. Therefore, the control of deflections of flexural members (i.e., beams) under the service load is getting more importance.

The effects of deflections are classified under four broad headings as under :

1. Sensory acceptability

A personal judgement leads to sensory acceptability. It depends upon the social background of the person using the structure. The sagging of beams and the drooping of the cantilever are the **visual effects.** The vibrations due to dynamic loads of live or wind load are the **tactile effects** (which are perceptible by touch). Noise from the vibrations are **auditory**

effects (which are related to the sense of hearing). These effects are all **sensory effects** (which may be realized by senses). It is difficult to establish deflection limits on sensory acceptability. These effects vary with personal opinion.

2. Serviceability of structures

The intended use of roof surfaces is to drain water off the floor. The floors are kept plane, Members support sensitive equipment. It is easy to define deflection limits for the control of deflections.

3. Effect on non-structural elements

The walls, partitions, ceilings, doors, windows, ventilators, ...etc. are the non-structural elements. The deflections of beams must not exceed specified limits. In which case, these are likely to damage the walls, partitions and ceilings. The movement of operating elements (doors, windows and ventilators) should not be prevented by the deflections. Depending upon the type of non-structural elements and the methods of their installations, the deflection limits are specified.

4. Effects on structural elements

The deflections cause instability (e.g., in arches, shells or long columns) and change in the stress system (such as, change in the bearing area due to rotation at the ends of a beam. Deflection cause dynamic effects that increase stresses (i.e., the resonant vibrations due to moving loads) are the effects on the structural elements.

The behaviour of the structures is assumed for the purpose of design. The limits specified for the deflections such that the deflections do not alter the assumed behaviour of the structure. In case, it is possible, the effects of deflections on the structural behaviour may also be included in the design of beams.

14.3 METHODS OF CONTROLLING DEFLECTIONS

Deflections in the structural component members may be normally vertical displace-ments or horizontal displacements or both. There are two methods of controlling deflections.

1. Method of sufficient stiffness

The deflections may be controlled by ensuring that the structural component members have sufficient stiffness to limit the deflections at the service loads. The method is indirect. However, it is simple and satisfactory in many cases when the spans, loads, load distributions, members' sizes and proportions fall in the usual range. Limits are specified to control deflection at the service loads as under :

(i) Limiting vertical deflection

The deflection of a member is calculated without considering the impact factor or dyna-mic effect of the loads causing the deflection. The deflection of a member shall not be such as to impair the strength and efficiency of the structure and lead to damage to finishings.

Generally, the maximum deflection should not exceed (1/325) of the span. This limit may be exceeded in cases where greater deflection would not impair the strength or efficiency of the structure or lead to damage to finishings. The deflections of the beams may be decreased by increasing the depth of beams, decreasing the span, providing greater end restraint or by any other means.

(ii) Limiting horizontal deflection

At the caps of columns in simple storey buildings, the horizontal deflection due to lateral force should not ordinarily exceed (1/325) of the actual length '*l*' of the column. This limit may be exceeded in cases where the greater deflection would not impair the strength and efficiency of the structure or lead to damage to finishings.

According to AISC specifications, the deflections of beams and girders for live load and plastered should not exceed (1/360) of the span.

2. Method of actual calculations of deflections

It is only in exceptional cases that the actual deflections are calculated. Whenever, it becomes necessary then, the actual deflections are calculated. These calculations are compared with the limiting values specified in the codes.

14.4 LIMITATIONS OF VERTICAL AND HORIZONTAL DEFLECTIONS IN CRANE STRUCTURES

Limitations of vertical and horizontal deflections for electrically and hand operated overhead crane structures are as under:

1. Limitation of vertical deflection

When the crane structure is subjected to vibration or shock impulses, it may be desirable to maintain reasonable deflection limits such as will produce a stiff structure less apt to vibrate and shake appreciably. For example, the excessive deflection in crane runway supporting girder will lead to uneven up and down motion of the crane as it proceeds. The impact stresses in such cases would be increased.

In case of crane runway girder, the maximum vertical deflection under dead and imposed loads as per the recommendations of IS : 800-1984, shall not exceed the following values :

(i) Where the cranes are manually operated and to similar loads (L/500)

(ii) Where the cranes are overhead travelling and operated electrically upto 500 kN (L/500)

(iii) Where the electric overhead travelling cranes operated over 500 kN (L/1000)

(iv) Other moving loads, such as charging cars, etc. (L/600)

Where, L is the span of crane runway girder.

2. Limitation of horizontal deflection

The horizontal deflection at the column cap level of columns supporting crane runway girders in the building shall not exceed as may be specified by the purchaser.

For the plates of normal proportions

$J = 0.32\, b \,.\, t^3$ and therefore

For the flanged sections, symmetrical about the minor axes

$$x = (h_s)\left(\frac{\Sigma\,(b.t + h_w.t_w)}{\Sigma\,(bt^3 + h_w t_w^3)}\right)$$

where, b is the width of each plate element, and t, the thickness of these elements

(iv) **V-slenderness factor** (found from the following expression

$$V = \left[\left(4N\,(1-N)\frac{1}{20}\left(\frac{\lambda}{x}\right) + \psi\right)^{1/2} + \psi\right]^{1/2}$$

where, N is the ratio

$[I_{cf} / (I_{cf} + I_{tf})]$ where,

I_{cf} is the second moment of the area of the compression flange about the minor axis of the section, and

I_{tf} is the second moment of the area of the tension flange about the minor axis of the section.

***ψ* is the mono symmetric index obtained as follows :**

For I and T sections with lipped flanges (Determined as follows)

ψ = 0.8 (2N – 1) (1 + D_L / D) for N > 0.5

ψ = 1.0 (2N – 1) (1 + D_L/2D) for N < 0.5

Example 14.1 : Check the adequacy of beam-section ISMB 450, @ 0.724 kN/m for deflection. Use limit state of serviceability.

Note : The beam-section ISMB 450, @ 0.724 kN/m has been proposed as a trial section has been checked for bending strength in Example 11. and shear strength in Example 12. and it has been checked against web-buckling and web-crippling in Example 13.1.

New this beam section ISMB 450, @ is to be checked for deflection. Using limit state of serviceability.

Step 1. For limit state of serviceability

For dead load, load factor, γ_{fL} = 1.00

For live load, load factor, γ_{fL} = 1.00

Dead load, D.L. = 19.32 kN/m

Live load, L.L. = 12.00 kN/m

Total load, TDL = (19.32 + 12) = 31.32 kN/m

Effective span, ES = 7.30 m

Moment of inertia of the section about strong axis, I_{xx} = 30,390.80 x 10^4 mm^4

Step 2 : Maximum (actual deflection)

The beam is simply supported at both the ends and it carries uniformly distributed load. Maximum (actual) deflection due to service load occurs at the centre.

$$y_{max} = \left[\frac{5}{384} \times \frac{wL^4}{E.I_{xx}}\right]$$

$$= \left[\frac{5}{384} \times \frac{31.32 \times (7.30 \times 1000)^4}{2 \times 10^5 \times 30{,}390.80 \times 10^4}\right]$$

$$= 19.05 \text{ mm.}$$

Step 3 : Allowable deflection

$$= \left(\frac{1}{200} \times \text{effective span}\right)$$

$$= \left(\frac{7.30 \times 1000}{200}\right) = 36.50\text{mm}$$

Note : Actual deflection, 19.05 mm is less than the allowable deflection 36.50 mm. Hence, the beam section is adequate for **limit of serviceability.**

Problems

14.1. In Problem 11.2, check the design for deflections.

14.1. In Problem 11.4, check the design for deflections.

14.3. In Problem 11.6, check the design for deflections.

14.4. In Problem 11.7, check the design for deflections.

14.5. In Problem 11.8, check the design for deflections.

15

LIMIT STATE OF COLLAPSE IN (FLEXUR) BENDING AND SHEAR, STRENGTH OF LATERALLY UNSUPPORTED (UNRESTRAINED) DESIGN OF STEEL BEAMS (SB-IV)

15.1 INTRODUCTION

The laterally unsupported beams are also termed as laterally unrestrained beams. The symmetrical beams even if loaded in the plane containing principal axis may bend out of the plane of load under certain conditions. The rolled steel beams and channels have large moment of interia about xx-axis. I_{xx} (i.e. the strong principal axis). Such beams have maximum flexural rigidity about xx-axis (EI_{xx}). In these sections, the material is distributed in the flanges (i.e, away from the centroid). The flexural rigidity (i.e., resistance to in-plane bending increases. The shapes of beams are made economical in the manner. Such shapes have less resistance to bending and torsion in the direction of yy-axis (i.e., weak axis) about **one-sixth,** $I_{yy} = I_{xx} / 6$ approxi. The beams may bend in the weak direction if it is not held or supported in a direction normal to the weaker axis. The beam is weakened further if a horizontal load is applied in the direction normal to yy-axis. Design of most of the beams are done by ensuring an adequate degree of safety against overall buckling. Overall instability for beams is complicated. A perfectly straight beam loaded exactly in the plane of web at the elastic critical load fails suddenly. The beam deflects sideways and twists about its longitudinal axis as shown in Fig. 15.1. The bending of the beam is accompanied by twisting. This phenomenon is termed as **lateral buckling** or **lateral torsional buckling**. This tendency further increases when the supported length of compression flange increases.

This influence gets magnified on the application of further moment or any lateral load. Eventually, the beam may fail at moment less than the full moment capacity of the section. However, the beams are never perfectly straight not homogeneous and not loaded exactly in the plane that is assumed for the design. The beams which do not have effective lateral restraint deform in the lateral direction are known as **'Lateral unsupported beams'**.

The bending of beams is accompanied by twisting. This phenomenon is termed as **lateral buckling** or **lateral torsional buckling.** When the torsional resistance and moment of inertia of any beam for the weak axis are small as compared with the moment of inertia about the strong axis, the beam may buckle laterally wheather the beams are symmetrical or not. The buckling tendency is maximum when the stress attains the value of critical stress. The

compression flange buckles side-ways or laterally and it is also accompanied by twisting. The tendency of lateral buckling of compression flange also increases as the ratio (I_{xx}/I_{yy}) increases. This tendency further increases when the supported length of compression flange increases.

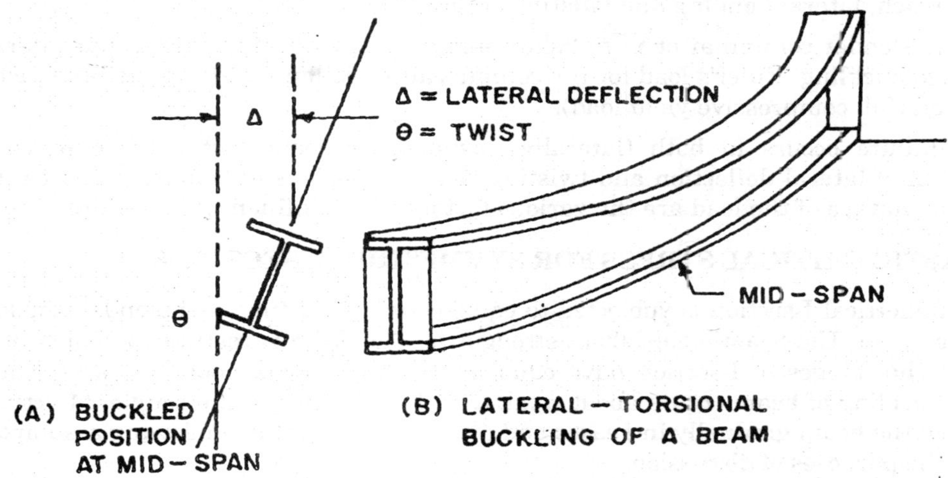

Fig. 15.1. Lateral torsional buckling of a beam

There are many factors which influence lateral stability of beams in actual structure. Imperfections in the beam (i.e. initial bow and initial twist in the beam section (if any), accidental eccentricity of loading and premature yielding due to the presence of residual stresses) are the various such factors. These factors result in unequal (i.e., different) stresses in the compression and tension flanges.

In order to avoid lateral buckling due to a load in the plane of one of the principal axis of the member, the beam should have enough torsional stiffness to resist torsional forces induced and the beam should be laterally supported in the direction perpendicular to the weaker principal axis. The compression flange of beam tends to buckle if not held in line by attached construction. Sometimes, the special conditions may require that the beams may be loaded in the plane of the web without continuous or intermittent lateral support at sufficient frequent intervals.

The bending stress in compression for sectional shapes with I_{yy} smaller than I_{xx} is moment of inertia of the whole section about the axis lying the plane of bending. yy axis, and I_{xx} is moment of inertia of the whole section about the axis normal to the plane of bending, xx-axis) is reduced for lateral buckling in proportion to elastic critical stress, σ_{cb}.

15.2 RATIONAL ANALOGY BETWEEN LATERALLY UNSUPPORTED BEAMS AND COLUMNS

The analogy between the compression flange of a **laterally unsupported beam** and **a column** is presented to show that general behaviour for lateral buckling between them is very similar. Laterally unsupported beams and axially loaded long (slender) columns is very similar. Both show similar characteristic due to buckling (i.e, bending).

A simply supported beam at its both the ends, which is laterally unsupported (except at supports). The beam is of small length and carries incremental transverse load. The compression flange of the beam subjected to bending may be considered as a column, subjected to axial load. This beam deflects downward, first in the direction of load. When the value of bending moment reaches its critical value, the compression flange buckles laterally. As a result of which, lateral bending and twisting occurs.

A long slender column after initial shortening in the axial direction, bends laterally and twist due to buckling. Euler's load for the column causes collapse (failure) before reaching the squash load (full compressive yield load).

The failure occurs in both (laterally unsupported beam and long columns) by the combination of lateral deflection and twisting. The proportions of beams, support conditions and the the nature of the load are the various factors which influence the failure.

15.3 ELASTIC CRITICAL STRESS FOR SYMMETRICAL SECTIONS

A symmetrical I-section is subjected to equal couple, M acting at the ends. It is loaded in the strong plane. The beam bends about strong axis. The beam is restrained against torsion at the ends. The flanges of I-section have equal moment of inertia about yy- axis. The lateral torsional buckling of beam occurs about weak-axis. It occurs when the couple, M acting at the two ends of the beam gradually increases and attains critical value. Following assumptions are made for the purposes of discussion.

Initially, there is no distorsion in the beam. The behaviour of beam remains elastic. The yielding does not take place. Externally, lateral and torsional loads are not applied. There are no residual stresses across the cross-section of the beam. The beam is simply supported vertically and laterally at its ends. Pure torsional resistance under uniform torsion and the warping torsional resistance are two distinct characteristics of the beam section which fails in lateral torsional buckling. The critical value of moment, M_{cr} is found as follows.

15.4 LATERAL TORSIONAL BUCKLING OF BEAMS WITH COMPACT CROSS-SECTIONS

An ideally perfect straight beam loaded exactly in the plane of web fails at the elastic critical load. The beam deflects sideways and twists about the longitudinal axis. This type of response may be seen in laboratory tests. Although the basic theory gives sufficient descri-ption of the behaviour of beams tested under very carefully controlled laboratory conditions, it does not provide for the several of factors which influence the lateral instability of beams in actual structures, Initial bow and initial twist in the beam-section, accidental eccentricities of the loading and premature yielding due to the presence of residual stresses. Therefore, whilst elastic buckling theory helps in the identification of controlling parameters of the problem, proper use must also be made of representative test data for satisfactory establishment of the design rules.

It is seen clearly from the experiments that the closely spaced beams with restrained spans may attain fully plastic moment, M_p. The beams having unrestrained long spans fail by elastic lateral torsional instability at moments which are very close to M_{cr} (the theoretical critical value).

The ratio of these two quantities may be used as a measure of beams proneness to lateral torsional collapse. The effective slenderness may be found as $(M_p/M_{cr})^{1/2}$ for lateral torsional buckling. The behaviour of beam shows following three regions when the test data are plotted as shown in Fig. 15.2.

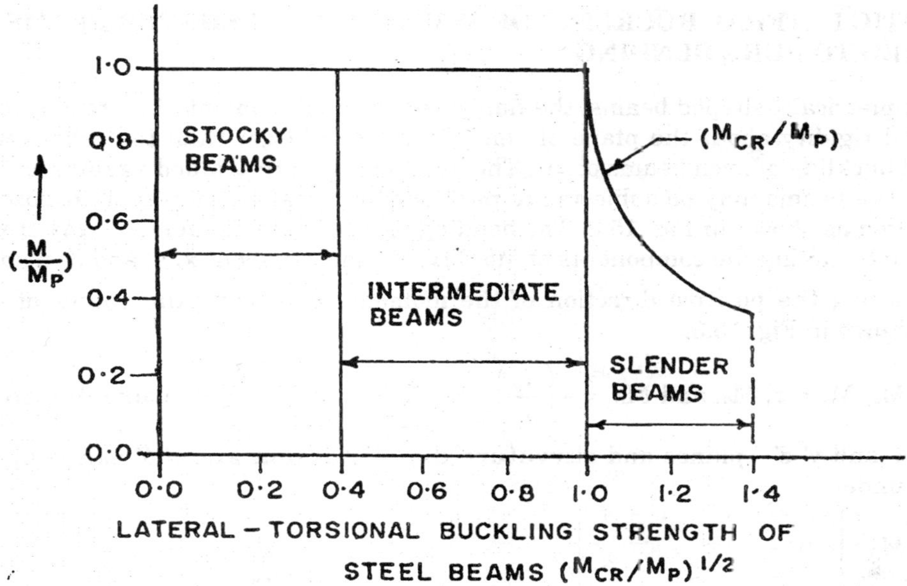

Fig. 15.2.

Lateral torsional buckling strength of steel beams. $(M_p/M_{cr})^{1/2}$ Fig. 15.2.

1. Stocky-beams : The beams for which plastic hinge action is possible are known as stocky-beams. [i.e., when $(M_p/M_{cr})^{1/2} < 0.4$], then, the value of bending moment may reach the value, M_p.

2. Beams having intermediate slenderness: The beams which fail through the combined effects of plasticity and instability at moments less than either M_p or M_{cr}, [i.e., $0.4 < (M_p/M_{cr})^{1/2} < 1.2$].

3. Slender beams : The beams which fail at moments approaching M_{cr}, [i.e., $(M_p/M_{cr})^{1/2} > 1.2$].

It is simply assumed that M_{cr} corresponds to the theoretical elastic critical moment, M_{cr} in the above described explanation for the particular beam under consideration. The theoretical elastic critical moment is a complex function of a number of parameters. The geometry of cross-section of beam, bending and torsional properties and its span are the most important parameters. The type of restraint provided in the lateral plane and the pattern of moment are influenced by the conditions of support provided in the transverse plane.

It is not necessary to check the resistance to lateral torsional buckling seperately. The members may be considered as laterally supported in the following cases.

(i) Bending is about weak axis of the section.

(ii) Section is hollow (rectangular/tubular) or solid bars

and (iii) In case of major axis bending, λ_{LT} non-dimensionless ratio, which one is given by $\lambda_{LT}^2 = \left(\sigma_y / \sigma_{cr.b}\right)$

15.5 ELASTIC LATERAL BUCKLING OF SYMMETRICAL I-SHAPED BEAMS SUBJECTED TO PURE BENDING

In symmetrical I-shaped beams, the flanges have equal moment of inertia about yy-axis. The flexural rigidity about the plane of bending is much larger than that about weak-axis. The lateral buckling of such beams occur. The beams may be restrained against torsion at the ends only. The beams may be subjected to pure bending, that is, couples M at both the ends are considered as shown in Fig. 15.3. The bending and twisting moments at any cross-section are obtained by taking the components of Mo (Mx, M_y and Mz about x, y, and Z, respectively).

Considering the positive direction of the moments and using the values of cosines of angles as shown in Fig. 15.3.

Mx = M_o, M_y = ϕ, M_o, and Mz $= -\left(\frac{d_u}{d_z}\right)$ M_o. There shall be **curvature due to bending** in both x' z' and y' z' - planes and **curvature due to torsion about Z-axis, which may be** written as under.

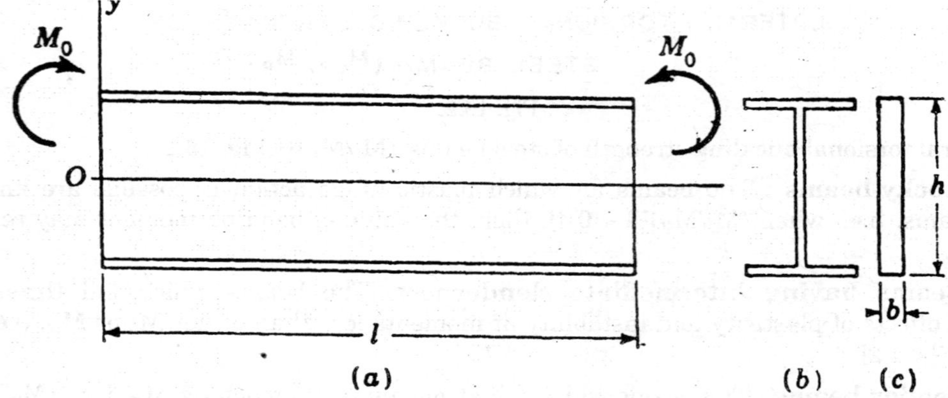

Fig. 15.3

$$E \,.\, I_x \,.\left(\frac{d^2v}{dz^2}\right) = M_x = M_o \qquad \text{(i)}$$

$$\text{and } E.I_x \left(\frac{d^2V}{dz^2}\right) = M_o = 0 \qquad \text{(ii)}$$

where, v is the displacement of the centroid in y-direction. Also the curvature in the x' z' plane is

$$EI_y \left(\frac{d^2u}{dz^2}\right) = M_y = \phi M_o \qquad \text{(iii)}$$

$$E.I_y \left(\frac{d^2u}{dz^2}\right) = M_y = 0 \qquad \text{(iv)}$$

where, u is the displacement of the centroid in the x-direction.

Differential equation for torsion of I-shaped beam is

$$M_z = GJ\left(\frac{d\phi}{dz}\right) - E.Cw\left(\frac{d^3\phi}{dz^3}\right) + \left(\frac{du}{dz}\right).M_o = 0 \quad \text{(v)}$$

Note : When the beam buckles slightly, the torsional component of M_g is the proportional to the slope in the x-z plane

$M_z = -\left(\frac{du}{dz}\right), M_o,$ which gives torsional differential equation as above (v)

In the above two expressions (ii) and (iv) two assumptions are inherent. Both of which relate to the assumptions of small deformations.

It is assumed that I_x and I_y are equal to I_x and I_y, respectively and so also I_x is large as compared to I_y. Therefore, the displacement, v in the plane of bending does not influence torsional function, ϕ Differentiating the expression (v) with respect to Z.

$$C.\left(\frac{d^2\phi}{dz^2}\right) - C_1\left(\frac{d^4\phi}{dz^4}\right) + \left(\frac{d^2\phi}{dz^2}\right).M_o = 0 \quad \text{(vi)}$$

since, $\quad EI_y.\left(\frac{d^2u}{dz^2}\right) = \phi.M_o$

or $\quad \left(\frac{d^2u}{dz^2}\right) = \left(\frac{\phi M_o}{EI_y}\right) \quad$ (vii)

Substituting this in the expression (vi), (d^2u/dz^2) may be eliminated.

$$C_1.\left(\frac{d^4\phi}{dz^4}\right) - C.\left(\frac{d^2\phi}{dz^2}\right) - \left(\frac{\phi M_0^2}{EI_y}\right) = 0 \quad \text{(viii)}$$

$$\left(\frac{d^4\phi}{dz^4}\right) - \left(\frac{C}{C_1}\right)\left(\frac{d^2\phi}{dz^2}\right) - \left(\frac{\phi}{C_1}\right)\left(\frac{M_o^2}{EI_y}\right) = 0 \quad \text{(ix)}$$

Let (C/C_1) be equal to (2α) and $\beta = \left(\frac{1}{C_1}\right).\left(\frac{M_o^2}{EI_y}\right)$, then the expression (ix) be written as under

$$\left(\frac{d^4\phi}{dz^4}\right) - 2\alpha.\left(\frac{d^2\phi}{dz^2}\right) - \beta.\phi = 0 \quad (15.1)$$

General solution of this differential Eq. (15.1) may be written as under

$\phi = A_1 . \sin(mz) + A_2 . \cos(mz) + A_3 . e^{+nz} + A_4 . e^{-nz}$

where m and n are positive, real quantities, defined by the relations

$m^2 = -\alpha + (\alpha^2 + \beta^2)^{1/2}$ and

$n^2 = +\alpha + (\alpha^2 + \beta^2)^{1/2}$

These constants of integration, A_1, A_2, A_3 and A_4 may be evaluated from the end conditions of the beam. It is assumed that the ends of beam cannot rotate about Z-axis, but these ends are free to warp. Then,

$$\textbf{At z = 0 and z = } \boldsymbol{l}\textbf{,} \quad \phi = \left(\frac{d^2\phi}{dz^2}\right) = 0$$

At z = 0, it is concluded that $A_2 = 0$ and $A_3 = -A_4$.

The angle of twist, ϕ may be represented in the form

$$\phi = A_1 . \sin(mz) + A_3 \left(e^{+nz} - e^{-nz}\right) \tag{15.2}$$

$$\text{or } \phi = A_1 . \sin(mz) + 2A_3 \sin h(nz) \tag{15.3}$$

$$\text{or } \phi = A_1 . \sin(mz) - 2A_4 \sin h(nz) \tag{15.4}$$

At z = *l*, ϕ = 0, using this condition,

$$A_1 . \sin\ (mz) - 2A_4 \ \sin h\ (nz) = 0 \tag{15.5}$$

$$A_1 . m^2 \sin\ (ml) + 2A_4 . n^2 \sin h\ (nl) = 0 \tag{15.6}$$

$$\sin\ (ml) . [n^2 . \sin h\ (nl) + m^2 \sin h\ (nl)] = 0 \tag{15.7}$$

Since, m and n are non-zero and positive quantities, it is concluded that

$\sin\ (ml) = 0$ (15.8) A

From Eq. 15.6, $A_4 = 0$

Therefore, the form of buckling of the beam is given by

$$\phi = A_1 . \sin\ (mz) \tag{15.8)B}$$

For the fundamental buckling mode. m = 1. This shows that the beam buckles in the shape of sine-wave. Smallest value of m which satisfies Eq. 15.

$$m^2 = \left(\pi^2 / l^2\right) \tag{15.9}$$

that is,

$$-\alpha + \left(\alpha^2 + \beta^2\right)^{1/2} = \left(\pi^2 / l^2\right) \tag{15.10}$$

Substituting, $\alpha = (C/2. C_1)$ and $m = (1/C_1) . \left(M_0^2 / EI_y\right)$ and solving for the critical value of the moment, M_o

$$\left(M_o\right)_{cr} = \left(\frac{\pi}{l}\right)\left[EI_y . C\left(1 + \left(\frac{C_1}{C}\right)\left(\frac{\pi^2}{l^2}\right)\right)\right]^{1/2} \tag{15.11}$$

This expression for the critical moment (applied moment) may be represented in the form

$$\left(M_o\right)_{cr} = \gamma_1 . \left(\frac{1}{l}\right)\left(EI_y . C\right)^{1/2}$$

where, γ_1 is a dimension less factor as defined by

$$\gamma_1 = \pi\left[1 + \left(\frac{C_1}{C}\right)\left(\frac{\pi^2}{l^2}\right)\right]^{1/2} \tag{15.12}$$

Substituting, $2\alpha = (GJ/E.C_w)$ and $\beta = (M_o / E^2C_w . I_y)$, it may be written as follows.

$$\frac{M_0^2}{E^2.C_w.I_y} + \left(\frac{GJ}{2EC_w}\right) = \left(\frac{\pi^2}{l^2} + \frac{GJ}{2EC_w}\right) \tag{15.13}$$

$$\left(\frac{M_0^2}{E^2.C_w.I_y}\right) = \left(\frac{\pi^2}{l^2} + \frac{GJ}{2EC_w}\right)^2 - \left(\frac{GJ}{2E.C_w}\right)^2 \tag{15.14}$$

$$M_0^2 = E^2.C_w.I_y\left[\frac{\pi^4}{l^4} + 2.\frac{1}{2}\frac{GJ}{EC_w}.\frac{\pi^2}{l^2} + \left(\frac{GJ}{2E.C_w}\right)^2 - \left(\frac{GJ}{2E.C_w}\right)^2\right] \tag{15.15}$$

$$M_0^2 = E^2.C_w.I_y\left[\frac{\pi^4}{l^4} + \frac{\pi^2}{l^2}.\frac{GJ}{E.C_w}\right] \tag{15.16}$$

$$M_0^2 = \left(\frac{\pi^4}{l^4}.E^2.C_w.I_y + \frac{\pi^2}{l^2}.\frac{GJ}{E.C_w}.E^2.C_w.I_y\right)$$

$$M_0^2 = \left(\frac{\pi^2}{l^2}\right)\left[\left(\frac{GJ}{E.C_w}\right).E^2.I_y.C_w + \left(\frac{\pi^2}{l^2}.E^2.C_w.I_y\right)\right] \tag{15.17}$$

$$M_0^2 = \left(\frac{\pi^2}{l^2}\right)\left[\left(\frac{GJ}{E.C_w}\right).E^2.I_y.C_w + \left(\frac{\pi^2}{l^2}.E^2.C_w.I_y\right)\right] \tag{15.18)A$$

$$M_0^2 = \left(\frac{\pi^2}{l^2}\right)\left[\left(\frac{GJ}{E.C_w}\right).\left(E^2.I_y.C_w\right) + \left(\frac{\pi^2}{l^2}\right)\left(E^2.C_w.I_y\right)\right] \tag{15.18)B$$

$$M_0^2 = \left(\frac{\pi^2}{l^2}\right)\left[\left(EI_y\ GJ\right) + \left(\frac{\pi^2}{l^2}\right)\left(EI_y.E.C_w\right)^2\right]$$

$$M_0^2 = \left(\frac{\pi^2}{l^2}\right)\left[\left(\left(E.I_y.GJ\right)^2\right)^{1/2} + \left(\left\{\left(\frac{\pi^2}{l^2}\right).\left(EI_y.E.C_w\right)\right\}^2\right)^{1/2}\right]$$

$$M_0^2 = \left(\frac{\pi^2}{l^2}\right)\left[(\text{Torsional Resistance})^2 + (\text{Warping Resistance})^2\right]$$

where, Torsional resistance = $(EI_y\ GJ)^2$

$$\text{Warping resistance} = \left[\left(\frac{\pi}{l}\right)^2 \left(EI_y\ E.C_w\right)^2\right]$$

$$M_0^2 = \left(\frac{\pi^2}{l^2}\right)\left(E.I_yGJ\right)\left(1 + \frac{\pi^2}{l^2}\frac{E.C_w}{GJ}\right) \tag{15.19}$$

Solving for $M_o = M_{cr}$

$$M_{cr} = \left(\frac{\pi}{l}\right)\left(E.I_y.GJ\right)^{1/2}\left(1+\frac{\pi^2}{l^2}.\frac{E.C_w}{GJ}\right)^{1/2} \tag{15.20}$$

Dividing by the section modulus Z, it results

$$\sigma_{cr} = \left(\frac{M_{cr}}{Z}\right) = \left(\frac{\pi}{l}\right)\left(\frac{\left(EI_y.GJ\right)^{1/2}}{Z}\right).\left(1+\frac{\pi^2}{l^2}.\frac{E.C_w}{GJ}\right)^{1/2}$$

$$\sigma_{cr} = \frac{\left(EI_y\,GJ\right)^{1/2}}{Z.l}.C_4 \tag{15.21}$$

where, C_4 is a symbol used in SSRC Guide

$$C_4 = \pi\left(1+\frac{\pi^2}{\lambda^2 l^2}.\frac{E.C_w}{GJ}\right) \tag{15.22}$$

$$\text{Let } \lambda^2 = \left(\frac{GJ}{E.C_w}\right) \tag{15.23}$$

$$\text{then, } \left(\frac{1}{\lambda}\right)^2 = \left(\frac{E.C_w}{G.J.}\right) \tag{15.24}$$

From Eq. 15.18B

$$M_0^2 = \left(\frac{\pi^2}{l^2}\right)\left[\left(EI_y.GJ\right)+\left(\frac{\pi^2}{l^2}\right)\left(EI_y\right)\left(E.C_w\right)\right] \tag{15.18)C}$$

In case of a simply supported prismatic members with symmetrical cross-section, the elastic lateral moment, M_{cr} may be written from Eq. 15.18 (C) as under : $\left(M_{cr}^2 = M_0^2\right)$

$$M_{cr}^2 = \left(\frac{\pi^2.EI_y}{(L_{LT})^2}\right)\left[GJ+\frac{\pi^2.EC_w}{(L_{LT})^2}\right] = \beta_b.Z_p.\sigma_{cr.b} \tag{15.18)D}$$

It is the expression as given in IS : 800-2007

where, $\sigma_{cr.b}$ is the critical stress in bending of non-slender rolled steel sections in Eq. (15.18)D. It may be approximately calculated. The values as given in Table 14 (IS : 800-2007) has been prepared using the following expression

$$\sigma_{cr.b} = (M_{cr} / Z_p)$$

$$\sigma_{cr.b} = \frac{1.1\pi^2 E}{\left(L_{LT}/r_y\right)^2}\left[1+\frac{1}{20}\left(\frac{L_{LT}/r_y}{h_f/t_f}\right)^2\right]^{1/2} \tag{15.18) E}$$

Following simplified expression may be used in the case of prismatic members made of standard rolled I-section and welded doubly symmetrical I-sections for calculating the elastic lateral buckling moment.

$$M_{cr} = \left(\frac{\pi^2 EI_y h_f}{2L_{LT}^2}\right)\left[1+\frac{1}{20}\left(\frac{L_{LT}/r_y}{h_f/t_f}\right)^2\right]^{1/2} \qquad (15.18)\ F$$

where,

J = Torsional constant $= \sum\left(b_i t_i^3/3\right)$ for open sections

C_w = warping constant

I_y, r_y = Moment of inertia and radius of gyration, respectively about the weaker axis

L_{LT} = Effective length for lateral torsional buckling

h_f = Centre to centre distance between flanges; and

t_f = Thickness of the flange

As per the recommendations of IS : 800-2007, resistance to lateral torsional buckling need not be checked seperately (i.e, member may be treated as laterally supported in the following cases :

(i) Bending is about the minor axis of the section

(ii) Section is hollow (rectangular/tubular) or solid bars, and

(iii) In case of major axis bending, λ_{LT} is less than 0.4.

where, λ_{LT} is non-dimensional slenderness ratio for torsional buckling. It is to note that for stock beams, (M_p/M_E) is less than or equal to 0.4.

In case, M_x is less than or equal to ($M_{R.LTB}/m_{LT.B}$), where,

M_x = Maximum moment about major strong axis

$M_{R.LTB}$ = Lateral torsional buckling moment of resistance.

$m_{LT.B}$ = Equivalent uniform moment factor for lateral torsional buckling.

Note : Suffixes, L, T and B are used to represent lateral, torsional and buckling.

In addition, M_x must not exceed the moment capacity of cross-section about the major axis. $M_{R.LTB}$ in found as the product of bending strength, (σ_b) and the section modulus appropriate to the class of the cross-section. For the large majority of the rolled steel sections, this will be the plastic modulus (Z_{px}). The values σ_b in terms of the **equivalent slenderness**, λ_{LT} which one is **defined** as

$$\lambda_{LT} = \left(\frac{\pi^2 E}{\sigma_y}\right)^{1/2} \cdot \left(\frac{M_p}{M_{cr}}\right)^{1/2} \qquad (15.25)$$

That is, the product of the quantity used as the abscissa as shown in Fig. 15.2 and a constant for given grade of steel. The limiting values of λ_{LT} for which s_b may be adopted as σ_y leading to $M_{R.LTB}$ = M_p, which may be re-presented as λ_{ELO}. Provided lateral bracing is employed at a spacing not exceeding $\lambda_{E.LO}$ (No allowance for failure by lateral torsional buckling is necessary).

$m_{LT.B}$ in Eq. 15.1 recognizes the fact that uniform, single curvature moment is the most severe arrangement in terms of lateral stability. For this case, $m_{LT.B}$ is **unity.**

Alternatively, for the arrangement in which the beam ABCD is loaded only at point of effective lateral restraint, producing in unrestrained length subjected only to unequated moments, a reduced value may be used.

An appropriate formula that produces values of M_{LTB} is given by

$$m_{LT.B} = (0.57 + 0.33\,\beta + 0.10\,\beta^2) \qquad (15.26)$$

Less than or equal to 0.43.

In this, β is the ratio of smaller to the larger end moment on the segment between points of lateral restraint such that (1 is more than or equal to β, which one is more than or equal to –1.0). However, $\lambda_{LT.B}$ is always calculated on the basis of uniform moment (β = 1.0)

The value of β is positive for opposite moments at the ends (single curvature bending) and negative for moments for the same nature (double curvature bending as shown in Fig. 15.4.

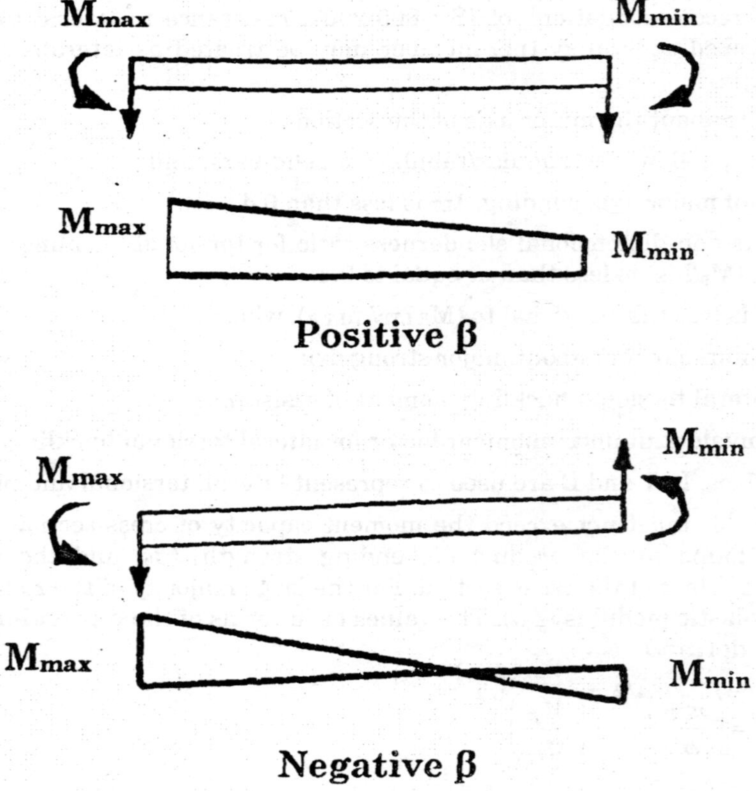

Fig. 15.4. Non-uniform distribution of bending moment.

For a particular case of β, the value of M at which the elastic instability occurs may be expressed as a ratio m involving the value of M_{cr} for the segment (i.e that is, the elastic critical moment for β = 1.00).

"m" is usually referred to as equivalent moment factor. Fig. 15.4 shows the relationship (M_{cr}/M)

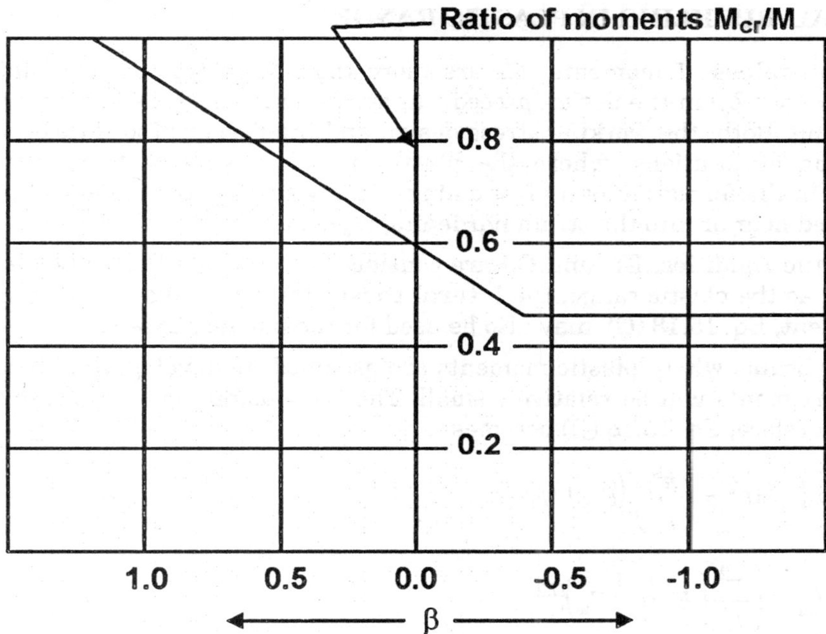

Fig. 15.5. 'm' Factor for equivalent uniform moment.

Fig. 15.4 shows that m = 1.0 for uniform moment and m is less than 1.0 for non-uniform moment. Therefore, beam with variation of moment over the unsupported length is less vulnerable to lateral stability as compared to that subjected to uniform moment. Its value is a measure of the intensity of the actual pattern of moments as compared with the basic case. In many cases, its value is dependent only on the shape of the moment diagram and some examples in table which presents some examples.

A good estimate of the critical moment due to actual load may be found using the value of m in the expression

$$M = (1/m)\,(M_{cr}) \tag{15.27}$$

This approximation helps in predicting the buckling of the segments of a beam, which is loaded through the transverse members preventing local lateral deflection and twist. Each segment is treated as a beam with unequal end moments and critical moments may be determined from the relationship Eq. (15.27). The critical moment of each segment may be found and lowest of them would give a conservative approximation to the actual critical moment.

It may be noted here that the values of m apply only when the point of maximum moment occurs at one end of the segments of the beams with uniform cross-section and equal flanges.

In all other cases, m = 1.00 for intermediate values, of β can be obtained from Eq. 15.26 or may be interpolated from Fig. 15.5. The local strength at the more heavily stressed end may be checked against moment capacity, M_p.

M_{max} shall be less than or equal to M_p.

15.6 LATERAL BUCKLING IN PLASTIC RANGE

When the values of moments, M_o are more than $M_y = (\sigma_y \,.\, Z)$ and upto the plastic moment, $M_p = (\sigma_y \,.\, Z_p)$ in the design procedures are considered, there is distinct possibility of lateral buckling. Both, the working stress design and the design procedures inherently need lateral bracing at locations, where the plastic hinges are likely to form in the failure mechanism. On the formation of a plastic hinge at any section, the extreme fibres of a beam will be strained near or into the strain hardening region.

In case, the rigidities, EI_y and GJ are considered to include the values in the inelastic range as well as the elastic range, the lateral torsional buckling equilibrium equations for pure moment, Eq. 15.18 (C) may also be used for the plastic range.

Since for beams where plastic moments are assumed to develop, the distances between lateral support points will be relatively small. The term using torsional rigidity, GJ may be neglected. Therefore, Eq. 15.18 (C) becomes

$$M_0^2 = M_{cr}^2 = \left(\frac{\pi^4}{l^4}\right)\left(EI_y\right).\left(EC_w\right) \tag{i}$$

$$M_{cr} = \left(\frac{\pi^2}{l^2}\right) E\,.\left(I_y\,.\,C_w\right)^{1/2} \tag{15.28)A}$$

Since, M_{cr} attains M_p, substitute $M_p = (\sigma_y \,.\, Z_p)$ for M_{cr} and $(C_w = I_y \,.\, h^2/4)$ and $(I_y = A\, r_y^2)$

$$M_E = \left(\frac{\pi^2}{l^2}\right).\,E\,.\left(I_y\,.\,\frac{I_y\,.\,h^2}{4}\right)^{1/2}$$

$$M_E = \left(\frac{\pi^2}{l^2}\,.\,E\,.\,I_y\,.\,\frac{h}{2}\right)$$

$$M_E = \left(\frac{\pi^2}{l^2}\right).\,E\,.\,A.r_y^2\,.\,\frac{h}{2}$$

$$\left(\frac{l}{r_y}\right)^2 = \left(\pi^2 E\right)\left(\frac{1}{M_E}\right)\left(\frac{Ah}{2}\right)$$

Since $M_p = \left(\sigma_y\,.\,Z_p\right) = \sigma_y\left(\frac{Ah}{2}\right)$

and $\left(\frac{Ah}{2}\right) = \left(\frac{M_p}{\sigma_y}\right)$, **then**

$$\left(\frac{l}{r_y}\right)^2 = \left(\frac{\pi^2 E}{2\sigma_y}\right)\left(\frac{M_p}{M_E}\right) \tag{15.28)B}$$

Substituting $\left(\frac{l}{r}\right) = \lambda_{E.LT}$

$$\lambda_{E.LT}^2 = \left(\frac{\pi^2 E}{\sigma_y}\right) \cdot \left(\frac{M_p}{M_E}\right) \tag{15.29}$$

where, $\lambda_{E.LT}$ is known as **equivalent slenderness ratio** and it is defined by Eq. (15.29).

The strain in the extreme fibre will reach or already entered the range of the strain hardening. Then, the hardening modulus of elasticity, E_{st} shoul be used instead of the modulus of elasticity E. Then, Eq. 15.29 may be written as

$$\lambda_{E.LT}^2 = \left(\frac{\pi^2 . E_{st}}{\sigma_y}\right) \left(\frac{M_p}{M_E}\right) \tag{15.30}$$

For the **stocky beam** (viz, plastic beam section) for which, the ratio (M_p/M_E) should be either less than or equal to 0.4. As such Eq. 15.30 may be written

$$\lambda_{E.LT}^2 = 0.4 \left(\frac{\pi^2 E}{\sigma_y}\right) \tag{15.31}$$

Perry's coefficient for lateral torsional buckling

(i) For rolled steel section (RSS)

$n_{LT} = a_b \left(\lambda_{E.LT} - \lambda_{L.O}\right)$

but $n_{LT} > 0$ (Greater than zero)

(ii) For welded steel section (WSS)

$n_{LT} = 2\, a_b \,.\, \lambda_{LO}$, but

n_{LT} shall be less than a equal to $2\, a_b \left(\lambda_{E.LT} - \lambda_{LO}\right)$

n_{LT} more than or equal to $a_b \,.\, \left(\lambda_{E.LT} - \lambda_O\right)$

n_{LT} more than or equal to zero

where, λ_{LO} = Limiting equivalent slenderness

a_b = Constant. It is taken as 0.007

15.7 DESIGN BENDING STRENGTH OF LATERALLY UNSUPPORTED BEAMS CONTROLLED BY LATERAL TORSIONAL BUCKLING

In case of a prismatic members with symmetrical cross-sections in simply supported beams the elastic lateral moment M_{cr}, in Art 15.6, Eq. 15.18 (C) was expressed as Eq. 15.18 (D)

$$M_{cr}^2 = \left(\frac{\pi^2 EI_y}{L_{LT}^2}\right)\left[GJ + \left(\frac{\pi^2 . E.C_w}{L_{LT}^2}\right)\right] = \beta_b . Z_p . \sigma_{cr.bd} \tag{15.18D}$$

From IS : 800-2007, Clause 8.2.2

β_b = 1.0 for plastic and compact section

= (Z/Z_p) for semi-compact section

Z_p, Z = Plastic modulus and elastic section modulus with respect to extreme compression fibre

$\sigma_{cr.bd}$ = Design bending compression stress obtained as under

$\sigma_{cr.bd} = \chi \cdot \sigma_y/\gamma_{msy}$

χ = Bending stress reduction factor to account for the lateral torsional buckling given by

$$\chi = \frac{1}{\left\{\phi_{LT} + \left[\phi_{LT}^2 - \lambda_{LT}^2\right]^{1/2}\right\}} <= 1.0 \tag{15.32}$$

$$\phi_{LT} = \frac{1}{2}\left[1 + \alpha_{LT}\left(\lambda_{LT} - 0.2\right) + \lambda_{LT}^2\right] \tag{15.33}$$

where, α_{LT}, the imperfection parameter

α_{LT} = 0.21 for rolled steel section

α_{LT} = 0.49 for welded steel section

Non-dimensional slenderness ratio, λ_{LT} is given by

$$\lambda_{LT} = \left(\frac{\beta_b Z_p \cdot \sigma_y}{M_{cr}}\right)^{1/2} \tag{15.34}$$

which should be less than or equal to

$$\lambda_{LT} = \left(\frac{1.2\, Z_p \cdot \sigma_y}{M_{cr}}\right)^{1/2} = \sqrt{\frac{\sigma_y}{\sigma_{cr.bd}}} \tag{15.35}$$

where,

M_{cr} = Elastic critical moment

$\sigma_{cr.bd}$ = Bending compressive stress corresponding to elastic lateral moment (IS 800-2007)

From the express (i) Art 15.7,

Elastic lateral torsional buckling moment has been expressed as per IS : 800-2007, Clause 8.2.2.1

$$M_{cr} = \left(\frac{\pi^2 EI_y}{(L_{LT})^2}\right)\left[GJ + \left(\frac{\pi^2 E.C_w}{(L_{LT}^2)}\right)\right]^{1/2} = \beta_b \cdot Z_p \cdot \sigma_{cr.b} \tag{15.36}$$

In the beginning, which has been treated as

$$M_{cr} = \beta_b \cdot Z_p \cdot \sigma_{cr.bd} \tag{15.37}$$

The values of $\sigma_{cr.bd}$ for non-slender rolled steel sections were calculated and given in Table 14 in IS : 800-2007 (Clause 8.2.2.1)

$$\sigma_{cr.bd} = \frac{1.10\,\pi^2.E}{\left(L_{LT}/r_y\right)^2}\cdot\left[1+\frac{1}{20}\left(\frac{L_{LT}/r_y}{h_f/t_f}\right)^2\right]^{1/2} \qquad (15.38A)$$

Following simplified expression may be used in case of prismatic members made of standard rolled steel I-sections and welded doubly symmetric I-sections for determining M_{cr} (IS : 800-2007, Clause 8.2.2.1)

$$M_{cr} = \frac{\pi^2\,EI_y\,h_f}{2\,L^2_{LT}}\left[1+\frac{1}{20}\left(\frac{L_{LT}/r_y}{h_f/t_f}\right)\right]^{1/2} \qquad (15.38)B$$

For calculating the elastic lateral buckling moment, M_{cr}, following simplified expression may be used in case of prismatic members made of standard rolled steel I-sections and welded doubly symmetric I-section.

where,

J = Torsional constant = $\sum\left(\frac{1}{3}b\,.t^3/3\right)$ for open section

C_w = Warping constant

I_y, r_y = Moment of ineria and the radius of gyration, respectively about the weaker axis

L_{LT} = Effective length for lateral torsional buckling

h_f = Centre to centre distance between flanges, and

t_f = Thickness of flange

Because of the different types of structural actions involved, the analysis for the lateral torsional buckling is very complicate. For the design purpose, the basic theory of elastic lateral stability cannot be used directly because the formulae for the elastic critical moment, M are too complicate for the routine and there are limitations to their extension in the ultimate stage.

Alternatively

A simple method of determining the buckling resistance of beams may be described as under :

Along the lines similar to BS : 5950 (Part I) and analogus to the Perry-Robertson method for columns, the buckling resistance moment may be found from the expression

$$(M_E - M_b)\,(M_p - M_b) = n_{LT}\,M_E\,.\,M_b \qquad (15.39)$$

which may be expressed in the determinant form as below

$$\begin{vmatrix} (M_E - M_b) & (-M_b) \\ (-n_{LT}\,M_E) & (M_p - M_b) \end{vmatrix} = 0 \qquad (15.40)$$

where,

M_E = Elastic critical moment

M_p = Plastic moment capacity

M_b = Moment of resistance in lateral torsional buckling

n_{LT} = Perry's coefficient for lateral torsional buckling Expanding Eq. 15.40

$$M_E . M_p - M_E . M_b - M_b M_p + M_b^2 - n_{LT} M_E . M_b = 0 \quad \text{(ii)}$$

Rearrange in the terms of this expression (ii)

$$M_b^2 - \left[M_p + (1 + n_{LT})M_E\right] M_b + M_p M_E = 0 \quad \text{(iii)}$$

Expression (iii) is a quadratic equation. There are two roots of this expression. Smaller-root of this expression gives the value of moment of resistance for lateral torsional buckling of laterally unsuppored beams. From **Hindu method (i.e, by using shridhar Acharaya's formula)**, the roots of above expression may be written as below :

$$M_b = \left(\frac{M_p + (1 + n_{LT})M_E}{2}\right) \pm \frac{1}{2}\left(\left\{M_p + (1 + n_{LT})M_E\right\}^2 - M_p M_E\right)^{1/2} \quad \text{(iv)}$$

Expression (iv) may be written as under

$$M_b = \phi_B \pm \left[\phi_B^2 - M_E M_P\right]^{1/2} \quad (15.41)$$

where,

$$\varphi_b = \frac{1}{2}\left(M_p + (1 + n_{LT} M_E\right) \quad (15.42)$$

Eq. (15.41) defined ϕ. From Eq. (15.36)

$$\left(M_b - \phi_B\right) = \pm\left[\phi_B^2 - M_E . M_R\right]^{1/2} \quad \text{(v)}$$

Squarring both the sides of the expression (v)

$$\left(M_b . \phi_B\right)^2 = \left[\phi_B^2 - (M_e)\right] \quad \text{(vi)}$$

Re-arranging the terms of this expression

$$(M_E . M_p) = \left[\phi_B^2 - \left(M_b - \phi_B^2\right)\right] \quad \text{(vii)}$$

Expanding the right side for difference of two squares

$$(M_E . M_P) = \left[\phi_B - (\phi_B) + M_b\right]\left[\phi_B + \phi_B - M_b\right] \quad \text{(viii)}$$

$$M_E . M_p = M_b . \left(2\phi_B - M_b\right) \quad \text{(ix)}$$

$$M_b = \left(\frac{M_E . M_P}{\left(2\phi_B - M_P\right)}\right) \quad \text{(x)}$$

Substituting the value of M_b in the denominator of the expression (ix)

$$M_b = \left(\frac{M_E . M_P}{\phi_B + \phi_B - \phi_B + \left[\phi_B^2 - M_E M_P\right]^{1/2}}\right) \quad \text{(xi)}$$

$$M_b = \left(\frac{M_E . M_P}{\phi_B + \left[\phi_B^2 - MEM_P\right]^{1/2}}\right) \quad (15.43)$$

and

$$\phi_\beta = \left(\frac{M_P + (1+n_{LT}).M_E}{2} \right) \quad (15.44)$$

As defined above

M_E = Elastic critical moment

$$M_P = (\sigma_y \,.\, Z_P / \gamma_{msy}) \quad (15.45)$$

n_{LT} = Perry's coefficient, similar to column buckling

Z_P = Plastic modulus of section

Based on above concept, BS : 5950 : Part I uses a curve in order to simplify the analysis. Similar to column curves, the bending strength of the beam is expressed as a function of slenderness, (λ_{LT})

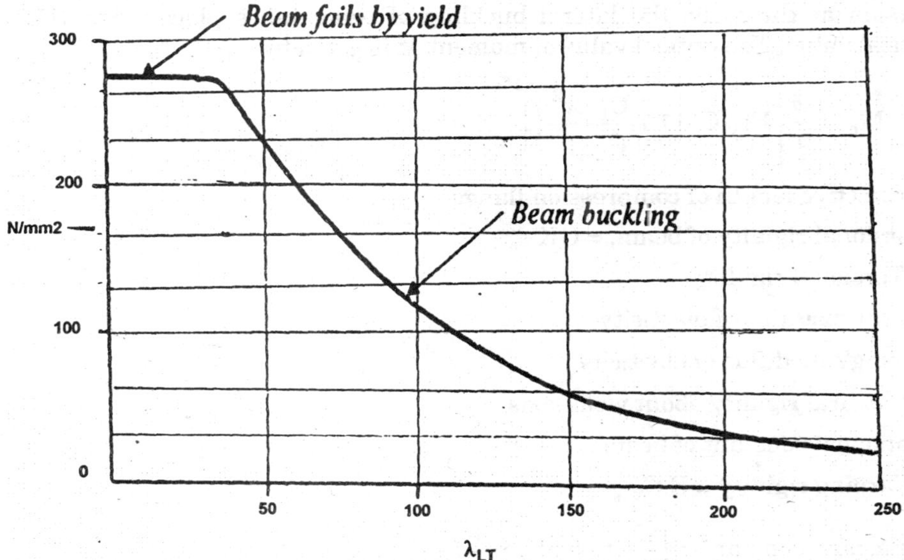

BENDING STRENGTH FOR ROLLED SECTIONS OF DESIGN 275 N/mm2 ACCORDING TO BS: 5950

Fig. 15.6

15.8 LATERAL-TORSIONAL BUCKLING FOR BEAMS OF DIFFERENT CROSS-SECTIONAL SHAPES

The elastic critical moment, M_{cr} has been found for doubly symmetric I-sections (which are open sections). Channel sections and angle sections are open sections, which are symme-tric about only one axis, Whereas, the equal and unequal angle open sections are unsymme-trical sections.

Circular and box-sections are closed sections and these are symmetrical about both the axes.

In **open sections**, the material remains distributed in the flanges and kept at a distance from the centre of gravity, C.G. The moment of inertia, I obtained is maximum as far as possible and as such, these sections provide moment of resistance against in-plane bending. The connections of such members with other adjacent members are convenient. For laterally restrained beams, the bending strength about strong axis is kept principal design consideration. At the same time, these sections have comparatively smaller moment of inertia about weak-axis. These sections are weak in bending about weaker axis and twisting about longitudinal axis.

Closed-sections have large torsional stiffness (about one-hundred times that of open- sections) Hollow circular tubes are most efficient shapes for torsional resistance. The connections of such sections are inconvenient. After development of welding, circular tubes are very commonly used as purlins in roof trusses. Tubular trusses are becoming popular and are used commonly now a days.

The flanges have equal moment of inertia about yy-axis

If symmetrical I-beam is subjected to couples M, at the ends and beam is restrained against torsion at the ends, the lateral buckling of beam takes place when the value of M attains critical value. The critical value of moment M is given by

$$M_{cr} = \frac{\pi}{l}\left[EI_y\, C. \left(1+\frac{C_1 . \pi^2}{C_2 . l^2}\right)\right]^{1/2} \qquad (15.46)$$

where, l = Effective length of compression flange

C = Torsional rigidity of beam, = GK

GK = Torsional rigidity

G = Shear modulus of elasticity,

E = Young's modulus of elasticity

EI_y = Flexural rigidity about weak-axis

K = Torsional constant of beam

C_1 = Warping rigidity = $E.C_w$

C_w = Warping constant = $\left(\frac{1}{24} t_f\, b_f^3\, h^2 = \frac{1}{2} I_f . h^2\right)$

$= \frac{1}{4} I_y . h^2$ (considering $I_f = \frac{1}{2} I_y$)

$E.C_w$ = Warping rigidity

I_f = Moment of inertia of compression flange about yy-axis of the girder

I = Moment of inertia of beam about yy-axis at the point of maximum bending moment

b = Width of flange of I-beam

t_f = Thickness of flange

h = Distance between the centre of gravity of compression flange and centre of gravity of tension flange. It may be taken approximately as depth of section minus thickness of flange, ($h - t_f$).

Substituting the values of C and C_1 in Eq. 15.46

$$M_{cr} = \frac{\pi}{l}\left[E.I_y.C + \frac{C_1}{C}\left(E.I_y.C\frac{\pi^2}{l^2}\right)\right]^{1/2} \quad (15.47)$$

$$M_{cr} = \frac{\pi}{l}\left[E.I_y.G.K. + \frac{\pi^2}{l^2}.E.I_y\,E.\frac{1}{4}I_y.h^2\right]^{1/2} \quad (15.48)$$

$$M_{cr} = \frac{\pi}{l}\,(EI_y.GK)^{\frac{1}{2}}\left[1 + \frac{\pi^2}{l^2}\left(\frac{EI_y.h^2}{4\,G.K}\right)\right]^{1/2} \quad (15.49)$$

$$M_{cr} = \frac{\pi}{l}\left[EI_y\,GK.\left(1 + \frac{\pi^2}{l^2}\frac{EI_y\,h^2}{4GK}\right)\right]^{1/2} \quad (15.50)$$

By rearranging the terms and taking the terms (π^2 EI_y . h^2) /4GK common from the parenthesis, Eq. 15.50 is written as follows :

$$M_{cr} = \left(\frac{\pi}{l}\right)(EI_y.GK)^{1/2}\left(\frac{\pi^2\,EI_y.h^2}{l^2\,4\,GK}\right)^{1/2}\left[\frac{l^2}{\pi^2}.\frac{4GK}{EI_y\,h^2} + 1\right]^{1/2} \quad (15.51)$$

$$M_{cr} = \frac{\pi^2}{2}\left(\frac{EI_y.h}{l^2}\right)\left[1 + \frac{1}{\pi^2}\frac{4GK.l^2}{EI_y.h^2}\right]^{1/2} \quad (15.52)$$

$$M_{cr} = \frac{\pi^2}{2}\,\frac{EI_y.h}{l^2}.\left[1 + \frac{1}{\pi^2}\frac{4GK.l^2}{EI_y.h^2}\right]^{1/2} \quad (15.53)$$

From Eq. (15.50) it may be observed that the critical bending moment depends not so much on the material, since value of E is constant for all steel, but on torsional rigidity, GK and elastic stiffness, EI_y for the bending in the plane perpendicular to the load.

Non uniform torsional buckling and lateral buckling occur together. Torsional and flexural stiffness of the beam are connected with the torsional and lateral components of buc-kling deformations. In Eq. 15.53, the second term (under square-root) is warping resistance. It adds to the resistance of the beam for short deep girders, its value is large. Long-shallow girders have small warping stiffness. Then, the elastic critical moment, M_{cr} is given by

$$M_{cr} = \frac{\pi}{l}\,(EI_y.GK)^{1/2} \quad (15.54)$$

The value of torsional rigidity GK is small for I-section consisting of thin plates [as K = (Σ(1/3) b.t^3)]. Both the terms under square-root contribute comparable. However, the light-gauge sections have most of the resistance to torsional deformation from the warping. The effect of length of beam is also considerable for deep and short beams $\left(\pi^2\,E.C_w / l^2\,G.K.\right)$ is large. It shows more warping resistance. Its value is small in case for long and shallow beams.

The elastic critical moment M_{cr} has been determined for a beam having I-section. This beam is simply supported and carries equal and moments. It is treated as a standard case. In practical situation, the cross-section of beam, the conditions at the support, loads acting over the beam ...etc. differ. Necessary changes are to be done for the design purpose.

A simply supported condition provides lowest lateral restraint. As the value of M_{cr} obtained is also lowest. Other restraint conditions provide greater values than this value. Lateral **bending, twisting and warping** are three types of deformations, which occur in later buckling. As such there are various types of support conditions possible. The effective length of the compressive flange takes into consideration, the different types of support.

Substituting, $G = 0.4\ E$, $K = (0.9\ b_f \cdot t_f^{\ 3})$ and $I_y = (2 \times t_f . b_f^{\ 3}/12)$

where, t_f and b_f are the thickness and width of the flange, respectively.

$$M_{cr} = \frac{\pi^2}{2} \cdot \frac{EI_y\,h}{l^2}\left[1 + \frac{1}{\pi^2} \times \frac{4 \times 0.4E \times 0.9\, b_f\, t_f^3}{E \times 2 \times t_f \times b_f^2/12} \cdot \frac{l^2}{h^2}\right]^{1/2} \quad \text{(i)}$$

$$M_{cr} = \frac{\pi^2}{2} \cdot \frac{EI_y\,h}{l^2}\left[1 + \frac{8.64}{\pi^2} \times \frac{t_f^2}{b_f^2} \cdot \frac{l^2}{h^2}\right]^{1/2} \quad \text{(ii)}$$

For an I-section and I-section of wide flanges, the approximate radii of gyrations are given by $r_y = 0.22\ b$ (viz. $b = 4.545\ r_y$) and $r_y = 0.25\ b$ (viz. $b = 4.000\ r_y$), respectively. As an approximation $b_f = 4.2\ r_y$), then

$$M_{cr} = \frac{\pi^2 . EI_y\,h}{2\,l^2}\left[1 + \frac{8.64\, t_f^2}{\pi^2 \times 17.64\, r_y^2} \cdot \frac{l^2}{h^2}\right]^{1/2} \quad \text{(iii)}$$

$$M_{cr} = \frac{\pi^2 . EI_y\,h}{2\,l^2}\left[\left(1 + \frac{1}{20}\left(\frac{l\, t_f^2}{r_y . h}\right)\right)\right]^{1/2} \quad \text{(iv)}$$

The elastic critical stress for bending for beams with I_y smaller than I_x, is given by $(\therefore M_{cr} = f_{cb} . Z_{xx})$

$$\sigma_{cb} = \frac{\pi^2\, EI\, h}{2 . Z_{xx}\, l^2}\left[1 + \frac{1}{20}\left(\frac{l . t_f}{r_y . h}\right)^2\right]^2 \quad \text{(v)}$$

The value of modulus of elasticity, E is taken as 2×10^5 N/mm² (MPa) and $I_y = A . r_y^2$. Substituting these values in the expression (v)

$$\sigma_{cb} = \frac{9.8526 \times 10^5\ A . h}{(l/r_y)^2\, Z_{xx}}\left[1 + \frac{1}{20}\left(\frac{l . t_f}{r_y . h}\right)\right]^2 \quad \text{(vi)}$$

The value of $(A.h/Z_{xx})$ for I-section may be taken as 2.688 approximately. Then,

$$\sigma_{cb} = \frac{26.5 \times 10^5}{(l/r_y)^2}\left[1 + \frac{1}{20}\left(\frac{l . t_f}{r_y . h}\right)^2\right]^2 \quad \text{(15.55)}$$

The value of critical stress, C_s is increased by 20 per cent for rolled steel beams, channels and plate girders provided.

$(t_e/t_w) > 2$, and $(d_1/t_w) > 85$

where, l = Effective length of compression flange

r_y = Radius of gyration about yy-axis of the gross-section of the whole girder at the point of maximum bending moment

h = Overall depth of the girder, at the point of maximum bending moment

t_f = Effective thickness of the compression flange

= k_1 x Mean thickness of the horizontal portion of the compression flange at the point of maximum bending moment.

k_1 makes the allowance for reduction in thickness and breadth of flanges between points of effective lateral restraint and depends on ψ the ratio of the total area of both flanges at the point of least bending moment to the corresponding point of greater moment between such points of restraints.

The values of k_1 have been given in Table 15.1.

TABLE 15.1. VALUES OF K_1 FOR BEAMS WITH CURTAILED FLANGES (As per IS : 800-1984)

ψ	1.0	0.9	0.8	0.7	0.6	0.5	0.4	0.3	0.2	0.1	0.0
k_1	1.0	1.0	1.0	0.9	0.8	0.7	0.6	0.5	0.3	0.3	0.3

Note : The flanges shall not be reduced in width to give a value ψ lower than 0.25 where the value of ψ calculated for compression flange alone is smaller than that when both the flanges are combined, this smaller value of ψ shall be used.

The values of critical bending stress f_{cb} are increased by 20 per cent for all rolled beams and channels and built-up beams for which (T/t_w) is not greater than 2 and (d_1/t_w) is not greater than $(1344/\sqrt{f_y})$ where, T is the mean thickness of the compression flange and d_1 is the clear depth of the web. The effective thickness of compression flange is equal to k_1 times the mean thickness of the horizontal portion of the compression flange at the point of maximum bending moment. The mean thickness of flange may be found by dividing area of horizontal portion of the compression flange by width of flange.

In order to provide safety against lateral buckling, allowable bending compressive stress σ_{bc} is reduced in proportion to critical bending stress f_{bc} at which lateral buckling of compression flange occurs.

The critical bending stress in compression for section symmetrical about the xx-axis may also be found from the following expression.

$$\sigma_{cb} = \frac{\pi}{Z_x \cdot l}\left[\frac{EI_y GK}{\mu}\left(1+\frac{C_1}{C}\frac{\pi^2}{l^2}\right)\right]^{1/2} \qquad (15.56)$$

where, $\mu = \left(\frac{I_x - I_y}{I_z}\right) \left(I_x - I_y\right)/I_x$

I_x = Moment of inertia of whole section about xx-axis

The values C_1 and C are substituted, then

$$\sigma_{bc} = \frac{\pi}{Z_x \cdot l}\left[\frac{EI_y\, GK}{\mu}\left(1 + \frac{\pi^2}{l^2}\frac{EC_w}{GK}\right)\right]^{1/2} \qquad (15.57)$$

The values of E, G and C_w mentioned above may be substituted. For I-section, Eq. (15.57) reduces to

$$\sigma_{cb} = \frac{4.094 \times 10^5}{Z_z}\left[I_y\, K\left(1 + 12.3\frac{I_f \cdot h^2}{Kl^3}\right)\right]^{1/3} \text{ N/mm}^2 \qquad (15.58)$$

For channels, Eq. (15.58) gives conservative values since value of warping constant for channels is more than that for I-sections.

The values of torsional constant 'K' for Indian Standard rolled steel beams and Indian Standard rolled steel channel sections are given in Appendix E in IS : 800-1962. For the sections composed of approximately rectangular sections

$$K = \Sigma\ (b' t^3/3)$$

where, b' and t' are the width and average thickness of the rectangular components of the section, respectively.

For box members, conservative values of 'σ_{cb}' are obtained. The value of K is given by

$$K = \left(\frac{4A^2}{S/t}\right) \text{ approximately}$$

where, A = Total enclosed area of section

S = Length along the periphery along which t is constant

t = Thickness of box member

t = Effective thickness (in the case of curtained flanges)

$$K = \frac{2b^2 \cdot d^2\, t}{(b + d)} \text{ approximately}$$

where, b = Width of box members

d = Depth of box members

For a plate or flat in bending in a plane parallel to its surface, the value of elastic critical stress in bending is given by

$$\sigma_{bc} = \left(\frac{4.094 \times 10^5 \cdot t^2}{lD}\right) \text{N/mm}^2 \qquad (15.59)$$

where, t is Thickness of plate or flat and D is depth of the plate

l is Effective length of part in compression.

Elastic critical stress

The elastic critical stress in bending may be calculated by an **elastic flexural torsional buckling analysis.** In case, this analysis is not carried out, the elastic stress in bending, f_{cb} for the beams and plate girders with the moment of inertia about axis of minimum strength (viz. yy-axis), I_{yy} smaller than that about axis of maximum strength (viz. xx-axis) I_{xx} may be determined from the expression derived as follows:

A paper titled as the basis for design of beams and plate girders in the revised BS : 153 (British Standards) structural paper 48, proceedings of the Institution of Civil Engineers, London, August 1956 was published by Kerensky, D.A., Flint, A.R., and Brown, W.C. Professor Kerensky gave the expression for **critical bending moment** for beams and plate girders as under :

$$M_{cr} = \frac{\pi}{l}(EI_y . GK)^{1/2}\left[\left(1+\frac{\pi^2}{l^2}\frac{EI_y . h^2}{4GK}\right)^{1/2} + (2\lambda-1)\frac{\pi h}{2.l}\left(\frac{EI_y}{GK}\right)^{1/2}\right] \quad \text{(i)}$$

where, λ is the ratio of moment of inertia of compression flange about yy-axis, I_f to the moment of inertia of the whole section about yy-axis I_y. $(2\lambda - 1)$ may be replaced by k_2. For the values of λ = 1 or 0, k_2 is equal to + 1.0 and -1.0. above expression (i) may be written as below [as expression (ii) derived in Eq. 15.15].

$$M_{cr} = \frac{\pi^2 EI_y . h}{2.l^2}\left[1+\frac{1}{20}\left(\frac{l\,t_f}{r_y . h}\right)^2\right]^{1/2} + k_2\left(\frac{\pi^2}{2}.\frac{EI_y . h}{l^2}\right) \quad \text{(ii)}$$

$$M_{cr} = \frac{\pi^2 DI_y . h}{2.l^2}\left[1+\frac{1}{20}\left(\frac{l\,t_f}{r_y . h^2}\right)\right]^{1/2} + k_2\left(\frac{\pi^2}{2}.\frac{EI_y . h}{l^2}\right) \quad \text{(iii)}$$

where I_y' is the modified moment of inertia. It is equal to $k_1 . I_y$, (I_y is the moment of inertia of the whole section about yy-axis at the section of maximum bending moment).

Since, $M_{cr} = \sigma_{cb}.Z_{xx}$, the critical bending stress is given by

$$\sigma_{cb} = \frac{\pi^2 EI_y . h}{2.Z_{xx}\, l^2}\left[1+\frac{1}{20}\left(\frac{l\,t_f}{r_y . h}\right)^2\right]^{1/2} + k_2\left(\frac{\pi^2 EI_y . h}{2.Z_{xx}\, l^2}\right) \quad (15.60)$$

This expression (iv) may be written as below : (as Eq. 15.61 has been derived).

$$\sigma_{cb} = \frac{26.5\times10^5 . k_1}{(l/r_y)^2}.\left[1+\frac{1}{20}\left(\frac{l\,t_f}{r_y . h}\right)^2\right]^{1/2} + k_2\left(\frac{26.5\times10^5 \times k_t}{(l/r_y)^2}\right) \quad (15.61)$$

where the moment of inertia of the tension flange about yy-axis exceeds that of the compression flange, Eq. 15.55 may be written as below :

$$f_{cb} = \left[\frac{26.5\times10^5 \times k_1}{(l/r_y)^2} \left\{ 1 + \frac{1}{20}\left(\frac{l\,.\,t_f}{r_y\,.\,D} \right)^2 \right\}^{1/2} + k_2 \frac{26.5\times10^5 \times k_1}{(l/r_y)^2} \right] \frac{C_2}{C_1} \text{ N/mm}^2 \qquad (15.62)$$

$$f_{cb} = k_1\,.\,[X + k_2\,.\,Y]\,.\,C_2 / C_1 \qquad (15.63)$$

IS : 800-1984 recommends Eq. 15.63 for calculating elastic critical stress in bending

where $$X = Y \left[1 + \frac{1}{20}\left(\frac{l.T}{r_y\,.\,D} \right) \right]^{1/2} \text{ N/mm}^2 \text{ (MPa)} \qquad (15.64)$$

$$Y = \left(\frac{26.5\times10^5}{(l/r_y)^2} \right) \text{ N/mm}^2 \text{ (MPa)} \qquad (15.65)$$

k_1. l, r_y, D and T symbols as used in Table 15.1(A) to (F) have been defined above.

k_1 = a coefficient. It allows for the inequality of the flanges. It depends on ω. ω is the ratio of moment of inertia of the compression flange alone to that of the sum of the moments of inertia of the flanges, each calculated about its own axis parallel to the y-y of the girder, at the point of maximum bending moment.

The values of co-efficient, k_2 for different values of ω are given in Table 15.2.

TABLE 15.2. VALUES OF k_2 FOR BEAMS WITH UNEQUAL FLANGES (As per IS : 800-1984)

w	1.0	0.9	0.8	0.7	0.6	0.5	0.4	0.3	0.2	0.1	0.0
k_2	0.5	0.4	0.3	0.2	0.1	0.0	-0.2	-0.4	-0.6	-0.8	-0.10

c_1 = lesser distance to the extreme fibre of the section

c_2 = maximum (greater distance to the extreme fibre of the section

I_y = moment of inertia of the whole section about the axis lying in the plane of bending (yy-axis); and

I_x = moment of inertia of the whole section about the axis normal to the plane of bending (xx-axis).

(a) For flanges of equal moment of inertia

$w = 0.5,\ k_2 = 0$

(b) For tees and angles bending in the plane of web with the xx- axis (neutral axis) nearer the extreme fibre of compression zone.

$w = 1.0,\ k_2 = 0.5$

(c) The bending stress in the leg when loaded with the flange or table in compression shall not exceed 0.66 σ_y. When loaded with the leg in compression, the permissible bending stress shall be calculated from Eq. 15.62 and Eq. 15.63 with k_2 = 1.0 and T as the thickness of leg.

The values of X and Y are given in Table 15.5 for appropriate values of (D/T) and (l/r_y).

The values of elastic critical stress in bending, f_{cb} may be increased by 20 per cent when (T/t_w) ratio is not greater than 2.0 and (d_1/t_w) is not greater than $1344/(f_y)^{1/2}$ where d_1 is the depth of web and t_w is the thickness of web.

When the bending of beams occur about axis of minimum strength (yy-axis), the maximum permissible bending stress in tension σ_{bt} or in compression σ_{bc} in beams should not exceed 0.66 σ_y, where σ_y is the yield stress of structural steel used.

The principal distinction between the design of laterally supported beams and laterally unsupported beams is the method of dealing with the bending in the laterally unsupported beams. The allowable bending stress in compression is reduced where necessary, depending upon the beams susceptibility of buckling. The reduction in allowable bending stress in compression provides adequate safety against the **lateral torsional buckling.** The reduction in allowable stress increases with the increase of the unsupported length of compression flange to the width of flange ratio and depth of beam to the thickness of flange ratio.

15.9 EFFECTIVE LENGTH OF COMPRESSION FLANGE FOR LATERAL TORSIONAL BUCKLING

The concept of effective length incorporates the various types of support conditions. For the beam with simply supported end-conditions and no intermediate lateral restraint to the compression flange and where each end of beam is restrained against torsion the effective length is equal to the actual length between the supports. When the large lateral and torsional restraint is provided at the supports, the effective length is less than the actual length and alternatively this length becomes more when there's less retraint. The effective length factor would indirectly account for the increased lateral and torsional rigidities provided by end restraints. In the laterally unsupported beams, bending compressive stress is reduced in proportion to the critical bending stress, s_{bc}. Which can be calculated from the effective length, L_{LT} and using other properties. Lateral bending of flange occurs with twisting and warping which depends on the effective length and effective lateral supports.

1. **For simply supported beams and girders where each end of the beam is restrained against torsion** of span length L, where there is no lateral restraint to the compression flange, but each end of the beam is restrained against torsion, the effective length of the compression flange for lateral torsional buckling, L_{LT} is to be used are given in Table 15.2.

TABLE 15.2(A). EFFECTIVE LENGTH OF COMPRESSION FLANGE FOR SIMPLY SUPPORTED BEAMS, L_{LT}, FULLY RESTRAINED AGAINST TORSION

S.NO.	CONDITIONS OF WARPING RESTRAINT AT SUPPORT	LOADING NORMAL	CONDITIONS DE-ESTABILIZING
(1)	(2)	(3)	(4)
(i)	Both flanges fully restrained	0.70L	0.85L
(ii)	Compression flange fully restrained	0.75L	0.90L
(iii)	Both flanges fully restrained	0.80L	0.95L
(iv)	Compression flange partially restrained	0.85L	1.00L
(v)	Both Flanges not restrained against warping	1.00L	1.20L

The destablizing factors shown in Tables 15.2 (A) & (B) depend upon.

(i) Level of application of transverse loads.

(ii) Influence of type of loading.

(iii) Loading applied at points of lateral restraints.

(iv) The lateral supports should provide torsional restraint, which may be provided as under:

(a) By providing web or flange clats in the end connections.

(b) By providing bearing stiffners acting in conjuction with the bearing of the beam.

(c) By providing lateral end frames or any other external support to the end of the compression flange.

(d) By fixing the ends of the beam into the walls.

Note: The end restraint element shall be capable of safely resisting in addition to wind and other applied external forces, a horizontal force acting at the bearing in a direction normal to the compression flange of the beam at the level of the centroid of the flange and having a value qual to or not less than 2.5 percent of the maximum force occuring in the flange.

TABLE 15.2(B). EFFECTIVE LENGTH OF COMPRESSION FLANGE FOR SIMPLY SUPPORTED BEAMS, L_{LT} PARTIAL RESTRAINED AGAINST TORSION

S.NO.	CONDITIONS OF WARPING RESTRAINT AT SUPPORT	LOADING NORMAL	CONDITIONS DE-ESTABILIZING
(1)	(2)	(3)	(4)
(vi)	Both flanges fully restrained	1.0L + 2D	1.2 L + 2D
(vii)	Both flanges not restrained against warping	1.2 L + 2D	1.4 L + 2D

NOTES : 1. Torsional restraint prevents rotation about the longitudinal axis.

2. Warping restraint prevents rotation of flanges.

3. D is the overall depth of the beam.

2. In simply supported beams with intermediate lateral restraint againt lateral torsional buckling : The effective length for lateral torsional buckling to be used, L_{LT} shall be taken as the length of relevant segment in between the lateral restraint. The effec-tive length shall be equal to 1.2 times. The length of the relevant segmetn in between the lateral restraint Restraint against torsional rotation at supports in these beams may be provided by:

(i) Web or flange cleats, or

(ii) bearing stiffeners acting in conjunction with the bearing of the beam, or

(iii) lateral end frames or external supports providing lateral restraint to the compression flanges at both the ends, or

(iv) their being, built-up into walls

3. **For beams, which are provided with members** giving effective lateral restraint to the compression flange at intervals along the span, in addition to the end torsional restraint required, the effective length for the torsional buckling shall be taken as the distance, centre-to-centre of the restraint members in the relevant segment under normal

loading conditions and 1.2 times this distance, where the load is not acting on the beam at the shear centre and is acting towards the shear centre so as to *have deestabilzing* effect during lateral torsional buckling deformation.

4. **For cantilever beams of projecting length** L, the effective length, L_{LT} to be used shall be taken as recommended in IS : 800-2007 in Table 16 for different conditions.

5. **Where a member is provided intermediate supports** to improve the lateral buckling strength, these restraints should have sufficient strength and stiffness to prevent lateral movement of the compression flange of that point, relative to the end supports. These intermediate lateral restraints should be either connected to an appropriate bracing system capable of transferring the restraint force to the effective lateral support at the ends of the members, or should be connected to an independent robust part of the structure capable of transferring the restraint forces. Two or more parallel members requiring such lateral restraint shall not be simply connected together assuming mutual dependence for the lateral restraint.

6. **The intermediate lateral restraints** should be connected to the member as close to the compression flange as practicable. Such retraints should be closer to the shear centre of the section. However, if torsional restraint preventing relative, rotation between the two flan ges is provided, the intermediate lateral restraint may be connected at any appropriate level.

7. **For beams which are provided with members giving lateral restraint at intervals along the span** the effective lateral restraint, shall be capable of resisting a force of 2.5% of the maximum force in the compression flange taken as divided equally between the points at which the restraint members are provided. Further, each restraint point should be capable of resisting one percent of the maximum force in the compression flange.

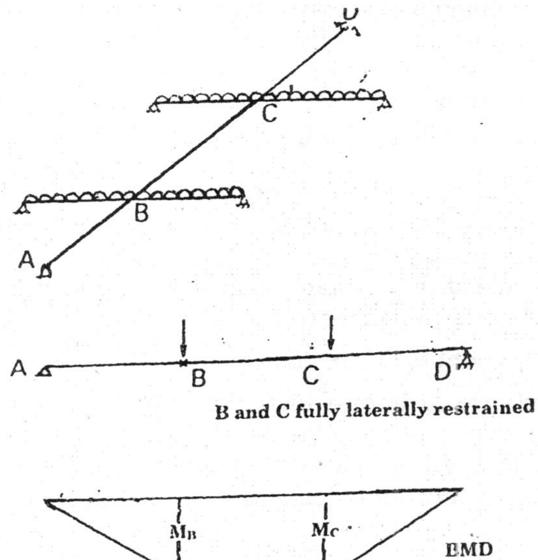

BEAM LOADED AT POINTS OF EFFECTIVE LATERAL RESTRAINT

Fig. 15.7.

8. In a series of such beams, with solid webs, which are conncted together by the same system of restraint members, the sum of the restraing forces shall be taken as 2.5 percent of the maximum force in the flange in one beam only.

9. In case of the series of latticed beams, girders or roof trusses which are connected together by the same system of restraint members the sum of the restraining forces required shall be taken as 2.5 percent of the maximum force in the compression plus 1.25 percent of this force for every member of the series other than the first, upto a maximum total of 7.5 percent.

10. Purlins adequately restrained by sheeting need not be normally checked for the restraing forces required by rafters, roof trusses or portal frames that carry the predominately roof loads provided, there is bracing of adequate stiffnes in the plane of rafters or roof sheeting which is capable of acting as a stress skin diaphragm.

11. In case of beams with double curvature bending, adequate direct lateral support to the compression flange in the hogging moment may be provided as given above for simply supported beam. The effect of support to tension (top) flange in the hogging moment region on lateral restraint to the compression flange may be considered as per special literature.

15.10. USE OF `m' FACTORS IN DESIGN

The shape of the moment diagram influence the lateral stability of a beam. A beam designed for uniform moment loading will be un-necessarily conservative. In order to account for the non-uniformity of moments, a modification of the moment may be made based on a comparison of the elastic critical moment for the basic case. This can be done in the following two ways.

1. Use of equivalent moment value (M = m. Max)

M_{max} is the larger of the two end moment, M_{max} is to be checked against (M_b/M). This method is more suitable for the cases where the loads are applied only at points of effective lateral restraint. Here, λ_{LT} is always found on the basis of uniform moment (β = 1.00) and an allowance for the actual shape of the moment diagram is made by providing following (design check) using an equivalent uniform moment.

M_x less than or equal to ($M_{R.LTB}/m_{LT.B}$).

where, $M_{R.LTB}$ is the resistance to lateral torsional buckling moment, M_x is the maximum moment about the strong axis, and $m_{LT.B}$ is the equivalent uniform moment factor for the lateral torsional buckling. In addition, M_x must not exceed the moment capacity of the cross-section about the strong axis equivalent slenderness ratio, λ_{LT} is obtained as described in the subsequent section.

Note : In certain situations, the maximum moment occurs within the span of beam.

The reduction in stiffness due to yielding would result in a smaller lateral buckling strength. The prediction according to above method based on the pattern of moments would not be conservative. Here, the second method is more appropriate. Fig. 15.8 shows the method.

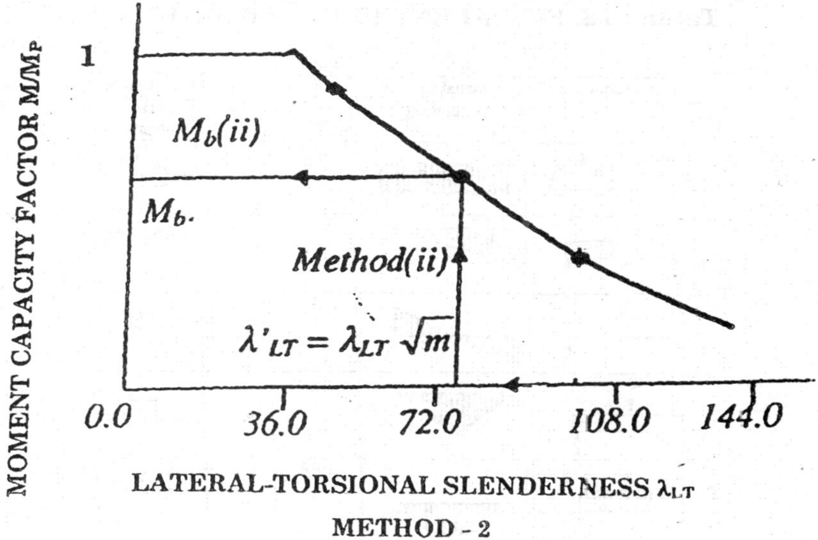

Fig. 15.8.

(ii) Using an effective slenderness ratio $\lambda_{LT} = \lambda_{LT}(m)^{1/2}$

λ'_{LT} is the effective lateral torsional slenderness ratio and λ_{LT} is the lateral torsional slenderness ratio.

The design capacity (bending strength), M_b is written by using the idea of lateral torsional slenderness λ_{LT} as below

$$\left[\frac{M_b}{M_p}\right] = f\left[\frac{1}{\lambda_{LT}^2}\right], \text{ where}$$

M_p is the fully plastic moment. The values of `m' factors for different beams and loads are given in Table 15.3.

15.11 SLENDERNESS CORRECTION

When the maximum moment occurs away from a braced point (e.g, when the beam is uniformly loaded in the span), a modification to the slenderness ratio, λ_{LT} may be made. Allowable critical stress is determined for an effective slenderness $n.\lambda_{LT}$, where, n is the slenderness correction factor. Its values for few cases have been tabulated in Table 15.4.

For the design purpose, one of the above two methods is used. If suitable values are adopted for m and n, both the methods yield identical results. The differences arises only in the way in which the correction is made. **In the second method** (i.e, in the n-factor method) the slenderness is reduced to take the advantage of the effect of the non-uniform moment.

While **in the first method**, (i.e, m-factor method), the moment is checked against lateral moment capacity M_b is reduced from M_{Max} to M by the factor-m. It is always safeto use m = n = 1. The design is based on the uniform moment case. In any situation, either m = 1 or n = 1 (i.e, only one method is used).

Table 15.3. EQVIALENT UNIFORM MOMENT

Beam and loads	Actual bending moment	M_{max}	m	Equivalent uniform moment
M M		M	1.0	
M		M	0.57	
M M		M	0.43	
W		Wℓ/4	0.74	
W		$W\ell^2/8$	0.88	
W W, ℓ/4, ℓ/4		Wℓ/4	0.96	

Lateral torsional slenderness ratio, λ_{LT} already defined for a particular material (i.e, E and σ_y are already known. These may be considered as a product of constant and $(M_P/M_{cr})^{1/2}$, λ_{LT} is called as the newly defined slenderness ratio.

Note : 1 : Buckling resistance (bending strength), M_b is always less than the critical moment, M_{cr}. This method is more conservative specially for low values of λ_{LT}.

Note 2 : When the maximum occurs within the span of the beam, this method is more appropriate.

Fig. 15.9. Shows second method

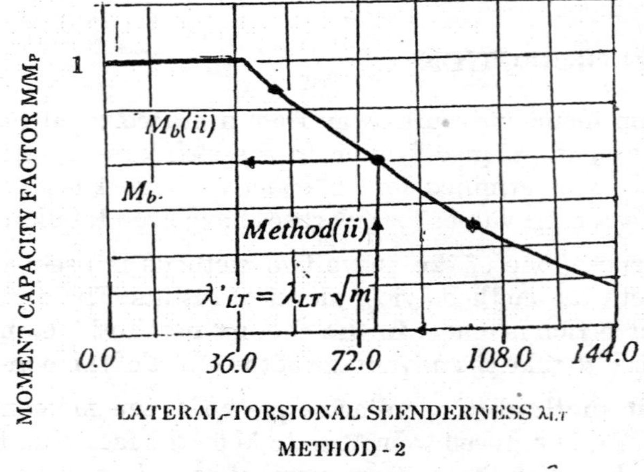

Fig. 15.9.

In this method a correction factor, n is applied to the slenderness ratio, λ_{LT} and the design strength is found for n. λ_{LT} where n = $\sqrt{M}$. This slenderness correction factor is explained in the subsequent section.

Both these methods, as represented by Fig. 15.7 and Fig. 15.8. may be compared

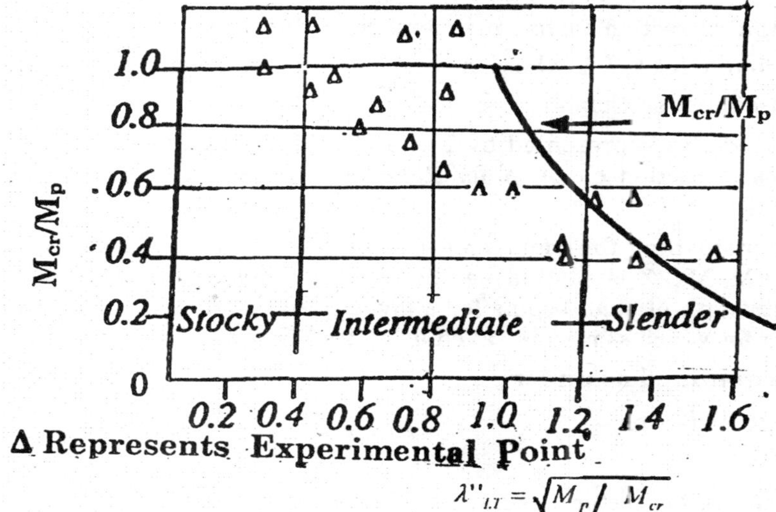

THEORETICAL ELASTIC CRITICAL MOMENT

Fig. 15.10.

The values of slenderness connection factor, `h' for the different load patterns are given in Table 15.4 attached here with.

15.12 DETERMINATION OF EQUIVALENT SLENDERNESS RATIO, λ_{LT}

The lateral torsional buckling resistance moment, M_b is given by the product of bending stress, s_b and the plastic modulus of the section

$$M_b = (\sigma_b \,.\, Z_p) \tag{i}$$

where, σ_b = Bending strength permitting for the susceptibility to lateral torsional buckling

Z_p = Plastic modulus of the section

It may be noted that $\sigma_b = \sigma_y$ for low values of slenderness of the beams and the value of s_b drops as shown in Fig. (15.10), as the beam becomes longer and the beam slenderness found as given below, increases. This behaviour is similar to columns. The slenderness ratio, λ_{LT} for the beams is given by

$$\lambda_{LT} = \left(\frac{\pi^2 . E}{\sigma_y}\right)^{1/2} \left(\frac{M_P}{M_E}\right)^{1/2} \tag{15.}$$

Lateral torsional slenderness ratio, λ_{LT} already defined for a particular material (i.e, E and σ_y are already known. These may be considered as a product of constant and $(M_p/M_{cr})^{1/2}$, λ_{LT} is called as the newly defined slenderness ratio.

Note : Buckling resistance (bending strength), M_b is always less than the critical moment, M_{cr}. This method is more conservative specially for low values of λ_{LT}.

Note : When the maximum moment occurs within the span of the beam, this method is more appropriate.

Fig. 15.9 show second method

In this method, a correction factor, n is applied to the slenderness ratio, λ_{LT} and the design strength is found for n.λ_{LT} where $n = \sqrt{m}$. This slenderness correction factor is explained in the subsequent section.

Both these methods, as represented by Fig. 15.7 and Fig. 15.8 may be compared. The values of slenderness correction factor, n' for the different load patterns are given in Table 15.4 attached here with.

Note : It is to note that the right side of Eq. 15. is the product of $(M_p/M_E)^{1/2}$ and $(\pi^2E/\sigma_y)^{1/2}$. In case, $(M_p/M_E)^{1/2}$ is plotted as the abscissa and the (M/M_p) as the ordinate torsional buckling strength of steel beams for a given grade of steel has been obtained. The limiting values of λ_{LT} for which s_b may be taken as s_y (for $\lambda_{LT} < 0.4$) leading to $(M_b = M_p)$

Value of λ_{LT} is obtained as under

$$\lambda_{LT} = (n, u, v)\ \ (l_e / r_y)$$

where,

n = slendernes correction factor (assumed as 1.0 unity)

u = Buckling parameter (assumed as 0.9 for rolled steel beams and 1.00 for the other sections

v = slenderness factor. It is dependent on the proportion of the yielding as the beam is restricted to supports, consequently, it results in a small reduction in the laterall buckling strength. In order to avoid overstressing at one end, an additional check, $(M_{max} < M_p)$ should also be satisfied. For preliminary assessment v is taken as 1.00 (unity)

For λ_{LT}, n, u, v and $\lambda = (l/r_y)$ are found as under:

(i) Slenderness correction factor, n

(i) n is found from the consideration of loading and the restrained conditions of the member. For members of uniform cross-section throughout the length, may be noted from Tables 15.5A, (B) and (C). For the members of varying cross-section throughout their length, the text is dealt at present.

(ii) **Buckling parameter, u and torsional index, x (A) For flanged sections,** symmetrical about the minor axis.

$$u = \left(\frac{4\, Z_P^2 \cdot \gamma}{A^2\, h_S^2}\right) \text{ and } \qquad \text{(ii)}$$

$$x = 0.566\, h_s \cdot (A/J)^{1/2} \qquad \text{(iii)}$$

Table 15.4. Slenderness Correction Factor, "n"

Slenderness correction factor, *n*			
Load pattern	Actual bending moment	*n*	Equivalent uniform moment
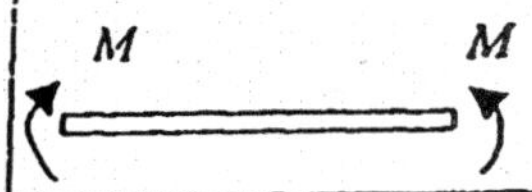 M M		1.0	
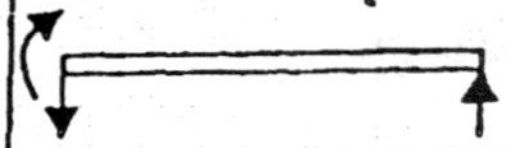	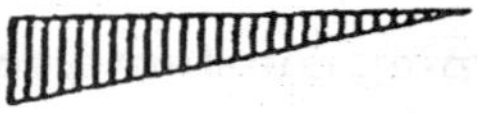	0.77	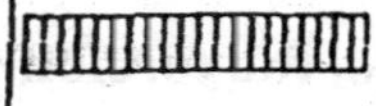
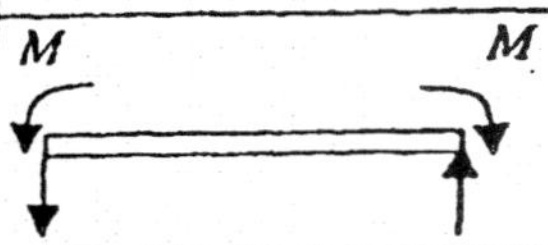 M M	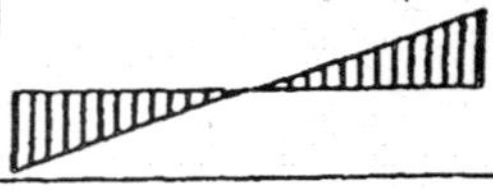	0.65	
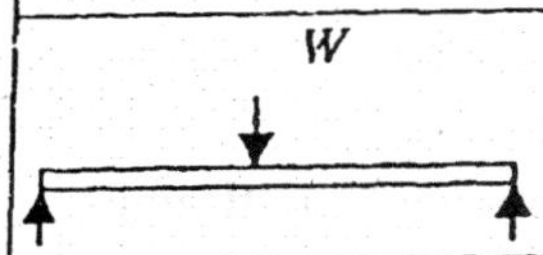 W		0.86	
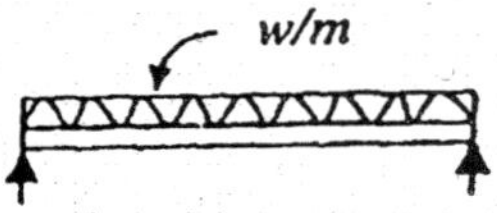 w/m		0.94	
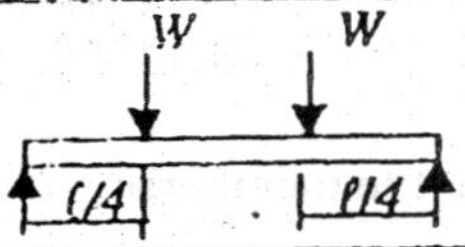 W W, l/4, l/4		0.94	
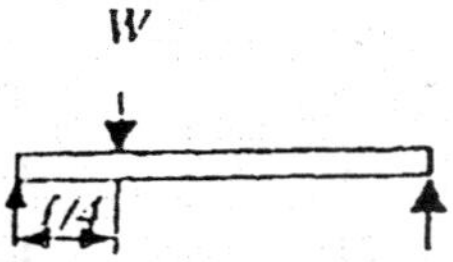 W, l/4	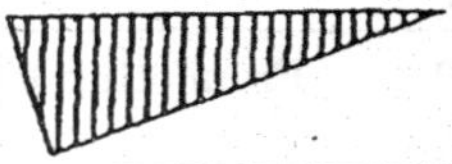	0.94	

(B) For the flanged sections, symmetrical about the major axis

$$u = \left(\frac{I_y . Z_P^2 . \gamma}{A^2 . C_w}\right)^{1/2} \quad \text{and}$$

$$x = 1.132 \left(\frac{A . C_w}{I_y . J}\right)^{1/2} \qquad \text{(iv)}$$

For rolled steel sections, u = 0.9 (Nethercot) and for all other sections, u = 1.00

where,

Z_p = Plastic modulus of the section about major axis

$\gamma = (1 - (I_y/I_x)$

I_x = Second moment of area about the major axis

I_y = Second moment of area about the minor axis

C_w = Warping constant

J = Torsional constant

h_s = Distance between the shear centres of the flange and

b_1, b_2 = widths of the flanges

(iii) J and C_w, the torsional and the warping constants, respectively

These are calculated from the approximately expressions

$$J = \frac{1}{3}\left(b_1\, t_1^3 + b_2\, t_2^3 + h_w . t_w^3\right)$$

$$C_w = \frac{t_1 . t_2 . b_1^3 . b_2^3 . h_s}{12\left(b_1^3\, t_1 + b_2^3\, t_2\right)}$$

where, b_1 and b_2 are the widths of flanges, t_1 and t_2 are the thicknesses of the flanges

t_w = thickness of web

h_s = Distance between the shear centres of the flanges

h_w = Depth of web

TABLE 15.5 (A). VALUES OF M AND N FACTORS FOR MEMBERS OF UNIFORM SECTION MEMBERS LOADED BETWEEN ADJACENT RESTRAINT

Description	Deestabilizing loads on the members			
	No		Yes	
	m	n	m	n
(i) Sections with equal flanges	1.0	From other tables	1.0	1.0
(ii) Sections with unequal flanges	1.0	1.0	1.0	1.0

TABLE 15.5 (B). VALUES OF M AND N FACTORS FOR MEMBERS OF UNIFORM SECTIONS. MEMBERS NOT LOADED BETWEEN ADJACENT LATERAL RESTRAINT

Description	Deestabilizing loads on the members			
	No		Yes	
	m	n	m	n
(i) Sections with equal flanges	From other tables	1.0	1.0	1.0
(ii) Sections with unequal flanges	1.0	1.0	1.0	1.0

Table 15.5 (C). CANTILEVERS WITHOUT INTERMEDIATE LATERAL RESTRAINT

	m	n	m	n
	1.0	1.0	1.0	1.0

For the plates of normal proportions

$J = 0.32\,b\,.\,t^3$ and therefore

For the flanged sections, symmetrical about the minor axes

$$x = (h_s)\left(\frac{\sum(b.t + h_w.t_w)}{\sum(bt^3 + h_w\,t_w^3)}\right)$$

where, b is the width of each plate element, and t, the thickness of these elements

(iv) v-slenderness factor (found from the following expression

$$v = \left[\left(4N(1-N) + \frac{1}{20}\left(\frac{\lambda}{x}\right) + \psi\right)^{1/2} + \psi\right]^{1/2}$$

where, N is the ratio

$\left[I_{cf} / (I_{cf} + I_{tf})\right]$ where,

I_{cf} is the second moment of the area of the compression flange about the minor axis of the section, and

I_{tf} is the second moment of the area of the tension flange about the minor axis of the section.

ψ is the mono symmetric index obtained as follows :

For I and T sections with lipped flanges (Determined as follows)

$\psi = 0.8(2N-1)(1 + D_L/D)$ for N > 0.5

$\psi = 1.0(2N-1)(1 + D_L/2D)$ for N < 0.5

where, D_L is the depth of lip and D is the overall depth of the section.

Note : when the flange is not lipped $D_L = 0.$, then

$\psi = 0.8(2N-1)$ for N > 0.5 and

$\psi = 1.0(2N-1)$ for N < 0.5

Alternatively

Slendereness factor, v (As described by Nethercot)

It is dependent on the proportions of yielding. It is restricted to the supports. Consequently, it results in a small reduction in the lateral buckling strength. In order to avoid over-stressing at one end, at additional check ($M_{max} < M_p$) should be satisfied.

For situations, where the maximum moment occurs away from a braced point (e.g, when the beam is uniformly loaded in the span), a modification to the slenderness, l_{LT} may be used.

Allowable critical stress, is determined for an effective slenderness (n. l_{LT}), where, n is the slenderness correction factor as given in Table 15.

For the design purpose, one of the above methods either (i) the moment correction factor method (m - method) or

(ii) slenderness correction factor method (n-method) may be used.

In case, suitable values are taken for m and n, both the methods give identical results. The difference arises only in the way in which the correction is made.

In **the n-factor method** the slenderness is reduced to take the advantage of the effect of the non-uniform moment.

Whereas, in the **m-factor method,** the moment to be checked against lateral/moment capacity, M_b is reduced from M_{max} to M by the factor, m. It is always safe to use m = n = 1 thereby the design is based on uniform moment.

In any situation, either (m = 1) or (n = 1), that is, any one method should be used.

15.13 DESIGN PROCEDURE FOR LATERALLY UNSUPPORTED BEAMS

Except for the stocky beams, the main design consideration for all laterally unsupported beams is **instability**. In assessing the load carrying capacity of the beams the value of M_{cr} is important. The important of instability and as a result, the controlling mode of failure is shown by the non-dimensional slenderness $(M_p/M_{cr})^{1/2}$

Adequacy of a given beam section over a span (with specified support conditions) carrying given load is checked) following the procedure as given below.

Step 1: The geometrical properties of given beam section are noted from ISI Handbook No. 1

Step 2 : The beam section is classified/typed by calculating the actual ratios (0.5 b_f/t_f) for the flanges and $[(D - 2h_2)/t_w]$ for the web and comparing them with the specified requirements for the outstand (i.e, buckling) as per IS 800-2007. Whether the section is plastic, or compact is decided.

Step 3 : Slenderness ratio, λ_{LT} is found

$$\lambda_{LT} = (n,\ u,\ v),\ \ (l_e/r_y)$$

Step 4 : Bending stress, σ_b is noted corresponding to calculated value of slenderness ratio, λ_{LT} as above.

Step 5 : Buckling resistance moment, M_b is obtained as below

$$M_b = (\sigma_b \,.\, Z_p)$$

Step 6 : Design (factored/ultimate) moment M_{max} is calculated. For the given beam and support conditions and the loads, [e.g, for the simply supported beam carrying uniformly distributed design (factored/ultimate) load

$$M_{max} = (w.\ l^2/8)$$

In case, M_{max}, design (factored) moment is less than the buckling resistance moment, M_b [i.e, $M_{max} < M_b$], the given beam section is adequate against lateral torsional buckling.

There are many differences between the assumptions made in the theory and the actual characteristics of the beam. At tc same time, the use of theoretical expressios is complicate. Theoretical prediction is elastic. It gives an upper bound for the true strength of the member.

In case, a non-dimensional plot with $(M_p/M_{cr})^{1/2}$ as absciss a and (M/M_p) as the ordinate is plotted as shown in Fig. 15.8 and Fig. 15.9 (where, M_p is the plastic moment capacity of the section, and M is the failure moment). It indicates clearly te lateral torsional behaviour of beam.

The use of such a curve as being representative of the actual test data may be validated by conducting the experiments on the beams.

A good design may be obtained by calculating the value of λ_{LT} and then, σ_b more accurately. **The design bending strength** of laterally unsupported beam as controlled by lateral torsional buckling, M_d may be found by using the relation.

$$M_d = (\beta_b \, \sigma_{bd} \, . \, Z_p)$$

Note : Equivalent uniform moments for the beams, and the segments of the beams between lateral supports may be obtained to find their relative instability. The lateral instability is checked for an equivalent moment ($M = m. \, M_{max}$), where, m is the equivalent factor.

In case, M_b is greater than M, the section proposed is satisfactory.

At the severely stressed locations, the local strength should be checked against development of M.

Example 15.1. A simply supported WB 400, @ 0.667 kN/m carries a uniformly distributed load (dead load 14 kN/m plus live load 12 kN/m over an effective span of 5.80 m. This beam is connected with the flanges of columns using double web cleat angles at both the ends. Check the adequacy of this beam against lateral torsional buckling.

Solution: Given beam section is simply supported in a vertical plane at its both the ends. The beam is fully restrained against lateral deflection at its both the ends. However, there is no restraint against rotation in plane.

Step 1 : Geometrical properties of WB 400, @ 0.667 kN/m

From ISI Handbook No. 1

Depth of the section, $D = 400$ mm

Width of the flange, $b_f = 200$ mm

Thickness of the flange, $t_f = 13$ mm

Thickness of the web, $t_w = 8.6$ mm

Moment of inertia

about strong axis, $I_{xx} = 23{,}426.7 \times 10^4$ mm^4

about weak axis, $I_{yy} = 1388 \times 10^4$ mm^4

Modulus of section,

about strong axis, $Z_{xx} = 1171.30 \times 10^3$ mm^3

about weak axis, $Z_{yy} = 138.80 \times 10^3$ mm^3

Radius at the root of flange, $r_1 = 13$ mm

Depth upto the root of flange, $h_2 = 29.75$ mm

Step 2 : Classification/type of given I-section

For the flange

Actual ratio, $(0.5b_f/t_f) = \left(\dfrac{0.5 \times 200}{13}\right) = 7.69$

Yield stress for steel, $\sigma_y = 250$ N/mm^2

Yield stress ratio

$\varepsilon = (250/s_y)^{1/2} = (250/250)^{1/2} = 1.00$

Requirement = (8.9 e) = (8.9 x 1.0) = (8.9)

Actual ratio 7.69 < Requirement 8.9

For the web

Actual ratio. $(\therefore\ h_2 = 29.75\ mm)$

$$\left[(D - 2h_2)/t_f\right] = \left[\frac{400 - 2 \times 29.5}{8.6}\right] = 39.59$$

Requirement

(83 x ε) = (83 x 1.0) = 83

Actual ratio 39.59 < Requirement 83

Given beam section is **Plastic.**

Step 3 : Check for lateral torsional buckling

(i) Effective length of compression flange (Effective span) = 5800 mm

Slenderness ratio of the beam about minor axis

Radius of gyration, r_y = 40.4 mm

$$\lambda = \left(\frac{l_e}{r_y}\right) = \left(\frac{5800}{40.4}\right) = 143.56$$

(ii) Overall depth of the beam, D = 400 mm

Thickness of flange, t_f = 13 mm

Torsional index, $(D/t_f) = \left(\dfrac{400}{13}\right) = 30.77$

Slenderness factor, v, depends on the proportion of the flanges and torsional index (D/t_f)

(iii) For equal flanges, v = 0.78 and the slenderness ratio (λ = 143.56)

(iv) Slenderness correction factor n = 1.00 (assumed)

Buckling parameter, u = 0.9

Equivalent slenderness of the beam

$\lambda_{LT} = (n.\ v.\ u.)\ .\ (l_e/r_y)$

$\lambda_{LT} = (1.0 \times 0.78 \times 0.9)\ .\ (143.56) = 100.78$

Step 4 : Bending stress factor for λ_{LT} = 100.78

σ_b = 122.6 N/mm^2

Elastic modulus of section, $Z_{xx} = 1171.3 \times 10^3$ mm^3

Assuming the shape factor for I-section, S = 1.12

Plastic modulus of section

$Z_p = (S. Z_{xx}) = (1.12 \times 1171.3 \times 10^3)\ mm^3$

$= 1311.856 \times 10^3 \times mm^3$

Step 5 : Resistance against buckling

$M_b = (\sigma_b \times Z_p)$

$$= \left(\frac{122.6 \times 1311.856 \times 10^3}{1000 \times 1000}\right)$$

$= 160.83$ kN-m

Step 6 : Design/factored/ultimate moment

Effective span = 5800 mm

Self weight of beam = 0.60 kN/m

Factored dead load = 1.35 (14 + 0.6) = 19.71 kN/m

Factored live (imposed load

= (1.50 x 12) = 18 kN/m

Total, w = (19.71 + 18) = 37.710 kN/m

Maximum design (factored) moment

$$M_{fd} = \left(\frac{wl^2}{8}\right)$$

$$M_{max} = \left(\frac{37.710 \times 5800^2}{8 \times 1000}\right) = 158.570\ kN - m$$

$M_{max} = 158.570$ kN-m is less than the moment of resistance of the beam section, $M_b = 160.83$ kN-m

Example 15.2 A beam section MB 450, @ 0.724 kN/m have been provided at a spacing of 3 m to carry an uniformly distributed load. Find the maximum dead load plus live load which may be supported by this beam section. Take the effective span for the beam 7.30 m.

Step 1. Geometrical properties of MB 450, @ 0.724 kN/m

From ISI Handbook No. 1

Overall depth of section, D = 450 mm

Width of the flange, $b_f = 150$ mm

Thickness of the flange, $t_f = 17.4$ mm

Thickness of web, $t_w = 9.4$ m

Moment of inertia

About strong axis, $I_{xx} = 30{,}390.8 \times 10^4\ mm^4$

About weak axis, $I_{yy} = 834.7 \times 10^4\ mm^4$

Modulus of section

About strong axis, $Z_{xx} = 1350.7 \times 10^3\ mm^3$

About weak axis, Z_{yy} = 111.2 x 10^3 mm^3

Radius at the root of flange, r_1 = 15 mm

Depth upto the root of flange, h_2 = 35.4 mm

Step 2 : Classification/type of section

Yield stress for steel, σ_y = 250 N/mm^2

Yield stress, ratio

$$\text{Epsilon, } \varepsilon = \left(\frac{250}{250}\right)^{1/2} = 1.0$$

For the flange

$(0.5\ b_f\ t_f) = (0.5 \times 150/17.4) = 4.3$

Requirement ratio

$(8.9\ \varepsilon) = (8.9 \times 1.0) = 8.9$

Actual ratio 4.3 < Requirement ratio 8.9

For the web

$$[(D-2h_2)/t_w] = \left[\left(\frac{450 - 2 \times 35.4}{9.4}\right)\right] = 40.3404$$

Requirement ratio

$83\ \varepsilon = (83 \times 1.00) = 83$

Actual ratio 40.3404 < Requirement ratio 83

MB 450, 0.724 kN/m

beam section is a **Plastic section**

Step 3 : Resistance/strength of beam section

Section modulus of beam section

Z_{xx} = 1350.7 x 10^3 mm^3

Z_{yy} = 111.2 x 10^3 mm^3

For I-section, shape factor,

S = 1-12 (assumed)

Plastic modulus of section

$Z_p = (S \times Z_{xx})$

$= (1.12 \times 1350.7) \times 10^3 = 1512.784 \times 10^3\ mm^3$

Resistance of beam section (i.e, moment capacity)

$$M = \left(\frac{\sigma_y \cdot Z_p}{\gamma_{msy}}\right) = \left(\frac{250 \times 1512.784 \times 10^3}{1000 \times 1.15}\right) = 328.866 \text{ kN-m}$$

Moment capacity/resistance of the beam section must be more than or at the most equal to moment due to factored/design moment due to due to external load/action.

$M_{fd} = M_{Resistance}$

Step 4 : Design moment due to factored (external) load, w kN/m

$$M = \left(\frac{w.l^2}{8}\right) = \left(\frac{w \times 7.30 \times 7.30}{8}\right)$$

$$w_{fd} = \left(\frac{328.866 \times 10^6 \times 8}{7.3 \times 7.3 \times 1000 \times 1000}\right) = 49.370 \text{ kN/m}$$

The factored dead load + live load (1.35 D.L. + 1.50 L.L.) = 49.370 kN/m

Which may be supported as service (characteristic) load shall be (DL + L.L.)

15.14 COMPUTER PROGRAMME FOR ANALYSIS AND DESIGN OF LATERALLY UNSUPPORTED BEAMS

To-day, it is `Computer age'. High speed and larger memory computers are avilable and these are used in every step of practical life.

One category of the persons prepare flow-charts and then, they write computer programme's. While other category of the persons write computer programmes and then, they prepare flow-charts. For the laterally unsupported beams, a computer programme has been written and its is named as CLUSB.

This computer programme deals with to find the adequacy of the beam section for the proposed beam which one is laterally unsupported.

Flow-chart for CLUSB

Brief step by step description of above programme has been given with respect to this programme Fig. 15.11 (A-D) shows flow chart. Every step (block) includes many computer instructions.

Dead load and live load and effective span, self-weight and the radius of gyration, are read. Depth of the beam section and its width, thickness of web, thickness of flange, yield stress and depth of the root of section h_2 are also read. In addition to these, the modulus of section and moment of inertia about strong and weak both.

The beam cross-section classification/type is found the axes are read.

Actual ratios for flange ($0.5\ b_f/t_f$) and web ($(D - 2h_2)/t_w$) are calculated. These ratios are compared with those specified in IS: 800-2007 for the requirement. Accordingly, the beam section is marked and noted as plastic, compact or semi-compact.

Then, the programme proceeds to check lateral torsional buckling and the equivalent slenderness ratio λ_{LT} = (n.u.v) and (l_e / r_y) is calculated.

$$\text{Slenderness ratio} = \left(\frac{\textit{Effective length}}{\textit{Radius of gyration, about weak axis}}\right)$$

is found. Torsional index is found. The values of n, u and v are read. Corresponding to slenderness ratio, λ_{LT} bending stress, σ_b is noted.

Plastic modulus of section is calculated as shape factor = 1.12 assumed as multiplied by Z_{xx}. The resistance to buckling (i.e, moment capacity/strength of beam section is obtained.

BRST = $(\sigma_b . Z_p/1000)$.

Factored/design load W_{fd} = (1.35 D.L. + 1.5 L.L.) kN per unit length is obtained. Maximum bending moment due to the external load/action is calculated

$M_{max} = (w_{fd} . le^2/8)$

In case, the resistance of the beam section (i.e, its strength) is more than the maximum bending moment, then, it is declared that the beam section is adequate.

INPUT-OUTPUT STATEMENTS FOR CLUSB (i.e. To Check adequacy of a given beam section laterally unsupported).

1. Input statements:

Following are the input data used in programme CLUSB

1. READ*, WDL, WLL, EL, WSB, RY

Dead load, WDL = 14.0 kN./m

Live-load, WLL = 12.0 kN/m

Length of the compression flange, EL = 5.80 m

Self-weight of the beam, WSB = 0.60 kN/m

Minimum radius of gyration, RY = 40.40 mm

2. READ&*, DTH, WTH, TW, TFL, SIGY, H2

Depth of I-beam section, DTH = 400 mm

Width of beam section, WTH = 200 mm

Thickness of the web, TW = 8.60 mm

Thickness of the flange, TFL = 13.0 mm

Yield stress of grade of steel, SIGY = 250 N/mm^2

Radius at the root of flange, H2 = 29.75 mm

3. READ* = ZXX, ZYY

Modulus of the section about x-axis, ZXX = 1171.30 cm^2

Modulus of the section about y-axis, ZYY = 138.80 cm^2

4. READ*, AIXX, AIYY

Moment of inertia of beam section about x-axis, AIXX = 23426.70 cm^2

Moment of inertia of beam section about x-axis, AIXX = 1388.0 cm^4

5. READ*, EN, U, V

EN, U and V are the factors used to modify slenderness ratio for lateral tension buckling, EN = 1.00, U = 0.90 and V = 0.78.

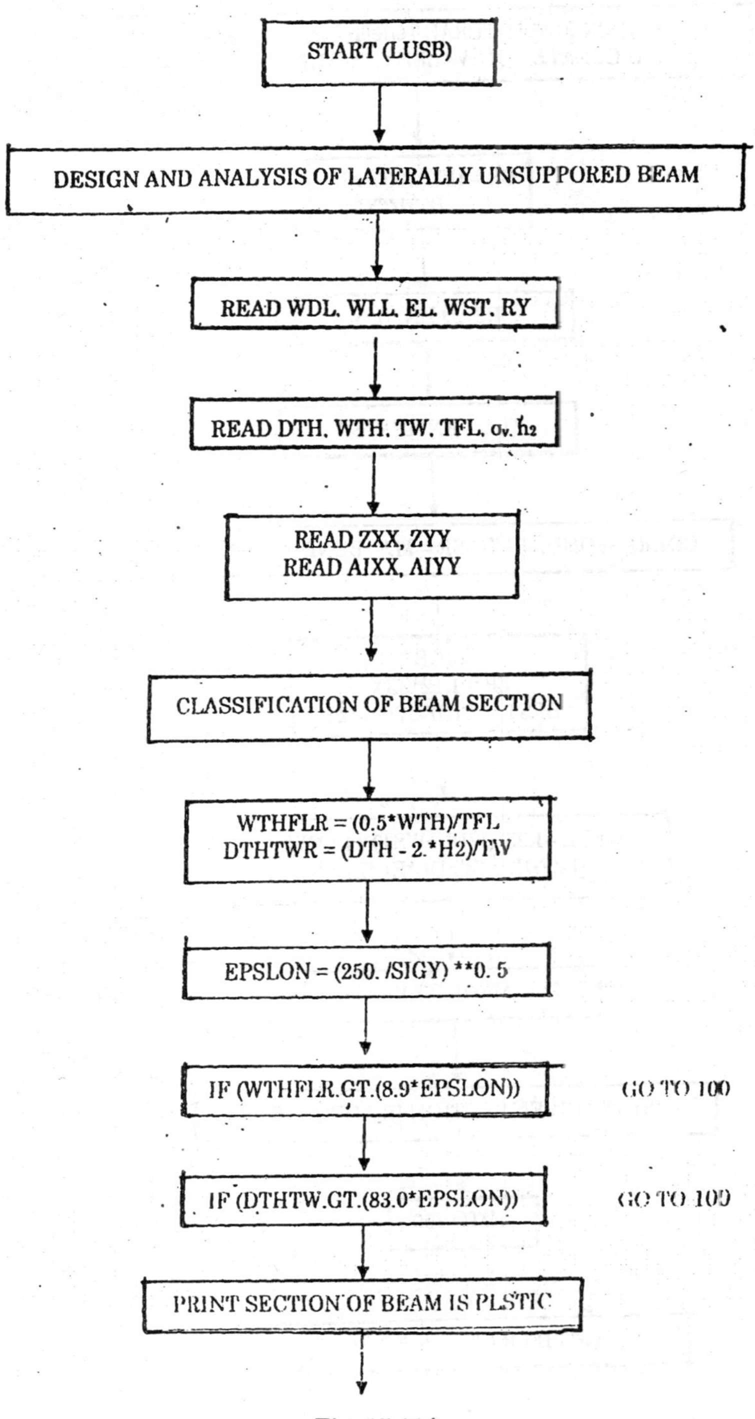
START (LUSB)
DESIGN AND ANALYSIS OF LATERALLY UNSUPPORED BEAM
READ WDL. WLL. EL. WST. RY
READ DTH. WTH. TW. TFL. σv. h2
READ ZXX, ZYY
READ AIXX, AIYY
CLASSIFICATION OF BEAM SECTION
WTHFLR = (0.5*WTH)/TFL
DTHTWR = (DTH - 2.*H2)/TW
EPSLON = (250. /SIGY) **0. 5
IF (WTHFLR.GT.(8.9*EPSLON))
GO TO 100
IF (DTHTW.GT.(83.0*EPSLON))
GO TO 100
PRINT SECTION OF BEAM IS PLSTIC

Fig. 15.11A.

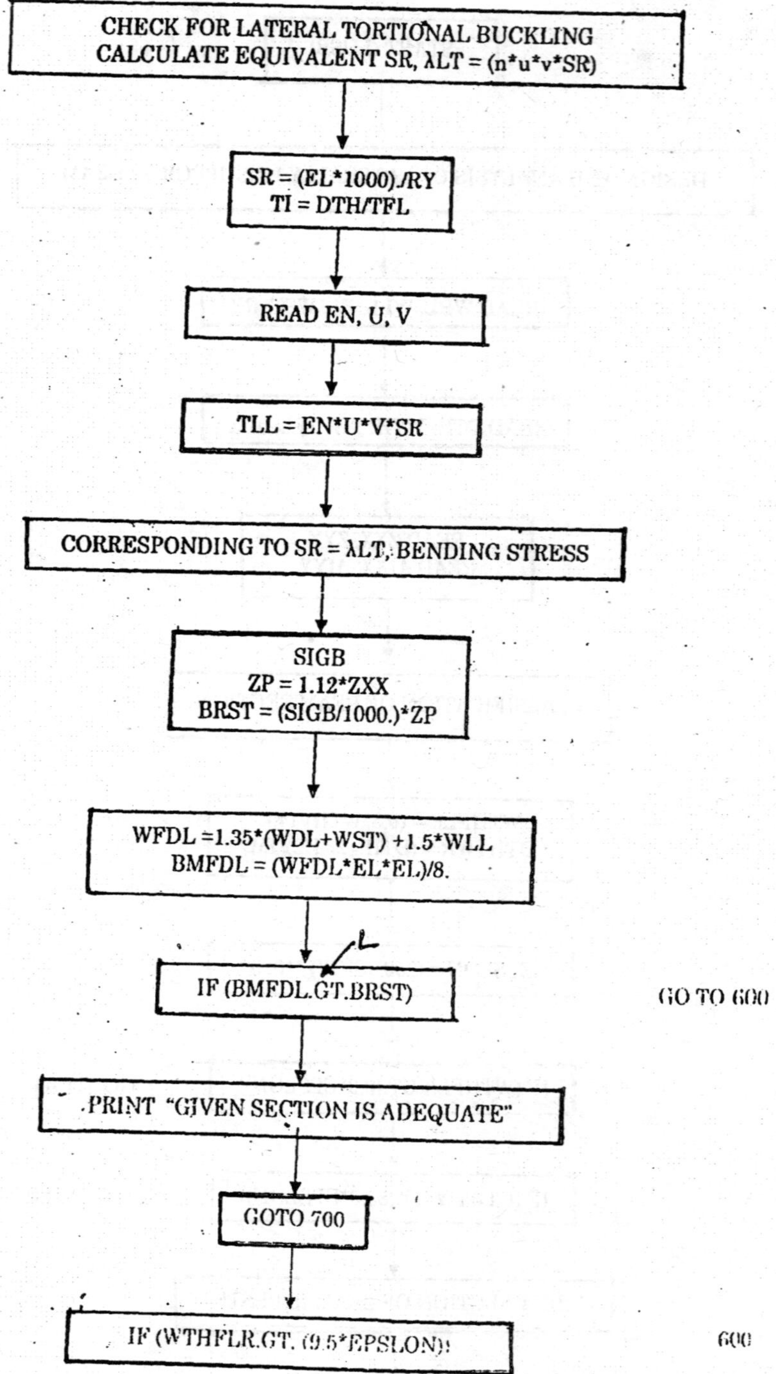
CHECK FOR LATERAL TORTIONAL BUCKLING
CALCULATE EQUIVALENT SR, λLT = (n*u*v*SR)
SR = (EL*1000)./RY
TI = DTH/TFL
READ EN, U, V
TLL = EN*U*V*SR
CORRESPONDING TO SR = λLT, BENDING STRESS
SIGB
ZP = 1.12*ZXX
BRST = (SIGB/1000.)*ZP
WFDL =1.35*(WDL+WST) +1.5*WLL
BMFDL = (WFDL*EL*EL)/8.
IF (BMFDL.GT.BRST)
GO TO 600
PRINT "GIVEN SECTION IS ADEQUATE"
GOTO 700
IF (WTHFLR.GT. (9.5*EPSLON))
600

Fig. 15.11B

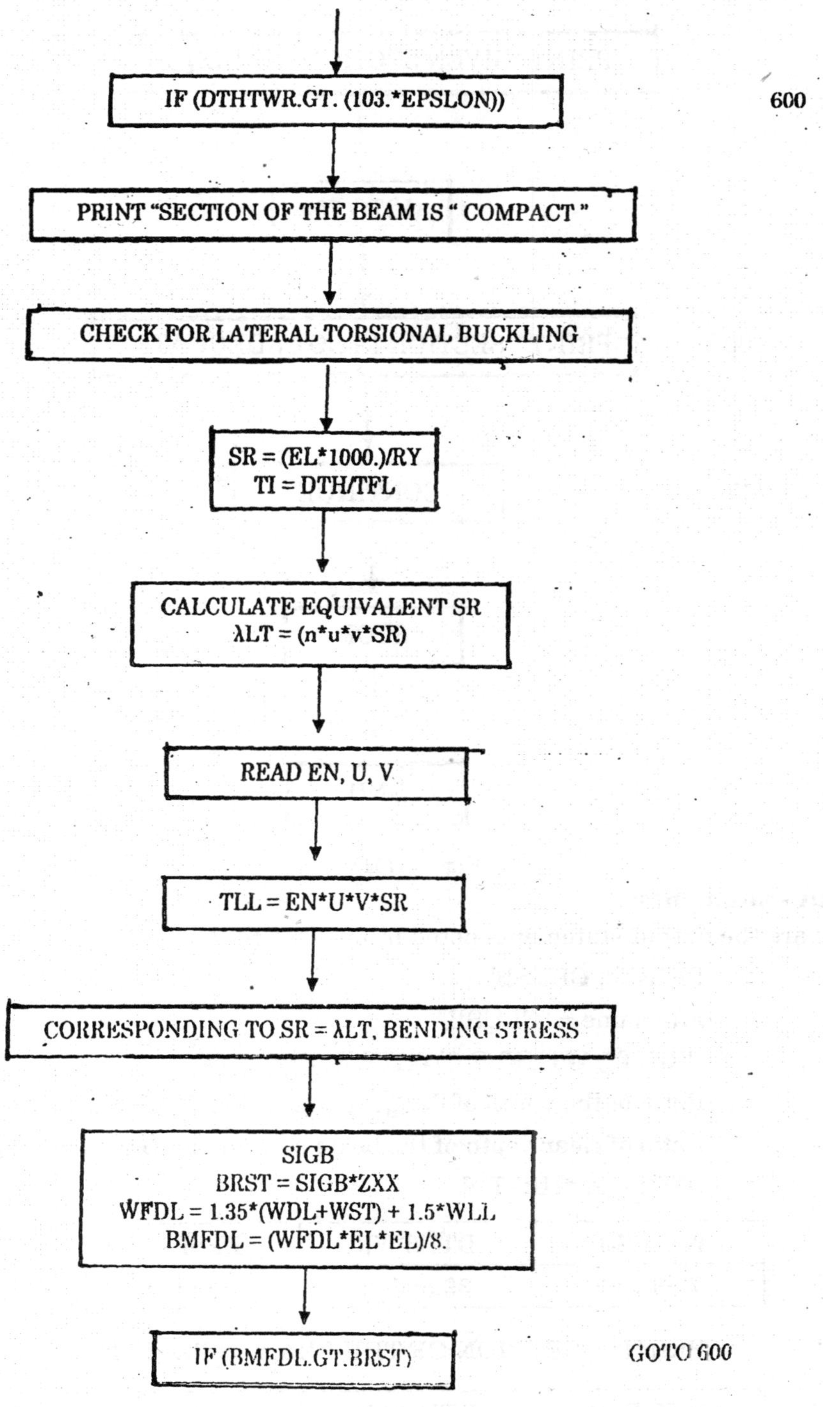
IF (DTHTWR.GT. (103.*EPSLON))
600
PRINT "SECTION OF THE BEAM IS " COMPACT "
CHECK FOR LATERAL TORSIONAL BUCKLING
SR = (EL*1000.)/RY
TI = DTH/TFL
CALCULATE EQUIVALENT SR
λLT = (n*u*v*SR)
READ EN, U, V
TLL = EN*U*V*SR
CORRESPONDING TO SR = λLT, BENDING STRESS
SIGB
BRST = SIGB*ZXX
WFDL = 1.35*(WDL+WST) + 1.5*WLL
BMFDL = (WFDL*EL*EL)/8.
IF (BMFDL.GT.BRST)
GOTO 600

Fig. 15.11C

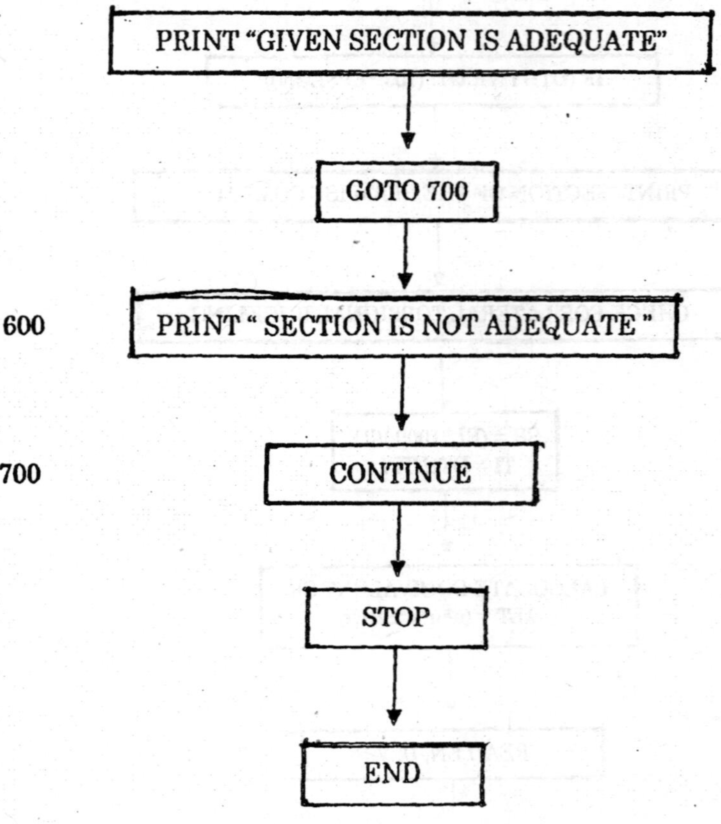

Fig. 15.11D

II. Output statements:

Following are the output statements noted from CLUSB.

1. PRINT*, CLUSB"

   File name = "CLUSB"

2. PRINT*, WTHFLR, DTHTWR, EPSLON"

   Ratio of the width of flange/its thickness, WTHFLR = 0.5 (b_f/t_f)

   Ratio of clear depth of the beam section and thickness of the web, (DTH – 2. *HB)/TW

WTHFLR	DTHTWR	EPSLON
7.692307	39.59302	1.00000

3. PRINT*, "SECTION OF THE BEAM IS PLASTIC"

WTHFLR	DTHTWR	EPSLON
—	—	—

4. PRINT*, "SECTION OF THE BEAM IS COMPACT"

5. PRINT*, BMFDL BRST"

6. PRINT*, BMFDL, BRST

Bending moment for factored design load, BMFDL (kN-m)

Resistance of the beam section in bending, BRST (kN-m)

BMFDL (kN-m)	BRST (kN-m)
158.5706	160.833

In case BRST is greater than BMFDL, it is declared.

7. PRINT*, "BEAM SECTION IS ADEQUATE".

PROBLEMS

15.1. A beam of simple framed building is made up of LB 550, @ 0.863 kN/m (b_f = 190 mm, t_f = 15 mm, A = 10.97 mm², t_w = 9.9 mm) and carries a total load of 400 kN uniformly distributed over the entire span. It is connected to the flange of column at ends of section WB 600, @ 0.481 kN/m b_f = 200 mm, t_f = 10 mm, A = 6133 mm², t_w = 7.4 mm). Using 20 mm ϕ rivets, design a suitable connection for the joints. Draw a neat sketch of the joint.

REFERENCES

1. David, A. Nethercot `Limit State Design of Structural Steel work' Third Edition (Based on revised BS : 5950 : Part 1, 2000), second Edition published in 1991 by Chap and Hall Reprinted 1994, 1995, 1996.
2. R.S. Narayanan and A.W. Beeby `Introduction to Design for Civil Engineers'. Simultaneously published in the USA and Canada by SPON PRESS (London and New York), First published in 2001 by Spon Press, 11 New Feter Lane, London, EC 4P 4 EE. 29 West 35th Street, New York, NY 10001.
3. Rangachari Narayanan, V. Kalyanaraman *et al.* (INSDAG) `Teaching Resource on structural steel Design', Vol.: 1, Organised by Institute for Steel Development And Growth (INSDAG) Delhi College of Engineering, Delhi, (June-July, 2001), Sponsored by Ministry of Steel, Government of India.
4. Draft for the approval for IS : 800 - (new revised code yet to be issued, circulated by Bureau of Indian Standards (BIS), Manak Bhawan, New Delhi.
5. A.B. Clark and S.H. Coverman - Structural Steel Work (Limit State Design), First Published in 1987, by Chapman and Hall Ltd. 11 New Fetter Lane, London, EC 4P 4EE, 29 West 35th Stress, New York 10001.
6. Adams, P.F. Kulak, G.L. and Gilmor, M. (1990) - ``Limit State Design in Structural Steel", 4th Edition, Canadian Institute of Steel Construction.
7. British Standards Institution - ``BS : 5950, Part - 1, Structural Use of Steel-work in Buildings", British Standards Institution, London, 1985.

8. AISC-LRFD. -``Load and resistance Factor Design Specification for Structural Steel Buildings". American Institute of Steel Construction (AISC), Chicago, III, 1993.

9. British Standards Institutions - (1996, 1997, 1998) BS : 6399 : Parts 1, 2 and 3. ``Design Loading for Buildings".

10. British Steel (1996). - ``The Prevention of Corrosion on structural Steel-work".

11. Ramchandra and Virendra Gehlot - ``Design of Steel Structures - 1" Twelfth Edition, 2007, published by M/s Scientific Publishers (India), 5-A, New Pali Road, Jodhpur.

12. Ramchandra and Virendra Gehlot -``Limit State Design of Concrete Structures" Third Ed., 2007 published by M/s Scientific Publishers (India), 5-A, New Pali Road, Jodhpur.

13. Eurocode 3. - ``Design of Steel Structures", British Standards Institute 1992.

14. Ramchandra- ``Stub Column Tests on Indian Standard H-Beams". A Dissertation (M.E.) submitted to University of Jodhpur, August, 1970.

15. Virendra kumar Gehlot (M.E. Student) - A Seminar Report submitted to Head of Department of Structural Engineering, MBM. Engineering College, JNV University of Jodhpur, Jodhpur

16. Virendra Kumar Gehlot. M.E. Thesis `Limit State Design of Steel Beams', Department of Structural Engineering, MBM Engineering College, Jodhpur.

16

LIMIT STATE OF COLLAPSE IN BENDING AND SHEAR DESIGN OF BUILT-UP (COMPOUND) BEAMS (SB-V)

16.1 INTRODUCTION

Built-up beams are the fabricated members interconnected by using more than one element, which acts as a single beam. A crane gantry girder is an example of a built-up. An I-section with its web vertical and a channel section with its web horizontal are connected together.

The built-up beams are also termed as **compound beams or compound girders**. The built-up beams are used when the span, load and corresponding bending moment are of such magnitudes that single rolled steel beam section becomes inadequate to provide required section modulus. The built-up beams are also used when rolled steel beams are inadequate for limited depth. In building construction, the depth of beam is limited by the space provided by the architect's drawings. The beams of small depth do not provide required section modulus. Therefore, plates are attached to the beams. The strength of rolled steel beam is increased by adding plates to its flanges. Which is therefore, one method of forming built-up section. The other method is to compound a number of rolled steel sections, themselves. The built-up beam sections commonly used are showing in Fig. 16.1. The built-up sections shown in Fig. 16.1 (A) and (B) are used for heavy loads and small spans. The built-up section shown in Fig. 16.1 (C) is also used for very heavy loads and small spans. The built-up section shown in Fig. 16.1 (D) is used for light loads and large spans. In this built-up section, the neutral axis passes in between the two flanges in contact. The bending stresses in these two flanges are very small. These two flanges in contact practically do not provide any moment of resistance. Hence, such section is provided where deflection is a major consideration. The built-up sections shown in Fig. 16.1 (E) and (F) are provided for the gantry girders. The top flange in these sections have been strengthened. This increases the resistance of the section for the lateral loads. In built-up beams in general cover plates are attached to the flanges of I-section. The depth between flanges cannot be increased. The material in compression flange cannot be utilised fully. As such the built-up beams remain uneconomical. Neglecting the moment of inertia of the attached cover plates about their own axes the moment of inertia increases approximately A . $(d/2)^2$ for every plate attached to main beam section.

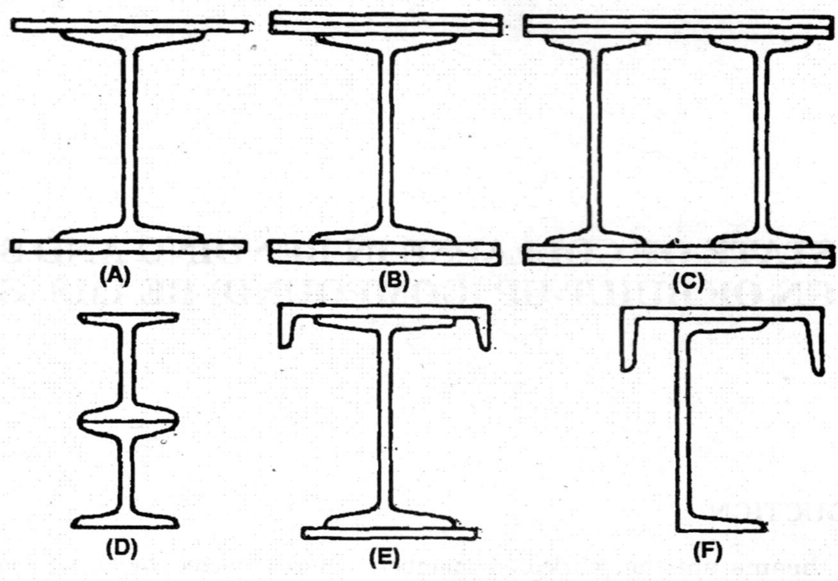

Fig. 16.1. Built-up beam sections.

In the riveted built-up beams, the area of tension flange is reduced by rivet holes. The deduction is made for rivet holes in tension flange while determining moment of inertia and section modulus of the built-up beam section. In actual practice, it is usual to make allowance for rivet holes in the both flanges. The rivets connecting flange plates are staggered so that each flange is reduced by one rivet hole as shown in Fig. 16.2.

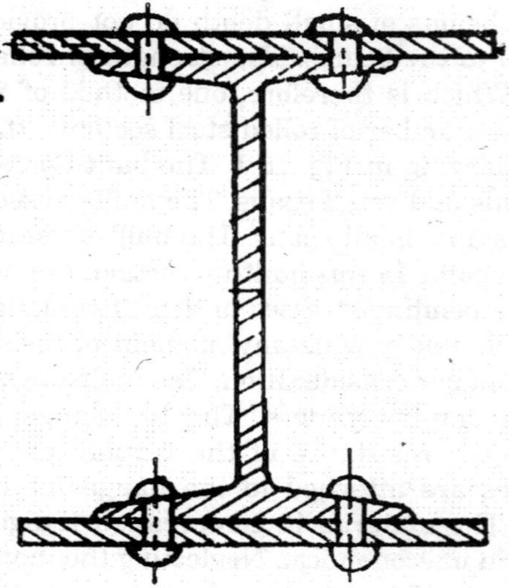

Fig. 16.2.

The curtailment of cover plates of built-up beams is similar to curtailment of cover plates of plate girder. It is discussed in chapter - 18 (Design of plate girders).

16.2. DESIGN OF BUILT-UP BEAMS

The built-up beams are designed by trial and error method. The built-up section is selected and it is checked for the stresses. The built-up beams with solid webs are proportioned in general on the basis of the moment of inertia of the gross cross- section with neutral axis taken at the centroid of the section.

The following are the usual steps in design of built-up beams :

Step 1. The effective span and load required to be carried by the built-up beam are known. The bending moment M and maximum shear F in built-up beam are calculated.

Step 2. In the beginning, the value of yield stress f_y, for structural steel to be used is assumed. The permissible bending stress σ_{bc} is assumed as 0.66 f_y. The required section modulus, Z for the beam section is calculated

Required section modulus, $Z = (M/\sigma_{bc})$,

where, σ_{bc} is allowable bending compressive stress.

For laterally supported beams, allowable bending compressive stress is equal to the allowable bending tensile stress. For laterally unsupported beams, allowable bending compressive stress is determined by the condition of lateral restraint and the section of built-up beams.

From ISI Handbook No. 1, a trial section for beam is adopted having the modulus of section, Z about 25 to 50 percent in excess of that required. The geometrical properties of beam section, are noted. The permissible bending compressive stress corresponding to the geometrical dimensions of beam section and restraint conditions is found from IS : 800-1984. As a check, the section modulus, Z needed corresponding to actual permissible stress is found and it is shown that it is less than the section modulus of the beam provided.

When the depth of beam is not limited, then the rolled steel beam is selected for the maximum section modulus. The plates are provided for the difference of section modulus (i.e., between that required and that provided by the rolled steel beam). When the depth of beam is limited, then, the usual practice is to select from ISI Handbook No. 1, the strongest rolled steel beam that will allow for necessary thickness of cover plates, at top and bottom. The area of cover plates required is then found by trial and error method. The moment of inertia of trial plated beam section is found, and then checked. It involves a considerable amount of work. The required area of cover plates may be found from the following guidance, which will result in satisfactory selection of the required area of cover plates at the first trial.

The moment of inertia of any area about any axis is the second moment of that area about that axis.

Let the moment of inertia of built-up plated beam (symmetrical section) with one cover plate attached with each flange be given by

$$I = \left[I_{beam} + 2.A_p \cdot \left(\frac{h}{2}\right) \cdot \left(\frac{h}{2}\right) \right] \quad \text{(i)}$$

where, I_{beam} is the moment of inertia of the rolled beam section, A_p is the area of cover plates in one flange and h is the distance between the centroids of the top and bottom flange plates.

The distance between the centroids of the top and bottom flange plates, h may be adopted equal to the depth of rolled steel beam itself, since the thickness of flange plates is small compared with the depth of beam section.

From the expression (i)

$$\left(\frac{I}{(h/2)}\right)=\left(\frac{I_{beam}}{(h/2)}+2\,A_p\cdot\left(\frac{h}{2}\right)\left(\frac{h}{2}\right)\right) \quad \text{(ii)}$$

In comparsion to the momnt of inertia, it is convenient to use the modulus of sections.

$$\text{or } Z_{reqd} = (Z_{beam} + A_p.h)$$

$$\text{or} \quad 2A_p = \left(\frac{Z_{regd} - Z_{beam}}{h/2}\right) \quad \text{(iii)}$$

where, $Z_{reqd} = Z_{read} = \left(\frac{I}{(h/2)}\right)$ and $Z_{beam} = \left(\frac{I_{beam}}{h/2}\right)$

$2A_P$ = Area of two cover plates (one plate) attached with each flange.

The expression (iii) gives the area of cover plates in one flange, when there are no rivet holes in the plates. In the riveted built-up beams, an allowance of about 13 percent of the area of cover plates found is added. The required area of cover plates in one flange is then provided by one or two plates and preferably not more than three plates. The rolled steel beam and cover plates provide the required section modulus. The width of cover plates and thickness are decided with the restrictions of the outstand.

In case, only one cover plate is to be attached than it is connected with the compression flange. However, in case thickness of this cover plate exceeds 1.5 times, the thickness of flange, it is more efficient to attach cover plates with both the flanges. These cover plates are checked individually for slenderness and outstands to avoid local buckling.

Step 3. From steel section tables, select a suitable section. The outstand of flange plates that is their projection beyond the outer line of connection shall not exceed the relevant values given in IS : 800-1984 in (a), (b) and (c) below. The area of excess flange shall be neglected while calculating the effective geometrical properties of the section.

These values have also been expressed in terms of yield stress of the structural steel as it has been done in recommending the various permissible stresses and other.

(a) Flanges and plates in compression with the unstiffened edges : $256t/(f_y)^{1/2}$ subjected to a maximum of 16 t.

(b) Flanges and plates in compression with the stiffened edges : 20 t to the innermost face of the stiffening.

(c) Flanges and plates in tension 20 t.

Step 4. The built-up section is checked for bending stresses in the extreme fibres. The gross moment of inertia of the built up section is found on the basis of gross cross-section of the built-up section with neutral axis assumed at the centroid of the section. The maximum tensile stress in bending is found from the maximum compressive stress in bending by increasing the maximum compressive stress in bending in the ratio of gross area to the net area of tension flange. The effective sectional area for the compression flange in the gross area with deductions for excessive widths of the plates as discussed for compression members and for open holes occurring in a plane perpendicular to the direction of stress at the section being considered. The effective sectional area for the tension flange is the gross area with deductions for the rivet holes.

Actual bending compressive stress

$$\sigma_{bc.cal} = \left(\frac{\text{Max Bending Moment}}{\text{Gross Moment of Inertia}}\right) \times \left(\begin{array}{c}\text{Distance of extreme fibre in} \\ \text{compression from neutral axis}\end{array}\right)$$

$$\sigma_{bc.cal} = (M/I_{xx})\, y_1$$

The actual bending tensile stress $\sigma_{bt.cal}$ is determined by increasing the actual bending compressive stress in the ratio of gross area of the tension flange to the net area of the tension flange.

$$\sigma_{bt.cal} = \sigma_{bc.cal} \times \left(\frac{\text{Gross area of the tension flange}}{\text{Net area of the tension flange}}\right)$$

The actual bending stresses should not exceed allowable bending stresses.

Step 5. The built-up section is checked for shear force and deflection.

Example 16.1. A simply supported beam is to support a uniformly distributed load 70 kN/m excluding weight of the beam, over a clear span of 8 metres. Design a plated rolled steel beam if MB 500, @ 0.869 kN/m and 10 mm thick plates are only available. The compression flange of the beam is laterally restrained.

Design :

Step 1 : Effective span of beam

Clear span of beam = 8

Provide 300 mm wide bed blocks on each side to support the beam

Effective span = Centre to centre of bearing = 8.30 m

Step 2 : Total uniformly distributed load

Uniformly distributed load on the beam = 70 kN/m

Assume self-weight of beam = 1.5 N/m

Total uniformly distributed load = 71.50 kN

Step 3 : Bending moment

The maximum bending moment, M occurs at the centre,

$$M = \left(\frac{w.l^2}{8}\right) = \left(\frac{71.50 \times 8.30 \times 8.30}{8}\right) = 615.7 \text{ m-kN}$$

Let, the value of yield stress for the structural steel to be used be 250 N/mm² (MPa)

Step 4 : Permissible bending stress

The maximum permissible bending stress in compression and that in tension are equal for the laterally supported beam and it may be assumed as

$$\sigma_{bc} = (0.66 \times 250) = 165 \text{ N/mm}^2$$

Step 5 : Section modulus required

$$Z = \left(\frac{M}{\sigma_{bc}}\right) = \left(\frac{615.7 \times 10^6}{165}\right) = 3731.51 \times 10^3 \text{ mm}^3$$

Step 6 : Properties of trial section

From the steel section tables, for MB 500, @ 0.869 kN/m

$I_{xx} = 45218.3 \times 10^4\ mm^4$, $t_w = 10.2$ mm, $t_f = 17.2$ mm

$Z_{xx} = 1808.7 \times 10^3\ mm^3$, h = 500 mm,

$b_f = 180$ mm, $r_y = 35.2$ mm, $h_2 = 37.95$ mm

Provide two cover plates 340 mm x 10 mm on each flange.

Provide 22 mm diameter rivet to connect flange flanges.

Step 7 : Check for fibre stress

Gross area of tension flange

$= (180 \times 17.2 + 2 \times 340) = 9932\ mm^2$

Net area of tension flange

$= [9932 - 23.5\ (17.2 + 20)] = 9057.8\ mm^2$

Gross moment of inertia of built-up section

$I_{xx} = [45218.3 + 2 \times 2 \times 2 \times 34 \times (26)^2] \times 10^4 = 137418.3 \times 10^4\ mm^4$

Actual bending compressive stress

$$\sigma_{bc.cal} = \left(\frac{615.7 \times 10^6 \times 270}{137418.3 \times 10^4}\right) = 120.973\ N/mm^2$$

Actual bending tensile strength

$$\sigma_{bc.cal} = 120.973 \times \left(\frac{9932}{9057.8}\right) = 132.648\ N/mm^2$$

Actual bending stress in compression and that in tension do not exceed

$(0.66 \times 250) = 165\ N/mm^2$ (MPa)

Provide MB 500, @ 0.869 kN/m with two cover plates 340 mm x 10 mm on each flange as shown in Fig. 16.4.

Example 16.2 : Propose a trial section for a simply supported rolled steel beam section with one cover plate attached with each flange to carry a uniformly distributed load 50 kN/m inclusive of dead load plus self weight plus imposed (live) load. The effective span of beam is 8 m use limit state method of design. The depth of beam should not be more than 500 mm. Take value of yield stress for the structural steel to be used 250 N/mm². The compression flange of the beam is laterally supported by floor construction.

Design : Effective span of beam = 8 m Yield stress for the structural steel to be used is 250 N/mm². The compression flange of the beam is laterally supported by floor construction.

Step 1 : The beam is to carry a uniformly distributed service load = 50 kN/m. It is dead load plus self-weight imposed load. Partial factor of safety for loads, $\gamma_{fL} = 1.5$. The load is uniformly distributed load.

Design (factored/ultimate) load

$w_{Fd} = (1.5 \times 50) = 75$ kN/m

Step 2 : Design (factored/ultimate) moment occurs at the centre

$$M_{fd} = \left(\frac{w.L^2}{8}\right) = \left(\frac{75 \times 8 \times 8}{8}\right) = 600 \text{ kN} - \text{m}$$

Step 3: Moment of resistance/strength of beam section required.

$$MR = \left(\frac{\sigma_y . Z_p}{\gamma_{ms}}\right)$$

where, Z_p is the plastic modulus of the section.

$$\left(\frac{\sigma_y . Z_p}{\gamma_{ms}}\right) = \left(600 \times 10^3 \times 10^3\right) \text{ N} - \text{mm}$$

$$Z_p = \left(\frac{600 \times 10^3 \times 10^3 \times 1.15}{250}\right) \text{mm}^3$$

$$Z_P = 2760 \times 10^3 \text{ mm}^3$$

Step 4 : Elastic section modulus Let the shape-factor. S be 1.12, since, it shall be an I-section (Assumed). Total elastic section modulus required.

$$Z_{xx} = \left(\frac{Z_p}{S}\right) = \left(\frac{2760 \times 10^3}{1.12}\right) = 2464.286 \times 10^3 \text{ mm}^3$$

Step 5 : Overall depth of beam-section should not exceed 500 mm. Let one cover plate 8 mm thick shall be attached with each flange Depth of beam section shall not be more than h = (500 – 2 × 8) = 484 mm

Step 6 : Let the depth of beam section

h = 450 mm, and

Let it be ISWB 450, @ 0.794 kN/m Section modulus provided by this section.

$$Z_{beam} = 1558.1 \times 10^3 \text{ mm}^3$$

Width of flange, t_f = 200 mm

Section modulus to be provided by two plates (one plate shall be attached with each flange)

$$Z_{PL} = (2464.286 \times 10^3 - 1558.1 \times 10^3)$$

$$Z_{PL} = 906.186 \times 10^3 \text{ mm}^3$$

Step 7 : Let width of cover plate be 300 mm (i.e, larger than the width of flange) Projection of plate on each side

$$= \frac{1}{2} \times (300 - 200) = 50 \text{ mm}$$

Area of one plate be A_{PL} Section modulus contributed by these two plates about neutral axis

$$2\,A_{PL} \times \left(\frac{450}{2}\right) = 906.186 \times 10^3$$

$$A_{PL} = \left(\frac{906.186 \times 10^3}{450}\right) = 2013.746\ \text{mm}^2$$

Width of each plate = 300 mm

Thickness of each plate = $\left(\frac{2013.746}{300}\right)$ **mm = 6.712 mm**

Provide 8 mm thick plate 300 mm width with projection from the flange 0.5 (300 - 200) = 50 mm

Trial beam section shall be

ISWB 450, @ 0.794 kN/m. with one cover-plate 300 mm x 8 mm attached with each flange as shown in Fig. 16.3.

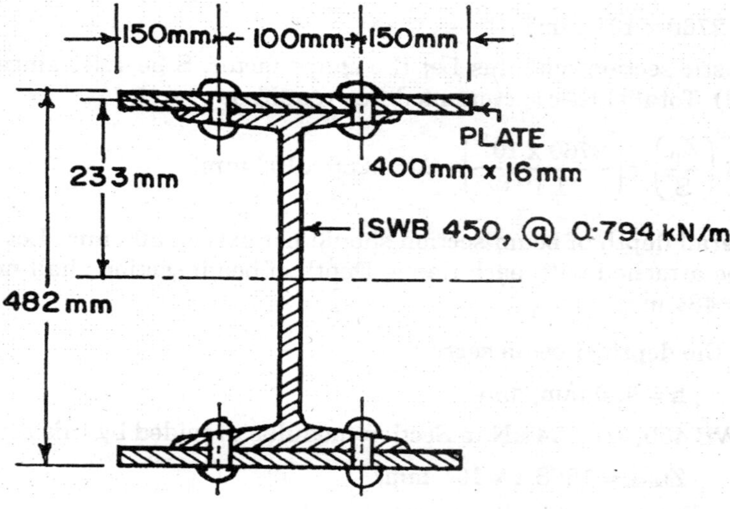

Fig. 16.3.

Example 16.3 : Trial built-up beam section ISWB 450, @ 0.794 kN/m with one cover plate 300 mm x 8 mm attached with each flange as shown in Fig. 16.3, as in Example 16.2 is adequate for flexure (bending).

Step 1 : Geometrical properties of proposed ISWB 450, @ 0.794 kN/m in Example 16.2, from ISI Handbook No.1,

Overall depth of the section, D = 450 mm

Width of flange, b_f = 200 mm

Thickness of flange, t_f = 15.4 mm

Thickness of web, t_w = 9.2 mm

Moment of inertia of beam -section

About strong-axis, $I_{xx} = 35057.6 \times 10^4\ \text{mm}^4$

About weak-axis, $I_{yy} = 1706.7 \times 10^4$ mm^4

Modulus of beam-section

About strong axis, $Z_{xx} = 1558.1 \times 10^3$ mm^3

About weak axis, $Z_{yy} = 170.7 \times 10^3$ mm^4

Radius at the root of the flange, $r_1 = 14.0$ mm

Depth upto the root of flane $h_2 = 33$ mm

One cover plate 300 mm wide x 8 mm thick is attached with each flange. Projection of cover-plate beyond the flange on each side

$$= \frac{1}{2}(300 - 200) = 50 \text{ mm}$$

Step 2 : Classification/type of the beam-section

Yield stress for the steel proposed

$$\sigma_y = 250 \text{ N/mm}^2$$

Yield stress ratio

$$\text{Epsilon } \epsilon = \left(\frac{250}{\sigma_y}\right)^{\frac{1}{2}} = \left(\frac{250}{250}\right)^{\frac{1}{2}} = 1.00$$

Criterion for flange

Actual ratio $(0.5\ b_f/t_f)$ = (0.5 x 200 x 15.4) = 6.49

Requirement of flange

Ratio (8.9 ε) = (8.9 x 1.0) = 8.9

Actual ratio 6.49 is less than 8.9 requirement

Criterion for the web

$$\text{Actual ratio } \left[\frac{D - 2t_f}{t_w}\right]$$

$$= \left(\frac{450 - 2 \times 15.4}{9.2}\right) = \left(\frac{419.2}{9.2}\right) = 45.56$$

Requirement for the web

Ratio (83 ε) = (83 x 1.0) = 83

Actual ratio 45-56 is less than the requirement 83.

Section ISWB 450, @ 0.794 kN/m is classified as **Plastic-section**

Projection beyond the flange/thickness of plate = $\left(\frac{50}{8}\right) = 6.25$ less than 12 t = 12 x 8 = 96 mm

Local buckling does not occur.

Step 3 : Resistance/moment capacity of the built-up beam

Elastic section modulus of beam section, Z_{xx} = 1558.1 x 10^3 mm^3

Contribution of two cover plate (neglecting moment inertia about own axis)

$$\left(2A_{PL} \times \frac{h}{2}\right) = \left(\frac{2 \times 300 \times 8 \times 450}{2}\right)$$

$$= 3 \times 8 \times 45 \times 10^3\ mm^3 = 1080 \times 10^3\ mm^3$$

Total elastic section modulus of built-up beam

Z_{xx} = (1558.1 + 1080) x 10^3 mm^3

= 2638.1 x 10^3 mm^3

Assuming shape factor, S = 1.12

Since, it is an I-section.

Plastic modulus of the built-up beam, Z_p = (1.12 x 2638.1) x 10^3 mm^3

= 2954.67 x 10^3 mm^3

Resistance/strength of beam section

$$= \left(\frac{\sigma_y \times Z_p}{\gamma_{ms}}\right)$$

$$= \left(\frac{250 \times 2954.67 \times 10^3}{1.15 \times 10^3 \times 1000}\right)$$

$$= \left(\frac{730.667}{1.15}\right) = 642.32\ kN-m$$

Design (factored/ultimate) moment due to external action = 600 kN-m

Design (factored/ultimate) moment due to external action 600 kN-m is less than resistance/strength moment capacity) of built-up beam 642.32 kN-m. Hence, the beam-section in **adequate** for flexure (i.e, bending).

Example 16.4 Trial built-up beam section, ISWB 450, @ 0.794 kN/m with one cover plate 300 mm x 8 mm attached with each flange as shown in Fig. 16.3, as in Example 16.1 is adequate for shear.

Step 1. From Example 16.1,

Effective span, l = 8 m

Design (factored/ultimate) load w_{fd} = 75 kN/m

Design (factored) shear force due to external load (action)

$$V_{fd} = \left(\frac{w.l}{2}\right) = \left(\frac{75 \times 8}{2}\right) = 300\,kN$$

Shear is resisted by the web Overall depth of web, t_w = 9.2 mm

Area of web in shear

A_{vw} = (D x t_w)

$= (450 \times 9.2) = 4140 \text{ mm}^2$

Step 2 : Shear strength/resistance of the beam section

$$\sigma_y = 250 \text{ N/mm}^2, \sigma_{yw} = \frac{\sigma_y}{\sqrt{3}} = 0.6\sigma_y$$

$$S_{vd} = \left(\frac{\sigma_{yw} \times A_{vw}}{\gamma_{msy}}\right) = \left(\frac{0.6\ \sigma_y \times A_{vw}}{\gamma_{msy}}\right)$$

$$= \left(\frac{0.6 \times 250 \times 4140}{1.15 \times 1000}\right) = 540 \text{ kN}$$

Note : Design (factored) shear force due to external load (action) 300 kN is less than the shear strength/resistance 540 kN.

($V_{fd} < V_{vd}$) is satisfied.

The beam section is adequate for shear requirement.

Example 16.5. A simply supported built-up consists of MB 500, @ 0.869 kN/m and tw 10 mm thick cover plates are attached to each flange. This built-up beam supports the load over a clear span 8 m. Find the maximum uniformly distributed load (Dead load + Self-weight + Imposed (live) load) which may be placed over this beam.

Step 1 : Let the width of supports be 300 mm at each end.

Effective span of beam = 8.30 m. This built-up beam is shown in Fig. 16.4.

Step 2 : Geometrical properties of given beam section. MB 500, @ 0.869 kN/m from ISI Handbook No.1

Depth of beam section, D = 500 mm

Width of flange, b_f = 180 mm

Thickness of flange, t_f = 17.3 mm

Thickness of web, t_w = 10.2 mm

Moment of inertia of the beam section

About strong axis, $I_{xx} = 45218.3 \times 10^4 \text{ mm}^4$

About weak axis, $I_{yy} = 1369.8 \times 10^4 \text{ mm}^4$

Modulus of beam section

About strong axis, $Z_{xx} = 1808.7 \times 10^3 \text{ mm}^3$

About weak axis, $Z_{yy} = 152.2 \times 10^3 \text{ mm}^3$

Radius upto the root r_1 = 17 mm

Depth upto the root, h_2 = 37.90 mm

Two 10 mm thick x 340 mm wide cover plates are attached with each flange.

Step 3 : Neglecting the moment of inertia of plates about their axes. Moment of inertia of any plate area A_{pL} about any axis is second moment of this area about that axis.

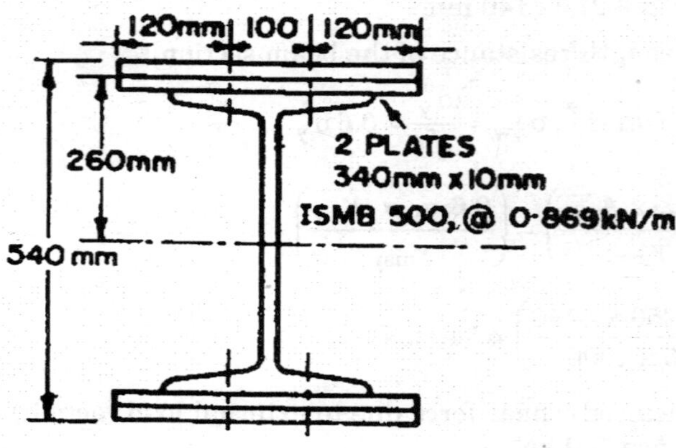

Fig. 16.4

Similarly, modulus of section of any plate area A_{PL} about any axis is first moment of area about that axis.

Area of one plate, A_{PL} = (10 x 340) = 3400 mm²

There are four plates (2 plates in each flange)

Section modulus of beam-section about xx-axis, Z_{xx} = 1808.7 x 10³ mm³

Approximately, the contribution to section modulus of each plate about neutral axis = $\left(A_{PL} \cdot \frac{h}{2}\right)$

One plate (3400 x 250) = 850 x 10³ mm³

Second plate (3400 x 250) = 850 x 10³ mm³

Third plate (3400 x 250) = 850 x 10³ mm³

Fourth plate (3400 x 250) = 850 x 10³ mm³

Total elastic section modulus

(1808.7 + 4 x 850) x 10³ mm³

Z_{xx} = (1808.7 + 3400) x 10³ = 5208.7 x 10³ mm³

Step 4 : Plastic modulus of the built-up section

Since, built-up section is of I-section shape-factor, S = 1.12 (assumed)

Plastic modulus of given built-up section

Z_p = (1.12 x 5208.7 x 10³) mm³ = 5833.744 x 10³ mm³

Step 5 : Resistance/strength of given built-up section

$$MR = \left(\frac{\sigma_y \times Z_p}{\gamma_{msy}}\right)$$

$$= \left(\frac{250 \times 5833.744 \times 10^3}{1.15 \times 1000 \times 1000}\right) = 1260.21 \text{ kN-m}$$

Step : Let the factored design) load be w_{fd} kN/m. This is uniformly distributed load. The maximum bending moment occurs at the centre Design (factored) moment

$$M_{fd} = \left(\frac{w.\,l^2}{8}\right)$$

$$= \left(\frac{w_{fd}\ \text{x}\ 8.30\ \text{x}\ 8.30}{8}\right) \text{kN} - \text{m} = 8.61\ w_{fd}\ \text{kN-m.}$$

Step 6. Design (factored) moment may be at the most equal to the resistance/strength of beam section. Therefore434441

$$8.611\ w_{fd} = 1260.21$$

$$w_{fd} = \left(\frac{1260.21}{8.611}\right) = 146.34\ \text{kN/m}$$

Dead load + imposed (live) load + Self weight (Service-load)

$$= \left(\frac{46.34}{1.5}\right) = 97.56\,\text{kN/m}$$

16.3 LINTELS

The brick masonry or stone masonry over doors, windows and other openings in walls is supported by beams. Such beams are known as *steel lintel* or simply as *lintels*. The various types of lintels are shown in Fig. 16.5.

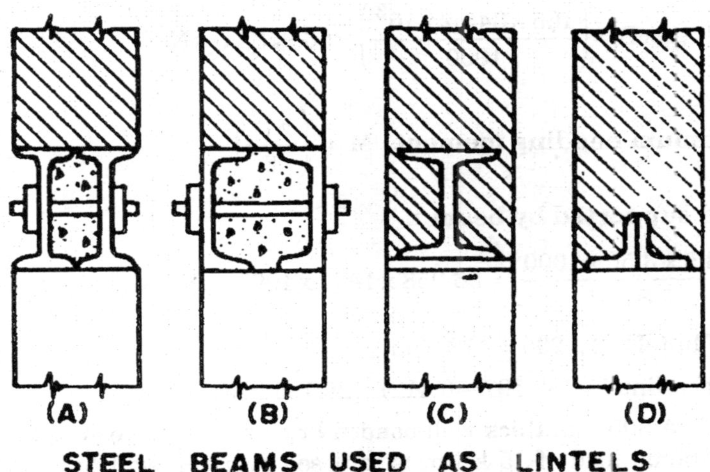

Fig. 16.5.

It is considered that the lintel supports the masonry contained in an equilateral triangle as shown in Fig. 16.6. The masonry should have well bond and should have an arch action. The masonry should extend at least half the effective span to either side of the opening. The masonry above the lintel should be upto a height of 1.25 times the effective span. Then, the load is considered as uniformly distributed load.

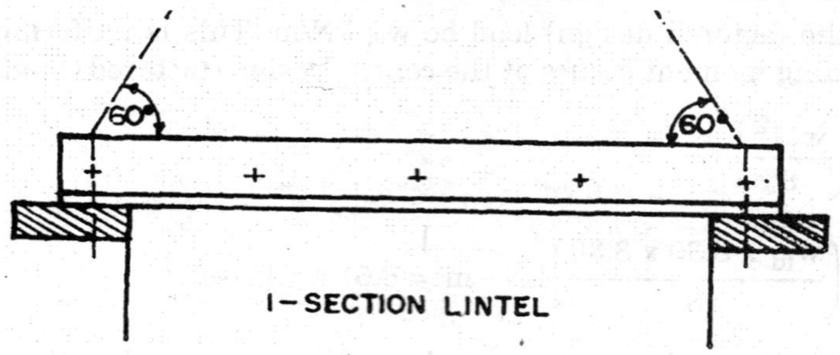

Fig. 16.6.

When the height of masonry above the lintel is not sufficient it is considered to support whole masonry about it. The floor or roof loads which are immediately above the opening, are considered as separate load to be supported by the lintels.

Example 16.6. A lintel consist of two LC @ 0.206 kN/m connected by pipe and bolt separator as shown in Fig. 16.5 (B). The effective span is 3.50 m. Determine the maximum uniformly distributed load for the lintel.

Solution :

Step 1 : Properties of given section

Two LC 200, @ 0.206 kN/m are used as a lintel.

From the steel section tables, the section modulus of two LC 200, @ 0.206 kN/m

$Z_{xx} = (2 \times 172.6 \times 10^3) = 345.2 \times 10^3 \text{ mm}^3$

Step 2 : Moment of resistance of the lintel

$$M_r = (\sigma_{bc} \times Z) = \left(\frac{165 \times 345.2 \times 10^3}{1000}\right) = 56.958 \text{ m-kN}$$

Step 3 : Maximum bending moment, $M = \left(\frac{wl^2}{8}\right)$

Step 4 : Load supported by beam

$$\left(\frac{w \times 3.5 \times 3.50 \times 1000}{8}\right) = 56.958 \times 1000 \times 1000, \quad w = 37.197 \text{ kN/m}$$

Self-weight of lintel = 2 x 206 = 412 N/m.

Load imposed on lintel = (37.197 – 0.412) = 26.785 kN/m

Example 16.7. A 300 mm thick well bonded brick well is to be supported over an opening of 2.75 m. An additional load of 15 kN/m is imposed over the wall from the roof. Arch action exists. Design a suitable section for the lintel. Weight of brick masonry = 21.60 kN/m³.

Design :

Step 1 : Effective span of lintel

Width of the opening = 2.75 m

Provide 150 mm wide bed blocks to support the lintel.

The effective span of lintel is the distance between centre to centre of bearings

$$l = (2.75 + 0.15) = 2.90 \text{ m}$$

Step 2 : Load supported by beam

Weight of brick masonry in equilateral triangle

$$\left(\frac{1}{2} \times 2.90 \times 2.90 \times \frac{\sqrt{3}}{2} \times 0.30 \times 21.60\right) = 23.598 \text{ kN}$$

Additional load imposed over the lintel from roof, = (2.90 x 15.00) = 43.50 kN

Self-weight of lintel = 2.85 kN

Total uniformly distributed load, (23.598 + 43.50 + 2.85) = 69.948 kN

Step 3 : Bending moment

The maximum bending moment, M occurs at centre

$$M = \left(\frac{wl^2}{8}\right) = \left(\frac{69.948 \times 2.90}{8}\right) = 25.356 \text{ kN-m}$$

Step 4 : Section modulus required

$$Z = \left(\frac{25.56 \text{ x } 1000 \text{ x } 1000}{165}\right) = 153.67 \times 120^3 \text{ mm}^3$$

Step 5 : Trial section

From steel section tables, try two JC 175, @ 0.112 kN/m

Section modulus provided = (2 x 82.3 x 10^3) = 164.6 x 10^3 mm^3

Thickness of web, t_w = 3.6 mm, and Depth of section = 175 mm

Moment of inertia of section, I_{xx} = 719.9 x 10^4 mm^4

16.4 JACK ARCH

Jack-arches are constructed with bricks laid on edge, springing from and supported on the lower flanges of rolled steel beams. The spacing of rolled steel beams shall be generally 1.0 to 1.4 metres, but shall never exceed 2.0 metres. The beams are designed to support the portion of floor coming over them. Jack arches are arches of small rise. Rise of the arch is kept about 1/8th to 1/12th of the span. The intermediate beams receive thrust from arches on either side and thrust is balanced. The end beams receive side thrust from the arch. The rods are provided to resist the side thrust of end beams. The tie rods are property anchored with stout mild steel nut and washers. The details of jack arch floor construction are shown in Fig. 16.7.

Specification. ISI recommends in IS : 2118-1962 the following specifications for the design and construction of jack arch floor or roof :

1. The deflection of the beams shall not exceed 1/480 of the span.
2. Thickness of arch shall be not less than 100 mm.
3. The tie rods shall be not less than 12 mm in diameter.

4. The tie rods shall be placed generally not less than 75 mm above the bottom of the joist. The spacing of the tie rods shall not exceed 20 times the width of rolled steel beams.

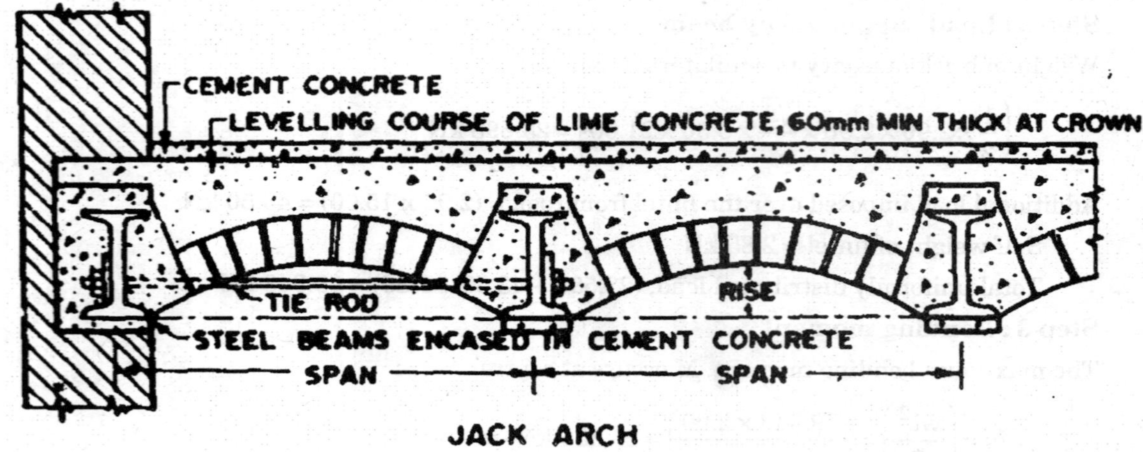

Fig. 16.7.

The tie rods are designed to resist side thrust found as follows :

$$T = \left(\frac{WT}{8R}\right)$$

where, F = Side thrust per metre width of arch

W = Live and dead load per metre width of the arch

L = Span of the arch and R = Rise of the arch.

Design of jack arch consists of design of

(i) thickness of arch, (ii) the rods, and; (iii) rolled steel beams.

Example 16.7. Design a jack arch roof for a room of 6 metres x 4.5 metres. Total uniformly distributed load including self weight of the roof is 9.750 kN/square metre. Allowable compressive stress is masonry is 0.3 N/mm².

Design of Jack arch :

Step 1 : Load supported

Provide rolled steel beams on span of 4.5 m, at a regular interval of 1.5 m.

Span of jack arch = 1.5 m

Provide rise of the arch = $\left(\frac{1}{12} \times \text{span}\right) = \left(\frac{1}{12} \times 1.5 \times 1000\right) = 125$ mm

Load per metre width for one span of arch, W = (1.2 x 1 x 9.75) = 14.625 kN/m

Step 2 : Side thrust from the arch

$$T = \left(\frac{WL}{8R}\right) = \left(\frac{14.625 \times 1.50 \times 1000}{8 \times 125}\right) = 21.938 \text{ kN}$$

Let *t* be the thickness of arch at the crown, A be the area of arch ring per metre width and σ_c be the allowable compressive stress in masonry A x σ_c = T

(1000 x t x 0.3) = (21.938 x 1000), t = 73.125 mm

Provide minimum thickness = 100 mm

Step 3 : Rolled steel beam

Clear span of beam = 4.5 m

Assume bearing of 300 mm on either sides of beam

Effective span of the beam is the distance between centre of bearing, = 4.80 m

Uniformly distributed load, w = 4.625 kN/m

Step 4 : Maximum bending moment, M occurs at the centre.

$$M = \left(\frac{wl^2}{8}\right) = \left(\frac{14.625 \times 4.8 \times 4.8}{8}\right) = 42.12 \text{ m-kN}$$

Step 6 : Section modulus required

$$Z = \left(\frac{42.12 \times 1000 \times 1000}{165}\right) = 253.273 \times 10^3 \text{ mm}^3$$

Step 7 : Properties of rolled steel beam

From steel section table, try LB 250, @ 0.279 kN/m

Thickness of web t_w = 6.1 mm, Depth of section h = 250 mm

Width of section b = 125 mm

Section modulus Z_{xx} = 297.4 x 10^3 mm^3

Moment of inertia I_{xx} = 3717.8 x 10^4 mm^4

Moment of inertia of beam required

$$\left(\frac{3717.8 \times 13.238 \times 10^4}{10}\right) = 4860 \times 10^4 \text{ mm}^4$$

Provide LB 275, @ 0.330 kN/m

Step 8 : Design of tie rods

Spacing of the tie rods shall not exceed 20 times width of the beam

(i.e., 20 x 140) = 2800 mm

Provide the rods at 2.0 m. Thrust resisted by tie rod

= (21.938 x 2.0) = 43.876 kN

Allowable tensile stress in tie rods upto 20 mm diameter

= (0.6 x 250) = 150 N/mm²

$$\text{Net area required} = \left(\frac{43.876 \times 1000}{150}\right) = 292.51 \text{ mm}^2$$

Provide tie rods 25 mm in diameter (ISRO 25).

16.5 FILLER JOISTS

The steel beams of light sections are termed as joists. The joists placed in plain cement concrete are called *filler joists*. When filler joists are continuous over more than two supports, then, these are called *continuous fillers*. Whenever possible, continuous fillers may extend over three spans. The joists are supported by main steel beams as shown in Fig. 16.8.

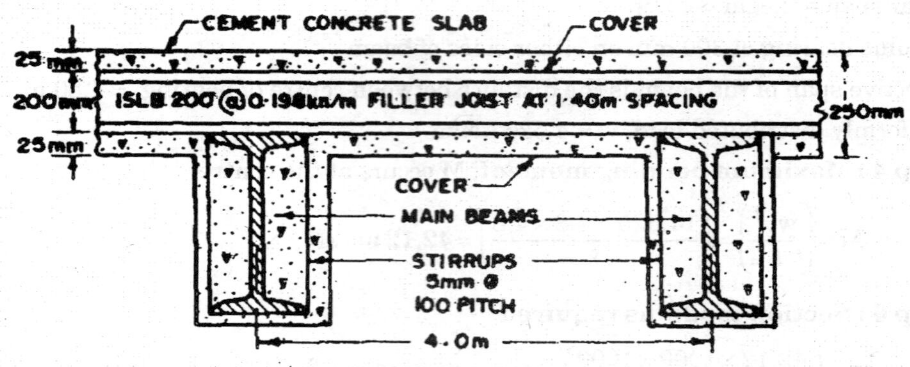

Fig. 16.8

The joists in continuous fillers are made over supports in such a manner to preserve considerable measure of continuity. The joints are staggered so that full continuity of the fillers occurs in 50 per cent of those crossing any support, such a continuity being evenly displaced.

When filler joists spans between two main steel beams only, then, these are called *discontinuous fillers*. Discontinuous filler joists are shown in Fig. 16.9. These are connected to main beams by means of cleat angles. Overall depth of construction in case of discontinuous filler joist is small than that of continuous filler joist construction.

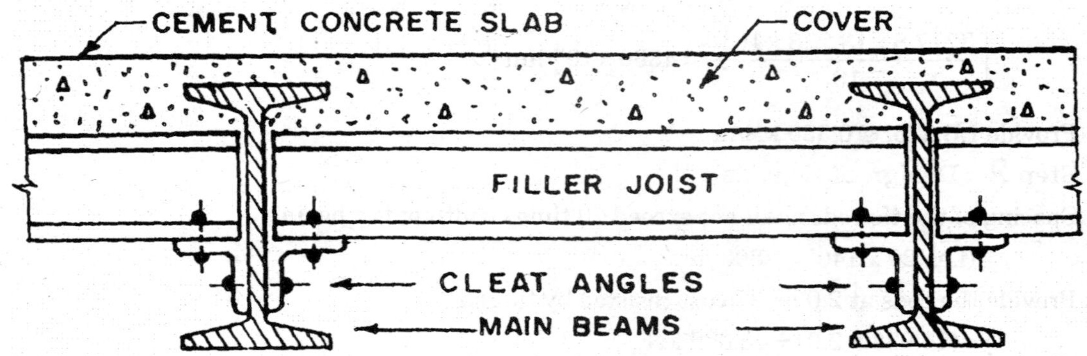

DISCONTINUOUS FILLER

Fig. 16.9.

The bottom of filler joists remains flush with cement concrete slab.

The design of filler joists is done as per following specifications.

1. Bending moment. The bending moments on slabs of which filler joists form part, shall be calculated to satisfy the following :

(a) As slabs spanning continuously over supports and subjected to those combinations of dead and live loads producing the maximum positive and negative moment.

(b) In the case of three or more approximately equal spans of continuous filler joists described above, as slabs designed for uniformly distributed loading, satisfying moment value given in Table 16.1.

TABLE 16.1. BENDING MOMENT VALUES FOR FILLER JOISTS

Locations	Values of B.M.
1. Near Middle of end span	+ $wL^2/10$
2. At support, next to end span	- $wL^2/10$
3. At middle of intermediate span	+ $wL^2/12$
4. At other interior supports	- $wL^2/12$

where w is the dead load plus load per unit length of span and L is the span, centre to centre of supports.

Note : For single spans, the slab shall be assumed to be freely supports.

The spans are considered approximately equal when the longest span does not exceed the shortest span by more than 15 percent.

2. Moment of resistance. The moment of resistance of the slab shall be calculated from the section properties of the filler joists only except that where the filler joists are completely embedded in a solid concrete slab, the moment of resistance of slab may alternatively be calculated as a composite reinforced concrete section.

3. Spacing. The spacing of filler joist centre concrete slab is given in Table 16.2 unless the concrete is reinforced to span as slab or function as an arch between the filler joists.

Table 16.2. SPACING OF FILLER JOISTS

Imposed load per m	Volume of n
1. Upto 2.50 kN	9
2. Above 2.50 kN upto 5.0 kN	8
3. Above 5.0 kN upto 7.50 kN	7
4. Above 7.50 kN upto 10 kN	6
5. Above 10 kN	5

Where a slab of concrete is designed to function as an arch between the filler joist, the thrust from the arch section shall be taken up by steel tie or by other means so as not to cause appreciable increase of stress in the joists. The concrete shall have a working strength of not less than 15 N/mm^2 at 28 days.

4. Thickness of concrete. Where the underside of the concrete is arched between the filler joists, the thickness at the crown shall be not less than 50 mm. The thickness of structural concrete over hollow blocks shall nowhere be less than 30 mm when the filler joists exceed 450

mm centre to centre and nowhere less than 50 mm when the spacing of filler joists is greater than 450 mm centre to centre.

5. Bending stress. **The bending stress in filler joists other than those designed as part of reinforced concrete slab, shall not exceed (165 + 0.62 t) N/mm², where t is an allowance, the value of which is equal to the thickness in mm of the structural concrete cover to the compression flange of the filler joists, provided that the allowance t is applied only in cases where the filler joists are embedded at least flush with underside of their bottom flanges in a solid concrete slab throughout and that any cover less than 25 mm and in excess of 75 mm shall be neglected.**

In cases where the underside of the slab is flush with the bottom flanges of the filler joists, the allowance t shall not apply in respect of support moments; nevertheless if the top flange is covered, the allowance may be made in calculating to the resistance sagging moment.

6. Shear stress. **The shear stress shall be calculated as taken entirely on the filler joists and stress shall not exceed the allowable stress in shear, i.e., 0.4 f_y.**

7. Bearing stress. **The bearing stress shall be calculated as taken entirely on the filler joist and stress shall not exceed the allowable stress in bearing i.e., 0.75 f_y.**

8. Span depth ratio. **The span of filler joist centre to centre of supports, shall not exceed 35 times the depth from the underside of the joist to the top of the structural concrete or 12 times this depth in the case of cantilever fillers.**

16.6 ENCASED BEAMS

When a steel beam is encased in cement concrete throughout the entire length, it is called encased beam. The cased beam is shown in Fig. 16.10.

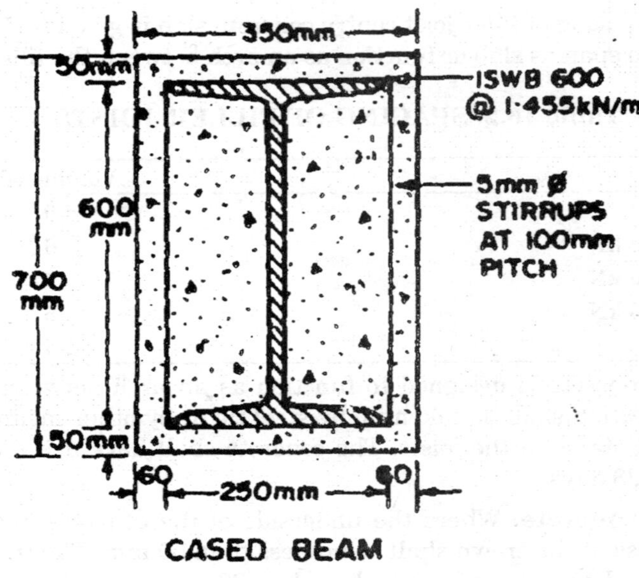

Fig. 16.10.

A steel beam is embedded in cement concrete for two purposes. It may be necessary to make a steel frame building fire resisting. The beams may be encased in cement concrete for architectural requirement. As per IS : 800-1984 beams may be designed as encased beams when the following conditions are fulfilled :

(a) The section is of single web and I-shape or of double open channel formed with the web not less than 40 mm apart.

(b) The over-all dimensions of the steel section do not exceed 750 mm x 450 mm overplating where used, the larger dimensions being measured parallel to web (as per IS : 800-1962).

(c) The beam is unpainted and it is solidly encased in ordinary dense concrete, with 10 mm aggregate (unless solidity can be obtained with a large aggregate) and of grade designation M 15, min (IS : 800-1978).

(d) The minimum width of solid casing is equal to (b_o + 100) mm where b_o is the width of the steel flanges in mm.

(e) The surface and edges of the flanges of the beam have a concrete cover not less than 50 mm.

(f) The casing is effectively reinforced with steel wire of at least 5 mm diameter and the reinforcement shall be in the form of stirrups or binding at not more than 150 mm pitch, and so arranged as to pass through the centre of the covering to the edges and soffit of the lower flange.

Design of encased beam

The steel beam section shall be considered as carrying the entire load but allowance may be made for the effect of concrete on the lateral stability of the compression flange. This allowance should be made by assuming for the purpose of determining the permissible stress in compression, that the equivalent moment of inertia, I_y about yy-axis is equal to Ar^2_y, where A is the steel section and r_y may be taken as 0.2 (b_o + 100 mm). The width of flange b_o, is taken in mm. The other properties required for calculating critical bending stress for Eq. 11. may be taken as for uncased section. The permissible bending stress so determined shall not exceed 1.5 times that permitted for the uncased sections.

It is noted that this does not apply to beams and girders having a depth greater than 1000 mm or a width greater than 500 mm or to box sections.

Example 16.8. A filler joist is continuous over three equal spans. The distance between centre to centre of main beams is 4 m. The floor consists of 250 mm thick plain cement concrete slab. Unit weight of plain cement concrete is 24 kN/m^3. The live load on floor is 5 kN/m^2. Design filler joist.

Design :

Step 1 : Load supported by joist

Distance between centre to centre of main beam = 4 m

Thickness of plain cement concrete slab = 250 mm

Assume, thickness of top cover = 25 mm

Thickness of bottom cover = 25 mm

Therefore, thickness of structural concrete slab = 200 mm

Live load on floor = 5 kN/m^2

Therefore, spacing of filler joist centre to centre

n x thickness of structural slab = 7 x 20 = 1400 mm

Dead load

Weight of plain cement concrete inclusive of cover at top and bottom

$$\left(\frac{24 \times 4 \times 1.400 \times 200}{1000}\right) = 26.88 \text{ kN}$$

Self-weight of filler joist assuming 0.200 kN/m

= (0.200 x 4) = 0.800 kN

Total dead load = (26.81 + 0.80) = 27.680 kN (say = 27.70)

Live load

Total live load = (5 x 4 x 1.40) = 28 kN

Total dead load plus live load

= (28 + 27.70) = 55.70 kN

Step 2 : Maximum bending moment

The maximum bending moment occurs at middle of end span

$$\left(\frac{55.70 \times 4}{10}\right) = 22.28 \text{ kN - m}$$

Allowable bending stress

(165 + 0.62 t) = (165 + 0.62 x 25) = 180.5 N/mm^2

Step 3 : Required section modulus for filler joist

$$Z = \left(\frac{22.28 \times 10^6}{180.5}\right) = 126.32 \times 10^3 \text{ mm}^3$$

From ISI Handbook No. 1, try LB 200 @ 0.198 kN/m

Z_{xx} = 169.7 x 10^3 mm^3

Step 4 : Check

Span of filler joist = Centre to centre of main beams (4 m)

Depth from the underside of the joist to the top of the structural concrete

= 225 mm

$$\text{Span depth ratio} = \left(\frac{4 \times 100}{22.5}\right) = 17.75$$

< 35. Hence design is satisfactory.

Provide LB 200, @ 0.1908 kN/m filler joist.

5 mm in diameter stirrups are provided at 100 mm pitch.

Complete design is shown in Fig. 16.10.

Example 16.9. An encased beam carries total uniformly distributed load of 40.00 kN/m. The effective span of beam is 8 m. The ends of beams are restricted against torsion, and not restrained against lateral bending. Design the encased beams.

Design :

Step 1 : Load supported by beam

Total uniformly distributed load is 40 kN/m

Step 2 : Maximum bending moment, M occurs at the centre

$$M = \left(\frac{40.0 \times 8 \times 8}{8}\right) = 32 \text{ kN-m}$$

Step 3 : Maximum shear force, F occurs at the support]

V = 160.00 kN

Step 4 : Selection of trial section

Select an uncased rolled steel beam section such that the section of beam provides required section modulus corresponding to the allowable bending compressive stress for the uncased section for the given span.

Select WB 600, @ 1.451 kN/m

$\sigma_{bc} = (0.66 \times 250) = 165$ N/mm^2

Step 5 : Section modulus required

$$Z = \left(\frac{32.00 \times 1000 \times 1000}{165}\right) = 193.94 \times 10^3 \text{ mm}^3$$

Step 6 : Properties of trial section

From ISI Handbook No. 1 for WB 600, @ 1.451 kN/m

Sectional area A = 18486 mm^2 and Depth of section = 6000 mm

Thickness of flange t_f = 23.6 mm and Width of flange b = 250 mm

Section modulus Z = 3854.2 x 10^3 mm^3

Section modulus provided is greater than section modulus required. Hence, safe.

For encased beam, $r_y = 0.2\ (b_o + 100)$ mm

= 0.2 (250 + 100) = 70 mm

Moment of inertia of cased beam

$I_y = Ar^2_y = (18486 \times 70^2) = 9020 \times 10^4$ mm^4

Distance between c.g. of flanges = (600 – 23.6) = 576.4 mm

PROBLEMS

16.1 Two LB 600, @ 0.995 kN/m joists at 300 mm with one 36 mm thick plate, 600 mm wise on each flange is used as a simply supported beam on a span on 12 m. Find the maximum uniformly distributed load per metre run which can be allowed on this built-up girder. Neglect rivet holes.

16.2 Design a simply supported plated rolled steel beam section to carry a uniformly distributed load 36.00 kN/m inclusive of self-weight of beam. The effective span of beam is 7.20 m, the depth of beam is limited to 500 mm. The beam is restrained against lateral deflection.

16.3 A freely supported beam spanning 8 m is made up of an I-section MB 500, @ 0.869 kN/m with two plates in each flange of 200 mm x 12 m each. Calculate the maximum uniformly distributed load the beam can carry if the compression flange is fully restrained. Assumed the flange connections to be made with 18 mm ψ rivets. Show calculations for the curtailment of the outermost plate in each flange.

16.4 A plated steel beam of span 8 metres effective is to carry a uniformly distributed load of 20 kN per metre inclusive of its own weight. Design a suitable plated beam including minimum spacing of rivets using any one of the three section listed below. Plate of 10 mm thickness is available in any width. The flange of beam should not be wider than 300 mm. Assume that 32 mm diameter rivets are used for the connection.

16.5 A beam has an effective span of 10 m and carries a uniformly distributed load of 100 kN. Design the cross-section, of the beam if its depth is limited to 45 mm.

16.6 The cross-section of a built-up beam consists of an LB 325 with a flange plate of 1200 mm x 10 mm provided at the top as well as at the bottom. It uses 18 mm diameter rivets.

It is simply supported over an effective span of 6 m. Calculated the maximum uniformly distributed load the beam can carry. If the beam is laterally supported throughout its length along the compression flange. Consider bending moment only.

17

LIMIT STATE FOR COLLAPSE IN BENDING AND SHEAR DESIGN OF CRANE GANTRY GIRDERS

17.1 INTRODUCTION

A crane is a machine used for moving/shifting heavy loads. While, gantry means a structure used to support travelling cranes. A crane gantry girder is a good example of a built-up beam.

The overhead travelling cranes are used in factories and workshops to lift the heavy materials, equipments etc., and to carry them from one place to the other. These cranes are either hand operated or electrically operated. The crane consists of bridge spanning the bay of the shop. A trolley or crab is mounted on the bridge. The trolley moves along the bridge. The bridge as a whole moves longitudinally on rails provided at the ends. The rails on either side of the bridge rest on crane gantry girders. The rails, supporting the crane and resting over gantry girders are shown in Fig. 17.1(A) and in partial plan in Fig. 17.1(B). The gantry girders are the girders which support the loads transferred through the travelling (moving) wheels of the cranes.

As it is seen in Fig. 17.1, a channel-section is placed over an I-section with 15 web kept horizontally. The flanges remain vertical. The moment of inertia, I_{xx} of the channel section gets added to that of I-section. It reduces and resists lateral (i.e., horizontal) deflection. The complete beam acts as **laterally supported beams.**

The effect of cranes to be considered under the imposed loads shall include the following :

1. The vertical loads from the crane.
2. The eccentricity effects induced by the vertical loads, and the impact factors.
3. Lateral (surge) thrust across the crane rail, and
4. Longitudinal horizontal thrust along the crane rail.

The crane loads to be considered shall be as indicated by the customer. In the absence of any specific indications, the load combinations shall be as follows :

(a) The vertical loads with full impact from loaded crane or two cranes in case of tandum operation together with the vertical loads, without impact, from as many loaded cranes as may be positioned for maximum effect along with maximum horizontal thrust (surge) from one crane only or two cranes in case of tandum operation;

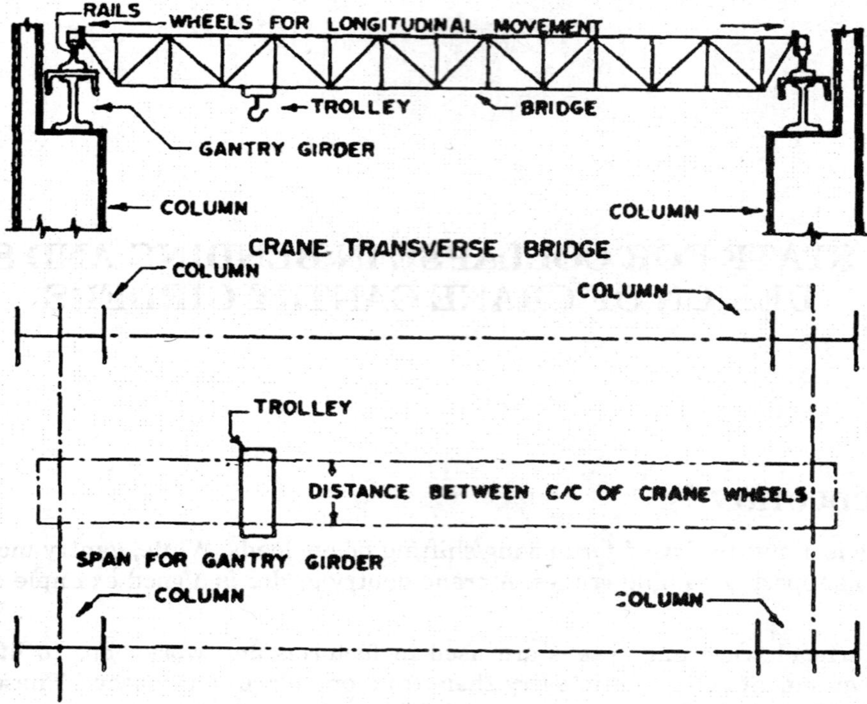

Fig. 17.1.

(b) For the multibay multicrane gantries the loads specified above, subject to consideration of cranes in maximum of any two bays of the building cross-section;

(c) The longitudinal thrust on a crane track rail shall be considered for the maximum of two loaded cranes on the track; and

(d) The lateral thrust (surge) and the longitudinal thrust acting respectively across and along the crane rail shall not be assumed to act simultaneously. The effect of both the forces, shall however be investigated separately.

While investigating the effect of earthquake forces the resulting effect from dead loads of all the cranes parked in each bay positioned for the maximum effect shall be considered.

The crane runaway girders supporting bumpers shall be checked for bumper impact loads.

The stress developed due to secondary effects such as handling, erection, temperature effects, settlement of foundations shall be appropriately added to the stresses calculated from the combination of loads are recommended by IS: 800-1984. The total stresses thus calculated shall be within the permissible limits.

While considered the simultaneous effects of vertical and horizontal surge loads of cranes, the permissible stresses for the design of crane gantry girders and their supporting structures may be increased by 10 per cent.

Where the wind load is the main acting on the structure, no increase in the permissible stresses is allowed.

The gantry girder is to be designed for the following additional loads given in Table 17.1 as per IS : 875-1984.

TABLE 17.1.

	Type of Load	Additional load
1	VERTICAL LOAD	
	(a) for electric overhead cranes	25 percent of maximum static wheel loads
	(b) for hand operated cranes	10 percent of maximum static wheel loads.
2.	HORIZONTAL FORCES TRANSVERSE TO RAILS	
	(a) for electric overhead cranes	10 percent of weight of trolley and weight lifted on the crane.
	(b) for hand operated cranes	5 percent of weight of trolley and weight lifted on the crane.
3.	HORIZONTAL FORCE ALONG THE RAILS	5 percent of the static wheel loads.

The gantry girder is designed on the assumption that either of the horizontal forces transverse to rails or along the rails act at the same time as the vertical loads including the impact load. The horizontal forces act at the rail level. The gantry girder is subjected to bending in vertical plane as well as in horizontal plane along with twisting. The design calculations are simplified by providing a channel at the top flange of the girder, and neglecting the bottom flange for transverse load computations. The transverse loads are comparatively small and this simplification in design calculations does not result in serious error. The design example discussed here is for simply supported gantry girder, the compression flange is subjected to horizontal and vertical loads. The horizontal forces act in the plane of the top flange and in the direction perpendicular to it. Therefore, the top flange is strengthened by providing a channel section as shown in in Fig. 17.5 to take the horizontal forces in the plane of this flange. The channel section provides flange areas to resist bending in horizontal plane due to horizontal forces acting in transverse direction. It increases moment of inertia about yy-axis. The flanges of channel section resist the bending in the horizontal plane. The bending of the crane gantry girder occurs about vertical axis as well as about the horizontal axis of the member. The actual bending stresses for bending of the girder in the vertical and horizontal planes are computed. The combined bending compressive stress is taken as the sum of the two calculated fibre stresses. The combined bending compressive stress should be less than or equal to the allowable bending compressive stress.

$$\left(\sigma_{bc.x.cal} + \sigma_{bc.y.cal}\right) < \sigma_{bc}$$

where, $\sigma_{bc.x.cal}$ = Actual bending compressive stress in vertical plane

$\sigma_{bc.y.cal}$ = Actual bending compressive stress in horizontal plane

σ_{bc} = Allowable bending compressive stress.

For detailed classification of cranes, the code of practice for design, manufacture, erection and testing (structural portion of cranes and hoists IS : 807-1976 may be referred.

The additional load for overhead travelling cranes has been given in Table 17.1 as the percentage of the maximum static wheel load. The maximum wheel load is the reaction on a wheel due to the total load given by weight of crane plus crab and the lifted load.

The term **'maximum static wheel load'** needs explanation. The minimum approach of crane hook is the distance from the gantry girder upto which the crab or trolley may approach. This extreme position of the crab with respect to the span of crane gives the maximum reaction on one of the gantry girders. This maximum reaction is distributed equally among the crane wheels. The reaction on each wheel is termed as the maximum static wheel load.

The allowable bending compressive stresses for bending in horizontal plane is equal to the allowable bending stresses in tension. The allowable bending compressive stress for bending in vertical plane is reduced in proportion of critical stress in bending.

IS : 800-1984 recommends that the allowable stress in axial tension, axial compression, and bending stresses, and allowable stresses for rivets are increased by 10% for the design of gantry girder for the combination of vertical and horizontal loads as discussed above. This increase in allowables stresses is not in addition to that allowed for erection loads with or without wind or seismic forces.

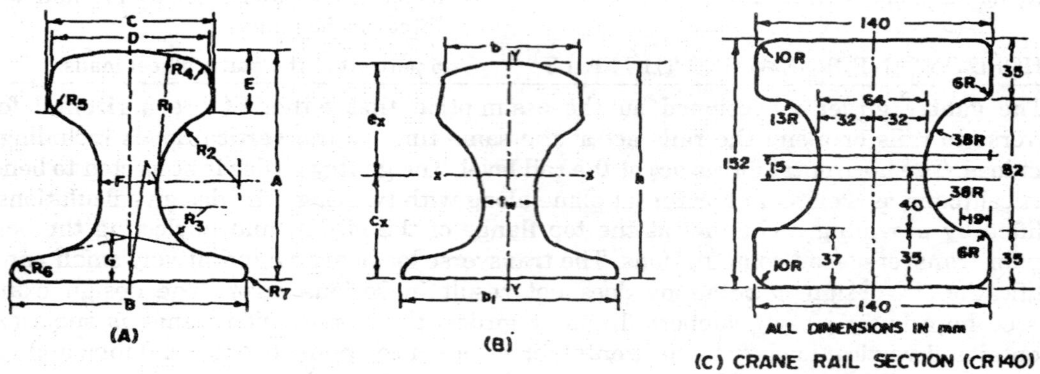

Fig. 17.2.

The bridge of a crane as a whole moves longitudinally on rails provided at the ends and mounted on the crane gantry girders. The sections of rails are known as crane rails sections. These sections are designated by the letter ISCR followed by the head width of the rail section in millimeters. The dimensions of crane rail sections are given in Table 17.2 and are shown in Fig. 17.2(A) as per IS : 3443-1966 (Indian Standard Specifications for Crane Rail Sections). The sectional properties of the crane rail sections are given in Table 17.3. The dimensions corresponding to Table 17.3 are shown in Fig. 17.2(B). The dimensions of crane rail section ISCR 140 are shown in Fig. 17.2(C).

TABLE 17.2. DIMENSIONS OF CRANE RAILS AS PER IS : 3443-1996

Designation	Dmension in mm													
	A	B	C	D	E	F	G	H	R1	R2	R3	R4	R5	R6	R7
CR50	90	90	55	50	25	20	20	9.70	300	26	18	6	6	5	1.5
CR60	105	105	65.5	60	27.5	24	22	9.82	350	32	20	6	6	5	1.5
CR80	139	130	87	80	35	32	26	9.75	400	44	26	8	8	6	1.5
CR100	150	150	108	100	40	38	30	11.25	450	50	30	8	8	8	2.0

TABLE 17.3. SECTIONAL PROPERTIES OF CRANE RAIL SECTION AS PER IS : 3443-1966

Designation	Area	Weight	Head width (b)	Bottom width (b)	Overall height of section (h)	Wcb thick-ness (Narro-west point) (tw)	cx	ex	Moment of Inertia Ixx	Iyy	$Z_{x1} = \left(\frac{I_{xx}}{c_x}\right)$	$Z_{x2} = \left(\frac{I_{xx}}{c_x}\right)$	$Z_y = \left(\frac{I_{yy}}{b_1/2}\right)$
(1)	(2)	(3)	(4)	(5)	(6)	(7)	(8)	(9)	(10)	(11)	(12)	(13)	(14)
	mm²	N/m	mm	mm	mm	mm	mm	mm	cm⁴	cm⁴	cm³	cm³	cm³
CR50	3802	298.5	50	90	90	20	43.2	46.8	357.54	111.42	82.76	76.40	24.76
CR60	5099	400.3	60	105	105	24	48.3	56.7	654.60	195.88	135.52	115.45	37.31
CR80	8113	638.9	80	130	130	32	64.3	65.7	1547.4	482.39	240.65	235.52	74.21
CR100	11332	889.6	100	150	150	38	76.0	74.0	2864.73	940.98	376.94	387.12	125.46
CR141	15580	1220	100	140	152	64	76.0	76.0	3934.0	1855.00	517.00	517.00	265.00

In selecting a crane gantry track rail for a given application on the basis of the crane manufacture's specified maximum wheel load (and the wheel diameter) and by working to a table of maximum permissible wheel loads as recommended by some authority for a range of rails, it is desirable to remember that the maximum wheel load occurs rarely in service. The maximum wheel load is reached wen the load on the hook of the crane is at full rated capacity and when the trolley is at the limit of its transverse motion. These two conditions rarely occur together. Actual wheel loads in service conditions (and with the crane on its long travel motion) normally lie in the range of 50 to 75 per cent of the specified possible maximum wheel load. The various formulae are used in determining maximum wheel load. One such formula in use in U.K. for steel mill cranes converted to metric unit is as follows :

$$W = (4.90\ a.D/1000)$$

where, W = Maximum permissible wheel load in kN

a = Rail head width in mm, and

D = Wheel tread diameter in mm

The crane gives the satisfactory services when the gantry girder is properly designed and the following matters are ensured in its subsequent maintenance :

(i) The laid track is reasonably straight. The deviation should not be more than 5 mm for every 10 m length of long travel track.

(ii) The track span dimension is fairly constant. The variation should not be more than 10 mm on either side.

(iii) The ends of the rails at the joints are held in close and hard contact except where expansion joints are necessary in long gantries which themselves have definite expansion joint details.

Where rails and wheels do not give a satisfactory life it will generally the consequence of shortcomings in one or more of (i), (ii), and (iii) and/or defect in the crane wheel bearing, or end carriage which make the wheels out of alignment in plan or displaced from the vertical in elevation. These defects and also shortcomings in (i) and (ii) above cause excessive wear of the wheel flanges and also on the sides of rail head. Poorly designed and poorly maintained rail joints will not normally adversely affect the life of the wheels or even the wheel bearings if these later are suitably lubricated plain bush type.

However, bed rail joints shorten considerably the life of antifriction bearings in crane-wheels. Also bad rail joints shorten the life of the rails. Quite often poorly designed and poorly maintained rail joints deteriorate to a point where it is deemed desirable to replace along lengths of rails in which there is no defect whatsoever, except at the joints.

Limitations of vertical and horizontal deflections for electrically operated overhead crane structures have been described in chapter.

Example 17.1. Design a simply supported gantry girder to carry one electric over head travelling crane.

Crane capacity = 300 kN

Weight of crane excluding trolley = 190 kN

Weight of trolley = 100 kN

Minimum approach of crane hook = 1.2 metres

Distance between centres of crane wheel = 3.5 metres

Distance between centres of crane wheel = 18 metres

Span of gantry girder = 6 metres

Weight of rail section = 0.300 kN/m

Height of rail section = 75 mm

Design :

A crane gantary girder is shown in Fig. 17.5. It consists of I-section. A channel section is placed with its web horizontally. I-section and channel section both jointly act as a built-up section. The moment of inertia of a channel section gets added to that about yy-axis. The crane gantry girder is designed to lift and to move and shift live load euqal to its full capacity.

Step 1 : Maximum wheel load

Weight of trolley (i.e., capacity of crane) + lifted load = (300 + 100) = 400 kN

The weight of crane (excluding trolley) 190 kN acts as uniformly distributed live load as shown in Fig. 17.3.

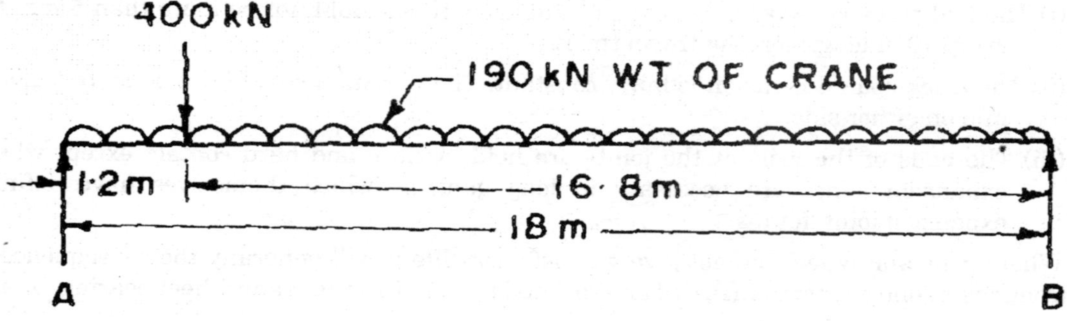

Fig. 17.3.

The vertical reaction on each wheel of crane would be maximum, when trolley is at nearest distance to trolley girder as shown in Fig. 17.3 that is minimum hook approach distance given as 1.2 m.

Take moment about B, then reaction at A

$$R_A = \frac{1}{18}\left[400 \times 16.8 + 190 \times \frac{18}{2}\right] = 468 \text{ kN}$$

This vertical load at one end of the crane bridge is transferred to the gantry girder through two wheels.

Maximum vertical load on each wheel of crane, = (1/2 x 468) = 234 kN

The characteristic (service) loads are shown in Fig. 17.4.

Step 2 : Maximum bending moment (due to D.L. + L.L. + I.L.)

The maximum bending moment in the gantry girder under a moving load occurs when the line of action of that load and c.g. of the loads are at equal distance from the centre of span. That is,

$$EC = CF = 0.875 \text{ (Fig. 17.4)}$$

The reactions at the supports A and B at service (characteristic level) are as follows :

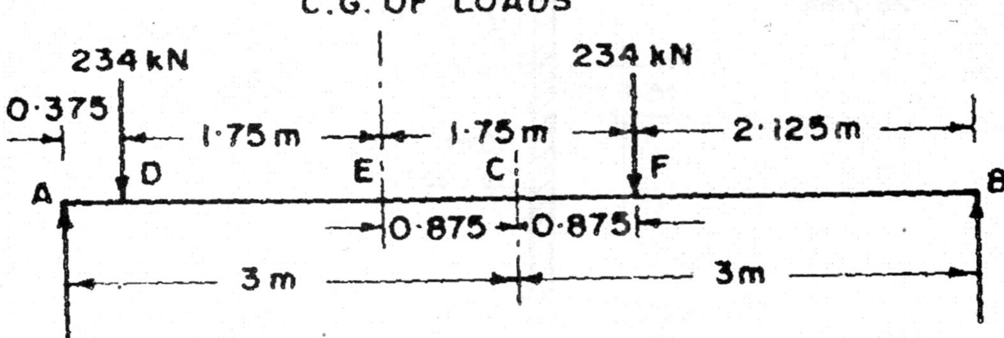

Fig. 17.4.

$$R_A = 234 \times \frac{1}{6}\left[(6 - 0.375) + 2.125\right] = 302.23 \text{ kN}$$

R_B = (2 x 234 – 302.23) = 165.77 kN

The reactions at the support A and B at ultimate [factored (design)] Level

$$R_A = \gamma_{FL}.(302.23) = (1.5 \times 302.23) = 453.345 \text{ kN}$$

$$R_B = \gamma_{FL}.(165.77) = (1.5 \times 165.77) = 248.655 \text{ kN}$$

Total = 1.5 x 46 = 702 kN

Factored maximum bending moment due to moving load occurs at F that is 1.5 x R_B x FB

FB = 2.125 m

M_F = 1.5 (165.77 x 2.125) = 1.5 x 352.3 kN = 528.75 kN-m

Add 25 per cent impact moment viz., 1.5 x 88.1 = 132.15 kN-m

(1) **Live load moment** = 1.5 x (352.3 + 88.1) = 1.5 x 440.4 kN-m = 660.6 kN-m

Assume self-weight of the girder as 2 kN/m

Weight of rail section is 0.300 kN/m, Total dead load = 2.3 kN/m

Factored Maximum bending moment due to dead load = γ_{fL} × dead load

$$1.35 \times \left(\frac{wl^2}{8}\right) = \left(\frac{1.35 \times 2.3 \times 6 \times 6}{8}\right) = 1.35 \times 10.35 = 13.97\ kN - m$$

(2) **Dead load moment** = 13.97 kN-m

(3) **Total vertical moment** = 1.5 (440.4) + 1.35 (10.35) = (660.6 + 13.97) = 674.57 kN-m

Yield stress for the steel, σ_y = 250 N/mm²

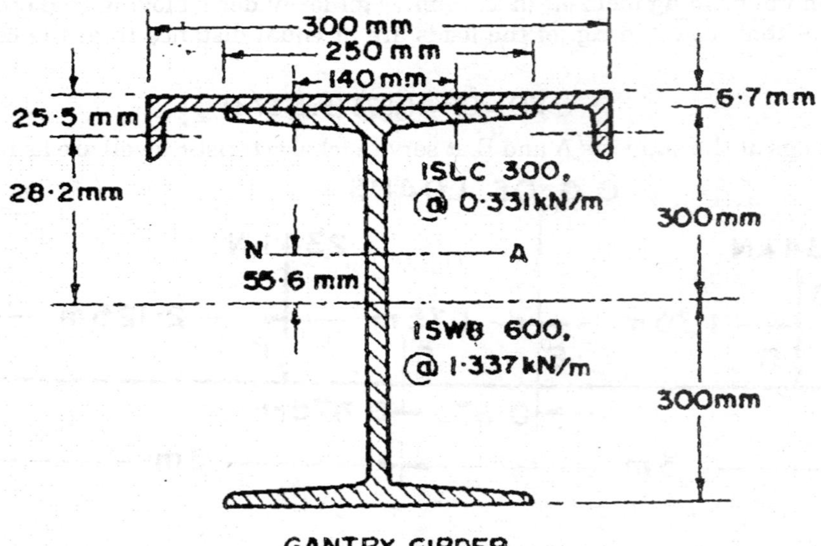

Fig. 17.5

Resistance of the beam section

$$= \left(\frac{\sigma_y \times Z_p}{\gamma_{msy}}\right) = \left(\frac{250 \times Z_p}{1.15}\right)$$

$$\left(\frac{250 \times Z_p}{1.15}\right) = \left(674.57 \times 10^3 \times 10^3\right) \text{ N-mm}$$

Plastic modulus required for the built-up section

$$Z_P = \left(\frac{674.57 \times 1.15 \times 10^3 \times 10^3}{250}\right)$$

$$= \left(\frac{775.76 \times 10^3 \times 10^3}{250}\right)$$

$= 3103.07 \times 10^3 \text{ mm}^3$

Total elastic section modulus required. Assuming the shape factor,

S = 1.12

$$Z_{xx} = \left(\frac{3103.07 \; x \; 10^3}{1.12}\right) = 2770.6 \times 10^3 \; mm^3$$

Let a channel section LC 300 a 0.331 kN/m be proposed to be provided over an I-section.

Section modulus of the channel section about strong axis

$= 403.2 \text{ x } 10^3 \text{ mm}^3$

Section modulus required for I-section

$= (2770.6 - 403.2) \text{ x } 10^3 \text{ mm}^3$

$= 2367.4 \text{ x } 10^3 \text{ mm}^3$

Provide ISWB 600, @ 1.337 kN/m

Section modulus of this I-section

$= 3540.0 \text{ x } 10^3 \text{ mm}^3$

Section modulus of the channel section

$= 403.2 \text{ x } 10^3 \text{ mm}^3$

Total section modulus provided ($3943.2 \text{ x } 10^3$), which is more than required $= 3103.7 \text{ x } 10^3 \text{ mm}^3$

Step 3 Horizontal (transverse/lateral) force at service load level

Horizontal force (transverse/lateral) to the rail

10 percent of (weight of trolley + load lifted

$$= \frac{10}{100} x \; (300 +)00 \;\; = \;\; 40 \text{ kN}$$

Horizontal force transverse to the rail on each wheel = 20 kN

Horizontal reaction at support A. Fig. 17.4 and Fig. 17.6.

It is calculated by proportion of reaction at support A due to vertical load Fig. 17.4.

$$= \left(\frac{20 \; x \; 302.33}{234}\right) = 25.83 kN$$

Horizontal reaction at support B

$= (40 - 25.83) = 14.17$ kN

Factored value of reaction at

Support A = (1.5 x 25.83) = 38.745 kN

Factored value of reaction at support B = (1.5 x 14.17) = 21.255 kN

Total = 60.000 kN

Step 4:Bending moment due to horizontal/lateral forces (i.e, in horizontal plane)

Moment in the horizontal plane due to external action (load)

= Reaction at support B x Distance BF

= (21.255 x 2.125)

= 45.167 kN-m

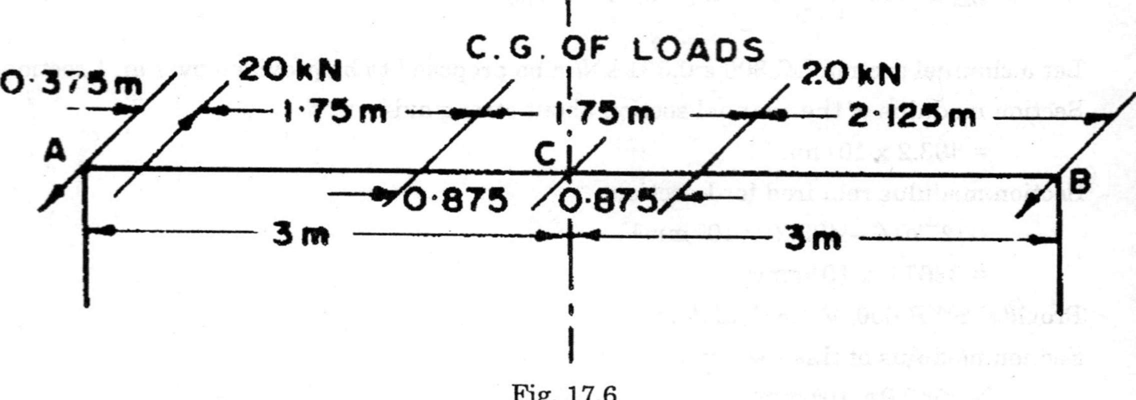

Fig. 17.6

Step 5 : Resistance/strength of the built-up crane girder section

Elastic section modulus of I-section ISWB 600, @ 1.337 kN/m about week axis = 376.2 x 10^3 mm^3

Elastic section modulus of channel section ISLC 300, @ 0.331 about strong axis = 403.2 x 10^3 kN/m

These two values are added together and jointly resist the horizontal (transversal lateral) deflection = (376.2 + 403.2) x 10^3 mm^3

= 779.4 x 10^3 mm^3

Resistance/strength of the built-up crane girder section to resist moment in the horizontal plane

$$= \left(\sigma_y \; x \; Z_P\right)$$

$$= \left(\frac{250 \; x \; 1.12 \; x \; 779.4 \; x \; 10^3}{1000 \; x \; 1000}\right) = 218.232 \text{ kN-m}$$

Horizontal moment due to external action (force) 45.167 kN-m is much less than the resistance/strength of built up crane girder section. Hence, crane girder is adequate.

Step 6 : Check for shear strength for the built-up crane girder

Maximum shear force due to vertical loading at service load level, Fig. 17.4

$$= \left(234 + 234 \; x \; \frac{2.5}{6.0}\right) kN = 331.5 \text{ kN}$$

Add 25 percent for impact = 82.75 kN

Total shear due to vertical loading = 414.25 kN

Factored shear force = (1.5 x 414.25) kN = 621.375 kN

Dead load shear = $\left[\frac{(1337 + 331.5) x6}{2 x 1000}\right] kN$ = 5.0855 kN

Factored dead load shear

V_{fd} = (1.35 x 5.0855) kN = 6.865 kN

Total Factored shear force

(621.375 + 6.865) = 628.240 kN

Step 7 : Shear resistance

Web of channel section resists shear

Depth of web = 300 mm

Thickness of web = 6.7 mm

Area resisting shear = (6.7 x 300) = 2000 mm²

Yield stress for web = ($\sigma_{yw}/\sqrt{3}$) = 0.6 σ_y

Shear resisted = $\left(\frac{0.60 x\ 250\ x\ 2000}{1000}\right) = 300\ kN$

Web of I-section resists shear

Depth of web = 600 mm

Thickness of web = 11.2 mm

Area of web = (600 x 11.2) = 6270 mm²

Shear resistance = $\left(\frac{0.6x\ 250\ x\ 6720}{1000}\right)$ = 1008 kN

Total shear resisted = (1008 + 300) = 1308 kN

Factored shear due to external action 628.24 kN is less than shear resistance of the crane girder 1308 kN. Hence, the crane girder is adequate for shear.

PROBLEMS

17.1 Design a gantry girder to carry an electric overhead travelling crane to suit following data :

Crane capacity ...200 kN
W.t. of crab alone ...70 kN
Wt. of crane ...150 kN
Minimum approach of crane hook ...1.2 mm
Dist. between centres of crane wheels ...3.5 m
Dist. between cranes of gantry girders ...15.0 m
Span of gantry girder ..7m

Wt. of rail section ...0.300 kN/m
Height of rail section ...80 mm

17.2 A gantry girder is provided for mill building to carry an electric overhead travelling crane with the following data as shown in Fig. 17.7.

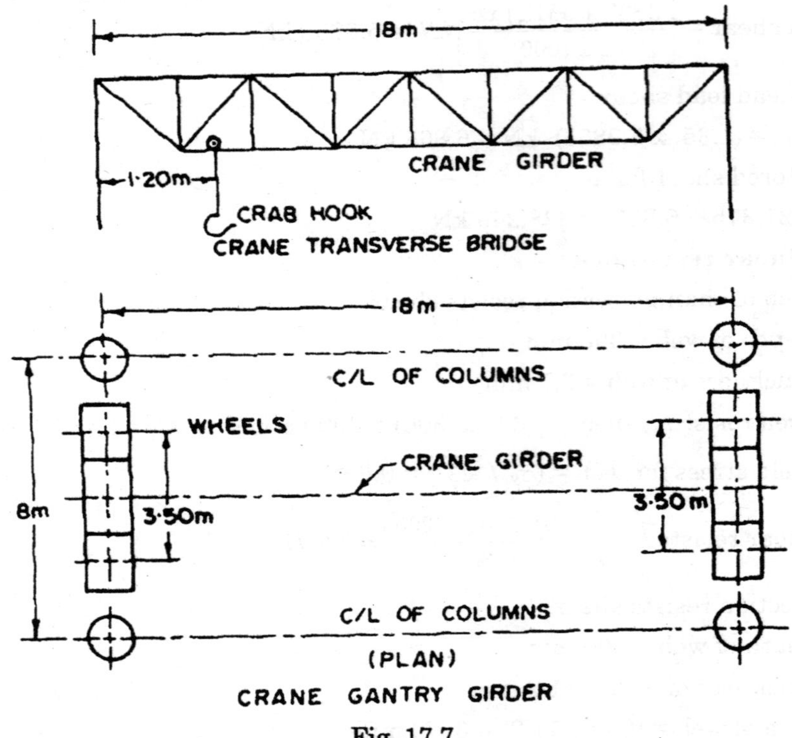

Fig. 17.7

(i) Centre to centre distance between columns (span of gantry girder)= 8 metres

(ii) Centre to centre distance between gantry girders = 18 metres

(iii) Crane capacity = 300 kN

(iv) Self-weight of crane hook = 30 kN

(v) Self-weight of crane hook = 30 kN

(vi) Allowable approach of crane hook from the vertical axis of the gantry girder = 1.2 metres

(vii) Distance between centres of wheels moving on the gantry girder and supporting the crane girder = 3.5 metres

(viii) Weight of rail section attached on the top of the gantry girder = 0.300 kN/metre.

Determine :

(a) The maximum B.M. and S.F. produced by the vertical loads

(b) The maximum B.M. and S.F. produced by the lateral loads

(c) The maximum longitudinal force exerted along the rails.

Also design the section of gantry girder.

17.3 A gantry girder is composed of an MC 550 x 150 mm, @ 1.037 kN/m and a channel MC 300 x 90 mm, @ 0.358 kN/m placed on the top of the joist with its flange down. Calculate the maximum flange and web stresses in the girder section, given the following :

(i) Effective span of gantry girder = 7.3 m

(ii) Effective span of crane girder between centres of gravity of rails = 17 m

(iii) Distance between the pair of carriage wheels moving on each gantry rail = 3.6 m

(iv) Capacity of overhead crane = 150 kN

(v) Total weight of crane girder excluding the crab = 180 kN

(vi) Weight of crab = 50 kN

(vii) Weight of gantry girder inclusive of rail = 1.80 kN/m

(viii) Lateral load due to horizontal surge = 10 per cent of lifted weight and weight of crab

(ix) Impact for vertical wheel loads = 25 per cent

(x) Closest position of crane hook from the centre of gantry rails = 1m.

(b) What would have been the effect on the vertical moment and shear diagrams if an allowance for longitudinal load were also taken.

18

Part - VIII

DESIGN OF PLATE GIRDERS

LIMIT STATE OF COLLAPSE IN BENDING DESIGN OF PLATE GIRDERS (P.G.-I)

18.1 INTRODUCTION

A plate girder is a beam built-up from the plate elements to achieve more efficient arrangement of material than it is possible with rolled steel beams. Plate girders are economical where spans are long enough to allow saving in cost by proportioning for the particular requirements. Plate girders may be riveted, bolted, or welded.

When the span and load combination is such that the rolled steel beams become insufficient to furnish the requirement, and built-up beams become uneconomical, then plate girders are used. The built up beams are used where overall depth is limited. As such the use of rolled steel beams, built-up beams and plate girder is a step by step approach for the increase in loads and the spans. The object of design is to achieve overall economy, which involves the cost of fabrication in addition to the cost of material. The cost of fabrication is more for built-up beams as compared to beams, and it is still higher for plate girders as compared to both. Attempt is made to provide deep sections for economy as regards material and cost of fabrication.

The plate girders are deep structural members subjected to heavy transverse loads. They consist of plates and angles riveted together. Plates and angles form an I-section. They are used in building constructions and also in bridges. The plate girders are economically and efficiently used for span upto about 45 m in building construction. The depth of plate girder may range upto 5 m or more, 1.5 m to 2.5 m depths are very common. Common sections used for plate girders are shown in Fig. 18.1. The plate girders have large moment of resistance (i.e., strength of the section) in bending. It is larger than the beams and built-up beams and somewhat smaller than the truss girders.

During the period 1870-1900, riveted plate girders consisting of angles connected to a web plate, with or without cover plates, were extensively used in the United States on spans from about 15 m to 45 m.

Fig. 18.1 (A) shows the simplest form of a plate girder. In case, this simple section cannot furnish sufficient flange material, additional plates are riveted to outstanding legs of angles as shown in Fig. 18.1(B) and (C). When number of cover plates becomes excess, then the section of plate girder is modified as shown in Fig. 18.1(D). In case of deck girders of rail-road bridges, sometimes a form as shown in Fig. 18.1(E) is used. Sometimes for long span bridges carrying heavy loads, heavy sections shown in Fig. 18.1 (F) and (G) are used. Section as shown in Fig. 18.1 (H) is adopted for crane runway girders.

Sometimes in building construction, the depth of a girder is limited to provide necessary head room. Limitations of head room require the depth of girder much less than the economical depth. It becomes necessary to provide large web area to resist shear forces. In such cases, sections as shown in Fig. 18.1 (J) and (K) are often used in which two or more webs are provided. These are called box girders. In addition to shear, webs resist bending moment to a considerable extent.

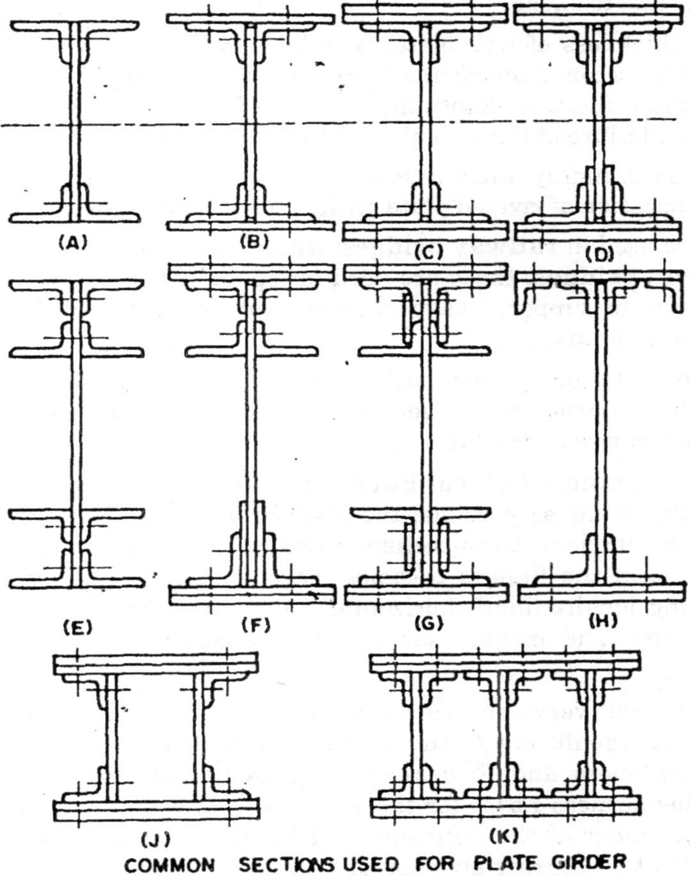

Fig. 18.1 (1). Common section of plate girder.

In the past, the plate girders were fabricated and constructed by riveting and rivets were used. Therefore, these are called **'Riveted Plate Girders'**. In field, the plate girders are fabricated and constructed by bolting and bolts are used. Rivets and bolts necessiate the flange angles to connect the web to the flange angles.

Rivets, riveted connections and riveted plate girders had been used since long. Afterwards, welded and welded joints have been established, welded plate girders are being used, which have been described in the subsequent articles.

In the design of a plate girder, two flange plates are connected to a web plate to form an I-section. The depth of web (i.e, the depth of a plate girder) is kept large. The complete cross-section is made efficient and economical. The flange plates are kept apart as far as possible from the neutral axis. As the depth of plate girder increases, the flange plates resist large in-plane bending moments. For the given bending moments, the axial forces and the respective stresses (compressive and tensile) reduce for the larger depth. The flanges are properly designed so that the **local buckling** may not occur (specially in the compression flange).

However, the larger depth of a plate girder needs smaller area of web plate. Deep web plates having large depth needs thin web plate, which resists shear. In thin web plates, **web buckling** may occur due to shear.

Therefore, the thickness of web plate, t_w is kept so as to satisfy the requirements of minimum thickness as recommended in IS : 800-2007. Accordingly, the vertical transverse stiffeners and horizontal (i.e, the longitudinal) stiffeners are designed properly and carefully so that the premature failure of the web plate does not occur.

In the recent years, highly automatic workshops have been established. As a result of which the cost of fabrication of riveted plate girders have been reduced significantly.

Plate girders are used in **railway bridges**, where the loads are heavy, and spans (15 m to 40 m). In **highway bridges**, these are used for spans (24 m to 46 m). Plate girders are also used in buildings to support heavy concentrated loads specially to build large halls without intermediate columns.

At present, the plate girders are built using mainly three plates (two plates for the flanges and one plate for the web) together by welding. Welded plate girdes have been described lateron in a seperate chapter.

It is important to consider (i) **local buckling** of the compression flange and (ii) **shear buckling** of the web. As far as possible, the large area of material is kept away from the neutral axis to have efficient cross-section to resist in-plane bending. The moment of resistance obtained from the flanges becomes large. These flanges have compressive and tensile stresses acting longitudinally/horizontally. The axial compressive and tensile force develop in the flanges. The magnitudes of these forces reduce as the depth of girder increases.

The web plate is kept very thin. However, it is seen that the premature failure due to web buckling in shear should not occur. These thin webs are stiffened (strengthened) by means of vertical stiffeners and if, necessary, horizontal stiffeners are provided on the compression side. Slenderness ratios for the web (depth of web, d_w to the thickness of web, t_w) control the requirement of the stiffeners and buckling of web. Slender webs (i.e, d_w/t_w ratio is more) buckle at comparatively smaller vales of applied shear loading. Webs having span/ depth ratio less than 12 results in deep girders.

Design of plate girders become economical when deep thin webs are provided. Deep thin webs need thorough knowledge in stability especially plate stability. Stiffeners are provided to increase the load carrying capacity.

Plate girders may fail by yielding of the tension flange and the buckling of the compression flange, local buckling of flanges or lateral torsional buckling or vertical buckling of web.

The strength of plate girders depends upon post-buckling strength of web. Normally near the supports and near the neutral axis, the values of shear becomes high. Diagonal compression may cause buckling of web.

Design of a plate girder requires consideration of buckling of the flanges and/or the web, it is convenient to consider the various posible cases in terms of interaction between moments and the shear forces as shown in Fig. 18.1(2) Fig. 18.1(B).

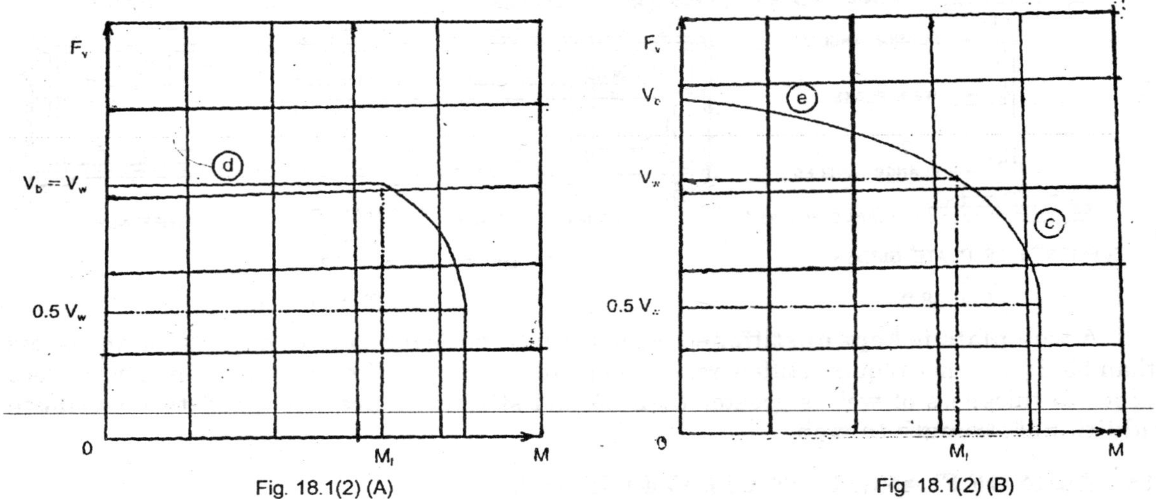

Fig. 18.1 (2) (A) & (B).

1. **For zero shear**, the moment capacity is simply the product of the design strength of the material, s_y times the appropriate section modulus (i.e, plastic modulus of the flanges about neutral axis)

2. In case, **the web is assumed to carry shear only, that is, all moment is assumed to be taken by the flanges** Fig. 18.1(A).

3. In case, the applied shear does not exceed 50 percent of the shear resistance of the web v_{WR}, then, the moment capacity of the section may be obtained as in 1, may be developed

4. For the combination of sher more than 0.5 V_{WR}, and the moment more than M_f, interaction should be considered as marked 'C' in Fig. 18.1(B).

Slender members results in efficient and economical design. The post buckling capacity of the web (i.e, the ability of the girder to resist transverse loads considerably in excess of the load at which the web buckles under shear) may result/prove advantageous.

Therefore, efforts may be made to design a girder of high strength to weight ratio by incorporation the **post buckling strength of the web** in the method of design may be used.

18.2 ELEMENTS OF A PLATE GIRDER

The vertical plate of the plate girder is termed as **web plate**. The angles connected at the top and bottom of the web plate are known as **flange angles**. The horizontal plates connected with the flange angles are known as **flange plates** or **cover plates**. The web plate, flange angles and flange plates are shown in Fig. 18.2. The bearing stiffeners, intermediate stiffeners and horizontal stiffeners used with the plate girder are shown in Fig. 18.3.

The cross section of a plate girder consists of (i) a compression flange (ii) a tension flange (iii) a web plate. These are designed as per the recommendations and specifications described in IS: 800-2007.

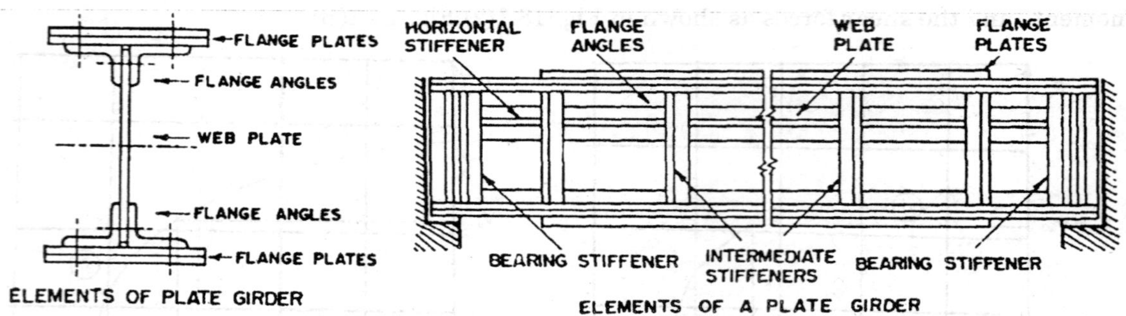

Fig. 18.2

Fig. 18.3. Plate girder

A web plate is kept unstiffened when the ratio of clear depth to thickness of web is less than 85. It does not require stiffeners. A web plate is called **stiffened,** when the ratio of clear depth to thickness of web is greater than 85 and **stiffeners are provided to contribute additional strength to web.**

18.3 AREA OF FLANGES OF A PLATE GIRDER

When a plate girder is carrying transverse load acting downward, its top flange is in compression and bottom flange is in tension. The area of flange comprises of areas of flange angles, flange plates (if flange plates are attached) and part of the web plate. This part of web plate sandwiched between flange angles also resist bending moment. The part of web plate which is available for flange area is computed as below. The moment capacity of the plate girder is found as under.

Bending moment, $M = \sigma_y \cdot (I_{xx}/y_{max})$

where, I_{xx} = Gross moment of inertia of plate girder

σ_y = Allowable bending stress in compression

y_{max} = Vertical distance of extreme fibre of section from neutral axis

The gross moment of inertia of plate girder

$$I_{xx} = \left[2A_f\left(\frac{d_e}{2}\right)^2 + \frac{t_w \cdot d_e^3}{12}\right] \text{ approximately} \qquad \text{(i)}$$

where, A_f = Gross area of flange, d_e = Effective depth, t_w = Thickness of web

The moment of inertias of flange plates and flange angles about their own axes are neglected. Then

$$M = \left(\frac{\sigma_y \times 2A_f \times d_e^2}{(d_e/2) \times 4}\right) + \left(\frac{\sigma_y \times t_w \times d_e^3}{(d_e/2) \times 12}\right) \qquad \text{(ii)}$$

$$M = \left(A_f \times \sigma_y' \times d_e\right) + \left(t_w \times d_d\right) \times \left[\sigma_y \times \frac{d_e}{6}\right] \quad \text{(iii)}$$

$$M = \left(A_f \times \sigma_y \times d_e\right) + \left(A_w \times \sigma_y \times \frac{d_e}{6}\right) \quad \text{(iv)}$$

$$M = \left(A_f + \frac{1}{6} A_w\right) \times \sigma_y \times d_e \quad \text{(18.1A)}$$

The effective depth is considered approximately equal to the depth of web, where, A_w is the area of web.

In the riveted plate girder, the vertical stiffeners are connected by using the rivets. The splice plates are also connected to the web by the rivets. As such rivet holes are made in the web. In the tension flange, net area is used. The net area is found after making deductions for the rivet holes. It is assumed that 25 mm diameter holes are made for rivets at 100 mm spacing. Then the net area of web is three-fourth of the gross-area. Therefore, net area of tension flange

$$A_{fnet} = \left(A_f + \frac{3}{4} \times \frac{1}{6} \times A_w\right) = \left(A_f' + \frac{1}{8} A_w\right)$$

where, A_f = Net area of flange excluding web.

From this, it is seen that approximately 1/6th web area acts as flange area if the gross area is considered, and 1/8th web area (on account of reduction in flange area for rivet/bolt holes) acts as flange area if the net area is considered. The portion of web which acts as flange, is termed as **web equivalent**.

The effective sectional area of compression flange shall be the gross area with deductions for excessive width of plates as it was specified for compression members and open holes occurring in a plane perpendicular to the direction of stress at the section being considered.

The effective sectional area of tension flanges shall be **the gross-area** with the deduction for holes where the rivets are also used.

The bending (flexural) strength of a plate girder is based on the yielding of the tension flange or otherwise buckling of the compression flange, which one is decided either by (i) local buckling or (ii) lateral torsional buckling. When the flanges are plastic, compact or semi-compact (i.e, (d_w / t_w) is more than 67ϵ, where ε (epsilon) is the yield stress ratio $(250/\sigma_y)^{1/2}$.

Moment resstance capacity of the plate girder after the flanges have been yielded is calculated from the following. Since, it is assumed that the flanges resist bending moment.

$$M_c = \left(\sigma_y \,.\, Z_{pf}\right) / \left(\frac{1}{\gamma_{msy}}\right)$$

where, σ_y is the yield stress in N/mm² for the grade/quality of steel. Z_{pf} is the plastic modulus of the flanges about the transverse axis of the plate girder-section.

γ_{msy} is the partial safety factor for the material strength. For steel, γ_{msy} = 1.15.

Note : As it is known that the modulus of section is calculated by considering first moment of the area.

Since, it is assumed that the bending moment is completely resisted by the flanges, the forces in the flanges may be considered equal (equal to $\sigma_y b_f t_f$)

Moment resisted

M = (Force x lever arm)

$$M = \sigma_y \ (b_f \ t_f). \ d_w$$

$$(b_f \ t_f) = \left(\frac{M}{\sigma_y . d_w} \right)$$

Gross cross-sectional area of the section

A_{gross} = Area of two flanges + area of web

A_{gross} = $(2 \ b_f \ t_f) + (d_w \ . \ t_w)$

$$A_{gross} = \left(\frac{2M}{\sigma_y \ . \ d_w} \right) + (d_w \ . t_w)$$

Let, $d_w = k \ t_w$ (i.e, slenderness of web $(d_w/t_w) = k$). Then

$$A_{gross} = \frac{2M}{\sigma_y \ . \ (k \ . t_w)} + k. t_w$$

Diffentiating with respect to t_w and equating to zero.

$$\frac{d \ A_{gross}}{d \ t_w} = \left(\frac{-2M}{\sigma_y \ . \ K. t_w^2} + k \right) = 0$$

$$\left(\frac{2 \ M}{\sigma_y \ k^2} \right) = t_w^2, \quad t_w = \left(\frac{2M}{\sigma_y \ . \ k^2} \right)^{1/2}$$

Optimum value of depth

$$d_w = \left(\frac{M.k}{\sigma_y} \right)^{1/3} \qquad (18.1B)$$

Therefore, when the area of one flange is equal to area of web, it gives optimum value (Reference : Martin and Purkins, 1992) Salmon and Johnson, 1996 extended this study. Following expression was suggested for the area needed for one flange

$$A_f = \left| (M / \sigma_y . d_w) - (A_w / 6) \right|$$

where, A_w is the area of web plate. This expression may be used for initial size.

Initial size of a plate girder

The **depth of plate** girder is main important size for the plate girder. The stiffness and the bending resistance is controlled by the depth of plate girder. The design of the plate girder should satisfy both the requirements for (i) strength, and deflection. Overall depth of girder D may be proposed as the ratio of overall depth/equivalent span for the following.

(i) For plate girders in buildings

Ratio (D/L) = (1/15 to 1/25)

(ii) For highway bridges

Ratio (D/L) = (1/12 to 1/18) and

(iii) For railway bridges

Ratio (D/L) = (1/10 to 1/15)

Note : The equivalent span length is defined as the length between the points of adjacent contraflexures under uniformly distributed load.

In case, the top flange is attached with a reinforced/composite concrete deck/girder or to an orthotropic steel deck, above proposed values of ratios (D/L) may be reduced. In case, the launching procedure is used for the erection, the depth of girder may be kept constant. For unequal spans for a continuous girder, the depth of plate girder is kept varied for the economic reasons.

Commonly the **thickness of web** is proposed 6 mm to 8 mm keeping the corrosion in view. For the bridges of large depth a minimum thickness of 8 mm to 10 mm may be suggested. Keeping in view the handling stresses.

At present, as far as possible, the tendency is to avoid stiffeners contrary to the practice in the past. Therefore, the cost of fabrication and time are saved.

For span 4 m to 6m, the **thickness of web** is kept constant. For the steel of grade with yield stress, $\sigma_y = 250$ N/mm^2,

Ratio (d_w/t_w) is proposed less than 360. For the steel of grade with yield stress, $\sigma_y = 360$ N/mm^2,

Ratio (d_w/t_w) is proposed less than 240

Maquoi and Skaloud, 2000 suggested that this ratio is still kept less in case, fatigue strength controls the design

For the stiffened web, the recommendations specified in IS : 800-2007 may be followed.

For Non-composite girders,

the requirements for strength and rigidity control the design of **flanges.** The width of flange plate may be proposed 0.3 times the depth of section.

(ii) For the composite girders, it is proposed as 0.2 *times* the depth of section.

The flanges are kept plastic compact or semi-compact. Accordingly the ratio (b_f/t_f) is kept in view, so that the local buckling is avoided before attaining yield.

For the grade of steel $\sigma_y = 410$ N/mm^2, (b_f/t_f) ratio is kept about 24. The width of flange is kept about 1000 m and thickness about 80 mm to 100 mm. The width of tension flange is kept about 1.30 *times* larger.

Transverse stiffeners are attached symmetrical on both the faces. Web stiffeners are welded continuously to the web. Panel aspect ratio is kept 1.2 to 1.6.

For the end panel designed without using tension field is kept about 0.6 to 1.0.

18.4 DEPTH OF A PLATE GIRDER

The depth between outer surfaces of flanges is termed as **over-all depth** `$d_o$' or depth of the plate girder. In general, the depth of plate girder is kept 1/10 to 1/12 of span. The distance between c.g. of compression flange and c.g. of tension flange is known as **effective depth** `d_e' of the plate girder. The distance between vertical legs of flange angles at the top and at the bottom is known as **clear depth** `d' of plate girder, in case horizontal stiffeners are not used. The overall depth, effective depth and clear depth of plate girder are shown in Fig. 18.4. The depth of plate girder is an essential factor for the design of plate girder. Sometimes, structural requirements limit the depth of plate girder. When the depth of plate girder is less than 750 mm, then, such girders are known as **shallow plate girders**. The shallow plate girders resemble rolled steel beams. When the depth of plate girder is more than 750 mm or when the depth is atleast eight times the depth of vertical leg of the flange angles, then, such girders are known as **deep plate girders**.

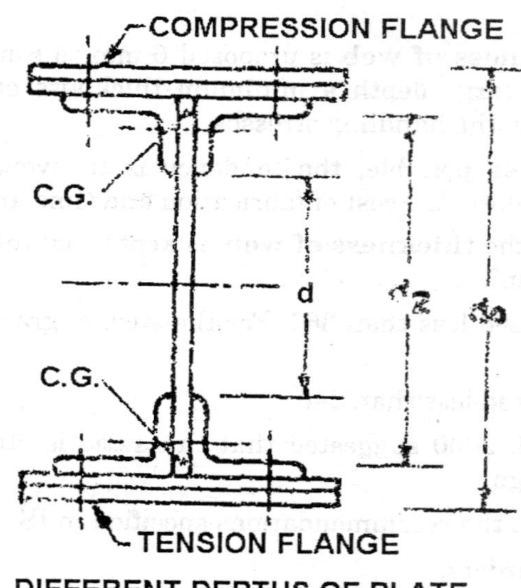

DIFFERENT DEPTHS OF PLATE GIRDER

Fig. 18.4.

The variation in cross-sectional area of the plate girder is to be checked as a function of the depth of the web-plate is calculated, which will give minimum area.

Thomas C. Shedd [Structural Design in Steel, New York, John Wiley & Sons, Inc, 1934, Chapter 3] Boris Bresier, T.Y. Lin and John Scalzi [Design of Steel Structures, second edition, New York, John Wiley & Sons, Inc, 1968, pp 497-554] and Omer W. Blodgette [Design of Welded structures, cleveland, Ohio, James F. Lineoln, Arc Welding Foundation, 1966] Charles G. Schilling [`Optimum Proportions for I-shaped Beams, Journal of structural Division, ASCE, 100, ST 12, (December, 1974, pp 2385-2401] have provided extended treatment of **optimum proportioning**. The average gross- area of the plate girder for the complete span may be expressed as under.

$$A_{gross} = 2C_1 A_f + C_2.A_w \qquad (18.2)$$

Where, C_1 is a factor to consider for reducing flange size at zones of lower bending moment than the maximum bending moment C_2 is a factor to consider for reducing web thickness at zones of reduced shear ($A_w = d_w . t_w$). Substituting in Eq. 18.2.

$$A_f = [M/(\sigma_{ave} . d) - A_w/6]$$

$$A_{gross} = 2C_1\left(\frac{M}{\sigma_{ave} . d} - \frac{d_w . t_w}{6}\right) + C_2\,(d_w . t_w)$$

The differentiation of A_{gross} with respect to depth of web and equating the same to zero gives condition for the minimum average gross-area. Therefore, $(\partial A_{gross}/\partial d) = 0$. There arises following three cases :

1. **There is no restriction for depth. The desired ratio of (d_w/t_w) may be adopted. Let K = (d_w/t_w)**

$$A_{gross} = 2C_1\left(\frac{M}{\sigma_{ave} . d} - \frac{dw^2}{6\,K}\right) + C_2 . \frac{dw^2}{K}$$

$$\left(\frac{d\,A_{gross}}{dd_w}\right) = \frac{-2C_1 . M . \sigma_{ave}}{(\sigma_{ave})^2\,(d_w)^2} - \left(\frac{4\,C_1\,d_w}{6K}\right) + \frac{2C_2 . d_w}{K} = 0$$

$$-6\,C_1\,M.K - 2\,C_1\,(d_w)^3 . \sigma_{ave} + 6\,C_2\,(d_w)^3\,\sigma_{ave} = 0$$

Therefore, $$d = \left(\frac{3\,M\,C_1\,K}{\sigma_{ave}\,(3\,C_2 - C_1)}\right)^{1/3} \quad (18.3)$$

In case, the reduction in section in zones of lower stress is neglected, then, $C_1 = C_2 = 1$. Then,

$$d_w = \left(\frac{3\,M . K}{2 . \sigma_{ave}}\right)^{1/3} \quad (18.4)$$

Using $C_1 = C_2 = 1$ and substituting (M/σ_{ave})

$$A_{gross} = \frac{4\,d_w^2}{3K} - \left(\frac{d_w^2}{3K}\right) + \left(\frac{d_w^2}{K}\right) = \left(\frac{2d_w^2}{K}\right) \quad (18.5)$$

From which, the mass of the plate girder per metre may be estimated, using the fact that unit weight of steel is 7840 kg/m²

$$A_{gross} = 1.72\left(\frac{M^2}{\sigma_y^2 . K}\right)^{1/3} \quad (18.6)$$

The units for M and σ_{ave} are in kN and mm respectively. It is to note that, in case, the stiffeners are attached, the mass of plate girder will increase by five to ten percent.

2. **The thickness of web plate is kept minimum and t_w is treated as constant. Then**

$$\left(\frac{\partial A_{gross}}{\partial d}\right) = \left(\frac{-2C_1 . M . \sigma_{ave}}{\sigma_{ave}^2 . d_w^2}\right) - \left[\frac{C_1 t_w}{3}\right] + (C_2 t_w) = 0$$

$$d_w = \left[\frac{6 C_1 M}{\sigma_{ave} . t_w (3C_2 - C_1)}\right]^{1/2} \quad (18.7)$$

In case, $C_1 = C_2 = 1$ (unity)

$$d_w = \left(\frac{3 M}{\sigma_{ave} . t_w}\right)^{1/2} \quad (18.8)$$

The mass of plate girder per metre

$$= 0.0181 \left(\frac{M . t_w}{\sigma_{ave}}\right)^{1/2} \quad (18.9)$$

mm units are used for the variable.

3. **There, is a heavy shear, which A_w controls the area of web and A_w is considered as constant.**

$$A_{gross} = 2C_1 \left(\frac{M}{\sigma_{ave} . d_w} - \frac{A_w}{6}\right) + C_2 . A_w \quad (18.10)$$

In case, the depth of web, d_w is maximum, it gives the minimum gross area, A_{gross}

In case, the same variety of steel (i.e., having same yield stress and other constituents), the value of C_1 may vary from 0.70 to 0.90, when used with the maximum positive moment, 0.85 to 0.90 is the usual range.

The value of C_2 is not as likely to vary except on the continuous structures where it might be 1.05 when used with maximum positive moment or 0.95 when used with maximum negative moment. In order to avoid complexity of evaluating C_1 and C_2 for the continuous structures, C_1 and C_2 may be adopted unity.

18.5 ECONOMICAL DEPTH OF A PLATE GIRDER FOR WORKING STRESS METHOD

The economical depth of the plate girder is the depth which gives the minimum weight of plate girder. The economical depth of plate girder is found as below : Let,

M = Maximum bending moment

d_e = Effective depth of plate girder (Depth of web plate)

t_w = Thickness of web plate

σ_b = Allowable bending stress

l = Length of plate girder

ρ = Unit weight of steel

Area of flange $A_f = M / (\sigma_b . d_e)$

When two or more cover plates are used in plate girder in each flange, then cover plates are curtailed. The depth of plate girder varies as the cover plates are curtailed. The part of total weight of plate girder (comprising of flanges, web plates, splice plates, stiffeners and filler plates) varies with the variation in the depth of plate girder. As such, the average section of flange is approximately reduced to 80%. The weight of stiffeners is approximately 60% of web plate. The total weight of plate girder

W = Average weight of flanges + weight of web plate inclusive of stiffener

$$W = \left[2 \times 0.80 \times \frac{M}{\sigma_b \times d_e} \times l \times \rho\right] + [1.6 \times t_w \times d_e \times l \times \rho] \quad \text{(i)}$$

$$W = \left[1.60 \times \frac{M}{\sigma_b \times d_e} \times l \times \rho\right] + [1.6 \times t_w \times d_e \times l \times \rho] \quad \text{(ii)}$$

In case, the portion of web plate in resisting the bending moment is considered for, then, the total weight of plate girder is given by

$$W = \left(1.60 \frac{M}{\sigma_b \times d_e} \times l \times \rho\right) - \left(2 \times \frac{1}{8} \times t_w \times d_e \times l \times \rho\right) + (1.6 \times t_w \times d_e \times 1 \times \rho) \quad \text{(iii)}$$

Differentiating the expression (iii) for W with respect to d_e

$$\left(\frac{d_w}{d\,d_e}\right) = -\left(1.60 \frac{M}{\sigma_b \times d_e^2} \times l \times \rho\right) + (1.6 \times t_w \times l \times \rho) - \left(\frac{1}{4} \times t_w \times l \times \rho\right) \quad \text{(iv)}$$

For the economical depth, $[dw/d(d_e)] = 0$

$$\therefore\ d_e = 1.1 \left(\frac{M}{\sigma_b \times t_w}\right)^{1/2} \quad \text{(18.11)}$$

In case the portion of web plate in resisting the bending moment is not considered, then the total weight of plate girder is given by the expression (i) Differentiating the expression (ii), and equating to zero, the economical depth of plate girder works out to be $(M/\sigma_b \cdot t_w)^{1/2}$. When the span of plate girder is small, then the flange plates may not be curtailed, and the section is uniform throughout the span, For such plate girder, the economical depth works out to 1.21 $(M/\sigma_b \cdot t_w)^{1/2}$.

The weight of plate girder varies 1 per cent for 10 per cent variation in the depth of the girder. In practice, economical depth of plate girder is assumed 15 to 20 percent less than that computed.

18.6 SELF-WEIGHT OF A PLATE GIRDER (WORKING-STRESS METHOD)

From Art. 18.5, the total weight of plate girder

$$W_1 = \left[1.60 \frac{M}{\sigma_b\, d_e} \times l \times \rho + 1.60\, t_w \times l \times \rho\right] \quad \text{(i)}$$

The economical depth of plate girder, from Eq. 18.11

$$d_e = 1.1\left(\frac{M}{\sigma_b \cdot t_w}\right)^{1/2}$$

$$\therefore t_w = 1.21\left(\frac{M}{\sigma_b \cdot d_e^2}\right) \text{ Therefore,} \qquad (ii)$$

$$W_1 = \left[\left[1.60\frac{M}{\sigma_b\, d_e} \times l \times \rho\right] + \left(1.60 \times 1.21\frac{M}{\sigma_b\, d_e^2} \times d_e \times l \times \rho\right) - \left(\frac{1}{4} \times 1121 \times \frac{M}{\sigma_b\, d_e^2} \times d_e \times l \times \rho\right)\right] \qquad (iii)$$

$$W_1 = \left[\left[1.60\frac{M}{\sigma_b \cdot d_e} \times l \times \rho\right] + \left[1.35 \times 1.21\frac{M}{\sigma_b \cdot d_e} \times l \times \rho\right]\right] \qquad (iv)$$

$$W_1 = 3.2\left(\frac{M \cdot l\, \rho}{(\sigma_b \cdot d_e)}\right) \qquad (v)$$

The expression (v) is for the weight of plate girder with gross- area of the flanges without making deductions for rivet holes. The weight of plate girder is about 10 to 20% greater than that given by the expression (v). The unit weight of steel p is 78450 N/m³. Therefore, the weight of plate girder is given by

$$W_1 = \left(\frac{1.2 \times 3.2 \times 78450 \times M\, l}{100 \times 100 \times 100 \times \sigma_b \times d_e} \times 100 \times 10\right) = \left(\frac{300 \times M\, l}{\sigma_b \cdot d_e}\right)$$

where, W_1 = Weight of the plate girder in N

M = Bending moment in N-mm, l = Span of the plate girder in metre

σ_b = Bending stress in N/mm², d_e = Effective depth of plate girder in mm.

The expression (vi) is useful when the bending moment M and the depth, d_e are known. Therefore, this expression is used to check the self-weight of the plate girder.

For all practical purpose, self-weight of girder is taken as

$$W_1 = (W.\ l/300) \qquad (18.12)$$

where, W = Total superimposed load and l = Span of plate girder in metres

The expression for self-weight of plate girder is approximate since it is assumed that web plate resists shear only and bending moment is resisted by flanges.

The self-weight w in kN/m may be assumed as (W/200), whose, W is the total factored load in kN.

18.7 IMPACT (COEFFICIENT OF DYNAMIC AUGMENTATION : CDA)

The **dynamic action** of the moving load on a structure is called impact.

The impact is caused due to sudden application of load, unbalanced driving wheel of locomotives, uneven track, wheel tyres worn out of truth, etc. The stresses and strains (deformations) due to sudden application of load are twice than those due to gradual application of load simultaneously and the actual increase in deformation is generally less than twice that due to gradually applied load because of damping effect of other construction. Therefore, the effect of impact is considered less than 100 percent. The effect of impact depends upon nature of moving load, speed of moving load, type of structure, loaded length of structure etc. Provision for effect of impact is made by impact allowance or impact load. The effect of impact is generally expressed as the percentage of the moving load. Impact load is determined as a product of **impact factor** and live (moving) load. The impact factor is specified by different authorities separately for different type of moving loads and different type of structures. Indian Railway Board has specified for steel girders for single track span, for metre gauge and broad gauge impact factor as under.

$$\textbf{Impact factor } i = \left(\frac{20}{14 + L}\right) > 1.00 \qquad (18.13)$$

$$\text{Coefficient of dynamic augmentation CDA} = \left(0.15 + \frac{8}{6 + L}\right) \text{ not more than.}$$

where, L is the loaded length of span in metres.

Instead of impact factor, a new term `CDA' **coefficient of Dynamic Augmentation"** has been used in the revised `**Bridge Rules**' used for the Railway Bridges. Formerly, an impact factor was used.

18.8 ASSUMPTIONS FOR DESIGN OF PLATE GIRDER

The plate girders are designed by trial and revised until the actual bending stresses are less than or equal to the permissible bending stresses. The stiffeners are attached as needed, until the shear stress is less than or equal to the allowable shear stress. The web plate is designed on the basis of economical depth of the web. The initial design of flange size is made by the flange area method and it is finally checked by moment of inertia method. The moment carried by the flange is

$$M_f = \left(A_f \cdot \sigma_b' \left(d_o - t_f\right)\right) \qquad \text{(i)}$$

and the moment shared by the web is

$$M_w = \left(\sigma_b'' \cdot Z_w\right) = \left(\sigma_b'' \, t_w \, d_e^2/6\right) \qquad \text{(ii)}$$

The total moment resisted by the plate girder

$$M = (M_f + M_w) \qquad \text{(iii)}$$

However, to simplify, the following assumptions are made for the design of the plate girder : BS : 5950 recommends that the design may be based on any of the following assumptions.

1. The moment is resisted by the flanges and shear is resisted by the web only (shear is considered as uniformly distributed over the web).
2. The moment is resisted by the entire section while the web is designed to resist shear and the longitudinal stresses due to bending only.
3. A combination of (1) and (2) above by approximating a percentage of the shear to the web and remaining to the entire section. This method is rarely used.

A mathematical simplification and a good and simple understanding of the load transit mechanism is given by the **first assumption**.

18.9 ALLOWABLE BENDING STRESS (FOR WORKING STRESS METHOD)

The maximum bending compressive stress was calculated for the plate girders bending about the axis of maximum strength (xx-axis), $\sigma_{bc.cal}$ shall not exceed the maximum permissible bending compressive stress, σ_{bc} in N/mm² (MPa) obtained by the following formula:

$$\sigma_{bc} = 0.66\left(\frac{\sigma_{cb}\cdot\sigma_y}{\left[(\sigma_{cb})^n + (\sigma_y)^n\right]^{1/n}}\right)$$

where, σ_{cb} = Elastic critical stress in bending

σ_y = Yield stress in the steel in N/mm² (MPa); and n = A factor assumed as 1.4.

The expression to calculate the elastic critical stress, σ_{cb} has been given in chapter 6 `Design of beam'.

The values of σ_{bc} as derived from the above formula for some of the Indian standard structural steels have also been given in Table 6.2 in Chapter 6. Design of beam'. Same values as given in Table 6.2 are used for the plate girders.

18.10 DESIGN OF WEB PLATE (PRELIMINARY DESIGN FOR A TRIAL SECTION)

In the design of web plate of a plate girder, the economical depth and thickness of web plate are determined. The depth of web plate depends on the bending moment developed due to the external loads acting over plate girder. Approximately, the economical depth of web plate is given by the following formula :

$$\textbf{Economical depth of web plate} = 1.1\left(\frac{M}{\sigma_{bc}\times t_w}\right)^{1/2} \quad (18.14)$$

where, M = Maximum bending moment

σ_{bc} = Allowable bending stress, t_w = Thickness of web plate.

For the economy, the depth of web plate selected should be such that wet, plates are manufactured and are available for that depth. The depth of plate girder varies from 1/10th and 1/12th the span. The thickness of web plate should be such that maximum shear stress in the web does not exceed maximum allowable shear stress and average shear stress in the web does not exceed average allowable shear stress. The thickness of web plate should also provide necessary bearing area for rivets connecting web plate and flange angles, so that rivets are provided at proper spacing. A minimum thickness of 6 mm is adopted to provide for corrosion. The thickness of web plate is fixed keeping in view the buckling of web in shear, and shear along with bending.

The economical depth of web plate may be approximately determined by the use of **Rawter and Clark formula** also.

$$\textbf{Economical depth of web plate} = K\left[\frac{M}{\sigma_{bc}}\right]^{1/3} \quad (18.15)$$

where, K = A parameter, which is adopted as 4.5 for the riveted plate girder

M = Maximum bending moment in N-mm

σ_{bc} = Allowable bending stress in N/mm^2 (MPa).

After deciding the depth of plate girder, the webs are designed to resist the shear considering the requirements of minimum thickness. The effective sectional area of webs in the full depth of web plate multiplied by the thickness of web. In case, the web is designed using the tension field method the weld should be able to transfer for, p_{yw} the tension field tensile stress which are used for the preliminary design and to decide a trial section.

18.11 DESIGN OF FLANGES (PRELIMINARY DESIGN PROCEDURE FOR A TRIAL SECTION)

In the riveted plate girder, the flange portion consists of flange plates, flange angles and that part of the web which is between the flange angles. The effective sectional area of **compression flange** is the gross area with deductions for excessive width of plates and for open holes, occurring in a plane perpendicular to the direction of stress at the section. The effective area of **tension flange** is the gross area with deduction of holes.

In practice, the gross-area of compression flange and tension flange are kept equal. The tension flange of the plate girder is designed on the net area basis and the corresponding gross area is provided for compression flange. After selecting thickness and depth of web plate, flanges are designed by the following methods.

1. Flange area method, 2. Moment of inertia method

1. Flange area method. The trial section of plate girder is determined by flange area method. The flange area method is also termed as **approximate method.** The bending stress in compression flange and in tension flange are assumed uniformly distributed as shown in Fig. 18.5 apart from triangular stress distribution. This assumption is satisfactory, so long as the depth of the flange is small in compression compared as to overall depth of the plate girder. It is a simplified method.

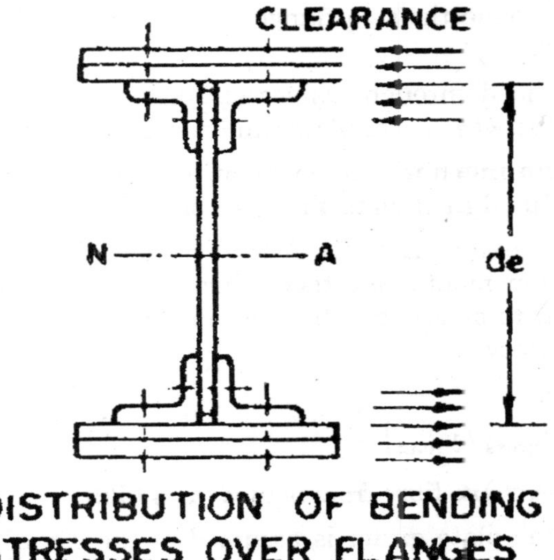

Fig. 18.5.

The forces in the compression flange `C' and in the tension flange `T' are equal and act at the c.g. of flanges in opposite direction. They form a couple which resists the bending moment.

$$M = (C \times d_e) \qquad \text{(i)}$$

$$M = (T \times d_e)$$

$$M = (A_f \times \sigma_b' \times d_e)\text{, therefore,} \qquad \text{(ii)}$$

$$A_f = (M/(\sigma_b \times d_e)) \qquad \text{(18.15A)}$$

where, M = Maximum bending moment

$\acute{\sigma}_b$ = Average bending stress

d_e = Effective depth of plate girder

A_f = Net area of tension flange.

In the design of net area of tension flange, a large part of the area of the flange is provided by the flange angles and the number of flange plates is kept to a minimum. Preferably not less than one-third net area of the tension flange is provided by angles. One-eighth area of web plate acts as flange area. The remaining flange area is provided by the flange plates. It is also possible that the flange of a plate girder consists of only flange angles and web equivalent and without flange plates. In the exposed situations, where the flange angles are used at least one plate over the top flange angles shall extend for the full length of the girder. The number of flange plates in each flange should not be more than three. When more than three plates are used, then, the length of rivets connecting flange plates and flange angles becomes large. The large length of rivet is subjected to bending in addition to shearing and bearing. As such the rivet grip becomes less effective. Where two or more flange plates are used taking rivets shall be provided, if nessary and comply with the requirements. The net area of flange provided should be little more than that required. Corresponding gross area to the net area of tension flange is adopted for compression flange. The unequal angle sections are used for flange angles, and the long legs are kept horizontal. The outstand of flange plates, (i.e., their projection beyond the outer line of connections to flange angles) is provided as recommended for the compression to flange angles.

In order to transmit load through flanges and not by direct bearing through web plate clearance of 5 mm is kept between cover plates and web plate as shown in Fig. 18.5.

2. Moment of inertia method. The moment of inertia method of design of plate girder section is **BIS method of design** of plate girder section.

This is also known as **exact method**. A trial section of plate girder, designed by flange area method is checked by moment of inertia method. The moment of resistance of trial section is computed. If it is equal to or greater than the maximum bending moment; then the trial section designed is satisfactory.

Moment of resistance,

$$\text{M.R.} = \sigma_b \times \left(I_{gross} / y_{max}\right) \qquad \text{(i)}$$

where, y_{max} = Distance of extreme fibre from neutral axis (d/2).

The neutral axis of the plate girder is assumed at the c.g. of gross area of plate girder. However, the net moment of inertia of plate girder may be calculated and used to determine the bending stress. The gross moment of inertia of the plate girder section is found as follows :

$$I_{gross} = \left[2I_f + 2A_{f1} \cdot \left(\frac{d}{2}\right)^2 + \frac{1}{12} t_w \, d^3\right] \qquad \text{(ii)}$$

where, A_{f1} = Flange area excluding web equivalent.

It is to note that the difference between the overall depth, the effective depth and the depth of the web is very small. Therefore, the depth of web and the effective depth have been assumed to be equal to the depth of plate girder section. The moment of inertia of flange plates about their own axes, I_f is very small. Hence, I_f is neglected. Therefore, the gross moment of inertia is given by

$$I_{gross} = \left[\frac{1}{2} A_{f1} \cdot d^2 + \frac{1}{12} t_w \, d^3\right] \qquad \text{(iii)}$$

or $$I_{gross} = \left(A_{f1} + \frac{1}{6} A_w\right) \frac{d^2}{2}$$

The moment of resistance is found by substituting this in the expression (i)

$$M.R. = \frac{\sigma_{bc}}{(d/c)} \left(A_{f1} + \frac{1}{6} A_w\right) \frac{d}{2} = \sigma_{bc} \cdot \left(A_{f1} + \frac{1}{6} A_w\right) . d \qquad \text{(iv)}$$

When the moment of resistance is equal to or greater than the maximum bending moment, then, the trial section designed is satisfactory. The moment of resistance of the plate girder section may also be written as

$$M.R. = \sigma_{b1} \left(A_{f1} + \frac{1}{8} A_w\right) . d \qquad \text{(v)}$$

In case, σ_b and σ_b'' in the expression (i) and (ii) are taken equal and $(d_o - t_f)$ is adopted as d_e, then, the bending moment resisted by flanges, then

$$M_f = (M - M_w)$$

Substituting the values of these moments, the compression and the tension flanges are checked for bending stresses in extreme fibre.

Equating (v) and (vi),

$$\sigma_{bt} \cdot \left(A_{f1} + \frac{1}{8} A_w\right) . \, d = \sigma_{bc} \cdot \left(A_{f1} + \frac{1}{6} A_w\right) . d$$

$$\sigma_{bt} \cdot A_{net} = \sigma_{bc} \cdot A_{gross} = \sigma_{bc} \times \left(A_{gross} / A_{net}\right)$$

then $$M.R. = \sigma_b' \left(A_f + \frac{1}{6} t_w . d_w\right) . \, d_e = \sigma_b' \left(A_1 + \frac{1}{6} A_w\right) . \, d_e \qquad \text{(vi)}$$

$$\left(A_f \times \sigma_b' \times d_e\right) = \left(M - \sigma_b' . \frac{t_w . d_e^2}{6}\right) \qquad \text{(vii)}$$

The bending stress in the extreme fibre in compression flange is given by

$$\sigma_{bc.cal} = \left(\frac{M}{I_{gross}} \times y_{max}\right)$$

The bending stress in the extreme fibre in tension flange is computed by increasing σ_{bc} in proportion of gross area and net area of flanges.

The bending stress in the extreme fibre in tension flange

$$\sigma_{bt.cal} = \sigma_{bc.cal} \times \left(\frac{\text{Gross area of flange}}{\text{Net area flange}}\right) \quad (18.4)$$

σ_{bt} should be less than allowable bending stress.

The ratio of depth of web, dw/thickness of web, tw (i.e dw/tw) decides the identity of laterally supported plate girder. In case (dw/tw) is more than 67ε, then the plate girders are used where, epsilon, is the ratio of yield stress = $(250/\sigma_y)^{1/2}$

18.12 CURTAILMENTS OF FLANGE PLATES : (LENGTH OF FLANGE PLATES)

When the plate girder is subjected to loading, then, the maximum bending moment occurs at one section usually. When the plate girder is simply supported at the ends, and subjected to the uniformly distributed load, then, maximum bending moment occurs at the centre. The values of bending moment decrease towards the supports. The flanges of plate girder or the plate girder section is designed to resist the maximum bending moment. The flange area designed to resist the maximum bending moment is not required at other sections. The flange area is reduced as the bending moment decreases. The reduction in flange area is done by curtailing the flange plates. Each flange plate is curtailed or made shorter than the neighbourhood flange plate nearer to the flange angles. The curtailed length of the flange plates suits the variation in the bending moment. It gives economy as regards the material and cost. The actual lengths of flange plates curtailed are found as follows :

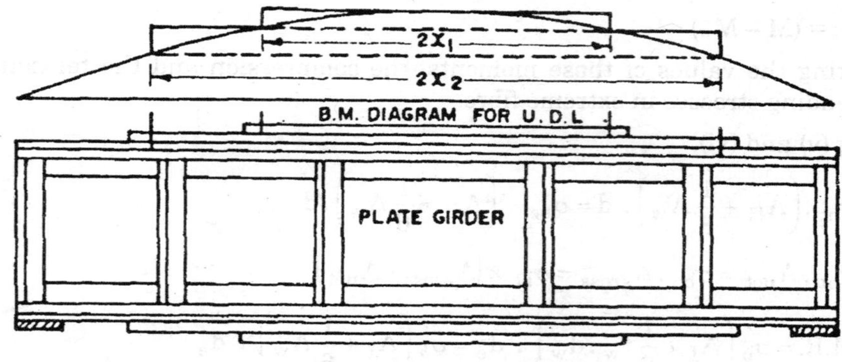

Fig. 18.6. Curtailment of flange plates

When the plates girder is carrying uniformly distributed load, the bending moment diagram has parabolic curve.

Let, A_f = Total flange area including web equivalent

$A_1, A_2 = A_n$ = Area of individual cover plates, counted from outer side

$2x_1, 2x_2...., 2x_n$ = Theoretical length of individual plates as shown in Fig. 18.6.

l = Span of plate girder.

The flange area is given by

$$A_f = M/(\sigma_{bc} \times d_e) \tag{i}$$

From this, it is seen that flange area varies as the moment. As such, flange area also varies following parabolic curve. The effective depth of plate girder is assumed as constant.

$$\text{Then,} \quad \left(\frac{A_1}{A_f}\right) = \left(\frac{4x_1^2}{l^2}\right) \tag{ii}$$

$$x_1 = \frac{1}{2}\left(\frac{A_1}{A_f}\right)^{1/2}, \quad x_2 = \frac{1}{2}\left(\frac{A_1 + A_2}{A_f}\right)^{1/2} \tag{iii}$$

$$x_n = \frac{1}{2}\left(\frac{A_1 + A_2 + ... A_n}{A_f}\right)^{1/2} \tag{18.15}$$

Alternatively. Eq. (18.15) may also be found by equating the bending moment at any section to the moment of resistance of the section. The maximum bending moment for the uniformly distributed load of intensity w per unit length over the simply supported plate girder

$$M_{max} = \left(\frac{wL^2}{8}\right) \tag{iv}$$

The moment of resistance of the section resisting the maximum bending moment

$$M.R_{max} = \left(A_{f1} + \frac{1}{6}A_w\right).\sigma_{bc}.d \tag{v}$$

where, A_{f1} = Area of flange excluding web equivalent

Equating the expressions (iv) and (v)

$$\left(A_{f1} + \frac{1}{6}A_w\right).\sigma_{bc}.d = \left(\frac{wL^2}{8}\right) \tag{vi}$$

In case, one flange plate is curtailed from the top, two flange plates are curtailed from the top and n flange plates are curtailed from the top, then, the expressions similar to the expression (vi) may be written as follows :

$$\left[\left(A_{f1} + \frac{1}{6}A_2\right) - A_1\right] \sigma_{bc}.d = \left(\frac{wL^2}{8} - \frac{w(2x_1)^2}{8}\right) \tag{vii}$$

$$\left[\left(A_{f1} + \frac{1}{6}A_w\right) - (A_1 + A_2)\right] \sigma_{bc}.d = \left(\frac{wL^2}{8} - \frac{w(2x_2)^2}{8}\right) \tag{viii}$$

$$\left[\left(A_{f1} + \frac{1}{6}A_w\right) - (A_1 + A_2 + ..).+ A_n\right] \sigma_{bc}.d = \left(\frac{wL^2}{8} - \frac{w(2x_n)^2}{8}\right) \tag{ix}$$

By dividing the expressions (vii), (viii) and (ix), respectively, by the expression (vi)

$$\left[1-\frac{A_1}{A_{f1}+\frac{1}{6}A_w}\right]=\left\{1-\frac{(2x_1)^2}{l^2}\right\} \tag{x}$$

$$\left(1-\frac{A_1+A_2}{A_{f1}+\frac{1}{6}A_w}\right)=\left\{1-\frac{(2x_1)^2}{l^2}\right\} \tag{xi}$$

$$\left(1-\frac{A_1, A_2+\ldots+A_n}{A_{f1}+\frac{1}{6}A_w}\right)=\left\{1-\frac{(2x_1)^2}{l^2}\right\} \tag{xii}$$

$$\text{Therefore,}\quad x_1=\frac{l}{2}\left(\frac{A_1}{A_{f1}}\right)^{1/2} \tag{xiii}$$

$$x_2=\frac{l}{2}\left(\frac{A_1+A_2}{A_{f1}}\right)^{1/2} \tag{xiv}$$

$$x_n=\frac{l}{2}\left(\frac{A_1+A_2+\ldots.+A_n}{A_{f1}}\right)^{1/2} \tag{xv}$$

where, $A_f=\left(A_{f1}+\frac{1}{6}A_w\right)$

The computations are done for the curtailment of flange plates in tension flange. Net areas of plates and flanges are used in calculations. The curtailment of plates in compression flange is done at the same positions as that in tension flange, for economy in fabrication.

Graphical method of curtailment of flange plates

For other system of loading, other than that of uniformly distributed load, analytical expression for the theoretical lengths of cut-off for the flange plates becomes complicated. In such a situation, the graphical method of curtailment of flange plates is adopted. The graphical method is suitable for all types of loading including the system of uniformly distributed load. In the graphical method, the bending moment diagram for the given system of loading is drawn to the scale. Let the bending moment diagram for the given system of loading be as shown in Fig. 18.7. The maximum section of the plate girder is designed for the maximum value of the bending moment DD_1.

The moment of resistance of the section resisting the maximum bending moment from expression (v) above.

$$\text{M.R.}=\left(A_{f1}+\frac{1}{6}A_w\right)\sigma_b\,.\,d$$

$$\text{or}\quad \text{M.R.}=\left(A_{fa}+A_1+A_2+\ldots+A_n+\frac{1}{6}A_w\right)\sigma_b\,.\,d$$

$$\text{or} \quad M.R. = \left[\left(A_{fa} + \frac{1}{6}A_w\right) + A_1 + A_2 + \ldots + A_n\right] \sigma_b \cdot d \qquad \text{(xvi)}$$

where, A_{fa} = Area of the flange-angles

$A_1, A_2 \ldots, A_n$ = Area of individual cover plates, counted from outer side

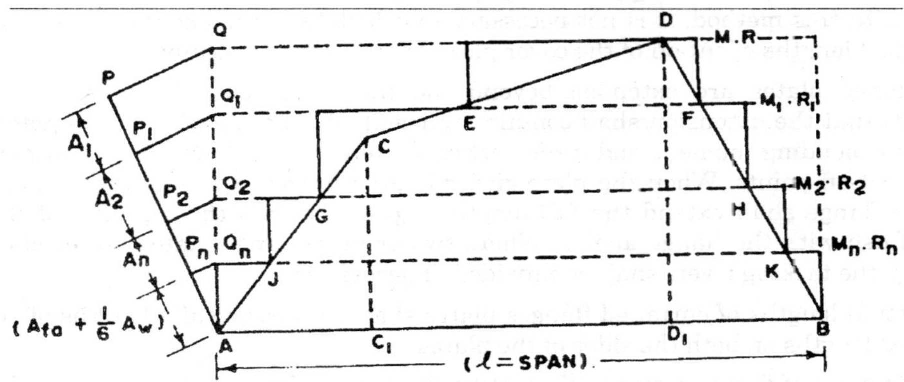

Fig. 18.7

The expression (xvi) gives moment of resistance of the section. The moments of resistance of the section with one cover plate curtailed, with two cover plates curtailed, with n cover plates curtailed are given by the following expressions

$$M_1R_1 = \left[\left(A_{fa} + \frac{1}{6}A_w\right) + A_2 + \ldots + A_n\right] \sigma_b \cdot d \qquad \text{(xvii)}$$

$$M_2R_2 = \left[\left(A_{fa} + \frac{1}{6}A_w\right) + \ldots + A_n\right] \sigma_b \cdot d \qquad \text{(xviii)}$$

$$M_nR_n = \left[\left(A_{fa} + \frac{1}{6}A_w\right)\right] \sigma_b \cdot d \qquad \text{(xix)}$$

The moments of resistance of the sections, M_1R_1, M_2R_2, ... M_nR_n are found and then plotted on the same bending moment diagram as shown in Fig. 18.7. The horizontal lines intersecting with bending moment diagram given the theoretical lengths of curtailment of plates, EF, GH, ... and JK.

Alternatively, from the expression (xvi) it is seen that the total moment of resistance comprises of those due to flange angles and equivalent of web, i.e., $\left(A_{fa} + \frac{1}{6}A_w\right)\sigma_b\, d$, due to one cover plate, i.e., $A_1\ \sigma_b$ d, due to second cover plate, i.e., $A_2\ \sigma_b$ d due to nth cover plate, i.e., $A_n\ \sigma_b$ d. It is to note that the moments of resistance of different flange elements are proportional to their respective areas.

A line AQ is drawn perpendicular to AB at A to represent the moment of resistance of the maximum section, (AP = DD_2, maximum bending moment). A line AP is drawn at any inclination from A. On this line, AP, AP_n, P_2P_n, P_1P_2 and PP_1 are marked equal to

$\left(A_{fa} + \frac{1}{6} A_w\right)$, A_n, A_2 and A_1, respectively to some scale. The points P and Q are joined. Then, from P_1, P_2, P_n lines P_1Q_1, P_2Q_2 and P_nQ_n are drawn parallel to PQ. The horizontal lines Q_1E, Q_2GH and Q_nJK are drawn through Q_1, Q_2 and Q_n, respectively. The intersections of these horizontal lines with the bending moment diagram give the theoretical lengths of cut-off as EF, GH and JK. In this method, it is not necessary to calculate the moments of resistance. Thus, the theoretical lengths of cut-off of the cover plates are found graphically.

The flange plates are extended beyond the theoretical cut-off points. IS : 800-1984 recommends that the extension shall contain sufficient rivets to develop in the plate, the load calculated for bending moment and girder action (taken to include the curtailed plate) at the theoretical cut-off points. When the plate girder is used in the exposed situation, at least one plate of top flange shall extend the full length of girder, unless the top edge of the plate is machined flush with the flange angles. Where two or more flange plates are used on the one flange, then, the tacking rivets shall be provided, if necessary.

The actual lengths of curtailed flanges plates shall be theoretically lengths of cut-off plus the extended lengths on both the sides of the plates.

The outstand of flange plates with unstiffened edges, that is, their projection beyond the outer line of connections to flange angles shall not exceed the values specified in the design of compression members.

18.13 CONNECTION OF FLANGE ANGLES TO WEB (WORKING STRESS METHOD/RIVETED PLATE GIRDER)

The flange angles of plate girder are connected to the web by sufficient rivets to transmit the maximum horizontal shear force resulting from bending moments in the girder, combined with any vertical loads which are directly applied to the flanges.

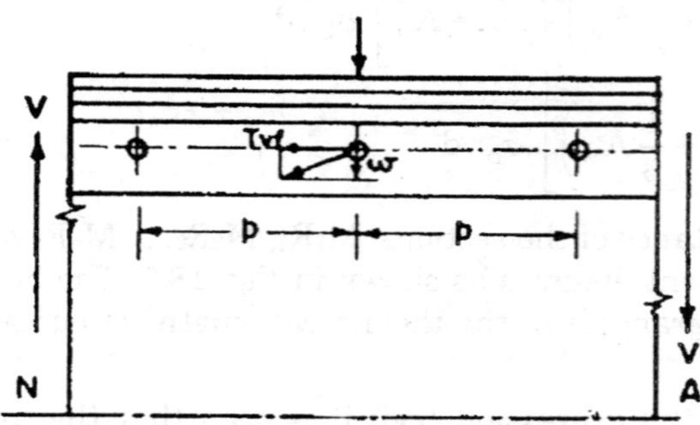

Fig. 18.8.

The design of connection of flange angles to web depends upon horizontal shear. A vertical component of load is added in the horizontal shear. The rivets connecting flange angles to web resist the resultant of vertical component of load and horizontal shear as shown in Fig. 18.8. The pitch of rivets is found as under :

The shear stress at any point of the cross-section is given by

$$\tau_{vf.cal} = (VQ/(I.b)) \tag{18.16}$$

where, $\tau_{vf.cal}$ = Calculated shear stress

V = Shear force at the cross-section

Q = Static moment about the neutral axis of the portion of cross- sectional area beyond the location at which stress is determined

I = Moment of inertia of section

b = Width of section, at which stress is determined.

Shear force per unit depth, $\tau_{vf.cal} = \left(\frac{VQ}{I.b} \times b \times 1\right) = \left(\frac{V.Q}{I}\right)$ (i)

Moment of section, $I = \left(2\,A_f \times \frac{d_e}{2} \times \frac{d_e}{2}\right)$ (ii)

and $Q = \left(A.\bar{y} = A_f \times \frac{d_e}{2}\right)$ (iii)

Shear force per unit depth, $\tau_{vf.cal} = \left[\frac{V \times A \times \left(\frac{d_e}{2}\right)}{2A_f \times \left(\frac{d_e}{2}\right) \times \left(\frac{d_e}{2}\right)}\right] = \left(\frac{V}{d_e}\right)$ (iv)

The shear stress per unit depth is accompanied by horizontal shear per unit length of equal intensity. The horizontal shear per unit length is therefore, given by

$$\tau_{vf.cal} = \frac{V}{d_1} \times \frac{A_{f1}}{\left(A_{f1} + \frac{1}{6}.A_w\right)} \quad (18.17)$$

where, A_{f1} is the gross area of flange excluding web equivalent. The resultant of horizontal shear per unit length and vertical component of load `w' is given by

$$= \left[(\tau_{vf.cal})^2 + (w)^2\right]^{1/2} \quad (vii)$$

The resultant force per pitch length should be less than or equal the rivet value.

Then, $p \times \left[(\tau_{vf.cal})^2 + (w^2)\right]^{1/2} = R$

$$p = \left(\frac{R}{\left[(\tau_{vf.cal})^2 + (w^2)\right]^{1/2}}\right) \quad (viii)$$

Substitute the value of $\tau_{vf.cal}$. The pitch of rivets connecting flange angles to web for loaded flange is given by

$$p = R \Bigg/ \left[\left(\frac{V}{d_e} \times \frac{A_{f1}}{\left(A_{f1} + \frac{1}{6}A_w\right)}\right)^2 + (w)^2\right]^{1/2} \quad (18.18)$$

The pitch of rivets connecting flange angles to web for unloaded flange (w = 0) is given by

$$p = R / \left[\frac{V}{d_e} \times \frac{A_{f1}}{\left(A_{f1} + \frac{1}{6} A_w\right)} \right] \tag{18.19}$$

where, A_{f1} is net area of tension flange. The bottom flange is in tension and it is unloaded flange.

Where a load is directly applied to the top flange, it is considered as dispersed uniformly at an angle of 45° to the horizontal.

18.14 CONNECTION OF FLANGE PLATES TO FLANGE ANGLES

The flange plates of the plate girder are connected to the flange angles by sufficient rivets.

The design of connection of flange plate to flange angles is designed for the horizontal shear. The horizontal shear per unit length (for tension flange)

$$t_{vf.cal} = \frac{V}{d_e} \times \left(\frac{A_1 + A_2 + \ldots + A_n}{A_{f1} + \frac{1}{8} A_n} \right) \tag{i}$$

where, $A_1, A_2, \ldots A_n$ = Areas of flange plates, counting from outside.

The pitch of rivets for tension flange, if rivets are to be provided in one straight line, is given by

$$p = \frac{R \times d_e}{V} \times \left(\frac{A_{f1} + \frac{1}{8} A_w}{A_1' + A_2' + \ldots + A'} \right) \tag{18.20A}$$

For the compression flange $p = \dfrac{R \times d_e}{V} \times \left(\dfrac{A_{f1}' + \frac{1}{6} A_w}{A_1' + A_2' + \ldots + A'} \right)$ (18.20B)

But the rivets are provided in two parallel lines, as such rivets are provided at double the pitch computed above.

18.15. DESIGN OF A PLATE GIRDER SECTION

The plate girders are designed following the procedure as described below.

The support conditions, the clear span, the dead load and the imposed (live) load. and the purpose (for buildings, for highway bridges and for railway bridges), for which the plate girders shall be used are the given parameters.

1. Total factored load, W applied to the plate girder is calculated in kN.

$$W = (\gamma_{fLD} \,.\, D.L. + \gamma_{fLL} \,.\, L.L)$$

= (1.35 D.L. + 1.5 LL) kN

2. Self weight of plate girder is assumed

3. Factored (ultimate/design) bending moments and factored (ultimate/design shear force are calculated.

4. The optimum depth of web'd of the plate girder may be found using the expression

$$d = (M.k. / \sigma_{yf})^{1/3}$$

5. The thickness of web plate, t_w is found from the expression

$$t_w = (M/ (\sigma_{yf}\ k^2))^{1/3}$$

where, k is the ratio of depth of web d_w and thickness of web, t_w

$$k = (d_w/t_w)$$

Ratio (d_w/t_w) is kept less than or equal to $345\ (\varepsilon_f)^2$, so that buckling of flange does not occur. For serviceability requirement ratio (d_w/t_w) is kept less than 200. For unstiffened web, the ratio (d_w/t_w) is kept less than 67.

Where, Yield ratio (epsilon)

$$\varepsilon_f = (250/\sigma_{yf})^{1/2}$$

σ_{yf} = yield stress for the steel used for the flanges

The requirements for the minimum thickness of web, t_w should be satisfied as recommended in IS : 800-2007.

6. **Design of the flanges**

(i) Width of flange, b_f is assumed as $b_f = 0.3$ *times* depth of web

(ii) **Thickness of flange, t_f**

Either the **plastic** or **Compact** sections are used. For requirement of

(a) Plastic section $(b_f/t_f) <= 8.4$

(b) Compact section $(b_f/t_f) = 9.4$

7. The design bending strength of the plate girder cross-section, M_d should be more than the factored (ultimate/design) bending moment due to external load

$$(M_d > M_{fd})$$

8. The design for shear strength has been described in chapter 19.

Example 18.1 A mild steel plate girder simply supported at two ends has an effective span of 20 m. It carries a dead load 50 kN/m and uniformly distributed live load 50 kN/m longer than the span.

Design the maximum section of the plate girder without using tension-field action . Allow for impact Use limit state of design method. Take yield stress for steel for the flanges and web, σ_y = 250 N/mm². Use `Simple-post-critical method'.

Design :

Dead load, D.L = 50 kN/m

Live load, L.L = 50 kN/m

Effective span, L = 20 m

Step 1. Factored Dead load = (1.35 x 50) = 67.5 kN/m

Factored live load = (1.50 x 50) = 75 kN/m

Total factored load = 142.5 kN/m

Assume self-weight of the plate girder

$$w = \left(\frac{W}{200}\right) = \left(\frac{142.5}{200}\right) \times 20 = 14.25\ kN/m$$

Revised dead load = (50 + 14.25) = 64.25 kN/m

Revised factor DL = (1.35 x 64.25) = 86.3775 kN/m

(including self-weight)

Factored live load = 75 kN/m

Coefficient of dynamic augmentation

CDA to allow for impact =

$$= \left(0.15 + \frac{8}{6+}\right) L = \left(0.15 + \frac{8}{6+20}\right)$$

$$= 0.4577\ \text{kN}$$

Factored live load (augmented)

= (1.4577 x 75) = 109.327 kN/m

Total load = (109.327 + 86.3775) = 195.70 kN/m

Step 2: Factored (Ultimate/design) moment

Factored load = 195.70 kN/m, Effective span = 20 m

Load is uniformly distributed

$$M_{fd} = \left(\frac{wL^2}{8}\right) = \left(\frac{195.70 \times 20 \times 20}{8}\right) = 9785\ kN-m$$

Step 3 : Factored (ultimate/design) shear force

$$V_{fd} = \left(\frac{wL}{2}\right) = \left(\frac{195.70 \times 20}{2}\right) = 1957\ kN$$

Step 4 : Optimum depth of plate girder

Yield stress for steel for the flanges and web plate σ_y = 250 N/mm²

$$\text{Yield ratio, (epsilon) } \varepsilon = \left(\frac{250}{250}\right)^{1/2} = 1.00$$

From IS : 800-2007, for minimum thickness of web, in case, the transverse and horizontal stiffeners shall be used. Then,

Ratio $K = \left(\frac{d_w}{t_w}\right) = 250 \times \varepsilon$

$\therefore K = (250 \times 1.0) = 250$

Optimum depth of plate girder

$$d = \left(\frac{M.k}{\sigma_y}\right)^{1/3} = \left(\frac{9785 \times 10^6 \times 250}{250}\right)^{1/3}$$

$$= (9.785^{1/3} \times 10^{3/3} \times 10^{6/2}) = 2138.88 \text{ mm}$$

The depth of web may be proposed as

$d_w = 2000$ mm.

Step 5 : Thickness of web, t_w

$\sigma_y = 250$ N/mm², $k = 250$

$$\text{Thickness of web, } t_w = \left(\frac{M}{\sigma_y . k^2}\right)^{1/3}$$

$$= \left(\frac{9785 \times 10^6}{250 \times 250 \times 250}\right) = \left(\frac{9785^{1/3} \times 10^{6/3}}{250}\right)$$

$$= \left(\frac{2.1388 \times 1000}{250}\right) = 8.5552 \; mm$$

Thickness of web, t_w is proposed as 10 mm.

Step 6. Width of flange, b_f

$b_f = (0.3 \times d_w) = (0.3 \times 2000) = 600$ mm

Width of flange, b_f is proposed as 600 mm

Step 7 : Thickness of flange, t_f

For welded steel plastic - section

(0.5 Width/thickness) ratio = 8.4

Thickness of flange, t_f

$$t_f = \left(\frac{0.5 \, b_f}{8.4}\right) = \left(\frac{0.5 \times 600}{8.4}\right) = 35.714 \; mm$$

Thickness of flange, t_f is proposed as 36 mm

Step 8 : Moment capacity/bending strength of the section

From IS : 800-2007, clause 8.2.1.2,

the design bending strength

$$M_d = \left(\beta_b \cdot Z_p \cdot \frac{\sigma_y}{\gamma_{msy}}\right)$$

β_b = 1.0 for Plastic and compact section

$$M_d = (1.0 \times Z_p \times \frac{250}{1.15}) = 217.397\ Z_p$$

Plastic modulus of section

$$Z_p = 2\, b_f \cdot t_f \left(\frac{2036 - 36}{2}\right) + \frac{1}{4}\, t_w\, d_w^2$$

$$= \left(\frac{2 \times 600 \times 36 \times 2000}{2}\right) + \left(\frac{1}{4} \times 10 \times 2000^2\right)$$

$$= (21600 \times 2000 + 10 \times 10^6)$$

$$= 53.2 \times 10^6\ mm^3 = 53200 \times 10^3 \times mm^3$$

Design bending strength

$$M_d = \left(\frac{217.397 \times 53.2 \times 10^6}{10^6}\right) = 11565.52\ \text{kN-m}$$

M_{fd} = 9785 kN-m

Since, M_{fd} is less than M_d, the proposed section is adequate for the bending strength.

For the shear strength, the section is checked in chapter 19. Design of plate girder section is shown in Fig. 18.9.

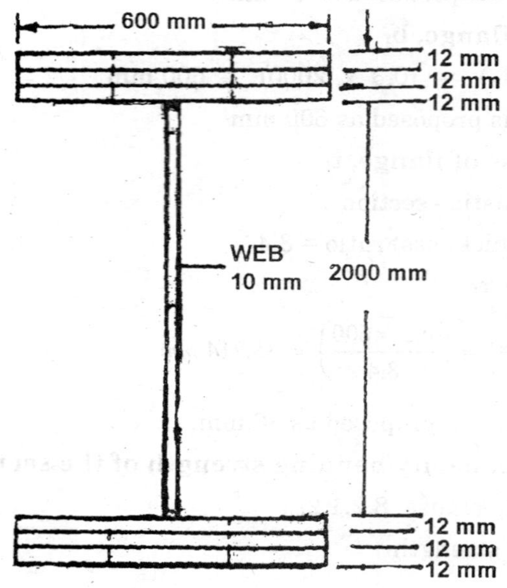

Fig. 18.9

PROBLEMS

18.1 A plate girder of span 18 m is simply supported at its both the ends. The girder is to carry a uniformly distributed load of 90 kN/m inclusive of its own weight. Design the maximum section of the girder. The depth of web is not to exceed 1600 mm. Allow for impact. Use simple post critical method of limit state design fcr bending.

18.2 The effective span of a plate girder is 16 m. The girder is simply supported at its both the ends. It carries a uniformly distributed load of 100 kN/m inclusive of self-weight Design the central section for its bending capacity. Allow for the impact, Use simple post critical method of limit state design for bending.

19

LIMIT STATE OF COLLAPSE IN SHEAR DESIGN OF PLATE GIRDERS (P.G.-II)

19.1 INTRODUCTION

The web plate of a plate girder is kept very thin to get maximum economy as regards its weight. These are made strong and stable by providing different types of stiffeners. which include load carrying stiffeners, bearing stiffeners, intermediate transverse stiffeners and longitudinal stiffeners. In addition to these, diagonal and torsion stiffeners are also provided, wherever necessary. The web plates of plate girder are designed to resist shear considering the requirements of its minimum thickness.

The factored design shear force V_{fd}, shall satisfy the requirement

$$(V < = V_d) \tag{19.1}$$

where, V_{fd} is the design strength

$$V_d = \left(\frac{V_n}{\gamma_{msy}}\right) \tag{19.2}$$

where, γ_{msy} is the partial safety factor for the material strength and V_n is the nominal strength of a cross-section, which may be controlled by plastic shear resistance or strength of the web controlled by shear buckling.

The nominal plastic shear resistance under pure shear may be calculated from

$$(V_n = V_p) \tag{19.3}$$

where,

$$V_p = \left(\frac{\sigma_{yw} \cdot A_v}{\sqrt{3}}\right) \tag{19.4}$$

σ_{yw} = yield strength of the web,

A_v = Area in shear

The shear area may be found as under.

(1) For I and channel sections

(i) For bending about major axis

Hot-rolled $A_v = (h_w \, . \, t_w)$ (19.5A)

Welded $A_v = (d_w \, . \, t_w)$ (19.5B)

(ii) **For bending about the minor axis**

Hot rolled or welded

$A_v = (2\,b \, . \, t_f)$ (19.5C)

(2) Rectangular hollow section having uniform thickness (RHS)

(i) Loaded parallel to depth (h)

$$A_v = \left(\frac{A.\,h}{b + h}\right) \quad (19.5D)$$

(ii) Loaded parallel to width (b)

$$A_v = \left(\frac{A.\,b}{b + h}\right) \quad (19.5E)$$

(3) Circular hollow section having uniform thickness (CHS)

$$A_v = \left(\frac{2A}{\pi}\right) \quad (19.5F)$$

(4) Plates and solid-bars

$$A_v = A \quad (19.5G)$$

where,

A = Cross-sectional area

b = Overall breadth of the tubular section, bread of I-section flanges

d = Clear depth of web between flanges

h = Overall depth of the section

t_f = Thickness of flange

t_w = Thickness of web

Note: Fasteners holes need not be considered for in plastic design (shear strength) calculations provided that

$$A_{vn} > = \left[\left(\frac{\sigma_y}{\sigma_u}\right)\left(\frac{\gamma_{msu}}{\gamma_{msy}}\right) A/0.9\right] \quad (19.5H)$$

In case, A doesnot satisfy the above condition, the effective shear area may be taken as that satisfying the above limit. Block-shear failure criteria may be verified at the end connections.

Resistance to shear buckling

Resistance to shear buckling shall be verified as specified in IS : 800-2007, when

(i) For a web, **without stiffeners** (i.e, for unstiffened web)

$(d_w/t_w) > 67\ \varepsilon$ and (19.6A)

(ii) For a web **with stiffeners** (i.e, for stiffened web)

$(d_w/t_w) > 67\ \varepsilon\ (K_v/5.35)^{1/2}$ (19.6B)

where, K_v is the shear coefficient and $\epsilon = (250/\sigma_y)^{1/2}$, yield stress ratio

Depending upon the ratio, (d_w/t_w) where, t_w is the thickness of web, the resistance to shear buckling is verified. The minimum thickness of web, $t_{w.min}$ is first decided depending upon its requirements.

Resistance to shear buckling is found by the methods recommended by IS : 800-2007 for its determination.

19.2 WEB PLATE SUBJECTED TO PURE SHEAR

When a simply supported plate girder subjected to uniformly distributed load, then, near the supports, it carries pure shear.

When the web plate is subjected to pure shear uniformly distributed along the edge as shown in Fig. 19.1, then, the **critical shear stress is given by the expression given below**

$$\tau_c = \left(\frac{\pi^2 \cdot E}{12\,(1-\mu^2)\ (b/t_w)^2} \cdot K_s\right) \qquad (19.17)A$$

where, μ = Poisson's ratio, K_s = Constant for the plate subjected to pure shear

(a/b) = Aspect ratio and a = Length of the plate

b = Width of the plate and t_w = Thickness of the plate

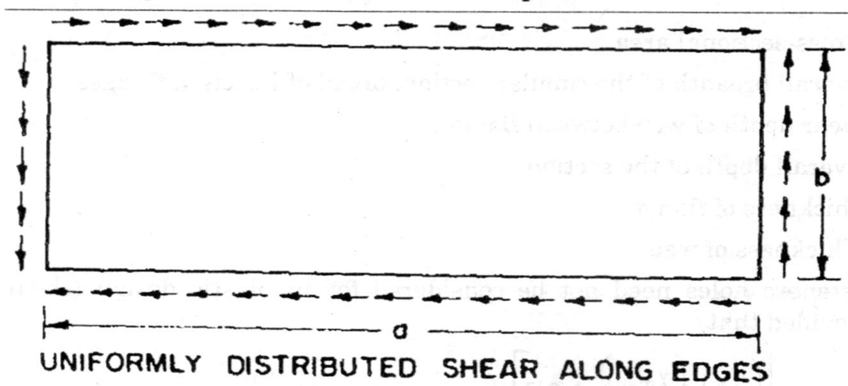

Fig. 19.1.

The value of constant, K_s depends upon aspect ratio (a/b) and the support conditions for the edges of plate. The concept of deep plate girders was developed by considering the importance of this aspects ratio. [Reference: S. Timoshenko and S. Woinowsk-Kriger Theory of Plates and shells, 2nd Ed., New York, 1959. McGraw-Hill Company Inc. (79-82)].

Plate simply supported along all four edges (i.e. displacement prevented are but rotation about edges unrestrained)

The values of constant K_s are given by the following expressions :

$$K_s = \left[5.34 + \frac{4.00}{(a/d)^2}\right](\text{for } a/d > 1) \text{ wide when the panels are wide} \tag{i}$$

$$K_s = \left[4.00 + \frac{5.34}{(a/d)^2}\right](\text{for } a/d < 1) \text{ for webs with closely spaced transverse stiffeners.} \tag{ii}$$

When the ratio of depth of web, and its thickness (d_w/t_w) of a plate girder is more than 85 then the vertical (i.e, transverse) stiffeners are attached with the web plate. The plate girders are referred as **transversely stiffened**. A long span plate girder will be divided into number of web panels. Each panel will carry separate combinations of bearing moments and shear forces. Those panels which are closed to the supports, carry large shear forces and those panels which are close to the centre, carry large bending moments. These panels are studied in two steps.

For the purpose of design, it is desirable to keep Eqs. 19.17 (A), (B) and (CD) interms of depth of girder, d (unsupported depth/height of web) and the spacing of the transverse stiffeners, a. Two cases [(a/d) > 1 and (a/d) < 1] shall be considered.

In the web plate for the plate girder, the dimension a may always be selected as the larger dimension. Therefore, the aspect ratio (a/b) will be greater than unity. From the expression (i) for $a/d = \infty$, the value of second term becomes zero, and the value of K_s is 5.34. For a/b = 1, the value of K_s is equal to [5.34 + (4.00/1)] = 9.34. For different values of the aspect ratio, the values of K_s can be found from the expression (i). The values of K_s for some values of the aspect ratio have been given in Table 19.1. When transverse stiffeners are provided only at supports 9 K_s = 5.34.

Buckling is defined as sudden bending, warping, curling or crimpling of the elements under compressive stresses.

Web buckling. The web plate of a simply supported plate girder may be considered as subjected to pure shear close to the support, to pure bending at the centre, and to combined shear and bending at the intermediate positions, when subjected to uniformly distributed load. The web plates used are deep and thin. The web plate may fail by buckling.

TABLE 19.1

a/b	1.0	1.2	1.4	1.6	1.9	2.0	3.0	5.0	∞
Ks	9.34	8.12	7.38	6.90	6.58	6.34	5.78	5.36	5.34

As the values of aspect ratio increase from 5.0 to ∞, the change in value of K_s is insignificant. For unstiffened web, value of K_s can be assumed equal to 5.34 for all practical purposes. For structural steel value of Poisson's ratio is 0.3 and Young's modulus of elasticity E is 2.11 x 10^5 N/mm^2. Substituting these values in above expression Eq. (18.17) reduces to

$$\sigma_{cr} = \left(\frac{\pi^2 \times 2.11 \times 10^5 \times 5.34}{12\,(1 - 0.3 \times 0.3)(b/t_w)^2}\right) N/mm^2$$

$$\sigma_{cr} = \left(\frac{10.05 \times 10^5}{(b/t_w)^2}\right) \text{N/mm}^2 \qquad (19.1)$$

The value of width of web plate may be substituted equal to clear depth of web (viz., d)

$$\sigma_{cr} = \left(\frac{10.05 \times 10^5}{(d/t_w)^2}\right) \text{N/mm}^2 \qquad (19.2)$$

It is seen that the critical shear stress in the web is inversely proportional to ratio (d/t_w). When the value of critical shear stress reaches **the value of yield stress in shear,** (viz. (0.6 × 250) = 150 N/mm²), then.

$$\left(\frac{d}{t_w}\right)^2 = \left(\frac{10.05 \times 10^5}{150}\right)$$

$$\left(\frac{d}{t_w}\right)^2 = 6700\ . \quad \left(\frac{d}{t_w} = 81.85\right) \text{(approximately)} = (82) \qquad (19.3)$$

For values of σ_y other than 250 N/mm², it shall be $\left(\frac{d}{t_w}\right)\left(\frac{\sigma_y}{250}\right)^{1/2} = 82$

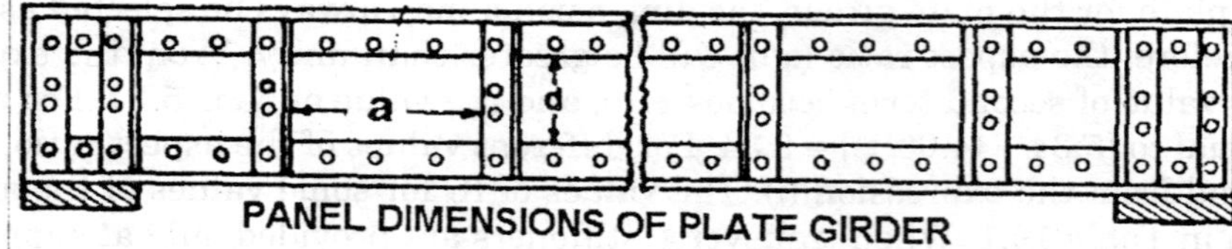

Fig. 19.2. Verticals show transverse stiffeners verticals at the ends show bearing stiffeners.

When the ratio d/t_w is less than 82, the failure of web plate takes place by yielding. When the ratio of d/t_w is more than 82, the buckling of web takes place. Therefore the allowable shear stress is decreased. As d/t_w ratio increases, the value of critical shear stress decreases rapidly. The value of critical shear stress may be increased by decreasing the aspect ratio. When the aspect ratio is decreased, the value of constant K_s increases. The aspect ratio may be decreased by reducing the length a, by providing transverse stiffeners on the web plate, as shown in Fig. 19.3. The value of constant K_s increases as the length a (i.e., the spacing between the transverse stiffeners, c) reduces. The value of critical stress increases. When the spacing between the transverse stiffeners c, becomes less than the clear depth of web, d, then, value of critical stress further increases since, b is the smaller dimensions in Eq. 19.17 and it is universally proportional to square of (b/t_w) ratio. Sub-dividing a simply supported plate of length, a, by sufficient rigid transverse stiffeners of spacing c, is shown in Fig. 19.2 smaller panels are formed. The smaller panels may be considered approximately as simply supported along the four edges. The buckling strength of web plate is released, and economical design of web may be done.

For simply supported edges, the minimum value of K_b is 23.9. The critical stress for bending is given by

$$\sigma_{cr} = \left[\frac{45.4 \times 10^5}{(d/t_w)^5}\right] \text{N/mm}^2 \tag{19.5}$$

For the plate clamped at the longitudinal edges, the critical stress for bending is given by

$$\sigma_{cr} = \left[\frac{75.2 \times 10^5}{(d/t_w)^5}\right] \text{N/mm}^2 \tag{19.6}$$

The critical stress for combined bending and shear lies in between these two values. For combined bending and shear, the web plate may be checked by calculating the equivalent stress, $\sigma_{e.cal}$ due to co-existent bending (tension or compression) and the shear stresses obtained as below. $\sigma_{e.cal}$ shall not exceed the value, $\sigma_e = 0.9\,\sigma_y$ (the maximum permissible equivalent stress) :

$$\sigma_{e.cal} = \left[\sigma_{bt.cal}^2 + 3\,\tau_{vm.cal}^2\right]^{1/2} \tag{19.7A}$$

$$\sigma_{e.cal} = \left[\sigma_{bc.cal}^2 + 3\,\tau_{vm.cal}^2\right]^{1/2} \tag{19.7B}$$

The theoretical values of critical stress for the buckling of web have been determined by small-deflection buckling theory making the following assumptions :

(i) The thickness of web plate is small compared with its surface dimensions.

(ii) The plate is perfectly straight before loading.

(iii) The plate deflections are small compared with the plate thickness.

The small-deflection buckling theory provides a safe but often over conservative design. When the web plate is adequately supported then, it will carry stresses substantially exceeding theoretical buckling stresses.

19.3 REQUIREMENTS FOR MINIMUM THICKNESS OF WEB

IS : 800-2007 recommends the following requirement for the web, which shall be satisfied for the design of a section.

1. Requirements for the serviceability

(i) When transverse stiffeners are not provided (i.e, the web is unstiffened)

(a) When the web is connected to the flanges along both longitudinal edges

(i) $(d_w/t_w) \; < \; = 200\,\varepsilon$ (19.7)A

(ii) When the web is connected to the flanges along one longitudinal edge only

$(d_w/t_w) \; < \; = 90\,\varepsilon$

where, epsilon, ratio of yield stress $\varepsilon = (250/\sigma_y)^{1/2}$ (19.7)B

(b) When only transverse stiffeners are provided (i.e, for the stiffened web)

(a) when the webs are connected to the flanges along both longitudinal edges)

(i) When ($3d_w >= a >= d_w$)

$(d_w/t_w) <= 200\ \varepsilon$ (19.8)A

(ii) When ($0.74\ d_w <= a < d_w$)

$(a/t_w) <= 200\ \varepsilon_w$ (19.8)B

(iii) when ($a < d_w$)

$(d_w/t_w) <= 270\ \varepsilon_w$ (19.8)C

(iv) When $a >= 3d_w$, the web shall be considered as **unstiffened.**

(c) When the transverse stiffeners and longitudinal stiffeners at one level only are provided 0.2 d from the compression flange

(i) (When $2.4\ d >= a > d_w$)

$(d_w/t_w) <= 250\ \varepsilon_w$ (19.9)A

(ii) When $0.74\ d_w <= a <= d_w$

$(a/t_w) <= 250\ \varepsilon_w$ (19.9)B

(iii) When $a < 0.74\ d$

$(d_w/t_w) <= 340\ \varepsilon_w$ (19.9)C

(d) When the second longitudinal stiffeners (located/attached at neutral axis)

$(d_w/t_w) <= 400\ \varepsilon_w$ (19.9)D

where,

d_w = depth of web

t_w = Thickness of web

a = Spacing of the transverse stiffeners

ε_w = yield stress ratio of web $(250 / \sigma_{yw})^{1/2}$, and

σ_{yw} = yield stress of the web.

2. Requirements for the buckling of compression flange

The web thickness shall satisfy the following requirement in order to avoid buckling of compression flange into the web.

(a) When transverse stiffeners are not provided (i.e, the web is **unstiffened)**

$(d_w/t_w) <= 345\ \varepsilon_f^2$ (19.10)A

(b) When the transverse stiffeners are provided (i.e, when the web is **stiffened)**

(i) When $a >= 1.5\ d_w$

$(d_w/t_w) <= 345\ \varepsilon_f^2$ (19.10)B

(ii) When $a < 1.5\ d_w$

$(d_w/t_w) <= 345\ \varepsilon_f$ (19.10)C

where,

d_w = Depth of the web

t_w = Thickness of the web

a = spacing of the transverse stiffeners

ε_f = Yield stress ratio of the web $(250 / \sigma_{yw})^{1/2}$

σ_{yw} = yield stress o. web

To ensure that the web will not buckle under normal service, conditions, above criteria have been specified. It is seen that the web is strong enough so that the flange will not buckle into the.

A thinner web may be used in case the ratio (a/d) is less than one and the web panel has higher strength.

19.4 SHEAR CAPACITY OF WEB PLATES (WORKING STRESS METHOD)

The shear capacity of web plate depends upon the ratio of depth of web, d_w and the thickness of web, t_w (i.e., ratio d_w/t_w). The web plates are kept **unstiffened** (i.e., stiffeners are not attached to the web) and the web plates are kept **stiffened** (i.e., stiffeners are attached to the web). The shear capacity of the web plates are calculated depending on whether it is unstiffened or stiffened. The web plate is designed accordingly.

1. Unstiffened web plate

The shear capacity of web without intermediate stiffeners depends on its resistance against buckling due to shear. The depth of web is kept limited to $d_{w.lim}$

$$d_{w.lim} = [V_{cr} / (T_{cr} \cdot t_w)] \tag{19.3}$$

The values of critical stress in shear, τ_{cr} is recommended in code BS : 5950 for the webs which are too slender. It is decided by considering slenderness parameter, λ_w calculated as below

$$\lambda_w = (0.6\, \tau_{y.web} / \tau_{e.cr})^{1/2} \tag{19.4}$$

where, $\tau_{e.cr}$ is the elastic critical shear strength. It is determined on the values of aspect ratio, (a/d) and ratio (d_w/t_w). $\sigma_{y.web}$ is the design strength of the web a is the spacing of the stiffeners (stiffener, may be bearing stiffeners) and d is the depth of web.

1. For aspect ratio (a/d) $\leq$ unity

The elastic critical stress is given by

$$\tau_{e.cr} = \left[0.75 + 1 / (a/d)^2\right] . [1000/ (dw/tw)]^2 \tag{19.5}$$

2. For aspect ratio (a/d) > unity

The elastic critical stress is given by

$$\tau_{e.cr} = \left[1 + 0.75/(a/d)^2\right] [1000 / (dw \,.\, tw)^{2}] \tag{19.6}$$

Code of practice BS : 5950, Part 1 recommends the following for design purpose the values of elastic critical stress corresponding to the different ranges of slenderness values of web, λ_w

(1) $(\lambda_w < 0.8)$ $\qquad \tau_{cr} = 0.6\,\sigma_{y.w}$ (19.7)

(2) $(0.8 < \lambda_w < 1.25)$, $\tau_{cr} = 0.6\,\sigma_{y.w}\,[1 - 0.8\,\lambda_2 - 0.64]$ (19.8)

(3) $(\lambda_w > 1.25)$, $\qquad \tau_{cr} = \tau_{e.cr}$ (19.9)

In general, τ_{cr} depends, on design stress of web, $\sigma_{y.w}$. For thin slender web, $\tau_{cr} = \tau_{e.cr}$

2. Stiffened web (i.e. vertical stiffeners are attached with the web)

A plate girder is to be designed to carry given load economically and efficiently. Then the depth of the plate girder is to be large. Deep plate girders provide large moment of inertia (as the depth between the flanges becomes large). A web plate is provided between two flanges to act as one unit. The web plate is to resist the shear force. Thick web plate increases the weight of the plate girder and therefore, the thick web plates are considered unnecessary. A thin web plate is provided and the stiffeners are attached with the web plate. It is known as **stiffened web plate.** The strength of web plate is therefore, increased.

The web plates of plate girders are designed by limiting the shear capacity to shear buckling strength. This procedure is not economical. It does not consider the mobilisation of the additional shear capacity. The shear resistance is increased in the following manner.

The resistance to buckling is improved due to decrease in aspect ratio (a/d). **Tension field action** (i.e., diagonal tension in the web) develops. Then, the web resists significantly more stress than the elastic critical strength of web in shear.

The behaviour of plate girder is like N-truss due to diagonal tension in the web plates and stiffeners behave as short struts. When the web plate yields and plastic hinges develop in the top flange and in the bottom flange.

For sway mechanism, two plastic hinges are formed at the top and two plastic hinges are formed at the bottom.

The resistance to shear buckling is determined as under :

$$V_{buckling} = \left[\tau_{buckling} + \tau_f\,(k_f)^{1/2}\right](d_w\,t_w) \quad \therefore\ 1/\sqrt{3} = 0.5778 = 0.6 \text{ (Approximately)}$$

$$V_{buckling} < (0.6\,\sigma_y \,.\, d_w \,.\, t_w) \tag{19.10}$$

First term within parenthesis above consists of elastic critical stress, τ_{cr} and tension field strength of the panel. That is,

$$\tau_{buckling} = \tau_{e.cr} + \frac{V_{buckling}}{2\left[(a/d) + \left(1 + \frac{a}{d}\right)^2\right]^{1/2}} \tag{19.11}$$

where,
$$V_{buckling} = \left[\left(\sigma_{yw}^2 - 3\,T_{e.cr} + \phi_t^2\right) - \phi_{tw}\right] \tag{19.12}$$

$$\phi_{tw} = 1.5\,\tau_{cr} \,/\left[1 + \left(\frac{a}{d}\right)\right]^{1/2} \tag{19.13}$$

where, a is the spacing of the vertical stiffeners, d is the depth of the plate girder, t_w is the thickness of web and $\sigma_{y.w}$ is the design stress for the web.

$$\sigma_{yw} = \frac{1}{\sqrt{3}}\,\sigma_y = 0.6\,\sigma_y \text{ (approximately)}$$

Term $T_f\ (k_f)^{1/2}$ in Eq. 19.10, represents the flanges' contribution to the post buckling strength. Its value depends on the plastic moment capacity of the flanges. $M_{p.fL}$.

The flanges resist diagonal tension. In case, the flanges attain their ultimate capacity, then, the plastic hinges develop. The plastic moment capacity of the flanges, $M_{p.fL}$ and that of the web, $M_{p.web}$ are related by the factor, k_f, (it is shown subsequently).

The shear strength is also dependent on the flanges

$$\tau_{flange} = 0.6\, \sigma_{y.w} \left[4\sqrt{3} \left(V_{buckling} / \sigma_{y.w} \right)^{1/2} \sin\left(\frac{\theta}{2}\right) \right] \tag{19.14}$$

where, $\theta = \tan^{-1}(d/a)$

When, the plate girder resist pure shear, in that case,

$$M_{p.fl} = 0.5\, b\, \tau^2\, \sigma_{y.fl} \tag{19.15}$$

Due to presence of overall bending moment, the contribution of flanges to shear will be decreased by the longitudinal shear, σ induced, due to overall bending moment by the factor $(1 - f/\sigma_{y.f})$.

At maximum moment region, *f* approaches $\sigma_{y.fl}$. Nearly, this factor reduces to zero. There, the flanges do not contribute to resistance due to shear. Their contribution becomes neglible. The plastic moment capacity of web may be obtained as below.

$$M_{p.web} = 0.25\, (dw)^2 \,.\, t_w \,.\, \sigma_{yw} \quad \text{and} \tag{19.16}$$

$$k_f = 0.25 \left(M_{p.fl} / M_{p.web} \right) \tag{19.17}$$

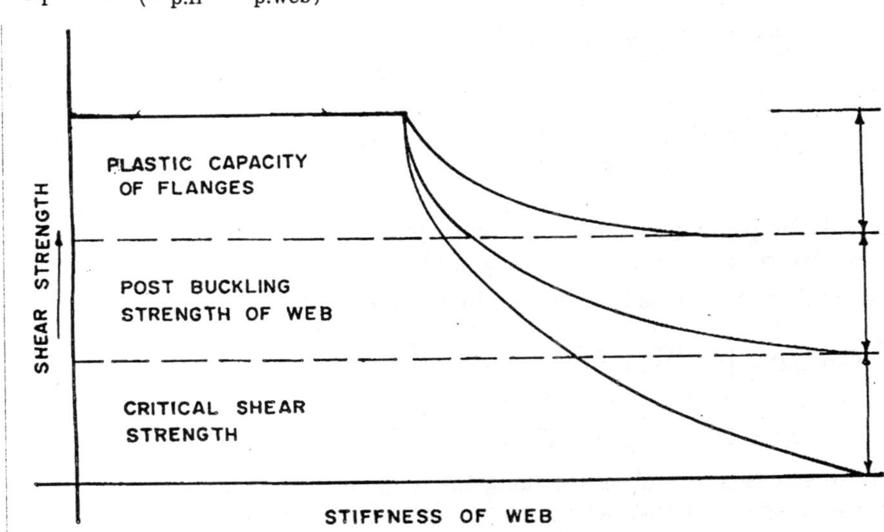

Fig. 19.3. Variation of shear strength with stiffness of web.

The critical shear strength, post buckling strength due to diagonal tension sufficiently resisted by the vertical stiffeners and the moment capacity of the flanges are shown in Fig. 19.3.

To develop diagonal tension in the end panel, sufficient anchor force F_{anchor} is needed.

$$F_{anchor} = (0.75\, d_w).\, t_w \left[1 - \frac{\tau_{cr}}{0.6\,\sigma_y}\right]^{1/2} \tag{19.18}$$

The end panel is designed for the anchor force, F_{anchor}. It imposes additional force on the end post. There, it has to be stout. In simple design, it is assumed that it is limited to V_{cr}, so that diagonal tension does not develop in the end panel. In such a case, the end panel behaves as a beam spanning between the flanges to resist shear and bending moment resulted due to anchor force, F_{anchor} produced by diagonal tension in last but one panel.

This procedure is conservative. The post buckling strength of end panel is not utilized where the shear force is maximum. The aspect ratio (a/d) of end panel becomes less than those of other panels. The compressive forces due to bearing and the moment due to diagonal tension, $M_{d.tf}$ in the last but one panel is used to design end stiffener.

For economical design, the diagonal tension is used to design the end panel. The bearing stiffener at the end and the end post are designed for combination of stresses in the flanges, $M_{d.tf}$. It needs stout stiffener. Two stout stiffeners are provided instead of one.

19.5 DIAGONAL BUCKLING OF A WEB PLATE

The diagonal buckling of web occurs when the ratio of clear depth to thickness of web (d/t_w) exceeds 85. It occurs because of diagonal compression as shown in Fig. 19.4. A small element of the web plate is subjected to the bending stress σ_b acting horizontally and shear stress, τ_{vf} acting vertically and which one is also accompanied by complimentary shear stress, τ_{vf} acting horizontally. The principal stress which small be acting on this element is under :

$$\sigma_1 = \frac{\sigma_b}{2} + \left[\left(\frac{\sigma_b}{2}\right)^2 + (\tau_{vf})^2\right]^{1/2} \tag{i}$$

$$\sigma_2 = \frac{\sigma_b}{2} - \left[\left(\frac{\sigma_b}{2}\right)^2 + (\tau_{vf})^2\right]^{1/2} \tag{ii}$$

At the end, the plate girder is simply supported, where, the bending moment and thereby bending stress are zero. Therefore, $\sigma_1 = \tau_{vf}$ and $\sigma_2 = \tau_{vf}$ (numerically). One principal stress is compressive, while the other is tensile. The inclination of the principal stress is given by

$$\tan 2\theta = \left(\frac{-\tau_{vf}}{(\sigma_b/2)} = \frac{-\tau_{vf}}{\text{zero}}\right) = \infty$$

$$2\theta = 90^\circ \qquad \therefore\ \theta = 45^\circ$$

Both these stresses σ_1 and σ_2 act along the diagonals inclined 45° with the horizontal line. The compressive stress acting along the diagonal causes buckling of the web plate and phenomenon is known as **diagonal buckling**. In this phenomenon, the diagonal compression and the diagonal tension of equal intensities, acting mutually at right angles are set up in the web at the neutral axis at an inclination of 45° with horizontal. At the neutral axis, the intensity of diagonal compression is equal to that of shear. Intensity and inclination of diagonal compression and of diagonal tension are changed by bending compressive stress above the neutral axis, and bending tensile stress below the neutral axis.

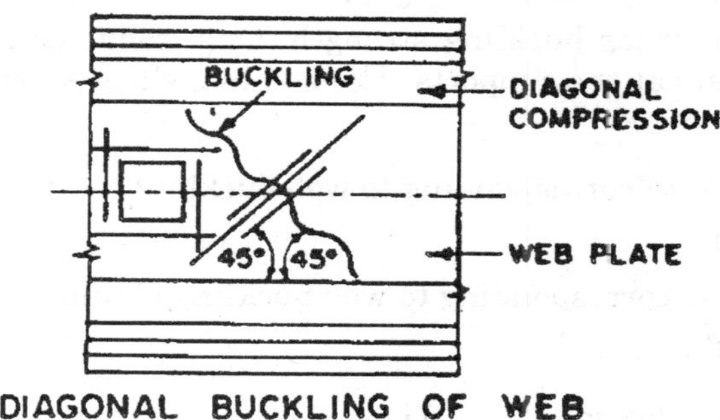

Fig. 19.4.

The strength of web plate is increased by providing stiffeners. When the stiffeners are provided, and diagonal buckling of web occurs, the flanges and vertical stiffeners form imaginary panels in the plate girder. Each panel of the plate girder behaves like a panel of a Pratt truss. The flanges behave like the top and bottom chords of the truss and the web plate constitutes the web members. The vertical stiffeners act as struts and diagonal strips of web act as ties as shown Fig. 19.5. The buckling strength of a plate girder web that has regularly spaced transverse stiffeners is increased by its capacity because of truss lie behaviour. The diagonal carrying tension are considered as effective. The contribution of those diagonals carrying compression to the shear resistance of the web is neglected (although each would, in fact, contribute its small buckling load). Therefore, the shear resistance between any two transverse stiffeners of an adequate stiffened web is the sum of the shear buckling strength of the web and the vertical component of the web yield strength in direct diagonal tension. This latter is sometimes known as **tension field strength.**

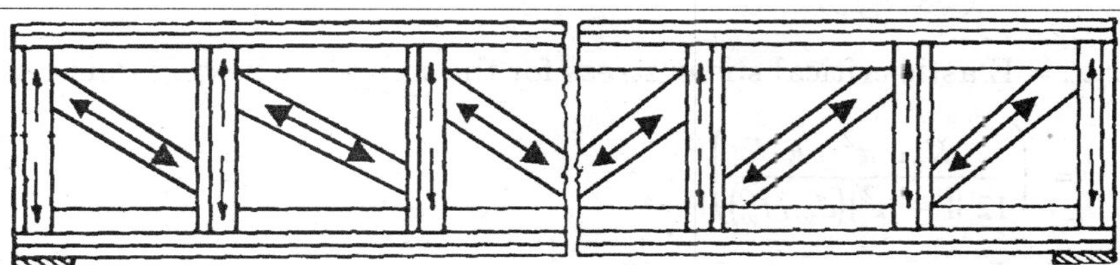

TRUSS LIKE BEHAVIOUR OF PLATE GIRDER (TENSION FIELD)

Fig. 19.5.

In the verticals, the arrows represent compression and in diagonals these show tension (internal system of marking)

19.6 DESIGN METHODS FOR DETERMINATION OF SHEAR BUCKLING RESISTANCE

IS : 800-2007 recommends that the nominal shear strength, V_n of webs of the plate girders with or without intermediate stiffeners as controlled by buckling may be found by using one of the following methods.

1. Simple post-critical method

A general method which may be used for the design of webs of I-section girders with or without intermediate transverse stiffeners is referred as `simple post-critical method.' This method is based on the shear buckling strength. Only requirement is that the web should have transverse stiffeners at the supports. The nominal shear strength is given by

$$V_n = V_{cr} \tag{i}$$

where, V_{cr} is the shear force corresponding to web buckling

$$V_{cr} = (A_v . \tau_b) \tag{19.11)A}$$

τ_b is the shear stress corresponding to web buckling obtained as under

(i) When $\lambda_w <= 0.8$

$$\tau_b = \left(\frac{\sigma_{yw}}{\sqrt{3}}\right) = (0.577)\,\sigma_{yw} = 0.6\,\sigma_{yw} \tag{19.11)B}$$

(ii) When $0.8 < \lambda_w < 1.2$

$$\tau_b = [1 - 0.8\,(\lambda_w - 0.8)]\left(\frac{\sigma_{yw}}{\sqrt{3}}\right) \tag{19.11)C}$$

(iii) When $\lambda_w >= 1.2$

$$\tau_b = \left(\frac{\sigma_{yw}}{\sqrt{3}\,.\,\lambda_w^2}\right) \tag{19.11)D}$$

where, λ_w is a non-dimensional slenderness ratio for the web for shear buckling stress. It is found as

$$\lambda_w = \left(\frac{\sigma_{yw}}{\sqrt{3}\;\tau_{cr.e}}\right)^{1/2},\ and \tag{19.11)E}$$

$\tau_{cr.e}$ = Elastic critical shear stress for the web plate having Poisson's ration, μ

$$= \left(\frac{K_v\,.\,\pi^2\,.\,E}{12\,(1-\mu^2)(d_w/t_w)}\right) \tag{19.11)F}$$

$K_v = 5.35$ when the transverse stiffeners are attached only at supports

(a) For $(c/d) < 1.0$

$$K_v = 4.0 + 5.35\ /\ (c/d)^2 \tag{19.11)G}$$

(b) For $(c/d) >= 1.0$

$$K_v = 5.35 + 4.0\ /\ (c/d_w)^2 \tag{19.11)H}$$

where, d_w is the depth of web and c = a = spacing of the transverse stiffeners.

For the expressions (iii), (iv) and (v) a plot for τ_b as an ordinate and and λ_w as an abscissa was suggested by Baseler (1961) and then by Rock and Skaloud (1971) A Table 19.2. For the different values of shear buckling strength, τ_b and for the various values of stiffeners spacing ratio (c/d_w) and depth to thickness ratio (d_w/t_w) has been given.

For stock webs where the yield controls and for the slender webs where buckling controls, a linear transformation concept was used. It may be seen from Fig. 19.6 that the expressions suggested by code of practice IS : 800-2007, neglects the post-buckling strength of the web.

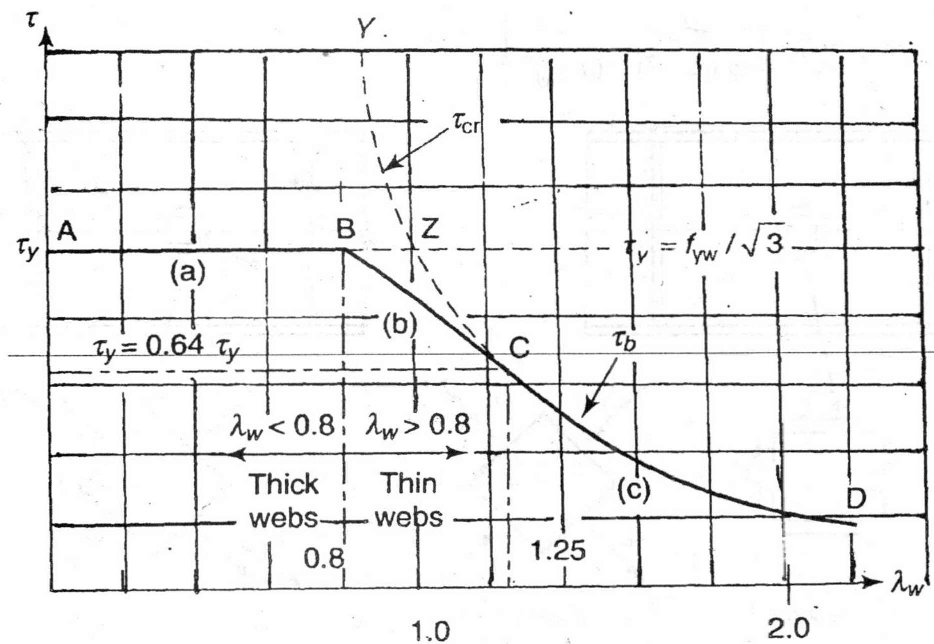

Fig. 19.6.

2. Tension field method

This method is based on pre-buckling of web (i.e, elastic buckling or before/while the web buckles) and post-shear buckling) (i.e, inelastic buckling) after buckling of the web in shear) and behaviour of web while/after failure. That is complete structural behaviour. Tension field method has been described subsequently.

19.7 BEHAVIOUR OF WEB PLATE SUBJECTED TO PURE SHEAR

The structural behaviour of web plate subjected to shear may be understood by considering the effect of complementary stresses. It is seen that diagonal compression and diagonal tension develop in the panels near the supports. This structural behaviour may be studied by considering different stages of the web plate (i) Pre-buckling (i.e, before buckling) (ii) Post-buckling (i.e, after the buckling (iii) Failure (collapse) of web plate.

Pre buckling (elastic buckling) behaviour (Before buckling) under pure shear

An element E inside a plate in the end panel, carries shear stresses, τ_{sh}. It is accompanied by complimentary shear stress, τ_{sh} as shown in Fig. 19.7. This along AC. Corresponding shear stress in the plate becomes critical shear stress, τ_{cr}. The value of τ_{cr} may be obtained from classical theoy of stability, in case, the boundary conditions are known. It is difficult to establish exact boundary conditions of the web plate due to restraint by the flanges and the stiffeners. The boundary conditions may be assumed simply supported, conservatively. In such a case, the critical shear stress is given by

$$\tau_{sh.cr} = Ks\left[\frac{\pi^2 . E}{12\left(1-\mu^2\right)} \frac{1}{\left(d/t_w\right)^2}\right] \tag{19.12}$$

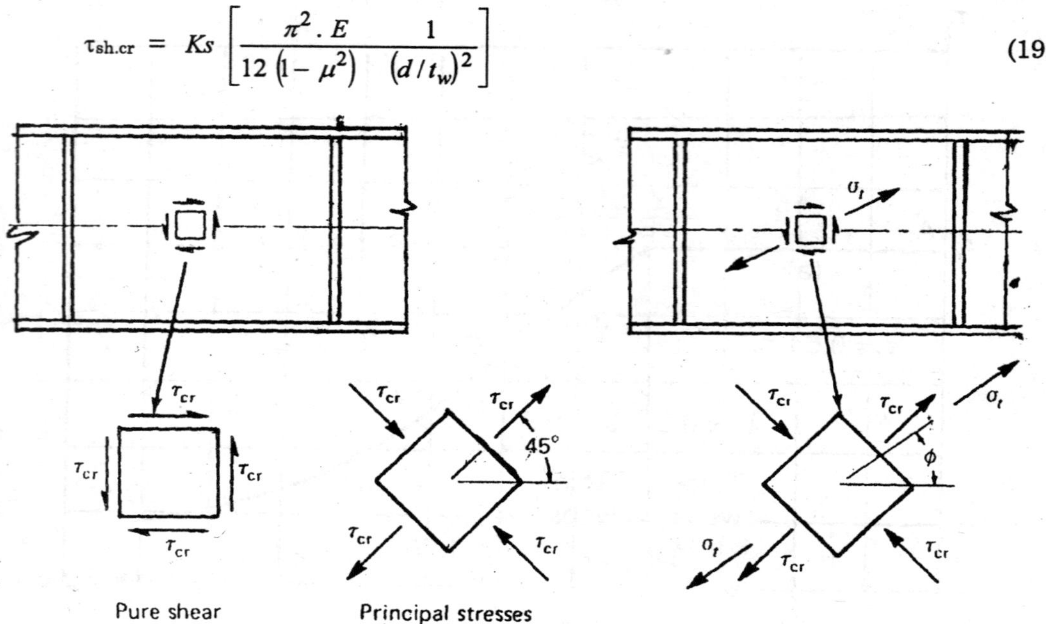

Fig. 19.7

Eqs. 19.17 (B) and (C) define the values of K_s. The values of $\tau_{sh.cr}$ for the different values of aspect ratio (a/d) have been listed here in Table 19.2.

Table 19.2. VALUES OF τ_{cr} IN N/mm²

Ratio (d_w/t_w)	Aspect ratio, (a/d)			
	1.0	1.5	2.0	5.0
100	169	129	115	Buckling does not control
125	108	83	73	-do-
150	75	57	51	204
175	55	42	37	150
200	42	32	29	115
225	33	25	23	91
250	27	21	18	72

Note : E = 2 x 10^5 N/mm², μ = 0.3, π =3.1412

When the value of (d_w/t_w) is sufficiently small, (d_w/t_w < 85) τ_{cr} increases above $\tau_{v.y} = \sigma$ and the web yields in shear before buckling.

2. Post-buckling behaviour (i.e. Inelastic or buckling)/ (strength after*) under pure shear

A plate stiffened by the stiffeners on the sides and by flanges at top and bottom has considerable post-buckling. The material of web plate in plate-girders is to be used efficiently. For which thin webs are necessary. It results in buckling of the web-plate at low shear stress Beyond the elastic critical stress, the diagonal AC is not able to resist any more compression. The web plate ceases to resist any further increase in compression stresses.

Further increase in compressive stresses may be considered as if, it resolved into compents (vertical and horizontal).

(i) Vertical component is transferred to and supported by the vertical stiffeners

(ii) The horizontal component is transferred to and resisted by the flanges. The flanges bend inwards due to the pull exerted by the tensile membrane stresses in the web on further in crease of loads, the tensile membrane stress in web applies greater pull on the flanges, till the stress in the web reaches the yield value of web.

The diagonal tension is resisted by the web only.

This complete behaviour is very similar to (analogus to) a Pratt truss as shown in Fig. 19.5.

Vertical stiffeners act as vertical members which carry compression some portion of web acts as diagonals in each panel, which carry tension. This ability of plate girder to act as truss girder was recognized by Baselar as early as 1898.

The diagonal tensile forces are carried by membrane action of web. It is referred as **tension-field action.**

The web panels remains anchored (surrounded) by the boundaries)

(i) at top and bottom, it is supported by the flanges and

(ii) on the either side, adjacent transverse stiffeners.

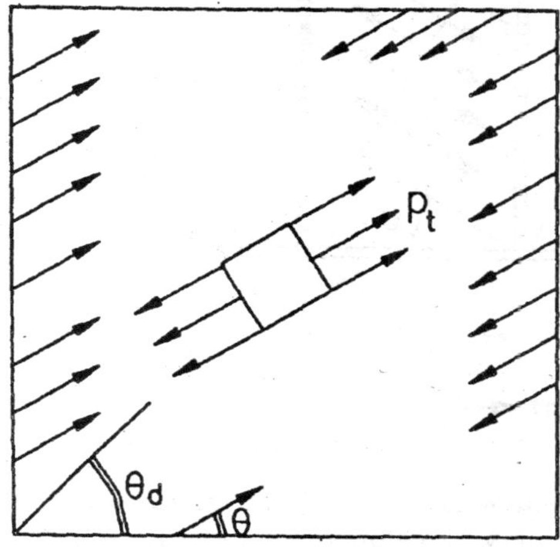

Fig. 19.8. Post buckled behaviour.

At this stage, the angle of inclination of the membrane stress, θ is not known. It is shown in Fig. 19.8.

The ultimate shear strength, in general, may be expressed as sum of following two strengths

1. Buckling strength of web in the elastic stage plus
2. Post-buckled strength, V_{tf} from tension-field action. Therefore,

$$V_u = (V_{cr} + V_{tf}) \tag{19.13)A$$

The buckling strength both elastic and inelastic may be written as

$$V_{cr} = \tau_y \,.\, (d_w \,.\, t_w) \,.\, \left(\frac{\tau_{cr}}{\tau_y}\right) \quad (19.13)B$$

$$V_{cr} = \tau_y \,.\, (d_w \,.\, t_w) \,.\, C_v \quad (19.13)C$$

Where, $C_v = (\tau_{cr} / \tau_v)$ is a non-dimensionalised constant.

Note: Intermediate stiffeners help (i) to develop **tension-field action** and (ii) to improve the shear capacity of the web.

Intermediate stiffeners do not support applied loads. Bearing stiffeners provided at the points of concentrated loads protect the web from direct compressive loads.

Concentrated loads acting on the top flange may produce yielding of web, crippling of web and the side sway buckling when the compression in web causes buckling in the lateral direction. To protect the web to deflect laterally, longitudinal stffeners are provided.

The shear strength, V_{tf} from the tension-field action in the web develops in a **band** of tensile forces, that occurs after the web has buckled under the diagonal compression.

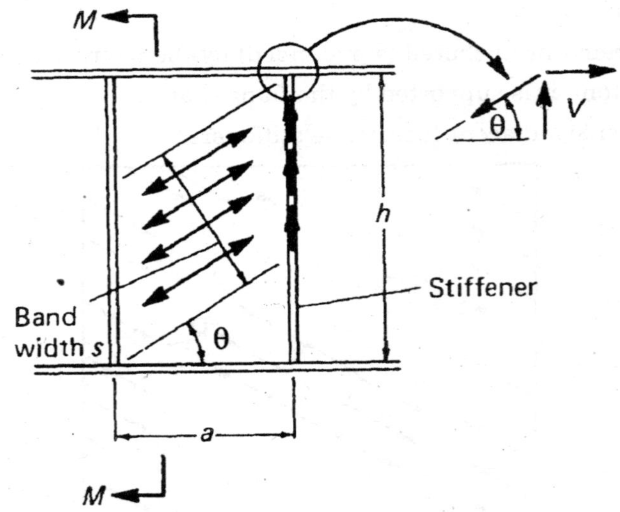

Fig. 19.9.

Let σ_t be the tensile stress which develops in the membrane in the web at an angle, *t*. In case, such tensile stresses may develop over the full depth of the web, then, the total diagonal tensile force, T would be as below

$$T = \sigma_t \,.\, (d_w \,.\, t_w) \,.\, \cos\theta \quad (19.14)A$$

The vertical component of which is the shear force, v, given by

$$V = \tau \,.\, \sin\theta = \sigma_t \,.\, (d_w \,.\, t_w) \,.\, \cos\theta \,.\, \sin\theta \quad (19.14)B$$

Tension-field action actually develop only over a band-width, such that the vertical component may be transferred to the vertical stiffeners which are designed to carry the necessary compressive force. Let the band-width be s_{bw} as shown in Fig. 19.9.

The memberane tensile force contributory to one stiffener is σ_t. $(S_{bw} \,.\, t_w)$ and the shear force developed by the compression in the stiffener is

$$V_{tf} = \sigma_t \,.\, (d_w \,.\, t_w) \,.\, \sin\theta \qquad (19.15)A$$

where, θ = Angle of inclination, which provides maximum component from the **tensile field.**

From the geometry, Fig. 19.10

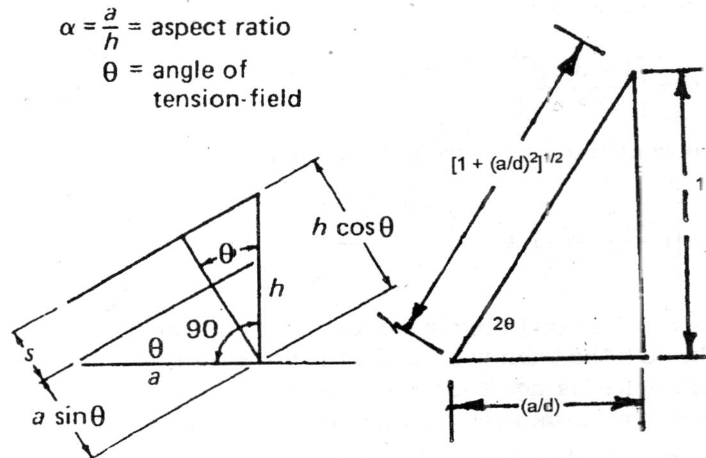

Fig. 19.10.

band-width

$$s_{bw} = (d.\cos\theta - a\,.\,\sin\theta) \qquad (19.15)B$$

where, a = Spacing of the stiffeners substituting in the expression

$$V_{tf} = \sigma_t \,.\, t_w\,[(d_w \,.\, \cos\theta - a.\ \sin\theta)\,.\,\sin\theta] \qquad (19.15)C$$

$$V_{tf} = \sigma_t \,.\, t_w\,[d_w \,.\, \cos\theta \,.\, \sin\theta - a\ \sin^2\theta] \qquad (19.15)D$$

Since, $\sin 2\theta = 2 \sin\theta\ \cos\theta$

$$\frac{1}{2}\sin 2\theta = \sin\theta.\ \cos\theta$$

Therefore,

$$V_{tf} = \sigma_t \,..\, t_w \left[\frac{d_w}{2}\,.\,\sin 2\theta - a\sin^2\theta\right] \qquad (19.16)A$$

For the maximum V_{tf}, $(d\ V_{tf} / d\theta)$ should be zero. Differentiating

$$\left(\frac{d\,V_{tf}}{d\theta}\right) = \sigma_t \,.\, t_w \,.\left[\frac{d_w}{2}.\,2\cos\theta - 2\,a\ \sin\theta\,\cos\theta\right] = 0 \qquad (19.16)B$$

$$\sigma_t \,.\, t_w \left[\frac{d_w}{2}.\,2\cos\theta - a\ \sin 2\theta\right] = 0 \qquad (19.16)C$$

$$\tan 2\theta = \left[\frac{d_w}{a}\right] = \left[\frac{1}{a/d_w}\right] \tag{19.16)D}$$

From the trigonometry,

$$\sin 2\theta = \left[\frac{1}{1+(a/d)^2}\right]^{1/2} \tag{19.16)E}$$

$$\text{since,} \quad \sin^2\theta = \left[\frac{1-\cos 2\theta}{2}\right]^{1/2}$$

The maximum contribution, V_{tf} from the tension-field is found by substitution

$$V_{tf} = \sigma_t \cdot \left[\left(\frac{dw \cdot t_w}{2}\right)\left(1+\left(\frac{a}{d}\right)^2\right)^{1/2} - \frac{a}{d}\right] \tag{19.16) F}$$

It is not practical to use (xv) directly, since the shear contribution from the part of the section) that cuts through the triangles outside the band-width s_{bw} must be added. The state of stress in these triangles is not known. It needs an alternate approach to find the total shear, V_{tf}, when the optimum value of the angle, θ is reached.

Baselor used an **alternate approach** to cut a free body as in Fig. 19.11.

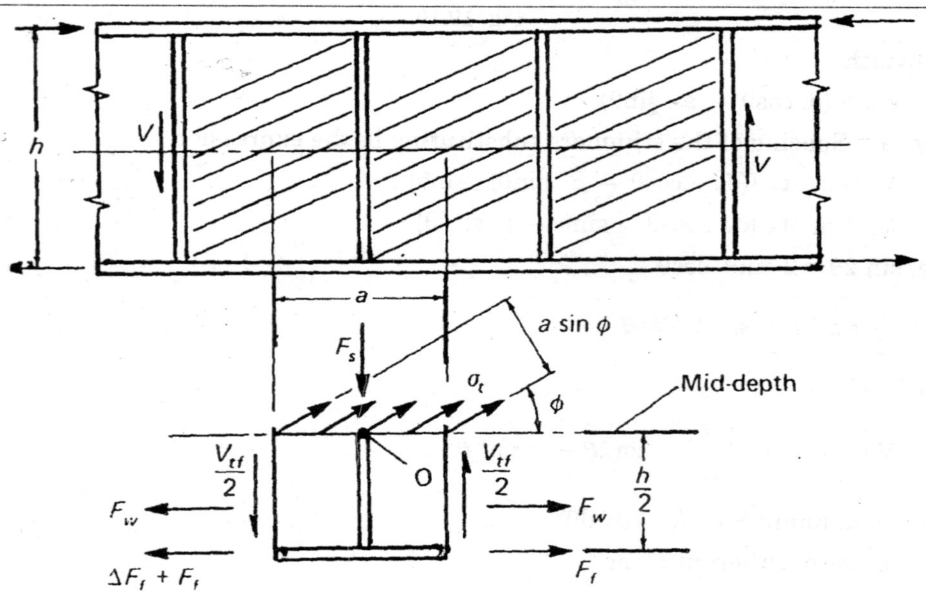

Fig. 19.11.

The section is taken vertically mid-way two adjacent stiffeners and horizontally at mid-depth. This provides an approach to the **tension-field**, where, the state of stress is known and the shear resultant on each side (vertical face) equals $\left(\frac{1}{2}V_{tf}\right)$ from symmetry.

Shear strength contribution from tension-field action

Using the free body of Fig. 19.11, and considering equilibrium from the horizontal forces

$$F_{HF} = (\sigma_t . (t_w . a) . \sin\theta) . \cos\theta \tag{19.17)A}$$

$$= \sigma_t . \left(\frac{t_w . a}{2}\right) . \sin 2\theta \tag{19.17)B}$$

since $(\sin\theta . \cos\theta = \frac{1}{2}\sin 2\theta)$

Considering the equilibrium in the direction of rotation

$$F_{Hf} . \frac{1}{2} d_w - \frac{1}{2} v_{tf} . a = 0 \tag{19.17)C}$$

$$\text{Therefore, } F_{Hf} = \left(\frac{V_{tf} . a}{d_w}\right) \tag{19.17)D}$$

Substituting for F_{Hf} in the expression (ii)

$$\left(\frac{V_{tf} . a}{d_w}\right) = \sigma_t . \left(\frac{1}{2} t_w . a\right) . \sin 2\theta \tag{19.17)E}$$

Substituting the value of sin2θ, from Fig. 19.17F

$$V_{tf} = \sigma_t . \left(\frac{1}{2} d_w . t_w\right)\left[\frac{1}{(1 + (a/d_w))^2}\right]^{1/2} \tag{19.17)F}$$

3. Behaviour of web at failure (collapse)

A new load carrying mechanism develops along the principal tensile direction, after the web has buckled along the direction of the principal compressive stress. This new mechanism is referred as **tension-field action.** This resists any further increase in the shear load. This new mechanism remains anchored at the boundaries (i) along top and bottom flanges, and (ii) along the either side of web stiffener members Fig. 19.8

At this stage, the total stress in the web plate consists of (i) applied critical shear stress, $\tau_{sh.cr.e}$ when the web plate buckled, and (ii) the post-buckled membrane tensile stress, p_t due to the tension field action, as shown in Fig. 19.12

By resolving the stresses in the direction (i) along, and (ii) perpendicular to the direction to the angle of inclination, θ

$$p_\theta = \tau_{sh.cr.e} + p_t \tag{19.18)A}$$

$$p_{(\theta + 90°)} = -\tau_{sh.cr.e} . \sin 2\theta \tag{19.18)B}$$

$$\tau_\theta = -\tau_{sh.cr.e} . \cos 2\theta \tag{19.18)C}$$

After reaching the stresses in web to yield, the value of the stress may be obtained by the von-Mises yield criterion

$$p_{\theta}^2 + p_{(\theta+90)}^2 - p_{\theta} \cdot p_{(\theta+90°)} + 3\,\tau_{\theta}^2 = \sigma_{yw}^2 \tag{19.19}$$

where, σ_{yw} = Tensile yield stress of web

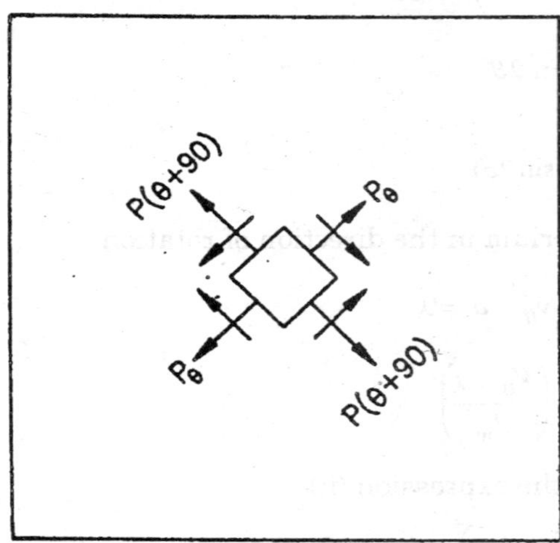

Fig. 19.12. State of stress in the web in the post buckled stage.

Substituting the values of p_{θ}, $p_{(\theta+90°)}$ *and* τ_{θ} from the expressions (i), (ii) and (iii) in the expression (iv)

$$p_{yt} = -\frac{3}{2}\,\tau_{sh.cr.e}\sin 2\theta + \left[\sigma_{yw}^2 + \left(\frac{9}{4}\sin 2\theta - 3\right)\tau_{sh.cr.e}^2\right]^{1/2} \tag{19.20)A}$$

The expression (vi) may be written in the non-dimensionalised form as below

$$\left(\frac{p_{yt}}{\sigma_{yw}}\right) = (-1)\left(\frac{\sqrt{3}}{2}\,\frac{\tau_{sh.cr.e}}{\tau_{yw}}\right)\sin 2\theta + \left[1 + \left(\frac{\tau_{sh.cr.e}}{\tau_{yw}}\right)^2\left(0.75\,\sin^2\right)2\theta\;-1\right]^{1/2} \tag{19.20)B}$$

where, $\tau_{sh.cr.e}$ is the elastic critical shear stress and $\tau_{yw} = \left(\sigma_{yw}/\sqrt{3}\right)$

This may be written as under

$$\frac{p_{yt}}{p_{yw}} = \left[\left(1 - \frac{\tau_{sh.cr.e}}{\tau_{yw}}\right)\left(1 - 0.75\right)\sin^2 2\theta\right]^{1/2} - \left(\frac{\sqrt{3}}{2}\left(\frac{\tau_{sh.cr.e}}{\tau_{yw}}\right).\sin 2\theta\right) \tag{19.20)C}$$

Once, the web yields, four plastic hinges are formed in the flanges as shown in Fig. 19.13, permitting shear mechanism, The failure (collapse) of the girder occurs. The plastic moment capacity of the flange plate shall be M_{pt}

$$M_{pt} = (\sigma_y \;.\; Z_{pf})$$

where, Z_{pf} is the plastic modulus of the flanges about neutral axis.

Porter *et al.* (1975) suggested a **tension field** which consists of a single band. It was an improvement of the model suggested earlier by Rocky and Skalound (1971). Porter *et al.*

(1975) and Rocky *et al.* in shear have shown, both experimentally and theoretically, that the plate girders fail in shear, when a section of the web plate yield and the plastic hinges form in the flanges. Four plastic hinges form in the flanges. This allows a shear mechanism. In the above studies, the model as suggested by **Porter *et al.* (1975)** has been considered. This model is often referred as **Cardiff model**. The effect of the bending stiffeners of the flanges on the width of the diagonal tension has been considered. Shear model as proposed by Porter *et al.* (1975) provides a fairly good correlation between theory and test results.

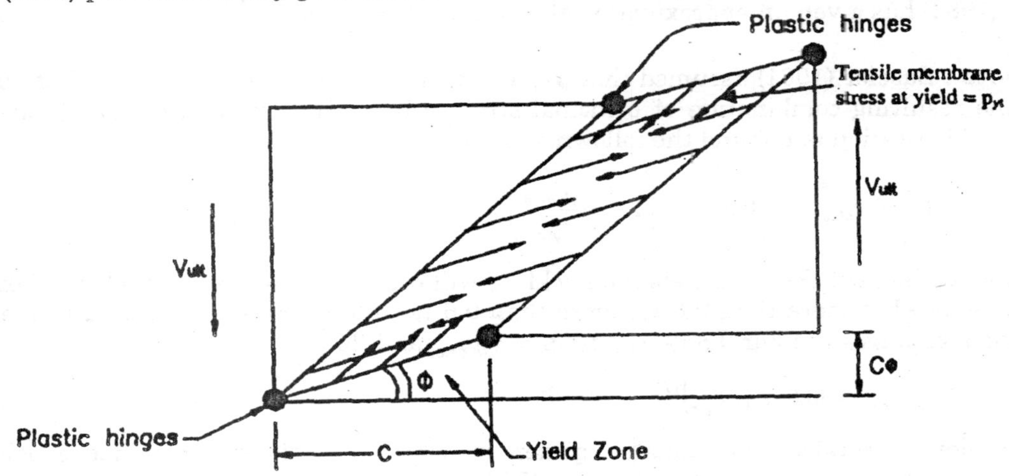

Fig. 19.13. Collapse of the panel.

Rockey *et al.* (1978) have proved that the total shear load at failure, V_s may be found from the following expression

$$V_s = t_{sh.cr.e}\,(d_w \,.\, t_w) + f_t \,.\, t_w \,.\, \sin^2\theta\,(d_w \cot\theta - a + S_t) + 4\,(M_{pf} / s_t) \qquad (19.21)A$$

$$\text{Plastic moment capacity of the flange} = \sigma_y \,.\, \left(\frac{b_f . t_f^2}{4}\right)$$

s_t = Distance between adjacent plastic hinges, and

f_t = Tensile field strength

It may be observed from that the expression (viii) has been found by the summation of three components, namely

(i) the strength of web buckling,

(ii) the yield strength of post-buckling of the web plate and

(iii) the plastic moment capacity of the flanges

A non-dimentionalised form of the expression may be written using the yield strength of the web plate, $V_{yw} = \sigma_{yw}\,(d_w \,.\, t_w)$

The position of the plastic hinges in the flanges, s_t or s_c may be located by considering the equilibrium of the flange and a free-body of a portion of web as per Rockey *et al.* 1973).

$$s_t = \frac{2}{\sin\theta}\left(M_{pf} / (f_t . t_w)\right)^{1/2} \qquad (19.22)A$$

Substituting the value of σ_t in the expression (viii) and dividing throughout by V_{yw}

$$\left(\frac{V_s}{V_{yw}}\right) = \left(\frac{\tau_{sh.cr.e}}{\tau_{yw}}\right) + \sqrt{3}\ \sin^2\theta\left(\cot\theta - (a/d)\right) + 4\sqrt{3}\ (\sin\theta)\left(\frac{f_t}{\sigma_{yw}}\ .\ \frac{M_{pf}}{d_w^2 t_w\ .\ \sigma_{yw}}\right)^{1/2} \quad (19.23)$$

This expression may be solved, in case θ is known. By a series of parameteric studies, Evans (1983) has given an approximate value of θ as $\theta = \frac{2}{3}\tan^{-1}\left(\frac{d}{a}\right)$. (19.24)

Note : Baselor (1961) assumed that p_{yt}, and $\tau_{sh.cr.e}$ are additive (as if f_{yt} acted at 45° and used the resulting combination of principal stresses in a linear approximation of the von-Mises yield criterion and found the following expression for shear stress

$$\tau_s = \tau_{sh.cr.e} + 0.5\ \sigma_{yw}\left(1 - \frac{\tau_{sh.cr.e}}{\tau_{yw}}\right).\ \sin\theta \quad (19.25)$$

It was assumed that the inelastic buckling will occur if. $\tau_{sh.cr.e}$ exceeds 0.8 τ_{yw} [being in tension somewhat more than 0.6 τ_{yw} more than 0.6 τ_{yw} in compression and defines inelastic buckling stress $\tau_{sh.cr.e}$ as For 0.8 τ_{yw} <= $\tau_{cri} \leq = \tau_{sh.cr.e}$ as 1.25 τ_{yw}]

$$\tau_{sh.cr.e} = \left(0.8\ \tau_{ri}\ .\ \tau_{yw}\right)^{1/2} \quad (19.26)$$

Baselor's formula overestimates girder in which the flanges are not capable of supporting lateral load from the tension field. Because, it gives the shear strength for complete **tension field** instead of the limited band-width s_{bw} as observed in the tests (Galumbos, 1998).

Note: Only plastic sections are capable to form plastic hinges. In case, the sections are classified as compact, semi-compact, and/or slender, these sections cannot develop plastic hinges in the flange. Therefore, this method is **not applicable** such sections.

Forces carried by the stiffeners may be determined by considering the equilibrium in the vertcal direction.

$V_s = \left[\sigma_t\ .\ (t_w\ .\ a)\ \sin\theta\right].\ \sin\theta$ and substituting for $\sin^2\theta$ from geometry of a right angled triangle having an angle equal to θ.

$$V_s = \sigma_t\ .\left[\left(\frac{a.\ t_w}{2}\right)\right]\left[1 - \frac{(a/d_w)}{\left(1 + (a/d_w)^2\right)^{1/2}}\right] \quad (19.27)$$

Substituting for s_t from the expression

$\sigma_t = (1 - C_v)\ F_y$

$$F_s = \left[F_y\ (1 - \phi_v\ .\ F_y\ .\left(\frac{a.t_w}{2}\right)\right]\left[1 - \frac{(a/d_w)}{\left(1 + (a/d_w)^2\right)^{1/2}}\right] \quad (19.28)$$

This represents the force received at ultimate shear strength to accommodate **tension-field action.**

19.8 TENSION FIELD ACTION METHOD FOR DETERMINATION OF RESISTANCE /STRENGTH OF WEB IN SHEAR BUCKLING

A plate girder with intermediate stiffeners ($1.0 <= (c/d_w) <= 3.0$) in addition to the transverse stiffener at supports, the end posts provide anchor for the tension field possesses considerable post buckling shear strength due to **tension-field action**. It is also necessary that ratio of the spacing of the transverse stiffeners, C and the depth of web [c/d is >= 1.0.

This method may be used for a certain range of girders only and gives efficient designs of the plate girders, since, this method takes the advantage of the post-buckling reserve of buckling strength.

There are three components which contribute to evaluate shear strength

(i) Primary (main) buckling strength of the web plate called as critical buckling strength of the web plate

(ii) Tension-field action of the web plate

(iii) Plastic moment capacity of the flanges, $M_{pf} = (\sigma_y \,.\, Z_{pf})$ (19.29)

A successful model for **tension field action** for the plate girders used in civil engineering structures was formulated by Baseler in 1961. First Rode (1916) provided physical explanation for the very shearing capacity of the slender plate girder webs, Wagner in 1929 developed a pure meaning for complete **tension field theory** for air craft structures.

Baseler's approach (1961, 1963) did not consider the flexural rigidity of the flanges. Therefore, true anchorage of the tension field along the flanges could not be reflected.

A number of researchers (e.g, Clark and Sharp, 1971; Steinhardt and Schroter, 1971) verified tension-field theory.

IS : 800-2007 recommends that the nominal shear resistance, V_n is found from ($V_n = V_{tf}$), where

$$V_{tf} = \left|\tau_b \,.\, A_v + 0.9 \;\; w_{tf} \,.\, t_w . f_v \;\; \sin\theta\right| \quad (19.30)A$$

τ_b = shear stress corresponding to buckling strength of the web plate

f_y = Yield strength of the tension field obtained from

$$f_v = \left[\sigma_{yw}^2 - 3\right]\tau_b^2 + \psi^{2\;1/2} - \psi, \quad (19.30)B$$

$$\psi = 1.5 \,.\, \tau_b \,.\, \sin 2\theta \quad (19.30)C$$

θ = Inclination of the tension field

$$\theta = \tan^{-1}\left[\frac{d_w}{c}\right] \quad (19.30)D$$

w_{tf} = Width of the tension-field

$$w_{tf} = d.\cos\theta + (C - S_c - S_t)\sin\theta \quad (19.30)E$$

s_c, s_t = Anchorage lengths of the tension field along the compression and tension flanges, found from

$$s = \left(\frac{2}{\sin\theta}\right)\left[\frac{M_{fr}}{\sigma_{yw} . t_w}\right] \quad (vi)\ (19.)$$

where,

M_{fr} = Reduced plastic moment capacity of the respective flange plate (disregarding any edge stiffener) after considering the axial force, P_f in the flange, due to overall bending and any external axial force in the cross-section, and M_{fr} is determined as below

$$M_{fr} = 0.25\ b_f\ .\ t_f^2\ .\ \sigma_{yf}\ \left[1 - \left\{P_f / \left(b_f\, t_f\, \sigma_{yf} / \gamma_{mf}\right)\right\}\right] \tag{19.31}$$

b_f, t_f = Width and thickness of the relevant flange, respectively

σ_{yf} = Yield stress of the flange

σ_{yw} = Yield stress of the web

d_w = depth of the web

c = Spacing of the stiffeners

BS: 5950-1: 2000. (British code of practice) presents an expression for the evaluation of τ_b

For 0.8 < λ_w < 1.25

$$\tau_b = \left[1 - 0.625\ \left(\lambda_w - 0.8\right)\right]\left(\sigma_{yw} / \sqrt{3}\right),\ and \tag{19.32)A}$$

For λ_w > = 1.25

$$\tau_b = 0.9\ \sigma_{yw} / \left(\sqrt{3}\ \lambda_{yw}^2\right) \tag{19.32)B}$$

This permits for a higher value of shear stress for thin webs as shown in Fig. 19.14. These expressions are used for determining the value, τ_b.

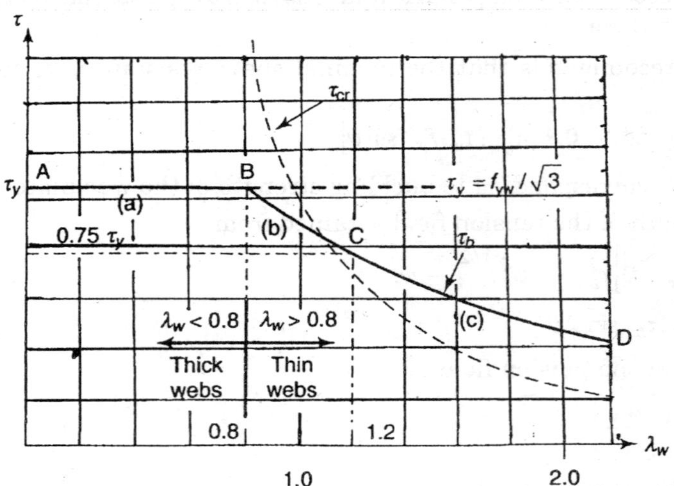

Simple post-critical shear strength as per BS: 5950

Fig. 19.14.

19.9 DESIGN OF END PANELS

The web plate of a plate girder is divided into interior panels and the end panels. The interior panels (e.g., panel A) is designed using tension field action. The provisions specified in IS : 800-2007 may be followed

The vertical components of the tension-field are resisted by the stiffeners. Tension field of interior panel (e.g., third panel) is balanced by the tension field of neighbouring panel (e.g., second panel from left side). These interior panels are anchored by the respective neighbou-ring panels.

Where as, the ed panel does not have anchorage on the left-side (and so also the anchorage has to be provided by an end stiffener. This end panel is specially to resist the bending induced by the tension-field.

It is just possible that the tension field may not normally be developed in an end panel . In case of end panel, the horizontal components of the tension fields as shown in Fig. 19.15 cannot be balanced. As such the end panels are designed by using only either of following two methods.

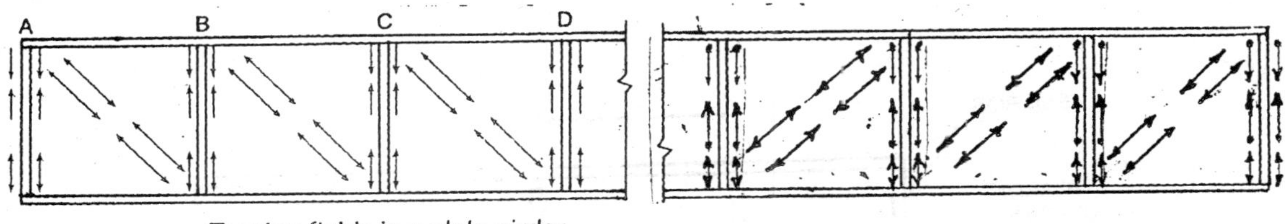

Tension fields in a plate girder

Fig. 19.15.

1. **Simple post-critical method** as per the provisions given in IS : 800-2007 apparently by not using tension-field action.

Additionally, the end panel along with the stiffeners should be checked as a beam spanning between the flanges to resist a shear force, R_{tf} and a moment, M_{tf} due to tension field forces. Further end stiffener should be capable of resisting the reaction plus a compressive force due to the moment, M_{tf}.

2. Tension-field action

The end panels of the web plate of a plate-girder may also be designed using tension field action following the provisions as specified by IS : 800-2007.

Additionally, it should be provided with an **end post** which may consist of (i) a single stiffener, or (ii) Double stiffeners, which satisfies the provisions specified in IS : 800-2007

(i) Single stiffener

The top of the end post should be rigidly connected to the flange using full strength welds.

The end-post should be capable of resisting the reaction plus a moment from the anchor forces equal to (2/3) M_{tf} due to tension field forces.

Where, M_{tf} is found from the provisions of IS : 800-2007 and described earlier.

The width and thickness of the end post are not to exceed the width and thickness of the flange.

(ii) **Double stiffeners** : The end post should be checked as beam spanning between the flanges of the plate girder and capable of resisting a shear force R_{tf} and a moment, M_{tf} due to tension field forces evaluated as specified in IS : 800-2007.

In Fig. 19.16 (A) at the left support **single stiffener** has been shown, while in the Fig. 19.16B left support, the **double** stiffener has been shown.

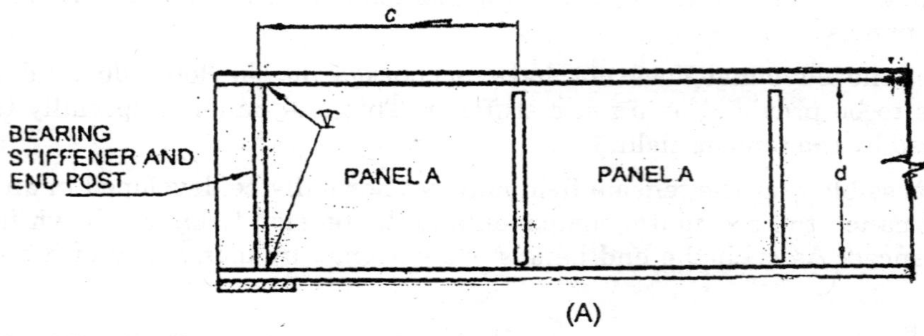

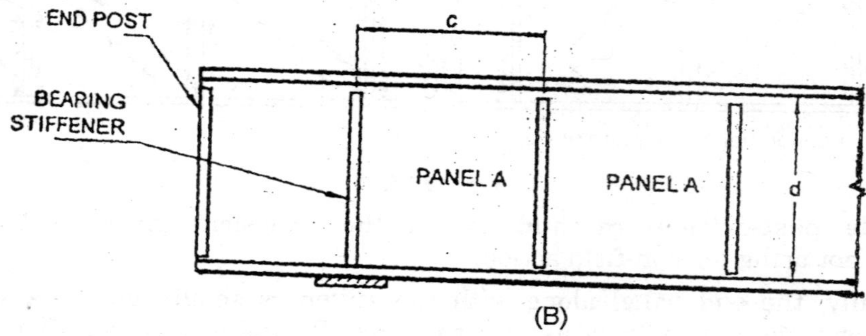

Fig. 19.16

Note : This is for the purpose of sketch, however, same type of stiffeners are provided as both the sides

Anchor forces : The longitudinal shear R_{tf}, develops from the anchor of tension field from the end panel which acts on the adjacent inner panel are evaluated as below :

$$R_{tf} = \frac{1}{2} Hq \quad and \tag{19.33)A}$$

$$M_{tf} = \frac{1}{10} Hq \cdot d, \quad where$$

$$Hq = 1.25\ V_p \left[1 - \frac{V_{cr}}{V_p}\right]^{1/2} \tag{19.33)B}$$

$$V_p = \left(\sigma_y \cdot \frac{d_w \cdot t_w}{\sqrt{3}}\right) \text{ and} \tag{19.33)C}$$

d_w = Depth of web

If the actual factored shear force, V in the panel using tension field approach is less than the shear strength, V_{tf} as described earlier, then, the values of Hq may be reduced by the ratio $\left|(V - V_{cr})/(V_{tf} - V_{cr})\right|$ where, V_{tf} is the basic shear strength for the panel utilising tension field action, and V_{cr} is the critical shear strength for the panel designed utilizing tension-field.

19.10 MODIFIED WEB BUCKLING STRESS

The web buckling stress in the presence of bending stresses gets modified. The modified web buckling stress may be found from the following interaction formula [Gaylord *et al.* 1992].

$$\left[\frac{\tau_{cr.m}}{\tau_{sh.cr.e}}\right]^2 + \left[\frac{f_{mbc}}{f_{cr.b}}\right] = 1 \qquad (19.34)\text{A}$$

where, $\tau_{cr.m}$ is the modified buckling stress in shear

$\tau_{sh.cr.e}$ = Elastic critical shear stress in the web (may be carrying pure shear)

f_{mbc} = Compressive bending stress in the extreme fibre at the mid-panel due to bending moment

$f_{cr.b}$ = Buckling for the plate given by

$$f_{cr.b} = K_b \cdot \left[\frac{\pi^2.E}{12\left(1-\mu^2\right).\left(d_w/t_w\right)^2}\right] \qquad (19.34)\text{ B}$$

$$= 23.9\left[\frac{\pi^2.E}{12\left(1-\mu^2\right).\left(d_w/t_w\right)^2}\right] \qquad (19.34)\text{C}$$

Buckling coefficient K_b depends on the aspect ratio (a/d) of the plate and number of buckles along the plate. For the long plates, simply supported on all the four edges, K_b = 23.9. The limiting value of (d_w / t_w) is

$$(d_w / t_w)\ (\sigma_y / 250)^{1/2} = 131.4 \qquad (19.34)\text{D}$$

For the plates, simply supported on the loaded edges and clamped on the other edges, K_b = 39.6. The limiting value of (d/t_w) is

$$(d_w/t_w)\ (\sigma_y/250)^{1/2} = 169.2$$

Note: The transverse stiffeners becomes ineffective unless more closely spaced than (0.67 d_w). Where as, the longitudinal stiffeners remain effective in changing the buckled shape.

19.11 WEBS SUBJECTED TO PURE BENDING

Fig. 18.1(2) represents the interaction between bending and shear effects. A high bending moment region exists beyond the point C. In this region, the failure may occur by bending. The compressive flange carries almost all the compressive stress, since the web subjected to compressive stress has already buckled. Therefore, full plastic moment, M_p cannot develop in the cross-section of the plate girder.

Having adequate lateral support, the failure occurs by inward collapse of the compression flange at an applied ultimate moment, M_u. It is approximately equal to the moment necessary to develop first yield in the extreme fibres of the compression flange.

However, due to the influence of buckling of web, its value is reduced. The ultimate moment capacity may be calculated by using Coopers (1971) formula. Cooper suggested the following formula in 1971

$$\left(\frac{M_u}{M_y}\right) = 1 - 0.0005 \left(\frac{A_w}{A_f}\right) \left[d - 5.7\right] \left(E/\sigma_{yf}\right)^{1/2} \text{ is more than or equal to } (M_f / M_y) \quad (19.35)$$

where, M_y is the bending moment necessary to develop yield in the extreme fibre of the flange neglecting web buckling (i.e, the web is assumed fully effective).

M_u is the moment needed to develop yield in the extreme fibres of the flange

Corresponding stresses in the web will be less than the yield (Point D in the interaction diagram. The ordinate at the point D on the curve may be evaluated approximately from

$$\left(\frac{V_D}{V_c}\right) = \left(\frac{M_p - M_u}{M_{pw}}\right)^{1/2}$$

where, M_p is the fully plastic moment capacity of the complete cross-section, M_{pw} is the plastic moment of resistance of the web alone. It is obtained from

$$M_{pw} = (0.25\, \sigma_{yw} \,.\, (d_w \,.\, t_w)) \,.\, d_w$$

Therefore, the complete interaction diagram may now be drawn.

Reduction of plastic moment capacity of the flanges

Due to bending moments, high axial forces are produced in the flanges. The plastic moment capacity of the flanges are reduced for the axial forces. Their effects are considered. The reduced capacity of the flanges from the theory of plasticity

$$M'_{pf} = M_{pf} \left[-1 \left(\frac{p_f}{p_{yst}}\right)^2\right]$$

where, p_f is the average axial stress for the portion of flange between the hinges.

19.12 STRUCTURAL BEHAVIOUR OF FLANGES AND WEBS (OF A PLATE GIRDER)

Plate girders may have flanges which are weak/very strong and webs which are very thick/very slender. The behaviour of plate girder depends upon the rigidity offered by these elements. The behaviour of a plate girder in these several cases may be studied.

1. Flages of a plate girder which are weak.

In case, the flanges of a plate girder are very weak, then the plastic moment capacity, M_{pf} would be small. In such a case, the tension-field action is supported completely by the vertical stiffeners and it is not supported by the flanges.

Therefore,

$$\left(\frac{V_s}{V_{yw}}\right) = \left(\frac{\tau_{sh.cr.e}}{\tau_{yw}}\right) + \sqrt{3}\, \sin^2\theta \left(\cos\theta - \frac{a}{d_w}\right) \frac{p_{yt}}{p_{yw}} \quad (19.36)$$

The tension-field action remains anchored completely on the transverse stiffeners as shown in Fig. 19.17.

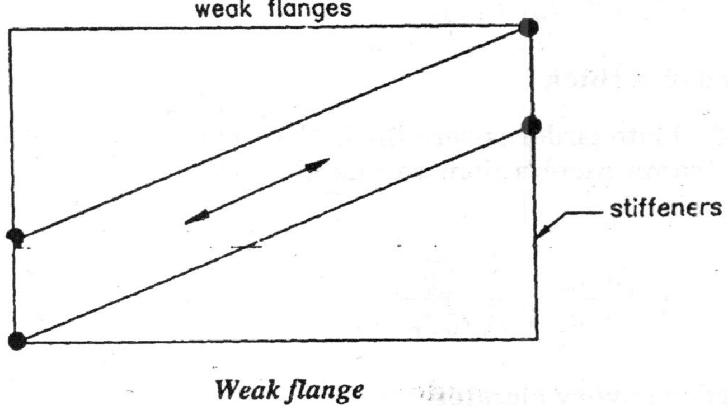

Weak flange

Fig. 19.17.

2. Flanges of a plate girder which are very strong

In such a case, it is seen that the distance between the plastic hinges, s_t increases. The plastic hinge forms at a distance away from the end panel. When, C = a, the hinges will develop at the four corners which appears like a **picture frame type mechanism** as shown in Fig. 19.18. The angle of inclination of the tension field θ = 45°. This type of shear failure is suggested by Narayanan (1983). The ultimate shear in such a case is given by

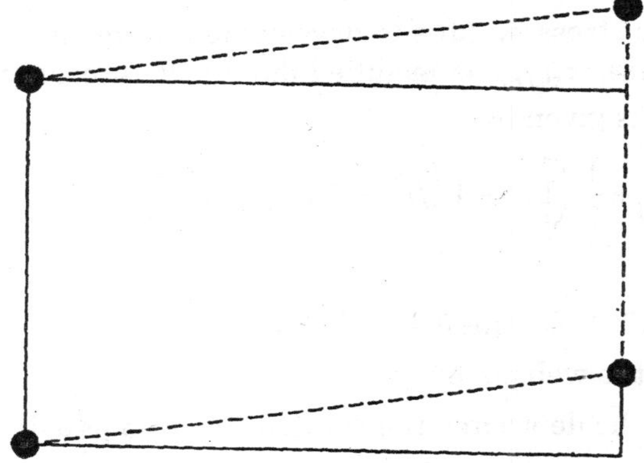

Picture frame mechanism of strong flange

Fig. 19.18.

$$\left(\frac{V_s}{V_{yw}}\right) = \frac{1}{4}\cdot\left(\frac{\tau_{sh.cr.e}}{\tau_{yw}}\right)$$

$$+ \frac{\sqrt{3}}{2}\left(1 - \frac{1}{4}\left(\frac{\tau_{sh.cr.e}}{\tau_{yw}}\right)^2\right)^{1/2}$$

$$+ \; 4\sqrt{3} \,.\, \left(\frac{d_w}{a}\right) \,.\, \left(\frac{M_{pf}}{d_w^2 \, t_w \,.\, \sigma_{yw}}\right) \tag{19.37}$$

3. Web of a plate girder is thick

When the web of a plate girder is very thick, the yielding occurs before buckling failure. However, **a picture frame mechanism** will develop with $\tau_{sh.cr.e} = \tau_{yw}$. The shear force at failure is given by

$$\left(\frac{V_s}{V_{yw}}\right) = 1 + 4\sqrt{3}\left(\frac{d_w}{a}\right)\left(\frac{M_{pf}}{d_w^2 \, t_w \,.\, p_{yw}}\right) \tag{19.38}$$

4. Web of a plate girder is very slender

For a very slender web, the elastic buckling stress shall be very small. The contribution **of tension-field action** becomes very significant. The value of p_{yt} will become very large. The general expression for the shear strength remains valid. Angle θ may be found as

$$\theta = \frac{2}{3} \tan^{-1}\left(\frac{d_w}{a}\right) \tag{19.39}$$

For the crane gantary girders the method of design of tension field cannot be used.

Modified membrane stress for yield of web

The tensile membrane stress p_{yt} (due to development of tension field action) is added to elastical critical shear stress, $\tau_{sh.cr.e}$ is modified due to the presence of bending stress. The modified membrane stress is given by

$$p_{yt.m} = \left(-\frac{1}{2}A\right) + \frac{1}{2}\left(\left[A^2 - 4\left(f_b^2 + 3\,\tau_{crm}^2 - p_{yw}^2\right)\right]^{1/2}\right) \tag{19.40)A}$$

where,

$$A = 3\,\tau_{crm} \,.\, \sin 2\theta + f_b \,.\, \sin^2\theta - 2\,f_b \,.\, \cos^2\theta \tag{19.40)B}$$

f_b = Bending stress in the web panel.

This varies over both, the depth and the width of the web panel.

The value of p_{ytm} depends on the value of f_b. It has to be calculated at various locations to get the resultant.

19.13 LOAD AND BEARING STIFFERS

The load bearing stiffeners (simply known as bearing stiffeners) are attached with the web plate of the plate girder to avoid local bending failure of the flange, local crippling or buckling and crushing of web (web crippling) due to concentrated loads (ao also the points of supports, i.e. reaction). Those stiffeners which get loads, and which are used to prevent local buckling of the web due to concentrated loads. The bearing stiffeners are provided under concentrated loads and at the points of supports. The bearing stiffeners strengthen the web and transmit heavy concentrated loads or reactions to the flanges of plate girder and from the flanges to the web plate. When these are provided at ends. these are also termed as end **bearing stiffeners.** They should not be joggled or crimped. They are provided straight as shown in Fig. 19.18. The bearing stiffeners extend between the full depth between flange to flange and have a tight bearing against the flange transferring the load. The filler plates of thickness equal to the thickness of flanges angles are inserted between web plate and stiffeners.

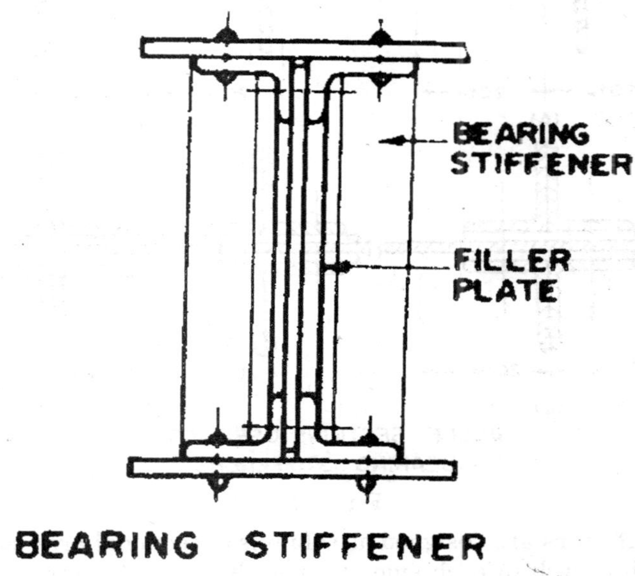

Fig. 19.19.

The various types of bearing stiffeners used are shown in Fig. 19.19. The angle sections in the end bearing stiffeners are arranged with reference to the point of/points of maximum pressure on the flanges. In the fixed bearing (plate bearings and cast-iron shoe bearings), the maximum pressures are likely to occur at the outer and inner edges of the sole plate (the plate attached to the bottom flange of plate girder at the support). At such plates, the arrangements of the angles as shown in Fig. 19.19(B) and (D) are suitable. In the pin or centre-bearing, the maximum pressure occurs at the middle of bearing. Therefore, the arrangement of the angles as shown in Fig. 19.9(C), as an equivalent cruciform shaped (a cross-shaped strut) (*i.e.*, in the form of cluster of four angles is suitable. Sometimes, two angles (one on either side of the web plate) are attached to the free or outer edge of the web, and four angles in the form of a cluster are arranged at the point of maximum pressure.

The end of bearing stiffeners are machined or ground, and are fitted tightly between top and bottom flange angles. They are provided symmetrically about the web. The outstanding width of the bearing stiffener is kept such as to extend as nearly as practicable upto the outer edges of flange angles. The bearing stiffeners are initially selected on the basis of bearing area required. The intensity of bearing pressure should not exceed $\sigma_p = 0.75\ \sigma_y$ N/mm². The bearing areas of outstanding legs are taken clear of root of the flange angles as shown in Fig. 19.20. The load bearing capacity calculated using the effective area should be greater than or equal to the applied load or the reaction.

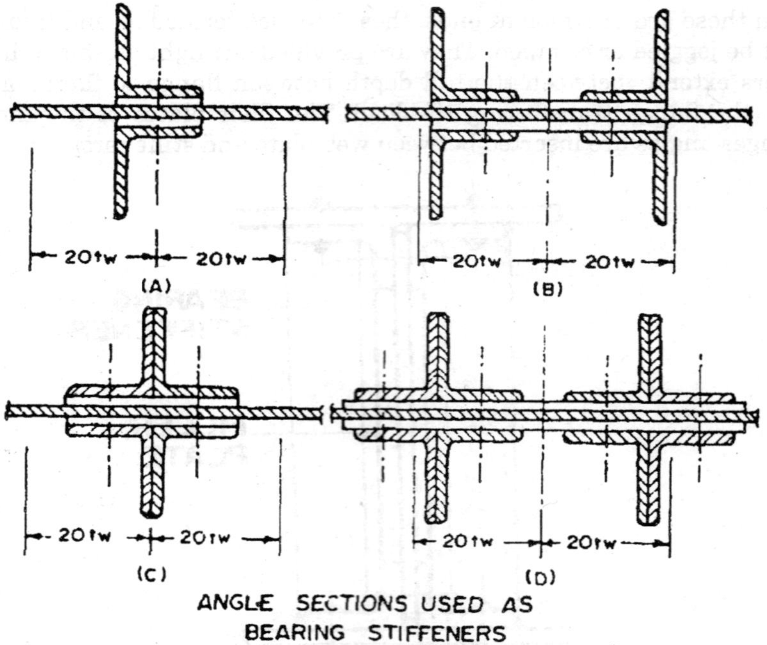

Fig. 19.20.

The bearing stiffeners are designed as columns of sectional area consisting of stiffeners together with a length of web in each side of centre line of stiffeners as shown in Fig. 19.20 equal to 20 times the thickness of web, if available. This area is used to determine the radius of gyration and to check the column stresses. The effective length is assumed, 0.7 times actual length (length between top and bottom flanges) length of the stiffeners because of secured connection with the web. The radius of gyration is computed about the centre line of the web. The rivets connecting stiffeners to the web under concentrated loads are designed to transmit these loads. At supports the rivets connecting stiffeners to the web are designed to transmit the reaction. The filler plates are used in between the stiffeners and the web plate. The thickness of filler plate is equal to the thickness of the flange angles. The rivets are connected through the filler plates. The rivets carrying calculated shear stress through packing greater than 6 mm thick are increased above the number required by normal calculations by 2.5% for each 2 mm thickness of packing. For double shear connections packed on both sides, the number of additional rivets required are determined from the thickness of thicker packing. The additional rivets are preferably be placed in an extension of the packing. When torsional restraint at the ends of plate girder is provided only by stiffeners, then the moment of inertia of the stiffeners about the centre line of the web shall not be less than

$$\left(\frac{D^3T}{250} \times \frac{R}{W}\right)$$

where D = Overall depth of the girder

T = Maximum thickness of compression flange

R = Reaction on the bearing and W = Total load on girder.

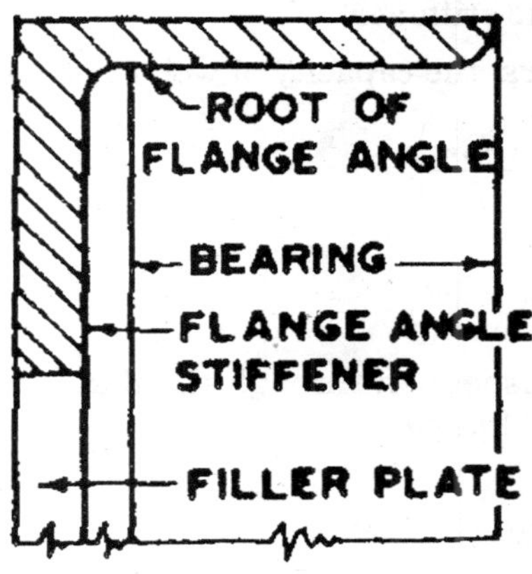

Fig. 19.21.

The load carrying capacity of the bearing stiffeners as a column should be greater than or equal to the applied load or the reaction.

19.14 DESIGN OF LOAD CARRYING STIFFENERS AND BEARING STIFFNERS

The buckling strength, F_{cdw} of **the load carrying stiffeners** is found as follows :

(i) Area of cross-section is calculated as $(b_1 + n_1).\ t_w$, where

b_1 = Width of stiff bearing on the flange

n_1 = Dispersion of load through the web at 45°, to the level of half the depth of the cross-section.

(ii) The effective length for determining the slenderness ratio has been mentioned in Art. 19.

The buckling strength is calculated about the axis parallel to the web using column curve in IS : 800-2007 for compression.

The buckling resistance of the load carrying stiffeners, f_{xd}, should be more than the external load or reaction, F_x.

The requirement is as under for the buckling

$(F_{xd} > F_x)$ check

Where, the stiffeners also act as an intermediate stiffener, it should be checked for the effects of combined loads

Load carrying web stiffeners should also be of sufficient size that the bearing strength of the stiffener, F_{psd} given below is not less the load transferred, F_x. It is the **requirement for the bearing check.**

$$F_{pcd} = \left(\frac{A_q \cdot \sigma_{yq}}{0.8\ \gamma_{msy}}\right) .> = F_x$$

where,

F_x = External load or reaction

A_q = Area of the stiffener in contact with the flange

σ_{yq} = Yield stress of the stiffener

For **the bearing stiffeners**, the capacity of web at its connection to flange

$$F_w = \left(\frac{(b_1 + n_2)\, t_w \,.\, \sigma_{yw}}{\gamma_{msy}}\right)$$

where,

b_1 = stiff bearing length

n_2 = Length obtained by dispersion through the flange to web junction at a slope of (1 : 2.5) to the plane of the flange

t_w = Thickness of web

f_{yw} = Yield stress of the web

Bearing stiffeners should project nearly as much as the overhange of the flange through which the load is transferred.

19.15 INTERMEDIATE STIFFENERS

The intermediate stiffeners are used for the economical design of the web plate of the plate girder.

The intermediate stiffeners are used to avoid diagonal buckling of the web. Depending upon the ratio of clear depth to the thickness of web (d/t_w), vertical stiffeners, or vertical and horizontal stiffeners are provided throughout the length of the girder. In Art. 18.4, the diagonal buckling of web, it is seen that a complex and variable stress pattern is set up in the web due to combined bending moment and shear. The inclined compressive stresses are set up in the direction at right angles to the compressive stresses and at the neutral axis, the inten-sities of the stresses are numerically equal. The direction of inclined compressive stresses varies from 45° at the neutral axis to a considerably flatter angle at the edge of the flange. Due to this, the web plate tends to buckle in one direction and tends to straighten in the right angle direction. The intermediate stiffeners prevent the web plate from buckling under the inclined compressive stress. The vertical intermediate stiffeners divide the web plate into small panels. These panels are supported along the lines of stiffeners. The resistance of web plate to buckling is measurably increased. The intermediate vertical stiffeners also have a second function. When the vertical stiffeners are fitted against the top and bottom flanges, then, they maintain the original 90° angle between the flanges and the web. When the dimensions of web are very large, then the panel dimensions are reduced by providing the horizontal stiffeners on the compression side of the web.

Unless the vertical stiffeners are closely spaced, these are not sufficient in increasing the resistance to buckling in bending. The horizontal stiffeners placed on the compression zone of the web, effectively, increase its resistance to buckling due to bending. A combination of both vertical and horizontal web stiffeners is often used depending upon (d/t_w) ratio. When the web plate is thoroughly stiffened, it becomes effective upon permissible stresses in shear.

19.16 VERTICAL (TRANSVERSE) STIFFENERS

The vertical stiffeners are also termed as **transverse stiffeners**. The vertical stiffeners are provided throughout the length of the girder when the thickness of web is less than the limits specified for the minimum thickness of the web plate. Intermediate transverse stiffeners are used to improve the buckling strength of a slender web due to shear. These stiffeners remain effective even after the web has bucked. These stiffeners provide anchorage for the tension field. So also prevent the flanges moving towards one another. They are joggled or crimpled as shown in Fig. 19.22. They may be provided straight. In that case, filler plate of thickness equal to that of flange angles is inserted between the stiffener and web plate. They are fitted tightly between outstanding legs of top and bottom flange angles. When two angles are used for the vertical stiffeners, they are provided on either side of web. When single angle is used for the vertical stiffeners, then they are placed alternately on opposite sides of the web.

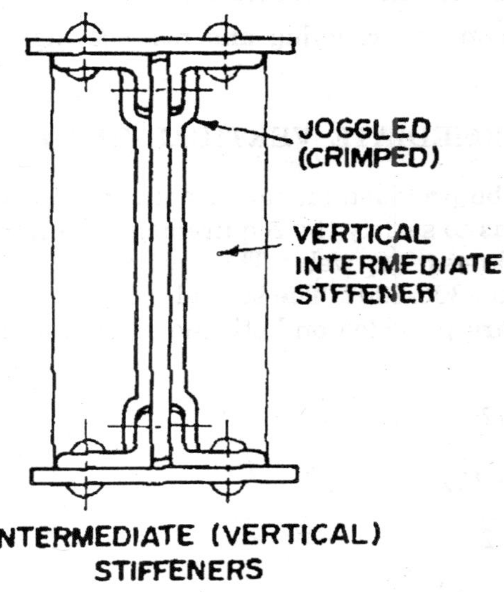

Fig. 19.22.

The vertical stiffeners are provided at spacing not greater than 1.5 d and not less than 0.33 d, where, d is the distance between flanges angles or where there are no flange angles, the clear distance between flanges ignoring fillets. When the horizontal stiffeners are provided, d shall be taken as the clear distance between the horizontal stiffener and the tension flange (farthest flange) ignoring fillets. The vertical stiffeners divide the web plate into number of panels of dimensions c and d as shown in Fig. 18.6. The panels are supported along the lines of stiffeners. The greater unsupported clear dimension of web panel should not be greater than 270 times the thickness of web, and the lesser unsupported clear dimension of the same web panel should not be greater than 180 times the thickness of web. These limiting values of panel dimensions have been adopted by keeping the aspect ratio a/b as 1.5 and to avoid the web buckling. The spacing of stiffeners is kept less near the supports, where the shear force is more. The spacing in increased towards the centre since the shear force decrease towards the centre. Actual average shear stress in the web should not exceed the allowable average shear stress in the stiffened web.

The length of outstanding leg of vertical stiffener may be taken equal to 1/30 of the clear depth of girder plus 50 mm. The outstand of stiffener from the web shall not be more than $256.t/\sigma_y^{1/2}$ (where, t is the thickness of the section, so long as the outer edge of each stiffener is not continuously stiffened, or 12 t for the flat sections. The length of connected leg of vertical

stiffener should be sufficient to accommodate the rivets connecting the stiffener to the web. The moment of inertia *I* of the stiffener selected should not be less than

$$I < \left(1.5\, d^3\, t_w^3 / c^2\right)$$

where, *I* = The moment of inertia of a pair of stiffener about the centre line of web, or of a single stiffener about the face of the web.

t_w = The minimum required thickness of the web.

c = The maximum permitted clear distance between vertical stiffeners.

These stiffeners are non-load carrying and non-bearing stiffeners. These are called as stability stiffeners.

19.17 DESIGN OF INTERMEDIATE VERTICAL (TRANSVERSE) WEB STIFFENERS

These stiffeners may be provided on one or both the sides of web. The spacing between these stiffeners is kept so as to satisfy the requirement of minimum thickness, which onward depends upon the thickness of web. The outstand is kept so as to satisfy the recommendations as per IS : 800-2007. The second moment of aream, I_s about the centre-line of the web in case, these are provided on both the sides of web, otherwise about the face of web, such that

(i) **In case (c/d) > = $\sqrt{2}$**

$$I_s >= 0.75\ (d_w\, t_w^3)$$

(ii) **In case (c/d) < $\sqrt{2}$**

$$I_s >= 1.5 \left(\frac{d_w^3\, t_w^3}{c^2}\right)$$

In case, c = $\sqrt{2}\ d_w$ from the expression (ii) $c^2 = 2\ d_w^2$.

Substituting for I_s = $\left(\frac{1.5\, d_w^3\, t_w^3}{2\, d_w^2}\right)$ = 0.75 dw. ^3_w as in expression (i)

where,

d_w = Depth of the web

t_w = Minimum required web thickness for the spacing using tension field

c = Actual spacing of the stiffener

These are checked for the bearing check. for force in a stiffener

$$F_q = \left(\frac{V - V_{cr}}{\gamma_{msy}}\right), \text{ so also}$$

F_q <= Fqd, where

F_{qd} = Design resistance of the intermediate stiffeners

V = Factored shear force adjacent to the stiffener and

V_{cr} = Shear, buckling resistance of web panel designed without using tension field

In additon to above check these stiffeners should satisfy following interaction expression

$$\left(\frac{Fq - Fx}{F_{qd}}\right) + \left(\frac{F_x}{F_{xd}}\right) + \left(\frac{Mq}{M_{yd}}\right) <= 1$$

In case, (Fq < Fx), then (Fq − Fx) should be taken as zero

F_q = Force in the stiffener

F_{qd} = Design resistance of these stiffeners corresponding to buckling about an axis parallel to web

F_x = External load or reaction at the stiffener

F_{xd} = Design resistance of a load carrying stiffener corresponding to buckling about axis parallel to the web

M_q = Moment on the stiffener due to eccentricity applied load and the transverse load, if any, and

M_{yq} = Yield moment capacity of the stiffener based on its elastic modulus about its centroidal axis parallel to web.

19.18 CONNECTION OF INTERMEDIATE STIFFENERS TO WEB

The stiffeners are connected to web as to withstand a shear between each component of the stiffener and the web (in kN/mm) of not less than $t_w^3/(5b_s)$, where b_s is the **outstanding width of the stiffener**, in mm and t_w is the web thickness in mm.

For the stiffeners subjected to external loading, the shear between the web and the stiffener due to such loading has to be added to the above value.

Stiffeners not subjected to external loads or moments may terminate clear of the tension flange and in such a situation, the distance cut short from the line of weld should not be more than $4t_w$.

19.19 HORIZONTAL (LONGITUDINAL) STIFFENERS

The horizontal stiffeners are also termed as ***longitudinal stiffeners***. The buckling resista-nce of the web is increased by providing longitudinal stiffener. When the thickness of web is less than the limits specified (for minimum thickness of web plate), are provided on the web at a distance from compression flange equal to 2/5th of distance of compression flange to the neutral axis i.e., at (1/5)th x d, where d is the clear depth of the plate girder. The stiffener is close to the crest in the buckles of unstiffened plate. The value of K_s of 129.4 may be obtained. The moment of inertia *I*, of the horizontal stiffener should not be less than $4c_1 . t_w^2$. [Reference: Allen and Bulson, 1980].

where, c_1 is the actual distance between the transverse (vertical stiffeners.)

When the thickness of web is less than the limits specified, second horizontal stiffeners (single or double) in addition to vertical stiffeners and horizontal stiffeners described above the provided at the neutral axis of the girder. The moment of inertia '*I*' of this horizontal stiffeners should not be less than $d^2.tw^3$. The moment of inertia is computed about the line as described for vertical stiffeners. The outstand of the stiffeners from the web shall not be more than 256.t/ $\sigma_y^{1.2}$ (where, *t* is the thickness of the section) or 12 t for the flat sections. d_2 is twice the clear distance from the compression flange angles/plates to neutral axis.

The requirements for moment of inertia for vertical stiffeners are based on the theoretical and experimental work. Timoshenko used a value of 0.3 in the expression of moment of inertia for the vertical stiffeners. Whereas Moore, R.L. proposed a value of 1.33 based on his investigation on effectiveness of the stiffeners on shear resistance of plate girder webs. Moore's this value was found to be adequate by ultimate load test, but it has been raised to 1.5. The expressions for moment of inertia of horizontal stiffeners are based on theoretical work and it is due to Bleich, F.

Horizontal web stiffeners may extend between vertical stiffeners but need not to be continuous over then same stiffeners may perform more than one function. These should be designed to comply with the requirements of all the functions.

Connection of intermediate stiffeners to web plate. The intermediate vertical stiffeners and horizontal stiffeners (when they are not subjected to external load) are connected to the web by rivets to withstand shear force not less than

$$\left(125\, t_w^2/h\right) kN/m$$

where, h is the outstand of stiffener in mm and t_w is the thickness of web in mm.

Same stiffeners may perform more than one function. These should be designed to comply with the requirements of such functions.

External Forces on intermediate stiffeners. In case, the vertical loads are acting eccentric or transverse forces are acting on the vertical intermediate stiffeners causing bending moment and shear, the moment of inertia of the stiffeners mentioned above shall be increased as under :

(a) **The vertical load acting eccentric** with respect to the vertical axis of the web causing bending moment

$$\text{Increase of } I = \left(\frac{150\, MD^2}{E.t_w}\right) \times 10^4 \text{ mm}^4$$

(b) **The lateral loading of stiffener**

$$\text{Increase of } I = \left(\frac{0.3\, VD^3}{E.t_w}\right) \times 10^4 \text{ mm}^4$$

where, M = Applied bending moment in kN-m

D = Overall depth of girder in mm

E = Young's modulus of elasticity, 2.047 x 105 N/mm² (MPa)

t_w = Lateral force to be taken by the stiffener and deemed to be applied at the compression flange of girder in kN.

19.20. IMPORTANT PARAMETERS NECESSARY FOR DESIGN OF STIFFENERS

Following parameters are important, and necessary which should be known and considered for the design of various stiffeners

1. Stiff bearing length
2. Outstand of the stifener from the face of the web
3. Eccentricity of loading with respect to the centroid of the stiffener

1. **Stiff bearing length** of any element, b_f is defined as that length which cannot deform appreciable in bending. It is obtained by considering the dispersion of the load through steel bearing at 45°, through a solid material such as the bearing plates, flange plates as shown in Fig. 19.23 (A), (B) and (C).

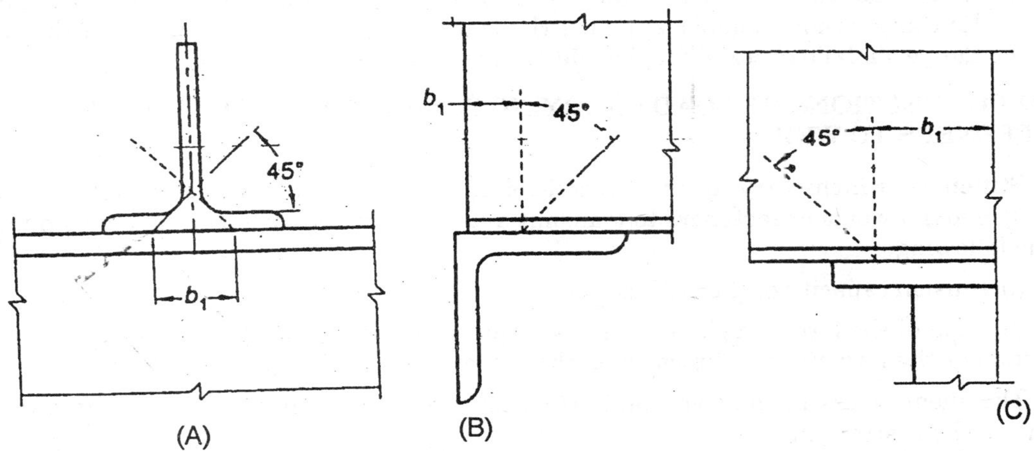

Fig. 19.23.

2. **Outstand of web stiffeners :** Unless the outer edge of the stiffener is continuously stiffened, the outstand from the face of the web should not exceed 20 tq. ε. (whereas, tq is the thickness of the stiffener).

When the outstand of web is between 14 tq.ε and 20 tq. ε, then the stiffener designed should be on the basis of a 14 tq. ε (where, ε epsilon is the yield stress ratio $(250/\sigma_y)^{1/2}$.

3. **Eccentricity of loading with respect to the centroid of the stiffener :** In case, a load or a reation is acting eccentrically to the centre line of the web (i.e, the centroid of the stiffener does not lie on the centre line of the web, the resulting eccentricity of load should be considered in the design of the stiffener.

19.21 BUCKLING RESISTANCE OF STIFFENERS

The buckling resistance, F_{qd} is based on the design compressive stress, f_{cd} of a strut (curve as in IS : 800-2007). **The radius of gyration** is taken about the axis parallel to the web. The effective section is full area or the core area of the stiffener together with an effective length of web on each side of the centreline of the stiffeners limited to 20 times the thickness of the web. The design strength used should be the minimum value obtained for buckling about the web or the stiffener.

The effective length for the intermediate transverse stiffeners used in calculating the buckling resistance, F_{qd} should be taken as 0.7 times the length, L of the stiffener.

The effective length for the **load carrying web stiffeners** used to find the buckling resistance, F_{qd}, assumes that the flange through which the load or reaction is applied is effectively restrained against lateral movement relative to the other flange, and it is taken as

(i) KL = 0.7 L, when the flange is restrained against rotation (Clamped/fixed) in the plane of the stiffeners by other structural member

(ii) KL = L, when the flange is not restrained, where,

L is the length of the stiffener.

In case, the load or reation is applied to the flange by a compression member, then unless effective lateral restraint is provided at this point, the stiffener should be designed as part of the compression member applying the load, and the connection between the column and the flange shall be checked for the effects of strut action.

19.22 CONNECTIONS OF LOAD CARRYING STIFFENERS AND BEARING STIFFENERS TO WEB

Stiffeners, which carry loads or reactions applied through a flange is connected to the web by enough welds or fasteners to transmit a design force which one is taken as minimum of the following.

(i) Tension capacity of the stiffener or

(ii) Sum of the forces applied at the two ends of the stiffener when these act in the same direction or the larger of the forces when these are acting in the opposite direction.

The shear stress in the web due to the design force transmitted by the stiffener is less than the shear strength of web.

Stiffeners resisting **the tension** should be connected to the flange with continuous welds or non slip fasteners.

Stiffeners resisting the **compression may either** be fitted against the loaded flange or be connected by continuous welds or non-slip fasteners. These stiffeners are to be connected to both the flanges in case.

(i) the load is applied directly over the support

(ii) the stiffener is an end stiffener of a web.

(iii) the stiffener also acts as a torsional stiffener.

When the shear is the controlling element for the longitudinally stiffened plate girder, a collapse mechanism (similar to one described previously) provided the stiffeners remained rigid upto failure [Reference : Narayanan 1983; Evans and Tang 1986]. Once a tension field develops, it spreads over ther complete depth of the girder. Therefore, equations expressed mathematically earlier are also valid. However, $\tau_{sh.cr.e}$ values are increased due to smaller panel dimension.

In the code of practice AS : 4100-1998 of Australia recommends that

$$I_s >= 4\ d_w\ t_w^3 \left[1 + \frac{4A_s}{d_w t_w}\left(1 + \frac{A_s}{d_w t_w}\right)\right]$$

where, A_s is the area of the stiffener.

Linear buckling theories are commonly used in the design of longitudinal stiffeners. The minimum value of stiffener rigidity is established. It is ensured that the buckling is limited to individual sub-panels of the web.

19.23 BEHAVIOUR OF TRANSEVERSE STIFFENERS AT ULTIMATE LOAD

Tension field produced in the web is supported by the flanges (at the top and bottom) and stiffeners (on the sides) which define its boundary. The web stiffeners help to obtain ultimate capacity of the plate girder. The buckling stresses are increased. During the post-buckling, these stiffeners support the tension field in the web. The flanges are prevented to move/pulled towards each other. The members of the boundary must be able to support the forces imposed due to tension field.

Initially, the transverse stiffeners are not subjected to axial load. After the buckling of the web plates, the axial loads act on the transverse stiffeners, which increase regularly as the membrane tension field in the web plate. [Rockey *et al.,* 1981]. The stiffeners possess sufficient rigidity so that the stiffeners remain straight. The web buckling is kept restricted to the individual panel.

A stiffener CD is shown in Fig. 19.24 attached to the web between two adjacent panels. **Shear failure mechanism** have developed in both these panels. For this transverse stiffeners, it is a most critical case of loading. Forces acting over the stiffeners have been identified by attaching a suffix, ss. Subscript ff is used to identify the forces in the flanges.

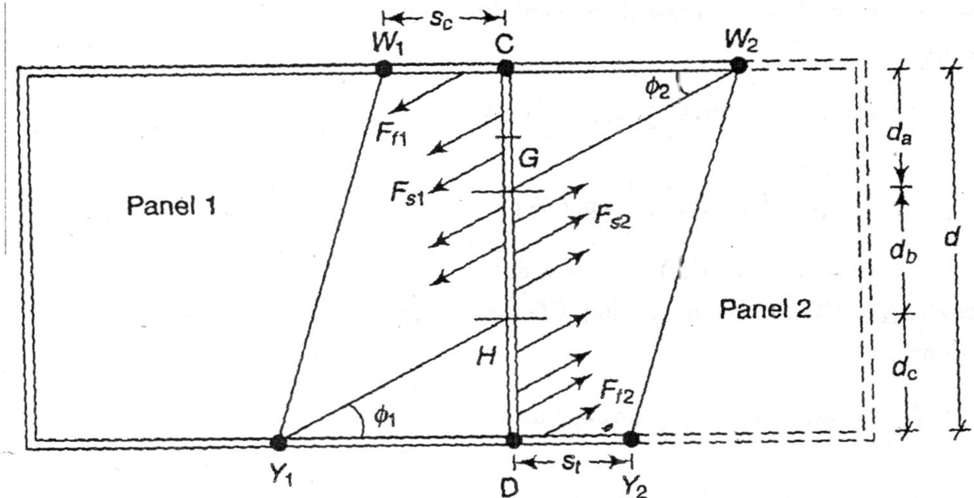

Fig. 19.24. Force imposed on transverse stiffeners by tension field

On the left top side of the stiffener resultant load acting on the portion W_1C, force F_{ff1} is inclined at an angle θ_1 (measured from horizontal).

On the right bottom side of the stiffener, resultant load acting on the portion $D^{1/2}$ force, F_{ff2} is also inclined at an angle θ_2 (measured from horizontal)

The vertical components of these two forces, V_c and V_D tend to pull the flanges towards each other.

From Fig. 19.20, components V_c and V_D applied by the flange, W_1C and $D^{1/2}$ to the stiffener are

$$V_c = -F_{ff1} . \sin\theta_1 = -p_{yt} . (t_w . s_t). \sin\theta_1 . \cos\theta_1$$

$$V_D = +F_{ff2} . \sin\theta_2 = +p_{ytw} . s_t . \sin\theta_2 . \cos\theta_2$$

where, $F_{ff1} = (p_{yt} . t_w . s_c) \cos\theta_1$ and

Negative sign shows that this load is applied in the direction from C to D (i.e, downward) and the positive sign shows that this load is applied in the direction from D to the horizontal components C (i.e, upward).

Moreover, the loading imposed directly upon the stiffener by web tension may be divided into three zones (i) Top part CG is subjected to a pull to the left by the left hand panel

(ii) Bottom part HD is similarly subjected to a pull to the right by the right hand side panel.

(iii) middle part GH is similarly subjected to the left and to the right (these two horizontal component of forces are practically balancing each other. As such this middle part GH remains practically unloaded by the tension field.

The vertical component of forces on the portions CG and DH are respectively found as

$$V_c = -p_{yt} . (t_w CG) . (\sin\theta_1) . (\cos\theta_1)$$

$$V_D = -p_{yt} . (t_w DH) . (\sin\theta_2) . (\cos\theta_2)$$

Where, p_{yt} is the tensile stress developed due to tension field action per unit area

That is,

$$V_c = -p_{yt} . (t_w . CG)\left(\frac{1}{2}\sin 2\theta_1\right) \text{ and}$$

$$V_D = -p_{yt} . (t_w . DH)\left(\frac{1}{2}\sin 2\theta_2\right)$$

$(t_w . CG)$ and $(t_w . DH)$ are the areas

Since, height CG is in compression, CG = s_c and height DH is in tension, DH = s_t

Therefore,

$$V_c = -p_{yt} . (t_w . s_c)\left(\frac{1}{2}\sin 2\theta_1\right) \text{ and}$$

$$V_D = -p_{yt} . t_w . s_t\left(\frac{1}{2}\sin 2\theta_2\right)$$

Where, s_c and s_t are the heights of band-width

Note : s = (CG + DH + GH) = Total band-width.

Forces acting over the stiffeners

Therefore,

Knowing the ultimate shear loading and the geometry of failure mechanism, the forces acting over the stiffeners may be obtained from above expressions.

Note : From the experimental studies Rockey in 1981 and Puthali in 1979 observed that the actual behaviour of the stiffeners is somewhat different from the simple model presented above (which was developed/seen earlier) along following lines

1. A portion of the web plates on each side of the stiffener acts is resisting the axial load, inspite of the fact that the web has theoretically yielded due to tension field. The effective cross-section is in the form of a T-section or a cruciform, in case, the stiffeners have beens provided on both the sides of the web plate. as shown in Fig. 19.25.

Rockey *et al.* 1981 pointed out that when a width of 40 t_w of the web is assumed to act with the stiffener, the design procedure suggested by them are in agreement with the test data.

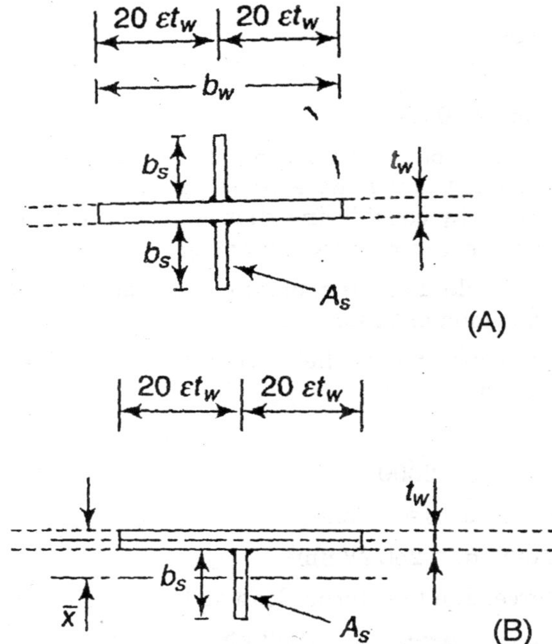

Fig. 19.25 (A&B)

2. The simple theory developed previously for the behaviour of the stiffener assumes that the loading applied on the stiffener is concentric.

However, the eccentricity of loading cannot be avoided practically.

The stiffener carries both axial load and bending moment arised due to eccentric and imperfections present in the stiffener. Load (eccentricity = $\bar{x}$ from the centroidal axis)

Horne 1979 gave an approximate expression to find the capacity of the stiffener,

$$\left(\frac{M}{M_{ps}}\right) = 1.0 \left(\frac{P_{ys} \cdot t_s \left(b_s - \bar{x} + 0.5\ t_w\right)^2}{M_{ps}} \quad \frac{P}{P_s}\right)$$

where,

$$\bar{x} = \left(\frac{(b_s t_s - b_w t_w)}{2\,t_s + (t_w/2)}\right)$$

where,

M_{ps} = Plastic moment capacity in the absence of loads

P_s = Squash load (i.e, full axial yield load)

t_s = Thickness of stiffener

p_{ys} = Design stress of stiffener

t_w = thickness of web

b_s = Width of the stiffener

b_w = Width of the web

(which may be assumed as 40 t_w)

A girder subjected to axial loads and accompanying moment, which are less the limit specified above may be designed using the expressions described earlier. The procedure is simplified by the design codes by considering that the compressive force in the stiffener is contant and full depth of web as a compression member.

Example 19.1. In Example 18.1, the plate-girder section has been designed for the bending strength (i.e, for the moment capacity)

Check the plate girder section for its shear strength (i.e, for shear capacity). **Use simple post-critical method Design :**

From **Example 18.1**

Depth of web-plate, d_w = 2000 mm

Thickness of web-plate, t_w = 10 mm

Yield stress for steel, σ_y = 250 N/mm^2

Factored shear force, due to external load (action), V_{fd} = 1957 kN

Step 1 : Factored shear force, V_{fd} = 1957 kN

As per IS : 800-2007, clause 8.4, for shear, the factored shear force, V_{fd} due to external load (action) should be less than or equal to the design shear strength, (It is requirement)

$$V_{fd} <= V_d = \left(\frac{V_n}{\gamma_{msy}}\right)$$

γ_{msy} = Partial safety factor for the strength of steel against failure = 1.15.

Step 2 : Nominal plastic shear-resistance

Reference may be made in IS: 800-2007 in order to calculate **resistance to shear buckling**

$$\text{Ratio}\left(\frac{d_w}{t_w}\right) = \left(\frac{2000}{10}\right) = 200$$

This ratio has been kept large, and the web-plate has been kept thin so that the design of stiffeners may be illustrated. Jointly, a complete design may be explained.

(i) when $(d_w/t_w) <= 67\,\varepsilon = 67$, then

the web plate does not need stiffeners. However,

(ii) When $(d_w/t_w) > 67\,\varepsilon \left(\frac{K_v}{5.35}\right)^{1/2}$, the web plate needs stiffeners.

In Example 18.1 $(d_w/t_w) = 200$

It is more. There the web plate needs intermediate stiffeners.

Note : Simple post critical method (i.e, tension-field action is not to be used)

Provide 8 panels, @ spacing of the trasnverse stiffeners c = 2500 mm

Ratio $(c/d_w) = (2500/2000) = 1.25$

Shear buckling coefficient for $c/d_w > 1.0$

$$K_v = 5.35 + (4 / 1.25^2)$$

$$= (5.35 + 2.56) = 7.91$$

Step 3 : Non dimensional slenderness ratio, λ_w for the web

$$\lambda_w = \left(\frac{\sigma_{yw}}{\sqrt{3}\ \tau_{cr.e}}\right)^{1/2}, \textit{ where}$$

Elastic critical shear stress for the web

$$\tau_{cr.e} = \left(\frac{K_v \cdot \pi^2 E}{12(1-\mu)(d/t_w)^2}\right)$$

Poisson's ratio, $\mu = 0.3$

Substituting the values

$$\tau_{cre} = \left(\frac{7.91 \times 9.85\ \ 96 \times 2 \times 10^6}{12(1-0.3)\,200 \times 200}\right) = 46.42 \text{ N/mm}^2$$

Average shear stress

$$\tau_{av} = \left(\frac{1957 \times 1000}{2000 \times 10}\right) = 97.85\ N/mm^2$$

$$\lambda_w = \left(\frac{\sigma_{yw}}{\sqrt{3} \cdot \tau_{cr.e}}\right)^{1/2} = \left(\frac{250}{1.732 \times 46.42}\right)^{1/2} = (3.109)^{1/2}$$

$$= 1.7633$$

For $\lambda_w >= 1.2$

Shear stress corresponding to web buckling

$$\tau_b = \left(\frac{\sigma_{yw}}{\sqrt{3}\,.\lambda_w^2}\right) = \left(\frac{250}{1.732 \times 1.7633^2}\right) = 46.427 \text{ N/mm}^2$$

Note : Values of critical shear stress $\tau_{cr.e}$ = 46.72 N/mm² and shear stress corresponding to web buckling are practically equal

Shear force corresponding to buckling

$$V_{cr} = \left(\frac{\tau_b \times d_w \times t_w}{1000}\right) kN = \left(\frac{46.427 \times 2000 \times 10}{1000}\right) = 464.27\ kN$$

As per IS : 800-2007, clause 8.7.1.1 (a), intermediate stiffeners are needed to improve the buckling strength of web due to shear

Example 19.2 In Example 18.1, the reaction at each support is 1957 kN Design bearing and load carrying stiffeners at the supports.

Design : From Example 18.1, the reaction at each support = 1957 kN

Assume width of supports 300 mm

Step 1 (i) Minimum stiff bearing length

Provided by the support, $b_1 = \left(\frac{300}{2}\right) = 150\ mm$

(ii) Dispersion length (1 : 2. 5)

Thickness of tension flange = 36 mm

$\eta_2 = (2.5 \times 36) = 90$ mm

Step 2 : Local capacity of the web

$$F_w = \left(\frac{(b_1 + \eta_2)\,.\,t_w\,\sigma_{yw}}{\gamma_{msy}}\right)$$

$$= \left(\frac{(150 + 90) \times 10 \times 250}{1.15 \times 1000}\right) = 521.739 kN$$

Local capacity of web, F_w = 521.739 kN is much less than the reaction at supports 1957 kN

Hence, at both the supports, the load carrying stiffeners are needed. As per IS : 800-2007, clause 8.7.1.2 and 8.7.1.3.

Effective section = Core area of stiffener + effective length of the web on either side of the centre line of the stiffener (= 20 t_w)

Step 3 : Load carrying stiffener

Try a pair of flats = 250 mm x 16 mm

Yield stress ratio, (epsilon)

$\varepsilon = (250 / 250)^{1/2} = 1.0$

Maximum outstand of the stiffener

= (20 x 16 x 1.0) = 320 mm

= (14 x 16 x 1.0) = 224 mm

Core area should be reduced to 2 x 224 mm x 16 mm

Local buckling resistance

Area = (2 x 224 x 16 + 20 x 16 x 10)

= (7168 + 3200) = 10368 mm²

Moment of inertia about axis of web

$$I_{xx} = \left[\left(\frac{16 \times 448^3}{12}\right) + \left(\frac{20 \times 16 \times 16^3}{12}\right)\right]$$

$199.887 \times 10^6 + 1.09 \times 10^6) = 120.977 \times 10^6\ mm^4$

Radius of gyration

$$r_{xx} = \left(\frac{120.977 \times 10^6}{10.360 \times 10^3}\right)^{1/2} = (116.773)^{1/2} = 108.06\ mm$$

$$\lambda = \left(\frac{0.7 \times 2000}{108.06}\right) = 12.95$$

From Table 9c (IS : 800 - 2007)

$\lambda = 10, \quad \sigma_{cd} = 227\ N/mm^2$

$\lambda = 20, \quad \sigma_{cd} = 224\ N/mm^2$

$$\lambda = 12.95, \quad \sigma_{cd} = (227 - \frac{2.95 \times (227 - 224)}{10}) = 226.15\ N/mm^2$$

Buckling resistance of the stiffener

$$= \left(\frac{226.15 \times 10{,}360}{1000}\right) = 2342.55\ kN$$

> 1957 kN

Hence, the load carrying stiffener is safe.

Example 19.3 : In Example 18.1, the section for the plate girder has been proposed for the bending strength (moment capacity)

Check the end-panel for its shear strength (shear capacity) without using tension field action.

Design :

Step 1 : Anchor forces (Check for shear)

From IS : 800-2007, Clause 8.5.3

Resultant longitudinal shear, R_{tf} and a moment, M_{tf}

$$R_{tf} = (Hq/2) \text{ and } M_{tf} = \left(\frac{Hq.\,d}{10}\right)$$

where, $H_q = 1.25\ V_p \left(1 - \frac{V_{cr}}{V_p}\right)^{1/2}$

$$V_p = \left(\frac{\sigma_y \cdot d_w \cdot t_w}{\sqrt{3}}\right)$$

d_w = Depth of web

t_w = Thickness of web

$$= \left(\frac{250 \times 2000 \times 10}{1000 \times 1.732}\right) = 2886.836\ kN$$

From Example 19.2

V_{cr} = Local capacity of the web

= 521.739 kN

$$H_q = 1.25 \times 2886.836 \left(1 - \frac{521.739}{2886.836}\right)^{1/2}$$

$$= 3608.545 \quad (1-1807)^{1/2}$$

$$= 3608.545 \times (0.8196)^{1/2}$$

$$= (3608.545 \times 0.9051) = 3266.221\ kN$$

$$R_{tf} = (H_q / 2) = \left(\frac{3266.221}{2}\right) = 1633.11 kN$$

$$A_v = (d_w.\ t_w) = (2000 \times 10) = 20{,}000\ mm^3$$

$$V_n = \left(\frac{\sigma_{yw} \cdot A_v}{\sqrt{3}\ \gamma_{msy}}\right) = \left(\frac{250}{1.732} \times \frac{20000}{1000}\right)$$

$$= 2886.836\ kN > 1957\ kN$$

The end panel is safe to carry shear due to anchoring forces

Step 2 : Anchoring forces (check for the moment capacity)

From IS : 800-2007, Clause 8.5.3

$$M_{tf} = \left(\frac{H_q\ .d}{10}\right) = \left(\frac{3266.221 \times 2000 \times 1000}{10 \times 1000 \times 1000}\right)$$

$$= 326.622\ kN\text{-}m$$

$$\text{Let } y = \left(\frac{c}{2}\right) = \left(\frac{Spacing\ of\ stiffener}{2}\right)$$

$$= \left(\frac{2500}{2}\right) = 1250\ mm$$

$$I = \left(\frac{1}{12} \times 10 \times (2500)^3\right)$$

$$I = \left(\frac{1}{12} \times 10 \times 15.625 \times 10^9\right) = 1.302 \times 10^{10}\ mm^4$$

From theory of simple bending

$$\left(\frac{M_q}{I}\right) = \left(\frac{\sigma_y}{y}\right)$$

$$M_q = \left(\frac{I \times \sigma_y}{y \times \gamma_{msy}}\right) = \left(\frac{1.302 \times 10^{10} \times 250}{1250 \times 1.15 \times 10^6}\right)$$

$$= \left(\frac{1.302 \times 10^6}{1250 \times 1.15}\right) = 905.74 \quad kN-m$$

M_{tf} = 326.622 kN-m

Hence, the end panel is safe to carry moment.

Example 19.4 In Example 18.1, the section of a plate girder has been designed for the bending strength. (i.e, for the moment capacity.)

In Example 19.1, the section of plate girder has been checked for its shear strength (i.e, for the shear capacity). It has been seen that intermediate stiffeners are needed to improve the buckling strength of web due to shear

Design and check (i) the bearing stiffener as load carrying member (ii) the intermediate stiffeners

Design

Stiff bearing length, b_1 = 0 (assumed)

Local (load) capacity of the web,

F_w = 521.739 kN

Step 1 : Factored shear force at 2500 mm from the support, from Example 18.1

$$= \left(1957 - \frac{195.70 \times 2500}{1000}\right) kN$$

= (1957 – 489.25) = 1467.75 kN

Bearing stiffener is designed for

= (1467.75 – 521.739) = 946.011 kN

Bearing capacity of the stiffener alone

$$= \frac{250}{1.15} \times \left(\frac{2 \times 224 \times 16}{1000}\right)$$

$$= \left(\frac{250 \times 7.168}{1.15}\right) = 1550.26\ kN > 1467.25\ kN$$

Therefore, the stiffener is safe against buckling.

Step 2 : (ii) Intermediate stiffener at one panel distance (i.e, 2500 mm) from the support, is the most critical intermediate stiffener

Ratio (Spacing of the stiffener to the depth of web) $\left(\frac{c}{d_w}\right) = \left(\frac{2500}{2000}\right) = 1.25$

In case, (c/d_w) is less than $\sqrt{2}$

From IS : 800-2007, clause 8.7.2.4

$$\text{Minimum stiffeners, } I_s = \left(\frac{1.5\ d_w^3\ t_w^3}{c^2}\right)$$

$$= \left(\frac{1.5 \times (2000)^3 \times (10)^3}{1000 \times 2.500 \times 2.500 \times 1000}\right) = \left(\frac{1.5 \times 8 \times 10^{12}}{6.25 \times 10^6}\right) = 1.92 \times 10^6\ mm^4$$

Provide moment of inertia more than I_s required = $1.92 \times 10^6\ mm^4$

Provide two flats 100 mm x 10 mm

$$I_s \text{ (provided)} = \left[\left(\frac{10 \times 210^3}{12}\right) - \left(\frac{10 \times 10^3}{12}\right)\right] = 7.714 \times 10^6\ mm^4$$

I_s (provided) = $7.714 \times 10^6\ mm^4$

more than I_s (required) = $1.92 \times 10^6\ mm^4$

Therefore, the proposed size of stiffeners is satisfactory. The proposed size is checked for outstand and buckling.

Step 3 Check for outstand

Size of the stiffeners = 100 mm x 10 mm

As per IS : 800-2007, clause 8.7.1.2,

Outstand should not exceed

$20\ t_q.\ e = (20 \times 10 \times 1.0) = 200$ mm

Outstand of the core section

$14\ tq.\ e = (14 \times 10 \times 1.0) = 140$ mm

Outstand of the proposed stiffeners is b_s = 100 mm is less than 140 mm. Therefore, the outstand requirement is satisfactory.

Step 4 : Check for buckling

IS : 800 - 2007, clause 8.7.2.5

Force in the stiffener

$$Fq = \frac{(V - V_{cr})}{\gamma_{msy}} <= F_{qd}$$

where,

F_{qd} = Design resistance strength of the intermediate stiffener

V = Factored shear force adjacent to the stiffener

V_{cr} = Shear buckling resistance of the web panel designed without using the tension field action.

Shear force at 2500 mm from the support = 1467.75 kN

Local bearing capacity of the web = 521.739 kN

$$F_q = \left(\frac{1467.75 - 521.739}{1.15}\right) = \left(\frac{946.011}{1.15}\right) = 822.618\ kN$$

Step 5 : Buckling resistance of intermediate stiffeners

Size of stiffeners = 2 x 100 mm x 10 mm

Area of stiffeners =(2 x 100 x 10)

= (2 x 100 x 10) + Available portion of web

= 2000 + (20 + 10 + 20 x 10) x 10 mm²

= (2000 + 4100) = 6100 mm²

Moment of inertia of the stiffener about centre line of web

$$= \left[\left(\frac{10 \times 210^3}{12}\right) + \frac{(20 + 20 + 10)10^3}{12} \times -\left(\frac{10 \times 10^3}{12}\right)\right] mm^4 = 7.717 \times 10^6 mm^4$$

Radius of gyration

$$r_x = \left(\frac{7.717 \times 10^6}{6.100 \times 10^3}\right)^{1/2} = \left(\frac{7.717}{6.100} \times 10^3\right)^{1/2}$$

$$= (1.265 \times 10^3)^{1/2} = 35.568 \text{ mm}$$

Effective length = (0.7 x 2000) = 1400 mm

Slenderness ratio

$$\lambda = \left(\frac{1400}{35.568}\right) = 39.36$$

$$\sigma_y = 250\ N/mm^2,\quad \lambda = 39.36$$

From IS : 800-2007, Table 9 (C)

(Buckling curve)

$$\sigma_{cd} =$$

$\lambda = 30$, $\sigma_{cd} = 211$ N/mm²

$\lambda = 40, \ \sigma_{cd} = 198 \text{ N/mm}^2$

$$\lambda = 39.36, \quad \sigma_{cd} = \left[198 + \frac{0.64 \times 13}{10}\right] = 198.08 \text{ N/mm}^2$$

Buckling resistance

$$= \left(\frac{198.08 \times 6100}{1000}\right) = 1207.8 \; kN > 822.618 \text{ kN}$$

Hence, satisfactory in buckling. Stiffener is not subjected to any external load.

Note : Example 19.1, 19.2, 19.3 and 19.4 have been worked-out and illustrated by using **simple post critical method**

Examples in `Chapter 23, `Design of Welded Plate Girders' have been worked out and illustrate by using **`Tension-field action method'**

19.24 WEB PLATES SUBJECTED TO COMBINED BENDING AND SHEAR

In a simply supported plate girder carrying uniformly distributed load, at cross-sections in between supports and the centre, the web-plate is subjected to combined bending moment and shear force. The cross-sections near the centre will be subjected to predominant bending moment and small shear force. Where, the cross-sections near the supports, will be subjected to predominant shear force and small bending moment.

The ultimate capacity is obtained by the interaction between the influences of bending moment and shear forces as shown in Fig. 19.26.

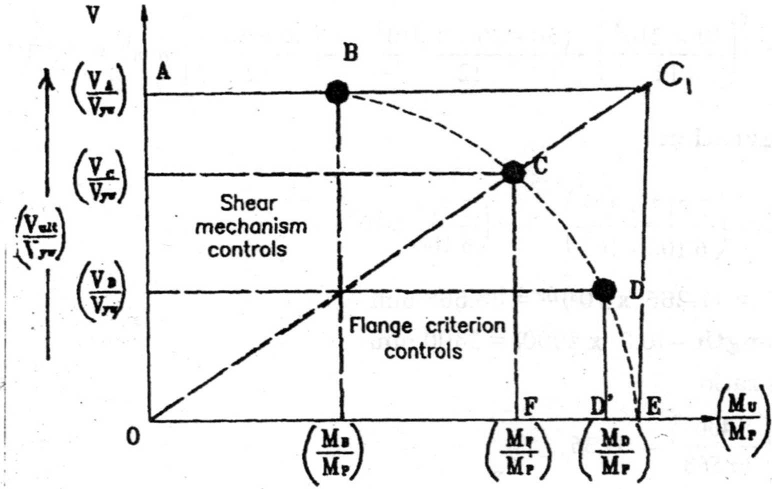

Fig. 19.26 (A)

The shear force capacity is plotted as an ordinate along y-axis and the bending moment capacity is plotted as an abscissa along x-axis. The co-existence values of shear and bending moment which may be sustained are seen as any point in the interaction diagram. The vertical ordinates are non-dimensional values use V_{yw} (yield shear of the web) and the horizontal values use M_p (fully plastic moment resistance of the section. The portio of curve between points A and C is the zone in which the girder will fail by predominant shear (i.e, shear mechanism)

Firstly, the stresses due to bending moment combined with the stresses due to shear give low buckling load.

Secondly, in the post-buckling range, the bending stresses influence the values of the tension field membrane for developing yield in the web.

Lastly, the axial forces due to bending of plastic moment of the flanges. Due to combined effects of the bending moment and shear force, the states of stress become more complicate.

A simple and effective design procedure may be developed by making some assumptions on the interaction of moment and shear.

The vertical ordinate at A represents the shear capacity which reduces gradually due to the presence of co-existent bending moment. Beyond the point C, the value of bending moment is high. The failure may commence by the collapse of flanges by one of the following manner.

(i) By yielding of flange material

(ii) By inward buckling of the compression flange, or

(iii) By lateral buckling of the flange

The line OC represents distinct change in failure criterion. The left side of line OC represents shear failure and the right side of line OC represents failure in bending (in flexure)

In general, the bending failure may commence, when the applied moment, M_f is approximately equal to the plastic moment, M_p provided by the flange plates only, neglecting the contribution from the web.

$$M_F = M_p = \sigma_{yf} \, . \, b_f \, . \, t_f \, (D + t_f)$$

where,

b_f = Width of flange

t_f = Thickness of flange

D = Depth of web

σ_{yf} = Design stress of the flange material.

In the portion ABC, Fig. 19.26B, the presence of additional bending moment needs the consideration of the following three factors :

(i) The reduction in the web buckling stress due to the presence of bending stresses

(ii) The reduction of plastic moment capacity of the flanges due to the presence of the axial stresses due to the bending stresses in the flanges.

(iii) The influences of bending stresses on the value of membrane stress needed causing yield in the web.

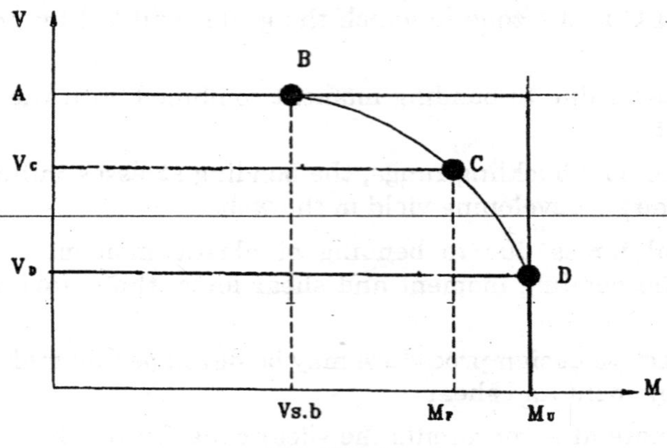

Fig. 19.26B

PROBLEMS

19.1 In Problem 18.1, the plate girder section is to be designed for the bending strength Check the plate girder for its shear strength. Use simple post critical method of `limit state design.'

19.2 In Problem 18.1, the plate girder section is to be designed for the bending strength. Design bearing and load carrying stiffeners at supports.

19.3 In Problem 18.1, the section for the plate girder is to be designed for the bending strength. Check the end panels for its shear strength.

19.4 In Problem 19.1, the section has to be checked for the shear strength. In case, the section needs intermediate stiffeners to improve the buckling strength of web due to shear, design and check the bearing stiffeners as load carrying member and (ii) the intermediate stiffeners

REFERENCES

1. Narayanan R. `Plated Structures, Stability and Strength' chapter one ``longitudinally and transversely reinforced plate girders (by Evans, H.R.) Elsevier, Applied Science Publishers, London, 1983

2. Owens, D.R.G. Skaloud, M. and Rockey, K.C. `Ultimate load behaviour of longitudinally reinforced Web Plates IABSE Memories 1970, Zurich 113-148.

3. Cooper, P.B. The ultimate bending moment for the plate girder. Proceedings of IABSE Collocum, London, 1971.

4. Mele M and Pathali, R. University of Rome Reports CECAR/136, 1979.

5. Rockey, K.C., Evans H.R. and Porter D.M. `A design method for predicting collapse behaviour of plate girders Proceedings of the Institution of Civil Engineers 1978, pp 85-112.

6. Rockey, K.C. and Skaloud, M. `Influence of flange stiffeners on load carrying capacity of Webs in shear. Proceedings of the Eighth Congress IABSE, New York.

7. Basler, K. 1961. `Strength of Plate Girders Under Combined Bending and Shear, ASCE, Journal Structural Division, October, Vol. 87

8. Porter, D.M., Rockey, K.C. and Evans, H.R. 1975. `The Collapse Behaviour of Plate Girders Loaded in Shear, structural Engineer 53 (8), 313-325.

9. Rockey, K.C. and Skaloud, M. 1972. `The Ultimate Load Behaviour, of Plate Girders in shear', Structural Engineer, 5011

10. Cooper, P.B. 1967. `Strength of longitudinally stiffened plate Girders' ASCE Journal of Structural Division, 93 (ST2), 419-452.

11. Basler, K. `Strength of Plate Girders in Shear, Transactions, ASCE 128, Part II (1963-719)

12. Basler, K. `Strength of plate Girders in Bending

13. Ballio, G. and Mizzolani, 1983. Theory and Design of steel structures chapman and Hall, London

14. Galumbos, T.V. (ed.) 1990. Guide to Stability Criteria for metal structures, 5th Edition John-Wiley and Sons, Inc New York, 911pp.

15. Basler, K. `Strength of Plate Girder in shear', Transactions, ASCE, 128, Part II (1963), 683-719 (Also as Paper No : 2967, Journal of Structural Division, ASCE, October, 1961).

16. Basler, K. `New Provisions for plate Girder Design', Proceedings, AISC, National Engineering Conference, New York, 1961, (650-74)

17. Basler, K., `Strength of plate Girders in Bending', Transactions, ASCE, 128, Part II (1963), 655-686, (Also as Paper No. 2913, Journal of Structural Division, ASCE, August, 1961).

18. Basler, K., `Strength of plate Girders, Under Combined Bending and Shear', Journal of structural Division, ASCE, 87 ST7 (October, 1961), 181-197.

19. Basler, K. and Thurlimann, Bruno. `Plate Girder Research., Proceedings AISC, National Engineering Conference, 1959.

20

WEB SPLICE AND SPLICE IN FLANGE ELEMENTS OF RIVETED PLATE GIRDERS (PG-III)

(Note : Working stress method has been used to explain web and flange splices)

20.1 INTRODUCTION

A joint in the web plate provided to increase its length is known as **web splice**. The plates are manufactured upto a limited length. When the maximum manufactured length of plate is insufficient for full length of a plate girder, web splice becomes essential. It also becomes essential when the length of plate girder is too long to handle conveniently during transportation and erection.

As far as possible, web splices may be located at sections where excess flange areas are available. The excess flange areas are available at sections prior to the curtailment of flange plates. Preferably, splices may be located under stiffeners. The stiffeners provide additional strength to the splice. The splices should not be located at the sections, where maximum bending moments occur. In case, only one splice may suffice for full length of the girder, it may be located at such sections.

The splices and cut-cuts for service ducts in the webs should preferably not located at points of maximum shear force and the heavy concentrated loads. The web splices shall preferably be made with complete penetration butt welds.

Where additional plates are needed to increase the strength of the web, these shall be placed on both the sides of web. These plates shall be equal in thickness.

The proportion of shear force assumed to be resisted by these plates shall be limited by the magnitude of horizontal shear which these plates may transfer to the flanges through the fastening. These fastenings shall be carried up to the points at which the flanges without the additional plates in sufficient.

The gap between the web plates and the flange plates shall be kept to a minimum and for fillet welds shall not exceed 1 mm at any point before welding.

The effective cross-sectional area of the web of plate girder shall be taken as the full-depth of the web-plate multiplied by the thickness.

The cross-section of the splice plates is normally kept 5 prevent larger than the flange elements that are spliced.

The connections are required at least 5 percent more load than the load at the spliced portion or at least 50 percent of the effective strength of the spliced portion.

The web splices are designed to resist the shears and moments at the spliced sections. The splice plates are provided on each side of the web. There are following three types of web splices, which are commonly used :

1. Rational splice 2. Moment splice 3. Shear splice.

20.2 WEB SPLICE (RATIONAL SPLICE)

This type of web splice is shown in Fig. 20.1. The stresses are transmitted directly in this type. This type is most satisfactory than other two types. The splice plates A as shown in Fig. 20.1 are provided between flange angles. A clearance of 6 mm is left between splice plates and flange angles. The splice plates B are provided for portion of web underneath the flange angles. If sufficient excess flange area is available at the splice section, the splice plates B need not be provided. These plates are designed for shear and moment, which would be resisted by the portion of the web, if the web was not spliced. The rivets are provided at uniform spacing in this type. The pitch of rivets connecting splice plates to the web is found as under :

Vertical shear per unit depth, $\tau_{v.cal} = (V/d_e)$ (i)

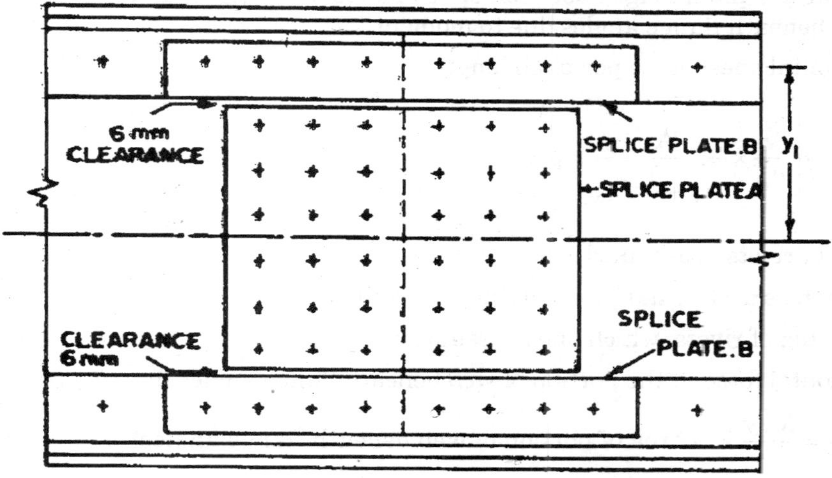

Fig. 20.1.

The bending stress upto the level of rivets connecting flange angles,

$$\sigma_{b.cal} = \left(\frac{M}{I}\right) . y_1 \quad \text{(ii)}$$

where, M = Bending moment at the splice section

I = Moment of inertia of the girder

y_1 = Distance upto the level of rivets connection flange angles from neutral axis.

The horizontal force per unit length d ue to moment = $(\sigma_{bc} . t_w)$

If the rivets are provided in one vertical row, and p is the pitch of rivets, the resultant of vertical and horizontal forces per pitch should not exceed the rivet value, R.

$$\text{Then,}\quad p \times \left[(\sigma_{b.cal} \times t_w)^2 + \left(\frac{V}{d_e}\right)^2\right]^{1/2} = R$$

$$\therefore\ p = R / \left[(\sigma_{b.cal} \times t_w)^2 + \left(\frac{V}{d_e}\right)^2\right]^{1/2} \quad (20.1)$$

If the computed pitch of rivets is less than minimum pitch, rivets are provided in two or three vertical rows. The rivets are provided at spacing of twice pitch computed above, if rivets are provided in two vertical rows. The rivets are provided at spacing of three times the pitch computed above if rivets are provided in three vertical rows. The thickness of splice plates A is kept equal to half the thickness of web, but not less than 6 mm. The width of splice plates A is kept sufficient to accommodate the rivets.

The area of portion of web beneath the flange angles is found. The splice plates B provide this area. The rivets connecting splice plates B to the flange angle area designed to resist horizontal shear per pitch length between splice plates and angles, and horizontal force in the portion of web beneath flange angles due to moment.

The horizontal shear force per pitch length

$$P_1 = \left(\frac{V}{d_e} \times \frac{A_f}{A_f + \frac{1}{6}A_w} \times p\right)$$

The pitch of rivets is assumed

A_f = Gross area of flange excluding web equivalent

p = Pitch of rivets, which one is assumed

The horizontal force in the portion of web beneath flange angles due to moment

$$P_2 = \frac{M.y}{I} \times \text{ (Area of portion of web beneath flange angles)}$$

where, y is the distance of rivets connecting splice plates B to the flange angles from the neutral axis.

If n is the number of rivets required,

$$n.\ R = n.\ P_1 + P_2,\ n = P_2/(R - P_1) \quad (19.2)$$

The rivets connecting splice places B to flange angles are provided at close spacing, so that their length width parallel to length is small.

Example 20.1. The bending moment and shear force at a particular section of a plate girder are 5760 kN-m and 1080 kN respectively. The cross-section of plate girder an that section is as shown in Fig. 20.2. Design the web splice.

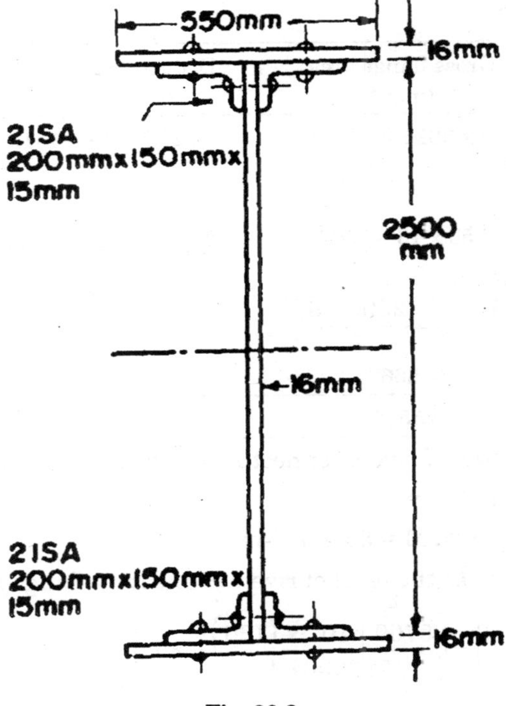

Fig. 20.2.

Design :

Step 1 : Effective depth of plate girder

Gross flange area = 25517 mm²

Net flange area = 21688 mm²

Let $\bar{y}$ be the distance of compression flange from top

$$\bar{y} = \left[\frac{2 \times 50.25\,(1.6 + 3.27) + 1.6 \times 55 \times 0.8}{2 \times 50.25 + 1.6 \times 55}\right] \times = 32.1 \text{ mm}$$

Effective depth of the plate girder

$d_e = (253.2 - 2 \times 3.21) \times 10 = 2467.8$ mm

Step 2 : Gross moment of inertia of the plate girder about xx- axis.

$$I_{xx} = \left[\frac{1}{12} \times 1.6 \times (250)^3 + 4 \times 969.2 + 4 \times 50.25 \times (125 - 3.72)^2 + 2 \times 55 \times 1.6 \times 125.8^2\right]$$

$\times\ 10^4$ mm^4

$I_{xx} = 7829002.1 \times 10^4$ mm^4

Step 3 : Area of flanges

The short legs of flanges are connected with the web the web plate. From ISI Handbook No. 1, the rivets are provided at 90 mm from top of the flange angles.

FLANGE AREA

Description	Gross flange area (mm^2)	Deduction for rivet hole (mm^2)	Net flange area (mm^2)
Flange Angles 2 ISA 200 mm x 150 mm x 15 mm	2 x 5025 = 10050	4 x 23.5 x 15 = 1410	8640
Flange plate 550 mm x 16 mm	550 x 16 = 8800	2 x 23.5 x 16 = 752	8048
Equivalent web area	$\frac{1}{6}A_w = \frac{2500 \times 6}{6}$ = 6667		$\frac{1}{8}A_w = \frac{2500 \times 6}{8}$ = 5000
Total	25517		21688

The distance upto the level of rivets connecting flange angles to web

(1250 – 90) = 1160 mm

Step 4 : Bending moment, M = 5670 kN-m

Step 5 : Bending stress at the level of rivets connecting flange angles

$$\sigma_{bc.cal} = \left(\frac{M}{I_{xx}} \cdot y_1\right) = \left(\frac{5760 \times 10^6 \times 1160}{7829002.1}\right) = 85.3\ N/mm^2$$

Thickness of web = 16 mm

Step 6 : Design of riveted connection

Use 22 mm diameter power driven rivets

Strength of rivets in double shear, $\left(2 \times \frac{\pi}{4} \times \frac{(23.5)^2 \times 100}{1000}\right) = 86.70\ kN$

Strength of rivet in bearing, $\left(\frac{23.5 \times 16 \times 300}{1000}\right) = 112.8\ kN$

Rivet value, R = 86.70 kN

Let p be the pitch of rivets, in case rivets are used in one row

$$p = \frac{R}{\left[(\sigma_{bc.cal} \times t_w)^2 + \left(\frac{V}{d_e}\right)^2\right]^{1/2}}$$

$$p = \frac{86.70 \times 1000}{\left[(85.2 \times 16)^2 + \left(\frac{1080 \times 1000}{2467.8}\right)^2\right]^{1/2}} = 60.49\ mm$$

Provide rivets in three rows at 165 mm pitch, using 13 rivets in each row.

Depth of splice plates A = (2200 – 2 x 6) = 2188 mm

Thickness of **splice plate** A = $\frac{1}{2}t_w = \frac{1}{2}\times 16 = 8\text{ mm}$

Width of splice pates A to accommodate rivets = 423.0 mm

This provides distance between adjacent rows of rivets = 3d and edge distance = 1.5 d, where d is the gross diameter of rivets used.

Step 7 : Splice plates B

Area of web plate under flange angles = (150 x 16) = 2400 mm²

Provide 130 mm wide plates on both sides

$$\text{Thickness of plates} = \left(\frac{2400}{130\times 2}\right) = 9.23\text{ mm}$$

Provide two plates 130 mm x 10 mm

The horizontal shear force per pitch length between flange angles and splice plates

$$P_1 = \left(\frac{V}{d_e}\times\frac{A_f}{A_f + \frac{1}{6}A_w}\times p\right)$$

Assume pitch of rivets p as 60 mm (p > 2.5 d minimum pitch)

$$p_1 = \left(\frac{1080\times 1000}{2467.8}\times\frac{18850}{1850+6667}\times\frac{60}{100}\right) = 19.40\text{ kN}$$

The horizontal force in the portion of web beneath the flange angles due to moment

$$P_2 = \frac{M}{I}.\,y\times(\text{Area of web beneath flange angles})$$

or $$= \left(\frac{5760\times 10^6}{7829002.1}\times\frac{1160\times 2400}{10^4\times 1000}\right) = 204.83\text{ kN}$$

Number of rivets required with 60-mm pitch, from Eq. 19.2.

$$n = \left(\frac{P_2}{R_1 - P_1}\right) = \left(\frac{204.50}{(86.70-19.40)}\right) = 4$$

The length of splice plates (i.e., width parallel length of plate girder) is 480 mm

The complete web splice is shown in Fig. 20.3.

20.3 WEB SPLICE (MOMENT SPLICE)

This type of web splice is shown in Fig. 20.4. There are four moment plates, (two on each face), marked as splice plates A and two shear plates (one on each face), marked as splice plate B. It is assumed that moment plates resist moment resisted by web, and shear plates resist shear resisted by web. In fact, each set of plate resist shear as well as moment. But in case of deep girders, shear resisted by splice plates A is small compared with plates B. Similarly the moment resisted by splice plates B is small compared with splice plates A. This type of splice may be used for girders about 2 metres deep. A clearance of 6 mm is provided between splice plates A and flange angles, and between splice plates A and splice plates B. The web splice (moment splice) is designed as under :

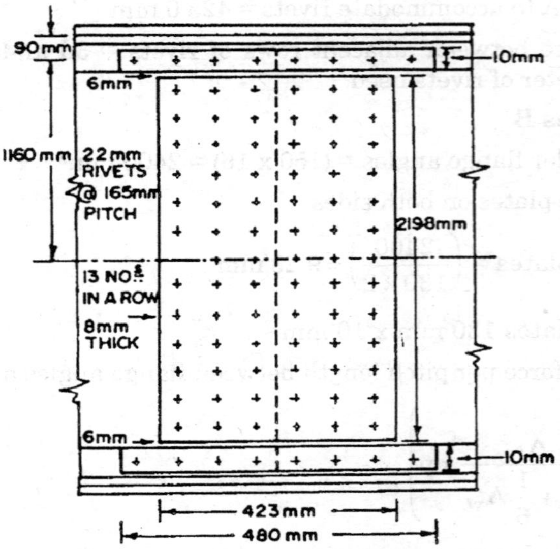

Fig. 20.3.

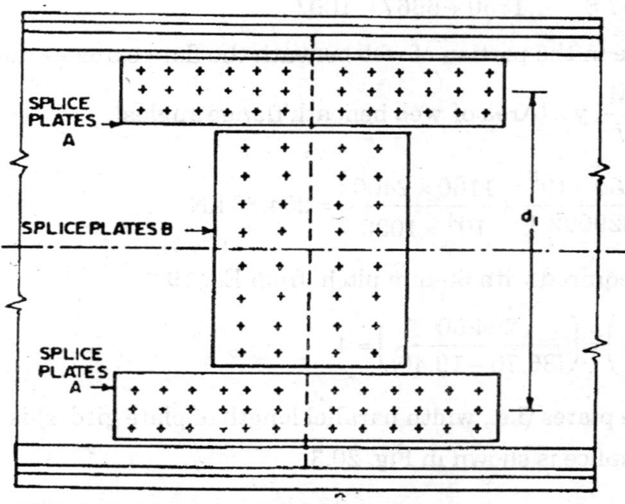

Fig. 20.4.

The moment resisted by the web plate is as under

$$M_w = M_w = \left(\frac{I_w}{I} \times M\right) \quad (20.3)$$

where, I_w = Gross moment of inertia of the web

I = Gross moment of inertia of the girder

M = Bending moment at the web splice section

Moment plates

The moment of resistance of four moment plates (splice plates A) is equal to moment resisted by web.

The moment of resistance of four moment plate plates is $A_s.\ \sigma_1 . d_1$

where, A_s = Net area of two moment plates

d_1 = Distance between centre to centre of splice plates A (Moment plates)

σ_1 = Bending stress at the centre of splice plate A, and assumed uniform in these plates.

Therefore, $M_w = A_s.\ \sigma_1 . d_1$ (i)

$$\sigma_{b1} = \left(\frac{d_1}{d_w} \times \sigma_{bc}\right) \quad \text{(ii)}$$

where, $\sigma_{bc.cal}$ is the bending stress at the extreme fibre of the web plate

$$\sigma_{bc.cal} = \left(\frac{M}{A_n . d_w}\right) \quad \text{(iii)}$$

$$\therefore M_w = \left(A_s . d_1 . \frac{d_1}{d_w} \sigma_{bc.cal}\right)$$

$$A_s = \left(\frac{M_w . d_w}{\sigma_{bc.cal} . d_1^2}\right) \quad (20.4)$$

Let t_w be the thickness of the moment plate

$$t_s = \frac{1}{2} . \frac{A_s}{(d_s - n_1)d} \quad (20.5)$$

where, d_s = Depth of moment plate

n_1 = Number of rivets in one vertical row

d = Diameter of rivets.

The horizontal force in the two moment plates A

= $(A_1 \times \sigma_{b1})$

The number of rivets required to connect moment plates A to the web plate on each side of web splice is given by

n = $(A_s.\ \sigma_{b1} / R)$ (20.6)

where, R = Rivet value.

Shear plates. The shear plate B resist shear at the web splice section. The combined thickness of these plates is designed to resist shear at web splice section. The width of the splice plates B is kept sufficient to accommodate rivets. The number of rivets, required to resist shear

$$n = (V/R)$$

where, V is the shear at the web splice and R is the rivet value.

These rivets are provided on each side of the web splice.

Example 20.2. Design the web splice given in Example 20.1.

Design :

Step 1 : Moment resisted by web plate

Gross moment of inertia of plate girder, $I_{xx} = 7829002.1 \times 10^4 \text{ mm}^4$

Gross moment of inertia of web plate $I_w = 2083334.3 \times 10^4 \text{ mm}^4$

The bending moment resisted by the web plate from Eq. 20.3.

$$M_w = \left(\frac{I_w}{I} \times M\right) = \left(\frac{20833333.3}{7829002.1} \times 5760\right) = 1535 \text{ kN-m}$$

The depth of web plate, d_w is 2500 mm, Net flange are $A_{fn} = 21688 \text{ mm}^2$

Step 2 : Moment plates

The depth of moment plates is kept sufficient to accommodate rivets in 3 rows. 22 mm diameter rivets are used at 100 mm spacing centre to centre and 50 mm edge distance.

The distance between centre to centre of moment plates,

$$d_1 = (2500 - 2 \times 152 - 2 \times 150 - 2 \times 6) = 1888 \text{ mm}$$

The bending stress at the extreme fibre of web plate,

$$\sigma_{bc.cal} = \left(\frac{5760 \times 1000 \times 1000}{21688 \times 2500}\right) = 106.23 \text{ N/mm}^2$$

The net area of two moment plates

$$A_s = \left(\frac{M_w \cdot d_w}{\sigma_{bc.cal}\ d_1^2}\right) = \left(\frac{1535 \times 10^6 \times 2500}{106.23 \times 1888^2}\right) = 10134.38 \text{ mm}^2$$

The thickness of moment plates as per Eq. (20.1)

$$t_s = \frac{1}{2}\left(\frac{A_s}{d_s - n_1\, d}\right) = \frac{1}{2} \times \left(\frac{10134.38}{(300 - 3 \times 23.5)}\right) = 22.08 \text{ mm}$$

Provide 18 mm thick plates.

Step 3 : Design of riveted connections

Strength of power driven rivets in double shear,

$$\left(2 \times \frac{\pi}{4} \times \frac{(23.5)^2 \times 100}{1000}\right) = 86.70 \text{ kN}$$

Strength of rivets bearing, $\left(23.5 \times 18 \times \frac{300}{1000}\right) = 126.9$ kN

Rivet value, R = 86.70 kN.

Number of rivets required to connect moment plates to web

$$\text{Bending stress } \sigma_{b1} = \left(106.23 \times \frac{1800}{2500}\right) = 80.22 \text{ N/mm}^2$$

$$\text{From Eq. (20.6), } \left(\frac{A_s \times \sigma_{b1}}{R}\right) = \left(\frac{18 \times 2\,(300 - 3 \times 23.5)}{86.70}\right) \times \frac{80.22}{1000} = 7.64$$

Provide 12 rivets on each side of web splice section

The distance between rows of rivets is 70 mm

Edge distance = 35 mm, and Width of moment plates = 560.0 mm

Step 3 : Shear plates

6 mm clearance is kept between moment plates and shear plates.

Depth of shear plates = (2500 – 2 x 150 – 2 x 300 – 4 x 16) mm = 1536 mm

Shear force at web splice = 1080 kN

Allowable shear stress = 0.4 f_y = 0.4 x 250 = 100 N/mm²

Thickness of two shear plates, $\left(\frac{1080 \times 1000}{1576 \times 100}\right) = 7.03$ mm

Thickness of one shear plates = 3.515 mm

Provide 6 mm thick shear plate

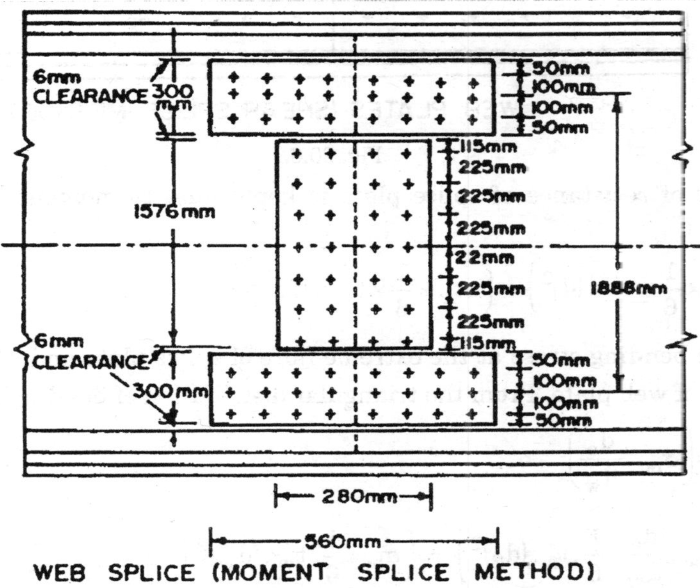

Fig. 20.5

Step 4 : Design of riveted connection

Number of rivets required (1080/86.70) = 12.45

Provide 14 rivets in two rows at 225 mm pitch.

Width of splice plate = 280.0 mm

The distance between rows of rivets in 70 mm

Edge distance = 35 mm

The complete web splice of the web plate is shown in Fig. 20.5.

20.4 WEB SPLICE (SHEAR SPLICE)

This type of web splice is shown in Fig. 20.6. The splice plates are provided between flange angles. A clearance of 6 mm is left between splice plates and flange angles. The web splice (shear splice) is designed as under :

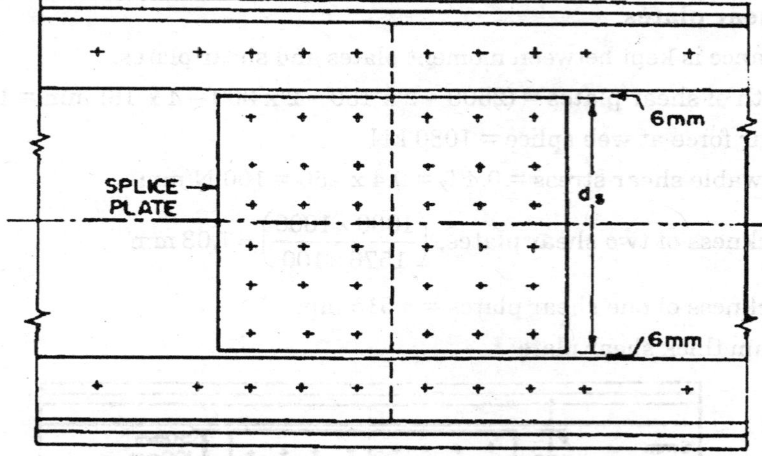

Fig. 20.6.

The moment of resistance of splice plate is kept equal to moment of resistance of web plates

$$\left(\sigma_{b1} \times \frac{1}{6} \times t_s \times (d)^2\right) = \left(\sigma_{bc} \times \frac{1}{6} \times t_w \times (d_w)^2\right)$$

where, σ_{b1} is the bending stress at the extreme fibre of splice plate and the bending stress at the extreme fibre of web plate. From the triangular distribution of bending stress,

$$\sigma_{b1} = \left(\sigma_{bc} \cdot \frac{d_e}{d_w}\right)$$

$$\therefore \left(\sigma_{bc} \cdot \frac{d_e}{d_w} \cdot \frac{1}{6} \cdot t_s \cdot (d_2)^2\right) = \left(\sigma_{bc} \cdot \frac{1}{6} \cdot t_w \cdot (d_w)^2\right)$$

$$\text{Thus} \quad A_s = A_w \left(\frac{d_w}{t_s}\right), \quad \text{where} \tag{20.7}$$

A_s and d_s = Area and depth of splice plates

A_w and d_w = Area and depth of web plates

Total thickness of splice plates

$$t_s = \left(\frac{A_s}{b_s}\right) = \left(\frac{\text{Area of the splice plate}}{\text{With of the splice plate}}\right)$$

The width of splice plate is kept sufficient to accommodate the rivets.

The splice plates are designed to resist shear and moment which would be resisted by web, if web was not spliced

$$M_w = \left(\frac{I_w}{I} \times M\right)$$

where I_w = Gross moment of inertia of the web

I = Gross moment of inertia of the girder

M = Bending moment at the splice section.

The vertical shear `V' at the splice section is shown in Fig. 20.7(A). It acts at an eccentric distance from c.g. of rivets, connecting splice plates to the web. It is considered equivalent to a vertical force `V' and an eccentric moment $M_e = V \times e$ as shown in Fig. 20.7(B).

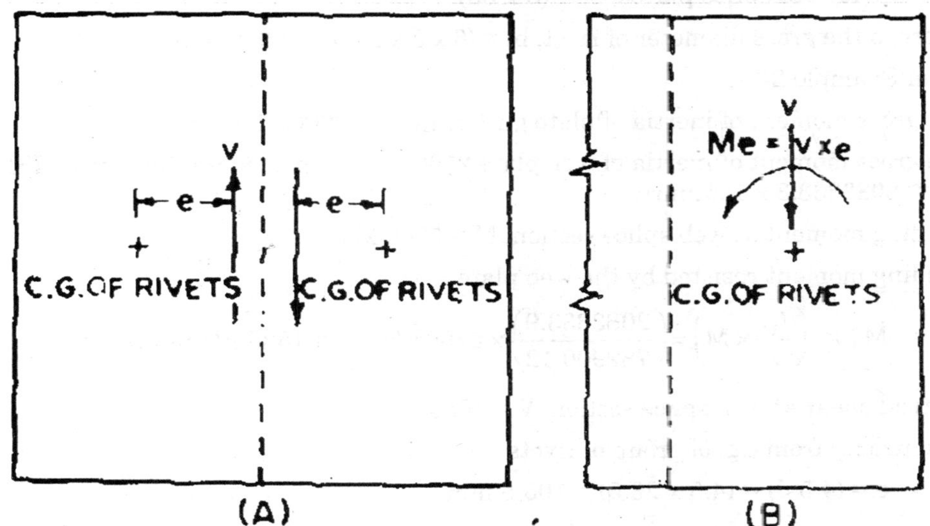

Fig. 20.7

The splice plates resist a total moment `M_s'.

$$M_s = (M_e + M_w)$$

The rivets connecting splice plates to the web are designed to resist a vertical force `V' and a moment `M_e'. This becomes design of eccentric riveted connections.

Example 20.3. Design the web splice given in Example 20.1.

Design :

Step 1 : Area of splice plate

The depth of web plate, d_w is 2500 mm

Depth of splice plates, d_e = (2500 – 2 x 150 – 2 x 6) = 2188 mm

Thickness of web plate, t_w = 1.6 mm

Area of web plate, $A_w = d_w . t_w$ = (2500 x 16) = 40000 mm²

Area of splice plate required, From Eq. 20.7.

$$A_s = A_w \left(\frac{d_w}{d_s}\right)^2 = 40000 \times \left(\frac{2500}{2188}\right)^2 = 52221\,\text{mm}^2$$

Step 2 : Total thickness of splice plates required

$$t_s = \left(\frac{A_s}{d_s}\right) = \left(\frac{52221}{2188}\right) = 23.887\text{ mm}$$

Provide two splice plates 12 mm thick one on each side of web plate. The width of the splice plate is kept to accommodate three rows of rivets on either side of splice section.

Step 3 : Width of splice plate

Use 22 mm power driven rivets

Width of splice plates, b_s = (6 x 3d)

where, d the gross diameter of rivet, b_s = (6 x 3 x 23.5) = 423.0 mm

From Example 20.1,

The gross moment of inertia of plate girder, I_{xx} is 7829002.1 x 10⁴ mm⁴

The gross moment of inertia of web plate 2500 mm deep x 16 mm thick from ISI Handbook No. 1, I_w = 2083333.3 x 10^4 mm^4

Bending moment at web splice section, M = 5760 kN-m

Bending moment resisted by the web plate

$$M_w = \left(\frac{I_w}{I} \times M\right) = \left(\frac{2083333.9}{782900.12}\right) \times 5760\,\text{kN-m} = 1535\text{ kN-m}$$

Vertical shear at web splice section, V = 1080 kN

Eccentricity from c.g. of group of rivets

e = (4.5 d) = (4.5 x 23.5) = 105.5 mm

$$\text{Eccentric moment, } M_e = \left(\frac{1080 \times 105.5}{1000}\right) = 113.4\,\text{kN-m}$$

Splice plates resist a total moment

$M_s = (M_e + M_w)$ = (113.4 + 1535) = 1648.4 kN-m

Step 5 : Design of riveted connections

Group of rivets resist a moment, = 1648.4 kN-m

and a vertical force = 1080 kN

Strength of power driven rivet in double shear,

$$\left(2 \times \frac{\pi}{4} \times (23.5)^2 \times \frac{100}{1000}\right) = 86.70 \text{ kN}$$

Strength of power driven rivet in bearing, $\left(23.5 \times 16 \times \frac{300}{1000}\right) = 112.8 \text{ kN}$

Rivet value, R = 86.70 kN

Let n be the number of rivets to be provided in one row. The rivets are provided in 3 rows. Assume pitch of rivets p in each row at 90 mm.

$$n = \left(\frac{6M_s}{(3p) \times R}\right)^{1/2} = \left(\frac{6 \times 1648.4 \times 1000}{3 \times 90 \times 86.70}\right)^{1/2} = 20.55$$

Provide 24 rivets in each row.

Step 6 : Check for forces in rivets

The rivets are arranged in a narrow strip. The vertical distance to the extreme rivet, is as below.

y_n = 1035 mm

$\Sigma y^2 = 6. [4.5^2 + 13.5^2 + 22.5^2 + 31.5^2 + 40.5^2 + 49.5^2$

$+ 58.5^2 + 67.5^2 + 76.5^2 + 85.5^2 + 94.5^2 + 103.5^2]$ x 100

= 6 x (46557.20 x 100) mm

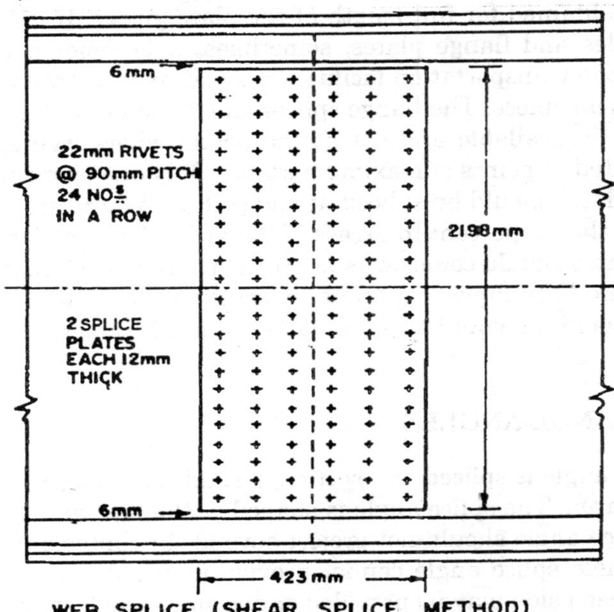

WEB SPLICE (SHEAR SPLICE METHOD)

Fig. 20.8

The force due to direct shear in each rivets is as under

$$F_1 = \left(\frac{1080}{24 \times 3}\right) = 15 \text{ kN}$$

Force in an extreme rivet resisting the moment

$$F_2 = \left(\frac{M.y_n}{\Sigma y^2}\right) = \left(\frac{1648.4 \times 1000 \times 1000 \times 1035}{6 \times 46557.20 \times 100 \times 1000}\right) = 61 \text{ kN}$$

$$\cos\theta = \frac{70.5}{\sqrt{(70.5^2 + 103.50^2)}} = 0.068$$

Resultant of two forces

$$F = \left(F_1^2 + F_2^2 + 2F_1\, F_2 \cos\theta\right)^{1/2}$$

$$= ((15)^2 + (61)^2 + 2 \times 15 \times 61 \times 0.068)^{1/2}$$

$$= 63.90 \text{ kN} < 86.70 \text{ kN, rivet value.}$$

Hence, the design of rivet is satisfactory.

The complete web splice is shown in Fig. 20.8.

20.5 FLANGE SPLICES

A joint in the flange element provided to increase the length of flange angle or flange plate is known as flange splice.

The flange splices should be avoided as far as possible. In general, the flange angles and flange plates can be obtained for full length of the plate girder. In spite of availability of full length of flange angles and flange plates, sometimes, it becomes necessary to make flange splices, for example, the transportation facilities may not permit transportation of plate girder for the entire span as one piece. The flange splices should be located at the section where some excess of flange area is available and not at the points where web splice is done. It should preferably not be located at points of maximum stress. Only one element of the flange, (*viz.,* one flange angle or one plate) should be spliced at one section. Where splice plates are used, their area shall be not less than 5 percent in excess of the area of flange element spliced. The centre of gravity of splice plates should coincide as nearly as possible with that of the element spliced. The rivets are designed for force 5 percent in excess to force in the element spliced. The strength of splice should be equal to or greater than 50 percent of the strength of splice elements.

20.6 SPLICE OF FLANGE ANGLES

When one flange angle is spliced at a section, a single splice angle may be provided in case it provides sufficient area. The splice angle is shaped at the heel and made to fit in the fillet of flange angle. The splice angle should not project beyond the flange angles specially in exposed positions. When a single splice angle cannot provide required cross-sectional area, then two splice angles, one on each side may be provided or one splice angle on one side and one plate on the other side may be provided, Fig. 20.9 shows splices of flange angles of three types.

In Fig. 20.9(A), a single splice angle is used for flange angle splice, and splice angle is in direct contact with the flange angle. Force from the flange angle shifts to splice angle on the same side of web. As a result of this, horizontal shear between flange angle and web plate is not affected. The splice and is connected with the flange angle with the same rivets and same spacing as used for connecting flange angle with the web. In order to minimize the length of splice angle, the rivets may be provided at less spacing. The length of splice is kept sufficient to accommodate rivets required to transmit force from flange to splice angle and back. In Fig. 20.9(C), a single splice angle and a splice plate is used for flange angle splice. The plate is not in direct contact with the element spliced. The part of the force from flange angle shifts to opposite side of the web through the web and unspliced flange angle. As a result of this, horizontal shear between flange angle and web is affected. It requires additional rivets for connections. The force in the flange angle is assumed to be distributed to the elements of the splice in proportion to their cross-sectional area. The strength to be transmitted by rivets connecting splice plate is equal to force in splice plate plus due to horizontal shear. Therefore,

= ($n.R_s$ = P + Horizontal shear) where,

n = Number of rivets required to connect splice plate on each side of splice

P = Force to be carried by splice plate

R_s = Strength of rivet in single shear

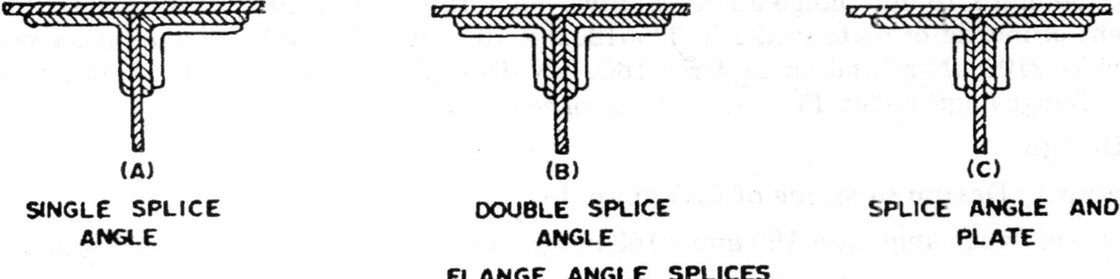

Fig. 20.9

The strength of rivets is considered single shear because P is force to be carried in one plane only.

The horizontal shear per unit length, $\tau_{vf.cal} = (V/d_e)$

The horizontal shear per unit length in one plane $\frac{1}{2}\tau_{vf.cal} = \left(\frac{1}{2}\frac{V}{d_e}\right)$

The horizontal shear per unit length in one plane, $F_s = \left(\frac{1}{2}\frac{V}{d_s} \times n \,.\, p\right)$ where,

p = Pitch of rivets

V = Shear force at the splice section

d_e = Effective depth of the girder

$$n.\, R_s = \left(P + \frac{1}{2}\frac{V}{d_s} \times n \,.\, p\right)$$

$$n = \frac{p}{\left(R_s - \frac{1}{2}\frac{V}{d}.p\right)} \tag{20.8}$$

20.7 SPLICE OF FLANGE PLATES

In case, it becomes necessary to splice an outer flange plate, a splice plate of same cross-section as the plate is provided. The length of splice plate is kept sufficient to accommodate necessary number of rivets. The strength of rivet is found in single shear. The splice plate is in direct contact. Therefore, the force to be transmitted by splice plate is not affected.

In case, it becomes necessary to splice an inner flange plate, the splice to be located at the theoretical cut-off of the next outer plate. The outer plate is extended. This serves as a splice plate. Splice of an inner plate is also done by providing an extra plate, which is placed outside of all flange plates. In this, splice plate is not in direct contact with the flange plate splices. The area of cross-section of splice plate is kept equal to the cross-sectional area of flange plate cut. The rivets may be designed to take full strength of flange plate cut and the shearing stress due to transmission of flange increment is neglected.

Example 20.4. A plate girder is composed of a 1600 mm x 12 mm web plate, 4 ISA 150 mm x 150 mm x 18 mm flange angles and two 500 mm x 12 mm thick flange plates. The gross moment of inertia of plate girder is 2355120.2 x 10^4 mm^4. The web of bending moment at a section is 2100 kN-m and shear force 1600 kN. One flange angle is spliced at this section. Design flange angle splice. Provide a single splice angle.

Design :

Step 1 : Design of splice of flange angle

Size of flange angle ISA 150 mm x 150 mm x 18 mm (ISA 150 150, @ 0.399 kN/m)

Use 22 mm diameter power driven rivets for connections

Net area of flange angle (5079 – 2 x 235 x 18) = 4233 mm^2

∴ Add 5 per cent = 212 mm^2, Total area required = 4445 mm^2

From ISA 200 mm x 200 mm x 25 mm (ISA 200 200, @ 0.736 kN/m) cut a splice angle 130 mm x 130 mm x 25 mm. The splice angle is cut and shaped to clear root of flange angle, and it does not project beyond flange angle.

Gross area of splice angle

[13 x 2.5 x (13 – 2.5) x 2.5] = 58.75 x 100 mm^2

Deduct 2 rivet holes

2 x 23.5 x 25 = 1175 mm^2

Net area provided by splice angle, (5875 – 1175) = 4700 > 4445 mm^2

Strength of power driven rivet in single shear, $\left(\frac{\pi}{4} \times (23.5)^2 \times \frac{100}{1000}\right) = 43.35$ kN

Strength of rivets in bearing, $\left(23.5 \times 12 \times \frac{300}{1000}\right) = 84.6$ kN

Rivet value = 43.35 kN

Net flange area = 16501 mm² (See table below)

Gross flange area = 19358 mm²

The stress at the level of c.g. of flange angles

$$\sigma_b = \left(\frac{2100 \times 10^6}{2355120.2 \times 10^4}\right) \times \left(\frac{(812 - 12 - 43.8) \times 19358}{16501}\right) = 78.1 \text{ N/mm}^2$$

Force in flange angle, $\left(\frac{78.1}{1000} \times 4445\right) = 347.15 \text{ kN}$

Number of rivets required, $\left(\frac{347.15}{43.35}\right) = 8.008$

Step 3 : Area of flanges

FLANGE AREA

Description	Gross flange area mm²	Deduct rivet hole mm²	Net flanges area mm²
Flange angle 2 ISA 150 mm x 150 mm x 18 mm	2 x 5071 = 10158	4 x 23.5 x 18 = 1692	8466
Flange plate 500 mm x 12 mm	500 x 12 = 6000	2 x 23.5 x 16.5 = 565	5635
Equivalent web area	$\frac{1}{6} \times 1600 \times 12 = 3200$	–	$\frac{1}{8} \times 1600 \times 12 = 2400$
	19358	–	16501

Provide 10 rivets on each side of flange angle, 5 rivets are provided in each leg at 60 mm spacing. The length of splice angle = 60 x 10 = 600 mm

The complete flange splice is shown in Fig. 20.10.

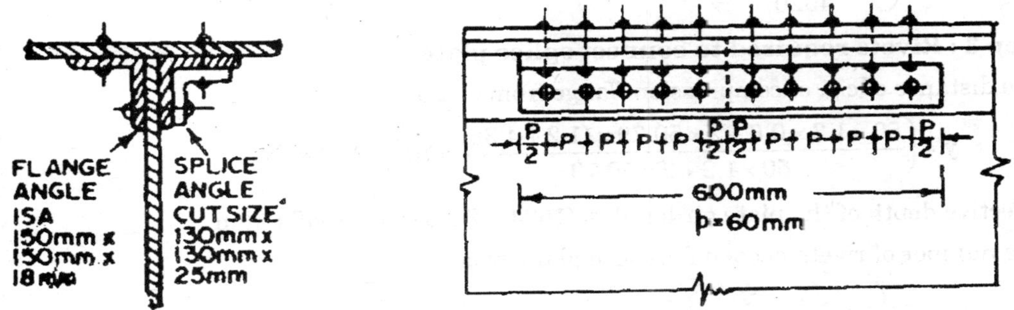

SINGLE ANGLE SPLICE (FLANGE ANGLE SPLICE)

Fig. 20.10

Example 20.5. Design flange angle splice in Example 20.1. Provide a splice angle and a rectangular plate.

Design :

Step 1 : Design of splice angle, From Example 20.4

Net area of flange angle ISI 150 mm x 150 mm x 18 mm (ISA 150 150, @ 0.399 kN/m) = 4233 mm²

Add 5 percent = 212 mm²

Total area of splice elements required = 4445 mm²

Provide ISA 130 mm x 130 mm x 15 mm (ISA 130 130, @ 0.289 kN/m) as splice angle

The splice angle does not project beyond flange angle

New area provided by splice plate

(36.81 – 2 x 2.35 x 1.5) x 100 = 29.76 x 100 mm²

Step 2 : Design of splice plate`

Area required from splice plate, = (4445 – 2986) = 1459 mm²

Provide 120 mm x 16 mm splice plate

Net area provide by splice plate = (120 x 16 – 23.5 x 16) = 1544 mm²

Total area provide by splice elements, = (2976 + 1544) = 4520 mm²

From Example 20. , the stress at the level of c.g. of flange angle, $\sigma_b = 78.1\ N/mm^2$

Force to be transmitted by flange angle, = (78.1 x 4445) = 347.15 kN

Force transmitted by splice angle, $\left(\dfrac{347.15 \times 2976}{4520}\right) = 228.56\ kN$

Rivets required to connect splice angle, $n = \left(\dfrac{228.56}{43.35}\right) = 5.27$

Provide 6 rivets on each side of flange splice, 3 rivets in each leg at 6 mm spacing. The force transmitted by splice plate

$$P = \left(\frac{347.15 \times 1544}{4520}\right) = 118.584\ kN$$

Step 3 : Rivets required to connect splice plate

The distance of c.g. of compression flange from outside

$$\bar{y} = \left(\frac{50 \times 1.2 \times 0.6 + 2 \times 50.79 \times (1.2 + 4.38)}{50 \times 1.2 + 2 \times 50.79} \times 10\right) = 37.1\ mm$$

Effective depth of the plate girder, d_e = (1624 – 2 x 37.1) = 1549.8 mm

The number of rivets needed for splice plate, from Eq. 20.

$$n = P/\left(R_s - \frac{1}{2} \times \frac{V}{d_e} . p\right)$$

$$n = \frac{118.584}{\left(43.35 - \frac{1}{2} \times \frac{1600 \times 1000}{1594.8} \times 600\right)} = 12.11$$

Provide 14 rivets on each side of flange splice. The length of splice plate. = 28 x 1680 mm

The complete flange splice is shown in Fig. 20.11.

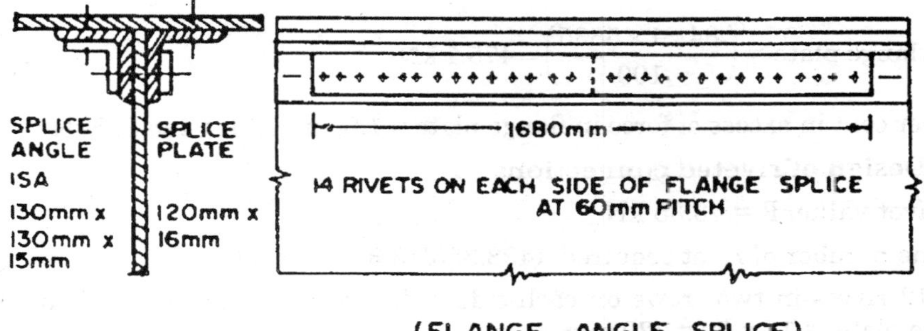

Fig. 20.11

Example 20.6. It is required to splice flange plate only in Example 20.4. Design splice for flange plate.

Design :

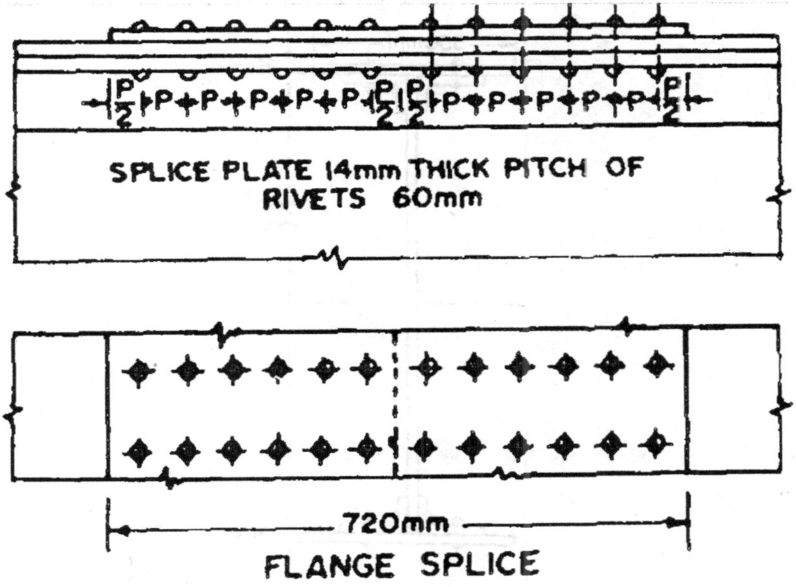

Fig. 20.12.

Step 1 : Force in flange plate

From Example 20.4, net area of flange plate, = 5635 mm²

Add 5 per cent = 282 mm²

Net area required by splice element, = 5917 mm²

Provide 50 mm x 14 mm splice plate

Net area provided = (50 x 1.4 – 2 x 2.35 x 1.6) x 100 = 6245 mm² > 5917 mm²

The stress in spliced flange plate, $\left(\dfrac{2100 \times 10^6 \times 19558}{2355120.2 \times 10^4 \times 16501}\right) = 84.31 \text{ N/mm}^2$

Force in flange plate $= \left(\dfrac{84.31 \times 5635}{100}\right) = 475.1 \text{ kN}$

Force 5 per cent in excess of force in flange plate = 1.05 x 475.1 = 498.855kN

Step 2 : Design of riveted connections

Rivet value, R = 43.35 kN

The number of rivet required, (498.855/43.35) = 11.50

Provide 12 rivets in two rows on each side of flange plate splice at 6 mm spacing. The length of splice plate, = 12 x 60 = 720 mm

The complete flange plates splice is shown in Fig. 20.12.

Example 20.7 A web plate of a plate girder is 2500 mm x 16 mm. Each flange plate of the girder is 550 mm x 16 mm. The cross-section of the plate girder is shown in Fig. 20.13. Design the web splice. The bending moment and the shear force at a particular section are 5200 kN-m. Design the section of a welded plate girder use Limit State method.

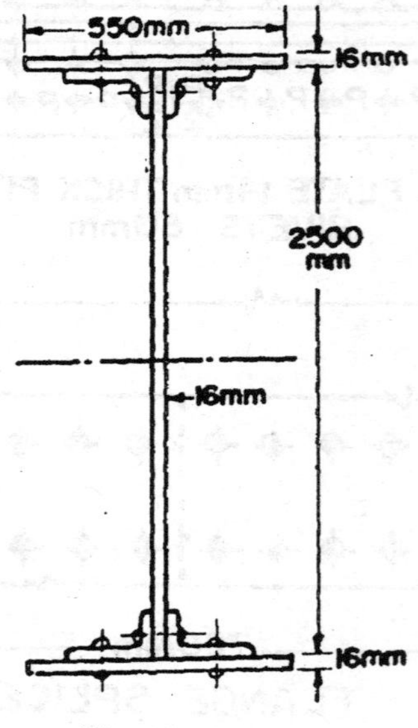

Fig. 20.13.

Design:

The bending moment, M = 5200 kN-m

The shear force, V = 1000 kN

Size of web plate d_w x t_w = 2500 mm x 16 mm

Step 1. Factored bending moment $M_{fd} = \gamma_{fL} . M$

(1.5 x 5200) = 7800 kN-m

Factored shear force, $V_{fd} = \gamma_{fL}. V$

(1.5 x 1000) = 1500 kN

Depth of the web plate = 2500 mm

Step 2 : Gross moment of inertia, I_{xx}

about xx-axis

$$I_{xx} = \left[\frac{1}{12} \times 1.6 \times (250)^3 + 2 \times 55 \times 1.6 \times 125.8^2\right] \times 10^4 \text{ mm}^4$$

$$= [2083.33 \times 10^3 + 2785.312 \times 10^3] \times 10^4 \text{ mm}^4$$

$$= 4868645.64 \times 10^4 \text{ mm}^4$$

Overall depth of the web-plate

$$= 2500 + 2 \times 16 = 2532 \text{ mm}$$

Thickness of web plate = 16 mm

Step 3 : Weld (Double - V groove butt-weld) may be used.

Effective throat thickness

$t_{tw} = t_w$ = Thickness of web plate

t_{tw} = 16 mm

Step 4 : Shear capacity of weld per mm length

$$\tau_{sd} = \left[\frac{1 \times 16 \times}{1000}\left(\frac{250}{\sqrt{3}}\right) \times \left(\frac{1}{1.15}\right)\right] = 2.01 \; kN/\; mm$$

Length of weld needed to resist the factored shear force

$$\left(\frac{V_{fd}}{\tau_{sd}}\right) = \left(\frac{1500 \times 1000}{2.01 \times 1000}\right) = 746.268 \; mm$$

It is less than the depth of web plate 2500 mm is more than 746.268 mm

Step 5 : Design bending strength

(i) $M_{sd}.1 = \left(1.2 \,.\, \frac{Z_e \, \sigma_y}{\gamma_{msy}}\right)$, *where*

Z_e = Elastic section modulus of the section

$$Z_e = \left(\frac{I_{xx}}{y}\right) \qquad y = \frac{1}{2}(depth \;\; of \;\; girder)$$

$$y = \left(\frac{1}{2} \times 2532\right) = 1266 \; mm$$

$$Z_e = \left(\frac{4868645.64 \times 10^4}{1266}\right) = 38456.92 \times 10^3 \; mm^3$$

$$M_{sd}.1 = \left(\frac{1.2 \times 38456.92 \times 10^3 \times 250}{1000 \times 1.15}\right)$$

$$= (10032 \times 10^6) = 10032 \text{ kN-m}$$

$$> 7800 \text{ kN-m}$$

(ii) $M_{sd2} = \left(\beta_b \cdot Z_p \cdot \sigma_y / \gamma_{msy}\right)$

Note : The mechanical properties of the weld-metal are always assumed to be larger than the elements joined.

PROBLEMS

20.1. It is required to a design a triple plate, i.e., with moment and shear plates web splice for a plate girder where the shear is 480 kN and moment 54,00 kN-m. The size of web plate is 1800 mm x 10 mm. The available clear distance between toes of flange angles is 1520 mm. The total moment of inertia of the girder section is 466 x 10^8 mm^4.

Use 22 mm diameter power driven rivets.

20.2. Design a suitable web splice for a plate girder at a section at which the bending moment is 5000 kN-m and shear force is 1000 kN. The plate girder is built-up from a web plate 2500 mm deep x 16 mm thick and two ISA 200 mm x 150 mm x 15 mm (ISA 200 150, @ 0.394 kN/m) angles and a 600 mm x 16 mm cover plate in each flange. The rivets used are of 22 mm diameter. Draw a sketch of the splice. The long legs of the flange angles are placed horizontally.

21

WELDED JOINTS SUBJECTED TO AXIAL LOAD

21.1 INTRODUCTION

The welding is one of the methods of connecting the structural members. In the welding, a metallic link is made between the structural members. The **weld** is defined as a union (a strong metallurgical bond) between two pieces of metal at faces rendered plastic or liquid by heat or by pressure or both. A typical weld is shown in Fig. 21.1. This shows the various zones of the weld. In the **fusion zone**, the portion of the structural members has been fused (i.e. made plastic or liquid by heat). In the weld metal zone, the portion of structural members to be connected and filler metal are melted or made plastic in making a weld. The portion of the structural members to be connected and filler metal melted or made plastic is termed as weld metal. The metal added in the welding is termed as filler metal. In the heat-affected zone, the parent metal is metallurgically affected by the heat of welding.

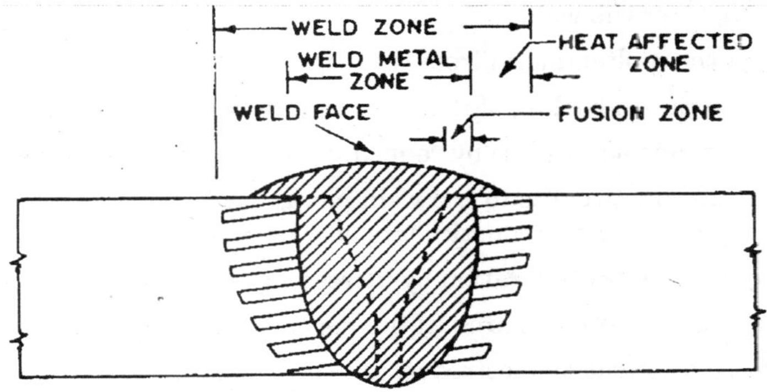

VARIOUS ZONES OF A TYPICAL WELD

Fig. 21.1. Various zones of a typical weld.

The welding process in which the weld is made between metals in a state of fusion without hammering or pressure is known as **fusion welding**. The welding process in which the weld is made by sustained pressure while the surfaces to be united are plastic is known as pressure welding. The welding is done by several processes. These processes are listed below :

1. **(A) Fusion welding. (i) Arc welding** (a) Metal arc welding (Shielded/or Unshielded); (b) Carbon arc welding (Shielded/or Unshielded)

1.(B) Fusion welding. (ii) Gas welding (a) Oxy-Acetylene welding (b) Oxy-Hydrocarb welding

2. Pressure welding (i) Resistance welding (ii) Pressure welding

3. Thermit welding. 4. Union melt welding

In general, the process of ``metal arc welding" is used for the structure steel connectio The procedure for metal arc welding is covered by IS : 823-1964.

In the metal arc welding, the electrodes are used. When considering electrodes, it necessary always to bear in mind the fact that heated metal combines chemically with oxyg and nitrogen from the atmosphere forming oxides and nitrides, which result in brittle p quality weld metal. For this reason, it is essential to provide some means of preventing t atmosphere from reacting the hot weld area. This can be done either by shrouding the ar with an inert gas or by the use of a suitable flux. **The flux** is fusible material used in weldi or oxygen-cutting to dissolve and facilitate removal of oxides and other undesirable substanc

The metal arc welding was originally done with bare electrodes consisting of a piece wire or rod of the same compositions as to the metal to be welded. For the reasons mention above, the welds were of poor quality and thus bare electrodes are now rarely encounter Bare wire is of course used for automatic, welding, but in this case, the weld area is protect from the atmosphere by powdered flux or inert gas.

Most modern electrodes are of the coated or covered type consisting of a metal core w surrounded by a thick coating applied by extrusion, winding or other process. The success welding operation depends on the composition of coating, which is varied to suit differe conditions and metals. The coating on the electrodes play the following function :

1. To stabilise the arc and enable the use of alternating current (if necessary)
2. To flux away any impurities present on the surface being welded.
3. To form a slag over the weld to
   (a) Protect the molten metal from the atmospheric contamination,
   (b) Slow the rate of cooling of the weld and to reduce chance of brittleness, and
   (c) Provide a smooth surface by reducing the ripples caused by welding operations.
4. To add certain constituents to the weld metal to compensate for the loss of any volat alloying elements or any constituents lost by oxidation.
5. To speed up the welding operation by increasing the rate of melting.

The various types of electrodes are : (i) the normal penetration electrodes,

(ii) the deep penetration electrodes, (iii) the automatic electrodes,

(iv) shielded arc electrode, and (v) low hydrogen electrodes.

The electrodes are used to suit the different conditions. The electrodes have the followi types of flux covering. The numbers indicating the digit refer to type of covering.

Digit (type of covering)

1. Having a high cellulose content.
2. Having a high content of titania and producing a fairly viscous slag.

3. Containing an appreciable amount of titania and producing a fluid slag.

4. Having a high content of oxides/or silicates of iron and manganese and producing an inflated slag.

5. Having an high content of iron oxide and/or silicates and producing a heavy solid slag.

6. Having a high contents of calcium carbonate and fluoride.

7. Having a high content of iron powder.

8. A covering of any type not classified above.

The code number of covered electrodes shall consist of seven essential and one or two ditional symbol, each indicating a specific characteristic of the electrode. The essential nbol shall consist of the letter `M' indicating the suitability of the electrode for metal arc lding and six digits in Hindu-Arabic numerical each denoting the following :

(a) Type of covering material (mentioned above)

(b) Welding position in which the electrode may be used.

(c) Welding current conditions.

(d) Ultimate tensile strength of deposited metal,

(e) Percentage elongation in tensile test of disposited metal, and

(f) Impact test values of deposited metal.

The essential symbols are followed by letter P or A or both, denoting the electrode for deep netration, or automatic welding or for both purposes. The size of electrode is designated by e diameter of the core wire in mm.

It became possible to provide the perfect rigid joints in the rigid frame structures for the signers with the advent of welding. There is no loss of cross-section of a member in a welded nt. In riveted joint, the cross-sectional area is reduced due to rivet holes. The welded joints e capable to transmit axial forces, shears and moments.

.2 ADVANTAGES OF WELDING

The use of welding is increasing rapidly because of its several advantages over riveting. ne advantages of welding are as under :

1. There is silence in the process of welding.

When the riveting is done, the rivet head is prepared by hammering. The hammering uses noise in the process of riveting.

2. There is safety of welding operator in the welding.

The riveting is done cold as well as hot. While hammering, rivets may fly away and injure e persons. When the hot riveting is done, the rivets are made red hot and handed over to ace them in position. During transactions, the red hot rivets may fall down and strike the orkers.

3. The welding may be done quickly in comparison to the riveting.

4. The welded joints have better appearance than riveted joints.

5. The welded joints are more rigid than the riveted joints.

6. The welding is more adaptable than riveting.

In the complicated structures, riveting may be extremely difficult or not possible but the welding may be done easily. For circular steel pipes and tube connections, riveting is practically impossible, but welding may be adopted conveniently. The welding is also convenient for additions in the existing structures.

7. The welded joints have more efficiency than riveted joints. The efficiency of a joint is the ratio of strength of the joint to the strength of the solid plate. In the welded joints, holes are not made as it is done in riveted joints. The strength of joint even in a tension member may be obtained equal to the strength of solid member. As such, in welded joint efficiency of joint may be achieved equal to 100 per cent. In the riveted joint, the efficiency of a joint in general is about 80 per cent only.

8. The welding is economical in comparison to the riveting.

9. Welding allows direct transfer of stresses.

10. The joints are made air and water tight welding joins plates. Welding is ideal for oil-storage tanks and ships.

In the welding connecting angles, gusset plates, splice plates and rivet head are not required. Thus, it reduces the weight of structure also. In the riveted connections, the minimum size is governed by the diameter of the rivet besides the requirement of force to be transmitted. Even if a smaller size of a member is sufficient to transmit the force, but the minimum size of member is to be used, which may provide minimum edge distance and minimum pitch for connecting rivets. In the welding, in absence of such restrictions, small size members may be used.

Gusset and splice plates are not needed. There is saving in time in detailing, fabrication and erection in field.

In the welding, the cost of labour is less than the riveting. The cost of painting and other maintenance is also less.

Overlapping of mild steel bars at joints in reinforced cement concrete structures is not necessary, if the bars are welded.

21.3 DISADVANTAGES OF WELDING

Besides the several advantages, there are disadvantages of welding. The disadvantages of welding are as under :

1. The members are likely to distort in the process of welding.

Uneven heating and cooling are caused due to welding. As a result of this, the distortion of members may be take place.

2. A welded joint fails earlier than riveted joint, if the structure is under fatigue stresses.

3. There is a greater possibility of brittle fracture in welding than the riveting.

4. The inspection of welded joint is more difficult and more expensive than the riveted joint.

The inspection of riveted joint may be done simply by tapping with a hammer. For inspection of a welded joint, a stethoscope and a light hammer is required. Some of the defects in a weld may be detected by this. The inspection of welded joint may also be done by the magnetic power method, radiography by means of X- rays, Gamma rays or by trepanning.

5. A highly skilled person is required in the welding than in the riveting.
6. The welded joints are over rigid.
7. Under fatigue, welded joints are highly prone to cracking.

21.4 TYPES OF WELDED JOINTS

The various types of welded joints are as follows :

1. Butt weld, 2. Fillet weld, 3. Slot weld and Plug weld.
4. Spot weld, 5. Seam weld, 6. Pipe weld.

21.5 BUTT WELD

A butt weld is also termed as **groove weld.** The butt weld is used to joint structural members carrying direct compression or tension. It is used to make tee-joint and butt-joint. Various types of butt joint are as under. These are named depending upon shape of the grove made for welding.

(i) Square butt weld. A square butt weld is a weld in the preparation of which the fusion faces lie approximately at right angles to the surfaces of the components to be joined and are substantially parallel to one another. The square butt weld is shown in Fig. 21.2.

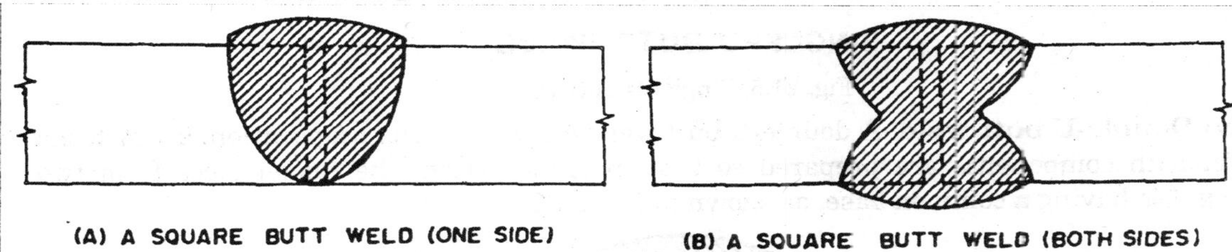

Fig. 21.2 (A&B)

(ii) Single-V butt weld. A single-V butt weld is a weld in the preparation of which the edges of both components are prepared so that in the cross-section, the fusion faces form a `V' as shown in Fig. 21.3.

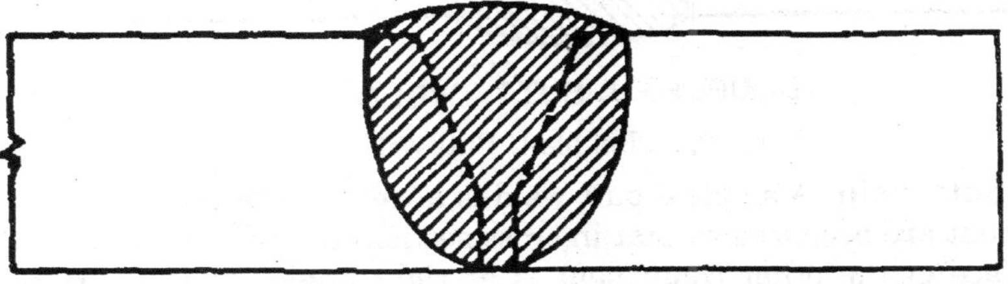

Fig. 21.3. Single-V butt weld.

(iii) Double-V butt weld. A double-v butt weld is a weld in the preparation of which the edges of both components are double beveled so that in cross-section, the fusion faces from two opposing V's as shown in Fig. 21.4.

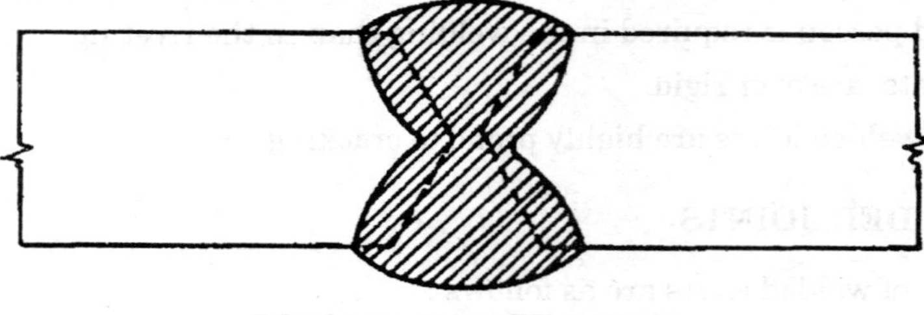

Fig. 21.4. Double-V butt weld.

(iv) Single-U butt weld. A single-U butt weld is a weld in the preparation of which the edges of both components are prepared so that in cross-section, the fusion faces form a `U' as shown in Fig. 21.5.

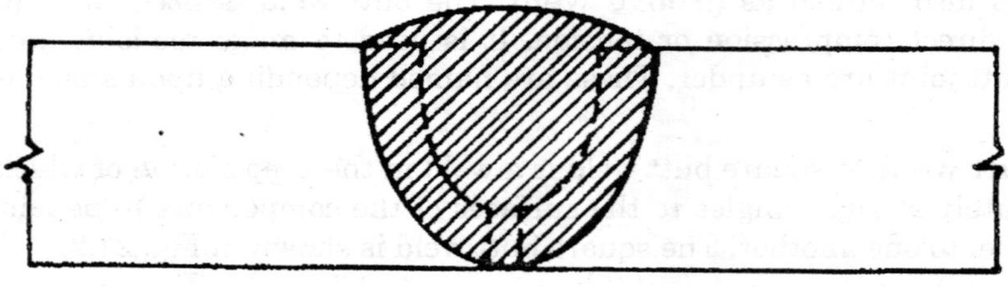

Fig. 21.5. Single -U butt weld.

(v) Double-U butt weld. A double-U butt weld is a weld in the preparation of which the edges of both components are prepared so that in cross-section, the fusion faces form two opposing `U's having a common base, as shown in Fig. 21.6.

Fig. 21.6. Double-U Butt weld

(vi) Single-J butt weld. A single-J butt weld is a weld in the preparation of which the edges of one component are prepared so that in cross-section, the fusion face is in the form of a `J' and the fusion face of the other component is at right angles to the surface of the first component as shown in Fig. 21.7.

(vii) Double-J butt weld. A double-J butt weld is a weld in the preparation of which the edges of one component are prepared so that in cross-section, the fusion face is in the form of two opposing `J's and the fusion face of the other component is at right angles to the surfaces of the first component as shown in Fig. 21.8.

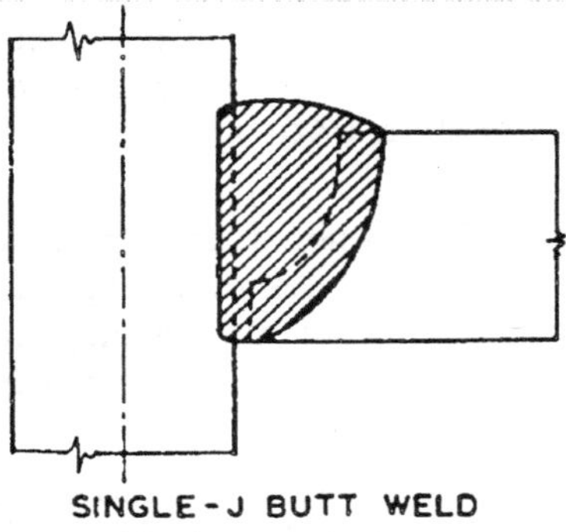

Fig. 21.7. Single-J butt weld.

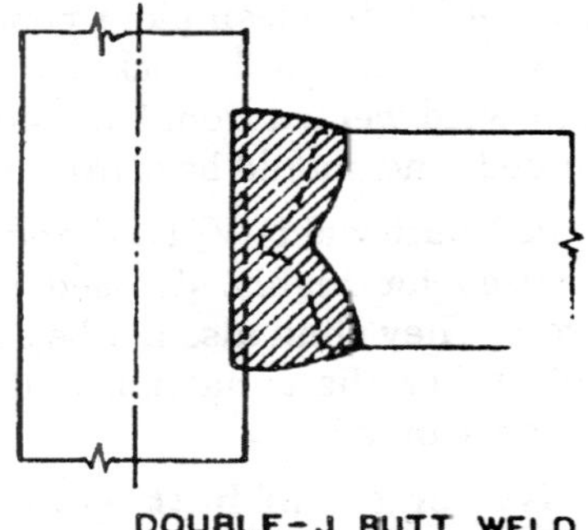

Fig. 21.8. Double -J butt weld.

(viii) Single-bevel butt weld. A single-bevel butt weld is a weld in the preparation of which the edge of one component is bevelled and the fusion face of the other component is at right angles to the surfaces of the first component as shown in Fig. 21.9.

(ix) Double-bevel butt weld. A double-bevel butt weld is a weld in the preparation of which the edges of one component are double bevelled and the fusion face of the other component is at right angles to the surfaces of the first components as shown in Fig. 21.10.

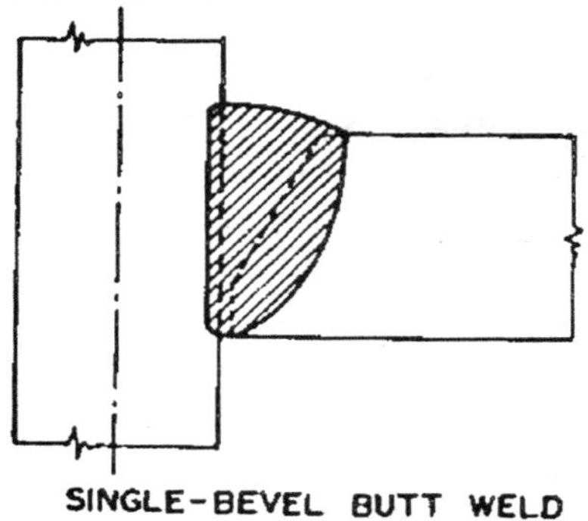

Fig. 21.9. Single-Bevel butt weld.

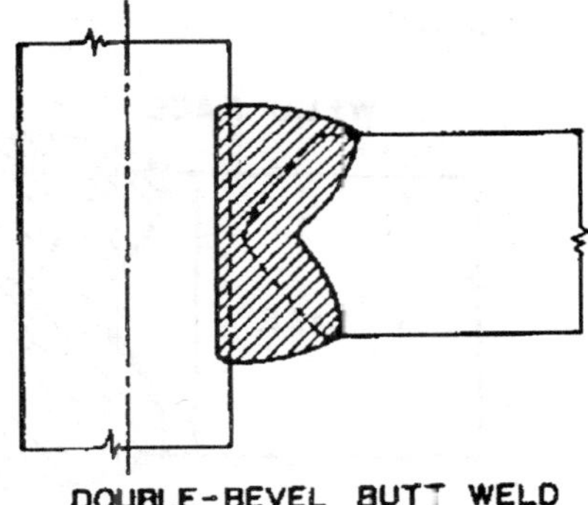

Fig. 21.10. Double -Bevel butt weld.

Specifications of the butt weld

(i) Size of butt weld. The size of a butt weld is specified by the **effective throat thickness.**

The effective throat thickness in case of complete penetration butt weld is taken as the thickness of thinner part joined. The effective throat thickness of T or L butt-joints is equal to the thickness of abutting part. The double-V, double-U, double-J and double-bevel butt welds are the examples of complete penetration butt weld.

The effective throat thickness in case of incomplete penetration butt weld is taken as 7/8th of the thickness of the thinner part joined. But for the purpose of stress calculation, a required effective throat thickness not exceeding 5/8th of the thickness of thinner part joined should be used. An incomplete penetration butt-weld is also termed as unsealed single butt- weld. Single-V, single-U, single-J and single-bevel butt joints are the examples of incomplete penetration butt weld. In incomplete penetration butt weld, the weld metal is not deposited intentionally through the full thickness of the joint. The unwelded portion in incomplete penetration butt weld welded from both sides shall not be greater than 1/4th of the thickness of thinner part joined, and should be central in the depth of the weld.

The unsealed butt welds V, U, J and bevel types and incomplete penetration butt welds should not be used for highly stressed joints and joints subjected to dynamic, repeated or alternating forces. They shall also not be subjected to a bending moment about the longitudinal axis of the weld other than that normally resulting from the eccentricity of the weld metal relative to the parts joined.

(ii) **Effective length of butt weld.** The effective length of butt weld is the length for which the continuous full size weld of the specified size (throat thickness) of the weld exists but not less than *times* size of the weld.

(iii) **Effective area of butt weld.** The effective area of a butt weld is taken as the product of the effective throat thickness and the effective length of butt weld.

(iv) **Reinforcement.** The extra metal deposited proud of the surface of the parent metal as shown in Fig. 21.11 is called reinforcement. This reinforcement is provided to provided sufficient surfaces convexity and to ensure full effectiveness at the joint. This requires a minimum practical surface convexity of 1.0 mm. This reinforcement should not exceed 3.0 mm. This is not considered as part of throat thickness. This reinforcement may also be removed if a flush surface is desired.

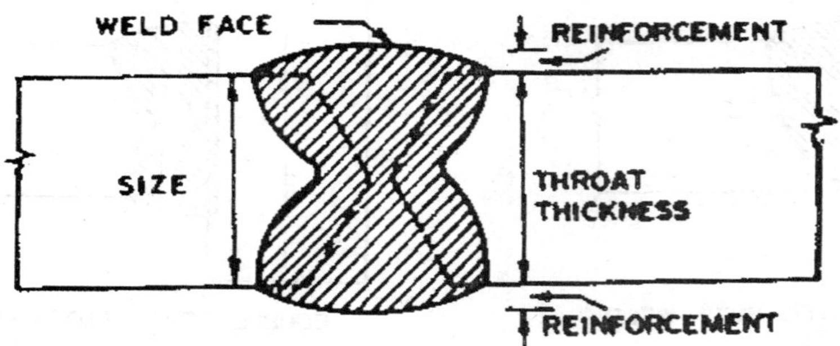

Fig. 21.11.

Intermittent butt welds are used to resist shear only. The effective length of intermittent weld is taken not less than four times the thickness of thinner part joined. The longitudinal space between the effective lengths of welds is taken not more than 16 times the thickness of thinner part joined. The intermittent weld is not used in position subjected to dynamic, repetitive and/or alternate stresses.

When the structural members of unequal thickness are butt welded and difference in thickness of members exceeds 25 per cent of the thinner part or 3.0 mm in metal arc welding and 6.0 mm or more in oxy acetylene welding, the thicker part is bevelled so that the slope of the surface from one part to the other is not steeper than one in five as shown in Fig. 21.12 (A). Where this arrangement is not practicable, the weld metal should be built-up at the junction with the thicker part to dimension at least 25 per cent greater than that of the thinner part (in metal are welding) as shown in Fig. 21.12 (B), alternatively, the weld metal should be built-up to the dimensions of thickener members as shown in Fig. 21.12 (C).

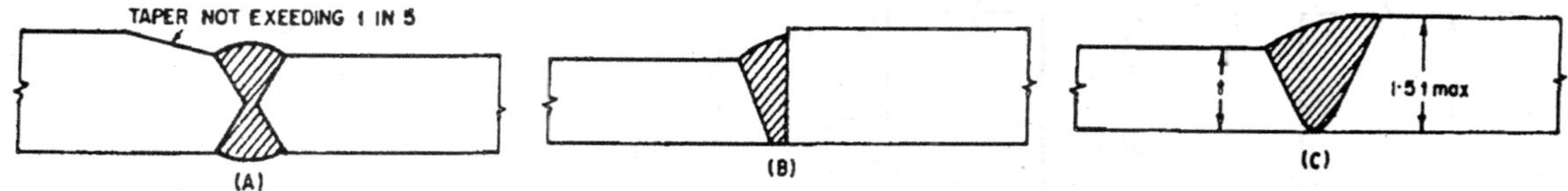

Fig. 21.12.

In case of complete penetration butt weld, generally, design calculations are not necessary, as these will usually provide the strength at the joint equal to the strength of the member connected.

Design strength of butt weld: Butt welds shall be treated as parent metal with a thickness equal to the throat thickness and the stresses shall not exceed those allowed in the parent metal.

21.6 FILLET WELD

A fillet weld is a weld of approximately triangular cross-section joining two surfaces approximately as right angles to each other in lap joint, or tee-joint. A fillet weld is shown in Fig. 21.13 (A).

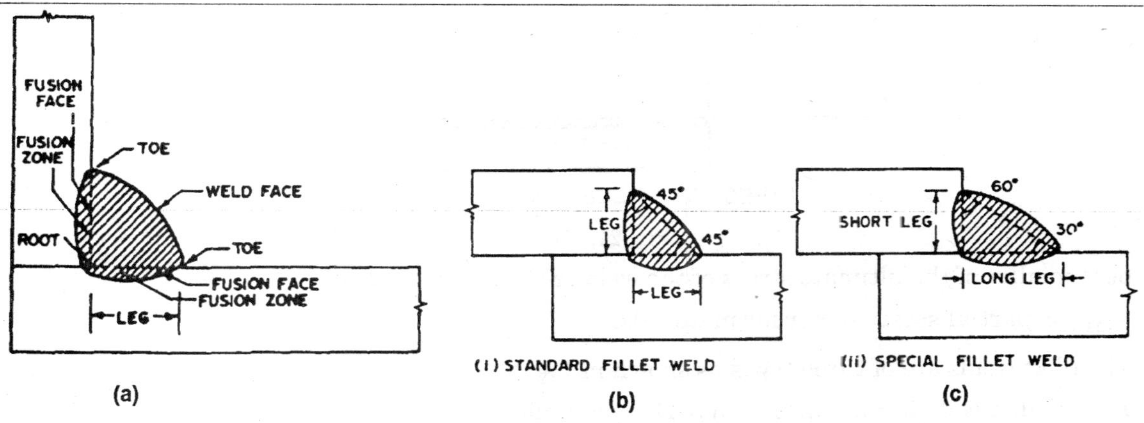

Fig. 21.13.

When the cross-section of fillet weld is 45-degree, isosceles triangle as shown in Fig. 21.13 B (i), it is known as standard fillet weld. The standard 45-degree fillet welds are generally used. When the cross-section of the fillet weld is 30-degree and 60-degree triangle as shown in Fig. 21.13 (B) (ii), it is known as a special fillet weld.

A fillet weld is termed as concave fillet weld or convex fillet weld or mitre fillet weld depending on the weld face is concave or convex or approximately flat as shown in Fig. 21.14, respectively.

A fillet weld is termed as `normal weld or deep penetration fillet weld depending upon the depth of penetration beyond the root is less than 2.4 mm or (2.4 mm or more), respectively.

Design strength of a fillet weld f_{wd} shall be based on it throat area. It is given by

$$\sigma_{wd} = (\sigma_{wn}/\gamma_{msw}), \text{ where } (\sigma_{wn} = \sigma u/\sqrt{3})$$

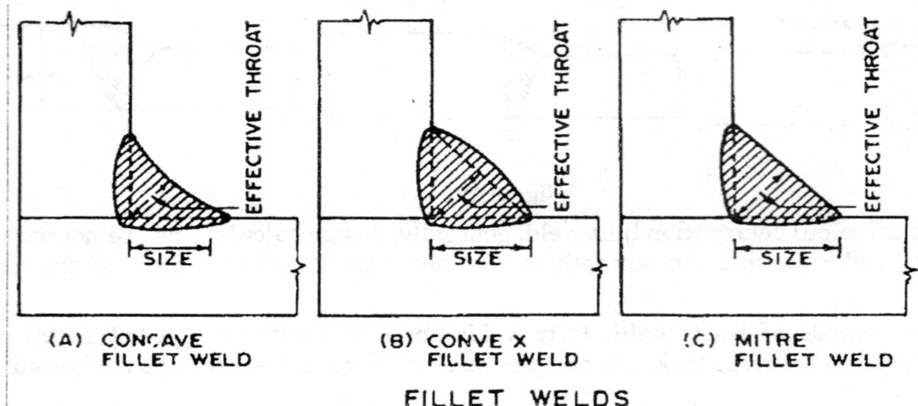

Fig. 21.14. Fillet weld

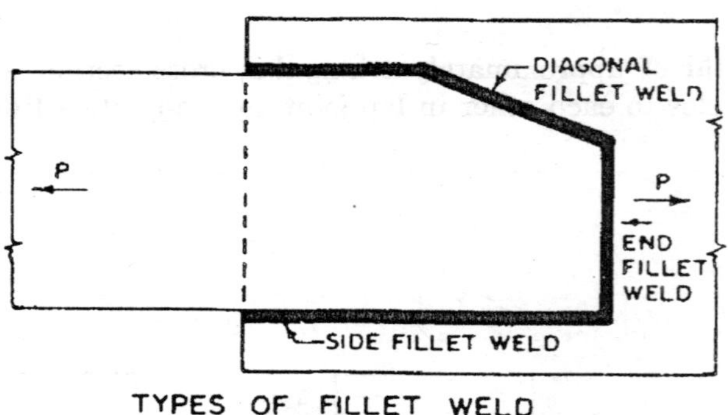

Fig. 21.15.

σ_u = smaller of the ultimate stress of the weld or the parent metal

γ_{msw} = partial safety factor for materials.

The fillet welds are of three types as shown in Fig. 21.15.

1. Side fillet weld 2. End fillet weld 3. Diagonal fillet weld.

1. Side fillet weld is a fillet weld stressed in longitudinal shear, i.e., a fillet weld, the axis of which is parallel to the direction of these applied load. It is also termed as longitudinal fillet weld.

In case, the maximum length l of the side welds transferring shear along its length exceeds 150 times the throat size of weld. The reduction in weld strength as per long joint should be considered.

2. End fillet weld is a fillet weld stressed in transverse shear, i.e., a fillet weld, the axis of which is at right angles to the direction of the applied load. It is also termed as transverse fillet weld.

3. Diagonal fillet weld is fillet weld the axis of which is inclined to the direction of the applied load.

A fillet weld is placed on the sides or end of the base metal and it is subjected to shear along with tension or compression and usually bending.

Long joints: When the length of weld joint, l of a splice or end connection in a compression or tension element is more than 150 t_t, the design capacity of weld shall be reduced by the factor.

where, $\beta_{tw} = (1.2 - 0.2\, l_j / 150\, t_t) <= 1.0$

l_j = length of the joint in the direction of force transfer, and

t_t = throat size of weld.

Specifications of fillet weld

(i) Size of fillet weld. The size of normal fillet weld is specified as minimum leg length of a convex or mitre fillet weld or 1.414 times the effective throat thickness of a concave fillet weld. The size of deep penetration fillet weld is specified as minimum leg length plus 2.4 mm. The length of leg is the distance from the root to the toe of a fillet weld, measured along the fusion face as shown in Fig. 21.13 (A). The minimum size of single run fillet weld is adopted as per Table 21.1 as recommended in IS : 816-1969.

TABLE 21.1. MINIMUM SIZE OF OF FIRST RUN OR OF A SINGLE RUN FILLET WELD (As per IS: 800-2007)

Thickness of thicker part		Minimum size mm
Over (mm)	Upto and including (mm)	
–	10	3
10	20	5
20	32	6
32	50	8 First run 10 mm size of fillet

Note 1. When the minimum size of the fillet weld is greater than the thickness of the thinner part, the minimum size of the weld should be equal to the thickness of thinner part. The thicker part shall be adequately preheated to prevent cracking of the weld.

Note 2. Where the thicker part is more than 50 mm thick, special precaution like preheating will have to be taken.

(ii) Effective throat thickness. The effective throat thickness of a fillet weld is the perpendicular distance from the root to the hypotenuse of the largest isosceles right-angled triangle that can be inscribed within the weld cross-section as shown in Fig. 20.14. The effective throat thickness of a fillet weld shall not be less than 3 mm and shall generally not exceed 0.7 t and 1.0 t under special circumstances where t is the thickness of thinner part.

$$\text{Effective throat thickness} = \frac{1}{\sqrt{2}} \times (\text{Size of weld}) = (0.7 \times \text{Size of weld})$$

In general, for the purpose of stress calculations, the effective throat thickness = K. Size of weld, where, K is a constant.

The value of constant K for different angles between fusion faces is adopted as per Table 21.2 as recommended in IS : 816-1969.

TABLE 21.2. VALUE OF K FOR DIFFERENT ANGLES BETWEEN FUSION FACES (As per IS: 800-2007)

Angle between fusion faces	Value of constant K
60° - 90°	0.70
91° - 100°	0.65
101° - 106°	0.60
107° - 113°	0.55
114° - 120°	0.50

A fillet weld is not used for joining parts, if the angle between fusion faces is greater than 120° or less than 60°.

(iii) Effective Length. The effective length of the weld is the length of the weld for which the specified size and throat thickness i.e., correctly proportioned cross-section of the weld, exists. It is taken as the actual length minus twice the size of weld, since the specified size and throat thickness do not exist at the ends. The effective length of the weld is shown on the drawings. In practice the actual length of weld is made equal to the effective length shown on the drawing plus twice the weld size but not less than four *times* the sie of weld.

The effective length of fillet weld **should not be less than** four times the size of the weld.

When the ends are returned, as shown in Fig. 21.16 then the ends should be carried continuous around the corners for distance not less than twice the size of weld. This should be applied particularly to side and top fillet welds in tension.

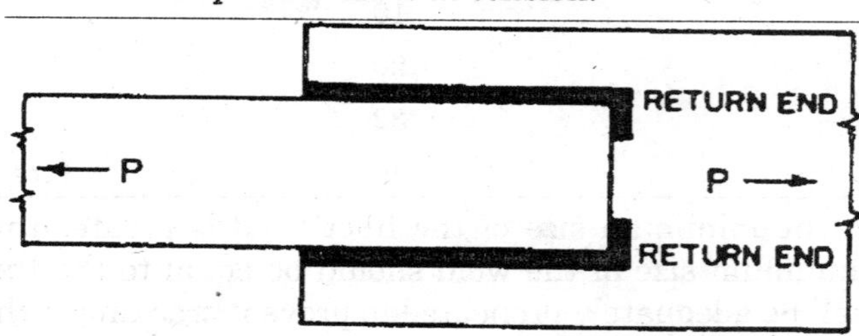

Fig. 21.16

(iv) Effective area The effective area of a fillet weld is taken as the product of length or of weld of effective length and effective throat thickness.

(v) End returns: Fillet welds terminating at the ends or sides of parts should be returned continuously around the corners for a distance of not less than twice the size of weld, unless it is not impracticable. It is important on the tension end.

Intermittent fillet welds

The intermittent fillet welds may be used to transfer calculated stress across a joint when the strength required is less than that developed by continuous fillet weld of the smallest allowable size for the thickness of the parts joined. Any section of intermittent fillet welding shall have an effective length of not less than four times the weld size with a minimum of 40 mm except for plate girder stiffeners.

The clear spacing between the effective lengths of intermittent fillet welds carrying calculated stress shall not exceed the following number of times the thickness of the thinner plate joined and shall in no case be more than 200 mm.

(i) 12 times for compression, and (ii) 16 times for tension the thinner plate jointes.

The longitudinal fillet welds at the ends of built-up members shall have an effective length of not less than the width of the component part joined unless end transverse welds are used, in which case, the sum of the end longitudinal and end transverse welds shall be not less than twice the width of the component part.

The chain intermittent welding is to be preferred to staggered intermittent weld. Where the staggered intermittent welding is used, the ends of the component parts shall be welded on both the sides.

In a line of intermittent fillet welds, the welding should extend to the ends of parts connected; for welds staggered about two edges, this applies generally to both the edges.

The intermittent fillet welds are not used in the case of main members of structures directly exposed to weather. It shall not be used in position subjected to dynamic repetitive or alternating stresses.

Fillet weld applied to the edge of a plate or section

1. Where a fillet weld is applied to the **square edge** of a part, the specified size of the weld should generally be of least 1.5 mm less than the edge thickness in order to avoid washing down of the exposed as shown in Fig. 21.17A.

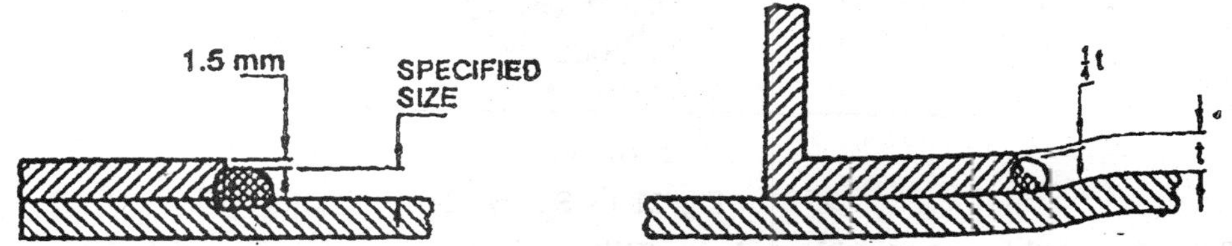

Fig. 21.17A.

2. Where the fillet held is applied to the rounded toe of a rolled section, the specified size of the weld should generally not exceed (3/4)th of the thickness of the section at the toes as shown in Fig. 21.17B.

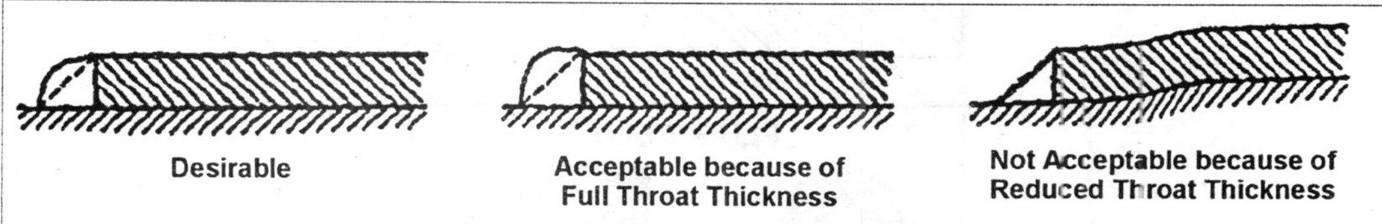

Fig. 21.17B.

3. Where the size specified for a fillet weld is such that the parent metal will not project beyond the weld, no melting of the outer cover or covers shall be allowed to occur to such an extent as to reduce the throat thickness as shown in Fig. 21.18.

4. When fillet welds are applied to the edges of a plate or section in members subjected to dynamic loading, the fillet weld shall be of full size with its leg length equal to the thickness of the plate or section with the limitations specified as in (3).

5. End fillet weld, normal to the direction of force shall be unequal size with a throat thickness of the part as shown in Fig. 21. The difference in thickness of the welds shall be negotiated at a uniform slope.

21.7 SLOT WELD AND PLUG WELD

The slot weld is the weld used to join two touching contiguous components by a fillet weld round the periphery of a slot in one component so as to join it to the surface of the other component exposed through slot as shown in Fig. 21.18.

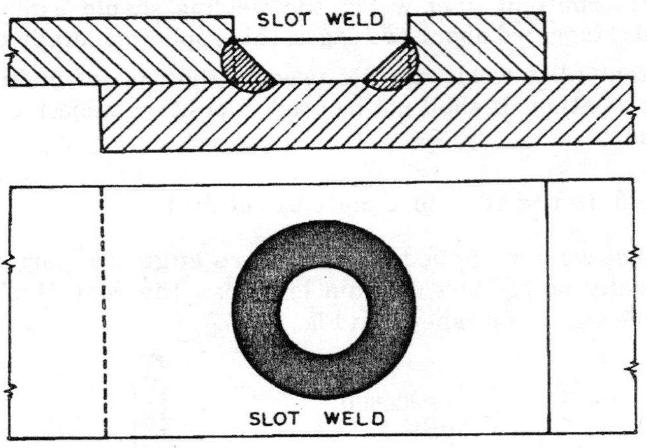

Fig. 21.18. Slot weld (A)

The plug weld is the weld made by filling in a hole in one part with filler metal so as to join it to the surface of contiguous part exposed through a hole as shown in Fig. 21.19.

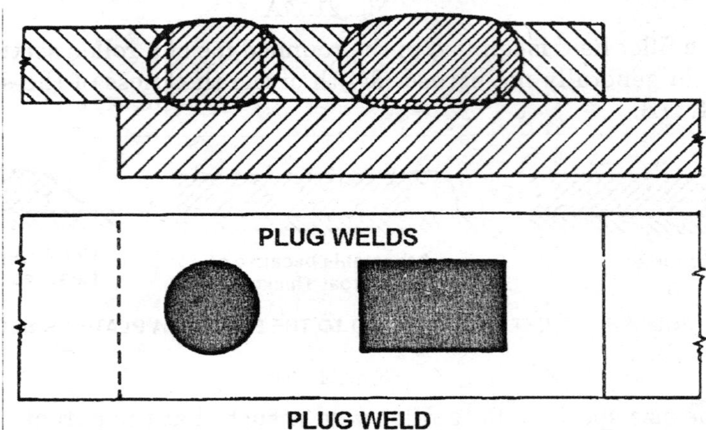

Fig. 21.18. Plug weld (B)

In case where there is insufficient space to accommodate the required length of fillet weld, plug weld or slot weld is used, to provide extra strength. These are also used in addition to the fillet weld when plates are welded. These welds balance the stresses in the plates and avoid buckling.

Specifications

The dimensions of slot or hole should comply with the following limits in terms of the thickness of the part in which the slot or hole is made :

(i) The width of diameter should not be less than three times the thickness or 25 mm whichever is greater.

(ii) The corners at the enclosed ends of slots should be rounded with a radius not less than 1-1/2 times the thickness or 12 mm whichever is greater.

(iii) The distance between the edge of the part and the edge of the slot or hole or between the adjacent slot or holes should not be less than twice the thickness and not less than 25 mm for holes.

(iv) The effective area of plug weld shall be considered as the nominal area of the hole in the plane of facing surface. The plug welds shall not be designed to carry stress.

1. Slot welding

The spot welding is done by the process of resistance welding. In a resistance welding process, a weld is produced by the heat obtained from resistance to the flow of electric current through the work parts held together under pressure by electrodes. The size and shape of the individually formed welds are limited primarily by the size and contour of the electrodes.

In case, slot welds are provided then, these should be set back behind the beginning of the welds on the edges of the plate. It is ensured that the effective area of the plate is not reduced. Due to deposition of slots, stress concentration does not occur.

2. Seam welding

The seam welding is also done by the process of resistance welding. The welding in which the pressure is applied continuously, and current impulsively to produce a linear weld, the work piece being between two electrodes wheels or between an electrode wheel and electrode bar. The electrode wheels apply the pressure and may be rotated continuously or stopped during the passage of the current.

3. Pipe welding

The pipe welding is used in pipe manufacture, in pipe fabrication and in pipe joint. The steel pipe is manufactured by five different processes (*viz.,* butt weld, lap weld, electric weld, hammer weld and seamless weld).

21.8 IMPERFECTIONS (DEFECTS) IN WELD

The welding should be done carefully so that there are no defects or imperfections in it. The strength of an imperfect or a defective weld is less than that of a perfect designed weld. Imperfect weld remains weak and unsafe. Therefore, necessary precautions should be taken to avoid imperfections in welding. The various imperfections in welding may be classified in two categories :

1. Surface imperfections in the weld or adjacent metal.
2. Internal imperfections in the weld or adjacent metal.

1. Surface imperfections in weld

The surface imperfections are generally removed in final dressing of the surface of weld. Sometimes, surface imperfections cannot be eliminated completely. The following are main surface imperfections in the weld or adjacent metal :

1. Edge of the plate melted off,
2. Exposed inclusion.
3. Exposed porosity,
4. Incompletely filled groove.
5. Overlap.
6. Undercut.

1. Edge of the plate melted off. This type of imperfection occurs in a fillet weld, the free edge of plate may be melted off as shown in Fig. 21.20. This may happen because the free edge may not be sufficient built-up.

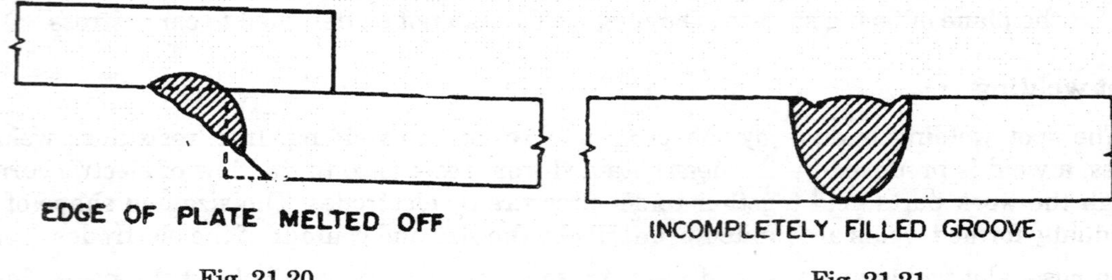

Fig. 21.20

Fig. 21.21

2. Exposed inclusion. A slag or any other foreign matter may be entrapped at the surface. This type of defect is called exposed inclusion.

3. Exposed porosity. A group of gas pores may remain at the surface of weld. These gas pores may not be completely removed while final dressing of the surface of weld.

4. Incompletely filled groove. This type of defect occurs in butt weld. A continuous or an intermittent channel may remain in the face of a butt-weld as shown in Fig. 21.21. As a result of this, the thickness of throat of the weld remains less than that of the parental metal.

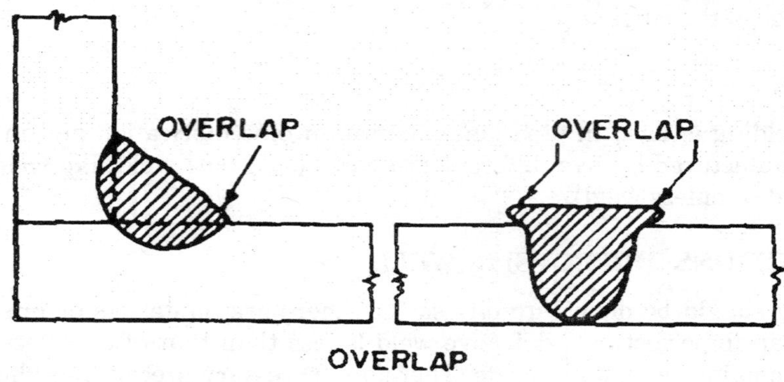

Fig. 21.22

5. Overlap. When the overflow of weld metal takes place over the surface of weld and fusion of the parent metal does occur, then, this defect is caused at toe or root of the weld as shown in Fig. 21.22.

6. Undercut. A groove may form in the base metal along the toe of a weld because of excess heat of arc in the metal arc welding. The groove may remain unfilled by weld metal as shown in Fig. 21.23. This defect is known as undercut. This defect may be removed in final dressing of the surface of the weld by depositing weld metal.

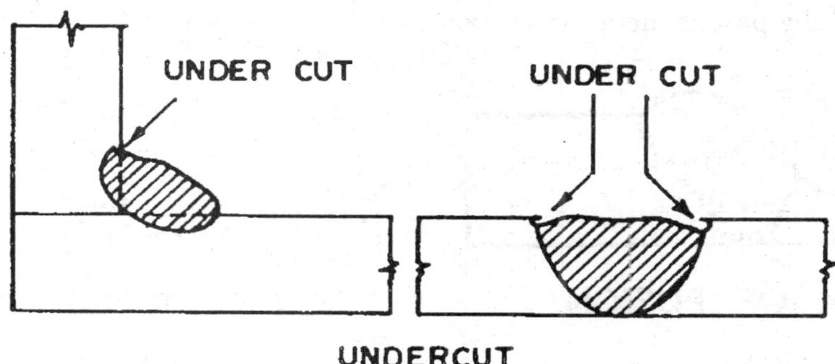

Fig. 21.23. Undercut

2. Internal imperfections in weld

The internal imperfections in the weld or adjacent metal cannot be removed. The following are main internal imperfections in the weld :

1. Gas pore 2. Inclusion
3. Lack of fusion 4. Incomplete penetration 5. Porosity.

1. Gas pore. While welding, gas may be entrapped in the weld. As a result of this, a small cavity occurs in the weld as shown in Fig. 21.24. The cavities upto 1.6 mm in diameter included in the defects are known as gas pore.

2. Inclusion. While welding, gas or other foreign matter may be entrapped in the weld. This effect is known as inclusion. The inclusion may appear generally in the form as large isolated inclusions, clusters of small inclusion or line inclusion. This defect is more irregular in shape than gas cavity.

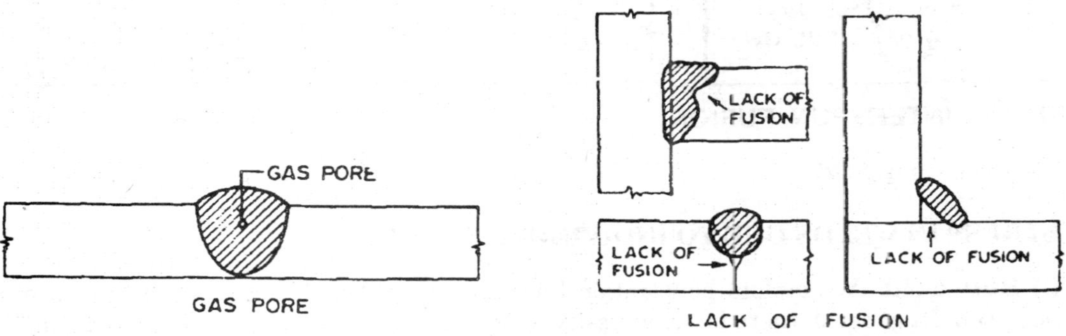

Fig. 21.24. Fig. 21.25

3. Lack of fusion. The lack of union in a weld is known as lack of fusion. The lack of union in a weld may remain between weld metal and parent metal, or between parent metal and parent metal or between weld metal and weld metal as shown in Fig. 21.25.

The lack of fusion may occur in the following forms :

(i) Lack of side fusion. The lack of union between weld metal and parent metal at a side of a weld outside the root is called lack of side fusion. It is shown in Fig. 21.25.

(ii) Lack of root fusion. The lack of union between weld metal and parent or between adjacent faces of the parent metal at the root is called lack of root fusion. It is shown in Fig. 21.26.

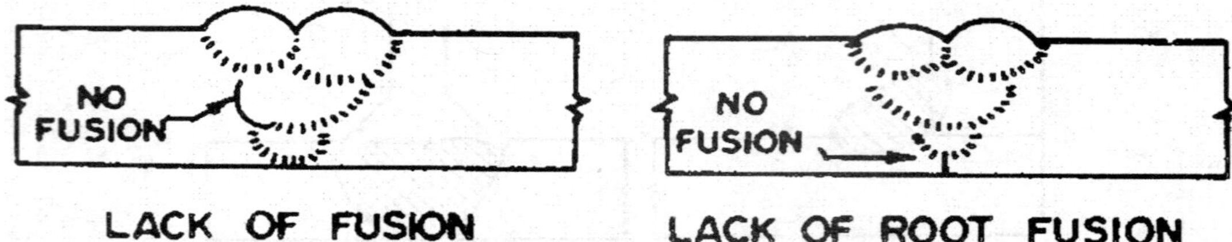

Fig. 21.25 Fig. 21.26

(iii) Lack of inter-run fusion. In a multi-run weld, the lack of union between adjacent runs of weld metals is called lack of inter-run fusion. It is shown in Fig. 21.27.

4. Incomplete penetration. In case of multi-run weld, small fissures may form by previous runs or run. These small fissures may usually occur at the toes of the underlying runs or run.

These small fissures may remain unfilled. At such positions, a gap may occur. This defect is known as incomplete inter-run penetration. It is shown in Fig. 21.28.

Sometimes, at the root of a groove, the gap or groove angle is insufficient. As a result of this, weld metal fails to fill root and a gap is left. Such defect is known as incomplete roof penetration.

5. Porosity. A group of gas pores may occur in a weld. Such defect is known as porosity in weld. Depending upon size of pores, it is called fine or medium or coarse. The porosity may occur as clusters or chains or it may be scattered.

Fig. 21.27 Fig. 21.28

21.9 STRESS IN WELDS DUE TO INDIVIDUAL FORCES

(i) Butt weld. A complete penetration butt weld provides strength equal to that of the member itself. In general, there is no necessity to determine stress in the weld. There is also no necessity to determine the size of weld so long as the weld metal possesses the physical properties of the plate. In case of incomplete penetration, the stress in direct compression or tension or shear force alone is given by

$$p = (P/A),\ (P/l_w \cdot t_e) \text{ where,} \qquad (21.1)$$

P = Pull or thrust to be transmitted by the weld

l_w = Effective length of butt weld

t_e = Effective throat thickness

Stress in bending $\sigma = (M/I) \times y$, where (21.2)

M = Bending moment at the weld section

I = Moment of inertia of weld group

y = Distance of the fibre of weld under consideration

Stress in torsion $\tau_s = (T/J) \times r$, where (21.3)

τ_s = Shear stress due to torsion

T = Torsion at the weld section

r = Distance to the fibre of weld under consideration

J = Torsional moment of inertia

Stress in beam shear $\tau_s = \left(\frac{FA\,\bar{y}}{I.b}\right)$, where, (21.4)

τ_s = Shear stress in the weld

I = Moment of inertia of weld group

b = Width of the weld at the weld section under consideration

A_y = Statistical moment of area of weld above the fibre under consideration.

The above expressions for stresses in butt weld in direct compression or tension, or bending or torsion or in beam-shear are based upon simple analysis, and are satisfactory for purpose of design.

(ii) Fillet weld. The stress in a fillet weld is considered as shear on the throat regardless of the direction of the applied load. In design, it is convenient to use longitudinal weld. The distribution of stress in weld is shown in Fig. 21.29. The stress in weld near the ends are high.

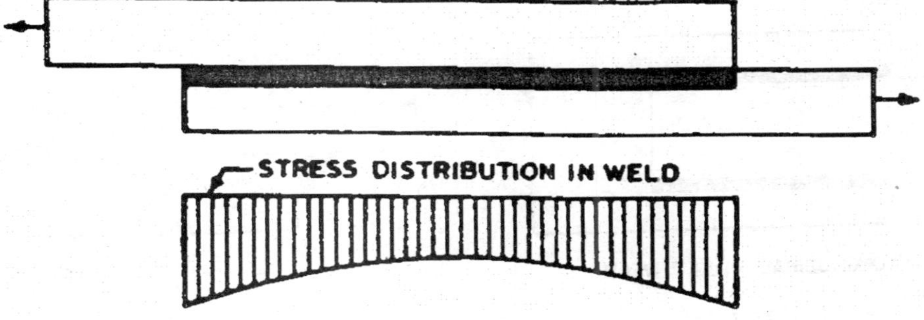

Fig. 21.29.

For the structural welds, the heavily coated electrodes are generally used. The heavily coated electrodes produce more ductile weld. Because of plastic deformation near the ends, the shear stress in the weld becomes more or less uniform before failure. The following formulae are considered sufficient and satisfactory for purpose of design :

1. Direct stress formula, $(\sigma = P/A) = (P/l_w \cdot t_w)$

2. Flexure formula, $(\sigma_b = M/I \times y)$

3. Torsion formula, $(\tau = (T/J) \times r)$

4. Beam-shear formula $\tau_s = \left(\frac{F.A\bar{y}}{I.b}\right)$

1. Direct stress formula. It is assumed that when the weld is subjected to direct compression or tension, the stress in weld is obtained by dividing the load by effective throat area. The effective throat area is product of effective length and effective throat thickness of weld. The heavily coated electrodes are used. These electrodes produce ductile weld. The ductility of weld allows equalization of stress in such weld before failure. Therefore, the use of direct stress formula is satisfactory for purpose of design.

Fig. 21.30 shows fillet welds joining two plates in lap joint. The loads acting are eccentric by distance *t*. Along with direct stress, welds are subjected to bending and shear.

Fig. 21.31 shows fillet weld joining two plates at right angles. The resisting loads P/2 are eccentric by the distance t/2 on either side of the plate. Along with direct stress, welds are also subjected to the bending and shear.

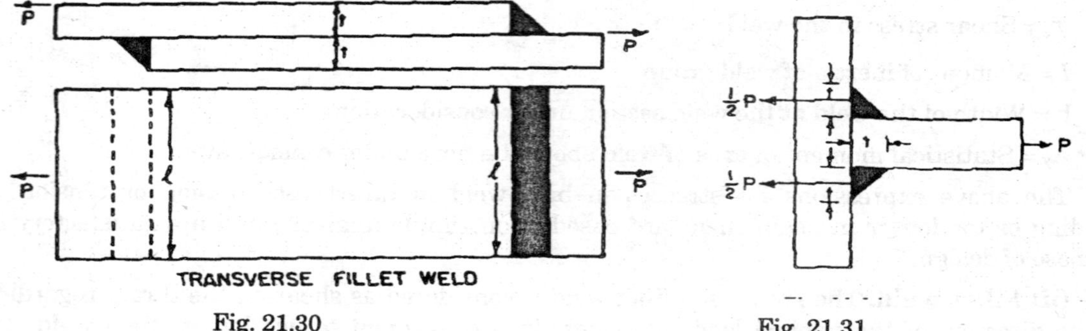

Fig. 21.30

Fig. 21.31.

Fig. 21.32 shows fillet welds joining two plates in lap joint. The longitudinal fillet weld have been used.

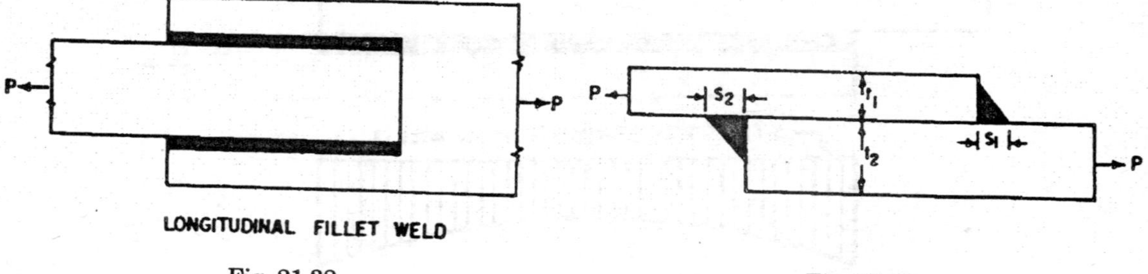

Fig. 21.32

Fig. 21.33.

It is common practices to treat welds shown in Fig. 21.30, 21.31 and 21.32, stressed uniformly.

In Fig. 21.30, direct stress

$$\sigma = \left(\frac{P}{A}\right) = \left(\frac{P}{\left(1/\sqrt{2}\right).2l_w.s_w}\right)$$

where, l is the effective length of weld and s is the size of weld

In Fig. 21.31, direct stress

$$\sigma = \left(\frac{P}{A}\right) = \left(\frac{P}{\left(1/\sqrt{2}\right).2l_w.s_w}\right)$$

In Fig. 21.32, direct stress

$$\sigma = \left(\frac{P}{A}\right) = \left(\frac{P}{\left(1/\sqrt{2}\right).2l_w.s_w}\right)$$

The direct stresses in all these three cases are equal in intensities. Since the allowable stress in shear is less, it controls the design.

In Fig. 21.33, the thickness of two plates connected by transverse fillet weld in lap joint are not equal. The strengths of weld per unit length are equal. Therefore, the load between welds is spread between plates is in proportion to their respective thickness.

$$\text{Load in plate A}, P_1 = \left(\frac{P.t_w}{\left(t_{1_w} + t_{2_w}\right)}\right)$$

where, t_1 is the thickness of plate A and t_2 is the thickness of plate B

$$\text{Stress in weld, } \sigma = \left(\frac{P}{\left(1/\sqrt{2}\right).l\,s_1}\right) = \left(\frac{P_{t1}}{\left(t_1 + t_2\right)}.\frac{\sqrt{2}}{l.s_1}\right) \quad (21.5)$$

where, l is the effective length of weld and s_1 is the size of weld in plate A

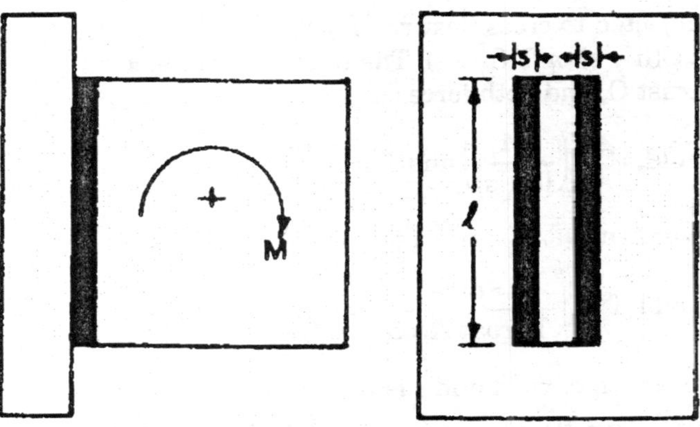

Fig. 21.34. Longitudinal flexure

$$\text{Load in plate B}, P_2 = \left(\frac{P_2.t_2}{\left(t_1 + t_2\right)}\right)$$

$$\text{Stress in weld, } \sigma = \left(\frac{P_2}{(1/\sqrt{2})l.s_2}\right) = \left(\frac{P.t_1}{(t_1+t_2)}\right).\frac{\sqrt{2}}{l.s_2} \quad (21.6)$$

where, s_2 is the size of weld in plate B.

2. Flexure formula. Fig. 21.34 shows fillet weld subjected to longitudinal flexure. The throat section is subjected to critical stress. The failure will occur at this section. The moment M, is resisted by width of weld as $2.s/\sqrt{2}$ and depth of weld as 'l'. The bending stress

$$\sigma = \frac{M}{I}.y = \left[\frac{M}{\frac{1}{12}\left(2.\frac{s}{\sqrt{2}}\right).\left(l_w^3\right)}\right].\frac{1}{2} = \left[\frac{6M}{\frac{1}{\sqrt{2}}.l_w^2\ 2s}\right] \quad (21.7)$$

Fig. 21.35 shows fillet welds subjected to cross flexure. The foot of weld is subjected to maximum shear stress and maximum tensile stress.

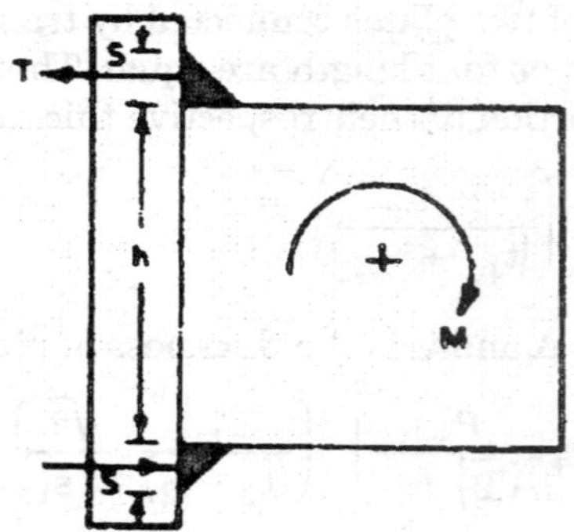

Fig. 21.35. Cross flexure

The fillet weld subjected to cross flexure may be analysed by steel theory. The moment M_1 is resisted by a couple C(h + s) or T (h + s). The top weld is subjected to tension T, while bottom weld is subjected to thrust C, and both forces are equal.

$$\text{Force in weld, } C = \left(\frac{M}{(h+s)}\right) \text{ and } T = \left(\frac{M}{(h+s)}\right)$$

where, (h + s) is the lever arm.

$$\text{Stress in weld. } f = \left(\frac{\text{Force}}{\text{Throat Area}}\right) = \left(\frac{M}{(h+s).(s/\sqrt{2}).1}\right) \quad (21.8)$$

where, l is the effective length of weld and s is the size of weld.

3. Torsion formula. Fig. 21.36 shows a circular pipe connected to a plate by a fillet weld. The weld is subjected to torsion, T. The throat of weld is subjected to critical stress. The shear stress due to torsion in weld, τ is equal to (T/J).r.

$$J = \pi d\left(\frac{s}{\sqrt{2}}\right) = \left(\frac{d}{2}\right)^2 = \left(\frac{\pi.s.d^3}{4\sqrt{2}}\right) \qquad (\because\ r = d/2)$$

$$\therefore\ \tau = \left(\frac{T.(d/2)}{\pi.s.d^3/4\sqrt{2}}\right) = \left(\frac{2\sqrt{2}\,T}{\pi.s.d^2}\right) \quad (21.9)$$

TORSION OF CIRCULAR FILLET WELD

Fig. 21.36. Torsion circular fillet weld

It is seen that the torsional formula is exact for fillet weld of circular section.

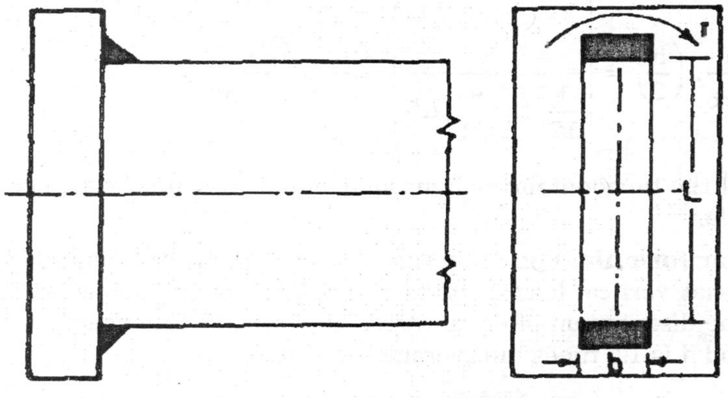

TORSION OF WIDELY SPACED SHORT FILLET WELB

Fig. 21.37. Torsion of widely spaced short fillet weld

Fig. 21.37 shows a bar of rectangular section connected to a plate by fillet welds. The fillet welds are short and are widely spaced. The welds are subjected to torsion, T. The torsion is resisted by a resisting couple.

$$\text{Force in each weld} = \left(\frac{\text{Torsion}}{\text{Lever Arm}}\right) = \left(\frac{T}{L}\right)$$

Shear stress in weld

$$\tau_s = \frac{T}{L}.\left(\frac{1}{(s/\sqrt{2}).b}\right) = \left(\frac{T\sqrt{2}}{L.s.b}\right) \quad (21.10)$$

Fig. 21.38 shows a plate of rectangular cross-section. It is connected to another plate by two adjacent long and narrow fillet welds. The welds are subjected to torsion T. The polar moment of inertia J is approximately equal to moment of inertia I_{xx}. The distance from centre of rotation (e.g., of weld group) to the extreme fibre of weld is approximately equal to half the depth of weld.

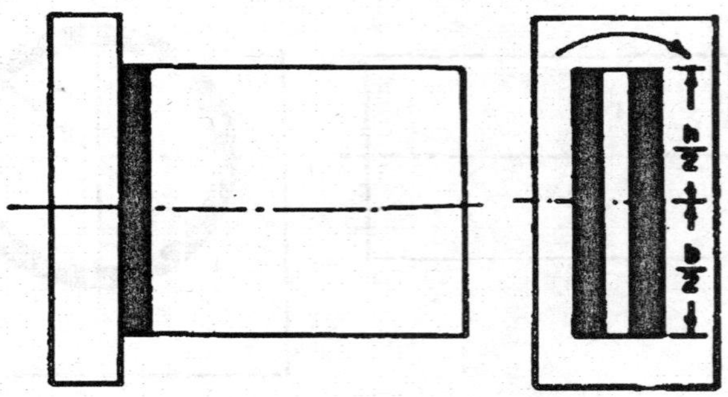

Fig. 21.38. Torsion of adjacent fillet weld

Therefore, shear stress in such weld

$$f_s = \frac{T}{I_{xx}} \cdot \left(\frac{h}{2}\right) = \left[\frac{1}{\frac{1}{12} \cdot 2\left(\frac{s}{\sqrt{2}}\right) h^3}\right] \cdot \frac{h}{2} = \left(\frac{3\sqrt{2}T}{s \cdot h^2}\right) \tag{21.11}$$

It is seen that the torsional formula for long and narrow fillet weld becomes equivalent to the flexural formula.

4. Beam shear formula. Fig. 21.39 shows a rectangular bar connected to a plate by fillet weld. The bar carries vertical load at its free end. The weld is subjected to shear stress and bending stress. The distribution of shear stress is parabolic. The shear stress is maximum at the neutral axis and it is 1.5 times the average shear stress.

The shear stress, $t_s = \left(\frac{FA\,\bar{y}}{I \cdot b}\right)$

Statical moment of area of the weld above neutral axis about xx- axis

$$A.\bar{y} = 2 \times \left(\frac{s}{\sqrt{2}}\right) \times \left(\frac{h}{2}\right) \times \left(\frac{h}{4}\right) = \left(\frac{s.h^2}{4\sqrt{2}}\right)$$

$$I_{xx} = \frac{1}{12} \cdot 2\left(\frac{s}{\sqrt{2}}\right)\left(h^3\right) = \frac{1}{6\sqrt{2}} s.h^3$$

Width of weld at xx = $\left(\frac{2s}{\sqrt{2}}\right)$

$$\tau_{s(max)} = \left[\frac{F \cdot \frac{sh^2}{4\sqrt{2}}}{\frac{1}{6\sqrt{2}} sh^3 \left(\frac{2s}{\sqrt{2}}\right)}\right] = \left(\frac{3}{2} \times \frac{F}{\left(\frac{2s}{\sqrt{2}} . h\right)}\right) = 1.5 \ f_s \ (\text{average})$$

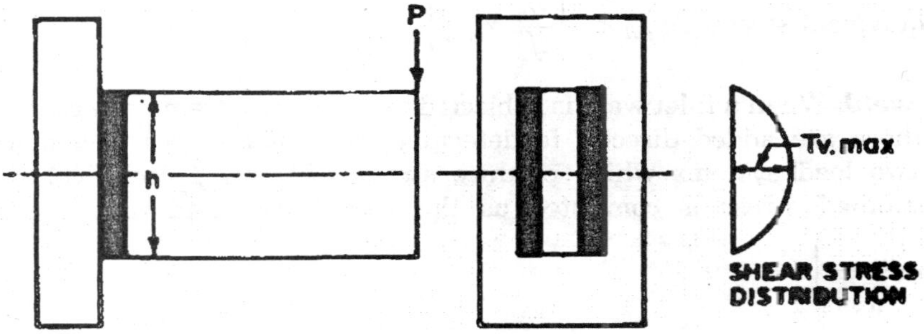

Fig. 21.39.

21.10 COMBINED STRESSES IN WELDS

1. Butt weld. A structural material may be subjected to any possible combined stress. When two structural members are connected by butt weld, it takes the place of equivalent amount of base metal. Therefore, the butt weld may also be subjected to any possible combined stress. In butt weld, the allowable stresses for compression, tension and shear are different. Therefore, maximum values of each stress are found to determine the critical stress. When butt weld is subjected to tensile or compressive stress (axial and/or bending) along with direct shear stress, the welds are proportioned so that quantity.

$$\left[\frac{p_s}{P_s}\right]^2 + \left[\frac{p_{t/c}}{p_{t/c}}\right]^2$$ does not exceed unity. Where,

p_s = Actual shear stress in the weld

P_s = Maximum permissible shear stress in the weld,

p_{vc} = Actual tensile/compressive stress (Axial or bending) in the weld.

P_{uc} = Maximum permissible tensile/compressive stress (axial or bending) in the weld.

When the butt weld is subjected to following combined stresses, then, the combined stress shall not exceed allowable stress for the parent metal or equivalent stress, $\sigma_{e.cal}$

(A) For shear combined with bending compressive or tensile.

$$\sigma_{e.cal} = \left[\sigma^2_{bt.cal} + 3\ \tau_{vm.cal}\ 2\right]^{1/2} \quad (21.12A)$$

or $$\sigma_{e.cal} = \left[\sigma_{bc.cal}\ 2 + 3\ \tau_{vm.cal}\ 2\right]^{1/2} \quad (21.12B)$$

(B) For the combined bearing, bending and shear stresses

$$\sigma_{e.cal} = \left[\sigma_{bt.cal}{}^2 + \sigma^2_{p.cal} + \sigma_{bt.cal}\ \sigma_{p.cal} + 3\ \tau^2_{vm.cal}\right]^{1/2} \quad (21.12C)$$

or $$\sigma_{e.cal} = \left[\sigma_{bt.cal}^{2} + \sigma_{p.cal}^{2} + \sigma_{bc.cal}\,\sigma_{p.cal} + 3\,\tau_{vm.cal}^{2}\right]^{1/2} \tag{21.12D}$$

where, $\sigma_{bt.cal}$; $\sigma_{bc.cal}$; $\tau_{vm.cal}$ and $\tau_{vm.cal}$ are the numerical values of the co-existing bending (tensile or compressive), shear and bearing stress, respectively. $\sigma_{e.cal}$ should not exceed $\sigma_{e.max}$ permissible equivalent stress. $\sigma_{e.cal} <= \left(\frac{\sigma_u}{\sqrt{3}}\gamma_{msw}\right)$

2. Fillet weld. When a fillet weld is subjected to two load systems, which produce collinear stress, then these are added directly to determine combined stress. When a fillet weld is subjected to two load systems which produce stresses in two perpendicular direction, the maximum combined stress is computed as the vector sum and that should not exceed $\sigma_{e.cal} <= \left(\frac{f_u}{\sqrt{3}\,.\,\gamma_{msw}}\right)$.

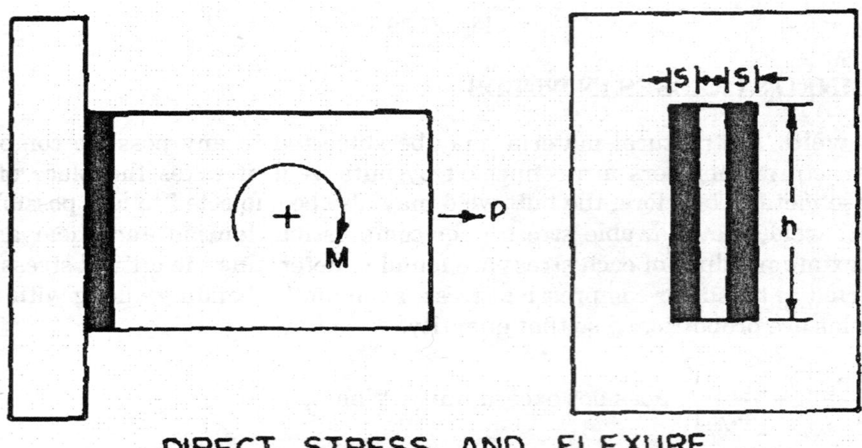

Fig. 21.40. Direct stress and flexure

Fig. 21.40 shows a rectangular bar connected to plate by fillet welds. The fillet weld is subjected to direct stress due to load P, and bending stress due to moment M. The two load systems produce collinear stresses. The combined stress in weld in found by adding these stresses directly.

$$\sigma_{b(max)} = \left(\frac{P}{A} + \frac{M_y}{I_{xx}}\right) = \left[\frac{P}{s\frac{S}{\sqrt{2}}h} + \frac{M.\frac{h}{2}}{2.\frac{1}{2}.\frac{S}{\sqrt{h}}h^3}\right]$$

or $$\sigma_{s(max)} = \left(\frac{P\sqrt{2}}{2s.h} + \frac{3\sqrt{2M}}{s.h^2}\right) \tag{21.13}$$

Fig. 21.41 shows rectangular bar connected to a plate by fillet welds. The weld is subjected to an eccentric load. The load system produces shear stress and bending stress. The stresses produced are collinear. The combined stress is obtained by two stresses

$$f_{s(max)} = \left(\frac{P}{A} + \frac{M.y}{I_{xx}}\right) = \left(\frac{P}{2\frac{s}{\sqrt{h}}h} + \frac{P\left(L+\frac{h}{2}\right).\frac{h}{2}}{2\times\frac{1}{12}\times\frac{s}{\sqrt{2}}h^3}\right)$$

$$f_{s(max)} = \left(\frac{P\sqrt{2}}{2s.h} + \frac{3\sqrt{2}.P.(L+h/2)}{s.h^2}\right) \tag{21.14}$$

CROSS SHEAR AND FLEXURE

Fig. 21.41. Cross shear and flexure

Fig. 21.42 shows a circular bar connected to a plate by fillet weld around the periphery of the bar. The bar is subjected to tangential load P at its free end. The weld is subjected to shear due to torsion and cross shear. The shear stress due to torsion is uniform around the entire periphery. At the upper end of the bar, the two stresses are collinear and the combined stress is found by adding them directly.

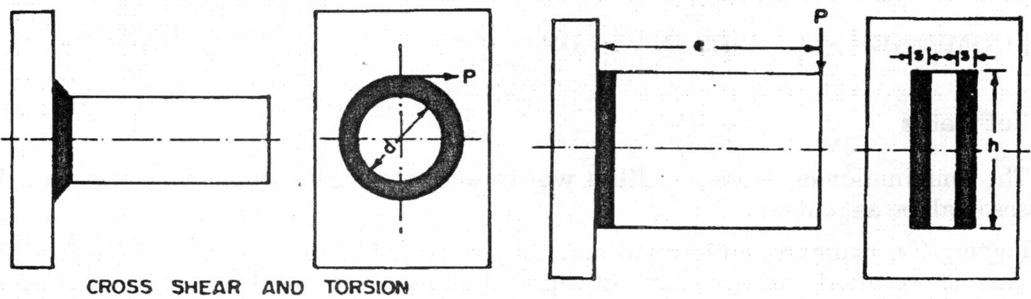

Fig. 21.42 Fig. 21.43

Fig. 21.43 shows a rectangular bar connected to a plate by fillet welds. The bar is carrying a vertical load, P, at its free end. The weld is subjected to shear stress and bending stress. The stresses are acting in two perpendicular directions. The combined stress is obtained by vectorial sum of two stresses.

Shear stress, $\tau_s = \frac{P}{A} = \left(\frac{P}{2:\left(s/\sqrt{2}\right)h}\right) = \left(\frac{P\sqrt{2}}{2h}\right)$

The maximum bending stress

$$\sigma_b = \left(\frac{My}{I}\right) = \left(\frac{P.e.(h/2)}{2.\frac{1}{12}.\frac{s}{\sqrt{2}}h^3}\right) = \left(\frac{P.e(3\sqrt{2})}{s.h^2}\right)$$

The combined stress, $\sigma = \left[\tau_s^2 + \sigma_b^2\right]^{1/2}$

When the fillet weld is subjected to following combined stress, then the equivalent stress, $\sigma_{e.cal}$ is calculated, which should not exceed (σ_y/γ_{msw}).

(A) Combined bending and shear in fillet weld

$$\sigma_{e.cal} = \left[\sigma_{bt.cal}^2 + 3\,\tau_{vm.cal}^2\right]^{1/2} \quad (21.14A)$$

$$\sigma_{e.cal} = \left[\sigma_{bc.cal}^2 + 3\,\tau_{vm.cal}^2\right]^{1/2} \quad (21.14B)$$

Eqs. (21.14 A) and (21.14 B) are used when the stress $(\sigma_{bt.cal}/\sigma_{bc.cal})$ and $\tau_{vm.cal}$ are mutually perpendicular to each other.

(B) Combined axial and shear in fillet weld

$$\sigma_e = \left[\sigma^2 + 1.8\,\tau^2\right]^{1/2} \quad (21.14\ C)$$

where, σ is the axial stress in the weld, σ_{bt} σ_{bc} and τ have been defined in (21.12 A), (21.12 B) and (21.12 C), respectively.

21.11 COMBINED STRESSES IN WELDS

1. Fillet welds

The combination of stresses in **fillet welds** and **butt welds** when subjected to combined stresses shall be as under :

Factors (for example, multi-axial state of stress, variation in the values of yield stress, presence of residual stresses and strain hardening) complicate the analysis of weld behaviour. However, the analysis of average stresses in the throat of weld is simplified in many cases.

Consider two plates as shown in Fig. 21., which have been connected perpendicular to each other by using two fillet welds. Let l_w be the effective length of weld and t_{tw} be the throat thickness of the weld. Let s_1 and s_2 be the sizes of weld and these are not equal (s_1 be the smaller size and s_2 be the larger size) and the inclination be $\alpha = \tan^{-1}(s_2/s_2)$. Each weld trnasfers longitudinal shear, v_L and transverse shears V_{tx} and V_{ty} between the plates.

Trahair, N.S., Bradform, M.A. and Nethercot, D.A., 2001 (The Behaviour and Design of Steel Structures to BS : 5950, 3rd Edition, Spon Press London, pp. 448) expressed the average normal stress, f_w and shear stress, τ_w in terms of above forces as below :

$$\tau_w . l_w\ t_{tw} = V_{tx} . \sin\alpha + V_{ty} . \cos\alpha \quad \text{(i)}$$

$$\tau_w . l_w . t_{tw} = \left[\left(V_{tx} . \cos\alpha - V_{ty}\ \sin\alpha\right)^2 + V_L^2\right]^{1/2} \quad \text{(ii)}$$

It is to note that local stresses in the weld arising from effects of bending, shear lag, and stress concentrations together with longitudinal normal stress induced by compatibility between the weld and end plate have been neglected in the design.

In common practice to assume that the **static strength of weld** is attained by the average throat stresses, f_w and τ_w alone.

Owens, G.W. and Cheal, B.D. 1989 (Structural steelwork Connections, Butterworths, London, pp. 330) stated that the **ultimate strength of the weld,** f_{uw} reaches

When,

$$f_{uw} = \beta\left(f_w^2 + \gamma\ \tau_w^2\right)^{1/2} \quad \text{(iii)}$$

γ = A parameter, its value varies from 1.8 and 3.0. However, they decided to use $\gamma = 3$.

Note : In this expression (iii), the interactive resistance to rupture (ultimate) has been related and not to onset of yielding.

β = A function of parent metal and weld metal strength.

In case, the design strength, f_{uw} is related to the parent metal strength, then, the lower the parent metal strength, the lower would be the value of β. Since, the ratio of weld metal strength to the parent metal with increases with increasing parent metal strength. Owens and Cheal, 1989 gave the typical values of β for the different values of range of yield, σ_y in N/mm²

For Range of yield	Values of β
$0 < \sigma_y < 240$	0.70
$240 < \sigma_y < 280$	0.80
$280 < \sigma_y < 340$	0.85
$340 < \sigma_y < 400$	1.00

Considering β = 1.00 (Unity), the expression (iii) for equivalent stress may be written as

$$f_e = f_{uw} = \left(f_w^2 + 3\ \tau_w^2\right)^{1/2} \quad \text{(iv)A}$$

$$\left[\left(f_{uw}\right) . l_w . t_{tw}\right]^2 = \left(f_w^2 + 3\ \tau_w^2\right) \quad \text{(iv)B}$$

From the expressions (i) and (ii)

$$f_w = \left(\frac{1}{l_w . t_w}\right)\left[V_{tx} . \sin\alpha + V_{ty} . \cos\alpha\right] \textit{and} \quad \text{(v)A}$$

$$\tau_w = \left(\frac{1}{l_w . t_{tw}}\right)\left[\left(V_{tx} . \cos\alpha - V_{ty} . \cos\alpha\right)^2 + V_L^2\right]^{1/2} \quad \text{(vi)A}$$

$$\left[\tau_w . \left(l_w . t_{tw}\right)^2\right] = \left[\left(V_{tx}\ \cos\alpha - V_{ty} . \sin\alpha\right)^2 + V_L^2\right]$$

Substituting for f_w and τ_w in the (vi)B expression (iv)

$$\left(f_{uw} \cdot l_w \cdot t_{tw}\right)^2 = \left[\left(V_{tx} \cdot \sin\alpha + V_{ty} \cdot \cos\alpha\right)^2 + V_L^2\right] \text{ and} \tag{iv)C}$$

$$3\left(\tau_w \cdot l_w \cdot t_{tw}\right)^2 = 3\left[\left(V_{tx} \cdot \cos\alpha - V_{ty} \cdot \sin\alpha\right)^2 + V_L^2\right] \tag{iv)C}$$

Therefore,

$$\left(f_{uw} \cdot l_w \cdot t_{tw}\right)^2 = 3\left[\left(V_{tx}^2 + V_{ty}^2 + V_L^2\right)\right] - 2\left(V_{tx} \sin\alpha + V_{ty} \cdot \cos\alpha\right)^2 \tag{vii}$$

Neglecting the second term on right hand side

$$\left(\frac{f_{uw}}{\sqrt{3}}\right) = \left(\frac{V_R}{l_w \cdot t_{tw}}\right) \tag{viii}$$

In code of Practice, IS : 800-2007, weld shear strength has been replaced the design strength ($f_{uw}/\sqrt{3}\,\gamma_{msy}$) and considers f_u as the smaller of the ultimate stress

Note: The weld metal always has a higher strength. Therefore, the parent metal strength is used in the expression suggested in IS : 800-2007.

As such, the design strength of weld as per IS : 800-2007 may be found as under

$$P_{dw} = \left(\frac{f_u}{\sqrt{3}\,\gamma_{ms}}\right) \cdot \left(l_w \cdot t_{tw}\right)$$

where, $t_{tw} = 0.7 \cdot S$ (taking $s_1 = s_2 = s$)

2. Butt-Welds

The check for the combination of stresses in butt-welds need not be carried out provided that

(i) the butt-welds are axially loaded, and (ii) in single and double bevel welds,

The sum of normal and shear stresses does not exceed the design normal stress, and the shear stress does not exceed 50 percent of the design shear stress.

In case, the butt-welds are subjected to bearing, stress, σ_{br}, bending stress, σ_b and the shear stress, τ_w under the most unfavourable conditions of loading, the **equivalent stress,** f_e as found from the following expression as per IS : 800-2007, σ_e shall not exceed the values allowed for the parent metal

$$\sigma_e = \left(\sigma_b^2 + \sigma_{br}^2 + \sigma_b \cdot \sigma_{br} + 3\,\tau_w^2\right)^{1/2}$$

Where, σ_e = Equivalent stress

σ_b = Calculated stresses due to bending in N/mm²

σ_{br} = Calculated stress due to bearing in N/mm², and

τ_w = shear stress in weld

Where a packing is welded between two members and is less then 6 mm thick, or is too thin to allow provision of adequate welds or to prevent buckling, the packing shall be trimmd flush with the edges of the element subjected to the design action and the size of welds along the edges shall be increased over the required size by an amount equal to the thickness of the packing. Otherwise, the packing shall extend beyond the edges and shall be fillet welded to the pieces between which it is fitted.

Combined stresses in welds, when welded joints are constructed with mild steel conforming to IS : 226-1962 as parent metal and with electrodes conforming to IS : 814-1974 are adopted as per Table 21.3 are recommended in IS : 816-1969.

TABLE 21.3. STRESSES IN WELDS

Note. IS 816-1969 is yet to be revised in S.I. units. As such the values of the maximum permissible stresses have been converted from those given in m.k.s. units.

S.No.	Kind of stresses	Max. permissible values
1	Tension on section through throat of butt weld	σ_y/γ_{msw}
2	Compression on section through throat of butt weld	σ_y/γ_{msw}
3	Fibre stresses in bending	
	(a) tension	σ_y/γ_{msw}
	(b) compression	σ_y/γ_{msw}
4	Shear on section through throat of butt and fillet welds	$\sigma_y/(\sqrt{3}\,\gamma_{msw})$
5	Plug welds	$\sigma_y/(\sqrt{3}\,\gamma_{msw})$

The maximum permissible value of stresses of shear and tension are reduced to 80 percent of those given in Table 21.3, in case, the welding in done at site, (field).

When the effects of wind or earthquake forces are considered, then, maximum permissible values of stresses are increased by 25 per cent.

It is to note that maximum permissible stresses given in Table 21.3 are same as for the parent metal (mild steel IS : 226-1962).

Example 21.1 Two plates 16 mm thick are joined by

(i) a double - U butt-weld; (ii) a single - U butt-weld

Determine the strength of welded joint in tension in each case. Effective length of weld, l_w is 150 mm. Yield strengt of steel, σ_y = 250 N/mm^2. Take partial safety factor for the material strength γ_{ms} = 1.15

Solution :

(i) **Double - U butt weld** (complete penetration weld)

Effective throat thickness, t_t = 16 mm

Effective length of weld, l_w = 150 mm

Step 1 : Yield stress of steel, σ_y = 250 N/mm^2

Partial safety factor. γ_{msw} = 1.15

$$\text{Design yield stress, } \sigma_{yd} = \left(\frac{250}{1.15}\right) = 217.4\ N/mm^2$$

$$\text{Design strength} = \left(\frac{217.4 \times 16 \times 150}{1000}\right) = 521.76\ kN$$

(ii) **Single-V butt weld**

Effective throat thickness

$$t_t = \left(\frac{5}{8} \times 16\right) = 10\ mm$$

Effective length of weld = 150 mm

Step 2 : Design strength = $\left(\sigma_{yd} \times l_w \times t_{tw}\right)$

$$= \left(\frac{217.4 \times 10 \times 150}{1000}\right) = 326.1\ kN$$

Example 21.2: In a truss girder of a bridge as shown in Fig. 21.44 connected to the gusset plate by fillet weld. Determine the strength of the weld. The size of the fillet weld is 6 mm.

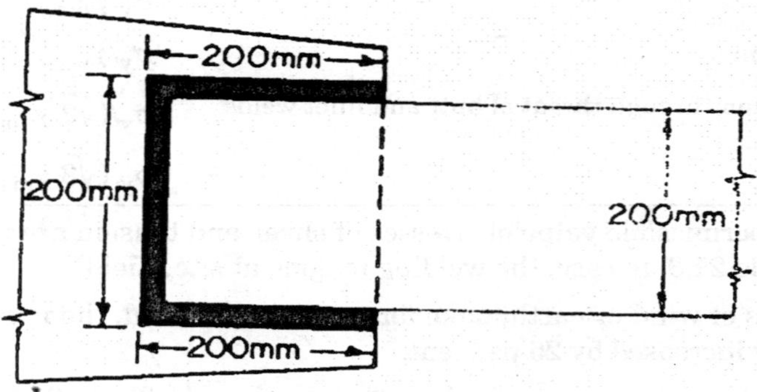

Fig. 21.44.

Solution : Size of fillet weld = 6 mm

Effective throat thickness, t_t

$t_t = (0.7 \times 6) = 4.2$ mm

Effective length of weld

$l_w = (200 + 200 + 200) = 600$ mm

Step 1 : Yield stress of steel, σ_y = 250 N/mm²

Partial safety factor, γ_{msw} = 1.15

$$\text{Design stress, } \sigma_{yd} = \left(\frac{250}{1.15}\right) = 217.4\ N/mm^2$$

Step 2 : Design strength

$$= \left(\frac{217.4 \times 4.2 \times 600}{1000}\right) = 547.848\ kN$$

Example 21.3 : In Example 21.2, the pull to be transmitted by the tie is 300 kN. Determine the necessary overlap of the tie.

Solution : Pull to be transmitted by the tie = 300 kN (While in service)

Size of fillet weld, s_w = 6 mm.

Step 1 : Yield stress of steel, σ_y = 250 N/mm²

Partial safety factor, γ_{msw} = 1.15

Design stress, $\sigma_{yd} = \left(\frac{250}{1.15}\right) = 217.4 N/mm^2$

Effective throat thickness

t_{tw} = (0.7 x 6) = 4.2 mm

(i) Pull transmitted by end fillet weld

$$= \left(\frac{217.4 \times 200 \times 4.2}{1000}\right) = 182.616\ kN$$

(ii) Let l_w be the necessary overlap for the weld. The pull transmitted by the side fillet welds (two welds)

$$= \left(\frac{2 \times 217.4 \times l \times 4.2}{1000}\right) = 1.82616\ kN$$

Step 2 : Factored pull to be transmitted

= (1.5 x 300) = 450 kN

Necessary overlap

$182.616 + 1.82616\, l_w = 450$ kN

$$l_w = \left(\frac{267.384}{1.82616}\right) = 146.418\ mm$$

Necessary overlap of the tie is 146.418 mm

Example 21.4. The web plate of a built-up welded I-section is 200 mm x 12 mm and the flange plates are 100 mm x 12 mm. The size of fillet weld is 6 mm. Compute the maximum shear force that may be allowed at any section.

Solution :

Step 1 : Moment of inertia of the builtup section about xx-axis

$$I_{xx} = \frac{1}{12}\left[10 \times 22.4^3 - 12.0 \times 20^3\right] \times 10^4\ mm^4 \quad = 1366.186 \times 10^4\ mm^4$$

Step 2 : Intensity of shear stress (at the weld section)

$$\tau_s = \left[\frac{V\,.\,A\,\bar{y}}{I_{xx}\,.\,(2t_t)}\right]$$

where,

$A\,.\,\bar{y}$ = moment of the area above the weld section about xx-axis

V = shear force at the section

t_t = Effective throat thickness of one weld

$$\tau_s = \left[\frac{V \times 100 \times 12 \times (100 + 6)}{1366.186 \times 10^4 \times 2 \times 0.7 \times 6}\right] N/mm^2$$

$$= (V / 902.2)$$

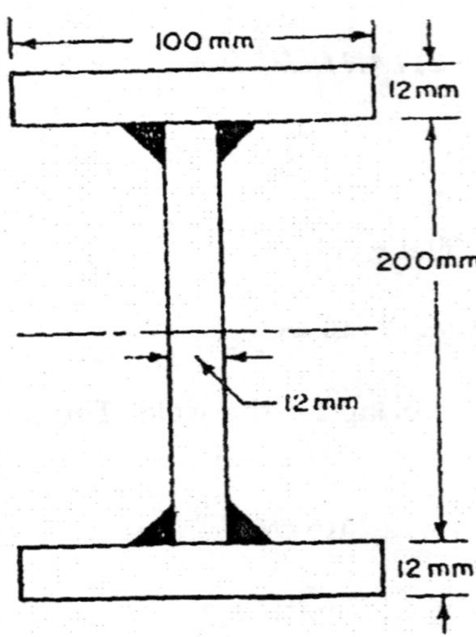

Fig. 21.45

Step 3 : Allowable shear in the weld

$$= \left(\frac{\sigma_y}{\sqrt{3}}\right) = 0.60\ \sigma_y\ approximately$$

$$V_{fd} = \left(\frac{0.60 \times 250}{1000} \times 902.2\right) = 135.33\ kN$$

$$V = \left(\frac{135.33}{1.5 \times 1.15}\right) = 78.45\ kN\ at\ service.$$

Example 21.5: Design a suitable longitudinal fillet weld to connect the plates as shown in Fig. 21.47 and transmit a pull equal to the full strength of the plate. Yield stress in the weld is 250 N/mm² and tensile stress in the plate is σ_y = 250 N/mm². The plates are 10 mm thick.

Design :

Yield stress of steel in plates = 250 N/mm²

Yield stress in the weld, σ_{yw} = 250 N/mm²

Partial safety factor for the strength of steel and steel, γ_{ms} = 1.15

$$\text{Design stress, } \sigma_{yd} = \left(\frac{250}{1.15}\right) = 217.4\ N/mm^2$$

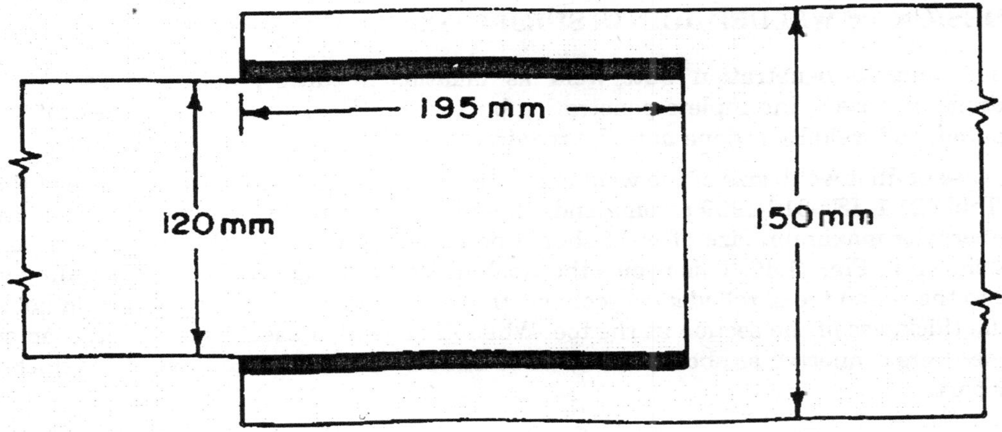

Fig. 21.46

Thickness of plate 150 mm width = 10 mm

Thickness of plate 120 mm width = 10 mm

Step 1 : Size of weld

The maximum size of weld required for thickness upto 19 mm is 5 mm. The maximum size of weld is limited by the thickness of plate (Thickness of plate - 1.5 mm).

$$= (10 - 1.5) = 8.5 \text{ mm}$$

Provide 6 mm fillet weld

Step 2 : Pull transmitted by 1 mm weld

$$= \left(\frac{217.4 \times 1 \times 0.7 \times 6}{1000}\right) = 0.913 \ kN$$

Step 3 : Tensile strength of plate (smaller width)

$$= \left(\frac{250 \times 120 \times 10}{1000}\right) = 300 \ kN$$

Step 4 : Necessary length of weld

Side welds are provided on two sides. Therefore,

$$= \left(\frac{300}{0.913} \times \frac{1}{2}\right) = 164.29 \ mm$$

Provide 165 mm longitudinal welds on each side.

Check :

(i) Length of weld 165 mm is more than the perpendicular distance 120 mm between them (Welds)

(ii) Spacing between the welds 120 mm.

21.12 DESIGN OF WELDED JOINTS SUBJECTED TO AXIAL LOAD

The complete penetration butt weld as mentioned earlier, does not require design calculations. In case of incomplete penetration butt weld, effective throat thickness of the weld is computed, and welding is done upto the required length.

In case of fillet weld, size of the weld is fixed keeping in view the minimum size of the weld as per Table 21.1. IS : 816-1969 recommends that when fillet weld is applied to the square edge of member, the maximum size of weld should be less than the edge thickness by at least 1.5 mm, as shown in Fig. 21.46. This avoids the washing down of edges of weld. When fillet weld is applied to the round toe of rolled steel sections, the maximum size of the weld should not exceed 3/4 of the thickness of the section at the toe. When fillet weld is used for lap joint, then overlap of the members connected as shown in Fig. 21.46 should not be less than five times thickness of thinner part.

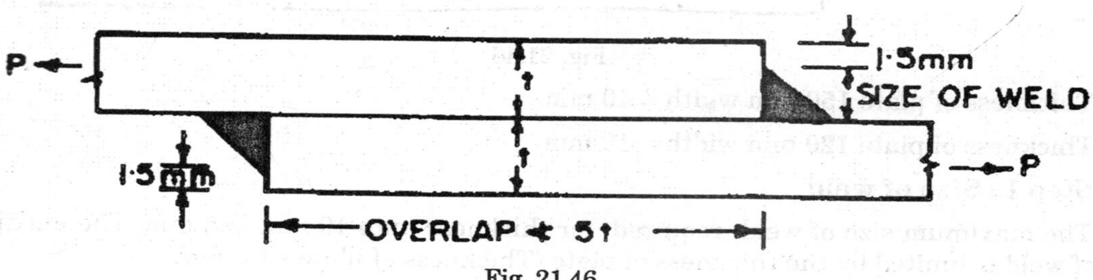

Fig. 21.46

The strength of the fillet weld is determined per mm length for the size of the weld adopted. The effective length of the weld is then computed for the full or thrust to be transmitted by the weld. In case, only side fillet welds are applied, the length of the each weld should not be less than perpendicular distance between them, and spacing between them shall not be more than 16 times the thinner part.

The design procedure for the design of fillet weld is as under :

1. Depending upon the thickness of plates to be connected, the size of fillet welds is assumed. The effective throat thickness t_{tw} = (0.7 . s)

2. The strength of 1 mm (one mm) of the weld as found.

3. Effective length of the weld, l_w is obtained as below :

External factored load is equal to the effective length of weld, l_w multiplied by the design strength of 1 mm weld. Therefore, the effective length of weld is determined.

This effective length may be provided either parallel to the load axis (as longitudinal fillet welds) or perpendicular to the load axis (as transverse fillet welds)

Both the welds are treated as if these are stressed equally. In case, the length exceeds 150 t_{tw}, the design capacity is reduced as per code IS : 800-2007.

4. In case, only longitudinal fillet welds are provided, then a check is applied. It is seen that the length of each longitudinal fillet weld is more than the perpendicular distance between them.

5. End returns are provided to keep the continuity of the weld. In that case, length of end returns are kept equal to twice the size of weld.

Example 21.6 : Design a suitable longitudinal fillet welds to connect the plates as shown in Fig. 21.47, and to transmit a pull equal to the full strength of thin plates. Yield stress of steel, σ_y for the plates is 250 N/mm². The plates are 10 mm and 12 mm thick, respectively.

Design : Yield stress of steel, $\sigma_y = 250$ N/mm²

Partial safety factor for the strength of material, $\gamma_{msy} = 1.15$

Design stress, $\sigma_{yd} = \left(\frac{250}{1.15}\right) = 217.4 N/mm^2$

Step 1 : Size of weld

The maximum size of weld required for thickness upto 19 mm is 5 mm. At the same time, it is limited by the thickness of the thinner plate, (i.e, (10-1.5)) = 8.5 mm Provide 6 mm size of fillet weld

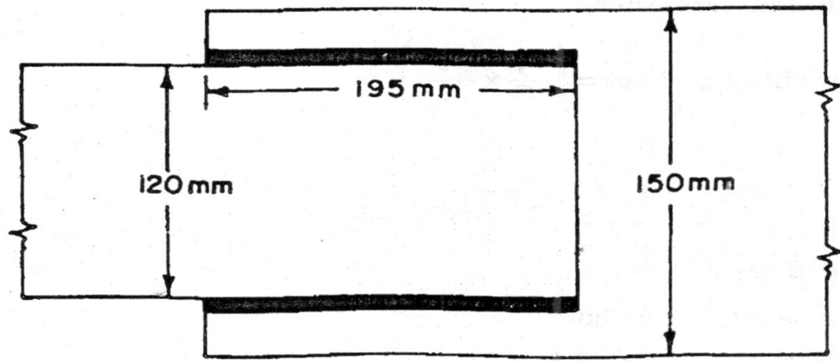

Fig. 21.47

Step 2 : Pull transmitted by 1 mm weld

$$= \left(\frac{217.4 \times 1 \times 0.7 \times 6}{1000}\right) = 0.913 \ kN/mm$$

Step 3 : Strength of thinner plate

$$P = \left(\frac{250}{1.15} \times \frac{10 \times 120}{1000}\right) = 260.869$$

Step 4 : Effective length of weld needed

$$l_w = \left(\frac{260.869}{0.913}\right) = 285.727 \ mm$$

$$\frac{1}{2}(l_w) = 142.86 \text{ mm} \quad \text{(Provide 145 mm on each side.)}$$

Example 21.7 : Two plates 120 mm x 10 mm are over-lapped and connected together by transverse fillet weld in lap joint to transmit pull equal to full strength of the plate. Design the suitable welding. Yield stress of steel for the plates is 250 N/mm²

Design : Yield stress in the steel, $\sigma_y = 250$ N/mm²

Partial safety factor for the strength of material, $\gamma_{msy} = 1.15$

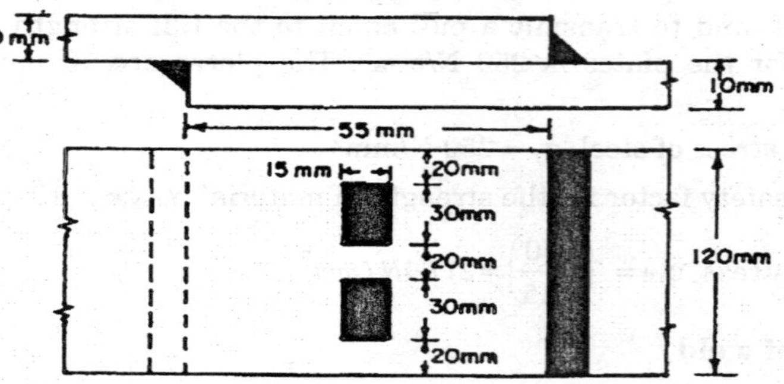

Fig. 21.48

Note : Transverse fillet welds provided in the direction perpendicular to the pull (load) transmitted fail in transverse shear

$$\text{Design shear stress } \sigma_d = \left(\frac{\sigma_y}{\sqrt{3}} \times \frac{1}{\gamma_{msy}}\right)$$

$$\sigma_d = \left(\frac{250}{\sqrt{3} \times 1.15}\right) = \left(\frac{217.4}{1.732}\right) = 125.51 \; N/mm^2$$

Step 1 : Size of weld

Minimum size of weld = 5 mm

Maximum size of weld = (10 - 1.5) = 8.5 mm

Provide 6 mm size of fillet welds

Fillet welds are provided at the two ends

Total length of two welds = (2 x 120)

= 240 mm

Step 2 : Pull transmitted by 6 mm weld

$$P_d = \left(\frac{125.51 \times 240 \times 0.7 \times 6}{1000}\right) = 126.514 \; kN$$

Step 3 : Maximum pull that may be transmitted

$$P = \left(\frac{250}{1.15} \times \frac{120 \times 10}{1000}\right) = 260.87 \; kN$$

Step 4 : To transmit the pull equal to the full strength of plate, provide additional weld. 4 plug welds are provide d as shown in Fig. 21.48

Provide two rectangular 4 plug welds 30 mm x 15 mm

Step 5 : Strength of plug welds

$$P_{pw} = \left(\frac{4 \times 125.51 \times 30 \times 15}{1000}\right) = 2 \times 112.959 \; kN$$

Step 6 : Total pull transmitted

= (126.514 + 2 x 112.959) kN

= (126.514 + 225.918) = 352.432 kN > 260.87 kN

Hence, satisfactory.

Example 21.8 A tension member consists of two MC 225, @ 0.259 kN/m. The channels are connected to either side of a gusset plate 12 mm thick. Design the welded joint to develop the full strength of the tension member. The develop is limited to 400 m. Yield stress in steel, σ_y = 250 N/mm². The partial safety factor, γ_{msy} = 1.15.

Design : Yield stress in steel, σ_y = 250 N/mm²

Partial safety factor for the strength of the material, γ_{msy} = 1.15

$$\text{Design stress, } \sigma_d = \left(\frac{250}{1.15}\right) = 217.4\ N/mm^2$$

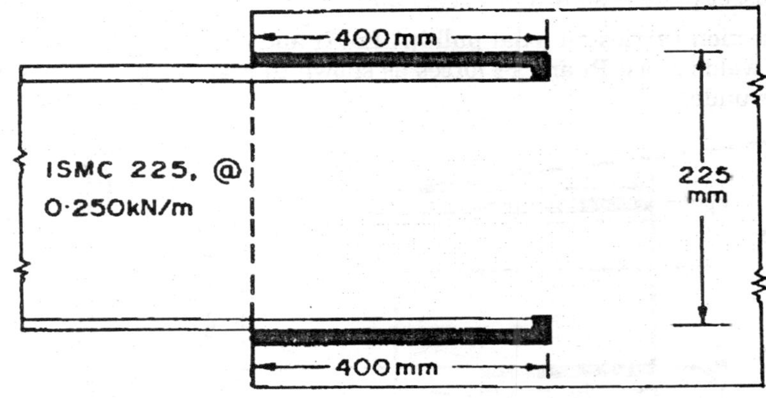

Fig. 21.49.

Step 1 : Properties of Section, MC 225, @ 0.259 kN/m

From ISI Handbook No. 1

Thickness of web = 6.4 mm

Thickness of flange, t_f = 12.4 mm

Cross-sectional area, A = 3301 mm²

Step 2 : Tensile strength of each channel

$$P_c = \left[\frac{217.4 \times 3301}{1000}\right] = 717.64\ kN$$

Provide 6 mm weld

Step 3 : Strength of weld per one mm length

$$P_w = \left[\frac{217.4 \times 1 \times 0.7 \times 6}{1000}\right] = 0.913\ kN/m$$

Step 4 : Total length of fillet weld necessary to connect one channel section

$$l_w = \left[\frac{717.64}{0.913}\right] = 786.024\ mm$$

The overlap of channel is limited to 400 mm.

Step 5 : Overlap as shown in Fig. 21.49

Total length, l_w = (400 + 400) = 800 mm > 786.024 mm

21.13 DESIGN OF WELDED JOINTS (UNSYMMETRICAL SECTIONS SUBJECTED TO AXIAL LOAD)

In order to avoid the effect of eccentricity, the load is applied along the neutral axis of unsymmetrical sections (e.g., angle-section, tees). When the unsymmetrical sections are connected by welding, then, fillet welding is so applied that centre of gravity of the welds lies on the line of action of action of the load : (i.e., it coincides with the neutral axis). An angle section connected to a gusset plate is shown in Fig. 21.50.

The angle section carries an axial pull `p'. Let l_1 and l_2 be the lengths of welds applied at the sides. These welds resist P_1 and P_2 forces as shown in Fig. 21.50. The lengths l_1 and l_2 may be determined as under.

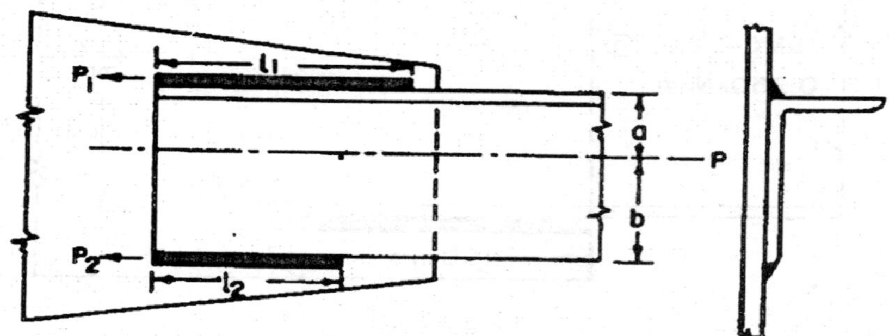

Fig. 21.50

Equating the moment of forces about the line of action of P_2

$$P_1.(a+b) = P.b,\quad P_1 = \frac{P.b}{(a+b)} \tag{21.15}$$

Values of a and b, are noted from ISI Handbook No. 1 for the given section.

Equate the moment of forces about the line of action of P_1

$$P_2.(a+b) = P.a,\quad P_2 = \frac{P.a}{(a+b)} \tag{21.16}$$

Let, τ_{vm} be the maximum permissible stress in the fillet weld, and s be the size of the weld.

The strength of weld per mm length = $(\tau_{vm} \times 0.7\ s)$

$$l_1 = \left(\frac{P_1}{\tau_{vm} \times 0.7\ s}\right) = \left(\frac{P.b}{\tau_{vm} \times 0.7\ s\,(a+b)}\right) \tag{21.17}$$

$$l_2 = \left(\frac{P_2}{\tau_{vm} \times 0.7\, s}\right) = \left(\frac{P.a}{\tau_{vm} \times 0.7\, s\,(a+b)}\right) \tag{21.18}$$

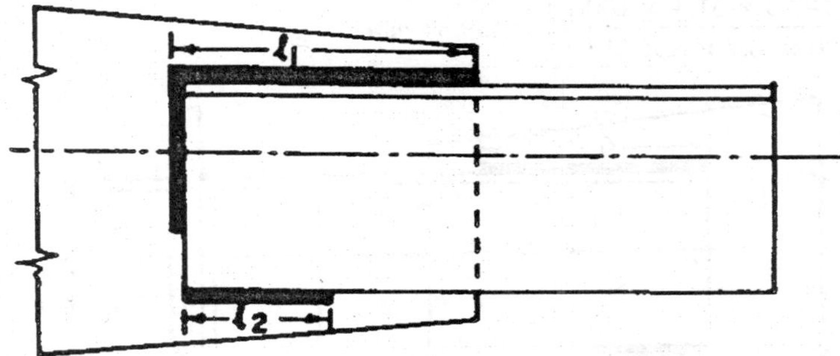

Fig. 21.51.

The lengths of the welds should further satisfy the specifications of the filler weld. The total length of the weld required to apply is computed by dividing the load transmitted by the strength of the weld per mm length. In case the lengths available at the sides become insufficient or to make the joint more effective, end fillet weld is also applied as shown in Fig. 21.51. In such cases, the lengths of welds l_1 and l_2 to be applied at the sides are computed for the load to be transmitted less the strength of end fillet weld.

Example 21.8 A tension member of a roof truss consists of 2 ISA 90 mm x 60 mm x 10 mm. The tension member carries a pull of 250 kN. The angles are connected to either side of a gusset plate 10 mm thick. Design the fillet welded connected. Yield strength of steel, σ_y = 250 N/mm². Partial safety factor for the strength of material = 1.15. Load factor, γ_{fL} = 1.5

Design : Yield stress for steel, σ_y = 250 N/mm²

Partial safety factor, γ_{msy} = 1.15

$$\text{Design stress, } \sigma_y = \left(\frac{250}{1.15}\right) = 217.4\ N/mm^2$$

Pull to be transmitted = 250 kN

Factored pull, P_{fd} = (1.5 x 250) = 375 kN

Step 1 : Size of weld

The size of weld should not exceed (3/4)th thickness of the rolled steel section at the toe.

$$= \left(\frac{3}{4} \times 10\right) = 7.5\ mm$$

Let the size of weld be 6 mm to be provided on both the sides each angle section is to transmit half the factored pull by each angle

$$P = \left(\frac{1}{2} \times 375\right) = 187.5\ kN$$

From Eq. 21.7, let the length of welds be l_{w1} at the top and l_{w2} on the bottom on the angles. From ISI Handbook No. 1, for ISA 90 mm x 60 mm x 10 mm, a = 28.7 mm and b = 61.3 mm

$$l_{w1} = \left[\frac{P.b \times 1000}{(a+b) \times 0.7 \times 6 \times 217.4}\right]$$

$$= \left[\frac{187.5 \times 61.3 \times 1000}{90 \times 0.7 \times 6 \times 217.4}\right] = 139.86\ mm$$

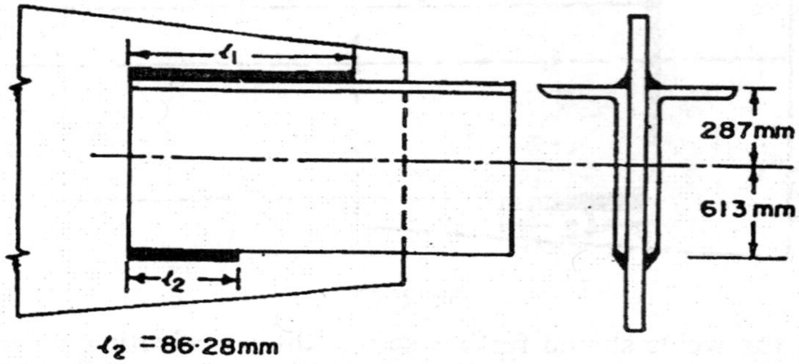

Fig. 21.52.

From Eq. 21.18

$$l_{w2} = \left[\frac{P.a \times 1000}{90 \times 0.7 \times 6 \times 217.4}\right]$$

$$= \left[\frac{187.5 \times 28.7 \times 1000}{90 \times 0.7 \times 6 \times 217.4}\right] = 65.483\ mm$$

The effective length of welds are shown in Fig. 21.52. For the actual lengths of the welds, twice the size of weld are added.

Example 21.9 In a roof truss, a tension member ISA 110 mm x 110 mm x 8 mm carries a pull of 210 kN. This tension member is connected with the gusset plate 8 mm thick. Design the welded joint. Apply end fillet weld symmetrical to the centroidal axis. Yield stress in steel, $\sigma_y = 250$ N/mm². Partial safety factor for the strength of material $\gamma_{msy} = 1.15$ and partial safety factor for loads, $\gamma_{fL} = 1.5$.

Design : Yield stress for steel, $\sigma_y = 250$ N/mm²

Partial safety factor, $\gamma_{msy} = 1.15$

Design stress, $\sigma_d = \left(\frac{250}{1.15}\right) - 217.4\ N/mm^2$

Factored load/pull to be transmitted

(1.5 x 210) = 315 kN

Step 1 : Size of welds

In case of rolled steel section, the size of weld at the toe should not exceed (3/4)th the thickness 8 mm

$$= \left(\frac{3}{4} \times 8\right) = 6\,mm$$

Maximum size of weld connecting vertical leg of the angle with gusset

$(8.0 - 1.5) = 6.5$ mm.

Provide a uniform size of weld 6 mm. The distance of axis from top is 30 mm and from bottom it is 80 mm. (Noted from ISI HB No. 1)

Provide an end fillet weld of effective length = (30 + 30) = 60 mm

Step 2 : Strength of end fillet weld

$$P_{EFW} = \left[\frac{217.4 \times 0.7 \times 6 \times 60}{1000}\right] = 54.785\ kN\ ,\ (315 - 54.785) = 260.275\ kN$$

Step 3 : Effective length of welds

From Eq. 21.17

$$l_{w1}\ \text{(at top)} = \left(\frac{260.275 \times 80 \times 1000}{110 \times 0.7 \times 6 \times 217.4}\right) = 207.31\ \text{mm}$$

From Eq. 21.18

$$l_{w2}\ \text{(at bottom)} = \left(\frac{260.275 \times 30 \times 1000}{110 \times 0.7 \times 6 \times 217.4}\right) = 77.74\ \text{mm}$$

The effective lengths of welds at the top and bottom are shown in Fig. 21.53.

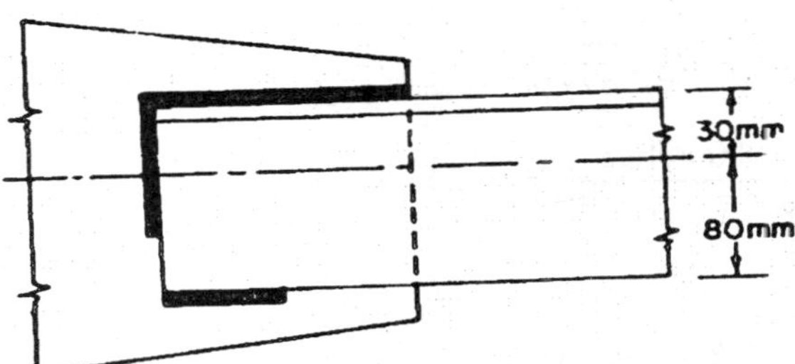

Fig. 21.53

Example 21.10 A joint in a welded truss is shown in Fig. 21.54. Design the welded joint Yield stress of steel is 250 N/mm² Partial safety factor for the strength of material is 1.15.

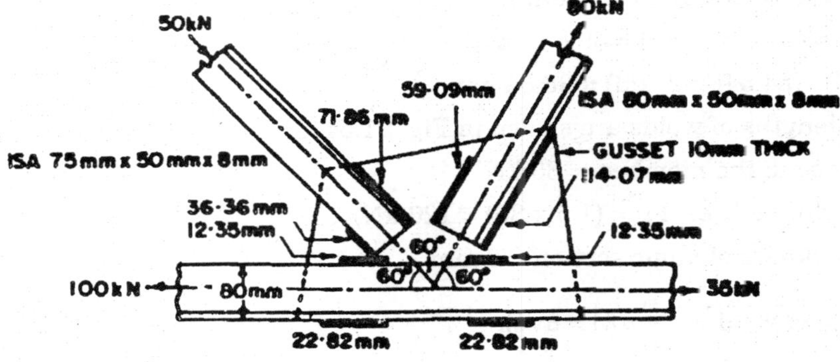

Fig. 21.54.

Design : Yield stress of steel, $\sigma_y = 250$ N/mm²

Partial safety factor for the strength of material, $\gamma_{msy} = 1.15$

$$\text{Design stress, } \sigma_y = \left(\frac{250}{1.15}\right) = 217.4\ N/mm^2$$

Step 1 : Size of weld

Force in one member = 50 kN

Factored forced, F_{fd} = (1.5 x 50) = 75 kN

Size of weld at the toe of rolled steel-section should not exceed (3/4)th thickness = $\left(\frac{3}{4}\times 8\right) = 6\ mm$

Fillet weld of 6 mm size is provided as shown in Fig. 21.54.

From ISI Handbook No. 1, ISA 75 mm x 50 mm x 8 mm (ISA 70 for 50, @ 0.074 kN/m)

a = 25.2 mm, and b = 49.8 mm

(a + b) = 75 mm

Step 2 : Effective length of weld

From Eq. 21.17,

$$l_{w2} = \left[\frac{P.b}{\sigma_d \times 0.7 \times s\,(a+b)}\right]$$

$$= \left[\frac{75 \times 1000 \times 49.8}{217.4 \times 0.7 \times 6 \times 75}\right] = 54.54\ mm$$

From Eq. 21.18

$$l_{w1} = \left[\frac{P.a}{\sigma_d \times 0.7\ s \times (a+b)}\right]$$

$$= \left[\frac{75 \times 25.2 \times 1000}{217.4 \times 0.7 \times 6 \times 75}\right] = 27.6\ mm$$

Step 3 : Actual lengths of welds

l_{w1} = (27.6 + 2 x 6) = 39.6 mm

l_{w2} = (54.54 + 2 x 6) = 66.54 mm

Effective lengths of welds are shown in Fig. 21.54

Force in the member = 80 kN

Factored force, F_{fd} = (1.5 x 80) = 120 kN

Thickness of angle section = 8 mm

$$\text{Size of weld} = \left(\frac{3}{4}\times 8\right) = 6\ mm$$

From ISI Handbook No. 1, for ISA 80 mm x 50 mm x 8 mm (ISA 80, 50, @ 0.077 kN/m)

a = 27.2 mm, and b = 52.7 mm

(a + b) = 80 mm

Step 4 : Effective length of weld

From Eq. 21.17

$$l_{w1} = \left[\frac{P \, . \, b}{\sigma_d \times 0.7 \times s \times (a+b)}\right]$$

$$= \left[\frac{120 \times 52.7 \times 1000}{217.4 \times 0.7 \times 6 \times 80}\right] = 86.575 \ mm$$

From Eq. 21.18

$$l_{w2} = \left[\frac{P \, . \, a}{\sigma_d \times 0.7 \times s \times (a+b)}\right]$$

$$= \left[\frac{120 \times 27.2 \times 1000}{217.4 \times 0.7 \times 6 \times 80}\right] = 44.684 \ mm$$

Step 5 : Actual lengths of welds

l_{w1} = (56.575 + 2 x 6) = 98.575 mm

l_{w2} = (44.684 + 2 x 6) = 56.684 mm

The effective lengths of welds are shown in Fig. 21.54

Horizontal member is a continuous tension member. It carries a pull of 100 kN on one side and 35 kN on the other side. The welds are designed for the difference of forces.

(100 – 35) = 65 kN

Factored value = (1.5 x 65) = 97.5 kN

The welding is done on both the sides Forces in one side = $\left(\frac{1}{2} \times 97.5\right) = 48.75 \ kN$

From ISI Handbook No. 1, for ISA 80 mm x 50 mm x 10 mm (ISA 80, 50, @ 0.094 kN/m

a = 28.1 mm, b = 51.9 mm

(a + b) = 80 mm.

Size of weld = $\left(\frac{3}{4} \times 10\right) = 7.5 \ mm$

Use 6 mm size for weld

Step 6 : Effective lengths of welds

From Eq. 21.17

$$l_{w1} = \left[\frac{P\,.\,b}{\sigma_d \times 0.7 \times s \times (a+b)}\right]$$

$$= \left[\frac{48.7 \times 51.9 \times 1000}{217.4 \times 0.7 \times 6 \times 80}\right] = 34.6\ mm$$

From Eq. 21.18

$$l_{w2} = \left[\frac{P\,.\,a}{\sigma_d \times 0.7 \times s \times (a+b)}\right]$$

$$= \left[\frac{48.7 \times 28.1 \times 1000}{217.4 \times 0.7 \times 6 \times 80}\right] = 18.734\ mm$$

Step 7 : Actual length of welds

$l_{w1} = (34.6 + 2 \times 6) = 46.6$ mm

$l_{w2} = (18.734 + 2 \times 6) = 30.734$ mm

Effective lengths of the welds have been shown in Fig. 21.54.

PROBLEMS

21.1 Two plates 14 mm thick are joined by (i) a double V butt weld, (ii) a single-V butt weld. Determine the strength of the welded joint in tension in each case. The effective length of weld is 200 mm.

21.2 Design a suitable side fillet weld to connect two plates 100 mm x 10 mm and 120 mm x 12 mm, and to transmit pull equal to the full strength of thin plate.

21.3 Two plates 100 mm x 10 mm are connected in a lap joint by means of end fillet weld. Design suitable weld to transmit pull equal to full strength of the plate.

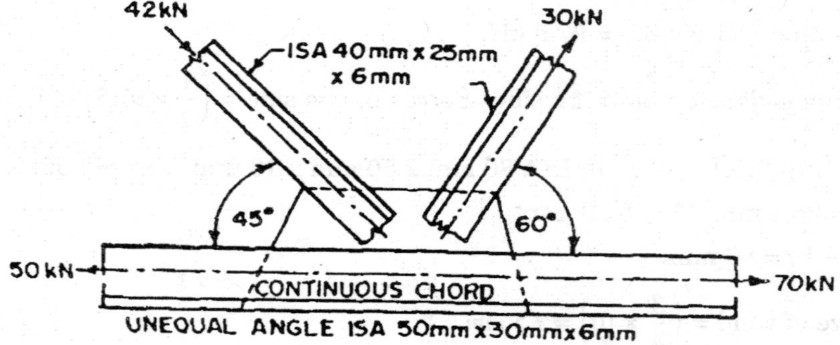

Fig. 21.55.

21.4 A truss joint is as sketched in Fig. 21.55. Design and detail the welds connecting the members to the gusset.

22

WELDED JOINTS SUBJECTED TO ECCENTRIC LOAD

22.1 INTRODUCTION

When the c.g. of group of the weld does not lie on the line of action of applied load, then the welded joint is subjected to eccentric load. Similar to the riveted bracket connections, the welded joints subjected to eccentric load are also of two types as under :

1. The c.g. of group of the weld lies in the plane of line of action of the applied load.
2. The c.g. of group of the weld does not lie in the plane of line of action of the applied load.

Type 1. Fig. 22.1 shows a fillet welded bracket connection. A vertical load, P is acting at a distance `e' from the c.g. of the group of the weld.

The eccentric load `P' is considered equivalent to direct load `P' passing through c.g. of the group of the weld and a twisting moment `P x e'.

Let a weld of uniform size be applied throughout, and t be the effective throat thickness.

The direct shear stress in the weld

$$p_s = \left(\frac{P}{(2b + d) \times t}\right)$$

The stress in the weld doe to twisting moment is maximum in the weld at the extreme distance from the c.g. of group of the weld and acts in the direction perpendicular to the radius vector. The maximum stress due to twisting

$$p_b = (P.\ e.\ r/I_{zz})$$

where, r = Distance to the extreme weld from the c.g. of weld group

I_{zz} = Polar moment of inertia, and $I_{zz} = (I_{xx} + I_{yy})$

The vector sum of two stresses is given by

$$p = [(p_s)^2 + (p_s)^2 + 2p_s p_s \cos\tau]^{1/2} \qquad (22.1)$$

The resulting stress p, is not to exceed the maximum permissible stress in the weld. In this type of eccentric welded joint, the stresses p_s and p_b are not normal to each other. As such, the combined stress is determined by using Eq. 22.1.

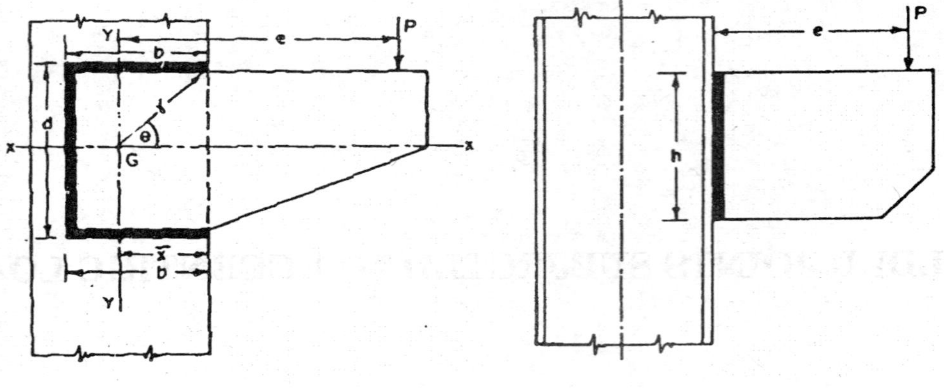

Fig. 22.1 Fig. 22.2

The thickness of gusset plate (bracket plates) should not be less than the size of the fillet weld. It should also be thicker than the thickness of flange of the column section.

Type 2. In this type, c.g. of group of weld does not lie in the plane of action of the applied load. In such cases, butt weld can be applied or filled weld can also be used.

Fig. 22.2 shows a bracket connection in which complete penetration butt weld has been applied.

The load P is acting at an eccentric distance `e'. The weld is subjected to direct shear stress as well as bending stress. Let t be the effective throat thickness of the weld.

Direct shear stress, $p_s = (P/(h\ .\ t)$

where, h = Effective depth of the weld.

The actual bending stress at the extreme distance of the weld

$$p_{ve} = \left(\frac{M}{Z}\right) = \left(\frac{M}{\frac{1}{6}t.h^2} = \frac{6M}{t\,h^2} = \frac{6P.e}{t.h^2}\right)$$

where, p_{ve} = Actual tensile/compressive bending stress in the weld,

For the combined stress in the butt weld, the expression

$$\left[\frac{P_s}{p_s}\right]^2 + \left[\frac{p_{t/e}}{p_{t/e}}\right]^2 \text{ should not exceed unity.}$$

For the purpose of design, the effective depth weld is computed from the following approximate depth formula

$$d = \left[\frac{6M}{t \times P_{t/e}}\right]^{1/2} \quad (22.2)$$

where, P_{te} is the maximum permissible tensile or compressive bending stress in the weld.

The effective throat thickness of the weld is know, knowing the thickness of the member transmitting the load. The butt weld so designed is checked for combined stresses as above.

Fig. 22.3 shows a bracket connection in which fillet weld has been applied on both the sides of the plate.

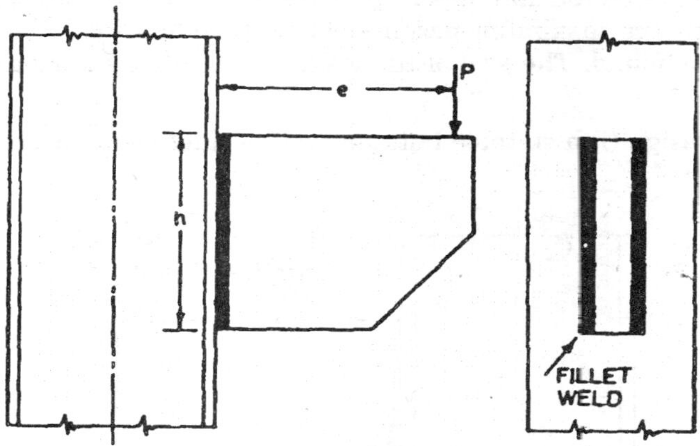

Fig. 22.3

The fillet welds are subjected to direct shear stress and bending stress. Let t be the effective throat thickness of the weld.

The direct shear stress in the weld, $\tau_v = \left(\frac{P}{2t.h}\right)$

The bending stress in the weld, $p_b = \left(\frac{M}{Z} = \frac{6P.e}{2.t.h^2}\right)$

The combined stress in the **fillet weld** is obtained by the vector sum of these stresses

$$p = \left[(\tau_v)^2 + (p_b)^2\right]^{1/2} \qquad (22.3)$$

The failure is due to critical stress across the throat of weld (at 45° to the weld leg since weld is 45° weld).

For the purpose of the design, size of the weld is fixed depending upon thickness of the member transmitting the load. The effective throat thickness of the weld is thereby known.

For fillet welds subjected to combined normal and sehar stress, the equivalent stress σ_e shall satisfy the following:

$$\sigma_e = \left(\sigma_n^2 + 3\,\tau_s^2\right)^{1/2} <= \frac{\sigma_u}{\sqrt{3}\,\gamma_{msu}}$$

σ_u = ultimate stress 410 N/mm²

$\gamma_{msu} = 1.25$

The depth of fillet weld applied on both the sides is obtained by

$$h = \left[\frac{6M}{2.t.P_s}\right]^{1/2} \qquad (22.4)$$

where, P_s is the maximum permissible shear stress in the weld.

In case, the depth of weld is limited, the combined stress in the weld for shear and bending is computed assuming the effective throat thickness of the fillet weld. The combined stress in the fillet weld is not to exceed maximum permissible shear in the weld. Equating the combined stress in the weld with the maximum permissible shear stress in the weld, effective throat thickness of the weld is found. The size of the weld is computed knowing the effective throat thickness of the weld.

Example 22.1. Design (i) a suitable butt weld, (ii) a fillet weld for the bracket connection as shown in Fig. 22.4.

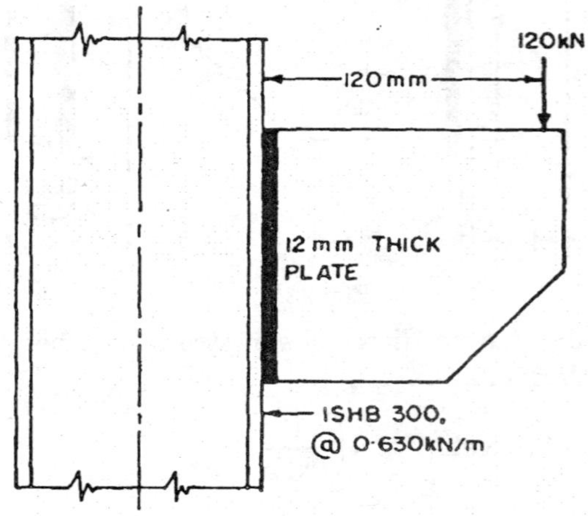

Fig. 22.4

Allowable stress on the butt weld in bending is 157.5 N/mm².

Design : (i) Butt weld

Step 1 : Depth of weld (Factored load = (1.5 × 120) = 180 kN)

Provide a double V butt weld. The effective throat thickness of weld is 12 mm.

Let h be the approximate depth of the weld. From Eq. 22.2, $h = \left[\dfrac{6M}{t \cdot P_{t/c}}\right]^{1/2}$

where, M = Bending moment and t = Effective throat thickness of weld

P_{vc} = Allowable tensile or compressive stress in the weld in bending

$$h = \left[\frac{6 \times 180 \times 1000 \times 120}{12 \times 157.5}\right]^{1/2} = 213.81 \text{ mm}$$

Try 230 mm depth of the weld

Step 2 : Stresses in weld

Direct shear stress in the weld, $\tau_v = \left(\dfrac{120 \times 1000}{230 \times 12}\right) = 43.478 \text{ N/mm}^2$

Actual bending stress, $P_{t/c} = \left(\dfrac{6 \times 120 \times 1000 \times 120}{12 \times 230^2}\right) = 136.11 \text{ N/mm}^2$

Check

Step 3 : For combined stress in the butt weld,

$$\left[\frac{P_s}{p_s}\right]^2 + \left[\frac{P_{t/e}}{p_{t/e}}\right]^2 = \left[\frac{43.478}{110}\right]^2 + \left[\frac{136.1}{157.5}\right]^2 = 0.903 < 1$$

Hence safe.

Step 4 : Check for equivalent stress

For butt weld, the equivalent stress $\sigma_{e.cal}$ due to coexisting bending stress (tension or compression) and shear stress is given by

$$\sigma_{e.cal} = \left[\sigma^2_{bc.cal} + 3\,\tau^2_{vm.cal}\right]^{1/2} = \left[(136.1)^2 + 3 \times (43.478)^2\right]^{1/2} = 155.54 \text{ N/mm}^2$$

As per IS: 816-1969, the equivalent stress $\sigma_{e.cal}$ shall not exceed the values allowed for the parent metal. For the parent metal (structural steel having yield stress, f_y as 250 N/mm²), the equivalent stress, σ_e (as per IS : 800-1984) is given by

$$\sigma_e = (0.9 \times 250) = 225 \text{ N/mm}^2 : (\sigma_{e.cal} \ngtr \sigma_e)$$

Hence, the design is safe. Provide 230 mm depth of double V butt weld.

(ii) Fillet weld

Equivalent stress should not exceed = $\left(\dfrac{\sigma_u}{\sqrt{3} \times 1.25}\right) = 410/(1.732 \times 1.25) = 189.376 \text{ N/mm}^2$

Step 1 : Depth of weld : Factored load = (1.5 x 120) = 180 kN

Apply 8 mm fillet weld on each side of plate.

Effective throat thickness of fillet weld, (0.7 x 8) = 5.6 mm

$$\text{Design stress} = \left(\frac{410}{\sqrt{3}} \times \frac{1}{1.25}\right) = 189.376 \text{ N/mm}^2$$

Strength of 1 mm fillet weld, $\left(\dfrac{189.376 \times 5.6 \times 1}{1000}\right) = 1.0605 \text{ kN}$

Approximate depth of fillet weld applied on two faces

$$h = \left[\frac{6 \times 180 \times 120}{2 \times 1.0605 \times 1}\right]^{1/2} = 247.19 \text{ mm}$$

Try 300 depth for fillet welds.

Step 2 : Stresses in welds

Direct shear stress in fillet weld, $\tau_v = \left(\dfrac{1.5 \times 120 \times 1000}{2 \times 300 \times 5.6}\right) = 1.5 \times 35.71 = 53.565 \text{ N/mm}^2$

Bending stress in fillet weld, $p_b = \left(\dfrac{1.5 \times 6 \times 120 \times 1000}{2 \times 5.6 \times 300^2}\right) = 1.5 \times 85.71 = 128.565 \text{ N/mm}^2$

Step 3 : Check for combined stress in fillet weld

p = 1.5 [35.71² + 85.71²]$^{1/2}$, 1.5 × 92.85 = 139.275 N/mm² < 189.376 N/mm². **Hence safe.**

Step 4 : Check for equivalent stress

For fillet weld, the equivalent stress $\sigma_{e.cal}$ due to co-existing bending stress and shear stress is given by

$$\sigma_{e.cal} = \left[\sigma^2_{e.cal} + 1.8\ \tau^2_{vm.cal}\right]^{1/2} = \left[85.71^2 + 1.8 \times 36.71^2\right]^{1/2} = 98.19 \text{ N/mm}^2$$

$$\not> \left(\sigma_e = 189.376 \text{ N/mm}^2\right)$$

Hence, the design is safe. Provide 300 mm depth of 8 mm fillet weld.

Example 22.2. Plates have been connected with the flanges of an I-section by applying 8 mm fillet weld as shown in Fig. 22.5. Compute the maximum load which may be placed at a distance of 100 mm from the flanges.

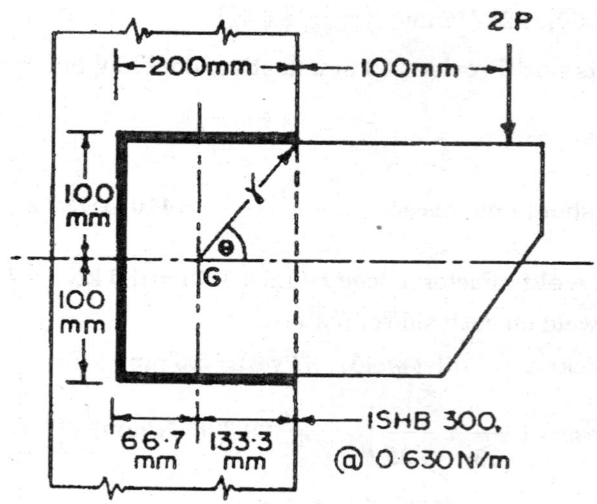

Fig. 22.5

Solution :

Step 1 : Size of weld

Consider one face, Size of weld = 8 mm

Effective throat thickness = (0.7 x 8) = 5.6 mm

Step 2 : Properties of welds

Let $\bar{x}$ be the distance of centroid 'G' of weld group from left hand edge of the plate

$$\bar{x} = \left[\frac{(2 \times 200 \times 5.6 \times 100) + (200 \times 5.6 \times 0)}{(2 \times 200 \times 5.6) + (200 \times 5.6)}\right] = 66.7 \text{ mm}$$

Moment of inertia of weld group about xx-axis

$$I_{xx} \doteq [2 \times 20 \times 0.56 \times 100 + \frac{1}{-} \times 0.56 \times 20^3] \times 10^4 \text{ mm}^4$$

Moment of inertia of weld group about yy-axis

$$I_{yy} = \left[\frac{1}{12} \times 2 \times 0.56 \times 20^3 + 2 \times 20 \times 0.56 \, (20 - 6.67)^2 + 20 \times 0.56 \times 6.72^2 \right] 10^4 \text{ mm}^4$$

$$= 1493.2 \times 10^4 \text{ mm}^4$$

Polar moment of inertia of weld group

$$I_{xx} = \left(I_{xx} + I_{yy}\right) = (4480 + 1493.2) = 5973.2 \times 10^4 \text{ mm}^4$$

Distance to the extreme weld from the centroid of weld group

$$r = [100^2 + 133.2^2]^{1/2} = 166.64 \text{ mm}, \cos\theta = (13.33/16.65)$$

Let *2P* be the maximum load which can be placed on the bracket.

Load transmitted by each face = P

Step 3 : Stresses in welds

Direct shear stress $\tau_v = \left(\frac{P}{(2 \times 200 + 200) \times 5.6}\right) = \left(\frac{(0.0297 \text{ P})}{100}\right) \text{N/mm}^2$

Twisting moment resisted by the weld group, T = P x (133.3 + 100) = 233.3 P N-mm

Maximum shear stress due to twisting, $p_b = \left(\frac{233.3P \times 166.64}{5973.2 \times 10^4}\right) = \left(\frac{0.0650 \text{ P}}{100}\right)$

Combined stress in the weld group, From Eq. (22.1),

$$p = \left(\frac{P}{100}\right)$$

$$\frac{P}{100} \times \left[(0.0297)^3 + (0.0650)^2 + 2 \times 0.0297 \times 0.0650 \, (13.33/16.65)\right]^{1/2} = \left(\frac{0.0905P}{100}\right) \text{N/mm}^2$$

Step 4 : Stress in weld

$$\sigma_u / (1.732 \times 1.25) = 410 / (1.732 \times 1.25) \text{ N/mm}^2$$

$$= 189.376 \text{ N/mm}^2$$

Step 5 : Check for combined stress

Combined stress should not exceed $\left(\frac{\sigma_u}{\sqrt{3} \times 1.25}\right) = \left(\frac{410}{1.732} \times \frac{1}{1.25}\right) = 189.236 \text{ N/mm}^2$

$$\therefore \left(\frac{0.0905P}{100}\right) = 189.376$$

$\therefore$ P = 209.255 kN, and 2P = 418.51 kN

Maximum load which may be placed is (418.51/1.50) = 275.03 kN

Example 22.3. A bracket plate is welded to the flange of a column as shown in Fig. 22.6. Calculate the size of the weld required.

Solution :

Step 1 : Properties of welds

Let t be the effective throat thickness of the weld group and x be the distance of its centroid from left hand edge of the plate.

$$\bar{x} = \left(\frac{2 \times 180 \times t \times 90 + 250 \times t \times 0}{2 \times 180 \times t + 250 \times 5}\right) = 53.1\ \text{mm}$$

Moment of inertia of weld group above xx-axis

$I_{xx} = [2 \times 18 \times t \times 12.5^2 + 1/12 \times t \times 25^3] \times 10^4 = 6942t \times 10^4\ \text{mm}^4$

Moment of inertia of weld group about yy-axis

$I_{yy} = [1/12 \times 2 \times t \times 18^3 + 2 \times 18 \times t\,(9 - 5.31)^2 + 25t \times 5.31^2] \times 10^4$

$= 1884.5t \times 10^4\ \text{mm}^4$

Polar moment of inertia of weld group,

$I_p = (6942 + 1884.5)t \times 10^3 = 8826.5t \times 10^3\ \text{mm}^4$

Distance to the extreme weld from the centroid of weld group

$r = [(125)^2 + (126.9)^2]^{1/2}$

$= 178.8$ mm

$\cos\theta = (126.9/178.8) = 0.71$

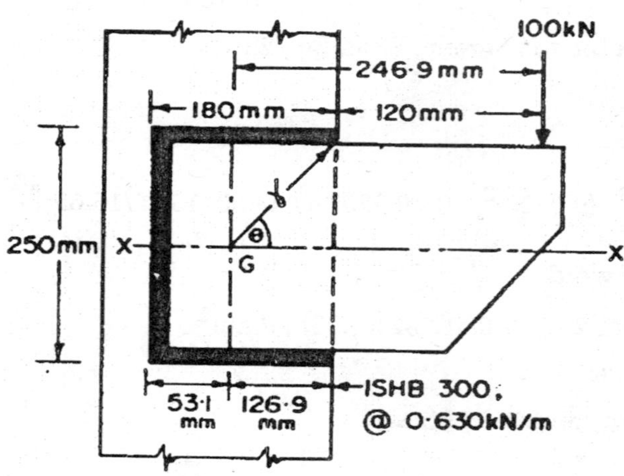

Fig. 22.6.

Step 2 : Stress in welds

Direct shear stress

$$\tau_v = \left(\frac{1.5 \times 100 \times 1000}{(2 \times 180 + 250).\,t}\right) = \left(\frac{1.5 \times 163.93}{t}\right)\ \text{N/mm}^2$$

Maximum shear stress due to twisting,

$$p_b = \left(\frac{1.5 \times 100 \times 1000 \times 246.9 \times 178.8}{8826.5\, t \times 10^3}\right) N/mm^2 = \left(\frac{1.5 \times 500.15}{t}\right) N/mm^2$$

Step 3 : Combined stress in the weld group, from Eq. 22.9,

$$P = 1.5\left[\left(\frac{163.93}{t}\right)^2 + \left(\frac{500.15}{t}\right)^2 + 2 \times \left(\frac{163.93}{t}\right)\left(\frac{500.15}{t}\right) \times 0.73\right]^{1/2} = \left(\frac{1.5 \times 627.25}{t}\right) N/mm^2$$

Combined stress in the fillet weld should not exceed

$$\sigma_u / \left(\sqrt{3} \times \gamma_{msu}\right) = \left(\frac{410}{1.732 \times 1.25}\right) = 189.376\ N/mm^2$$

$$\therefore t = \left(\frac{1.5 \times 163.93}{189.376}\right)^2 = 3.31 \times 1.5 \text{ mm size of weld required} = \left(\frac{1.5 \times 3.31}{0.7}\right) = 7.0975 \text{ mm}$$

Provide 5 mm fillet weld.

Example 22.4. Calculate the size of the weld required for the joint cutting used as a bracket loaded as shown in Fig. 22.7. The flange welds are of double the size of welds.

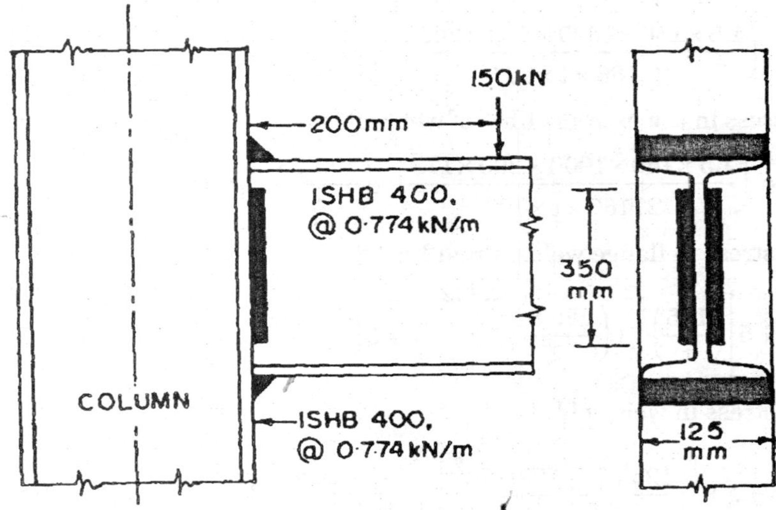

Fig. 22.7

Solution :

Step 1 : Properties of welds

Let 2 t be the effective throat thickness of flange welds.

Effective throat thickness of web weld = t mm

Moment of inertia of weld group about xx-axis

I_{xx} = [2 x 12.5 x 2t x 20^2 + 2 x 1/12 x t x 35^3] x 10^4

= (27166 x t x 10^4) mm^4

Step 2 : Stresses in welds

1. Direct shear stress in weld

Direct load shared by welds in the proportion to the areas of welds.

Total area of welds = (2 x 125 x 2t + 2 x 350 x t) = 1200 t mm²

Load shared by flange welds = $\left(\frac{500\text{ t}}{1200\text{ t}}\times 1.5\times 150\right) = 1.5\times 62.50\text{ kN} = 93.75\text{ kN}$

Load shared by web weld = 1.5 x 87.50 kN = 131.25 kN

Shear stress in the flange weld, $p_s = \left(\frac{1.5\times 62.50\times 1000}{2\times 125\times 2t}\right) = \left(\frac{1.5\times 125}{t}\right)\text{ N/mm}^2$

2. Bending stress in welds

The bending stress in the welds in proportion to the distance from the neutral axis of weld group.

Bending stress in the weld, $p_b = \left(\frac{M}{I_{xx}}\right).y$

$$p_b = \left(\frac{1.5\times 150\times 1000\times 200\times 200}{27166\times t\times 10^3}\right) = \left(\frac{1.5\times 220.86}{t}\right)\text{ N/mm}^2$$

Bending stress in the extreme fibre of web welds

$$p_{b1} = \left(\frac{1.5\times 150\times 1000\times 200\times 175}{27166\times t\times 10^3}\right) = \left(\frac{193.26}{t}\right)\text{ N/mm}^2$$

Combined stress in flange welds. From Eq. 22.

$$p = 1.5\left[\left(\frac{125}{t}\right)^2 + \left(\frac{220.86}{t}\right)^2\right]^{1/2} = 1.5\left(\frac{253.78}{t}\right)\text{ N/mm}^2$$

Combined stress in web weld

$$p_1 = 1.5\times\left[\left(\frac{125}{t}\right)^2 + \left(\frac{193.26}{t}\right)^2\right]^{1/2} = \left(\frac{1.5\times 230.16}{t}\right)\text{ N/mm}^2$$

Combined stress in the weld should not exceed $\sigma_u/\left(\sqrt{3}\times 1.25\right)$ = 410 /(1.732 $\times$ 1.25) = 189.376 N/mm^2,

$$\therefore\ t = \left(\frac{1.5\times 253.78}{189.376}\right) = 1.34\times 1.5 = 2.0\text{ mm},$$

Size of web welds = $\left(\frac{2.01}{0.7}\right) = 2.87\text{ mm}$.

Adopt 5 mm weld. Size of flange welds

2s = 2.87 mm. Adopt 2 x 5 = 10 mm weld.

Example 22.5. A bracket plate is welded to the flange of column as shown in Fig. 22.8. The bracket plate carries a load of 250 kN at a distance of 500 mm from the flange of column. Provide intermittent fillet weld on both sides of the bracket plate. Design the welded connection.

Design :

Step 1 : Number of intermittent fillet welds

The number of weld lengths is found in the same manner as the number of rivets necessary to resist the moment caused. From Art. 2.17, eccentric riveted connections, (design of bracket connections type 1) relation

$$n = \left[\frac{6\,M\,(n-1)}{p\,.\,R\,.\,n}\right]^{1/2}$$

may be used. Factor $\frac{(n-1)}{n}$ is assumed to be nearly unity, therefore

$$n = \left[\frac{6M}{p\,.\,R}\right]^{1/2}$$

The intermittent fillet welds are provided on both the sides of bracket plate. Therefore,

$$n = \left[\frac{6M}{2p\,.\,R}\right]^{1/2}$$

where, n = Number of weld lengths in one line and p = Strength of one weld length.

Let the size of weld for 12 mm thick bracket plate be 8 mm.

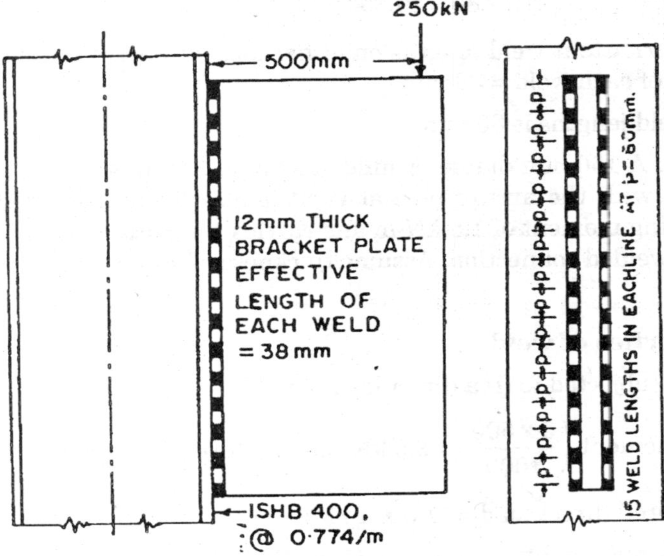

Fig. 22.8

The effective length of intermittent weld should be four times the size of weld or 38 mm whichever is more.

Therefore, effective length of weld = 38 mm

$$\text{Design stress in weld} = \frac{\sigma_u}{\sqrt{3} \times \gamma_{msu}}$$

$$\text{Factored force} = \left(\frac{410}{1.732 \times 1.25}\right) = 189.376 \text{ N/mm}^2$$

$$= (1.5 \times 125) \; kN$$

Step 2 : Spacing between welds

Assume pitch of welds, i.e., distance between centre to centre of adjacent intermittent welds p = 80 mm

$$\text{Clear spacing between the effective length of welds,} = \left(80 - 2 \times \frac{38}{2}\right) = 42 \text{ mm}$$

The clear spacing does not exceed 16 t for compression and 24 t for tension.

$$\text{Factored moment at weld section} = \left(1.5 \times 250 \times \frac{500}{100}\right) = 1.5 \times 125 = 187.5 \text{ kN-m}$$

$$\text{Strength of one 38 mm weld, R} = \left(0.7 \times 8 \times \frac{38 \times 189.376}{1000}\right) = 40.303 \text{ kN}$$

Number of weld lengths in one line

$$n = \left[\frac{6M}{2P.R}\right]^{1/2} = \left[\frac{1.5 \times 6 \times 125 \times 1000}{2 \times 80 \times 40.303}\right]^{1/2} = 13.21$$

Provide 15 intermittent weld lengths on either face of bracket plate as shown in Fig. 22.8, the effective length of each weld = 38 mm

The pitch of weld lengths is 80 mm.

Example 22.6. A 150 mm diameter mild steel pipe 0.50 metre long is welded to a vertical plate 12 mm thick, with the axis of pipe at right angles to the face of the plate. The pipe is subjected to a twisting moment of 1.5 kN-m and carries a vertical load of 5 kN acting at the end of pipe. Design the welded connection. Assume thickness of pipe 6 mm.

Design :

Step 1 : Properties of weld

The welding is subjected to (i) a direct load = 5 kN.

(ii) a bending moment $\left(\frac{5 \times 500}{1000}\right) = 2.5$ kN-m and (iii) a twisting moment = 1.5 kN-m

(i) Factored direct load = (1.5 x 5) kN

(ii) Factored bending moment = (1.5 x 2.5) kN-m

(iii) Factored twisting moment = (1.5 x 1.5) kN-m

Let *t* mm be the effective throat thickness of weld.

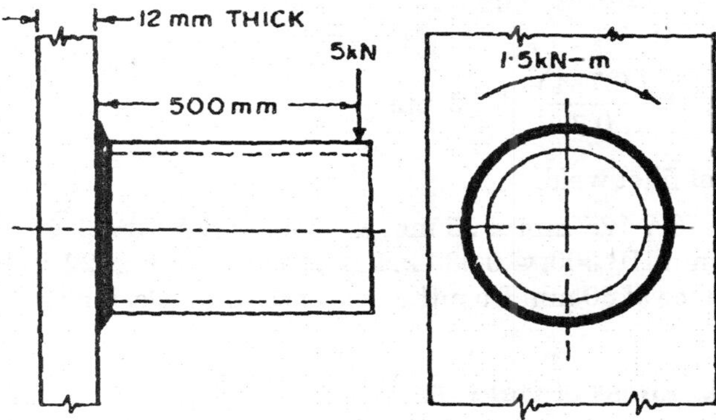

Fig. 22.9.

The approximate polar moment of inertia of weld group, $I_{xx} = \Sigma \delta a \times r^2$

where, δa is the elementary area of weld and r is the distance from the centre

$$I_{xx} = (\pi rt \text{ x } r^2 = 2\pi r^3 t) = (2\pi \text{ x } 75^3 \text{ x t x } 10^3) = 2640t \text{ x } 10^4 \text{ mm}^4$$

$$\text{Moment of inertia of weld about xx-axis, } I_{xx} = \frac{1}{2} I_{xx} = 1320\, t \times 30^4 \text{ mm}^4$$

Step 2 : Stresses in welds

Shear stress in the weld due to twisting

$$p_{tw} = \left(\frac{1.5 \times 1.5 \times 10^6 \times 75}{2640\, t \times 10^3}\right) = \left(\frac{1.5 \times 42.6}{t}\right) \text{N/mm}^2$$

Stress in the weld due to bending

$$p_b = \left(\frac{1.5 \times 2.5 \times 10^6 \times 75}{1320 \times t \times 10^3}\right) = \left(\frac{1.5 \times 142}{t}\right) \text{N/mm}^2$$

The stress in the weld due to shear in maximum at the centre. But average shear stress is considered. Average shear stress remains uniform.

$$p_s = \left(\frac{1.5 \times 5 \times 1000}{2\pi \times 75t}\right) = \left(\frac{1.5 \times 10.62}{t}\right) \text{N/mm}^2$$

Combined stress in the weld group at the top

$$1.5 \times \left[\left(\frac{10.62}{t}\right)^2 + \left(\frac{142}{t}\right)^2 + \left(\frac{42.6}{t}\right)^2\right]^{1/2} = \left(\frac{1.5 \times 148.63}{t}\right) \text{N/mm}^2$$

The combined stress in the fillet weld is not to exceed maximum permissible stress in shear

$$= \left(410/\sqrt{3} \times 1.25\right) = 189.376 \text{ N/mm}^2$$

$$\therefore t = \left(\frac{148.53}{189.376}\right) = 0.784 \text{ mm}$$

Step 3 : Size of weld $\left(\frac{0.784}{0.7}\right) = 1.12 \text{ mm}$

Provide 3 mm size of fillet weld.

Example 22.7. An ISA 125 mm x 95 mm x 12 mm (ISA 12595, @ 0.196 kN/m) is welded with the flange of a column HB 300, @ 0.630 kN/m as shown in Fig. 22.10. The bracket carries a load of 100 kN at a distance of 60 mm from the face of column. Design the bracket connection.

Design :

Step 1 : Throat thickness of weld

Width of flange of column HB 300, @ 0.630 kN/m

b = 250 mm

The horizontal fillet welds of 250 mm length each, are provided as shown in Fig. 22.10.

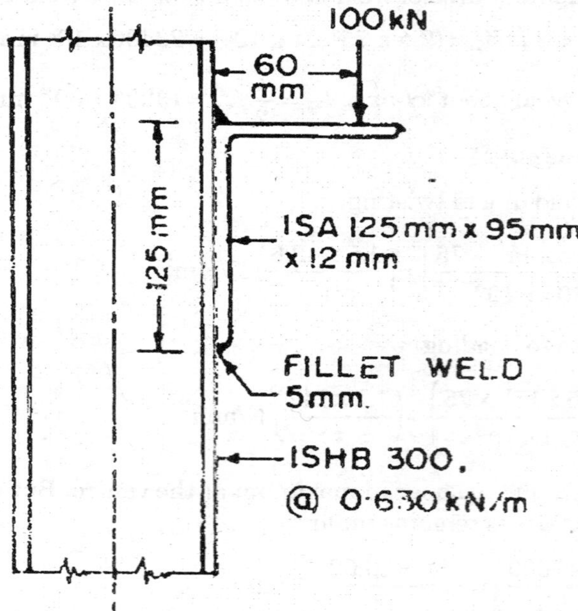

Fig. 22.10

Let t mm be the throat thickness of weld. The weld is subjected to direct shear and bending.

Step 2 : Stresses in welds

Shear force in weld

$$\tau_{vm.cal} = \left(\frac{1.5 \times 100 \times 100}{2 \times t \times 250}\right) \text{N/mm}^2 = \left(\frac{1.5 \times 200}{t}\right) \text{N/mm}^2$$

Moment about the faces of column

$= (1.5 \times 100 \times 100 \times 60/10^6) = 1.5 \times 0.6$ kN-m

The bending moment is resisted by a couple. The weld at the top is subjected to pull and that at the bottom is subjected to thrust. This is a case of discontinuous weld i.e. cross flexure

Stress in weld, $f_s = \left(\dfrac{M}{(h+s).(s/\sqrt{2}).1}\right) = \left(\dfrac{1.5 \times 0.6 \times 10^6}{125 \times t \times 250}\right)$

(h + s) is taken as h only, $f_t = \left(\dfrac{28.8}{t}\right)$ N/mm^2

Step 3 : Resultant stress in the weld, $f = \left[f_s^2 + f_t^2\right]^{1/2}$

$189.376 = \dfrac{1}{t}\left[(200)^2 + (28.8)^2\right]^{1/2}$, $t = \dfrac{1}{189.376}\left[(200)^2 + (28.8)^2\right]^{1/2} = \left(\dfrac{28.8}{189.376}\right) = 0.152\ \text{mm}^2$

Size of weld, $s = \left(\dfrac{0.152}{0.7}\right) = 0.217$ mm

The size of weld is very small. Provide 5 mm weld as shown in Fig. 22.10.

PROBLEMS

22.1 A welded bracket is of the design shown in Fig. 22.11. Design the welded connections completely.

22.2 An ISA 100 mm x 75 mm x 10 mm (ISA 100 75, @ 0.130 kN/m) is welded with the flange of a column HB 350, @ 674 N/m as shown in Fig. 22.12.

The bracket carries a load of 80 kN at a distance of 40 mm from the face of column. Design the bracket connection.

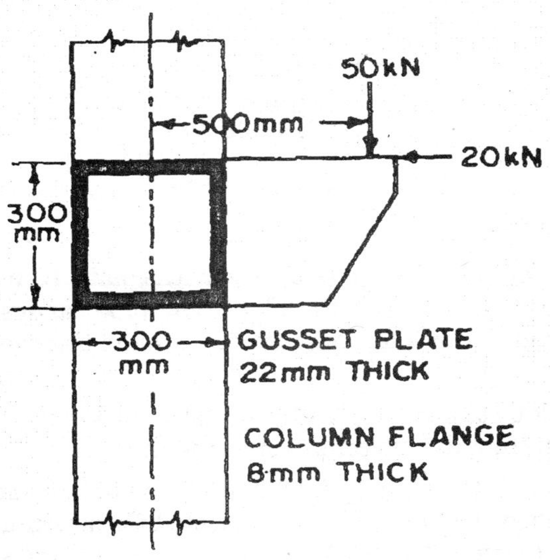

Fig. 22.11

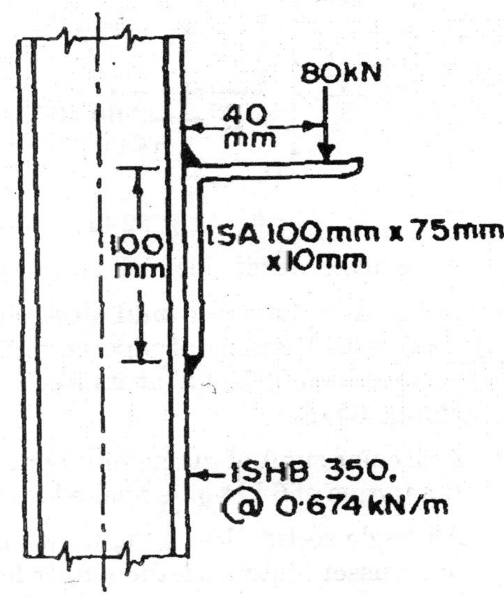

Fig. 22.12

22.3 Fig 22.13 shows an I-section bracket 350 mm x 140 mm connected to a steel stanchion by 12 mm fillet welds. The weld lengths are indicated on the figure. Determine the load P which the bracket can support if the stress in the welds is not to exceed 110 N/mm².

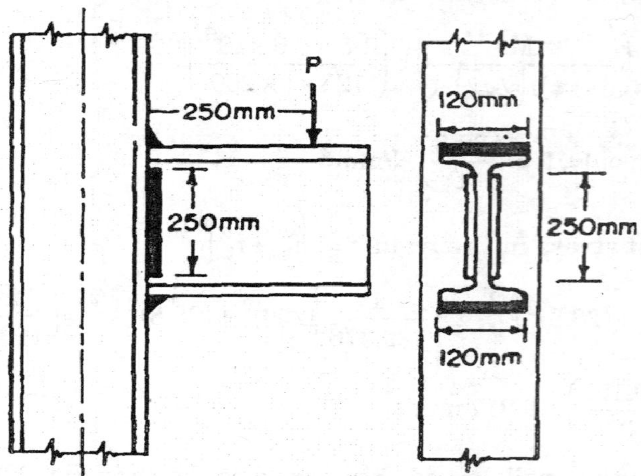

Fig. 22.13

22.4 A bracket is loaded as shown in Fig. 22.14. Design the welded connection.

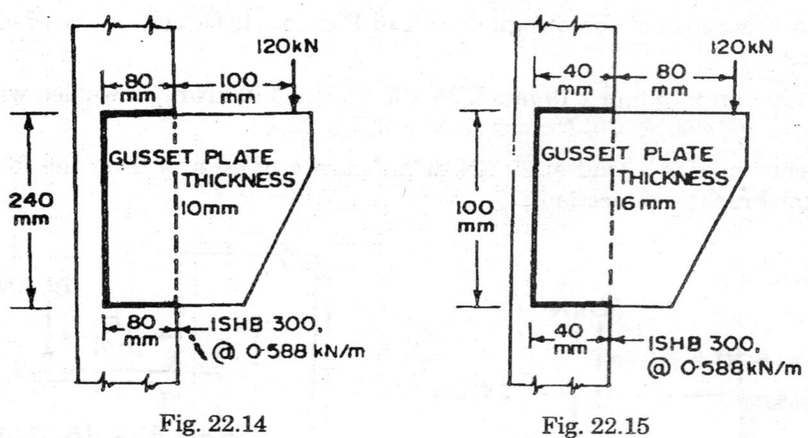

Fig. 22.14 Fig. 22.15

22.5 A welded bracket is loaded as shown in Fig. 22.15. Design the suitable weld.

22.6 A 100 mm diameter mild steel pipe 0.40 m long is welded to a vertical plate 10 mm thick with the axis of pipe at right angles to the face of the plate. The pipe carries a vertical load of 10 kN at its free end. Design the welded connection. Assume thickness of pipe 6 mm.

22.7 A circular shaft of diameter 100 mm is welded to a rigid plate by a fillet weld 6 mm size. If a torque of 6 kN-m is applied to the shaft, find the maximum stress in the weld.

22.8 An angle section ISA 40 mm x 25 mm x 6 mm (ISA 4025, @ 0.028 kN/m) is to be joined to a gusset plate with the longer leg attached. The joint is to be designed for maximum load carrying capacity of the angle iron in tension. The thickness of the gusset plate is 8 mm. Design the welded joint.

22.9 Two angles of ISA 90 mm x 90 mm x 8 mm (ISA 9090, @ 0.108 kN/m) transmit at tensile force cf 250 kN. The angles are connected one on each side of a gusset plate 10 mm thick by welding. Design the joint using 6 mm weld. The c.g. line for each angle

$C_{xx} = C_{yy} = 28.7$ mm.

22.10 Design the bracket connection (Fig. 22.16) assuming that the loads shown are for one gusset plate only.

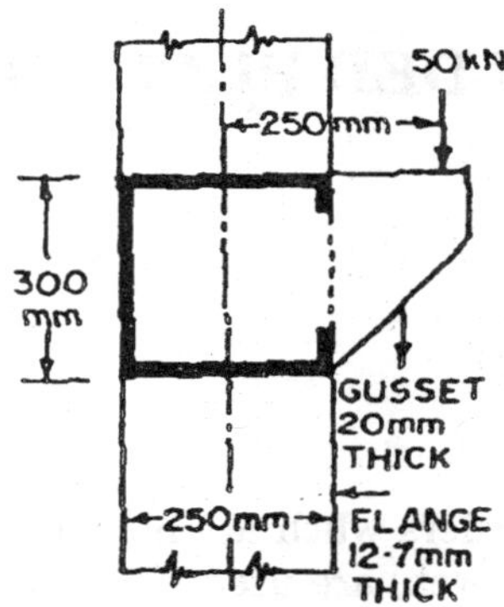

Fig. 22.16

22.11 Design a single angle tension member of a roof truss carrying an axial load of 100 kN. The member is connected to a gusset by welds on end and sides. Design the welded connection.

22.12 A bracket web plate 12 mm thick transmits a load of 200 kN at an eccentricity of 350 mm to flange of a column. Design the fillet welds on either side of web.

23

DESIGN OF WELDED PLATE GIRDERS (WPG)

23.1 INTRODUCTION

At present, the welded plate girders whch are built mainly using three plates (two plates for the flanges and one plate for the web) are used. These plates are kept together by welding. The flange plates and the web plate are fabricated in the workshops. Numerically controlled welding machines are used for fabrication.

In the welded plate girders, whole of the section is effective in resisting the loads. Therefore, it is efficient than the riveted plate girder. Fig. 23.1(A) and (B) show the sections of welded plate girder commonly used in section. Fig. 23.1(C) shows the elevation of a plate girder with bearing stiffeners at the ends and intermediate stiffeners. In the welded plate girder, flange angles are not used. Thc flange plates are directly welded to the web of the girder as shown in Fig. 23.1(A). The cut sections from wide flange beams are also used as shown in Fig. 23.1(B). The stacked flange plates are not used for welded plate girders because of multiple welding involved, but it is possible to vary the cross-sectional area by flanking the flange plate required for maximum bending moment with one or more successively thinner plates, butt-welded end to end. In welded plate girder, it is not necessary to make an allowance for rivet holes to determine the net section for tension flange. There is saving in the section of tension flange. Generally, the welded plate girders weigh less than riveted plate girders. The saving may vary from 5 to 15 per cent depending on the governing specifications. The welded plate girder is economical in material and cost. The section of welded plate girder is proportioned by the same principles as in the case of riveted plate girder. Instead of rivets, welds are used for connections.

23.2 DESIGN OF WEB PLATE

The depth of plate girder is adopted as 1/8 to 1/12 of the span of girder. The web plate is designed for maximum shear. The web plate designed should also satisfy the requirements of local buckling. The economical depth of web is given by

$$d_e = 5[M/f_b]^{1/3} \tag{23.1)A}$$

where, M is the maximum bending moment and f_b is the maximum permissible bending stress.

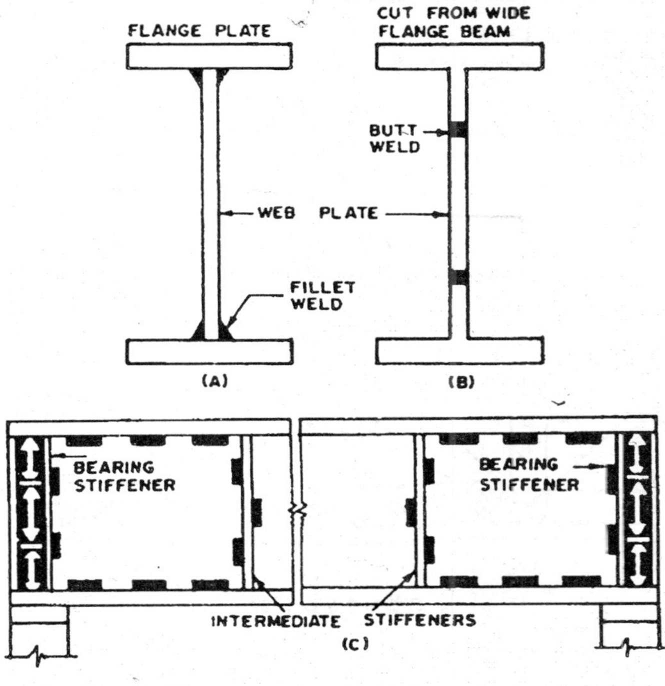

Fig. 23.1

23.3 DESIGN OF FLANGE PLATES

The area of flange is determined by approximate flange area method and then it is checked by moment of inertia method. Allowance is not made for rivet holes to determine the area of tension flange. The gross area of flange is effective to resist the moment. The flange includes (1/6)th the web area in the tension flange. In order to avoid the local buckling, the outstanding of flange plates, with unstiffened edges, that is, their projection beyond the face of the web shall not exceed 16 t for the flanges in flexural compression and 20 t for the flanges in flexural tension, where, t is the thickness of the plate.

In general, one plate is used in each flange. The thickness of flange plate is maximum at the section, where the bending moment is maximum. The value of bending moment reduces towards the end. Therefore, the maximum thickness of flange plate provided is not needed throughout the entire span. Either the thickness or width of flange plate may be reduced as the value of bending moment decreases. For different lengths of plate girder, either different thicknesses are used or width of flange is reduced by tapering it. The plates are joined by butt weld at the junction of two plates and the flange is made continuous. When the difference in thickness of two plates exceeds 25% thickness of thinner plate or 3.2 whichever is more, the thickness of thick plate is reduced to that of thin plate at the butt joint. This is done by providing a slope in the thicker plate. This slope should not be greater than one in five. This is shown in Fig. 23.2(A). In case, the thick plate cannot be reduced, the weld metal is built-up 25% greater than the thickness of thinner plate as shown in Fig. 23.2(C). The thickness of thicker plate in such case should not be greater than 50% the thickness of thinner plate.

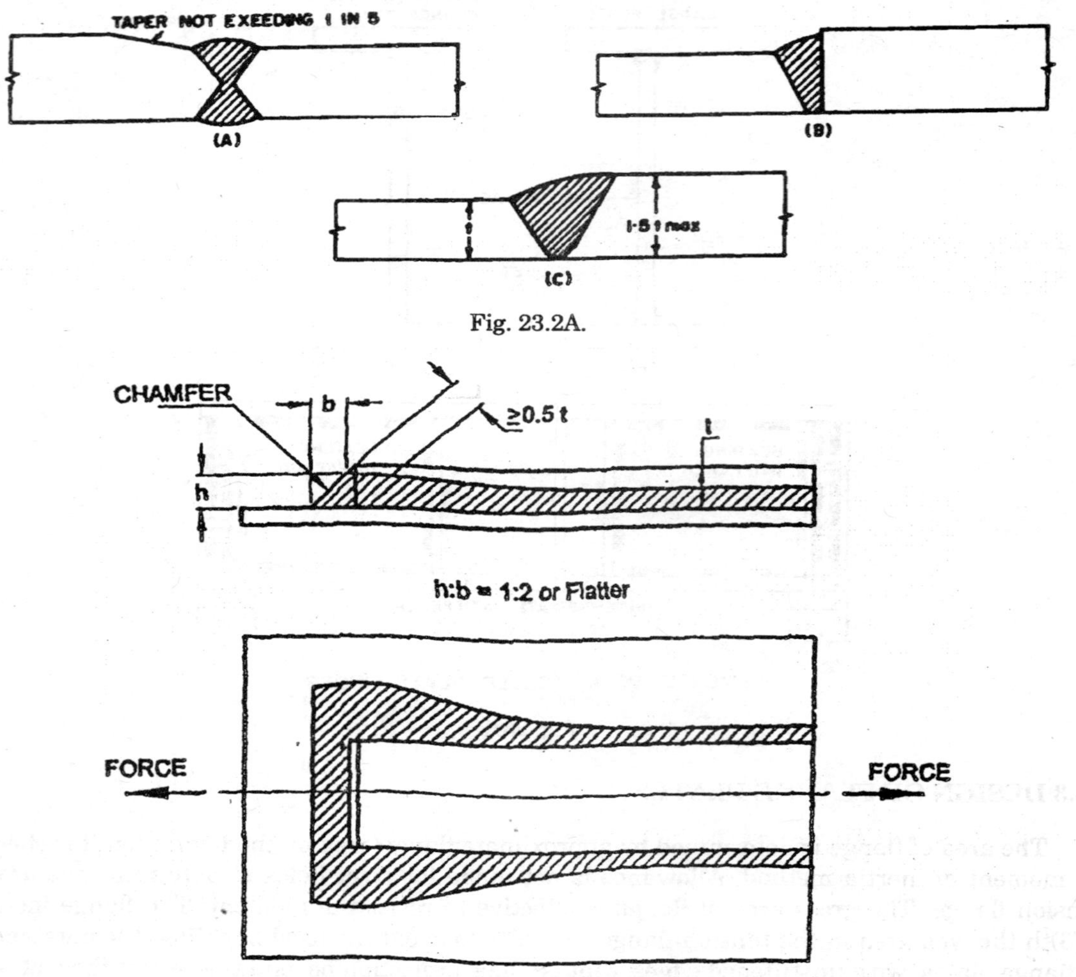

Fig. 23.2A.

Fig. 23.2B.

Butt-welds: Check for the combined stresses in butt-welds need not be carried out provided that

(i) butt-welds are axially loaded, and (ii) in single and double bevel welds, the sum of normal and shear stresses does not exceed the design normal stress, and the shear stress does not exceed 50 percent of the design shear stress.

Combined bearing stress, σ_{br}, bending stress, σ_b (compressive or tensile) and shear stress, τ under the most unfavourable conditions of loading in butt welds, the equivalent stress σ_e as obtained from the following expression, shall not exceed the values allowed for the parent metal

$$\sigma_e = \left(\sigma_b^2 + \sigma_{br}^2 + \sigma_b \cdot \sigma_{br} + 3\tau_s\right)^{1/2} \qquad (23.1)B$$

23.4 DESIGN OF CONNECTION OF FLANGE PLATE AND WEB

The welds connecting flange plates with the web plate may be applied continuously or intermittently on both the sides of web. When continuous weld is used, the size of fillet weld is changed in different portions of girder. The fillet weld is designed for horizontal shear per unit length. The horizontal shear per unit length for the loaded flange is given by

$$f_{sh} = \left[\frac{V \cdot A\bar{y}}{I_{xx} \cdot (2t)} \times (2t \times 1)\right] \quad \therefore \quad f_{sh} = \left[\frac{V \cdot A\bar{y}}{I_{xx}}\right] \quad \text{where,} \qquad (23.2)$$

V = Shear force at the section

$A\bar{y}$ = Moment of the area above the horizontal section considered about xx-axis (neutral axis)

t = Effective throat thickness of the weld

I_{xx} = Moment of inertia of the welded plate girder about xx-axis (neutral axis).

Let w be vertical load per unit length over the loaded flange. The combined stress per unit length is given by

$$f = [(f_{sh})^2 + w^2]^{1/2} \qquad (23.3)A$$

For the unloaded flange, w is zero.

For Fillet welds

When subjected to combined normal and the shear stress, as per IS: 800-2007, the equivalent stress, σ_e shall satisfy

$$\sigma_e = \left(\sigma_n^2 + 3\,\tau^2\right)^{1/2} <= \left(\frac{\sigma_u}{\sqrt{3} \cdot \gamma_{msu}}\right) \qquad (23.3)B$$

where σ_n = Normal stresses compression or tension due to axial force or bending moment

τ = shear stress due to shear force

$$\text{Design stress } \sigma_d = \left(\frac{\sigma_u}{\sqrt{3} \times \gamma_{msu}}\right) = \left(\frac{410}{1.732 \times 1.25}\right) = 189.376 \text{ N/mm}^2$$

The size of fillet weld to be applied should be such that the strength of weld per unit length is equal to the combined stress per unit length. If s is the size of fillet weld in mm, strength of weld per mm length

$$0.707 \times (2s) \times 189.376 = 267.777\ s \text{ Newton}$$

Therefore, $267.777\ s = f_{sh}$, $\quad s = (f_{sh} / 267.777)$

The size of welds should not be less than minimum size of weld required as per Table 21.1.

In case intermittent fillet welds are provided, the pitch of weld is found similar to that or rivets connecting flanges with the web. The minimum size of weld depends upon thickness of thicker plate to be connected. The thickness of plate is maximum near the centre where the bending moment is maximum. Near the supports, the thickness of the plate is small, but shear force is maximum. A suitable uniform size of weld is selected for the entire span.

The effective length of intermittent fillet weld should not be less than four times the size of weld or 38 mm, whichever is more. The clear spacing between the weld should not be greater than 12 t or 200 mm for compression flange and 16 t or 200 mm for tension flange, where t is thickness of thinner plate.

If R is the strength of one weld, the pitch of weld for loaded flange

$$p = R/(f_{sh})^2 + w^2]^{1/2} \quad (23.4)$$

For the unloaded flange, w = 0.

The weld is provided on both faces, therefore, the weld is provided at a distance equal to twice the pitch calculated. The size of intermittent weld adopted used to be large. It is economical to use continuous fillet weld for connections.

The gap between the web plate and the flange plates shall be kept to a minimum, and for fillet welds shall not exceeds 1 mm at any point before welding. The dispersion of loads through the flange is assumed at 30° to the horizontal.

Note : Complete text related to shear strength of Design of plate-girders (specially Web plate of aplate-girders) have been described in `Chapter 19'. Examples 19.1, 19.2, 19.3 and 19.4 concerning the design of various elements of plate-girder have been worked-out and illustrated in Chapter 19 by using simple post critical method.

However, Examples dealing with design of various elements of plate-girder have been worked-out and illustrated in this `Chapter 23' by using **tension field action method.**

23.5 DESIGN OF BEARING STIFFENER

The stiffeners consists of flat plates instead of angles as in case of riveted plate girder as shown in Fig. 23.1(C). The ends of plates are machined to provide direct bearing so that the load is transmitted by bearing. The welds connecting the stiffeners to the web are designed to transmit full load or support reaction. Besides this, the bearing stiffeners are designed similar to those in riveted plate girder.

23.6 INTERMEDIATE STIFFENERS

The stiffeners consist of flat plates instead of angles as incase of riveted plate girder as shown in Fig. 23.1(C). The outstand of stiffeners of flat sections shall be not more than 12 t, where t is the thickness of flat section used. The intermediate stiffeners are designed similar to those in riveted plate girder. The weld connecting stiffeners and web are designed to resist horizontal shear (1.5 t_w^2/h) kN/m length, where t_w is the thickness of web in mm and h is the outstand of the stiffener in mm. The minimum size of weld depends upon the thickness of stiffener or thickness of web.

In case, intermittent welds are used, the instance between the effective lengths of any two welds, even if staggered on opposite side should not exceed 16 times the thickness of stiffener nor 300 mm. In case intermittent welds are provided on one side of stiffener only or both sides but taggered of where single plate stiffeners are butt welded to the web, the effective length of each weld should not be less than 10 times the thickness of stiffener. In case, intermittent welds are provided in pairs on both sides, the effective length of each weld should not be less than four times the thickness of the stiffener.

Example 23.1 A welded plate girder simply supported at its two ends has an effective span of 28 m. It carries a uniformly distributed load of 25 kN per metre. Design the maximum section of welded plate girder.

Design : The effective span of welded plate girder is 28. The plate girder carries a uniformly distributed load, w of 25 kN/m.

Self-weight of girder

$$w_1 = \left(\frac{wL}{300}\right) = \left(\frac{25 \times 28}{300}\right) = 2.33 \text{ kN}$$

Total uniformly distributed load

$$w = (25 + 2.33) = 27.33 \text{ kN/m}$$

Step 1: Maximum bending moment, M and maximum shear force, F

Factored load, $W_{fd} = (1.5 \times 27.33) = 41$ kN/m

Factored maximum bending moment

$$M_{fd} = \left(\frac{41 \times 28 \times 28}{8}\right) = 4018 \text{ kN}-\text{m}$$

Factored maximum shear force

$$V_{fd} = \left(\frac{41 \times 28}{2}\right) = 574 \text{ kN}$$

Step 2: Optimum depth of web plate

Yield stress for the steel for the flanges and web plate, $\sigma_y = 250$ N/mm^2

$$\text{Yield ratio (epsilon) } \varepsilon = \left(\frac{250}{\sigma_y}\right)^{1/2}$$

$$\varepsilon = \left(\frac{250}{250}\right)^{1/2} = 1.0$$

In case, the longitudinal (i.e, horizontal) stiffeners are not provided, the maximum (d_w/t_w) ratio may be kept less than 360 for $\sigma_y = 250$ N/mm^2.

Then, let this ratio

$$K = \left(\frac{d_w}{t_w}\right) = (250 \times \varepsilon)$$

$$K = (250 \times 1.0) = 250$$

Optimum depth of plate girder

$$d = \left(\frac{MK}{\sigma_y}\right)^{1/3} = \left(\frac{4018 \times 250 \times 10^6}{250}\right)$$

$$= (4018 \times (100)^3)^{1/3} = 1589.7786 \text{ mm}$$

Adopt the depth of web, d_w = 1200 mm

For a large depth of the web plate, the thickness of web plate, t_w be adopted as 10 mm

Step 3 : Design of flange plates

The width of flange plate may be taken as 0.3 times the depth of girder

$$b_f = (0.3 \times 1200) = 360 \text{ mm}$$

Adopt the width of flange plate as 500 mm

For preliminary size, the width of thickness ratio for Fe 410 steel be kept to 24

$$t_f = \left(\frac{500}{24}\right) = 20.8 \text{ mm.}$$

Provide thickness of flange plate

$$t_f = 32 \text{ mm}$$

Provide 2 x 500 mm x 32 mm flange plates (one at the top in compression and one at the bottom in tension) and 1200 mm x 10 mm web plate as shown in Fig. 23.3.

Step 4 : Gross-moment of inertia of the plate girder section about xx-axis

$$I_{xx} = \left(\frac{1}{12} \times 10 \times 1200^3 + 2 \times 32 \times 500 \times 616^2\right)$$

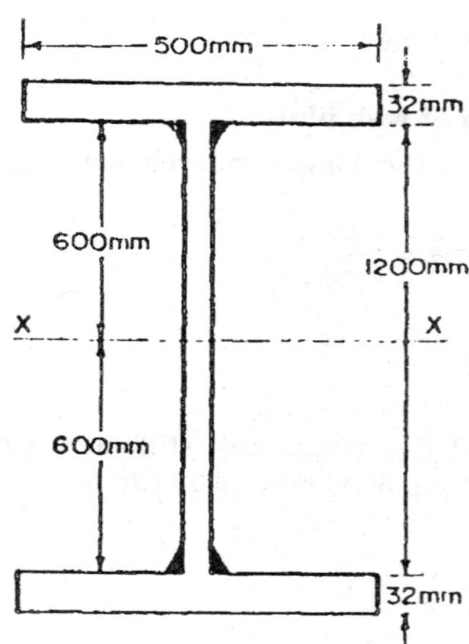

Fig. 23.3.

$$= \left(\frac{17280 \times 10^6}{12} + 320 \times 37.9456 \times 10^6\right)$$

$$= \left(1440 \times 10^6 + 12142.592 \times 10^6\right)$$

$$= 1358259.2 \text{ x } 10^4 \text{ mm}^4$$

Step 5 : Elastic section modulus, Z_E

$$Z_E = \left(\frac{I_{xx}}{y_{max}}\right) = \left(\frac{1358259.2 \times 10^4}{\frac{1200}{2} + 32}\right) mm^3$$

$$= \left(\frac{1358.2592 \times 10^3 \times 10^4}{632}\right) = 21491.4 \times 10^3 \times mm^3$$

Step 6 : Moment capacity of the plate girder section.

$$M_d = \left(\frac{Z_E \times \sigma_y}{\gamma_{msy}}\right) = \left(\frac{21491.4 \times 250 \times 10^3}{1.15 \times 10^6}\right)$$

$$= \left(\frac{2149.14 \times 2.50 \times 10^3 \times 10^3}{1.15 \times 10^6}\right) = 4672.04 \text{ kN} - \text{m}$$

> 4018 kN-m

M_d = 4672.04 kN-m is more than M_{fd} = 4018 kN-m.

Step 6(B) : Moment capacity of plate girder section

Design bending strength as per IS : 800-2007, Clause 8.2.1.2

$$M_d = \left(\beta_b \frac{\times Z_p \times \sigma_y}{\gamma_{msy}}\right)$$

β_b = 1.00 for plastic and compact section

Plastic modulus of section

$$Z_p = 2\, b_f\, t_f\, (D - t_f) + \frac{1}{4} t_w\, d^2$$

$$= \left[2 \times 500 \times 32\,(1264 - 32) + \frac{1}{4} \times 10 \times 1200^2\right]$$

$$= \left[3.2 \times 10^4\,(1232) + \frac{1}{4} \times 1440 \times 10^4\right]$$

$$= (3942.4 + 360) \times 10^4 = 43024.0 \times 10^3 \text{ mm}^3$$

$$M_d = \left(\frac{1.0 \times 43024 \times 10^3 \times 250}{1.15 \times 10^6}\right)$$

$$= \left(\frac{4302.4}{1.15} \times 2.50\right) = 9353.04 \text{ kN-m}$$

M_d = 9353.4 kN-m is much greater than M_{fd} = 4018 kN-m.

Step 7 : Shear capacity

. As per IS : 800-2007, clause 8.4, the factored design shear force V_{fd} due to external load should be less than or equal to design shear strength

$V_{fd} <= V_d$ (V_{fd} = 574 kN-m)

Nominal plastic shear resistance

As per IS : 800-2007, clause 8.4.1

$$V_d = \left(\frac{A_v\ \sigma_{yw}}{\sqrt{3} \times \gamma_{msy}}\right),\ A_v = (d_w t_w)$$

$$A_v = (1200 \text{ x } 10) = 12000 \text{ mm}^2$$

$$V_d = \left(\frac{12000 \times 250 \times 10}{1.732 \times 1.15 \times 1000}\right) = 15061.75 \text{ kN}$$

Plate girder section is safe in shear

Step 8 : Reduction of flange plate

The thickness of flanges may be reduced to 20 mm.

Moment of inertia of welded plate girder with 2 x 500 mm x 20 mm flange plates

$$I_{xx} = \left[\frac{1}{12} \times 1 \times 120^3 + 2 \times 2 \times 50 \times 61^2\right] \times 10^4$$

$$= 888200 \text{ x } 10^4 \text{ mm}^4$$

Elastic-section modulus

$$z_E = \left(\frac{I_{xx}}{\frac{1200}{2} + 20}\right) = \left(\frac{888200 \times 10^4}{620}\right) \text{ mm}^3$$

$$= 14325.8 \text{ x } 10^3 \text{ mm}^3$$

Moment of resistance

$$M_{dr} = \left(\frac{Z_E \times \sigma_y}{\gamma_{msy}}\right)$$

$$\left(\frac{14325.8 \times 10^3 \times 250}{1.15 \times 10^6}\right) = 3114.3\ kN - m$$

Expression (equation of parabolic bending moment diagram with left hand support as origin is

$$y = k\,.\,x\,(L - x)$$

Maximum bending moment 4018 kN-m

occurs at centre (y_{max} = 4018 kN-m)

At x = 14 m, 4018 = k x 14 x 14

$k = 20.5 \qquad \therefore y = 20.5\,x\,(L - x)$

At (y = 3114.3) = 20.5 x (28 – x), x = 7.36 m

The thickness of flange plates 32 mm may be reduced to 20 mm, at 7.56 m from left hand support.

Alternatively : Instead of reducing the thickness of flange plates, it is convenient to reduce width of flange plates. The tapering width is provided as shown in Fig. 23.4. Equation of parabolic bending moment diagram is

y = 20.5 x (L – x)

At x = 7.50 m

y = (20.5 x 7.5 x 20.5)

= 31.51.875

The bending moment at x = 7.5 m

M_2 = 3151.875 kN-m

Let b be the width of flange at x = 7.5 m, equating the moment of resistance width of flange may be calculated which works out to be 240 mm.

Note: The width of flange may be found at different sections. The flange plate may be tapered on its both the edges from both the ends, such that the taper accommodates the required width at those sections.

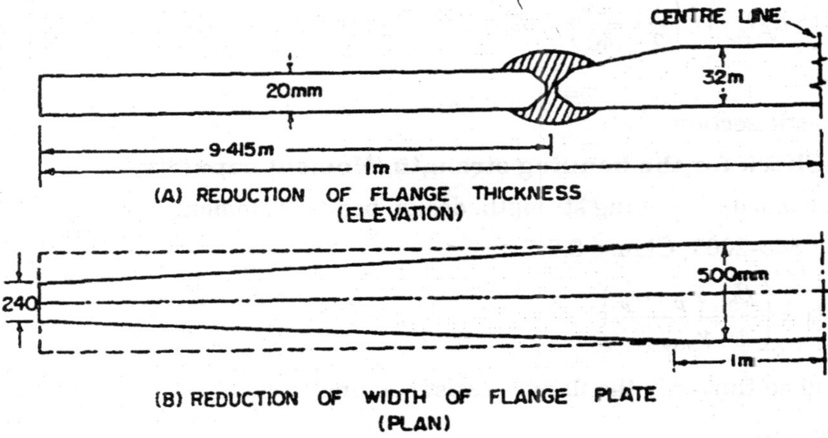

Fig. 23.4.

Example 23.2 : The web plate of a 2000 mm deep welded plate girder is 10 mm thick. Two flanges are 500 mm wide and 40 mm thick. The effective span of the girder is 28 m. Total load per one metre include of its own weight is 270 kN/m. Check the given section whether section is (i) adequate for its bending strength, and (ii) adequate for its shear strength.

Check : Given load = 27 kN/m

Effective span is 28 m

The girder is simply supported at its both the ends.

Step 1 : Factored load

Total uniformly distributed load inclusive of self-weight, w = 27.0 kN/m

Factored load, w_{fd} = (1.5 x 27.0) = 40.5 kN/m

Step 2 : Factored shear force

$$V_{fd} = \left(\frac{40.5 \times 28}{2}\right) = 567 \text{ kN}$$

Step 3 : Factored bending moment :

Load is uniformly distributed

Effective span = 28 mm.

Maximum bending moment occurs at the centre. Factored bending moment

$$M_{fd} = \left(\frac{w_{fd} \times l_e^2}{8}\right)$$

$$= \left(\frac{40.5 \times 28 \times 28}{8}\right) = 3969 \text{ kN} - \text{m}$$

Ratio of $\left(\frac{\text{Width}}{\text{Thickness}}\right)$ of flanges

$$\left(0.5\, b_f / t_f\right) = \left(\frac{1}{2} \times \frac{500}{40}\right) = 6.25$$

< 8.4

It is a plastic section.

Step 4. Check for the bending strength /Moment capacity,

Moment capacity, bending strength, design bending moment

From IS : 800-2007, Clause 8.2.1.2

$$M_d = \left(\frac{\beta_b \,.\, Z_p \,.\, \sigma_p}{\gamma_{msy}}\right)$$

It is assumed that only the flanges resist bending moment

$\beta_b = 1.0$

Plastic modulus of the section

$$Z_p = \left(\frac{2\, b_f\, t_f \left(D - t_f\right)}{2}\right)$$

$$= \left(\frac{2 \times 500 \times 40 \times (2080 - 40)}{2}\right) \text{mm}^3$$

$$= \left(\frac{20{,}000 \times 2040}{2}\right) = 40.8 \times 10^6 \ mm^3$$

$$M_d = \left(\frac{1.0 \times 40.8 \times 10^6 \times 250}{1.15 \times 10^6}\right) = 8869.56\ kN-m$$

The factored bending moment, M_{fd} = 3969 kN-m is less than the moment capacity/design bending moment/bending strength M_d = 8869 kN-m.

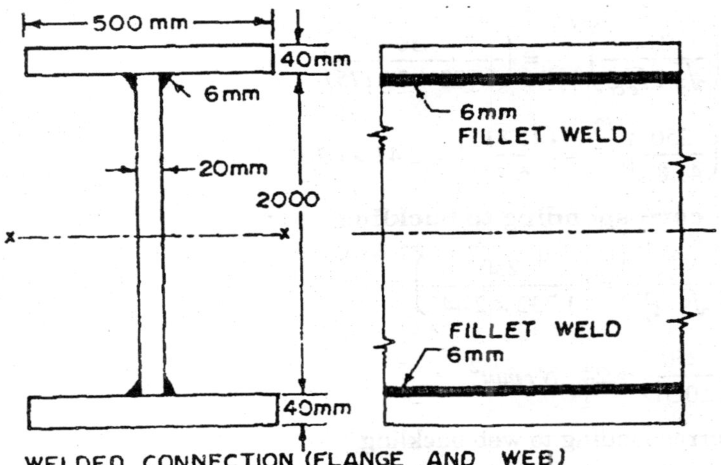

Fig. 23.5.

Hence, the given section is adequate for the given load.

Fig. 23.5 show the cross-section of given welded plate-girder.

Step 5 : Check for shear strength

As per IS : 800-2007, Clause 8.6.1

Minimum thickness of web

(i) For serviceability requirement

$$\left(\frac{d_w}{t_w}\right) = \left(\frac{2000}{10}\right) = 200\ \varepsilon = 200\ (\therefore \varepsilon = 1)$$

As per IS : 800-2007, clause 8.6.1.2 (A)

(ii) To avoid flange buckling due to compression

$$\left(\frac{d_w}{t_w}\right) = \left(\frac{2000}{10}\right) = 200 < 345\ \varepsilon_f^2 \qquad (\therefore \varepsilon_f = 1.00)$$

Consider the simple post-critical method

Nominal shear strength of $V_n = V_{cr}$

K_v = 5.35 when the transverse stiffeners are provided only at supports. The elastic critical stress for the web plate, Poisson's ratio = 0.3

$$\tau_{e.cr} = \left(\frac{k_v \cdot \pi^2 \times E}{12\left(1-\mu^2\right)\left(d/t_w\right)^2}\right) \quad \because (\pi = 3.14)$$

$$= \left(\frac{5.35 \times 9.8596 \times 2 \times 10^5}{12\left(1-0.3^2\right)(200)^2}\right) = 24.175\ N/mm^2$$

For shear buckling stress

Non-dimensional web slenderness ratio

$$\lambda_w = \left(\frac{\sigma_{yw}}{\sqrt{3}\ .\ \tau_{e.cr}}\right)^{1/2} = \left(\frac{250}{1.732 \times 24.175}\right)$$

$$= \left(\frac{250}{41.87}\right)^{1/2} = \left(\frac{15.811}{6.47}\right) = 2.44 > 1.20$$

Shear stress corresponding to buckling

$$\tau_{ke} = \left(\frac{\sigma_{yw}}{\sqrt{3}\ \lambda_w^2}\right) = \left(\frac{250}{1.732 \times 2.442}\right)$$

$$= \left(\frac{250}{10.31}\right) = 24.\ \ N/mm^2$$

Shear force corresponding to web buckling

$$V_{cr.bkf} = (d_w\ .\ t_w\ .\ \tau_{bkl})$$

$$= \left(\frac{2000 \times 10 \times 24.248}{1000}\right) = 484.96\ kN.$$

The web-plate needs intermediate transverse stiffeners.

The compression flange is laterally restrained. The rotation at supports is not allowed. Therefore, it is not necessary to check for lateral-torsional buckling.

Example 23.3 : In Example 23.2, Design suitable weld connections between the web plate and the flange plates (i) Provide continuous weld, and (ii) Provide intermittent weld.

Design : Properties of the plate-girder

Flange plates = 500 mm wide x 40 mm thick

Web plate = 2000 mm deep x 10 mm thick.

Area of flange plates (one flange plate)

A_f = (500 x 40) = 20,000 mm²

Moment of this area about xx-axis

(Neutral axis) $\bar{y} = 1020\ mm.$

$$A\ .\ \bar{y} = (2000 \times 1020) = 20.4 \times 10^6\ mm^3$$

Step 1 : Factored load

Total uniformly distributed load inclusive of self-weight = 27 kN/m

Factored load, w_{fd} = (1.5 x 27) = 40.5 kN/m

Step 2 : Factored shear force

$$V_{fd} = \left(\frac{40.5 \times 28}{2}\right) = 567\ kN$$

Factored load on compression flange per 1 mm length

$$= \left(\frac{40.5 \times 1}{100}\right) = 0.0405\ kN/mm$$

Step 3 : Moment of inertia of the plate girder

$$I_{xx} = \left[\frac{1}{12} . t_w . d_w^3 + 2 \times b_f . \times t_f \times \bar{y}^2\right]$$

$$= \left[\frac{1}{12} \times 1 \times 200^3 + 2 \times 50 \times 4 \times 102^2\right] \times 10^4\ mm^4$$

$$= \left[66.7 \times 10^4 + 400 \times 10404\right] \times 10^4\ mm^4$$

$$= 4827000 \times 102^4\ mm^4$$

Step 4 : Horizontal shear per 1 mm length

$$\tau_{sh} = \left(\frac{V_{fd} . A \bar{y}}{I_{xx} . 1}\right)$$

$$= \left(\frac{567 \times 20.4 \times 10^6}{4827000 \times 10^4}\right) = 239.627\ N/mm$$

$\tau_{sh} = 237.627$ N/mm, w = 0.0405 N/mm^2

Step 5 : Combined stress for 1 mm length

$$\sigma_{comb} = \frac{1}{1000}\left[0.239^2 + 0.0405^2\right]^{1/2} = 0.0586\ kN/mm$$

Step 6A : Continuous fillet weld, size

$$\text{Design stress, } \sigma_d = \left(\frac{\sigma_u}{\sqrt{3} \times 1.25}\right)$$

$$\sigma_d = \left(\frac{410}{1.732 \times 1.25}\right) = 189.376\ N/mm^2$$

Size of weld

$$s = \left(\frac{0.0586 \times 1000}{2 \times 0.7 \times 1 \times 189.376}\right) = 0.221\ mm$$

Provide 6 mm fillet weld for the flange plate and the web plate on both the sides as shown in Fig. 23.6.

Step 6B : Intermittent fillet weld : size

Let 6 mm size of weld be used Effective length of weld

= 4 times or 38 mm (whichever is more)

= (4 x 6) or 38 mm

= 24 mm or 38 mm.

Adopt effective length of weld as 38 mm.

Strength of weld provided on both the sides of the web-plate.

$$F_{dw} = \left(\frac{2 \times 38 \times 0.7 \times 6 \times 189.376}{1000}\right) = 60.45 \text{ kN}$$

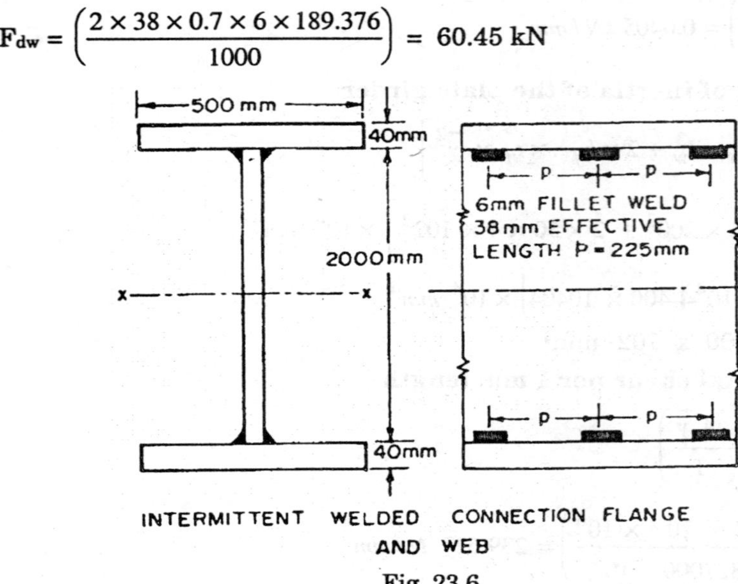

Fig. 23.6

Spacing of weld

$$= \left(\frac{60.45}{0.0586}\right) = 1031.57\,mm$$

Allowable clear spacing

= (12 x 10) = 120 mm

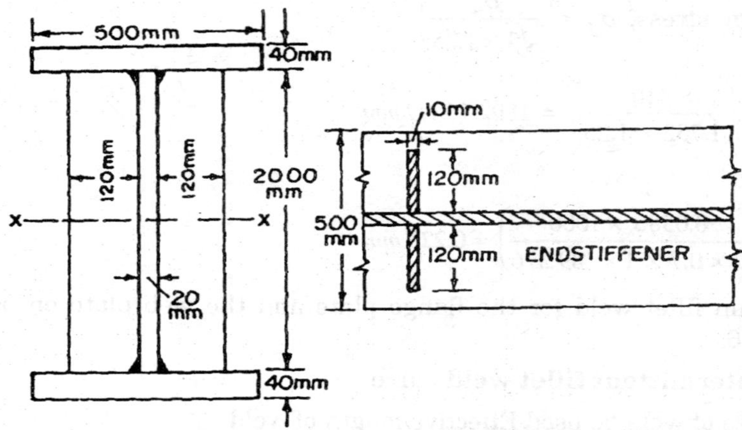

Fig. 23.7.

Provide intermittent fillet weld at clear spacing 225 mm. The intermittent fillet weld of 6 mm size is provided at 225 clear spacing as shown in Fig. 23.7.

Example 23.4: In Example 18.1, factored bending moment, M_{fd} = 9785 kN-m and factored shear force, V_{fd} = 1957 kN due to external load

Redesign the plate girder with intermediate and end stiffeners.

Use simple Post critical buckling method.

Design :

Factored bending moment, M_{fd} = 9785 kN-m.

Factored shear force, V_{fd} = 1957 kN

Size of flange plates as in **Example 18.1** (Width of flange b_f = 600 mm and thickness of flange, t_f = 36 mm)

In Example 18.1, the design bending strength (i.e, moment capacity) M_d = 11565.52 kN-m. It is more than the factored bending moment M_{fd} = 9785 kN-m)

In this **Example 23.4,** size of the flange plates is same as in **Example 18.1,** the size of the web plate (i.e, depth of web plate proposed, d_w = 2000 mm, thickness of web plate, t_w = 10 mm). In this Example 23.4, use of web plate of 10 mm.

Step 1 : Check for shear :

As IS : 800- 2007, Clause 8.4.1 and 8.4.2.2 (A) Nominal plastic shear resistance

$$V_n = V_p = \left(\frac{A_v \cdot \sigma_{yw}}{\sqrt{3}}\right)$$

where, A_v = area in shear and σ_{yw} and strength of the material of web

Let c = clear spacing between the intermediate stiffeners be 2500 mm

d_w = Depth of web = 2000 mm

Effective span = 20 m

$$\text{Number of panels} = \left(\frac{20 \times 1000}{2500}\right) = 8$$

Ratio (c/dw) = (2500/2000) = 1.25

First web panel near the end is the most critical panel for the maximum shear

$$\text{Ratio}\left(\frac{d_w}{t_w}\right) = \left(\frac{2000}{10}\right) = 200$$

(i) **For serviceability requirement** for the minimum thickness of web

$$\text{Ratio } (d_w/t_w) <= 200\ \varepsilon \qquad (\because \varepsilon = 1.00)$$

$$= 200$$

Requirement is satisfied.

(ii) Clear spacing between the stiffeners

$$c > d_w \ (\because C = 2500\ mm,\ d_w = 2000\ mm)$$

$$\left(\frac{d_w}{t_w}\right) = \left(\frac{2000}{10}\right) = (200 \times 1.0) = 200$$

However, $\left(\frac{d_w}{t_w}\right)$ $200 < 345\ \varepsilon^2 \quad (\because\ \varepsilon = 1.00)$

As per IS : 800-2007, Clause 8.6.1.2 the buckling of compression flange will not occur.
Since (C/d_w) = (2500/2000) = 1.25 is more than 1.00

$$K_v = \left(5.35 + \frac{4.0}{(C/d_w)^2}\right) = \left(5.35 + \frac{4.0}{(1.25)^2}\right)$$

$$= (5.35 + 4/1.5625) = (5.35 + 2.56) = 7.91$$

Poisson's ratio for steel = 0.3

Average shear stress

$$\tau = \left(\frac{V_{fd}}{d_w . t_w}\right) = \left(\frac{1957 \times 10^3}{2000 \times 10}\right) = 97.85\ N/mm^2$$

Elastic critical shear stress for the web plate

$$\tau_{ecr} = \frac{K_v . \pi^2\ E}{12\left(1 - \mu^2\right)(d/t_w)^2}$$

$$= \left(\frac{7.91 \times 9.8596 \times 2 \times 10^5}{12\left(1 - 0.3^2\right)\ 200 \times 200}\right) = 35.71\ \text{N/mm}^2$$

Non-dimensional slenderness ratio for shear buckling stress for web

$$\lambda_w = \left(\frac{250}{\sqrt{3}\ \tau_{e.cr}}\right)^{1/2} = \left(\frac{250}{1.732 \times 35.71}\right)^{1/2}$$

$$= 2.01 > 1.2$$

$$\tau_b = \left(\frac{\sigma_{yw}}{\sqrt{3} \times 2.01^2}\right) = \left(\frac{250}{1.732 \times 4.04}\right) = 35.728\ N/mm^2$$

It is note that $\tau_{e.cr}$ and τ_b are practically equal.

Step 2 : Shear force corresponding to buckling

$$V_{cr} = (\tau_b \times d_w \times t_w)$$

$$= \left(\frac{35.728 \times\ 2000 \times 10}{1000}\right) = 714.56\ kN\ <\ 1957\ \text{kN}$$

The web plate needs intermediate stiffeners. As per IS: 800-2007, clause 8.7.1.1 (A), the intermediate transverse stiffeners are to be provided to improve the buckling strength of a slender web.

Check End Panel (Shear capacity)

Step 3 : Anchor forces

As per IS : 800-2007, Clause 8.5.3

$$H_q = 1.25\left(1 - \frac{V_{cr}}{V_p}\right)^{1/2}, \ where$$

$$V_p = \left(\frac{\sigma_y}{\sqrt{3}} . d_w \ t_w\right)$$

$$= \left(\frac{250}{1.732} \times \frac{2000 \times 10}{1000}\right) = 2886.836 \ kN$$

$$V_{cr} = 714.56 \text{ kN}$$

$$H_q = 1.25 \ (V_p) \ \left(1 - \frac{V_{cr}}{V_p}\right)^{1/2}$$

$$= 1.25 \ x \ 2886 \ \left(1 - \frac{714.56}{2886.836}\right)^{1/2}$$

$$= 3130.296 \text{ kN}$$

Resultant longitudinal shear, R_{tf}

$$R_{tf} = \left(\frac{1}{2} \times H_q\right) = \frac{1}{2} \times (3130.296) = 1565.148 \ kN$$

Area $A_v = (d_w . t_w) = (2000 \text{ x } 10) = 2 \text{ x } 10^4 \text{ mm}^2$

$$V_n = \left(\frac{\sigma_{yw}}{\sqrt{3}} \times \frac{A_v}{\gamma_{msy}}\right) = \left(\frac{250}{1.732} \times \frac{2 \times 10^4}{1.15 \times 1000}\right) = 2510.29 \text{ kN} > 1565.148 \text{ kN } (R_{tf})$$

End-panel is adequate to carry the shear due to anchoring forces.

Step 4 : Moment capacity

$$M_{tf} = \left(\frac{H_q . d_w}{10}\right)$$

$$= \left(\frac{3130.296 \times 2000}{10}\right)\left(\frac{10^3}{10^6}\right)$$

$$= \left(\frac{6260.5192}{10}\right) = 626.06 \ kN - m$$

$$y = \left(\frac{C}{2}\right) = \left(\frac{2500}{2}\right) = 1250 \ mm$$

$$I = \left(\frac{1}{12} \times t_w \times C^3\right) = \left(\frac{1}{12} \times 10 \times 2500^3\right)$$

$$= 13020.833 \text{ x } 10^6 \text{ mm}^4$$

$$M_q = \left(\frac{I}{y} \times \frac{\sigma_y}{\gamma_{msy}}\right) = \left(\frac{13020.833 \times 10^6 \times 250}{1250 \times 1.15}\right)$$

$$= \left(\frac{10.417 \times 250 \times 10^6}{1.15}\right) = 2264.49\ kN-m$$

$> M_{tf} = 626.06$ kN-m

The end panel is adequate to carry the bending moment due to anchor forces

Step 5 : Intermediate stiffeners

As per IS : 800-2007, clause 8.7.2.4

Intermediate stiffener next to the bearng stiffener at support is the most critical transverse (intermediate) stiffeners.

$$\text{Ratio}\left(\frac{\text{Clear spacing of stiffeners}}{\text{Depth of web}}\right) = \left(\frac{c}{d_w}\right)$$

$$\left(\frac{c}{d_w}\right) = \left(\frac{1250}{2000}\right) = 0.625 < \sqrt{2}$$

(i) **Moment of inertia** (i.e, second moment of area, I_s about the centre of the web, keeping the stiffeners on both the sides of web,

$$I_s > = 1.5\left(\frac{d_w^3\, t_w^3}{C_3^2}\right)$$

$$= 1.5\left(\frac{2000^3 \times 10^3}{1250 \times 1250}\right) = 768 \times 10^4 \times mm^4$$

Provide two flats = 2 x 120 mm x 10 mm

Moment of inertia provided

$$I_s = \frac{1}{12} \times 10 \times (120 + 120 + 10)^3 - \left(\frac{1}{12} \times 10 \times 10^3\right)$$

$$= 1302 \times 10^3 \times 10^4 \times mm^4 > 768 \times 10^4\ mm^4$$

The stiffeners have stiffeness more than the required. Hence, the stiffeners are satisfactory.

(ii) **Check for outstand** (Clause 8.7.1.2)

Outstand of the stiffeners = b_f = 120 mm

$14\, t_q\, \varepsilon = (14 \times 10 \times 1.0) = 140$ mm.

$b_s < 14\, t_q \,.\, \varepsilon$ satisfied.

(iii)A Check for buckling (Clause 8.7.2.5)

$$F_q = \left(\frac{V_{fd} - V_{cr}}{\gamma_{msy}}\right)$$

V_{fd} = Factored shear force

V_{cr} = Shear buckling resistance of the web panel

$$V = (1957 - 60 \times 2) = 1837 \text{ kN}$$

$$V_q = \left(\frac{1837 - 714.56}{1.15}\right) = \left(\frac{1122.44}{1.15}\right) = 976.03\ kN$$

(iii)B Buckling resistance (Claue 8.7.1.5)

Effective length of web on each side of web of the centreline of the stiffeners limited to 20 times web thickness

$$20\ t_w = (20 \times 10) = 200 \text{ mm}$$

Effective section is the full area or core area of stiffener + Portion of web

$$\text{Area} = [(240 \times 10) + (2 \times 200 \times 10)]\ mm^2$$

$$= (2400 + 4000) = 6400 \text{ mm}^2$$

Moment of inertia

$$I_{xx} = \left(\frac{1}{12} \times 10 \times 250^3\right) + \left(\frac{400 \times 10^3}{12}\right) - \left(\frac{10 \times 10^3}{12}\right) = 1305.33 \times 10^4 \text{ mm}^4$$

Radius of gyration

$$= \left(\frac{1305.33 \times 10^4}{6400}\right)^{1/2} = \left(\frac{36.129 \times 100}{8 \times 10}\right)$$

r_x = 45.16 mm

Effective length = 0.7 x 2000 = 1400 mm

$$\lambda = \left(\frac{1400}{45.16}\right) = 31$$

From Table 9(c) of IS : 800-2007 for σ_y = 250 N/mm² (buckling curve, c)

$\lambda = 31$, σ_{cd} = 202 N/mm²

$$\text{Buckling resistance, } F_{qd} = \left(\frac{202 \times 6400}{1000}\right) = 1292.8\ kN$$

$> F_q$ = 976.03 kN

The stiffener is safe against buckling

(iv) **Interaction expression** (clause 8.7.2.5)

In addition to above the stiffeners should satisfy the following interaction expression

$$\left(\frac{F_q - F_x}{F_{qd}}\right) + \left(\frac{F_x}{F_{xd}}\right) + \left(\frac{M_q}{M_{yq}}\right) <= 1$$

F_q = Force in the stiffener = 973.03 kN

F_{qd} = Design resistance of an intermediate web stiffener corresponding to buckling about an axis parallel to the web = 1292.8 kN

In case $F_q < F_x$, $(F_q - F_x)$ should be taken as zero.

F_x = External load or reaction at the stiffener = 0.

F_{xd} = Design resistance = F_{qd} = 1292.8 kN

M_q = Moment on the stiffener due to eccentrically applied load and transverse load if any = 0.

M_{qd} = Yield moment capacity of the stiffener based on its elastic modulus

$$\left(\frac{F_q - F_x}{F_{qd}}\right) = \left(\frac{973.03 - 0}{1292.8}\right) + \left(\frac{0}{1292.8}\right) + 0 = 0.7526 < 1.00$$

Hence, satisfactory.

Note : The intermediate stiffener should extend to the compression flange but not necessarily be connected to it. Stiffener not subjected to external load or moment may be terminated at a distance of about 4tq (4 x 10 = 40 mm) from the tension flange. In case, there is moving load, the stiffeners may be connected to compression flange and just extend it upto the bottom flange.

Step 6 : Design of end stiffeners (stiffener at supports)

Reaction at support = 1957 kN

Force F_m due to M_{tf}

$$F_m = \left(\frac{M_{tf}}{c}\right) = \left(\frac{626.06 \times 10^6}{1250 \times 10^3}\right) = 500.848 \ kN$$

Total compression force

F_c = Reaction at support + F_m

= (1957 + 500.848) = 2457.848 kN

According to IS : 800-2007, Clause 8.7.5.2

Area of the stiffener is A_q. The bearing strength of the stiffener

$$F_{psd} = \left(\frac{A_q \cdot \sigma_{yq}}{0.8\ \gamma_{msy}}\right)$$

$$A_q = \left(\frac{0.8\ \gamma_{msy}}{\sigma_{yq}} \cdot F_{psd}\right)$$

$$= \left(\frac{0.8 \times 1.15 \times 2457.848}{250} \times 10^3\right) mm^2 = 9044.9 \ mm^2$$

Provide stiffeners of two flats of size 2 x 250 mm x 20 mm

Area provided = 10,000 mm^2

(i) **Check for out stand**

b_s = 250 mm < 20 t_q = 20 x 20 = 400 mm

14 $t_q \varepsilon$ = (14 x 20 x 1.0) = 280 mm > b_s

As per criterion, satisfactory for outstnad

(ii) **Check for buckling (Clause 8.7.1.5)**

For simplicity, neglect the buckling resistance of the web.

$$I_x = \left(\frac{20}{12} \times 510^3\right) - \left(\frac{1}{12} \times 20 \times 10^3\right) \text{ mm}^4$$

$= (1.66 \times 132.651 \times 100 \times 10^4) - (0.16 \times 10^4)$

$= (22020.066 \times 10^4 - 0.16 \times 10^4) = 22019.906 \times 10^4 \text{ mm}^4$

Effective area = (2 x 250 x 20) = 10,000 mm²

Radius of gyration

$$r_x = \left(\frac{22019.906 \times 10^4}{10{,}000}\right)^{1/2} = 148.39 \; mm$$

Flange is restrained against rotation and lateral deflection

$l_e = (0.7 \times 2000) = 1400$ mm

$$\lambda = \left(\frac{1400}{148.39}\right) = 9.4345$$

From Table 9 (C) of IS : 800-2007 for $\sigma_y = 250$ N/mm²

For $\lambda = 10$, $\sigma_{cd} = 227$ N/mm²

Buckling resistance of the stiffener

$P_{qs} = (\sigma_{cd} \times A_e)$

$$= \left(\frac{227 \times 2 \times 250 \times 20}{1000}\right) = 1135 \; kN < 2457.848 \text{ kN}$$

It needs additional load bearing stiffener.

Example 23.5 : In Example 18.1, factored bending moment, M_{fd} = 9785 kN-m and factored shear force, V_{fd} = 1957 kN due to external load.

Redesign the plate-girder with intermediate and end stifeners. Use **tension field action method.**

Design :

Factored bending moment, M_{fd} = 9785 kN-m

Factored shear force, V_{fd} = 1957 kN

Size of web plate

Depth of web = 2000 mm

Thickness of web = 10 mm

Size of flange plates

Width of flange plates, b_f = 600 mm

Thickness of flange plates, t_f = 36 mm

It is assumed that the bending moment is resisted by the flange plates.

Step 1 : Check for shear

Consider that there are eight (08) panels c = 250 mr each. First panel near the end stiffener is the most critical panel of web.

Using **Tension field method** as per IS : 800-2007, Clause 8.4.2.2 (B). Ratio of (actual spacing/depth of web) = (c/d_w) = (2500/2000) = 1.25 > 1.00. Tension field method may be applied.

From **Example 23.4,** τ_{cr} = 35.71 N/mm²

τ_b = 35.728 N/mm²

Shear force, corresponding to buckling

V_{cr} = 714.56 kN

Factored shear force

V_{fd} = 1957 kN

Nominal shear resistance

$V_n = V_{tf}$

$$V_{ft} = \left|\left(\tau_b \cdot d_w \cdot t_w\right) + 0.9\, w_{tf}\, t_w \cdot \sigma_y \sin\theta\right| \;<=\; V_P$$

Yield strength of the tension field

$$\sigma_{ytf} = \left[\sigma_{yw}^2 - 3\,\tau_b^2 + \psi^2\right]^{1/2} - \psi$$

θ = Inclination of tension field, $\psi = 1.5\,\tau_b \cdot \sin 2\theta$

$$= \tan^{-1}\left(\frac{d_w}{c}\right)$$

$$\theta = \tan^{-1}\left(\frac{2000}{1250}\right) = \tan^{-1}(1.6) = 57.9946$$

$$\psi = 1.5 \times 35.728 \times \sin(2 \times 57.9946) \simeq 48.17$$

Axial force in the flange, due to overall bending

$$P_{a.fl} = \left(\frac{9785 \times 10^3}{2000 + 36}\right) = \left(\frac{9785 \times 10^3}{2036}\right) = 4805.99\ kN$$

Reduced plastic moment of the respective flange plate

$$M_{pr} = \frac{1}{4} b_f \cdot t_f^2 \cdot \sigma_{yf} \cdot \left[1 - \left(\frac{P}{\left(\dfrac{b_f t_f\ \sigma_{yf}}{\gamma_{msy}}\right)}\right)^2\right]$$

$$M_{pr} = \frac{1}{4} \times 600 \times 36^2 \times 250 \left[1 - \left(\frac{4805.99}{\left(\dfrac{600 \times 36 \times 250}{1.15}\right)}\right)^2\right]$$

$M_{pr} = 190.38 \times 10^6$ N-mm

Anchorage length of tension field along the flange ($s = s_c = s_t$)

$$S = \frac{2}{Sin\theta}\left[\frac{M_{pr}}{\sigma_y . t_w}\right] <= C$$

$$= \frac{2}{\sin(57.9946)}\left[\frac{190.38\times10^6}{250\times10}\right]^{1/2} = 650.85 \text{ mm} < 1250 \text{ mm}$$

Width of tension field

$W_{ft} = d \cos\theta - (c - s_c - s_t) \sin\theta$

$= 2000 \cos(57.996) - (1250 - 2 \times 650) \sin(57.996)$

$= 1102.2$ mm

Yield strength of tension field

$$\sigma_{ytf} = \left[\sigma_{yw}^2 - 3\,\tau_b^2 + \psi^2\right]^{1/2} - \psi$$

$= [250^2 - 3 \times 35.728^2 + 48.17^2]^{1/2} - 48.17$

$= 198.79$ N/mm^2

$$V_{tf} = \left[d_w . t_w . \tau_b + 0.9\, w_{tf}\, t_w \sin\theta\right]$$

$$= \left[2000\times10\times35.728 + 0.9\times1102.2\times10\times\sin(57.996)\right]$$

$= V_{np}$

$$V_{tf} = \left[2\times3.5726\times10^5 + 0.9\times0.11022\times 10^5\times0.8488\right]$$

$= 10^5\,[7.1456 + 0.0842]$

$= 7.22979 \times 10^5 / 10^3 = 7.22974 \times 100$ kN

$= 722.974$ kN

Note : The shear strength is increased from 714.56 kN to 722.974 kN due to the **tension field action.**

Nominal shear strength $V_n = V_{tf}$

$V_n = 722.974$ kN < 1957 kN

The shear strength shall be increased by providing end stiffeners

Step 2 : Design of end panel using Tension field action

As per IS : 800-2007, Clause 8.5.2, the end panel shall be provided with an **end-post,** which may consist of a single or double stiffener

(i) Double stiffener shall be provided

It is checked as a beam spanning betwen the flanges of the plate girder and capable of resisting a shear force, R_{tf} and a moment, M_{tf} due to the tension field forces as given in Clause 8.5.3.

(ii) Anchor forces due to tension field action (clause 8.5.3)

(A) Resultant longitudinal shear

$$R_{tf} = \left[\frac{1}{2} \cdot H_q\right], \textit{ where}$$

$$H_q = \left[1.25\, V_p \left(1 - \frac{V_{cr}}{V_p}\right)^{1/2}\right]$$

From Clause 8.4.1

$$V_{d.p} = \frac{1}{\sqrt{3}}\left(A_v \cdot \sigma_{yw}\right)$$

$$= \frac{1}{1.732}\left(\frac{2000 \times 10 \times 250}{1000}\right) = 2886.84 \text{ kN}$$

Actual factored shear force V_{fd} =

V_{fd} = 1957 kN

As per clause 8.5.3, in case the actual factored shear force V_{fd} is less than the shear strength, then H_q may be reduced by the ratio

$$\left(\frac{V - V_{cr}}{V_{tf} \times V_{cr}}\right)$$

Hq = 3130.296 kN

R_{tf} = 1565.148 kN

V = 1837 kN, V_{cr} = 714.56 kN

V_{tf} = 722.974 kN, V_{cr} = 714.56 kN

$$\text{Above ratio} = \left(\frac{1837 - 714.56}{722.974 \times 714.56}\right) = \left(\frac{1122.44}{51.66 \times 10^4}\right) = 0.00217$$

$$\text{Reduced Hq} = \left(3130.296 - \frac{0.217}{100} \times 3130.296\right) = 3123.5 \text{ kN} > V_{fd}\ 1957 \text{ kN}$$

Hence, the end panel is safe in shear

(B) **Moment capacity of end panel**

Mq = 2264.49 kN-m > M_{tf} = 626.06 kN-m

(From Example 23.4)

The end panel is safe

(C) (i) **Design of stiffeners**

Load bearing stiffener at support reaction at support = 1957 kN

From Example 23.4

F_m = 500.848

Required A_q = 9044.9 mm²

Provide A_q = 10,000 mm² > Aq required

Two flat plates = 2 x 250 mm x 20 mm

(C) (ii) Check for outstand (Clause 8.7.1.2)

Outstand, b_s = 250 mm

$14\ t_q \cdot \varepsilon_q = (14 \times 20 \times 1.0) = 280$ mm

b_s = 250 mm is less than 280 mm

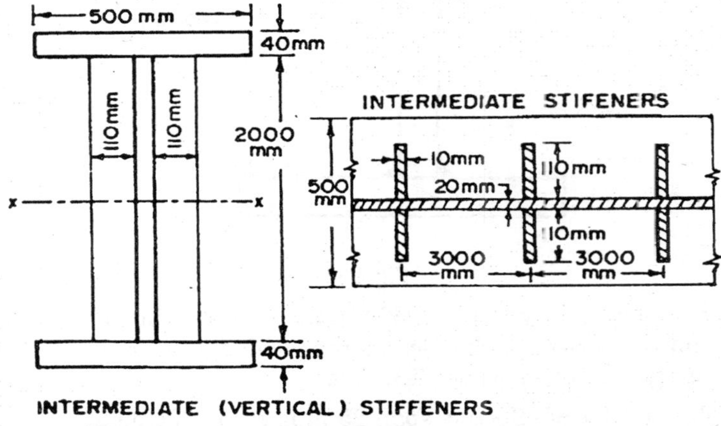

INTERMEDIATE (VERTICAL) STIFFENERS
(WELDED PLATE GIRDER)

Fig. 23.8.

(C) (iii) Check for buckling (Clause 8.7.1.5)

$$I_x = \left(\frac{20 \times (280+10)^3}{12}\right) - \left(\frac{20 \times 10^3}{12}\right) = 4048.57 \times 10^4 \text{ mm}^4$$

Effective area = (2 x 250 x 20) = 10,000 mm²

Radius of gyration

$$r_y = \left(\frac{4048.57 \times 10^4}{10000}\right)^{1/2} \text{ mm}$$

$$= \left(\frac{63.628 \times 100}{100}\right) = 63.628\ mm$$

The compression flange is restrained against lateral deflection and rotation. Effective length (from Example 23.4), l_e = 1400 mm

$\lambda = (1400/63.628) = 22$

PROBLEMS

23.1 A welded plate girder is simply supported at both the ends. The effective span of plate girder is 16 m. It carries a uniformly distributed load of 30 kN/m of exclusive of its own weight. Design the maximum section of welded plate girder, end bearings and intermediate stiffeners.

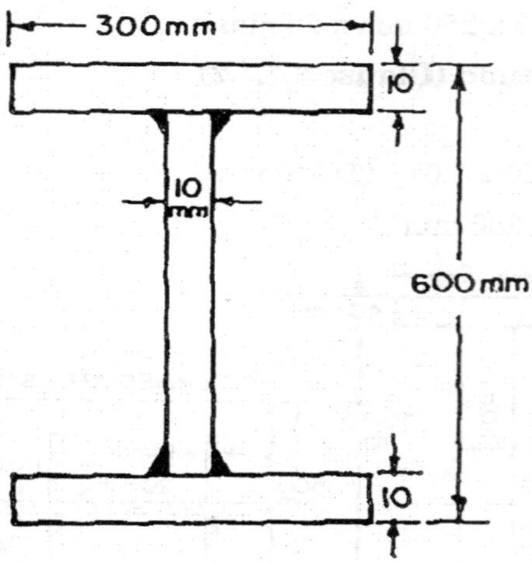

Fig. 23.9.

23.2 An I-section is built-up by welding 600 mm x 15 mm (thick) web plate and two flange plates 150 mm x 12 mm (thick). Design the welded joint to develop full strength of the section. Adopt safe stresses as per I.S. code.

23.3 A welded plate girder of effective span 25 m is made of web 2000 mm deep and 20 mm thick and flanges 500 mm wide and 30 mm thick. It carries a uniformly distributed load of 30 kN/m including its own weight. Design the intermittent weld connection between the web and the flange.

23.4 A welded plate girder has a uniform section as shown in Fig. 23.9. The plate girder carries a uniformly distributed load of 40 kN/m over the entire span of 25 m. The girder is simply supported at both the ends. Design the welded connection between flanges and web plate.

23.5 A 12 m span welded plate girder is subjected to a uniformly distributed load of 30 kN/m along with a concentrated load of 150 kN at 4 m from one of the supports. Design

(i) the cross-section of the plate girder assuming the web to be 6 m,m thick.

(ii) the welded joint for connecting the flange plates with the web.

(iii) the web stiffeners including bearing stiffeners at supports

REFERENCES

1. IS: 800-2007, General Construction Steel, Code of Practice (Third Revision) Published by Bureau of Indian Standards Manak Bhawan, 9, Bahadur Shah Zafar Marg, New Delhi-2.

2. N. Subramanian, Design of Steel Structuress' Published by Oxford, University Press, 2008, India.

3. R. Narayanan, `Teaching Resources on Structural Steel Design', Institution of Steel Development and Growth.

24

DESIGN OF WELDED BEAM CONNECTIONS

24.1 INTRODUCTION

As defined in Chapter 21, welding is formation of a metallic link between two the structural members. Welding is defined as joining two pieces of metal at faces made plastic or liquid by heat or by pressure or by both, so that fusion occurs.

The welded beam connections are similar to the riveted connections in many respects. The welded beam connections are generally more compact and free than the riveted connections. The beams are connected to beams or columns at their ends by direct fillet or butt welds, or through the plates and angles. The welded beam connections are of four types :

1. Direct welded connections
2. Welded framed connections
3. Welded seat connections
4. Moment-resistant welded connections.

This chapter deals with the design and detailing of requirements for the connections between the members, which consist of components such as cleats, gusset plates, brackets, connecting plates by welds.

These connections are designed so as to be consistent with the assumptions made in the analysis of the structures and comply with the requirements specified. These connections shall be capable of transferring the calculated design actions (externmal loads)

Where the members are connected to the surface of a web or the flange of a section, the ability of the web or the flange to transmit the applied forces locally should be checked and where necessary local stiffening shall be provided.

24.2. DIRECT WELDED CONNECTIONS

The beams are connected to beams or columns at their ends by direct fillet welds or butt welds. The fillet welds are applied on both the sides of the web of a beam as shown in Fig. 24.1. The web is connected directly to the face of supporting member. The fillet welds are designed for vertical reaction at the end of beam. The restraining moment existing at the weld section is not considered. As far as possible, the fillet welds are applied near the top. The minimum size of fillet weld is kept equal to the thickness of web or 7 mm whichever is more. The welds, web of beam and element of supporting member are checked for shear. Let V be the vertical reaction at the end of beam. Then, for shop welds.

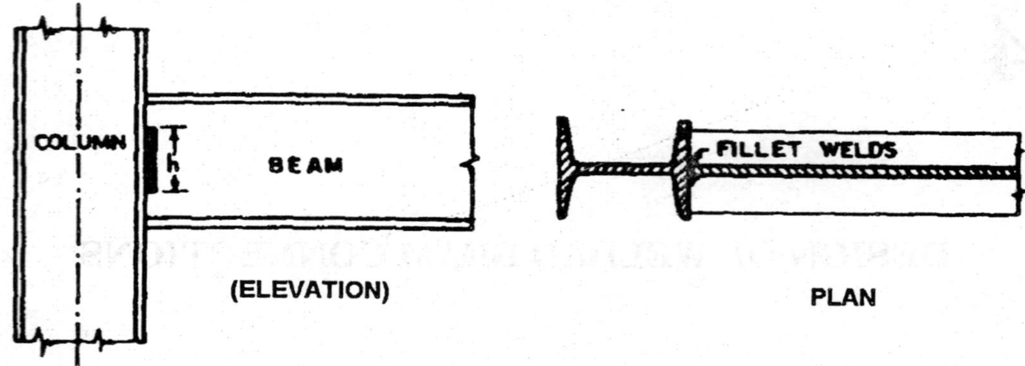

Fig. 24.1.

(A) Design stresses in fillet welds

Design strength of a **fillet weld** f_{wd} is based on its throat thickness.

$f_{wd} = (f_{wn} / \gamma_{mw})$, where

$f_{wn} = \left(\sigma_u / \sqrt{3}\right) =$ smaller of the ultimate stress of the weld or the parent metal, and

γ_{mw} = Partial safety factor (1.25)

(B) Design stresses in butt welds

Butt welds are treated as parent metal with a thickness equal to the throat thickness, and the stresses shall not exceed those allowed in the parent metal.

$$\text{The shear stress, } \left(\frac{V}{2 \times 0.7\, s.h}\right) < \left(\sigma_u / 1.25\right) \text{N/mm}^2 \qquad (24.1)$$

The rotation of end of beam takes place. This rotation may take place by yielding of web or yielding of weld. The size of weld is made larger than the minimum size necessary, so that yielding will occur in the web plate. In order to provide adequate flexibility, the depth of weld should not be greater than one-half to two-third the depth of beam, and size of the weld is kept at least four-fifth the thickness of web. However, size of weld equal to two-third the thickness of web is considered effective to determine depth of weld.

When the beams are connected to beam or columns at their ends directly or columns at their ends directly by butt weld, a backing strip is used as shown in Fig. 24.2. The backing strip is welded with the web in the shop. Then the web of beam is connected to the supporting member by butt weld.

The extra weld metal is deposited on the butt weld. This provides extra strength to butt weld. The depth of butt weld is kept small as far as possible in order to allow end rotation. The butt weld is designed to resist the vertical reaction at the end of beam. If V be the vertical reaction, then, the shear stress in butt weld.

$$\left(\frac{V}{d \times t_w}\right) < \tau_v \left(\sigma_u / 1.25\right) \text{N/mm}^2$$

The direct fillet weld or butt weld needs preparation of accurate length of beam. The rotation of end of beam takes place by yielding of web. Therefore, as far as possible, these types of connections are not used.

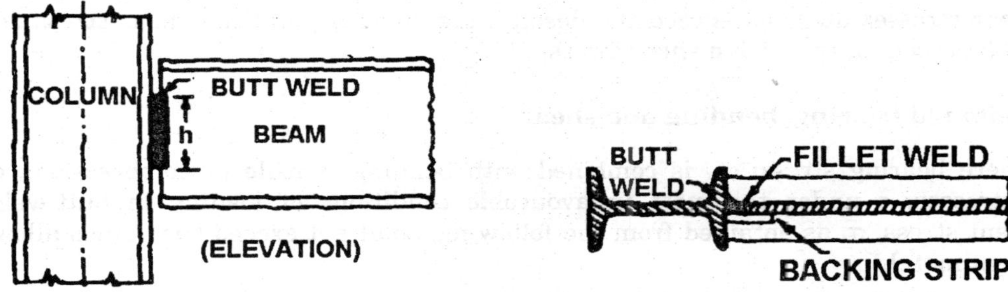

Fig. 24.2.

(A) Stresses due to individual forces

When subjected to either compressive or tensile or shear force alone, the stress in the weld

$$\sigma_a \text{ or } \tau = \left(\frac{P}{t_t \cdot l_w}\right), \text{ where}$$

σ_a = calculated normal stress due to axial force, in N/mm²

τ = shear stress, in N/mm²

P = Force transmitted (Axial force)

V = Shear force

t_t = Effective throat thickness of weld, in mm

l_w = Effective length of weld

(B) Combined stresses in fillet welds

When the fillet welds are subjected to a combination of normal and shear stresses, the equivalent stress σ_e shall satisfy the following

$$\sigma_e = \left(\sigma_a^2 + 3\tau^2\right)^{1/2} <= \left(\frac{\sigma_u}{\sqrt{3} \times 1.25}\right), \text{ where}$$

σ_a = Normal (axial) stress, compression or tension due to axial force or bending moment

τ = shear stress due to shear force or tension

The check for the combination of stresses need not be done for

(i) side fillet welds joining cover plates and flange plates and

(ii) fillet welds where sum of normal and shear stresses does not exceed σ_{wd} (design stress in weld = σ_u / $(\sqrt{3} \times 1.25)$

(C) Combined stresses in butt welds

Check for the combination of stresses in butt-welds need not be carried out provided that (i) butt welds are axially loaded and (ii) in single and double bevel welds, the sum of normal and shear stresses does not exceed the design normal stress, and the shear stress does not exceed 50 percent of the design shear stress.

(D) Combined bearing, bending and shear

Where bearing stress, σ_{br} is combined with bending (tensile or compressive), σ_b and shear stresses, τ under the most unfavourable conditions of loading in butt-welds the equivalent stress, σ_e as obtained from the following, shall not exceed the values allowed for the parent metal.

$$\sigma_e = \left(\sigma_b^2 + \sigma_{br}^2 + \sigma_b \cdot \sigma_{br} + 3\tau^2\right)^{1/2}, \text{ where}$$

σ_e = Equivalent stress

σ_b = calculated stress due to bending, in N/mm^2

σ_{br} = Calculated stress due to bearing in N/mm^2

τ = shear stress in N/mm^2

Example 24.1 A MB 500, @ 0.869 kN/m transmits an end reaction of 130 kN to the flange of a column MB 250, @ 0.510 kN/m. The fillet welds are applied directly on both the sides of the beam, Design the connections.

Design :

Step 1. Size of weld

From ISI Handbook No. 1 for MB 500, @ 0.869 kN/m beam

Thickness of web, t_w = 10.2 mm.

For MB 250, @ 0.510 kN/m column.

Thickness of flange, t_f = 9.7 mm

Size of fillet weld = (4/5) x 10.2 = 8.16 mm

Provide 9 mm fillet weld

Step 2 : Factored shear force

Beam transmits an end reaction to the flange of the column.

V = 130 kN

Factored shear force, V_{fd} = (1.5 x 130) = 195 kN

Design stress in the weld

$$\sigma_{dw} = \left(\frac{\sigma_u}{\sqrt{3} \times 1.25}\right) = \left(\frac{410}{1.732 \times 1.25}\right) = 189.376 \text{ N/mm}^2$$

Throat stress in the weld

$$t_t = (0.7 \times 9) = 6.3 \text{ mm}$$

Strength of 1 mm length of weld

$$= \left(\frac{189.376 \times 1 \times 6.3}{1000}\right) = 1.143 \text{ kN/mm}$$

Fillet weld is provided on both the sides.

$$\text{Depth of weld, h} = \left(\frac{195}{2 \times 1.143}\right) = 85.302 \text{ mm}$$

Provide 9 mm weld, 90 mm deep on both the sides of the web, as shown in Fig. 24.2.

24.3 WELDED FRAMED CONNECTIONS

The direct fillet weld or butt weld needs preparation of accurate length of beam. The gap between end of beam and supporting member must be short. For fillet weld, the end of beam must remain completely flush with the element of supporting member. Butt weld may be provided even if the gap is upto the extent of 3 mm. But, when beams are manufactured for ordered length, the rolling mill allows a tolerance of 12 mm either way. The gap may be greater at one flange of beam than at the other. The beam end may be prepared by milling, cold sawing or flame cutting. All these operations are fairly expensive. To avoid this difficulty, welded framed connections are used. In the welded framed connections, two plates or two angles are used. The plate connection as shown in Fig. 23.3 may be considered as a special case. The plates make the connection somewhat flexible, than the direct welded connection.

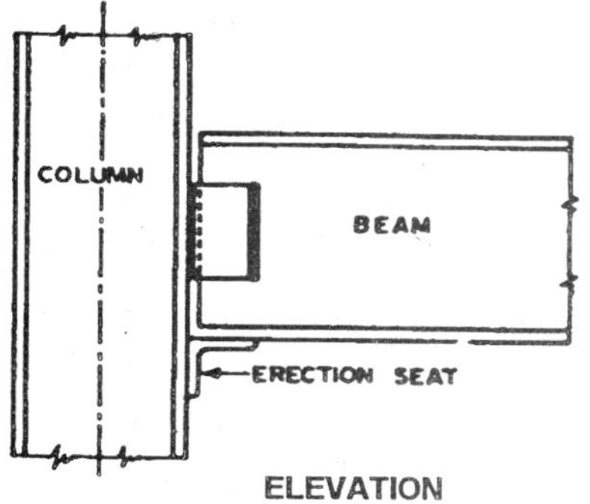

ELEVATION

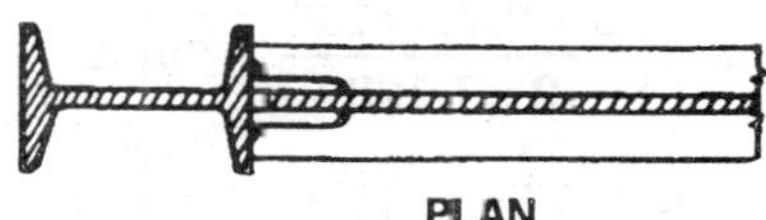

PLAN

Fig. 24.3.

One plate is used on one side of web of the beam and other plate is attached on the other side. For connection, one plate is welded in the shop with the beam and other plate is welded in the shop with the element of supporting member. When both the members are kept in position, erection bolts are inserted and welding in done. To allow rotation at the end of beam, the length of plates is limited to two-third the depth of beam. The size of weld connecting plates to the web of beam is kept equal to four-fifth of thickness of web.

The welds connecting plate to the supporting member are designed to resist the vertical reaction V, at the end of beam, Whereas, the welds connecting plates to the web of beam are designed to resist vertical reaction, V, moment, V.*l*, where *l* is distance between two welds. The length of plates is kept about 5 mm. The plates may be thicker than the welds connecting plates and web of beam by 1.6 mm. In that case, the plates will also be safe, if the welds connecting plates and web of beam are safe.

Example 24.2 In Example 24.1, two plates are used for connection on both the sides of web of a beam. Design the welded connections.

Design : From Example 24.1, the vertical reaction at the end of beam transmitted = 130 kN

Two plates 50 mm long are provided for the welded frame connection

Step 1 : Moment to be resisted

$$M = \left(\frac{130 \times 50}{1000}\right) = 6.5 \text{ kN} - \text{m}$$

Step 2 (A). Factored shear force to be transmitted = End reaction

$$V_{fd} = (1.5 \text{ x } 130) = 195 \text{ kN-m}$$

Step 2 (B). Factored moment to be resisted

$$M_{fd} = (1.5 \text{ x } 6.5) = 9.75 \text{ kN-m}$$

For the web thickness. 10.2 mm,

9 mm fillet welds are adopted.

Step 3 : Strength of 1 mm weld (Weld value)

1. Weld connecting plates and web in the shop :

$$\text{Design stress, } \sigma_d = \left(\frac{410}{1.732 \times 1.25}\right) = 189.376 \text{ N/mm}^2$$

$$F_{w1} = \left(\frac{0.7 \times 9 \times 189.376}{1000}\right) = 1.193 \text{ kN/mm}$$

Fillet welds are provided on both the sides of the web

Depth of weld

$$d = \left(\frac{6 \times 9.75 \times 10^3}{2 \times 1.193}\right)^{1/2} \text{mm}$$

$$= (15.66 \text{ x } 10) = 156.6 \text{ mm}$$

Try 180 mm deep welds

Vertical shear per one mm length

$$\tau_v = \left(\frac{195 \times 1000}{2 \times 180}\right) = 541.66 \text{ N/mm}$$

Horizontal shear per 1 mm length

$$\tau_H = \left(\frac{6 \times 9.75 \times 10^6}{2 \times 180 \times 180}\right) \text{N/mm}$$

$$= \left(\frac{58.5 \times 10^6}{2 \times 3.24 \times 10^4}\right) = 902.77 \text{ N/mm}$$

Resultant shear

$$\tau_s = \left(541.66^2 + 902.77^2\right)^{1/2} \text{ N/mm}$$

$$= \left|(29.34 + 81.5)^{1/2} \times 100\right| = 1052.8 \text{ N/mm}$$

Strength of one mm weld = 1193, N/mm is more than resultant shear. Hence, safe.

2. Weld connecting plate to the supporting member in the field

Let the size of weld be s. Then

$$0.7 \times s \times (0.1 \times 189.376) \times 2 \times 180 = 195 \times 1000$$

$$s = \left(\frac{195 \times 1000}{0.7 \times 18.9376 \times 360}\right) = 4.086 \text{ mm}$$

Provide fillet welds of size 6 mm. Thickness of plates = (9 + 1.6) = 10.6 mm

Provide 12 mm x 50 mm plates

The connections of two framing angles are similar to the riveted frame connections. The welded framed connections with two framing angles are more flexible than other type of welded connections. The bending of outstanding legs of the angles provides flexibility to the connection. During the erection, the erection bolts are provided through the connection angles.

A welded framed connection is shown in Fig. 24.4. In order to provide flexibility to the connections, thickness of angle is kept less than the size of weld. For economical consid-eration, the length of leg of angle connecting the web of beam should be short. The welds are usually returned to the end of beam on this leg, as shown in Fig. 24.4. This welding is done in the shop. This will now be referred as shop weld. The outstanding legs are connected to the supporting member in the field and and the weld will be referred as field weld. The field welds are returned at the top for short distance.

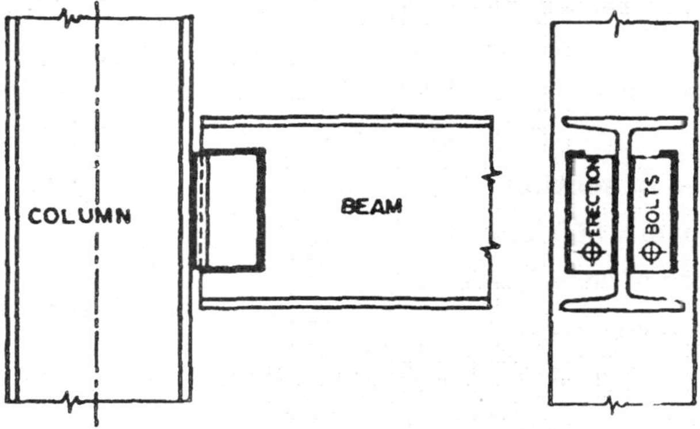

Fig. 24.4.

Fig. 24.5 shows a beam with two frame angles attached to it. The beam is assumed simply supported at the face of outstanding legs of connecting angles. Therefore, the field welds are not subjected to any moment. Both the angles are connected individually to the web of beam. These angles are not connected together. Each field weld is subjected to the vertical reaction R/2 as shown in Fig. 24.5. Therefore, each shop weld is subjected to vertical reaction R/2, and a twisting moment, (R/2) l_2.

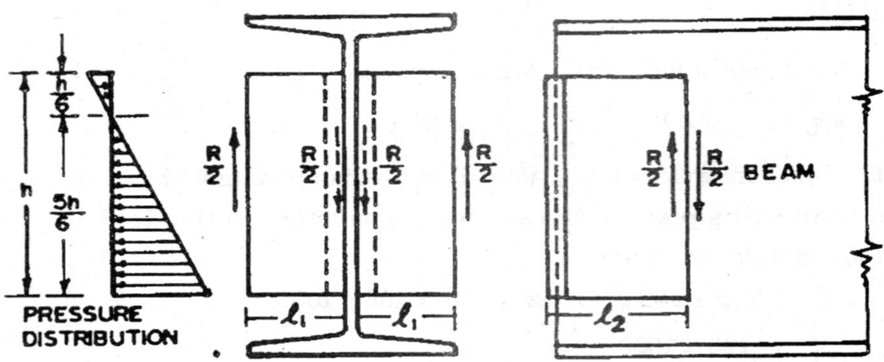

Fig. 24.5.

Each angle is also subjected to a moment (R/2). l_1. Though this moment is resisted by both shop weld and field weld, it is assumed that this moment is resisted by field weld only. This moment is resisted by bearing pressure between angles and web at the top and between angles and weld for the remaining length.

It is assumed that the neutral axis is situated at h/6 from the top of angles as shown in pressure distribution diagram, Fig. 24.5. The maximum horizontal shear in the field may be obtained as under :

$$\left(\frac{1}{2}\tau_v \cdot \frac{5}{6}h\right) \cdot \frac{2}{3}h = \frac{R}{2} \cdot l_l\ ,\ \tau_v = \frac{9}{5} \cdot \left(\frac{R.l_1}{h_2}\right)$$

The resultant of horizontal shear and shear due to vertical reaction R/2 is found by vector sum. The resultant shear stress must be less than the allowable shear. The depth of angle is kept about one-half to two-third the depth of the beam, 2 ISA 90 mm x 60 mm are used for framing angles. The thickness of angles is determined after designing the welds.

Example 24.3. In Example 24.1, two framing angles are used as shown in Fig. 24.6 for connections on both the sides of web of beam. Design the welded connections.

Design :

Step 1 : Properies of section

Depth of beam = 500 mm

Half of depth of beam = 250 mm

(2/3) depth of beam = 333.3 mm

The depth of angles is adopted between **one-half** to **two-thirds** the depth of beam

Assume depth of framing angles = h = 300 mm

Let 2 ISA 90 mm x 60 mm are used for the framing agles.

Step 2 : Design of field weld

Vertical reaction on each weld

$$\left(\frac{R}{2} = \left(\frac{130}{2}\right) = 65\ kN\right)$$

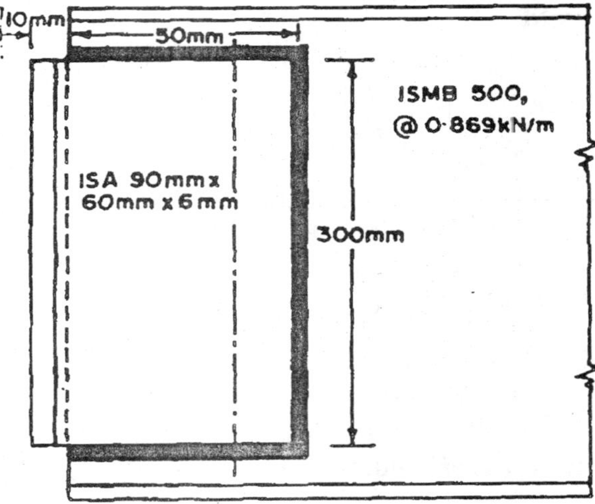

Fig. 24.6.

Horizontal shear per one mm length

$$\tau_{v.sh} = \left(\frac{9}{5} \times \frac{R \times l_1}{h^2}\right) \text{ N/mm}$$

$$\tau_{v.sh} = \frac{9}{5} \times \left(\frac{130 \times 1000 \times 90}{300 \times 300}\right) = 234 \text{ N/mm}$$

Vertical shear per one mm length

$$\tau_{v.sv} = \left(\frac{65 \times 1000}{300}\right) = 216.6 \text{ N/mm}$$

Resultant shear stress for 1 mm length

$$= \left(\tau^2_{v.sh} + \tau^2_{v.sv}\right)^{1/2} = \left(234^2 + 216.6^2\right)^{1/2} = 320 \text{ N/mm}$$

Factored resultant shear stress

per one mm length

= (1.5 x 320) = 480 N/mm

Let s be the size of field weld Design stress for wled

$$= \left(\frac{\sigma_u}{\sqrt{3}} \times \frac{1}{1.25}\right) = \left(\frac{410 \times 1}{1.732 \times 1.25}\right) = 189.376 \text{ N/mm}^2$$

Strengh of one mm weld

0.7 s x 1 x (0.8 x 189.376) = 480

$$s = \left(\frac{480}{0.7 \times 0.8 \times 189.376}\right) = 4.526 \text{ mm}$$

Provide 6 mm size for field fillet weld

2. Design of shop weld

Let t_t be thoat thickness.

Depth of weld = 300 mm

$$\bar{x} = \left(\frac{2 \times 50 \times \frac{50}{2} \times t_t}{2 \times 50 \times t_t + 300 \times t_t} \right) = 6.25 \text{ mm}$$

Therefore, twisting moment

$$M = \frac{195}{2} \times (60 - 6.25) = 5240.625 \text{ kN} - \text{m}$$

Step 3 : Properties of weld

Let t_t be the throat thickness of weld.

Moment of inertia about xx-axis

$$I_{xx} = \frac{1}{12}\left[t_t \times 30^3 + 2 \times 5 \times t_t \times 15^2\right] \times 10^3 \text{ mm}^4 = 4500 \, t_t \times 10^3 \text{ mm}^4$$

Moment of inertia about yy-axis

$$I_{yy} = \left[30 \times t_t \times 0.625 + 2 \times \frac{1}{12} \times t_t \times 5^3 + 2 \times 5 \times t_t \, (2.5 - 0.625)^2\right] \times 10^3 \text{ mm}^4$$

$$= 51.4 \; t_t \text{ x } 10^3 \text{ mm}^4$$

$$I_p = (4500 \, t_t + 51.4 \, t_t) = 4551.4 \, t \text{ x } 10^3 \text{ mm}^4$$

Step 4 : Stresses in welds

1. Horizontal shear in the weld

$$\tau_{v.sh} = \left(\frac{5240.625 \times 10^3 \times 150}{4551.41 \times 10^3 \times t_t} \right) = \left(\frac{172}{t_t} \right) \; N / mm^2$$

2. Vertical shear in the weld

$$t_{v.sv} = \left(\frac{5240.625 \times 10^3 \times 43.75}{4551.3 \times 10^3 \; t_t} + \frac{195 \times 50}{2 \times 50 \times t_t + 300 \; t_t} \right) = \left(\frac{216}{t_t} \right) \; N / mm^2$$

3. Resultant shear stress

$$= \left(\frac{216^2}{t_t} + \frac{172^2}{t_t} \right)^{1/2}$$

$$= \frac{1}{t_t}(46656 + 29584)^{1/2} = \left(\frac{276.115}{t_t} \right) \text{N/mm}^2$$

$$\frac{276.115}{t_t} = 189.376, \quad t = 1.458 \; mm$$

Provide 6 mm size for shop welds. The thickness of angles is kept equal to the size of weld (i.e, = 6 mm)

24.4 WELDED SEAT CONNECTIONS

When a beam is connected to a beam or a column by means of an angle at the bottom of the beam, which is shop welded to the supporting member and an angle ISA 100 mm x 100 mm x 6 mm at top which is field welded, the connections are known as seated connections. The top angle provides stability to the beam. The welded seat connections provide some flexibility as the welded framed connections. The welded seat connections provide satisfac-tory and efficient support to the beam. Similar to the riveted connections, the welded seat connections are also unstiffened and stiffened.

Unstiffened welded seat connections. The unstiffened welded seat connections are shown in Fig. 24.7. In Fig. 24.7, one beam is supported at the flange of a column and other beam is supported at the web. The unstiffened seat connections are used to support light loads. The vertical welds are used to connect seat angles. These welds are returned at the top of seat angles. The return ends of weld are not considered to carry-loads. The top angles are welded along the toss by 6 mm welds. Therefore, it remains free to bend and provides flexibility to the connection.

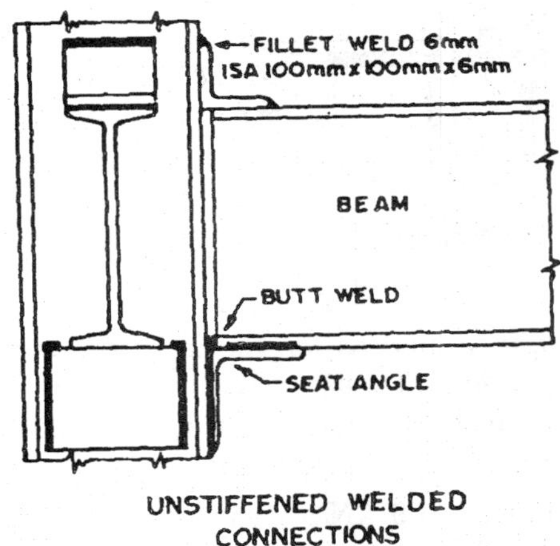

Fig. 24.7.

The bearing length `b' of seat angle is determined in the manner seat similar to that of the riveted connection.

$$\text{The bearing length, } b = \left(\frac{V}{\sigma_b \cdot t_w} - \sqrt{3} \cdot h_2\right) \cdot \nleftarrow \left(\frac{1}{2} \frac{1}{\sigma_b \cdot t_w}\right)$$

The end reaction `V' is considered as uniformly distributed on the outstanding leg of seat angle over the length `b'. A clearance of 10 mm is provided between the end of beam and column. The critical section is assumed at the edge of filler. The distance of end reaction from critical section of the outstanding leg of seat angle is determined, and bending moment `M' is found at the critical section. The length of seating angle is assumed equal to width of the flange of beam. The thickness of seating angle is found by equating moment of resistance to the moment at the critical section.

The neutral axis for vertical welds is assumed to be situated at mid-height. The horizontal shear per mm and vertical shear per mm of weld are determined and combined resultant shear per mm length should be less than or equal to allowable shear per mm of welds.

Stiffened welded seat connections. The design of stiffened seat connections is somewhat simpler than that of riveted connections, In the stiffened welded seat connections, two plates welded together forming a tee-section are used. A split-beam section is also used to form a seat in stiffened welded seat connections. The stiffened welded seat connections are shown in Fig. 24.8. The thickness of stiffening plate is kept equal to thickness of web of the beam. The thickness of seat plate is kept equal to flange thickness. The stiffening plate is welded along both the vertical sides. The horizontal welds are provided along the under-side of the seat plate. The length of each horizontal weld is kept one-fifth to one-half the length of vertical leg. The horizontal welds and vertical welds form a double angle section as shown in Fig. 24.8.

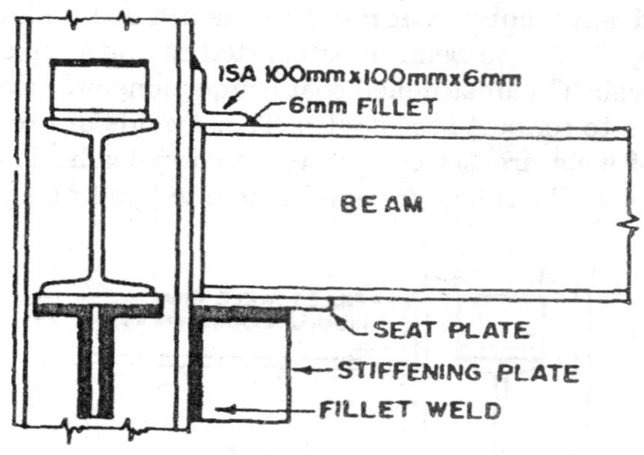

STIFFENED WELDED SEAT CONNECTIONS

Fig. 24.8.

The bearing length `b' is determined as for the unstiffened welded seat connections. For the stiffened seat connections the eccentricity of beam reaction is more than the unstiffened one. The position of beam reaction is assumed to lie at the mid-point of bearing length. The bearing length is measured from outer end of seat plate.

Example 24.4. A MB 500, @ 0.869 kN/m transmits an end reaction of 130 kN to the flange of a column HB 250 @ 0.510 kN/m. Design an unstiffened welded seat connection.

Design : Beam transmits an end reaction

$V = 130$ kN

Factored value = (1.50 x 130) = 195 kN

Design stress, $\sigma_d = \left[410/\left(\sqrt{3} \times 1.25\right)\right] = 189.376 \text{ N/mm}^2$

Step 1 : Bearing length of seat connection

From ISI HB No. 1, for MB 500, @ 0.869 kN/m

Width of flange, $b_f = 180$ mm

Thickness of web, $t_w = 10.2$ mm.

Thickness of flange, $t_f = 17.2$ mm

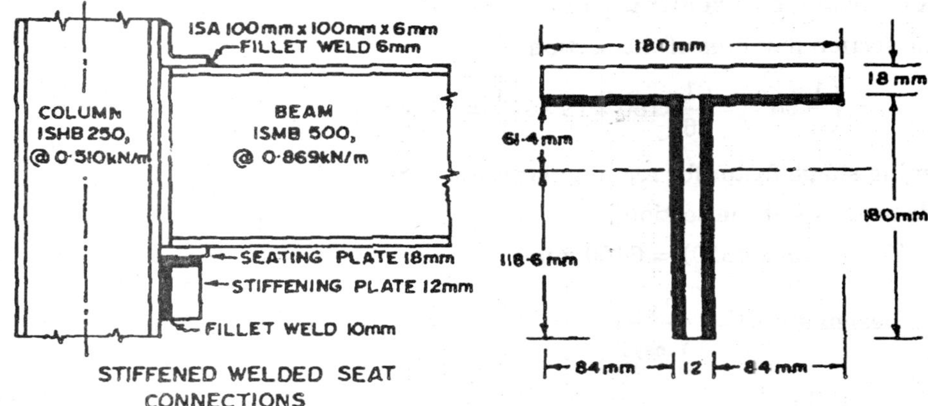

Fig. 24.9. Fig. 24.10.

Depth upto root of flange, h_2 = 37.95 mm

Bearing length

$$b = \left[\frac{V}{\sigma_d \cdot t_w} - \sqrt{3}\,.\,h_2\right]$$

$$= \left(\frac{195 \times 1000}{189.376 \times 10.2} - \sqrt{3} \times 37.95\right) = 35.22 \text{ mm}$$

But b should not be less than

$$= \left(\frac{1}{2} \times \frac{V}{\sigma_d \cdot t_w}\right) = \frac{1}{2} \times \left(\frac{195 \times 1000}{189.376 \times 10.2}\right) \text{mm}$$

$$= \frac{1}{2} \times (100.95) = 50.475 \text{ mm}$$

Let the bearing length of seat angle,

b be 55 mm

Try seating angle of size ISA, 150 mm x 115 mm x 12 mm (ISA 150, 115, @ 0.238 kN/m)

The long leg is kept vertical. Radius at the root, r_1 = 11.0 mm

Distance of end reaction from the critical section

$$= \left[10 + \frac{1}{2} \times 33.7 - (12 + 11)\right] \text{mm}$$

$$= [(26.85 - 23)] = 3.85 \text{ mm}$$

Step 2 : Factored Bending moment at the critical section

= (3.85 x 195) = 750.75 kN-mm

Width of flange (beam), b_f = 180 mm

Length of seating angle = 180 mm

Design strength of seating angle

Elastic section modulus of the section

$$Z_E = \left(\frac{1}{6}\,bd^2\right) = \left(\frac{1}{6}\times 180 \times 12 \times 12\right) = 4320 \text{ mm}^3$$

Assuming shape factor for rectangular section, S = 1.50

Plastic modulus of the section

$$Z_p = (1.50 \times 4320) = 6480 \text{ mm}^2$$

$$\text{Design stress} = \left(\frac{\sigma_y}{\gamma_{msy}}\right)$$

$$= \left(\frac{250}{1.15}\right) = 217.39 \text{ N/mm}^2$$

$$\text{Design strength} = (\sigma_d \times Z_p)$$

$$= \left(\frac{217.39 \times 6480}{1000}\right) = 1408.68 \text{ kN} - \text{mm}$$

Design strength is more than the factored bending moment. Hence, the seating angle is safe.

Step 3 : Design of weld

Length of vertical weld = 150 mm

Vertical shear per one mm weld

$$= \left(\frac{195 \times 1000}{2 \times 150}\right) = 650 \text{ N/mm}$$

Distance of end reaction from the weld

$$= \left(\frac{550}{2} + 10\right) = 37.5 \text{ mm}$$

Bending moment at the weld

$$= \left(\frac{195 \times 37.5}{1000}\right) = 7.3125 \text{ kN} - \text{m}$$

Horizontal shear per one mm length

$$= \left(\frac{7312.5 \times 1000}{2 \times \frac{1}{6} \times 3 \times 150 \times 150}\right) = 97.5 \text{ N/mm}$$

Resultant shear stress per one mm

$$\tau_s = \left(650^2 + 97.50^2\right)^{1/2} \text{ N/mm}$$

$= (4225 \times 100 + 45.06 \times 100)^{1/2}$

$= 653.45$ N/mm

Let, s be the size of weld. Then

$0.7\, s \times 1 \times 189.376 = 653.45$

$s = 4.93$ mm

Provide 6 mm fillet weld and ISA 150 mm x 115 mm x 12 mm

(ISA 150, 115, @ 0.238 kN/m seating angle. ISA 100 mm x 100 mm x 6 mm (ISA 100, 100 @ 0.092 kN/m) is used at the top and 6 mm weld is applied at the toes.

Example 24.5 A MB 500, @ 0.869 kN/m beam transmits an end reaction of 240 kN to the flange of a column section HB 250, @ 0.510 kN/m. Design stiffened seat connections.

Design: End reaction transmitted, V = 240 kN factored value = (1.5 x 240) = 360 kN

From ISI Handbook No. 1, the geometrical properties, of beam and column

For MB 500, @ 0.869 kN/m

Width of flange, $b_f = 180$ mm

Thickness of flange, $t_f = 17.2$ mm

Thickness of web, $t_w = 10.2$ mm

Depth upto root of radius, $h_2 = 37.95$ mm

Step 1 : Bearing length of seat connection

$$\text{Design stress, } \sigma_d = \left(\frac{\sigma_u}{\sqrt{3} \times 1.25}\right) \text{N/mm}^2$$

$$= \left(\frac{410}{\sqrt{3} \times 1.25}\right) = 189.376\ N/mm^2$$

$$b = \left(\frac{V}{\sigma_b \cdot t_w} - \sqrt{3}\,.\,h_2\right) \text{mm}$$

$$= \left(\frac{360 \times 1000}{189.376 \times 10.2} - \sqrt{3} \times 37.95\right) \text{mm}$$

$$= (186.371 - 65.731) = 120.64 \text{ mm}$$

The bearing length, b should not be less than

$$b = \left(\frac{1}{2} \times \frac{V}{189.376} \times \frac{}{10.2}\right) \text{mm}$$

$$= \frac{1}{2} \times \left(\frac{360 \times 1000}{189.376 \times 10.2}\right) = 93.371\ mm$$

The bearing length may be provided, b = 130 mm

Provide a clearance of 9.36 mm between the end of beam and the flange of the column

Width of seat plate = (120.64 + 9.36) = 130.00 mm

Thickness of a seat plate

Thickness of flange, (beam), t_f = 17.2 mm

Provide 18 mm thick and 130 mm long and 180 mm wide seat angle

Thickness of stiffening plate

Thickness of web (beam), t_w = 10.2 mm

Provide 12 mm thick stiffening plate. The distance of end reaction from outer end of seat plate.

$$\left(130 - \frac{1}{2} \times 93.371\right) = 83.31\ mm$$

Step 2 : Factored bending moment

M_{fd} = (360 x 83.31) = 29993.22 kN-mm

Step 3 : Properties of weld

$$\bar{y}_1 = \left(\frac{2 \times 180 \times 90}{2 \times 180 + 168}\right) \text{mm}$$

$$= \left(\frac{180 \times 180}{528}\right) = \left(\frac{32400}{528}\right) = 61.36\ mm$$

$$\bar{y}_2 = (180 - 61.36) = 118.64\ mm$$

Moment of inertia

$$I_{xx} = \left[2 \times \frac{1}{12} \times 18^3 + 2 \times 18\ (11.86 - 9)^2 + 16.8 \times 6.136^2\right] \times 10^4\ mm^4$$

= 1898 x 10^4 x mm^4

Step 4 : Stresses in weld

Vertical shear per one mm of weld

$$\tau_{v.sv} = \left(\frac{360 \times 1000}{2 \times 180 \times 168}\right) = 5.95\ N/mm$$

Horizontal shear per one mm of weld

$$\tau_{v.sh} = \left(\frac{29993.22 \times 1000 \times 6.136}{1898 \times 10^4}\right) \times 10 = 96.96 \text{ N/mm}$$

Since the connection will not allow failure of weld in thrust y = 61.36 mm

Step 5 : Resultant shear stress per one mm

$$\tau = \left(5.95^2 + 96.96^2\right)^{1/2} \text{ N/mm}$$

$$= (35.40 + 9401.24)^{1/2} = 97.142 \text{ N/mm}$$

Step 6 : Size of the weld

Let the size of weld be s

$$0.7 \times s \times 1 \times 189.376 = 97.142$$

$$s = \left(\frac{97.142}{0.7 \times 189.376}\right) = 0.732\ mm$$

Provide 10 mm x 6 mm (ISA 100, 100, @ 0.092 kN/m is used at top. A complete welded stiffened seat connected designed in shown in Fig. 24.9.

24.5 MOMENT RESISTANT WELDED CONNECTIONS

When the ends of beam are subjected to moments and shears, moments resistant welded connections as shown in Fig. 24.11 are used. The end moments of beam are resisted by top plate and bottom flange. When the moment of beam, say, on the left side of connection is clockwise, the top plate clockes tensile force and the bottom flange of beam carries compressive force. The forces are equal in magnitude, and opposite in direction. These forces form a resisting couple. When the direction of moment is anticlockwise, these elements of connections carry equal forces of opposite sign. The top plate is butt welded to the flange of column. The butt weld needs a 5 mm minimum gap between plate and the flange of column. It requires a backing strip. The backing strip is tack welded to the flange. The length of top plate is kept large. The entire length of top plate is not welded. A length equal to one-half to one times the width of the plate is kept unwelded. This portion of plate is kept unwelded so that if yielding occurs, it occurs in this, rather than in weld. This provides ductility to the connections. The size of fillet weld should be such that the length of side welds remains about one-half to one times the width of plate. In case, the moment resistance connections are made with the web of column, the top plate is flared. This allows side welds to the column flange.

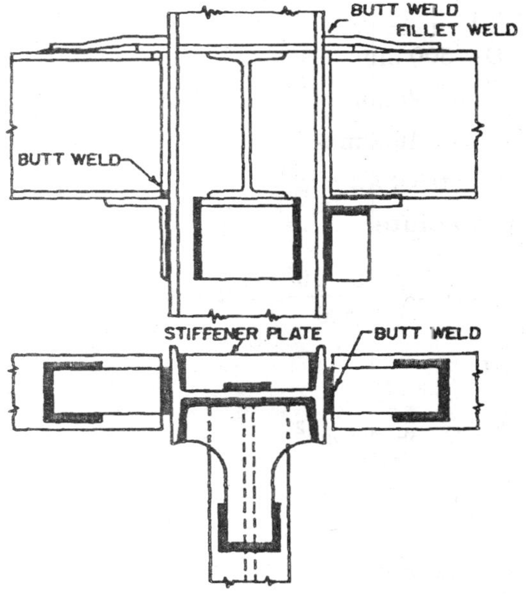

Fig. 24.11.

The welded seat connections are designed to resist the vertical end reactions of beam. This welded seat connections are kept unstiffened or stiffened depending upon the magnitude of end reactions.

Fig. 24.12 shows moment resistant welded connections for beams to beams. In case, the top flanges of two beams to be connected are at the same level of supporting beam, then a plate is welded as shown in Fig. 24.12(A). In case, the top flange of two beams to be connected are not at the same level as that of the flange of supporting beam, then, the top plates are butt welded to the web as shown in Fig. 24.12(B). As an alternative to this a slot may be cut in a channel to clear the flange of supporting beam and it is connected to the flanges as shown in Fig. 24.12(C).

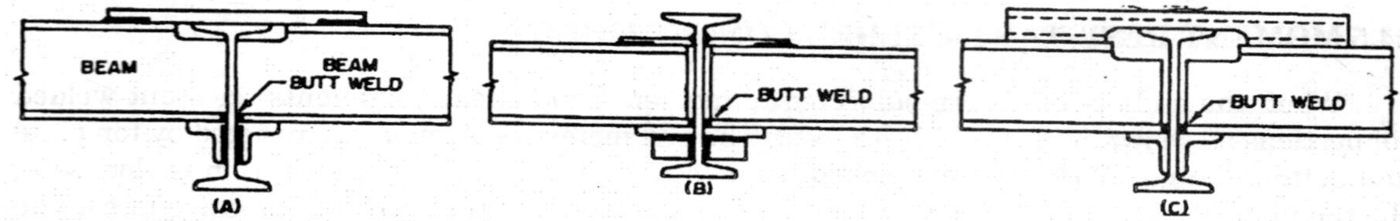

Fig. 24.12.

Example 24.6 : A MB 500, @ 0.869 kN/m transmits an end reaction of 130 kN and an end moment of 120 kN-m to the flange of the column HB 250, @ 0.510 kN/m Design the welded connections.

Design : Factored value of end reaction

$V_{fd} = (130 \times 1.5) = 195$ kN

Factored bending moment

$M_{fd} = (120 \times 1.5) = 180$ kN

Geometrical Properties of beam section

From ISI Handbook No. 1

Depth of beam, D = 500 mm

Width of flange, $b_f = 180$ mm

Thickness of web, $t_w = 10.2$ mm

Thickness of flange, $t_f = 17.2$ mm

Step 1 : Thickness of top plate

$$\text{Force in the top plate} = \left(\frac{180 \times 10^3}{500}\right) = 360\ kN$$

Design stress for top plate

$$= \left(\frac{\sigma_y}{1.15}\right) = \left(\frac{250}{1.15}\right) = 217.39\ N/mm^2$$

Area required for the top plate

$$= \left(\frac{360 \times 1000}{217.39}\right) = 1656\ mm^2$$

Width of flange = 180 mm

Adopt width of top plate = 160 mm

Thickness of top plate required

$$= \left(\frac{1656}{160}\right) = 10.35\ mm$$

Provide 160 mm x 12 mm top plate top plate is connected to the flange of column 250, @ 0.510 kN/m 12 mm thick stiffener plates are provided, so that the flange of column does not deform.

Step 3 : Fillet weld

For 12 mm thick top plate, adopt 12 mm size of weld

Strength of one mm fillet weld

$$= (0.7 \times 10 \times 1 \times 189.376)\ N/mm$$

= 1325.632 N/mm

Length of fillet weld

$$= \left(\frac{360 \times 1000}{1325.632}\right) = 271.568\ mm$$

Provide 120 mm side fillet welds on each side and 160 mm and fillet weld total length of weld provided is 400 mm. The length of unwelded portion is kept equal to the width of plate (i.e, 160 mm). Therefore, the total length of top plate is 290 mm.

The bottom flange of beam may be directly welded the flange of column. The welded unstiffened seat connections for end reaction 130 kN designed in Example 24.4 are adopted. The connections designed are shown in Fig. 24.13.

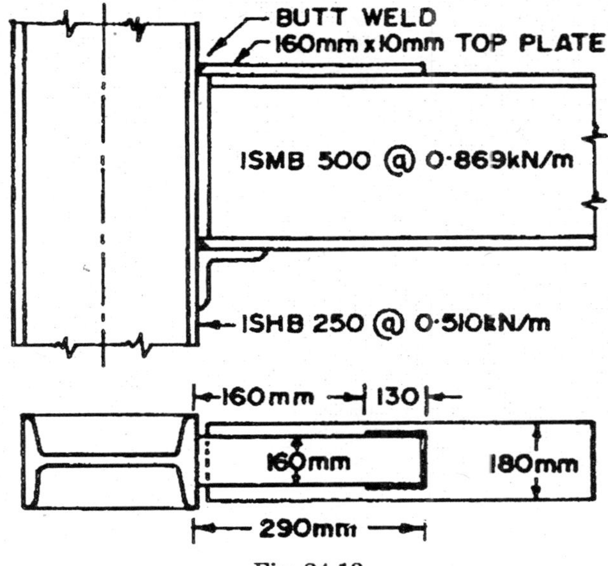

Fig. 24.13.

PROBLEMS

24.1 A LB 300, @ 0.377 kN/m transmits an end reaction of 80 kN to the flange of a column HB 200, @ 0.400 kN/m. The fillet welds are applied directly on both the sides of web of the beam. Design the connections.

24.2 In Problem 24.1 if two plates are used for connections on both the sides of web of beam, then, design the welded connections.

24.3 In Problem 24.1, if two framing angles are used for connections on both the sides of web of a beam, then, design the connections.

24.4 A LB 400, @ 0.569 kN/m transmits and end reaction of 110 kN to the flange of a column HB 350, @ 0.674 kN/m. Design unstiffened welded seat connections.

24.5 A HB 450, @ 0.653 kN/m transmits an end reaction of 220 kN to the flange of a column HB 300, @ 0.630 kN/m. Design stiffened welded seat connections.

24.6 A MB 300, @ 0.442 kN/m transmits an end reaction of 11 kN and an end moment of 80 kN-m to the flange of a column HB 300, @ 0.630 kN/m. Design the welded connections.

REFERENCE

1. IS: 800-2007, `General Construction in Steel Code of Practice' (Third Revised Edition) Published by Bureau of Indian Standard Manak Bhawan, 9- Bahadur Shah Zafar Marg, New Delhi 110 002.

25

DESIGN OF ROOF TRUSSES

25.1 INTRODUCTION

Truss girders are the frame-works which usually comprise of straight struts (stressed in compression) and ties (stressed in tension) and sometimes both in response to moving (dynamic) loads.

Truss girders are designed to be stiff, even if, all the members are considered to be pin-jointed. This is obtained by the inclusion of sufficient number of triangles. Even if, all the joints are fixed rather than pinned, the triangulation provides main contribution to the stiffeness triangles provide rigid shapes which may be made with pinned-joints. With given one set of three sides of a triangle, there is only one set of angles. There is only one possible triangular shape above or below the base line.

As it is known that for three sides of a triangle a, b and c and angles opposite to them are A, B and C, then

$$\left(\frac{a}{\sin A}\right) = \left(\frac{b}{\sin B}\right) = \left(\frac{c}{\sin C}\right) = \text{constant}$$

and sum of three angles of a triangle (A + B + C) = 180°. Above relation is satisfied by only one set of angles for a given one set of lengths a, b and c.

When the beams are subjected to bending, then, bending stress varies from zero at the neutral axis to maximum towards the extreme fibres. The different fibres in a cross-section of the beams are subjected to different intensity of stress. The stress distribution is triangular. The entire section of the beam is not utilised.

Where as the bending of trusses is quite distinct from the beams. In the truss-girders, the complete section of each member is subjected to uniform stress. The strength of the member is fully utilized. The bending action is provided by elongation in the tension members and the shortening in the compression members (i.e, by changes in the lengths of the members of the truss girder). It results in deflection of joints in the truss girders.

Compared with a simple beam, a truss girder looks very complicated. Whereas members are found, it is simple, specially if the joints are pin-jointed.

The roof trusses are the frame structures in which separate straight members are so arranged and connected at their ends that the members form triangles. The axes of members meeting at one joint intersect at a common point. The riveted joints used for the connections of the members are considered to act as pin-joints. The external loads are applied to the joints of the truss so that the members carry direct forces only. In case all the members lie in one plane, the trusses are called planar trusses. When beams are subjected to bending, then bending stress varies from zero at the neutral axis to maximum towards the extreme fibres. The different fibres in a cross-section of the beams are subjected to different intensity of stress. The entire section of beam is not utilized. Whereas the bending of trusses is quite distinct from the beams. In roof trusses, the entire section of each member is subjected to uniform stress. The strength of member is fully utilized. The bending action in roof trusses is provided by elongation and shortening of the members of the truss. It further results in deflection of truss. The forces in various members are either compressive or tensile. The members carrying compressive forces in a roof truss are called **struts**, and those carrying tensile forces are called **ties**. In a roof truss, members are so arranged that the length of members in compression are small while the lengths of members in tension are long. As far as possible, it is seen that the length of a member is not more than 3 metres. The members of a truss are classified as **main members** and **secondary members**. The main members are the structural members which are responsible for carrying and distributing the applied loads and stability of the truss. The secondary members are the structural members which are provided for stability and or restraining the main members from buckling or similar modes of failure.

The roof trusses are used at places which require sloping roofs. The sloping roofs are necessary at places where rainfall is more and at places where snow fall occurs. The roof trusses are also used in many single storeyed industrial buildings, workshops, godowns, ware-houses, residential buildings and schools, where large column free spaces are required for operational purposes. The roof trusses are suitable for relatively light loads and large spans. The roof trusses are also used to span the distance between walls or end supports when the least dimension of the building becomes large and unsuitable and economical to span with simple beams or joists, columns and girder. The roof trusses have the advantages of permitting a wider variety of roof shapes and greater unobstructed interior floor area at less cost. The trusses are specially advantageous where greater depth is required to satisfy the condition of adequate stiffness.

The primary function of a roof truss is to support the roof covering, external loads carried by roof covering and ceiling load if any. These loads are transmitted as reactions to the walls or to the supporting stanchions. In general, the external loads are applied at the joints of truss. Sometimes, it becomes necessary to apply loads at intermediate points. In such cases, the members are subjected to bending in addition to direct forces.

25.2 TYPES OF ROOF TRUSSES

The various types of roof trusses used are shown in Fig. 25.1. The suitable spans for these trusses have also been shown for each type of truss. King post truss shown in Fig. 25.1 (A) is a wooden truss. It can also be built of a combination of wood and steel. **Queen post** truss shown in Fig. 25.1(B) is also a wooden truss. **Howe** trusses had been named after Willium Howe. Howe trusses shown in Fig. 25.1(C) are made of combination of wood and steel. The vertical members are tension members and made of steel. **Pratt** trusses shown in Fig. 25.1(D) are made of steel. The Pratt trusses had been named after Thomas W. and Caleb Pratt. The diagonal members near the top are highly stressed because of their sharp inclination. These are less economical than **fink** trusses. The vertical members are compression members and diagonal members are tension members.

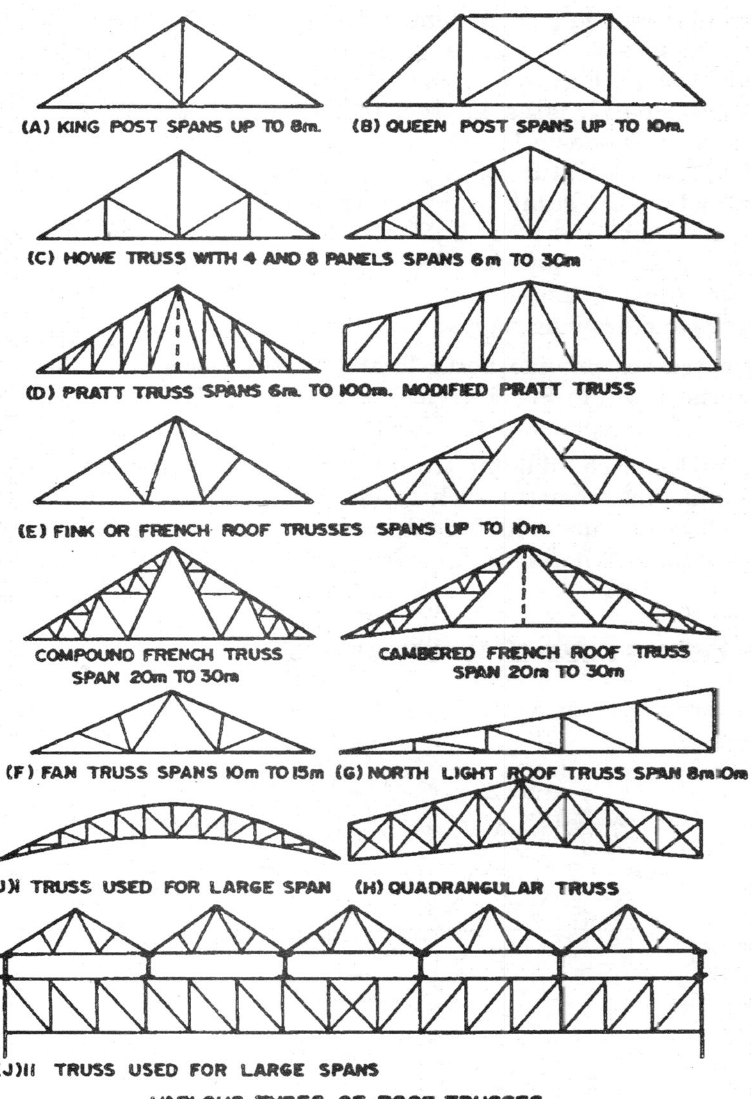

Fig. 25.1. Various types of trusses

The modified shape of a Pratt roof truss is also used. **Fink roof trusses** shown in Fig. 25.1(E) are made of steel. They are very economical form of roof trusses. The lengths of compression members are small in these types of trusses. Fink trusses are also called **French roof trusses**. The fink truss had been named after its originator Albert Fink. A Fink truss is constructed by drawing perpendicular to and from the centre of the top-left and right hand top chords and extending until they meet the lower chord. The remaining two web members are drawn from these intersections to the peak as shown in Fig. 25.1(E). The number of panels in half the truss may be increased to four by sub-dividing each panel into 2, as shown in Fig. 25.1(E). This process may be repeated again resulting in eight panels as shown in Fig. 25.1 (E). A disadvantage of this type of truss is that the number of panels can be increased only by doubling the previous number. A modification of fink truss that permits greater flexibility in number of panels is the **fan truss. Fan roof trusses** shown in Fig. 25.1(F) are made of steel. Fan roof trusses are form of fink roof trusses. In fan trusses top chord is divided into small lengths, in order to provide supports for purlins which would not come at joints in trusses shown in Fig. 25.1(E). The web numbers do not intersect the top chord at right angles. This feature of fan truss may be used or combined with the fink truss. The number of panels in the **combined fink-fan truss** may be made six instead of four (i.e., in between four and eight). The sky light may be mounted on top of most of trusses when desired. For unsymmetrical layouts, the natural light is received by using **saw tooth trusses** or **north light trusses**. The roof trusses shown in Fig. 25.1(G) are used for industrial buildings and drawing rooms.

Such type of trusses provide north light. These trusses are called **North light roof** trusses. The trusses shown in Fig. 25.1(H) are used for roofs. The trusses shown in Fig. 25.1 (J) are used for very large span such as railway sheds and auditoriums.

In general, the lower chord of the truss is kept straight. In large rooms, series of trusses are used one behind the other. In such cases, trusses with straight chords appear to sag. Then, the lower chord is cambered. It is done for better appearance. As a result of this, the depth of truss is reduced. Therefore, the cost of trusses arrangement will be economical and pleasing.

25.3 VARIOUS TERMS USED IN ROOF TRUSSES

The various terms used in roof trusses are given below and are shown in Fig. 25.2.

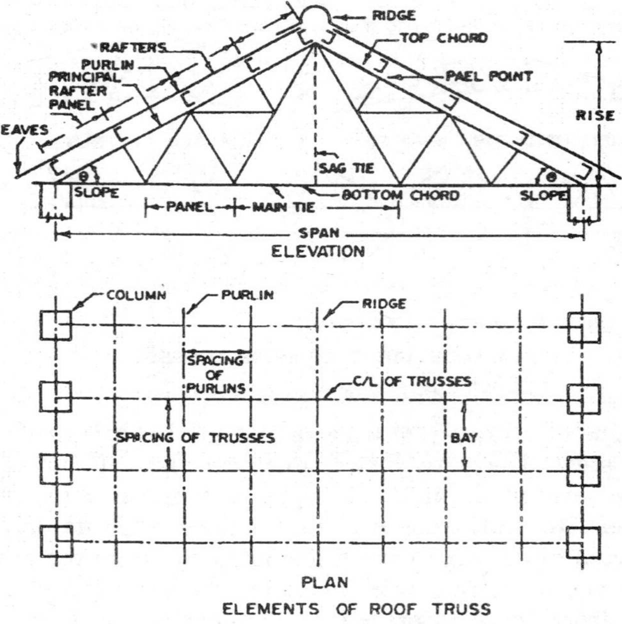

Fig. 25.2.

1. **Top chord**. The top chord is defined as the uppermost line of members extending from one-support to the other and that passes through the peak of a truss. The top chord is also known as the upper chord of the roof truss.

2. **Bottom chord**. The bottom chord is defined as the lower most line of members of a truss extending from one support to the other. The bottom chord is also known as the **lower chord** of the roof truss.

3. **Span**. The span of a roof truss is defined as the distance centre to centre of supports. The span of a roof truss is decided by the dimensions of area to be kept free of columns.

4. **Rise.** The rise of a roof truss is defined as the distance from the highest point to the line joining supports.

5. **Pitch.** The pitch of a symmetrical truss is defined as the **ratio of rise to the full span.** Where the roofs are to carry snow loads in addition to wind load, a pitch of 1/4 is most common and economical. In cases where the roof has to carry snow loads in addition to wind loads. Where the snow loads do not occur, lower pitches upto 1/6 are suitable. Lower pitches are advantaeous since the wind pressure on the roof is reduced. It corresponds to a slope of 1 in 2 or an angle of 26 1/2°. The pitches 1/3, 1/5 and that corresponding to an angle of inclination of 30° are also commonly employed. The wind force is less when the pitch of roof truss is less.

6. **Slope**. The slope of a symmetrical truss is defined as **ratio of rise to half the span.** The minimum slope depends to a great extent on the type of roof covering material. The slope provided must be sufficient to drain off the rain water without allowing leakage at the joints of covering material. The steep slopes remain better for architectural effect. But the wind pressure becomes large on such roofs. It will need strong roof trusses and large quantity of roof material. The pitches of symmetrical trusses used, their respective slope and the angles of inclination with the horizontal are given in Table 25.1 below :

TABLE 25.1

Pitch	Slope	Angle of inclination with horizontal
1/2	1	45°00'00"
1/3	2/3	33°41'00"
1/4	1/2	20°34'00"
1/5	2/5	21°48'00"
1/6	1/3	18°26'00"
1/8	1/4	14°02'00"

The slope of a roof is defined as the tangent of the angle that the plane of the roof makes with horizontal; or the tangent of the angle between the top and bottom chords of the roof trusses provided the bottom chord is horizontal. The slope of the roof therefore is not equal to the pitch and great care must be exercised to see that two terms are not used synonymously. The slope of the roof is equal to twice the numerical value of pitch in all the cases whether truss is symmetrical or unsymmetrical.

7. **Purlins**. The purlins are secondary structural members subjected to transverse loads and rest on the top chords of roof trusses. The purlins support the sheathing that carries roof

covering or roof covering directly. The purlins are horizontal beams spanning between the two adjacent trusses.

Purlins are beams which rest on roof trusse's and support the sloping roof. Purlins distribute the external load to the strusses. Purlins also serve as the flange bracing for the rafters. The weight of purlins varies from 10 per cent to 25 per cent and may be equal to or even more than the weight of the trusses.

Channels, angle, tube sections and cold formed C-or Z sections are commonly used as purlins. Cold-formed C-, Z purlins may be economical than hot rolled purlins for spans 5 m to 8 m. Angles and channels without sag bars may be economical upto 5 m and tubes upto 6 m. Open web joints are used as purlins, when the span exceeds 12 m.

Purlins are placed over the main rafters of the roof trusses. These are placed bending of the top chords of the roof trusses. For the larger trusses, it is more economical to place purlins at closer intervals. In general, purlins are placed forms 0.6 m to about 2 m.

Maximum safe span of the roof covering and glazing sheets decides the spacing of purlins when these are placed directly on purlins

For corrugated G.I. Sheets	1.5 m to 1.75 m
corrugated A.C. sheets	1.4 m
for 6 mm thickness	1.4 m
for 7 mm thickness	1.6 m

In case, spans are large and the configuration of turss is such that the it is not possible to place purlins at panel points, the purlins are placed between panel points. In such cases, the upper chords carry bending moments additionally. The weight of trusses may be increased by 10 percent to 15 percent.

The most desirable depth to the span ratio is (1/24). All purlins are designed in accordance with the requirements, for uncased beams and the limitations of bending stress based on lateral instability of the compression flange and the limiting deflections as specified in IS : 800-2007.

For simply supported purlins, the maximum bending moment may be taken as ($wL^2/8$). As per IS: 800-2007, the advantage may be taken of the continuity of the purlins over supports. In that case, the bending moment may be taken as ($wL^2/10$). The bending about the two axes should be determined separately and checked according to the biaxial bending requirement.

During erecting angle, channel, or I-sections purlins, it is desirable that these are erected over the rafter with their flanges facing up slope. In such a case, the twisting moment does not cause instability.

8. **Sub-purlins**. The sub-purlins are secondary system of purlins resting on the rafter. The sub-purlins are spaced to support the tiles or slate coverings.

9. **Rafters**. The rafters are beams and rest on the purlins. The rafters support the sheathing. These may support sub-purlins directly. These are called common rafters to distinguish from principal rafter.

10. **Sheathing**. The sheathing are covering of boards or reinforced concrete. The sheathings are supported on purlins or rafters. The sheathings provide support for the roof covering.

11. **Panel.** The panel is defined as a distance between two adjacent joints in a principal rafter of a roof truss. It is also defined as the distance between the two adjacent purlins.

12. **Bay.** The bay is defined as the distance between the adjacent trusses.

13. **Spacing of purlins.** The spacing of purlins is defined as the distance between two adjacent panel points, if purlins are placed at panel points only. It depends upon type of roof covering material and on the slope of the truss. The spacing of purlins should be less than or equal to the safe span of roof covering material if these are placed directly over purlins. In case, the roof covering material is supported on common rafters, then, purlins must be placed at any desired spacing. As far as possible, the purlins must be placed at the panel points so that in the top chord of truss, there is no bending moment due to purlin load. Generally, the spacing of purlins varies from 2 m to 3m.

14. **Spacing of trusses.** The spacing of trusses is decided by local conditions and requirements. The positions of doors, windows and columns, also control the spacing of trusses. It is desirable to have a uniform spacing of trusses. It is desirable to have a uniform spacing of trusses in one portion or in a wing of a structure, so that as many trusses as possible may be made identical.

15. **Ridge line.** The ridge line or a ridge is a line joining the vertices of the trusses.

16. **Eaves.** The bottom edges of an inclined roof surface or a pitched roof is termed as eaves.

17. **Sag tie.** As shown in Fig. 25.2, a tie member is provided to join the peak of truss and middle tie member. The length of middle tie member used to be large. The deflection of this member due to self weight may be large. When a sag tie is provided, it decreases the deflection of middle tie member.

18. **Principal rafter.** The top chord members as shown in Fig. 25.2, are called principal rafters. The principal rafter carries compressive force.

25.4 ECONOMICAL SPACING OF ROOF TRUSSES

Columns are spaced depending upon the functional requirements. Roof trusses are spaced depending upon the spacing of column. However, in general, the spacing of trusses should be such that the cost of roofing is minimized.

For a roof structure, the most economical spacing of trusses depends on many factors. It is not possible to give a simple formula. The economic spacing of trusses depends upon the weight per square metre of floor area of trusses, purlins columns, roof covering etc. It also depends on relative cost of each type of material. In case, the spacing of trusses is large, the cost of purlins will be large. If the spacing of trusses is kept small, the cost of purlins is small, but the cost of trusses will be large. Therefore, the spacing of trusses should be kept such that the overall cost of roof structure as a whole is minimum. The economic spacing of trusses is defined as the spacing which makes the cost of trusses, purlins, columns, roof covering etc., a minimum.

Let s, be the spacing of trusses; t, be the cost of trusses; p, be the cost of purlins; and r, be the cost of roof covering material.

The cost of trusses, t, is inversely proportional to the spacing of trusses, s. Therefore,

$$t = (c_1/s) \qquad \text{(i)}$$

The size of purlins depends on the bending moment. The bending moment carried by purlin varies as the square of the span of purlin. Hence, the cost of purlins, p, is directly proportional to the square of spacing of trusses. Therefore,

$$p = (c_2 \,.\, s^2) \tag{ii}$$

The cost of roof covering materials, r, is directly proportional to the spacing of trusses. Therefore,

$$r = (c_3 \,.\, s) \tag{iii}$$

where, c_1, c_2 and c_3 are constants

The overall cost of roof structure $X = (t + p + r)$

$$\text{Therefore, } X = \left(\frac{-c_1}{s} + c_2 \,.\, s + c_3 \,.\, s\right) \tag{iv}$$

For, economic spacing of the trusses, the overall cost of these items should be minimum

$$\text{Therefore, } \left(\frac{dX}{ds}\right) = 0,\ \left(\frac{dX}{ds}\right) = \left(\frac{-c_1}{s^2} + 2\,c_2 \,.\, s + c_3\right) \tag{v}$$

$$\left(-\frac{C_1}{S} + 2\,C_2 \,.\, S + C_3 \,.\, S\right) = 0, \text{ or } (-t + 2p + r) = 0 \text{ or } (t = r + 2p) \tag{25.1}$$

Therefore, for economic spacing of trusses, the cost of trusses should be equal to twice the cost of purlins plus cost of roof covering material. In this expression the factor *s*, spacing of trusses does not occur. Therefore, it is not possible to determine the spacing directly from this expression. This expression provides only a check to verify that the spacing provided is economical or not.

The economical spacing will usually be kept between 3 m to 5 m for spans upto 15 m and 4.5 m to 7.5 for spans 15 m to 30 m.

The spacing of long span trussses (longer than such about 45 m) should be considered carefully spaced at 12 m to 15 m, rather than 6m to 7.5 m.

25.5 ROOF COVERING MATERIALS

The following materials are used as roof covering materials :

1. Slates. The slates are most durable roof covering material. The slates are not affected by heat, rain or other atmospheric actions. The slates are heavy in weight. The slates are attached to battens by slating nails. The slating nails are made of copper or non-rusting material. A minimum pitch of 1/4 is required for roof trusses using slates as roof covering materials. The slopes of 27° and 33° are provided for 500 mm x 250 mm size and 400 mm x 200 mm size of slates, respectively.

2. Tiles. The tiles are used as roofing material as substitute for slates. The tiles are manufactured from asbestos, clay and many compositions. The tiles are available in market in various forms. Tiles are cheap in comparison to slates. The tiles are not durable as slates. A minimum pitch of 1/4 is required for roof trusses using tiles as roof covering materials.

3. Lead. The lead-sheets were formerly used as roof covering materials specially on flat roofs. The lead is heavy in weight. The lead is also a costly material. The lead sheets are very durable. But these are not suitable for steep slopes.

4. Zinc. The zinc is also one of the material used for roof covering. The zinc sheets are comparatively light in weight and cheap in cost in comparison to lead. Under atmospheric action, the corrosion and decay of zinc sheets may take place. The zinc sheets can be used on steep slopes.

5. Glass. The glass sheets are manufactured in various sizes and thickness. The glass sheets are mainly used for light from sloping roof.

6. Galvanized corrugated iron sheets. The galvanized steel sheets are made by galvanizing black sheets, rolled from good quality low-carbon mild steel, free from cracks, pittings, blisters, laminations, twists, scales and other surface defects. The black sheets used for galvanizing are annealed. The galvanizing is carried out by the hot-dip process by first pickling the black sheets, and then dipping them in bath of molten zinc, at a temperature suitable to produce a complete and uniform adhesive coating of zinc. The zinc coating contains not less than 97.5 percent of pure zinc. The galvanizing corrugated sheets used for roof covering should be free from twist or buckle and should have uniform corrugations, true in depth and pitch and parallel to the sides of sheets.

The depth of corrugation is 19 mm (nominal), and pitch of corrugations is 75 mm. The number of corrugations in a galvanised corrugated sheet is 8 or 10. The nominal overall width of galvanized corrugated sheets measured between the crown of outside corrugations in case of 8 corrugations per sheet, is 750 mm before corrugations and 600 mm after corrugation and that in case of 10 corrugations per sheet, (pitch of corrugations 75 mm, depth = 19) is 900 mm before corrugation and 800 mm after corrugations. The galvanized sheets are manufactured in lengths of 1.8 m, 2.2 m, 2.5 m, 2.8 m and 3.0 m. The dimensions of galvanized corrugated sheets are given as per IS : 277-1962.

Thickness of sheets 0.63, 0.8, 1.0, 1.25 and 1.6 mm. For roof side overlap : $1\frac{1}{2}$ to 2 corrugations and end overlap of 150 mm are used. The spacing of purlins is controlled by length of sheet thickness and applied loading. Approximate section modulus $Z = \frac{4}{5}$ bdt.

These sheets are used for slopes of roof 1 in 4 and above. These sheets are very light in weight, and can be fixed to purlins by J-bolts. The holes in these sheets should be made on ground. Bitumen washers are used below. G.I. washers to make the joint water-tight. The decay of galvanized sheets starts rapidly at holes due to atmospheric action. The decay of these sheets takes place due to atmospheric action as soon as the galvanizing has worn off. The life of galvanized corrugated sheet is comparatively short.

7. Asbestos cement sheets. A.C. sheets for roof covering are available in two varieties, namely, '**Asbestos cement corrugated sheets**' and '**Asbestos cement Trafford sheets**'. A.C. sheets are good insulators against temperature. A.C. sheets do not decay because of atmospheric action. These are not perishable. The asbestos cement corrugated sheets are shown in Fig. 25.3 (A) and asbestos cement trafford sheets are shown in Fig. 25.3 (B). The trafford sheets are also termed as **semi-corrugated sheets**. A.C. sheet are available in 1.75 m, 2.0 m, and 3 m lengths. These are manufctored in 6 mm and 7 mm thickness. The maximum permissible spacing of purlins for 6 mm sheet: 1.4 m and for 7 mm sheet 1.6 m.

Both the varieties of A-C, sheets are available in lengths of 1.50 m, 1.75 m, 2 m, 2.25 m, 2.50m, 2.75 m and 3 m. Both the sheets are manufactured in nominal thickness, T, as 6 mm and 7 mm. For corrugated sheets, the pitch P is 146 mm, and depth of corrugation D, is 48 mm. The overall width B of these sheets is 1.50 m and effective width C is 1.01 m. For trafford sheets, the pitch P is 338 mm and depth of corrugation D is 45 mm. The overall width B of the trafford sheets is 1.01 m and effective width is 1.014 m. The dimensions of these A.C. sheets are given as per IS : 459-1962.

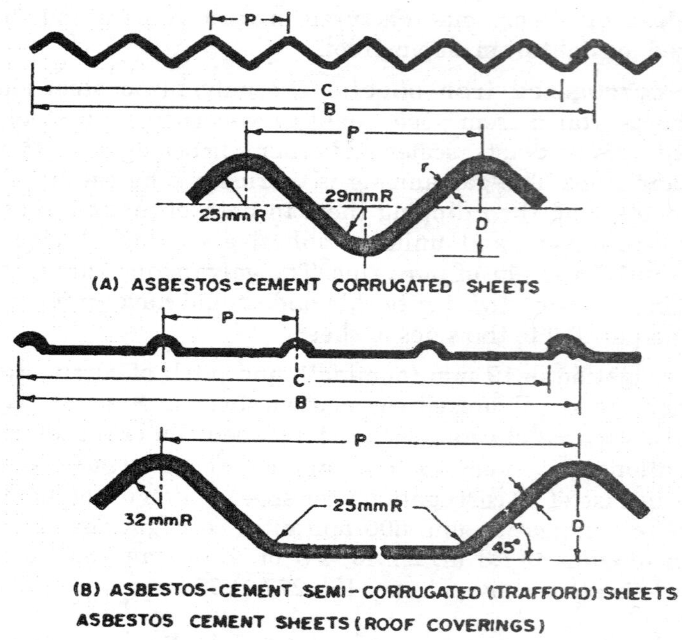

Fig. 25.3. Asbestos cement sheets

The asbestos cements sheets are used for 30° slope and above. A.C. sheets are fixed with purlins with J-bolts. The holes are drilled on the top of corrugations. The bitumen washers are placed under G.I. washers in order to make it water-tight.

8. Corrugated aluminium sheets. The corrugated aluminium sheets find extensive application where use of aluminium is justified due to its corrosive resistance, lightness and thermal efficiency. The dimensions of corrugated aluminium sheets may be noted from IS : 1254-1965. The corrugated aluminium sheets are known as general purpose sheets. The thickness of sheet is kept not less than 0.45 mm. The widths available are 650 mm and 800 mm and lengths, 1800, 2400 and 3600 mm.

25.6 LOADS ON ROOF TRUSSES

The roof trusses are subjected to dead load and live load. In addition to these loads, the roof trusses are also subjected to some special loads such as ceiling or suspended floors or heavy machinery.

1. Dead load. In Chapter 2, loads etc., in Art. 2.2, dead load has been described in general. Dead load on roof trusses includes the weight of roof covering, the weight of purlins, the weight of bracing and the self weight of trusses.

(i) Weight of roof covering. The unit weights of buildings materials have been given in Table 2.1. It includes the weights of asbestos cement corrugated and semi-corrugated sheets, C.G.I. sheets, tiles, glass and slates. The weights of these materials are given in kN per square metre of plan areas.

(ii) Weight of purlins. The weight of purlins is assumed as 0.070 to 0.150 kN per square metre of plan area.

(iii) Weight of bracings. The weight of bracings is assumed as 0.015 kN of plan area.

(iv) Weight of trusses. For the design of roof trusses, the weight of truss is assumed. The weight of truss varies with the span, and the rise of truss, the spacing of trusses, the type of roof covering material, the geographical situation of the roof structure. It is not possible to have an exact formula which includes all these factors. The self weight of truss is a small part of the total design loads for the roof truss. The self weight of truss may be assumed as 0.090 to 0.150 kN per square metre of plan area. The self weight of truss can also be found by empirical formula given below :

The self weight of truss in kN per square metre of plan area

$$w = \frac{1}{100}\left(\frac{l}{3} + 5\right) \text{kN/m}^2 \qquad (25.2)$$

where, l is the span of truss in metres.

This formula is applicable for roof trusses with C.G.I. sheets and pitch equals to 1 to 4 and for 4 m spacing of trusses. For other spacing of trusses, a proportionate value for self weight is determined.

The weight of trusses w in N/m² of area covered, depending upon the design loads q in N/m² may be found from the following formulae for the spans mentioned:

$$\text{Span } L = 18 \text{ m}, \; w = \frac{1}{100}\left(2.2 + \frac{q}{12.5}\right) \text{kN/m}^2 \qquad \text{(i)}$$

$$\text{Span } L = 24 \text{ m}, \; w = \frac{1}{100}\left(2.78 + \frac{q}{5.42}\right) \text{kN/m}^2 \qquad \text{(ii)}$$

$$\text{Span } L = 30 \text{ m}, \; w = \frac{1}{100}\left(4.44 + \frac{q}{3.47}\right) \text{kN/m}^2 \qquad \text{(iii)}$$

$$\text{Span } L = 36 \text{ m}, \; w = \frac{1}{100}\left(5.27 + \frac{q}{2.1}\right) \text{kN/m}^2 \qquad \text{(iv)}$$

The weight of steel roof trusses may also be estimated from the following expressions. These expressions include the plan area covered by the steel roof trusses. The weight of riveted steel trusses are found as follows :

$$\text{For sheeted roof trusses. } w = \frac{1}{100}(4.88 + 0.075 \times A) \text{kN/mm}^2 \qquad \text{(v)}$$

$$\text{For partly glazed roof trusses, } w = \frac{1}{100}(4.88 + 0.088 \times A) \text{kN/mm}^2 \qquad \text{(vi)}$$

where, A = plan area in m²

The weight of welded roof trusses area approximated as follows :

$$\text{For sheeted roof trusses, } w = \frac{1}{100}(5.37 + 0.053 \times A) \text{kN/m}^2 \qquad \text{(vii)}$$

For partly glazed roof trusses, $w = \frac{1}{100}(5.37 + 0.064 \times A)\, kN/m^2$

2. Live loads. In Chapter 2, the loads etc. : Art 2.3 live loads have been described in general. The imposed (live) loads on various types of roofs other than wind load and snow load, as per IS : 875-1987, for roofs with slopes upto and including 10 degrees, is adopted as 1.5 kN per square metre of plan area where access is provided to roof. The minimum live load measured on plan shall be 3.75 kN uniformly distributed over any span of one metre width of the roof slab and 9.0 kN uniformly distributed over the span in the case of all beams. Where the access is not provided, except the maintenance, live load on roofs is adopted as 0.750 kN per square metre of plan area. In the case, the minimum live load measured on plan shall be 1.9 kN uniformly distributed over any span of one metre width of the roof slab and 4.50 kN uniformly distributed over the span in the case of beams. The live load for sloping roof with slopes greater than 10 degrees is adopted as 0.75 kN per square metre of plan area less 0.020 kN/m^2 for every degree increase in slope over 10 degrees subjected to minimum of 0.400 kN/m^2 per square metre of plan area.

The live loads for curved roofs with slopes at springing greater than 10 degrees shall be $(0.75 - 52\, r^2) \times kN/m^2$,

where, $r = (h/l)$

h = the height of the highest point of the structure measured from springing

l = the chord width of the roof if singly curved and shorter of the two, if doubly curved.

This is subjected to minimum of 0.400 kN/m^2. The live load also includes snow load and wind load.

3. Snow loads. The snow loads depends upon geographical situation of roof structure. It depends upon latitude of the place and atmospheric humidity. If a roof is subjected to snow load, it should be designed for the actual load due to snow or for the live load mentioned above whichever is more severe. The actual load due to snow will depend upon the shape of roof and its capacity to retain the snow and each case shall be treated on its own merits. In the absence of any specific information, the bending due to the collection of snow may be assumed to be 2.5 N/m^2 per mm depth of snow. The possibility of total or partial snow load should be considered, that is, one half of the roof fully loaded with the design snow load and the other half loaded with half the design snow load. In the case of roof with slopes greater than 50°, snow load may be disregarded. Where, however, there are possibilities of formation of snow pockets, these should be taken into accounts.

4. Wind loads. The wind load is one of the most important loads that an engineer has to deal with and is also one of that is most difficult to evaluate properly. The magnitude of wind pressure depends on **wind velocity** and the **shape of the structure**. The magnitude of wind velocity varies with the geographical location of the structure and the height of the structure.

Sir Isaac Newton established the relationship between the wind pressure and the wind velocity and is known as **Newtonian theory**. When a stream of air flows around the object, then, the wind creates a pressure on the nose of object. The pressure developed may be determined from the consideration of energy involved. According to **Bernoullis theorem**, the sum of energies at all points.

$$\text{Therefore, } (K.E. + P_1.E_1 + P_2.E_2) = \text{Constant} \qquad \text{(v)}$$

where, K.E. = Kinetic energy, which a mass M possesses, $Wv^2/2g$

W = Weight of that mass

g = Acceleration due to gravity

v = Linear velocity of the wind

$P_1.E_1$ = Potential energy, W.h

h = Height above the datum plant

$P_2.E_2$ = Pressure energy, W.p/w

P = Pressure of thc fluid (wind)

W = Weight this pressure can lift

w = Unit weight of the fluid (wind)

The expression may be written in terms of the expression for these energies. Therefore,

$$\left(\frac{W.v^2}{2g} + W.h + \frac{W.p}{w}\right) = \text{Constant} \qquad \text{(vi)}$$

The expression (vi) is divided by W, $\therefore \left(\frac{v^2}{2g} + h + \frac{p}{w}\right) = \text{Constant}$ (vii)

The sum of three terms in the expression (vii) is known as **total head.** It is represented by H. These terms are known as velocity head, gravity head and pressure head, respectively.

$$H = \left(\frac{v^2}{2g} + h + \frac{p}{w}\right). \text{ or } H.w = \left(\frac{w.v^2}{2g} + w.h + p\right) \qquad \text{(viii)}$$

The expression (viii) gives total energy, $E = \left(\frac{w.v^2}{2g} + w.h + p\right)$ (ix)

The unit of total energy is now N per sq. metre. The energy per unit volume given by the expression (ix) remains constant. The energy at the nose of object and the stagnation point (the point where wind velocity is zero) are equal

$$\left(p_o + \frac{w.v_o^2}{2g} = p_s\right) \qquad \text{(x)}$$

where, p_o = Pressure at the nose of object

p_s = Pressure at the stagnation point of object

v_o = Velocity at the nose

From the expression (x), $(p_s - p_o) = \left(\frac{w.v_0^2}{2g}\right)$ (xi)

When the wind is brought to rest, then it results in rise in pressure $(p_s - p_o)$. It is termed as impact pressure. (kinetic pressure) of the air and is represented by q. Hence,

$$q = \left(\frac{w \cdot v_0^2}{2g}\right) = \left(\frac{w \cdot v^2}{2g}\right) \qquad \text{(xii)}$$

When the temperature and the barometric pressure change, then, the weight of wind per unit volume also changes. The weight of wind at sea level at a temperature of 17°C is 0.347 N. At 0°C, the weight of wind is 0.365 N. It increases with the decrease of temperature. The value of acceleration due to gravity may be adopted at 9.81 m/sec². The expression (xii) may be written as

$$\therefore q = (K/v^2) \qquad \text{(xiii)}$$

where, K is the constant or a co-efficient. It takes into account the unit of wind velocity in km per hour.

The basic wind pressure p is adopted equal to q x C_D and the value of K is taken as 0.06 in IS : 875-1964. Therefore,

$$p = (0.06 \times v^2) \times 10 \text{ N/m}^2 = 0.6\, v^2 \text{ N/m}^2 \qquad \text{(xiv)}$$

where, C_D is the shape factor.

The velocity of wind (wind speed) varies with the height above ground elevation. The relationship showing the variation of wind is as follows :

$$\frac{v_1}{v_2} = \left(\frac{H_1}{H_2}\right)^{0.225} \qquad \text{(xv)}$$

where, v_1 is the velocity of wind (wind speed) in km per hour at height H_1 in metres and v_2 is the velocity of wind in km per hour at height H_2 in metres.

The wind velocities (wind speed) are measured by meteorological observatories with the help of instrument known as anemometers. These instruments are kept at standard height of 10 m in the different parts of our country and the wind velocities (speeds) are measured at this height.

Basic wind speed : The **basic wind speed,** V_b is the wind speed measured in a 50 year return period. The basic wind speed is based on peak gust velocity averaged over a short interval of time of about 3 (three) seconds and it corresponds to mean heights above ground level in an open terrain (category 2).

TABLE 25.2. BASIC WIND SPEEDS IN m/sec (As per IS : 875 (Part 3) - 1987)

City/Town	Speed (m/sec)	City/Town	Speed (m/sec)	City/Town	Speed (m/sec)
Agra	47	Dehradun	47	Mysore	33
Ahmadabad	39	Delhi	47	Nagpur	44
Ajmer	47	Durgapur	47	Nainital	47
Almora	47	Gangtok	47	Nasik	39
Amritsar	47	Gauhati	50	Nellore	50
Asansol	47	Gaya	39	Panjim	39
Aurangabad	39	Gorakhpur	47	Patiala	47

Bahraich	47	Hyderabad	44	Patna	47
Bangalore	33	Imphal	47	Pondicherry	50
Barauni	47	Jabalpur	47	Portblair	44
Bareilly	47	Jaipur	47	Pune	39
Bhatinda	47	Jamshedpur	47	Raipur	39
Bhilai	39	Jhansi	47	Rajkot	39
Bhopal	39	Jodhpur	47	Ranchi	39
Bhubaneshwar	50	Kanpur	47	Roorkee	39
Bhuj	50	Kohima	44	Rourkela	39
Bikaner	47	Kurnool	39	Simla	39
Bokarao	47	Lakshadweep	39	Surat	44
Bombay	44	Lucknow	47	Tiruchirrapalli	47
Calcutta	50	Ludhiana	47	Trivendrum	39
Calicut	39	Madras	50	Udaipur	47
Chandigarh	47	Madurai	39	Vadodara	44
Combatore	39	Mandi	39	Varanasi	47
Cuttak	50	Mangalore	39	Vijaywada	50
Darbhanga	55	Moradabad	47	Visakapatnam	50
Darjeeling	47				

Fig. 25.4 shows basic maximum wind speed in m/sec (peak gust velocity are ranged over a short time interval of about 3 seconds duration) map of our country, as applicable to 10 m height above mean ground level for different zones. As per IS : 875 (Part 3 - wind loads) - 1987, six wind zones have been formed which correspond to basic wind speed of 55, 50, 47, 44, 39 and 33 metre per second, respectively, as shown in Fig. 25.4. The basic wind speeds for some important cities/towns are given in Table 25.2.

Note: Earlier wind pressure maps (one giving winds of shorter duration and another excluding winds of shorter duration) have been replaced by a single wind map (Fig. 25.4).

It is to note that the occurrence of a tornado is possible in virtually any part of India. These are particularly more severe in the northern parts of our country. The recorded number of these tornados is too small to assign any frequency. The devastation caused by a tornado is due to exceptionally high winds about its periphery, and the sudden reduction in atmospheric pressure at its centre, resulting in an explosive outward pressure on the elements of the structure. The regional basic wind speeds do not include any specific allowance for tornado. It is the usual practice to allow for the effect of tornado unless special requirements are called for as in the case of important structures, such as, nuclear power reactors and satellite communication towers.

It is to further note that the total number of cyclonic storms that have struck different sections of east and most coasts are included in Fig. 25.4 based on available records for the period from 1877 to 1982. The figures above the lines (between the stations) indicate the total number of severe cyclone storms with or without a core of hurricane winds (speeds above 87 kmph) and the figures in the brackets below the lines indicate the total number of cyclone storms. Their effect on land is already reflected in the basic wind speeds specified in Fig. 25.4. These have been included only as an additional information.

Fig. 25.4.

The cyclonic storms gradually reduce in speed as they approach the sea coast. The cyclonic storms generally extend upto 60 km inland after striking the sea coast. Their effect on land is already reflected in basic wind speed specified in Fig. 25.4. The influence of wind speed **off the coast** upto a distance of about 200 km may be taken as 1.15 times the value on the nearest coast in the absence of any definite wind data.

25.7 DESIGN WIND SPEED

The basic wind speed, V_b for any site shall be noted from Fig. 25.4 and it is modified to include the effect of risk level, terrain roughness, height and size of the structure, and local topography. The **design wind speed**, V_z may be mathematically expressed as under. The design wind speed is the wind speed for which the structure is designed.

$$V_z = (k_1 \,.\, k_2 \,.\, k_3) \,.\, V_b \text{ where,} \qquad (25.3)$$

V_z = design wind speed at any height z in m/sec,

k_1 = risk coefficient (probability factor),

k_2 = terrain, height and structure size factor, and k_3 = topography factor

Above factors k_1, k_2 and k_3 have been described in IS : 875 (Part 3) - 1987, which are as follows : It is to note that the **design wind speed** upto 10 m height from the mean ground level shall be considered constant.

1. Risk coefficient (probability factor), k_1

The basic wind speed is based on 50 years mean return period. Fig. 25.4 gives basic wind speeds for terrain (category 2) as applicable at 10 m ground level. The design life of a structure is based on the **function aspect** and **importance of a structure**. A structure is designed to withstand the **frequency distribution** of the wind over that period. The values of risk coefficients based on probability concept are given in Table 25.3.

TABLE 25.3. RISK COEFFICIENTS FOR DIFFERENT CLASSES OF STRUCTURES IN DIFFERENT WIND SPEED ZONES (As per IS : 875 (Part 3) - 1987)

	Class of structures	Mean Probable life years	k_1 factors for basic wind speeds (m/s)					
			33	39	44	47	50	55
1	Temporary structures	5	0.82	0.76	0.73	0.71	0.70	0.67
2	Structures presenting a low degree of hezard to life and property	25	0.94	0.92	0.91	0.90	0.90	0.89
3	All general buildings and structures	50	1.0	1.0	1.0	1.0	1.0	1.0
4	Important building and structures	100	1.05	1.06	1.07	1.07	1.08	1.08

Note : (It is to note that the temporary sheds and structures include those used during construction operations (e.g., form work and falsework), structures during construction stages and boundary walls. Isolated towers in wooded areas, farm buildings other than the residential buildings present a low degree of hezard to life and property in the event of failure. Hospitals, communication buildings/towers, power plant structures are important buildings and structures).

The risk factor k_1 is based on statistical concepts which take into consideration of the degree of reliability needed and period of time in years during which these will be exposed to wind, (i.e., life of the structures). Whatever wind speed is adopted for the design purposes, there is always a probability (however small) that it may be exceeded in a storm of exceptional voilance; the greater the period of years over which these will be exposed to wind, the greater is the probability.

Higher return period ranging from 100 years to 1000 years (implying lower risk level) in the association with greater period of exposure may have to be selected for exceptionally important structures, such as, nuclear power reactors and satellite communication towers.

The expression given below may be used in such cases to estimate k_1 factors for different periods of exposure and chosen probability of exceedance (risk level). The probability level of 0.63 is normally considered sufficient for the design of buildings and structures against wind effects and the values of k_1 corresponding to this risk level are given below :

$$k_1 = \left(\frac{X_{N.P}}{X_{50.0.63}}\right) \tag{25.4}$$

$$k_1 = \left(A - B\left[\log - \frac{1}{N}\log\left(1 - P_n\right)\right]\right) / (A + 4B) \tag{25.5}$$

where,

N = mean probable design life of a structure in years.

P_N = risk level in N consecutive years (probability that the design wind speed is exceeded atleast once in N successive years), normal value is 0.63.

($X_{50,\,0.63}$) = extreme wind speed for (N = 50 years and P_N = 0.63)

The values of A and B coefficients for different basic wind speed zones.

Values of	Wind Zones of speeds (m/sec.)					
	33	39	44	47	50	55
A	83.2	84.0	88.0	88.0	88.8	90.8
B	9.2	14.0	18.0	20.5	22.8	27.3

2. Terrain, height and structure size factor, k_2

Terrain category means the characteristics of the surface irregularies of an area which arise from natural or constructed features. The terrain categories are numbered in the increasing order of roughness. The effect of obstructions in the surrounding of the structure creates the turbulence and roughness, which also depends upon the direction of wind. The terrain is classified into following four categories :

(a) Category 1 ; The exposed open terrain with few or no obstruction and in which the average height of any object surrounding the structure is less than 1.5 m above ground surface. It is to note that open sea coasts and flat plains having no stress are included in this category.

(b) Category 2 : The open terrain with scattered obstructions having heights usually between 1.5 m to 10 m above ground surface. It is to note that air fields, open parklands and undeveloped sparsely built-up out skirts of towns and suburbs are included in this category. This category may also include open land adjacent to sea coast due to roughness of large sea waves at high winds.

(c) Category 3 : The terrain having numerous closely spaced obstructions (e.g. building structures upto 10 m in height with or without a few isolated tall structures).

It is to note that well wooded areas, and shrubs, towns and industrial areas full or partially developed are included in this category. It is likely that the next higher category than this will not exist in most design situations and that selection of a more severe category will be deliberate. The particular attention must be given to performance of obstructions in areas affected by fully developed tropical cyclones. The vegetation which is likely to be blown down or defoliated cannot be relied upon to maintain the conditions of category 3. Where such situation may exist, either an intermediate category with velocity multipliers midway between the values for category 2 and 3 given or category 2 should be selected having due regard to local conditions.

(d) Category 4 : The terrain with numerous large high closed spaced structures. It is to note that large city centres, (generally with obstructions above 25 m and well developed industrial complexes) are included in this category.

The terrain and building sizes classified into following three classes which depend on their sizes.

(i) Class A : The structures and/or their components such as cladding, glazing, roofing etc. having maximum dimensions (horizontal/vertical) is less than 20 m above ground surface).

(ii) Class B : The structures and/or their components such as cladding, glazing, roofing etc. having maximum dimension (horizontal/vertical) between 20 m and 50 m above ground surface.

(iii) Class C : The structures and/or their components such as cladding, glazing, roofing etc. having maximum dimension (horizontal/vertical) more than 50 m above ground surface.

Table 25.4 gives multiplying factor, k_2 by which the basic wind speed noted from Fig. 25.4 must be multiplied to get the wind speed at different heights, in each terrain category for different sizes/classes of buildings/structures.

Depending on the direction of wind, the terrain category used in the design of a structure may vary. The velocity profile for a given terrain category does not develop to full height immediately with the commencement of that terrain category. It develops gradually to height h_x which increase with the **fetch** or upward distance. Fetch is the distance measured along the wind from a boundary at which a change in the type of terrain occurs. The developed height, h_x in metres may be noted from Table 25.5.

TABLE 25.4. VALUES OF FACTORS, k_2 (as per IS 875 (Part 3) - 1987)

Height (m)	Terrain categories and classes of buildings											
	(1)			(2)			(3)			(4)		
	A	B	C	A	B	C	A	B	C	A	B	C
10	1.05	1.03	0.99	1.00	0.98	0.93	0.91	0.88	0.82	0.80	0.76	0.67
15	1.09	1.07	1.03	1.05	1.02	0.97	0.97	0.94	087	0.80	0.76	0.67
20	1.12	1.10	1.06	1.07	1.05	1.00	1.01	0.98	0.91	0.80	0.76	0.67
30	1.15	1.13	1.09	1.12	1.10	1.04	1.06	1.03	0.96	0.97	0.93	0.83
50	1.20	1.18	1.14	1.17	1.15	1.10	1.12	1.09	1.02	1.10	1.05	0.95
100	1.26	1.24	1.20	1.24	1.22	1.17	1.20	1.17	1.10	1.20	1.15	1.05
150	1.30	1.28	1.24	1.28	1.25	1.21	1.24	1.21	1.5	1.24	1.20	1.10
200	1.32	1.30	1.26	1.30	1.28	1.24	1.27	1.24	1.18	1.27	1.22	1.13
250	1.34	1.32	1.28	1.32	1.31	1.26	1.29	1.26	1.20	1.28	1.24	1.16
300	1.35	1.34	1.30	1.34	1.32	1.28	1.31	1.28	1.22	1.30	1.26	1.17
350	1.37	1.35	1.31	1.36	1.34	1.29	1.32	1.30	1.24	1.31	1.27	1.19
400	1.38	1.36	1.32	1.37	1.35	1.30	1.34	1.31	1.25	1.32	1.28	1.20

450	1.39	1.37	1.33	1.38	1.36	1.31	1.35	1.32	1.26	1.33	1.29	1.21
500	1.40	1.38	1.34	1.39	1.37	1.32	1.36	1.33	1.28	1.34	1.30	1.22

Note : Intermediate values may be obtained by linear interpolation, if desired. It is permissible to assume constant wind speed between two heights for simplicity. The height mentioned above is the height of the structure.

TABLE 25.5. FETCH LENGTH AND DEVELOPED HEIGHT, h_x (As per IS : 875 (Part 3) - 1987)

Fetch length x km	Developed height, h_x (m) (Terrain category)			
	1	2	3	4
0.2	12	20	35	60
0.5	20	30	35	95
1	25	45	80	130
2	35	65	110	190
5	60	100	170	300
10	80	140	250	450
20	120	200	350	500
50	180	300	400	500

For the structures of heights more than the developed height h_x as noted from Table 25.5, the velocity profile may be obtained in accordance with the less or least rough terrain.

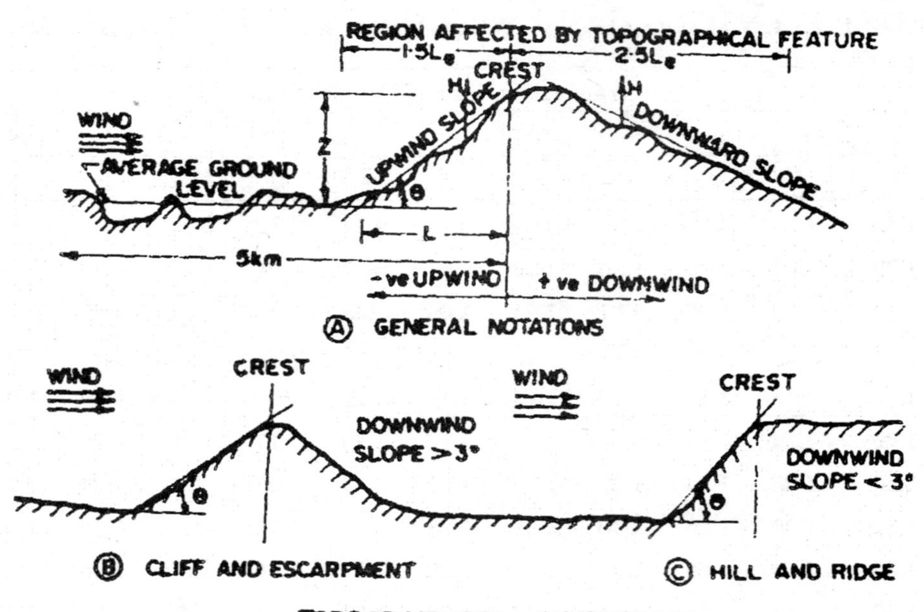

Fig. 25.5.

3. Topography factor k_3. The basic wind speed, V_b as given in Fig. 25.4 takes into consideration the general level of site above sea level. This does not allow for local topographic features such as hills, valleys, cliffs, escarpments, or ridges which may significantly influence the wind speed in their surrounding. Near the summits of hills or crests of cliffs, escarpments or ridges, the wind gets accelerated, where as near the foot of cliffs, steep escarpments or ridges, the wind gets decelerated. The wind is parallel to the ground in the planes. The wind is assumed to be normal to the face of the face of the structure. The topography factor k_3 is given by

$$k_3 = (1 + C\,s) \qquad (25.6)$$

$$\text{For } 3^\circ < \phi < 17^\circ \quad C = 1.2\,.(z/L)\;.\; L_e = L \qquad (25.7, 25.8)$$

$$\text{For } \phi > 17^\circ \quad C = 0.36\,,\; L = (z/0.3,) \text{ where} \qquad (25.9, 25.10)$$

ϕ = upward slope of the ground in the wind direction

S = a factor taken from Figs 25.5 and 25.6

L = actual length of the upward wind slope in the direction of wind

z = effective height of the crest (feature)

Le = effective horizontal length

The effect of topography becomes significant at a site when the upward wind slope, ϕ is more than 30°, and below that, the value of k_3 may be taken equal to unity. The value of k_3 lies in the range of 1.0 to 1.36 for slopes greater than 3°. From Eqs. (25.8) and (25.10), it is seen that the values of factors s depend upon the effective horizontal length, L_e. For the design of all the structures within the effected zone (1.5 L_e on the windward side and 2.5 L_e on the leeward side) as shown in Fig. 25.5, this coefficient has to be considered.

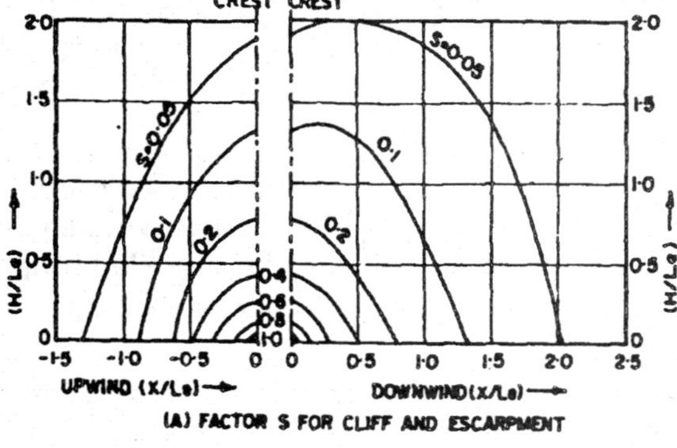

(A) FACTOR S FOR CLIFF AND ESCARPMENT

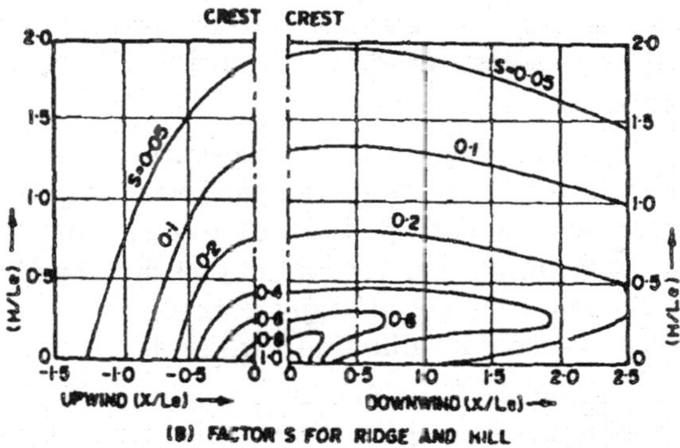

(B) FACTOR S FOR RIDGE AND HILL

Fig. 25.6.

25.8 DESIGN WIND PRESSURES

The design wind pressure, p_z depends upon the basic wind speed, V_b, the height of structure above ground level, the terrain categories, the local topography, the aspect ratio (viz., length and breadth of the building/structure, (i.e., size of the structure), the shape of structure and the solidity ratio or openings in the structures.

The design wind pressure at any height mean ground level shall be obtained from the following expression

$$p_z = 0.6 (V_z)^2 \text{ N/m}^2 \quad (25.11)$$

$$p_z = 0.0006 (V_z)^2 \text{ kN/m}^2 \quad (25.12)$$

where, V_z is the design wind speed in m/sec at height z. This coefficient 0.6 (or 0.0006) (in S.I. units) in the above expressions depends on a number of factors and mainly on the atmospheric pressure and air temperature.

It is to note that wind pressures and wind forces (loads) acting over the structures shall be calculated for the building as a whole, the individual structural elements as roofs and walls and the individual cladding units (glazing and their fixing).

25.9 WIND FORCES (LOADS) ON BUILDINGS/STRUCTURES

For a particular surface or part of the surface of a building, the **pressure coefficients** are always given. The wind load acting normal to a surface is determined by multiplying the **area of that surface** or its appropriate portion by the **pressure coefficient,** C_p and the **design wind pressure** at the height of the surface from the ground. The averages values of the these pressure coefficients (**external pressure coefficients** and **internal pressure coefficient**) for some shapes of buildings have been given in subsequent sections.

The average values of pressure coefficients are given for critical wind directions in one or more quadrants. In order to calculate the maximum wind load on the building, the total wind load should be found for each of the critical directions shown from all the quadrants. In case, the considerable variation of pressures occurs over a surface, it has been subdivided and the mean pressure coefficients have been mentioned for each of its several parts.

In addition, the areas of high local suction (negative pressure concentration) frequently occurring near the edges of walls and roofs are separately noted. These coefficients for the local effects should only be used for calculation of forces on these local areas influencing roof sheeting, glass panels, individual cladding units including their fixtures. These should not be used for calculating forces on entire structural elements such as roof walls or structure as a whole.

It is to note that the pressure coefficients have been found from measurements on models in wind tunnels, and the great majority of data available has been used in conditions of relatively smooth flow. In case of rectangular buildings, sufficient field data are available. For these buildings, turbulent flow has been used. The design of wall glazing and cladding has become a source of major concern in recent years. The damage to glass may be hazardous and cause considerable financial losses.

In Chapter 2, loads etc. Art. 2.4, the wind load has been described. In order to determine the wind load on roofs and other inclined surfaces, the direction of wind is assumed as horizontal. The roof surfaces are also inclined to the horizontal. The normal component and tangential component of wind on inclined roof surfaces may be found. It is to note that these components of horizontal wind pressure cannot be found by resolving. The tangential component of wind pressure is small and it is neglected. The normal component of wind pressure can be found from empirical formula.

In the beginning in 1829, **Col. Duchemin**, a French Army Officer derived an empirical formula from experiments. This empirical formula bears his name. The Duchemin's formula is

$$p_n = p.\left(\frac{2\sin\alpha}{1+\sin^2\alpha}\right) \quad (25.13A)$$

where, p is the wind pressure on surface in the direction of the wind.

p_n = Component of wind pressure normal to the inclined roof surface.

α = Angle which the inclined roof surface makes with the horizontal direction of wind. The angle α is shown in Fig. 25.6.

$$p_v = p_n \,.\, \cos\alpha, \text{ or } p_v = p.\left(\frac{2\sin\alpha\cos\alpha}{1+\sin^2\alpha}\right) \quad (25.13B)$$

$$\text{and } p_h = p_v \,.\, \sin\alpha \text{ or } p_h = p.\left(\frac{2\sin\alpha\cos\alpha}{1+\sin^2\alpha}\right) \quad (25.13C)$$

It is considered that 1.5 kN/m² wind pressure on surface normal to the direction of wind is sufficient for roof structure in an exposed position. For the roof structures, in protected position, 1.00 to 1.25 kN/m² of wind pressure is sufficient.

The wind pressures are expressed in terms of basic wind pressures described in Art. 2.4. The design wind load on the roof surface is determined as the resultant effect of **internal air pressure** and **external air pressure**.

25.10 WIND LOAD ON INDIVIDUAL MEMBERS

In order to find the wind load on individual structural elements/members (e.g., roofs and walls, and individual cladding units and their fittings, it is essential to consider the pressure difference between the opposite faces of such elements or units. For cladding, it is therefore, necessary to know the **internal pressure** as well as the **external pressure**. The wind load, W_L acting in a direction normal to the individual structural element or cladding unit may be determined from the following expression.

$$W_L = (C_{pe} - C_{pi}) \,.\, A \,.\, p_d \text{ where,} \quad (25.13D)$$

C_{pe} = external pressure coefficient

C_{pi} = internal pressure coefficient

A = surface area of structural element or cladding unit, and

p_d = design wind pressure

The **external pressure coefficient** and **internal pressure coefficients** are described in the subsequent sections.

It is note that the external pressure coefficient is positive if it is acting outside to inside. However, the internal air pressure is treated positive if it is acting inside to outside.

25.11 EXTERNAL PRESSURE COEFFICIENTS

Average external pressure coefficients shall be considered for walls, pitched roofs, monoslope roofs and canopy roofs, as described hereunder :

1. Walls

The average external pressure coefficients for the walls of clad buildings of rectangular plan as shown in Fig. 25.7 are given in Table 25.6.

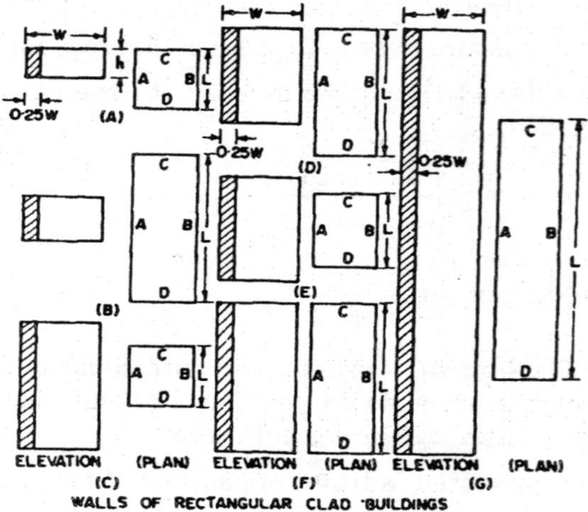

Fig. 25.7.

2. Pitched roofs of rectangular clad buildings

The average external pressure coefficients and pressure concentration coefficients for pitched roofs of rectangular clad buildings as shown in Fig. 25.8 are noted from Table 25.7, where, the pressure coefficients are not given, the average coefficients shall be used.

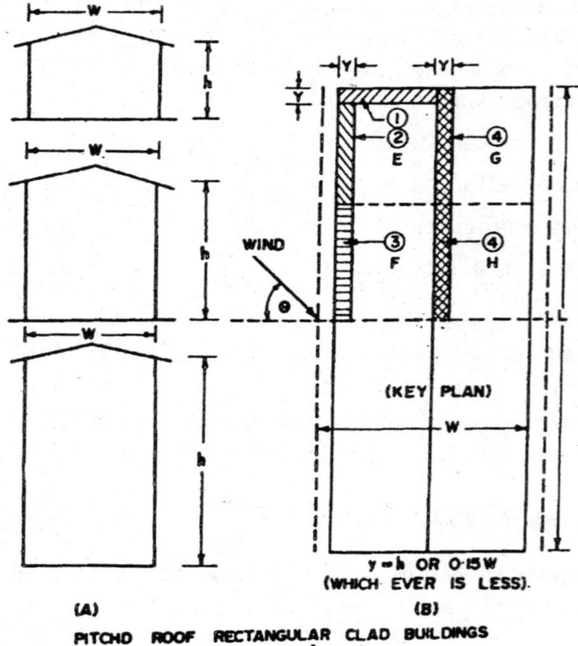

Fig. 25.8.

It is to note that the pressure concentration is assumed to act outward (suction pressure) at the ridges, eaves, cornics and 90 degree corners of roofs. This pressure concentration is not included with the net external pressure while calculating overall loads.

TABLE 25.6. EXTERNAL PRESSURE COEFFICIENTS, C_p FOR WALLS OF RECTANGULAR BUILDINGS (As per IS : 875 (Part 3) - 1987)

Building Height Ratio	Building plan Ratio	Wind Angle θ	C_{pe} for surfaces A	B	C	D	Local, C_{pe} Shaded area
(h/w)< 0.5}>	1 < (L/W) < 1.5	0°	+ 0.7	− 0.2	−0.5	− 0.5	− 0.8
	Fig. 25.7(A)	90°	− 0.5	− 0.5	+ 0.7	− 0.2	
	1.5 < (L/W) < 4.0	0°	+ 0.7	− 0.25	− 0.6	− 0.6	− 1.0
	Fig. 25.7(B)	90°	− 0.5	− 0.5	+ 0.7	− 0.1	
0.5 (h/w) < 1.5	1 < (L / W) < 1.5	0°	+ 0.7	− 0.25	− 0.6	− 0.6	−1.1
	Fig. 25.7(C)	90°	− 0.6	− 0.6	+ 0.7	− 0.25	
	1.5 < (L/W) <4.0	0°	+ 0.7	− 0.3	+ 0.7	− 0.7	−1.1
	Fig. 25.7(D)	90°	− 0.5	− 0.5	+ 0.7	− 0.7	
1.5(h/w) < 6	1 < (L/W) <1.5	0°	+ 0.8	− 0.25	− 0.8	− 0.8	−1.2
	Fig. 25.7(E)	90°	−0.8	− 0.8	+ 0.8	− 0.25	
	1.5 < (L / W) < 4.0	0°	+ 0.7	− 0.4	− 0.7	− 0.7	−1.2
	Fig. 25.7(F)	90°	− 0.5	− 0.5	+ 0.8	− 0.1	
(h/w) > inf	(L/W) = 1.5	0°	+ 0.95	− 1.85	− 0.9	− 0.9	−1.2
	Fig. 25.7 (G)	90°	− 0.8	− 0.8	+ 0.9	− 0.85	
	(L/W) = 1.0	0°	+ 0.95	− 1.25	− 0.7	− 0.7	−1.25
	Fig. 25.7 (G)	90°	− 0.7	− 0.7	+ 0.95	− 1.25	
	(L / W) = 2	0°	+ 0.85	− 0.75	− 0.75	− 0.75	−1.25
	Fig. 25.7 (G)	90°	− 0.75	− 0.75	+ 0.85	− 0.75	

Note : h is the height to caves or parapet, L is the greater horizontal dimension of building and, W is the lesser horizontal dimension of building.

TABLE 25.7. EXTERNAL PRESSURE COEFFICIENTS, C_{pe} FOR PITCHED ROOF OF RECTANGULAR CLAD BUILDINGS (As per IS : 875 (Part 3) - 1987)

Building height ratio	Angle of Roof	Wind Angle, θ		Wind Angle θ		Local Coefficients			
	α	EF	GH	EF	GH	1	2	3	4
	0°	−0.8	− 0.4	− 0.8	− 0.4	− 2.0	− 2.0	− 2.0	
	5°	− 0.9	− 0.4	− 0.8	− 0.4	− 1.4	− 1.2	− 1.2	− 1.0
(h/w)<0.5	10°	− 1.2	− 0.4	− 0.8	− 0.6	− 1.4	− 1.4		− 1.2
	20°	− 0.4	− 0.4	− 0.7	− 0.6	− 1.0			− 1.2
	30°	0	− 0.4	− 0.7	− 0.6	− 0.8			− 1.1

	45°	+ 0.3	− 0.5	− 0.7	− 0.6				− 1.1
	60°	+ 0.7	− 0.6	− 0.7	− 0.6				− 1.1
	0°	− 0.8	− 0.6	− 1.0	− 0.6	− 2.0	− 2.0	− 2.0	
	5°	− 0.9	− 0.4	− 0.9	− 0.6	− 2.0	− 2.0	− 2.0	− 1.0
0.5 (h/w) <1.5	10°	− 1.1	− 0.6	− 0.8	− 0.6	− 2.0	− 2.0	− 1.5	− 1.2
	20°	− 0.7	− 0.5	− 0.8	− 0.6	− 1.5	− 1.5	− 1.5	− 1.0
	30°	− 0.2	− 0.5	− 0.8	− 0.8	− 1.0			− 1.1
	45°	+ 0.2	− 0.5	− 0.8	− 0.8	− 0.8			
	60°	+ 0.6	− 0.5	− 0.8	− 0.8	− 0.8			
	0°	− 0.7	− 0.6	− 0.9	− 0.7	− 2.0	− 2.0	− 2.0	
	5°	− 0.7	− 0.6	− 0.8	− 0.8	− 2.0	− 2.0	− 1.5	− 1.0
0.5 (h/w) <1.5	10°	− 0.7	− 0.6	− 0.8	− 0.8	− 2.0	− 2.0	− 1.5	− 1.2
	20°	− 0.8	− 0.6	− 0.8	− 0.8	− 1.5	− 1.5	− 1.5	− 1.2
	30°	− 1.0	− 0.5	− 0.8	− 0.7	− 1.5			
	40°	− 0.2	− 0.5	− 0.8	− 0.7	− 1.0			
	50°	+ 0.2	− 0.5	− 0.8	− 0.7				
	60°	+ 0.5	− 0.5	− 0.8	− 0.7				

Note : h is the height to eaves or parapet and w is the lesser horizontal dimension of the building. Where the local coefficients are not given the overall coefficients apply. For hipped roofs, the local coefficients for the hip ridge may be conservatively taken as the appropriate ridge value.

TABLE 25.8. EXTERNAL PRESSURE COEFFICIENTS, C_{le} FOR MONO - SLOPE ROOFS FOR RECTANGULAR CLAD BUILDINGS WITH (h/w) < 2 (As per IS : 875 (Part 3) - 1987)

Angle of roof, a degrees	Angle of wind θ										Local C_{fe}					
	0°		45°		90°		135°		180°							
	H	L	H	L	H&L	H&L	H	L	H	L	H1	H2	L1	L2	He	Le
5	-1.0	-0.5	-1.0	-0.9	-1.0	-0.5	-0.9	-1.0	-0.5	-1.0	-2.0	-1.5	-2.0	-1.5	-2.0	-2.0
10	-1.0	-0.5	-1.0	-0.8	-1.0	-0.5	-0.8	-1.0	-0.4	-1.0	-2.0	-1.5	-2.0	-1.5	-2.0	-2.0
15	-0.9	-0.5	-1.0	-0.7	-1.0	-0.5	-0.6	-1.0	-0.3	-1.0	-1.8	-0.9	-1.8	-1.4	-2.0	-2.0
20	-0.8	-0.5	-1.0	-0.6	-0.9	-0.5	-0.5	-1.0	-0.2	-1.0	-1.8	-0.8	-1.8	-1.4	-2.0	-2.0
25	-0.7	-0.5	-1.0	-0.6	-0.8	-0.5	-0.3	-0.9	-0.1	-0.9	-1.8	-0.7	-0.9	-0.9	-2.0	-2.0
30	-0.5	-0.5	-1.0	-0.6	-0.8	-0.5	-0.1	-0.6	0	-0.6	-1.8	-0.5	-0.5	-0.5	-2.0	-2.0

Note : These values apply to length (w/z) from windward end and
** these values apply to remaining length
h is the height of eaves at lower side
L is the greater horizontal dimension of a building and W is he lesser horizontal dimension of the building.

3. Mono-slope roofs of rectangular clad buildings

The average pressure coefficients and pressure concentration coefficient for mono-slope (lean-to) roofs of rectangular clad buildings as shown in Fig. 25.9 may be noted from Table 25.8.

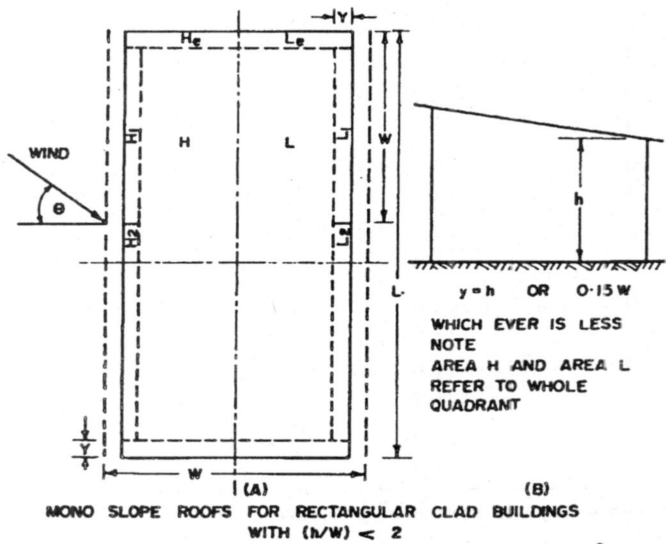

Fig. 25.9.

4. Canopy roofs with [(L/4) (h/w) 1 and 1 (L/W) 3]

The pressure coefficients are given in Table 25.9 and 25.10 separately for mono-pitch and double-pitch canopy roofs (as shown in Figs. 25.10 and 25.11, respectively) such as open-air parking garages, shelter areas, outer areas, railway platforms, stadiums and theatres. These coefficients take into consideration the combined effect of the wind exerted on and under the roof for all wind direction; the resultant is to be taken normal to the canopy.

In case the local coefficients overlap, the greater of the two given values should be considered. However, the effect of partial closure of one side and/or both sides, such as those due to trains, buses and stored materials shall be foreseen and considered.

TABLE 25.9. PRESSURE COEFFICIENTS FOR MONO-SLOPES FREE ROOFS (AS PER IS : 875 (PART 3) - 1987)

Angle of roof (degrees)	Solidity ratio	Maximum (largest+ve) and Minimum (argest-ve) Pressure coefficients			
		Overall coefficients	Local coefficients		
			1	2	3
0	All values of ϕ	+ 0.2	+ 0.5	+ 1.8	+ 1.1
5		+ 0.4	+ 0.8	+ 2.1	+ 1.3

10		+ 0.5	+ 1.2	+ 2.4	+ 1.6
15		+ 0.7	+ 1.4	+ 2.7	+ 1.8
20		+ 0.8	+ 1.7	+ 2.9	+ 2.1
25		+ 1.0	+ 2.0	+ 3.1	+ 2.3
30		+ 1.2	+ 2.2	+ 3.2	+2.4
0	0	– 0.5	– 0.6	– 1.3	– 1.4
	1	– 1.0	– 1.2	– 1.8	– 1.9
5	0	– 0.7	– 1.1	– 1.7	– 1.8
	1	– 1.0	– 1.6	– 2.2	– 2.3
10	0	– 0.9	– 1.5	– 2.0	– 2.1
	1	– 1.3	– 2.1	– 2.6	– 2.7
15	0	– 1.1	– 1.8	– 2.4	– 2.5
	1	– 1.4	– 2.3	– 2.9	– 3.0
20	0	– 1.3	– 2.2	– 2.8	– 2.9
	1	– 1.5	– 2.6	– 3.1	– 3.2
25	0	– 1.6	– 2.6	– 3.2	– 3.2
	1	– 1.7	– 2.8	– 3.5	– 3.5
30	0	– 1.8	– 3.0	– 3.8	– 3.6
	0	– 1.8	– 3.0	– 3.8	– 3.6

Note : For mono-pitch canopies, the centre of pressure should be taken to act at 0.3 W from the windward edge.

The **solidity ratio** ϕ is equal to the area obstruction under the canopy divided by the gross area under the canopy, both areas normal to wind direction ($\phi = 0$ represents a canopy with no obstruction under neath) and $\phi = 1$ represents the canopy fully blocked with contents to the downward eaves).

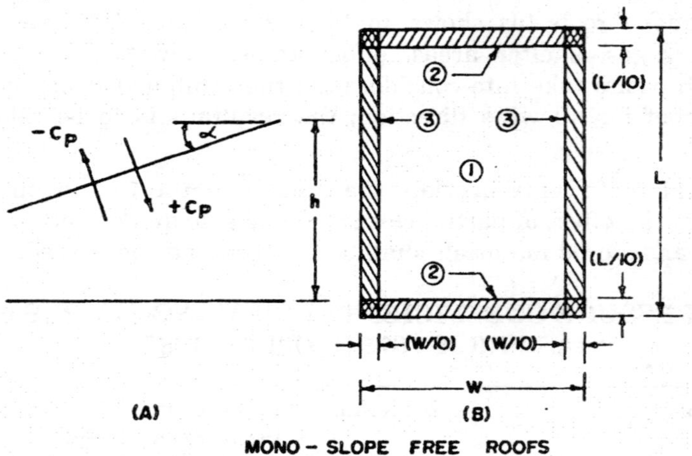

Fig. 25.10.

The values of C_p for intermediate solidities may be linearly interpolated between these two extremes and apply upwind of the position of maximum blockage only. Downwind of the position of maximum blockage, the coefficients for $\phi = 0$ may be used.

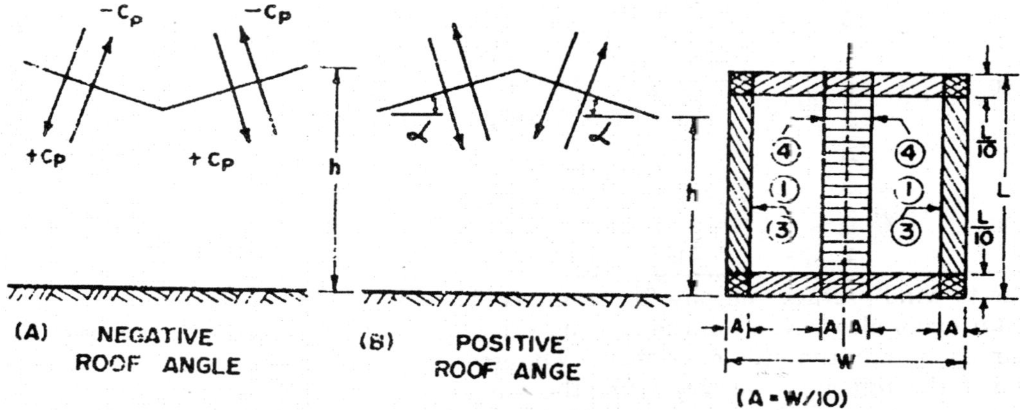

Fig. 25.11.

TABLE 25.10. PRESSURE COEFFICIENTS FOR FOR FREE STANDING DOUBLE - SLOPED ROOFS (As per IS : 875 (Part 3) - 1987)

Angle of roof (degrees)	Solidity ratio	Maximum (largest +ve) and Minimum (argest -ve) Pressure coefficients				
		Overall coefficients	Local coefficients			
			1	2	3	4
– 20	All values of ϕ	+ 0.7	+ 0.8	+ 1.6	+ 0.6	+ 1.7
– 15		+ 0.5	+ 0.6	+ 1.5	+ 0.7	+ 1.4
– 10		+ 0.4	+ 0.6	+ 1.4	+ 0.8	+ 1.1
– 5		+ 0.3	+ 0.5	+ 1.5	+ 0.8	+ 0.8
+ 5		+ 0.3	+ 0.6	+ 1.8	+ 1.3	+ 0.4
+ 10		+ 0.4	+ 0.7	+ 1.8	+ 1.4	+ 0.4
+ 15		+ 0.4	+ 0.9	+ 1.9	+ 1.4	+ 0.4
+ 20		+ 0.6	+ 1.1	+ 1.9	+ 1.5	+ 0.4
+ 25		+ 0.7	+ 1.2	+ 1.9	+ 1.6	+ 0.5
+ 30		+ 0.9	+ 1.3	+ 1.9	+ 1.6	+ 0.7
– 20	0	– 0.7	– 0.9	– 1.3	– 1.6	– 0.6
	1	– 0.9	– 1.2	– 1.7	– 1.9	– 1.2
– 15	0	– 0.6	– 0.8	– 1.3	– 1.6	– 0.6
	1	– 0.8	– 1.1	– 1.7	– 1.9	– 1.2
– 10	0	– 0.6	– 0.8	– 1.3	– 1.5	– 0.6
	1	– 0.8	– 1.1	– 1.7	– 1.9	– 1.3
– 5	0	– 0.5	– 0.7	– 1.3	– 1.6	– 0.6
	1	– 0.8	– 1.5	– 1.7	– 1.9	– 1.4
+ 5	0	– 0.6	– 0.6	– 1.4	– 1.4	– 1.1
	1	– 0.9	– 1.3	– 1.8	– 1.8	– 2.1
+ 10	0	– 0.7	– 0.7	– 1.5	– 1.4	– 1.4
	1	– 1.1	– 1.4	– 2.0	– 1.8	– 2.4

+ 15	0	– 0.8	– 0.9	– 1.7	– 1.4	– 1.8
	1	– 1.2	– 1.5	– 2.2	– 1.9	– 2.8
+ 20	0	– 0.9	– 1.2	– 1.8	– 1.4	– 2.0
	1	– 1.3	– 1.7	– 2.3	– 1.9	– 3.0
+ 25	0	– 1.0	– 1.4	– 1.9	– 1.4	– 2.0
	1	– 1.4	– 1.9	– 2.4	– 2.1	– 3.0
+ 30	0	– 1.0	– 1.4	– 1.9	– 1.4	– 2.0
	1	– 1.4	– 2.1	– 2.6	– 2.2	– 3.0

Due to wind pressure on any fascia and to friction over the surface of the canopy, the horizontal loads also act in addition to the pressure forces normal to the canopy. For any wind direction, only the greater of these two forces need be considered. Fascia loads should be found on the area of the surface facing the wind, using a force coefficient of 1°.

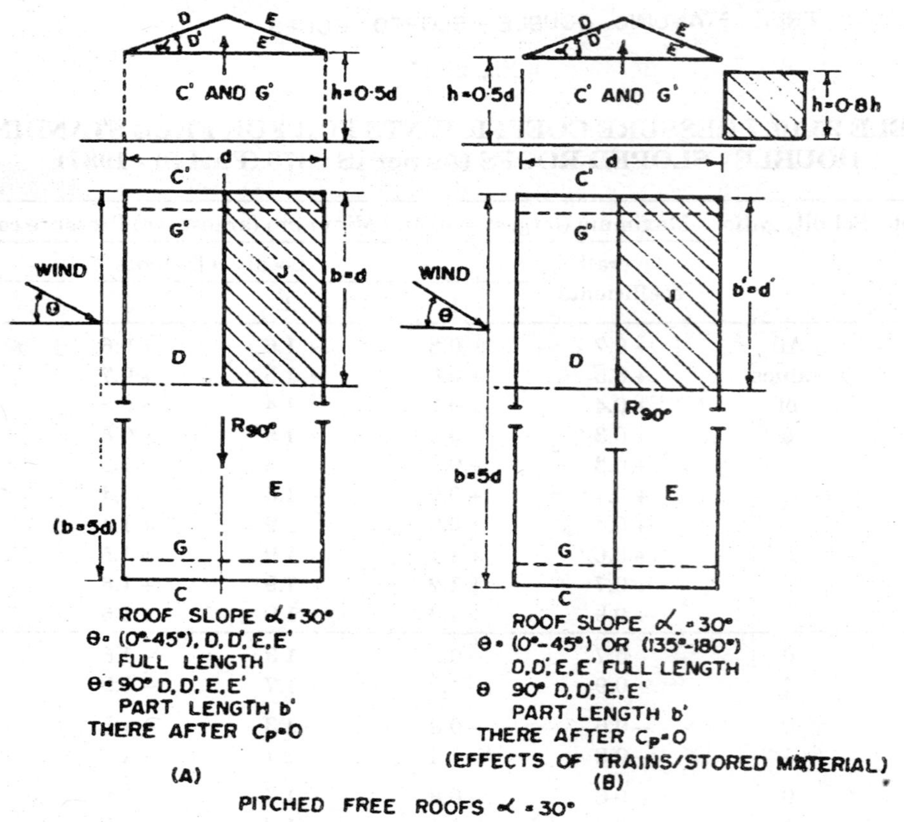

Fig. 25.12.

For pitched and toughed free roofs for some specific cases for which aspect ratios and roof slopes have been specific as shown in Fig. 25.12 (A), and (B), Fig. 25.13 (A) and (B) and Fig. 25.24 (A) and (B) the **internal and external pressure coefficients** may be noted from Tables (25.11), 25.12, 25.13 (A), 25.13 (B), 25.14 (A) and 25.14 (B), respectively. However, while using these Tables, any-significant departure from it should be investigated carefully. No increase shall be made for the local effects except as indicated.

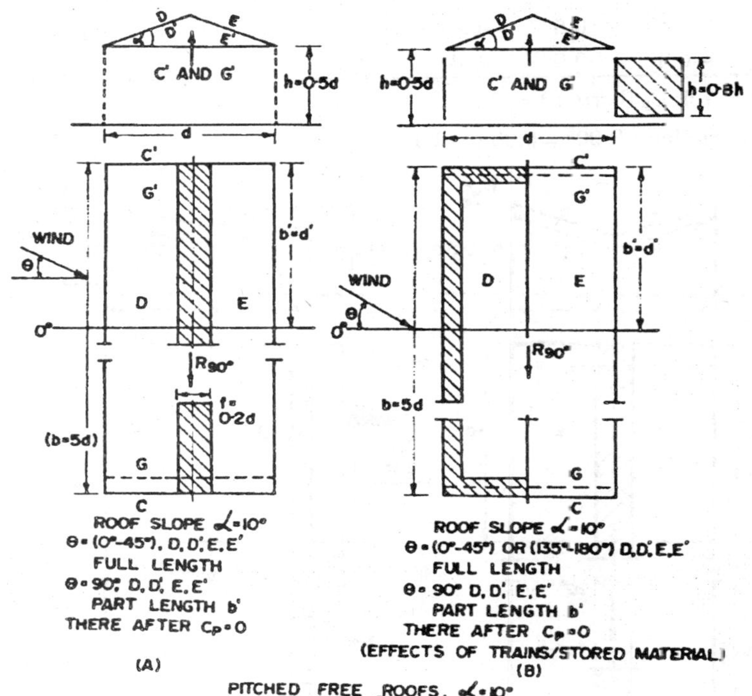

Fig. 25.13.

TABLE 25.11. PRESSURE COEFFICIENTS (TOP AND BOTTOM) FOR PITCHED ROOFS, $\alpha = 30°$ (As per IS : 875 (Part 3) - 1987)

Angle of wind θ	Pressure coefficients, C_p							
	D	D'	E	E'	End surfaces			
					C	C'	G	G'
0°	0.6	-1.0	-0.5	-0.9	–	–	–	–
45°	0.1	-0.3	-0.6	-0.3	–	–	–	–
90°	-0.3	-0.4	-0.3	-0.4	-0.3	0.8	0.3	0.4

For j : C_p (top) = -1.0; C_p (bottom) = -0.2

Tangentially acting friction : R 90® = 0.05 p_d . bd

TABLE 25.12. PRESSURE COEFFICIENTS (TOP AND BOTTOM) FOR PITCHED FREE ROOFS, $\alpha = 30°$ WITH EFFECTS OF TRAIN OR STORED MATERIAL (As per IS : 875 (Part 3) - 1987)

Angle of wind θ	Pressure coefficients C_p							
	D	D'	E	E'	End surfaces			
					C	C'	G	G'
0°	+ 0.1	+0.8	-0.7	+0.9	–	–	–	–
45°	– 0.1	+0.5	-0.8	+0.5	–	–	–	–

90°	– 0.4	-0.5	-0.4	-0.5	-0.3	0.8	0.3	0.4
180°	– 0.3	-0.6	-0.4	-0.6	–	–	–	–

45° For j : C_p (top) = -1.5; C_p (bottom) = -0.5

Tangentially acting friction : R 90° = 0.05 p_d . bd

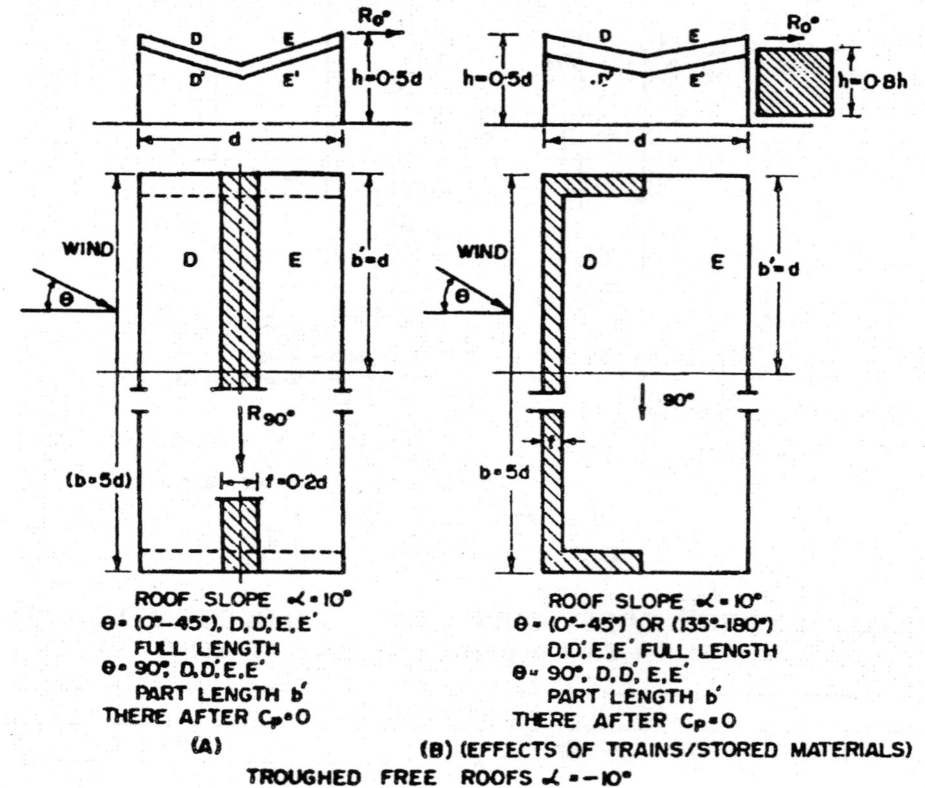

Fig. 25.14.

TABLE 25.13 (A). PRESSURE COEFFICIENTS (TOP AND BOTTOM) FOR PITCHED ROOFS, α = 10° (As per IS : 875 (Part 3) - 1987)

Angle of wind θ	Pressure coefficients C_p							
	D	D'	E	E'	End surfaces			
					C	C'	G	G'
0°	-1.0	+0.3	-0.5	-0.2	–	–	–	–
45°	-0.3	+0.1	-0.3	-0.2	–	–	–	–
90°	-0.3	0	-0.3	0	-0.4	0.8	0.3	-0.6

0° For f : C_p (top) = - 1.0; C_p (bottom) = +0.4

0° – 90° Tangentially acting friction : R 90° = 0.01 p_d . bd

TABLE 25.13 (B). PRESSURE COEFFICIENTS (TOP AND BOTTOM) FOR PITCHED ROOFS, α = 10° WITH EFFECTS OF TRAIN OR STORED MATERIAL (As per IS : 875 (Part 3) - 1987)

Angle of wind θ	Pressure coefficients C_p							
	D	D'	E	E'	End surfaces			
					C	C'	G	G'
0°	-0.3	+0.8	-0.6	+0.7	–	–	–	–
45°	-0.5	+0.4	-0.3	+0.3	–	–	–	–
90°	-0.3	0	-0.3	0	-0.4	+0.8	+0.3	-0.6
180°	-0.4	-0.3	-0.6	-0.3	–	–	–	–

0° For f : C_p (top) = -1.6; C_p (bottom) = +0.9

0° - 180° Tangentially acting friction : R 90° = 0.01 p_d . bd

TABLE 25.14 (A). EXTERNAL PRESSURE COEFFICIENTS FOR TROUGHED FREE ROOFS, α = 10° (As per IS : 875 (Part 3) - 1987)

Angle of wind θ	Pressure coefficients C_p			
	D	D'	E	E'
0°	+ 0.3	- 0.7	0.2	- 0.9
45°	0	- 0.2	0.1	- 0.3
90°	-0.1	+ 0.1	-0.1	0.1

TABLE 25.14 (B). PRESSURE COEFFICIENTS (TOP AND BOTTOM) FOR TROUGHED FREE ROOFS, α = 10° WITH EFFECTS OF TRAIN OR STORED MATERIALS (As per IS : 875 (Part 3) - 1987)

Angle of wind θ	Pressure coefficients C_p			
	D	D'	E	E'
0°	-0.7	0.8	- 0.6	0.6
45°	-0.4	0.3	- 0.2	0.2
90°	-0.1	0.1	- 0.1	0.1
180°	-0.4	-0.2	-0.6	-0.3

25.12 INTERNAL PRESSURE COEFFICIENTS

The internal air pressure in a building depends upon the **degree of permeability** of the structure (cladding) to the flow of air. The internal air pressure may be positive or negative (that is, a positive pressure or suction) depending upon the direction of flow of air in relation to the openings in the buildings. Four types of building having different degrees of permeability and internal pressures shall be considered.

1. Building having a small degree of permeability

In the case of buildings where the cladding permits the flow of air with openings not more than about 5 percent of the wall area but where there are no large openings, it is necessary to consider the possibility of the internal pressure being positive or negative. Following two design conditions shall be examined.

1. A buildings with an internal pressure coefficient of +0.2, and
2. A buildings with an internal pressure coefficient of -0.2.

Design wind pressure on roofs

The internal pressure coefficient is algebraically added to the external pressure coefficient and the analysis which indicates more distress of the member shall be adopted. In most situations, a simple inspection of the sign of external pressure will at once indicate the proper sign of the internal pressure coefficient to be taken for design.

The term **normal permeability** relates to the flow of air commonly afforded by the structures (claddings). It includes the flow of air not only through open windows and doors, but also through the slits round the closed windows, doors, ventilators, chimneys and through the joints between the roof coverings, the total open area being less than 5 percent of area of the walls having the openings.

2. Buildings having medium degree of permeability

Buildings with medium openings may also show either positive or negative internal pressure depending upon the direction of wind. The buildings with medium openings between about 5 to 20 percent of wall area shall be designed for an internal pressure coefficient of +0.5 and again with an internal pressure coefficient of -0.5. The analysis shall be done for maximum distress of the members.

3. Buildings having large degree of permeability

Buildings with large openings, that is, openings large than 20 percent of the wall area shall be designed once with an internal pressure coefficient of +0.7 and again with an internal pressure coefficient of -0.7. The analysis shall be carried out for maximum distress on the members.

4. Buildings having high degree of permeability

Buildings with one open side or having opening exceeds 20 percent of wall area may be assumed to be subjected to internal positive pressure or suction (internal negative pressure) similar to those for buildings with large openings. The buildings with one sided openings are shown in Fig. 25.15. The values of internal pressure coefficients with respect to the direction of wind are also indicated.

The values of internal pressure coefficients for all the four about categories of buildings have been noted briefly in Table 25.15.

TABLE 25.15. INTERNAL PRESSURE COEFFICIENTS, C_{pi}
(As per IS : 875 (part 3) - 1987)

S.No.	Categories of buildings	C_{pi}	
1	Buildings with small degree of permeability (openings > percent)	+0.2	-0.2
2	Buildings with medium degree of permeability (openings 5 to 20 percent)	+0.5	-0.5
3	Buildings with large degree of permeabilities (opening 20 percent)	+0.7	-0.7
4	Buildings with one open side or (opening 20 percent)	Fig. 25.15	

[Note. For (B/L) equal to unity, average values may be used.]

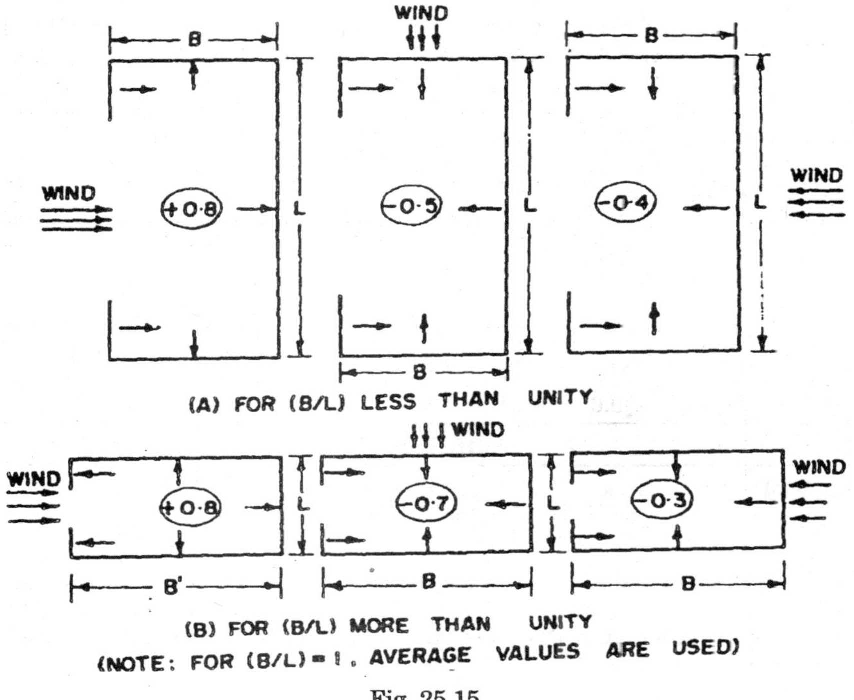

Fig. 25.15.

25.13 EXTERNAL PRESSURE COEFFICIENTS FOR SOME OTHER ROOFS

The external pressure coefficients for the following more types of roofs are adopted as specified in IS : 875 (Part 3) - 1987.

1. Pitched and saw-tooth roofs of multi-span building

For the pitched and saw-tooth roofs of multispan buildings, as shown in Figs. 25.16 and 26.17 (see page) the external average pressure coefficients and pressure concentration coeffic-ients may be noted from Tables 25.16 and 25.17, respectively provided all the spans are equal and the height to eaves shall not exceed the span.

2. Pressure coefficients on overhangs from roofs

The pressure coefficients on the top overhanging portion of the roof are taken to be the same, as that of the nearest top portion of the non-overhanging portions of the roofs. The pressure coefficients for the underside surface of the overhanging portions are taken as follows and shall be taken as positive in case the overhanging portion is on the windward side.

Overhangs	External pressure coefficients, C_{pc}
(a) sloping downward	1.25
(b) horizontal	1.00
(c) sloping upward	0.75

TABLE 25.16. EXTERNAL PRESSURE COEFFICIENTS, (C_{pe}) FOR PITCHED ROOFS OF MULTISPAN BUILDINGS (ALL SPANS EQUAL) WITH (h w) (As per IS : 875 (Part 3) - 1987)

Angle of Roff α	Angle of wind θ	First span		First Interspan		Other intermediate		End span		Local coefficients	
(degrees)	(degrees)	a	b	c	d	m	n	x	z	1	2
5	0	– 0.9	– 0.6	– 0.4	– 0.3	– 0.3	– 0.3	– 0.3	– 0.3	– 2.0	– 1.5
10		– 1.1	– 0.6	– 0.4	– 0.3	– 0.3	– 0.3	– 0.3	– 0.4	– 2.0	– 1.5
20		– 0.7	– 0.6	– 0.4	– 0.3	– 0.3	– 0.3	– 0.3	– 0.5	– 2.0	– 1.5
30		– 0.2	– 0.6	– 0.4	– 0.3	– 0.2	– 0.3	– 0.2	– 0.5	– 2.0	– 1.5
45		+ 0.3	– 0.6	– 0.6	– 0.4	– 0.2	– 0.4	– 0.2	– 0.5	– 0.2	– 1.5

DISTANCE

Angle of roof α	Angle of wind θ	h_1	h_2	h_3
upto 45°	90°	– 0.8	–0.6	– 0.2

Note : It is to note that when the angle of wind θ is 0°, the horizontal forces due to frictional drag are allowed for in the above values. And when the angle of wind is 90°, the above values are allowed for frictional drag as specified in IS : 875 (Part 3) 1987.

TABLE 25.17 EXTERNAL PRESSURE COEFFICIENTS, (C_{pe}) FOR SAW-TOOTH ROOFS OF MULTI-SPAN BUILDINGS (ALL SPAN EQUAL) WITH (h w) (As per IS : 875 (Part 3) - 1987)

Angle of wind θ (degrees)	First span		First Interspan span		Other intermediate		End span		Local coefficients	
	a	b	c	d	m	n	x	z	1	2
0	+ 0.6	– 0.7	– 0.7	– 0.4	– 0.3	– 0.2	– 0.1	– 0.3	– 2.0	– 1.5
180	– 0.5	– 0.3	– 0.3	– 0.3	– 0.4	– 0.6	– 0.6	– 0.1	– 2.0	– 1.5

DISTANCE

Angle of wind θ	h_1	h_2	h_3
90° 270°	– 0.8	–0.6	– 0.2

Note : It is to note that when the angle of wind, θ is zero, the horizontal forces due to frictional drag are allowed for in the above values. And when the angle of wind, θ is 90°, the frictional drag is allowed as recommended in IS : 875 (Part 3) - 1987.

It is to note that for the over hanging portions on the sides to the than the windward side, the average pressure coefficients on adjoining walls may be used.

Combined roofs and roofs with a sky-light

The average external pressure coefficients for combined roofs with a sky light as shown in Figs. 25.18 and 25.19 may be noted from Tables 25.18 and 25.19 respectively.

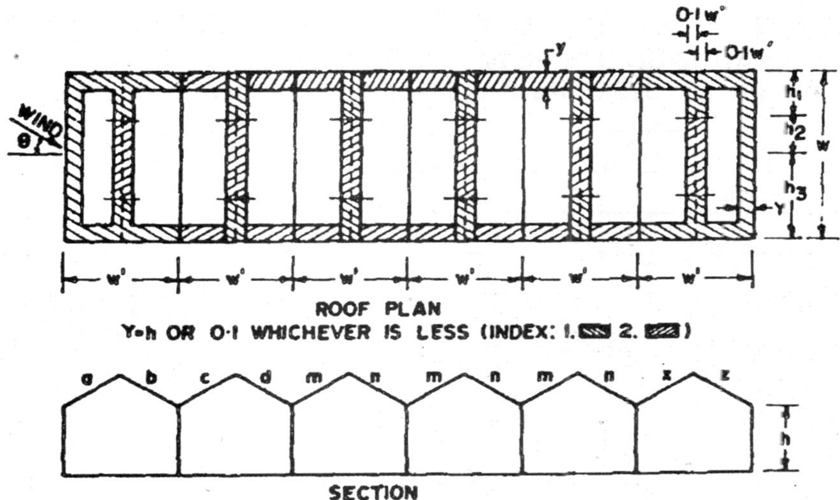

Fig. 25.16.

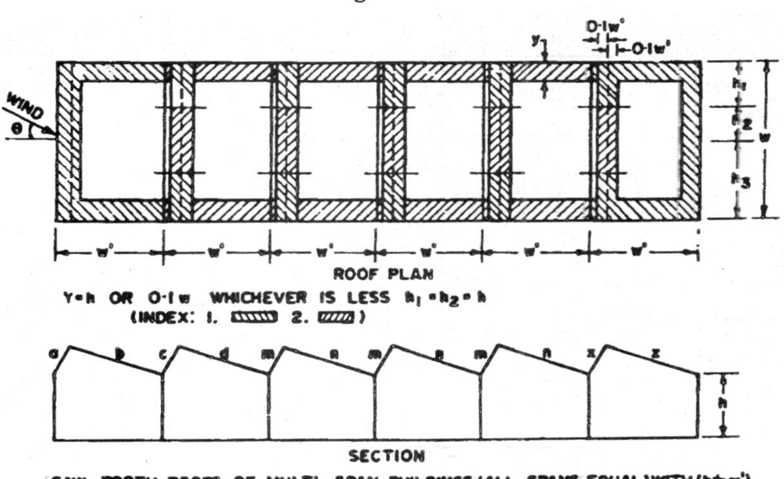

Fig. 25.17.

TABLE 25.18. EXTERNAL PRESSURE COEFFICIENTS, C_{pe} FOR COMBINED ROOFS AND ROOFS WITH SKY-LIGHT (As per IS : 875 (Part 3) - 1987) Values of C_{pe}

Portion	Direction 1	Direction 2
a	From Fig. 25.18	
b	$C_{pe} = -0.5(h_1/h_2) < 1.5$ $C_{pe} = -0.7\ (h_1/h_2) > 1.5$	– 0.4
c and d	See Table 9	
e	See Art 25.13(2)	

TABLE 25.19. EXTERNAL PRESSURE COEFFICIENTS, C_{pe} FOR COMBINED ROOFS AND ROOFS WITH SKY-LIGHT (As per IS : 875 (Part 3) - 1987) Pressure coefficients, C_{pe}

$b_1 > b_2$		$b_1 < b_2$	
Portions		Portions	
a	b	a	b
− 0.6	+ 0.7	Table 25.18	Table 25.18

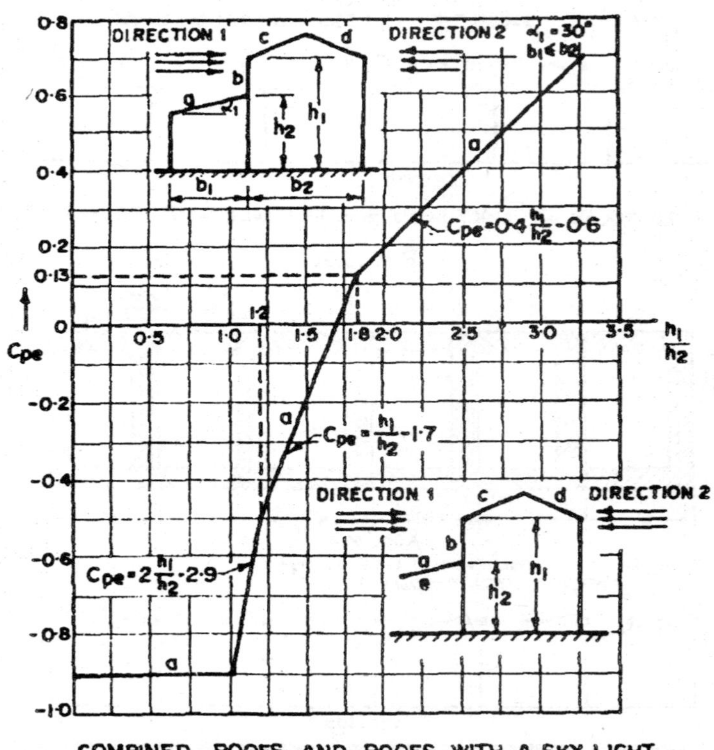

Fig. 25.18.

25.14 COMBINATION OF LOADS ON ROOF TRUSSES

A judicious combination of loads on roof trusses specified in Art. 25.6, keeping in view the possibility of (i) their action together and (ii) their disposition in relation to other loads and the severity of stresses or deformation caused by the combination of various loads is necessary to ensure by the combination of various loads is necessary to ensure the required safety and economy in the design of roof trusses.

The combination of loads on roof trusses is used to determined the maximum stresses in the members of a truss. The following combination of loads are used :

(i) Dead load + snow load, if any.

(ii) Dead load + partial or full live load whichever causes the maximum stresses in the members.

(iii) Dead load + wind load + internal pressure.

(iv) Dead load + wind load + internal suction.

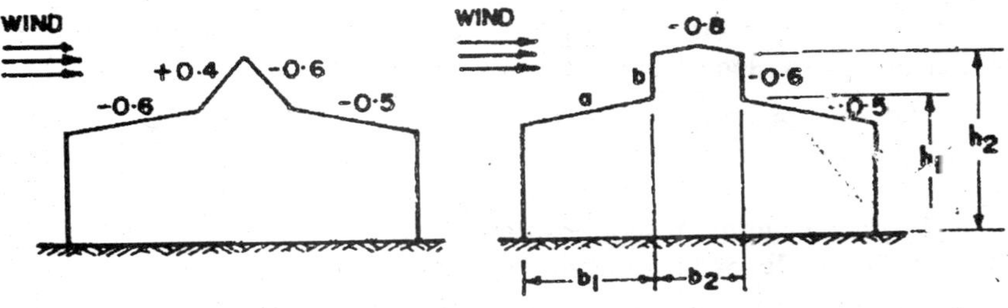

COMBINED ROOFS AND ROOFS WITH A SKY LIGHT

Fig. 25.19.

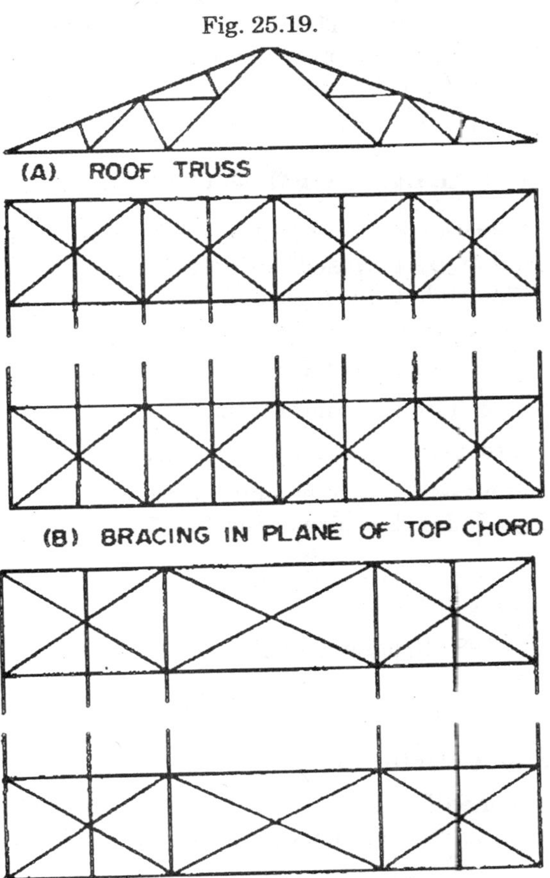

BRACING OF ROOF TRUSSES

Fig. 25.20.

25.15 BRACING OF ROOF TRUSSES

The bracing of roof trusses is done to provide a stiff rigid structures so that structure is not subjected to vibrations due to rolling loads of machinery. In building structures, the roof trusses are supported on masonry walls. The purlins at ends also rest on end masonry walls. For such structures, lateral bracing is not necessary. The purlins provide sufficient lateral support. The roof trusses are also securely fixed with anchor bolts in the masonry walls.

When the roof trusses are supported over steel columns, then lateral bracing should be provided for stability against lateral forces. In absence of lateral bracing, the structure will collapse in a high wind storm. The structure may collapse due to stress and vibrations due to rolling loads. The lateral bracing provided should be such that it transmits lateral forces directly as possible to the walls or foundations of the buildings.

The roof trusses should be braced in the place of top chord and also in the plane of bottom chord. The bracings should also be provided in the plane of columns. The bracing is provided in every third or fourth panel. The bracing of a roof truss in top and bottom chords is shown in Fig. 25.20. The bracings are providied in such a manner that they diagonals form angles about 45° with the loads to be carried. The bracings may consist of single angles sections.

25.16 ASSUMPTIONS IN ANALYSIS OF ROOF TRUSSES

The analysis of roof trusses consist of two parts. Part one consists of determination of loads and reactions. The other part consists of determination of the internal forces in the members of roof trusses. The following assumptions are made for determining the forces (stresses) in the members :

1. The roof truss is not restrained by the reactions.
2. The axes of members meeting at a joint intersect at a common point.
3. The riveted joints in roof trusses act as frictionless hinges.

Example 25.1. A roof truss-shed is to be built in Jodhpur city area for an industrial use. Determine the basic wind pressure. The size of shed is 18 m x 30 m.

Solution :

Step 1 : Basic wind speed

From IS : 875 (Part 3) - 1987 (from wind zone map of the country or Table 25.2, the basic wind speed in Jodhpur.

$$V_d = 47 \text{ m/sec} \qquad \text{(i)}$$

Step 2 : Risk coefficient (factor - k_1)

The design life for roof shed for industrial use may be assumed 50 years. From IS : 875 (Part 3) -1987 or Table 25.3, the risk coefficient (factor - k_1), for $V_b = 47$ m/sec,

$$k_1 = 1.00 \qquad \text{(ii)}$$

Step 3 : Terrain, height and structure size factor (factor - k_2)

The terrain is Jodhpur city industrial area. Therefore, the area is of type category 3. The size of shed shall be 18 m x 30 m. Therefore, the maximum horizontal dimension is between 20 m and 50 m. Therefore, the roof shed structure is of type class B. In category 3, the terrain with numerous closely spaced obstructions having the size of building structures upto 10 m in height with or without a few isolated tall structures, therefore, from IS : 875 (Part 3) 1987 or Table 25.4.

$$k_2 = 0.98 \qquad \text{(iii)}$$

Step 4 : Topography factor, k_3

Near the roof-shed, the ground is assumed to be plain. Therefore, the topography factor from Eq. 25.6,

$$k_3 = (1 + C \,.\, S)$$

where $C = (Z/L) = 0$, $k_3 = 1.00$ (iv)

Step 5 : Design wind speed (Eq. 25.3)

$$V_z = k_1 \,.\, k_2 \,.\, k_3 \,.\, V_b$$

$$V_z = (1 \times 0.98 \times 1 \times 47) = \mathbf{46.06\ m/sec.}$$

Step 6 : Basic wind pressure (from Eq. 25.12)

$$p_z = 0.0006 \times V_z^2 \text{ kN/m}^2$$

$$= 0.0006 \times (46.06)^2 = \mathbf{1.2729\ kN/m^2}$$

Example 25.2. An industrial roof shed of size 20 m x 30 m is proposed to be constructed at Mangalore near a hillock of 160 m and the slope is 1 in 2.8. The roof shed is to be built at a height of 120 m from the base of the hill. Determinc the design wind pressure on the slope. The height of roof shed shall be 12 m.

Solution :

Step 1 : Basic wind speed

From IS : 875 (Part 3) - 1987 or Table 25.2, the basic speed V_b is 39 m/sec.

Step 2 : Risk coefficient (factor - k_1)

For 50 years of design life of the structure and all general buildings for V_b = 39 m/sec, risk coefficient.

$$k = 1.00 \qquad \text{(i)}$$

Step 3 : Terrain factor, k_2

Near the hillock side, the terrain shall be open with well scattered obstructions having heights generally between 1.5 m to 10 m. Therefore, the terrain is of type category 2. The size of roof shed is 20 m x 30 m. Therefore, it is of type class B (since maximum horizontal dimension is between 20 m and 50 m). The height of roof-shed shall be 12 m. Therefore, the terrain, height and structure size factor, from IS : 875 (Part 3) - 1987 or Table 25.

$$k_2 = \left[0.98 + \frac{(1.02 - 0.98) \times 2}{5}\right] = 0.996$$

Step 4 : Topography factor k_3

The effective height of the hill, z is 160 m.

The slope is 1 in 2.8. Therefore, $\theta = \tan^{-1}(1/2.8) = 19.654°$

Length of upward slope L = (160 x 2.8) = 448 m

For $\theta > 17°$, effective horizontal length of the hill

$$L_e = (Z/0.3) = (160/0.3) = 533.33 \text{ m}$$

For $(\theta = 19.654) > 17°$, from IS : 875 (Part 3) - 1987, Appendix C.

$$C = 0.36 \text{ and } k_3 = (1 + C \cdot S)$$

Let x be the horizontal distance of the building from the crest. It is measured positive towards leeward side and negative towards the windward side.

$$x = -(L-120) = -(448 - 120) = -328 \text{ m}$$

$$\frac{x}{L_e} = \left(\frac{-328}{533.33}\right) = -0.615$$

$$\frac{H}{L_e} = \left(\frac{12}{533.33}\right) = 0.0225$$

From Fig. 25.6 (B), for $(x/L_e) = -0.615$ and $(H/L_e) = 0.0225$

$$S = 0.2$$

From Eq. 25.6, for (B) = 19.654°) > 17, c = 0.36

$k_3 = (1 + C \cdot S)$, therefore,

$$k_3 = (1 + 0.36 \times 0.2) = 1.072$$

Step 5 : Design wind speed (from Eq. 25.3)

$$V_z = k_1 \cdot k_2 \cdot k_3 \cdot V_b$$

$V_z = (1 \times 0.996 \times 1.072) \times 39 =$ **41.641 m/sec**

Step 6 : Design wind pressure (from Eq. 25.12)

$$p_z = 0.0006\, V_z^2 \text{ kN/m}^2$$

$$p_z = (0.0006 \times 41.641^2) = 1.04 \text{ m/sec}$$

Example 25.3. A hospital building of size 50 m x 100 m and a height 10 m is proposed to be built at Poona on a hill top. Determine the design wind pressure on the building. The height of hill is 320 m with a slope of 1 in 4. The hospital is proposed at a distance 120 m from the crest on the downward slope.

Solution :

Step 1 : Basic wind speed

From IS : 875 (Part 3) - 1987 or Table 25.2, the basic wind speed at Poona is 39 m/sec.

Step 2 : Risk coefficient (factor - k_1)

For important buildings and structures, such as hospitals for which the design life be assumed 100 years, the risk coefficient from IS : 875 (Part 3) - 1987 or Table 25.3, for $V_b = 39$ m/sec

$$k_1 = 1.06 \qquad \text{(i)}$$

Step 3 : Terrain, height and structure size factor, k_2

The height of hospital shall be 10 m. The open terrain with well scattered obstruction having heights generally between 1.5 m to 10 m belongs to category 2. The size of hospital dimension is more than 50 m. As such, it belongs to class (C). From IS : 875 (Part 3) - 1987 for Table 25.4, for 10 m height

$k_2 = 0.93$ (ii)

Step 4 : Topography factor k_3

The effective height of the hill, Z is 320 m. The slope is θ = 1 in 4

$\theta = \tan^{-1}(1/4) = 14.036°$

Length of upward slope, L = (320 x 4) = 1280 m

Effective horizontal length of the hill, for (θ = 14.036°) < 17°,

$Le = L = 1280$ m

Value of C for $\theta = 17°$, $C = 1.2 \times \left(\frac{Z}{L}\right) = \left(\frac{1.2 \times 320}{1280}\right) = 0.3$ (iv)

Height of the hospital building shall be 10 m. Let x be the distance from the crest, (then, x = + 120 m). Non-dimensional factors

$$\left(\frac{X}{L_e}\right) = \left(\frac{120}{1280}\right) = +0.09375 \text{ and } \left(\frac{H}{L_e}\right) = \left(\frac{10}{1280}\right) = +0.0078$$

From Eq. 25.6 (B), the value of S for ridge and hill, S = 1.00 (v)

From Eq. 25.6, $k_3 = (1 + C . S) = (1 + 0.3 \times 1.00) = 1.3$ (vi)

Step 5 : Design wind speed (from Eq. 25.3)

$V_z = k_1 . k_2 . k_3 . V_b$

V_z = [(1.06) . (0.93) . (1.3) . (39)] = **49.98 m/sec**

Step 6 : Design wind pressure (from Eq. 25.12)

$p_z = 0.0006 . V^2_z$

$p_z = 0.0006 \times (49.98)^2$ = **1.4988 kN/m²**

Example 25.4 : A communication lower of 80 m height is proposed to be built over hill top of height 520 m with a gradient of 1 in 5. The horizontal approach distance is 2.8 km from the level ground. The lower is proposed at Abu mount. Determine the design wind pressure.

Solution :

Step 1 : Basic wind speed : From IS : 875 (Part 3) - 1987, or Table 25.2, the basic wind speed at Abu mount (wind zone 5),

$V_b = 39$ m/sec (i)

Step 2 : Risk coefficient (factor - k_1)

For important buildings and structures such as the communication towers, the design life may be assumed 100 years. From IS : 875 (Part 3) - 1987, or Table 25.3, for $V_b = 39$ m/sec,

$k_1 = 1.06$ (ii)

Step 3 : Terrain, height and structure size factor, k_2

The height of communication tower is 80 m. The structure shall be having maximum vertical dimension more than 50 m. The structure belongs to class (C). The exposed open terrain with few or no obstruction and in which the average height of any object surrounding the structure is less than 1.5 m. The terrain belongs to category-1. The terrain factor

$$k_2 = \left[1.14 + \frac{(1.20 - 1.14) \times 30}{50}\right] = 1.176$$

Step 4 : Topography factor, k_3

The effective height of the hill, Z is 520 m. The slope is 1 in 5

$$\theta = \tan^{-1}(1/5) = 11.31°$$

The length of upward slope, L is 2.800 km, that is, 2800 m. The effective length of hill,

for $\theta = 11.31° < 17°$, Le = L = 2800 m

Value of C for $\theta = 17°$

$$C = 1.2 \text{ x } \left(\frac{Z}{L}\right) = \left(\frac{1.2 \times 520}{2800}\right) = 0.222857$$

For H = 80 m, and x = 0,

$$\left(\frac{x}{L_e}\right) = 0 \text{ and } \left(\frac{H}{L_e}\right) = \left(\frac{80}{2800}\right) = 0.02857$$

From Fig. 25.6 (B), or IS : 875 (Part 3) - 1987,

S = 1.00

$k_3 = (1 + C . S)$

$k_3 = (1 + 0.222857 \text{ x } 1) = 1.222857$

Step 5 : Design wind speed (from Eq. 25.3)

$V_z = k_1 . k_2 . k_3 . V_b$

$V_z = (1.06) . (1.176) . (1.222857) \text{ x } (39)) = 59.45$ m/sec

Step 6 : Design wind pressure (from Eq. 25.12)

$p_z = 0.0006 . v_z^2$ kN/m^2

$p_z = (0.0006 \text{ x } 59.45^2) = 2.121$ kN/m^2

25.17 DESIGN OF ROOF TRUSSES

The following are usual steps in the design of roof trusses :

Step 1 : The type of truss which suits the requirements of span and roof covering to be used is selected.

Step 2 : The general proportion for the type of truss selected is fixed as discussed in Art. 25.3. The line diagram of roof truss is drawn.

Step 3 : The dead and live loads acting on the roof truss are determined as discussed in Art. 25.6. The wind load on roof truss depending upon slope of the truss and other conditions is found as discussed in art. 25.6. Thus, various loads acting on the roof truss are estimated.

Step 4 : The purlins are ordinary provided at the panel points. The purlin section is designed as discussed in Art. 25.22.

Step 5 : After the design of purlins, the loads acting at the panel points of the roof truss are determined.

Step 6 : The forces in various members of the roof truss are determined for various combinations of loads as discussed in Art. 25.14. The forces in roof truss may be found analytically or graphically by drawing force (stress) diagram. A design table for forces in various members of the roof truss is prepared. In this design table, the forces in various members for different combinations of loads are tabulated. The design forces for various members of the roof truss are noted.

25.18 DESIGN OF MEMBERS OF A ROOF TRUSS

Rolled steel single or double angle sections, tee-sections, hollow circular, square or rectanglar sections are used in the roof trusses. For long span roof trusses, heavy rolled steel sections (for example I-sections and channel sections) are used. In case of long-span bridge truss girders built-up I-sections, channels and plates are used.

Access to the surface is provided for inspection, cleaning and repainting during service. It is also an important consideration for using built-up sections. For highly corrosive atmosphere, fully closed welded box sections are used. Hollow sections are used with their ends fully closed. The maintenance cost is reduced and the durability of trusses is increased.

Depending upon a particular combination of loads, forces in members are found (whether compressive or tensile) for some other combination of loading, the reversal of forces may occur. Therefore, the members are designed for both maximum compression and maximum tension.

The principal rafter is designed as a continuous strut. While other compression members are designed as discontinuous strut. Lower chord members are designed as a continuous tie.

Step 7 : The compression members of roof truss are designed as discussed in Chapter 6. The principal rafters of roof truss are designed as continuous struts as discussed in Art. 6.18. Other compression members, have their length in between two joints and these are not continuous. Such members are designed as discontinuous struts as discussed in Art. 6.18. The double angle sections will be designed for principal rafter and single angle sections are designed as far as possible for other members. The unequal angle sections are selected for double angle sections. The long legs of angles sections are connected together. The equal angle sections are selected for single angle section. The angle sections smaller in size than 50 mm x 50 mm x 6 mm will not be used. The members are so designed that their slenderness ratios are not greater than 180. The effective length of compression members in roof truss is adopted as 0.7 to 1.0 times the distance between centre to centre of intersections of longitudinal axes of members at the joint.

Step 8 : The tension members are designed as discussed in Chapter 5. The double angle sections are used for all tension members except those which are lightly stressed. The single angle sections are objectionable. The single angle sections have a tendency to twist the truss and produce eccentric forces at the joint. The long legs are kept outstanding. The angle sections smaller in size than 50 mm x 50 mm x 6 mm will not be used. If reversal of force takes place in tension members, their slenderness ratio should be less than 350.

Step 9. The members meeting at a joint should have their centroidal axes inter-secting at one point so as to avoid eccentricity effect. The joint are designed for the forces in the members. When the members are continuous over a joint, the forces on both sides of the joints act in opposite directions. The rivets are provided for the difference of their values. The rivets are placed symmetrical to the joints. The rivet of nomical diameter less than 16 mm will not be used. The thickness of gusset plate at any joint will not be less than 6 mm.

Step 10. The lateral bracings for roof trusses are provided if necessary.

Step 11. Design of end support. The various elements used at the end support of truss are shown in Fig. 25.21. The design of end supports consists of (i) design of shoe angle, (ii) design of bearing plate, and (iii) design of anchor bolt.

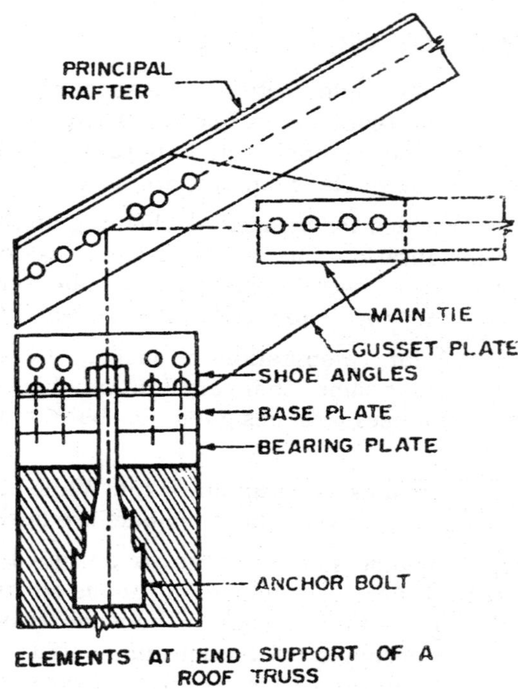

Fig. 25.21.

(i) Design of shoe angles. The shoes angles are designed to accommodate number of rivets necessary to transmit maximum end reactions. The number of rivets are found by dividing maximum reaction by rivet value. If the number of rivets works out small, then, rivets are provided on the basis of practical considerations. Two shoe angles are used. The shoe angles are connected with the gusset plate. The horizontal legs of shoes angles are connected with base plate. The number of rivets for connecting horizontal legs of shoe angles is kept equal to the number of rivets used for connection of vertical legs. The nominal size of 75 mm x 50 mm x 6 mm may be used for the shoe angles.

(ii) Design of bearing plate. The bearing plate is provided to distribute end reaction so that the bearing pressure on masonry or concrete does not exceed the allowable bearing pressure. Let R be the maximum end reaction. The area of bearing plate required

$$A = (R/\sigma_p)$$

where, σ_p is the allowable bearing pressure in masonry or concrete and it is taken as 4 N/mm^2.

The width of bearing plate b be kept equal to twice the length of horizontal leg of shoe angle plus the thickness of gusset plate.

The length of bearing plate.

$$l = \left(\frac{\text{Area of bearing plate}}{\text{Width of bearing plate}}\right)$$

If σ_{p1} is actual bearing pressure in masonry or concrete, then σ_{p1} should be less than σ_p.

The thickness of bearing plate t, is designed to resist the bending moment M, at xx caused due to actual bearing pressure acting at the bottom of bearing plate as shown in Fig. 25.22. The moment of resistance of bearing plate.

$$M_1 = (\sigma_{bs} \times Z) = \left(\sigma_{bs} \times \frac{1}{6} \times x \times t_1^2\right) \qquad (\because M = M_1)$$

$$\text{Therefore, } t_1 = \left(\frac{6M}{x \times \sigma_{bs}}\right) \tag{25.14}$$

where, Z = modulus of section of bearing plate

σ_{bs} = Allowable bearing pressure in steel bearing plate = 185 N/mm² (MPa)

t = (t_1 – thickness of angle)

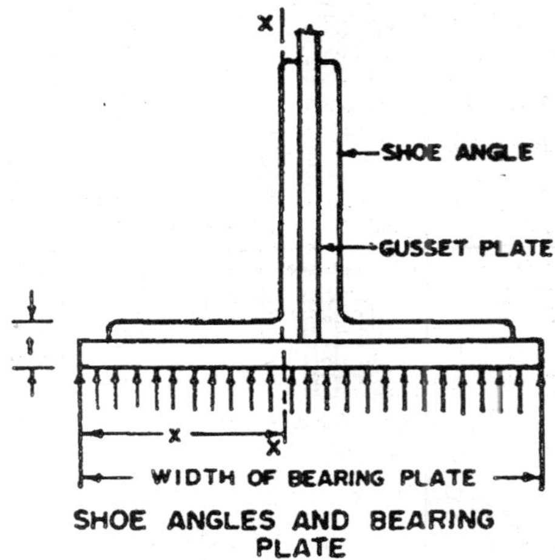

Fig. 25.22.

A base plate is provided under the bearing plate. The thickness of base plate is kept equal to the thickness of bearing plate.

(iii) **Design of anchor bolt**. The anchor bolts are designed to resist net uplift pressure acting on the roof truss.

One end of the truss is kept fixed, while slots of elliptical shape are provided at the other end. The elliptical slots are provided in the horizontal legs of those angles and in the bearing plate. This provides allowance for the expansion or contraction of the truss. The expansion or contraction of truss may take place due to rise or fall of temperature. The bearing plate can slide over the base plate.

Step 12. After the design of all the elements of roof truss, a complete drawing is prepared. The complete drawing is necessary for fabrication. The complete drawing is also useful for estimating the weight of truss.

25.19 TYPICAL DETAILS OF ROOF TRUSSES

In Fig. 25.23 details at the shoe joint of a roof truss have been shown. The roof trusses have horizontal movement at their support due to expansion and contraction of its members under change of temperature. In general provision for horizontal movement at support is made at one support either by providing a sliding plate as shown in Fig. 25.23 (A) or by providing a roller as shown in Fig. 25.23 (C). The joints in roof trusses are so designed that lines of action of forces meet at joint to avoid effect of eccentricity. There is eccentricity in the connections shown in Fig. 25.23 (A), (B) and (C). If these connections are made as shown in Fig. 25.23 (G), (H) and (J); this eccentricity may be avoided. Fig. 25.23 (B), (D) and (F) show trusses supported on columns. When the roof trusses are supported over columns, provision for horizontal moment is not made.

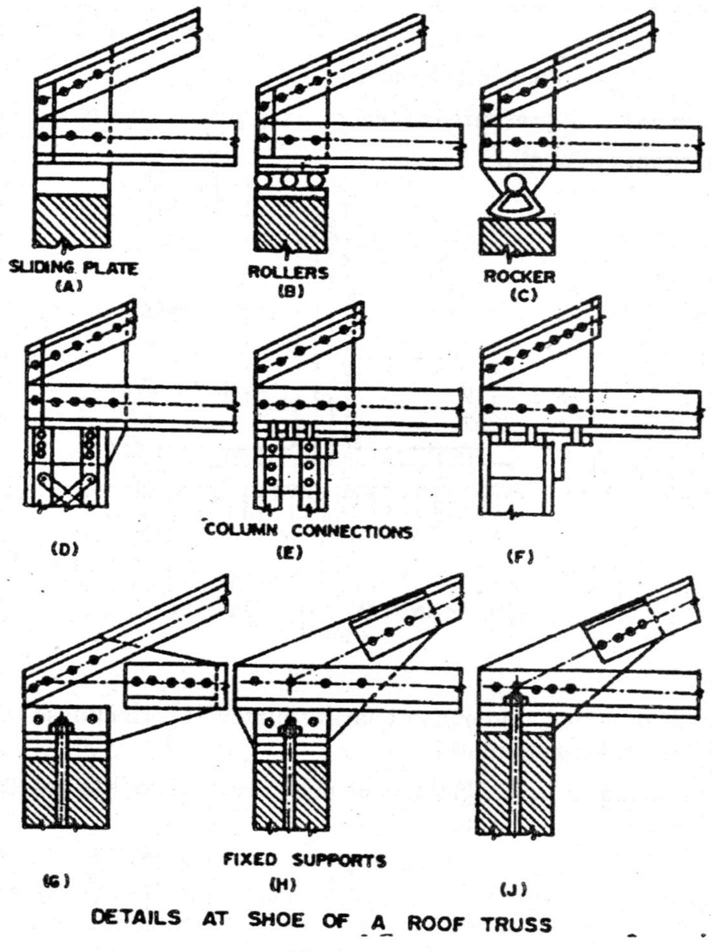

Fig. 25.23

The typical details at the ridge and eves of roof trusses are shown in Fig. 25.23. A.C.C. sheets are shown as roofing material over the trusses. A.C.C. sheets are connected to the angle purlins by hook bolts. The bitumen washers are provided under G.I. washers to make the joint tight.

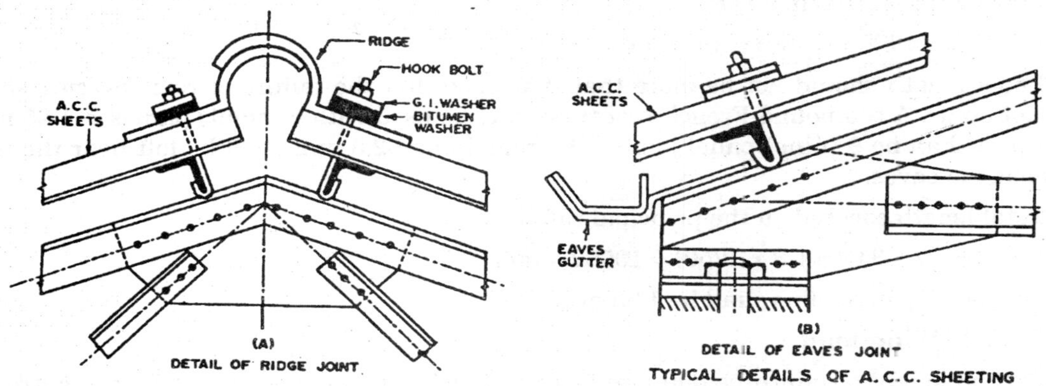

Fig. 25.24.

Fig. 25.25 shows the details of northlight glazing.

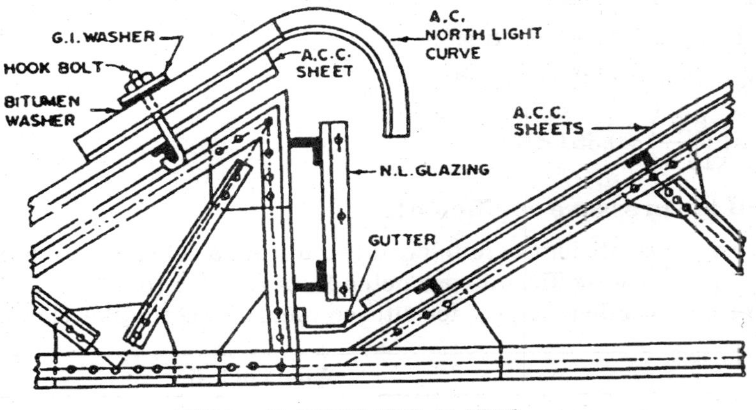

Fig. 25.25.

Example 25.5. A roof truss-shed is to be built in Jodhpur city area for an industrial use. The size of shed shall be 18 m x 30 m. Determine the basic wind pressure [Example 25.1).

Roof trusses of 18 m span and 4.5 rise shall be provided at 4 m spacing centre to centre. The height of cladding or eave level is 6 m. Determine the wind load parallel to be ridge causing a total outward normal wind pressure on the interior purlins, for (i) small permeability (ii) medium permeability and (iii) high permeability.

Solution :

Step 1 : From Example 25.1

The basic wind speed, V_b is 47 m/sec. The design wind speed (as worked out), V_z is 41.641 m/sec. The design wind pressure, p_z is 1.04 kN/m^2.

Step 2 : Geometry of truss

Span of the roof-trusses to be used is 18 m its rise shall be 4.5 m. Slope of the roof

$$\theta = \tan^{-1}(4.5/9) = 26.565° \quad \text{(i)}$$

Length of the principal rafter

$$= [(4.5)^2 + (9)^2]^{1/2} = 10.06 \text{ m} \quad \text{(ii)}$$

Panel length should not be more than 1.4 m, so that the failure of roof covering sheets does not occur. A compound French roof turss may be used with 8 (eight) panels @ 1.25 m, so that there shall be 9 (nine) panel joints. The remaining 62.3 mm shall be left near the ridge and near the eaves.

Total length covered on the principal rafter

$$= (2 \times 31.15 + 8 \times 1250) = 10062.3 \text{ mm}$$

The spacing of purlins shall be 1.25 m.

Step 3 : Wind load

Let the height of plinth be 600 mm and the projection of eave girder behond centre line be 400 mm. the eave height of the roof shed above ground level.

$$h = (6 + 0.600) = 6.6 \text{ m}$$

Gross-width of the roof shed

$$w = (18 + 2 \times 0.400) = 18.8 \text{ m}$$

Ratio of height to width of roof shed

$$\left(\frac{h}{w}\right) = \left(\frac{6.6}{18.8}\right) = 0.351 < 0.5$$

(A) External air pressure coefficients

The wind pressure coefficients are to be adopted from a set (h/w) < 0.5 and for shape θ = 26.565°. The wind pressure coefficients are noted from IS : 875 (part 3) - 1987, or from Table 25.7 (External pressure coefficients, C_{pe} for pitched roofs of rectangular clad buildings.

Side of pitched roof	Wind angle θ			
	θ = 0°	*	θ = 90°	*
Windward	– 0.1374	– .8687	– 0.7	–
Leeward	– 0.4	– 1.1344	– 0.7	

(*Local effects)

(B) Internal pressure coefficients : Internal air pressure in a building depends upon the degree of permeability of cladding to the flow or air.

(i) For small permeability $C_{pi} = \pm 0.2$

(ii) For medium permeability $C_{pi} = \pm 0.5$

(iii) For high permeability $C_{pi} = \pm 0.7$

Note : The internal pressure coefficient is added algebraically to the external pressure coefficient and the worst combination shall be considered.

Step 4 : Total wind load parallel to ridge

(i) For small permeability

External + internal pressure coefficients $(C_{pe} - C_{pi}) = [-0.7 - (+0.2)] = 0.9$

Total wind load $W_w = (C_{pe} - C_{pe}) . p_d.A$ $W_w = -0.9\ p_d. A$

(ii) For medium permeability

External + internal pressure coefficients $(C_{pe} - C_{pi}) = [-0.7 - (+0.5)] = -1.2$

Total wind load $W_w = -1.2\ p_d . A$

(iii) For high permeability

External + internal pressure coefficients $(C_{pe} - C_{pe}) = [-0.7 - (+0.7)] = -1.4$

Total wind load $W_w = -1.4\ p_d.A$

Note : p_d (design wind pressure) is 1.04 kN/m². A (effective exposed area, that is, spacing of purlins multiplied by the length of purlin is (1.25 x 4) = 5 m². Therefore, total wind load parallel to ridge for

(i) **Small permeability**

$W_w = [(-0.9) . (1.04) . (5)] = -$ **4.68 kN**

(ii) **Medium permeability**

$W_w = [(-1.2) . (1.04) . (5)] = -$ **6.24 kN**

(iii) **high permeability** :

$W_w = [(-1.4) . (1.04) . (5)] = -7.28$ kN

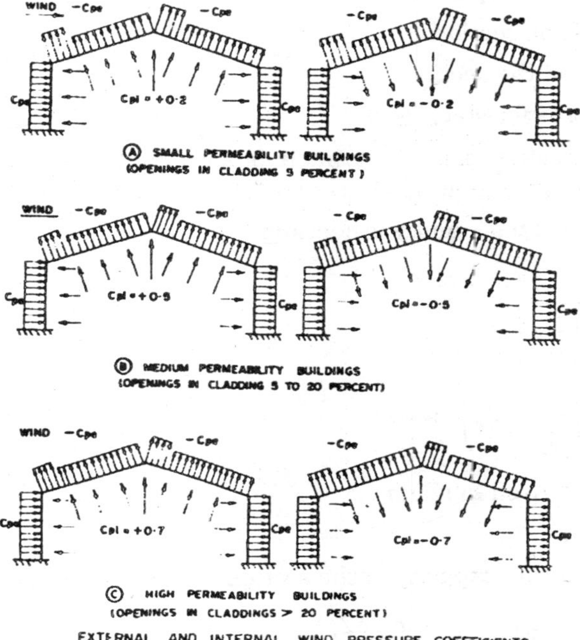

Fig. 25.26

Figs. 25.26(A), (B) and (C) show external air pressure coefficients, C_{pe} for pitched roofs, and internal air pressure coefficients (positive and negative) for small permeability buildings, medium permeability buildings and high permeability buildings, respectively.

25.20 DESIGN OF CHANNEL AND I-SECTION PURLINS

Design procedure for the design of channel and I-section purlins as under.

1. Centre to centre distance of adjacent roof-trusses gives the span of the purlins.

2. Dead load for the roof covering material and imposed live load and the wind load are calculated.

The components of these loads in the direction perpendicular and parallel to the sheet are obtained. These loads are multiplied by the respective values of partial safety factors for loads, γ_{fL}

3. Using the values of factored loads, the values of factored bending moments ($M_{fd.z}$ and $M_{fd.y}$ and the factored shear forces ($V_{fd.z}$ and $V_{fd.y}$) are calculated.

4. The value of plastic section modulus of the section required is determined,

$$Z_{pZ} = \left(\frac{M_{fd.Z} \times \gamma_{msy}}{\sigma_y}\right) + 2.5\left(\frac{d}{b}\right) \times$$

$$\left(\frac{M_{fd.Z} \cdot \gamma_{msy}}{\sigma_y}\right), \text{ where}$$

γ_{msy} = Partial safety for steel 1.15

d = depth of the section

b = Breadth of the section

σ_y = yield stres for steel used

5. Check for the section classification. It is seen that the section is adequate for the bending strength. The requirement is M_{fd} should be less than the bending strength M_{ds}

6. Shear capacity for axes Z and y is checked

$$V_{dZ} = \left(\frac{\sigma_y \quad A_{vz}}{\sqrt{3} \cdot \gamma_{msy}}\right)$$

$$V_{dy} = \left(\frac{\sigma_y}{\sqrt{3} \cdot \gamma_{msy}}, A_{vy}\right), \text{ where}$$

$A_{v.z} = (h_w \cdot t_w)$ Area in shear

$A_{v.y} = 2\, b_f \cdot t_f$

7. Calculate the Lending capacity of the section for both the axes.

$$M_{dz} = \left(\frac{\sigma_y \cdot Z_{pz}}{\gamma_{msy}}\right) <= 1.2\left(\frac{\sigma_y \cdot Z_{e.z}}{\gamma_{msy}}\right)$$

$$M_{dy} = \left(\frac{\sigma_y \cdot Z_{py}}{\gamma_{msy}}\right) <= r_f \left(\frac{\sigma_y \cdot Z_{e.y}}{\gamma_{msy}}\right)$$

In this secon expression 1.2 is replaced by γ_f. In the y-direction, the shape-factor (Z_p / Z_e) is move than 1.2.

In case factor 1.2 is used, the onset of yielding (under unfactored load) cannot be prevented.

8. Inter-action equation is used to check local capacity

$$(M_z/M_{dz}) + (M_y + M_{dy}) <= 1.0$$

where M_{dz} is lateral torsional strength of member.

9. It is seen that the deflection is within allowble limits.

10. The bottom flange of the purlin will be under compression (since, it is laterally unsupported under combined wind load and dead load.

The lateral-torsional buckling cpacity of the section has be calculated and checked overall buckling check is to be made.

25.21 EMPIRICAL DESIGN OF ANGLE PURLINS

BS : 5950-2000 specifies empirical rules for the design of purlins. As such, the procedure has been titled as empirical design. Hot rolled angle sections are used for the purlins. Angle sections are unsymmetrical sections about both the axes. Following are the general rules for the empirical design of purlins.

(i) Unfactored loads are considered in the design, that is, partial safety factor for loads, g_{fL} is taken as unity.

(ii) The span of purlins should not exceed 6.5 m centre to centre of main supports

(iii) Where the purlins span only one bay, each purlin should be connected by at least two fasteners (bolts) at each end

(iv) In case, the purlins are continuous over two or more spans with staggered joints in adjacent lines of member at least one end of any single bay members should be connected by not less than two bolts.

In addition to the above rules following specific rules have been given.

(i) The roof-slope should not exceed 30°

(ii) The loading on the purlins should be substantially uniformly distributed. Other types of load should not be more than 10 percent of total roof load.

The elastic modulus, Z of a purlin section about its axis parallel to the plane of the cladding should not be less than Wp L/1800 mm³, where Wp is the total unfactored load on one span (N) due to dead and imposed load or dead load minus wind load and L is the span in mm.

The dimension D in mm of the angle purlin perpendicular to the plane of the cladding should not be less than (L/45), and the dimension B in mm' parallel to the plane of cladding should not be less than (L/60).

25.22 BI-AXIAL BENDING IN PURLINS

In general, the beams are subjected to transverse loading. The plane of bending (or plane of loading) coincides with one of the principal axes. The neutral axis of the section coincides with other principal axis.

Where as, in case of purlins, the line of action of resultant of the vertical loads and normal component of wind load do not coincide with any of the principal axes. The plane of loading for different cross-sections used for the prulins is shown in Fig. A.1 (Appendix-A). In such cases, the resultant of loads remain, inclined with respect to the principal axes of the cross-section. In case of purlins, **bi-axial bending** occurs. The purlins are subjected to such a loading condition, the the bending occurs about both the axes (i.e, the strong axis and weak-axis). Therefore, biaxial bending occurs in purlins. One of the two components passes through shear centre of the cross-section.

Note: Appendix B deals with `shear centre'. The shear centre is defined as that point through which when the loads act, the twisting or torsion does not occur in the beam.

The shear centre remains located on an axis of symmetry. For a cross-section of a structural member which has two axes of symmetry, the shear centre and the centroid of the cross-section coincide.

When one of the component of loads acts through the shear centre, simple bending occurs about two perpendicular directions. One component of load acting perpendicular to Z-direction, it produces bending about y-direction. And that component which acts perpendicular to y-direction, produces bending about Z-direction. In case, the bending occurs about Z-axis, only, then

$$(M_z/M_{dz}) <= 1.0, \text{ where}$$

M_z is the factored bending moment about the z-z axis and

M_{dz} is the design bending strength about the z-z axis

Similarly in case, the bending occurs about y-axis only then

$$(M_y/M_{dy}) <= 1.0, \text{ where}$$

M_y is the factored bending moment about the y-y axis and

M_{dy} is the design bending strength about the y-y axis.

When the bending occurs about both the zz-axis and yy-axis. then the interaction expression needs that the sum of the two ratios is less than or equal to 1.0. That is

$$(M_z/M_{dz}) + (M_y/M_{dy}) <= 1.0$$

It is to note that any shape bent about its weak axis cannot buckle in the other direction, so lateral buckling is not a limit state for bending about weak axis

Purlins subjected to bi-axial bending may be designed only by trial and error

In 1985 Linder and Gietzelt presented a design method for beams subjected to bi-axial bending. In 1990, Sarma and Lakshmanan provided a method for channel section and tube purlins under bi-axial bending.

For plastic and compact section

In above stated interaction expression,

for $M_{dz} = \left(\dfrac{\sigma_y \,.\, Z_{p.z}}{\gamma_{msy}}\right)$ *and*

$M_{dy} = \left(\dfrac{\sigma_y \,.\, Z_{p.y}}{\gamma_{msy}}\right)$ *may be*

substituted. Then

$$\left(\frac{M_z \cdot \gamma_{msy}}{\sigma_y \,.\, Z_{pz}}\right) + \left(\frac{M_y \cdot \gamma_{msy}}{\sigma_y . Z_{p.y}}\right) <= 1.0$$

Multiplying this expression by Z_{pz}

$$\left(\frac{M_z \cdot \gamma_{msy}}{\sigma_y}\right) + \left(\frac{M_y \cdot \gamma_{msy}}{\sigma_y}\right)\left(\frac{Z_{pz}}{Z_{py}}\right) <= Z_{pz}$$

A fairly close design may be found by knowing an approximate range of (Z_{pz}/Z_{py}), Using the properties of beam and neglecting the effect of web, this quantity (Z_{pz}/Z_{py}) may be approximately found. Using b_f = Width of flange and d = depth of beam

$$\left(\frac{Z_{pz}}{Z_{py}}\right) = \frac{(I_{pz})(b_f/2)}{(I_{py})(d/2)} = \frac{2\, b_f\, t\, (d/2)^2\, (b_f/2)}{(d/2)\,(2)\,(1/12)\,(t\, b_f^3)} = \left(\frac{3d}{b_f}\right)$$

By increasing the coefficient of (d/b_f) from 3 to 5), the error in the approximation due to neglecting the moment of inertia of web is offset closely.

In 1992, Gaylord et al suggested that this coefficient 3.5 may be replaced by a factor 2.5 for limit state method. Then

$$Z_{pz} = \left(\frac{1}{\sigma_y}\right)\left[\left(M_z \cdot \gamma_{msy}\right) + \left(M_y \cdot \gamma_{msy}\right) \times \left(2.5 \times \frac{d}{b_f}\right)\right]$$

There is no allowance for lateral torsional buckling. Therefore, some trials are needed.

IS : 800-2007, section 9 deals with member subjected to combined forces. Clasue 9.3 deals with the members subjected to axial force and the bending moment.

Under combined axial force and bending moment, the section strength as controlled by the materal failure and the member strength as controlled by buckling failure shall be checked in accordance with clause 9.3.1. Section/clause 9.3.1 deals with **Section strength**. According to clause 9.3.1.1 **for plastic and compact section**

$$\left(\frac{M_y}{M_{ndy}}\right)^{\alpha_1} + \left(\frac{M_z}{M_{ndz}}\right)^{\alpha 2} <= 1.0$$

Conservatively, the following expression may also be used under combined axial force and bending moment

$$\left(\frac{P}{P_d}\right) + \left(\frac{M_y}{M_{dy}}\right) + \left(\frac{M_z}{M_{dz}}\right) <= 1.0 \quad \text{where}$$

M_y, M_z = Factored bending moments applied about the minor and the major axis of the cross-section, respectively.

M_{ndy}, M_{ndz} = Design reduced bending strength under combined axial force and the respective uniaxial moment acting alone.

P = Factored applied axial force (Tension, T or Compression, C)

P_d = Design strength in Tension, T_d as obtaned from 6 or in compression due to yielding given by

$$P_d = \left(\frac{\sigma_y \cdot Ag}{\gamma_{msy}}\right)$$

M_{dy}, M_{dz} = Design strength under corresponding moment acting alone (clause 8.2)

Ag = Gross area of the cross-section.

α_1, α_2 = Constants given in Table 17, IS : 800-2007, and

γ_{msy} = Partial factor of safety in yielding.

Example 25.6 Design an I-section purlin to support galvanized corrugated iron sheet roof. The purlins are 1.25 m apart over roof trusses spaced 5 m centre to centre. The roof surface has an inclination of 30° to the horizontal. The weight of corrugated iron sheet is 0.1331 kN/m². The weight of fixtures etc 0.053 kN/m². The design wind pressure for the medium permeability is 1.20 kN/m² (outward) parallel to the ridge.

Design :

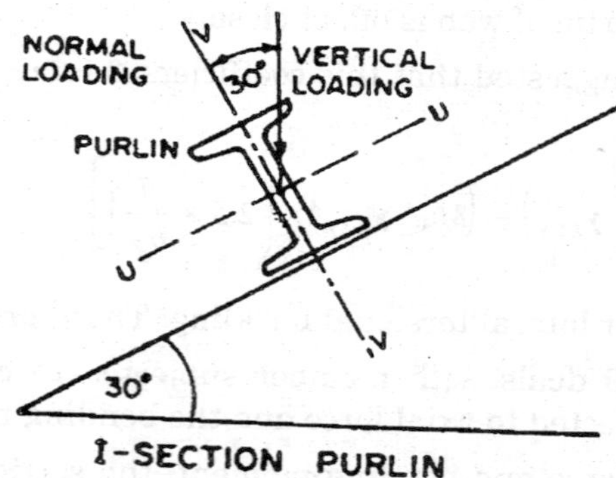

Fig. 25.27

Step 1 : Dead load on purlin per metre length

Weight of galvanized corrugated iron sheets = (0.1331 x 1.25 x 1) = 0.1664 kN/m

Extra load for overalap and fixtures = (0.053 x 1025 x 1) = 0.0663 kN/m

Self-weight of purlins (assumed = 0.120 kN/m

Total dead load = 0.3527 kN/m = 0.360 kN/m

Component of dead load normal to roof = (0.360 x cos 30°) = $\left(\frac{0.360 \times \sqrt{3}}{2}\right)$

= 0.312 kN/m

Component of dead load parallel to the roof

$$= (0.360 \text{ x } \sin 30^\circ) = \left(\frac{0.360 \times 1}{2}\right) = 0.180 \text{ kN/m}$$

Factored dead load normal to roof

= (0.312 x 1.32) = 0.412 kN/m

Factored dead load parallel to the roof

= (0.180 x 1.32) = 0.238 kN/m

Step 2 : Imposed (live) load

The imposed (live) load for sloping roof for slope more than 10°

$$[0.75 - (30 - 20)(0.020)] = 0.35 \; kN/m^2$$

It is subjected to a minimum of 0.4 kN/m²

Live load (0.4 x 1.25 x 1) = 0.5 kN/m

Component of live load normal to the roof

$$(0.5 \text{ x } \cos 30^\circ) = \left(\frac{0.5 \times \sqrt{3}}{2}\right) = 0.433 \; kN/m$$

Component of live load parallel to the roof

$$(0.5 \text{ x } \sin 30^\circ) = \left(\frac{0.5 \times 1}{2}\right) = 0.25 \; kN/m$$

Factored value live load normal to the roof

= (1.5 x 0.433) = 0.650 kN/m

Factored value live load parallel to the roof

= (1.5 x 0.25) = 0.375 kN/m

Step 3 : Wind load on purlin

Given design pressure for medium permeability (parallel to ridge) is 1.20 kN/m² (outward or negative). That is, it acts upward.

Step 4 : Combination of loads (Factored values)

(i) Dead load + live load

Normal to the roof

= (0.412 + 0.650) = 1.062 kN/m

Parallel to the roof

= (0.238 + 0.375) = 0.613 kN/m

(ii) Dead load + live load + Wind load

Parallel to the ridge

[0.238 + 0.613 + 1.2 x 1.20] = 2.291 kN/m

Step 5 : Bending moment and shear force

(i) Bending moment (Factored values)

Normal to the roof

$$M_{yfd} = \left(\frac{\omega.L^2}{10}\right)$$

$$= \left[\left(\frac{0.412 \times 5}{10}\right) \times 5 + \left(\frac{0.650 \times 5}{9}\right) \times 5\right] = (1.03 + 1.80) = 2.83 \text{ kN-m}$$

Parallel to the roof

$$M_{z.fd} = \left[\left(\frac{0.238 \times 5}{10}\right) \times 5 + \left(\frac{2.291 \times 5}{10}\right) 5\right] \text{ kN-m}$$

$$= (0.595 + 7.275) = 7.87 \text{ kN-m}$$

(ii) Shear force (Factored values)

Normal to the roof

$$V_{y.fd} = \left[\left(\frac{0.412 \times 5}{2} + \frac{0.50 \times 5}{2}\right)\right] = 1.155 \ kN$$

Parallel to the roof

$$V_{z.fd} = \left[\left(\frac{0.238 \times 5}{2} + \frac{2.291 \times 5}{2}\right)\right] = 6.3225 \ kN$$

Step 6 : Assuming a trial I-section

LB 125, @ 0.119 kN/m

From ISI HB No. 1, geometrical properties

Sectional area, A = 1512 mm²

Depth of section, h = 125 mm

Width of flange, b_f = 75 mm

Thickness of flange, t_f = 6.5 mm

Thickness of web, t_w = 4.4 mm

Elastic section modulus Z_{xx} = 65.1 x 10^3 mm³

Elastic section modulus Z_{yy} = 11.6 x 10^3 mm³

Radius at the roof r_1 = 8 mm

Depth of web, d_w = h – 2 (t_f + r_1)

Assuming shape factor 1.12 = 125-2 (6.5 + 8) = 96 mm

Plastic section modulus, $Z_{p.xx}$ = (1.12 x 65.1) = 72.9 x 10^3 mm³

Plastic section modulus, $Z_{p.yy}$ = 1.12 x 11.6 = 12.992 x 10^3 mm³

Plastic modulus of section required

$$= \frac{M_z}{\sigma_y} \times \gamma_{msy} + 2.5 \times \left(\frac{d}{b}\right)\left(\frac{M_y}{\sigma_y} . \gamma_{msy}\right)$$

$$= \frac{1.15}{250}\left[M_z + 2.5 \times \left(\frac{d}{b}\right) M_y\right]$$

$$= \frac{1.15}{250}[7.87 + 11.79] \times 10^6 \ mm^3$$

$= 90.436 \ x \ 10^3 \ mm^3$

Elastic section modulus required = $(90.436 / 1.12) \times 10^3 \ mm^3 = 80.714 \times 10^3 \ mm^3$

Select ISLB 150, @ 0.142 kN/m

Elastic section modulus provided.

$Z_{xx} = 91.8 \times 10^3 \ mm^3$, which one is more than required.

Step 7 : Section classification

For ISLB 150, @ 0.142 kN/m

Width of flange, $b_f = 80$ mm

Thickness of flange, $t_f = 6.8$ mm

Thickness of web, $t_w = 4.8$ mm

Depth of web, $d_w = h - 2\ (t_f + r_1)$

$[150 - 2\ (6.8 + 9.5)] = 117.4 \ mm$

Ratio $(0.5\ b_f/t_f) = (0.5 \ x \ 80 / 6.8) = 5.88$

< 9.4

Ratio $(d_w / t_w) = (117.4/4.8) = 24.458$

< 29.3

Section is plastic.

Step 8 : Shear capacity of section

Area of Web, $A_v = (150 \ x \ 4.8) = 720 \ mm^2$

$$V_d = \left(\frac{\sigma_y}{\sqrt{3}\ \gamma_{msy}} \times A_v\right) = \left(\frac{250}{1.732 \times 1.15} \times \frac{720}{1000}\right) = 90.37 \ kN$$

$0.6\ V_d = (0.6 \times 90.37) = 54.22 \ kN > 6.3225 \ kN$

Note : In purlin design, the shear capacity is usually large.

Step 9 : Moment capacity of section

For ISLB 150, @ 0.142 kN/m

$Z_{xx} = 91.8 \times 10^3 \ mm^3$

$Z_{yy} = 13.8 \times 10^3 \ mm^3$

Assuming shape factor 1.12

$$M_{dz} = \left(\frac{Z_{pz} \cdot \sigma_y}{\gamma_{msy}}\right) = \left(\frac{1.12 \times 91.8 \times 10^3 \times 250}{1.15 \times 1000}\right)$$

$= 22.35$ kN-m > 7.87 kN-m

$$M_{d.y} = \left(\frac{Z_{py} \cdot \sigma_y}{\gamma_{msy}}\right) = \left(\frac{1.12 \times 13.8 \times 10^3 \times 250}{1.15 \times 1000}\right)$$

$= 3.36$ kN-m > 2.83 kN-m

Step 10 : Overall member strength

For bi-axial bending

$$\left(\frac{M_z}{M_{dz}}\right) + \left(\frac{M_y}{M_{yd}}\right) <= 1.0$$

$$\left(\frac{7.87}{22.95}\right) + \left(\frac{2.83}{3.36}\right) = 1.19$$

Slightly more than 1.0.

Example 25.7, In Example 25.6, design a channel section purlin.

Design : From Example 2.5, from step 1

Step 1 : Total dead load (Components of dead load)

(i) Normal to roof

Factored value

$$= (0.360 \text{ x } \cos 30°) = \left(\frac{0.360 \times \sqrt{3}}{2}\right) = 0.312\ kN/m$$

= (0.312 x 1.32) = 0.412 kN/m

(ii) Parallel to roof

$$= (0.360 \text{ x } \sin 30°) = \left(\frac{360 \times 1}{2}\right) = 0.180\ kN/m$$

Factored value

= (0.180 x 1.32) = 0.238 kN/m

From Example 25.6, from **Step 2**

Step 2 : Imposed (live) load (component of loads)

(i) Normal to the roof

$$= (0.5 \text{ x } \cos 30°) = \left(\frac{0.5\sqrt{3}}{2}\right) = 0.433\ kN/m$$

Factored value

= (1.5 x 0.433) = 0.650 kN/m

(ii) Parallel to the roof

$$= (0.5 \text{ x } \sin 30°) = \left(\frac{0.5 \times 1}{2}\right) = 0.25\ kN/m$$

Factored value

$= (1.5 \times 0.25) = 0.375$ kN/m

Step 3 : Wind load on purlin

Given design pressure for medium permeability (parallel to ridge) is 1.20 kN/m^2 (outward or negative). That is acts upward.

Step 4 : Combination of loads (Factored values)

(i) Dead load + live load

Normal to the roof

$= (0.412 + 0.650) = 1.062$ kN/m

Parallel to the roof

$= (0.238 + 0.375) = 0.613$ kN/m

(ii) Dead load + live load + wind load

Parallel to the ridge

$[0.238 + 0.613 + 1.2 \times 1.20] = 2.291$ kN/m

Step 5 : Bending moment and shear forces

(i) Factored bending moment (normal to roof)

$$M_{y.fd} = \left(\frac{w.L^2}{10}\right)$$

$$= \left[\left(\frac{0.415 \times 5}{10}\right) \times 5 + \left(\frac{0.650 \times 5}{9}\right) \times 5\right]$$

$= (1.03 + 1.80) = 2.83$ kN/m

(ii) Factored bending moment (Parallel to roof)

$$M_{z.fd} = \left[\left(\frac{0.238 \times 5}{10}\right) \times 5 + \left(\frac{2.291 \times 5}{10}\right) \times 5\right]$$

$=[(0.5951 + 7.275)] = 7.87$ kN-m

(iii) Factored shear force (normal to the roof)

$$V_{y.fd} = \left[\left(\frac{0.412 \times 5}{2}\right) + \left(\frac{0.50 \times 5}{2}\right)\right] = 1.155\ kN$$

(iv) Factored shear force (Parallel to the roof)

$$V_{z.fd} = \left[\left(\frac{0.238 \times 5}{2}\right) + \left(\frac{2.291 \times 5}{2}\right)\right] = 6.3225\ kN$$

Step 6 : Assuming a trial channel section

ISMC 125, @ 0.127 kN/m

Frɔm ISI Handbook No.1, geometrical properties,

sectional area, A = 1619 mm^2

Depth of section, h = 125 mm

Width of flange, b_f = 65 mm

Thickness of flange, t_f = 8.1 mm

Thickness of web, t_w = 5 mm

Elastic section modulus, $Z_{xx} = 66.6 \times 10^3\ mm^3$

Elastic section modulus , $Z_{yy} = 13.1 \times 10^3\ mm^3$

Radius at the root, r_1 = 9.5 mm

Depth of web, $d_w = [h - 2(t_f + r_1)] = [125 - 2x(8.1 + 9.5)] = 89\ mm$

Step 7 : Plastic section modulus required

$$= \left[\left(\frac{M_{pz} \times \gamma_{msy}}{\sigma_y}\right) + 2.5\left(\frac{d}{d}\right)\left(\frac{M_{py} \times \gamma_{msy}}{\sigma_y}\right)\right]$$

$$= \left[\frac{M_{pz} \times 1.15}{250} + 2.5 \times \left(\frac{125}{65}\right)\left(\frac{M_{py} \times 1.15}{250}\right)\right]$$

$$= \frac{1.15}{250}\left[M_{pz} + 2.5 \times \left(\frac{125}{65}\right) M_{py}\right] \times 10^6$$

$$= \frac{1.15}{250}[7.87 + 2.5 \times 1.923 \times 2.83] \times 10^6$$

$$= \frac{1.15}{250}[7.87 + 13.605] \times 10^6 = \left(\frac{1.15}{250} \times 21.475 \times 10^6\right) = 98.785 \times 10^3\ mm^3$$

For ISMC 125, @ 0.127 kN/m³

$M_p = 77.2 \times 10^3\ mm^3$

For ISLC 150, @ 0.114 kN/m

$M_p = 106.2 \times 10^3 \times\ mm^3$

Geometrical properties for ISLC 150, @ 0.144 kN/m, from ISI Handbook No. 1,

Sectional area, A = 1836 mm²

Depth of section, h = 150 mm

Width of flange, b_f = 75 mm

Thickness of flange, t_f = 7.8 mm

Thickness of web, t_w = 4.8 mm

Modulus of section, $Z_{xx} = 93 \times 10^3\ mm^3$

Modulus of section, $Z_{yy} = 20.2 \times 10^3\ mm^3$

Radius at the root, r_1 = 8 mm

Moment of inertia, $I_{xx} = 697.2 \times 10^4\ mm^4$

Moment of inertia, $I_{yy} = 103.2 \times 10^4\ mm^4$

Step 8 : Section classification

Ratio (b_f / t_f) = (75/7.8) = 9.6 < 15.7 mm

Depth of web, $d_w = [h - 2(t_f + r_1)]$

= [150 - 2 x (7.8 + 8)] = 118.4 mm

Ratio (d_w/t_w) = (118.4/4.8) = 24.66 mm

< 42

Section is semi-compact section.

Step 9 : Shear capacity of the section

In z-direction

$$V_{dz} = \left(\frac{\sigma_y}{\sqrt{3}\,\gamma_{msy}}\right) \times \left(h \times t_w\right)$$

$$= \left(\frac{250}{1.732} \times \frac{1}{1.15}\right) \times \left(\frac{150 \times 4.8}{1000}\right) = 90.370\ kN$$

V_d = (0.6 x 90.370) = 54.22 kN > 6.3225 kN

In y - direction

$$V_{dy} = \left(\frac{\sigma_y}{\sqrt{3} \times \gamma_{msy}}\right) \times \left(\frac{2 \times 75 \times 7.8}{1000}\right)$$

$$= \left(\frac{250}{1.732 \times 1.15}\right)\left(\frac{150 \times 7.8}{1000}\right) = 146.85\ kN \ \text{>1.155 kN}$$

Note : Shear capacity is usually more than shear force for the purlins.

Step 10 : Bending strength of the section

$$M_{dz} = \left(\frac{\sigma_y . Z_{pz}}{\gamma_{msy}}\right) = \left(\frac{250}{1.15} \times \frac{106.2 \times 10^3}{1000 \times 1000}\right) = 23.086\ kN - m\ > 7.87\ \text{kN-m}$$

$$\left(\frac{Z_{pz}}{Z_{xx}}\right) = \left(\frac{106.2 \times 10^3}{93 \times 10^3}\right) = 1.1419$$

Z_{py} = (1.1419 x Z_{yy})

= (1.1419 x 20.2) = 23.06 x 10^3 mm^3

$$M_{dy} = \left(\frac{\sigma_y . Z_{py}}{\gamma_{msy}}\right) = \left(\frac{250 \times 23.06 \times 10^3}{1.15 \times 10^3 \times 10^3}\right)$$

= 5 kN-m > 2.83 kN-m

Step 11 : Overall strength of member

Following interaction expression should be satisfied

$$\left(\frac{M_z}{M_{dz}}\right) + \left(\frac{M_y}{M_{dy}}\right) <= 1.0$$

$$\left(\frac{7.87}{23.086}\right)+\left(\frac{2.83}{5}\right)=(0.341+0.566) \;=\; 0.907<1.0$$

Step 12 : Check for deflection

$$\text{Deflection, y} = \left(\frac{5\,\omega\, l^4}{384\; EI_4}\right)$$

$$y \;=\; \left(\frac{5\times 0.433\times 5000^4}{384\times 2\times 10^5\times 697\times 10^4}\right)=2.527\; mm$$

Allowable deflection

$$=\;\frac{1}{180}\times(5000)=27.77\; mm$$

Hence, the section is safe.

Example 25.8 Design an angle section purlin for a trussed roof from the following data.

Span of roof truss = 12 m

Spacing of roof trusses = 5 m

Spacing of purlins along the slope of roof truss = 1.2 m

Slope of roof truss = 1 vertical to 2 horizontal

Wind load on roof surface = 1.04 kN/m²

Vertical load from roof sheeting etc = 0.20 kN/m²

Design

Step 1 : Slope of roof truss

Slope of roof truss = 1 vertical to 2 horizontal

$$\tan\theta=\frac{1}{2},\theta\;=26.565^{\circ}<30^{\circ}$$

Angle iron (section) purlin may be designed according to BS : 5950 - 2000

Step 2 : Vertical load on purlin per one metre length

Vertical load from roof sheeting etc.

(0.200 x 1.2 x 1) = 0.240 kN/m

Assume self-weight of purlin = 0.120 kN/m

Total vertical load = 0.360 kN/m

Step 3 : Wind load on roof surface normal to roof

(1.04 x 1.2 x 1) = 1.248 kN/m

Step 4 : Total load (assumed to act normal to roof)

(1.248 + 0.360) = 1.608 kN/m

Step 5 : Maximum bending moment

$$M = \left(\frac{WL}{10}\right) = \left(\frac{1.068 \times 5 \times 5000}{10}\right) = 4.02 \text{ kN-m}$$

It is assumed that the vertical loads are resisted by zz-axis and the horizontal component is resisted by sheeting. Design strength, $\sigma_y = 250 \text{ N/mm}^2$

Step 6 : Required section modulus

$$Z_{ez} = \left(\frac{4.02 \times 10^6}{250}\right) = 16.08 \times 10^3 \text{ mm}^3$$

Step 7 : Size of angle section

Lengh of leg perpendicular to cladding

$$= \left(\frac{5000}{45}\right) = 111.11 \text{ mm}$$

Lengh of leg parallel to cladding

$$= \left(\frac{5000}{60}\right) = 83.33 \text{ mm}$$

From ISI Handbook No. 1,

Select ISA 125 mm x 95 mm x 6 mm (ISA 125, 95, @ 0.129 kN/m)

Modulus of section provided

$Z_{ex} = 23.4 \times 10^3 \text{ mm}^3$

Provide ISA 125 mm x 95 mm x 6 mm

(ISA 125, 95 @ 0.129 kN/m) size for the angle section purlin.

Note : Deflection need not be checked in this case.

Example 25.9 : A 10 m x 40 m godown is to be constructed. The steel roof trusses will be used for roofing. The trusses will be supported over masonry walls 300 mm thick. Galvanized corrugated iron sheets will be used for covering. Propose a suitable type of roof truss.

The basic wind pressure is 1.04 kN/m², and there is no snowfall. Determine the load at each point.

Design :

Step 1 : Preliminary geometry of truss

Clear spacing between the masonry walls = 10 m

Thickness of masonry = 0.30 m

Centre to centre spacing between the masonry walls = 10.30 m

Therefore, span of roof truss = 10.30 m

For galvanized corrugated iron sheet,

provide pitch of roof truss = 1/4

$$\text{Rise of roof truss} = \frac{1}{4} \times \text{span} = \left(\frac{1}{4} \times 10.30\right) = 2.575 \text{ m}$$

Let θ be slope of roof truss for pitch of 1/,4

$$\tan\theta = \frac{1}{2} \quad \therefore \quad \theta = 26.565° = 26°34'$$

Length along the sloping roof

$$= \left[\left(\frac{10.30}{2}\right)^2 + (2.575)^2\right]^{1/2} = 5.76 \text{ m}$$

The length along the sloping roof may be divided into six equal panels.

Length of each panel = (5.76/6) = 0.960 m (not > 1.40 m)

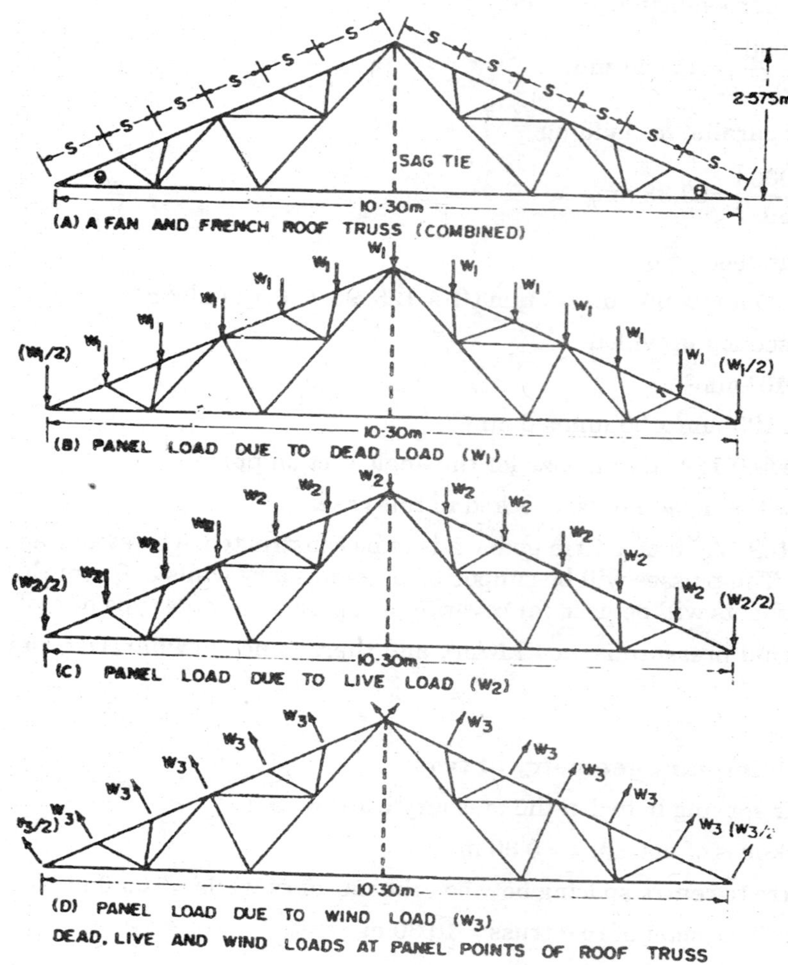

Fig. 25.28.

Therefore, for above proportions, a French roof truss as shown in Fig. 25.28 (A) may be selected. These trusses may be provided at 4 m spacing. 11 trusses will be provided over the entire length of godown. The purlins will be provided at each panel point.

Step 2 : Load at each panel point is determined as under :

1. Dead load

Weight of galvanized corrugated iron sheets = 0.133 kN/m^2

Weight of purlins (assumed) = 0.150 kN/m^2

Weight of bracing (assumed) = 0.015 kN/m^2

Self weight of roof truss

$$w = \frac{1}{100}\left(\frac{l}{3}+5\right) = \frac{1}{100}\left(\frac{10.30}{3}+5\right) = 0.084 \text{ kN/m}^2$$

Total dead load = (0.133 + 0.150 + 0.015 + 0.084) = 0.382 kN/m^2

Spacing of trusses = 4 m

Panel length in plan = 0.960 cos (26° 34') = 0.858 m

Load at each intermediate panel due to dead load, w_1

(0.382 x 4 x 0.858) = 1.311 kN

Load at end panel due to dead load ($w_1/2$) = 0.656 kN.

2. Live load

Access is not provided for roofing except for maintenance. Slope of roof truss

= 26° 34′ = 26.565°

∴ Live load = (0.750 – 0.020 x 16.565)= 0.4187 kN/m^2

Load over each intermediate panel due to the live load, w_2

(0.4187x 4 x 0.960) = 1.61 kN

Load at each end panel due to live load ($w_2/2$) = 0.804 kN

3. Wind load

Slope of roof truss = 26.565°

Intensity of design wind pressure = 1.4 kN/m^2

For pitched roof, the height of eave, and the width of cladding, w are also necessary data. Let the height of plinth be 600 mm and the protection of eave girder beyond the centre line be 400 mm. The gross width of roof-shed.

w = (10 + 2 x 0.400) = 10.80 m assumed.

The height of eave of the roof-shed above ground level.

h = (3.18 + 0.60) = 3.78 m assumed.

Ratio of height to the width of roof-shed $\left(\frac{h}{w}\right) = \left(\frac{3.78}{10.8}\right) = 0.35 < 0.5$

(A) External air pressure coefficients – The wind pressure coefficients are to be adopted from a set (h/w) < 0.5 and for slope θ = 26.565°. The wind pressure coefficients are noted from IS: 875 (Part 3) - 1987 or from Table 25.7 (External pressure coefficients, C_{pe} for pitched roofs of rectangular clade buildings)

Side of pitched roof	Wind angle θ			
	θ = 0°	*	θ = 9°	*
Windward	–0.1374	– 0.8687	– 0.7	–
Leeward	– 0.4	– 1.1344	– 0.7	–

(B) Internal pressure coefficients : The godown shall be a building (cladding) having large opening. Let the large opening be in 40 m length and direction of wind be in the direction perpendicular to this length having the opening (*i.e.*, as shown in Fig. 25.29).

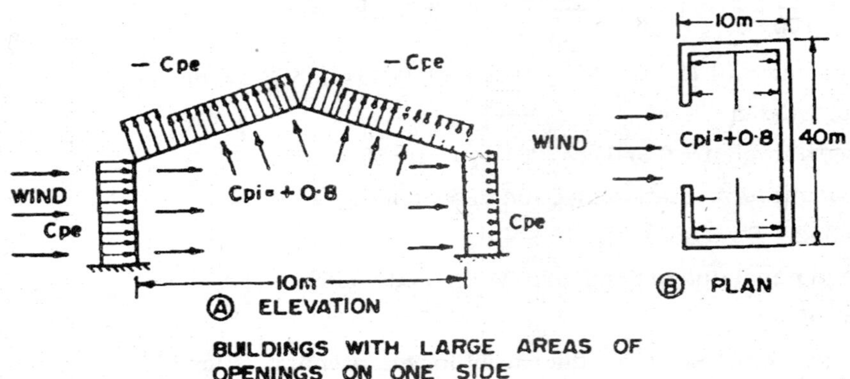

Fig. 25.29.

Total wind load C_{pe} = – 0.8 parallel to the ridge

For buildings having large. opening

$$\left(C_{pe} - C_{\pi}\right) = \left[-0.7 - (+0.8)\right] = -1.5$$

Total wind load (let the spacing of roof trusses be 4 m)

$$W_3 = (C_{pe} - C_{pi}) \,.\, p_d \,.\, A, \; W_3 = -1.5 \,.\, p_d \,.\, A$$

$$W_3 = -1.5 \times 1.04 \times (0.960 \times 4) = 5.99 \text{ kN}$$

$$\left(\frac{W3}{2}\right) = 2.995 \text{ kN}$$

The dead load, live load and wind load at each panel point of truss are shown in Fig. 25.1 (B), (C) and (D) respectively.

Note. For the design of various members, the stresses in members can be found for each loading either graphically or analytically. The maximum force in each member may be found for the combination of loads whichever is severe. The members may be designed for maximum force in them.

Example 25.10. At the apex joint of a Fink roof truss, four members AB, AC, AD and AE meet so that angles BAC = EAD = 30° and CAD = 60°. The length of members between the panel points are AB = AE : 170 kN compression and AB = AD : 100 kN tension. Design the members and detail the joint A. Length of each panel = 0.960 m.

Design :

Force in AB = AE = 170 kN compression.

AB and AE are rafter members in Fink roof truss. The rafter members in a truss are continuous struts. Panel length should not be more than 1-40 m.

Length of members AB = AE = 0.960 m

Effective length AB = AE = 0.85 x 0.960 m, (l = 0.816 m)

Factored load compression = (170 x 1.5) = 255 kN

Factored load, tension = (100 x 1.5) = 150 kN

Step 1A : Design of compression members

AB and AE are the members of principal rafters of the roof-truss, for which a pair of angle sections are used. Try 2 ISA 90 mm x 60 mm x 8 mm (2 ISA 90, 60 @ 0.890 kN/m. Long legs are kept back to back. 6 mm thick gusset plates shall be used for the connections. From ISI Handbook No. 1, **Geometrical properties** (Two angles back to back, Table 22)

Sectional area provided = 2274 mm²

Radius of gyration, r_{xx} = 28.4 mm

Radius of gyration, r_{yy} = 24.5 mm

(Distance between back to back of angles 6 mm)

Minimum radius of gyration = 24.5 mm

Step 2 A : Section classification, σ_y = 250 N/mm²

Yield stress ratio (Epsilon) $\varepsilon = \left(\frac{250}{250}\right)^{1/2} = 1.0$

Ratio $\left(\frac{90}{8}\right) = 11.25, \left(\frac{60}{8}\right) = 7.5 < 15.7$

The section is a semi-compact section.

Step 3A : Slenderness ratio

$$\lambda_y = \left(\frac{960}{24.5}\right) = 39.18 < 180$$

The slenderness ratio is within the permissible limits

Step 4A : Design compressive stress, σ_{cd}

From Table 9C (IS : 800-2007)

$\lambda = 30, \quad \sigma_{cd} = 211$ N/mm²

$\lambda = 40, \quad \sigma_{cd} = 198$ N/mm²

$$\lambda = 39.18, \quad \sigma_{cd} = \left[198 + \left(\frac{211-198}{10}\right) \times 0.82\right] \text{ N/mm}^2$$

$$= (198 + 1.066) = 199.066 \text{ N/mm}^2$$

Step 5A : Capacity of member in axial compression

$$P_{td} = \left(\frac{199.066 \times 2274}{1000}\right) = 452.67\ kN \quad > 255 \text{ kN}$$

Hence, the section is safe in axial compression

Tension capacity

$$P_{td} = \left(\frac{250 \times 2274}{1.15 \times 1000}\right) = 494.347\ kN$$

Hence, the section is safe in axial tension. Though, a smaller section may be selected, this section may be used to take care of handling and erection stresses.

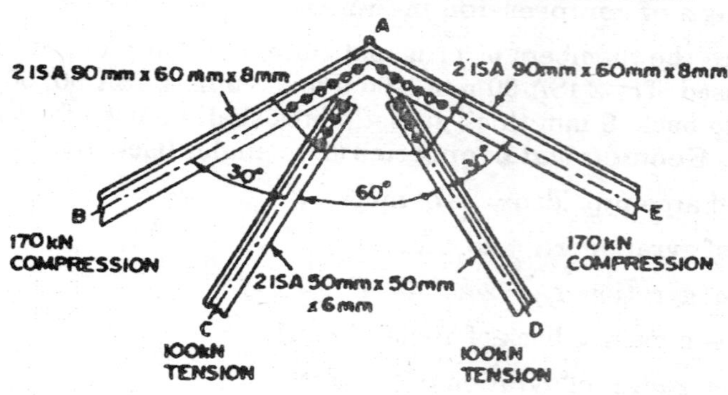

Fig. 25.30.

Step 1B : Design of tension member

Force in the member = 100 kN

Factored force in the member = 150 kN

Try 2 ISA 50 mm x 50 mm x 6 mm (2 ISA 50, 50, @ 0.045 kN/m). Angle sections are provided on both the sides of a gusset plate.

From ISI Handbook No. 1, Table 22 (Properties of two angles back to back)

Sectional area, A = 1136 mm^2

Radius of gyration, r_{xx} = 15.1 mm

Radius of gyration, r_{yy} = 23.1 mm

Step 2 B: Section classification

(b/t) = (50/6) = 8.33 < 8.4

The section is a plastic section

Step 3 B: Slenderness ratio = $\left(\frac{960}{15.1}\right) = 63.576$

Step 4B : Design compressive stress

From Table 9C, (IS : 800-2007)

For $\lambda = 60$, $\sigma_{cd} = 168$ N/mm^2

$\lambda = 70$, $\sigma_{cd} = 152$ N/mm^2

$$\lambda = 63.576, \qquad \sigma_{cd} = \left[168 - \left(\frac{168-152}{10}\right) \times 3.576\right] \text{ N/mm}^2$$

$$= (168 - 5.72) = 162.28 \text{ N/mm}^2$$

Step 5 B : Capacity of member in (axial compression)

$$P_{cd} = \left(\frac{162.28 \times 1136}{1000}\right) = 184.35 \; kN$$

Tension capacity

Two angle sections are used back to back. Load is applied axially. Load carrying capacity in tension.

$$P_{td} = \left(\frac{250}{1.15} \times \frac{1136}{1000}\right) = 246.956 \; kN > 150 \text{ kN}$$

Note : In case, a single angle section is used, then the load is applid through one-leg. The procedure for design shall be adopted for flexural torsional buckling strength of single angle through one leg. In that case Equivalent slenderness ratio λ_e is found.

$$\lambda_e = \left[k_1 + k_2 \, \lambda_{yy}^2 + k_3 \, . \, \lambda_{\phi\phi}^2\right]^{1/2}$$

Values of k_1, k_2 and k_3 are adopted from Table 12, IS : 800-2007

$$\lambda_{vv} = \frac{(L / r_{vv})}{\varepsilon \left(\dfrac{\pi^2 E}{250}\right)^{1/2}}, \quad and \; \lambda_{\phi\phi} = \frac{b_1 + b_2}{\varepsilon \left(\dfrac{\pi^2 E}{250}\right)^{1/2} . \, 2t}$$

Welded joint shall be done lengths of welds are found.

Design is shown in Fig. 25.30

Example 25.11: The sliding end of a roof truss rests on 450 mm brick wall through a concrete bearing pad. The maximum normal reaction on the bearing is 125 kN. The principal rafter is inclined at 30° to the main tie which one is horizontal. In case the panel length of the principal rafter is 1.38 m, design the rafter, the tie and the sliding joint. Also design the anchor. bolt for a pull of 7.50 kN.

Design :

Normal reaction on bearing = 125 kN

Force in the principal rafter (compression)

$$= (125 \text{ x } \sec 60°) = (125 \text{ x } 2) = 250 \text{ kN}$$

Factored force = (1.5 x 250) = 375 kN

Force in tie member (Tension)

$$= (250 \, . \, \cos 30°) = \left(250 \times \frac{\sqrt{3}}{2}\right) = 216.50 \; kN$$

Factored force = (1.5 x 216.50) = 324.76 kN

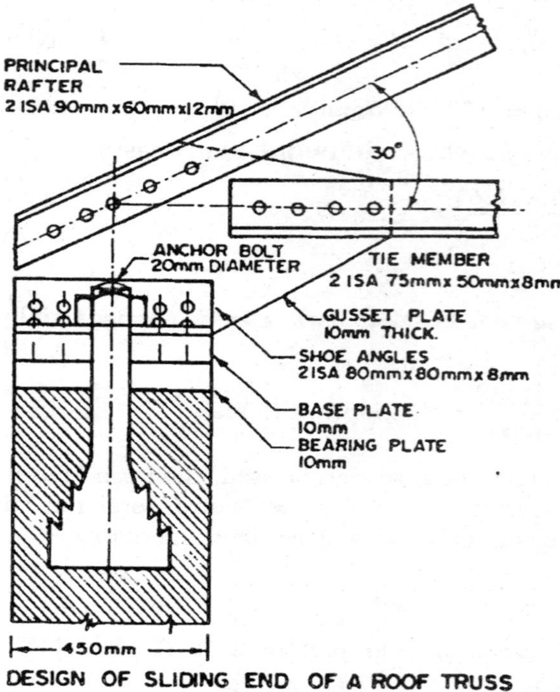

Fig. 25.31.

Step 1 : Design of compression member

Principal rafter is a compression member. It is a continuous strut

Panel length = 1.38 m

Effective length = (0.85 x 1.38) = 1.173 m

Principal rafter shall consist of 2 angle sections.

Let the trial section be 2 ISA 90 mm x 60 mm x 12 mm (2 ISA 90, 60 @ 0.130 kN/m) 10 mm thick gusset plate shall be used From ISI Handbook No. 1, from Table 22)

Properties of two angles back to back

Sectional area, A = 3314 mm²

Radius of gyration, r_{xx} = 27.9 mm

Radius of gyration, r_{yy} = 27.0 mm

(Distance back to back 10 mm)

Maximum slenderness ratio

$$\lambda = \left(\frac{1173 \times 1000}{27}\right) = 43.44$$

Step 2 : Design compression stress

From IS : 800 - 2007, Table 9(C),

For $\sigma_y = 250$ N/mm²

$\lambda = 40, \quad \sigma_{cd} = 198$ N/mm²

$\lambda = 50, \quad \sigma_{cd} = 183$ N/mm²

$$\lambda = 43.44, \quad \sigma_{cd} = \left[198 - \left(\frac{198-183}{10}\right) \times 3.44\right] \text{N/mm}^2$$

$$\sigma_{cd} = (198 - 5.16) = 192.84 \text{ N/mm}^2$$

Step 3 : Axial force capacity in compression

$$P_{cd} = \left(\frac{192.84 \times 3314}{1000}\right) = 639.07\ kN > 375 \text{ kN}$$

Provide 2 ISA 90 mm x 60 mm x 12 mm (2 ISA 90, 60 @ 0.131 kN/m) for the principal rafter.

A smaller size angle section may be tried.

Step 4 : Design of tension member

Factored force in main tie = 324.76 kN

Try 2 ISA 75 mm x 50 mm x 8 mm (2 ISA 75, 50, @ 0.074 kN/m)

From ISI Handbook No. 1, Table 22 (two angles back to back)

Properties of two angles back to back

Sectional area, A = 1876 mm²

Radius of gyration, r_{xx} = 14 mm

Radius of gyration, r_{yy} = (with distance back to back = Thickness of gusset plate 10 mm) = 38.3 mm

Step 5 : Section classification

$$\left(\frac{b}{t}\right) = \left(\frac{75}{8}\right) = 9.375 < 15.7$$

$$\left(\frac{b}{t}\right) = \left(\frac{50}{8}\right) = 6.25$$

Angle section is a semi-compact sectio

Step 6 : Axial force capacity in tension

$$P_{td} = \left(\frac{\sigma_y}{\gamma_{msy}} \times \text{Sectional area}\right)$$

$$= \left(\frac{250}{1.15} \times \frac{1876}{1000}\right) = 407.826 \text{ kN} > 324.76 \text{ kN}$$

It is safe in tension.

Provide 2 ISA 75 mm x 50 mm x 8 mm (2 ISA 75, 50, @ 0.074 kN/m) as shown in Fig. 25.31 kN

Step 7 : Design of joint

Welded joints are used to connect principal rafter and the main tie member to the gusset plate

Step 8 : Design of sliding joint

Maximum normal reaction at bearing = 125 kN

Its factored value = (1.5 x 125) = 187.5 kN

Welded joint is used to connect shoe angles with gusset plate and so also to connect show angles with the base plate.

1. **Bearing plate**

Factored value of normal reaction = 187.5 kN

Length of the base plate = 450 mm

Width of the bearing base plate

(80 + 80 + 10) = 170 mm

Bearing pressure on concrete bearing pad

$$= \left(\frac{187.5 \times 1000}{450 \times 170}\right) = 2.451\ N/mm^2$$

It is less than allowable bearing pressure

Consider one mm strip of base plate. Factored Bending moment about xx as shown in Fig. 25.24

$$M_{fd} = \left(\frac{2.451 \times (80-8)^2}{2}\right) = 6352.94\ N-mm$$

Design strength of base plate in bending

$$= \left(\frac{\sigma_y}{\gamma_{msy}} \times Z_p\right) = \left(\frac{250}{1.15} \times \frac{1}{4} \times 1 \times t^2\right) = 54.35\ t^2\ \text{N-mm}$$

$54.35\ t^2 = 6352.94$, $t = (116.894)^{1/2}$

$\therefore t = (10.8\ \text{mm} - 8\ \text{mm}) = 2.81\ \text{mm}$

Provide 6 mm thick base plate,

450 mm x 170 mm x 6 mm bearing plate below the base plate. An elliptical hole is kept on each side of shoe angles and the base plate, sothat the bearing plate may slide over bearing plate.

2. **Anchor bolt** Pull in anchor bolt = 7.50 kN

Factored value = (1.5 x 7.50) = 11.25 kN

Design stress in axial tension in anchor bolt = $\left(\frac{250}{1.15}\right) = 217.39\ N/mm^2$

Area required at the root of the thread = $\left(\frac{11.25 \times 1000}{217.39}\right) = 51.75\ mm^2$

Anchor bolts shall be provided having area more than 51.75 mm^2 on each side of show angles. A complete sliding joint is shown in Fig. 25.31.

PROBLEMS

25.1 It is proposed to construct a shed in Jodhpur. The average height of roof above ground level is 12 m. Determine the wind load for the following :

(i) When degree of permeability is small.

(ii) When shed is having normal permeability.

(iii) When shed is having large opening.

25.2 A gable with corrugated asbestos roofing is supported on steel trusses of span 9 m, rise 2.25 m, spaced at 3 m c/c. The dead load is 1.220 kN/m^2 normal to roof. Taking purlins one at each end of the slope and the others equally spaced, design an unequal angle purlin with long leg perpendicular to the slope of roof.

25.3 Design an angle iron purlin for a roof with the following data :

Span of truss = 8 m

Spacing of trusses = 4 m c/c

Pitch of truss = 1/4

Spacing of purlins along the slope of truss = 1.20 m c/c

Roof covering = Asbestos sheets

Basic wind pressure = 1.04 kN/m^2.

25.4 A 10 m x 50 m godown is to be coverted by roof trusses. Propose a suitable type of roof truss for the purpose and design the various members of the truss.

25.5 The forces in the members of the roof truss at joint A, AB compression 500 kN and AD tension 600 kN. Design the members AB and AD and the joint A as shown in Fig. 25.32. Sketch the joint A including shoe detail.

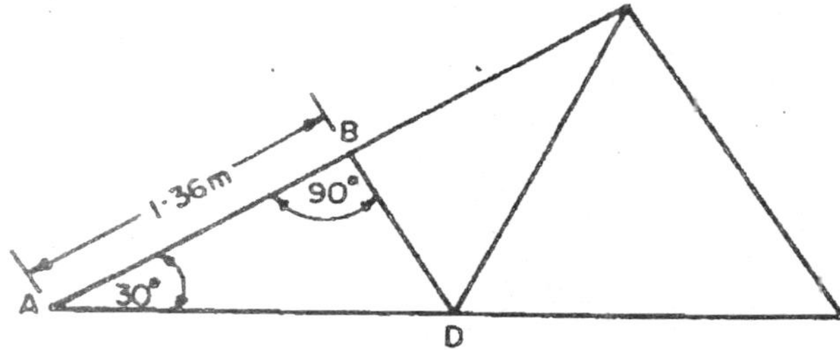

Fig. 25.32.

25.6 In steel roof truss, the vertical reaction at the left hand support is 500 kN. The principal rafter at this support is inclined at an angle of 30° to the horizontal. Design the main tie, the principal rafter, and the riveted joint. The panel length of the principal rafter may be assumed to be 1.20 m. Draw a fully dimensioned neat sketch.

25.7 For a steel roof with a asbestos cement sheet covering design an I-section purlin from the following data :

Spacing of trusses = 4 m

Spacing of purlins of principal rafter = 1.32 m

Inclination of principal rafter to the main tie = 30°

Wind pressure of roof = 1.000 kN/m^2.

REFERENCES

1. IS: 800-2007 General Construction in Steel-Code of Practice (Third revision) published by Bureau of Indian Standards Manak Bhawan, 9, Bahadur Shah Zafar Marg, New Delhi 110 062.
2. BS: 5950-2000. British Standard, Structural use of steel work Building code of practice for design of simple and continuous construction: Hot rolled section. Britissh Standard Institution.

26

DESIGN OF ROUND TUBULAR TRUSSES AND CONNECTIONS

26.1 INTRODUCTION

The round tubular and hollow sections are used as structural components since long in plane and space frames. Tubular (round) sections have equal moment of inertia and their corresponding radius of gyration about all the axes. Welding of the members may be done by using complete penetration groove weld, fillet weld or any combination of welds. The welding is done to ensure that the yielding and re-distribution may occur in the base metal. This permits the weld to develop the required deign load before reaching the ultimate breaking strength. A large number of tubular structures have been constructed in the past. The tubular sections have been used for columns, compression members, tension members, roof trusses, mill buildings, air plane hangers, cross bracings and beams. The tubular sections are also used for single storey industrial buildings, ware-houses, and shopping centres where long span column free areas are required. The tubular sections are used to advantages in structures designed for material handling equipments (e.g. a bridge, derrick and tower cranes) where weight savings may be very substantial economic consideration. The tubular sections are effectively used in large space frame, lattice structures for arenas, stadium exhibition halls where appearance as well as weight become an important design consideration. The masts and transmission towers are other examples where tubular sections are utilised effectively. In the past, the use of tube was hampered because of connection details. A major contribution to the solution of this problem has undoubtedly been the development of fully automatic oxyacetylence tube cutting machines, which not only cut tubes to fit flat surfaces but also cut them to fit cylindrical surfaces such as tube as well. The machine cuts the tubes to the correct profile with a bevelled edge to simplify the welding process at the joint.

The round tubular sections have as much as 30 to 40 per cent less surface area than that of an equivalent rolled steel shape. Therefore, the cost of maintenance, cost of painting, fire proofing and/or other protective coatings reduce considerably. The moisture and dirt do not collect on the smooth external surface of the tubes. Therefore, the possibility of corrosion also reduces. The ends of tubes are sealed. As a result of this, the interior surface is not subjected to corrosion. The interior surfaces do not need any protective treatment.

The use of round tubular members is becoming more generally adopted for structures. The tubes are of special interest to the architects from an effective viewpoint and to the engineers from a structural effectiveness viewpoint.

The tubular sections have more torsional resistance than the other sections of the equal weights. The tube sections have a higher frequency vibrations under dynamic loading than the other sections including the solid round one. The round tubular sections offer less resista-nce to wind. Some of the codes allow a one-third reduction in wind load compared to equiva-lent projected area. IS : 875-1984 recommends a shape factor of 0.7 for round tubular sections.

The tubular sections are most efficient sections which are adoptable to many different situation. The tubular sections cannot be surprised in its efficiency by other sections.

26.2 ROUND TUBULAR SECTIONS

The round tubular sections are manufactured from steel which when analysed in accordance with the methods in IS : 228, Methods of Chemical Analysis of Pig Iron, Cast Iron and Plain Carbon and Low Alloy Steels shall show not more than 0.06 percent sulphur, not more than 0.06 per cent phosphorus. The plain carbonsteel tubes as specified in IS : 1161-1963 are used for the structural purpose. The sizes and properties of steeltubes for structural purpose are given in Tables 26.1(A) and (B) as per IS : 1161-1963. These tubes are of following types :

1. HFW : Hot finishes welded. 2. HFS : Hot finished seamless.
3. ERW : Electric resistance or induction butt welded.

The hot rolled carbon steel structural tubes are made in round shapes from a continuous welding or seamless process. In the continuous weld process coils of steel called skelp are welded and to end to form a continuous band of steel and passed through furnaces. As the skelp exists from the furnace, it is formed into a round pipe and welded by pressure. The pipe is then placed into sketch reducing mill, where, it is brought to the desired diameter and wall thickness.

In seamless tubes, a solid round bar of redetermined size is treated and them pierced by a mandrel while rotating at high speed. The tube is then placed to other rolling operations which bring it to the proper diameter and wall thickness.

The structural members consisting of round steel tubes are designed as per specification given in IS : 806-1957 (Code of Practice for Use of Steel Tubes in General Building Construction).

TABLE 26.1 (A). SIZES AND PROPERTIES OF STEEL TUBES FOR STRUCTURAL PURPOSES

Nominal Bore	Outside diameter	Class	Thickness	Weight	Area of X-section	Internal volume	Surface		M.I.	Z	r	r^2
							Ext.	Int.				
mm	mm		mm	N/m	mm²	mm³/m	mm²/m	mm²/m	m⁴	cm³	cm	cm²
1	2	3	4	5	6	7	8	9	10	11	12	13
15	21.3	Light	2.0	6.2	121	23500	66900	54300	0.57	0.54	0.69	0.47
		Medium	2.6	12.1	153	20400		50600	0.68	0.64	0.67	0.45
		Heavy	3.2	14.4	182	17400		46800	0.77	0.72	0.65	0.42
20	26.9	Light	2.3	14.1	178	39100	84500	70100	1.36	1.36	0.87	0.76
		Medium	2.6	15.7	198	37000		68200	1.48	1.48	0.86	0.75

		Heavy	3.2	18.9	238	32000		64400	1.70	1.70	0.85	0.71
25	33.7	Light	2.6	20.1	254	63800	105900	89500	3.00	1.84	1.10	1.22
		Medium	3.2	24.2	307	58500		85800	3.60	2.14	1.08	1.18
		Heavy	4.0	29.5	373	51900		80700	4.19	2.49	1.06	1.12
32	42.4	Light	2.6	25.7	325	108700	133200	116900	6.43	3.05	1.41	1.99
		Medium	3.2	31.1	394	101800		113000	8.99	4.24	1.36	1.93
		Heavy	4.0	38.1	483	91900		108100	8.99	4.24	1.36	1.86
40	40.3	Light	2.9	32.7	414	141900	151700	133500	10.70	4.43	1.61	2.59
		Medium	3.2	35.9	453	137900		131500	11.59	4.80	1.60	2.56
		Heavy	4.0	44.1	550	127600		126300	13.77	5.70	1.57	2.47
50	60.3	Light (1)	2.9	41.4	523	233300	189400	171200	21.59	7.16	2.03	4.13
		Light (2)	3.2	45.4	574	228200		169300	23.47	7.18	2.02	4.09
		Medium	3.6	50.7	641	221500		166800	25.37	8.58	2.01	4.03
		Heavy	4.5	61.7	788	206700		161200	30.90	10.20	1.98	3.92
65	76.1	Light	3.2	58.0	733	381600	239100	219000	48.78	12.8	2.58	6.66
		Medium	3.6	64.9	820	372800		216500	54.01	14.2	2.57	6.59
		Heavy	4.5	79.2	1010	353600		210800	65.12	17.1	2.54	9.48
80	88.9	Light	3.2	68.1	862	534600	279300	259200	79.21	17.8	3.03	9.19
		Medium	4.0	84.3	1070	514000		254200	96.34	21.7	3.00	8.03
		Heavy	5.0	103.0	1320	488900		247900	116.04	26.2	2.97	8.83

(where M.I. is moment of inertia, Z is modulus of section, r is the radius of gyration and r^2 square of radius of gyration)

TABLE 26.1 (B). SIZES AND PROPERTIES OF STEEL TUBES FOR STRUCTURAL PURPOSES

Nominal Bore	Outside diameter	Class	Thickness	Weight	Area of X-section	Internal volume	Surface		M.I.	Z	r	r^2
							Ext.	Int.				
mm	mm		mm	N/m	mm^2	mm^3/m	mm^2/m	mm^2/m	m^4	cm^3	cm	cm^2
1	2	3	4	5	6	7	8	9	10	11	12	13
90	101.6	Light	3.6	87.6	1110	699900	319200	296600	133.2	26.2	3.47	12.02
		Medium	4.2	97.0	1230	688100		294100	146.3	28.8	3.45	11.98
		Heavy	5.0	119.0	1520	659000		287800	177.5	34.9	3.42	11.70
100	114.3	Light	3.6	99	1250	900900	359100	336500	192.0	33.6	3.92	15.33
		Medium	4.5	121	1550	870900		330800	234.3	41.0	3.89	15.10
		Heavy	5.4	145	1850	841300		335200	274.5	48.0	3.85	14.86
110	127.0	Light	4.5	122	1730	1093600	399000	370700	325.3	51.2	4.33	18.78
		Medium	5.0	150	1920	1075100		367600	357.1	56.2	4.32	18.64
		Heavy	5.4	162	2060	1060500		865100	382.0	60.2	4.30	18.52
125	139.7	Light	4.5	149	1910	1341700	438900	410600	437.2	62.6	4.78	22.87
		Medium	5.0	166	2120	1321200		407500	480.5	68.8	4.77	22.71
		Heavy	5.4	179	2280	1305000		405000	514.5	73.7	4.75	22.58
135	152.4	Light	4.5	164	2090	1615100	478800	450500	572.2	75.1	5.23	27.37

		Medium	5.0	182	2320	1592500		447400	629.5	82.6	5.21	27.19
		Heavy	5.4	195	2490	1574800		447900	674.5	88.5	5.20	27.05
150	165.3	Light	4.5	178	2270	1913800	518700	490400	732.6	88.7	5.78	32.27
		Medium	5.0	197	2510	1889400		487300	806.6	97.7	5.66	32.07
		Heavy	5.4	212	2710	1869900		484700	864.7	105.0	5.65	31.92
150	168.3	Light	4.5	181	2320	1993100	528700	500500	777.2	92.4	5.79	33.56
		Medium	5.0	201	2570	1968100		497300	855.8	102.0	5.78	33.36
		Heavy (1)	5.4	217	2760	1948300		494800	917.7	109.0	5.76	33.21
		Heavy (2)	6.3	253	3210	1904000		489100	1053.4	125.0	5.73	32.85
175	193.7	Light	5.0	233	2960	2650400	608500	577100	1320	136	6.67	44.54
		Medium	5.4	250	3190	2627400		574600	1417	146	6.66	44.36
		Heavy	5.9	273	3480	2598700		571600	1536	159	6.64	44.13
200	219.1	Light	5.0	264	3360	3434000	688300	650000	1918	176	7.57	57.33
		Medium	5.6	294	3760	3894700		653100	2142	195	7.55	57.02
		Heavy	5.9	310	3950	3375100		651300	2247	205	7.54	56.86
225	244.5	Heavy	5.9	342	4420	4252900	768100	731100	3149	258	8.44	71.21

The members subjected to both bending and axial stresses shall be so proportioned that the quantity $[(f_a'/f_a) + (f_b'/f_b)]$ does not exceed unity.

where, f_a' = Calculated axial stress

f_a' = Axial load divided by appropriate area of member

f_a = Permissible stress in member for the axial load

f_b' = Calculated bending stress in the extreme fibre

f_b = Permissible bending stress in the extreme fibre

When the bending occurs about both the axes of the members, then f_b shall be taken as

$$f_b = [(f_{bx})^2 + (f_{by})^2]^{1/2} \tag{26.1}$$

where f_{bx} and f_{by} are the two calculated unit fibre stresses. The equivalent stress, σ_e due to co-existing bending shear stresses shall not exceed the value given as per IS : 806-1957. (Note : These values have been converted in S.I. units).

The equivalent stress σ_b is obtained from the following formula

$$\sigma_e = \left[\sigma_b^2 + 3\,\tau_s^2\right]^{1/2} \tag{26.2}$$

where, σ_b is the calculated bending stresses at the point under consideration and τ_s is the calculated actual shear stress at the point under consideration.

26.3 TUBE COLUMNS AND COMPRESSION MEMBERS

The round tubular sections provide the most efficient cross-sectional shape for the columns and compression members having lateral restraint in all directions normal to the axis of member. The diameter of such a member should be as large as possible with the additional requirement that (d/t) (see Fig 26.1), should be small enough to assure that pressure failure by local buckling will not occur.

The local buckling strength of very short perfect tubes depends primarily on (L/d) ratio. In case of the medium tubes, the local buckling strength is primarily a function of (d/t) ratio. For long tubes, the local buckling failure should be avoided so that the general column buckling is the primary problem. The medium length tubes are most frequently used in practice and the theoretical local buckling stress based on small deflection theory for solution of **Donell's equation**.

$$\sigma_b = C.E./(R.t) \tag{26.3}$$

where, $C = [3(1-v^2)]^{-1/2}$ (0.6 when v = 0.3) and (26.4)

R and t are defined in Fig. 26.1.

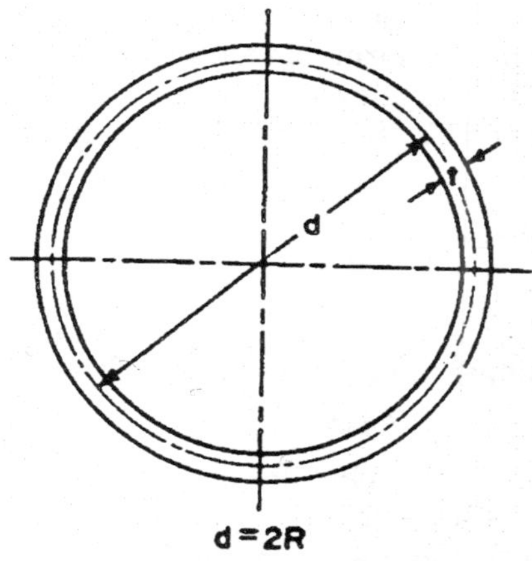

Fig. 26.1.

The medium length tube solution defined by Eq. 26.3 is applied for value of Z greater than 2.85, whereas Z is given by Batdorf

$$Z = \frac{L^2}{R.t}\left[\left(1 - v^2\right)\right]^{1/2} \tag{26.5}$$

where, Z is the limiting constant for medium length tubular columns.

The experiments show that as (d/t) increases, failure usually occurs at a stress increasing lower than that obtained from Eq. 26.3. This is because residual stresses and/or small irregularities in shape have an extremely detrimental effect on the compressive strength of round tubular section. The value of constant C depends to some degree on the tube manufacturing process and should be based on the recommendations made by the particular manufacturer.

In the design of steel tubular columns and compression members the attention must be paid to two points : **crinkling and heat treatment**. The yield strength of mild steel may be considerably reduced by any heat treatment which it receives such as welding. On this account, the precautions should be taken to prevent heat treatment or the strength must be taken as that of the annealed material.

For the round tubular columns and compression members, the effective length of member is adopted as given in Table 6.2, chapter 6. The limits specified for maximum slenderness ratio in Table 6.2 are also applicable for the tube members. In addition to this, the members should satisfy the minimum thickness requirement. The steel tubes used for construction exposed to the weather shall be not less than 4 mm thick and for construction not so exposed shall be not less than 3.2 mm thick when the structures are not readily accessible for maintenance, the minimum thickness shall be 5 mm. Not withstanding the above, the thickness of tube subjected to compressive stress shall not be less than that given by the formula

$$t = 0.865\ [D]^{1/3} \tag{26.6}$$

where, D is the outside diameter of tube in mm. The ends of the members are sealed.

26.4 CRINKLING OF TUBULAR SECTIONS

When a steel tubular member is subjected to compression, then, the tube may crinkle (i.e., the walls of the tubes may cave in and form folds after the manner of concertina). These folds may be circular, oval or polygonal and they may occur after or before the longitudinal stress reaches the yield point. Professor R.V. Southwell investigated the matter and deduced the following formula analytically

$$p = E.\frac{t}{R}\left[\frac{m^2}{3\left(m^2-1\right)}\right]^{1/2} \tag{26.7}$$

where, p = Stress causing the collapse, and t = Thickness of the tube

R = Mean radius of tube and 1/m = Poisson's ratio.

The stress which will produce the lobbed failures is $[(K^2 - 1)/(K^2+1)]$ times that given by Eq. 26.7, where K is the number of lobes formed. This formula applies only if p is less than the yield stress. Thus, for a mild steel tube having a yield point of 315.0 N/mm^2 the formula will only apply when (t/R) ratio is less than about (1/400).

26.5 TUBE TENSION MEMBERS

The effectiveness of a member subjected to axial tension depends on its cross-sectional area, grade of steel and its method of connection. Therefore, a round tube or any rolled steel shape with the same net cross-sectional area and the same material would have the same equivalent resistant to the force. The tubular section do not have any advantage as tension member. The tubular sections have higher cost of production than other rolled steel sections.

26.6 TUBULAR ROOF TRUSSES

The design of roof trusses has been discussed in Chapter 26. The roof trusses are subjected to external loads. The external loads. The external loads, in general, are applied at the joints. Therefore, the different members of the roof trusses are subjected to axial compressive and tension forces only. The tubular compression members have been discussed in Art. 26.4. The tubular tension members have been discussed in Art. 26.6. The members of tubular trusses are generally joined by welding. A tubular roof truss is shown in Fig. 26.2.

In order to facilitate the erection of the trusses, the complete truss is splitted in different parts (viz., AEGB, DFHC and JK). These parts are the assembled after the erection. These parts are then connected by bolts and nuts. The arrangements for bolt connections made are as shown in Fig. 26.2 at J, K and L. The rectangular pieces of plates are welded to the projected end of the parts of the truss. The bolts are placed in the holes in both the plates and then, the nuts are tightened. The different parts of the truss are then assembled and a complete truss is built. The rectangular pieces with holes are also welded with the principal rafters at the joints and these pieces are kept projected as shown in Fig. 26.2 at E, K, G and H. These projected pieces are used to provide tubular purlins. The tubular purlins are provided with holes at the ends with internal threads. The tubular purlins are placed from one side. The studs are used for the connections of the purlins. The studs are placed through the holes of projected plates and tightened in the internally threaded ends of the purlins. The purlins are placed and fixed with the other ends of the studs. Similar arrangements are also made for tubular bracings with the bottom tie of the truss as shown in Fig. 26.2 at B and C.

In addition to the type of truss shown in Fig. 26.2, the tubular trusses of various other types are also made with suitable arrangements for the erection.

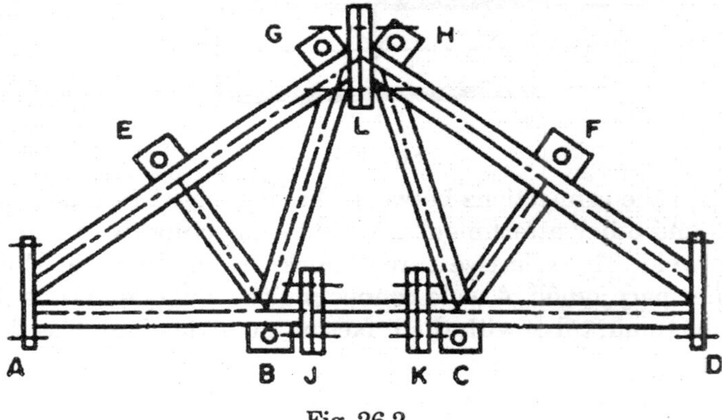

Fig. 26.2.

26.7 JOINTS IN TUBULAR TRUSSES

The welding is generally adopted for connections in tubular steel construction. The welded connections provide rigid joints. The rigid joints results in saving in materials and greater overall economy. The actual condition of rigidity should be taken into consideration while designing these types of joints.

The welded connections in the tubular structures may however, be designed on a simple design basis comparable with IS : 800. In such cases, the secondary stresses may be neglected in the design of roof trusses except where the axes of the members do not meet at a point where there is such eccentricity, the effects of the eccentricity should also be considered.

In joints in compression members, the ends of the members are faced for complete bearing over their whole area. The welding and joining materials are kept sufficient to retain the members accurately in place and to resist all forces other than direct compression including those arising during transit, unloading and erection.

Where such members are not faced for complete bearing, the welding and joining material shall be sufficient to transmit all the forces to which the joint is subjected. Wherever possible the joints shall be proportioned and arranged so that the gravity axes of the members and the joints are in line, so as to avoid eccentricity.

Initially, the tube members in tubular trusses were welded to the gusset plates as shown in Fig. 26.3.

The welded joint with the gusset plates in the tubular trusses were found to be unsatisfactory for severe load conditions. The difficulty with this type of connection arises from the concentrated stress flow between the tube walls and the gusset plate even though the gusset plate is capable of transferring the forces effectively.

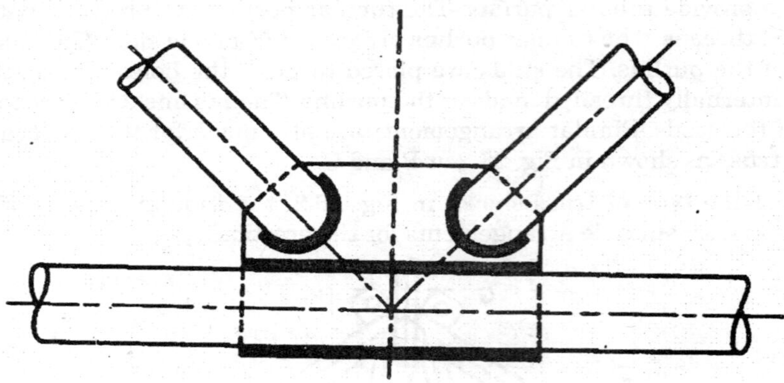

Fig. 26.3.

As far as possible, tube connections between the members are made directly tube to tube without gusset plates and other attachments. A weld connecting two tubes end to end shall be full penetration butt weld. The effective throat thickness of the weld shall be taken as the thickness of the thinner part joined. A weld connecting the end of one tube (branch tube) to the surface of another tube (main tube) with their axes at an angle of not less than 30° shall be one of the following :

I. A butt weld throughout. II. A fillet weld throughout.

III. A fillet-butt weld, the weld being a fillet weld in one part and a butt weld in another with a continuous change from the one form to the other in the intervening portions.

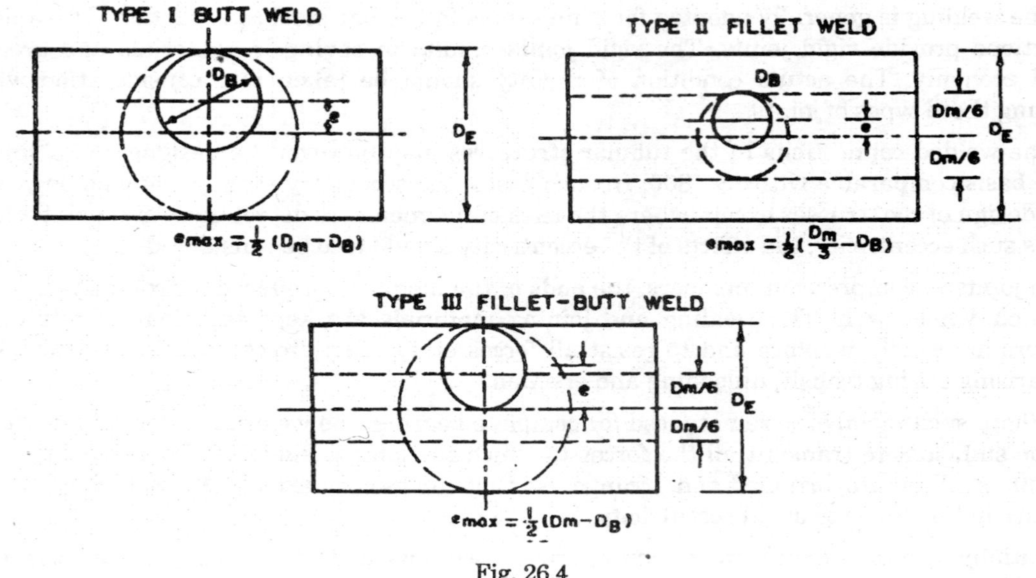

Fig. 26.4

Type I may be used whatever the ratio of the diameter of the tubes joined provided complete penetration is secured either by the use of the backing material, or by deposing a sealing run of metal on the back of joint or by some special method of welding. Type I weld is shown in Fig. 26.4. When Type I is not employed then, Type II should be used whether the diameter of branch tube is less than one third of the diameter of the main tube. Type II weld is shown in Fig. 26.4. When the diameter of branch tube is equal or greater than one-third of the diameter of the main tube, then, the Type III weld as shown in Fig. 26.4 is used.

For the purpose of stress calculation, the throat thickness of the butt weld portion shall be taken as the thickness of the thinner part joined. The throat thickness of the fillet weld and filletbutt weld is taken as the minimum effective throat thickness of the fillet or fillet butt weld.

A weld connecting the end of one tube to the surface of another with the axes of the tubes intersecting at an angle less than 30° shall be permitted only if ade-quate efficiency of the junction has been shown.

The length of curve of intersection of the tube with another may be determined from the following formula :

$$L = \left(a + b + 3\sqrt{a^2 + b^2}\right), \text{ where, } a = [d/2], \text{ cosec } \theta \quad (26.8)$$

For intersection with a tube : $b = \left[\dfrac{d}{3} \cdot \dfrac{3 - (d/D)^2}{2 - (d/D)^2}\right]$ (26.9)

Alternatively, $b = (D/4).\ \phi$ (26.10)

where, $\sin(\phi/2) = (d/D)$ (26.11)

where ϕ is measured in radius

For intersection with flat plates : $b = (d/2)$ (26.12)

where θ = Angle between branch and main tubes

d = Outside diameter of branch tube

D = Outside diameter of main tube.

The above parameters of tubes are shown in Fig. 26.5.

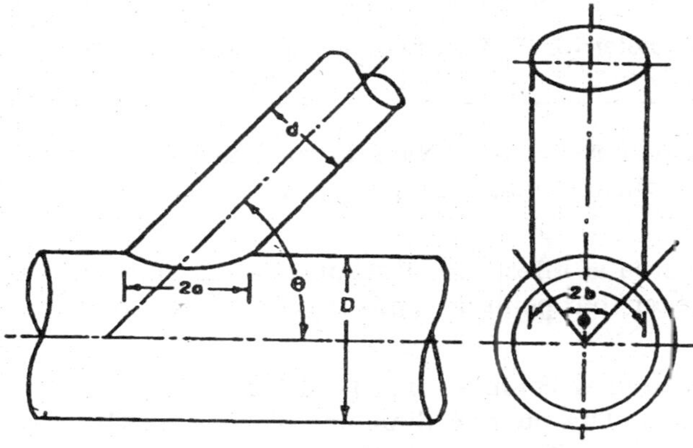

Fig. 26.5

26.8 DESIGN OF TUBULAR BEAMS

The tensile and compressive stresses in bending in the extreme fibres of tubes shall nor exceed the values given in Table 26.4. The maximum shear stress in tubes in flexure calculated by diving the total shear by an area equal to half the net cross- sectional area of the tube shall not exceed the values given in Table 26.5.

When the tubular steel beam rests on abutment or other supporting member, it shall be provided with a shoe adequate to transmit the load to the abutment and to stiffen the end of tube.

Where a concentrated load is applied to a tubular member transverse to its length or the effect of load concentration is given by the intersection of triangular truss member, consideration shall be given to the local stresses set up and the method of application of the load and stiffening shall be provided as necessary to prevent the local stresses from being excessive. The increase in the intensity of local bending stresses caused by concentrated loads particularly marked if either the diameter of connected member or the connected length of a gusset or the like is small in relation to the diameter of the tubular member to which it is connected.

The deflection of a member shall not be such as to impair the strength, efficiency or appearance of the structure or lead to damage to fittings and finishings. Generally, the maximum deflection should not exceed 1/325 of the span for simply supported beams. This requirement may be deemed to be satisfied if the bending stress in compression or in tension does not exceed 3153 (D/l) N/mm², where D is the outer diameter of the tube in mm and l is the effective length of the beam in mm.

26.9 DESIGN OF TUBULAR PURLINS

The requirement mentioned in Art. 26.9 regarding limiting deflection may be waved in the design of simple tubular purlins provided the following requirements are satisfied:

1.	Name of end fixing	Simply supported	Effectively continuous
2.	Maximum value of section modulus (mm³)		
	(a) IS : 1161 grade St. 35 (Yst. 22)	(WL/11200) x 10^3	(WL/16800) x 10^3
	(b) IS : 1161 grade St. 40 (Yst. 25)	(WL/13230) x 10^3	(WL/19840) x 10^3
3.	Minimum outside diameter IS: 1161 grade St. 35 or St. 40 (mm) Yst. 32)	(L/40) x (10)	(L/70) x (10)

where, W is the total distributed load in N on the purlins arising from dead load and snow load excluding wind, and L is the distance in mm between the centres of the steel principals or other supports.

A purlins is considered as effectively continuous at any intermediate point of support, if it is actually continuous over that point or it has there a joint able to provide a fixing moment of not less than WL/2.

Example 26.1: A tubular column consists of IS: 1161 grade St 35 steel. The column is hinged at both the ends. The outside diameter of tube is 219.1 mm. With internal bore diameter 200 mm heavy class. The weight of one m length of tube is 0.310 kN. The length of column is 4.5 m. Determine safe load carrying capacity of the column.

Solution : Column is hinged at both the ends.

Length of column = 4.5 m.

Effective length, l_e = (1 x 4.5) = 4.5 m

Outside diameter of tube section = 219.1 mm @ 0.310 kN/m.

Tube section belongs to heavy class.

Internal bone diameter = 200 mm

Radius of gyration of tube section, r = 75.4 mm

Cross-sectional area (from IS : 806 – 1957), A = 3950 mm^2

Step 1 : Slenderness ratio

Slenderness ratio of the column

$$\lambda = \left(\frac{4.5 \times 1000}{75.4}\right) = 59.68 \quad < 180$$

Step 2 : Design compression stress, σ_{cd}

From IS : 800-2007, Table 9 (C) for σ_y = 250 N/mm^2

For $\lambda = 50$, $\sigma_{cd} = 183$ N/mm^2

$\lambda = 60$, $\sigma_{cd} = 168$ N/mm^2

$$\text{For } \lambda = 59.68, \quad \sigma_{cd} = \left(168 + \left(\frac{168 - 168}{10}\right) \times 0.32\right) \text{ N/mm}^2$$

$$= (168 + 0.448) = 168.448 \text{ N/mm}^2$$

Step 3 : Design strength of column

$$P_{cd} = \left(\frac{168.448}{1.15} \times \frac{3950}{1000}\right) = 578.58\, kN$$

Safety load carrying capacity at (serivce load)

$$= \left(\frac{578.58}{1.5}\right) = 385.72\ kN$$

Example 26.2 : The principal rafter in a round hollow tubular truss carries a maximum force of 184 kN. A tension member meeting at right angles to the principal rafter carries a force of 200 kN. Design the member using IS : 1161 grade st. 35 steel for the tube. The panel length along the principal rafter is 1.28 m.

Design : Principal rafter consists of a round hollow tube section

force in compression = 184 kN

Force in tension member = 200 kN

Factored force in compression member

= (1.5 x 184) = 276 kN

Factored force in tension member

= (1.5 x 200) = 300 kN

Panel length along the rafter = 1.28 m

Effective length = (0.85 x 1.28) = 1.088 m

Step 1 : Design of compression member

Yield stress, $\sigma_y = 250\ N/mm^2$

This needs a trial section, for which, followings may be kept in view

(i) Effective length of the compression member

For the member of principal rafter,

Effective panel length used to be small < 1.40.

Looking the section table (As per IS : 806 - 1957), it is natural that a heavy section shall be required for heavy force and vice-versa. Therefore, the value of radius of gyration shall be more. Likely value of slenderness shall be small, and the corresponding value of design stress shall be somewhat about (250/1.15 = 217.39 N/mm²)

The product of design stress, σ_{cd} and the cross-sectional area of trial section will suggest axial force capacity

Slenderness ratio	Design compressive stress, N/mm²
0	(250/1.15) = 217.39 N/mm²
10	(227/1.15) = 197.39 N/mm²
20	(224/1.15) = 194.78 N/mm²
30	(211/1.15) = 183.478 N/mm²
40	(198/1.15) = 172.174 N/mm²

Step 2 : Trial section

Try tube section with outer diameter 114.3 mm (class-heavy), with nominal bone diameter 100 mm

Cross-sectional area, A = 1850 mm²

Radius of gyration, r = 38.5 mm

$$\text{Slenderness ratio, } \lambda = \left(\frac{1.088 \times 1000}{38.5}\right) = 28.26$$

Step 3 : Design compressive stress, σ_{cd}

For $\lambda = 20, \quad \sigma_{cd} = 194.78\ N/mm^2$

$\lambda = 30, \quad \sigma_{cd} = 183.478\ N/mm^2$

$$\lambda = 28.26, \quad \sigma_{cd} = \left(183.478 + \left(\frac{194.78 - 183.478}{10}\right) \times 1.74\right) N/mm^2$$

$$= [(183.478 + 1.97)] = 185.44\ N/mm^2$$

$$P_{cd} = \left(\frac{185.44 \times 1850}{1000}\right) = 343.07\ kN \quad > 276\ kN$$

$$\text{Force while in service} = \left(\frac{343.07}{1.5}\right) = 228.7\ kN \quad > 184\ kN$$

Hence, satisfactory.

Note : Still, a smaller section may be tried.

Thickness of tube 5.4 mm is more than 4 mm (minimum thickness required). Provide tube of outside diameter 114.3 mm (class-heavy), @ 0.145 kN/m

Step 4 : Design of tension member

Design stress for axial tension

$$\sigma_{td} = \left(\frac{\sigma_y}{Y_{msy}}\right) = \left(\frac{250}{1.15}\right) = 217.39\,\text{N/mm}^2$$

Factored force in tension = 200 kN

$$\text{Area required for the tube} = \left(\frac{200 \times 1000}{217.39}\right) = 920\ \text{mm}^2$$

Provide tube with outer diameter 76.1 mm (class-heavy) with nominal bone diameter 80 mm.

Cross-sectional area provided = 1010 mm^2

> 920 mm^2

Thickness of tube 4.5 mm is more than minimum thickness 4 mm required for exposed situation.

Example 26.3 : Design the welded joint for the tubular truss members in Example 26.2

Step 1 : Length of curve of intersection

Principal rafter is the main tube member. From Example 26.2, the outside diameter, D is 114.3 mm (class-heavy)

Step 2 : Parameter a and b

The outside diameter of tension member (it is a branch tube member)

d = 76.1 mm (Class-heavy)

This tension member is meeting the principal rafter at right angles. Therefore, the angle between the main tube and branch tube, $\theta = 90°$

Therefore, (d/D) = (76.1/114.3) = 0.6658

$$a = \frac{d}{2}.\ \text{cosec}(\theta) = \left(\frac{76.1}{2} \times \frac{1}{1}\right) = 38.05\ \text{mm}$$

From Eq. 26.9,

$$b = \left[\frac{d}{3} \times \left(\frac{3 - \left(\frac{d}{D}\right)^2}{2 - (d/D)^2}\right)\right] \text{mm}$$

$$= \left[\frac{76.1}{3} \times \left(\frac{3 - 0.6658^2}{2 - 0.6658^2}\right)\right] = 41.82\ \text{mm}$$

Step 3 : Length of intersection curve

From Eq. 26.8, the length of intersection

$$L = a + b + 3\ (a^2 + b^2)^{1/2}$$

$$= 38.05 + 41.82 + 3\ (38.05^2 + 41.82^2)^{1/2} = 249.49 \text{ mm}$$

Step 4 : Design stress in weld in shear

$$\sigma_{dw} = \left(\frac{\sigma_u}{\sqrt{3}} \times \frac{1}{\gamma_{msu}}\right)$$

$$= \left(\frac{410}{1.732} \times \frac{1}{1.25}\right) = 189.376 \text{ N/mm}^2$$

Step 5 : Size of weld

Let t be the throat thickness of the fillet weld. Force to be transmitted 200 kN

Length of weld x Throat thickness x σ_{dw} = 200 kN

$$249.49 \times t \times 189.376 = 200 \times 1000$$

$$t = \left(\frac{200 \times 1000}{249.49 \times 189.376}\right) = 4.233 \text{ mm}$$

Size of weld required

$$0.7\ s = 4.233$$

$$s = 6.047 \text{ mm}$$

Provide 8 mm fillet weld

PROBLEMS

26.1 Two members meeting at a joint at 30° inclination in a tubular stress carry 120 kN compression and 36 kN tension. The length of compression member from centre to centre of joint is 2 m. The members are made of IS : 1161 grade St. 35 steel. Design the members.

26.2 Design the welded joint for the tubular truss in Problem 26.1.

APPENDIX-A

SHEAR CENTRE FOR A THIN-WALLED BEAM CROSS-SECTIONS

A.1 INTRODUCTION

A **beam** is defined as a structural member subjected to transverse load. The plane of transverse load i.e., transverse bending loads is parallel to the plane of symmetry of the cross-section and it passes through the **bending axis** of the beam and the **shear centre,** so that simple bending of beam occurs. The **bending axis** of a beam is the axis parallel to longitudinal axis through which the transverse bending loads pass in order that the bending of the beam shall not be accompanied by twisting of the beam. In simple bending, the transverse loads produce bending moments and shear forces in the beam at all sections of the beam. The **shear centre** (or **centre of twist**) for any transverse section of the beam is defined as the point of intersection of the bending axis and the plane of the, transverse section. In case a bar is subjected to a transverse twisting couple, then, there is a point in any transverse cross-section which will not be displaced, this point is known as the **centre of twist**. It is not necessarily the same point as the shear centre.

I-sections symmetrical about both the principal axes (i.e., axes of symmetry), Z-sections are the examples of beam sections, which are symmetrical about both the axes.

The shear centre is the location in a cross-section of a beam where, the torsion does not occur when flexural shears act in planes passing through that location.

It is complicate to locate the shear centre for all but simple shapes. Fortunately, most cross-sections have the shear centre at a convenient location. For the cross-sections of a beam having two axes of symmetry, one axis of symmetry and no axes of symmetry, the position of shear centre has been indicated in the description given for these cases below.

In case of pure bending, the plane of deflection curve coincides with the plane of the bending couples, provided these couples act in one of the principal planes of bending. There are following three types of beam cross-sections.

(1) Beam cross-section having two axes of symmetry : For such beam cross-section, the bending axis for a beam is the longitudinal centroidal axis of the beam and it passes through the shear centre. The shear centre for such a section coincides with the centroid of the beam section.

In case, the load is applied along the longitudinal centroidal axis sections of one symmetry axis will be subjected to torsion.

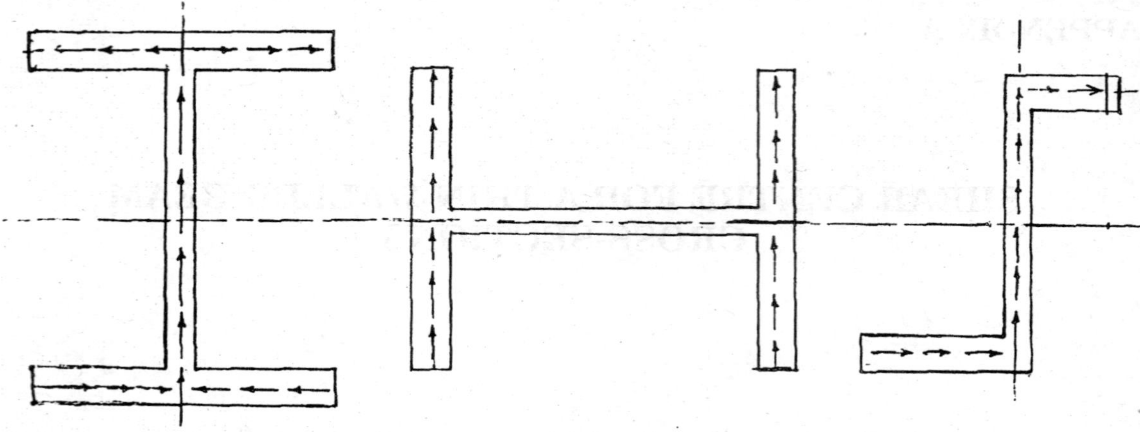

Fig. A.1. Beam cross-sections with two axes of symmetry
(Shear flow: Tension to compression flange)

2. Beam cross-sections having one axis of symmetry. (Fig. A.2) For such a beam cross-section, the shear centre does not coincide with the centroid of the beam section. The channel sections and tee-sections are the examples of sections which are symmetrical about one axis of symmetry. The shear centre lies on the other side of axis (which one is perpendicular to the axis of symmetry). In the channel section, it lies on the opposite side of the centroid. In the tee-section the centroid lies on the stem, where as, the shear centre lies in the flange.

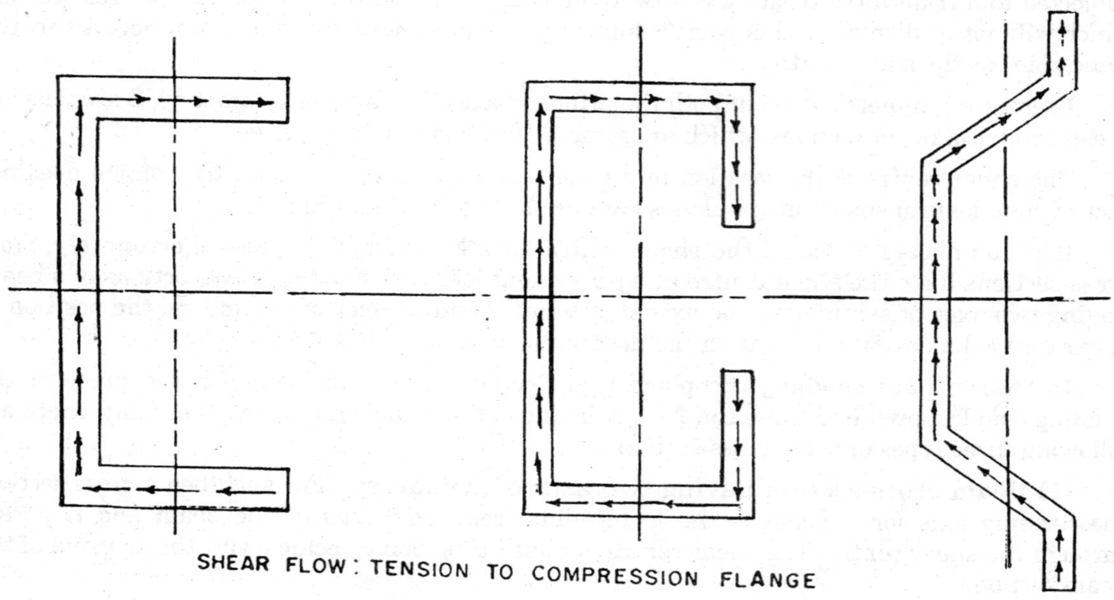

Fig. A.2(1). Shear flow: Tension to Compression Flange

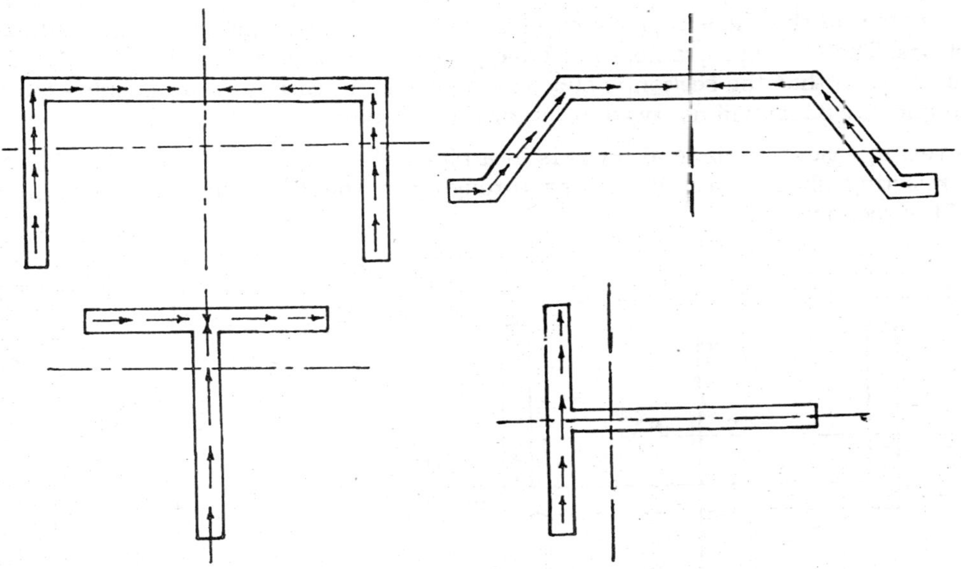

Fig. A.2(2).

(3) Beam cross-sections having no axis of symmetry : (Fig. A.3) In such a section, the shear centre does not lie on the longitudinal centroidal axis. The position of shear centre has to be located with respect to both the principal axes.

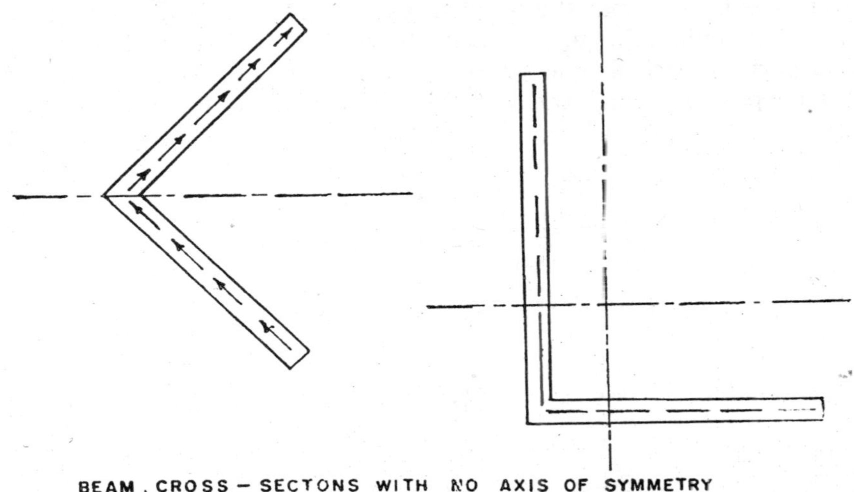

Fig. A.3.

The bending axis and the shear centre have special importance in beams having cross-sections composed of thin parts, which offer large resistance to bending but small resistance to torsion, for example, I-sections, channel sections, angle sections, and the various sections of beams made of thin plate materials (i.e., having rectangular sections).

It is to note that the shear centre for a cross-section of a beam depends only on dimensions of the sections. The shear centre does not have physical meaning unless the bending loads on the beams are such that the section is subjected to shear. In case, a beam is subjected to pure bending couple, the beam will not twist, as regards of the plane of loading.

However, in case, the plane of loads is in a plane of symmetry as shown in Fig. A.4, the resultant resisting shear force, V_R passes through the shear centre and also through the centroid of the section.

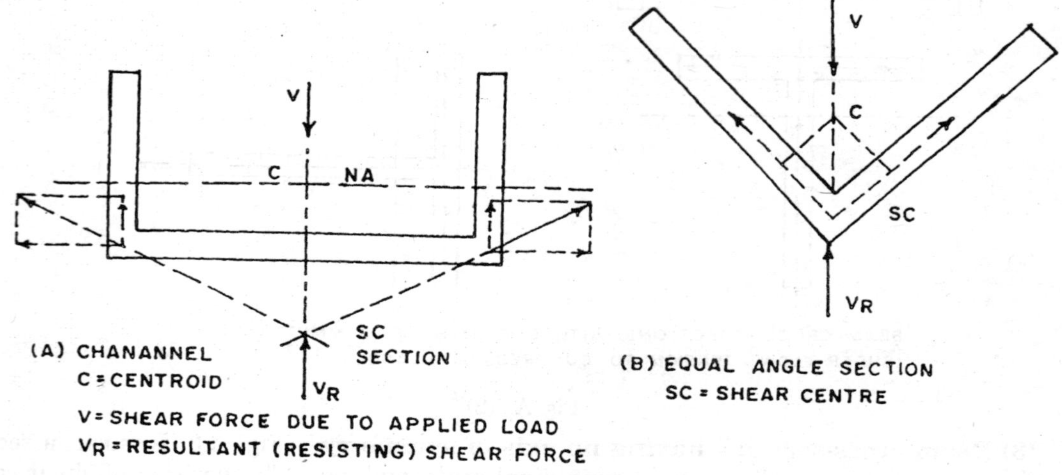

Fig. A.4.

Furthermore, it is to note that the I-section as shown in Fig. A.5, a lateral shear force is developed in each outstanding flange, just as in the channel section of Fig. A.5. But the shear forces in the two parts of each flange of the I-section are equal and opposite and hence, the resultant shear force passes through the centroid of the section. In this way, the centroid is also the shear centre.

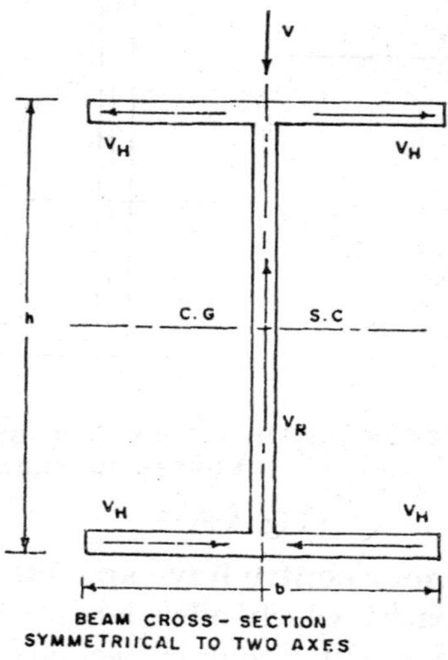

Fig. A.5(1).

It may be noted that, in case, the bending loads develop torsion, as it bends, the longitudinal flexural stresses in the beam are not given by the flexural formula.

$$\sigma_b = (M \cdot y/I) \quad \text{(i)}$$

Since, simple bending does not occur. The maximum longitudinal stress in the beam may be greatly in excess of that developed by loads of same magnitude, which act through the bending axis.

The shear centre is the location in the cross-section of a beam, where the torsion does not develop. The flexural shears act in planes passing through that location. In other words, the shear stresses developed due to loads applied through shear centre are zero,

$$\left(\text{i.e. } \int_0^n (\tau t) r \, ds = 0\right)$$

A straight cantilever beam is shown in Fig. A.5.2(A) with a channel section (which may be of an arbitrary shape). It is subjected to concentrated load W, that lies in the end plane (x = 0) of the beam cross-section. The vector representing W lies in a plane which forms angle, $0 < \phi < \pi$. It is taken positive when measured anti-clockwise in x-z plane as viewed from the positive x-axis. It is known as the **plane of the transverse load.** Let C.G. be chosen as the centroid of the beam cross-section at the right end of the beam (with the x-axis as the centroidal (i.e., longitudinal) axis of the beam (y,z)-axes are taken (arbitrary) so that the moment of inertia of the cross-section I_{zz}, I_{yy} and I_{yz} are calculated easily. These axes are taken to be principal axes and $I_{yz} = 0$.

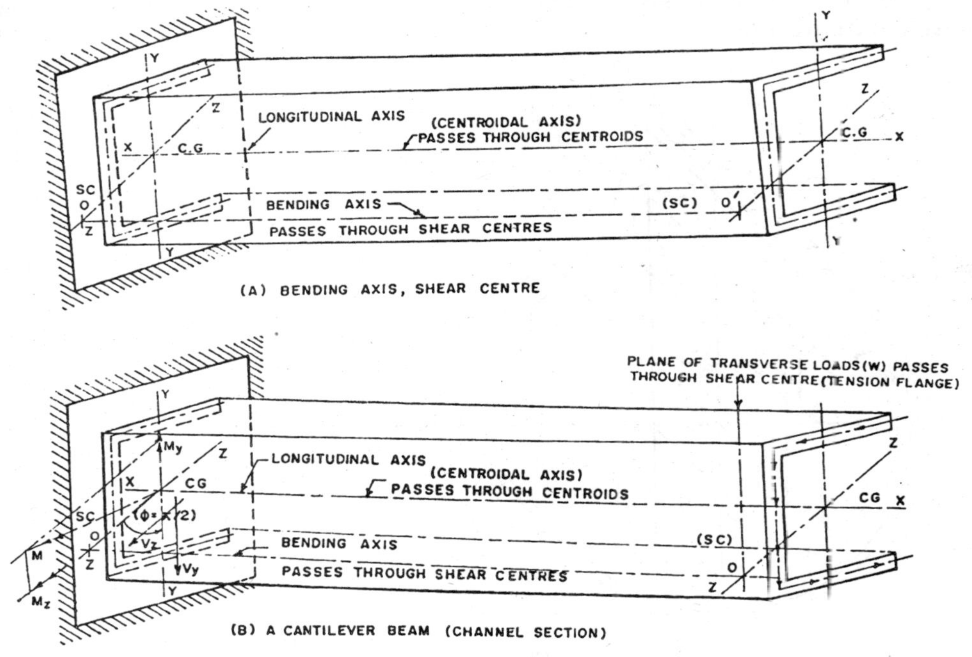

Fig. A-5(2).

Consider a cross-section of the beam at a distance x from the free end. The free body diagram of the part of beam to the free end of this section is shown in Fig. A. 5.2(B). For the equilibrium of this portion of the beam, a moment M, with its components of moments Mzz and

M_{YY}, shear components, V_Z and V_Y, and in general, a twisting moment T (with vector directed along the positive x-axis) must act on the cross-section at x. In case, the line of action of transverse load, W passes through a certain point (**shear centre**) in the cross-section then, T=0. In this description, it is assumed that the line of action of transverse load, W passes through the shear centre. Therefore, the twisting moment is zero and it is not shown in Fig. A.5.(B). It is to note that the transverse load, W needs V_y and V_z to be positive (directed along the positive (y–z) axes respectively). The component M_z is also directed along the positive, Z axis and since $\phi = \pi/2$, M_{yy} is negative (directed along the negative y-axis).

A line in the beam is particularly known as the **bending axis of the beam,** which one is parallel to the centroidal axis of the beam (which passes through the centroids of all of the cross-sections of the beam). Except for the special cases, the bending axis does not coincide with the centroidal axis.

The point of intersection of the bending axis with any cross-section of the beam, O (may be within or outside of the cross-section) is known as **shear centre** of the beam cross-section. As such the bending axis passes through the shear centres of all the cross-sections of the beam.

A.2 GENERAL BENDING THEORY

General bending includes **pure bending** of beams (in which there is no shear force at any cross-section and there is no twisting), **simple (ordinary) bending** of beams of symmetrical cross-sections and loaded symmetrical (in which, the shear force occurs at all cross-sections and since, the plane of loading passes through the shear centre, there is no twisting), and **unsymmetrical bending**.

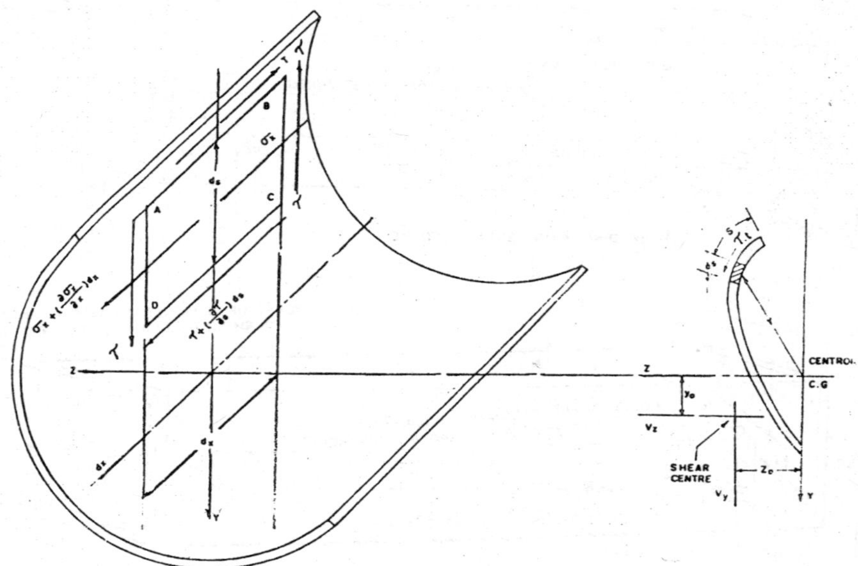

Fig. A-6.

Pure bending: A straight beam of uniform cross-sectional shape with no variation along the length as shown in Fig. A.6 is considered. A pure moment acts in the plane ABCD, which makes an angle ϕ with xz plane. The moments are represented by vectors normal to the plane of action (positive moment defined by using right-hand rule for rotation). It is assumed that beam is free from twisting.

A portion of the beam of length x as shown in Fig. A.4 (A) is examined. This portion of beam needs the following to satisfy the equilibrium on the free body

$$\Sigma F_x = 0, \ \int_A \sigma . dA = 0 \tag{A.1}$$

$$\Sigma M_z = 0, \ M_z = \int_A y . \sigma . dA \tag{A.2}$$

$$\Sigma M_y = 0, \ M_y = \int_A Z \ \sigma \ dA \tag{A.3}$$

The moments M_y and M_z are both positive according to the customary convention [as seen in Fig. A.4 (B)] of representing positive bending that which develops compression in the top portion of the beam. The subscripts for moment M shows the axis about which bending occurs; (i.e., the direction of the moment vector). The bending of beam xy plane only, in xz plane only and neither in xz nor in xy planes are considered, which are as follows :

1. Bending in the xy plane only

In case, the bending occurs in the xy-plane, the bending stress is proportional to distance y from the neutral axis (the distribution is linear and triangular). Therefore,

$$\sigma_b = a . y \tag{A.4}$$

From Eqs. A.1, A.2 and A.3

$$a_1 \int_A y \, dA = 0 \tag{A.5}$$

$$M_z = a_1 \int_A y^2 \, dA = a_1 . I_{zz} \tag{A.6}$$

$$M_y = a_1 \int_A y . z \ dA = a_1 . I_{YZ} \tag{A.7}$$

Eq. A.5 means that ZZ must be a centroidal axis. Then, the bending stress may be calculated as follows

$$\sigma_b = \frac{M_{Z.Y}}{I_{ZZ}} \text{ or } \quad \sigma_b = \frac{M_{y.y}}{I_{YZ}} \tag{A.8}$$

and the angle ϕ must be such that

$$\tan \phi = \left(\frac{M_z}{M_y}\right) = \left(\frac{I_{ZZ}}{I_{YZ}}\right) \tag{A.9}$$

It is not possible that an unsymmetrical beam section would be located is a plane making an angle 0 with xz plane and have bending occurring in the xy plane. Eq. A.5 also shows that in case a section is used with at least one axis of symmetry (for which S_A, y,z, dA = I_{yz} = 0), tan ϕ = ∞ , ϕ = 90°. If means that the loading and the bending, both occur in the x, plane.

It leads to one important conclusion that in case I_{yz} = 0, the bending occurs in the plane of loading for examples, on unsymmetrical shape such as angles or zees loaded in the xz or xy plane, the plane of bending will be different.

2. Bending in the xz-plane only. In case, the bending occurs in the xz plane, the bending stress is then proportional to z. Therefore,

$$\sigma_b = a_2\, z \tag{A.6)A}$$

From Eqs. A.1, A.2, and A.3

$$a_2 \int_A z\, dA = 0 \tag{A.6)B}$$

which means that y must be a centroidal axis. Also

$$M_Z = a_2 \int_A yz\ dA = a_2 \,.\, I_{xy} \tag{A.6)C}$$

$$M_y = a_2 \int_A z^2\ dA = a_2 \,.\, I_y \tag{A.7}$$

$$\text{and } \tan\ \phi = \left(\frac{M_Z}{M_Y}\right) = \left(\frac{I_{YZ}}{I_{yy}}\right)$$

In the case where $I_{yz} = 0$, $\tan \phi = 0$, that is the loading and bending both occur in the xz - plane.

3. Bending in neither xz plane nor in xy plane : It is a practical case. It considers **unsymmetrical bending** (already described in Appendix). When the bending of beam occurs in a plane other than the principal plane of section, the beam section is subjected unsymmetrical bending. The direction of neutral axis is not perpendical to the plane of bending. It is assumed that all the stresses are within elastic limit. Therefore, total bending stress is the sum of the bending stresses due to bending (as if the bending would have occurred separately), in each of xz-plane and xy plane.

$$\sigma_b = (a_1\, y + a_2 \,.\, z) \tag{A.8}$$

$$\text{and} \quad M_z = a_1 \,.\, I_{zz} + a_2\, I_{xy} \tag{A.9}$$

$$M_y = a_1\, I_{yz} + a_2\, I_{yy} \tag{A.10}$$

The values of a_1 and a_2 are determined by solving Eqs. A.9 and A.10. Substituting these values in Eq. A.8.

$$\sigma_b = \left(\frac{M_z \,.\, I_{yy} - M_y\, I_{yz}}{I_{zz} \,.\, I_{yy} - I_{yz}^2}\right)_y + \left(\frac{M_y \,.\, I_{zz} - M_z \,.\, I_{yz}}{I_{zz}\, I_{yy} - I_{yz}^2}\right) .\ z \tag{A.11}$$

It is to note that Eq. A.11 and Eq. A.13 are same. The difference in subscripts is due to notation of zz-axis and xx-axis.

It is further to note that equation Eq. A.11 is the general bending. The beam considered is straight. There is no variation along the length, in the cross-sectional shape. x-axis and y-axis are mutually perpendical centroidal axes. The bending stress is proportional to strain and the maximum value of bending stress is within the limit of proportionality.

The principal axes of a beam cross-section have been described in Appendix - B (B.2). The angle of inclination of neutral axis is given by (from Eqs. A.5, A.7 and Fig. A.1)

$$\tan\phi = \left(\frac{M_Z}{M_y}\right) \quad (A.12)$$

The bending stress at the neutral axis is equal to zero ($\sigma_b = 0$), as such.

$$-\left(\frac{z}{y}\right) = \left(\frac{M_Z\, I_{yy} - M_y\, I_{YZ}}{I_{ZZ}\,.\,I_{yy} - I_{yz}^2}\right).\left(\frac{I_{ZZ}\, I_{yy} - I_{yz}^2}{M_y\,.\,I_{ZZ} - M_z\, I_{yz}}\right) \quad (A.13)$$

Therefore, $-\left(\frac{z}{y}\right) = \left(\frac{M_Z\,.\,I_{yy} - M_y\, I_{YZ}}{M_y\,.\,I_{ZZ} - M_y\,.\,I_{YZ}}\right)$ (A.14)

Dividing both the numerator and the denominator of right side of Eq. A.14 by M_y

$$\tan\alpha = \left(\frac{(M_z/M_y)\, I_y - I_{yz}}{I_{zz} - (M_{ZZ}/M_y\ \ I_{YZ})}\right) \quad (A.15A)$$

Substituting A.13 $\tan\alpha = \left(\frac{I_{yy}\tan\phi - I_{YZ}}{I_{ZZ} - I_{YZ}\tan\phi}\right)$ (A.16A)

A.3. SHEAR STRESSES DUE TO BENDING OF THIN-WALLED OPEN CROSS- SECTIONS

A thin-walled open cross-section is shown in Fig. A.6(A), where yy and zz are the centroidal axes. Consider the equilibrium of an element ABCD (t.ds.dx). It is acted upon by bending stress, longitudinal stress, σ_x and shear stress, τ. These stresses develop due to bending moment. The shear stress, τ multiplied by the thickness, t (i.e., shear force per unit length) is called as **shear flow,** (τ.t). For the equilibrium in the direction, x

$$\frac{\delta(\tau.t)}{\delta s}.ds.dx + t\frac{\partial\sigma_x}{\partial x}.\partial_x = 0 \quad (A.15B)$$

or $\left(\frac{\partial(\tau.t)}{\partial s}\right) = -\left(\frac{t.\partial\sigma_x}{\partial x}\right)$ (A.16B)

Following three cases are considered.

1, In case, it is assumed that the moment is applied in the xy plane only (i.e., $M_y = 0$). Then, the bending stress (as in Eq. A.11).

$$\sigma_{bx} = \left(\frac{M_{ZZ}}{I_{YY}\,.\,I_{ZZ} - I_{yz}^2}\right)(I_y\,.\,y - I_{yz}\,.\,z)$$

$$\left(\frac{\partial\sigma_x}{\partial x}\right) = \left(\frac{\partial M_{zz}/\partial x}{I_{yy}\,.\,I_{zz} - I_{yz}^2}\right)(I_y\,.\,y - I_{zz}\,.\,z) \quad (A.17)$$

Since, $V_y = (\partial M_x/\partial x)$, and substituting this in Eq. A.16.

$$\left(\frac{\partial(\tau.t)}{\partial s}\right) = \left(\frac{-t.V_y}{I_{yy}\,.\,I_{zz} - I_{yz}^2}\right).(-I_y\,.\,y - I_{yz}\,.\,z) \quad (A.18)$$

To find (τ.t), integrate Eq. A.18 (for distances, from a free edge)

$$(\tau.t) = \left(\frac{-V_y}{I_{yy}.I_{ZZ} - I_{yz}^2}\right)\left[I_y \int_0^S y.t.ds - I_{yz}\int_0^S z.t.ds\right] \quad (A.19)$$

2. In case, it is assumed that the moment is applied in the xz plane only, (i.e., $M_z = 0$) the flexural stress due to bending as given by Eq. A.20.

$$\sigma_{bx} = \left(\frac{M_y}{I_{yy}.I_{zz} - I_{yz}^2}\right)\left(-I_{yz}.y + I_{ZZ}.x\right) \quad (A.20)$$

Differentiating (A.20)

$$\left(\frac{\partial \sigma_x}{\partial x}\right) = \left(\frac{\partial M_y/\partial x}{I_{yy}.I_{zz} - I_{yz}^2}\right)\left(-I_{yz}.y + I_{zz}.x\right) \quad (A.21)$$

Since, $V_z = (\partial M_x / \delta x)$ and substituting in Eq. A.21,

$$\left(\frac{\partial(\tau.t)}{\partial s}\right) = \left(\frac{-t.V_z}{I_{yy}\,I_{zz} - I_{yz}^2}\right).\left(-I_{yz}.y + I_{zz}.x\right) \quad (A.22)$$

To find (τ.t), integrating A.22 (for distance, s from a free edge)

$$(\tau.t) = \left(\frac{+V_z}{I_{yy}.I_{zz} - I_{yz}^2}\right).\left[I_{yz}\int_0^S y.t.ds - I_z\int_0^S z.t.ds\right] \quad (A.23)$$

3. In case, the moments are applied in both xy and xz-plants, then the desired shear stresses may be calculated by superimposing the results of Eq. A.19 and Eq. A.23

It is to be seen from Fig. A.6 that the equilibrium needs that the shear in y direction, V_y equals the components of (τ.t) in the y-direction added over the complete section. Similarly V_z equal the additon of (τ.t) components in the z-direction. The equilibrium in the angular direction must also be satisfied.

That is, the moment about the centroid of section, Fig. A.6(B)

$$\int_0^{s1} (\tau.t)r.ds = 0 \quad (A.24)$$

In case, the angular equilibrium is automatically satisfied when the shears due to bending act through the centroid, then, there shall be no torsion occurring simultaneously with bending.

A.4 SHEAR CENTRE AND SHEAR FLOW IN OPEN AND THIN CROSS- SECTIONS OF A BEAM

I-sections used as beams are open and thin cross-sections (in case), the thicknesses of flanges and web are thin). These sections are symmetrical about both the axes. Both the principal axes pass through centroid. Channel-sections and Tee sections used as beams are open and thin cross-sections (in case, the thicknesses of flanges and web are thin). These sections are symmetrical about one axis. Whereas, the other principal axis, passes through the centroid. However, the angle sections used as purlins (i.e., beams) are also open and thin sections, but these sections are not symmetrical about any of the axis.

When a beam carries a transverse load, it is subjected to bending moment, which varies along the axis of the member. Every cross-section of the beam carries external shear force along with the bending moment. For simply supported beam, the shear force is maximum at the supports. The shear force at the concentrated load is also large. Due to shear, the internal shear stresses are set up along with the bending stresses at all the cross-sections of the beam. The shear stress at any fibre of the cross-section of the beam is given by

$$\tau_v = \left(\frac{V.A\bar{y}}{I.t}\right) \tag{A.25}$$

where, τ_v is the shear force and V is the external shear force at the transverse cross-section of the beam.

Q = ($A\bar{y}$), static moment about the neutral axis of the area, A.

A = Area of portion of cross-sectional area beyond the location of fibre at which the shear stress is being determined.

$\bar{y}$ = Vertical distance of the fibre being considered from the neutral axis.

I = Moment of inertia of the cross-section of the beam about the neutral axis.

t = Thickness of web (i.e., width of fibre parallel to neutral axis in which the stress is found).

Eq. A.25 may also be written as under.

$$\tau_v = \left(\frac{V.Q}{I.t}\right) \tag{A.26A}$$

In case, the plane of loading (i.e., bending) does not pass through the shear centre, a torsional moment is developed that produces shear stresses due to torsion. It is assumed that the material is homogenous and the cross-section is uniform. There is no warping. From the theory of torsion.

$$\left(\frac{M_{tc}}{J}\right) = \left(\frac{q_s}{r}\right) \tag{A.26B}$$

$$q_s = \left(\frac{M_{TC}.r}{J}\right) \text{ and } \quad q_s = \left(\frac{V.e.t_f}{J}\right), \text{ where}$$

M_{Tc} = torsional moment (couple) = V.e'

V.e.' = External shear force and its eccentricity with respect to the shear centre

t_f = thickness of element, where shear stress is desired, (t = r)

J = Torsional constant, for a thin rectangle $\left(= \Sigma\, bt^3/3\right)$

(b/t) → 10 (section with webs and some flanges) (Reference: J.E. Bowels, P. 171)

$$J = \left(\frac{bt^3}{3} - 0.21\,t^4\right)$$

(b/t) → 4 (Stubby Flanges as for channels).

J = Contribution of web + two flanges

$$J = \frac{bt^3}{3} + 2\left(\frac{bt^3}{3} - 0.21\,t^4\right) \quad \text{(Since there is one web + two flanges)}$$

From Eq. A.26(B), it is seen that as long as the assumptions hold good, the torsional stress is proportional to the radial distance from the centre of twist.

The resultant of the internal shear stresses produce internal shear force, V_i at any cross-section. For the equilibrium of forces in the vertical direction, the internal shear force, V_i is equal in magnitude but opposite in the direction to the external shear force. Irrespective of the inclination of plane of loading (i.e., the distance from the centroid of the cross-section and its direction to the principal axes), there is a distinct point, O in the cross-section through which the internal shear force always, passes, (so that there is no twisting of the beam cross-section), that point is called **shear centre** of the cross-section of the beam. In case, the plane of loading passes through shear centre, the internal shear force, V_i and the external shear force, V become collinear, and their resultant is zero, the plane of loading does not pass through the shear centre, the internal shear force, V_i and the external shear force, V, act at a distance, a, and a twisting couple (V.a) is formed.

Some times Eq. A.25 may also be written as under for convenience.

$$(\tau_v \cdot t) = (V_y \cdot Q \cdot I) \tag{A.27}$$

The product of shear stress and thickness of web (i.e., width of fibre parallel to neutral axis) becomes shear force per unit length and it is referred as **shear flow** in the cross-section of the beam under consideration.

The shear flow (τ_v . t) seems like flow of a fluid within thin walled open section (i.e., within tubular section between the free surfaces of walls). The direction of shear flow is towards or away from the free tips on the tension side or compression side of the beam cross-section whichever, direction is consistent with the equilibrium in the vertical direction, which should be verified.

The direction of shear flow is from tension side to the compression side when looked at the section in the direction of increasing bending moment.

A.5 DETERMINATION OF SHEAR CENTRE FOR A GIVEN CROSS-SECTION

It is assumed that the plane of loading (*i.e.,* plane of bending) passes through the shear centre. So that the torsion does not occur when the shear stresses due to bending act on the plane through this point. Then, the cross-section of the beam may be analyzed for simple bending and shear using the following expressions

$$\sigma_b = \left(\frac{M_Z}{Z_{xx}}\right) + \left(\frac{2\,M_y}{Z_{yy}}\right)$$

$$\tau_v = \pm\frac{V.Q}{I_{zz}\,t} \pm \frac{V_y \cdot Q\,Y}{I_{yy}}$$

The shear centre for the given cross-section of a beam is located as follows :

1. In case, the cross-section of beam is symmetrical about an axis, then, the shear centre shall lie on the axis of symmetry. For the channel section, the shear centre lies on zz-axis, which one is the axis of symmetry.
2. In case, the cross-section of beam is symmetrical about both the axes, the shear centre coincides with the centre of gravity. For the I-section, which is symmetrical about both yy-axis and zz-axis, the shear centre coincides with the centre of gravity.
3. In case of (for example) unsymmetrical I-section, then, the centre of gravity shall be first located.
4. In case, the beam section consists of series of thin rectangles, and a common point, O, then, the shear centre coincides with this point.

A.6 SHEAR CENTRE FOR A CHANNEL-SECTION WITH PARALLEL FLANGES

A channel section with parallel flanges is shown in Fig. A.7. The position of a plane in which the vertical loads must act to produce simple bending with zz-axis as the neutral axis.

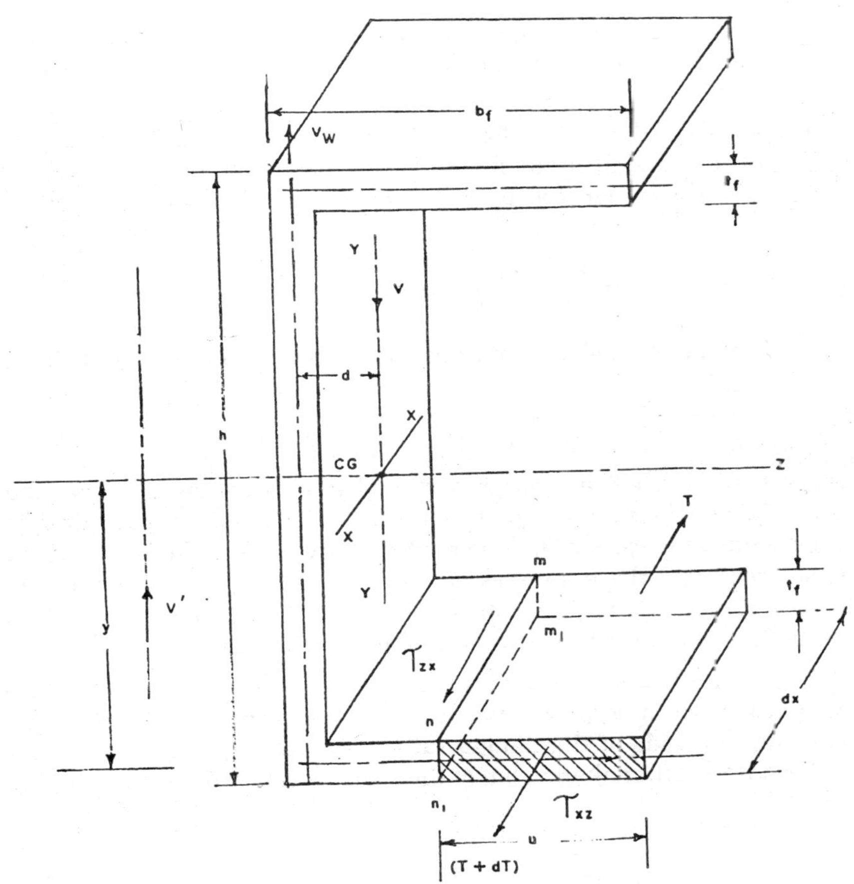

A CHANNEL – SECTION WITH PARALLEL FLANGES

Fig. A-7

For a channel section with the loading parallel to web, the total shear force carried by the web is equal to V_w, and that in the flanges produce, two equal and opposite horizontal forces. The horizontal shear stresses in the flanges are denoted by τ_{xz}. The subscript z in τ_{xz} indicates that shear stress is parallel to zz-axis and the subscript x indicates that the stress acts in plane perpendicular to xx-axis.

Consider an element, cut from the flange by two adjacent cross-sections, dx apart (along the length) and by a vertical plane $mnn_1 m_1$ parallel to web. In case the beam bends downward, the lower flange will be in tensile and the tensile forces T and (T + dI) act on the above element. Then, these tensile forces shall be equal to

$$T = \frac{M}{I_{ZZ}} \int y . dA \quad \text{and} \tag{i}$$

$$(T + dT) = \left(\frac{M + \frac{dM}{dx} dx}{I_{zz}} \right) \int y \, dA \tag{ii}$$

where, the integration extends over the shaded portion of the cross-section of the flange. The integral represents the moment of the shaded area with respect to the zz-axis (i.e., neutral axis).

The difference between the tensile forces [T – (T + dT)] is equal to the sum of the shear stresses, τ_{zx} acting over the side $mnn_1 m_1$. It is assumed that these stresses are uniformly distributed over this side and denoted by t_f, the thickness of flange. The shear stress, τ_{zx}, is found as below :

$$\tau_2 \, t_f . dx = (-dT) = -\left(\frac{dM}{dx} \right) \frac{dx}{I_{zz}} \int y \, dA$$

Since, (dM/dx) is equal to V_Y shear stress parallel to x-axis and perpendicular to z-axis,

$$\tau_{zx} = -\left(\frac{V_y}{I_z . t_f} \right) \int y dA \tag{A.28}$$

The moment of the shaded area is proportional to the distance, u from the edge of the flange; hence, τ_{zx} is proportional to u. The shear stress τ_{zx} is equal to τ_{zx}, (these shear stresses act on the two perpendicular faces of the element and are equal). The absolute values of these stresses are considered), and these stresses act horizontally at points along the line nn_i in the cross-section of the flange. Hence, the stresses, τ_{zx} are distributed non-uniformly over the cross-section of the flange, but these are proportional to the distance, u.

At the junction of the flange and web, the distribution of shear stresses is complicated. In approximate calculations, it is assumed that Eq. A.29 holds from u = 0 to u = b. The vertical distance between the centroids of the flanges (at the top and bottom) is h. It is seen that the first moment of the cross- sectional area, $b.t_f$ of the flange with respect to z-axis is $(b.t_f.h/2)$.

$$\therefore \int dA = b_f \, t_f$$

$$(\tau_{xz})_{max} = (\tau_{zx})_{max} = \frac{V_y}{I_{zz1} \cdot t_f} \int y . dA = -\left(\frac{V_y \, b . h}{2 \, I_{zz}} \right) \tag{A.29}$$

Resultant, V_z of the shear stresses, τ_{xz} distributed over the cross-sectional area b.t. of the flange are as shown in Fig. A.7.

$$V_Z = \left(\frac{V_y \cdot b_f\, h}{2\, I_{ZZ}}\right)\left(\frac{b_f\, t_f}{2}\right) = \left(\frac{V_y\, b^2 \cdot h \cdot t_f}{4\, I_{ZZ}}\right) \tag{A.30}$$

The sum of the shear stresses τ_{xz} over the cross-section of the upper flange shall evidently be an equal and opposite force. As such, the shear stresses over a channel section reduce to the forces as shown in Fig. A.8.

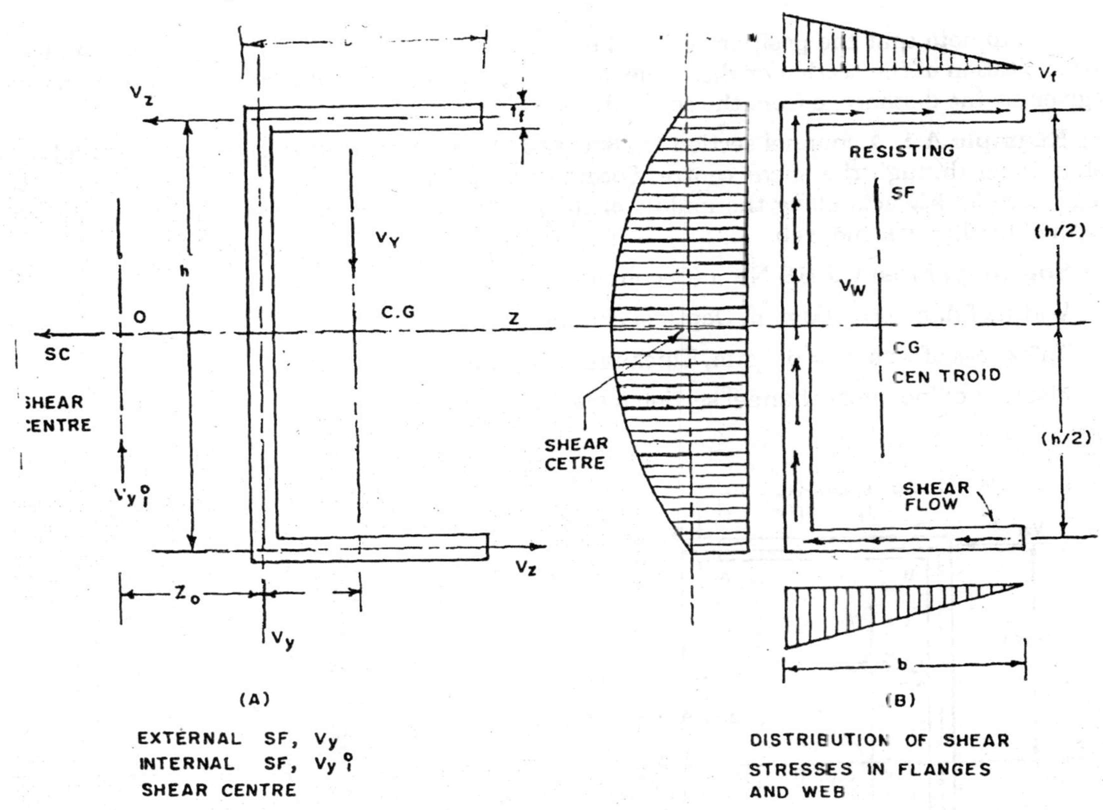

Fig. A-8

Taking the moment of these forces about the point, O, at a distance, e from the centre of web.

$$V \cdot e - V_{zf} \cdot h = 0$$

$$\therefore\ e = \left(\frac{V_{zf}\, h}{V_w}\right) = \left(\frac{b^2\, h^2\, t_f}{4\, I_{zz}}\right) \tag{A.31}$$

Therefore, it is seen that for simple bending about zz-axis representing the neutral axis the vertical plane of transverse loads must pass through the point O, which is known as the **shear centre.**

Moment of inertia of the channel section, about zz-axis.

$$I_{zz} = \left(\frac{1}{12} t_w . h^3 + 2 bt_f \left(\frac{h}{2}\right)^2\right) = h^2 \left(\frac{tw . h + 6 b.t_f}{12}\right)$$

Substituting this in Eq. A.31

$$e = \left(\frac{3 b^2 . t_f}{6 b_f . t_f + h . t_w}\right) \tag{A.32}$$

It is to note that the position of the shear centre is a unique location. It is independent of the orientation of the section or the plane of loading. The location of shear centre remains same as given by the distance, e from the centreline of the web.

Example A.1. A channel section LC 300, @ 0.331 kN/m is loaded in a plane parallel to the web passing through the shear centre. Locate the position of shear centre. In case, a vertical shear force 30 kN acts along the middle of the of the web. Also calculate shear force per unit length of the flanges and web.

Solution : From ISI HB No. 1 (Section tables)

Width of flanges = 100 mm, Depth of channel = 300 mm

Thickness of web, t_w = 6.7 mm, Thickness of flange, t_f = 11.6 mm

Moment of inertia of channel section = 6047.9×10^4 mm^4

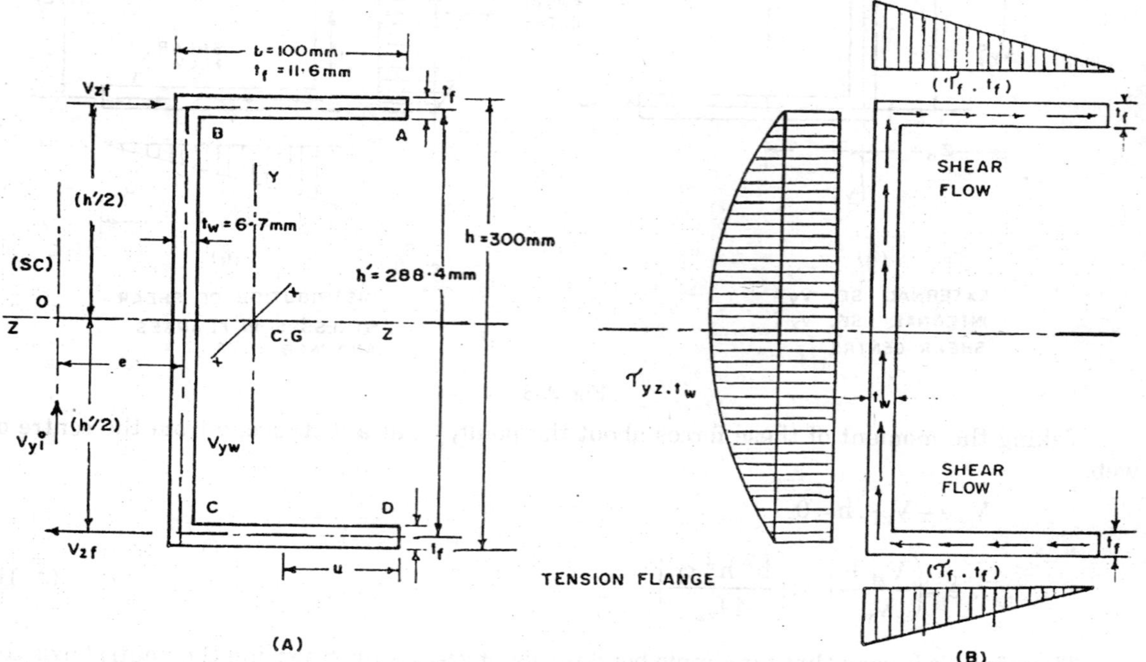

SHEAR FLOW SHEAR DISTRIBUTION, SHEAR CENTRE A THIN-WALLED OPEN CHANNEL SECTION

Fig. A-9

Consider the thin-walled open channel section as shown in Fig. A.9. The plane of loading is assumed parallel to yy-axis. The shear stress resultant due to bending in the top flange and bottom flange are V_{zf} and that in the web is V_{yw} as shown in Fig. A.9. ZZ-axis is the axis of symmetry. Therefore,

$$V_{zf} = V_{yw} \tag{i}$$

The shear stress at any point, at a distance, u from the free edge D, at a depth, y = (h'/2) from the neutral axis (zz-axis)

$$\tau = \frac{V_y}{I_{uu} \cdot t_f}\left(u \cdot t_f \cdot \frac{h}{2}\right) \tag{ii}$$

Therefore, total shear force in the bottom flange

$$V_{zf} = \left(\frac{-V_y}{I_{uu}}\right)\left(\int_0^b \left(u \cdot t_f \frac{h'}{2}\right)\right) du$$

$$V_{zf} = \left(\frac{-V_y}{I_{uu}}\right)\left[\frac{u^2}{2} \cdot t_f \cdot \frac{h'}{2}\right]_0^b$$

$$V_{zf} = \left(\frac{-V_y}{I_{uu}}\right) \cdot \left(\frac{b_f^2\, t_f \cdot h'}{4}\right) \tag{iii}$$

For the equilibrium of the vertical forces

$$V_y = V_{yi} = V_{yw} \tag{iv}$$

where, V_{yi} is the internal shear force acting at the shear centre, O as shown in Fig. A.9. Consider the moment of forces about the point, O

$$\left(V_{yw} \cdot e - V_{zf} \cdot h'\right) = 0 \tag{v}$$

Substituting the value of V_{zf} from the expression (iii)

$$V_{yw} \cdot e - h'\left(\frac{-V_y}{I_{uu}}\right) \cdot \left(b_f^2 \cdot \frac{t_f}{4} \cdot h'\right) = 0$$

$$e = \frac{-1}{I_{uu}} \cdot \left(\frac{b_f^2\, h^2 \cdot t_f}{4}\right) \tag{v}$$

Moment of inertia of the channel section about ZZ-axis (i.e., neutral axis)

$$I_{uu} = I_{ZZ} = \left(\frac{t_w\, h^3}{12} + \frac{b_f\, t_f\, h^2}{2}\right) \tag{vi}$$

Distance of the point, O (shear centre) from the middle line of web. (h'= 300 – 6.7)

$$e = \left(\frac{-3\, b_f^2 - t_f}{6\, b_f\, t_f + h'\, t_w}\right) = 29.33 \text{ mm} \tag{vi}$$

The negative sign for the shear centre indicates that it is on the opposite side of the centroid.

In case, a vertical shear force acts along the middle line of the web, the twisting couple developed shall be

M_{TC} = (27.08 x 30) = **834 kN-mm** = 0.834 kN-m

$$h' = \left[300 - 2\times11.6\frac{h'}{2}\right] = \mathbf{144.2} = 276.8 + 11.6 = 288.4$$

Area of each flange, A_f = (100 x 11.6) = 1160 mm²

Area of web, A_w = [(300 – 2 x 11.6) x 6.7]

= (276.8 x 6.7) = 1854.56 mm²

Moment of inertia of the channel section = 6047.9 x 10 mm²

Shear flow in each flange

$$\left(\tau_{zf}\ t_f\right) = \left(\frac{V}{I}\cdot\frac{A_f\,.\,h'}{2}\right) = \left(\frac{30\times1000}{6047.9\times10^4}\right)\times\left(\frac{1160\times288.4}{82}\right) = 82.97\ \text{N/mm}$$

$$\text{Shear flow in the web, } = \left(\tau_{zf}\,.\,t_f\right) + \left(\frac{V}{I_{zz}}\right)\cdot\left(\frac{A_w}{2}\times\frac{h'}{4}\right)$$

$$= 82.97 + \frac{30\times1000}{6047.9\times10^4}\times\frac{1854.6}{2\times4}\times288.4 = \mathbf{115.16\ N/mm}$$

Substituting the respective values

$$e = \left(\frac{-3\times100\times100\times6.7}{6\times100\times6.7+293.3\times11.6}\right) = \mathbf{-27.08\ mm}$$

From the expression (v), and substituting the respective values

$$e = -\frac{1}{I_{uu}}\left(\frac{b_f^2\,.\,h^2\,.\,t_f}{4}\right)$$

$$e = \frac{-1}{10^4\times6047.9}\times\left(\frac{100\times100\times293.3^2\times11.6}{4}\right) = \mathbf{-24.95\ mm}$$

$$\text{Moment of inertia of channel section, } I_{uu} = \left(\frac{1}{12}t_w\times h^3 + \frac{b_f\ t_f\ h'^2}{2}\right)$$

$$= \left(\frac{6.7\times300^3}{12} + \frac{93.3\times11.6\times293.3^2}{2}\right)$$

= (1507.5 x 10⁴ + 4653.8 x 10⁴) = 6161.3 x 10⁴ mm⁴

The difference between two values is due to difference between values of moment of inertias so also, the channel section is considered having rectangular elements. Whereas, in ISI HB. No.1 there is slope in the flanges of the channel section and the radius at the both at the junction.

Example A.2. A thin-walled open channel section with bent-lips 200 mm x 80 mm is 3.2 mm thick. The depth of lips is 25 mm. find the shear force per unit length in the lips, flanges and the web. Locate the position of shear centre. The properties of this channel section may be noted from IS : 811-1961 (Tables for light gauge sections).

Solution :

Total cross-sectional area = 1244.7 mm^2, Width of flange, b_f = 80 mm

Moment of inertia about zz-axis = 766.743 x 10^4 mm^4

Depth of section, h = 200 mm and Thickness $t_w = t_f = t_{lips}$ = 3.2 mm

A thin-walled open channel section with bent lips is shown in Fig. A.10. ZZ-axis is an orthogonal axis passing through the centroid, CG of the channel section. A shear force V_y acts parallel to the web of the channel section. V_{lip}, V_{zf} V_{yw} and V_{lip} show the direction of total shear flow in the elements FE, ED, DC, CB and BA as in Fig. A.10.

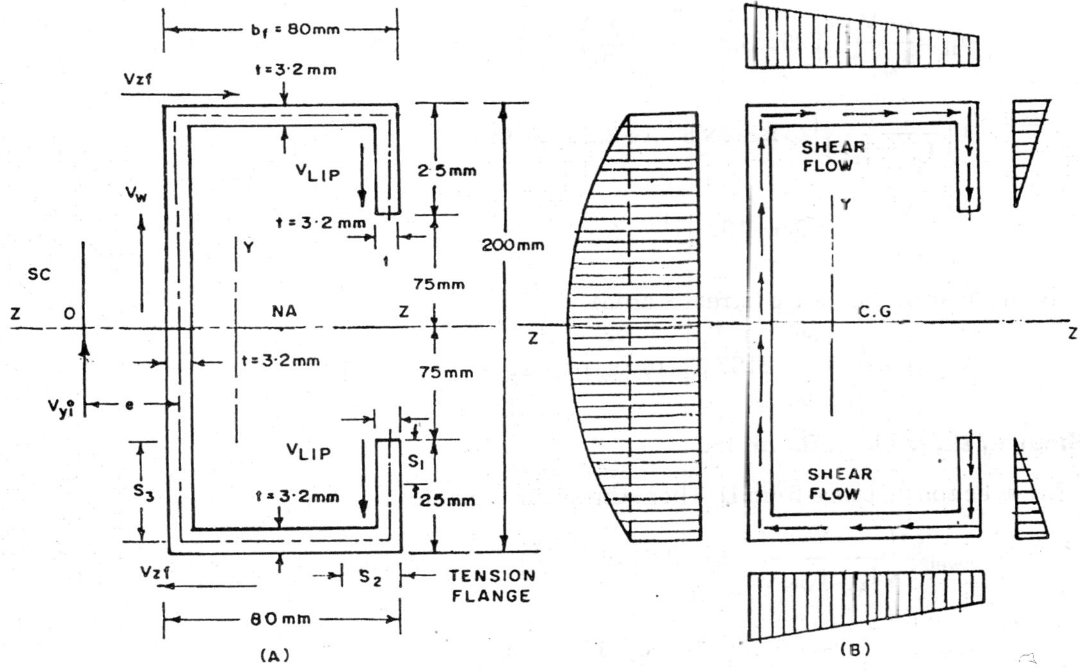

SHEAR FLOW, SHEAR DISTRIBUTION, SHEAR CENTRE (A THIN—WALLED OPEN CHANNEL SECTION WITH LIPS)

Fig. A-10

From symmetry V_{lip} (top) = V_{lip} (bottom), and V_{zf} (top) = V_{zf} (bottom)

The values of V_{lip}, V_{zf} and V_{yw} are found as under :

1. Shear in the lip element FE

Let s_i be the length of lip at bottom flange from F (free edge). Then, the shear stress in the lip element.

$$\tau_{lip} = \left(\frac{-V_y}{I_{zz}\cdot t_{lip}}\cdot A\,\overline{y}\right) = \left(\frac{-Y_y}{I_{zz}}\times\frac{1}{3.2}\right)\times\left[3.2\,s_1\left(75+\frac{1}{2}s_1\right)\right]$$

$$\left(\frac{-V_y}{I_{ZZ}}\right)\left[75\,s_1+\frac{1}{2}s_1^2\right] \quad \text{(i)}$$

Total shear in the lip (stress x area)

$$V_{lip} = \left(\frac{-V_y}{I_{zz}}\right)\int_0^{25}\left(75\,s_1+\frac{1}{2}s_1^2\right)t_{lip}\,.ds = -8.333\times10^4\left(\frac{V_y}{I_{ZZ}}\right) \quad \text{(ii)}$$

2. Shear in the flange element, ED

Let s_2 be the distance from E. Then, the shear stress in the flange element

$$\tau_{fl} = \frac{-V_y}{I_{zz}\cdot t_{fl}}\left[25\times3.2\times\left(75+\frac{25}{2}\right)+3.2\,s_2\times(75+25)\right]$$

$$= \left(\frac{-V_y}{I_{zz}\times3.2}\right)\left[25\times3.2\times87.5+3.2\,s_2\times100\right]$$

$$= \left(\frac{-V_v}{I_{zz}}\right)\left[2187.5+100\,s_2\right] \quad \text{(iii)}$$

Total shear in the flange (stress x area)

$$V_{fl} = \left(\frac{-V_y}{I_{ZZ}}\right)\int_0^{80}\left[2187.5+100\,s_2\right]t_{f1}\,ds_2 = -158.4\times10^4\left(\frac{V_y}{I_{ZZ}}\right) \quad \text{(iv)}$$

3. Shear in the web element, DC

Let s_3 be the distance from D. Then, the shear stress in the web

$$\tau_{web} = \left(\frac{-V_y}{I_{zz}\cdot t_{web}}\right)\left[25\times3.2\left(75+\frac{25}{2}\right)+80\times32\times100+3.2\times s_3\left(100-\frac{1}{2}s_3\right)\right]$$

$$\tau_{web} = \left(\frac{-V_y}{I_{zz}}\right)\left[25\times87.5+80\times100+100\,s_3-\frac{1}{2}s_3^2\right]$$

$$\tau_{web} = \left(\frac{-V_y}{I_{zz}}\right)\left[10187.5+100\,s_3-\frac{1}{2}s_3^2\right] \quad \text{(v)}$$

Total shear in the web element (shear stress x area)

$$V_{web} = \left(\frac{-V_v}{I_{zz}}\right)\int_0^{200}\left(10187.5+100\,s_3-\frac{1}{2}s_3^2\right)3.2\times ds_3$$

$$= \left(-\frac{865.312\times10^4\,V_y}{I_{ZZ}}\right) \quad \text{(vi)}$$

As a check, it may be noted that for the equilibrium of the vertical forces ($V_{web} = 2\ V_{lip} = V_y$). Take moment about the point, O (shear centre) and equate the algebraic sum of the moments to zero. It is to note further that [$V_{lip\,(bottom)} = V_{lip\,(top)}$] and [$V_{fl\,(bottom)} = V_{fl\,(top)}$]. Therefore,

$$2\ V_{lip}\ (e+80) + V_{fl} \times 200 - V_{web} \times e = 0$$

Substituting respective values

$$\frac{-V_y}{I_{zz}} \times 10^4\ [2 \times 8.33\ (e+80) + 158.4 \times 200 - 865.312\ e] = 0$$

$$e = \left(\frac{33012.8}{848.652}\right) = 38.9\ \text{mm}$$

Example A.3 A thin-walled open truffed channel with lips 10 mm thick is shown in Fig. A.11. Locate the position of shear centre.

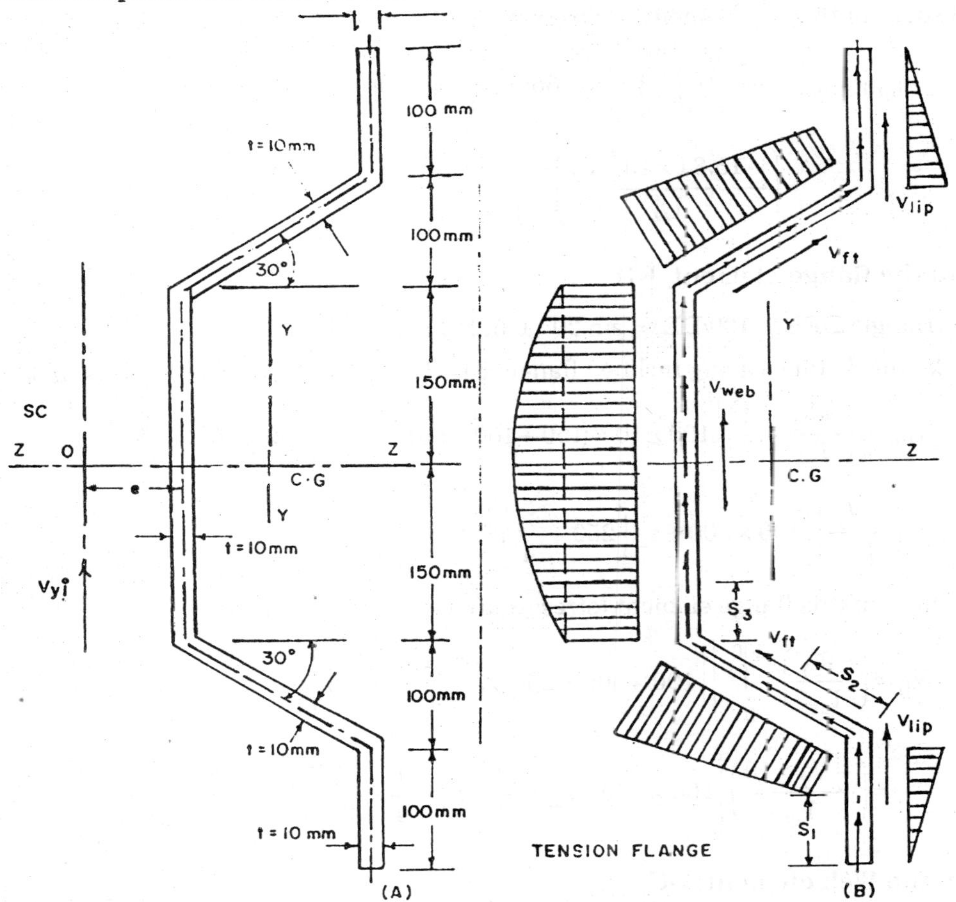

SHEAR FLOW SHEAR DISTRIBUTION, SHEAR CENTRE
A THIN – WALLED OPEN TRUFFED CHANNEL

Fig. A-11.

Solution : ZZ-axis is the axis of symmetry. The shear centre lies at point, O (shear centre) on this axis. The plane of loading is assumed to be parallel to the web. The directions of shear flows $V_{lip\,(bottom)}, V_{flange}, V_{web}, V_{flange}$ and $V_{lip(top)}$ is along FE, ED, DC, CB and BA, respectively as shown in Fig. A.11 from tension flange towards compression flange. From symmetry $V_{lip(bottom)} = V_{lip\,(top)}$ and $V_{fl.(bottom)} = V_{fl\,(top)}$. The resultant shear flows may be found as under.

(1) Shear in the lip element, FE

Let s_1, be the distance from F (free tip). Then, the shear stress

$$\tau_{lip} = \left(\frac{-V_y}{I_{ZZ}\,.\,t}\right)\left[S_1 \times t\left(350 - \frac{s_1}{2}\right)\right] \quad \text{(i)}$$

Total shear in this lip element = (Stress x area)

$$V_{lip\,(bottom)} = \left(\frac{-V_y}{I_{zz}}\right)\int_0^{100}\left[s_1\left(350 - \frac{1}{2}s_1\right)t\,.\,ds_1\right]$$

$$= \left(\frac{-15.84 \times 10^6\,V_y}{I_{zz}}\right) \quad \text{(ii)}$$

2. Shear in the flange element, ED

From triangle DEE_1, (100/DE) = sin 30°, 0.5, DE = 200 mm

Let s_2 be the distance along inclined flange, ED. Then, the shear stress in this flange

$$\tau_{lip} = \left(\frac{-V_y}{I_{zz}} \times \frac{1}{t}\right)\left[100 \times 10 \times (150 + 100 + 50) + 10\,s_2\left(150 + 100 - \frac{s_2}{4}\right)\right]$$

$$= \left(\frac{-V_y}{I_{zz}}\right)\left[100 \times 300 + s_2\left(250 - \frac{s_2}{4}\right)\right] \quad \text{(iii)}$$

Total shear in this flange element (stress x area)

$$V_{fl} = \left(\frac{-V_y}{I_{zz}}\right)\int_0^{200}\left[\left(100 \times 300 + 250\,s_2 - \frac{s_2^2}{4}\right)10\,ds_2\right]$$

$$= \left(\frac{-V_y \times 10}{I_{zz}}\right)\left[100 \times 300 \times s_2 + \frac{250}{2}s_2^2 - \frac{1}{4} \times \frac{s_2^3}{3}\right] = \left(-\frac{65.84 \times 10^6\,V_y}{I_{zz}}\right) \quad \text{(iv)}$$

3. Shear in the Web element, DC

Let s_3 be the distance from D. Then, the shear stress in this element

$$\tau_{web} = \left(\frac{-V_y}{I_{zz}\,.\,t}\right)\left[100 \times 10 \times 300 + 200 \times 10 \times 200 + s_3 \times 10 \times \left(150 - \frac{s_3}{2}\right)\right]$$

$$= -\frac{V_y}{I_{zz}}\left[7\times10^4 + 150\, s_3 - \frac{1}{2}s_3^2\right] \tag{v}$$

Total shear in the web element stress x area

$$V_{web} = \left(\frac{-V_y}{I_{zz}}\right)\int_0^{300}\left[\left(7\times10^4 + 150\, s_3 - \frac{1}{2}s_3^2\right)100\, ds_3\right] = \left(\frac{-172.5\times10^6\, V_y}{I_{zz}}\right) \tag{vi}$$

It is to note for the equilibrium of the vertical forces

$$2\, V_{lip} + 2\, V_{fl}\,\sin 30 + V_{web} = V_y$$

Take moment about the point, O (shear centre) at a distance, e from the middle of the web and equate the algegraic sum to zero.

$$2\, V_{lip}\,(e + 200\cos 30^\circ) + 2\, V_{fl}\times\sin 30^\circ\times e - 2\, V_{fl}\times\cos 30^\circ\times150) + V_{web}\times e = 0$$

$$2\left(\frac{-15.84\times10^6\, V_y}{I_{zz}}\right)\left(e + 200\times\frac{\sqrt{3}}{2}\right) + 2\left(\frac{-65.84\times10^6\, V_y}{I_{zz}}\right)\times\frac{1}{2}e$$

$$-2\left(\frac{-65.84\times10^6\, V_y}{I_{zz}}\right)\times\frac{\sqrt{3}}{2}\times150 + \left(\frac{-172.5\times10^6\, V_y}{I_{zz}}\right)\times e = 0$$

$$2\times(15.84)(e+173) + 2\left(65.84\times\frac{1}{2}\right)e - 2\left(65.84\times\frac{1.73}{2}\times150\right) + 172.5\, e = 0$$

$$31.68\, e + 31.68 \times 173 + 65.84\ \ e - (65.84 \times 173 \times 1.50) + 172\, e = 0$$

$$270.02\ \ e = 670.8 \times 173$$

$$e = \left(\frac{670.8\times173}{270.02}\right) = 42.98\text{ mm}$$

A.7 SHEAR CENTRE FOR A THIN-WALLED Z-BEAM CROSS-SECTION

A thin-walled Z-beam cross-section is shown in Fig. A.12. In case, it is assumed that the moment is applied in xy-plane only (i.e., $M_y = 0$). Then, the bending stress is given by Eq. A.11. Therefore,

$$\sigma_{bx} = \left(\frac{I_y\,.\,y - I_{yz}\,.\,Z}{I_{zz}\,.\,I_{yy} - I_{yz}^2}\right).\,M_{zz} \tag{i}$$

Consider a point at a distance s_1 (Fig. A.12 from free edge D. From Fig. A.12.

y = (h/2), and for the flanges

$$I_{yy} = \left(2\times\frac{1}{3}t\times b_f^3\right) = \frac{2}{3}t\, b_f^3 \tag{ii}$$

$$Z = \left(b_f - \left(\frac{1}{2}\right)s_1\right)$$ **and for the flanges product of inertia**

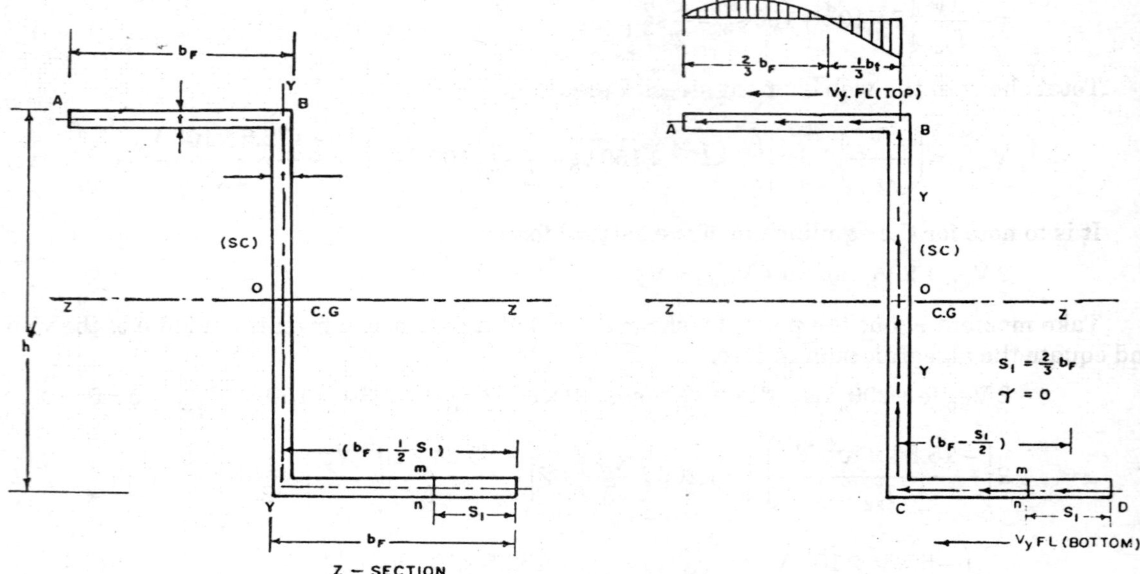

Fig. A-12

I_{yz} = (Area) (distance of cg. of complete flange from yy-axis) x (distance of c.g. from zz-axis)

$$= \left(b_f \cdot t \times \frac{b_f}{2} \times \frac{h}{2}\right)$$

Substituting above values in the expression (i)

$$\sigma_{bx} = \frac{h}{2} \cdot \left(\frac{2}{3} t \cdot b_f^3\right) + \left(b_f \cdot t \times \frac{bf}{2} \times \frac{h}{2}\right)\left(b_f - \frac{s_1}{2}\right) / \left(I_{zz} \cdot I_{yy} - I_{yz}^2\right) \quad \text{(iii)}$$

Shear stress, τ_{xz} from free edge D

$$\tau_{zz} = \left(\frac{V_y}{I_{zz} \cdot I_{yy} - I_{yz}^2}\right) \cdot \left(b_f^2\, h \cdot t \cdot s_1\right)\left(\frac{s_1}{4} - \frac{b_f}{6}\right) \quad \text{(iv)}$$

The expression (iv) shows the variation of the shear stress in the bottom flange. For $s_1 = 0$, $\tau_{xz} = 0$. Whereas, for $s_1 = b_f$, at point, C.

$$\tau_{xz} = \left(\frac{V_y}{I_{zz}\, I_{yy} - I_{yz}^2}\right) \cdot \left(\frac{b_f^4\, h \cdot t}{12}\right) \quad \text{(v)}$$

The shear stress τ_{xz} changes sign (i.e., it becomes zero) in between D and C, at $\left(\frac{s_1}{4} - \frac{b_f}{6}\right) = 0$. That is, $s_1 = \left(\frac{4}{6} b_f\right) = (2/3) \cdot b_f$ from the point D, free edge. The direction of shear stress in the top flange AB and the direction of shear flow may be found from the point of symmetry of Z-section about the centre of gravity, C.G. So also, the resultant shear flow [shear force per unit length = (τ_{xz} . t)] may be found from

$$V = \int (\tau . t . ds) = 0 . \quad \text{Therefore, } V_{fz} = \int_0^{b_f} \tau_{xz} (t. ds_1)$$

$$V_{fz} = \left(\frac{V_y}{I_{zz} . I_{yy} - I_{yz}^2} \right) . \left(b_f^2 . h . t^2 \right) \int_0^{b_f} s_1 \left(\frac{s_1}{4} - \frac{b_f}{6} \right) . ds_1 = 0 \qquad \text{(vi)}$$

Similarly, the resultant shear flow in the flange element CD, V_{fz} = 0. That is, $V_{fz.AB} = V_{fz.CD}$.

Therefore, $V_{fz.AB} = V_{fz.CD}$ = 0. The line of action of V_{web} coincides with the middle line of the web. The shear centre is located with the intersection of ZZ-axis. That is, the shear centre coincides with the centre of gravity of Z-beam cross-section.

A.8 SHEAR CENTRE FOR THIN WALLED OPEN I-SECTION WITH UNEQUAL FLANGES

The shear centre for a thin-walled I-section with unequal flanges positioned as shown in Fig. A.13 (i.e., with the web horizontal and the flanges vertical.

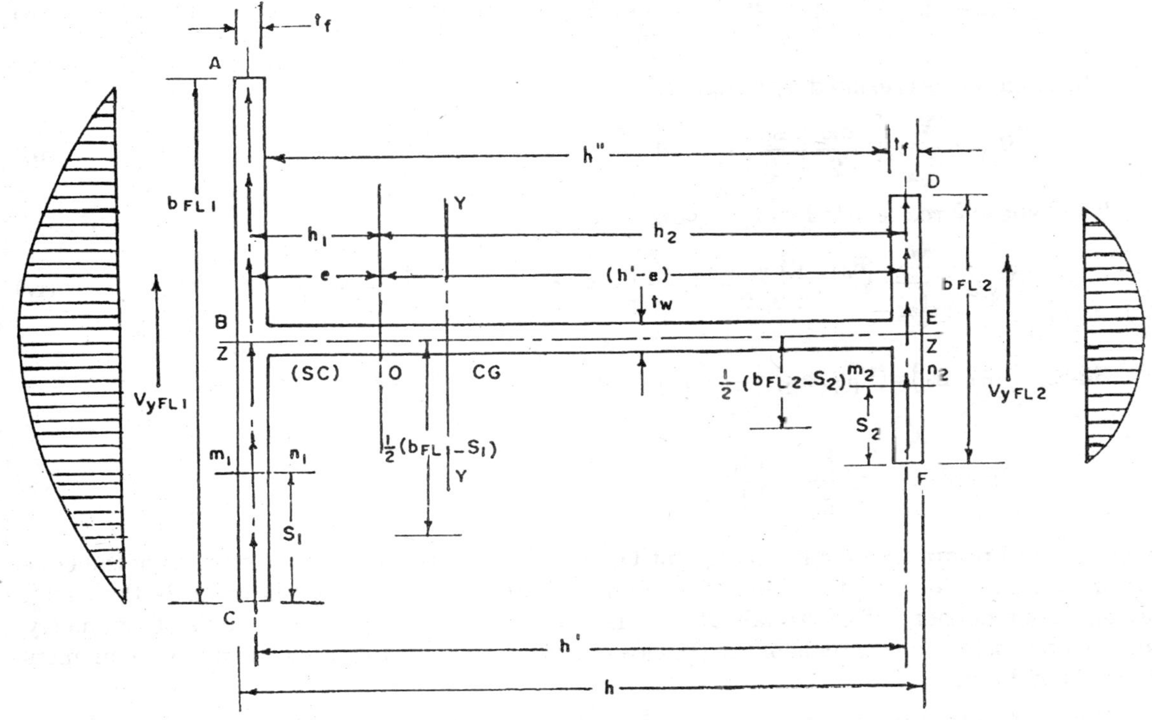

SHEAR DISTRIBUTION, SHEAR FLOW, SHEAR CENTRE A THIN-WALLED OPEN I-SECTION (UNEQUAL FLANGES)

Fig. A-13.

Let the transverse loads act in a plane perpendicular to zz-axis (i.e., vertically in a plane parallel to the flanges, which are vertical).

The shear stresses do not develop in the web (which one is horizontal). Since the product, $A.\bar{y}$ appearing in the expression to calculate the shear stresses in the numerator shall be zero. Total shear force, V_y shall be taken by the flanges

$$V_y = (V_{yfl} + V_{yf2})$$

where, V_{yfl} and V_{yf2} are the shear forces resisted by the flanges.

Let the flanges be considered as two separate beams and $I_{zz.1}$ and $I_{zz.2}$ shall be their moments of inertia, respectively. Their curvature and their deflection shall be equal. In case, the loads are distributed between the flanges in the ratio $I_{zz.1}$: $I_{zz.2}$. The shear stress in these flanges shall also be in the same ratio (the effect of shear forces on deflection is neglected). Let the loads are distributed in the ratio $I_{zz.1}$ and $I_{zz.2}$. Consider a section mn from bottom.

The shear stress at a distance resisted by are flange

$$\tau_{zy} = \frac{V_y}{I_{zz} . t_f}\left(\frac{b_{fl}}{2} - \frac{s_1}{2}\right) t_f , s_1 = \frac{V_y}{I_{zz}}\left(\frac{b_{fl} . s_1}{2} - \frac{s_1^2}{2}\right) \qquad \text{(i)}$$

Total shear stress resisted by this flange

$$V_{yfl1} = 2\frac{V_y}{I_{zz}} - \int_0^{b_{fl}/2}\left(\frac{b_{fl2} . s_2}{2} - \frac{s_2^2}{2}\right) ds_2 . t_f . = \left[\frac{V_y t_f}{I_{zz}} \times \frac{b_{f1}^3}{12}\right] = \left(\frac{V_{y1}}{I_{zz1}} \times I_{zfl_2}\right) \qquad \text{(ii)}$$

The shear stress resisted by other flange

$$V_{yf2} = \frac{V_y}{I_{zz}} . \left[\left(\frac{b_{f2}}{2} - \frac{s_2}{2}\right) s_2 . t_f\right] \qquad \text{(iii)}$$

Total shear force resisted by this flange

$$V_{yfl2} = \frac{V_y}{I_{zz}} \int_0^{b_{fl2}} \left(\frac{b_{f2} . s_2}{2} - \frac{s_2^2}{2}\right) ds_2 . t_f , \quad V_{yf2} = \left(\frac{V_y}{I_{zz}} \times I_{zfl_2}\right) \qquad \text{(iv)}$$

$$\text{Since, } (V_{fl1} . h_1) = (V_{fl2} . h_2) . \left(\frac{V_y}{I_{zz}} . I_{z.fl1} \times h_1\right) = \left(\frac{V_y}{I_{zz}} . I_z\, f_{12} \times h_2\right)$$

$$\text{Therefore, } \left(\frac{I_{z.fl_1}}{I_{z.fl_2}}\right) = \left(\frac{h_2}{h_1}\right) \qquad \text{(v)}$$

where, h_1 and h_2 are the distances of point O (i.e., shear centre) from the centroids of the cross sections (middle lines) of the flanges. The shear centre (i.e., point, O) is towards the flange having larger moment of inertia about zz-axis. The shear centre lies on the axis of symmetry. The distance between centroids of the flanges is divided into the inverse ratio as to the moment of inertia of flanges.

It is to note that in case as shown in Fig. A.14 (A), in which one of flanges vanishes. It may be assumed with sufficient accuracy that the point, O coincides with the centroid of the flange and that the transverse loads should act in the vertical plane through this point O, in order to have simple bending and the torsion is eliminated. It is also explained in next section.

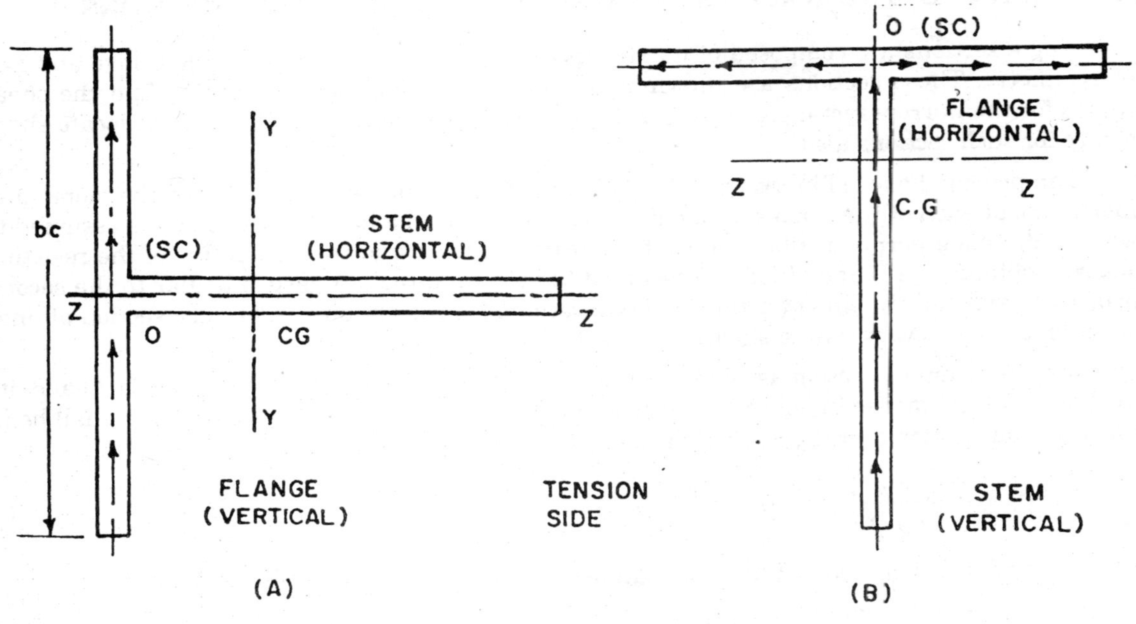

SHEAR FLOW, SHEAR CENTRE
THIN-WALLED TEE – SECTIONS

Fig. A-14

A.9 SHEAR CENTRE FOR THIN-WALLED TEE-SECTION

A tee-section may be positioned either as shown in Fig. A.14 (A) or as shown in Fig. A.14 (B).

In Fig. A.14 (A), the stem of tee-leg is horizontal while the flange is vertical. The transverse loads act in a plane perpendicular to zz-axis (i.e., in a plane parallel to the middle line of the flange).

The shear stresses do not develop in the stem (which one horizontal in Fig. A.14 (A). Since, the product $A\bar{y}$ appearing in the expression to calculate the shear stresses in the numerator, shall be zero. Total shear force shall be resisted by the flange (which is vertical, Fig. A.14(A). The point, O is the point of intersection of middle lines of the stem and the flange. The point, O shall be shear centre. It lies on the axis of symmetry. The transverse plane of loading (vertical plane) passes through the shear centre, so that there is no twisting.

In Fig. A.14 (B), the flange of tee-section is horizontal and the stem is vertical. YY-axis is the axis of symmetry. The transverse loads act in a vertical plane parallel to yy-axis. The shear force is resisted by the stem (which one is vertical) and so also by the flange (which one is horizontal). The point, O is the point of intersection of the middle lines of both the elements, and it is shear centre. It lies on the axis of symmetry (i.e., yy-axis, Fig. A.14 (B)). The transverse loads act in vertical plane which passes through the shear centre, so that, there shall be no twisting of the section.

A.10 SHEAR CENTRE FOR A THIN-WALLED EQUAL LEG ANGLE SECTIONS

Angle sections are composed of narrow rectangles but these sections do not have any axis of symmetry. These sections are unsymmetrical sections. The method used to find the shear centre for the narrow rectangles sections having axis of symmetry may be used to locate shear centre for such sections also.

The principal axes of inertia passing through of the section are located, and the moment of inertia about each of these axes is calculated. The plane of the loads on the beam is assumed to act in a direction perpendicular to one of the principal axes, and the action line of the resisting shear is obtained. A plane of load is assumed to act in a direction perpendicular to the second principal axis, and the line of action of the resisting shear is obtained. The intersection of lines of action of these two resisting shears locates the shear centre.

An equal angle section is shown in Fig. A.15(1). The length of each leg is l, and its thickness is t. In case of bending in a vertical plane, the shear stress, τ at points mn shall be in the direction as shown in Fig. A.15(1). The shear stress shall be equal to

$$\tau = \left(\frac{V_y}{I_{zz} \cdot t}\right) \int y \, dA \tag{i}$$

(Note : It is calculated in the same manner as in the case of channel section).

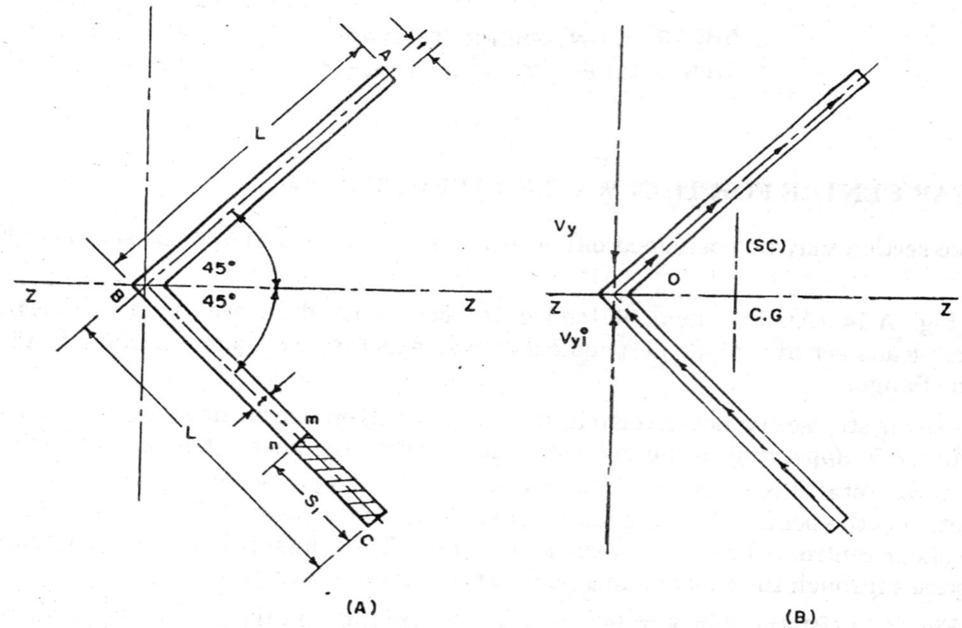

SHEAR FLOW, SHEAR CENTRE A THIN – WALLED EQUAL ANGLE SECTION

Fig. A-15(1)

In the expression (i) the integral sign represents the moment of the shadded area with respect to zz-axis. These shear stresses give a resultant force in the direction as shown in Fig. A.15(1)(B). The shear stress

$$\tau = \frac{-V_y}{I_{zz} \cdot t} \, y \, dA \; = \; \frac{-V_y}{I_{zz} \cdot t}\left(1 - \frac{u}{2}\right) \sin 45^\circ \, u.t \qquad \text{(i)}$$

Total shear force in the leg CB (stress x Area)

$$V_{BC} = \frac{-V_y}{I_{ZZ}} \int_0^l \left(1 - \frac{u}{2}\right) \sin 45^\circ \, u \, . \, t \, du = \left(\frac{-V_y}{I_{zz}} \, \frac{l^3 \, t}{3\sqrt{2}}\right) \qquad \text{(ii)}$$

A force of equal magnitude shall also be found for the upper leg. The resultant of these two forces V_{CB} and V_{BA} shall be equal to V_{yi} (internal shear force resisting the external shear force, V_y) as shown in Fig. A.15(B). It passes through, O (the point of intersection of middle lines of the legs). Point O is the shear centre for the equal angle section. It lies on the axis of symmetry zz-axis. The applied vertical load (external shear force, V_y) should pass through this point. There shall be no twisting moment.

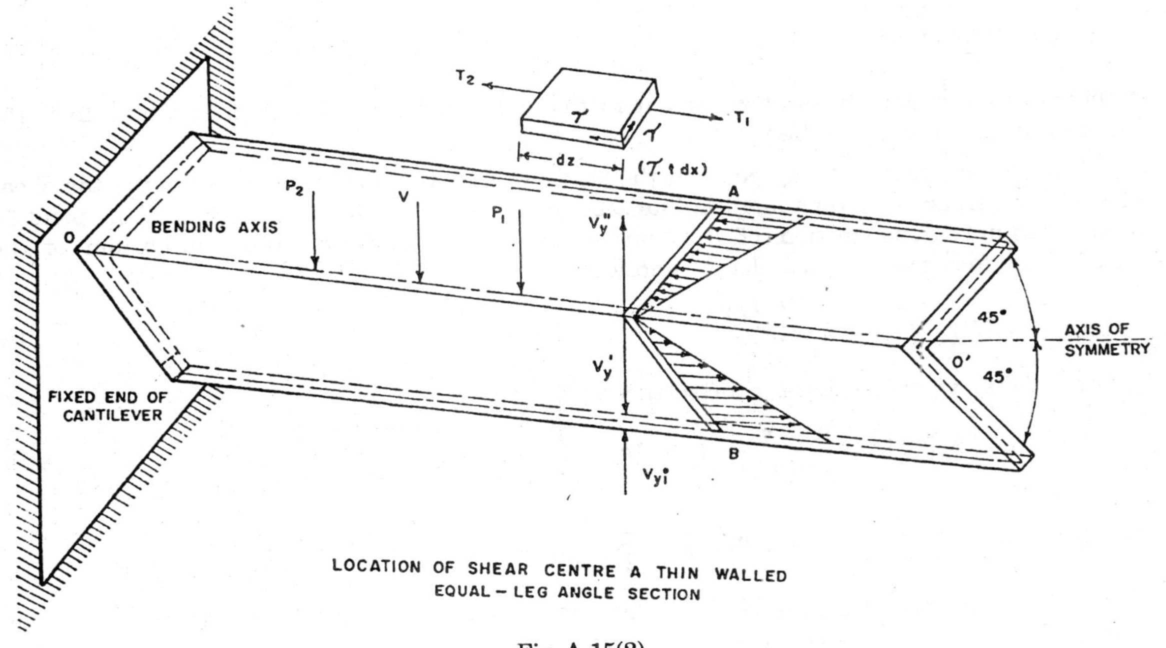

Fig. A-15(2)

A.11. LOCATION OF SHEAR CENTRE

Consider an equal-leg angle section used as a cantilever beam. It is subjected to transverse bending loads in a plane perpendicular to the axis of symmetry of the cross-section, as shown in Fig. A.15(2). The bending loads produce shear in the beam. The beam is prevented from twisting as it bends. The plane of loads contains the line of action of the resultant resisting shear force V_{yi} acting at each cross-section.

In case, the loads P_1 and P_2, etc. in Fig. A.15 (2) are replaced by their resultant V, and in case, V is resolved into a force V′ at 0 in any section AB and a couple composed of force V and V′′. It is done by introducing two equal and opposite collinear forces in of the plane of loads at 0 such that V' = V" = V.

The bending couple composed of V and V" is held in equilibrium by a resisting moment at any transverse section of the beam in accordance with the flexural formula, the bending stress in each long of the angle varying as shown at section AB.

Likewise, the vertical shear V' at the section is held in equilibrium by a resisting shear force V_{yi} at the section provided that V' is collinear with V_R.

The line of action of the resultant shear V_{yi} in any section such as AB in Fig. A.15 (2) passes through the point of intersection of its components V_1 and V_2, which for these thin sections lie along the centre lines of the upper and lower legs, respectively as shown in Fig. A.15 (2). Point O in Fig. A.15 (2) is the shear centre and the axis OO′, is the axis of bending.

V_1 and V_2 act along the centre lines of the legs of the angle and these are the two components of the resultant shear force V_{yi}.

The force V_1 is found by adding up the differential shear forces τ . dA (Fig. A.15(2)) over the complete area of the upper leg; i.e.

$$V_1 = \int \tau \, dA \qquad \text{(i)}$$

where, τ is the shear stress at any point in the cross-section of the upper leg, and dA is the differential area including the point.

The shear stress τ at any point in the transverse section is obtained by finding the shear stress at the same point on a longitudinal section of the beam. The shear stress at any point on a longitudinal plane section of a beam is found by considering the equilibrium of the longitudinal forces acting on a block. Therefore,

$$\tau = \left(\frac{V \, . \, A \, \bar{y}}{I \, . \, t}\right) = \left(\frac{V \, . \, Q}{I \, . \, t}\right) \qquad \text{(ii)}$$

where, V is the vertical shear for the section of the beam.

$$V_1 = \int \tau dA = \int_0^L \frac{V}{I_{zz} \, . \, t} A \, . \, \bar{y} \, . \, dA = \frac{V_y}{I_{zz} \, . \, t} \int_0^L A \, \bar{y} \, dA \qquad \text{(iii)}$$

$$A = t \, . \, (L - s) \, . \, ds \, . \, dA = tds \text{ , and } \bar{y} = \frac{\sqrt{2}}{2}\left(\frac{L+s}{2}\right), \text{ and}$$

where, s is measured along the leg from the point of intersection of the middle line of two legs, or from the outer corner of the legs, since the sections are thin. Hence,

$$V_1 = \left(\frac{\sqrt{2}}{4} \frac{V_y \, t}{I_{zz}}\right) \int_0^L \left(L^2 - s^2\right) ds$$

$$V_1 = \frac{\sqrt{2}}{6} \cdot \left(\frac{V_y \, . \, t \, . \, L^3}{I_{zz}}\right) = \frac{V_y \, . \, L^3 \, . \, t}{I_{zz} \, . \, 3\sqrt{2}} \qquad \text{(iv)}$$

Since, $I_{zz} = \left(t \, . \, L^3/3\right)$, the shear force V, on the area of the upper leg may be written as under

$$V_1 = \frac{\sqrt{2}}{2} \, . \, V_y \qquad \text{(v)}$$

The shear force, V_2 in the lower leg is obtained by above procedure and it is equal in magnitude to V_1 and it acts along the middle line of the lower leg, Fig. A.15(1). Forces V_1 and V_2 have inclination of 45° with the vertical. Since $V_1 = V_2 = (\sqrt{2}/2)\,V$, their resultant is a vertical force equal in magnitude to V.

Example A.4. A thin walled open I-section with unequal flanges is shown in Fig. A.16. Locate the position of shear centre.

Solution : ZZ-axis is the axis of symmetry. The shear centre lies on this axis. The shear force is resisted by the vertical flanges.

$$\left(V_{y.FL1} + V_{yFL.2} = V_y\right)$$

The shear stress in the web (which one is horizontal) and its middle line is the axis of symmetry (i.e., the neutral axis). For the portion of the web, product $(A.\bar{y})$ is zero. The normal shear force resisted by the web shall be zero, irrespective of the manner in which the bending moment varies. The directions of the shear force per unit length (i.e., shear flow) are also shown in Fig. A.16.

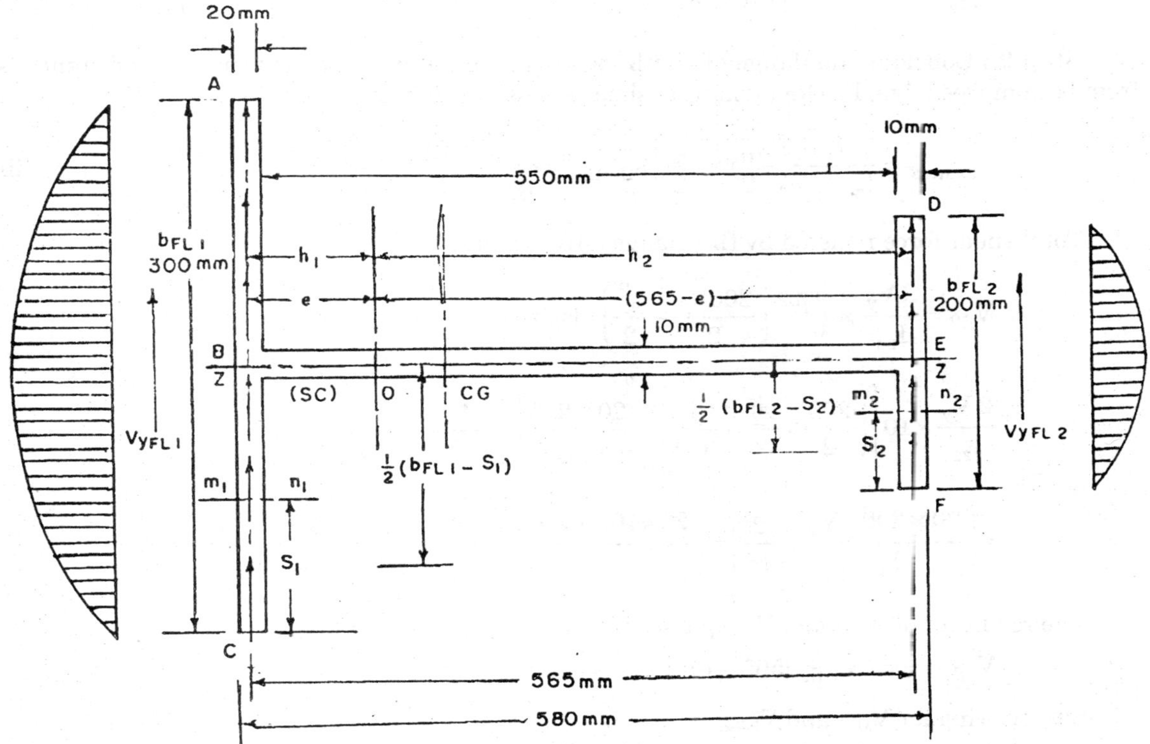

Fig. A-16

The distribution of shear stress in each flange is parabolic and are also in Fig. A.16. The position of the shear centre, the point is O is located as under.

Step 1. Consider the flange of width $b_{fl} = 300$ mm, and section mm at a distance s_1 from bottom free edge at C. The maximum shear stress in this flange.

$$\tau_{yzmax} = \left(\frac{V_y}{I_{zz}\,.\,t_{fl}}\right)\left(\frac{b_{fl}}{2} - \frac{s_1}{2}\right) t_{fl1}\,.\,s_1 = \left(\frac{V_y}{I_{zz}}\right)\left(\frac{300}{2} s_1 - \frac{s_1^2}{2}\right) s_1$$

$$\tau_{yz.max} = \frac{V_y}{I_{zz}}\cdot\left(\frac{300}{2} s_1 - \frac{s_1^2}{2}\right) \qquad \text{(i)}$$

Total shear force resisted by this flange (stress x area)

$$V_{yfl.1} = \frac{2\,V_y}{I_{zz}} \int^{bfl\,1/2} \left(\frac{300}{2} s_1 - \frac{s_1^2}{2}\right) ds_1\, t_{fl1} = \frac{2\,V_y}{I_{zz}} \times \left[\left(\frac{300}{2} \times \frac{s_1^2}{2} - \frac{s_1^3}{6}\right)\right] t_{fl1}$$

$$\frac{2\,V_y}{I_{zz}} \times 20 \times 300^3 \left[\left(\frac{1}{16} - \frac{1}{48}\right)\right] = 54 \times \frac{10^7}{I_{zz}} V_y \times \left(\frac{2}{48}\right) = 45 \times \left(\frac{10^6 \times V_y}{I_{zz}}\right) \qquad \text{(ii)}$$

Step 2 : Consider the flange of width. $b_{fl2} = 200$ mm and the section m_2n_2 at a distance, s_2 from bottom free edge, F. The maximum shear stress in this flange.

$$\tau_{yz.max} = \frac{V_y}{I_{zz}\, t_{fl2}}\left(\frac{b_{fl2}}{2} - \frac{s_2}{2}\right) s_2\, t_{fl2} = \frac{V_y}{I_{zz}}\left(\frac{200\, s_2}{2} - \frac{s_2^2}{2}\right) \qquad \text{(iii)}$$

Total shear force resisted by this flange (stress x area)

$$V_{fl2} = \frac{2\,V_y}{I_{zz}} \times \int_0^{bfl.2} \left(\frac{200\, s_2}{2} - \frac{s_2^2}{2}\right) ds_2\, t_{fl2}$$

$$\frac{2\,V_y}{I_{zz}} \times 10 \left[\left(\frac{200}{2} \times \frac{s_2^2}{2} - \frac{s_2^3}{6}\right)\right] = \frac{20 \times 200^3}{I_{zz}} \left[\left(\frac{1}{16} - \frac{1}{48}\right)\right]$$

$$= \left(\frac{20 \times 200^3\, V_y}{I_{zz}} \times \frac{2}{48}\right) = \left(\frac{20 \times 10^6 \times 2 \times 2^2 \times V_y}{I_{zz}} \times \frac{2}{48}\right) = \left(\frac{6.67 \times 10^6\, V_y}{I_{zz}}\right)$$

The resultant of $V_{y.fl1}$ and $V_{y.fl2}$ passes through the shear centre point O.

$$V_{yfl.1} \cdot e = V_{y.fl2}\,(565 - e)$$

Substituting for $V_{yfl.1}$ and $V_{y.fl2}$

$$\left(\frac{45 \times 10^6\, V_y}{I_{zz}}\right) \cdot e = \left(\frac{6.67 \times 10^6 \times V_y}{I_{zz}}\right), \; 45\, e = 6.67\,(565 - e)$$

$$e\,(45 + 6.67) = 6.67 \times 565, \quad e = \left[\frac{6.67 \times 565}{51.67}\right] \; \mathbf{72.935\ mm}$$

I. Alternatively

The distribution of shear stress for both the rectangular area of the flanges is parabolic. Maximum ordinate for the portion CB, of the flange AC

$$\tau_{CB} = \left(\frac{V_Y}{I_{ZZ}}\right)_{tfl} \times A\,\bar{y}$$

$$\tau_{CB} = \left(\frac{V_Y}{20 \times I_{ZZ}}\right)(20 \times 150 \times 75) = \left(\frac{225000\,V_y}{20 \times I_{ZZ}}\right) \quad \text{(i)}$$

Maximum ordinate for the portion FE, of the flange DF

$$\tau_{FE} = \left(\frac{V_Y}{10 \times I_{ZZ}}\right) \times (10 \times 100 \times 150) = \left(\frac{5000\,V_y}{I_{ZZ}}\right) \quad \text{(ii)}$$

The resultant of shear stress in the flanges AC and DF may be obtained as below

= (area of shear stress diagram) x thickness of flange

$$\therefore V_{Y.AC} = \left(\frac{225000}{20}\frac{V_y}{I_{ZZ}}\right) \times \left(\frac{2}{3} \times 300\right) \times 20 = \left(\frac{450 \times 10^5 \times V_y}{I_{zz}}\right) \quad \text{(iii)}$$

$$V_{Y.DF} = \left(\frac{50 \times V_y}{10 \times I_{ZZ}}\right) \times \left(\frac{2}{3} \times 200\right) \times 10 = \left(\frac{200}{3}\right) \times \frac{10^6\,V_y}{I_{ZZ}} = \left(\frac{6.67 \times 10^6 \times V_y}{I_{ZZ}}\right) \quad \text{(iv)}$$

The resultant of $V_{Y.AC}$ and $V_{y.DF}$ passes through the shear centre, O. Therefore,

$$V_{y.AC} \cdot e = V_{y.DF}\,(565 - e)$$

$$\left(\frac{4500 \times 10^6\,V_y}{I_{ZZ}}\right).e = \frac{6.67 \times 10^6 \times V_y}{I_{ZZ}}\,(565 - e)$$

$$(45 + 6.67).e = (6.67 \times 565), \quad e = \left(\frac{6.67 \times 565}{51.67}\right) = \mathbf{72.935\,mm}$$

II. Alternatively. Moments of inertias for the rectangular flanges are as under:

$$I_{ZZ.AC} = \left(\frac{1}{12} \times t_{f1} \times b_{f1.1}\right) = \left(\frac{1}{12} \times 20 \times 300\right) \text{mm}^4$$

$$= \left(\frac{540}{12} \times 10^6\right) = 45 \times 10^6\ \text{mm}^4 \quad \text{(i)}$$

$$I_{ZZ.DF} = \left(\frac{1}{12} \times 10 \times 200^3\right) = 6.67 \times 10^6\ \text{mm}^4$$

Let h_1 = e Fig. A.16, and h_2 = (565 – e), Fig. A.16

These distances of the point O, are inversely proportional to the moment of inertias of the respective flanges. Therefore,

$$\left(\frac{h_1}{h_2}\right)=\left(\frac{6.67\times10^6}{45\times10^6}\right),\quad (45+6.67)e=(6.67\times565)$$

$$e=\left(\frac{6.67\times565}{51.67}\right)=\mathbf{72.935\ mm}$$

Example A.5 : A channel-section with flanges sloping 30° with ZZ-axis (i.e., horizontal symmetrical axis) is shown in Fig. A.17 (A). Locate the shear centre for the cross-section.

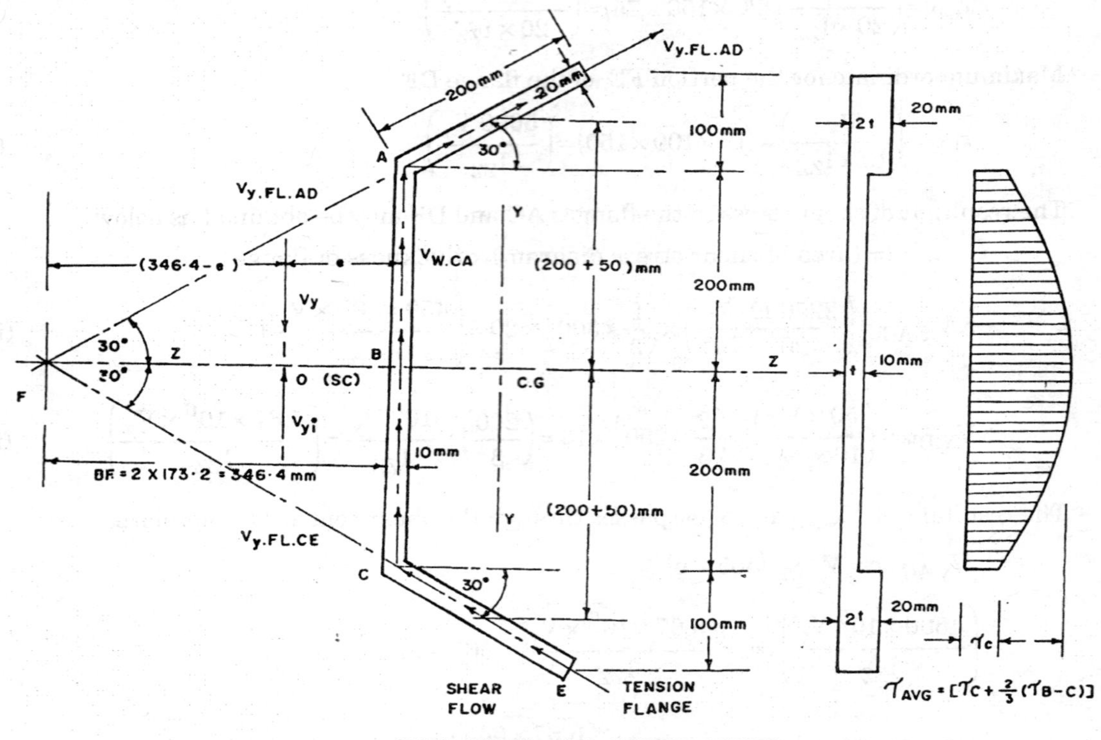

Fig. A-17

Solution :

For the simplicity, the actual cross-section may be approximated by the cross-section as shown in Fig. A.17(B). The moment of inertia about zz-axis for the cross-section, Fig. A.17(B) may be closely approximated to that the actual cross-section, Fig. A.17(A).

$$I_{ZZ}=\left(\frac{1}{12}\times20\times600^3\right)-\left(\frac{1}{12}\times10\times400^3\right)\ \text{mm}^4$$

$$= (360\times10^6 - 53.33\times10^6) = 306.67\times10^6\ \text{mm}^4$$

The shear flow forces $V_{y.FL.CE}$, $V_{YW.CA}$ and $V_{Y\ FL\ AD}$ are developed in the three elements of the cross-section (where, $V_{y\ .\ FL\ .\ CE} = V_{Y\ .\ FL\ AD}$). The magnitude of forces needs integration. Therefore, it is convenient to use point, F, Fig. A.17(A) as the centre for moments. The lines of action of shear flow forces $V_{y.FL.CE}$, $V_{YW.CA}$ both passes through the point, their moment about point, F shall be zero. As such, the magnitudes of these forces are not needed. The shear flow force from C to B to A varies parabolically the average shear flow in equal to the shear flow at C plus (2/3) the difference between the shear flow at B and the shear flow at C.

$$\tau_c = \left(\frac{V_y \,.\, A\,\bar{y}}{I_{zz}}\right) = \frac{V_y}{I_{zz}} \times (10 \times 200)(200+50) = \frac{V_y}{I_{zz}} \times (2000 \times 250) = \left(50 \times 10^4 \,\frac{V_y}{I_{ZZ}}\right)$$

$$\tau_B = \tau_C + \frac{V_y}{I_{zz}} \times (200 \times 10) \times (100) = \left[\left(50 \times 10^4 \,\frac{V_y}{I_{ZZ}}\right) + \left(\frac{V_y}{I_{ZZ}} \times 20 \times 10^4\right)\right] = \left(70 \times 10^4 \,\frac{V_y}{I_{ZZ}}\right)$$

Average shear stress

$$\tau_{ave} = \left[\tau_c + \frac{2}{3}(\tau_B - \tau_C)\right] = \left[50 \times 10^4 + \frac{2}{3}(70-50) \times 10^4\right] \frac{V_y}{I_{ZZ}}$$

$$= \left[\left(50 + \frac{2 \times 20}{3}\right)\right] \times 10^4 \,\frac{V_y}{I_{ZZ}} = \left(63.33 \times 10^4 \,\frac{V_y}{I_{ZZ}}\right)$$

$$V_{W.CA} = (\tau_{ave} \times 400) = \left(63.33 \times 400 \times 10^4 \,\frac{V_y}{I_{ZZ}}\right) = 253.32 \times 10^6 \left(\frac{V_y}{I_{ZZ}}\right)$$

The clockwise moment of V_y and the anti-clockwise moment of $V_{YW.CA}$ about the point, F (centre for moments) are equal.

$$(346.4 - e)\, V_y = 346.4.\ V_{YW.CA}$$

$$(346.4 - e)\, V_y = \frac{346.4 \times 253.32}{306.67 \times 10^6} \times 10^6 \,\frac{V_y}{I_{ZZ}}$$

$$(346.4 - e)\, V_y = \left(\frac{346.4 \times 253.32 \times 10^6}{306.67 \times 10^6} \times V_y\right)$$

$$(346.4 - e) = (346.4 \times 0.826),\ e = (-346.4 + 286.138)$$

$$\mathbf{e = 60.26\ mm}$$

A.12. SHEAR CENTRE FOR A THIN-WALLED SECTION (AN ARC OF A CIRCLE)

A thin-walled section (an arc of a circle subtending an angle 20° at the centre) is shown in Fig. A.18. The position of shear centre may be located as follows.

Consider an element $d\alpha$, subtending angular distance, α from the free edge, B. Angle subtended at the centre, C by an arc = D_1D_2

$$d\phi = D_1\, D_2 / R, \quad \therefore\ R.\, d\phi = D_1 D_2$$

Shear stress at the point, D.

$$\tau = \frac{V_y}{I_{zz}\,.\,t}(A\,.\,\bar{y}) = \frac{V_y}{I_{zz}\,.\,t}\left[(R\, d\phi\, t).\, R \sin(\theta - \phi)\right]$$

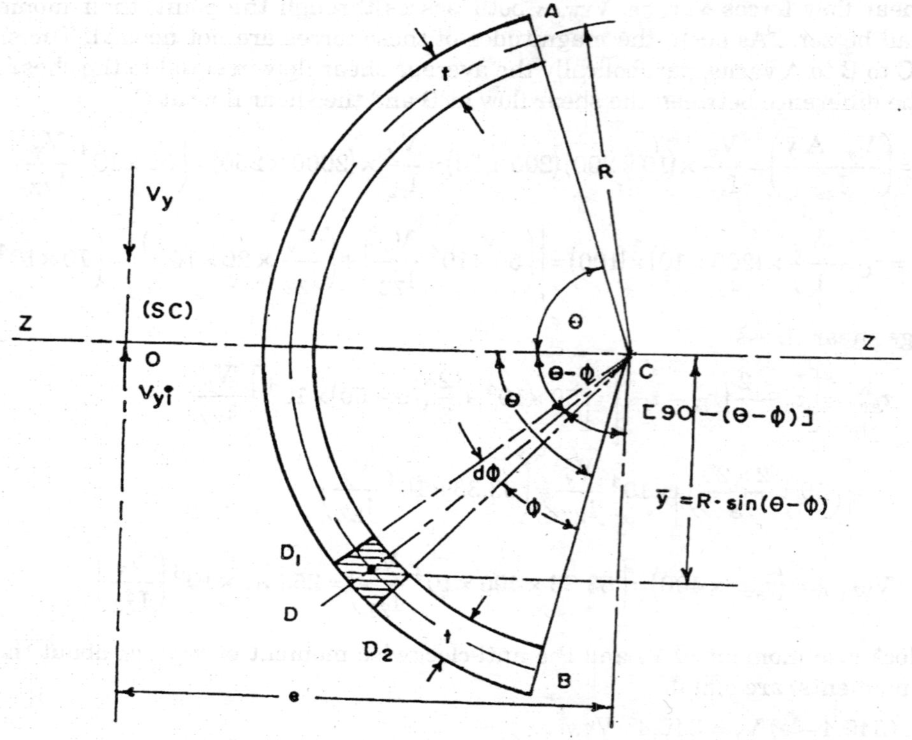

SHEAR CENTRE FOR A THIN-WALLED ARC OF A CIRCLE

Fig. A-18

Shear force (stress x area)

$$\left(\frac{V_y}{I_{ZZ}}\right)\int_0^{\alpha} R\,d\phi\,.\,R\sin(\theta-\phi).\,t = \left(\frac{R^2 t.\,V_y}{I_{zz}}\right)\left[\cos(\theta-\phi)\right]_0^{\alpha}$$

$$= \left(\frac{R^2\,t.\,V_y}{I_{ZZ}}\right)\left[\cos(\theta-\alpha)-\cos\theta\right] \quad \text{(i)}$$

Moment of inertia = (area) $(\bar{y})^2$

$$I_{ZZ} = 2\int_0^{\theta}(R.\,d\phi.\,t)\left[R.\sin(\theta-\phi)\right]^2$$

$$= R^3.\,t.\,(\theta-\sin\theta.\cos\theta) \quad \text{(ii)}$$

Substituting this in the expression (i)

$$V_{y1} = \left[\frac{R^2\,t\,V_y\left[\cos(\theta-\alpha)-\cos\theta\right]}{R^3.\,t\,(\theta-\sin\theta\cos\theta)}\right] \quad \text{(iii)}$$

$$V_{y1} = \left(\frac{V_y}{R}\right)\left(\frac{\cos(\theta - \alpha) - \cos\theta}{\theta - \sin\theta . \cos\theta}\right) \quad \text{(iv)}$$

The resultant of shear stress over complete section acting at the point O, (shear centre). Take moment about the centre, C

$$V_{yi} . e = \frac{V_y}{R} \int_0^{2\theta} . R . d\alpha . t . R \quad \text{(v)}$$

$$e = 2R\left(\frac{(\sin\theta - \theta\cos\theta)}{\theta - \sin\theta . \cos\theta}\right)$$

In case, the arc of the circle is in the shape of a semi-circle ($2\theta = \pi$), $\theta = \pi/2$), then, the distance of shear centre, O from C

$$e = (4R/\pi) \quad \text{(A.33)}$$

Example A.6 Locate the position of the shear centre for the cross section of a beam as shown in Fig. A.19.

Solution : Shear flow in the section is as shown in Fig. A.19.

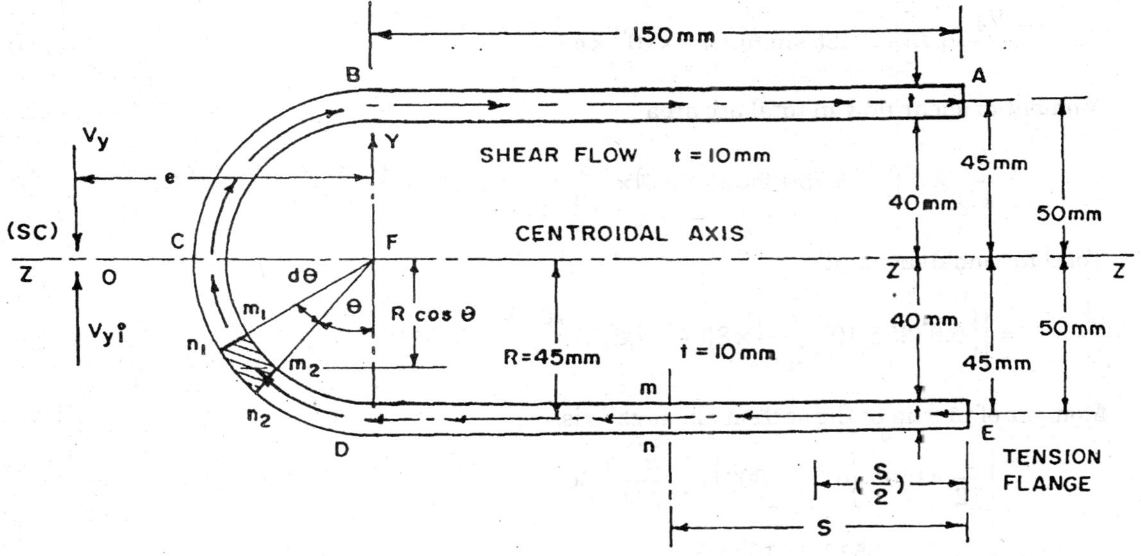

Fig. A-19

Step 1 : Shear force per unit length in the element ED.

Consider a section mn at a distance, s from E (free edge)

$$\tau_{ED} = \left(\frac{V_y}{I_{ZZ} . t}\right)[s . t \times 45] = \left(\frac{45\, V_y}{I_{ZZ}}\right) . s \quad \text{(i)}$$

shear force per unit length (i.e., shear flow in the complete element, ED (stress x area)

$$V_{Y.ED} = \frac{45\,V_y}{I_{ZZ}} \int_0^{150} s.ds.t_{ED} = \left(5.0625 \times 10^6\right)\frac{V_y}{I_{ZZ}} \quad \text{(ii)}$$

Step 2 : Shear force per unit length in the arc portion: Consider elementary area

= 10 x 45 d θ (iii)

Distance of it C.G. from neutral axis (i.e., zz-axis) = 45 cos θ (iv)

Moment of elementary area. = (10 x 45 dθ) (45 cos θ) (v)

$$\text{Total moment} = \int_0^{\theta} 10 \times 45^2 \times \cos\theta\, d\theta = \left(10 \times 45^2 \times \sin\theta\right) \quad \text{(vi)}$$

Shear flow in the elementary area

$$= \frac{V_y}{I_{ZZ}} \times \left(150 \times 10 \times 45 + 10 \times 45^2 \times \sin\theta\right) ds \times 10$$

$$= \frac{V_y}{I_{ZZ}} (6.75 + 2.025 \sin\theta) \times 4.5 \times 10^6\, d\theta \quad \text{(vii)}$$

Moment of shear flow in the elementary area about, F

$$\frac{V_y}{I_{ZZ}} (6.75 + 2.025 \sin\theta) \times 4.5 \times 10^6 \times 45 \quad \text{(viii)}$$

Moment of shear flow in total arc area

$$= \frac{V_y}{I_{zz}} \times 2 \int_0^{\pi/2} (6.75 + 2.025 \sin\theta) \times \frac{45^2}{10} \times 10^6 \times d\theta = \left(\frac{5114.29 \times 10^6 \times V_y}{I_{zz}}\right) \quad \text{(ix)}$$

Total torsional moment

$$= \left[\left(5.0625 \times 10^6 \frac{V_y}{I_{zz}}\right) \times 80 + 5114.2 \times \frac{10^6\, V_y}{I_{zz}}\right] = \left(5519.2 \times 10^6\right).\frac{V_y}{I_{zz}} \quad \text{(x)}$$

Moment of inertia of the section about zz-axis

$$= \left[\frac{1}{12} \times 150 \times \left(100^3 - 80^3\right) + \frac{\pi}{2 \times 64}\left(100^4 - 80^4\right)\right] = (9.55 \times 10^6)\ \text{mm}^4$$

$$= V_{y1} \times e = \frac{5519.2 \times 10^6 \times V_y}{9.55 \times 10^6}, \quad e = 577.93\ \text{mm}$$

Step 2 : The curved portion is an arc. of a circle of subtending an angle 180° at the centre, F. Consider the elementary area 5 x (R dt) = (5 x 45) d θ = 225 d θ.

The distance from horizontal axis (i.e., zz-axis which are is neutral axis)

= 45 cos θ (iii)

Elementary moment of area about neutral axis = $(225\, d\theta).(45 \cos\theta)$

$$\text{Total moment} = \int_0^{\theta = \frac{\pi}{2}} 20250 . \cos\theta . d\theta = (20250 \sin\theta)_0^{\pi/2} \quad \text{(iv)}$$

Shear flow in the elementary area = $\frac{V_y}{I_{ZZ}}[150 \times 10 \times 45]$

REFERENCES FOR SHEAR FLOW SHEAR CENTRES

1. Charles, G. Salmon, John, E. Johnson `Steel Structures Design and Behaviour' Second edition, 1980 (pp. 396-399).
2. Fred, B. Seely, Jamesh, O. Smith `Advanced Mechanics of Materials' Second edition, 1952 (pp. 97, 117, 122, 292).
3. Boris Bresler, T.Y. Lin, John B. Scalzi `Design of Steel Structures' Second edition, (pp. 292).
4. Stephen Timosphenko, `Strength of Materials' Part 1, Third edition (pp. 235-244).
5. Joseph E. Bowles, `Structural Steel Design', 1980 (p. 171).
6. G.H. Ryder, `Strength of Materials, Third edition 1961, (pp. 126-128).
7. Arthur, P. Boresi, Omar M. Side bottom, `Advanced Mechanics of Materials' Fourth edition, 1985 (pp. 3, 275-279, 316-351).
8. V.N. Vazirani, M.M. Ratwani, `Analysis of Structures' Vol. 1, 181 (pp. 569 - 583).
9. G.S. Pandit, S.P. Gupta, R. Gupta, `Theory of Structures, Vol. 1, 1999 (pp. 480-501).
10. Edwin H. Gaylord, Jr. Charles N. Gaylord, James E. Stallmeyer, `Design of Steel Structures' Third edition (pp. 381-385).

APPENDIX-B

UNSYMMETRICAL BENDING OF BEAMS

B.1 INTRODUCTION

The rolled steel sections and rectangular wooden beams are used as purlins in the roof trusses. The purlins are placed with their webs or sides perpendicular to the top chords of a roof truss as shown in Fig. B.1. The principal axes of rolled steel I-section, channel section, rectangular wooden section shown in Fig. B.1(A), (B) and (C) respectively are parallel to the sides of sections. Whereas Fig. B.1(D) shows rolled steel angle section. One leg of angle section is normal to the top chord of roof truss. It supports the normal component of loading. The principal axes of angle section are not parallel to its sides. In general, when a structural member is subjected to transverse loading, the plane of bending (or plane of loading) coincides with one of the principal axes. The neutral axis of the section coincides with other principals axes. Whereas in case of purlin, the line of action of resultant of vertical loads and normal component of wind load do not coincide with any of the principal axes. Therefore, the bending of purlins occurs in a plane other than the principal plane of section. Such a bending is known as **unsymmetrical bending**. The purlins are subjected to unsymmetrical bending.

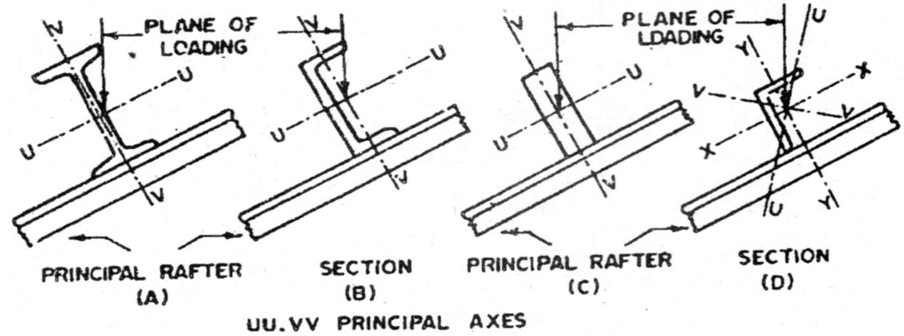

Fig. B.1

I-section shown in Fig. B.1 (A) and rectangular section shown in Fig. B.1 (C) have two axes of symmetry. The channel section shown in Fig. B.1 (B) has only one axis of symmetry. Whereas angle section shown in Fig. B.1 (D) has no axes of symmetry.

B.2 PRINCIPAL AXES OF SECTION

The moment of inertia of a section about any axis is defined as the second moment of an area about that axis. Fig. B.2 shows any plane section. The point O represents centre of gravity of the section. XX-axis and YY-axis represent two co-ordinates axes. The co-ordinates axes pass through this point O, and these are perpendicular to each other. (x, y) represent co-ordinates of any elementary area δA.

Then, $I_{xx} = (\Sigma y^2 \delta A)$, $I_{yy} = (\Sigma x^2 \delta A)$, $I_{xy} = (\Sigma xy \delta A)$, where,

I_{xx} = Moment of inertia of the section about xx-axis

I_{yy} = Moment of inertia of the section about yy-axis

I_{xy} = Product of inertia about xx-axis and yy-axis of section.

The product of inertia of a section is determined by taking moment of the area about one axis and then taking the moment of this moment about other axis.

Let UU and VV be the two perpendicular axes and these axes pass through point O as shown in Fig. B.2 (u, v) represent co-ordinates of the elementary area δA with reference to UU-axis and VV-axis.

Then, $I_{UU} = \Sigma v^2 \delta A$, $I_{VV} = \Sigma u^2 \delta A$, $I_{uv} = \Sigma uv\, \delta A$, where,

I_{UU} = Moment of inertia about UU-axis

I_{VV} = Moment of inertia about VV-axis

I_{UV} = Product of inertia about UU-axis and VV-axis.

The principal axes of a plane section are defined as two perpendicular axes in its plane passing through the centre of gravity of the section, such that the product of inertia is zero.

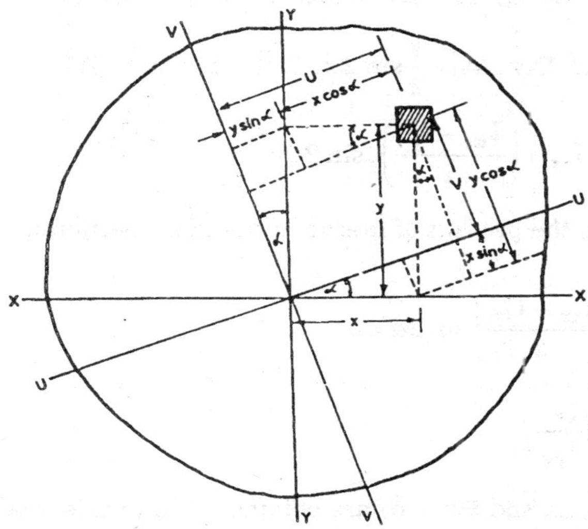

PRINCIPAL AXES OF A SECTION

Fig. B.2

If the product of inertia, I_{UV} is zero, then, UU-axis and VV-axis are known as **principal axes** of the plane section. The angle of inclination of principal axes with XX-axis and YY-axis is a as shown in Fig. B.2. If any plane section has an axis of symmetry, then, the product of inertia about that axis is zero. The axis of symmetry represents one of the principal axis. The other principal axis is at right angle to this axis of symmetry. If any plane section has two axes of symmetry, then both the axes of symmetry represent principal axes of the plane section. I-section and rectangular section are shown in Fig. B.1 (A) and Fig. B.1(C). UU-axis and VV-axis are two axes of symmetry. These axes are principal axes of the sections. The channel section shown in Fig. B.1 (B) has one axis of symmetry. This axis of symmetry UU is principal axis. The other principal axis VV is at right angle to this axis and passes through c.g. of section. Angle section shown in Fig. B.1(D) has no axis of symmetry. Such sections are called **unsymmetrical section**. The principal axes for such sections are located either by analytical method or by graphical method.

B.3. LOCATION OF PRINCIPAL AXES BY ANALYTICAL METHOD

The principal axes of a plane section shown in Fig. B.2 can be located by determining angle of inclination, α with UU-axis as below. The co-ordinates (u,v) of the elementary δ A with reference to UU-axis and VV-axis may be expressed in terms of co- ordinates (x, y) and α

$$u = (x \cdot \cos\alpha + y \cdot \sin\alpha) \text{ and } v = (y \cdot \cos\alpha - x \cdot \sin\alpha) \quad \text{(i)}$$

The product of inertia about UU-axis and VV-axis, $I_{UV} = \Sigma\, uv\, \delta A$ (ii)

$$= \Sigma (x \cdot \cos\alpha + y.\sin\alpha)(y.\cos\alpha - x.\sin\alpha)\,\delta A$$

$$= \left(\Sigma\, xy.\cos^2\alpha \cdot \delta A\right) - \left(\Sigma\, x^2 \cdot \sin\alpha\cos\alpha \cdot \delta A\right)$$

$$- \left(\Sigma\, xy.\sin^2 \cdot \alpha\,\delta A\right) + \left(\Sigma y^2 \cdot \sin\alpha\cos\alpha \cdot \delta A\right)$$

$$I_{UV} = \left(\cos^2\alpha - \sin^2\alpha\right)\Sigma xy \cdot \delta A + \sin\alpha \cdot \cos\alpha\left(\Sigma y^2 \cdot \sin\alpha\cos\alpha \cdot \delta A\right)$$

$$= \left(\cos^2 2\alpha \cdot \Sigma xy \cdot \delta A\right) + \frac{1}{2}\sin 2\alpha \cdot \left(\Sigma y^2 \cdot \delta A - \Sigma x^2 \cdot \delta A\right)$$

$$= \cos 2\alpha \cdot I_{xy} + \left(\frac{I_{xx} - I_{yy}}{2}\right) \cdot \sin 2\alpha \quad \text{(iii)}$$

For the principal axes, the product of inertia of the plane section about those axes is zero. Therefore, $I_{UV} = 0$

$$\cos 2\alpha \cdot I_{yy} + \left(\frac{I_{xx} - I_{yy}}{2}\right)\sin 2\alpha = 0 \quad \text{(iv)}$$

$$\tan 2\alpha = \left(\frac{-2\, I_{xy}}{I_{xx} - I_{yy}}\right) \quad \text{(B.1)}$$

From this, two values, (α and 90° + α) are obtained. UU-axis is inclined to XX-axis by an angle α, and VV-axis is inclined to the same axis by an angle (90° + α). Thus, two principal axis UU and VV of the plane section are located with reference to XX-axis.

The values of **principal moment of inertia** I_{UU} and I_{VV} can be found in terms of I_{xx}, I_{yy} and I_{xy} as below : $I_{UU} = \Sigma\, v^2 \cdot \delta A = \Sigma (y.\cos\alpha - x.\sin\alpha)^2\, \delta A$

$$\left(\Sigma\cos^2\alpha \cdot y^2\,\delta A\right) + \left(\Sigma\sin^2\alpha \cdot x^2\,\delta A\right) - \left(\Sigma\, 2\sin\alpha\cos\alpha \cdot xy\,\delta A\right)$$

$$= I_{xx} . \cos^2 \alpha . + I_{yy} . \sin^2 \alpha - I_{yy} . \sin 2\alpha \qquad (B.2)$$

$$= I_{xx} . \left(\frac{1+\cos 2\alpha}{2}\right) + I_{yy} . \left(\frac{1-\cos 2\alpha}{2}\right) - I_{xy} \sin 2\alpha$$

$$= \left(\frac{I_{xx}+I_{yy}}{2}\right) + \left(\frac{I_{xx}-I_{yy}}{2}\right) \cos 2\alpha - I_{xy} \sin 2\alpha$$

From value of tan 2 α, Eq. B.1 we have

$$\sin 2\alpha = \frac{-I_{xy}}{\left[\left(\frac{I_{xx}-I_{yy}}{2}\right)^2 + I_{xy}^2\right]^{1/2}} \quad \text{and} \quad \cos 2\alpha = \frac{\frac{1}{2}\left(I_{xx}-I_{yy}\right)}{\left[\left(\frac{I_{xx}-I_{yy}}{2}\right) + I_{xy}^2\right]^{1/2}}$$

Substituting the values of sin 2α and cos 2α,

$$I_{UU} = \left(\frac{I_{xx}+I_{yy}}{2}\right) + \left[\left(\frac{I_{xx}-I_{yy}}{2}\right)^2 + I_{xy}^2\right]^{1/2} \qquad (B.3)$$

Similarly, $I_{VV} = \sum u^2 \delta A = \sum(x . \cos \alpha + y . \sin \alpha)^2 \delta A$

$$= \left(\sum \cos^2 \alpha . x^2 \delta A\right) + \left(\sum \sin^2 \alpha . y^2 \delta A\right) + \left(\sum 2 \sin \alpha \cos \alpha . x . y . \delta A\right)$$

$$= I_{yy} \cos^2 \alpha + I_{xx} . \sin^2 \alpha + I_{sy} \sin 2\alpha \qquad (B.4)$$

$$= I_{yy} . \left(\frac{1+\cos 2\alpha}{2}\right) + I_{xx} . \left(\frac{1-\cos 2\alpha}{2}\right) + I_{xy} . \sin 2\alpha$$

$$= \left(\frac{I_{xx}+I_{yy}}{2}\right) - \left(\frac{I_{xx}-I_{yy}}{2}\right) \cos 2\alpha + I_{xy} \sin 2\alpha$$

$$I_{vv} = \left(\frac{I_{xx}-I_{yy}}{2}\right) - \left[\left(\frac{I_{xx}-I_{yy}}{2}\right)^2 + I_{xy}{}^2\right]^{1/2} \qquad (B.5)$$

Thus, the principal moment of inertias are

$$I_{uu} = \left(\frac{I_{xx}+I_{yy}}{2}\right) + \left[\left(\frac{I_{xx}-I_{yy}}{2}\right)^2 + I_{xy}{}^2\right]^{1/2} \qquad (B.6)$$

$$I_{VV} = I_{vv} = \left(\frac{I_{xx}+I_{yy}}{2}\right) - \left[\left(\frac{I_{xx}-I_{yy}}{2}\right)^2 + I_{xy}{}^2\right]^{1/2} \qquad (B.7)$$

If, the values of I_{UU} and I_{VV} are added, then,

$$(I_{UU} + I_{VV}) = (I_{xx} + I_{yy}) \qquad (B.8)$$

Thus, it is seen that the sum of moment of inertias about any two perpendicular axes of a plane section, passing through the centre of gravity of section remains unchanged.

B.4. LOCATION OF PRINCIPAL AXES BY GRAPHICAL METHOD

The principal axes UU and VV of a plane section shown in Fig. B.3 can also be located by graphical method. The values of principal moment of inertias I_{UU} and I_{VV} can also be found graphically. Following are the graphical methods for locating the principal axes and determining the values of I_{UU} and I_{VV}.

(i) Mohr's circle method, and (ii) Circle of inertia method.

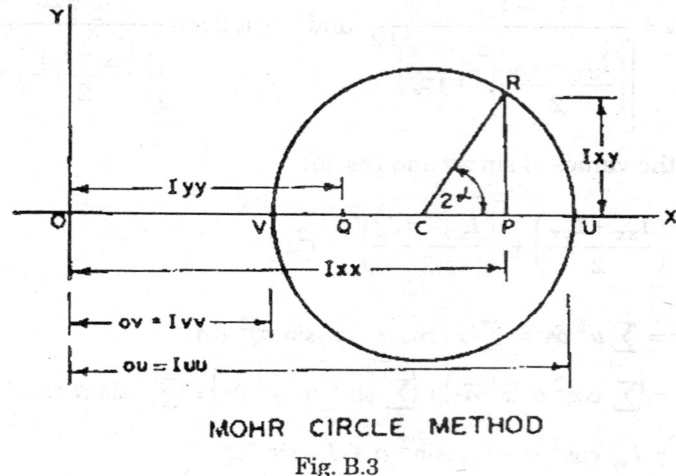

MOHR CIRCLE METHOD

Fig. B.3

1. Mohr's circle method

Let OX and OY be the two co-ordinate axes, passing through point O, as origin, as shown in Fig. B.3.

Plot OP = I_{xx} and OQ = I_{yy} on x-axis.

Draw PR = I_{xy} perpendicular to OX-axis at point P. PR is plotted upwards to OX-axis, if the value of I_{xy} is positive and it is plotted downwards to OX-axis, if its value is negative. The point C is obtained by bisecting the distance PQ. Joint C and R. With C as the centre, and CR as radius draw a circle as shown in Fig. B.3. The circle is known as **Mohr's circle**. The circle intersects OX-axis at two points U and V.

Distance PQ = $(I_{xx} - I_{yy})$. In the triangle PCR,

$$PC = \left(\frac{I_{xx} - I_{yy}}{2}\right),\ PR = I_{xy}\ ,\ CR = \left[\left(\frac{I_{xx} - I_{yy}}{2}\right)^2 + I_{xy}^{\ 2}\right]^{1/2}$$

$$\therefore \tan(\angle PCR) = \left(\frac{PR}{PC}\right) = \left(\frac{I_{xy}}{\frac{1}{2}(I_{xx} - I_{yy})}\right) = \left(\frac{2I_{xy}}{I_{xx} - I_{yy}}\right) = \tan(-2\alpha) \quad \therefore \angle PCR = 2\alpha$$

where, α represents the angle of inclination of principal axis UU of the plane section. The negative sign shown that angle is to be plotted in the negative direction, (i.e., clockwise). In Mohr's circle, if CR appears anti-clockwise to OX-axis, then angle is plotted clockwise. If CR appears clockwise to OX-axis, then, angle is plotted anti-clockwise.

Distance OU = (OC + CU) = (OC + CR) (CU = CK)

$$= \left(\frac{I_{xx} + I_{yy}}{2}\right) + \left[\left(\frac{I_{xx} + I_{yy}}{2}\right)^2 + I_{xy}{}^2\right]^{1/2} = I_{uu} \quad \text{(B.9)}$$

Distance OV = (OC – CV) = (OC – CR) (∴ CV = CR)

$$= \left(\frac{I_{xx} + I_{yy}}{2}\right) - \left[\left(\frac{I_{xx} - I_{yy}}{2}\right)^2 + I_{xy}{}^2\right]^{1/2} = I_{vv} \quad \text{(B.10)}$$

Thus, OU and OV in Mohr's circle method give principal moment of inertias I_{UU} and I_{VV}, respectively.

(ii) Circle of inertia method

Let OX and OY be the two co-ordinate axes passing through point, O, as origin as shown in Fig. B.4.

Plot OP = I_{xx} on y-axis and PQ = I_{yy} on x-axis.

Draw PR = I_{xy} perpendicular to CY-axis at point P. PR is plotted towards positive direction (i.e. right if I_{xy} is positive) and towards negative direction (i.e., left if I_{xy} is negative). The point C is obtained by bisecting the distance OQ. With C as the centre and OC as the radius, draw a circle as shown in Fig. B.4. The circle is known as **circle of inertia**. Join C and R and produce it on both sides. It intersects the circle at points U and V.

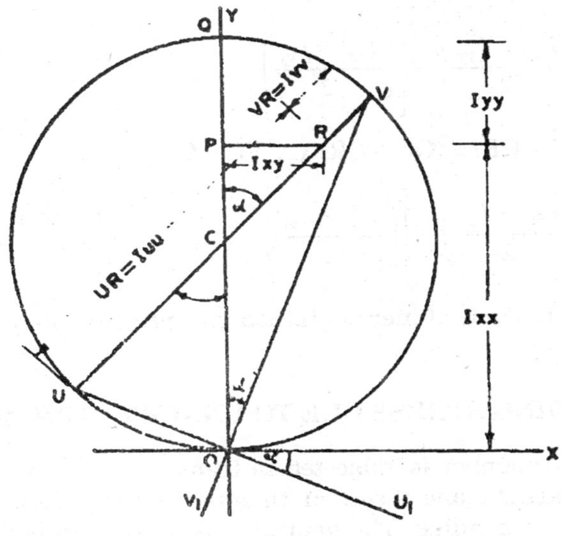

Fig. B.4

Distance OQ = (I_{xx} + I_{yy})

$$OQ = CQ = \left(\frac{I_{xx} + I_{yy}}{2}\right) = \text{Radius of circle}$$

Then, distance PC = (CQ + PQ). Therefore, $\left(\frac{I_{xx} + I_{yy}}{2} - I_{yy}\right) = \left(\frac{I_{xx} - I_{yy}}{2}\right)$

In Δ PCR,

$$CP = \left(\frac{I_{xx} - I_{yy}}{2}\right) \quad \text{and} \quad PR = I_{yy}$$

$$CR = \left[\left(\frac{I_{xx} - I_{yy}}{2}\right)^2 + I_{xy}^2\right]^{1/2}$$

$$\angle \tan\ PCE = \left(\frac{PR}{PC}\right) = \left(\frac{1}{\left(\frac{I_{xx} - I_{yy}}{2}\right)}\right)$$

$$\tan(\angle - PCR) = \frac{I_{xy}}{(I_{xx} - I_{yy})} \quad \therefore \angle PCR = 2\,\alpha$$

where, α represents the angle of inclination of principal axis UU of the plane section. The negative sign shown the angle is to be plotted in the negative direction, (i.e., clockwise), OU and OV give the directions of the principal axes UU and VV.

Distance UR = (UC + CR) = (OC + CR) (UC = OC)

$$= \left(\frac{I_{xx} - I_{yy}}{2}\right) - \left[\left(\frac{I_{xx} - I_{yy}}{2}\right)^2 + I_{xy}^2\right]^{1/2} = I_{uu} \qquad \textbf{(B.11)}$$

Distance VR = (VC – CR) = (OC – CR) (∴ VC = OC)

$$= \left(\frac{I_{xx} - I_{yy}}{2}\right) - \left[\left(\frac{I_{xx} - I_{yy}}{2}\right)^2 + I_{xy}^2\right]^{1/2} = I_{vv} \qquad \text{(B.12)}$$

Thus, UR and VR in circle of inertia method give principal moment of inertias I_{UU} and I_{VV}, respectively.

B.5 MAXIMUM BENDING STRESS DUE TO UNSYMMETRICAL BENDING

When a structural member is subjected to transverse loading such that the bending of beam occurs in the principal plane of section, then the neutral axis of the section of beam is perpendicular to the plane of bending. The neutral axis passes through the centroid of section. If the transverse loading is such that the bending of beam occurs in a plane other than the principal plane of section, the beam section is subjected to **unsymmetrical bending**. The direction of neutral axis is not perpendicular to plane of bending. The maximum bending stress in a beam-section subjected to unsymmetrical bending may be determined by following methods:

1. By resolving bending moment into two components along the principal axes.
2. By resolving bending moment in two components along any two perpendicular axes passing through the centroid of the section.
3. By locating the neutral axis of section.

1. By resolving bending moment into two components along the principal axes. Let the beam section is subjected to a bending moment, M. The plane of bending (i.e., the plane of loading) is inclined at θ with the principal axis-VV as shown in Fig. B.5. It is to note that angle θ is considered positive when measured in anti-clockwise direction from OV-axis.

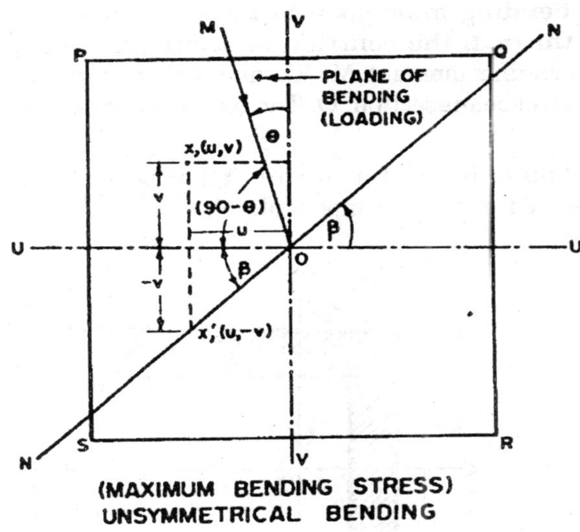

Fig. B.5

The bending moment, M may be resolved into two components along the principal axes UU and VV. The component of bending moment along UU-axis is M sin θ, and that along VV-axis M cos θ. After resolving the bending moment, into its two components, the theory of simple bending can be applied. For the component, M sin θ, the bending occurs in the principal plane passing through UU-axis, and VV-axis becomes a neutral axis. For the component M cos θ, the bending occurs in the principal plane passing through VV-axis and UU-axis becomes a neutral axis.

The bending stress at any point X, may be found by algebraic sum of the stress due to M. sin θ and M. cos θ. The bending stress at point X is given by

$$\sigma_b = \left(\frac{M.\cos g\theta}{I_{vv}} + \frac{M.\sin\theta}{I_{vv}} u \right), \text{ where} \tag{B.13}$$

I_{UU} = Principal moment of inertia of the section about UU-axis

Ivv = Principal moment of inertia of the section about VV-axis

u, v = Co-ordinates of point, X in the beam section with reference to principal axes UU and VV.

The co-ordinates (u,v) are assumed positive as shown in Fig. B.5. Due to M. cos θ the portion of beam section above UU- axis is subjected to compressive bending stress and that below UU-axis is subjected to tensile bending stress. Due to M. sin θ, the portion of beam section to the left of VV-section is subjected to compressive bending stress and that to the right of VV-axis is subjected to tensile bending stress. Thus, it is seen that point P is subjected to maximum tensile stress. The co-ordinates of point P and R are known. The maximum compressive and maximum tensile bending stresses in a beam section can be found by Eq. B.13.

The method is suitable for beam section which has at least one axis of symmetry. The axis of symmetry is also a principal axis. The points of extreme distances from principal axes may be seen by visual inspection. When a beam section has no axes of symmetry as shown in Fig. B.1 (D), it is not possible to locate the points which are at extreme distances by visual inspection, and this method is not convenient to determine the maximum bending stress for such sections.

2. **By resolving bending moment into two components along any two perpendicular axes, passing through the centroid of section.** Let the beam section shown in Fig. B.6 is subjected to a bending moment M. XX and YY are two perpendicular axes and pass through the centroid of the beam section, O. The bending stress at any point can be determined as under :

The plane of loading is inclined at θ with YY-axis as shown in Fig. B.6. The bending moment M may be resolved into two components M. cos θ and M. sin θ along YY and XX axes, respectively.

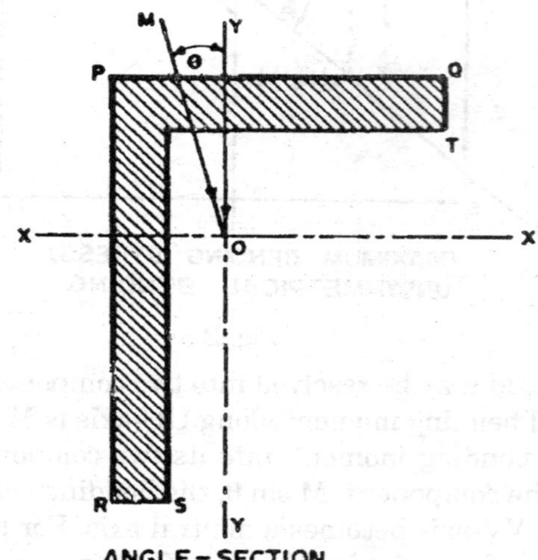

Fig. B.6

The bending of beam due to M. cos θ takes place about XX-axis, and that due to M. sin θ along YY and XX axes, respectively.

Then, M_{xx} = M. cos θ, and M_{yy} = M. sin θ (i)

where, M_{xx} is the component of bending moment, M, about XX-axis, and M_{yy} is the component of bending moment, about YY-axis.

The variation of bending stresses at any point whose co-ordinate (x, y), is linear then $\sigma_b = (a_1 . x + a_2 y)$ (ii)

where, σ_b = Bending stress at point, (x, y) and a_1, a_2 = Any constants

then $M_{xx} = \int \sigma_b . dA . y$ (iii)

where, dA = Any elementary area of beam section

$$M_{xx} = \int (a_1 x + a_2 y) y \, dA = a_1 \int xy \, dA + a_2 . \int y^2 \, dA$$

$$= (a_1 . I_{xy} + a_2 . I_{xy}) \quad \text{(v)}$$

where, I_{xx} is the moment of inertia of beam section about XX-axis and I_{xy} is the product of inertia of beam section about XX-axis and YY-axis.

$$\text{And } M_{yx} = \int \sigma_b \, dA \, x = \int (a_1 x + a_2 y) x \, dA \quad \text{(vi)}$$

$$= a_1 \int x^2 \, dA + a_2 \int xy . dA = (a_1 . I_{yy} + a_2 . I_{xy}) \quad \text{(vii)}$$

where, I_{yy} is the moment of inertia of beam section about YY- axis.

From (v) and (vii), we have

$$a_1 = \left(\frac{M_{yy} I_{xx} - M_{xx} I_{xy}}{I_{xx} I_{yy} - I_{xy}^2} \right), \text{ and } a_2 = \left(\frac{M_{xx} I_{yy} - M_{yy} I_{xy}}{I_{xx} I_{yy} - I_{xy}^2} \right) \quad \text{(viii)}$$

Therefore, substituting from (viii) and (ix) and values of a_1 and a_2 in (iii),

$$\sigma_b = \left(\frac{M_{yy} I_{xx} - M_{xx} I_{xy}}{I_{xx} I_{yy} - I_{xy}^2} \right) . x + \left(\frac{M_{xx} I_{yy} - M_{yy} I_{xy}}{I_{xx} I_{yy} - I_{xy}^2} \right) . y$$

$$\sigma_b = \left(\frac{M_{xx} I_{yy} - M_{yy} I_{xy}}{I_{xx} I_{yy} - I_{xy}^2} \right) . y + \left(\frac{M_{yy} I_{xx} - M_{xx} I_{xy}}{I_{yy} I_{xx} - I_{x}^2} \right) . x \quad \text{(B.14)}$$

Eq. (B.14) is specially useful for beam sections, in which the web and flanges or legs of angle sections are parallel to XX and YY axes. The values of bending stress at different corner points are found. The maximum bending stress in the beam section, is then be obtained.

3. By locating the neutral axes of beam section. When a beam section is subjected to unsymmetrical bending moment M, inclined at θ, with the principal axis VV as shown in Fig. B.5, then the neutral axis is not perpendicular to the plane of bending. The neutral axis of beam section may be located by

(i) Analytical method and (ii) Graphical method.

(i) Analytical method. The bending stress at any point X, is given by Eq. B.13. The bending stress at any point X_1 on the neutral axis is zero. Therefore, from Eq. (B.13),

$$\left[\frac{M . \cos\theta}{I_{UU}} . v + \frac{M . \sin\theta}{I_{VV}} . u \right] = 0 \quad \text{Therefore, } v = -u . \left(\frac{I_{UU}}{I_{VV}} \right) . \tan\theta \quad \text{(B.15)}$$

Eq. (B.15) is an equation of straight line, and represents the equation of neutral axis-NN of the beam section. From this equation it is seen that it passes through the centroid of the section O.

Let β be the angle of inclination of neutral axis-NN with the principal axis- UU. Then, from Fig. B.5, $\tan\beta = (-v/u)$ (i)

From Eq. (B.14), $-\left(\frac{v}{u}\right) = \left(\frac{I_{UU}}{I_{VV}}\right).\tan\theta$ (ii)

Substituting the value from this, we have, $\tan\beta = +\left(\frac{I_{UU}}{I_{VV}}\right).\tan\theta$ (B.16)

From Eq. (B.6), it is seen that angle β in general, is not same as θ. If the angle θ is zero, the bending moment M will be acting along the principal plane, and the neutral axis will become perpendicular to the plane of bending. The angle B will also be equal to θ in case

$\therefore (I_{UU} = I_{VV}) \therefore (\tan\beta = \tan\theta), \angle\beta = \angle\theta$

In such cases, the ellipse of inertia of the beam section becomes a circle. Any pair of two perpendicular axes passing through the centroid can be considered as principal axes. Therefore, the neutral axis remains perpendicular to the plane of bending.

UU and VV axes are the principal axes. Therefore, for product of inertia I_{UV} about UU-axis and VV-axis is zero. The moment of inertia of beam section about any axis inclined β with the axis-UU is obtained from Eq. B.2.

Therefore, $I_{NN} = (I_{UU}.\cos^2\beta + I_{VV}.\sin^2\beta)$ (B.17)

The angle of inclination of plane of loading with the neutral axis-NN, is equal to (90 – θ + β) or (90 + β – θ).

The angle of inclination of plane of loading with the line perpendicular to neutral axis-NN will therefore be equal to (β – θ).

The component of bending moment, M, with the axis, = M. cos (β – θ)

If, any point is at a maximum distance, Y_N, from the neutral, axis-NN. then, the maximum bending stress

$$\sigma_b = \left[\frac{M.\cos(\beta - \theta)}{I_{NN}}\right].Y_N \qquad \text{(B.18)}$$

The bending stress will be compressive or tensile depending upon the position of point relative to the neutral axis and direction of loading.

(ii) **Graphical method**. The radii of gyration about principal axes are known as **principal radii of gyration**. Let r_{UU} and r_{VV} be the principal radii of gyration about UU and VV principal axes of a beam-section. The moment of inertia about neutral axis I_{NN} may be found by determining the radius of gyration r_{NN} about it. The radius of gyration r_{NN} about neutral axis-NN is found graphically by constructing **momental ellipse**. The momental ellipses is also known as **ellipse of inertia**.

Draw UU and VV principal axes of a beam section passing through point O as shown in Fig. B.7. The point O represents centroid of beam-section. Plot OA = r_{UU} on UU-axis and OB = r_{VV} on VV-axis on some suitable scale. Draw two concentric circles with point O as centre and, OA and OB as radii, respectively. Draw MM at angle θ, with VV-axis. MM represents the plane of loading. Draw momental ellipse with OA as length of major axis and OB is length of minor axis as shown in Fig. B.7. Draw a line M`M' tangential to the momental ellipse such that it is parallel to MM. This line is tangent to the moment ellipse at point S. Join points O and S, and produce it on both sides.

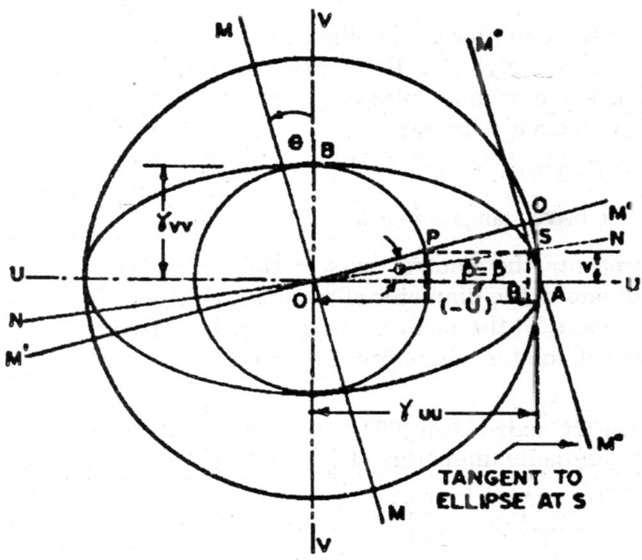

MOMENTAL ELLIPSE

Fig. B.7

This locates neutral axis NN of beam section. The radius of gyration r_{NN} about the neutral axis, NN is given by OS.

Then, $I_{NN} = A.r^2_{NN}$

where, A is the area of beam section. This can be proved as under :

Draw M'M' perpendicular to MM and passing through point O as shown in Fig. B.7. M'M' intersects the concentric circles at points P and Q, respectively. From point Q, draw a perpendicular QR to OA. From the point P draw a perpendicular to PS to QR.

Let (–u, v) be the co-ordinates of point S with reference to UU and VV principal axes. The co-ordinates of points S are as (–u, v). Since co-ordinates of point X in Fig. B.5 are assumed as (u, v) in the positive direction of axes, (i.e., u and v are positive in second quadrant). Then

$$-u = OR = OQ \,.\, \cos\theta = OA \,.\, \cos\theta = r_{UU} \,.\, \cos\theta \quad (OQ = OA)\,(\therefore OP = OB)$$

$$u^2 = r_{UU}^2 \,.\, \cos^2\theta$$

$$y = SR = PT = OP.\sin\theta = OB\,.\sin\theta = r_{VV}\,.\sin\theta$$

$$v^2 = \left(r_{vv}^2 \sin^2\theta\right)$$

From (i) and (ii), we have $\left(\dfrac{u^2}{r_{UU}{}^2} + \dfrac{v^2}{r_{VV}{}^2}\right) = \left(\cos^2\theta + \sin^2\theta\right)$

$$\therefore \left(\frac{u^2}{r_{UU}^2} + \frac{v^2}{r_{VV}^2}\right) = 1 \tag{B.19}$$

Eq. (B.19) represents the equation of an ellipse, with r_{UU} and r_{VV} as lengths of semi-major and semi-minor axes, respectively. Eq. B.19 also represents the locus of S. The ellipse represented by Eq. B.19 is same as momental ellipse which is drawn to find r_{NN}. The lengths of semi-major and semi-major axes are the same.

Let the angle SOA be β', $\tan\beta' = (-v/u)$

From Eq. (B.16) (i), we have, $\tan\beta = (-v/u)$, Therefore, $\beta' = \beta$.

Therefore, NN represents the neutral axis of the beam section. The line MM may be considered as direction of one of the diameter of momental ellipse. The line NN passes through middle point of MM and point, S, the point on tangent drawn parallel to MM. Line NN shows the direction of conjugate diameter, therefore, MM and NN form conjugate diameter of the ellipse.

The slope of MM with UU-axis is tan (90° + θ) or cot θ and that of NN with UU-axis is tan β. From the properties of conjugate diameters of an ellipse,

$$-\cot\theta \,.\, \tan\beta = \left(\frac{r_{UU}^2}{r_{VV}^2}\right) \tag{i}$$

$$\tan\beta = \left(+\frac{A\,.\,r_{UU}^2}{A\,.\,r_{VV}^2}\right).\tan\theta = \left(+\frac{I_{UU}}{I_{VV}}\tan\theta\right) \textit{ and } \therefore \tan\beta = (-v/u) \tag{ii}$$

Therefore, $$v = -u\left(\frac{I_{UU}}{I_{VV}}\right).\tan\theta \tag{B.20}$$

This relation obtained is same as Eq. B.15 and it is equation of the neutral axis. Further from Fig. B.7,

$(OS)^2 = (OR)^2 + (RS)^2 = (OQ.\cos\beta)^2 + (OP\,.\sin\beta)^2$ (i)

$(OQ = OA, OP = OB)$

$(OS)^2 = (OA\,.\cos\beta)^2 + (OB\,.\sin\beta)^2$

$= \left(r_{uu}^2 \,.\cos^2\beta + r_{vv}^2\,.\sin^2\beta\right)$ (ii)

Multiplying both sides by A, the area of cross-section of beam section, we have

$A.(OS)^2 = \left(A\,.\,r_{UU}^2\cos^2\beta + A\,r_{VV}^2\sin^2\beta\, I_{UU}\,.\cos^2\beta + I_{VV}\,.\sin^2\beta\right)$ (iii)

$I_{UU} = A\,r_{UU}^2$ *and* $I_{VV} = A.\,r_{VV}^2$ *and* $I_{uv} = 0$

From Eq. B.2, therefore, $A.(OS)^2 = I_{NN}$, $(OS)^2 = (I_{NN}/A) = r^2_{NN}$, $OS = r_{NN}$

After determining the value of radius of gyration graphically the moment of inertia I_{NN} about neutral axis is determined. The maximum bending stress at the extreme distance from the neutral axis may be found from Eq. B.18.

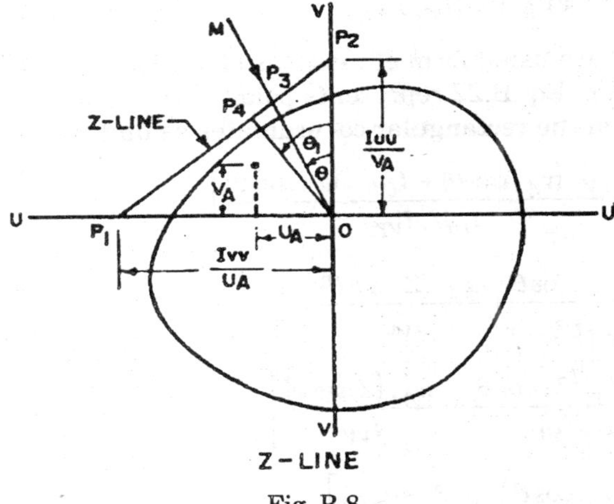

Fig. B.8

B.6 THE Z-POLYGON

The strength of a beam section in bending depends upon the modulus of section Z. Along with the geometrical properties of a beam section, the value of modulus of section Z also depends upon plane of bending (or plane of loading). The value of modulus of section Z, varies as the position of the plane of bending changes. The maximum and minimum values of modulus of section, Z, may be found by determining the corresponding positions of plane of bending.

Consider any point A, whose co-ordinates are (u_A, v_A) with reference to principal axes UU and VV. The beam section is subjected to bending moment M. The plane of bending is inclined at θ with VV-axis, as shown in Fig. B.8. From Eq. B.13 the bending stress at point is given by

$$\sigma_b = \left(\frac{M\cos\theta}{I_{UU}}\right).v_A + \left(\frac{M\sin\theta}{I_{VV}}\right).u_A = M\left[\frac{V_A.\cos\theta}{I_{UU}} + \frac{u_A.\sin\theta}{I_{VV}}\right]$$

$$\sigma_b = \left[\frac{I_{VV}.(v_A\cos\theta) + I_{UU}.(u_A\sin\theta)}{I_{UU}.I_{VV}}\right] = \left[\frac{M}{\dfrac{I_{UU}.I_{VV}}{I_{VV}.(v_A.\cos\theta) + I_{VV}.(u_A.\sin\theta)}}\right] \quad \text{(B.21)}$$

$$\therefore \sigma_b = (M/Z)$$

where, Z is the modulus of section of the beam section, and

$$Z = \left[\frac{I_{UU}.I_{VV}}{I_{VV}.(v_A.\cos\theta) + I_{UU}.(u_A.\sin\theta)}\right] \quad \text{(B.22)}$$

In Eq. B.22, I_{UU} and I_{VV} are principal moment of inertias of the beam section about UU and VV axes, respectively. For a given beam section, the value of I_{UU} and I_{VV} are constant. For a given point, A on the beam section u_A and v_A are also constant. Thus, it is seen that the value of Z depends upon θ only.

When $(\theta = 0°)$, $Z = (I_{UU}/u_A)$ (i)

And when $(\theta = 90^{\circ}), Z = (I_{VV} / u_A)$ (ii)

Expressions (i) and (ii) are usual form of modulus of section. For any given point of a given beam section, with θ variable, Eq. B.22 represents polar form of the equation of a straight line. Eq. B.22 may be expressed in the rectangular co- ordinates as under:

$$\frac{I}{Z} = \left[\frac{I_{VV} \cdot (v_A \cdot \cos\theta) + I_{UU} \cdot (u_A . \sin\theta)}{I_{UU} \cdot I_{VY}}\right] \quad \text{(iii)}$$

or
$$\frac{1}{Z} = \left[\frac{v_A . \cos\theta}{I_{UU}} + \frac{u_A . (Z . \sin\theta)}{I_{VV}}\right] \quad \text{(iv)}$$

or
$$\left[\frac{V_A \ (Z . \cos\theta)}{I_{UU}} + \frac{u_A . (Z . \sin\theta)}{I_{VV}}\right] = 1 \quad \text{(v)}$$

or
$$\left[\left(\frac{Z . \cos\theta}{I_{UU} / v_A}\right) + \left(\frac{Z . \sin\theta}{I_{VV} / u_A}\right)\right] = 1 \quad \text{(vi)}$$

Put Z. cos θ = v, and Z. sin θ = u

Then
$$\left[\left(\frac{u}{I_{VV} / v_A}\right) + \left(\frac{u}{I_{UU} / v_A}\right)\right] = 1 \quad \text{(B.23)}$$

This represents equation of a straight line, which is known as Z- line. The length of interception on UU-axis is (I_{VV}/u_A) and the length of interception on VV-axis as (I_{UU}/v_A).

Eq. B.23 can also be expressed as below :

$$\left(\frac{v}{I_{UU} / V_A}\right) = -\left(\frac{u}{I_{VV} / u_A}\right) + 1 \quad \text{(i)}$$

$$v = -u\left(\frac{u_A}{v_A}\right) . \frac{I_{UU}}{I_{VV}} + \left(\frac{I_{UU}}{v_A}\right) \quad \text{(ii)}$$

When the plane of bending passes through the point A

Therefore,
$$v = -u\left(\frac{I_{UU}}{I_{VV}}\right) . \tan\theta + \left(\frac{I_{UU}}{v_A}\right) \quad \text{(B.24)}$$

By comparing Eq. B.15 and Eq. B.24 for the plane of bending passing through given point, it is seen that Z-line is parallel to the neutral axis.

The Z-line represented by Eq. B.23 is plotted by taking OP, = (I_{UU}/u_A) and OP_2 = (I_{UU}/v_A) along UU-axis and VV-axis respectively. The points P_1 and P_2 are joined. The line P_1P_2 represents Z-line. The Z-line intersects the planet bending at P_3. OP_3 represents the modulus of section, Z. The bending stress at point A is given by

$$\sigma_b = (M / OP_3) \quad \text{(B.25)}$$

Let OP_4 represents perpendicular from point O, the centroid of beam section to the Z-line. The minimum distance from point O to Z-line is given by perpendicular distance OP_4. Then OP_4 represents the minimum value of Z. Let θ_1 be the angle of OP_4 measured from OV-axis. The bending stress at point A is maximum when the plane of loading is inclined at θ_1 to principal axis VV.

The maximum bending stress at the point A $\sigma_{b(\max)} = (M / OP_4)$ *where* $\theta = \theta_1$

The maximum distance of any point on Z-line from the centroid O, of the beam section, depends, upon the lengths of intercepts on UU-axis and VV-axis. If OP_2 is greater than OP_1, then OP_2 is the maximum distance. The maximum value of modulus of section Z is given by OP_2. The bending stress at point O is minimum. The plane of bending coincides with the principal axis VV.

Minimum bending stress at point A, $\sigma_{b(\min)} = (M / OP_2)$

where, $\theta = \theta$, $OP_2 = (I_{VV}/v_A)$

It is necessary to draw Z-lines for critical points, (the points which are at extreme distance from c.g. of beam section) in order to determine maximum bending stress. The Z-lines drawn for critical points form a polygon. This polygon is called **Z-polygon.** The values of Z for these critical points for a given plane of bending may be compared. The maximum bending stress occurs at a critical point having least value of Z. The Z-polygon for a rectangular beam section is drawn as below :

The principal axes UU and VV are drawn passing through point O, the centroid of rectangular beam. The rectangular beam section ABCD is drawn as shown in Fig. B.9. Let the width of section be `b' and depth of section be `h'. The rectangular beam section ABCD is drawn as shown in Fig. B.9. The rectangular section is symmetrical section. Therefore, co-ordinate axes XX and YY coincide with the principal axes of the section.

Z line for point A is drawn first. The co-ordinates of point A, lying in first quadrant are $u_A = (1/2)\, b . v_A = (1/2)\, h$.

$I_{UU} = I_{xx} = (1/12)\, bh^3$ and $I_{VV} = I_{yy} = (1/12)\, hb^3$

The length of intercepts of Z-line along UU-axis is

$$OP_1 = \left(\frac{I_{VV}}{u_A}\right) \text{ and } OP_1 = \left[\frac{(1/12)hb^3}{(1/2)b}\right] = \frac{1}{6}hb^2 \text{ and } Z_{yy} = Z_{VV} \quad (\therefore Z_{yy} = Z_{VV})$$

The length of intercepts of Z-line along VV axis is

$$OP_2 = \left[\frac{(1/12)bh^3}{(1/2)h}\right] = \frac{1}{6}bh^2 = Z_{xx} = Z_{UU} \quad (\therefore Z_{xx} = Z_{UU})$$

Eq. B.23 becomes
$$\left[\frac{u}{Z_{VV}} + \frac{v}{Z_{UU}}\right] = 1 \qquad \text{(B.26)}$$

On UU-axis, plot $OP_1 = \frac{1}{6} = hb^2$ and on VV-axis plot $OP_2 = \frac{1}{6} = bh^2$ to a suitable scale. Points P_1 and P_2 are joined. P_1P_2 represents Z- line for the point A. Similarly, Z-lines P_2P_3, P_3P_4 and P_4P_1 are drawn for points B, C and D respectively. $P_1P_2P_3P_4$ represents Z- polygon. The Z-polygon for a rectangular beam section is a rhombus.

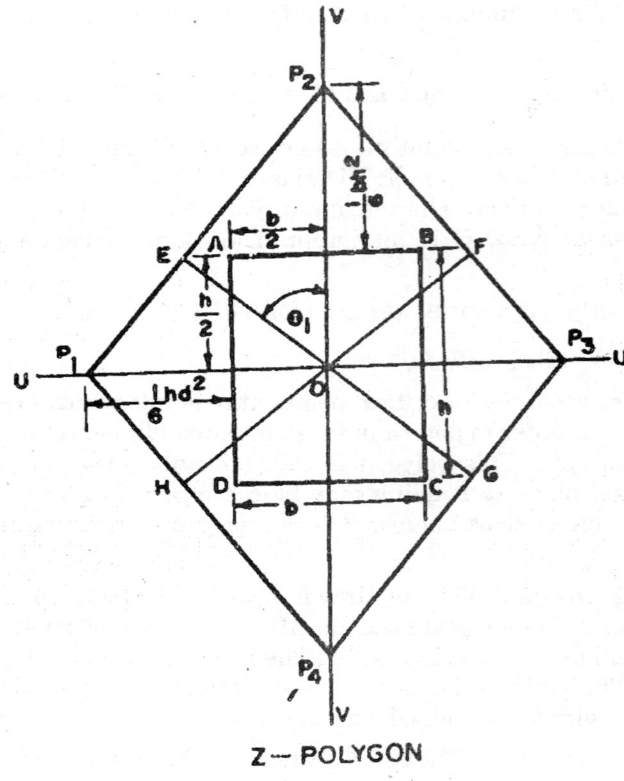

Fig. B.9

The position of plane of bending for maximum and minimum strength of a section can be found by inspection of Z-polygon. For example, for minimum strength of the section, the plane of bending is along VV-axis.

The value of Z is maximum along this line. Similarly, for minimum strength of the section, perpendicular OE, OF, OG and OH are drawn from point, O, to Z-lines as shown in Fig. B.9, the strength of beam section is minimum along EG and FH planes. The values of Z for plane bending along these planes is minimum. The inclination of plane EG is θ_1. The value of θ_1 is obtained by

$$\tan \theta_1 = \left(\frac{OP_2}{OP_1}\right) = \left(\frac{Z_{UU}}{Z_{VV}}\right) = \left(\frac{\frac{1}{6}bh^2}{\frac{1}{6}.\ hb^2}\right) = \frac{h,}{b} \qquad \therefore \theta_1 = \tan^{-1}\left(\frac{h}{b}\right) \tag{B.27}$$

B.7 DEFLECTION OF BEAMS UNDER UNSYMMETRICAL BENDING

The plane of bending or the plane of loading is inclined at an angle θ with principal axis-VV as shown in Fig. B.10. The bending moment M_1, acting along the plane of bending may be resolved into two components along VV-axis and UU-axis, respectively. The deflection of beam subjected to unsymmetrical bending may be found as under:

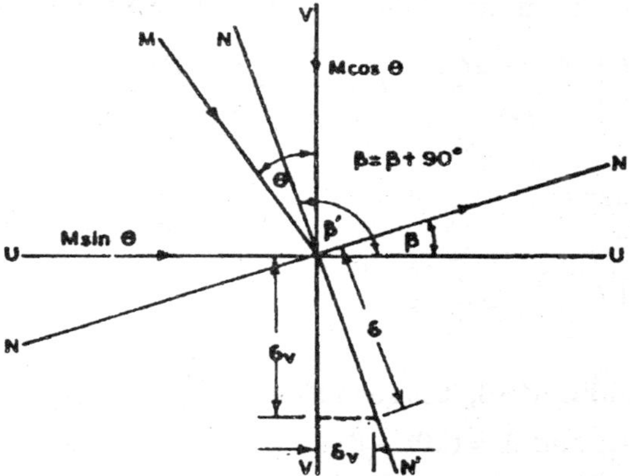

DEFLECTION OF BEAMS(UN SYMMETRICAL BENDING)

Fig. B.10.

From unit load method, the deflection of beam subjected to moment M_1, is given by

$$\delta = \int_o^L \left(\frac{M_1\, m_1}{EI}\right) = dx$$

where, m_1 = Moment due to unit load acting in the direction of desired deflection. Unit load is applied at the point where deflection is to be found

d_x = Elementary length of beam, measured along span of beam

L = Span of beam.

The deflection in the direction of VV-axis is given by

$$\delta_v = \int_o^L \left(\frac{M \cos\theta}{EI_{UU}}\right) .m_v \;. dx \qquad \text{(i)}$$

The deflection in the direction of UU-axis is given by

$$\delta_v = \int_o^L \left(\frac{M \sin\theta}{EI_{VV}}\right) .m_U \;. dx \qquad \text{(ii)}$$

The resultant deflection is found by adding these deflections vectorially. Therefore, resultant deflection

$$\delta = \left[(\delta_U)^2 + (\delta_V)^2\right]^{1/2} \qquad \text{(iii)}$$

$$\delta = \int_o^L \left(\frac{M\;m}{E}\right)\left[\left(\frac{\cos\theta}{I_{UU}}\right)^2 + \left(\frac{\sin\theta}{I_{VV}}\right)^2\right]^{1/2} \qquad \text{(iv)}$$

Since, $m_V = m_U = m$

$$\text{or } \delta = \int_o^L \left(\frac{M\;m}{E}\right)\left[\frac{{I_{VV}}^2 \cos^2\theta + {I_{UU}}^2 \sin^2\theta}{{I_{UU}}^2\, I_{VV}^2}\right]^{1/2} dx \qquad \text{(B.28)}$$

Let β' be the angle which the resultant deflection makes with the UU-axis.

$$\tan\beta' = \left(\frac{-\delta_U}{\delta_V}\right) = \int_0^L \left(\frac{M\ m\sin\theta}{EI_{VV}}\right) dx \,/\, \int_0^L \left(\frac{M\ m\cos\theta}{EI_{UU}}\right) dx$$

$$\text{or } \tan\beta = -\left(\frac{I_{UU}}{I_{VV}}\right)\tan\theta \qquad \text{(B.29)}$$

From Eq. B.16, $\tan\beta = +\left(\frac{I_{UU}}{I_{VV}}\right)\tan\theta$

By comparing Eq. B.29 and Eq. B.16, it is seen that

tan β' = tan (90° + β) and β' = (90° + β)

Therefore, it is seen that the resultant deflection occurs in the direction N'N', which is perpendicular to neutral axis-NN for any given plane of loading.

If the beam is subjected to uniformly distributed load w, then

$$\delta_{UU} = \frac{5}{384}\left(\frac{w\,.\sin\theta\,.\,L^4}{E\,I_{vv}}\right) \qquad \text{(i)}$$

$$\delta_{VV} = \frac{5}{384}\left(\frac{w\,.\cos\theta\,.\,L^4}{E\,.\,I_{UU}}\right) \qquad \text{(ii)}$$

The resultant deflection of the beam $\delta = \left[(\delta_{UU})^2 + (\delta_{VV})^2\right]^{1/2}$ (iii)

$$\delta = \frac{5}{384}\frac{wL^4}{E}\left[\frac{\sin^2\theta}{I^2_{VV}} + \frac{\cos^2\theta}{I^2_{UU}}\right]^{1/2} = \frac{5}{384}\left(\frac{wL^4\,.\cos\theta}{E\,.\,I_{UU}}\right)\left[1 + \left(\frac{I_{UU}}{I_{VV}}\tan\theta\right)^2\right]^{1/2}$$

From Eq. (B.16), $\tan\beta = \left(+\frac{I_{UU}}{I_{VV}}\tan\theta\right)$

$$\text{Therefore, } \delta = \frac{5}{384}\left(\frac{wL^4\,.\cos\theta}{E\,.\,I_{UU}}\right)\left[1 + \tan^2\beta\right]^{1/2} \qquad \text{(v)}$$

$$= \frac{5}{384}\cdot\left(\frac{wL^4}{EI_{UU}}\right).\cos\theta\,.\sec\beta = \frac{5}{384}\left(\frac{wL^4}{EI_{UU}}\right)\cdot\frac{\cos\theta}{\cos\beta}$$

$$= \frac{5}{384}\left(\frac{wL^4}{EI_{UU}}\right)\cdot\left(\frac{\cos\theta}{\cos\beta}\right)\cdot\left(\frac{\cos(\beta-\theta)}{\cos(\beta-\theta)}\right)$$

$$= \frac{5}{384}\left(\frac{wL^4}{EI_{UU}}\right)\cdot\left(\frac{\cos\theta.\cos(\beta-\theta)}{\cos\beta\left[\cos\beta\cos\theta + \sin\beta\sin\theta\right]}\right)$$

$$= \frac{5}{384}\left(\frac{wL^4}{EI_{UU}}\right)\cdot\left(\frac{\cos(\beta-\theta)\cos\theta}{\cos\beta\ \cos\beta.\cos\theta\left[1 + \tan\beta\tan\theta\right]}\right) \qquad \text{(vi)}$$

From Eq. B.16 $\tan\beta = \left(\frac{I_{UU}}{I_{VV}}\right).\tan\theta$

Value of tan θ is substituted in (vi)

$$\therefore \delta = \frac{5}{384}\left(\frac{wL^4}{EI_{UU}}\right)\left(\frac{\cos(\beta-\theta)}{\cos^2\beta.\left[1+\frac{I_{UU}}{I_{VV}}.\tan^2\beta\right]}\right)$$

$$\text{or} \quad \delta = \frac{5}{384}\left(\frac{wL^4}{EI_{UU}}\right)\left(\frac{\cos(\beta-\theta)}{\cos^2\beta.\left[\frac{I_{UU}\cos^2\beta + I_{VV}\sin^2\beta}{I_{UU}\,.\cos^2\beta}\right]}\right)$$

$$\delta = \frac{5}{384}\left(\frac{wL^4\cos(\beta-\theta)}{E\left[I_{UU}\,.\cos^2\beta + I_{VV}\,.\sin^2\beta\right]}\right) \qquad \text{(viii)}$$

From Eq. B.17, $I_{NN} = (I_{UU}\cos^2 B + I_{VV}\sin^2\beta)$

$$\therefore \delta = \frac{5}{384}\left(\frac{w.\cos(\beta-\theta)L^4}{E.I_{NN}}\right) \qquad \text{(B.30)}$$

In Eq. B.30, w cos (β – θ) is the component of resultant uniformly distributed load along the direction N'N' perpendicular to the neutral axis NN, for plane bending (or plane loading) inclined θ with VV-axis.

B.8 DESIGN OF PURLINS SUBJECTED TO UNSYMMETRICAL BENDING

The rectangular section, I-section, channel section and angle section are used for purlins of roof trusses. The rectangular section and I-section have two axes of symmetry. The channel section has at least one axis of symmetry. The axes of symmetry locate the principal axes in case of rectangular section, I- section and channel section.

All the purlins are designed in accordance with the requirements for uncased beams. However, the limitations of bending stress based on lateral instability of the compression flange and limitation of deflection are waived for the design of purlins. The calculated deflections should not exceed those permitted for the type of roof cladding used. In calculating the bending moment, the advantage may be taken of the continuity of the purlin over the supports. The bending stresses about the two axes should be determined separately and checked by the combined interaction equations.

The design of purlins, in cases where principal axes are located directly from axes of symmetry of a section, is done as follows :

The span of the purlin is centre to centre distances between adjacent trusses, (i.e., it is equal to the spacing of trusses).

The spacing of purlins should not exceed 1.4 m so that ACC sheets do not fail.

Step 1 : Roof sheets are supported by the purlins. These sheets transfer the loads to the purlins. The various loads acting over the purlins are weight of roof-sheets and fixtures, self-weight of the purlins, live load from the roof-sheets and the wind load. Except the wind loads, other loads are the gravity loads. These gravity loads act vertically. The vertical loads and wind load acting over the purlins are shown in Fig. B.11. The vertical loads act downward and the wind load acts normal to the sheets either inward or outward depending upon the wind pressure or suction. The purlins are connected to the rafters of the roof trusses. The major principal axis of the purlin-UU remains parallel to the principal rafter. Therefore, the vertical load remains inclined with the major principal axes. As a result of which unsymmetrical bending (that is, biaxial bending) of the purlins occur.

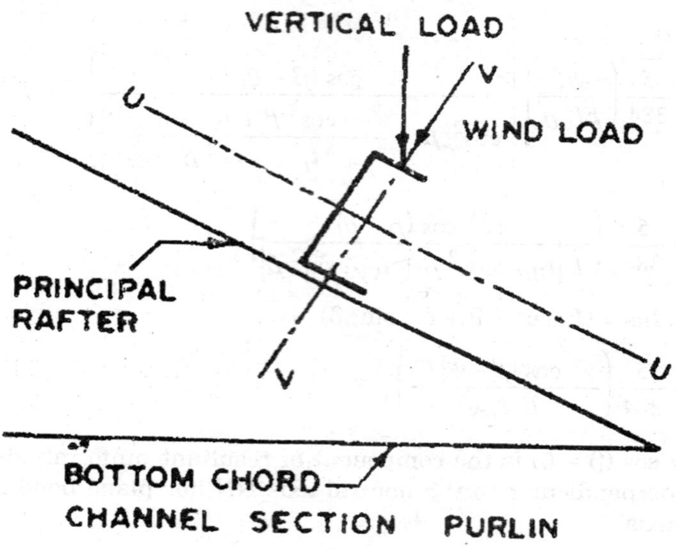

Fig. B.11.

The appropriate combination of loads on purlins are resolved into two components perpendicular to and parallel to the roof slope.

Step 2: The principal axes UU and VV of the section coincide with the co-ordinate axes XX and YY respectively. The maximum bending moments are found about these axes. Let the maximum bending moment about UU-axis (i.e., XX-axis) be M_{UU} and that about VV- axis (i.e., YY-axis) be M_{vv}.

Step 3 : The maximum bending stress at any point on a purlin section is given by

$$\sigma_b = \left(\frac{M_{UU}}{I_{uu}}\right) . \upsilon + \left(\frac{M_{VV}}{I_{VV}}\right) . u \qquad \text{(i)}$$

If M is the maximum bending moment caused due to resultant loading, and if a plane of bending (i.e., plane of loading) is inclined at θ then,

$$M_{UU} = M \cos\theta \text{ and } M_{VV} = M \sin\theta$$

Expression (i) is in the same form as Eq. B.13. In above expression

$$I_{UU} = I_{xx} \quad \text{and} \quad I_{VV} = I_{yy}$$

Expression (i) can be written as

$$\sigma_b = \left[\frac{M_{UU}}{\left(\frac{I_{UU}}{v}\right)} + \frac{M_{VV}}{\left(\frac{I_{VV}}{u}\right)} \right] = \left(\frac{M_{UU}}{Z_{UU}}\right) + \left(\frac{M_{VV}}{Z_{VV}}\right)$$

$$\text{or } \sigma_b = \frac{M_{UU}}{Z_{UU}} \cdot \left(1 + \frac{M_{VV}}{M_{UU}} \frac{Z_{UU}}{Z_{VV}}\right) \quad \text{(B.31)}$$

$$\text{or } Z_{UU} = \frac{M_{UU}}{\sigma_b} \left(1 + \frac{M_{VV}}{M_{UU}} \frac{Z_{UU}}{Z_{VV}}\right) \quad \text{(B.32)}$$

In Eq. B.32 it is to note that Z_{UU} and Z_{VV}, the moduli of section about UU-axis and VV-axis, respectively, are the same as Z_{xx} and Z_{yy}, the moduli of section about XX-axis and YY-axis, respectively. In Eq. B.32 suitable value of ratio (Z_{UU}/Z_{VV}) is assumed in order to determine, modulus of section Z_{UU}. For rectangular section, the value of (Z_{UU}/Z_{VV}) may be assumed as 5. For I-sections, this ratio may be assumed 5 to 7 for lightly loaded purlins, and 7 to 10 for heavily loaded purlins. For channel section, the value of (Z_{UU}/Z_{VV}) may be assumed as 8. From Eq. B.32 the required modulus of section Z_{UU} is determined by assuming σ_b equal to permissible bending stress. A trail section may be selected from ISI Handbook No. 1.

Step 4 : This trial section for purlins is checked for maximum bending stress. The maximum bending stress is determined from Eq. B.32. The maximum bending stress compressive or tensile in beam section should not exceed the maximum permissible bending stress in compression or in tension.

The weight of purlins as compared to the total weight of the steel structure may vary from 10 percent to 25 percent. The economic impact (i.e., cost), durability of roof, increase or decrease of the fixtures and the vibrations of roof sheets are the various factors, which should be kept in view while designing the purlins. The spacing of roof trusses may vary from 3 m to 12 m. The economical spacing of roof trusses depends on the various factors (e.g., type of soil, bearing capacity of soil, wind zone, inclination of the roof and the functional aspects). The various sections (e.g., channels, angles, tubes, Z- shape or cold formed, Warren or N-type truss, triangular or open section) may be used for the purlins. The channels or angles should be so placed that the accumulation of dust, moisture and water is minimum or avoided. The purlin sections should be connected with the principal rafters of the roof trusses in their correct orientations. For example, the flanges in case of channels may be kept sloping downward with the web normal to the rafter and upward. So also, in case of the angle sections, one leg is kept and connected normal to the rafter while the other leg is kept sloping downward. The durability, (that is, the life) of purlin reduces due to incorrect position (i.e., wrong orientation) of the purlin sections.

B.9. DESIGN OF ANGLE IRON PURLINS

The angle-section are most commonly used for purlins. The angle sections do not have any symmetric axis. Therefore it becomes essential to locate the principal axes of the angle-section. The principal axes may be located, if the geometrical properties of angle section are known. Depending upon the resultant load, the trial section is selected for purlins. Then, general procedure for design of purlins mentioned above may be followed.

If the slope of roof truss is not greater than 30° and the steel is conforming to Grades Fe 410-O or Fe 410-W, then, the angle purlins may be designed as an alternate to the general design procedure, as recommend by IS : 800-1984. However, the following requirements which are based on a maximum imposed load of 0.75 kN/m^2 are fulfilled.

Step 1 : The depth of angle purlin in the plane approximate to the incidence of the maximum load or maximum component of the load is not less than (1/45)th of the length of purlin.

Step 2. The width of angle purlin is not less than (1/60)th of the length of purlin.

Knowing the depth of purlin, and width of purlin, a trial angle selection is selected from ISI Hand-book No. 1.

Step 3. The maximum bending moment in a purlin is taken as

$$M_{max} = (WL/10)$$

where, W is total uniformly distributed load on the purlin including wind load and L is the distance centre to centre of the rafters or other supports of the purlins.

The loads are assumed as acting normal to the roof surface. The bending of the purlin about the minor axis is neglected. In calculating the bending moment, the advantage of continuity of the purlins over roof trusses has been taken.

Step 4. The maximum bending stress calculated for M_{max} should not exceed the maxi-mum permissible bending stress in compression or in tension. The design or purlins takes into account the wind load. Therefore, the permissible stress in bending in compression or tensile are 1.33 times the usual permissible values. (*viz.*, dead load plus live load, dead load plus half the live load plus the wind load or the dead load and wind load) shall be considered. It is to note that the dead load and live load cause biaxial bending, where as the wind load produces bending about major principal axis. The bending moments due to dead loads, due to live load and due to wind load are calculated.

These purlins may be treated as continuous beams supported over principal rafters of the roof trusses. The moments at the supports and mid-spans of the end and intermediate spans may be calculated as recommended in the code.

For end span (one end simply supported and other end continuous)

(a) Bending moment due to dead loads

$$\text{mid-span } M = +(w_d.L^2/16), \text{ and} \quad \text{(i)}$$

$$\text{support } M = (w_d.L^2/9) \quad \text{(ii)}$$

(b) Bending moment due to live loads

$$\text{mid-span } M = +(w_L.L^2/9), \text{ and} \quad \text{(iii)}$$

$$\text{support } M = (w_L.L^2/9) \quad \text{(iv)}$$

It is to note that for the intermediate spans, the bending moments are less than these values. It is recommended to use the value ($w_L.L^2/10$).

Following guide lines may be kept in view while selecting a particular section for the purlins. These guide lines will be advantageous in selecting the purlin sections.

1. The angle or channel sections may be used suitably and economically upto 5 m span without sag bars except in heavy snow zones.

2. The tubular sections may be used upto 6 m span. A minimum of 3 mm thickness of the tube is essential for the structural welding. In case, the thickness of tube is small, it will be rusted soon. As far as possible, tubes with thickness more than 3.5 mm may be used.

3. In case of snow zones, for angle sections and channels sections sag bars are used for purlins for 4 m and more spans the channel sections even sag bars for purlins of spans beyond 8 m are not suitable.

4. For purlins of spans 6 m and more, the triangulation or truss purlins may be used economically. However, their fabrication is slightly difficult.

The durability, maintenance aspect fabrications and the erection convenience should also be considered.

Example B.1. Locate the principal axes of ISA 200 mm x 100 mm x 10 mm (ISA 200 100, @ 0.228 kN/m). Determine principal moment of inertia and radii of gyration about the principal axes for this angle section.

Solution :

Step 1 : Geometrical properties of given section

From ISI Handbook No. 1, ISA 200 mm x 100 mm x 10 mm (ISA 200 100, @ 0.228 kN/m)

Sectional area, A = 2903 mm^2

$c_{xx} = 69.6$ mm, $c_{yy} = 20.1$ mm

$e_{xx} = 130.4$ mm, $e_{yy} = 79.9$ mm

$I_{xx} = 1210 \times 10^4$ mm^4, $I_{yy} = 209.2 \times 10^4$ mm^4

Step 2 : Product of inertia of section

The angle section is divided into two rectangle. Let A_1 and A_2 be the areas of these rectangles

$A_1 = 100 \times 10 = 1000$ mm^2 and $A_2 = (200 - 10) \times 10 = 1900$ mm^2

$x_1 = (50 - 20.1) = 29.9$ mm and $y_1 = (69.6 - 5.0) = 64.6$ mm

$x_2 = (20.1 - 5.0) = -15.1$ mm and $y_2 = (130.4 - 95) = 35.4$ mm

Product of inertia I_{xy} about xx-axis and yy-axis

$I_{xy} = (A_1.x_1y_1 + A_2.x_2y_2)$

$I_{xy} = [(10 \times 2.99 \times 6.46) + \{19 \times (-1.51) \times (3.54)\} \times 10^4$ mm^4

$= 295.3 \times 10^4$ mm^4

From Eq. B.1 $\tan 2\alpha = \left(\dfrac{-2I_{xy}}{I_{xx} - I_{xy}}\right) = \left(\dfrac{(-2 \times 295.3) \times 10^4}{(1210 - 209.2) \times 10^4}\right) = -0.59$

$\therefore 2\alpha = -31^\circ,\ \alpha = -15^\circ\ 30^\circ,\ \tan\alpha = -0.275$

(From ISI Handbook No. 1, $\tan\alpha = -0.27$)

The principal axes are as shown in Fig. B.12.

The negative sign indicates that angle a is to be measured clock- wise from xx-axis.

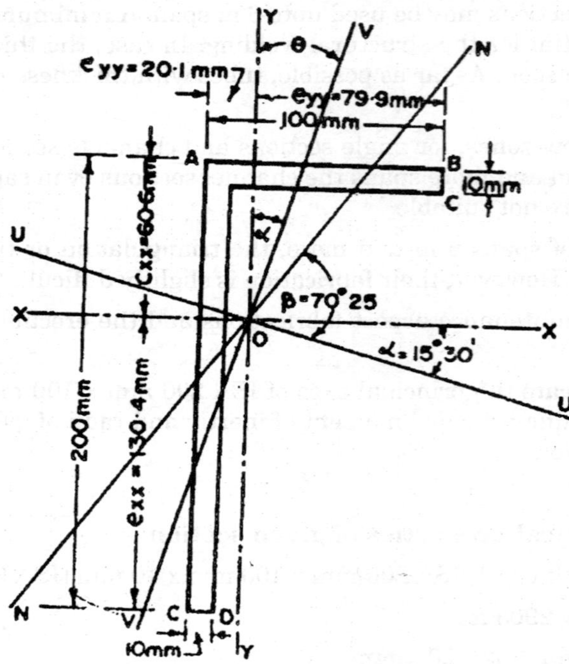

Fig. B.12

Step 3 : Principal moment of inertia of section (From Eq. B.6)

$$I_{UU} = \left(\frac{I_{xx}+I_{yy}}{2}\right) + \left[\left(\frac{I_{xx}-I_{yy}}{2}\right)^2 + I_{yy}{}^2\right]^{1/2}$$

$$= \left(\frac{1210+209.2}{2}\right) + \left[\left(\frac{1210-209.2}{2}\right)^2 + (295.3)^2\right]^{1/2} \times 10^4\ mm^4$$

I_{UU} = 1291.6 x 10^4 mm^4

(From ISI Handbook No. 1, I_{UU} = 1286.7 x 10^4 mm^4)

Step 4 : Radius of gyration of section

$$I_{VV} = \left(\frac{I_{xx}+I_{yx}}{2}\right) - \left[\left(\frac{I_{xx}-I_{yy}}{2}\right)^2 + I_{xy}{}^2\right]^{1/2}$$

$$= \left(\frac{1210+209.2}{2}\right) - \left[\left(\frac{1210-209.2}{2}\right)^2 + (209.3)^2\right]^{1/2} \times 10^4\ mm^4$$

= 127.6 x 10^4 mm^4

(From ISI Handbook No. 1, I_{VV} = 132.5 x 10^4 mm^4)

$$r_{UU} = \left(\frac{I_{UU}}{A}\right)^{1/2} = \left(\frac{1291.6 \times 10^4}{2903}\right)^{1/2} = 66.6\ mm$$

(From ISI Handbook No. 1, r_{UU} = 66.6 mm)

$$r_{VV} = \left(\frac{I_{UU}}{A}\right)^{1/2} = \left(\frac{127.6 \times 10^4}{2902}\right)^{1/2} = 21.0\ mm$$

(From ISI Handbook No. 1, r_{VV} = 21.4 mm)

The actual angle section is rounded at the toe and at the root. In above calculations, angle section is not rounded. As a result of this, there are small differences in the values calculated and the values noted from Handbook No. 1.

Example B.2. In Example B.1, the angle section is subjected to bending moment 6 kN-m acting in the vertical plane through the centroid of section. Determine the maximum bending stress induced in the angle section.

Solution :

Step 1 : Principal moment of inertias of section

From Example B.1,

$I_{UU} = 1291.6 \times 10^4\ mm^4$, $I_{VV} = 127.6 \times 10^4\ mm^4$

The plane of loading is vertical and passes through the centroid of the angle section. From Eq. B.16, for neutral axis

$$\tan\beta = \left(\frac{I_{UU}}{I_{VV}}\right)\tan\theta = \left(\frac{1291.6 \times 10^4}{127.6 \times 10^4}\right) \times \tan 15°30' = +2.81 \quad \therefore \beta = = +70°25'$$

The value of β is positive. It is plotted anti-clockwise from principal axis UU is shown in Fig. B.12 From Fig. B.12, it is seen that the point D is at extreme distance from the neutral axis.

Step 2 : Co-ordinates of farthest point D

$x_D = -(20.1 - 10) = -10.1$ mm

$y_D = 130.4$ mm, $\alpha = 15°30'$

$u_D = (-x_D \cos\alpha + y_D \sin\alpha)$

$= [-10.1 \times \cos(-15°\ 30') + (-1\ 30.4) \times \sin(15°\ 30')] = +25.1$ mm

$v_D = (y_D \cos\alpha \sin\alpha)$

$= [-130.4 \cos(-15°\ 30') - (10.1) \times \sin(-15°\ 30')] = -128.5$ mm

$M_{VV} = M.\cos\theta$ and $M_{UU} = M.\sin\theta$

Step 3 : Maximum bending stress at point D (From Eq. B.13),

$$\sigma_b = \left(\frac{M\cos\theta}{I_{UU}}\right), v_D + \left(\frac{M\sin\theta}{I_{VV}}\right), u_D$$

Both the components of bending stresses are additive

$$\sigma_b = \left(\frac{6\times10^6 \cos 15°30'}{1291.6\times10^4}\right)\times128.5 + \left(\frac{6\times10^6 \sin 15°30'}{127.6\times10^4}\right)\times 25.1\ N/mm^2 = 88.8\ N/mm^2$$

The maximum bending stress at point D is tensile.

(i) **Alternative.** The moment of inertia of the section about neutral axis from Eq. B.17.

$$I_{NN} = (I_{UU}\,.\cos^2\beta + I_{VV}\,.\sin^2\beta)$$

$$= [1291.6 \times \cos^2 70°\,25' + 127.6 \sin^2 70°\,25'] \times 10^4\ mm^4$$

$$= 250.3 \times 10^4\ mm^4$$

$$(\beta - \theta) = (70°\,25' - 15°\,30') = 54°\,55'$$

$$Y_N = (-v_D\,.\cos\beta - u_D\,.\sin\beta)$$

$$Y_N = (-128.5 \times 0.335 - 25.1 \times 0.942) = -66.6\ mm$$

From Eq. B.18, maximum bending stress at point D

$$\sigma_b = \left[\frac{M\,.\cos(\beta-\theta)}{I_{NN}}\right].Y_N = \left(\frac{6\times10^6 \times \cos 54°55' \times 66.6}{258.3\times10^4}\right) = 88.7\ N/mm^2$$

The maximum bending stress at the point D is tensile.

(ii) **Alternative**. The maximum bending stress at point D may be found by Eq. B.14. Eq. B.14 is useful specially for beam sections in which less of angle section are parallel to xx and yy axes. From Eq. B.14, the bending stress at point D,

$$\sigma_b = \left(\frac{M_{xx} I_{yy} - M_{yy}\,.I_{xy}}{I_{xx} I_{yy} - I_{xy}^2}\right).y_D + \left(\frac{M_{yy} I_{xx} - M_{xx}\,.I_{xy}}{I_{yy} I_{xx} - I_{xy}^2}\right).x_D,\ and\ \varphi = 0.$$

$M_{xx} = M \cos\ \varphi = 6 \times 10^6$ kN-m and $M_{yy} = M \sin\ \varphi$ = Zero.

$$a_2 = \left(\frac{M_{xx} I_{yy} - M_{yy} I_{xy}}{I_{xx} I_{yy} - I_{xy}^2}\right) = \left(\frac{6\times10^6 \times 209.2\times10^4}{[1210\times209.2 - (29.3)^2]\times10^8}\right) = 0.754$$

$$a_1 = \left(\frac{M_{yy} I_{xx} - M_{xx} I_{xy}}{I_{yy} I_{xx} - I_{xy}^2}\right) = \left(\frac{-6\times10^6 \times 295.3\times10^4}{[1210\times209.2 - (295.3)^2]\times10^8}\right) = -1.06$$

$$\sigma_b = (0.754\,y_D - 1.06\,x_D)$$

$$= [0.754\,(-130.4) - (1.06)\,(-10.1)] = -87.8\ N/mm^2.$$

The maximum bending stress at point D is tensile and it has been worked out from Eqs. B.13, B.14, and B.18. The values differ slightly.

Example B.3. LB 200, @ 0.198 kN/m is subjected to bending moment 12 kN-m. The plate of loading passes through centroid of beam section, and it is inclined 8° with the YY-axis in the anti-clockwise direction. Locate the neutral axis. Determine maximum bending stress induced in the beam section.

Solution :

Step 1 : Geometrical properties of section

From ISI Handbook No. 1, for LB 200, @ 0.198 kN/m

Depth of section, h = 200 mm, Width of section, b = 100 mm

$I_{UU} = I_{xx} = 1696.6 \times 10^4 \text{ mm}^4$

$I_{VV} = I_{yy} = 115.4 \times 10^4 \text{ mm}^4$

It is to note that I-section shown in Fig. B.13 is a symmetrical section.

The axes of symmetry XX and YY are principal axes UU and VV

From Eq. B.16, for neutral axis

$$\tan\beta = +\left(\frac{I_{UU}}{I_{VV}}\right)\tan\theta \qquad \text{or } \tan\beta = +\left(\frac{1696.6}{115.5}\right)\tan 8^\circ \quad \beta = 64^\circ\ 7'.$$

The value of β is positive. It is plotted anti-clockwise from principal axis-UU as shown in Fig. B.13.

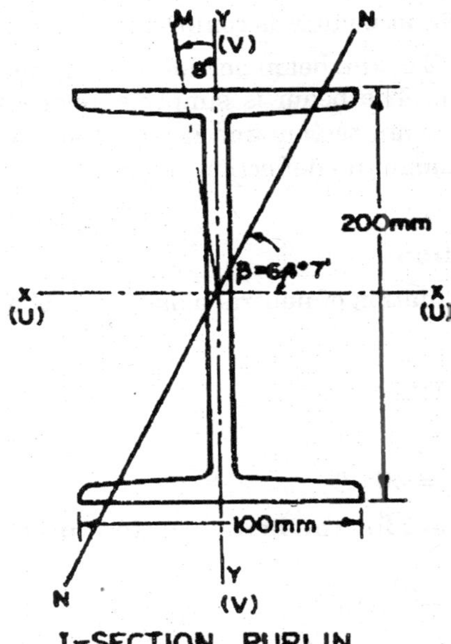

I-SECTION PURLIN

Fig. B.13

Step 2 : Moment of inertia of beam section about neutral axis (From Eq. B.17).

$I_{NN} = (I_{UU} \cdot \cos^2\beta + I_{VV} \cdot \sin^2\beta)$

or $I_{NN} = [1696.6 \cos^2 64^\circ\ 7' + 115.4 \times \sin^2 64^\circ\ 7'] \times 10^4 \text{ mm}^4$

$= 417.5 \times 10^4 \text{ mm}^4$

$(\beta - \theta) = [(64^\circ\ 7') - 8^\circ] = 56^\circ\ 7'$

Step 3 : Co-ordinates of points A and D are at extreme distance from the neutral axis

$y_N = [100 \cos\beta - (-50) \sin\beta]$

$= [(100 \cos 64^\circ\ 7' + 50 \sin 64^\circ\ 7')] = 88.6 \text{ mm}$

Step 4 : Maximum bending stress at point A (from Eq. B.18)

$$\sigma_b = \left[\frac{M \cdot \cos(\beta - \theta)}{I_{NN}}\right] \cdot y_N = \left(\frac{12 \times 10^6 \cos 56° 7'}{417.5 \times 10^4}\right) \times 88.6 = 141.8\ M/mm^2$$

The maximum bending stress at point A is compressive and that at point D is tensile.

Alternative. The maximum bending stress at point A, can be found by resolving bending moment along VV-axis and UU-axis. From Eq. B.13, bending stress at point A,

$$\sigma_b = \left(\frac{M \cos 8°}{I_{UU}}\right) \cdot v_A + \left(\frac{M \sin 8°}{I_{VV}}\right) u_A$$

or $$\sigma_b = \left(\frac{121 \times 10^6 \times 0.99 \times 100}{1696.6 \times 10^4}\right) + \left(\frac{12 \times 10^6 \times 1.39 \times 50}{115.4 \times 10^4}\right) N/mm^2 = 141.8\ N/mm^2$$

The maximum bending stress at point A is compressive and that at the point D is tensile.

Example B.4. In Example B.3, the beam section is carrying uniformly distributed load 6 kN/m. The effective span is 4 m. The beam is simply supported at both ends. The plane of loading passes through centroid beam section and it is inclined 8° with the yy-axis. Determine the magnitude and direction of maximum deflection. Take E = 2.04 x 10^5 N/mm².

Solution :

Step 1 : Properties of section

From Example 24.3, the inclination of neutral axis from UU-axis

$$\beta = +64° 7'$$

Moment of inertia about neutral axis

$$I_{NN} = 417.5 \times 10^4\ mm^4$$

$$(\beta - \theta) = [(64° 7') - 8°] = 56° 7'$$

Step 2 : Maximum deflection for the beam section (from Eq. B.)

$$\delta = \frac{5}{384} \cdot \left(\frac{w \cdot l^4 \cos(\beta - \theta)}{E \cdot I_{NN}}\right) mm$$

$$\delta = \frac{5}{384} \times \left(\frac{6 \times 10^4 \times 4 \times (4000)^4 \times \cos 56° 7'}{2.04 \times 10^5 \times 417.5 \times 10^4}\right) = 13.2\ mm$$

The maximum deflection occurs in the direction perpendicular to neutral axis NN. It occurs in the direction β measured from UU principal axis,

$$\beta = (\beta + 90°) = (64° 7' + 90°) = 154° 7'$$

Example B.5. Draw Z polygon for ISA 200 mm x 100 mm x 10 mm. The angle section is subjected to bending moment 6 kN-m acting in the vertical plane through the centroid of section. Determine the maximum bending stress induced in the angle section. Also determine the absolute maximum bending stress for the given moment and corresponding inclination of plane of loading with YY- axis.

Solution :

Step 1 : Geometrical proportion of given section

From ISI Handbook No. 1 for ISA 200 mm x 100 mm x 10 mm. (ISA 200 100, @ 0.228 kN/m)

Sectional area, A = 2903 mm^2

c_{xx} = 69.6 mm^2, c_{yy} = 20.1 mm

e_{xx} = 130.4 mm, e_{yy} = 79.9 mm

I_{xx} = 1210 x 10^4 mm^4, I_{yy} = 209.2 x 10^4 mm^4

I_{UU} = 1286.7 x 10^4 mm^4, I_{VV} = 132.5 x 10^4 mm^4

tan α = – 0.270, α = –15° 7'

It is to note that angle α is negative. It is plotted in clock-wise direction from XX-axis as shown in Fig. B.14. UU and VV are principal axes.

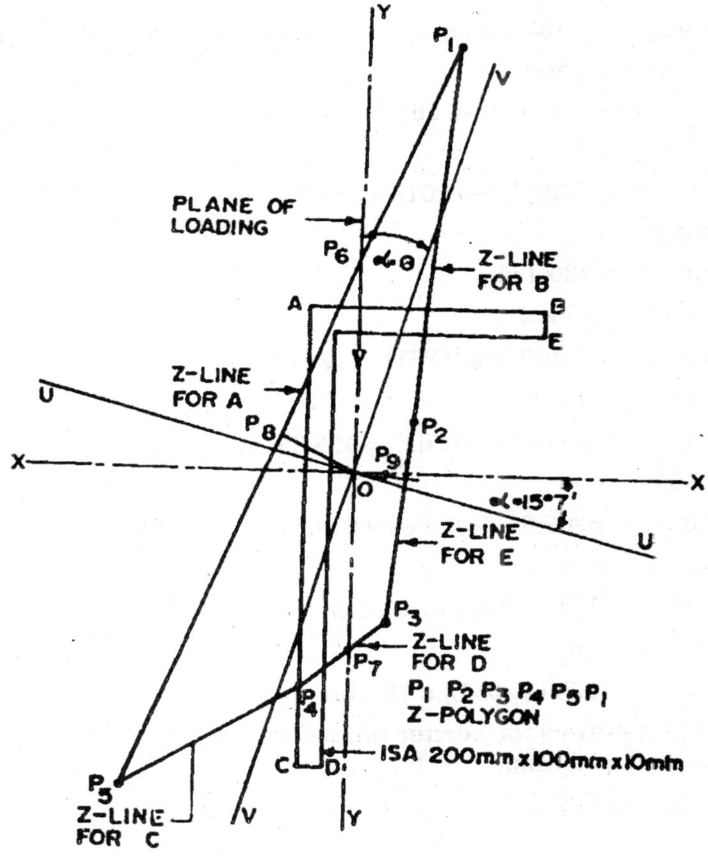

Fig. B.14

Step 2 : Co-ordinates of corner points with reference to UU and VV-axes.

(i) **Corner point A.**

x_A = –20.1 mm, y_A = + 69.6 mm

$u_A = (x_A . \cos \alpha + y_A . \sin \alpha)$

$= [(-20.1) \cos (-15° 7') + 69.6 \sin (-15° 7')] = -37.6$ mm

$v_A = (y_A . \cos \alpha - x_A . \sin \alpha)$

$= 69.6 \times 0.956 - (-20.1)(-0.261)) = 51.9$ mm

(ii) **Corner point B.**

$x_B = +79.6$ mm, $y_B = +69.6$ mm

$u_B = (x_B . \cos \alpha + y_B . \sin \alpha)$

$= (79.9 \times 0.965 - 69.6 \times 0.261) = +58.8$ mm

$v_B = (y_B . \cos \alpha - x_B . \sin \alpha)$

$= [(69.6 \times 0.965) - (79.9)(-0.261)] = 88.0$ mm

(iii) **Corner point C,**

$x_C = -20.1$ mm, $y_C = -130.4$ mm

$u_C = (x_C . \cos \alpha + y_C . \sin \alpha)$

$= [(-20.1)(0.965) + (-130.4)(-0.261)] = +14.7$ mm

$v_C = (y_C . \cos \alpha - x_C . \sin \alpha)$

$= [-130.4 \times 0.965 - (-20.1)(-0.261)] = -131.3$ mm

(iv) **Corner point D,**

$x_D = 10.1$ mm, $y_D = -130.4$ mm

$u_D = (x_D . \cos \alpha - y_D . \sin \alpha)$

$= [(-10.1)(0.965) + (-130.4)(-0.261)] = +24.4$ mm

$v_D = (y_D . \cos \alpha - x_D . \sin\alpha)$

$= [(-130.4)(0.965) - (-10.1)(-0.261)] = 128.6$ mm

(v) **Corner point E,**

$x_E = (+79.9$ mm, $y_E = (69.6 - 10)) = +59.6$ mm

$u_E = (x_E \cos \alpha + y_E \sin \alpha)$

$= 79.9 \times 0.965 + (59.6)(-0.261) = +61.4$ mm

$v_E = (y_E \cos \alpha - x_E \sin \alpha)$

$= [59.6 \times 0.965 - (79.9)(-0.261)] = 78.4$ mm.

Step 3 : Equation of Z-lines for corner points. It is to note that, now co-ordinates (u, v) are taken positive in second quadrant.

(i) **Corner point A,** from Eq. B.23,

$$\frac{u}{\left(\frac{I_{VV}}{u_A}\right)} + \frac{v}{\left(\frac{I_{UU}}{v_A}\right)} = 1, \quad \frac{u}{\left(\frac{132.5}{37.6}\right)} + \frac{v}{\left(\frac{1286.7}{51.9}\right)} = 1 \times 10^4$$

or $\quad v = (-7.06u + 24.8 \times 10^4) \qquad$ (i)

(ii) Corner point B

$$\frac{u}{\left(\frac{I_{VV}}{u_B}\right)}+\frac{v}{\left(\frac{I_{UU}}{v_B}\right)}=1,\ \frac{u}{\left(\frac{132.5}{-58.8}\right)}+\frac{v}{\left(\frac{1286.7}{88.0}\right)}=1\times10^4$$

or $v = (6.5u + 14.6 \times 10^4)$ (ii)

(iii) Corner point C

$$\frac{u}{\left(\frac{I_{VV}}{u_C}\right)}+\frac{v}{\left(\frac{I_{UU}}{v_C}\right)}=1,\ \frac{u}{\left(\frac{132.5}{-14.7}\right)}+\frac{v}{\left(\frac{1286.7}{-137.3}\right)}=1\times10^4$$

or $v = (1.08u - 9.78 \times 10^4)$ (iii)

(iv) Corner point D

$$\frac{u}{\left(\frac{I_{VV}}{u_D}\right)}+\frac{v}{\left(\frac{I_{UU}}{v_D}\right)}=1,\ \frac{u}{\left(\frac{132.5}{-24.4}\right)}+\frac{v}{\left(\frac{1286.7}{-128.6}\right)}=1\times10^4$$

or $v = (1.84\,u - 10.9 \times 10^4)$ (iv)

(v) Corner point E

$$\frac{u}{\left(\frac{132.5}{-61.4}\right)}+\frac{v}{\left(\frac{1286.7}{78.4}\right)}=1\ v=\left(7.46\,u+16.4\times10^4\right)$$

or $v = (7.46\,u + 16.4 \times 10^4)$ (v)

Step 4 : Z-polygon

(i) The intersection of Z-lines for points A and B locates point P_1 of Z-polygon.

Z-line for point A, $v = (-7.06u + 24.8 \times 10^4)$ [from Eq. (i)]

Z-line for point B, $v = (6.5u + 14.6 \times 10^4)$ [from Eq. (ii)]

Solving equations (i) and (ii),

Co-ordinates of P_1 are (75.4, 1948).

(ii) The intersection of Z-lines for points B and E locates point P_2 of Z-polygon.

Z-line for point B, $v = (6.5\,u + 14.6 \times 10^4)$ [from Eq. (ii)]

Z-line for point E, $u = (7.46\,u + 16.4 \times 10^4)$ [from Eq. (v)]

Solving equations (ii) and (v), co-ordinates of point P_2 are (–188, 240)

(iii) The intersection of Z-lines for points E and D, locates P_3 of Z-polygon.

Z-line for point E, $(7.464 + 16.4 \times 10^4)$

Z-line for point D, v = (–1.84u – 10.0 x 10^4) [from Eq. (iv)]

Solving equations (v) and (iv), co-ordinates of point P_3 are (–283, –580).

(iv) The intersection of Z-lines for points D and C, locates point P_4 of Z-polygon.

Z-line for point D, v = (–1.84u – 10.0 x 10^4) [from Eq. (iv)]

Z-line for point C, v = (–1.08u – 9.78 x 10^4) [from Eq. (iii)]

Solving equations (iv) and (iii), co-ordinates of point P_4 are (–28.9, 946.8).

(v) The intersection of Z-lines for points C and A, locates point P_5 of Z-polygon.

Z-line for point C, v = (–1.08u – 9.78 x 10^4) [from Eq. (iii)]

Z-line for point A, v = (–7.06u + 2.48 x 10^4) [from Eq. (i)]

Solving equations (iii) and (iv), co-ordinates of point P_5 are (57.6 – 1602).

The points P_1, P_2, P_3, P_4 and P_5 are plotted with reference to UU and VV principal axes. Join P_1 and P_2, P_2 and P_3, P_3 and P_4, P_4 and P_5 and P_5 and P_1. $P_1P_2P_3P_4P_{51}$ represents Z-polygon.

The bending moment M is acting in the vertical plane passing through the centroid of section. The vertical plane through the centroid intersects Z-line of point A at P_6 and Z-line of point D at P_7.

From Fig. B. , $OP_6 = 88 \times 10^3$ mm³, $OP_7 = 68 \times 10^3$ mm³

Therefore, point D is subjected to maximum bending stress

Step 5 : Maximum bending stress

$$\sigma_b = \left(\frac{M}{OP_7}\right) = \left(\frac{6 \times 10^6}{68 \times 10^3}\right) = 88.235\ N/mm^2$$

The bending stress at point D is tensile.

For absolute maximum bending stress draw perpendiculars OP_8 and OP_9 from centroid O. OP_9 gives minimum modulus of section. $OP_9 = 24 \times 10^3$ mm³. Absolute maximum bending stress for the given moment.

$$\sigma_b = \left(\frac{6 \times 10^6}{24 \times 10^3}\right) = 250\ \text{N/mm}^2$$

For absolute maximum bending stress, the inclination of plane of loading is 96° clockwise from vertical.

PROBLEMS

B.1. Locate the principal axes of ISA 90 mm x 60 mm x 10 mm [ISA 9060, @ 0.110 kN/m]. Determine the principal moment of inertias and radii of gyration about the principal axes for the angle section. Compare these values with those given in ISI Handbook No. 1. Other geometrical properties may be noted from handbook.

B.2. In Problem B.1, the angle section is subjected to bending moment 2 kN-m acting in the vertical plane through the centroid of section. Determine the maximum bending stress induced in the section.

B.3. An MC, 350, @ 0.421 kN/m is subjected to bending moment 8 kN-m. The plane of loading is inclined 10° with the YY-axis in anti-clockwise direction, and it passes through the centroid of channel section. Determine the maximum bending stress induced in the section. Locate the neutral axis.

B.4. In Problem B.3, the channel section is carrying uniformly distributed load 5 kN/m. The effective span is 5 m. The section is simply supported at both ends. The plane of loading is inclined 10° with YY-axis and it passes through the centroid of beam section. Determine the magnitude and direction of maximum deflection. Take, E = 2°04 x 10^5 N/mm^2.

B.5 Draw Z-polygon for MC, 350 @ 0.421 kN/m. Determine the maximum bending stress induced in the section for data given in Problem B.3. Also determine the absolute maximum bending stress for the given moments and corresponding inclination of plane of loading with YY-axis.

APPENDIX-C

BOLTS AND DESIGN OF BOLTED-CONNECTIONS

C.1 INTRODUCTION

Various types of connections are used for connecting the structural members. The choice of type of fasteners depends on connection strength required, space limitations of the connections, available technicians to fabricate and erect the structure, service conditions and finally the total cost of installation. These fasteners serve essentially the same function in transferring loads from one component to another. The connections are named after the type of fasteners used. In bolted connections, bolts and nuts are used. In tension or compression members, bolts may be used like rivets for end connections. The bolts may also be used for semi-rigid connections. The requirements for bolts as regards the pitch and minimum edge distance are same as those for the rivets.

Structural steel members are connected most easily and economically by using bolts in the field. Bolt connections are done whenever possible to join the members in the field. Even, when the welding becomes a part of the final field connection, bolts are needed to assemble and align the members. In erection, bolts are used whenever possible to transfer the load along with the weld.

This chapter deals with the design and detailing of requirements for the connections between the members, which consist of components such as cleats, gusset plates, brackets, connecting plates and welds.

These connections are designed so as to be consistent with the assumptions made in the analysis of the structures and comply with the requirements specified. These connections shall be capable of transferring the calculated designed actions (external loads).

Where the members are connected to the surface of a web or the flange of a section, the ability of the web or the flange to transmit the applied forces locally should be checked and where necessary local stiffening shall be provided.

Fully tensioned **friction grip bolts** may be designed to share the load with welding, providing the bolts are fully tightened to develop necessary pretension after welding.

SPECIFICATIONS

1. Clearances for holes and slot as regards their sizes are followed as specified as under. Bolts are located in standard size, over size, short slotted or long slotted hole.

(i) **Standard clearance hole :** Except where fitted bolts, bolts in low clearance or over size holes are specified, the diameter of standard clearance holes for bolts shall be as specified in Table B.1 as in IS : 800-2007.

(ii) **Over-size hole :** Holes of size larger than the standard clearance are given in Table 24.1 may be used in slip resistant connections and hold down bolted connections, only where specified, provided the over size holes in the outer ply is covered by a cover plate of sufficiently large size and thickness and having a hole not larger than the standard clearance hole (and hardened washer in slip resistant connections).

(iii) **Short and long slots :** Slotted holes of size larger than the standard clearance hole as given in Table C.1 may be used in slip resistant connectons and hold down bolted connections, only where specified, provided the over size holes in the outer ply is covered by a over plate of sufficiently large size and thickness and thickness and having a hole of size not larger than the standard clearance hole (and hardened washer in slip resistant connection).

TABLE C-1. Size of hole = Nominal diameter of the fastener + Clearance mm

1	2	3	4	5	6
1.	12-14	1.0	3.0	4.0	2.5 d
2.	16-22	2.0	4.0	6.0	2.5 d
3.	24	2.0	6.0	8.0	2.5 d
4.	Larger than 24	3.0	8.0	10.0	2.5 d

Note : Above numbers of columns in this Table C-1 are as under
1. Serial number; 2. Nominal size of fastener, of in mm; 3. Standard clearance in diameter with width of slot in mm; 4. Over-size clearance in diameter clearance in the length of the slot; 5. Short-slot; 6. Long-slot

3. Edge and end distances

(i) **Edge-distance** : The edge distance is the distance at right angles to the direction of the stress from the centre of a hole to the adjacent edge.

End-distance : The end distance is the distance in the direction of stress from the centre of a hole to the end of the element

(ii) **In slotted holes,** the edge and end distances should be measured from the edge or end of the material to the centre of its end radius or the centre line of the slot, whichever is smaller.

In over-size holes, the edge and end end distances should be takens as the distance from the relevant edge/end plus half the diameter of the standard clearance hole corresponding to the fastener, less the nominal diameter of the over-size hole.

(iii) **The minimum edge and end distances,** from the centre of any hole to the nearest edge of a plate shall not be less than 1.7 times the hole diameter in case of sheared or hand-flame cut edges; and 1.5 times the hole diameter in case of rolled, machine flame cut sawn and planed edges.

(iv) **The maximum edge-distance** to the nearest line of fasteners from an edge of any un-stiffened part should not exceed 12 t E, where

$\in = (250/\sigma_y)^{1/2}$ and t is the thickness of the thinner outer plate. This would not apply to fasteners inter connecting the components of back to back tension members. Where the members are exposed to corrosive influence the maximum edge distance shall not exceed 40 mm + 4t, where t is the thickness of thinner connected plate.

4. Minimum spacing

The distance between centre of fasteners shall not be less than 2.5 times the nominal diameter of the fastener.

5. Maximum spacing

(i) The distance, between the centres of any two adjacent fasteners shall not exceed 32 t or 300 mm, whichever is less, where, t is the thickness of the thinner plate.

(ii) The distance between the centres of two adjacent fasteners (pitch) in a line lying in the lying in the direction of stress, shall not exceed 16 t or 200 mm, whichever is less in tension member and 12 t or 200 mm whichever is less in compression members,where, t is the thickness of the thinner plate.

In the case of the compression members wherein forces are transferred through butting forces, this distance shall not exceed 4.5 times the diameter of the fasteners for a distance equal to 1.5 times for width of the member from the butting faces.

(iii) The distance between the centres ofany two consecutive fasteners in a line adjacent and parallel to an edge of an outside plate shall not exceed 100 mm + 4t or 200mm, whichever is less in compression and tension member, where, *t* is the thickness of the thinner outside plate.

(iv) When the fasteners are staggered at equal intervals and the not exceed 75 mm, the spacing specified above between the centre of fasteners may be increased by 50 percent, subject to the maximum spacing specified above.

6. Tacking fasteners

(i) In case of member cover in just above paragraph, when the maximum distance between centres of two adjacent fasteners as specified in that paragraph is exceed, tacking fasteners not subjected to calculated stress shall be used.

(ii) **Tacking fasteners** shall have spacing in a line not exceeding 32 times the thickness of the thinner outside plate or 300 mm whichever is less.

Where the plates are exposedto the weather,the spacing in a line shall not exceed 16 times the thickness of the thinner outside plate or 200mm, whichever is less.

In both the cases, the distance between the lines of the fasteners shall not be greater than the respective pitches.

(iii) All the requirements specified in just above paragraph generally apply to compression members, subjected to the stipulation in section 7 of IS : 800-2007 affecting the design and construction of compression members.

(iv) **In tension members** (see section 6 of IS: 800-2007) composed of two flats, angles, channels or tees in contact back to back or seperated back to back by a distance not exceeding the aggregate thickness of the connected parts, tacking fasteners with solid distance pieces should be provided at a spacing in line not exceeding 1000 mm.

(v) **For compression members** covered in section 7 of IS : 800-2007, tacking fasteners in a line shall be spaced at a distance not exceeding 600 mm.

7. Counter-shunk heads

For the counter shunk heads, one-half the depth of the counter sinking shall be neglected in calculating the length of the fastener in bearing in accordance to clause 10.3.3.

For fasteners in tension having counter-shunk heads, the tensile strength shall be reduced by 33.33 percent. No reduction is required to be made in shear strength calculations.

8. Bearing type bolt

(A) Effective area of bolt

(i) Since the threads may occur-in the shear plane, the area, A_e for resisting shear should normally be taken as the net tensile stress area, A_n of the bolt.

For bolts, where the tensile stress area is not defined, A_n shall be taken as the area at the root of threads.

(ii) Where it can be shown that the threads do not occur in the shear plane, A_e maybe taken as the cross-section area, A_s at the shank.

(iii) In the calculation of thread length, allowance should be made for tolerance and thread run-off.

(iv) A bolt subjected to factored shear force V_{fd} shall satisfy the conditon.

$V_{fd} = V_{db}$, where

V_{db} is the design strength of the bolt taken as the smaller of the value as controlled by shear V_{dsb} and bearing.

C.2 ADVANTAGES OF BOLTED CONNECTIONS

The advantages of bolted connections are as follows :

1. There is silence in preparing bolted connection.

   In riveting, hammering is done. The hammering causes noise in the riveting.

2. There is no risk of fire in bolted connection.

   The rivets are made red hot in riveting and there is risk of fire.

3. The bolted connections may be done quickly in comparison to the riveting.

4. Through the cost of bolts is more than the cost of rivets, the bolted connections are economical to use them than rivets because less persons are required for installation, and the work proceeds quickly.

5. The bolted connections facilitate the erection because of ease with which these connections can be done.

C.3 DISADVANTAGES OF BOLTED CONNECTIONS

The following are disadvantage of bolted connections :

1. If bolted connections become loose, their strength reduces considerably.
2. The unfinished bolts are not uniform in diameter and they have less strength.
3. The bolted connections have less strength when they are subjected to axial tension, because area at root of thread is less.
4. Generally, the diameter of hole is kept 1.6 mm more than the nominal diameter of black bolt. The bolt does not fill the hole and there remains a clearance in bolted connections.

C.4 BOLTS

A bolt is a metal pin with a head formed at one end and its shank is threaded at the other end in order to receive a nut. as shown in Fig. C.1.

The bolt for structural purposes are manufactured from mild steel and high strength steel. A bolt consists of head and body as shown in Fig. C.1. The body of a bolt is termed as shank. The bolts are manufactured to different lengths to suit different purposes. The size of bolt is expressed by the diameter of shank. The shank is threaded at one end. A nut can be provided on the threaded end of the bolt. The threaded portion of each bolt shall project through the nut at least by one thread. In all cases, when the full bearing area of the bolt is to be developed, the bolt shall be provided with a washer of sufficient thickness under the nut to avoid any threaded portion of the bolt being within the thickness or the part bolted together. The structural bolts are classified according to type of shank (unfinished or turned); material and strength (ordinary structural, or high strength steel); shape of head and nut (square or horizontal, regular or heavy duty); and pitch and fit of thread (standard, coarse or fine).

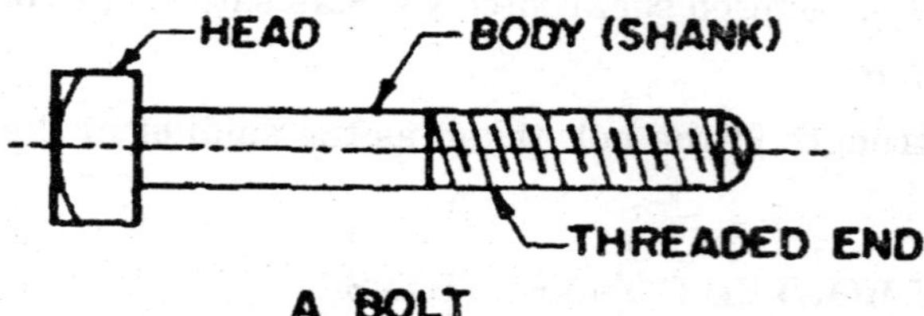

A BOLT

Fig. C.1

Table C.2. DESIGNATION SYSTEM FOR BLACK SQUARE BOLTS, SCREWS AND NUTS

Description			Designation
Fasteners	Thread size	Length mm	
Square bolts with square nut	M10	30	Square bolt M 10 x 30 N
Square bolt only	M10	30	Square bolt M10 x 30
Square screw	M10	30	Square screw M10 x 30
Square nut	M10	—	Square nut M 10
Square bolt with hexagonal nut	M10	30	Square bolt M 10 x 30 NH

The following are various types of bolts used for structural purposes :

1. Unfinished bolts

The finished bolts are also termed as ordinary bolts, common bolts, rough bolts and back bolts. The unfinished bolts are manufactured from low carbon steel. The unfinished bolts have square heads. The heads of unfinished bolts are made by forging. The shanks of unfinished bolts are not uniform in diameter. The unfinished bolts are designed for lower permissible stresses in shear and bending than the rivets and turned bolts. The unfinished bolts are not used for structural connections of members, which are subjected to shock or vibrations. The unfinished bolts are used for ordinary field work and light load. The unfinished bolts are used in light structures, secondary or bracing members, platforms, catwalks, purlins, girts, smaller trusses, and in similar applications in which the loads are small and static in nature. Such bolts are also used as temporary fitting up connectors. The unfinished or black bolts are used with square nuts. These are also used with hexagonal nuts. These bolts, screws and nuts are designated as illustrated in Table C.2 as per IS : 2585-1968.

A large portion of site connections are made with black bolts. From the point of view of the erector, this makes for simplicity as the erection can be completed in almost one opera-tion. The additional cost of special labour for riveting and the plant required is also provided.

The black bolts cannot, however, carry loads equal to those of a rivet with the same diameter. In such cases, the turned bolts are used.

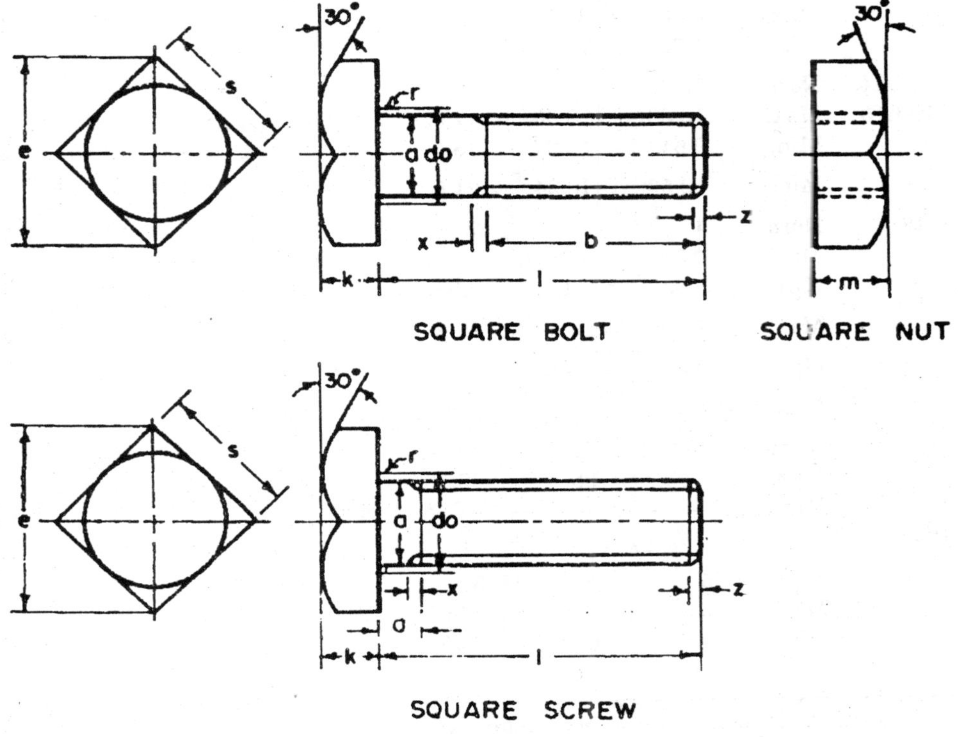

Fig. C.2

The various dimensions of bolts, screws and nuts are as shown in Fig. C.2. The dimensions are as per IS: 2585-1968 and have been given in Tables C.3 and C.4. The diameters range from 6 to 39 mm.

2. Turned bolts

Turned bolts are also termed as turned and fitted bolts. For these bolts, the holes are made slightly larger than the diameter of the bolts. These bolts are force fitted. Turned bolts are manufactured from hexagonal stock. The bars are turned to required diameters. The threads in turned bolts are usually cut with a die. If turned bolts are used in holes which are reamed, then turned bolts are equally satisfactory as rivets for many connections. The flat face of nut, inner face of head of bolt and both sides of washers are machined for better fitting. This type of bolt is used in connections, which require close-fitting bolts in drilled holes, such as in riveted constructions, where it is not possible to drive rivets satisfactorily. These bolts are sometimes useful in aligning mechanical equipment and structural members which require precise positioning.

TABLE C.3. DIMENSIONS FOR BLACK SQUARE BOLTS, SCREWS AND NUTS (as per IS : 2585-1968) (Diameter range M-6 to M-18)

	Size		M6	M8	M10	M12	M14	M16	M18
d		Norm.	6	8	10	13	14	16	18
		Max.	6.48	8.90	10.00	12.10	15.10	17.10	19.10
		Min.	5.70	7.64	9.64	11.57	13.57	15.57	17.57
s	<19-h14	Norm.	10	13	17	9	22	24	27
	>19-h15	Max.	10.00	13.00	17.00	19.00	22.00	24.00	27.00
		Min.	9.64	12.67	16.57	18.48	21.16	23.16	26.16
e		Min.	12.53	16.34	21.50	24.02	27.51	30.11	34.01
k	js 16	Norm.	4	5.5	7	8	9	10	12
		Max.	4.38	5.88	7.45	8.45	9.45	10.45	12.55
		Min.	3.62	5.12	6.55	7.55	8.55	9.55	11.45
da		Max.	7.2	10.2	12.2	15.2	17.2	19.2	21.2
r		Min.	0.25	0.4	0.4	0.6	0.6	0.6	0.6
a		Max.	4.0	4.5	5.0	6.0	7.5	7.5	9.0
b	b_1	Norm.	18	22	26	30	34	38	42
	b_2	Max.	—	28	32	36	40	41	48
	b_3	Min.	—	—	—	49	51	57	61
mfs 16		Norm.	5	5	8	10	11	13	15
		Max.	5.38	6.95	7.55	10.45	11.55	13.55	15.55
		Min.	4.62	6.05	8.45	9.55	10.45	12.45	14.45

Note 1. Sizes shown in brackets avoid second preference.
Note 2. Use b, for l<<130, b2 130<<K200 and b, for >>200

TABLE C.4. DIMENSIONS FOR BLACK SQUARE BOLTS, SCREWS AND NUTS (as per IS : 2585-1968) (Diameter range 20 to M-39)

Site			M20	(M22)	M24	(M27)	M30	(M33)	M36	M39
d		Nom.	20	22	24	27	30	33	36	39
d		Max.	21.50	23.30	25.30	28.30	31.30	34.60	37.60	40.60
		Min.	19.481	21.48	23.48	26.48	29.48	32.38	35.38	38.38
s	<19-114	Nom.	30	32	36	41	46	50	55	60
		Max.	30.00	32.00	36.00	41.00	46.00	50.00	55.00	60.00
	>19-h11	Min.	29.16	31.00	35.00	40.00	45.00	49.00	53.80	58.80
e		Min.	37.91	40.30	45.50	42.00	58.50	63.70	69.94	76.44
k	js16	Nom.	13	14	15	17	19	21	23	25
		Max.	13.55	14.55	15.55	17.55	19.65	21.55	23.65	25.65
		Min.	12.45	13.45	14.45	16.45	18.35	20.35	22.35	24.35
da		Max.	24.4	26.4	28.4	12.4	35.4	38.4	42.4	45.4
r		Min.	0.8	2.8	0.8	1.0	1.0	1.0	1.0	1.0
a		Max.	9.0	9.0	11	—	—	—	—	—
b	b_1	Nom.	46	50	54	60	66	72	78	84
	b_2	Max.	52	56	60	60	72	78	84	90
	b_3	Min.	65	69	73	60	85	91	97	100
m	js	Nom.	16	18	19	22	24	26	29	31
		Max.	16.55	18.55	19.55	22.65	24.65	26.65	29.65	31.80
		Min.	15.45	17.45	18.35	21.35	23.35	25.35	28.35	30.20

Note 1. Size shown in brackets are of second preference.
Note 2. Use b_1 for l 130, b_2 for 130 l 200, and b_3 for l 200.

The turned or black hexagonal bolts, nuts and lick nuts and screws, are designated as illustrated in Table C.5 per IS : 1363- 1967.

Table C.5. DESIGNATION SYSTEM FOR BOLTS, SCREWS, NUTS AND LOCK NUTS

Description			
Fastener	Size	Length	
Bolts with nuts and a lock nut	M 16	70	Hex. bolt M 16 x 70 NL
Bolts with a nut	M 16	70	Hex. bolt M 16 x 70 N
Bolt only	M 16	70	Hex. bolt M16 x 70
Screw	M 16	70	Hex. screw M 16 x 70
Nuts	M 16	—	Hex. nut M16
Lock nut	M 16	—	Hex. lock nut M 16
Double chamfered nut	M 16	—	Hex. nut DCM 16

The various dimensions of black hexagonal bolts, screws, nuts and lock nuts are as shown in Fig. C.3. These dimensions are as per IS : 1363-1967 and have been given in Tables C.6 and C.7. The diameters range from 6 mm to 24 mm.

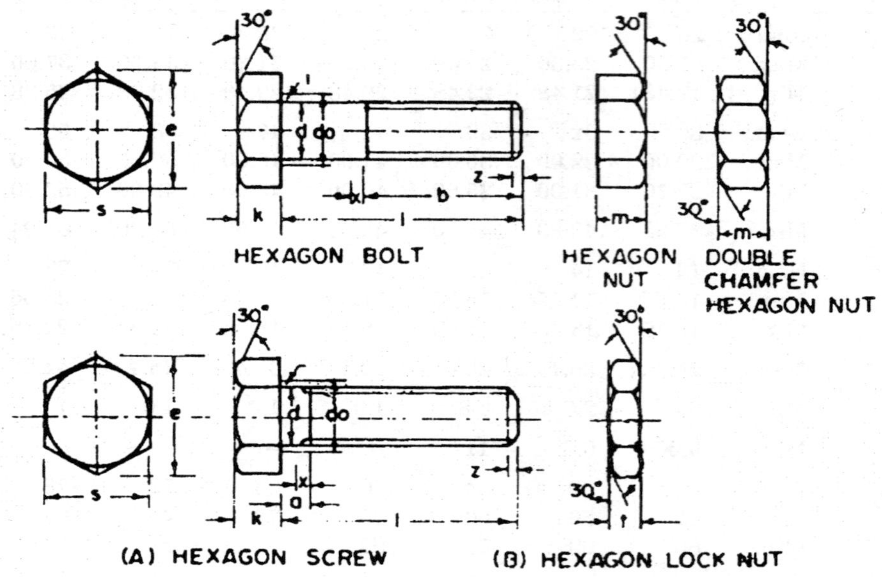

Fig. C.3.

TABLE C.6. DIMENSIONS FOR BLACK GRADE HEXAGONAL BOLTS, SCREWS, NUTS AND LOCK NUTS (Diameter range M-6 to M-18) as per IS : 1363-1967

Site			M6	(M7)	M8	(M10)	M12	(M14)	M16	(M18)
d		Nom.	6	7	8	10	12	14	16	18
		Max.	6.48	7.58	8.90	10.90	13.10	15.10	17.10	19.10
		Min.	5.70	6.44	7.64	9.64	11.57	13.57	15.57	17.57
s	<19-h14	Nom.	10	11	13	17	19	22	24	27
		Max.	10.00	11.00	13.00	17.00	19.00	22.00	24.00	27.00
	>19-h15	Min.	9.64	10.57	12.57	16.57	18.48	21.6	23.16	26.16
e		Min.	10.89	11.94	14.20	18.72	20.88	23.91	26.17	29.56
k	js16	Nom.	4.0	5.0	5.5	7.0	8.0	9.0	10	12
		Max.	4.38	5.38	5.88	7.45	8.45	9.45	10.45	12.55
		Min.	3.62	4.62	5.12	6.55	7.55	8.55	9.55	11.45
da		Max.	7.2	8.2	10.2	12.20	15.2	17.2	19.2	21.2
a		Max.	4.0	4.5	4.5	5.0	6.0	7.5	7.5	9.0
b	b_1	Min.	18	20	22	26	30	34	38	42
	b_2	Min.	—	—	28	32	36	40	44	48
	b_3	Min.	—	—	—	—	49	53	57	61

m	js16	Nom.	5.0	5.5	6.5	8.0	10	11	13	15
		Max.	5.38	5.88	6.95	8.45	10.45	11.55	13.55	15.55
		Min.	4.62	5.12	6.05	7.55	9.55	10.45	12.45	14.45
f	js16	Nom.	33.30	3	4	5	7	8	8	9
		Max.	2.70	3.30	4.38	5.38	7.45	8.55	8.55	9.45
		Min.		2.70	3.62	4.62	6.45	7.45	7.45	8.55
r		Min.	0.25	0.25	0.4	0.4	0.6	0.6	0.6	0.6

Note : 1. Sizes shown in brackets are of second preference.
Note 2. Use b, for l<130, for 130<200 and b3 for l<200.

Table C.7. DIMENSIONS FOR BLACK GRADE HEXAGONAL BOLTS, SCREWS, NUTS AND LOCK NUTS (Diameter range M-20 to M-30) as per IS : 1363-1967

Site			M20	(M22)	M24	(M27)	M30	(M33)	M36	M39
d		Nom.	20	22	24	27	30	33	36	39
		Max.	21.30	22.30	25.30	28.30	31.30	34.00	37.60	40.00
		Min.	19.48	21.48	23.48	26.48	29.48	32.38	35.38	38.38
s	<19-114	Nom.	30	32	36	41	46	50	55	60
		Max.	30.00	32.00	36.00	41.00	46.00	50.00	55.00	60.00
	>19-h15	Min.	29.16	31.00	35.00	40.00	45.00	49.00	53.80	58.80
e		Min.	32.95	35.03	39.55	45.20	50.85	55.37	60.79	66.44
k	js16	Nom.	13	14	15	17	19	21	23	25
		Max.	13.55	14.55	15.5	17.55	19.65	21.65	23.65	25.65
		Min.	12.45	13.45	15.45	16.45	18.35	20.35	22.35	24.35
da		Max.	24.4	6.4	28.4	32.4	35.4	18.4	42.4	45.4
a		Max.	9.0	9.0	11	—	—	—	—	—
	b_1	Nom.	46	50	54	60	66	72	78	84
	b_2	Max.	52	56	60	66	72	78	84	90
	b_3	Min.	65	69	73	79	85	91	97	103
m	js16	Nom.	16	18	19	22	24	26	29	31
		Max.	16.55	18.55	19.65	22.65	24.65	26.65	29.65	31.80
		Min.	15.45	17.45	18.35	21.35	23.35	25.35	28.35	30.20
f	js16	Nom.	9	10	12	12	12	14	14	15
		Max.	9.45	10.45	10.45	12.55	12.55	14.55	14.55	16.55
		Min.	8.55	9.55	11.45	11.45	13.45	13.45	13.45	15.45
r		Min.	0.8	0.8	0.8	1.0	1.0	1.0	1.0	1.0

Note : 1. Sizes shown in brackets are second preference.
Note 2. Use b1 for l<130, b2 for 130<l<200, and b3 for l<200.

C.5 HISTORY OF HIGH STRENGTH FRICTION GRIP BOLTS

C. Batho and E.H. Bateman ['Investigations on Bolts and Bolted Joints, Second Report of the steel structures Research Committee. London : His Majesty's Stationery Office, 1934] indicated the possibility of using high-strength bolts in steel-frame constructions in 1934. The slippage of the connected parts may be prevented by using these bolts having minimum yield strength 372 N/mm². Wilson, W.M. and Thomas F.P. ['Fatigue Tests on Riveted Joints', Bulletin 302, Engg Experiment Station, U. of Illinois, Urbana, Ill, 1938] in their follow-up tests subtantiated the earlier work. It was reported that high-bolts smaller in diameter than the holes in which these bolts were inserted had fatigue strength equal to that of well-driven rivets provided that the bolts were sufficiently pre-tensioned.

In 1947, Research Council on Riveted and Bolted structural Joints was formed. A.E.R. De Jonge ['Riveted Joints; a Critical Review of LIterature Covering their Development' New York : American Society of Mechanical Engineers, 1945] made extensive Bibliography in 1945.

Interest on the use of high-strength bolts in rail-road bridge maintenance developed in the American Railway Engineering Association (AREA) in 1948. A number of field tests installed, and in their confirmation, the adequacy of high-strength bolt connections was reported in the same year.

In `Symposium on High-Strength Bolts' Proceeding AISC, National Engineering Council in 1950, the use of High-strength bolts and a summary of research and behaviour was presented in 1950, the Research Council permitted the replacement of rivets with bolts on a one-to-one basis. It was conservatively assumed that it was necessary to transfer the load in all joints under service load conditions by friction. The factor of safety was kept high enough. A good fatigue resistance (i.e., no slip under varying stress or stress reversal consisting of many load cycles) was provided in every joint. In 1951, high strength bolts become on accepted load transfer method. Since then, their use have become increasingly popular, specially in tall buildings and in situations, where the loads, which are transferred are large.

Use of flat washers on 1:20 sloping surfaces was allowed by issuing a revision in 1954.

Use of impact wrenches for installing high-strength bolts was also allowed in this revision of the specifications. At the same time, it was also allowed to paint the surfaces in contact to create a bearing type connection (i.e., when the ultimate strength of the connection was to be based on the bolt in bearing against the side of hole.

William H. Munse in 1956 ['Research on Bolted Connections' Transactions, ASCE, 121 (1956), pp 1255-1266] reported the summary on bolt behaviour. It was concluded that in case, high-strength bolts are to be used as efficient and economical as possible, an initial-tension in the shank of bolt as high as practical must be induced in the bolts.

Additional research in 1960, made justified the increase in the minimum bolt tension. It was recognized that the bearing-type was ordinarily an acceptable substitute for a riveted-connection. It was also accepted that the connections having its design strength based on resistance to slip, known as **friction type connection** may only be necessary when direct tension acts on the holes or when the fatigue conditions are important.

In the absence of the torque wrench method for tightening the high-strength bolts, a turn-of-the nut procedure was also introduced in 1960. Two washers were required previously. It was reduced to one washer. One washer is sufficient under the element (head or nut) being turned, in case, the turn-of-the nut method of installation was used.

Research Council further liberalized in its specification. The requirement of washers was completely deleted ['Rivets and High Strength Bolts', A Symposium, Transactions, ASCE, 126, Part II (1961), 693-820] in 1962.

Research Council [Specifications for structural Joints using ASTM A325 or A490 Bolts, Research Council on Riveted and Bolted structural Joints of the Engineering Foundation, February, 04, 1976] made most significant change since, 1951.

Above changes included increased allowable capacities, explicit recognition of other than standard size holes for fasteners, and changes in the philosophy of design for friction-type and bearing-type connections. There was an object in making these changes. A more uniform reliability in connection performance for all types of loading and joints arrangements was provided. These changes were based on the Guide to Design Criteria for Bolted and Riveted Joints by John W. Fisher and John H.A. Struik [New York, John Wiley & Sons, Inc. 1974].

Research Council specifications involving new concepts and requirements relating to 1976 in the above complete chapter. These are also included in 1978 AISC specifications.

Table C.8. DIMENSIONS FOR HIGH STRENGTH (TENSILE) FRICTION GRIPS BOLTS (as per IS : 3757-1966)

	d	M12	M16	M20	M22	M24	M27	M30	M33	M36	M39
d1	Min.	20	25	30	34	39	44	48	53	58	63
b	12	30	38	46	50	54	60	66	72	78	84
		X	44	52	56	60	66	72	78	84	90
c	Nom.	0.4	0.6	0.8	0.8	0.8	0.8	1.0	1.0	1.0	1.0
e	Max.	25.4	31.2	36.9	41.6	47.3	53.1	57.7	63.5	69.3	75.0
k j6	Nom.	8	10	13	14	15	17	19	21	23	25
	Max.	8.45	10.45	13.55	14.55	15.15	17.55	19.65	21.65	23.65	25.65
	Min.	7.55	9.55	12.45	13.45	14.45	16.45	18.35	20.35	22.35	24.35
r	Max.	1.6	1.6	2.0	2.0	2.5	2.5	2.5	2.8	2.8	2.8
	Min.	1.2	1.2	1.5	1.5	2.0	2.0	2.0	2.2	2.2	2.2
e h15	Nom.	22	27	32	36	41	46	50	55	60	65
	Max.	22.00	27.00	32.00	36.00	41.00	46.00	50.00	55.00	60.00	65.00
	Min.	21.16	26.16	31.00	35.00	40.00	45.00	49.00	53.80	53.80	63.80

(1) For length upto 90 mm (2) For length over 90 mm upto 20 mm
Note. The dimensions d1 shall not exceed the actual width across the flat.
(1) For length upto 90 mm (2) For length over 90 mm upto 200 mm.
Note. The dimensions d_1 shall not exceed the actual width across the flat.

C.6. SHEAR CAPACITY OF BOLT

The design strength of bolt V_{dsb} as controlled by shear strength is given by (As per IS : 800-2007, clause, (10.3.3)

$$V_{dsb} = \frac{1}{\gamma_{mb}} \cdot (V_{nsb}), \text{ where}$$

V_{nsb} is the nominal shear capacity of a bolt, calculated as follows

$$V_{nsb} = \frac{\sigma_u}{\sqrt{3}}(n_n . A_{nb} + n_s \, A_{sb}), \text{ where}$$

σ_u = ultimate tensile strength of a bolt

n_n = Number of shear planes with threads intercepting the shear plane

η_s = Number of shear planes without threads intercepting the shear plane

A_{sb} = Nominal plain shank area of bolt, and

A_{nb} = Net shear area of the bolt at threads, may be taken as the area corresponding to root diameter at the thread.

(A) **For long joints:** When the length of the joint, lj of a spliceor end connection in a compression or tension element containing more than two bolts (that is, the distance between first and last rows of bolts in the joint, measured in the direction of load transfer) exceeds 15 d in the direction of load, **the nominal shear capacity,** V_{db} reduced by the factor β_{ij} given by

$$\beta_{ij} = 1.075 - lj/(200\ d)$$

but $$0.75 <= \beta_{ij} <= 1.0$$

$$= 1.075 - 0.005\ (lj/d)$$

Note : This provision does not apply when the distribution of shear over the length of joint is uniform, as in the connection of web of a section to the flange.

(B) **Large grip lengths :** When the grip length, l_g (equal to the total thickness of the connection plates) exceeds 5 times the diameter, d of the bolt, the **shear capacity,** shall be reduced by a factor β_{lg} given by

$$\beta_{lg} = 8\ d / (3d + lg) = 8/(3 + lg/d)$$

β_{lg} shall not be more than β_{ij}. The grip length, l_g shall in no case be greater than 8 d.

(C) **Packing plates** : The design shear capacity of bolts carrying shear through a packing plate in excess of 6 mm shall be decreased by a factor β_{pk} given by

$$\beta_{pk} = (1 - 0.0125\ t_{pk})$$

where, t_{pk} is the thickness of the thicker packing, in mm.

C.7 BEARING CAPACITY OF BOLT

At ultimate limit state (when subjected to shearing forces, a parallel shank friction grip bolt slips into bearing. Then, the bearing stresses between the bolt and the plate are required to be checked.

Design bearing capacity of the bolt/strength of a bolt on any plate v_{dpb} as controlled bybearing is given by (As per IS : 800-2007, Clause (10.3.4))

$$V_{dpb} = (V_{npb} / \gamma_{msp}),\ \text{where,}$$

V_{npb} = Nominal bearing strength of a bolt

$$= 2.5\ k_b\ d\ .\ t.\ \sigma_{up},\ \text{where}$$

k_b is smaller of (e/3do), [(P/3do) – 0.25]

(σ_{ub} / σ_{up}), 1.0

e, p = End and pitch distances of the fastener along bearing direction

d_o = diameter of the bolt hole (σ_{ub}), σ_{up} = Ultimate tensile stress of the bolt and ultimate tensile stress of the plate, respectively. Then γ_{msb} = 1.25

d = Nominal diameter of the bolt, and

t = Summation of the thicknesses of the connected plates experiencing bearing stress in the same direction, or if the bolts are countershunk, thickness of the plate minus one half the depth countersinking.

The bearing resistance (in the direction normal to the slots in slotted holes) of bolts in holes other than standard clearance holes may be reduced by multiplying the bearing resistance as above, V_{npb} by the factors given below :

(i) Over size and short slotted holes 0.7

(ii) Long slotted holes 0.5

C.8. TENSION CAPACITY OF BOLT

Due to the presence of holes, a plate in a joint may fail in tension through the weakest section. As such, the holes are kept in a longitudinal direction of the plate, so that the number of holes is equal to the number of rows across the width or staggered, so that the number of holes across the width become small. In the first case, the plate fails across the width.

A bolt subjected to a factored tensile force, $\tau_{fd.b}$ as per IS : 800-2007, Clause 10.3.5 shall satisfy

$T_{fd.b} <= T_{d.b}$, where

$T_{db} = T_{nb}/\gamma_{mb}$

T_{nb} = Nominal tensile capacity of the bolt calculated as under

$0.90\ \sigma_{ub} . A_n < \sigma_{yb} . A_{sb} . (\gamma_{mb} / \gamma_{msu})$

where,

σ_{ub} = Ultimate tensile stress of the bolt

σ_{yb} = Yield stress of the bolt

A_n = Net tensile stress area as specified in a appropriate Indian standard (for bolts where the tensile stress area is not defined, A_n shall be taken as the area at the bottom of the threads) and

A_{sb} = Shank area of the bolt

$\gamma_{msu} = 1.25$

C.9 COMBINED SHEAR AND TENSION IN BOLT

A bolt required to resist both design shear force (V_{sb}) and desgn tensile force (T_{db}) at the same time, as per IS : 800-2007, as per Clause 10.3.6, shall satisfy

$$\left[\frac{V_{sb}}{V_{db}}\right]^2 + \left[\frac{T_{fd.b}}{T_{db}}\right]^2 <= 1.0, \text{ where}$$

V_{sb} = Factored shear force acting on the bolt

V_{db} = Design shear capacity (Clause 10.3.2)

$T_{fd.b}$ = Factored tensile force acting on the bolt and

T_{db} = Design tension capacity (Clasue 10.3.5).

C.10 HIGH STRENGTH BOLTS

The high strength bolts are also called **high strength friction grip bolts**. Briefly, these are called **HSFG bolts**. These bolts are developed as a result of recent developments in the field of connections for the steel structures. These bolts are in use since 1938. Bureau of Indian Standards Institution has published a special publication SP : 6(4)-1969 (use of high strength friction bolts), which deals with the principles of design for such bolts. Use of high strength friction grip bolts has become very popular as a field fastener for structural connections. The term high strength friction grip bolts relates to bolts of high tensile steel. In these type of bolts, high initial pre-tension (i.e., usually high strength) is developed due to clamping force at the interfaces of elements being joined. Frictional resistance to slip between the plate surfaces subjected to clamping force oppose slip due to externally applied shear. These bolts are used along with high tensile steel nuts and hardened steel washer. These bolts are tightened to predetermined shank tension which are equal to proof load. As a result of this, the structural elements connected together are clamped between the nuts and the head of bolt. The load is transferred from one structural element to other by friction between the parts as shown in Fig. C.4 and not by shear in, and bearing on the bolts. These bolts do not allow slip to occur. When high strength friction grip bolts are used, all connected surface, including those adjacent to the bolts heads, nuts, or washers should be free of scale, burns, dirt and other foreign material that would prevent solid seating of parts. The contact surfaces of parts to be joined should be free of oil, paints, lacquer or galvanizing.

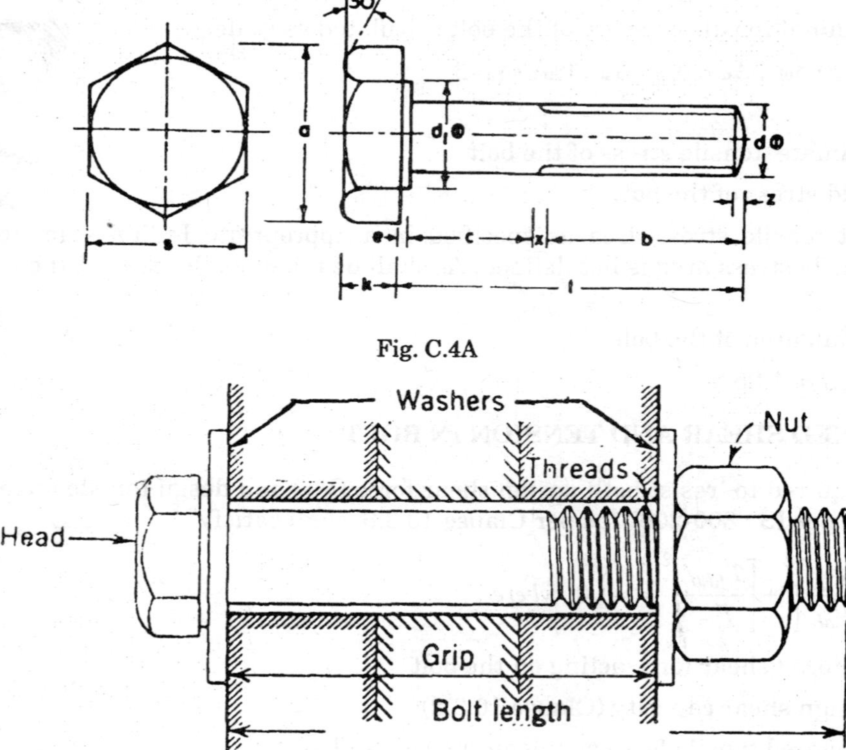

Fig. C.4A

Fig. C.4B

The high strength friction grip bolts have many advantages over rivets. Whereas the rivets are subjected to shear and bearing stresses, the bolts are subjected to uniform tensile stresses only. The high strength friction grip bolts have higher fatigue strength because there is no concentration of the stress in the hole. The bolts do not bear against the plates. Therefore, the uneven distribution of stress does not occur. These bolts are advantageously used in bridges and machine foundations subjected to vibrations. These bolts also simplify the problem of alterations and additions to structures as they can be assembled more easily than rivets. The various dimensions of high strength friction grip bolts are as shown in Fig. C.4. The dimensions are as per IS: 3757-1966 and have been given in Table C.7. The lengths of these bolts are from 30 mm to 80 mm with variation of 10 mm. These bolts are designated by expressing type of bolts, following by the size, the length and symbol representing the mechanical properties and IS number, e.g. friction grip bolts M 16 x 100 IS : 3757 : 10 K. The size of this bolt is M 16 (16 mm) and, the length of bolt is 100 mm. The bolt conforms to the mechanical properties of 10 K.

C.11 DEFINITIONS OF TERMS USED IN BOLTS AND BOLTING

1. **Nominal diameters of bolt.** The nominal diameter of a bolt is diameter of unthreaded shank of the bolt.
2. **Gross diameter of bolt.** The gross diameter of a bolt is the nominal diameter of the bolt.
3. **Gross area of bolt.** The gross area of a bolt is the nominal area of the bolt. It is the cross-sectional area of the bolt calculated from gross (or nominal) diameter of the bolt.
4. **Net area of the bolt.** The net area of a bolt is the area at the root of threaded part of cross-sectional area of the unthreaded part whichever is less. It is not cross-sectio- nal area of the bolt calculated from diameter at the root of threads (or core diameter).
5. **Diameter of the bolt hole.** The diameter of a bolt hole is the nominal diameter of the bolt plus 1.5 mm unless specified otherwise.

   The following definitions pertain to high strength friction grip bolts.
6. **Grip.** The grip of bolt is equal to total thickness of steel sections to be held together excluding washers.
7. **Ply.** The ply is equal to a single thickness of steel forming part of a structural joint.
8. **Length of bolt.** The length of bolt is equal to the distance from the underside of the bolt head t the extreme end of the shank, including any camber or radius.
9. **Effective interface.** The effective intersurface is a common contact surface between two load transmitting plies, excluding packing pieces, through which the bolt passes.
10. **Slip factor.** The slip factor is defined as ratio of the load per effective interface, required to produce slip in a pure shear joint to the total shank tension induced in the bolts.
11. **Factor of safety.** The factor of safety is the numerical value by which the load which would cause slip in a joint is divided to give the permissible working load on the joint.

C.12 BOLT, NUT AND WASHER ASSEMBLY

Fig. C.5 shows complete assembly of a bolt, nut and two washers. The complete assembly provides bolted connections.

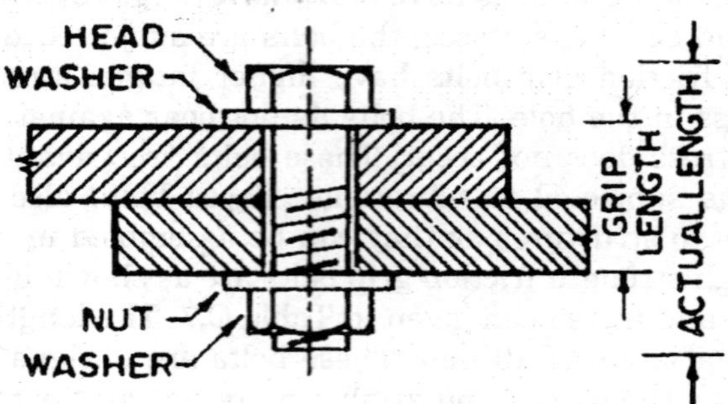

Fig. C.5. Bolt, nut and washers assembly

The washers are used below the nut and inner face of a head of bolt. The washers distribute the load to be transmitted over a large area. The washers provide smooth seating for the nut to turn on. The washers prevent the nut from cutting into the plates.

Two plates are connected by bolts. The combined thickness of plates connected is called grip length. The actual length of bolt is kept greater than the grip length. The nut and washers are placed in length which is in excess over the grip length. A small threaded portion remains projecting outside the nut.

There are four basic load transfer mechanisms, which have been critically developed for the bolt connections. The strength of bolt itself in either shear or tension may serve to limit the load that may be transferred. The strength of the bolted connections may also be a func-tion of the bearing of the bolt on the member being connected. High strength bolts are very commonly pretensioned, thereby creating a friction load between the two joined members.

(i) **Bolts in tension** - The strength of a bolt in tension is the product of nominal effective area of the bolt and the nominal strength of the bolt-material tension (i.e., permissible stress in tension in bolt).

(ii) **Bolts in shear** — The strength of a bolt in shear is the product of nominal effective area of the bolt (i.e., cross sectional area at the root of the threads) and the nominal strength of the bolt- material in shear.

C.20 SLIP RESISTANCE OF HSFG BOLTS

Clamping forces at the interface of the elements being joined develop due ot the initial pretension in the bolt.

Design for friction type bolt, in which slip is required to the limited, a bolt subjected only to a factored design shear force, $V_{fd.s}$ in the interface of connections at which slip cannot be tolerated, shall satisfy the following as per IS : 800-2007, Clause 10.4.3.

$$V_{fd.s} <= V_{d.sf}, \quad \text{where}$$

$$V_{fd.s} = (V_{ns.f} / \gamma_{mf})$$

$\gamma_{mf} =$ 1.15 for slip resistance at service load

V_{nsf} = Nominal shear capacity of a bolt as controlled by slip for friction type connection, calculated as follows

$V_{nsf} = \mu_f \cdot n_e \cdot K_n \; F_o$, where

μ_f = coefficient of friction (slip factor) as specified in Table 20, IS : 800-2007, ($\mu_f = 0.55$)

n_e = Number of effective interfaces offering frictional resistance to slip

K_n = 1.0 for fasteners in clearances holes

= 0.85 for fasteners in oversized and short slotted holes and for fasteners in long slotted holes loaded.

Perpendicular to the slot

= 0.7 for fasteners in long slotted, holes loaded parallel to the slot

γ_{mf} = 1.0 (In case, slip resistance is designed at service loads)

= 1.25 (In case, slip resistance is designed at ultimate load)

F_o = minimum bolt tension (Proof load) at installation and may be taken as $A_{nb} \cdot \sigma_o$

A_{nb} = Net area of bolt at threads, and

σ_o = Proof stress (= 0.70 σ_{ub})

Note : (1) V_{ns} may be evaluated at a service load or ultimate load using appropriate partial safety factors depending upon whether slip resistance is required at service load or ultimate load.

2. Provision for the long joints as in clause 10.3.3.1 (described) earlier also applies to friction grip connections. When the distance between the first and last rows of bolts in the joint, measured in the direction of the load transmitted exceeds 15 d in the direction of the load) by a factor β_{ij}, where

$\beta_{ij} = (1.075 - lj/(200d))$ but

$0.75 < \beta_{ij} <= 1.0$, where d is the nominal diameter of the bolt.

3. When friction type bolts are designed not to slip only under service loads, the design capacity at ultimate load may be calculated as per bearing type connections. (Clause 10.3.2 and 10.3.3)

4. The block shear resistance of the edge distance due to bearing force may be checked as per clause 6.4.

C.21 TENSION RESISTANCE OF HSFG BOLTS

For the parallel shank and waisted shank friction grip bolt, the tensile strength is similar to that of black bolts.

A friction bolt subjected to a factored tension force, T_{fd} shall be satisfy

$T_{fd} <= T_{df}$, where

$T_{df} = (T_{nf}/\gamma_{mf})$

T_{nf} = Nominal tensile strenght of the friction bolt, calculated as

$0.9\,\sigma_{ub} \cdot A_n <= \sigma_{yb} \cdot A_{sb}\,(\gamma_{msu}/\gamma_{msy})$

σ_{ub} = Ultimate tensile stress of the bolt

A_s = Net tensile stress area as specified in variousparts of IS : 1367 (for bolts, where the tensile stress area is not defined, A_n shall be taken as the area at the root of the threads)

A_{sb} = Shank area of the bolt and partial factor of safety

γ_{mf} = 1.25 as controlled by ultimate stress

= 1.15 as controlled by yield stress of bolt

C.22 COMBINED SHEAR AND TENSION

The interaction curve recommended for combined shear and tension for HSFG bolts (for which slip in the serviceability limit state is limited) is similar to that of black bolts.

Bolts in a connection which slip in the serviceability limit state shall be limited, which are subjected to a tension force, T and shear force, V, shall satisfy

$$\left(\frac{V_{s.fd}}{V_{df}}\right)^2 + \left(\frac{T_{fd}}{T_{df}}\right)^2 <= 1.0$$

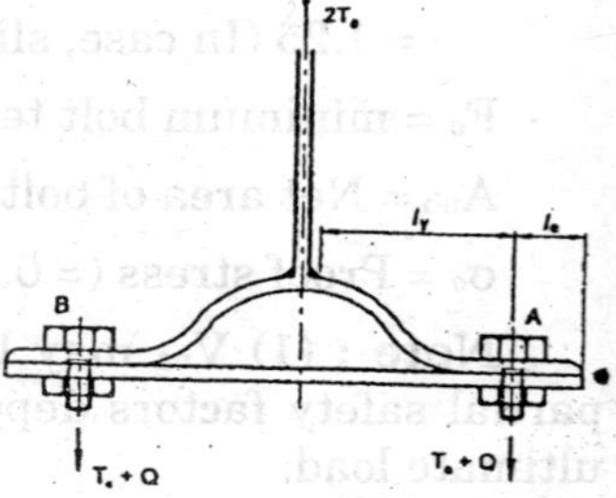

Fig. C-5(B). Combined prying force and tension

$V_{s.fd}$ = Applied factored shear at design load

V_{df} = Design shear strength

T_{fd} = Externally applied factored tension at design load, and

T_{df} = Design tension

C.23 PRYING FORCE, Q

Prying force is defined as an additional force developed in a bolt as a result of the flexing (bending) of a connection component (such as a beam end plate or leg of an angle). These additional forces induced in the bolts are mainly due to the flexibility of connected plates. These forces develop only when ends of the flanges are in contact due to the external load.

Prying force, Q has been illustrated in Fig. C.

Moment resisting beam - To- column connections commonly have areas in which the bolts are needed to transmit load by direct tension. In the design of such connections, such an additional force are considered.

In 1965, Douty and McGuire, in 1979, Agerskove, in 1983, Holmes and Martin and in 1984, Subramanian have worked in the field, so called prying action', Q

Many research scholars have studied in this field and have suggested expression to obtain the prying force developed in the bolt. (For example, in 1985, Astanesh, in 1987, Kulak et al, in 1989, Owens and Cheal, in 1985, 1992, Thornton). Expression as suggested by Owens and cheal, in 1989 has been recommended by IS : 800-2007.

Where prying force, Q is significant, it shall be calculated as given below and added to the tension in the bolt

$$Q = \left(\frac{l_v}{2l_e}\right)\left[T_e - \frac{\beta . \eta . \sigma_o . b_e . t^4}{27\, l_e . l_v^2}\right] \text{ where}$$

l_v = Distance from the bolt centre-line to the toe of the fillet weld or to half the root radius for a rolled section.

The prying force, Q acts at some intermediate point instead of at the edge of the flange, in case the distance from the bolt-line to the edge of the flange is comparitively large. In 1989, Owens and cheal suggested the following expression for minimum prying force and for minimum flange thickness design.

l_e = Distance between prying force and bolt centre-line and is minimum of either the end distance or the value given by

$$l_e = 1.1\,t\left(\frac{\beta\sigma_o}{\sigma_y}\right)^{1/2}, \textit{ where,}$$

β = 2 for non pre-tensioned bolt and

= 1 for pretensioned bolt

η = 1.5

b_e = Effective width of flange per pair of bolts

σ_o = Proof stress in consistent (kN, kN/mm²) and units.

t = Thickness of the end plate.

Second term in the above expression to find prying force is neglected (being very small). Then, it results in a simplified expression

$$Q = T_e \,.\, l_v/(2.\, l_e)$$

In case, it is assumed that the plastic hinges are located at the bolt line and the root, that is, when the thickness of flange is minimum, then, above simplified expression results.

BS : 5950 specifies that prying force may be found by reducing the bolt tensile strength by 20 percent.

C.24 BLOCK SHEAR FAILURE

It is seen in some research work in USA, that the angles, gusset plates and coped beam connections may fail due to block shear.

(Kulak and Grondin, 2000). Failure occurs in shear at a row of bolt holes parallel to the applied loads, accompanied by tensile rupture along a perpendicular face. A block of material tears out by the applied shear in this type of failure. The block shear strength T_{db} of a joint are calculated, from following two expressions [IS : 800-2007, Clause 6.4.1]

$$T_{db1} = \left(\frac{\sigma_y \,.\, A_{vg}}{\sqrt{3}\,\gamma_{msy}}\right) + \left(\frac{0.9\,\sigma_u\, A_{tn}}{\gamma_{msu}}\right)$$

and

$$T_{db2} = \left(\frac{0.9\,\sigma_u\, A_{vn}}{\sqrt{3}\,\gamma_{msu}}\right) + \left(\frac{\sigma_y \,.\, A_{tg}}{\gamma_{msy}}\right)$$

T_{db} is adopted as T_{db1} or T_{db2} whichever is small. Where

A_{vg} and A_{vn} = Minimum gross and net area respectively, in shear along a line of transferred force (along l_v)

A_{tg} and A_{tn} = Minimum gross and net areas, respectively, in tension from the hole to the toe of the angle or next last row of bolt in gusset plate (along l_t)

σ_u and σ_y = Ultimate and yield stresses of the material respectively and γ_{msy} and γ_{msu} = Partial safety factors for material resistance controlled by yielding (1.15) and controlled. By ultimate stress (1.25), respectively.

C.13 STRESSES IN BOLTS

Unfinished bolts are used in bolted connections in lap joint as shown in Fig. C.6 (A) and in butt joint as shown in Fig. C.6 (B). The plates are carrying an axial pull, P. The diameter of hole is 1.6 mm greater than the diameter of bolt. Firstly, there occurs a slip. Then plate elements come in contact with the bolts. The bolts are subjected to single shear in case of lap joint and double shear in case of butt joint. In addition to shear, bolts are also subjected to bearing.

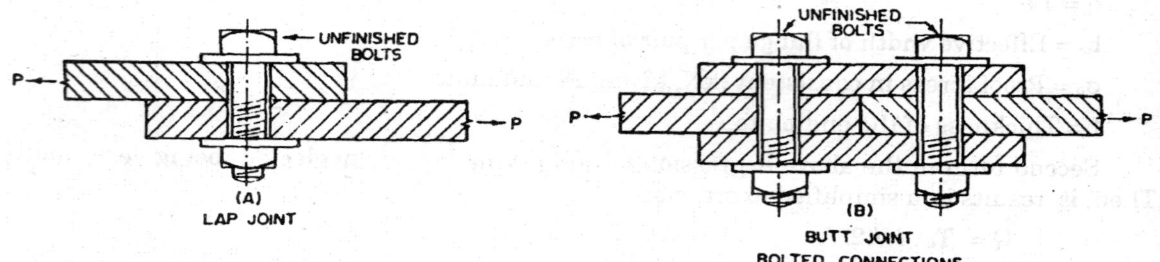

Fig. C.6. Bolted connections

When turned bolts are used, the holes are drilled and reamed. The turned bolts fit tightly in the holes as shown in Fig. C.7. The pull is transmitted directly. The bolts are subjected to shear and bearing.

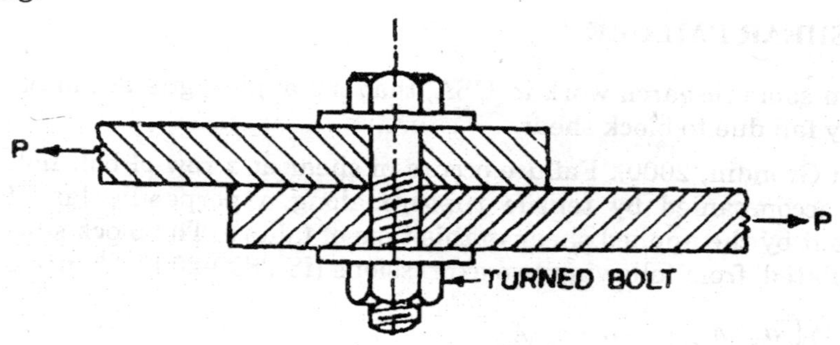

Fig. C.7. Bolted connection

Fig. C.7 shows a bolted connection. Pull acts along the longitudinal axes of bolts A, A. The bolts are subjected to tension. The tension in bolt is resisted by the cross-section of bolt at the root of thread (i.e., the net area of bolt).

Fig. C.8 shows bolted connections in which high strength friction grip bolts are used. There is no slip or movement between the connected parts in these type of connections. This type of connection is useful where rigid joints are required. The surfaces in contact must remain free from oil, grease, scale and paint. HSFG bolts are tightened by applying a torque to the head of bolt. These are tightened to shank tension equal to the proof load F. In case, any part will

remain loose, then slip will occur and the joint will act as an ordinary bolt joint. The plates are clamped together. As a result of this, friction uF, is developed between the plates as shown in Fig. C.9. The load is transferred by friction. These bolts do not act in shear or in bearing. When these bolts are tightened, longitudinal tensile stresses, and torsional shear-stresses are developed in the shank of bolt. The shank tension reduces slightly after completion of tightening. The torsion drops appreciably. Later on, the shank tension remains more or less constant for the whole of its working life, irrespective of whether, the connections remain loaded or unloaded.

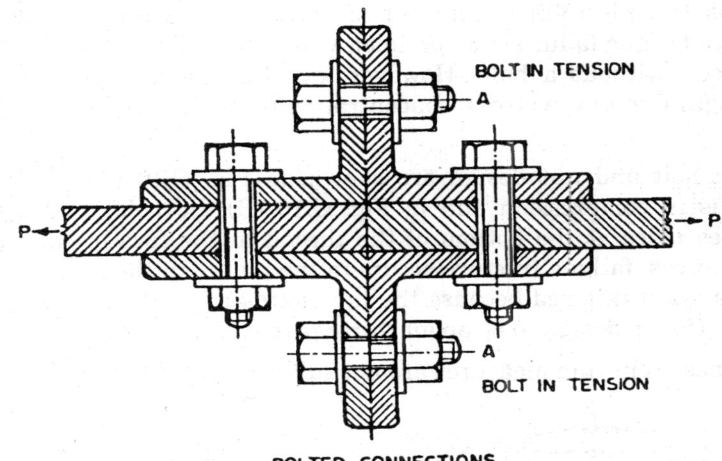

Fig. C.8. Bolted connections

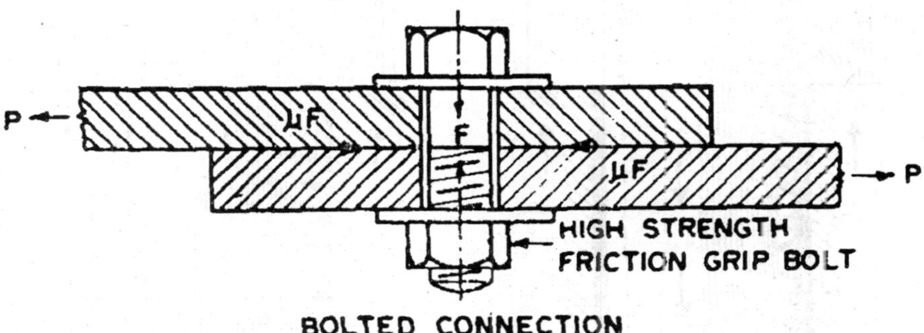

Fig. C.9. Bolted connections

There are three methods to see that correct shank tension has been reached in the bolt. These methods are as follows :

1. **Part turning.** In this method, the nut is screwed up tight and then tightened further by half to three-fourth of a turn depending on the length and diameter of bolt.

2. **Torque control.** In this method, a calibrated power operated or hand operated torque wrench is used to give a stated torque to bolt.

3. **Patented load-indicating bolts and washers.** In this, the nuts have projections that flatten out when the required tension is reached.

C.14 CONCENTRATIONS OF STRESS IN BOLTS

In machine parts, stress-concentrations are always present due to fillets, grooves, holes, key-ways, ...etc. and experience shows that most fatigue cracks in service begin at points of stress- concentration. Stress-concentration produce by sharp variations and these are specially damaging in the case of varying stresses.

In the cases of flat specimens with small circular-holes subjected to direct stress, the theoretical factor of stress-concentration is three (3). In the endurance tests, the magnitude of the peak stress is the controlling factor. It is expected that the tension compression load required to produce fatigue failure of a specimen with a hole, it would be about one-third of the load for a specimen without a hole. However, experiments showed that in this case, the reduction in strength due to the stress- concentration is small as compared with the calculated effect.

Two different bolt and nut design are shown, in Fig. C.10 (A) and Fig. C (B). In Fig. C.10(A), the nut is in compression and in Fig. C.10 (B) the bolt is in tension. High stress concentration takes place at the bottom of the thread in the face of the nut, and under the action of variable forces, fatigue fracture occurs in the plane. In the lip design, Fig 4.10(B), the peak stress is somewhat relieved because the lip is stressed in the same direction as the bolt. Fatigue tests show the lip design to be about thirty percent stronger.

Sometimes, these relieving measures are not sufficient to eliminate fatigue fractures.

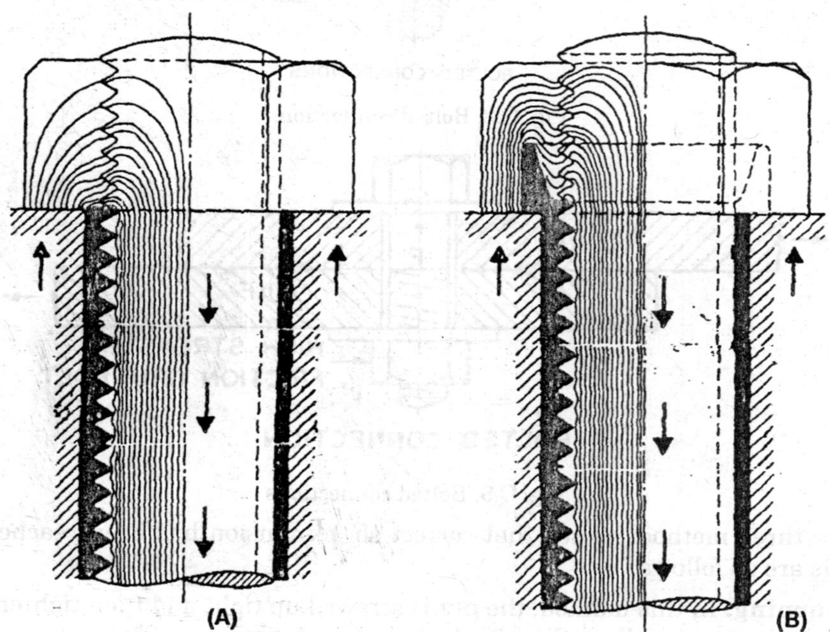

Fig. C.10. Stress concentration in bolts.

C.15 PROOF LOAD AND BOLT TENSION IN HIGH STRENGTH BOLTS

When the high strength bolts are installed, it is primary requirement to provide a sufficient pre-tension force. The pretension should be as high as possible without any chance of permanent deformation or failure of bolt. Bolt material exhibits a stress-strain (load-deformation) behaviour that has no yield point. Therefore, instead of directly using a yield stress, a so called proof-load is used. The proof-load is defined as the load obtained by multiplying the tensile stress area by a yield stress established by using either 0.2 percent off set strain or a 0.5 percent extension under the load. Tensile stress area may be determined by using following.

$$A_{TSA} = \left(\frac{\pi}{4}\right)[D - (0.9743.n)]^2$$

where,

A_{TSA} = Stress area in square mm

D = nominal diameter of bolt in inches, and

n = number of threads per mm

ASTM (American Society for Testing Materials have tabulated the values of this proof load for each diameter fastener using, for example, for 12 mm to 25 mm diameter bolts a strain offset value of 6.34 N/mm² and a length measurement value of 586 N/mm². The proof load stress is approximately 70 per cent to 80 percent of the minimum tensile strength for A325 and A490 bolts, respectively.

TABLE C.8. MINIMUM BOLT TENSION (As per AISC Table 1.23.5)

Bolt size (mm)	A 325 Bolt (kN)	A 490 Bolt (kN)
12.7	53	67
15.9	85	107
19.1	125	156
22.2	173	218
25.4	227	285
28.6	249	356
31.8	316	454
34.9	378	538
38.1	458	658

C.16 TRANSFER OF LOAD IN HSFG BOLT-CONNECTIONS

The high-strength bolt-behaves in a manner similar to the rivet. The bolt is installed either by the calibrated wrench method, or the turn-of-the-nut method or by the direct tension indicator method. The specified tension is large enough to give a frictional force capable of transferring the total load. The bearing of the bolt-shank against the side of the hole generally does not occur under the service load.

In case, there is over-load and once the frictional resistance becomes inadequate to transfer the load, the bearing against the side of the hole occurs.

In case, the failure of the connection is likely to happen soon, the frictional force does not affect the failure mode.

The strength of the plate, and the tensile strength of bolt and the shear strength of the bolt will control the strength of the connection.

C.17 DIFFERENCE BETWEEN NOMINAL STRESSES AND ACTUAL STRESSES

For many years, the design of connections has been based on the behaviour when the failure is likely to happen soon rather than on the behaviour at service loads (even though, the calculations are made on the service loads)

The failure of bolt connections may be in one of the following possible modes of failures.

1. Shear failure of bolt
2. Shear failure of the plate.
3. Bearing failure of the bolt.
4. Bearing failure of the plate.
5. Tensile failure of bolts.
6. Bending failure of bolts.
7. Tensile failure of plate.

Above modes are used to calculate the unit stresses. Since the behaviour assumed for the calculations does not occur, when the connection is behaving elastically, such stresses not real stresses, but instead merely serve as criteria of safety. Such stresses used in design calculations are known as nominal stresses.

The nominal stresses approach used in design focuses on the individual fastener capacity. It means, all fasteners of the the same size and material are treated as having equal strength in transmission of load. For example, there are two lines of rivets and there are eight rivets in one line, each fastener participates equaling and transmit one-sixteenth of the load in a lap-joint.

Once the friction is overcome and the fasteners act in the bearing the plate deformation between the fasteners would not actually be the same.

The nominal stress assumption means rigid plates, otherwise, the actual deformation on the fasteners could not be same.

The main reasons, why nominal stresses differ from actual stresses are as under.

1. The friction resistance to slip is neglected.
2. The deformation of plates is neglected.
3. The tensile stress concentrations at holes are neglected.
4. The shearing deformation of connectors is assumed proportional to the shearing stresses.
5. The shearing stresses are assumed uniform over connectors cross-sections.
6. The bearing stress is assumed uniform over a nominal control surface equal to connector diameter times plate thickness.
7. The bending of connectors is neglected.

It is difficult to obtain the actual stresses even if desired the actual stresses are depended on many factors often beyond the control of the designer, such as, poor alignment of holes, loose or unequal tightening eccentric loading not anticipated and poor construction.

C.18 FRICTION-TYPE CONNECTIONS

In case, slip resistance at service load is desired, the friction type connections are used.

The pre-tension force, T in the bolt equals the clamping force between the pieces being joined. The resistance to shear is a frictional force (μT), where, μ is the coefficient of friction.

The coefficient of friction or more properly, the slip coefficient, depends on the surface condition of the pieces being connected, with factors such as mill scale, oil, paint, or special surface treatments find the value of μ.

The range of μ has been found from about 0.2 to 0.6 depending on the surface conditions of the pieces being joined.

C.19 BEARING-TYPE CONNECTIONS

When the prevention of slip is not necessary, (that is, the slip may occur, under an occasional overload to bring the bolt shank into bearing against the side of hole) then, the bearing-type connections are used. For this type of the connections, the adequate strength of the connection is the sole criterion.

AISC specifications control the installation of high strength friction grip-bolts. In our country IS : 4000 specifications the installation of HSFG-bolts.

High-strength bolts are required to be installed in the same manner to obtain the same pre-tension, whether the connection is to be **bearing type or friction type.** The performance of bolts is at the service load is identical. That is, the service loads are transmitted by friction between the connected pieces.

The difference is performance is totally due to difference in factor of safety against slip. After the occurrence of slip under an occasional loading and for any subsequent loading, the load is transferred by friction in combination with bearing on the plate. As long as, the loading is static, and does not reverse the direction, such slip occurs only once. Thereafter, the bolt remains in `bearing against the material at the side of hole.

Note : The permissible stress in a bolt (other than a high strength friction grip bolt) of property class higher than 4.6 shall be those given in Table 2.10 multiplied by the ratio of its yield stress or 0.2 per cent proof stress or 0.7 times its tensile strength, whichever is less, to 235 N/mm^2 (MPa).

The calculated bearing stress of a bolt on the parts connected by it shall not exceed.

(i) the value of yield stress, σ_y for bolts in clearance holes, and

(ii) the value of 1.2 σ_y (σ_y is the yield stress) for close tolerance and turned bolts.

C.17 DESIGN OF HIGH STRENGTH FRICTION GRIP BOLTS

Indian Standard Institution has specified specifications for high tensile friction grip bolts in IS : 3757-1967 (High tensile friction grip fasteners for structural engineering purposes). The code of practice IS : 4000-1967 deals with assembly of structural joints using high tensile friction grip fasteners.

The connections made with high strength friction grip bolts may be subjected to shear, tension in the direction of the axis of bolt, combined shear and tension or repeated variation of stresses. The principles for design of these connections are as follows :

1. Shear connections

When the bolt connections are subjected to shear in the plane of the friction faces, the force in the members, P as shown in Fig. C.9 does not exceed the resulting frictional forces, that is, the maximum value of P may be expressed as follows :

$$P=\frac{1}{F.S}.(\mu.F) \tag{C.1}$$

where, μ is the coefficient of friction between the interfaces and F.S. is appropriate factor of safety. The coefficient of friction is also termed as slip factor. The slip factor is adopted as 0.45. For all loads except wind, the factor of safety is adopted as 1.4. When the effect of wind forces is considered, its value is adopted as 1.2. In case, the frictional forces are developed in N interfaces instead of one as shown in lap joint Fig. C.2 Eq. C.1 is modified as under

$$P=\left(\frac{1}{F.S}\times N.\mu.F\right) \tag{C.2}$$

The common contact surfaces subjected to forces in the opposite directions give the number of effective interfaces. The value of proof load F depends on the diameter and the proof stress of the material.

The friction grip bolts are manufactured as conforming to mechanical properties of 10 K or 8 G. The proof loads F for the different diameters of bolts as per IS : 3757-1966 have been given in Table C.9.

The values of proof loads for 10 K and 8 G bolts are based on 700 N/mm² and 600 N/mm², respectively.

TABLE C.9. PROOF LOADS FOR BOLTS

Diameter of bolts (mm)	Proof load (kN)	
	10 K bolts	8 G bolts
12	59.60	50.58
14	80.50	75.00
16	107.90	91.20
18	134.40	115.20
20	171.50	147.00
22	212.10	181.80
24	237.10	211.80
27	321.30	274.50
30	392.70	336.60
33	485.80	416.40
36	571.90	490.20
39	683.20	585.60

2. Tension connections

In case, the bolt connections are subjected to external tension in the direction of the axis of bolts, then, the clamping force is likely to be decreased. Very large values of external tension may separate interfaces and the clamping force may be decreased to zero and cause the failure of the connections. Therefore, the maximum external tension of any bolt should not exceed 0.6 of the proof load of the bolt used.

3. Combined shear and tension

In beam to column connections and in some bracket connections, the connections are subjected to combined shear and tension. IS: 4000-1967 specifies that the bolts are so proportioned that the following expression is satisfied.

$$\left[\left(\frac{\text{Calculated shear}}{\text{Slip factor}}\right)+1.2\times(\text{calculated tension})\right]=\left(\frac{\text{Proof load}}{\text{F.S.}}\right)$$

The effective clamping action of a bolt will cease when the externally applied tension reaches 0.6 of its proof load.

4. Repeated variation of stress

The nut and washer of the bolt remain in position by spring action due to tension in the bolt. As such these nuts and washers do not become loose even if these are subjected to repeated variation of stress. Also there is very little vibrations in the tension of the bolt. The fatigue strength of friction grip connections has been found to be 25 percent higher than the ordinary riveted joints. However, the external tension is limited to 0.5 of the proof load so that the separation of friction surfaces does not occur.

Example C.1 The uplift on wind ward side of a steel chimney is 276 N/mm of periphery at the base. Design the foundation bolts.

Design : Uplift force at the base of steel chimney = 276 N/mm

Factored force, P_{fd} = (1.5 x 276) / 1000

= 0.419 kN/mm.

Step 1 : Tensile strength of bolt

Design tensile stress in bolt

$$=\left(\frac{\sigma_y}{\gamma_{msy}}\right)=\left(\frac{250}{1.15}\right)=217.39\ N/mm^2$$

Use 37 mm diameter of bolts

Area at the root of threads = 840 mm^2

$$\text{Strength of bolt}=\left(\frac{217.39\times 840}{1000}\right)=182.60\ kN$$

Step 2 : Spacing of the bolts

$$=\left(\frac{182.60}{0.414}\right)=441.06\ mm$$

Provide 37 mm diameter bolts at 480 mm spacing along the periphery at the base of chimney.

Example C.2 Two plates have been connected by a double cover butt as shown in Fig. C.11 joint. The width of plate is 200 mm and the thickness of plate is 10 mm. Design bolted connected to transmit pull equal to the strength of plate.

Design :

Step 1 : Bolt Value

Let 20 mm diameter turned and fitted bolts are used. Two bolts are used in one row as shown in Fig. C.11.

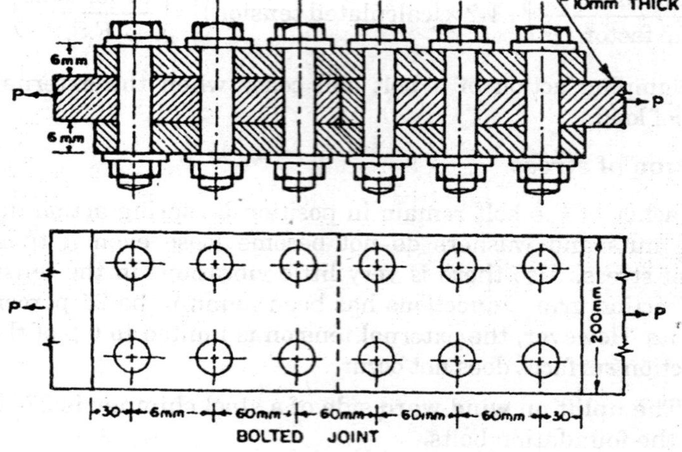

Fig. C.11.

Design strength of bolt is shear

$$\sigma_{cd} = \left(\frac{\sigma_y}{\sqrt{3}} \times \frac{1}{1.25}\right) = \left(\frac{410}{1.732} \times \frac{1}{1.25}\right) = 189.376\ N/mm^2$$

Strength of bolt in double shear

$$= \left(2 \times \frac{\pi}{4} \times 20^2\ \frac{\times 189.376}{1000}\right)$$

$$= \left(2 \times \frac{3.14}{4} \times 400 \times \frac{189.376}{1000}\right) = 118.93\ kN$$

Strength of bolt in bearing (Clause 10.3.4)

$$= \left(\frac{2.5\ k_b\ .\ d\ .\ t.\ \sigma_u}{1.25}\right) \qquad k_b = 1.0$$

$$= \left(\frac{2.5 \times 1 \times 20 \times 10 \times 410}{1.25 \times 1000}\right) = 164\ kN$$

When turned and fitted bolts are used then, holes are drilled and reamed. The diameter of hole is kept equal to bolt diameter.

Bolt value = 118.93 kN

Step 2 : Strength of plate in axial tension

Design stress, $\sigma_{td} = \left(\frac{\sigma_y}{1.15}\right) = \left(\frac{250}{1.15}\right) = 217.39\ N/mm^2$

Strength of plate in axial tension

$$= \left[217.39 \left(\frac{200 - 2 \times 20}{1000}\right) \times 10\right]$$

$$= \left[\frac{217.39 \times 160 \times 10}{1000}\right] = 347.824\ kN$$

Step 3 : Number of bolts

$$= \left(\frac{347.824}{118.93}\right) = 2.92$$

Provide 6 bolts 20 mm diameter on each side as shown in Fig. C.11.

Example C.3 Two plates of 16 mm thickness have been connected in a lap joint using high strength friction grip bolts. Design the joint so as to transmit a pull equal to the full strength of plate.

Design :

Step 1 : Strength of plate

$$\text{Design stress, } \sigma_{td} = \left(\frac{250}{1.15}\right) = 217.39\ N/mm^2$$

Two plates have been connected in lap joint. Number of interfaces, N is equal to one only. Let P be the pitch of the bolts. The strength of plate per pitch = p.

$$P = \left(\frac{P \times 16 \times 217.39}{1000}\right) = 3.478\ p\ kN$$

Step 2 : Force in one bolt

It is assumed that the plates are connected using four bolts only.

Force to be transmitted by one bolt

$$= \left(\frac{1}{4} \times 3.478\ p\right) = \frac{1.739}{2}\ p\ kN$$

$$\frac{1.739}{2}\ p = (N \times \mu \times F) \quad N = 1$$

$$F = \left(\frac{1.739 \times p}{2 \times 0.45}\right) = \frac{1}{2} \times 3.86\ p\ kN$$

It is assumed that the bolts are provided at P = 60 mm. Then

$$F = \left(\frac{3.86}{2} \times 60\right) = \frac{1}{2} \times 231.86\ kN$$

Step 3 : Bolt-value

Let the nominal diameter of bolt, d = 16 mm

Diameter of bolt hole = (16 + 2) = 18 mm

Bolts are in single shear

Design stress for bolt

$$\sigma_{db} = \left(\frac{410}{\sqrt{3}} \times \frac{1}{1.25}\right)$$

$$= \left(\frac{410 \times 1}{1.732 \times 1.25}\right) = 189.376\ N/mm^2$$

Area of bolt in shear

$$A_{vsb} = \frac{\pi}{4} \times (18)^2 = 254.34\ mm^2$$

Design shear strength of one bolt

$$= \left(\frac{189.376 \times 254.34}{1000}\right)$$

$$= 48.165 \text{ kN}$$

Bearing strength of plate (In case thicknesses of plates are unequal then thickness of thinner plate is to be used)

$$= \left(\frac{2.5\ k_b . d\ . t}{\gamma_{mb}}\right) . \sigma_u$$

k_b is smaller of following

End distance, e = 30 mm

$$(e/3do) = \left(\frac{30}{3 \times 18}\right) = 0.555$$

$$\left(\frac{p}{3do} - 0.25\right) = \left(\frac{60}{3 \times 18} - 0.25\right) = 0.86$$

$$\left(\frac{\sigma_{ub}}{\sigma_u}\right) = \left(\frac{400}{410}\right) = 0.9756$$

Therefore, $k_b = 0.555$

Bearing strength of plate

$$= \left(\frac{2.5 \times 0.555 \times 16 \times 18 \times 410}{1.25 \times 1000}\right).$$

$$= (2.5 \times 0.555 \times 94.467)$$

$$= 131.06 \text{ kN}$$

Bolt value = 48.165 kN

Step 4 : Number of bolts

$$= \left(\frac{231.86}{2 \times 48.165}\right) = 2.406$$

Provide 16 mm nominal diameter 4 (four) bolts to connect the given plates 18 mm thick in lap joint.

PROBLEMS

C.1. The uplift on a windward side of a steel chimney is 324 N/mm of periphery at the base. Design the foundation bolts.

REFERENCES

1. IS 2585 : Black square Bolts and Nuts and Black square screws.
2. IS: 3139 : Dimensions for screws Threads for Bolts and Nuts
3. IS: 3757-1952: High Tensile Friction Grip Fasters for Structural Engineering purposes.
4. IS: 4000-1967 : Assembly for structural Joints using High Tensile Friction Grip Fasterners.

APPENDIX-D

PINS AND DESIGN OF PINNED-CONNECTIONS

D.1 INTRODUCTION

In **pin-connections**, pins are used. Ordinarily, the bolts are used for temporary fastenings and erection. Occasionally bolts are used for permanent connections. When structural members are subjected to shock or vibrations, then, bolted connections are not used. When structures support machinery or rolling load,s then also, bolted connections are not used for beam and beam or column connections. The pin connections simplify the analysis of indeterminate structures. The pin connections are also occasionally used in buildings to carry extremely heavy column loads across auditoriums. The pin connections are used in hinged arches and few other structures. The pin connections acts as hinge where the moment is zero.

D.2 ADVANTAGES OF PIN CONNECTIONS

The following are the advantages of pin connections :

1. The pin connections simplify analysis of indeterminate structures. The moment at a pin-connection is zero.
2. The pin connections have freedom of rotation of connected members.
3. Only one pin is used in a pin-connection. Whereas in a rivet or bolted connection, more than one rivets or bolts are used.
4. The pin connections reduce secondary stress occurring at the joints.

D.3 DISADVANTAGES OF PIN CONNECTIONS

The following are the disadvantages of pin connections :

1. The pin cannot resist the longitudinal tension. Because, longitudinal tension produces friction which prevents free turning of pins. Free turning of pins is one of its principal function.
2. The pins and pin-holes require expensive machine work on them.
3. The pin connections have lack of rigidity. As a result of this, there is noise in bridges under traffic vibrations (specially in light structures).

D.4 PINS

Pins are the cylindrical pieces of steel inserted in holes which are aligned and join two pieces.

The pins for structural connections are manufactured from mild steel. The pins can freely turn in a pin-connections. The sizes of pins range from 9 mm diameter, used for connecting strap, iron bars, to railway bridge pins 330 mm or more in diameter.

The following are various types of pins used for structural connections :

1. Forged steel pin. Fig. D.1 (A) shows a forged steel pin used for structural connections. When forged steel pins are used, these are secured by means of a cotter pin or a tape pin, or a thin nut screwed upto shoulder on the end of pin. Fig. D.1 (B) and Fig. D.1 (C) show washer and cotter pin, respectively.

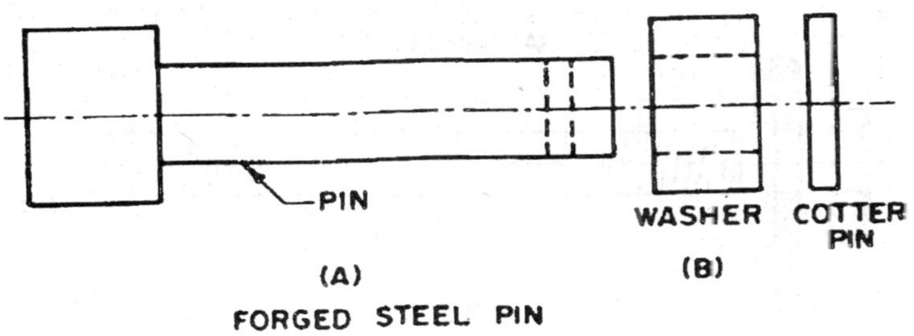

Fig. D.1.

2. Undrilled pins. Fig. D.2 (A) shows undrilled pins. Undrilled pins are used in bridges. Pilot nut, driving nut and holding nut shown, in Fig. D.2 (B), (C) and (D), respectively, are used for securing the pin when it is placed in position.

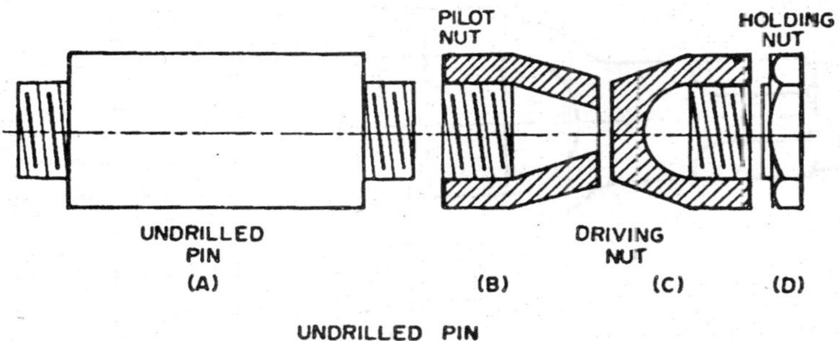

Fig. D.2.

3. Drilled pins. When a diameter of pin is more than 200 mm, then drilled pin as shown in Fig. D.3 (A) is used. A hole is drilled in this type of pin. Bolt and cap along with nut and washers as shown in Fig. D.3 (B), (C) and (D) are used for securing the pin, when it is placed in position.

D.5 PIN CONNECTION

In a pin-connected structure, the connection between the various members meeting at a joint is made by means of a cylindrical pin. One end of a bar is forged in the form of fork and hole is drilled in it. One end of other bar is forged and an eye is made. A hole is also drilled in it. The eye-bar can be inserted within the jaws of the fork. A forged steel pin is driven through the holes in them. The pin is secured by means of a cotter pin or a screw. The ends of both parts are made octagonal for good grip. Such type of pin connection is shown in Fig. D.4. This type of pin-connection is used in structures such as braced girders, links for suspension chains (chain-link cables of suspension bridges used for medium spans) etc.

Pin-connections are used to resist zero moments. These connections allow free rotations. In practice, the pins are machined to polished surface, so that the friction is minimum.

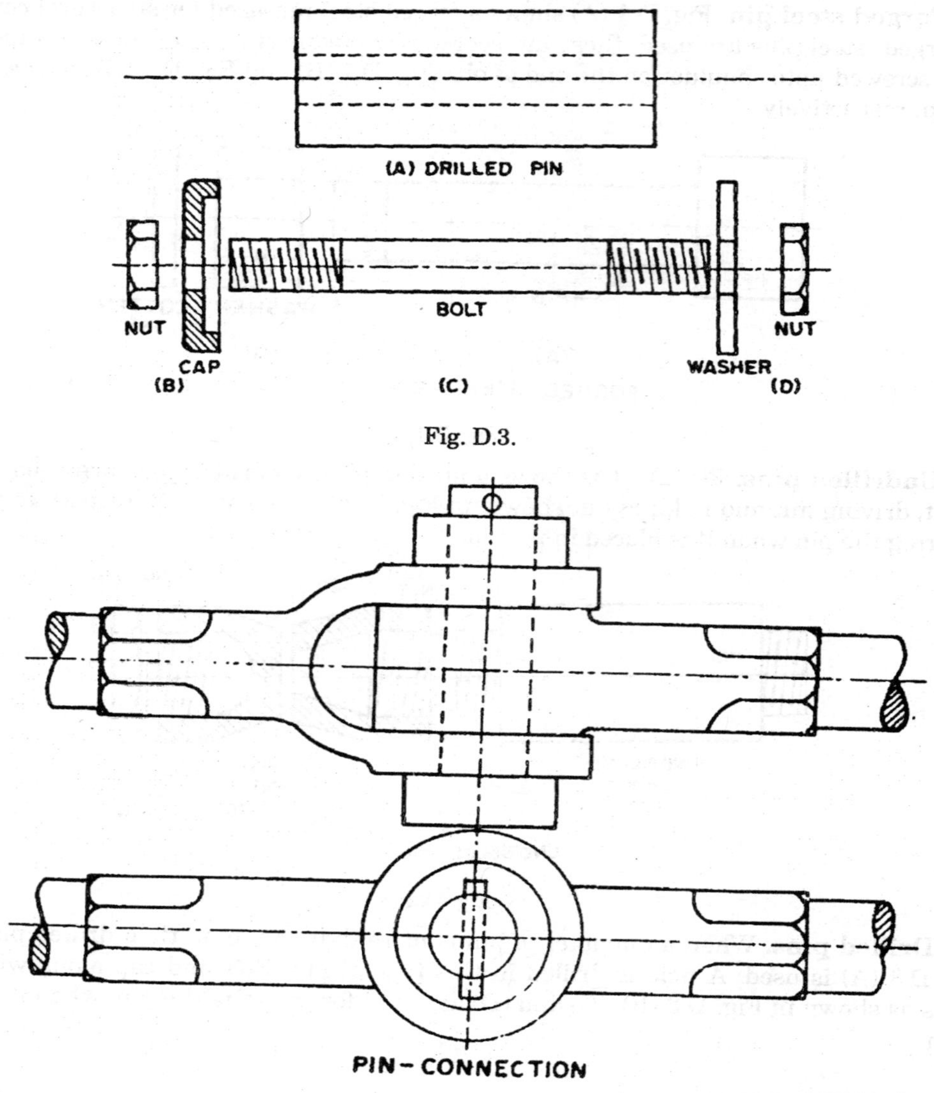

Fig. D.3.

Fig. D.4.

When two pairs of parallel eye bars are connected together then an undrilled pin is driven through the holes in them as shown in Fig. D.5. In such type of connection, a spacer is inserted between inner pair of eye bars. Such type of connections are used in structures such as truss bridge girders. A clearance of 3 mm is kept between adjacent elements to allow free rotation at the joint.

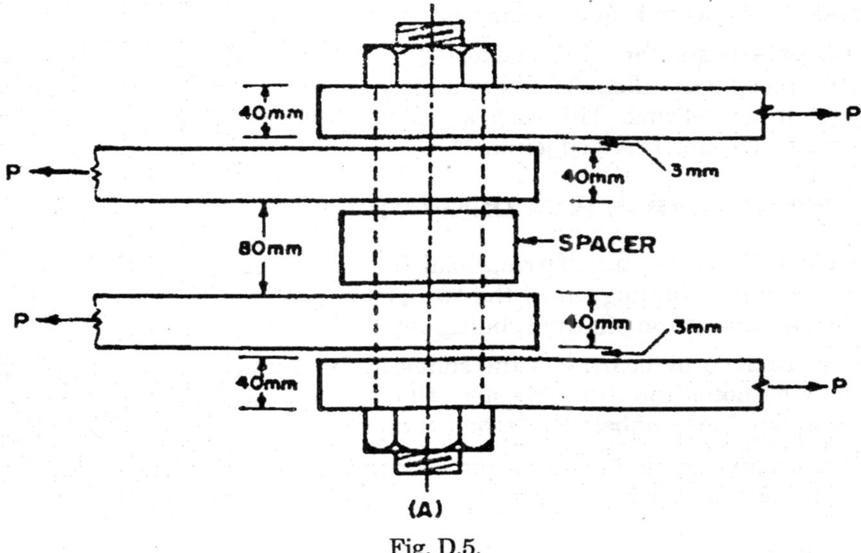

Fig. D.5.

D.6 STRESSES IN PINS

The pins are subjected to the following stresses in pin- connections :

1. Shear stress. A pin is fact, is a deep beam. The distribution of shear stress in a pin is the same as for beams of circular section. The maximum shear stress in a pin is

$$\tau_{vf(max)} = \left(\frac{V\,A\,\bar{y}}{I\,.\,b}\right)$$

where, V is shear force at the section, and A is area of cross- section above the fibre under consideration, and

$\bar{y}$ = Distance of c.g. of the above area from fibre under consideration

l = Moment of inertia of cross-section

b = Width of fibre under consideration.

For circular section of pin, the maximum shear stress occurs at the neutral axis. The maximum shear stress is, therefore,

$$\tau_{vf(max)} = \left[\frac{V\left(\frac{1}{2}\pi r^2\right)\left(\frac{4}{3}\frac{r}{\pi}\right)}{\left(\frac{\pi}{4}r^2\right)(2\pi r)}\right] = \frac{4}{3}\left(\frac{V}{\pi r^2}\right)$$

$$\tau_{vf\,(max)} = (4/3)\,.\,\tau_{vf(ave)}$$

where, = Radius of pin, $\tau_{v(ave)}$ = Average shear stress.

The allowable shear stress in pins may be adopted same as that for power driven shop rivets.

2. Bearing stress. The pins are also subjected to bearing stress. The allowable bearing stress in pins may be adopted same as that for power driven rivets.

3. Bending stress. Pins are subjected to bending stress. It may be necessary to proportions the pins for bending. The design of pin is likely to be governed by bending rather than by shear or by bearing. This is a significant difference between the design of a pin connection and the design of riveted joints.

D.7 TRANSFER OF LOAD IN PIN-CONNECTIONS

The simplest device for transferring load from one steel piece to another is with a pin inserted in holes which are aligned in the two pieces as shown in Fig. D.4 and Fig. D.5. The cotter pins shown prevent the pin from sliding out.

Load is transferred by bearing of the shank of the pin against the side of the hole. From Fig. D.5, it may be noted than the transfer of load is actually made by shear on the pin (the slight rotation of pin due to unbalanced moment is relatively negligible).

There is negligible friction between the connected pieces. Formerly, steel truss structures were actually connected by pins.

D.8 DESIGN OF PINS

Requirements as regards the design of pins have not been specified in IS 800. However, BS : 5950 (Part I) specifies following provisions, clause 6.5.3 for the design of pins. The capacity of pin-connections is obtained from (i) shear capacity (Clause 6.5.3.2), (ii) bearing capacity (clause 6.5.3.3) and (iii) bending capacity (clause 6.5.3.4)

1. **Shear capacity** : The shear capacity is found from the following

$$V_{d.p} = \left(\frac{\sigma_{up}}{\sqrt{3}} \times \frac{A}{\gamma_{msu}}\right)$$

$$= \left(\frac{\sigma_{up}}{1.732} \times \frac{A}{1.25}\right),$$

$$= (0.5774 \ \sigma_{up} A/1.25)$$

$$= (0.6 \ \sigma_{up} . \ A/1.25) \text{ where}$$

σ_{up} = Design strength of pin

A = Cross-sectional area of pin

2. **Bearing capacity** The bearing capacity is obtained from the following

$$P_{d.bp} = \left(1.2 \frac{\sigma_{yp}}{\gamma_{msy}} \times d.t\right), \text{ where}$$

σ_{yp} = Lower of the design strengths of the pin and the connected part.

d = Diameter of the pin

t = Thickness of the connected part

γ_{msy} = Partial safety factor for the strength of the pin = 1.15 (corresponding to σ_{yp})

3. **Bending capacity**: The bending moments on a pin is calculated on the assumption that the forces transmitted between the pin and the connected parts are uniformly distributed along the length in contact in each case.

The moment capacity of the pin is taken as

$$M_{d.b} = \left(1.2 \frac{\sigma_{yp} \cdot Z}{\gamma_{msy}}\right), \textit{ where}$$

σ_{yp} = Design strength

Z = Elastic section modulus of the pin ($Z = \pi\, d^3/32$)

γ_{msy} = Partial safety factor

$$M_{db} = \left(1.2 \times \frac{\pi}{32} \times d^3 . \frac{\sigma_{yp}}{\gamma_{msy}}\right)$$

$$\therefore d = \left[\frac{32}{1.2} \times \pi \times \frac{M_u}{\sigma_{yp}}\right]^{1/3}$$

$$d = \left[\frac{21.33 \times M_u}{\pi\, \sigma_{yp}}\right]^{1/3}$$

where, M_u is the ultimate bending moment (N-mm)

Example D.1 : Two pairs of parallel eye bars of size 150 mm x 40 mm are connected by a pin. The inner pair of eye-bar is spaced 80 mm apart. Design the pin connection.

Design : Two pairs of parallel eye bars are connected.

Size = 150 mm x 40 mm.

Spacing between inner pair = 80 mm

Step 1 : Pull transmitted by each eye bar

Pull transmitted by each pair of eye-bar is equal to shear force at x_1 or x_2

$$V_{ds} = \left(\frac{0.6 \times 250 \times 150 \times 40}{1000}\right) kN$$

V_{ds} = 900 kN

Step 2 : Maximum bending moment

M_{fd} = (900 x 43) = 38700 kN-mm

= 38.700 kN-m

Step 3 : Diameter of pin, d

$$d_{pin} = \left(\frac{21.33 \times 38700}{\pi \times 250} \times 1000\right)^{1/3}$$

$d_{pin} = (1051.555)(1000)^{1/3}$

$= (10.1689 \times 10) = 101.689$ mm

Let the diameter of pin to be used be 140 mm

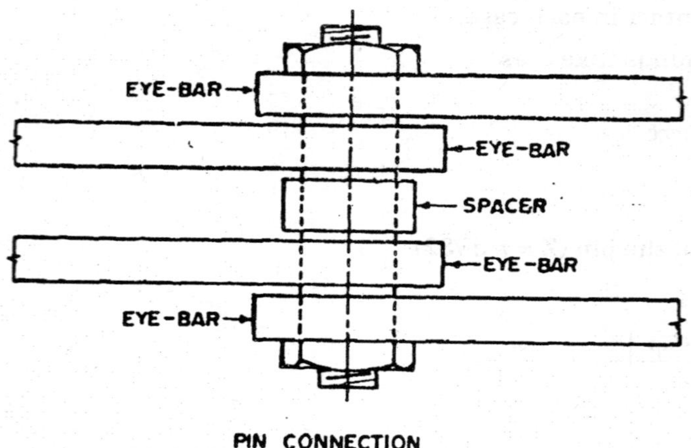

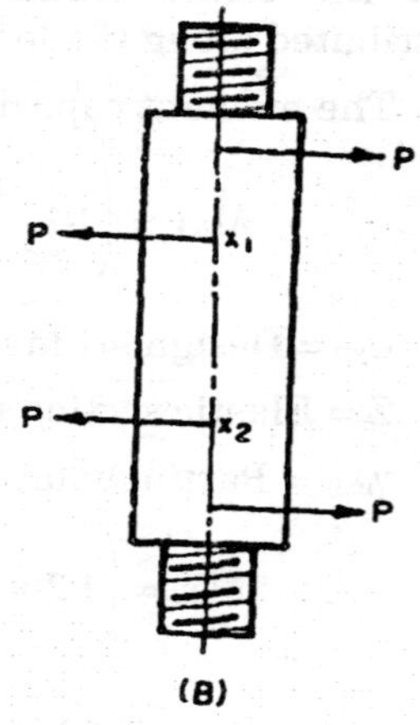

Fig. D.6.

Step 4 : Bending moment capacity

Elastic section modulus provided

$$Z = \left(\frac{\pi}{32} \times d^3\right)$$

$$= \left(\frac{3.14 \times 140^3}{32}\right)$$

$$= \left(\frac{3.14 \times 2744 \times 1000}{32}\right) = 269.255 \times 10^3 \text{ mm}^3$$

$$M_{db} = \left(\frac{1.2 \times 250 \times 269.255 \times 10^3}{1000 \times 1000 \times 1.15}\right) = 70.174 \text{ kN-m}$$

M_{db} is more than M_{fd}. Hence, it is satisfactory in bending.

Step 5 : Shear capacity

Cross-sectional area of pin (pin is cylindrical)

$$A = \left(\frac{\pi}{4} \times 140^2\right) = 15386 \, mm^2$$

Shear capacity of the pin

$$V_{dp} = \left(\frac{0.6 \times 410 \times 15386}{1000 \times 1.25}\right) = 3027.96 \text{ kN}$$

V_{dp} is more than V_{fd}. Hence, the pin is satisfactory in shear.

Step 6 : Bearing capacity

$$P_{dbp} = \left(\frac{1.2 \times 250}{1.15} \times \frac{140 \times 40}{1000}\right) kN = 1460.87 \text{ kN}$$

Hence, it is satisfactory in bearing

PROBLEMS

5.1. Two pairs of parallel eye bars of size 240 mm x 60 mm are connected by a pin. The inner pair of eye bar is spaced 100 mm apart. Design pin-connections.

REFERENCES

1. BS : 5950: Part 1, 2000 `British Standard Structural use of Steel-work Building', Code of Practice for the Design in simple and continuous construction, Hot Rolled Sections, Published by British Standard Institution.

2. N. Subramanian. `Design of Steel Structures' Published by Oxford University Press, YMCA Library Building, Jai Singh Road, New Delhi 110 001.

INDEX

DESIGN OF STEEL STRUCTURES-2

9th Revised and Enlarged Edition

Dr. Ramchandra & Virendra Gehlot

Eight edition of this book (July, 2006) is based on Bridge Rules (Adopted in 1941, Revised in 1964 and Reprinted in 1989), and IS: 800-1984 as amended in 1997. Authors have distributed present text in the edition in thirty two chapters [that is, in Four parts (1) Steel Bridges and Influence Lines Diagrams for axial forces for the members of different types of truss-girders, (2) Special Steel Structures (3) Analysis of Structures (specially, the method of tension co-efficients for determinate and indeterminate structures, (4) Aluminum structures). In order to emphasize that similar to various other subjects, this subject is also very vast. Therefore, space steel structures and stressed-skin steel structures have been described special features of this new-edition of this book may be mentioned as under (1) Historical development of different types of steel bridges details of some spans of longest spans of various types of steel bridges, (2) Design of Guyed Steel Chimneys (3) Instantaneous Centre of Rotation (ICR) and Plastic Analysis of Pitched slope (i.e., gable structure) and influences of axial forces and shear forces on the plastic moment of resistance of the member cross-sections.

CONTENTS

ISBN: 978-81-7233-644-8, 2016, 972 pages